AA002270

International Exhibition & Conference for Power Electronics, Intelligent Motion, Renewable Energy and Energy Management (PCIM Europe 2024)

Nuremberg, Germany
11 – 13 June 2024

Volume 4 of 5

ISBN: 978-1-7138-9966-2

Printed from e-media with permission by:

Curran Associates, Inc.
57 Morehouse Lane
Red Hook, NY 12571

Some format issues inherent in the e-media version may also appear in this print version.

Copyright© (2024) by Mesago Messe Frankfurt GmbH
All rights reserved.

Printed with permission by Curran Associates, Inc. (2024)

For permission requests, please contact VDE VERLAG GMBH
at the address below.

VDE VERLAG GMBH
Bismarckstr. 33
P.O.B. 12 01 43
10625 Berlin, Germany

Phone: +49 30 34 80 01 - 0
Fax: +49 30 34 80 01 - 9088

kundenservice@vde-verlag.de

Additional copies of this publication are available from:

Curran Associates, Inc.
57 Morehouse Lane
Red Hook, NY 12571 USA
Phone: 845-758-0400
Fax: 845-758-2634
Email: curran@proceedings.com
Web: www.proceedings.com

TABLE OF CONTENTS

VOLUME 1

KEYNOTE

K01   AI BETWEEN HYPE AND INDUSTRIAL-GRADE - THE IMPACT OF AI ON THE ENTIRE POWER ELECTRONICS LIFECYCLE.. 1
Rolf Hellinger

K02   INFRASTRUCTURE REQUIREMENTS FOR ELECTRIFIED HEAVY GOODS TRANSPORT IN GERMANY AND THE EU... 7
Martin Wietschel

K03   CHALLENGES AND SOLUTIONS TO POWER LATEST PROCESSOR GENERATIONS FOR HYPER SCALE DATACENTERS 15
Gerald Deboy

GAN RUGGEDNESS

OP001   AN IMPROVED ULTRAFAST DESATURATION-BASED PROTECTION SCHEME FOR GAN HEMT 19
Juncheng Lu

OP002   THE PERFORMANCE OF A GAN EMODE HEMT IN SURGE CURRENT SCENARIOS SUCH AS THE ACTIVE SHORT CIRCUIT.. 24
Dominik Nehmer

OP003   GATE RESISTANCE EFFECT ON SHORT-CIRCUIT ROBUSTNESS OF P-GAN HEMTS 34
Mohamed Lemine Dedew

ADVANCED PACKAGING TECHNOLOGIES

OP004   NEURAL NETWORK ASSISTED NUMERICAL SIMULATION BENCHMARKING FOR ELECTRIC VEHICLE THERMAL MANAGEMENT SYSTEM 40
Ekin Alp Bicer

OP005   RELATIONSHIP BETWEEN POROSITY IN CU SINTERED BONDING AND BONDING RELIABILITY... 49
Hideo Nakako

OP006   HIGH THERMAL DURABILITY OF THIN COPPER DIE-ATTACH LAYERS AND FINITE ELEMENT MODEL SIMULATION.. 56
Takaaki Eyama

THERMAL CYCLING RELIABILITY

OP007   THERMAL SHOCK TEST LIFETIME IMPROVEMENT WITH OPTIMIZED ADHESIVE STRENGTH BETWEEN EPOXY RESIN AND COPPER 62
He Kangjia

OP008　POWER CYCLING RELIABILITY AND FAILURE MODE ANALYSIS OF POL 67
Kenichi Koi

OP009　ACCELERATED POWER CYCLING OF GAN HEMTS USING SWITCHING LOSS
AND FAST TEMPERATURE MEASUREMENT 74
Wing Tai Leung

HIGH POWER CONVERTERS

OP010　CONTROL OF AN MMC-BASED HYBRID TRANSFORMER WITH STAR-POINT
VOLTAGE INJECTION 84
Rui Wang

OP011　PROTECTION AND CONTROL OF A DUAL MMC MEDIUM VOLTAGE SUPPLY 93
Max Dupont

OP012　STATION POWER ELECTRONICS CONVERTER WITH HIGH THERMAL
ENDURANCE TO POLE-TO-POLE SHORT CIRCUITS FOR LVDC DISTRIBUTION GRID 103
Frédéric Reymond-Laruina

GATE DRIVERS

OP013　SUPPRESSION OF OSCILLATIONS IN A SIC BRIDGE-LEG USING A CUSTOM
SINGLE-CHIP DIGITAL ACTIVE GATE DRIVER WITH 2×255 STRENGTH LEVELS 113
Qilei Wang

OP014　SIC MOSFET SHORT-CIRCUIT PROTECTION: A FASTER SOFT SHUT DOWN
METHOD FOR GATE DRIVERS 121
Julien Weckbrodt

OP015　PARAMETER IDENTIFICATION: GATE SENSOR FOR POWER TRANSISTOR
TOLERANCE COMPENSATION IN ADVANCED GATE DRIVER ICS 128
Christopher Wille

ADVANCED CONTROL TECHNIQUES ON ELECTRICAL DRIVES I

OP016　AN INNOVATIVE HIGH-SPEED TRACK RANGE RESTART STRATEGY FOR
PERMANENT MAGNET SYNCHRONOUS MOTOR 135
Anna Corbitt

OP017　STEADY-STATE ERROR REDUCTION OF REINFORCEMENT LEARNING BASED
INDIRECT CURRENT CONTROL OF PERMANENT MAGNET SYNCHRONOUS
MACHINES 140
Tobias Schindler

OP018　PERFORMANCE COMPARISON OF USING SHUNT-BASED AND INTEGRATED
CURRENT SENSING FOR SENSORLESS FIELD-ORIENTED CONTROL 150
John Emmanuel Tan

GAN CONVERTERS

OP019　DESIGN OF HIGH-POWER INVERTER WITH 12 PARALLEL GAN DEVICES 161
Takashi Sawada

OP020   OVER 99.7% EFFICIENT GAN-BASED 6-LEVEL CAPACITIVE-LOAD POWER
CONVERTER 167
   Stefan Mönch

OP021   CASCADED PRIMARY-SIDE-ONLY CONTROL OF A COMPACT 2 MHZ 500 W
WIRELESS POWER TRANSFER SYSTEM 174
   Tim Krigar

ADVANCED MATERIALS AND TECHNOLOGIES

OP022   POWER MODULE EVALUATION USING ULTRA HIGH HEAT DISSIPATION AND
HIGH HEAT RESISTANCE RESIN SHEET CONTAINING CARD HOUSE TYPE BORON
NITRIDE FILLER 180
   Ayano Imai

OP023   INVESTIGATING TEMPERATURE DEPENDENT WARPAGE IN METAL CERAMIC
SUBSTRATES FOR POWER ELECTRONICS DEVICES 190
   Benjamin Fabian

OP024   DEGRADATION MODE ANALYSIS OF DIFFERENT BONDING TECHNOLOGIES
OF SIC POWER SEMICONDUCTORS STRESSED BY ACTIVE POWER CYCLING 197
   Rasched Sankari

CHARGING STATION TECHNOLOGY

OP025   IMPLEMENTATION AND VERIFICATION OF A 50KW OPPORTUNITY WIRELESS
CHARGER DESIGN 205
   Carlos Costas Sos

OP026   PERFORMANCE EVALUATION OF SILICON-BASED 3-LEVEL VIENNA
RECTIFIER IN ISOPLUS SMPD PACKAGE 214
   Karsten Haehre

OP027   PERFORMANCE ANALYSIS OF A 25-KW SIC-BASED DUAL ACTIVE BRIDGE
CONVERTER BASED ON PARALLEL-CONNECTED DEVICES 222
   Francesco Porpora

MODELLING AND MONITORING

OP028   SEMICONDUCTOR CHIP MODELS ARE THE KEY FOR ENABLING VIRTUAL
DESIGN AND OPTIMIZATION WORKFLOWS OF POWER ELECTRONIC SYSTEMS 230
   Stefan Haensel

OP029   IMPROVED RESONANT FREQUENCY-BASED PARASITIC INDUCTANCE
ESTIMATION METHOD FOR SIC MOSFET HALF-BRIDGE CIRCUIT 238
   Hongpeng Zhang

OP030   FAST SIMULATOR WITH INVERTER TEMPERATURE ESTIMATION FOR
TRACTION EDRIVES IN VEHICLES SUBJECTED TO DRIVING CYCLES 248
   Simone Giuffrida

SOLID STATE TRANSFORMERS

OP031　A NEW FAMILY OF THREE-PHASE-UNFOLDER-BASED MVAC-LVDC SOLID-STATE TRANSFORMERS.. 254
Jonas Huber

OP032　VOLTAGE BALANCING OF A SPLIT-CAPACITOR IGCT 3L-NPC LEG FOR THE RESONANT DC TRANSFORMER.. 264
Renan Pillon Barcelos

OP033　COMPARATIVE ANALYSIS OF UNIDIRECTIONAL HIGH STEP-UP CONVERTERS FOR MEDIUM VOLTAGE APPLICATIONS 274
Stefan Subotic

ADVANCED CONTROL TECHNIQUES ON ELECTRICAL DRIVES II

OP034　STARTUP BEHAVIOR OF HARMONIC SUPPRESSION IN ELECTRICAL MACHINES USING ITERATIVE LEARNING CONTROL AND NEURAL NETWORKS........................ 284
Annette Mai

OP035　ANALYTICAL APPROACH OF THE VECTOR CURRENT CONTROL FLUX-WEAKENING STRATEGY FOR PERMANENT MAGNET SYNCHRONOUS MACHINES 290
Oriol Subirats Rillo

POWER ELECTRONICS FOR E-MOBILITY

OP036　INVESTIGATION ON DIRECT LIQUID COOLING DESIGN OF POWER MODULES WITH FLAT BASEPLATE FOR AUTOMOTIVE APPLICATION.. 298
Nobuhide Arai

OP037　A NOVEL APPROACH FOR AFFORDABLE ELECTRIC VEHICLES BASED ON DUAL 48V BATTERY SYSTEM WITH MULTI-FUNCTIONAL 3-LEVEL CONVERTER........................ 305
Radovan Vuletic

OP038　AN INNOVATIVE 3-LEVEL SOLUTION FOR AUTOMOTIVE APPLICATIONS: EMPACK.. 315
Pranav Panchal

OP039　GATED RECURRENT UNITS-ASSISTED STATE-SPACE MODELING FOR ELECTRIC VEHICLE TEMPERATURE PREDICTION 322
Xinyuan Liao

OP040　NOVEL BIDIRECTIONAL SINGLE-STAGE ISOLATED 600-V GAN M-BDSBASED SINGLE/THREE-PHASE-OPERABLE EV ON-BOARD CHARGER 330
Sven Weihe

ENCAPSULATION MATERIALS

OP041　APPLICATION-SPECIFIC INVESTIGATION OF INORGANIC POTTING MATERIAL IN DRIVE TRAINS 338
Soenke Fleck

OP042 THE INFLUENCE OF THE GLASS TRANSITION TEMPERATURE OF EPOXY MOLD COMPOUNDS ON THE RELIABILITY OF A SEMICONDUCTOR DEVICE 343
Stefan Schwab

OP043 CORROSION RESISTANT PACKAGING FOR POWER SEMICONDUCTOR MODULES - MODIFIED INSULATION MATERIALS FOR CONTAMINATED ENVIRONMENTS 351
Michael Hanf

OP044 INVESTIGATION OF INORGANIC ENCAPSULATION MATERIALS IN POWER ELECTRONIC SYSTEMS FOR HIGH POWER DENSITY APPLICATIONS 361
Stefan Behrendt

OP045 CHARACTERIZATION OF THERMALLY AGED SILICONE GELS FOR POWER SEMICONDUCTOR MODULES.. 369
Sonja Madloch

POWER QUALITY

OP046 A COORDINATED CONTROL OF HYBRID SINGLE-PHASE AC/DC MICROGRIDS BASED ON THE NATURAL HARMONIC INJECTION CONCEPT 378
Mehdi Baharizadeh

OP047 A HIGH-POWER DENSITY SIC BASED TP PFC WITH HIGH-FREQUENCY RIPPLE CANCELLATION LEG.. 383
Serkan Dusmez

OP048 HIGH FREQUENCY ACTIVE FILTER FOR AC-DC HIGH POWER CONVERTERS 390
Sarah Sifoune

OP049 LABORATORY SETUP FOR ACCURACY INVESTIGATION OF ELECTRICITY METERS AND MONITORS UNDER INDUSTRY-TYPICAL OPERATING CONDITIONS...................... 397
Matthias Schmidt

GRID CONNECTED CONVERTERS

OP050 REAL-TIME EVALUATION OF WEIGHTING FACTORLESS PREDICTIVE CONTROL OF LCL FILTER EQUIPPED GRID-SIDE CONVERTERS USING SORTING NETWORKS.. 403
Kristóf Bándy

OP051 RELAXED ROBUST CONTROL WITH PRAGMATIC SHORTAGE OF PASSIVITY FOR WIND, STORAGE AND PV POWER CONVERTERS 411
Sergio De Lopez Diz

OP052 AN EFFECTIVE DC VOLTAGE REGULATION OF ACTIVE FRONT-END RECTIFIER THROUGH MODEL PREDICTIVE CONTROL.. 419
Mobina Pouresmaeil

OP053 BI-DIRECTIONAL 11KW MULTI-LEVEL ACTIVE-NEUTRAL-POINT-CLAMPED AC-DC CONVERTER USING 600V/750V SI SUPER-JUNCTION AND SIC MOSFETS FOR HIGH-EFFICIENCY AND HIGH-DENSITY APPLICATIONS... 424
Mengxing Chen

OP054   A STUDY OF GRID-FORMING INVERTER CONTROL STRATEGY FOR FAULT-RIDE-THROUGH CAPABILITY 433
Hirofumi Uemura

PASSIVE COMPONENTS

OP055   FILM CAPACITORS FOR HIGH TEMPERATURE AC-DC INVERTER APPLICATIONS 440
Adel Bastawros

OP056   LOSS REDUCTION IN HF-TRANSFORMERS USING LAMINATED FERRITE E-CORES 447
Lukas Reißenweber

OP057   MULTIGAP TOROIDAL TRANSFORMER AND INDUCTORS FOR OVERCOMING FRINGING LOSSES IN HIGH FREQUENCY CONVERTERS 456
Pau Colomer

OP058   STUDY ON SAMPLE GEOMETRIES FOR FERRITE CHARACTERISATION IN THE MHZ RANGE 463
Till Piepenbrock

OP059   FEM-SUPPORTED AND NON-DESTRUCTIVE MAGNETIC CHARACTERIZATION METHOD FOR NON-LAMINATED STEEL 472
Stefan Tobler

DRIVES FOR HIGH DEMANDING APPLICATIONS

OP060   HIGHLY-COMPACT BEARINGLESS AXIAL-FLUX MOTOR FOR A PEDIATRIC IMPLANTABLE FONTAN BLOOD PUMP 480
Andreas Horat

OP061   A NOVEL PERMANENT MAGNET SYNCHRONOUS MOTOR DRIVE FOR REACTION WHEELS IN SATELLITES 490
Baris Colak

OP062   EXPLORING HIGH FREQUENCY OPERATION OF MOTOR DRIVES: PRACTICAL INSIGHTS ON EFFICIENCY AND LOSS 497
Asantha Kempitiya

OP063   HIGH POWER DENSITY SYSTEM DESIGN FOR GAN-BASED LV MOTOR DRIVES 502
Marco Cannone

OP064   DESIGN OF GAN TRANSISTOR BASED VARIABLE SPEED DRIVE INVERTER WITH OUTPUT VOLTAGE FILTERING 510
Kaspars Kroics

IGBT

OP065   THE 8TH GENERATION LV100 IGBT MODULE WITH HIGHER CURRENT RATING 518
Daichi Otori

LOSS REDUCTION BY LAMINATIGN FERRITE E CORES 525
Lukas Reißenweber

OP066   NEW PLANAR 4.5 KV SPLIT-GATE (SG) SI-IGBT DEVICE FOR IMPROVED
SWITCHING CHARACTERISTICS AND HIGH FREQUENCY OPERATION 534
Gaurav Gupta

OP067   4.5 KV DOUBLE-GATE REVERSE-CONDUCTING PRESS-PACK IEGT 543
Satoshi Yoshida

DEVICE CONCEPTS

OP068   EVALUATION OF A 3 KV POLARIZATION SUPERJUNCTION GAN HEMT 549
Alireza Sheikhan

OP069   MORE THAN 1200 V BREAKDOWN AND LOW AREA-SPECIFIC ON STATE
RESISTANCES BY PROGRESS IN LATERAL GAN-ON-SI AND GAN-ON-INSULATOR
TECHNOLOGIES 557
Richard Reiner

OP070   NOVEL 200 V MOSFET TECHNOLOGY PUSHES MOTOR DRIVE INVERTER
EFFICIENCY TO AN UNPRECEDENTED LEVEL 564
Mark Thomas

DEGRADATION MECHANISMS

OP071   MOISTURE ROBUST CHIP DESIGN - IMPROVED EDGE-TERMINATIONS FOR
HIGH LIFETIME UNDER HIGH HUMID CONDITIONS 571
Michael Hanf

OP072   METHOD FOR MEASURING THE INITIAL STATE OF A SOLDER JOINT
DELAMINATION IN A 3D PCB INTEGRATION ASSEMBLY OF SIC MOSFETS 581
Souhila Bouzerd

OP073   GENERIC LIFETIME MODEL FOR WIRE BONDS DEGRADATION IN IGBT
MODULES BASED ON A FRACTURE MECHANICS PARAMETER 589
Merouane Ouhab

ADVANCED CONVERSION CONCEPTS

OP074   MODULAR COAXIAL POWER CONVERTER FOR HIGH-DENSITY INTEGRATION
INTO MEDIUM-VOLTAGE CABLES 599
Mark Cairnie

OP075   CONTROLLED INDUCTOR BASED BCM BUCK CONVERTERS 608
Ziv Gellman

OP076   INFLUENCE OF VARYING COMMON MODE CHOKE SIZES ON THE
PERFORMANCE AND STABILITY OF AN ACTIVE EMI FILTER 615
Patrick Körner

PHOTOVOLTAIC SYSTEMS

OP077   A HIGH EFFICIENCY BATTERY CHARGER WITH MAXIMUM POWER POINT TRACKING FOR MAGNETIC ENERGY HARVESTERS 625
Antonio Miguel Munoz Gomez

OP078   SYMMETRIC FLYING-CAPACITOR BOOST CONVERTER FOR MEDIUM-VOLTAGE PHOTOVOLTAIC APPLICATIONS 635
Luis Alves Rodrigues

OP079   COMPARISON OF SI IGBT, SIC MOSFET AND ADJUSTABLE HYBRID SWITCH PV INVERTERS FOR DIFFERENT GEOGRAPHICAL LOCATIONS 645
Tanya Thekemuriyil

MODEL BASED SYSTEM ANALYSIS

OP080   OPTIMISING A POWER MODULE FOR ELECTRICAL AND THERMAL PERFORMANCE AND SYMMETRY USING EDA TOOLS 655
Wilfried Wessel

OP081   CONDUCTOR-BASED MODELING OF VOLTAGE DISTRIBUTION ALONG A SINGLE-TOOTH WINDING OF ELECTRICAL MACHINES 665
Hujun Peng

OP082   REDUCTION OF PWM HARMONICS WITH CARRIER PHASE SHIFTING IN A DUAL-STATOR PMSM WITH MAGNETIC COUPLED WINDINGS 672
Bünyamin Tekir

VOLUME 2

SIC DEVICES

OP083   THE NEW COOLSIC MOSFET 1200 V G2: ELECTRICAL PERFORMANCE AND COMPACT MODELLING 681
Andreas Huerner

OP084   PARALLELING SIC-POWER-MOSFET BODY DIODES UNDER HARSH SWITCHING CONDITIONS 690
Michael Rauh

OP085   3.3KV SBD-EMBEDDED SIC-MOSFET MODULE FOR TRACTION USE 699
Yoichi Hironaka

OP086   DEAD TIME OPTIMIZATION FOR HIGH POWER SIC MOSFET MODULE IN CONSIDERATION OF PARASITIC COMPONENTS 707
Pham Ha Trieu To

WBG RELIABILITY

OP087   PERFORMANCE INSTABILITY OF 650 V P-GAN GATE HEMT DEVICE UNDER TEMPERATURE-RELATED POSITIVE GATE BIAS STRESSES 717
Renze Yu

OP088 GATE OXIDE RELIABILITY OF CURRENT GENERATION 1.2 KV SIC MOSFETS UNDER STEP-WISE INCREASED GATE VOLTAGE.. 723
Roman Boldyrjew-Mast

OP089 AN ACCELERATED DYNAMIC GATE SWITCHING STRESS TEST CONCEPT OF SIC MOSFETS AT HIGH DRAIN-SOURCE VOLTAGE (HV-GSS) 731
Clemens Herrmann

OP090 SILICON CARBIDE POWER DEVICE USE IN SPACECRAFT AND AIRCRAFT 739
Akin Akturk

POWER ELECTRONICS FOR E-MOBILITY/ CONTROL

OP091 CURRENT RIPPLE REDUCTION BY COMBINATION OF SI IGBT AND SIC MOSFETS IN HEAVY DUTY FUEL CELL TRUCKS.. 745
Yavuz Gürlek

OP092 EVALUATION OF ACTIVE GATE DRIVERS WITH SWITCHABLE GATE RESISTORS AND INTERMEDIATE VOLTAGE LEVELS FOR SIC MOSFETS IN WLTC 754
Michael Frank

OP093 PERFORMANCE EVALUATION OF TCM-BASED, ZERO-VOLTAGE SWITCHING (ZVS) THREE-PHASE INVERTER FOR ELECTRIC VEHICLE DRIVE SYSTEMS 764
Khizra Abbas

OP094 A PARTIAL LOAD THREE-PHASE TRIANGULAR CURRENT MODE MODULATION CONCEPT WITH AN OPTIMIZED FILTER INDUCTOR FOR HIGH EFFICIENCY TRACTION DRIVES 774
Bhaskar Chatterjee

DC-DC CONVERTERS I

OP095 GAN VS SI SYNCHRONOUS RECTIFIER FOR LLC CONVERTER 784
Gokhan Sen

OP096 CO-SIMULATION DESIGN OF A GAN-BASED THREE-PHASE LLC CONVERTER WITH INTEGRATED THREE-PHASE MAGNETICS 791
Jhih-Cheng Hu

OP097 SWITCHING ASSISTING CIRCUIT IMPROVING THE EFFICIENCY OF DC-DC CONVERTERS BASED ON PIEZOELECTRIC RESONATORS.. 797
Ghislain Despesse

OP098 TRANSFORMER-BASED FIXED-RATIO RESONANT DC-DC CONVERTERS FOR 48V DATA CENTERS 803
Xufu Ren

PFC CONVERTERS

OP099 HIGH-DENSITY 3.3 KW GAN RECTIFIER FOR SERVER APPLICATIONS COMPRISING A 130 KHZ TOTEM-POLE PFC AND A 500 KHZ LLC.. 812
Manuel Escudero Rodriguez

OP100   ADDRESSING POWER SWITCH TECHNOLOGY SELECTION SI/SIC/GAN IN
HIGH EFFICIENCY ZVS-PFC RESONANT CONVERTERS 822
    Marco Torrisi

OP101   BUCK-TYPE CURRENT UNFOLDING CONVERTER WITH DISCONTINUOUS
CONDUCTION MODE IN ULTRA-LOW POWER-FACTOR OPERATION 831
    Tomoyuki Mannen

OP102   GAN BASED BI-DIRECTIONAL 6.6KW INTERLEAVED TOTEM-POLE PFC WITH
13KW/L POWER DENSITY AND HIGH EFFICIENCY 837
    Juncheng Lu

SIC MODULES

OP103   THE DESIGN OF A 2KV 1700A SIC MOSFET DUAL MODULE 843
    Jorge Mari

OP104   TECHNOLOGICAL APPROACHES TO HIGH-POWER DENSITY SIC POWER
MODULE FOR AUTOMOTIVE 849
    Takeshi Tokorozuki

OP105   EXTREMELY COMPACT SIC POWER MODULE FOR EV TRACTION INVERTERS
IN THE 250 KW CLASS 855
    Raffael Schnell

OP106   BENEFITS OF .XT INTERCONNECTION TECHNOLOGY FOR 3.3 KV XHP 2
MODULE WITH 3.3 KV COOLSIC MOSFET 863
    Matthias Bürger

ADVANCED COOLING

OP107   LARGE-AREA BONDING WITH LMEE: SUPPRESSION OF THE DEGRADATION
OF THE JUNCTION-TO-WATER THERMAL RESISTANCE IN POWER MODULES 870
    Yo Mochizuki

OP108   ACTIVE THERMAL CONTROL OF SIC MOSFETS UTILIZING TRANSIENT
THERMAL CHARACTERIZATION 875
    Varaha Satya Bharath Kurukuru

OP109   THERMAL MANAGEMENT SOLUTIONS BY ADDITIVE MANUFACTURING –
POWDER BED FUSION AND DIFFUSION BONDING 883
    Simon Jahn

OP110   ADVANCED PUMPED TWO-PHASE COLD PLATE FOR COOLING POWER
ELECTRONICS 888
    Elizabeth Seber

DC-DC CONVERTERS II

OP111   FEASIBILITY STUDY OF HIGH-POWER DENSITY ISOLATED CLLC DC-DC
INTERFACE WITH WIDE RANGE OF VOLTAGE/CURRENT REGULATION 893
    Oleksandr Husev

OP112   DC-BIAS REDUCTION IN HIGH-FREQUENCY DUAL ACTIVE BRIDGE DC-DC CONVERTERS THROUGH SLOW DC MEASUREMENTS 903
Patrick Lenzen

OP113   OPTIMIZED CURRENT SHARING TECHNIQUE FOR INTERLEAVED CLLC CONVERTERS FOR MINIMAL OUTPUT CURRENT DISTORTION 909
Martin Gendrin

OP114   PRIMARY-SIDE OUTPUT REGULATION PRINCIPLES IN DYNAMIC MULTI-MHZ INDUCTIVE POWER TRANSFER SYSTEMS AND ISOLATED DC/DC CONVERTERS 916
Ioannis Nikiforidis

SMART GRID

OP115   LOW VOLTAGE DC-GRIDS WITH GALVANIC ISOLATION: SYSTEM DISCUSSION, EFFICIENCY AND PERFORMANCE COMPARISON TO AC-FEEDING 926
Lukas Fräger

OP116   IMPLEMENTATION AND EXPERIMENTAL EVALUATION OF AN ADAPTIVE DC GRID CONTROLLER FOR DECENTRALISED GRID CONTROL 933
Steffen Menzel

OP117   DEMONSTRATING THE EFFECTIVENESS OF A DC SOLID-STATE CIRCUIT BREAKER'S FAST RESPONSE TIME 942
Ehab Tarmoom

OP118   MODELLING AND SIZING SENSITIVITY ANALYSIS OF A FULLY RENEWABLE ENERGY-BASED ELECTRIC VEHICLE CHARGING STATION MICROGRID 949
David A. Stone

MEASUREMENT TECHNIQUES AND METHODS

OP119   LED POWERED ROTOR TELEMETRY SYSTEM 958
Raphael Beyerle

OP120   'INFINITY GATE SENSOR': A DIFFERENTIAL MAGNETIC FIELD SENSOR FOR MEASURING GATE CURRENT OF SIC POWER TRANSISTORS 966
Yushi Wang

OP121   CHARACTERISING WIDE BANDGAP POWER MODULES: VALIDATING THE M-SHUNT CONCEPT FOR HIGH-POWER APPLICATIONS IN THE KILOAMPERE RANGE 976
Hauke Lutzen

OP122   CHARACTERIZATION OF POWER-MODULE PARASITICS: SUB-NANOSECOND LARGE SIGNAL PULSING VS. DOUBLE-PULSE TESTING 986
Gerhard Groos

STATISTICAL VARIATIONS IN THE PARASITIC CAPACITANCE OF A COIL 997
Kevin Talits

HIGH VOLTAGE SWITCHES

PP001   A 4.5 KV FAST RECOVERY DIODE PLATFORM FOR HIGH-CURRENT IGBTS 1002
Jan Vobecky

PP002   6.5 KV INNOVATIVE SILICON POWER DEVICE (I-SI) MODULE WITH HIGH POWER DENSITY AND LOW LOSS BY STORED CARRIER CONTROL 1007
Takashi Hirao

PP003   HIGH CURRENT DENSITY 4.5KV PRESSPACK IGBTS PUSH SOA LIMITS 1013
Hossein Davoodi

PP004   2.5KV IGBT MODULE WITH HIGH RELIABILITY FOR RENEWABLE APPLICATIONS 1018
Akiyoshi Masuda

PP005   NEW GENERATION 4.5KV IGCT AND FAST RECOVERY DIODE FOR RAILWAY POWER SUPPLY APPLICATIONS 1025
Umamaheswara Reddy Vemulapati

PP006   NEXT GENERATION 4.5 KV IGBT-ONLY STAKPAK MODULE WITH REDUCED LOSSES AND HIGH TEMPERATURE CAPABILITY 1031
Jeremy Jones

THERMAL MODELLING AND SIMULATIONS

PP007   FINITE ELEMENT ANALYSIS OF THE UPSCALING OF WARPAGE AND BIFURCATION HYSTERESIS LOOPS: FROM CU/SI DIE TO LARGE WAFERS 1039
Vincenzo Vinciguerra

PP009   MAXIMUM JUNCTION TEMPERATURE SIMULATION AND VALIDATION FOR THE HOT SPOT IN MULTI-CHIP SIC POWER MODULE 1046
Wonjin Dylan Cho

PP010   INTEGRATION OF CFD-SIMULATION RESULTS IN PLECS USING LOOKUP TABLES 1051
Simon Cepin

PP011   PCB ONLY THERMAL MANAGEMENT TECHNIQUES FOR EGAN FETS IN A HALF-BRIDGE CONFIGURATION 1057
Adolfo Herrera

HIGH POWER DENSITY DESIGNS

PP013   FROM 4X TO 3X STPAK – OPTIMIZATION FOR A MORE COMPACT EV TRACTION INVERTER SOLUTION... 1065
Vittorio Giuffrida

PP014   A MULTI-OBJECTIVE STRUCTURAL OPTIMIZATION METHOD BASED ON MULTI-PHYSICS SIMULATIONS FOR POWER MODULE 1072
Baihan Liu

PP015   HOLISTIC APPROACH TO MAXIMIZE LIFETIME AND POWER DENSITY IN HIGH POWER SEMICONDUCTOR MODULES 1077
Martin Schulz

PP016   REGULATED HIGH DENSITY SWITCH CAPACITOR TOPOLOGY 1082
Pierrick Ausseresse

PP017 SILICON INTERPOSER AS A SUBSTRATE FOR POWER MODULES WITH HIGH
POWER DENSITY AND SUPERIOR THERMAL PERFORMANCE 1087
   Ahmed Ammar

SPECIAL CONVERTER APPLICATIONS

PP018 ANALYTICAL MODELING AND STABILITY CHARACTERIZATION OF A
DAMPED VSCC CM ACTIVE EMI FILTER FOR SINGLE- AND THREE-PHASE AC-DC
APPLICATIONS 1092
   Timothy Hegarty

PP020 A REPETITIVE HIGH VOLTAGE NANOSECOND PULSE GENERATOR: FIRST
PROTOTYPE DESIGN AND TEST RESULTS 1101
   Serge Gavin

PP021 FREQUENCY SHIFT KEYED DUAL SIDE CONTROL OF INDUCTIVE POWER
TRANSFER: AN APPLICATION OF TALKATIVE POWER CONVERSION 1105
   Hamzeh Beiranvand

PP022 STUDY OF A MULTI-ACTIVE BRIDGE CONVERTER FOR A DOMESTIC
ELECTRICAL GRID 1113
   Abdennour Merrouche

INTEGRATION TECHNOLOGIES AND RELIABILITY DESIGN

PP023 FABRICATION DEVELOPMENT FOR GATE DRIVER EMBEDDED DOUBLE-
SIDED COOLING SIC POWER MODULE FOR ELECTRIC VEHICLE APPLICATION 1123
   Anna Corbitt

PP024 PRINTED CIRCUIT EMBEDDING OF PREPACKAGED 150V POWER MOSFETS IN
A PORTABLE WELDING APPLICATION 1128
   Thomas Gebhard

PP025 PROCESS CHALLENGES AND PROGRESS TOWARDS DIRECT CONNECTION OF
AUTOMOTIVE POWER MODULES (TMM) TO HEATSINK 1133
   Indrajit Paul

PP026 OPTIMIZING PCB STACKUPS FOR ENHANCED GAN TRANSISTOR
PERFORMANCE IN HIGH-POWER APPLICATIONS 1139
   Philipp Czerwenka

PP027 NEW GENERATION CERAMIC SUBSTRATES – KEY COMPONENTS FOR POWER
ELECTRONIC APPLICATIONS: PROCESSING AND CHARACTERIZATION 1147
   Stefanie Schindler

PP028 AI-ENHANCED VACUUM REFLOW OVEN: PRECISION CONTROL FOR
RELIABLE LARGE-AREA SOLDERING 1152
   Chih Hui Lee

PP030 CORROSION-COMPATIBLE DRIVE ELECTRONICS FOR ELECTRIC VEHICLES
AND INDUSTRIAL POWER MODULES 1158
   Tom Petzold

PP031 EVALUATING THE SAFETY ISOLATION OF THE PACKAGE IN AN INTEGRATED POWER DEVICE 1168
Thomas Anthony Capobianco

CONTROL METHODS I

PP032 FLEXIBLE CONTROL SYSTEM FOR MODULAR ONE-PHASE INTERLEAVED GAN-BASED TOTEM POLE PFC USING REAL-TIME HARDWARE 1174
Oleksandr Solomakha

PP033 A PEAK CURRENT MODE CONTROL METHOD FOR PFC 1180
Sean Yu

PP034 ADAPTIVE RESONANT CONTROLLER FOR A THREE-PHASE PFC CONVERTER FOR AN ON-BOARD CHARGE APPLICATION 1185
Rami Troudi

PP035 SYNTHESIS OF A FIELD ORIENTED CONTROL ALGORITHM BY USING TWO DIFFERENT POLE-ZERO COMPENSATION APPROACHES.................................1192
Marco Denk

PP037 AVERAGE CURRENT MODE CONTROL AND ITS LOOP DESIGN 1200
Niklas Schwarz

PP038 NOVEL POWER FEED-FORWARD REGULATION FOR DUAL STAGE PFC+DCDC CONVERTERS 1207
Alfredo Medina-Garcia

HIGH POWER AC-DC AND DC-AC CONVERTER

PP039 22 KW BI-DIRECTIONAL WALL-BOX CHARGER WITH 1200 V SIC MOSFET 1212
Sanbao Shi

PP040 DYNAMIC SWITCHING FREQUENCY SELECTION FOR EFFICIENCY OPTIMIZATION IN ON-BOARD CHARGER PFC STAGE BASED ON NOVEL SIC MOSFET POWER MODULE...1217
Giuseppe Aiello

PP041 DESIGN AND OPTIMIZATION OF SIC-BASED 11KW MOTOR DRIVE WITH HIGH EFFICIENCY 1222
Iris Liu

PP042 MODEL DESIGN DEVELOPMENT FOR FALSE TURN-ON CHARACTERIZATION IN SIC-BASED ACTIVE T-TYPE CONVERTER CONSIDERING ALL PARASITICS 1227
Amir Babaki

PP043 EFFICIENCY INVESTIGATIONS OF AN AUXILIARY RESONANT COMMUTATED POLE INVERTER..1233
Markus Zocher

PP044 A NOVEL HYBRID TWO-STAGE AC-DC CONVERTER WITH SOFT-SWITCHED CCM PFC STAGE FOR EVS CHARGING APPLICATIONS...............................1242
Lei Wang

PP045   A METHOD FOR TUNING LEAKAGE INDUCTANCE IN TRANSFORMERS 1249
Rosemary O'Keeffe

PP046   LOW COST HIGH DENSITY 300W/20V AC-DC CONVERTER ENABLED BY GAN
POWER ICS.. 1254
Tom Ribarich

PP047   25KVA GRID-TIED BI-DIRECTIONAL T-TYPE INVERTER WITH HIGH-
EFFICIENCY AND HIGH-POWER DENSITY USING SIC MOSFETS...................................... 1259
Tamanna Bhatia

PP048   COST-EFFECTIVE EFFICIENCY ENHANCEMENT IN AC-DC CONVERTERS: A
STUDY ACROSS THE FULL LOAD CYCLE 1264
Sebastian Gick

E-MOBILITY TRACTION I

PP049   NEXT GENERATION POWER MODULE WITH PARALLEL CONNECTED SIC
MOSFETS FOR BEV TRACTION INVERTERS.. 1272
Kohei Tanikawa

PP051   INVESTIGATION OF COMMON SOURCE FEEDBACK IN SIC POWER MODULES
REGARDING PERFORMANCE AND SHORT CIRCUIT ROBUSTNESS................................... 1277
Dominik Ruoff

PP052   HYBRIDPACK DRIVE POWER MODULES WITH SIC-MOSFET'S AND
MONOLITHIC RC- SNUBBER CHIPS FOR OPTIMIZED POWER DENSITY....................... 1283
Andre Uhlemann

PP053   ROBUST AUXILIARY POWER SUPPLY FOR EVS BASED ON INNOVATIVE
STI2GAN 650V IC.. 1289
Federica Cammarata

PP054   IMPACT OF VARIOUS SILICON DIODES ON THE HYBRID SWITCH INVERTER 1297
Michael Walter

PP055   ADVANCED PULSE SEQUENCE FOR SALIENCY-BASED HIGH-ACCURATE
ROTOR POSITION ESTIMATION OF RAILWAY TRACTION LOCOMOTIVE MOTORS 1307
Markus Vogelsberger

CONTROL TECHNIQUES

PP056   OPTIMIZED HALF-BRIDGE GATE-DRIVE WITH LOW TIME-SKEW FOR RC-
IGBTS AND SIC-MOSFET DEAD-TIME CONTROL 1315
Jan Fuhrmann

PP057   DESIGN OF A TRACTION INVERTER BASED ON PCB-EMBEDDED GAN
DEVICES 1322
Maurizio Tranchero

PP058   OPTIMIZING ELECTRIC VEHICLE PERFORMANCE WITH GAN DESIGN....................... 1330
Andrew Patterson

PP059   FAST ANALYTICAL CALCULATION OF THE MAGNETIC FIELD IN PERMANENT
MAGNET SYNCHRONOUS MACHINES WITH FLUX BARRIERS INCLUDING
SATURATION 1336
   Martin Ackermann

PP060   MODELING AND CONTROL OF LCL FILTERED 3L-VSCS IN INTERLEAVED
TOPOLOGY 1346
   Adeel Jamal

PP062   ENHANCING SAFETY AND EFFICIENCY FOR ISOLATED PLC I/O DESIGNS
WITH SPI DAISY CHAIN 1352
   Travis Lenz

VOLUME 3

PP063   COST-EFFECTIVE METHOD TO DISCHARGE DC LINK CAPACITORS WITH SIC
POWER MODULES... 1361
   Paul Kanatzar

POWER QUALITY

PP064   A STUDY ON CIRCULATION CURRENT IN PARALLEL OPERATION OF
TRANSFORMER LESS UPS 1368
   Koji Kato

PP065   DESIGN CHALLENGES AND CONSIDERATIONS FOR GATE DRIVERS OF SIC
MOSFETS AND THEIR TESTING.. 1374
   Niranjan Hegde

PP066   A PORTABLE EFFICIENCY CHARACTERIZATION SETUP FOR TECHNOLOGY
DEMONSTRATION OF POWER MODULES 1380
   Sebastian Tengvall

PP067   FAST EME CHARACTERIZATION OF BARE-DIE SIC MOSFETS 1385
   Robert Kragl

PP068   THEORETICAL COMPARISON OF COMPONENT-RELATED MEASUREMENT
METHODS OF PHOTOVOLTAIC INVERTERS FOR LONG-TERM TESTING.................................. 1393
   Niclas Reitz

DYNAMIC TRANSIENTS AND RELIABILITY OF HIGH-VOLTAGE SILICON & 4H-SIC
BIPOLAR JUNCTION TRANSISTORS UNDER AVALANCE AND SHORT-CIRCUITS 1402
   Mana Hosseinzadehlish

PP069   POWER CYCLING TEST OPTIMIZATION TOWARD RELIABILITY ASSESSMENT
OF SINTERED POWER MODULES.. 1410
   Robert Graham

PP070   REAL-TIME ESTIMATION AND SENSITIVITY ANALYSIS OF PARASITIC
CAPACITANCES IN ELECTRIC DRIVE SYSTEMS 1418
   Mohammadreza Bagheribavaryani

MODELLING AND TESTING

PP071 PARASITIC COMPONENT EFFECTS OF INTERNAL AND EXTERNAL PACKAGE
LEVEL ON SWITCHING PERFORMANCE OF SIC POWER MODULE 1428
Nguyen Nghia Do

PP072 A MULTI-PHYSICS ITERATIVE APPROACH FOR TEMPERATURE ESTIMATION
IN SIC POWER MODULE FOR ELECTRIC VEHICLE 1434
Stefano Orlando

PP073 VOLTAGE BALANCING METHOD FOR SERIES CONNECTION OF 50 SIC
MOSFETS 1441
Antoine Philippe

VOLTAGE BALANCING METHOD FOR SERIES-CONNECTION OF 50 SIC MOSFETS 1449
Antoine Philippe

PP074 A LABORATORY-SCALE MMC-BASED DC SYSTEM WITH RCP AND PHIL
SIMULATION CAPABILITIES 1457
Marc René Lotz

PP075 FILM CAPACITOR STANDARD SERIES DIGITALIZATION: ELECTROMAGNETIC
& THERMAL MODELLING IMPLEMENTATION IN CLARA WEB TOOL 1467
Fernando Aunon

PP076 ACCURACY EVALUATION AND PROPOSED DYNAMIC TUNING PROCEDURE
OF A COMPACT SIC SPICE MODEL 1475
Austin Curbow

PP077 INVESTIGATION OF USE-CASE-DEPENDENT MODELING APPROACH FOR
SWITCHED-MODE POWER CONVERTER FOR LVDC GRID EVALUATION 1485
Melanie Lavery

PP078 AVERAGED MODEL WITH BLOCKING CAPABILITY FOR SOLID-STATE
TRANSFORMERS 1495
Ahmed Meligy

ADVANCED COMPONENTS

PP080 SURFACANT-MODIFIED NANOCOMPOSITE THIN-FILM CAPACITORS 1504
Bartosz Gackowski

PP081 INCREASING ENERGY STORAGE CAPABILITIES OF POWDER CORES BY
ADAPTING THE WINDING AND THE USE OF FRINGING FLUX 1511
Paul Winkler

PP082 PEEC-BASED THERMAL MODELING OF PASSIVE COMPONENTS.................................. 1516
Sascha Langfermann

PP083 GALVANICALLY ISOLATED POWER SUPPLY FOR GATE DRIVERS IN HIGH
VOLTAGE APPLICATIONS 1523
Priyanka Ghosh

PP084 FABRICATION TECHNIQUE FOR NOVEL NANOCRYSTALLINE CORES WITH HIGH SATURATION POLARIZATION AND LOW LOSSES 1532
Merlin Thamm

PP085 EXCITATION-DEPENDENT TEMPERATURE BEHAVIOR OF THE QUASI-STATIC HYSTERESIS LOSS ENERGY DENSITY OF N87 FERRITE MATERIAL................................. 1538
Jeremias Kaiser

PP087 PASSIVE METHODS LIMITING LEAKAGE CURRENT IN METAL-OXIDE VARISTOR AS VOLTAGE CLAMPING DEVICE USED DC LOW VOLTAGE POWER ELECTRONICS-BASED CIRCUIT BREAKERS 1545
Kenan Askan

GAN DEVICES AND APPLICATIONS

PP088 ESD SOLUTIONS FOR 650V NORMALLY-OFF ALGAN/GAN HEMTS 1555
Thanh Hai Phung

PP089 A SIMULATIVE STUDY OF MEASUREMENT ERRORS DURING DOUBLE PULSE TESTING OF GAN DEVICES 1561
Severin Klever

PP090 PARALLEL CONNECTION OF GAN FETS: AN EXPERIMENTAL INVESTIGATION APPROACH... 1568
Marco Palma

PP091 REPETITIVE SHORT CIRCUITS ON 650 V GAN 1574
Adrien Lambert

PP092 COMPARISON OF SWITCHING LOSSES AND DYNAMIC ON RESISTANCE OF 600 V-CLASS GAN HEMTS.. 1584
André Thönnessen

PP093 PERFORMANCE EVALUATION OF DEADTIME AND GATE RESISTANCE FOR PARALLEL CONNECTED GAN HEMTS 1590
Junhyeok Jegal

PP094 REACHING BEYOND 1200V: LATERAL GAN HEMTS FOR HIGH-RELIABILITY EV AND INDUSTRIAL APPLICATIONS 1598
Kamal Varadarajan

SIC DEVICES AND TECHNOLOGIES

PP095 SMARTSIC 150 & 200MM ENGINEERED SUBSTRATE: INCREASING SIC POWER DEVICE CURRENT DENSITY UP TO 30%.. 1604
Eric Guiot

PP096 DYNAMIC TRANSIENTS IN HIGH-VOLTAGE SILICON AND 4H-SIC NPN BIPOLAR JUNCTION TRANSISTORS 1610
Mana Hosseinzadehlish

PP097 AN ADVANCED MULTI-ASPECT PERFORMANCE ANALYSIS OF PLANAR-GATE 1.2 KV SIC POWER MOSFETS 1613
Anja Katerina Brandl

PP098   SIC MOSFET DIE SORTING AND PARALLEL FOR OPTIMAL MODULE DESIGN 1621
        Zhong Ye

PP099   SIMULATION APPROACH FOR RADIATED ELECTRO-MAGNETIC FIELDS
         ESTIMATION ON ACEPACK DRIVE SIC POWER MODULE 1627
        Andrea Cusumano

CONTROL METHODS II

PP100   EXACT ANALYSIS OF CONTROL-TO-OUTPUT TRANSFER FUNCTIONS OF
         PWM-CONVERTERS - A COMPARISON OF TWO METHODS.. 1634
        Daniel Breidenstein

PP101   3-LEVEL FLYING CAPACITOR MULTILEVEL TOPOLOGY WITH DELTA-SIGMA
         MODULATION 1642
        Jannik Maier

PP102   MODEL BASED CONTROLLED POWER CONVERTER TEST PLATFORM......................... 1651
        Dawid Koczy

PP103   EDUCATIONAL HARDWARE TRAINER FOR TEACHING THE DUAL ACTIVE
         BRIDGE IN A DC GRID 1658
        Peter Van Duijsen

PP104   STUDY OF THE OPERATING PERFORMANCE OF A FCS-MPC-CONTROLLED
         MATRIX-CONVERTER FOR PMSM AT DIFFERENT FREQUENCY RATIOS 1664
        Robert Zipprich

PP105   ENHANCING REACTIVE POWER CAPACITY IN BATTERY-FED POWER
         CONDITIONING SYSTEMS 1673
        Lucas Araujo

PP106   PULSE SHARING: ACHIEVING HIGH EFFICIENCY AND EXCELLENT REGULA-
         TION IN MULTI-OUTPUT FLYBACK POWER SUPPLIES 1680
        Xingda Yan

PP107   RELIABILITY-OPTIMIZED SPACE VECTOR MODULATION (RO-SVM) FOR
         SEMICONDUCTORS LIFETIME ENHANCEMENT 1686
        Amin Rezaeizadeh

INTELLIGENT POWER MODULES

PP108   ANALYSIS AND OPTIMIZATION OF INTERNAL COUPLING INTERFERENCE IN
         INTEGRATED SIC POWER MODULE BASED ON DBC 1693
        Chenhang Zeng

PP109   MULTISPECTRAL ELECTROLUMINESCENCE SENSING OF SIC MOSFETS FOR
         JUNCTION TEMPERATURE AND CURRENT EXTRACTION.. 1703
        Lukas Ruppert

PP110   SIC-IPM FOR COMPACT AND ENERGY EFFICIENT LOW-POWER MOTOR
         DRIVES 1712
        Jongmu Lee

PP111   CONCEPT FOR A GAN-BASED INTELLIGENT MOTOR CONTROLLER WITH
INTEGRATED FAILURE PREDICTION FOR THE INVERTER AND THE DRIVE 1717
    Christoph Blechinger

PP112   INTRODUCING THE NEW 1200 V CIPOS MAXI IM817 INTELLIGENT POWER
MODULE FOR MOTOR DRIVE APPLICATIONS 1724
    Kihyun Lee

PP113   THERMAL PERFORMANCE OF INFINEON'S NEW 600 V CIPOSTM MICRO IM241
IPM FOR LOW POWER MOTOR DRIVE SYSTEMS WITHOUT HEATSINK 1732
    David Jo

INTRODUCING THE NEEW 1200 V CIPOSTM MAXI IM12BXXXC1 INTELLIGENT POWER
MODULE FOR MOTOR DRIVE APPLICATIONS 1737
    Kihyun Lee

INTELLIGENT GATE DRIVE UNITS

PP114   AN ADAPTIVE DEAD TIME CONTROL BASED ON SWITCH NODE VOLTAGE
DERIVATIVE 1745
    Lukas Knappstein

PP115   COUPLING COIL DESIGN AND POSITIONING OPTIMIZATION ON NEW HIGH
POWER SEMICONDUCTOR MODULE FOR FAST SHORT CIRCUIT DETECTION 1751
    Yannick Dumollard

PP116   ENABLING ACTIVE THERMAL CONTROL VIA AN ADAPTIVE MULTI-VOLTAGE
GATE DRIVER 1759
    Tianlong Albert

PP117   INNOVATIVE GATE DRIVE METHOD TRIC3 FOR MOTOR 1765
    Hisashi Sugie

PP118   A NEW CLASS OF SOLID STATE ISOLATORS ENHANCES THE RELIABILITY OF
SOLID STATE RELAYS 1770
    Wolfgang Frank

PP119   A SELF-DRIVING 3-LEVEL ACTIVE GATE DRIVER NETWORK TO CONTROL
THE SWITCHING SLEW RATE FOR SIC MOSFETS 1775
    Vin Loong Choo

E-MOBILITY TRACTION II

PP121   ANALYSIS OF LONG-TERM RELIABILITY OF SIC IN TRACTION INVERTER
CONSIDERING VTH INSTABILITY 1781
    Chi Zhang

PP122   EFFICIENT MAPPING OF ON-DEMAND DRIVE LOAD PROFILES ON INVERTER
STRESS 1788
    Zlatko Bosnjic

PP123   EV TRACTION INVERTER OPTIMAL DESIGN IS DOMINATED BY 3-LEVEL
ANPC 1797
    Timothé Delaforge

PP124 INTRODUCTION OF POWER SEMICONDUCTOR OPTIONS FOR AN EXCITER OF ELECTRICALLY EXCITED SYNCHRONOUS MOTOR 1804
Yeriel Bai

PP125 A NOVEL HIGH POWER DENSITY THREE PHASE TRACTION INVERTER ARCHITECTURE FOR ELECTRIC VEHICLE (EV) APPLICATIONS.................................... 1809
Yiyang Yan

PP126 A MODULAR DC-LINK CAPACITOR SOLUTION FOR THE MAIN POWERTRAIN INVERTER OF XEV 1814
David Olalla

PP127 FAULT IDENTIFICATION TESTING METHODS FOR A COMMERCIAL TRACTION INVERTER 1821
Anna Corbitt

PP128 SHORT CIRCUIT ROBUSTNESS FOR TRACTION INVERTERS FROM AN APPLICATION POINT OF VIEW 1828
Karl Oberdieck

INVESTIGATIONS OF PARTICULAR SIC DEVICE PHENOMENON

PP129 THE IMPACT OF THE DEADTIME ON THE STABILITY OF 1.2KV SIC MOSFET BODY DIODE UNDER HARD SWITCHING WITH SYNCHRONOUS RECTIFICATION..................... 1835
Mohammed Amer Karout

PP130 RC-DC SNUBBER IMPLEMENTATION FOR SUPPRESSION OF DIODE VOLTAGE PEAK AND RINGING IN A FULL SIC HALF-BRIDGE POWER MODULE 1844
Emanuela Alfonzetti

PP131 SUB-5 SECOND WIDE-BANDGAP POWER DEVICE CALORIMETRIC MEASUREMENTS UTILZIING OPTICAL SENSORS AND PELTIER ELEMENTS 1851
Ruben Schnitzler

PP132 SIC TRENCH MOSFETS IN AVALANCHE MODE WITH RC SNUBBER CIRCUIT............... 1858
Sebnem Tuncay

PP133 HIGH-FREQUENCY OSCILLATIONS IN SIC MOSFET POWER MODULES DURING TURN-ON SWITCHING TRANSIENT – ANALYSIS BASED ON SIMULATIONS AND MITIGATION METHODS.. 1865
Rajani Kumar Thirukoluri

PP134 A DYNAMIC CURRENT BALANCING METHOD USING FULL-COUPLED INDUCTORS IN PARALLELED GATE BRANCHES.. 1872
Jianwei Lv

PP135 QUANTITATIVE PERFORMANCE COMPARISON OF LARGE-FORMAT SIC MOSFET AND SI IGBT MODULES 1878
Arthur Boutry

THERMAL MANAGEMENT AND ADVANCED COOLING

PP136 SOLDER PREFORM TECHNOLOGY FOR IMPROVED THERMOMECHANICAL PERFORMANCE IN MOLDED POWER MODULE PACKAGE-ATTACH 1886
Joseph Hertline

PP138　EFFECT OF FLIP-CHIP DIE-ATTACH ON THE THERMAL BEHAVIOR OF POWER GAAS DIODES 1891
Felix Steiner

PP139　INFLUENCES OF SOLDER DELAMINATION ON THE THERMAL PERFORMANCE IN AUTOMOTIVE TRACTION MODULE 1896
Hansol Seo

PP141　DEVELOPMENT OF A PASSIVE CAPILLARY-PUMPED COOLING SYSTEM FOR HIGH-PERFORMANCE ELECTRONICS 1902
Justin Fey

PP143　ADVANCED COOLING OF POWER ELECTRONICS WITH COPPER COLD SPRAYED ALUMINIUM HEATSINKS & BUSBARS 1907
Michael Dasch

PP144　COLD PLATE DESIGN FOR COOLING LV100 SILICON CARBIDE POWER MODULE PACKAGING 1910
Wahid Cherief

PP145　AN IMPROVED DOUBLE-LAYER SPACER IN DOUBLE-SIDED COOLING POWER MODULE 1917
Linhao Ren

RELIABILITY TESTING

PP146　POWER CYCLING OF 1.7KV MULTI-CHIP POWER MODULES – SIC MOSFETS VS SILICON IGBTS 1923
Nick Baker

PP147　POWER CYCLING CAPABILITY OF DISCRETE SIC MOSFET DEVICES WITH DIFFERENT DESIGNS 1930
Luhong Xie

PP148　MODEL-BASED PARAMETER TUNING OF SEMICONDUCTOR DEVICES IN DC POWER CYCLING TEST 1936
Yi Zhang

PP149　INFLUENCE OF TRANSFER MOLDING ON THE RELIABILITY OF DCM SIC POW-ER MODULES 1942
Jacek Rudzki

PP150　DAMP HEAT BEHAVIOR OF HIGH HEAT CAPACITORS FOR APPLICATIONS IN ELECTRIC VEHICLES 1951
Adel Bastawros

PP151　INFLUENCE OF THE GATE VOLTAGE DURING ON-TIME ON THE POWER CYCLING CAPABILITY OF SIC MOSFETS 1955
Patrick Heimler

PP152　INVESTIGATION OF THE TEMPERATURE MEASUREMENT VIA VSD(T)-METHOD APPLIED TO PARALLELED SIC MOSFET CHIPS DURING POWER CYCLING 1964
Kevin Ladentin

PP153　APPROACHES OF TSEP MEASUREMENTS FOR POWER SEMICONDUCTORS 1969
Philipp Hauenschild

PP154 REALTIME JUNCTION TEMPERATURE ESTIMATION IN SIC POWER MODULES BASED ON MULTIPLE TSEP ACQUISITION 1978
Kevin Muñoz Barón

HIGH VOLTAGE WBG DEVICES

PP155 ENHANCED CURRENT MEASUREMENT APPROACH FOR NON-ISOLATED 6.5 KV SILICON CARBIDE MOSFETS 1987
Xinyuan Du

PP156 NEW 2KV SIC-MOS TECHNOLOGY FOR APPLICATION FIELDS IN THE INDUSTRIAL LANDSCAPE 1991
Igor Kasko

PP157 HIGH TEMPERATURE EXPERIMENTAL CHARACTERIZATIONS OF COSS OF 3.3 KV SIC MOSFET FOR MEDIUM VOLTAGE PV APPLICATIONS 1999
Paul Schmidt

PP158 IMPACT OF GATE CONTROL ON THE SWITCHING PERFORMANCE OF 3.3KV SBD-EMBEDDED SIC-MOSFET 2006
Junya Sakai

PP159 COMPARATIVE ASSESSMENT OF OVERLOADABILITY POTENTIAL OF 3.3 KV SI-IGBTS AND SIC-MOSFET POWER MODULES 2013
Muhammad Nawaz

PP160 IMPROVED RELIABILITY OF A 2200 V SIC MOSFET MODULE WITH AN EPOXY-ENCAPSULATED INSULATED METAL SUBSTRATE 2022
Hiroshi Kono

PP161 PARALLELING 3.3-KV/800-A RATED SIC-MOSFET MODULES – AN OPTIMIZATION METHOD 2028
Hiroyuki Irifune

PP162 PERFORMANCE ASSESSMENT OF 10 KV SIC MOSFET AND PIN DIODE IN 3L-NPC CONVERTER TOPOLOGY 2036
Renato Amaral Minamisawa

VOLUME 4

PP163 PERFORMANCE EVALUATION OF COOLSIC 2 KV SIC MOSFET DISCRETE IN 1500 V DC LINK SYSTEMS 2041
Ajith Kumar Sekar

PP164 A NEW 2.3 KV RATED SIC MOSFET MODULE WITH LOW-INDUCTANCE HIGH-POWER PACKAGE HPNC FOR 1500 VDC APPLICATIONS 2049
Junya Kawabata

PACKAGING AND INTERCONNECTION MATERIALS

PP166 MECHANISM FOR IMPROVING THE HEAT-RESISTANCE OF ADHESIVE INTERFACE IN FLEXIBLE PRINTED CIRCUITS 2053
Keita Suzuki

PP167   A SYSTEMATIC COMPARISON STUDY OF DIFFERENT BONDING
TECHNOLOGIES FOR SUBSTRATE ATTACHMENT OF POWER ELECTRONICS............................ 2060
  Lisheng Wang

PP168   STABILITY OF PRESSURE SINTERED INTERCONNECTS AS A FUNCTION OF
TEMPERATURE AND ENVIRONMENTAL CONDITIONS.. 2067
  Kentaro Yoshioka

PP169   THE EFFECT OF NANO-CU INTERCONNECTION MATERIALS ON THE
THERMOMECHANICAL PROPERTIES OF SIC DOUBLE-SIDED POWER MODULES 2074
  Suhang Wei

PP170   ALL-IN-ONE-SINTERING: DIE-ATTACH AND SUBSTRATE-ATTACH ON BARE
COPPER IN A PRESSURE ASSISTED SINTERING ONE-STEP PROCESS.. 2082
  Battist Rabay

PP171   SEQUENTIAL MANUFACTURING OF HIGHLY FUNCTIONALIZED THREE-
DIMENSIONAL CERAMIC COMPONENTS FOR POWER ELECTRONICS... 2088
  Lars Rebenklau

PP173   PARAMETRIC STUDY OF DAMAGE EVOLUTION IN SILVER SINTERED
LAYERS OF DOUBLE SIDED POWER ELECTRONICS MODULES OF ELECTRICAL
VEHICLES... 2094
  Saeed Akbari

DC-DC CONVERTER I

PP174   TRISTATE MODIFIED BOOST CONVERTER... 2104
  Johannes Gragger

PP175   COMPARATIVE EVALUATION OF THE CENTER TAPPED BOOST CONVERTER
TOPOLOGY 2112
  Bryan Radix

PP176   COMPARISON OF MULTI-LEVEL TOPOLOGIES TO REDUCE THE
COMPONENTS VOLTAGE STRESSES WHEN POWERED FROM INDUSTRIAL DC GRIDS..............2119
  Katharina Machtinger

PP177   HARD-SWITCHING HIGH-FREQUENCY GAN-BASED DC-DC CONVERTERS
WITH CONCOMITANT DATA TRANSMISSION FUNCTIONALITY 2128
  Abdelmoumin Allioua

PP178   EFFICIENT DESIGN OF HIGH-CURRENT, LOW-OUTPUT VOLTAGE DC-DC
CONVERTERS USING ARTIFICIAL INTELLIGENCE-BASED TOPOLOGY SELECTION
AND OPTIMIZATION 2138
  Thomas Harmand

HIGH POWER DC-DC CONVERTER I

PP180   A SIC BASED 60KW LLC CONVERTER WITH NOVEL TRANSFORMER DESIGN
FOR IMPROVING VOLTAGE BALANCE.. 2146
  Frank Wei

PP181 ANALYSIS OF INVERTER OPERATION MODES OF AN IGBT-BASED ZCS LLC CONVERTER FOR A 2 KW AUTOMOTIVE ON-BOARD DC-DC 2152
Daniel Urbaneck

PP182 DUAL OUTPUT HYBRID CONVERTER FOR 48 V DATA CENTERS: M-HSC.................... 2162
Simone Mazzer

PP183 3.6KW HIGH EFFICIENCY SIC-BASED HV/LV DC-DC CONVERTER FOR EVS 2167
Veera Bharath Chandra Reddy Gandluru

PP184 BIDIRECTIONAL DC-DC TOPOLOGIES COMPARISON FOR 800 V AUTOMOTIVE APPLICATIONS INTEGRATING 650 V GAN-ON-SI DEVICES.. 2175
Ilias Chorfi

PP185 ANALYSIS OF PHASE SHIELDING METHOD BASED ON ?-CR-Y THREE-PHASE INTERLEAVED LLC CONVERTER.. 2182
Jin Wen

PP186 22KW IMS-BASED BIDIRECTIONAL DC-DC CONVERTER USING SURFACE MOUNT SIC MOSFETS FOR OBCS 2185
Hamlin Wang

PP187 COMPARATIVE ANALYSIS OF DC-DC CONVERTERS FOR ELECTROLYZERS USING GEOMETRIC PROGRAMMING 2190
Tim McRae

PP188 DESIGN CONSIDERATION OF BI-DIRECTIONAL CLLLC RESONANT CONVERTER IN ENERGY STORAGE SYSTEMS 2200
Sheng-Yang Yu

SMART-GRID TECHNOLOGIES

PP189 ADAPTIVE FAST CHARGING SYSTEM WITH SECOND LIFE BATTERIES - AN OVERVIEW OF A RESEARCH PROJECT 2208
Lukas Böhning

PP190 PARALLEL OPERATION AND SYNCHRONIZATION OF MICROGRIDS BY USING THE THEVENIN THEOREM.. 2217
Marius Block

PP192 21 KA SOLID STATE DC BREAKER FOR SUPERGRID INSTITUTE'S HIGH POWER TEST FACILITY.. 2227
Christophe Conilh

PP193 DESIGN AND ANALYSIS OF A 50KW SIC-BASED ACTIVE FRONT END WITH A VERY SMALL LINE CHOKE FOR DC-GRIDS 2234
Raphael Otte

PP194 INVESTIGATION OF LOAD TRANSITIONS BETWEEN LOADED AND LOAD FREE CONDUCTOR SEGMENTS IN INDUSTRIAL CONDUCTOR SYSTEMS 2240
Jan-Niklas Koch

PP195 A METHOD TO CONTROL VOLTAGE AND POWER FLOW IN A DC GRID 2248
Peter Van Duijsen

ENERGY STORAGE SYSTEMS

PP196   CONSIDERATIONS ON A HIGH-CELL-COUNT CONVERTER-BASED BATTERY STORAGE SYSTEM WITH REDUCED COMMUNICATION EFFORT 2258
Paul Aspalter

PP197   STUDYING CONVERTORS FOR VOLTAGE EQUALIZATION IN ENERGY STORAGE SYSTEM WITH ACTIVE BMS 2268
Dimitar Arnaudov

PP198   CHALLENGES OF HIGH SIDE GATE DRIVER AND DISCONNECT MOSFET FOR BATTERY PROTECTION UNIT DURING START-UP, TURN-OFF AND OVER CURRENT EVENTS.. 2273
Niranjan Suravarapu Reddy

PP199   ELECTRIC INSULATION COORDINATION TO PREVENT ELECTRIC ARCS IN LITHIUMION BATTERIES 2278
Daniel Chatroux

PP201   BATTERY CHARGER WITH IMPEDANCE SPECTROSCOPY CAPABILITY FOR LI-ION CELLS.. 2286
Christian Branas

EMC

PP202   EFFICIENCY, VOLUME AND CO_2 EMISSIONS IMPACT IN A PFC CONVERTER WITH AN ACTIVE FILTER SOLUTION FOR OBC APPLICATION.. 2294
Kelly Ribeiro

PP203   ANALYTICAL AND EXPERIMENTAL VALIDATION COMMON MODE FEEDBACK LOOP FOR A THREE-PHASE_LEVEL VIENNA RECTIFIER... 2303
Daniel San Laureano Igartuburu

PP204   ROBUSTNESS OF FREQUENCY-DOMAIN TERMINAL MODELING OF ELECTROMAGNETIC INTERFERENCES IN STATIC CONVERTERS 2309
Mehyeddine Singer

PP205   STUDY OF EMI BEHAVIOR OF A 2-LEVEL GAN-INVERTER – SIMULATION AND MEASUREMENT.. 2316
Benedikt Kohlhepp

COMMON MODE CURRENTS IN RESONANT CIRCUITS GENERATED WITH A DELTA-SIGMA MODULATED VOLTAGE SOURCE INVERTER... 2326
Tobias Haas

PP206   ANALYSIS OF COMMON-MODE NOISE GENERATED DUE TO FAST-SWITCHING GAN DEVICES IN TOTEM-POLE PFCS.. 2334
Serkan Dusmez

PP207   CONDUCTED EMI FROM GAN-BASED 48V TO 12V DC-DC-CONVERTERS FOR AUTOMOTIVE APPLICATIONS.. 2342
Erik Kampert

ADVANCED DESIGN

PP208 APPLIED DESIGN AUTOMATION FOR FINDING FEASIBLE DESIGNS FOR HIGH-FREQUENCY PLANAR TRANSFORMERS 2350
Rando Raßmann

PP209 FREQUENCY DEPENDENT AREA PRODUCT METHOD 2359
Alfonso Martínez

HIGH RESOLUTION MIXED-SIGNAL PULSE WIDTH MODULATOR FOR HIGH-FREQUENCY DC-DC CONVERTERS 2364
Tim McRae

PP210 DESIGNING A CONTROL LIBRARY FOR GRID-FOLLOWING AND GRID-FORMING POWER INVERTERS 2370
Lars Lindner

PP211 INTELLIGENT OPTIMISATION OF A WIND TURBINE DIGITAL TWIN MODEL 2377
René Reimann

PP212 THERMAL TRANSIENT DIGITAL TWIN MODELLING FOR POWER CONVERTERS 2386
Xianghao Mo

PP213 A DIGITAL TWIN APPROACH TOWARD LIFETIME ANALYSIS AND PREDICTIVE MAINTENANCE OF POWER SEMICONDUCTORS FOR RAILWAY APPLICATION 2394
Emmanuel Batista

INDUCTORS

PP214 SATURABLE FERRITE CORE INDUCTORS IN LCL FILTERS OF THREE-PHASE VOLTAGE SOURCE INVERTERS 2400
Marius Kaufmann-Bühler

PP215 2D COPPER LOSS ANALYTICAL MODEL FOR PLANAR INDUCTOR COMBINING HIGH AND LOW PERMEABILITY MATERIALS 2408
Idriss Nachete

PP216 CNC-MANUFACTURED POWER INDUCTORS WITH EXCELLENT BANDWIDTH FOR MULTI-MEGAWATT CONVERTERS 2416
Thomas Kreppel

PP217 ANALYTICAL EVALUATION OF DIFFERENTIAL MODEL DC EMI FILTER INDUCTORS USING MATERIAL SATURATION COEFFICIENT 2425
Lukas Mueller

PP218 DESIGN AND PERFORMANCE EVALUATION OF AIR CORE INDUCTORS FOR VERY HIGH FREQUENCY POWER CONVERSION 2431
Florentin Salomez

PP220 IMPROVING MULTI-PHASE FERRITE MAGNETICS BY COUPLING FOR MV AND UPS CONVERTERS 2438
Michael Schmidhuber

E-MOBILITY CHARGING

PP221   22-KW BIDIRECTIONAL SINGLE-STAGE DIRECT-AC-AC POWER CONVERSION ON-BOARD CHARGER WITH HIGH-POWER-DENSITY IMPLEMENTATION................................... 2448
Oscar Lucia

PP222   BENCHMARKING DC FAST CHARGERS: A COMPARATIVE ANALYSIS OF POWER CONVERTER STRUCTURES FOR WIDE VOLTAGE RANGE 2453
Sadik Cinik

PP223   PERFORMANCE OPTIMIZATION OF SINGLE-PHASE ON-BOARD CHARGERS WITH RIPPLE PORT 2461
Davide Gottardo

PP224   A REDUCED-SENSOR MODULAR DUAL ACTIVE BRIDGE-BASED BATTERY CHARGING SYSTEM FOR ELECTRIC VEHICLES USING AN IMPROVED LINEAR EXTENDED STATE OBSERVER... 2469
Armel Asongu Nkembi

PP225   BIDIRECTIONAL NON-ISOLATED THREE-PHASE ONBOARD CHARGER WITH A LOW-VOLTAGE LOWER-PHASE OPERATION MODE... 2478
Steffen Frei

PP226   CONTROL OF A THREE-PHASE INDUCTIVE POWER TRANSFER SYSTEM BASED ON DD²Q COIL TOPOLOGY 2488
Nikola Mirkovic

PP227   COMPARISON OF TWO BIDIRECTIONAL 11KW 400V CLLC AND CLLLC RESONANT CONVERTERS FOR EV APPLICATIONS 2494
Hasan Mousavi Somarin

PP228   DYNAMIC WIRELESS CHARGING SYSTEM DESIGN FOR EXTRA-URBAN AREAS BASED ON RESONANT INDUCTIVE POWER TRANSFER 2503
Irene Maria Torres Alfonso

PP229   BIDIRECTIONAL ISOLATED 400-12V DC-DC CONVERTER WITH IMPROVED POWER DENSITY AND FULL-RANGE OPERAION FOR EV APPLICATIONS 2513
Oscar Lucia

HIGH POWER DC-DC CONVERTER II

PP230   GAIN OPTIMIZATION CONTROL METHOD FOR CLLLC RESONANT CONVERTERS UNDER PHASE SHIFT MODE 2518
Sean Yu

PP231   ANALYSIS OF COMMON AND SPLIT DC-BUS INTERLEAVED H-BRIDGE CONVERTERS FOR HIGH-CURRENT LOW-RIPPLE APPLICATIONS.................................... 2524
Bhavana Gudala

PP232   OPTIMAL FREQUENCY OPERATING POINTS FOR HYBRID SWITCHED CAPACITOR CONVERTERS AND LOSSLESS CURRENT SENSE METHOD 2532
Simone Mazzer

PP233 DESIGN AND TESTING OF A 250 KW 50 KHZ SIC-BASED HALF-BRIDGE-SERIES-RESONANT-CONVERTER 2538
Daniel Haake

PP234 30KW - 97% EFFICIENCY ISOLATED DC-DC CONVERTER WITH LARGE INPUT VOLTAGE RANGE BASED ON A BOOST DAB ASSOCIATION 2547
Jean-Jacques Huselstein

PP235 ANALYSIS OF A FULL-BRIDGE PUSH-PULL FORWARD DUAL ACTIVE BRIDGE DC-DC CONVERTER 2557
Gean Sousa

DC-DC CONVERTER II

PP236 SYMMETRICAL OPERATION OF FOUR CHANNEL RESONANT BOOST DC-DC CONVERTERS IN CONTINUOUS CONDUCTION MODE 2566
Kristóf Bándy

PP237 IMPACT OF MAGNETICS TOLERANCE ON THE POWER SHARING OF PARALLEL DUAL-OUTPUT PHASE-SHIFT FULL-BRIDGE CONVERTERS 2576
Riccardo Mandrioli

PP238 A BALANCING CONVERTER WITH SERIES CONNECTED MOSFETS FOR +/-700V BIPOLAR DC GRIDS 2583
Sachin Yadav

PP239 OPTIMIZATION AND DESIGN OF LOW-VOLTAGE AND HIGH-CURRENT POINT-OF-LOAD CONVERTER UNDER 48V BUS ARCHITECTURE 2591
Jiajia Guan

PP240 INTERLEAVED BOOST CONVERTER EFFICIENCY AND POWER DENSITY MODEL FOR ACTIVE AND PASSIVE COMPONENT DESIGN 2596
Damien Lemaitre

NOVEL AND ADVANCED SEMICONDUCTOR DEVICES

PP241 EVALUATION OF A HYBRID POWER SWITCH BASED ON TRENCH CLUSTERED IGBT AND SIC MOSFET 2606
Alireza Sheikhan

PP242 CONTRIBUTIONS FOR BUILDING BLOCKS FOR NORMALLY-OFF 650V GAN-ON-SI POWER INTEGRATED CIRCUITS 2612
Thanh Hai Phung

PP243 NEW BIDIRECTIONAL ASYMMETRIC HIGH VOLTAGE TVS (TRANSIENT VOLTAGE SUPPRESSOR) DIODE 2620
Boris Rosensaft

PP244 ISO247: HIGH PERFORMANCE CERAMIC BASED ADVANCED ISOLATED DISCRETE PACKAGE TO FULLY EXPLOIT THE ADVANTAGES OF SIC MOSFET 2627
Sachin Shridhar Paradkar

PP245 IMPACT OF CURRENT RIPPLE REDUCTION USING HIGH SWITCHING FREQUENCIES ON PMSM EFFICIENCY 2632
Jannik Fuchs-Gade

PP246   MAXIMIZING COST-EFFICIENCY IN ELECTRIC DRIVETRAINS: A SIC/SI
FUSION SWITCH APPROACH 2638
    Matthias Ippisch

ADVANCED CONTROL

PP247   CONCISE AND RELIABLE SIC MOSFET DRIVER CIRCUITS 2646
    Zhong Ye

PP248   ARTIFICIAL INTELLIGENCE ENHANCED RESOLVER SYSTEM FOR
AUTOMOTIVE TRACTION INVERTER APPLICATIONS BASED ON AURIX TC4X 2651
    David Zipperstein

PP250   MULTIFUNCTIONAL GRID MANAGER TOPOLOGY WITH CONFIGURABLE
OUTPUT 2657
    Peter Van Duijsen

PP252   CO2 FOOTPRINT OF MEDIUM VOLTAGE DC SOLID STATE
TRANSFORMER   2663
    Adriana Campos

SIC MOSFET

PP253   THERMO-ELECTRICAL ANALYSIS AND PERFORMANCE: A COMPARATIVE
STUDY BETWEEN MODULAR AND DISCRETE APPROACHES 2673
    Stefano Orlando

PP254   IMPACT OF PARAMETER SPREAD IN PARALLEL-OPERATED SIC MOSFETS
FOR HARD-SWITCHING CONVERSION 2680
    Andrea Piccioni

PP255   ASSESSMENT OF THE RDS,ON OF SIC MOSFET DIES THROUGH KELVIN WIRE
CONNECTION 2686
    Philipp Rehlaender

PP256   CHALLENGES IN SCALING SIC SINGLE-CHIP MEASUREMENTS TO
CORRESPONDING POWER MODULES 2693
    Hao Wang

PP257   SWITCHING PERFORMANCE EVALUATION OF HIGH-POWER 1.7 KV SIC
MOSFET MODULES USING A COMMON BUSBAR DESIGN 2700
    Sebastian Neira

PP258   CHARACTERIZING THE SWITCHING BEHAVIOR OF A 1.2 KV MIXED SIC JFET
AND MOSFET HALF BRIDGE 2708
    Tim Ringelmann

VOLUME 5

WBG HIGH FREQUENCY APPLICATION

PP259   PERFORMANCE EVALUATION OF THE PACKAGING OF SIC DIODES IN A 6.78
MHZ WIRELESS POWER TRANSFER SYSTEM 2718
    Ioannis Nikiforidis

PP260   VOLTAGE WAVEFORM GENERATION FOR SAWYER-TOWER COSS LOSS
MEASUREMENTS USING A HYBRID POWER CONVERTER 2724
    Malachi Hornbuckle

PP261   EVALUATION OF SIC DEVICES FOR OVER 500KHZ APPLICATION BASED ON
BUCK CIRCUIT 2730
    Minli Jia

PP262   LINEARIZATION OF DRAIN-SOURCE CAPACITANCES FOR ANTISERIAL
CONFIGURATED SIC MOSFETS IN HIGH FREQUENCY SOLID STATE SWITCHES 2737
    Lars Dresel

SIC RUGGEDNESS

PP263   EFFECTS OF NON-KILLER DEFECTS ON SIC MOSFET SHORT-CIRCUIT
RUGGEDNESS AND RELIABILITY 2745
    Sara Kuzmanoska

PP264   DYNAMIC REVERSE BIAS TEST: ELECTRO-THERMAL CHARACTERIZATION
OF SIC MOSFETS 2751
    Giuseppe Mauromicale

PP266   RADIATION HARDNESS OF SIC BASED INVERTERS BASED ON AN EV
MISSION PROFILE 2758
    Hadiuzzaman Syed

PP267   RAPID SHORT CIRCUIT PROTECTION USING DIDT DETECTION FOR SIC
POWER MODULES 2764
    Koki Samura

PP268   COMPARISON OF DYNAMIC GATE STRESS TEST RESULTS OF SIC MOSFETS 2769
    Mathias Gebhardt

PP279   EXTENDING SIC MOSFET SHORT-CIRCUIT WITHSTANDING TIME BY TWO-
LEVEL TURN-OFF GATE DRIVING 2778
    Kwokwai Ma

PP270   EXPERIMENTAL INVESTIGATIONS ON PARASITIC TURN-ON OF 1.2KV SIC
MOSFET DISCRETE DEVICES 2786
    Thanh-Toan Pham

PP271   BEHAVIOR MODELLING THE SHORT CIRCUIT CHARACTERISTICS OF SIC
MOSFETS USING COMPACT MODELS 2791
    Qing Sun

THERMAL CHARACTERIZATION

PP273   THERMAL ANALYSIS AND MODELLING OF CHARGING STATIONS FOR
ELECTRIC VEHICLES 2796
    Ruben Kopischke

PP274   JUNCTION TEMPERATURE MEASUREMENT OF A 3.3 KV SILICON CARBIDE
MOSFET POWER MODULE 2803
    Michael Gleissner

PP275 INNOVATIVE 3D POWER MODULE DEFAULTS DETECTION VIA THERMAL IMPEDANCE ANALYSIS AND SIMULATIONS 2811
Louis Alauzet

PP276 THERMAL CHARACTERIZATION OF AN AIR-COOLED PEBB BASED ON SIC MOSFET POWER MODULES 2819
Alexandre Marie

PP277 THERMAL BEHAVIOUR OF SIC MOSFET WITH PLANAR PACKAGING TECHNOLOGY 2826
Yijun Ye

RELIABILITY AND AVAILABILITY

PP279 IMPLEMENTING MODULE HEALTH MONITORING IN EV TRACTION INVERTERS 2831
Karol Rendek

PP280 RELIABILITY TESTS OF COPPER THICK-FILM SUBSTRATES FOR POWER ELECTRONIC APPLICATIONS... 2838
Henry Barth

PP281 POWER MODULE SOLUTIONS WITH IMPROVED RELIABILITY FOR ELEVATOR DRIVE APPLICATIONS 2843
Tiago Jappe

PP282 FAIL-OPERATIONAL LLC TOPOLOGIES WITH FAULT-TOLERANCE INTEGRATED REDUNDANT CAPABILITIES 2850
Aswathy M. Prince

PP283 THERMAL AND RELIABILITY OPTIMIZATION OF CLIPS IN SIC MOSFET POWER MODULES.. 2860
Zexiang Zheng

PP284 CONDITION MONITORING OF A GAN FULL-BRIDGE BY MEANS OF FORWARD VOLTAGE IN CONTINUOUS OPERATION.. 2866
Michael Vogt

PP285 A SIMPLE AND LOW COST OVERCURRENT PROTECTION SYSTEM BASED ON COMMERCIAL SHUNT FOR WIDE-BANDGAP DEVICES.. 2874
Emanuele Martano

PP286 SVM-BASED FAULT-TOLERANT CONTROL FOR A CASCADED H-BRIDGE MULTILEVEL CONVERTER UNDER MULTIPLE OPEN-CIRCUIT SWITCH FAULTS....................... 2880
Dong Xie

PP287 REVOLUTIONIZING MOBILITY: THE SECOND LIFE OF ONBOARD CHARGING SYSTEMS IN COMMERCIAL VEHICLES 2886
Ajay Krishna Voppu Muralikrishna

LOW VOLTAGE SWITCHES

PP288 A BEHAVIORAL TRANSIENT MODEL FOR IGBT DEVICE WITH ANTI PARALLEL FREEWHEELING DIODE.. 2893
Shiwu Zhu

PP289 PARAMETER EXTRACTION FOR AN ANN-ASSISTED IGBT MODEL IN
TRANSIENT SIMULATIONS 2901
  Huaiyuan Zhang

PP290 FABRICATION OF 600V RC-IGBT USING 300MM WAFER 2909
  Masaki Ueno

PP291 NEXT LEVEL OF POWER MODULE SOLUTION FOR PV C&I STRING INVERTER
WITH 1200V H7 TECHNOLOGY IN EASY3B PACKAGE 2914
  Tilo Poller

PP292 ANALYSIS OF MOSFET SWITCHING LOSSES IN RESONANT CONVERTERS
USING ELECTRICAL AND THERMAL MEASUREMENTS AND LOSS TRENDS WITH
MOSFET SIZE VARIATION 2921
  Alfio Scuto

PP293 OPTIMOS 6 135V FOR HIGH POWER MOTOR DRIVES 2930
  Kunal Jha

PP294 AUTO POWER-SOI: SHAPING THE FUTURE OF BATTERY MONITORING
TECHNOLOGY 2937
  Alex Lim

LIFETIME MODELLING AND CONDITION MONITORING

PP295 UNDERSTANDING THE IMPACT OF IEC60747-17 ON CAPACITIVE AND
MAGNETIC COUPLERS 2942
  Shu Ee Ong

PP296 PARIS LAW APPLIED TO WIRE BONDS DEGRADATION USING CRACK
GROWTH MEASUREMENT 2948
  Merouane Ouhab

PP297 CONDITION MONITORING TECHNIQUE OF POWER ELECTRONIC MODULES
VIA SQUARE-WAVE GATE SIGNAL EXCITATION 2956
  Isabel Austrup

PP298 STATISTICS-BASED LIFETIME SIMULATION ENVIRONMENT FOR POWER
MODULES INCORPORATING DEGRADATION MODELS 2963
  Karthik Debbadi

PP299 POWER CYCLING RESULTS FOR RELIABILITY STUDIES OF SIC-INVERTERS 2972
  Robert Keilmann

PP300 GAN CASCODE IN HIGH SPEED DRIVEN AIR COMPRESSORS FOR
AUTOMOTIVE FUEL CELLS 2981
  Florian Lippold

PP301 PROGNOSTIC ANALYSIS OF IGBT HEALTH: REAL-TIME ON-STATE VOLTAGE
PREDICTION THROUGH MACHINE LEARNING 2986
  Tanya Thekemuriyil

PP302 ROBUSTNESS ANALYSIS OF TEMPERATURE-SENSITIVE ELECTRICAL
PARAMETERS OF IGBTS 2995
  Laurids Schmitz

PP303   OBSERVATION OF THERMAL-RESISTANCE INCREASE OF DEGRADED IGBT
MODULES BY VCE (SAT) MEASUREMENT IN A CHOPPER CIRCUIT 3002
Kazunori Hasegawa

PULSE WITH MODULATION METHODS

PP304   MODULATION TECHNIQUE FOR REDUCED AC CONTENT OF THE DC LINK
CURRENT IN THREE-PHASE TWO-LEVEL INVERTERS 3007
Steffen Frei

PP305   COMMON MODE CURRENTS IN RESONANT CIRCUITS GENERATED WITH A
DELTA-SIGMA MODULATED VOLTAGE SOURCE INVERTER 3017
Tobias Haas

PP306   EVALUATION OF NEW MODULATION SCHEME FOR 3L-ANPC USING BOTH
CURRENT PATHS IN ZERO STATE 3020
Felix Eichler

PP307   AN INNOVATIVE SYNCHRONOUS RECTIFICATION METHOD FOR 11KW CLLC
CONVERTER 3029
Sanbao Shi

PP308   INTERLEAVED ASYNCHRONOUS DELTA-SIGMA MODULATION CONCEPT FOR
DYNAMIC POWER CONVERTERS 3034
Philipp Czerwenka

PP309   HIGH RESOLUTION MIXED-SIGNAL PULSE WIDTH MODULATOR FOR HIGH-
FREQUENCY DC-DC CONVERTERS 3042
Tim McRae

PP310   IMPLEMENTATION AND CONTROL OF OPTIMIZED PULSE PATTERNS FOR
SALIENT PERMANENT MAGNET SYNCHRONOUS MACHINES IN ELECTRIC VEHICLES........... 3045
Maximilian Hepp

PP311   A 3-LEG INTERLEAVED TP PFC WITH A 90° PHASE-SHIFTED ASYMMETRIC
LEG FOR REDUCED MAGNETICS.. 3060
Serkan Dusmez

PP312   FAULT-TOLERANT OPERATION ANALYSIS OF A FIVE-PHASE THREE-LEVEL
TNPC INVERTER FOR ELECTRIC AIRCRAFT PROPULSION SYSTEMS 3067
Chanuch Chaisakdanugull

AC-DC AND DC-AC CONVERTER

PP313   CCM TOTEM-POLE PFC FOR ULTRA-HIGH POWER DENSITY USB-PD
CHARGERS.. 3077
Manuel Escudero Rodruigez

PP314   COMPARISON OF HYBRID SI/SIC AND SIC TWO-LEVEL AND THREE-LEVEL
CONVERTERS FOR LOW-VOLTAGE LOW-POWER APPLICATIONS...................................... 3086
Tim Augustin

PP315   ANALYSIS OF ANALOGUE CURRENT AND FLUX BALANCING FOR THE DUAL-
ACTIVE-BRIDGE CONVERTER.. 3096
Christophe Basso

PP316 DESIGN AND OPTIMIZATION OF A SINGLE-STAGE PHOTOVOLTAIC
MICROINVERTER WITH INTEGRATED MAGNETICS 3103
Jin Wen

PP317 EXPERIMENTAL INVESTIGATION OF CLASS F INVERTER UNDER VARIOUS
LOAD CONDITIONS.. 3110
Baptiste Daire

PP318 ANALYSIS, MODELING, DESIGN, AND LIMITATIONS OF CURRENT INJECTION
BASED UPF RECTIFIER WITH SMALL DC-LINK CAPACITOR... 3118
Ramkrishan Maheshwari

PP319 HIGH-EFFICIENT ISOLATED AC-DC CONVERTER WITH CIRCULATING
CURRENT REDUCTION FOR AC ADAPTERS 3125
Hiroki Watanabe

PP320 A PHASE-LOCKED LOOP (PLL) BASED STRATEGY FOR ACCURATE BLANKING
TIMES IN BRIDGELESS TOTEM-POLE PFCS 3130
Sandu Tigira Tigira

PP321 CIRCULATING CURRENTS IN COUPLED MULTI-TERMINAL HYBRID AC-DC
GRIDS.. 3136
Fabian Herzog

ADVANCED CONVERTER TOPOLOGIES

PP322 COMPARISON OF 4500V STATE-OF-THE-ART XHP3 IGBT AND CONVENTIONAL
IHV IGBT FOR 3300V 3-LEVEL ANPC MEDIUM VOLTAGE DRIVES 3142
Martin Knecht

PP323 GENERALIZED SWITCHING SEQUENCE FOR VOLTAGE BALANCING IN A
FLYING CAPACITOR DC-DC CONVERTER WITH QUASI-2-LEVEL MODULATION 3150
Jose Andres Aguilar Croston

PP324 OPTIMIZATION-BASED SIZING OF A MODULAR MULTILEVEL CONVERTER
BASED ON 650 V GAN MODULES FOR NEW LVDC/MVDC GRIDS....................................... 3160
Gregoire Le Goff

PP325 A NOVEL THREE-PHASE LOW-SWITCH-COUNT AC-DC GRID CONVERTER
TOPOLOGY WITH GALVANIC ISOLATION.. 3169
Liska Steenbock

PP326 SINGLE-STAGE LED DRIVER BASED ON COUPLED INDUCTOR POWER
FACTOR CORRECTION AND LLC CONVERTER.. 3175
Alireza Ramezan Ghanbari

PP327 A INVERSE COUPLED DC-DC BOOST INDUCTOR WITH 2-KV SIC MOSFET
MODULE FOR 1500V SOLAR INVERTER MPPT.. 3181
Yusi Liu

PP328 ENVIRONMENTAL IMPACT OF MODULAR POWER ELECTRONICS SYSTEMS
CONSIDERING DIAGNOSTIC-DRIVEN UNIT REPLACEMENT 3187
Briac Baudais

POWER ELECTRONICS FOR RAILWAY APPLICATIONS

PP329　SWITCHING PERFORMANCE COMPARISON OF 3.3 KV SIC MOSFET AND SI IGBT POWER MODULES FOR RAILWAY TRACTION SYSTEMS 3197
Yue Zhao

PP330　COMPARISON OF THREE-LEVEL INVERTER TOPOLOGIES FOR MVDC REVERSIBLE RAILWAY SUBSTATIONS 3206
Luc Bimmel

PP331　CONTROL OF BIDIRECTIONAL POWER FLOW IN RAILWAY CATENARY OVERHEAD LINES.. 3213
Peter Van Duijsen

PP332　A RAIL TRACTION CONVERTER PLATFORM BASED ON POWER MODULE IMPLEMENTATIONS WITH 450 A, 600 A AND 800 A 3.3 KV IGBT MODULES.................................. 3221
Ekrem R. Gunes

PP333　COMPARISON OF SELECTED MEGAWATT-LEVEL TRACTION CONVERTER POWER MODULE IMPLEMENTATIONS IN TERMS OF COMMUTATION INDUCTANCE AND PRACTICALITY.. 3229
Abdulkerim Ugur

CURRENT RELATED TESTING

PP334　PITFALLS AND THEIR AVOIDABILITY IN THE DOUBLE-PULSE TEST 3237
Nikolas Förster

PP335　MODELING AND SIMULATION OF FLUXGATE BASED CURRENT SENSOR 3247
Yunus Çay

PP336　SIGMA-DELTA BASED CURRENT ACQUISITION WITH REDUCED SETTLING TIME 3256
Joschka Randerath

PP337　CHARACTERISATION OF WIDE-BANDGAP SEMICONDUCTORS IN DOUBLE PULSE TESTING USING OPTICALLY ISOLATED PROBES.. 3264
Lennart Hoffmann

PP338　NON-INVASIVE BATTERY CONDITION TESTING USING ELECTRICAL SIGNALS AND OSCILLOSCOPES.. 3269
Srikrishna N. H

PP339　INSTRUMENTATION REQUIREMENTS FOR FAST 130 V/NS SWITCHING OF 1700 V, 35 M? SIC MOSFETS 3276
Matthew Appleby

POWER ELECTRONICS FOR AEROSPACE APPLICATIONS

PP340　CONCEPTUALIZATION AND EXPERIMENTAL ASSESSMENT OF DESIGN ASPECTS FOR 3-LEVEL ANPC INVERTERS 3286
Lukas Radomsky

PP341   DESIGN OF A HIGH POWER DENSITY INVERTER AND FOC IMPLEMENTATION FOR UAVS 3296
Matthias Neuner

PP342   HIGHLY-INTEGRATED, FLEXIBLE POWER SOLUTION FOR AEROSPACE 5KVA – 20 KVA MOTOR DRIVE APPLICATIONS 3305
Alain Calmels

PP343   DATABASE-SUPPORTED PRELIMINARY DESIGN, SIMULATION AND EVALUATION OF POWER CONVERTERS IN ELECTRIC AIRCRAFT PROPULSION SYSTEMS 3315
Jeff Kugener

PP344   DESIGN AND ANALYSIS OF GATE-DRIVER FOR SIC-BASED INVERTER FOR MEGAWATT SCALE ALL ELECTRIC AIRCRAFT 3318
Jeff Kugener

MEASUREMENT TECHNIQUES AND METHODS

PP345   ADDRESSING TESTING CHALLENGES FOR POWER MODULES AND THREE-LEVEL INVERTERS 3328
Oleg Fotteler

PP346   CHARACTERIZATION OF THE BONDING QUALITY OF SILVER SINTERED COMPOUNDS BY MEANS OF LASER-INDUCED BREAKDOWN SPECTROSCOPY 3334
Yannick Bockholt

PP347   INVERTER-INTEGRATED MEASUREMENT OF THE FREQUENCY-DEPENDENT WINDING IMPEDANCE OF ELECTRIC MACHINES 3340
Christian Mühlfeld

PP348   COMPENSATION TECHNIQUES FOR BANDWIDTH-DISTORTED MEASUREMENTS OF FAST TRANSIENTS IN DOUBLE PULSE TESTS 3347
Christian Lottis

PP349   AN AERODYNAMIC LOAD MEASUREMENT TECHNIQUE FOR AUTONOMOUS AERIAL VEHICLES 3353
Mehmet Oguz Girgin

COMPENSATION TECHNIQUES FOR BANDWIDTH-DISTORTED MEASUREMENTS OF FAST TRANSIENTS IN DOUBLE PULSE TESTS 3358
Christian Lottis

PP350   A HIGH-BANDWIDTH MULTILEVEL COUNTER CIRCUIT FOR BEARING CURRENT EVALUATION 3364
Felix Schulte

TRANSFORMERS

PP351   CORE LOSS MODEL FOR CONSIDERING ANISOTROPY AND TEMPERATURE EFFECTS ON ELECTRICAL STEEL UNDER POWER ELECTRONIC CONDITIONS 3371
Michael Owzareck

PP353   CIRCULAR ECONOMY ORIENTED AND RECONFIGURABLE PLANAR
TRANSFORMER DESIGN FOR ISOLATED DC-DC CONVERTERS 3380
Fabian Groon

PP354   CONTROLLABLE MAGNETICS: VARIABLE TRANSFORMERS AND VARIABLE
INDUCTORS, THEORY – PRODUCTION – APPLICATION.. 3390
Florian Fenske

PP355   A THREE-PHASE INTERLEAVED LLC INTEGRATED TRANSFORMER USING
PCB WINDINGS FOR FUEL CELL DCDC CONVERTERS 3395
Jiajia Guan

PP356   TESTING THE PRIMARY-SECONDARY COIL COUPLING OF HIGH-FREQUENCY
TRANSFORMER IMPLEMENTED ON ETD AND TOROIDAL CORES 3400
Alexis Gioda

Author Index

PCIM Europe 2024, 11– 13 June 2024, Nuremberg

DOI: 10.30420/566262286

Performance Evaluation of CoolSiC™ 2 kV SiC MOSFET Discretes in 1500 V DC Link Systems

Syeda Qurat ul ain Akbar[1], Ajith Kumar Sekar[1], Jorge Cerezo[1]

[1] Infineon Technologies Austria AG, Austria

Corresponding author: Syeda Qurat ul ain Akbar, syedaquratulain.akbar@infineon.com
Speaker: Ajith Kumar Sekar, ajithkumar.sekar@infineon.com

Abstract

Higher DC link voltage, particularly 1500 VDC, is becoming popular for use in industrial applications such as energy storage systems and photovoltaic systems. Higher DC link voltage enables higher power levels while simultaneously reducing system losses. However, to achieve higher DC link voltage levels with the existing device technologies, multi-level converter topologies are required. This makes the systems complex. Infineon's new 2 kV discrete CoolSiC™ MOSFET presents an opportunity for the development of more efficient and simplified designs. This paper evaluates the performance of a simplified topology with the 2 kV CoolSiC™ MOSFET, comparing it with an alternative system employing 1200 V SiC devices to achieve a 1500 V DC link voltage. The simulations demonstrated that integrating the 2 kV CoolSiC™ MOSFET leads to improved performance of a simplified system with reduced component count and light weight design. The measurement data also validated the simulation outcomes, confirming the effectiveness of the 2 kV CoolSiC™ MOSFET.

1 Introduction

To increase power levels, photovoltaic systems have been transitioning towards higher system voltage, with 1500 V_{DC} becoming increasingly popular. This shift aims at reducing both power loss and system costs to make renewable energy more affordable [1-2].

For designing a solar inverter, 1500 V at DC link offers two options. The first option involves using a 3-level booster stage for DC-DC maximum power point tracking (MPPT) and a 3-level topology, e.g., active neutral-point-clamped (ANPC), for the DC-AC stage. Both stages can use 1200 V SiC devices to ensure safe and reliable system design. However, this approach is complex, and has a higher component count. The second option involves using a simplified 2-level topology with high voltage class devices. This approach can potentially be more efficient depending on the performance of the semiconductor devices used.

Discrete devices are generally chosen by designers for design flexibility, to optimize system costs, and to lower the overall cost of ownership. However, the commonly available discrete semiconductor devices of the highest voltage class, so far, are the 1700 V devices. Though using 1700 V

class MOSFETs in 1500 V solar inverter systems with a simplified 2-level topology may seem like a viable option, it is essential to consider the impact of cosmic radiation-induced failures [3]. Semiconductor devices are vulnerable to cosmic radiations. Their chance of failure increases drastically at blocking voltages that exceed 80% of their rated voltage [4]. Therefore, using 1700 V MOSFETs in 1500 V solar inverter systems with 2-level topology can lead to a significantly high failure rate.

These design challenges and reliability concerns can be resolved by Infineon's latest and the first in the market – the 2 kV CoolSiC™ MOSFET discrete.

The performance of a solar inverter utilizing the 2 kV CoolSiC™ MOSFET and diode was evaluated and compared with that of inverters utilizing 1200 V CoolSiC™ devices. System-level simulation results showed that a 2-level booster stage with a 2 kV CoolSiC™ had 20% lower losses than a 3-level booster stage implemented using 1200 V CoolSiC™ devices. Similarly, the 2-level DC-AC stage with 2 kV CoolSiC™ had 15% lower power loss than the 3-level ANPC stage implemented with 1200 V CoolSiC™ devices.

Furthermore, measurement results of DC-DC converters utilizing 2 kV and 1200 V CoolSiC™ devices confirmed the performance improvement achieved by the 2 kV device.

2 The 2 kV CoolSiC™ MOSFET

The latest 2 kV CoolSiC™ MOSFET is based on Infineon's first generation SiC MOSFET technology that utilizes the vertical trench cell structure. Across all voltage classes (650 V, 750 V, 1200 V and 2 kV), the M1H base technology design effectively combines low static and dynamic losses with exceptionally high gate oxide reliability, offering an optimal balance between performance and robustness [5-7]. This MOSFET structure also inherently exhibits a favorable capacitance ratio, characterized by a small Miller capacitance (C_{GD}) and a relatively large gate-source capacitance (C_{GS}). This characteristic allows for effective switching control with low dynamic loss [8], which is particularly essential for suppressing undesirable parasitic turn-on (PTO). In addition, the 2 kV CoolSiC™ offers humidity robustness over its entire operating lifetime. This has been validated through the 1000 hour HV-H[3]TRB reliability test [9].

2.1 Enhanced package design

The new 2 kV CoolSiC™ comes in a new discrete TO-247 PLUS-4-HCC package. This package provides enhanced creepage and clearance distance mandated by the IEC to meet the isolation requirements for higher switching frequencies and higher peak voltages to prevent any flashover or corona effects. The design of this new package effectively incorporates a pin-to-pin creepage and clearance of 14 mm and 5.4 mm respectively (see Fig. 1), ensuring robust high voltage insulation and reliable operation in high voltage and high frequency applications.

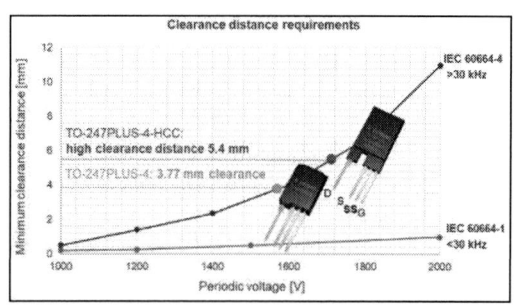

Fig. 1 The minimum clearance requirement for high frequency and high peak voltage as mandated by the IEC standard

The TO-247 PLUS-4-HCC package integrates an advanced chip interconnection technology, referred to as .XT, which utilizes the diffusion soldering method. Compared to traditional soft soldering methods, diffusion soldering considerably improves the junction-to-case thermal resistance of the device [10]. This optimizes system performance and improves the power dissipation capacity.

MOSFET	
R_{DSon} [mΩ]	TO-247PLUS-4-HCC
12	MYH200R012M1H
24	IMYH200R024M1H
50	IMYH200R050M1H
75	IMYH200R075M1H
100	IMYH200R075M1H

Table 1 Portfolio of 2 kV CoolSiC™ MOSFETs

The product portfolio comprises 2 kV CoolSiC™ MOSFETs and 2 kV CoolSiC™ Schottky diodes. The MOSFET and product portfolio is listed in Table 1. The diodes are offered in both TO-247 PLUS-4-HCC and TO-247-2-pin packages, as listed in Table 2.

2.2 Static conduction losses

The on-resistance of the device, which has a positive temperature co-efficient, dictates conduction losses during both forward (first quadrant operation) and reverse channel conduction (third quadrant operation or synchronous rectification). Figures 2 and 3, show the on-state voltage drop of IMYH200R024M1H for a gate-source ON voltage ($V_{GS(on)}$) of 15 V to 18 V and a gate-source OFF voltage $V_{GS(off)}$ of -2 V to 0 V at 25°C and 175°C, respectively.

Diode		
I_F [A]	TO-247PLUS-4-HCC	TO-247-2 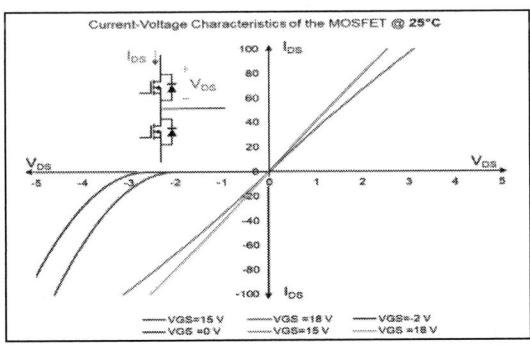
80	IDYH80G200C5	IDWD80G200C5
50	IDYH50G200C5	IDWD50G200C5
40	IDYH40G200C5	IDWD40G200C5
25	IDYH25G200C5	IDWD25G200C5
10	IDYH10G200C5	IDWD10G200C5

Table 2 Portfolio of 2 kV CoolSiC™ Schottky diodes

The curves for $V_{GS(off)}$ = -2 V represent conduction through the body diode. At $V_{GS(off)}$ = 0 V, the source-to-drain (V_{SD}) voltage drop decreases due to some contribution of the channel to the current. However, very low V_{SD} and linear characteristics are observed as soon as the channel is turned on by applying $V_{GS(on)}$. Consequently, the corresponding on-state voltage drop in the third quadrant significantly decreases. Therefore, to maintain low static losses in reverse conduction mode, synchronous rectification is strongly recommended. This can be done by activating the channel after an appropriate dead time.

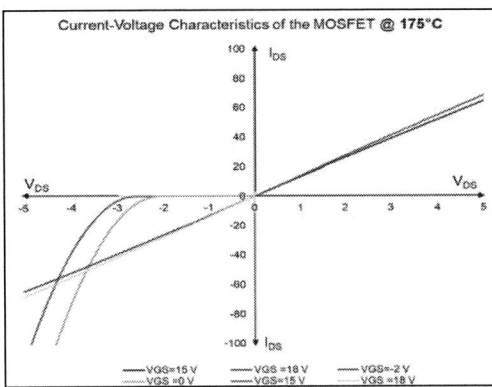

Fig. 3 Voltage drop across the channel during forward and reverse conduction for $V_{GS(on)}$ = 15 V and 18 V, conduction through the body diode for $V_{GS(off)}$ = -2 V and 0 V at 175°C

2.3 Driving voltage range

For CoolSiC™ MOSFETs, proper selection of gate source voltage is important as it significantly impacts device performance. A $V_{GS(on)}$ of 18 V is recommended because it yields a positive influence on conduction performance and reduces turn-on energy losses. Similarly, a negative $V_{GS(off)}$ ensures a secure and reliable OFF state and enhances the turn-off characteristics. Due to substantial advancements in gate oxide layer reliability and SiC MOSFET manufacturing process, their ability to withstand negative gate voltages within specified limits has significantly improved. The new 2 kV CoolSiC™ further expands this capability, allowing for a selection of negative gate-source voltages and ensuring safe operation even at -10 V [11].

2.4 Dynamic switching characteristics

In SiC MOSFETs, the contribution of turn-on energy loss (E_{on}) to total switching energy losses is higher than that of the turn-off energy loss (E_{off}). E_{off} represents the energy to charge the output capacitance of the device with lower dependence on temperature. Conversely, E_{on} has a more pronounced dependence on temperature. It is also influenced by the circuit configuration and the complementary switching device.

2.4.1 Body diode

In a half-bridge circuit configuration, implementing a dead time between the gate driving signals is

Fig. 2 Voltage drop across the channel during forward and reverse conduction for $V_{GS(on)}$ = 15 V and 18 V, conduction through the body diode for $V_{GS(off)}$ = -2 V and 0 V at 25°C

crucial to prevent shoot-through. During this period, the SiC body diode, which typically exhibits bipolar characteristics, conducts, and experiences reverse recovery when turning off. This can result in higher switching energy losses. Figures 4 and 5 depict the turn-on waveform of IMYH200R024M1H for different dead times at 25°C and 100°C, respectively. At 25°C, the increase in E_{on} is 13% and the reverse recovery energy (E_{rec}) is 26% and the dead time varies from 150 ns to 1 µs. However, at 100°C heatsink temperature, a significant increase in reverse recovery charge (Q_{fr}) is observed, leading to a 56% increase in E_{on}. This shows that at higher temperatures, higher currents, and for longer body diode conduction durations the impact of the reverse recovery charge becomes more prominent [12].

Consequently, in addition to higher conduction losses caused by the higher forward voltage drop in the body diode, switching energy losses and reverse recovery losses of the diode also increase due to prolonged body diode conduction.

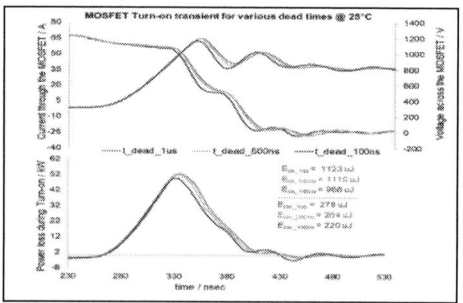

Fig. 4 Turn-on transient of IMYH200R024M1H for different dead time values at room temperature (25°C)

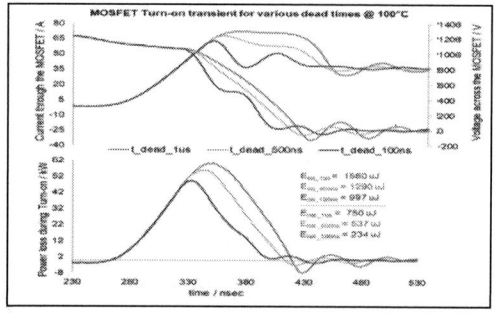

Fig. 5 Turn-on transient of IMYH200R024M1H for different dead time values at a heatsink temperature of 100°C

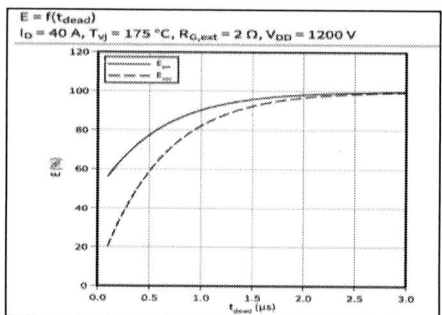

Fig. 6 E_{on} and E_{rec} of IMYH200R024M1H as a function of dead time at I_{ds} = 40 A, T_{vj} = 175°C, R_G = 2 Ω and 1200 V_{DC} link voltage.

The relationship between E_{on} and E_{rec} with dead time is given in the product datasheet and shown in Fig. 6. Under the nominal conditions specified in the datasheet, E_{on} can be reduced to less than 60% and E_{rec} to 20% of the maximum value for the shortest dead time duration. Therefore, careful consideration of the dead time is essential to ensure efficient and reliable operation, especially in applications with fast switching or higher efficiency requirements.

2.4.2 2 kV CoolSiC™ Schottky diode

The CoolSiC™ Schottky barrier diode (SBD), unlike the MOSFET's body diode, does not exhibit the reverse recovery effect, resulting in reduced temperature-dependent E_{on} energy loss in the MOSFET. Figure 8 illustrates the E_{on} losses of the MOSFET, switched against the SBD diode. Compared to the MOSFET's E_{on} in a half-bridge configuration, SBD diodes result in a 30-60% reduction in switching losses. A combination of the 2 kV CoolSiC™ Schottky diode and the 2 kV CoolSiC™ MOSFET is particularly useful in unidirectional converter topologies. For example, as the boost converter in the MPPT of a solar inverter as shown in Fig. 7.

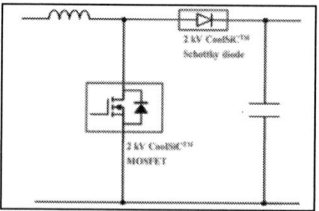

Fig. 7 A boost converter with 2 kV CoolSiC™ Schottky diode and MOSFET

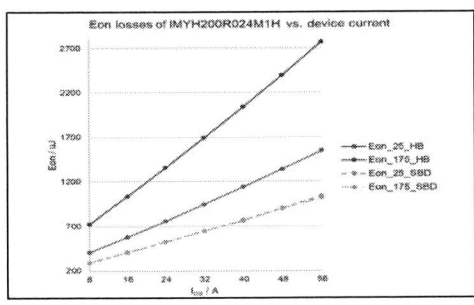

Fig. 8 Turn-on energy losses when switching against the body-diode in a half-bridge vs. switching against an SBD in a boost converter

3 Application performance

To demonstrate the performance of the new 2 kV CoolSiC™ MOSFET, system simulations in PLECS [13] and measurements on a DC-DC converter were conducted. Both the simulation and the measurements involved a comparison between the performances of a simplified system using 2 kV CoolSiC™ devices and a multi-level system using 1200 V CoolSiC™ devices.

3.1 Simulation analysis

The study involved the analysis of a 30 kW system, with simulations carried out for both the DC-DC and DC-AC stage.

3.1.1 The DC-DC boost stage for MPPT

Table 3 lists the simulation parameter values for the DC-DC boost stage in the MPPT of a solar inverter.

A 3-level, dual booster stage with 40 mΩ, 1200 V CoolSiC™ MOSFET (IMZA120R040M1H) and 20 A SBD (IDWD20G120C5) was simulated and compared with a 2-level booster stage with 24 mΩ, 2 kV CoolSiC™ MOSFET (IMYH200R024M1H) and 25 A SBD (IDYH25G200C5), both operating under the same system specifications as listed in Table 3. The schematics are presented in Fig. 9.

To compensate for the increased on-state resistance at higher junction temperature and higher switching energy losses of the 2 kV MOSFET, a lower $R_{ds(on)}$ of 24 mΩ was selected to replace the higher $R_{ds(on)}$ of 40 mΩ 1200 V CoolSiC™ MOSFET. Similarly, the 1200 V 20 A CoolSiC™ Schottky diode was replaced by a 2 kV 25 A CoolSiC™ Schottky diode, as the higher voltage class

has a higher forward voltage than the lower voltage class.

Parameters	Values
Input voltage	810 V
Output voltage	1500 V
Output power	30 kW
Switching frequency	40 kHz
Gate resistance	2 Ω
$V_{GS(off)}/V_{GS(on)}$	-2 V/18 V
$T_{heatsink}$	80°C
$R_{th(case-heatsink)}$	0.5 K/W

Table 3 Simulation parameters for the boost stage

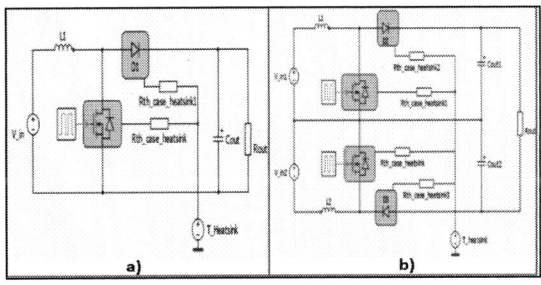

Fig. 9 **a)** 2-level booster stage with 2 kV CoolSiC™ **b)** 3-level dual booster stage with 1200 V CoolSiC™

Figure 10 shows the simulation results. The 2 kV system shows 20% lower total semiconductor loss compared to the 1200 V system. The overall part count (including semiconductor switches, inductors, and capacitors) within the DC-DC booster stage is also halved.

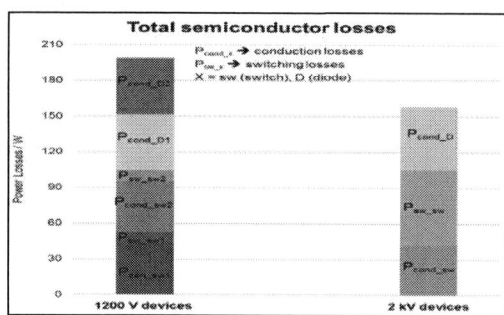

Fig. 10 Total semiconductor loss in a system utilizing 1200 V device vs. 2 kV devices

However, with respect to thermal performance within the same thermal management system, with an identical heatsink design and thermal impedance path between the device case and the heatsink, the 2 kV MOSFET was hotter than the 1200 V MOSFET, as shown in Fig. 11.

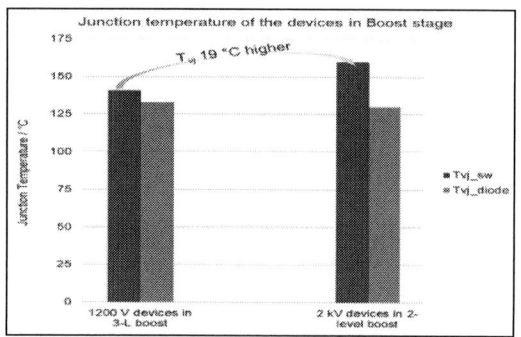

Fig. 11 Junction temperature of 1200 V vs. 2 kV devices in the DC-DC boost stage

To achieve the same power output using a 2 kV device, a better thermal management design is essential. The losses per device are considerably higher in a 2 kV system, unlike 1200 V systems where the losses are distributed among more components. The same cooling conditions and thermal path, therefore, naturally result in a higher device temperature in the 2 kV system.

3.1.2 The DC-AC inverter stage

Simulation parameters for the DC-AC inverter stage, including a 3-level ANPC implemented with IMZA120R040M1H and a simplified 2-level B6 implemented with IMYH200R024M1H, are listed in Table 4.

Parameters	Values
DC link voltage	1500 V
Output voltage (V_{rms})	600 V
Output power	10 kW
Power factor	1
Output line frequency	50 Hz
Switching frequency	30 kHz
Gate resistance	2 Ω
$V_{GS(off)}/V_{GS(on)}$	-2 V/18 V
$T_{heatsink}$	80°C
$R_{th(case-heatsink)}$	0.5 K/W
Dead time	150 ns

Table 4 Simulation parameters for the inverter stage

The schematics are shown in Fig. 12.

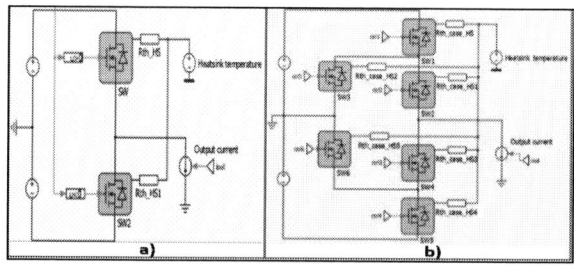

Fig. 12 **a)** 2-level B6 with 2 kV CoolSiC™ devices

b) 3-level ANPC with 1200 V CoolSiC™ devices

Figure 13 shows the simulation results. The 2-level B6 inverter employing 2 kV devices shows 15% lower total semiconductor device losses per phase than the 3-level ANPC using 1200 V devices.

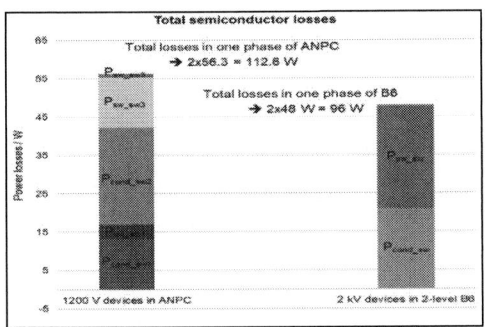

Fig. 13 Total semiconductor loss in one phase of an inverter utilizing 1200 V vs. 2 kV devices

3.2 Measurement analysis

For the measurement analysis, a 2 kV evaluation board from Infineon [14] was operated in the buck converter mode to maintain low output voltage. This mode was chosen over the boost converter mode to avoid any accidental voltage rise at the converter output and prevent the DC electronic load from exceeding the voltage limit, given the absence of a closed-loop control.

Fig. 14 Measurement setup of a 2 kV evaluation board operated as a buck converter

The cooling system comprised a 150 mm long LAM 3K heatsink with a 30 x 30 mm 12 V axial fan. Figure 15 shows the thermal resistance (R_{th}) of the heatsink [15].

The measurement used the same part numbers as those used in the simulation of the 30 kW boost converter (see Section 3.1.2). Two tests were conducted:

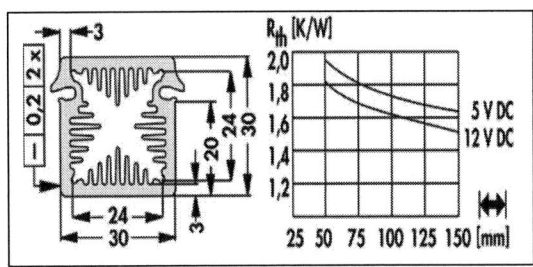

Fig. 15 The heatsink dimension and R_{th}

1. A buck converter with 2 kV CoolSiC™ MOSFET and Schottky diode: In this case the converter was tested for a maximum output power of 8.4 kW at a maximum input voltage of 1200 V (limited by the measuring equipment) and a switching frequency of 40 kHz. The fan operated at its maximum speed for 12 V V_{DC}, with an $R_{th\_heatsink}$ of 1.52 K/W as per the specification sheet.

2. A converter with 1200 V CoolSiC™ MOSFET and Schottky diode: In this case, only half the system was tested. It represented a single stage of the converter carrying half the power of the first test case. Additionally, the fan was switched off to raise the R_{th} of the heatsink. The resultant $R_{th\_heatsink}$ was approximately 1.75 K/W, i.e., a rise of more than 15%. The converter was then operated at half the power and DC link voltage, corresponding to 600 V input voltage and 4.2 kW of power.

The measurement results demonstrated that the converter with 2 kV devices achieved 98.5% efficiency, while the efficiency of converters with 1200 V devices, operating at half power and voltage, was 98.2%. The lower efficiency with the 1200 V CoolSiC™ MOSFETs can be attributed not only to semiconductor losses but also to the contribution of inductor losses and other passive elements to the circuit, which appeared to be more pronounced at lower voltages than at higher voltages despite the same ripple current in both systems, leading to lower efficiency.

A comparison between the thermal performance of the two systems revealed that the temperature of IMYH200R024M1H was approximately 100°C at a maximum load of 8.4 kW and 1200 V input voltage, while the temperature of IMZA120R040M1H was around 70°C at a load of 4.2 kW and 600 V input voltage, despite having 15% higher $R_{th\_heatsink}$. This indicates that the 2 kV system requires a better thermal management design to effectively cool down the device.

The difference in device temperature seen in the measurement and simulation results primarily arises due to the higher power loss ratio between the 2 kV and 1200 V systems in measurement.

This is a result of lower conduction losses at lower currents in the measurement, coupled with significantly higher switching losses in the 2 kV device. Consequently, the 2 kV device had a higher junction temperature. Implementing a cooling system designed for higher power systems can help reduce the temperature difference between the 1200 V and 2 kV devices.

Nevertheless, the measurement results validated the enhanced overall system efficiency and reduced component count of the 2 kV system, indicating a benchmark system benefit. These characteristics establish the 2 kV CoolSiC™ MOSFET as an attractive solution for applications to achieve higher system efficiency with lower component counts and decreased systems size and weight compared with state-of-the-art systems based on multi-level topologies.

4 Conclusion

Infineon's latest 2 kV CoolSiC™ in a discrete package meets the demand for enhanced design flexibility and system simplification. The innovative package technology, characterized by high creepage and clearance distance help ensure benchmarking reliability in systems with 1500 V DC link voltage. An evaluation of a solar inverter utilizing the 2 kV CoolSiC™ MOSFET and diode demonstrated considerable performance advantages over inverter designs employing 1200 V devices. The system-level simulations and measurements reinforced the superior performance of the 2 kV CoolSiC™, demonstrating 20% lower losses at the 2-level booster stage and 15% lower power losses at the 2-level DC-AC stage compared to their 1200 V counterparts. In essence, the 2 kV CoolSiC™ MOSFET in a discrete package represents a substantial step forward in achieving reduced component counts for better efficiency, and optimized system size and weight in high-voltage solar inverters. This development marks a significant milestone in the journey towards cost-effective and efficient photovoltaic systems.

References

[1] E. Serban, M. Ordonez, and C. Pondiche, "DC-Bus Voltage Range Extension in 1500V Photovoltaic Inverters," IEEE Journal of Emerging and Selected Topics in Power Electronics, vol. 3, Dec. 2015, pp. 901–917.

[2] Z. Corba, B. Popadic, V. Katic, B. Dumnic, and D. Milicevic, "Future of high power PV plants - 1500V inverters," International Symposium on Power Electronics, 2017.

[3] G. Soelkner, W.Kaindl, H.-J.Schulze, and G.Wachutka, "Reliability of power electronic devices against cosmic radiation-induced failure", Microelectronics Reliability, 2004.

[4] M. Slawinski, B. Sahan, and U. Jansen, "Evaluation of a NPC1 phase leg built from three standard IGBT modules for 1500 VDC photovoltaic central inverters up to 800 kVA," 18th European Conference on Power Electronics and Applications, Karlsruhe, Germany, 2016.

[5] D. Peters, R. Siemieniec, T. Aichinger, T. Basler, R. Esteve, W. Bergner, and D. Kueck, "Performance and Ruggedness of 1200V SiC - Trench – MOSFET," 29th International Symposium on Power Semiconductor Devices and ICs (ISPSD), 2017.

[6] D. Peters, T. Basler, B. Zippelius, T. Aichinger, W. Bergner, R. Esteve D. Kueck, and R. Siemieniec "The new CoolSiC™ Trench MOSFET Technology for Low Gate Oxide Stress and High Performance," PCIM Europe, Nuremberg, Germany, 2017.

[7] P. Friedrichs, "Reliability and robustness of SiC power devices – how to ensure the quality level established in the silicon world," The International Power Electronics Conference (IPEC-Himeji 2022 -ECCE Asia), 2022.

[8] D. Heer, D. Domes, and D. Peters, "Switching performance of a 1200 V SiC-Trench-MOSFET in a low-power module", PCIM Europe, Nuremberg, Germany, 2016.

[9] Infineon whitepaper: "How Infineon controls and assures the reliability of SiC based power semiconductors", 2020.

[10] M. Holz, J. Hilsenbeck, R. Otremba, A. Heinrich, P. Türkes and R. Rupp, "SiC Power Devices: Product Improvement using Diffusion Soldering", Materials Science Forum Vols. 615-617 (2009) pp 613-616, online at http://www.scientific.net

[11] Infineon Technologies AG application note: AN2018-09, "Guidelines for CoolSiC™ MOSFET gate drive voltage window".

[12] P. Sochor, et al., "Understanding the Turn-off Behavior of SiC MOSFET Body Diodes in Fast Switching Applications", PCIM Europe 2021.

[13] https://www.plexim.com/products/plecs

[14] Infineon Technologies AG application note: "EVAL-COOLSIC-2kVHCC, Evaluation Board for CoolSiC™ MOSFET 2000 V", 2023.

[15] https://www.fischerelektronik.de/web_fischer/en_GB/VA/LAM3K15012/datasheet.xhtml?branch=heatsinks

PCIM Europe 2024, 11– 13 June 2024, Nuremberg   DOI: 10.30420/566262287

A New 2.3 kV Rated SiC MOSFET Module with Low-Inductance High-Power Package HPnC for 1500 VDC Applications

J. Kawabata[1], S. Chen[1], Y. Kodaira[1], T. Uchida[1], T. Takaku[1], Y. Sekino[1], Y. Kusunoki[1], Y. Kobayashi[1], S. Ewald[2]

[1] Fuji Electric Co., Ltd., Japan

[2] Fuji Electric Europe GmbH, Germany

Corresponding author: Junya Kawabata, kawabata-junya@fujielectric.com
Speaker: Junya Kawabata, kawabata-junya@fujielectric.com

Abstract

An optimum combination of the newly developed 2.3 kV rated SiC MOSFET and the high power package: High Power next Core (HPnC) is introduced in this paper. The 2.3 kV SiC MOSFET module is designed for 1500 VDC applications and its excellent characteristics enables 2-level topology instead of 3-level NPC topology for the systems. In addition, the newly developed HPnC package, featured low-inductive inside structure, is suitable for fast switching devices represented by SiC MOSFET, and maximizes its performance. The simulation result of power losses in a 1500 VDC inverter with the 2.3 kV SiC MOSFET HPnC modules was 38% smaller than that of a conventional 3-level NPC with 1.2 kV Si IGBT modules, and the total footprint of the power modules was reduced by 68%. This result indicates that the developed module can realize a cost-effective power converter for 1500 VDC applications.

1 Introduction

In recent years, the global shift towards higher voltage levels, particularly in the realm of renewable energy and industrial applications, necessitates the continuous evolution of power electronic components and systems. New DC-voltage class of 1500 V has become the standard for several applications such as wind and photovoltaic energy converters, energy storage systems, UPSs and hydrogen electrolyzers, to enhance output power and improve transmission efficiency. For 1500 VDC applications, 3-level I-type NPC topology with series-connected 1.2 kV rated devices is a common solution. However, the 3-level I-type NPC requires at least three of half-bridge modules to implement the circuit, resulting in a large footprint, large commutation inductance and high on-state voltage due to the series connection of modules. We have developed a new voltage class 2.3kV SiC MOSFET module with a High Power next Core (HPnC) package, which is a new high power package to breakthrough these limitation. This module realizes a 2-level configuration in 1500 VDC applications, and it will contribute higher efficiency and simplification of the systems. This paper describes

the characteristics of the 2.3 kV SiC MOSFET device and the features of the HPnC package, and demonstrates the advantages of the 2.3 kV SiC MOSFET HPnC for 1500 VDC applications.

2 Development of 2.3 kV SiC MOSFET and HPnC Package

2.1 2.3 kV rated SiC MOSFET

A new voltage class 2.3 kV SiC MOSFET with a trench gate structure has been developed [1-5]. The SiC MOSFET has trench gate structures, and several technologies are applied to realize low $R_{DS(on)}$, such as thinning the drift layer, finer cell pitch, improving channel mobility and so on. By an innovative design optimization and quality improvement, the body diode of the MOSFETs can be utilized as a free-wheeling diode without anti-parallel SiC SBDs. To remove SiC SBDs, more die area can be ensured, and it enables further output current with additional SiC MOSFETs.

Figure 1 shows output characteristics comparison between 2.3 kV rated SiC MOSFET (2-level) and 1.2 kV Si IGBTs (3-level I-type NPC). Due to series

2049

connection of devices in the 3-level I-type NPC topology, the on-state voltage of Si IGBTs become very high. In contrast, the SiC MOSFET reduced on-state voltage by 0.9 - 1.2 V.

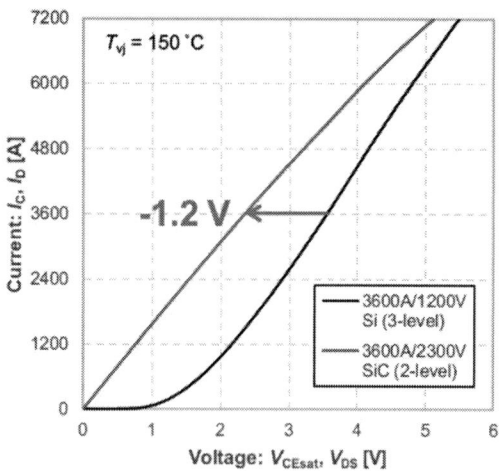

(a) Si IGBT and SiC MOSFET

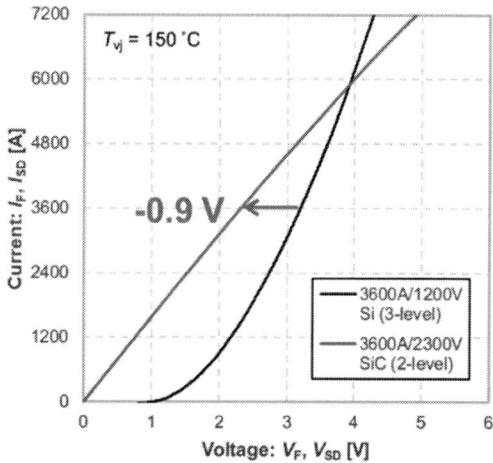

(b) Si FWD and SiC MOSFET (body diode)

Fig. 1 Static characteristics

Figure 2 shows turn-on and turn-off switching waveforms of the 2.3kV rated SiC MOSFET HPnC at the condition of DC 1500 V and rated module current 1200 A. Fast and stable switching are achieved, and the turn-off spike voltage is around 2.0 kV even at the full DC link voltage without any snubber capacitors.

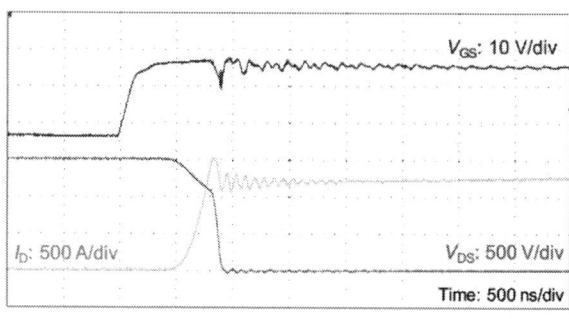

(a) Turn-on waveform

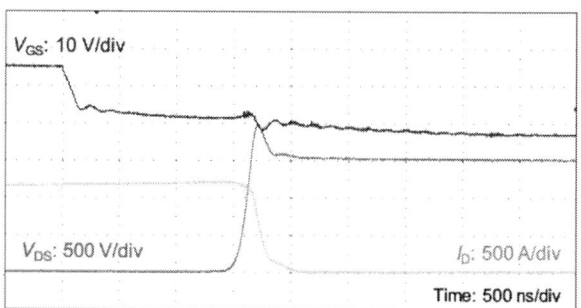

(b) Turn-off waveform

Fig. 2 Dynamic characteristics

2.2 New HPnC Package

In general, fast switching devices require low-inductance packages to suppress spike voltages and oscillations. Conventional packages have higher stray inductance, which causes snappy behaviors of fast switching devices. Newly developed HPnC package solve these problems of the conventional high-power device packages [6]. Figure 3 shows an outline appearance and the circuit configuration. Figure 4 shows internal terminal structure of the HPnC. The compact and symmetrical design of the package implements easy paralleling of the modules for further output power expansion of inverters. The laminated structure of P and N terminals inside the module reduced the stray inductance and achieves approximately 10 nH. In addition, a high thermal conductive substrate is applied to realize effective heat dissipation.

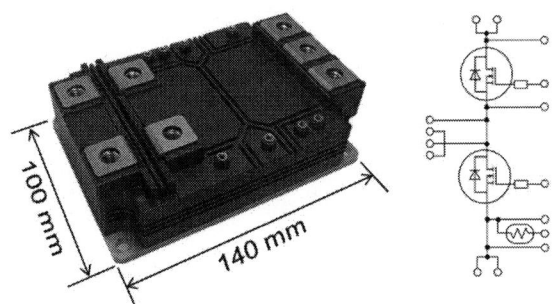

Fig. 3 New HPnC package outline

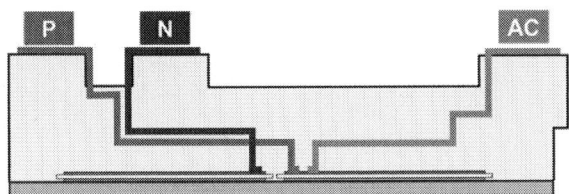

Fig. 4 HPnC internal terminal structures

The HPnC package is adapted to the 2.3 kV rated SiC MOSFET module. The module is a half-bridge configuration and the current rating is 1200 A.

The newly developed 2.3 kV SiC MOSFET HPnC enables a 2-level topology with a single half-bridge module for a 1500 VDC inverter instead of a 3-level topology. Figure 5 shows a configuration comparison of a 3-level I-type NPC circuit with conventional 1.2 kV rated modules and a 2-level circuit with a 2.3 kV HPnC module. The number of power modules is reduced and the total footprint size of the power modules is shrank by 68%. Therefore, the newly developed 2.3 kV HPnC contributes minimizing commutation inductance and increasing power density of the system.

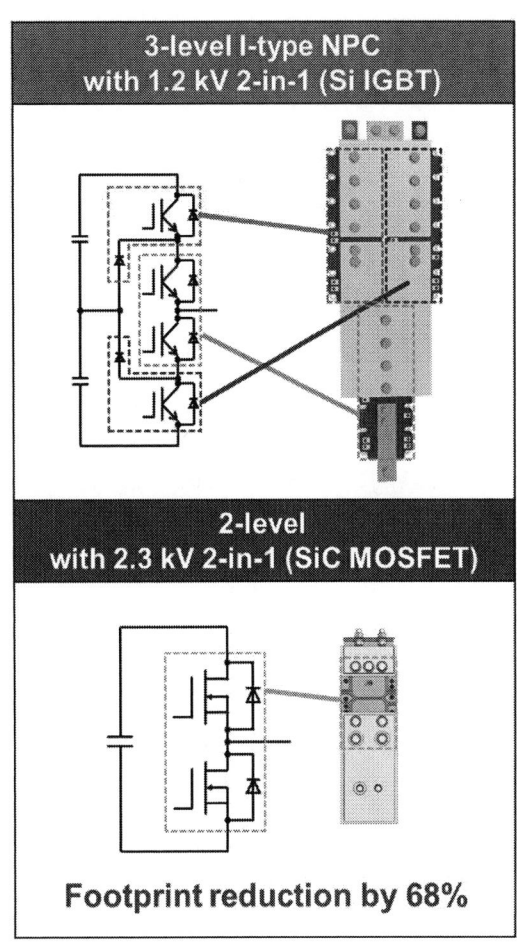

Fig. 5 Comparison of 3-level I-type NPC (top) and 2 level (bottom) configurations

2.3 Low Power Losses

To confirm advantages of newly developed 2.3 kV SiC MOSFET HPnC, power losses of 2.3 MW class 1500 VDC inverter was calculated and compared. For conventional 3-level configuration, 6 of 1800 A / 1.2 kV rated 2-in-1 modules per phase and total 18 modules were necessary. On the other hand, for 2.3 kV rated SiC MOSFET HPnC, 3 of 1200 A / 2.3 kV rated 2-in-1 modules per phase and total 9 modules were enough. Figure 6 shows a comparison of inverter power losses between a 3-level I-type NPC and 2-level inverters. Power losses of 2.3 kV SiC MOSFET HPnC is reduced by 13%. By SiC MOSFET, conduction losses are reduced by 58%, therefore even though the switching losses are increased due to higher carrier frequency, total power losses are lower.

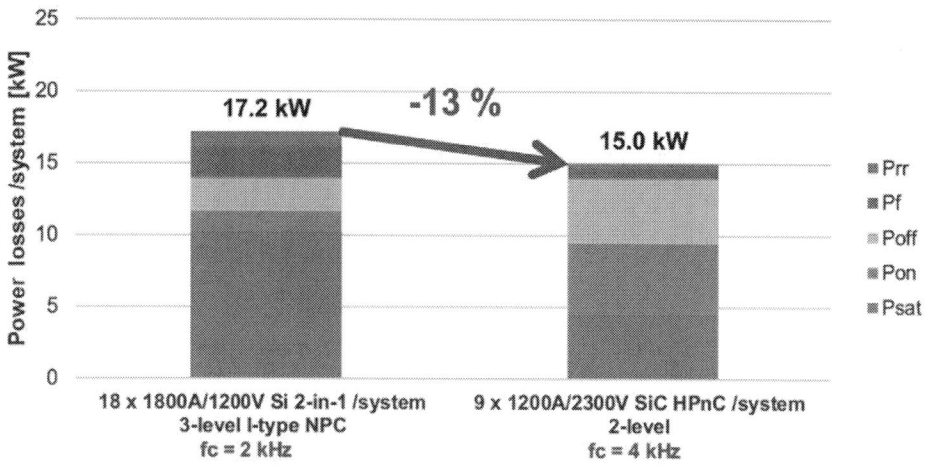

Fig. 6 Comparisons of power losses (left: Si 3-level, right: SiC 2 level)
Io = 1800 Arms, VDC = 1500 V, pf = 0.9, λ = 0.9, fo = 50 Hz, fc = 2 kHz (3-level), 4 kHz (2-level)

3 Conclusion

A new voltage class 2.3 kV SiC MOSFET with low-inductance new high-power package HPnC has developed. The 2.3kV SiC MOSFET HPnC is suitable for 1500 VDC applications and enables a simplified 2-level topology instead of a conventional 3-level I-type NPC topology with 1.2 kV devices. It has been demonstrated that a 2-level topology composed of this module achieves 13% reduction in power losses and 68% smaller footprint and lower commutation inductance compared to a 3-level NPC topology. It will contribute size and cost reduction and enhance reliability of systems.

References

[1] M. Chonabayashi et al., "All-SiC Module with 2nd Generation Trench Gate SiC-MOSFETs", Proceeding of PCIM Europe 2019, pp. 318-323.

[2] K. Okumura et al., "2nd- Generation SiC Trench Gate MOSFETs", FUJI ELECTRIC REVIEW, vol. 92, no. 4, pp. 225-227, 2019.

[3] Y. Sekino et al., "3.3 kV All-SiC Module with 1st Generation Trench gate SiC-MOSFETs for Traction Inverters", Proceeding of PCIM Europe 2020, pp. 98-103.

[4] B. Bradel et al., "Comparison of Dissipation Loss Reduction Rates of 1.2 kV and 1.7 kV All-SiC Modules against Si-IGBT Modules", Proceeding of PCIM Europe 2022, pp. 842-847.

[5] S. Chen et al., "Application of Newly-Developed 2.3kV Si and SiC Devices to Renewable Energy System", Proceeding of PCIM Europe 2023, pp. 696-700.

[6] Y. Sekino et al., "3.3 kV All SiC Module with 2nd Generation Trench gate SiC-MOSFETs for Traction", Proceeding of PCIM Europe 2022, pp. 584-590.

PCIM Europe 2024, 11– 13 June 2024, Nuremberg   DOI: 10.30420/566262289

Mechanism for Improving the Heat-Resistance of Adhesive Interface in Flexible Printed Circuits

Keita Suzuki[1] , Takahisa Manabe[1] , Hisae Oba[1] , Yuichi Aoyagi[1]

[1] NOK CORPORATION, Japan

Corresponding author: Yuichi Aoyagi, aoyagi.yuichi@jp.nokgrp.com
Speaker: Keita Suzuki, suzuki.21.keita@jp.nokgrp.com

Abstract

Flexible printed circuits (FPC's) have been utilized for various kinds of applications in electronics. Recently, due to the increasing demand for FPC's as power electronics components, they are required to withstand high temperatures. In this study we will explain the degradation mechanism of the adhesive interface between the conductive metal layer and the insulative polyimide base film at high temperatures. By chemiluminescence measurements, it has been proven that the copper layer which is attached directly to the fluoropolymer (FKM) adhesive layer causes the oxidation reaction of the adhesives when heated at high temperatures.

1 Introduction

1.1 Decarbonization in Mobility Industry

As environmental problems become more serious, it is becoming increasingly important to reduce emissions of greenhouse gasses such as carbon dioxide. The term "carbon neutralization" became much more popular within the last few decades in all industries [1,2]. Recently, using renewable energy is one of the prominent strategies to deal with the environmental problems [3].

Taking a look at the automotive industry, it is about to undergo a major transition due to carbon neutralization. The transition from fossil fuel-powered engine vehicles to electric vehicles powered by electricity generated from clean energy resources is taking place. Sales volume of eco-friendly mobilities such as electric vehicles (EV's) and plug-in hybrid electric vehicles (PHEV's) is expected to increase sharply, while that of gasoline vehicles will decline in the future [4].

EV's contain a large variety of power modules, and their circuit structures are completely different according to their intended functionality. As components of various shapes and sizes are being installed in electric vehicles, there is a need for electrical circuits that can be flexibly processed.

1.2 Flexible Printed Circuits

Flexible printed circuits (FPC's) are formed by sticking the thin conductive metal layer (e.g., copper) onto the insulative polyimide layer by using adhesive agents [5]. Their flexible, thin and light characteristics enable us to achieve simple wiring designs, and they are essential for creating small electronic devices such as smart phones, PC's, and game consoles.

Although the traditional applications of FPC's have been mainly focused on small devices, the main market of FPC's is now shifting towards the power electronics industry. Power trains are the assembly converting electricity to motion energy and contain many electric power modules. These power devices require frequent charge and discharge cycles, which can generate high-temperature heat emission. Thus, circuit materials are required, which can withstand extremely harsh conditions [6].

1.3 Heat-Resistance Issue of FPC's

In general, FPC's have a layer structure consisting of cover and base materials, conductive metals, and adhesives (Fig.1) [5]. Regarding heat-resistance properties, super engineering plastics such as polyimide can withstand high temperatures up to 400°C, while the melting point of the copper conductive layer is higher than 1000°C. These materials are unlikely to decompose under the environmental conditions of an EV's powertrain. On the other hand, there are various available materials for adhesives, and the high-temperature heat-resistance of the entire FPC will be greatly influenced by whether a heat-resistant adhesive is used or not. Adhesives play a key role in providing FPC's with their heat-resistance properties.

2053

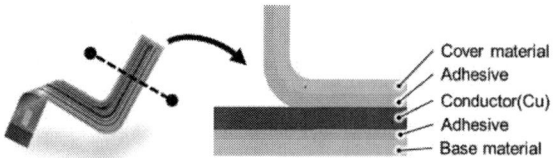

Fig. 1 Flexible circuits and its schematic drawing (cross section).

Fluororubber (FKM) and polyimides withstand up to 250°C for 3000 h [7,8]. Yonezawa *et. al.* have used a polymer made by mixing and copolymerizing a thermoplastic resin with polyimide as an adhesive [7]. This adhesive can be crosslinked by using crosslinking agents, and they show better heat durability than any other conventional adhesives.

Fluoropolymer also has an excellent heat-resistance property. Our group has previously shown an example of FKM-based adhesives. FPC's were heated in an oven at 200°C or 250°C for 3000 h, afterwards the peeling strength of FPC's was measured (Fig. 2a, 2b). While the sample largely kept the adhesive structure at 200°C for 3000 h, the sample lost adhesive strength at 250°C (Fig. 2c) [8].

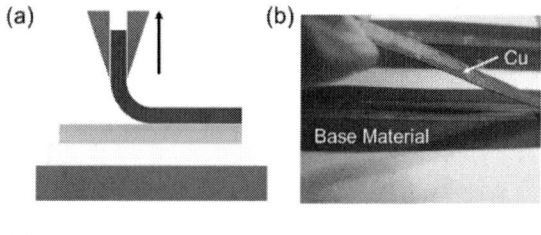

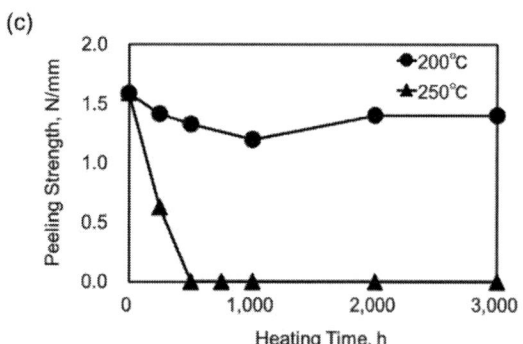

Fig. 2 Previous experimental results of our work. (a) Schematic drawing of peeling tests. (b) Image of a peeled FPC sample. (c) Peeling test results.

1.4 Oxidation of Adhesives

Polymer materials are often decomposed under long-term usage. This decomposition mechanism is underlying an automatic oxidation process which is caused by external factors such as heat or light. This process starts from generating radical species, which immediately reacts with O_2 in the atmosphere. Fig. 3a shows the degradation mechanism of a polymer main chain (described with R) [9]. The hydrogen atoms located at the polymer side chains leave and polymer radicals ($R \cdot$) are generated. Polymer radicals immediately react with oxygen, and peroxide radicals are generated. These peroxide radicals result in carbonyl groups. It was presumed that this mechanism of polymer oxidation may also occur in the present fluorinated adhesive.

1.5 Oxidation Mechanism by Metals

A specific phenomenon of accelerated oxidation is known in composite materials, such as FPC's, where the metal and adhesive polymer are in direct contact with each other [10]. Especially a catalytic oxidation reaction with Cu is very well known in the polymer industry (Fig. 4). The chemical reaction mechanism is as follows:

$$ROOH + Cu^+ \rightarrow RO \cdot + Cu^{2+} + OH^- \qquad (1)$$
$$ROOH + Cu^{2+} \rightarrow RO \cdot + Cu^+ + H^+ \qquad (2)$$
$$2\ ROOH \rightarrow \ RO \cdot + ROO \cdot + H_2O \qquad (3)$$

The mechanism mentioned above has been studied mainly in universal hydrocarbon polymers such as polyethylene [11]. On the other hand, the mechanism of accelerated oxidation in fluorinated adhesives in contact with copper is not clear. We hypothesized that this copper-accelerated oxidation phenomenon, copper damage, may also occur when FPC's are exposed to high temperatures.

1.6 Chemiluminescence

Mechanical measurements and spectroscopy have been conventional methods to analyze oxidation degradation of adhesives. The main method of mechanical measurement is a peeling test, which is an easy way to quantify the degree of oxidation [8]. However, it is impossible to understand the chemical mechanisms via mechanical measurements. To analyze the chemical mechanism usually IR spectroscopy methods are conducted for spectral analysis [12]. Both methods, however, require samples which are heated for thousands of hours, resulting in very long-term experiments. From an industrial point of view, the development of highly heat-resistant FPC's is an urgent task, therefore a more efficient method of degradation analysis is desired.

Therefore we focused on the chemiluminescence method to investigate the oxidation process [13]. Chemiluminescence (CL) is the emission of light which is observed when a molecule returns to the electronic ground state from an excited state. Usually, during polymer oxidation reactions, excited carbonyl species are generated with the addition of oxygen to the polymer main chain, and luminescence occurs with the deactivation of the carbonyl groups (Fig. 3a). Chemiluminescence emission is usually too weak to be observed. Tohoku Electronic Industrial CO. Ltd. has developed a device that can observe CL (Fig. 3b). In that equipment, the sample is heated in the furnace and photons are observed by a photomultiplier tube (PMT) which is cooled to -25°C [13]. Several examples to measure CL of general adhesive agents have been reported so far, and we took advantage of this method.

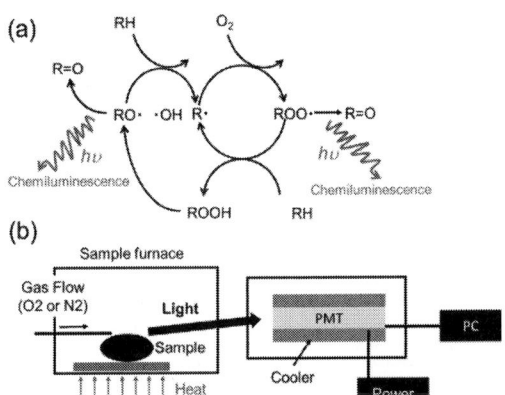

Fig. 3 (a) Radical reaction scheme of polymer oxidation. (b) Conceptual drawing of CL measurement.

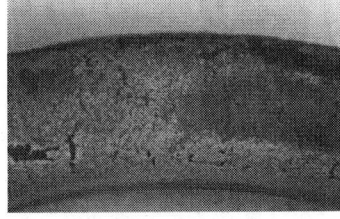

Fig. 4 Sample image of an oxidated ethylene-propylene-diene rubber (EPDM) material. The oxidation is caused by long-term exposure to copper ions in water.

1.7 Research Objective

In this study we clarify the mechanism of degradation process in the adhesive interface at high temperature. The chemical reaction which occurs between the conductive copper layer and the adhesive FKM layer is also discussed according to the results of CL measurements. Timescale can be a major issue when evaluating material degradation. To deal with this issue CL method is performed.

2 Experimental Section

2.1 Materials

2.1.1 Samples

As adhesive agents, FKM sheets are utilized. Copper thin films are purchased from Takeuchi Metal Foil & Powder Co., Ltd..

2.1.2 Preparation

A piece of copper layer (thickness: 50 um) was pasted to an FKM adhesive sheet (thickness: 10 um). The combined sheet was heated at 130°C under 2 MPa for 30 seconds, then further heated at 185°C under 2 MPa for 140 seconds. The sample was cooled to room temperature and then gradually heated at a rate of 1.5°C/min to 180°C, then kept at 180°C for 12 h. The sample is referred to as Cu-Ad (Fig. 5a). Also, the sample which consists of only adhesives was prepared for comparison (Ad, Fig. 5a), which went through the same preparation procedure as Cu-Ad. These sheet samples were cut into small pieces (10 mm × 10 mm) right before the CL measurements (Fig. 5b).

2.2 Methods

2.2.1 Measurements

CL was performed by LA-FS5 (Tohoku Electronic Industrial CO., Ltd., Miyagi, Japan) with a heating rate of 20 °C/min under oxygen or nitrogen atmosphere, from 50 °C to 350 °C. To Confirm that the results of CL measurements truly reflect the oxidation reaction of FKM, FT-IR spectra were recorded on a VIR-200 (JASCO Corporation, Tokyo, Japan) from 4000 to 650 cm^{-1} at a resolution of 4 cm^{-1} with a setup of diamond crystal ATR.

2.2.2 Analysis

Fig. 6 shows a conceptual drawing of typical CL spectrum. The intersection point is defined to be the initial oxidation temperature (IOT) which indicates how the sample is likely to be oxidized by O$_2$. IOT is determined by drawing the extension of background and the steepest slope line of a peak. The analysis to obtain IOT was achieved by THERMOKINETICS Software by AKTS.

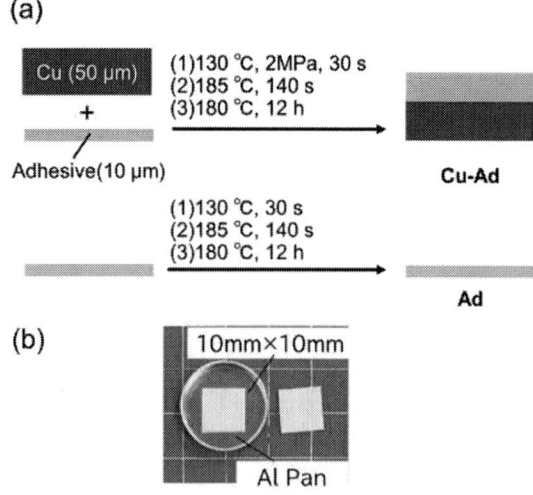

(b)

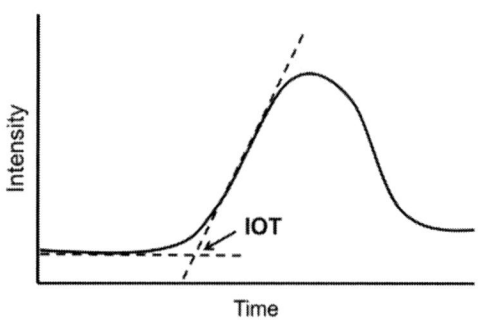

Fig. 5 (a) The preparation scheme of a Cu layer-attached adhesive sheet sample (Cu-Ad) and a neat adhesive sheet sample (Ad). (b) An image of CL sample.

Fig. 6 Conceptual drawing of determining initial oxidation temperature (IOT).

3 Results and Discussion

3.1 Oxidation Effect of Copper Layer

Fig. 7 shows the results of chemiluminescence measurements. As shown in the figure, the main oxidation peak of Cu-Ad appeared at 269°C, while that of Ad appeared at 341°C. Initial oxidation temperature (IOT) was also calculated for the spectra of each sample and the result is shown in Table 1. The IOT points of Cu-Ad and Ad are at 214°C and 288°C, respectively. Thus it is indicated that by attaching Cu to FKM adhesives, IOT was reduced by approximately 80°C. From these results, oxidation facilitation effect of copper toward FKM adhesives was clarified, and we could detect oxidative behaviour of copper in a short time experiment.
For only Cu-Ad, two distinct peaks are observed, while Ad only had one main oxidation peak. It is not possible to attribute these peaks from the measurement under oxygen, so we took another test to clarify the mechanisms.

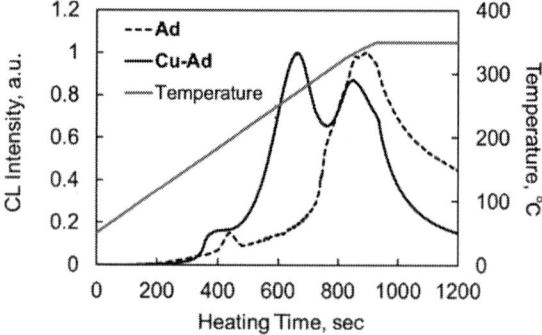

Fig. 7 CL spectra of Cu-Ad (solid line) and Ad (dashed line). The red line describes the profile of temperature.

	Main Peak, °C	IOT, °C
Ad	341	288
Cu-Ad	269, 335	214

Table 1 Comparison of the temperature at the main peak and IOT between Ad and Cu-Ad.

Fig. 8 shows the FT-IR spectra of the samples recorded before and after going through CL measurements. As shown in the figure, all the samples showed the vibration peaks of carbonyl groups ($v_{C=O}$, 1725 cm^{-1}) and fluorocarbon groups (v_{C-F}, 1110 cm^{-1}), respectively, indicating that the polymer main chain of FKM was oxidized and carbonyl groups were generated. This result supports the CL measurements, which implies the reaction between the O_2 molecules and FKMs. Furthermore, a broad C=O vibrational peak was identified in the Cu-Ad sample. This implies that when copper comes into contact with the adhesive, a reaction other than the usual oxidation mechanism is occurring. One possibility is the formation of carboxylate copper salts, but this needs to be confirmed in the future.

Table 2 compares the peak intensity ratios. Before heating, only weak carbonyl groups were observed in both samples. On the other hand, after heating, the intensity ratio of carbonyl groups to fluoroalkyl groups increased; Cu-Ad has a higher proportion of C=O, which is consistent with the CL measurements.

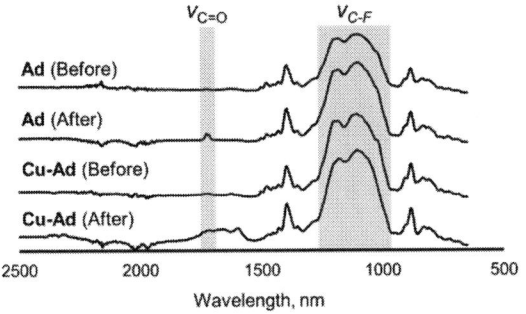

Fig. 8　FT-IR spectra of the samples recorded before and after CL measurements are achieved.

	$I_{C=O}$ / I_{C-F} (before)	$I_{C=O}$ / I_{C-F} (after)
Ad	0.079	0.15
Cu-Ad	0.055	0.20

Table 2　Comparison of the IR intensity rates of Ad and Cu-Ad. $I_{C=O}$ and I_{C-F} stand for the peak top intensity of $v_{C=O}$ (1725 cm^{-1}) and v_{C-F} (1110 cm^{-1}), respectively.

3.2 Analysis of oxidation peaks

To investigate the reason why the oxidation peaks were divided into two for Cu-Ad, a further measurement was carried out. Fig. 9 shows the CL spectra under O$_2$ gas flow and N$_2$ gas flow, respectively. Under N$_2$ gas flow, the CL intensity of peaks are relatively small compared to that under O$_2$ gas flow. One peak at lower temperature disappeared

under nitrogen atmosphere due to the absence of O$_2$. On the other hand, the other peak was still observed in the N$_2$ flow condition, which indicates that FKM polymers have another degradation pathway which requires no oxygen. From the CL spectra, at least two different degradation mechanism were estimated.

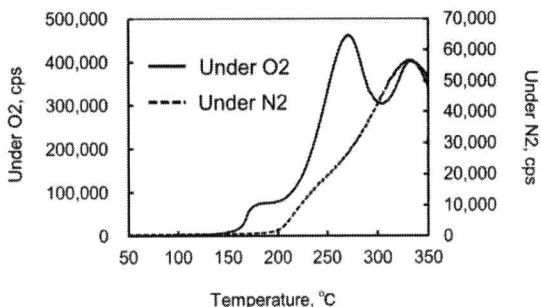

Fig. 9　CL spectra of Cu-Ad under O$_2$ gas flow (solid line) and under N$_2$ gas flow (dashed line), respectively.

3.3 Plausible mechanism of degradation of fluoropolymer adhesive

Although there are not so many reports on the degradation process of FKM polymers, some mechanisms have been suggested [14].

A general FKM is a copolymer which consists of vinylidene fuluoride (VdF) and hexafuluoro pentane (HFP). VdF has C-H bonds, and these C-H bonds can decompose to radical species. Thus, one plausible mechanism explanation is the same as the oxidation mechanism of general polymers also shown in Fig. 3a. C-F bonds can also decompose into radicals, but the bonding energy of C-F is much higher than that of C-H, so decomposition of C-F bonds is unlikely to occur (Fig. 10).

Another mechanism is the elimination of hydrogen fluoride. Both hydrogen and fluorine atoms depart from the main chain of the polymer, and then C=C double bonds are established. This reaction needs no oxygen, so it is considered that the oxidation peak which did not disappear under nitrogen atmosphere (see Fig.9) is attributed to this mechanism.

Fig. 11 shows the plausible mechanism of the degradation process of FKM adhesive in the FPC layer structure. FKM adhesives decompose at the copper-adhesive interface, resulting in decrease of adhesive mechanical strength. Copper ions (Cu$^+$, Cu^{2+}) are necessary in redox reactions of the polymer degradation process by copper, as mentioned before. It is estimated that the passive surface state film of copper layer (Cu_2O) is further oxidized to CuO at high temperature, and the generated Cu^{2+} ions gradually flow into the polymer adhesive layer. This ion leakage has yet to be confirmed experimentally and will be studied by using molecular dynamic simulations.

The automatic oxidation mechanism featuring copper reaction mechanism seems to be the main factor for FPC decomposition at extremely high temperature. At the same time, however, another process which proceeds without O_2 is observed by analysing chemiluminescence spectra under nitrogen atmosphere. Concerning this reaction, it is likely that hydrogen fluoride is eliminated which we clarify in the future.

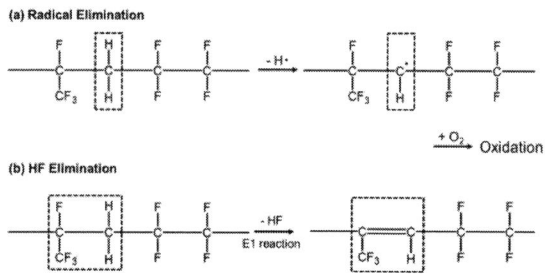

Fig. 10 Degradation chemical reactions suggested in previous works.

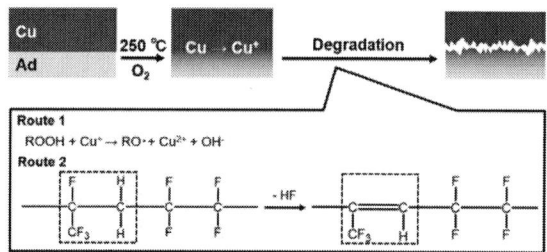

Fig. 11 The plausible mechanism of the degradation process of FKM adhesive in the FPC layer structure.

4 Conclusions

Herein we discussed the degradation process mechanism of FKM adhesive layers for FPC's at high temperature (250°C) by performing chemiluminescence measurements. It is revealed that the copper layer which is attached directly to the FKM adhesive layer acts as an oxidant against the adhesives, and facilitates the automatic radical oxi-

dation process of fluorocarbon structures. In addition, another degradation process with no O_2 molecules has been observed. We hope that our findings will lead to the development of FPC's which are able to withstand the high temperature coditions expected in power electronics environments.

5 Acknowledgement

The authors sincerely thank the members of Tohoku Electronic Industrial CO., Ltd., namely, Tetsu Sato, Itsuo Tanuma, Ryota Sameshima and Masahito Toyonaga for their generous support in chemiluminescence measurements and fruitful discussions.

References

[1] S. Fankhauser, S. M. Smith, M. Allen, K. Axelsson, T. Hale, C. Hepburn, J. M. Kendall, R. Khosla, J. Lezaun, E. M. Larson, M. Obersteiner, L. Rajamani, R. Rickaby, N. Seddon, T. Wetzer, *Nat. Clim. Change*, 2022, **12**, 15-21.

[2] L. Chen, G. Msigwa, M. Yang, A. I. Osman, S. Fawzy, D. W. Rooney, P. S. Yap, *Environ. Chem. Lett.*, 2022, **20**, 2277-2310.

[3] N. Z. Muradov, T. N. Veziroğlu, Int. J. Hydrogen Energy, 2008, **33**, 6804-6839.

[4] DENSO CORPORATION, https:www.denso.com/global/en/news/newsroom/2019/201912 09-51/, accessed in April, 2024.

[5] NIPPON MEKTRON, LTD., https://www.mektron.co.jp/company_e/company_profile_e/, accessed April 2024.

[6] K. Tsuruta, Denso Technical Review, 2011, **16**, 90-95.

[7] T. Yonezawa, S. Kaimori, M. Yamauchi, M. Kakimoto, Y. Ishii, Y. Uchita, *SEI Technical Review*, 2016, **188**, 108-111.

[8] Y. Yanagimoto, H. Oba, T. Manabe, Jp. Pat. 2022-173162, 2022.10.28 submitted.

[9] H. Yamamoto, S. Kawahara, *J. Soc. Rub Sci Tech., Jap.*, 2018, **91**, 99-104.

[10] Y. Otake, *The Society of Heating, Air-Conditioning and Sanitary Engineers of Japan*, 2006, **80**, 69-75.

[11] L. M. Gorghiu, S. Jipa, T. Zaharescu, R. Setnescu, I. Mihalcea, *Polym. Degrad. Stab.*, 2004, **84**, 7-11.

[12] Y. Aoyagi, PhD thesis, Leibniz University Hannover, 2018.

[13] T. Sato, R. Yamada, *J. Adhes. Soc. Jp*, 2019, **55**, 236-245.

[14] M. Akiba (2000), Gomu Erasutoma No Rekka To Jumyou Yosoku, Rubber Digest Co.

PCIM Europe 2024, 11– 13 June 2024, Nuremberg   DOI: 10.30420/566262290

A Systematic Comparison Study of Different Bonding Technologies for Large Substrate Attachment of Power Electronics

Lisheng Wang [1,2], Gert Rietveld [1,3], Raymond J. E. Hueting [1]

[1] University of Twente, The Netherlands
[2] E-Tronic (Guangzhou) Technology Co., Ltd., China
[3] VSL, The Netherlands

Corresponding author: Lisheng Wang, l.wang-9@utwente.nl
Speaker: Lisheng Wang, l.wang-9@utwente.nl

Abstract

Solder joints, silver (Ag) sintering, and transient liquid phase (TLP) bonding are widely used bonding technologies for packaging power modules. Each technology has advantages and limitations regarding reliability, thermal conductance and cost. So far, these technologies have not been systematically compared, and in this work we aim to fill this gap. To this end, we have performed shear strength tests on different solder, sintered Ag and TLP joint samples as a function of aging via thermal cycling and high-temperature storage tests. The joint microstructure and failure modes were analyzed using a scanning electron microscope with energy-dispersive X-ray spectroscopy. The results of our study show that sintered Ag has the maximum shear strength both before and after aging of around 70 MPa, while solder joints have the lowest shear strength that decreases from 50 MPa to about 35 MPa after thermal cycling and aging. For TLP bonding, the shear strength remains essentially stable at about 50 MPa after aging due to intermetallic compound growth. TLP bonding thus has a higher reliability than solder joints and is a cost-effective bonding alternative to Ag sintering.

1 Introduction

The substrate attachment between the substrate and heat spreader/sink plays a crucial role in power devices/modules to ensure adequate mechanical support, bonding and heat dissipation. Wide bandgap (WBG) semiconductors are expected to operate at higher temperatures than (conventional) silicon (Si) dies [1]. The higher the application temperature, the larger the interface thermal stress it will cause and the earlier failure could occur, which poses a challenge to the reliability of the bonding materials for substrate attachment [2].

Table 1 summarises the key properties of different substrate attachment bonding schemes [3–9]. Solder joints have been widely used in the power electronics industry for substrate attachments due to their reasonable cost, good wettability, high thermal conductance and low shear modulus [10]. However, solders are prone to fatigue failure due to the creep phenomenon [2, 11]. Silver (Ag) sintering is a promising bonding material for large substrate attachment due to its high reliability (10 to 100

Tab. 1: Summary of key properties of different substrate attachment bonding schemes. [3–8]

Materials properties	Solder [a]	Sintered Ag [b]	TLP [b]
CTE [ppm/K]	31	¯20	¯18
Young's modulus [GPa]	49	14-72	¯80
Shear strength [MPa]	41	60-90	55
Price [USD/gram]	0.2	7	2
Density [g/cm^3]	7.3	5.5	7
Estimated price [USD/($50\times50\times0.5$ mm^3)]	1.825	48.125	17.5

[a] $Pb_{95}Sn_5$ offers good reliability but is no longer suitable for commercial use in the automotive industry due to policy regulations. Due to its solid solution strengthening properties, the $SnSb_5$ (tin-antimony-based alloy) solder outperforms conventional lead-free solders (e.g., $Sn_{96.5}/Ag_{3.5}$) during thermal cycling and is the most widely used bonding material for substrate-attachment of power modules [9].

[b] The properties of sintered Ag and TLP strongly depend on the porosity ratio, here only the properties have been considered for porosity values above 80 %.

times higher than solder joints), and superior thermal conductance [12]. However, due to the large bonding area ($\geq 50 \times 50$ mm^2) and relatively large bonding layer thickness (≥ 100 μm), the substrate attachment material cost using Ag sintering can be very high. Therefore, a viable bonding scheme with a higher reliability than traditional solder joints, which is much less expensive than Ag sintering, should be developed. Transient liquid phase (TLP) bonding is a relatively new bonding layer technique that overcomes deficiencies in solder joints. In this case, a solid intermetallic compound (IMC) joint is formed by isothermal solidification reactions between a low-melting temperature phase (LTP) (e.g., tin/solder (Sn) and indium (In)) and high-melting temperature phase (HTP) (e.g., copper (Cu) and Ag) as the kinetics of liquid-solid diffusion is much faster than solid-state diffusion [8,13,14]. The main advantage of TLP bonding is that the melting point of the resulting IMC bonding is higher than that of traditional solder. Thus, it offers higher reliability as the creep resistance of the IMCs is improved compared to solder [8,13,14]. However, a systematic comparison in terms of reliability between solder, sintered Ag and TLP joints has not been reported so far, which is the objective of this study.

Therefore, after an explanation of the setup and approach in our experiments in Section 2, this study first investigates the reliability comparison of solder, sintered Ag and TLP focusing on thermal cycling (TC) and high-temperature storage (HTS) aging tests in Section 3.1. Then, to analyze our observations in the shear strength, the failure mode analysis and cross-sectional characterization are discussed in Section 3.2. Finally, in Section 3.3, microstructural evaluation of TLP bonding during thermal aging is analyzed, and a possible physical explanation for our observations is proposed.

2 Experiments

2.1 Preparation of joint samples

For the experiments, active metal brazed (AMB) blocks with electroless plated Ag layer were used as dummy devices [15] and aluminum (Al) blocks with a Ni(Nickel)/Cu/Ag [16] metal stack with dimensions of $17.4 \times 13.6 \times 3$ mm^3 were used as a substrate rather than Cu blocks to increase the coefficient of thermal expansion (CTE) mismatch to mimic the bonding of wide bandgap devices. A 100 μm thick SnSb$_5$ preform was used for this study. The same sintering Ag paste was used as reported in [15]. The TLP paste used in this work comprises microscale Cu particles dispended in an Sn$_x$Ag$_y$ (with an unknown composition) solder paste.

The AMB blocks were attached to Al blocks using the following process steps. Firstly, the sintering Ag and TLP wet paste are applied onto Al blocks by screen printing using a 300 μm thick 304 stainless steel mask. Then the sintering Ag and TLP wet paste are pre-dried in an air atmosphere at 80 °C and 120 °C for 60 min, respectively, to remove organic solvents inside the paste [16]. Next, AMB blocks are attached to the pre-dried paste. Finally, the pre-dried Ag and TLP paste are sintered in a vacuum sintering machine. The temperature (T) and time used for Ag sintering and TLP were set at 230 °C and 10 mins, and the pressure for Ag sintering and TLP were 15 MPa and 5 MPa, respectively [15].

2.2 Reliability testing and characterization

The shear strength of the different solder, sintered Ag and TLP joint samples was evaluated using a TRY MFM1200HF shear tester with a 500 kg die shear module. The die-shear speed and position height were set at 100 μm/s and 20 μm, respectively. Thermal Cycling (TC) and long-term high-temperature storage (HTS) aging tests were analyzed for the reliability assessment. The TC test was conducted at -50 °C to 150 °C per cycle per hour. The dwell times at -50 °C and 150 °C were both 10 minutes. The heating and cooling times were both 20 minutes. The HTS test was conducted at $T = 200$ °C in air for up to 1000 hours.

To analyze the failure modes and microstructural evaluation of the different bonding joints, the cross-section surfaces were scanned using a scanning electron microscope (SEM), model JSM-6390LA (JEOL, Japan), equipped with an energy-dispersive X-ray spectroscope (EDS). Conventional metallographic grinding and fine polishing procedures were used to prepare the cross-sections of the sintered Ag joints. Here, the grid and polishing papers used in the subsequent grinding and polishing phases had grits of 400, 1000 and 2000, respectively, and the grit sizes of the polishing papers were used in order from low to high. An ultrasonic cleaning process was used to remove the impurities on the polished surfaces as a final sample preparation step.

PCIM Europe 2024, 11– 13 June 2024, Nuremberg   DOI: 10.30420/566262290

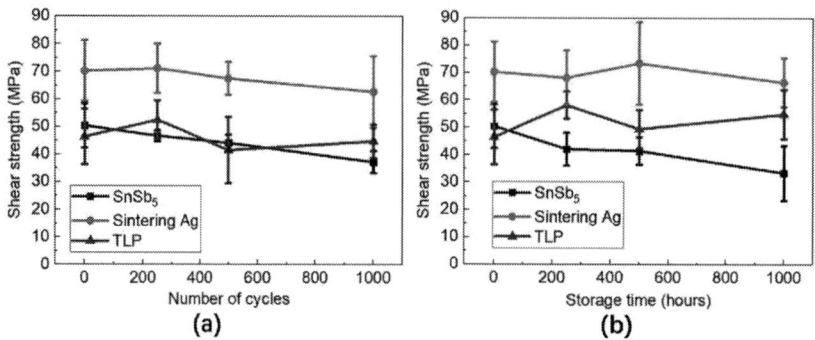

Fig. 1: Shear strength of solder, sintered Ag and TLP bonded dummy devices against (a) the number of TC cycles between -50 °C and 150 °C and (b) the HTS time at 200 °C.

3   Results and Discussion

3.1   Shear strength

Fig. 1 shows the shear strength of solder, sintered Ag and TLP-bonded dummy devices against the number of TC cycles and the HTS time. The shear strength of the solder joints steadily decreases with increasing number of TC cycles, from 50 MPa at the start to 37 MPa after 1000 cycles. Although there is also some decrease, the shear strength of sintered Ag bonding still remains above 63 MPa and is the highest compared to its counterparts. For more TC cycles, the shear strength of the TLP shows only some tiny fluctuations and is essentially stable near 46 MPa over 1000 TC cycles. As for HTS aging, the shear strength of the solder joints also shows a steady drop against the aging time from 50 MPa to 33 MPa after 1000 hours of aging. However, the shear strength of the Ag sintered joints essentially remains the same, at around 70 MPa, over the complete 1000 hours aging period. For the TLP bonding, although the shear strength shows some fluctuations, it shows a tendency to increase slightly from 46 MPa to 54 MPa after 1000 hours of HTS aging.

Metallization often plays a crucial role in exploiting the full potential of bonding materials [16]. Therefore the fracture surfaces of different substrate-attachment bonding materials are analyzed after shear strength tests, both before and after TC and HTS aging [16]. The results show that all failure modes of different materials are cohesive failures, which means that the adhesion between the metallization stacks and the Al substrate as well as that between the metal layers within those stacks is qualified. An important finding of our study is that the metallization we used in this work is no longer the bottleneck to exploit the full poten-

tial of the different bonding materials, so the shear strength represents the mechanical strength of the materials themself.

3.2   Cross-sectional characterization

As shown in the SEM picture of Fig. 2 (a), the solder tightly bonded the AMB and Al plate, and the interface is dense and void-free before aging. After 1000 hours of HTS aging, a strong delamination next to the interface between the solder and Cu is obvious in Fig. 2 (d, e). From the EDS points analysis results shown in Fig. 2 (e) it is deduced that voids are located at the CuSn IMC zone, and mainly at the Cu_3Sn layer [17]. Kirkendall voids arise due to the difference in diffusion rate between Cu and Sn [18]. In contrast, the solder layer shows a much thinner CuSn IMC layer (between the yellow lines in Fig. 2 (b, c)) after TC cycling. Cracks are located at the interface between the Cu and solder and are intermittently spread throughout the joint due to thermal stress resulting in the shear strength decrease shown in Fig. 1 (a).

As can be seen from Fig. 2 (f, g), most areas of the sintered Ag joints are dense and the void volume is in the nanoscale. The HTS aging hardly affected the sintered Ag joints which explains the shear strength results in Fig. 1 (b). However, mainly the edge of the joint shows damage, indicating an impact of thermal stress during TC aging, as shown in Fig. 2 (i). So the shear strength of sintered Ag joints decreases slightly after TC aging.

The existence of micro-sized voids and Cu particles is obvious for the TLP joints directly after bonding, see Fig. 2 (k), (l). The void size and content decrease after 1000 cycles of TC aging, shown in Fig. 2 (m). Horizontal cracks also appear at the edge of the joint after TC aging in Fig. 2 (n), while, as discussed in Section 3.1, the shear strength

2062

PCIM Europe 2024, 11– 13 June 2024, Nuremberg   DOI: 10.30420/566262290

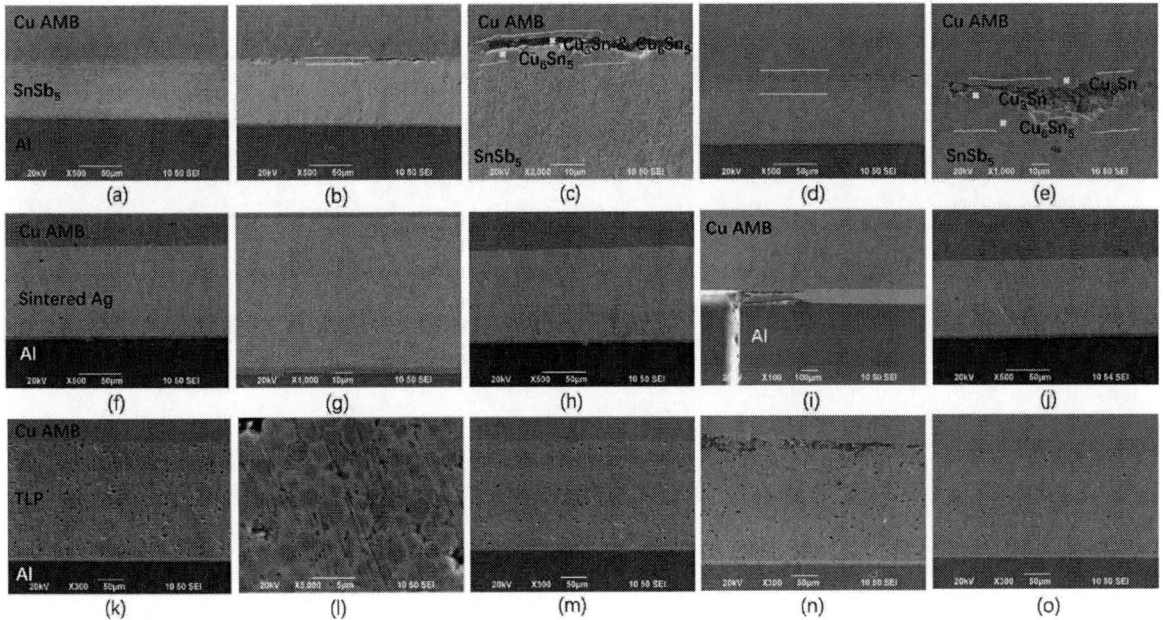

Fig. 2: Cross-sectional SEM images of (a) solder as bonded, (b) solder after 1000 cycles of TC aging, (c) zoomed in of (b), (d) solder after 1000 h of HTS aging, (e) zoomed in of (d), (f) sintered Ag as bonded, (g) zoomed in of (f), (h) sintered Ag after 1000 cycles of TC aging, (i) edge of (h), (j) sintered Ag after 1000 h of HTS aging, (k) TLP as bonded, (l) zoomed in of (k), (m) TLP after 1000 cycles of TC aging, (n) edge of (m), (o) TLP after 1000 h of HTS aging.

is stable near 45 MPa over 1000 TC cycles. The reason for this behavior will be discussed in Section 3.3. The void size and content are the lowest after 1000 hours of HTS aging as shown in Fig. 2 (o).

3.3 Microstructural evaluation of TLP bonding during thermal aging

The SEM cross-sectional images and EDS analysis of the TLP bonding layer indicate that the (clustered) Cu particles are surrounded by thin (white) SnAg solder layers which form interconnections between adjacent Cu particles, all over the TLP bonding layer between the top and bottom side of the substrate, as shown in Fig. 3 (a, b). The dividing boundary between Cu particles and the solder layer is obvious. In contrast after 1000 hours of HTS aging, as shown in Fig. 3 (i), necks are formed between adjacent Cu particles resulting in larger, irregular, other-than-round-shaped Cu particles. Moreover, the thickness and size of the white SnAg layers are much smaller than in the original case shown in Fig. 3 (a). EDS mapping images in Fig. 3 show that Cu and Sn atoms interdiffused after HTS aging. The EDS spectroscopy points analysis also indicates that the SnAg solder has fully been transformed into $SnAg_3$ IMC (white

layers) [19]. At the same time, mutual diffusion occurred between the Sn and Cu atoms and transformed into Cu_3Sn [14] (particles and necks) due to the thermally-driven HTS aging, as shown in Fig. 3 (j, k, l). It is inferred that the mutual diffusion effect leads to a reduction in the size of the white solder layers and voids. Since the thermal effect of TC aging is weaker than that of HTS aging, the mutual diffusion of Sn and Cu atoms is less, resulting in fewer IMCs and relatively larger void size and content in Fig. 2 (m) compared to that in Fig. 2 (o). The remaining (clustered) Cu particles in Fig. 3 (h) are obvious compared with Fig. 3 (l), confirming the weaker mutual diffusion effect during the TC aging process.

The obtained microstructural evolution results of TLP bonding during thermal aging can be better explained by a schematic illustration as in Fig. 4. First, the original metallurgical microstructure of the TLP layer is composed of SnAg solder and Cu particles. Submicron voids are formed between big Cu particles since there is insufficient solder to fill all gaps. After thermal aging, Ag, Cu and Sn atoms intermix and the IMC could be formed according to

2063

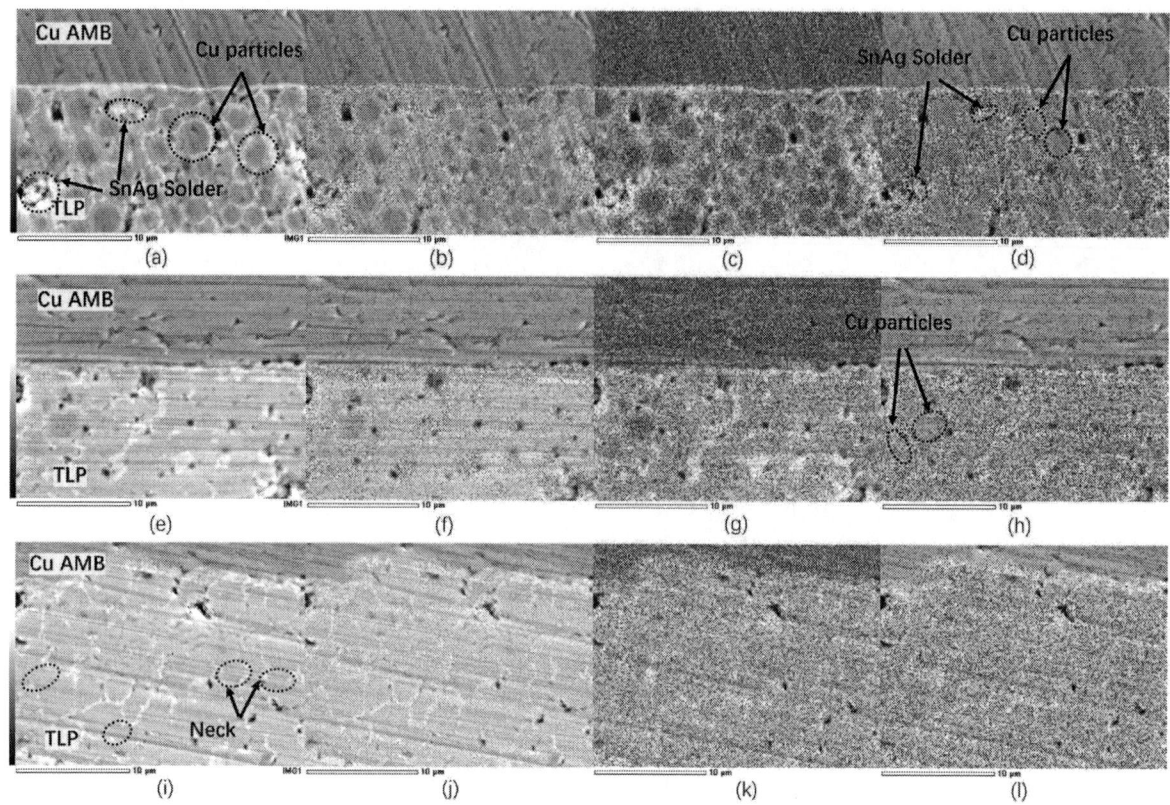

Fig. 3: SEM cross-sectional images and EDS mapping images of the TLP bonding layer: SEM image (a) directly after bonding, (e) after 1000 cycles of TC aging, and (i) after 1000 hours of HTS aging. In Figs. 3 (b, f, j) the green dots represent Ag, in Figs. 3 (c, g, k) red points are Cu and Figs. 3 (d, h, l) the purple points represent Sn.

the chemical reaction [14]:

$$(9x-3y)Cu + 3Sn_xAg_y \xrightarrow{\text{Thermal}} (3x-y)Cu_3Sn + ySnAg_3. \quad (1)$$

The Ag atoms diffuse and are enriched together to form a $SnAg_3$ IMC with Sn atoms, and then the spaces left by those Ag atoms are occupied by mobile Cu atoms to form a Cu_3Sn IMC resulting in larger grey spherical shape areas in Fig. 3 (i) compared with the Cu particles of Fig. 3 (a). Finally, for longer HTS times there will be more mutual diffusion, causing the solder layer to be fully transformed into $SnAg_3$ IMC (white layer) while the Sn and Cu atoms are fully transformed into Cu_3Sn (particles and necks).

Importantly, the shear strength of the TLP joints is significantly affected by the interfacial IMC. The shear strength of the Cu_3Sn IMC is larger than that of the SnAg solder [20], so although there are some cracks caused by thermal stress inside the TLP bonding joint after TC aging, the shear strength remains essentially stable due to the IMC

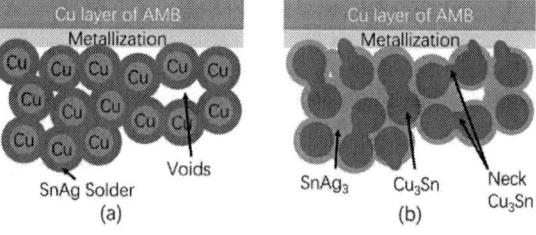

Fig. 4: Schematic cross-sectional illustration of the microstructural evolution for the TLP bonding: (a) as bonded and (b) after 1000 hours HTS aging at 200 °C.

growth, as shown in Fig. 1 (a). The improvement in mechanical properties introduced by the Cu_3Sn IMC leads to an increase in shear strength, as shown in Fig. 1 (b), since there is no thermal stress during HTS aging.

4 Conclusion

We have compared the reliability of three bonding techniques for (large-size) substrate attachment of power electronics by using 17.4×13.6 mm^2 Al blocks as substrate and AMB blocks as dummy devices. The results of our study show that sintered Ag joints have the maximum shear strength both before and after aging of around 70 MPa, while solder joints have the lowest shear strength that decreases from 50 MPa to about 35 MPa after thermal cycling and aging. For TLP bonding, the shear strength remains essentially stable at around 50 MPa after aging due to intermetallic compound growth. To gain insight into the microstructure and the failure modes of all bonding techniques, extensive SEM and EDS analysis was performed on cross-sections of the bonds. This work shows that silver (Ag) sintered joints outperform their bonding counterparts under study in shear strength both before and after aging. Solder joints appear prone to form Kirkendall voids and thermal stress cracks after aging, thereby having a minimal shear strength. Although some thermal stress cracks are inside the transient liquid phase (TLP) bonding, the shear strength remains essentially stable after aging most likely due to the formation of a Cu_3Sn IMC.

References

[1] K. Suganuma, "Interconnection technologies," In *Wide Bandgap Power Semiconductor Packaging*, K. Suganuma Eds., pp. 57-80, Woodhead Publ., 2018.

[2] L. Wang, W. Wang, R. J. E. Hueting, G. Rietveld and J. A. Ferreira, "Review of Topside Interconnections for Wide Bandgap Power Semiconductor Packaging," *IEEE Trans. Power Electron.*, vol. 38, no. 1, pp. 472–4990, 2023.

[3] W. Yu, H. Xie, L. Yin, J. Zhao, L. Xia, *et al.*, "Exceptionally high thermal conductivity of thermal grease: Synergistic effects of graphene and alumina," *International Journal of Thermal Sciences*, vol. 91, pp. 76–82, 2015.

[4] Y. He, "Rapid thermal conductivity measurement with a hot disk sensor: Part 2. Characteri-

zation of thermal greases," *Thermochimica Acta*, vol. 436, no. 1, pp. 130–134, 2005.

[5] M. F. M. Sabri, D. A. Shnawah, I. A. Badruddin, S. B. M. Said, F. X. Che, *et al.*, "Microstructural stability of $Sn_1Ag_{0.5}Cu_xAl$ (x= 1, 1.5, and 2 wt.%) solder alloys and the effects of high-temperature aging on their mechanical properties," *Materials Characterization*, vol. 78, no. 1, pp. 129–143, 2013.

[6] W. W. Sheng and R. P. Colino, "Materials," In *Power electronic modules: design and manufacture*, 1st ed. Boca Raton, FL, USA, CRC Press LLC, 2004.

[7] A. A. Wereszczak, D. J. Vuono, H. Wang, M. K. Ferber, and Z. Liang, "Properties of bulk sintered silver as a function of porosity," Oak Ridge National Lab. (ORNL), Oak Ridge, TN, USA, Tech. Rep., Jun. 2012.

[8] H. Shao, A. Wu, Y. Bao, Yue. Zhao and L. Liu, "Thermal reliability investigation of Ag-Sn TLP bonds for high-temperature power electronics application," *Microelectron. Rel.*, vol. 91, pp. 38–45, 2018.

[9] J. Booth, M. Varley, D. Slack, P. M. Croft, S. Jones, *et al.*, "High Reliability Large Area Substrate Solder Interconnect by Embedded Mesh Technique," in *Proc. Int. Exhib. Conf. Power Electron., Intell. Motion (PCIM), Renew. Energy Energy Manage.*, pp. 1432–1438, Nuremberg, Germany, May 16-18, 2017.

[10] R. Khazaka, L. Mendizabal, D. Henry and R. Hanna, "Survey of High-Temperature Reliability of Power Electronics Packaging Components," *IEEE Trans. Power Electron.*, vol. 30, no. 5, pp. 2456–2464, 2015.

[11] S. Gao, Z. Yang, Y. Tan, X. Li, X. Chen, *et al.*, "Bonding of Large Substrates by Silver Sintering and Characterization of the Interface Thermal Resistance," *IEEE Trans. on Industry Appl.*, vol. 55, no. 2, pp. 1828-1834, 2015.

[12] T. G. Lei, J. N. Calata, G. Q. Lu, X. Chen and S. Luo, "Low-Temperature Sintering of Nanoscale Silver Paste for Attaching Large-Area ($>$100 mm^2) Chips," " *IEEE Trans. Compon. Packag. Technol.*, vol. 33, no. 1, pp. 98–104, 2010.

[13] L.R. Billa, Y. Wang, T. Grant, X. Li, H. Neal, *et al.*, "Transient Liquid Phase Bond Reliability Evaluation of Die-attach for Power Module Packaging," in *24th Eur. Conf. on Power Electron. and App. (ECCE-EPE)*, , pp. 1—7, Hanover, Germany, Sep. 05-09, 2022.

[14] H. Tatsumi, A. Lis, H. Yamaguchi, T. Matsuda, T. Sano, *et al.*, "Evolution of transient liquid-phase sintered Cu–Sn skeleton microstructure during thermal aging," *Appl. Sci.*, vol. 9, pp. 1–12, 2019.

[15] L. Wang, Z. Lei, R. Liang, G. Rietveld and R. J. E. Hueting, "A new SiC power module assembly based on silver sintering bonding," in *25th Eur. Conf. on Power Electron. and App. (ECCE-EPE)*, pp. 1–8, Aalborg, Denmark, Sep. 04-08, 2023.

[16] L. Wang *et al.*, "Reliability of Ag sintered bonding of metal-plated Aluminum surfaces," *To be published*.

[17] S.E. Jeong, S.B. Jung, and J.W. Yoon, "A study of the growth rate of Cu-Sn intermetal-lic compounds for transient liquid phase bonding during isothermal aging," in *20th Electron. Packag. Tech. Conf. (EPTC)*, pp. 225—228, Singapore, Dec. 04-07, 2018.

[18] T. C. Chiu, K. Zeng, R. Stierman, D. Edwards and K. Ano, "Effect of thermal aging on board level drop reliability for Pb-free BGA packages," in *54th Electron. Comp. Tech. Conf.*, pp. 1256—1262, Las Vegas, NV, USA, June. 04, 2004.

[19] N, Saud, K.F. Ng, and R.M. Said, "Microstructure evolution and phase formation of transient liquid phase Sn/Ag solder alloy via multiple reflow soldering method," *AIP Conf. Proc.*, vol. 2347, No. 1, pp. 020193(1)–(9), 2021.

[20] B. S. Lee and J. W. Yoon, "Cu-Sn intermetallic compound joints for high-temperature power electronics applications," in *J. Electron. Mater.*, vol. 47, no. 1, pp. 430–5, 2018.

PCIM Europe 2024, 11– 13 June 2024, Nuremberg   DOI: 10.30420/566262291

Stability of Pressure Sintered Interconnects as a Function of Temperature and Environmental Conditions

Kentaro Yoshioka[1], Mutsuharu Tsunoda[1], Akihiro Mochizuki[1], Maurizio Fenech[2]

[1] MacDermid Alpha Electronics Solutions, Hiratsuka, Japan

[2] MacDermid Alpha Electronics Solutions, Langenfeld, Germany

Corresponding author: Kentaro Yoshioka, kentaro.yoshioka@MacdermidAlpha.com

Speaker: Kentaro Yoshioka, kentaro.yoshioka@MacdermidAlpha.com

Abstract

Performance and microstructural changes of pressure sintered silver die attach material are assessed following high temperature storage in different atmospheres and using different substrates. The aged joints are characterized using C-SAM and Electron Microscopy whereas the mechanical properties are assessed using Die Shear. The results show that oxygen has a strong influence on the residual strength of sintered joints and that the dense microstructure obtained by pressure sintering is stable following storage at 300°C for 500 hours under N_2 atmosphere. It is suggested that sintered joints have excellent stability in power modules packaged in high quality over-molding materials.

1 Introduction

Pressure silver sintering has become the de-facto process for realizing highly reliable die attach joints for automotive power electronics applications due to a number of positive attributes when compared to solder. Pressure sintered joints based on Argomax® nano-particle sinter technology have consistently shown to have a homogeneous and highly dense sintered microstructure that allows for enhanced reliability, a predictable lifetime, and unparalleled thermal transfer properties [1], [2], [3]. Going to a full sintered die attach solution allows the power module developer to comply with the EU commission´s RoHS directive [4] of avoiding Pb within electronic systems while allowing high Tj devices such as SiC based power modules to operate reliably and with maximum heat dispersion capacity. Sintering a ceramic substrate or a completely molded power module to a heatsink allows for the lowest stack Rth, allowing the power module developer to extract more power density out of a given system.

Implementing sintering technology at different levels in an automotive power module requires reliability testing based on guidelines such as the ECPE AQG 324 [5]. As pressure sinter technology is getting more widely adopted, special industrial applications could require the joint material to operate in conditions which are harsher than what standard automotive test recommendations would anticipate. High temperature exposure tests for pressure-less sintered assemblies have been conducted [6], [7], showing that sintered silver can be susceptible to atomic reorganization. In this study, the stability of a statically stressed highly dense pressure sinter joint that is subjected to high temperature under different environmental conditions is investigated.

2 Experiment

2.1 Sample preparation

Fig.1 shows the details and illustrations of the sample setup used in this study. For high temperature performance characterization, MacDermid Alpha Argomax® 5020 sinter paste was used to join SiC blank dies to bare Cu and Ni/Au plated AMB substrates. The paste was printed using a 150 μm stencil having a 15mm*15mm square aperture and subsequently dried at 130°C for 30 minutes in air. To obtain different sintered joint properties, a 7mm*7mm die and four 2.5mm* 2.5mm SiC dies were sintered using two different sintering conditions: 250°C,10MPa, 90 seconds and 300°C,20MPa,180 seconds, respectively. In addition, samples sintered on the Cu substrate were encapsulated in silicone gels to approximate practical conditions. Two commercially available two-component gels marketed for high heat durability sealing were used for encapsulation. The two-part component was thoroughly stirred manually, and vacuum defoaming was performed using a rotary pump until the gel was completely free of

2067

air bubbles. Gel-encapsulation was made in an aluminum cup that had been cleaned and air-baked at 350°C in advance, and vacuum defoaming was performed again until no air bubbles were visible. The gel was cured at 100°C for 40 minutes according to the recommended conditions.

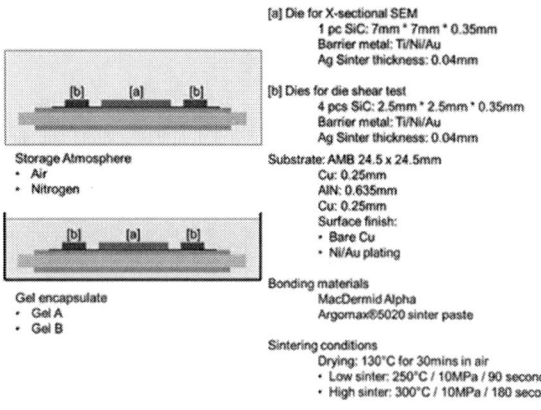

[a] Die for X-sectional SEM
1 pc SiC: 7mm * 7mm * 0.35mm
Barrier metal: Ti/Ni/Au
Ag Sinter thickness: 0.04mm

[b] Dies for die shear test
4 pcs SiC: 2.5mm * 2.5mm * 0.35mm
Barrier metal: Ti/Ni/Au
Ag Sinter thickness: 0.04mm

Substrate: AMB 24.5 x 24.5mm
Cu: 0.25mm
AlN: 0.635mm
Cu: 0.25mm
Surface finish:
• Bare Cu
• Ni/Au plating

Bonding materials
MacDermid Alpha
Argomax®5020 sinter paste

Sintering conditions
Drying: 130°C for 30mins in air
• Low sinter: 250°C / 10MPa / 90 seconds
• High sinter: 300°C / 10MPa / 180 seconds

Storage Atmosphere
• Air
• Nitrogen

Gel encapsulate
• Gel A
• Gel B

Fig.1 Sample conditions

2.2 Sample Testing and Analysis

High-temperature storage tests were conducted in air and N_2 atmospheres, with temperatures of 250°C and 300°C for 500 hours using a conventional electric oven. The full test matrix is presented in Table 1. The bonding state of the SiC dies were observed nondestructively using *Insight* constant-depth mode scanning acoustic microscope (C-SAM). Four 2.5mm*2.5mm SiC dies were subjected to die shear testing using a Nordson bond tester. The fracture surface was observed using an optical microscope. Following die shear testing, the sintered substrate was embedded in epoxy resin and prepared for cross section. In order to avoid damage to the observation area, the 7mm*7mm SiC die was cut using a wheel cutter and progressively ground and polished to the observation position. The observation area was finished by ion milling after polishing. Observations and elemental analysis were performed using a Hitachi High-Tech scanning electron microscope(SEM) equipped with Energy Dispersive X-ray Spectroscopy (EDS).

3 Results and discussions

3.1 Shear and Fractography for Nitrogen and Air exposed non-encapsulated samples and NiAu substrate finish

Fig.2 shows the die shear test results and the fracture surface of samples sintered at 250°C, 10MPa, 90 seconds on Ni/Au finished substrates before and after high temperature storage.

Substrate Finish	Sintering conditions Temp/Pressure/Time	Encapsulation	Storage condition	Storage Temperature
Cu	250°C/10MPa/90sec	None	Air	250°C
Cu	250°C/10MPa/90sec	None	Air	300°C
Cu	250°C/10MPa/90sec	None	N2	250°C
Cu	250°C/10MPa/90sec	None	N2	250°C
Cu	300°C/20MPa/180sec	None	Air	250°C
Cu	300°C/20MPa/180sec	None	Air	300°C
Cu	300°C/20MPa/180sec	None	N2	250°C
Cu	300°C/20MPa/180sec	None	N2	250°C
Ni/Au	250°C/10MPa/90sec	None	Air	250°C
Ni/Au	250°C/10MPa/90sec	None	Air	300°C
Ni/Au	250°C/10MPa/90sec	None	N2	250°C
Ni/Au	250°C/10MPa/90sec	None	N2	250°C
Ni/Au	300°C/20MPa/180sec	None	Air	250°C
Ni/Au	300°C/20MPa/180sec	None	Air	300°C
Ni/Au	300°C/20MPa/180sec	None	N2	250°C
Ni/Au	300°C/20MPa/180sec	None	N2	250°C
Cu	250°C/10MPa/90sec	Gel A	Air	250°C
Cu	250°C/10MPa/90sec	Gel A	Air	300°C
Cu	250°C/10MPa/90sec	Gel B	Air	250°C
Cu	250°C/10MPa/90sec	Gel B	Air	300°C

Table 1 Test matrix

The die shear strength of the sample stored at high temperature in the air showed a slight decrease as the holding temperature increased. On the other hand, the die shear strength of the samples stored at high temperatures in N_2 was increased. A significant spread in the die shear strengths is observed. This is mainly attributed to the fact that die fracture and explosive disintegration is often observed during testing due to the very high joint strength offered by pressure sintering.

Observation of the fracture surface showed no change in the samples stored at high temperatures in N_2. There was a clear difference between the outer and the inner part of the fracture surface of the sample stored at high temperature in the air, with the outer area being wider when stored at 300°C than at 250°C.

It is suggested that the improvement in die shear strength in the N_2-stored samples is due to further diffusion and consolidation of the initial sintered microstructure at high temperatures. The dynamics of consolidation are understood to be active also in the air stored samples, however any strength increase is countered by a progressive degradation of joint quality on the perimeter of the die.

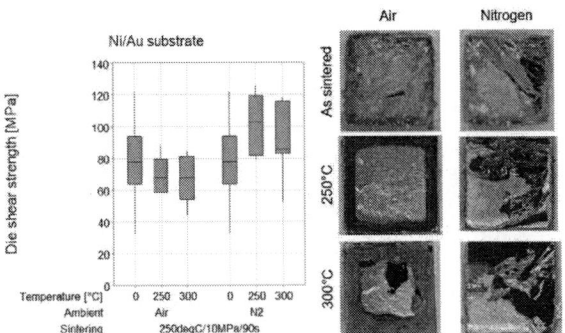

Fig.2 Ni/Au substrate low sinter

Fig.3 shows the die shear test results and the fracture surface of samples sintered at 300°C, 20MPa, 180 seconds on Ni/Au finished substrates before and after high temperature storage. It is understood that the higher initial die shear strength is due to the progress of the sintered structure due to the higher and longer sintering temperature, pressure, and time. The die shear strength after holding at high temperature was similar between low sinter and high sinter conditions. This suggests that high temperature exposure homogenizes sintered joints with a degree of independence from the original sintering conditions.

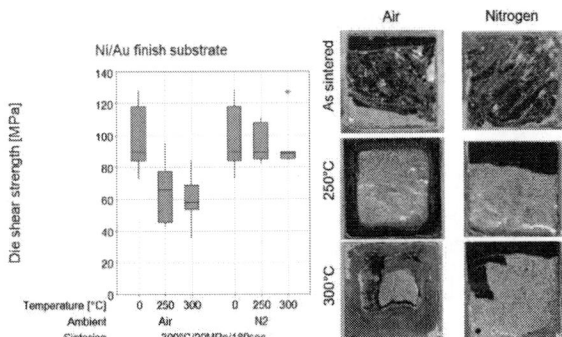

Fig.3 Ni/Au Substrate High sinter

3.2 Shear and Fractography for Nitrogen and Air exposed non-encapsulated samples and Cu finish

Fig.4 shows the die shear test results and the fracture surface of samples sintered at 250°C, 10MPa, 90 seconds on Cu substrates before and after high temperature storage. The initial die shear strength of the Cu substrate sample was lower than that of the NiAu substrate. NiAu plated substrates do not form immiscible oxides on their surface, whereas air exposed copper substrates produce a native oxide. As silver-to-copper solid state diffusion dynamics are highly influenced by the level of native

oxide, it is understood that the low sintering conditions employed in Fig4. are not sufficient to a fully diffused bond. Recovery in die shear strength is observed after high temperature storage, resulting in a shear strength that is equivalent the NiAu substrate sample. The color of fracture surface on the outside of the die of the sample stored in Air was reddish brown. It is considered that copper oxide was generated on the Cu substrate, similar to previous reports [8].

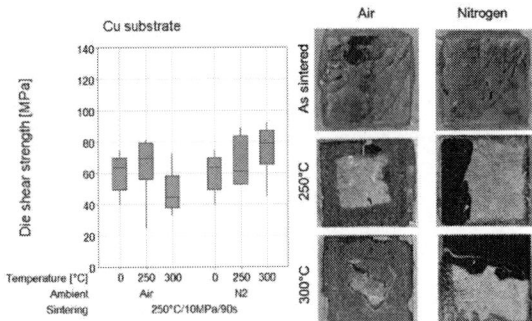

Fig.4 Cu substrate low sinter

Fig.5 shows the die shear test results and the fracture surface of samples sintered at 300°C, 20MPa, 180 seconds on Cu substrates before and after high temperature storage. It is suggested that the high initial die shear strength is due to the progress of the sintered structure due to the higher and longer sintering temperature, pressure, and time. After high temperature storage in air, the die shear strength decreased to the same level as low sinter due to the effect of copper oxidation. No significant changes were observed in die shear strength or fracture surface during high temperature storage in N_2.

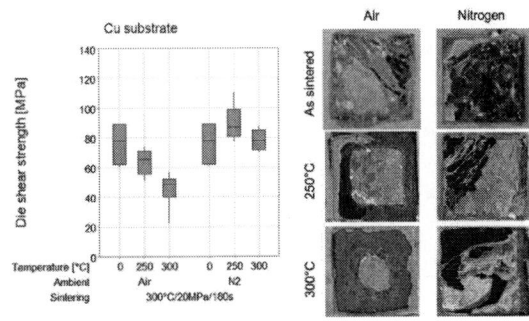

Fig.5 Cu substrate High sinter

3.3 Shear and Fractography encapsulated samples and Cu finish

Fig.6 shows the appearance of samples encapsulated with two types of gel before and after high-temperature storage at 250°C for 500h in air. None of the gels used in the experiment could withstand high-temperature storage at 300°C for 500 hours, and the entire gel was completely hardened and crushed into pieces. After high-temperature storage at 250°C for 500 hours, the surface of Gel A appeared to crack and disintegrate, however the inside maintained a gel state. Gel B did not show any fractures, and although it was slightly harder than its initial state, its surface maintained some elasticity as a gel.

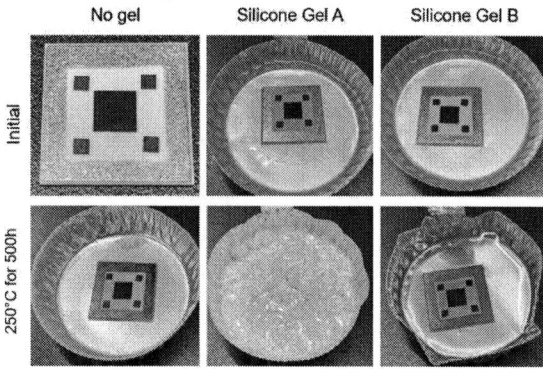

Fig.6 Apperance of gel encapsulate

Fig.7 shows the die shear strength and fracture surface of samples sintered at 250°C, 10MPa, 90 seconds on Cu substrate which are encapsulated with gels A and B before and after high temperature storage in air at 250°C for 500 hours. The die shear strength of the gel-encapsulated samples after high-temperature storage in Air and was significantly improved, consistently being higher than 110MPa.

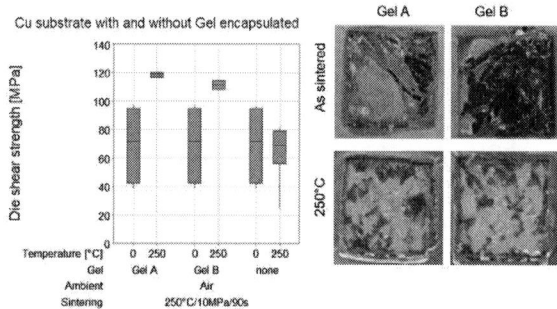

Fig.7 Gel encapsulate Cu substrat low sinter

Observation of the fracture surface showed that there was no change between the outside and inside of the die, and no color change was observed as seen in copper oxide. It is also believed that the

reason for the high and stable die shear strength is that no cracks occurred in any of the dies, allowing for the measuring equipment to read the true full strength of the sintered joint.

3.4 Cross Section and CSAM analysis

Figure 8 shows the initial cross-sectional microstructures of samples sintered at 250°C, 10MPa, 90 seconds and 300°C, 20MPa, 180 seconds, respectively. A dense sintered structure with evenly distributed sub-micron pores was obtained under both sintering conditions, but the sample sintered at 300°C, 20MPa, and 180 seconds had slightly larger voids that also exhibited a more rounded shape. This indicates that the sintered structure progressed into further interdiffusion and pore coalescence a more time and temperature was provided for the reaction to happen.

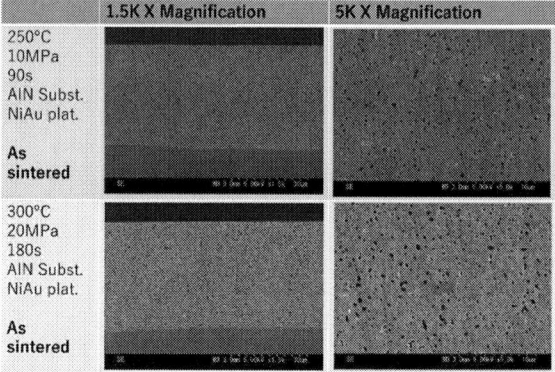

Fig.8 Sinter microstructure of samples with Low and High sinter conditions.

Fig.9 shows C-SAM images and cross-sectional SEI of a 7mm*7mm SiC die sample sintered with 250°C, 10MPa, 90 seconds on Ni/Au finish substrate after high-temperature storage at 300°C for 500h in Air and N_2, respectively. The cross-sectional SEM observation of the air-stored sample shows drastic coarsening, which is thought to correspond to the white area in the C-SAM image due to the reflection of ultrasonic waves from the relatively large voids. The voids are understood to be resulting from coarsening in the area starting from the periphery and going in several hundred micrometers into the die. SEM observation of N_2-stored sample shows that the shape of the vacancies became more rounded compared to initial one shown in Fig. 8, suggesting that a slight progression of sintering by simple atomic diffusion at high temperatures had occurred. The difference in microstructure changes after high temperature storage in Air and N_2 suggests that coarsening is largely influenced by atmospheric oxygen. This result is consistent with previous reports for pressure-less sintering [6], [7], albeit to a significantly lesser degree of crystallographic reorganization.

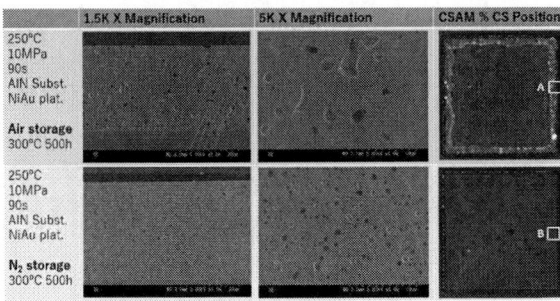

Fig.9 Sinter microstructure at the edge of the die for NiAu finished substrates as a function of storage conditions.

Fig.10 shows C-SAM images and cross-sectional SEI of a 7mm*7mm SiC die sample sintered with 250°C, 10MPa, 90 seconds on Cu substrate after high-temperature storage at 300°C for 500h in Air and N₂, respectively. SEM observation of air-stored sample shows a copper oxide layer connecting from the periphery along the substrate interface. As the silver exists between the substrate surface and the oxide layer, it is understood that a phenomenon of copper migration from the substrate diffusing inwards into the silver sintering zone is occurring. Voids at the interface between Cu substrate-silver sintered body also suggest inward diffusion of copper. Cu substrates do not show oxidation or drastic coarsening after high temperature storage in N₂ atmosphere.

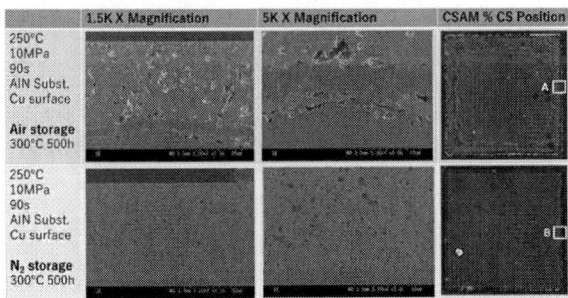

Fig 10 Sinter microstructure at the edge of the die for Cu finished substrates as a function of storage conditions.

Fig.11 shows the cross-sectional structure from the edge of the die to about 850 micrometers inside after high-temperature storage at 300°C for 500 hours in air. Near the edges of the die, the voids become coarser, and the copper oxide layer is thicker. From about 800 micrometers inward to

the edge of the die, no copper oxide layer is observed, and the size of the coarsening is similar to that of the sample after high temperature storage in N₂. This indicates that the oxygen necessary for drastic coarsening and the formation of copper oxide only progresses to this extent even during storage at extremely high temperatures of 300°C and 500 hours.

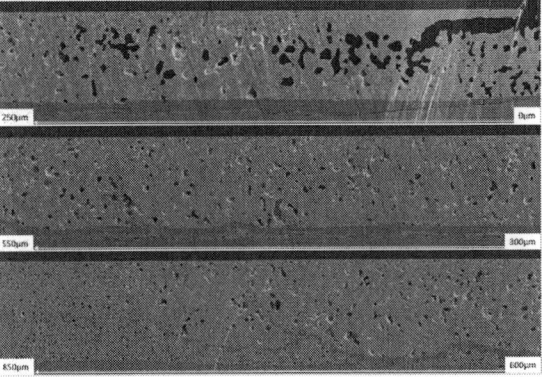

Fig 11 Microstructure coarsening and copper migration from the edge of the die down to 850µm into the center of the die. Storage at 300°C for 500h

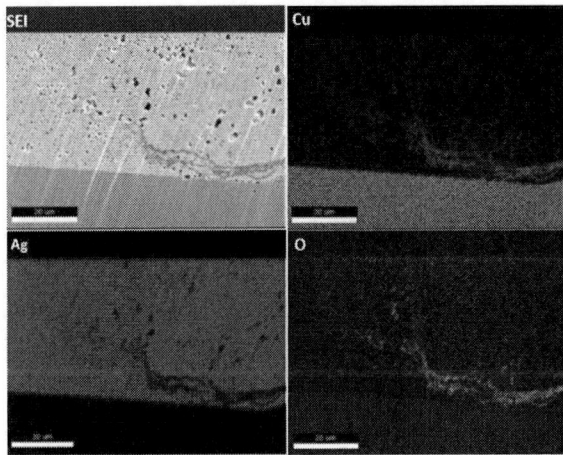

Fig 12 Elemental map showing migration of copper from substrate and precipitation into an oxide.

Fig.12 shows the result EDS mapping analysis at the end of the copper oxide layer. No oxygen was detected inside the end of the copper oxide layer. This indicates that oxygen enters only from the outer periphery of the die that is exposed to the atmosphere. Interestingly, copper and oxygen were simultaneously detected in the same location. Cu and Ag are consistently distinct and there is no evidence of intermetallics forming and Cu does not diffuse into the Ag inside the end of the copper oxide layer. This suggests that the behavior of Cu migration is mainly driven by the changing of the metal into stable copper oxide.

Fig.13 shows the results of cross-sectional structure after high-temperature storage at 250°C for 500 hours in air with and without silicone gel encapsulating. As shown in Fig.13(A), a copper oxide layer and drastic coarsening were observed in the structure of the sample without gel encapsulating. The extent of copper oxide layer and coarsening area was smaller compared to that after 300°C 500h high temperature storage in Air shown in Fig.11, but consistent with what was observed in Fig 3 and Fig 4 for 250°C storage conditions. As shown in Fig.13(B), the extent of the copper oxide layer and coarsening region was further reduced in the gel-encapsulated sample. Through this experiment, we were able to sufficiently reduce the influence of oxygen even with simple gel encapsulation at the laboratory level. This suggests that if a practical module encapsulation technique is used, there will be no major deterioration of the pressed silver sintered structure even in a high Tj environment.

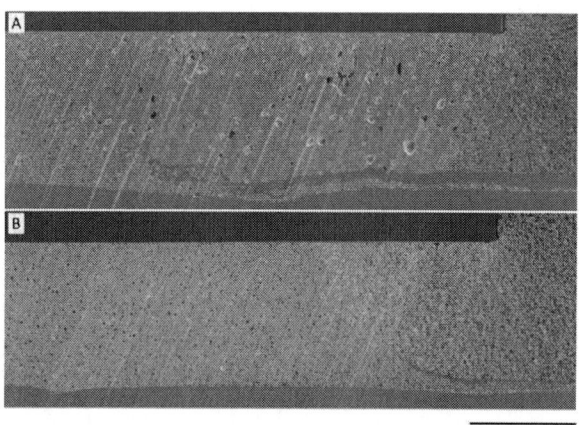

30μm

Fig 13 Microstructure coarsening and copper migration for air exposed die (A) and Gel encapsulated die (B). Storage at 250°C for 500h

Ellingham diagrams for silver, copper and other metals are shown in Fig.14. The diagrams explain well our observations of oxygen migration within a the sintered joint and favored precipitation of oxygen in copper rich areas. As can be seen from the diagram, the standard free Gibbs energy of silver oxide is near zero, which indicates that in the temperature range of our experiments, silver and silver oxide are in an unstable region where they can transform into either. Copper oxide is stable at the standard free Gibbs energy of the reaction between metal and metal oxide in the high-temperature storage temperature range of 250°C to 300°C

used in this study. This is consistent with the fact that production and growth of copper oxide layer near the interface in the Cu substrate sample after high temperatures storage in air. On the other hand, the instability of this silver-silver oxide is thought to be one of the reasons that not only solid-phase diffusion of Oxygen in Ag is prevalent, but also the drastic coarsening during high-temperature storage in air.

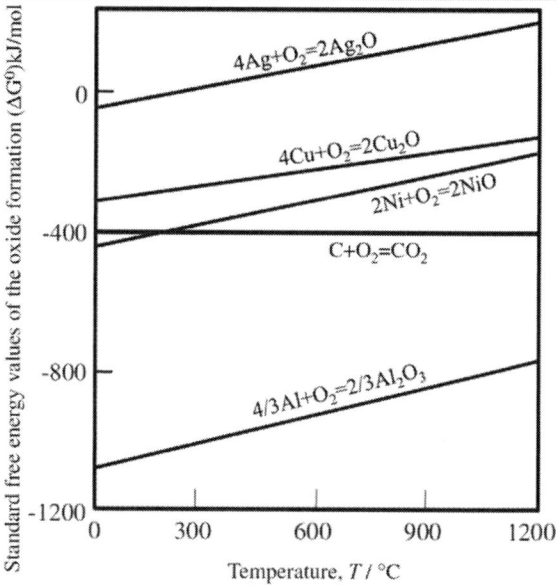

Fig.14 Ellingham diagrams for silver and various metals [9]

Of interest is that since the temperature range of 250°C to 300°C is similar to the sintering temperature range for general silver sintering, the Ag/O interdiffusion phenomenon may have a positive effect also on achieving good sintered bonding in a relatively short time in an air environment by accelerating solid-phase diffusion of silver during sintering.

4 Conclusions

The results of this study show that the dense microstructure obtained by pressure sintering i s stable at 300°C for 500 hours in an oxygen free environment. Dependencies of performanc e on surface plating is also investigated, wher e added high temperature stability could be o btained by having a passive underlayer such as nickel that prevents copper diffusion. Stabili ty in over-molded or gel encapsulated power module is expected to be ensured as long as the employed encapsulants do not allow access for oxygen. Over-all our studies indicate that basic environmental oxygen protection is sufficient to extract the full potential of silver sintering within the ensuring a dependable joint operating in the high Tj regime.

References

[1] F. Le Henaff, S. Azzopardi, E. Woirgard, T. Youssef, S. Bontemps, J. Joguet, "Lifetime Evaluation of Nanoscale Silver Sintered Power Modules for Automotive Application Based on Experiments and Finite-Element Modeling", IEEE Transactions on Device and Materials Reliability, vol.15, issue 3, pp. 326-334, 2015.

[2] Jianfeng Li, Christopher Mark Johnson, Cyril Buttay, Wissam Sabbah, Stéphane Azzopardi, "Bonding strength of multiple SiC die attachment prepared by sintering of Ag nanoparticles", Journal of Materials Processing Technology, vol.215, pp. 299-308, 2015.

[3] Jingru Dai, Jianfeng Li, Pearl Agyakwa, Christopher Mark Johnson, "Time-Efficient Sintering Processes to Attach Power Devices Using Nanosilver Dry Film", Journal of Microelectronics and Electronic Packaging, vol.14, issue 4, pp140-149, 2017.

[4] "DIRECTIVE 2011/65/EU OF THE EUROPEAN PARLIAMENT AND OF THE COUNCIL of 8 June 2011 on the restriction of the use of certain hazardous substances in electrical and electronic equipment", Official Journal of the European Union, 2011.

[5] "Qualification of Power Modules for Use in Power Electronics Converter Units (PCUs) in Motor Vehicles", ECPE Guideline AQG 324, 2018.

[6] Chuantong Chen, Chanyang Choe, Dongjin Kim, Zheng Zhang, Xu Long, Zheng Zhou, Fengshun Wu, Katsuaki Suganuma, "Effect of oxygen on microstructural coarsening behaviors and mechanical properties of Ag sinter paste during high-temperature storage from macro to micro", Journal of Alloys and Compounds, Vol. 834, 5, 155173, 2020.

[7] S.A. Paknejad, G. Dumas, G. West, G. Lewis, S.H. Mannan, "Microstructure evolution during 300 °C storage of sintered Ag nanoparticles on Ag and Au substrates", Journal of Alloys and Compounds, vol. 617, 25, pp. 994-1001, 2014.

[8] F. Yang, W. Zhu, W. Wu, H, Ji, C. Hang, M. Li, "Microstructural evolution and degradation mechanism of SiC–Cu chip attachment using sintered nano-Ag paste during high-temperature ageing", Journal of Alloys and Compounds, vol. 846, 15, 156442, 2020.

[9] L. S. Darken, R. W. Gurry, "Physical Chemistry of Metal", McGraw-Hill, pp. 342371, 1953

PCIM Europe 2024, 11– 13 June 2024, Nuremberg   DOI: 10.30420/566262292

The Effect of Nano-Cu Interconnection Materials on the Thermo-mechanical Properties of SiC Double-Sided Power Modules

Suhang Wei[1], Jiaxin Liu[1], Weishan Lv[2], Cai Chen[1], Yong Kang[1], Yue Wu[3], Zhipeng He[3]

[1] State Key Laboratory of Advanced Electromagnetic Technology, Huazhong University of Science and Technology, China

[2] School of Mechanical Science and Engineering, Huazhong University of Science and Technology, China

[3] State Key Laboratory of HVDC, Electric Power Research Institute, CSG, China

Corresponding author: Jiaxin Liu, liujx711@hust.edu.cn
Speaker: Suhang Wei, shwei@hust.edu.cn

Abstract

The high porosity and high coefficient of thermal expansion (CTE) of nano-silver limit its application in SiC double-sided power modules, and the easy electromigration of nano-silver reduces its reliability. Stress fluctuations and residual stress during the sintering process directly affect the yield limit and fatigue strength of the power module and determine the reliability of solder interconnection. Therefore, this paper uses nano-copper slurry with lower cost and better CTE matching as the interconnect material for SiC double-sided power modules to solve these problems. Through finite element simulation, we conducted thermo-mechanical stress simulations on three SiC double-sided power modules whose interconnect layer materials are nano-copper, nano-silver, and SAC305. The results show that during the sintering process, compared with nano-silver, the stress fluctuation of the three layers of nano-copper slurry is greatly reduced, and the average residual stress is reduced by 33.58%, 14.79%, and 37.22% respectively. The thin interconnect layer thickness of 50um makes the steady-state junction temperature of the nano-copper module lower and has a heat dissipation capability comparable to that of nano-silver. During the sintering heating and cooling processes, the overall average stress of the nano-copper module was 11.76% and 4.98% lower than that of the nano-silver module respectively, proving that nano-copper has an excellent ability to alleviate stress concentration and improve the reliability of SiC double-sided power modules.

1   Introduction

In recent years, SiC MOSFET power modules have gradually developed towards high power density, high frequency, small size, and high withstand voltage requirements. As device capacity and switching frequency increase, the power loss and operating junction temperature of semiconductor devices are higher. Therefore, reliable packaging of electrical, mechanical, and thermal management is particularly important. Due to melting point limitations, the maximum operating temperature of traditional tin-lead solder is lower than 300°C, which cannot meet the requirements of high-power density SiC double-sided packaging. Sintered nano-silver is gradually replacing traditional solders due to its ability to work in high-temperature environments [1]. Low-temperature sintering solders such as nano-silver are used to allow SiC MOSFET power modules to operate at higher junction temperatures, reduce porosity through sintering to make nanoparticles densely connected, provide better conductivity and reduce thermal stress caused by chip interconnection, and improve reliability.

However, shortcomings such as high porosity, high thermal expansion coefficient, and easy electromigration are not conducive to the use of nano-silver in SiC packaging. Moreover, the heat transfer path area of single-sided modules is small, with insufficient heat dissipation, and it is also difficult to work at high temperatures. Therefore, to alleviate the mismatch between solder and other packaging materials, this paper applies nano-copper paste to SiC MOSFET double-sided power modules and compares the thermo-mechanical properties with nano-silver and SAC305 (Sn96.5%-Ag3%-Cu0.5%). Compared with single-sided power modules, SiC double-sided power modules with multiple interconnection layers can achieve better heat transfer and show more significant stress transfer characteristics.

In this paper, a SiC double-sided power module for 1200V/100A is designed, and the pressureless sintering of nanoscale copper paste for chip attachment (5mm × 5mm) to the spacer layer and substrate, and spacer layer attachment to the substrate is studied. Considering that the main material of the substrate and spacer is copper, the Cu-Cu direct interconnection technology of nano-copper slurry and copper layer interconnection on DBC was used to conduct finite element simulation analysis on the transient heating and cooling process of the sintering of nano-copper slurry. Through simulation, the average stress and residual stress of nano-copper and nano-silver under the same sintering curve were compared and analyzed, and SAC305 was added for comparison analysis, which verified the high stability brought by the nano-copper slurry to the sintered joint during the processing stage. It is expected to bring higher reliability to double-sided SiC power modules under long-term high-temperature operation.

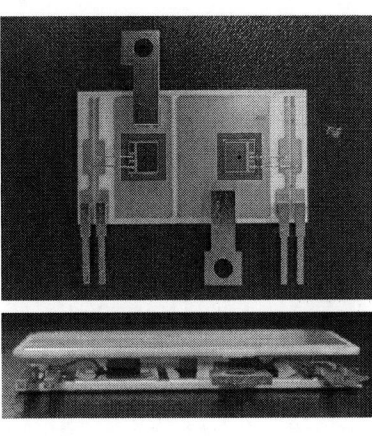

(c)

Fig. 1 (a) Three-dimensional diagram (b) Cross-sectional diagram in the double-sided power module (c) Physical picture

2 Module Manufacture Process

Nano-copper paste is printed onto the metalized ceramic substrate through a template. The 50um thick template is laser-cut to form two 5mm × 5mm holes on the bottom substrate, which together with the drive terminals and power DC+ and AC terminals form the first printing stencil aperture. Use a nickel-plated metal scraper to spread the copper paste evenly. A nano-copper interconnect layer is formed at the Die-DBC connection. The processing method of nano-silver paste is the same as that of nano-silver paste. It should be noted that since the flexibility and fluidity of nano-copper paste are lower than that of nano-silver paste, before stencil printing, the nano-copper paste needs to be taken out of the -5°C freezer in advance and stirred with a glass rod. Evenly, and let it sit for about 20 minutes to reduce the porosity of the nano-copper during subsequent sintering.

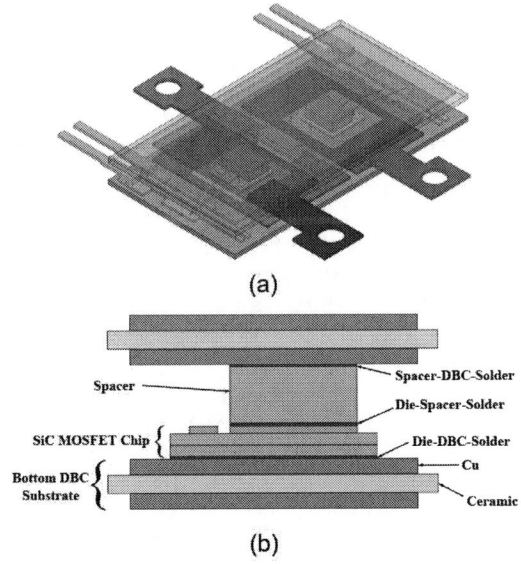

After stencil printing, place the chip and terminals vertically on the solder paste, and then put them into a vacuum reflow for 90 minutes of sintering. The same sintering curve is used for nano-copper and nano-silver, and SAC305 is used as the control group. Compared with nano-silver paste, nano-copper paste not only has excellent electrothermal properties but also has good electromigration resistance and low-cost benefits. However, nano-copper is prone to spontaneous oxidation in air, which increases the difficulty of practical application. Therefore, to strictly compare the impact of interconnect materials of SiC double-sided power modules on their sintering stability, it is necessary to vacuum and introduce nitrogen during the sintering process to ensure that the

Table 1 Comparison of thermodynamic properties of various parts of the module

Components	Materials	CTE $(10^{-6}/K)$	Thermal Conductivity (W/m·K)	Young's Modulus (MPa)	Passion ratio
Die	Cu	17	393	1.35×10^5	0.34
	SiC	4.2	223	4.48×10^5	0.17
Solders	Nano-silver	19	280	0.25×10^5	0.38
	Nano-copper	16.8	180	0.8×10^5	0.25
	SAC305	22	70	0.42×10^5	0.35
DBC	Cu	17	393	1.35×10^5	0.34
	Al2O3	6.8	28	2.8×10^5	0.23
Spacer	Mo70Cu30	7.5	190	2.3×10^5	0.31

sintering process is not affected by oxygen. This is different from the usual sintering of nano-silver paste in power modules.

The processing of double-sided power modules also requires the process of stencil printing of the top DBC, spacer placement, and top terminal sintering, which is the nano-copper interconnect layer of Spacer-DBC. The processed top DBC plate and bottom DBC plate are fixed with a jig. The bottom of the spacer of the top DBC is turned over and brushed with solder paste. This is the nano-copper interconnect layer of the Die-Spacer. After being vertically attached, a SiC double-sided power module is formed. The schematic diagram and physical picture of the SiC double-sided module is shown in Fig. 1(a), (b) and (c).

3 Result and Discussions

First, the double-sided SiC power module is modeled, and two Rohm 1200V/130A S4661 chips are selected to form a half-bridge structure. The lower copper layer of the top substrate and the upper copper layer of the bottom substrate are designed with discrete copper layer patterns to reduce the accumulation of thermal stress caused by the large-area copper layer, promote stress release, and provide electrical connection paths for the double-sided module. To further determine the overall temperature and stress distribution of the module, the driving gate, and auxiliary source terminals at both ends of the substrate cannot be ignored during the simulation process. The interconnect layers at the bottom of the chip, the top of the chip, and the top of the spacer are made of the same material and are connected with three different interconnect materials: nano-silver, nano-copper, and SAC305 to obtain three different modules.

Due to requirements such as anti-oxidation and anti-pollution, the surface of the chip usually contains a special coating. However, since the thickness of the coating is um or even nm level, it has little impact on die sintering process and long-term work stability, so the surface coating can be ignored in the simulation. The parameters of the SiC double-sided power module are shown in Table 1. It can be seen from the table that the CTE of nano-copper is closer to the SiC chip than nano-silver and SAC305. The purpose of using Mo70Cu30 as the spacer material is to further match the CTE of the SiC chip and reduce thermal stress. The CTE and thermal conductivity of nano-copper and bulk copper are different, and the impact of copper size on the simulation results needs to be additionally considered in the simulation.

3.1 Transient thermal analysis

The transient thermal simulation of Ansys Workbench finite element simulation is used to demonstrate the heat transfer capabilities of different solder materials. The power of the chip is set to 50W, and the convective heat transfer coefficient (htc) of the outer surface of the upper and lower DBC copper layers of the double-sided heat dissipation module is set to $1500W/m^2 \cdot °C$. During the sintering temperature rise, high-temperature stabilization stage and temperature drop stage of nano-silver and nano-copper, the temperature sampling intervals are set to 20s, 50s and 30s respectively, and the temperature changes of the sintering process are monitored with a moderate sampling frequency to ensure accuracy. while reducing simulation time. SAC305 uses a traditional reflow soldering curve, with a total sintering time of about 5 minutes and 30 seconds. The steady-state temperature distribution of modules with different materials is shown in Fig 2.

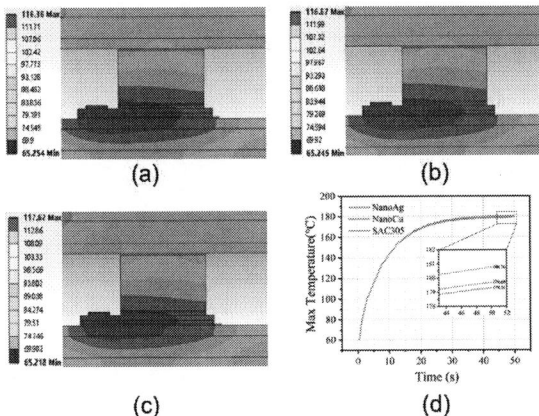

(a)　　　　　　　　　　　(b)

(c)　　　　　　　　　　　(d)

Fig.2 (a) Thermal temperature distribution of nano-silver (b) nano-copper (c) SAC305 in double-sided module (d) Transient maximum junction temperature curves of three materials

The results show that the thermal conductivity of the three interconnection layer materials differs by nearly 300%, but the difference in their steady-state Tj is very small, less than 5%. This is because the thickness of the interconnection layer is small. In the case of solid heat dissipation, the solder layer can speed up the transfer rate of heat flow and diffuse the heat flow path. However, the thermal resistance caused by the high thermal conductivity of low-temperature sintering materials such as nano-silver is relatively small. Compared with traditional lead-free solder, it is only reduced by 9%, and both are lower than 0.1 ℃ /W[2]. Therefore, the impact of the solder thermal resistance of the thin solder layer module on the transient junction temperature fluctuation and steady-state junction temperature is almost negligible.

In addition, the temperature distribution ranges around the chip are relatively similar, indicating that the nano-copper paste has excellent heat dissipation properties. In addition, most of the heat generated by the chip is taken away through the spacer of the copper-molybdenum alloy materials and DBC substrate of the double-sided module. They provide a large contact heat transfer area and have bidirectional heat dissipation flow. For SAC305, due to its low thermal conductivity, it has the highest steady-state temperature and the smallest heat dissipation range. The above results show that the heat transfer capability of nano-copper used in SiC double-sided power modules is similar to nano-silver and has better thermal conductivity than traditional SAC305 solder to meet packaging needs.

3.2　Sintering Stress Analysis

3.2.1　Sintering Process

The biggest advantage of nano-copper compared to nano-silver is its lower thermal expansion coefficient, which better matches the thermal stress of SiC chips. When the sintering process heats up, the drying process in this step volatilizes the organic solvent to form a dry nano-silver film; when the temperature exceeds the melting point of nano-copper, a sintering reaction occurs between nano-copper particles to form a high-density, highly conductive metal. The connection and sintering curve are shown in Fig. 4. The high-temperature and constant-temperature sintering time is 60 minutes, which provides enough time for the organic components wrapped on the surface of the slurry to fully volatilize and burn out, so that the nano-copper can be effectively interconnected, and the porosity can be reduced.

Since the thickness of the nano slurry layer is extremely small (about 50um), the main concern is the stress change during the sintering process. Therefore, the thermal stress caused by temperature changes during the pressureless sintering process of nano-interconnected materials was studied. The pressureless sintering process can avoid the decrease in chip strength and reliability caused by pressure sintering and improves the power density and heat dissipation capacity of SiC power modules[3]. The pressureless sintering process produces up to high shear strength on the silver-plated direct-bonded copper substrate to support the high-temperature operation of SiC power devices.

3.2.2　Anand model of nano-copper

Among the many constitutive relationships studied, the Anand model is widely used in solder simulation because of its accurate constitutive relationships. Moreover, the Anand model describes the macroscopic impedance to plastic flow inside the material, which can reflect the deformation behavior of viscoplastic materials related to strain rate and temperature, as well as the historical effects of strain rate, strain hardening, dynamic recovery, and other characteristics. Therefore, this paper uses the unified viscoplastic constitutive model Anand model [4] to simulate the uniaxial tensile inelastic deformation behavior of the material, considers the temperature dependence of thermal conductivity and thermal expansion coefficient, and analyzes the performance of different interconnect materials in SiC double thermo-mechanical properties of area power modules during sintering. The Anand model comparison of nano-copper and nano-silver is shown in Table 2.

Table 2 Anand models of three interconnection materials

Parameters	NanoAg	NanoCu	SAC305
$A(s^{-1})$	9.81	2.677	3700
$Q/R(k)$	5706.3	7146.478	11500
ξ	11	12	4
m	0.6572	0.884	0.47
s	67.389	53.874	7.72
n	0.00326	2.917×10^{-8}	0.0315
h_0	15800	210.351	70000
a	1	1	1.9
s_0	2.768	0.446	6.5

Nano-copper has a lower CTE and a higher Young's modulus, which can better transfer mechanical properties when temperature changes, reduce stress accumulation on adjacent copper, and accelerate stress transfer. In reliability studies of power modules, Von Mises stress simulations of various module layers are widely used to correlate with reliability measurement data. It was shown in [5] that the number of cycles to failure is inversely proportional to the maximum Von Mises stress. Therefore, this paper uses simulation to extract the stress changes during the sintering heating process, as well as the residual stress when a stable joint is formed after cooling, reflecting the stress accumulation of the module during the sintering process, thus serving as an evaluation method for the stability of the sintered joint formed by low-temperature sintering.

3.2.3 Sintering stress results

After conducting a transient thermal simulation on nano-copper, the thermal temperature distribution of the sintering process was applied as an excitation to the finite element stress simulation to explore the results of stress changes when the heat source accumulates. During sintering, we mainly focus on the stress fluctuation and residual stress during the sintering process. Larger stress fluctuations will affect the degree of densification and porosity of the connection between microscopic particles, thereby affecting the stability of the interconnection layer. Larger residual stress will cause the module to warp or twist, deform, or even crack under long-term operation.

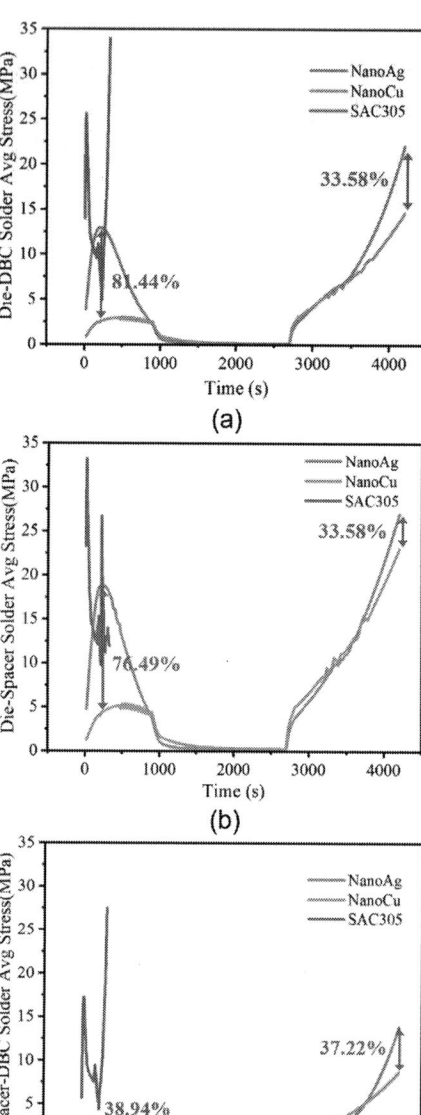

Fig. 3 Average stress changes during the sintering process of three interconnection layers under different materials (a) Die-DBC (b) Die-Spacer (c) Spacer-DBC

Fig.3 shows the average stress during the sintering process of three interconnect layers under different materials. The results show that the average stress and stress fluctuation of each interconnection layer of SAC305 is much higher than that of nano-silver and nano-copper as a whole, which shows that reflow soldering with traditional solder produces large stress changes. When the metal melts, the stress decreases, but the stress will rise sharply after solidification. Therefore, this interconnection method will bring greater porosity and greater stress on the interconnection layer during operation.

Compared with nano-silver, nano-copper has the largest stress drop in the heating and cooling stages of the Die-DBC layer, with a maximum drop of 81.44%. The stress drop in the Die-Spacer layer is also larger, and in the Spacer-DBC layer the decrease in stress is relatively small, but there is still a decrease of 37.22%. In addition, the stress fluctuation range of nano-copper is smaller because nano-copper has a lower CTE and a larger Young's modulus, which can alleviate the thermal stress caused by high temperatures in double-sided modules. The nano-copper interconnect layer closest to the heat source has the lowest average sintering thermal stress, which has a significant improvement effect on sintering densification and joint formation.

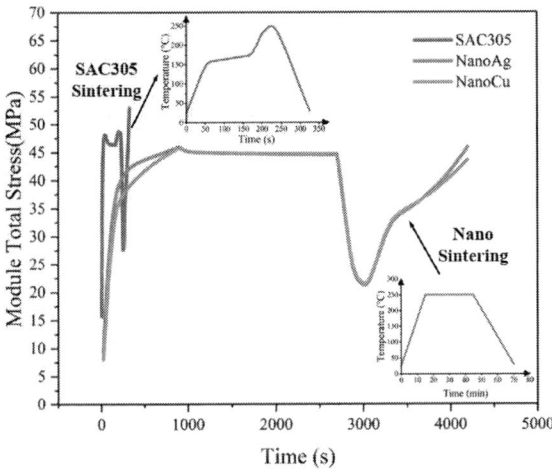

Fig.4 The overall average stress change curve of the module during the sintering process of different interconnection layer materials

Fig. 4 shows the average stress change of the module during the overall sintering process. The results show that during the heating process, the average stress of the nano-copper module is about 5MPa lower than that of nano-silver, and the stress is reduced by 11.76%, showing a good ability to resist temperature changes and maintain stress balance. When maintained at high temperatures, the average stress is almost the same, and nano-copper and nano-silver undergo solid phase transformation. The average stress of the nano-copper module is lower than that of the nano-silver module. The residual stress difference during the cooling process is 2.283MPa, and the

stress is reduced by 4.98% during sintering. This shows that the overall average stress of the nano-copper module during sintering is lower. When the sintered copper joint is formed, the performance in the module The generated stress brings more lasting reliability to the nano-copper module, and the feasibility of using nano-copper as a new interconnect material for SiC double-sided modules has been verified. Compared with nano-silver, nano-copper modules show lower stress during the sintering process, which can significantly improve the strength and reliability of the SiC double-sided module interconnection layer, reduce fatigue, fracture, and processing failure problems, and improve its reliability under temperature-changing conditions.

Fig.4 shows the results of residual stress after sintering of the double-sided power module. The double-sided power module used in Nano-copper shows lower stress in the SiC, drain, and power layers of the chip. The residual stress is lower than that of the double-sided power module used in nano-silver. The darker colors represent lower stress, and both nano-copper and nano-silver below the chip layer show lower stress.

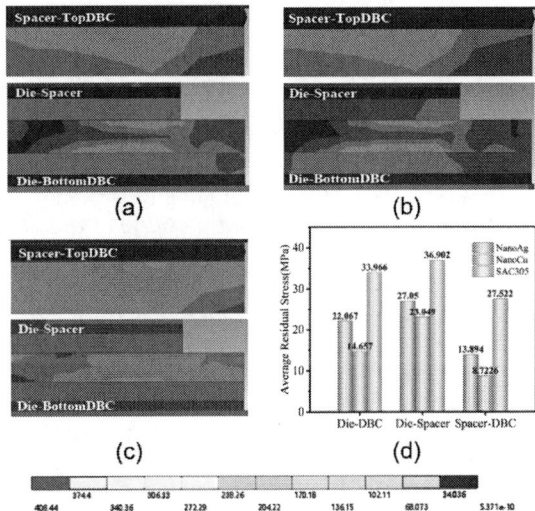

Fig.5 (a) Nano-silver (b) Nano-copper (c) SAC305 residual stress cross-sectional distribution diagram (d) Residual stress diagram of each layer

Fig. 5(d) shows that the residual stress of nano-copper slurry in Die-DBC, Die-Spacer, and Spacer-DBC layers is 33.58%, 14.79%, and 37.22% lower than that of nano-silver slurry respectively. The results show that Nano-copper has better CTE matching with SiC chips, upper and lower substrates, and spacers. The detailed explanation is as follows:

(1) The Die-DBC layer is connected in the Cu-Cu form of nano-copper particles and micron-copper particles on the DBC substrate. The CTE is consistent, so the residual stress of the nano-copper module is 33.58% lower than that generated by the nano-silver module, indicating that the nano-copper matching of CTE and DBC substrate can effectively reduce stress generation.

(2) The stress between the Die-DBC and Spacer-DBC layers is transferred through the SiC die and spacer. The stress changes of nano-copper, nano-silver, and SAC305 in these two layers are 40.49%, 37.04%, and 25.42% respectively. Nano-copper has the greatest stress reduction. This is because the CTE difference between the nano-copper module and the CTE of SiC, and the CTE of the Cu/Mo alloy spacer with nickel-plated surface layer is lower than that of nano-silver. It is shown that in the double-sided SiC power module, the stress of the nano-copper interconnect layer is more fully transmitted, reducing the accumulation of stress in the module.

(3) When cooling, the average stress of the module decreases, which means that the direction of the module changes from compressive stress to tensile stress during cooling, so the compressive stress decreases, and then the overall stress of the module increases because the tensile stress exceeds the compressive stress, and the stress reverses rise. During the cooling process, the stress change of the nano-copper module is 10.33% smaller than that of the nano-silver module, indicating that the CTE matching of the nano-copper interconnect layer is more suitable for the sintering connection of nanomaterials to provide a more stable sintering joint.

To sum up, the nano-copper interconnect layer, DBC substrate, and spacer are the mapping of the same material at different microscopic scales. When sintering, the CTE of the Cu-Cu interconnect layer is consistent, which greatly reduces the stress transfer of the chip heat to the adjacent connection layer, effectively alleviates the interconnect fatigue and stress product caused by CTE mismatch, and can significantly improve SiC thermomechanical properties of bifacial power modules.

4 Conclusions

This paper uses nano-copper slurry for SiC double-sided modules through Cu-Cu interconnection technology and compares it with nano-silver and SAC305. Finite element simulation results show that using nano-copper interconnect layers in SiC double-sided power modules can reduce the stress accumulation caused by temperature changes during the sintering process of the module, and after the formation of nano-copper stable joints, the overall cooling of the module brings the residual stress is low. Compared with nano-silver, the local stress analysis of the double-sided power module shows that the average residual stress of the nano-copper paste in the three interconnection layers of the Spacer-DBC, Die-Spacer, and Die-DBC is reduced by 33.58% and 14.79% and 37.22% respectively. The overall analysis of the double-sided power module shows that the total average stress is reduced by 11.76% and the total residual stress is reduced by 4.98%. Moreover, this paper also analyzes stress fluctuations and residual stress during the sintering process to illustrate their impact on the stability of sintered joints. Lower stress can bring higher reliability to the module and provide a theoretical basis for subsequent reliability experimental verification.

Overall, the low cost, low CTE, and good thermal conductivity of nano-copper can provide new ideas for interconnect layer materials for high-temperature applications of SiC double-sided power modules.

References

[1] Hansson, J., Nilsson, T. M. J., Ye, L., & Liu, J. (2017). Novel nanostructured thermal interface materials: a review. International Materials Reviews, 63(1), 22–45.

[2] G. Chen et al., "Transient Thermal Performance of IGBT Power Modules Attached by Low-Temperature Sintered Nano-silver," in IEEE Transactions on Device and Materials Reliability, vol. 12, no. 1, pp. 124-132, 2012.

[3] Liu at al., "Design and Characterizations of a Planar Multichip Half-Bridge Power Module by Pressureless Sintering of Nano-silver Paste," in IEEE Journal of Emerging and Selected Topics in Power Electronics, vol. 7, no. 3, pp. 1627-1636, 2019.

[4] Hu at al., "High temperature viscoplastic deformation behavior of sintered nano-copper paste used in power electronics packaging: Insights from constitutive and multi-scale modelling," in Journal of Materials Research and Technology, vol. 26, pp. 3183–3200,2023.

[5] C. Ding at al., "A Double-Side Cooled SiC MOSFET Power Module With Sintered-Silver

Interposers: I-Design, Simulation, Fabrication, and Performance Characterization," in IEEE Transactions on Power Electronics, vol. 36, no. 10, pp. 11672-11680, 2021.

PCIM Europe 2024, 11– 13 June 2024, Nuremberg   DOI: 10.30420/566262293

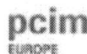

All-in-One-Sintering: Die-attach and Substrate-attach on Bare Copper in a Pressure Assisted Sintering One-step Process

Dr. Battist Rábay[1], Dr. Adrian Stelzer[1], Julien Hossain[1], Constanze Weber[2], Dr. Matthias Hutter[2], Dirk Buße[3], Alexander Dahlbüdding[3],

[1]Nano-join GmbH, Berlin, Germany; [2]Fraunhofer - Institut für Zuverlässigkeit und Mikrointegration, Berlin, Germany; [3]budatec GmbH, Berlin, Germany

Corresponding author: Dr. Battist Rábay, battist.rabay@nano-join.de

Abstract

Using silver sintering for die-attach in power modules is common these days. With increasing power densities, switching frequencies and therefore changing from Si to SiC puts even more mechanical and thermal stress on these components. As of today, sintering of AMB or DBC materials or even molded modules is still a challenge, which makes Sn-solder and thermal interface materials (TIM) still the best and cheapest alternative. In this work we will showcase a material and a process, which allows us the industry to sinter established module designs, die- and substrate level, in one step.

1 Introduction

The expectations in advanced materials for power module assembly in the automotive industry are constantly increasing,[1] not only in terms of performance and overall reliability, but even more challenging is the aspect of increasing costs. Changing from Si to SiC, DBC to AMB or from Sn-based interconnection materials to silver sintering, the price curve is constantly increasing.[2] Industry can only benefit from a combination from all new developments combined if the yield of production is high or better in comparison to soldering, the miniaturization of components enables manufacturers to keep material costs low or if process steps can be combined. Doing so, this ultimately must lead to a higher output in production and ideally to energy savings.

A logical combination of processes are the sintering steps of die-layer and substrate-layer.

Currently, the die-attach is performed via a pressure assisted sintering protocol.[2],[3] Followed by a molding and package procedure. To enable a heat path from top to bottom the module is in the second step interconnected with the cooling structure. The state of the art for this level which is used so far is either a thermal interface material (TIM) or a solder.[4] On using these two methods, the full potential of the material stack in terms of power throughput and heat dissipation cannot be unlocked. The in that stack generated bottleneck is the interconnect at the substrate level. The low price-point of TIMs and its ease of use for large areas, as no curing step is required, are the biggest selling points for that technology. The tradeoffs are a very low heat conductivity of up to 12 W/mK and the longevity of that technology as it suffers dramatically from the so-called "pump-out effect".[5] Soldering will lead to a higher thermal

2082

conductivity and mechanical stability of the assembled stack, but the process will inevitably cause voids below the substrate. Even if voiding is under control, the interconnect cannot compete with a sintered one in terms of performance and lifetime.[6]

Sintering of the substrate layer is getting more attention recently because of the limitations of above-mentioned materials. Nonetheless, sintering such large areas has its challenges not only from a price point of view. Delamination, warpage, and voiding are the most critical challenges to be solved for the package attach.[6]

As we presented in the past a material which can be used for large area sintering, we now established a process with this material which combines the pressure assisted sintering of the die-level and the substrate level in just one step. This will ultimately lead to a higher output, energy savings and a reduction of the total cost of ownership for fully sintered power modules. The results of this work will be presented in this paper.

2 Results and Discussion

In an earlier work we used the sintering paste NJ-Force from Nano-Join GmbH to attach three copper DBCs to a copper heat-spreader. The total area which was sintered in that study was around 3.000 mm². For this publication we are using the same dimensions and materials for our All-in-One-sintering approach (figure 1). To successfully accomplish the assembly of that stack we had to understand the factors of influence for each process step.

    1.) Printing of the paste on the substrate-level and die-level.

With such large structures and different material stacks there always comes a lot of warpage during the processing of these materials due their characteristics. In this material set, we had to deal with pre-bend copper baseplates and non-planar DBCs, which is a common starting position for such components. Printing paste on bend substrates can result in a non-homogeneous distribution of paste, which ultimately can lead to performance degradation over time of the sintered

interconnect. As of the limitations during that study we had to focus on stencil printing but other options such as "slit-nozzle dispensing" can help to improve the resulting paste pad.

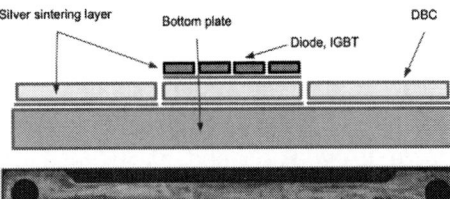

Figure 1: Demonstrator consists of copper baseplate, three copper DBCs, two Si-diodes and two Si-IGBTs.

We used a very simple stencil design, with just one opening and a thickness of 150µm. Which already

Figure 2: Printed and pre-dried silver paste on Cu-base plate.

led to very good and consistent results (figure 2). Pre-drying was carried out in N2-atmosphere for 20 min at 120°C in a standard box oven. The paste was dry but still had the required tackiness to the DBCs.

For the die-level the process was similar, with the only difference of the stencil thickness (100µm).

2.) Sintering conditions and parameters.

Finding the right sintering conditions for the substrate level alone and then combined for both levels took the most part of this work. For all sintering steps we used the SP300 from budatec GmbH.[7] This press has a two-chamber design, which allows the control of atmosphere, sintering temperature and pressure and cool down speeds. Being able to work on and track these crucial process parameters was key to generate a consistent quality of the resulting demonstrators. Sintering conditions for all the following results: 12 MPa at 275°C for 5 min in a nitrogen atmosphere.

The quality of the samples was evaluated by Fraunhofer IZM due to means of C-SAM analysis (figure 3).

Figure 3: Demonstrators for substrate attach after sintering with corresponding C-SAM below.

With slight variations, all these substrate level demonstrators showed a very homogeneous sintered silver layer.

As a next step we worked on the assembly of the All-in-One-sintered demonstrators. The main challenge was to have a uniform distribution of the pressure on the dies and the substrates at the same time. The SP300 uses a flat hard tool, which is very good for temperature control. To mimic the structure flexibility of a soft tool without losing the very good heat transmittance of the hard tool we

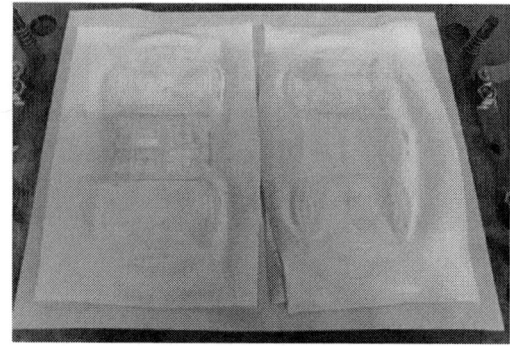

Figure 4: Mimicking a soft tool with two layers of teflon.

opted for thick teflon foil to resolve the challenge (figure 4).

With the same sintering parameters, we obtained the same quality for the substrate-level and for the die-level (figure 5).

Figure 5: Showcasing all sintered layers by C-SAM, left substrate and

To prove the performance of NJ-Force and this developed process thermal-shock-cycling was conducted at Fraunhofer IZM with a representative number of demonstrators.[8] A three chamber design was used, chamber one was at -55°C, chamber two at room temperature and chamber three at 125°C. One full cycle took 30 min, with a 15 min holding time in chamber 1 and 3. For such large structures a $\partial T=180°C$ is pretty challenging. After initial C-SAM every 250 cycles further C-SAM measurements had been conducted. Additionally, after passive cycling X-ray analysis was conducted on all samples, on two samples cross-sections had been performed to gain a deeper understanding of the constitution of the sintered interconnects after thermal cycling.

The evaluation of the damaged sintered layer per demonstrator is below 7% (die- and substrate-level combined) for the demonstrator with the "highest" degradation after 1000 cycles (figure 6).

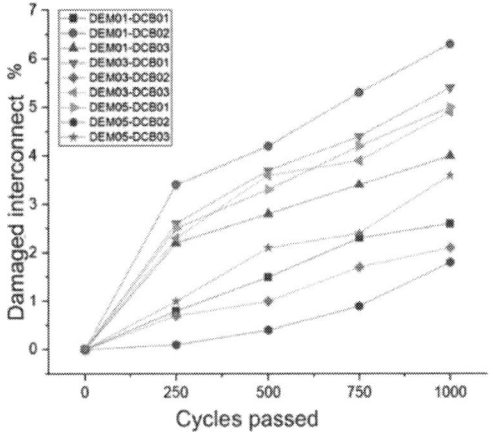

Figure 6: Correlation of damaged interconnect to cycles passed.

With silver sintering these studies are mainly conducted with silver plated components. As far as we know, with sintering on bare copper such results are unchallenged. Even with modern copper sintering pastes engineers must accept degradation of at least 10% for the die-attach.[9] The detailed C-SAM scans for each layer showed only minimal signs of degradation, the best looked even close to unchanged (figure 7).

Having no signs of degradation in these C-SAM images are only one side of the medal, to gain further insights on the toughness of these

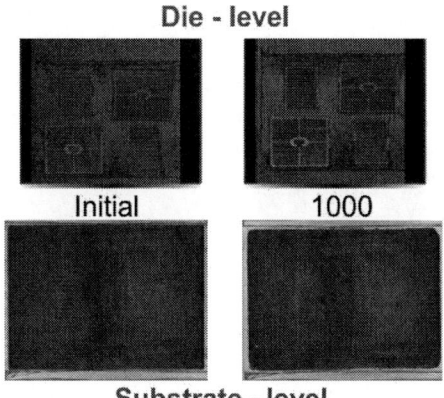

Figure 7: C-SAM before and after thermal shock.

interconnects, cross sections had been accomplished for some of the demonstrators after 1000 thermal-shock-cycles (figure 8). The images showcase the consistent and uniform density of the silver layers for both levels. Even after the conducted shock cycles no voiding or delamination had occurred. Only on the edges minimal damage was visible on some modules. As shown in figure 6, the overall degradation for all modules was very little. What also can be seen in these cross-sections is the very thin bond line thickness (BLT) after sintering. For the die-level 35µm and for the substrate-level 60µm can be detected. To obtain a best-in-class robustness with a thin BLT is unique, usually at least a twofold amount is necessary with other sintering materials to generate similar but not as good results, no matter if based on copper or silver. These unique

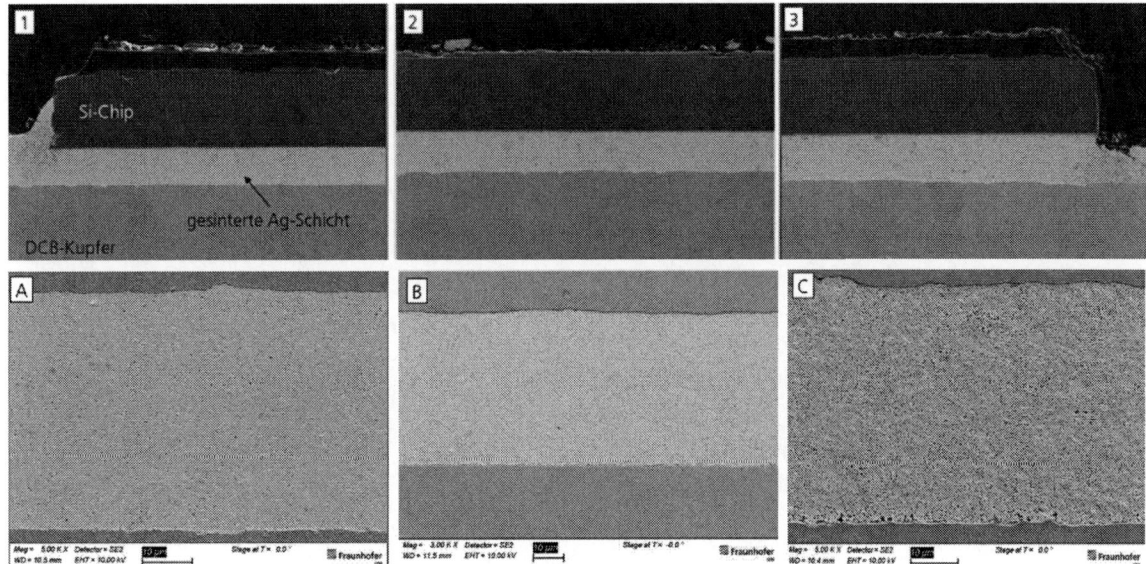

Figure 8: Cross-section after 1000 cycles for one demonstrator, die-level (top) and

selling points of this material and process completely reduces the price benefits of copper-based sintering materials. Especially, when it is taken into consideration that no special equipment or molding technologies are required for such a silver-based material.

3 Conclusion

As the design, materials and processes for power modules leaping their performance to new limits, the cost structure unfortunately becomes more and more a concern. Anyway, components such as AMB, SiC and sintering pastes are the materials of choice, they are cost intensive. Optimization of assembly processes, throughput and yield gaining importance as here the majority of costs can be saved. The All-in-One sintering process showcased in this work has the potential to reduce the total cost of ownership for such applications dramatically. Ideally, a usually two-step process, die sintering and substrate sintering, can become a one step process. Saving energy, consumables, and time by 50%. NJ-Force as the used material in this work shows a great flexibility from small to large surface areas but also for material stacks which are prone to warpage during sintering and stress test. Assembly was performed under moderate conditions. Performed analysis of the demonstrators before and after thermal shock cycling gave outstanding results as the data generated via cross sections and C-SAM show only little signs of degradation. All-in-One sintering will offer an efficient and cost friendly way to assemble completely sintered power modules for automotive applications in just one step.

4 Acknowledgements

We would like to thank our partners which contributed to this work: budatec GmbH for the use of their sintering equipment and Fraunhofer IZM for conducting thermal cycling and analysis of the assembled demonstrators. We also would like to thank the Federal Ministry of Education and Research and for funding the joint project "KoKo-Power", the BmBF and the Investitions Bank Berlin Brandenburg (IBB).

5 Reference

[1] a.) J. D. van Wyk and F. C. Lee, "On a Future for Power Electronics," in *IEEE Journal of Emerging and Selected Topics in Power Electronics*, vol. 1, no. 2, 2013, 59-72; b.) J. D. Van Wyk and F. C. Lee, "Power electronics technology at the dawn of the new millenium-status and future," *30th Annual IEEE Power Electronics Specialists Conference. Record. (Cat. No.99CH36321)*, 1999, 3-12, vol.1.

[2] a.) W. Sabbah, S. Azzopardi, C. Buttay, R. Meuret, E. Woirgard, "Study of die attach technologies for high temperature power electronics: Silver sintering and gold–germanium alloy", Microelectronics Reliability, Volume 53, Issues 9–11, 2013, 1617-1621; b.) M. Maruyama, R. Matsubayashi, H. Iwakuro, *et al.*, "Silver nanosintering: a lead-free alternative to soldering.", *Appl. Phys. A* 93, 2008, 467–470; c.) M. A. Khan, G. Simin, S. G. Pytel, A. Monti, E. Santi and J. L. Hudgins, "New Developments in Gallium Nitride and the Impact on Power Electronics," *2005 IEEE 36th Power Electronics Specialists Conference*, 2005, 15-26; d.) T. Paul Chow, High-voltage SiC and GaN power devices, Microelectronic Engineering, Volume 83, Issue 1, 2006, 112-122; F. Roccaforte, P. Fiorenza, G. Greco, R. Lo Nigro, F. Giannazzo, F. Iucolano, M. Saggio, "Emerging trends in wide band gap semiconductors (SiC and GaN) technology for power devices", Microelectronic Engineering, Volumes 187–188, 2018, 66-77; e.) C. Michele, A. Sitta, S. M. Oliveri, G. Sequenzia, 2021. "Silver Sintering for Silicon Carbide Die Attach: Process Optimization and Structural Modeling" *Applied Sciences* 11, no. 15: 7012. f.) B. Rabay and N. Papathanasiou, "Silver sintering: New Materials for die and substrate attach in high power applications," *MikroSystemTechnik Congress 2021; Congress*, Stuttgart-Ludwigsburg, Germany, 2021, pp. 1-3.

[3] a.) Y. Liu, H. Zhang, L. Wang, X. Fan, G. Zhang and F. Sun, "Effect of Sintering Pressure on the Porosity and the Shear Strength of the Pressure-Assisted Silver

Sintering Bonding," in *IEEE Transactions on Device and Materials Reliability*, vol. 18, no. 2, 2018, 240-246; b.) L. M. Chew, W. Schmitt, C. Schwarzer and J. Nachreiner, "Micro-Silver Sinter Paste Developed for Pressure Sintering on Bare Cu Surfaces under Air or Inert Atmosphere," *2018 IEEE 68th Electronic Components and Technology Conference (ECTC)*, 2018, 323-330; c.) W.S. Hong, M.S. Kim, C. Oh, "Low-Pressure Silver Sintering of Automobile Power Modules with a Silicon-Carbide Device and an Active-Metal-Brazed Substrate." *Journal of Elec Materi* 49, 2020, 188–195.

[4] A. Volke, M. Hornkamp, "IGBT modules (chapter 10.2.2)", (3rd edn), Infineon, 2017;

[5] a.) B. Wunderle, D. May, J. Heilmann, J. Arnold, J. Hirscheider, Y. Li, J. Bauer, M. Abo Ras, "A Novel Concept for Accelerated Stress Testing of Thermal Greases and In-situ Observation of Thermal Contact Degradation.", IEEE, 2018.

b.) S. Söhl, R. Eisele, "Impact of the Pump-Out-Effect on the thermal long-term behaviour of power electronic modules." *Microelectronics Reliability*, 2019, 100-101.

[6] B. Rábay, A. Stelzer, "Large surface area substrate attach in power module applications.", CIPS, 2022.; R. Dudek, R. Doring, M. Hildebrandt, S. Rzepka, S. Stegmeier, S. Kiefl, (2016). "Electro-thermal-mechanical analyses on stress in silver sintered power modules with different copper interconnection technologies.", *ESTC*, 2016, 1-8.

[7] budatec GmbH, Melli-Beese-Str. 28, 12487 Berlin, Germany.

[8] Fraunhofer-Institut für Zuverlässigkeit und Mikrointegration (IZM), Gustav-Meyer-Allee 26/ Gebäude 17, 13355 Berlin, Germany.

[9] a.) C. Schwarzer, L. M. Chew, M. Schnepf, T. Stoll, J. Franke, M. Kaloudis, „Investigation of Copper Sinter Material for Die Attach", *in Proceedings of SMTA International*, 2018; b.) S. K. Bhogaraju, A. Hanß, M. Schmid, G. Elger and F. Conti, "Evaluation of silver and copper sintering of first level interconnects for high power LEDs," 2018 7th Electronic System-Integration Technology Conference (ESTC), 2018, pp. 1; c.) Bhogaraju, Sri Krishna, Francesco Ugolini, Federico Belponer, Alessio Greci, and Gordon Elger. 2024. "Reliability of Copper Sintered Interconnects under Extreme Thermal Shock Conditions." *IMAPSource Proceedings* 2023 (EMPC): 25–29.

PCIM Europe 2024, 11– 13 June 2024, Nuremberg   DOI: 10.30420/566262294

Sequential Manufacturing of Highly Functionalized 3-Dimensional Ceramic Components for Power Electronics

Lars Rebenklau[1] , Henry Barth[1] , Paul Gierth[1] ,
[1] Fraunhofer Institute for Ceramic Technologies and Systems IKTS, Dresden, Germany

Corresponding author: Lars Rebenklau, lars.rebenklau@ikts.fraunhofer.de
Speaker: Lars Rebenklau, lars.rebenklau@ikts.fraunhofer.de

Abstract

The central task in standard and power electronics is the processing of electronic information and the communication with higher-level systems. In general, this requires hardware that is adapted to the task. This consists of electronic components, often a subcarrier and a housing. In addition, there are always other functional, ecological, and economic constraints. For example, increased functional density can be achieved by a 3-dimensional arrangement of electronic components or housing components. An effective way of achieving this is the use of additive ceramic manufacturing processes in combination with functionalization of these components by thick-film technology.

1   Introductions

In general perception, ceramic solutions are often discussed as economically disadvantageous due to the complex manufacturing technologies used and the associated high initial costs. However, ceramics have outstanding functional properties compared to other material systems. This is often due to their low weight, high corrosion stability, mechanical hardness, and suitability for use under high temperatures. Customized ceramics also permit economically competitive products. Temperature sensors in general, NTC and PTC sensors in particular, can be cited as striking examples of this. In addition, the use of ceramics as insulators, e.g. as substrate (DCB, AMB) or sensor housings, are common.

It is precisely at this point that current developments in the shaping of ceramic masses already allow the production of geometrically highly complex ceramic components. This is made possible using additive manufacturing technologies with the use of ceramic materials. During this shaping process, unfired ceramic materials are processed and sintered.

For selected applications, however, the focus is not only on the geometric shape of the structural element, but also on subsequent functionalization.

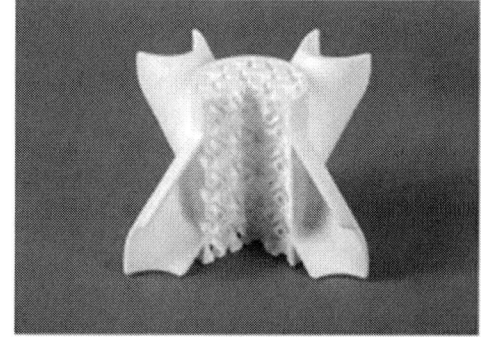

Fig. 1   3-dimensional ceramic with complex inner structure

These approaches are often associated with electrical metallization and sensor or actuator functions. This results in the need to "functionalize" the additively manufactured components with appropriate layer systems, for which the approach of multi-material printing is being pursued in various scientific papers. This means that different processing heads are used simultaneously to produce multi-material components. These are filled with different materials. In the case of ceramics, for example, this makes it possible to print both the unfired ceramic slurry and the electrically functional layers simultaneously. After printing these so called green bodies (non-sintered ceramics), the entire body is fired in thermal process, resulting in the desired multifunctional component.

2088

This paper presents an alternative approach based on ceramic materials, in which the multi-functional components are realized in two separate technological steps. In a first step, the ceramic component is manufactured, including sintering. In a second technological step, this ceramic component is functionalized.

For Al_2O_3 and other ceramics, the thick-film technology established in electronics and microsystems technology is suitable. With this technology, electrical functional layers or sensors can be printed and sintered onto the ceramic substrates. The electrical connection is realized using well established methods of packaging and interconnection technology in electronics.

The advantage of this approach is seen in the fact that the materials required for functionalization are industrially available. Extensive technological know-how also exists regarding both processing and further processing. This means that applications can currently be implemented. The main focus of this paper is not on additive ceramic manufacturing processes. The focus is on the functionalization of these elements. Regarding additive manufacturing on the basis of ceramics, reference is made to further literature [1].

2 Basics of thick-film technology

If one considers the possibilities of using ceramic materials for the realization of interconnection boards, it becomes clear that there are different technological routes. Each of them has specific advantages and disadvantages for individual requirements. To give an overview they are described in the following section. The most important structuring processes are:

- Assembly of metallic films using DCB/AMB technology
- Deposition of thin films using physical/chemical processes
- Processing of pasty materials by means of thick-film technology

2.1 DCB/AMB technology

The first two variants, DCB and AMB [2], address applications with very high line cross-sections, i.e. high layer thicknesses.

The main areas of application are in the field of power electronics with electronic components that have a low number of connections but very high current densities to be switched. To ensure the conductor cross-sections, flat copper foils with appropriate thicknesses are used. These foils are joined to the ceramic substrates either by sintering processes (DCB) or on the basis of special solders via brazing processes (AMB). Comparable to the standard printed circuit board technology, the structuring of the conductor tracks is carried out in the final step by means of subtractive etching processes. The achievable conductor geometries have heights corresponding to the film thickness (typical 300...500 µm) with minimally analog conductor widths. Fineline capability is not available. Processing is carried out on flat substrates.

2.2 Thin film technology

Diametrically opposed requirements can be solved by thin film structuring of conductors on ceramic substrates. Solutions are addressed for the contacting of high-pole components or high-frequency technology, which requires a very fine resolved conductor pattern. The target parameter is to ensure minimum conductor spacing and width.

Different vapor deposition or sputtering processes for different metallic or insulating materials can be used for layer generation. Layer deposition is generally performed on flat substrates. Structure generation is done by photo processes or etching processes.

2.3 Thick-film technology

The technology was developed in the 1980s and the main areas of application of thick-film technology were in the fields of industrial electronics and, from today's point of view, still comparatively limited applications in automotive electronics. Today's areas of application are in the fields of highly reliable electronics for harsh environmental conditions, sensor technology and special applications. Due to the current transformation of the automotive industry towards electric drives, power electronics and autonomous driving, future application areas for ceramic solutions are expected and are currently being researched.

Thick-film technology is based on the use of paste-like materials. These thick-film pastes are printed onto ceramic plates (substrates) by means of suitable printing techniques and then bonded in defined processes at temperatures of approx. 850°C.

In this context, the thick-film process means the production of the ceramic substrate with a conductor pattern of any complexity. The term hybridization is also mentioned in the literature in connection with thick-film technology. Hybridization means the further processing of this thick-film substrate with other technologies, e.g. the electroplating of termination pads as well as a component assembly. This is the combination of different technologies and thus hybridization.

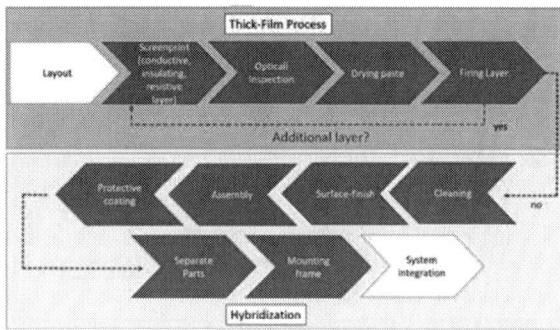

Fig. 2   Thick-film process flow (incl. hybridization)

2.4   Thick-film pastes

Basically, there are thick-film pastes as well as the final sintered thick-film layers with different electrical properties. There are thick-film pastes with the following properties:

- Conductive pastes based on metallic ingredients
- Resistive pastes based on metal oxides
- Insulation pastes with high glass content

In principle, thick-film pastes always consist of four main components.

- Active substances
- Inorganic binders
- Organic binders
- Solvents and auxiliaries

The solvents, organic binders and auxiliaries determine the printing behavior. They are removed after drying and firing. The inorganic binders serve to ensure adhesion of the active phase particles to each other and to the substrate. The active substance determines the electrical function of the paste. A distinction is made between conductive pastes, resistive pastes and insulating pastes. The construction of a thick-film circuit from several pastes requires that they are compatible with each other. The groups of compatible pastes offered by the paste manufacturer are called paste systems. They ensure processing according to uniform technology (e.g. the same firing profile), their chemical compatibility with each other and guarantee the properties of the individual pastes. Commercial paste systems consist of conductive, insulating and resistive pastes. In addition, pastes are also offered for special applications, e.g. sensor pastes or pastes for use on special substrates (steel, aluminum, glass). The conductivity of fired resistive and conductive pastes is characterized by the surface resistance R_F. It indicates the resistance between the opposite sides of a square of a printed layer. This definition is based on a defined layer thickness.

Apart from the electrical properties of the fired pastes and their compatibility, the flow behavior (rheology) of the pastes to be printed must meet the special requirements of screen printing. On the one hand, the paste pressed through the individual meshes during the printing process must bond to form a cohesive structure, and on the other hand, this structure should not flow further apart. This good flowability for a short time is achieved by the structural viscosity and thixotropy of the paste. Thixotropy is the reversible reduction in viscosity caused by constant shear stress. This flow behavior is specifically influenced by the selection of organic binders and solvents.

Conductive pastes

Pastes with precious metals such as gold or silver as the active substance are mainly used to produce conductive structures. The technological and electrical properties are specifically influenced by adding palladium or platinum.

Precious metal-free copper conductive pastes are an alternative. To avoid oxidation during firing, these pastes must be fired under inert gas (usually high-purity nitrogen). However, the advantage resulting from the precious metal savings is largely offset by the higher cost of producing the copper powder and by the necessary use of an inert gas.

The following requirements are placed on the fired conductive structures:

- adequate adhesive strength
- good electrical conductivity [R_F = (1.5...50) mΩ/sq.]

- solderability and bondability (if required)
- low tendency to diffusion and migration
- corrosion resistance

Resistor pastes

In resistor pastes, ruthenium oxide (RuO_2) and bismuth ruthenate ($Bi_2Ru_2O_7$) are mainly used as active substances. With these active substances, good electrical properties are achieved (temperature coefficient, long-term stability, noise) and a range of values of R_F between 10 Ω / square and 10 MΩ / square is covered.

The reproducible production of thick-film resistors with defined properties requires knowledge of all influences and exact adherence to the prescribed technological processing conditions. Most resistor pastes may be fired only once. When using several resistor pastes with different surface resistances, they are consequently printed and dried one after the other and then sintered together (co-firing process).

Dielectric and insulating pastes

The active substances of these pastes are special glass and ceramic frits which, in addition to the respective desired dielectric constant, have the highest possible insulation resistance. Insulation pastes can be divided into three groups:

- Insulation pastes with low dielectric constant for multilayer construction.
- Dielectric pastes with high dielectric constant for the manufacture of printed capacitors
- Low-sintering masking pastes for surface protection and solder stop

Insulation pastes with low dielectric constant are used for the realization of line crossings (cross over technology) and for the insulation of several conductive layers (multilayer). To keep unwanted capacitive coupling of the insulated conductive structures as low as possible, the active substance is optimized for minimum dielectric constant. (practical values 7...9). Insulation layers are printed at least twice. Multiple printing closes any pores (pin holes) and improves the tightness of the layer.

Dielectric pastes are used to manufacture printed capacitors. To keep the area required for such capacitors low, a higher dielectric constant is desired here. Systems of dielectric pastes realize graded dielectric constants up to about 2000. However, the capacitance of such printed capacitors varies greatly. Since no satisfactory balancing technology is known, only limited circuit tasks can be solved with such roughly tolerated capacitances. Therefore, chip capacitors are usually used in hybrid circuits.

Masking pastes were developed to protect printed resistors from environmental influences and improve their long-term stability and reliability. They consist of low-melting glasses and can therefore be sintered at temperatures as low as 500 °C. The already sintered resistors are then protected from environmental influences. At these temperatures, the already sintered resistors are only slightly affected. The resistors are adjusted through the cover. When masking pastes are printed, only the contact surfaces are kept free. Thus, the masking paste acts as a solder stop.

2.5 Screen printing of thick-film pastes and thermal processing

Various printing processes are available for the structured layer deposition of thick-film pastes. Examples include stamp printing, stencil printing and aerosol printing. However, the main printing process is screen printing of the thick-film pastes. The principle of screen printing with emulsion screen is shown in the next figure.

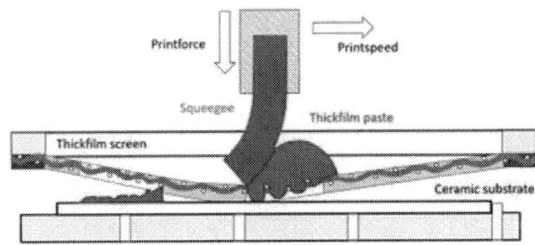

Fig. 3 Screen printing principle

For the stencil carrier to lift off from the substrate after printing, the printing form is placed at a distance above the substrate. This distance is called the take-off. For emulsion screens, it is (0.3...2) mm, depending on the screen mesh, frame size, screen tension and flow behavior of the paste. After starting the printing process, the squeegee is lowered and moved over the printing screen at a defined (adjustable) speed. It presses the printing paste on the stencil carrier through the open meshes of the screen fabric onto the substrate. This stretches the screen fabric and elastically deforms the squeegee. After passing over the set path length, the squeegee lifts off and is returned to its original position. The stencil carrier lifts out of the paste as a result of its pretensioning and rebounding and returns to its start position.

Screen printing is possible on flat substrates as well as on the surface of tubular substrates.

The printing is followed by thermal processes. First, the printed layers are dried. This is followed by sintering of the printed layers. An example profile with a firing temperature of 850 °C is shown below.

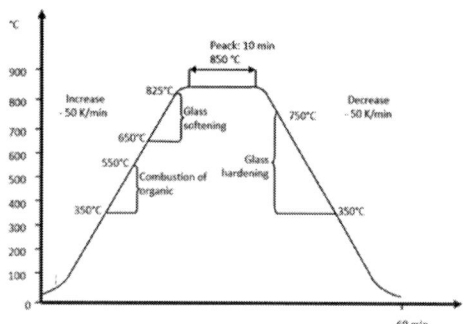

Fig. 4   Standard thick-film firing profile

2.6   Assembling

Conductive pastes are filled with metallic components. These have a low electrical resistance and are used to produce electrical conductors. Furthermore, the conductive pastes are used for connection assembly. Since the described assembly with further components or sensors is generally carried out by electronics technology assembly processes, the conductor traction materials must also permit corresponding assembly. Typical assembly methods for conductor pastes are:

- Soldering with soft solders and, for higher operating temperatures, soldering with brazing solders [3], [4].
- Thermosonic wire bonding with gold wires and ultrasonic bonding with aluminum or copper wires. The maximum wire diameters are approx. 50 µm for thermosonic bonding and up to 500 µm for ultrasonic bonding.
- The sintered assembly of power semiconductors based on nano-silver or -copper sintering pastes
- The use of electrically conductive or electrically insulating adhesives

Commercial thick-film pastes are designed for assembly. As a result, mounting on the conductor materials is possible using the above-mentioned processes.

Fig. 5   Assembled thick-film hybrid

However, it is important to note that the use of these standard assembly methods and materials limits the usable operating temperatures. Typical operating temperatures in the field of electronics are below 200 °C. Consequently, the commonly used material systems are also only designed for this temperature range.

3   Application example

The combination of the two technologies "additive manufacturing of ceramic components" and functionalization using thick-film technology opens a powerful toolbox. The geometries required for functionality are mapped using additive manufacturing. As this is based on the use of ceramic materials in this case, all the material-related advantages, such as high thermal conductivity, high insulation strength and temperature stability, can be utilized. Highly complex geometries are possible because of the additive manufacturing. Functionalization using thick-film technology allows access to the entire functional spectrum offered by thick-film pastes. This is combined with the excellent possibilities of various assembly processes.

In order to ideally combine the advantages of both technologies, applications in which a three-dimensional form must be combined with functionalization are therefore advantageous. The applications shown in the following therefore do not claim to be exhaustive, but serve to illustrate the possibilities:

The first example addresses the application of "active cooling" of components with power requirements. Various geometric substrate configurations are conceivable for this.

Figure 6 shows the principle of a pin-fin cooling structure. This is provided with electrical functional layers on the upper side, which allows components to be mounted.

Fig. 6 Principle image of an open pin-fin cooling structure

A consistent further development of this idea is to use a geometrically closed and actively cooled substrate. This has integrated cooling structures and is made of ceramic with appropriate thermal conductivity. Materials made of Al_2O_3 or AlN are possible. Using thick-film technology, functionalization with conductor tracks and assembling pads are possible. It is possible to mount active components on these metallizations. The next figure shows a corresponding sample. However, no electronic components were mounted here. In the sample shown here, the simulated power loss is applied using a thick-film heater. Concrete setups with active electronic components are currently being tested.

Fig. 7 Integrated cooling structure

(left: inner structure, right: functionalized on surface)

Other possible applications include ceramic and functionalized component- and/or sensor-housings. As previously described, the shaping process is based on additive manufacturing, including subsequent functionalization. The example shows the base plate of a three-dimensional ultra-compact sensor system.

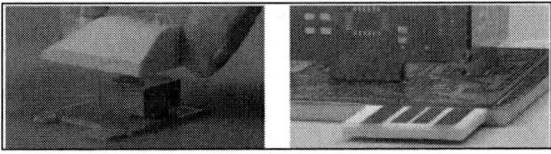

Fig. 8 Functionalized ceramic housing

4 Conclusions

The investigations discussed have shown that the technology of additive manufacturing of ceramic components can be ideally combined with that of thick-film technology. All functionalities that can be realized in thick-film technology can be transferred to complex ceramic components. The applications shown here are only exemplary solutions. The possibilities were presented extensively with a focus on the various thick-film pastes. It is also possible to package additional components on the functionalized ceramics. Packaging technologies have been developed to use the components at operating temperatures of up to 350 °C. Applications up to 600 °C are demonstrated [3].

The examples shown here are based on the use of Al_2O_3 ceramics. In principle, this work can also be carried out using AlN ceramics.

In addition, a wide variety of other applications have been successfully demonstrated.

Acknowledgements

Parts of the work were funded within the project "DynaCool" by the German Federal Ministry for Economic Affairs and Energy BMWi based on a resolution of the German Bundestag (FKZ: 16KN054345). In addition, the results of the Fraunhofer internal projects "TPC-LE" have been included in this presentation.

Literature
[1] A. Michaelis et. al.: "Advanced Manufacturing for Advanced Ceramics"; Procedia CIRP, Volume 95, Pages 18-22

[2] A. Pönicke et al."Aktivlöten von Kupfer mit Aluminiumnitrid- und Siliziumnitridkeramik"; Keramische Zeitschrift 06/2011

[3] L. Rebenklau, P. Gierth, H. Barth "High Temperature Packaging for Sensor Elements"; PCIM 2020

[4] L. Rebenklau "High Temperature Packaging for Sensors", PCIM 2019

PCIM Europe 2024, 11– 13 June 2024, Nuremberg   DOI: 10.30420/566262296

Parametric Study of Damage Evolution in Silver Sintered Layers of Double Sided Power Electronics Modules of Electric Vehicles

Saeed Akbari[1], Kooros Moabber[2], Konstantin Kostov[1], Mietek Bakowski[1], Jang-Kwon Lim[1], Klas Brinkfeldt[1]

[1] RISE Research Institutes of Sweden, Sweden

[2] Volvo Car Corporation, Sweden

Corresponding author:	Saeed Akbari, saeed.akbari@ri.se
Speaker:	Saeed Akbari, saeed.akbari@ri.se

Abstract

Double sided modules accommodating wide band gap (WBG) devices are increasingly used in electric vehicles owing to their lower thermal resistance and parasitic inductances. Compared with single sided modules having a single ceramic substrate, the mechanical constraint applied on the silver sintered bonding layers in double sided modules (with two ceramic substrates) poses a more challenging reliability issue. In this work, we develop a parametric model to investigate the effects of layout, geometry and material properties on damage distribution in silver sintered layers of double sided modules. Anand viscoplastic model was used to describe the inelastic deformation of sintered silver under power cycling. Equivalent inelastic strain accumulated in each power cycle was used as the damage parameter and failure criterion. The model enables parametric study of damage distribution in double sided modules, and help improve design for maximum reliability. Using this model, the effects of parameters such as spacer and die thicknesses were investigated in this study.

1. Introduction

Wide bandgap (WBG) power devices such as silicon carbide (SiC) and gallium nitride (GaN) operate at much higher power densities than the silicon counterparts. Important characteristics of WBG power devices such as high blocking voltage, high switching speed, and low ON-resistance make them an attractive candidate for high power density applications such as electric vehicles.

To exploit the full potential of WBG power devices, it is crucial to develop power modules with minimum parasitic inductances and maximum thermal efficiency. For these purposes, solutions such as double sided cooling [1-2] and embedding [3] are currently being investigated by academia and industry. However, reliability issues and early failures can limit further development of these novel approaches.

In single sided modules currently used for WBG devices, there is only one joining layer related to each chip, defined between the chip and the ceramic substrate. However, in double sided modules, there are two extra interconnection layers related to each chip, formed between chip and spacer as well as between spacer and top substrate. Additionally, in double sided modules more mechanical constraint is applied on interconnection layers by spacer and top substrates. These make reliability of double sided modules more challenging.

Crack propagation in interconnection layers is a predominant failure mode in double sided modules [1-2]. An example is shown in Fig. 1a, where a crack is observed in the solder layer between the copper post (spacer) and the power die after power cycling. Also, Fig. 1b shows crack growth in silver sintered layer of a double sided module.

Many studies have shown significant improvement of power cycling reliability in silver sintered modules compared with soldered ones [4-5]. For this reason, many power module manufacturers are replacing soldering by silver sintering. The higher reliability of silver joints compared with solder ones is associated with their much lower homologous temperature (ratio of operating temperature to the melting point). However, application of silver sintering is typically done at elevated pressures, which requires extra process steps.

To exploit the benefits of silver sintering fully, it is important to optimize the power module layout and select the spacer materials that minimize stress and damage in the silver sintered layers. For this purpose, we use a parametric finite element model to calculate damage distribution in sintered layers for a variety of joining layer thicknesses, spacer material properties, etc. This can help avoid unexpected failures in interconnection layers.

2. Finite Element Model

A subroutine was developed in Ansys Parametric Design Language (APDL) to create a two-dimensional coupled finite element model, schematically depicted in Fig. 2. The coupled thermomechanical simulation involved two analyses, each belonging to a different field. The thermal and structural fields were coupled by applying temperature results from thermal analysis as loads in structural analysis.

First, thermal analysis was performed to calculate temperature at each point. To this end, thermal element type Plane 55 was used to solve heat equation, and determine temperature distribution. This element has a single degree of freedom, temperature, at each node. A heat transfer coefficient of 4,000 W/mk was assumed on the model top and bottom surfaces.

Then, the outcome of thermal analysis, a result file with *.rth* extension, was exported as thermal load to the structural analysis. In this part, structural element type Plane 182 was used to capture deformation and thermal strains and stresses at each point. This element is defined by 4 nodes with two degrees of freedom at each node, including translations in x and y directions. At the end of the structural simulation, the damage was extracted at a path along the midplane of silver layers (X-axis in Fig. 2).

PCIM Europe 2024, 11– 13 June 2024, Nuremberg   DOI: 10.30420/566262296

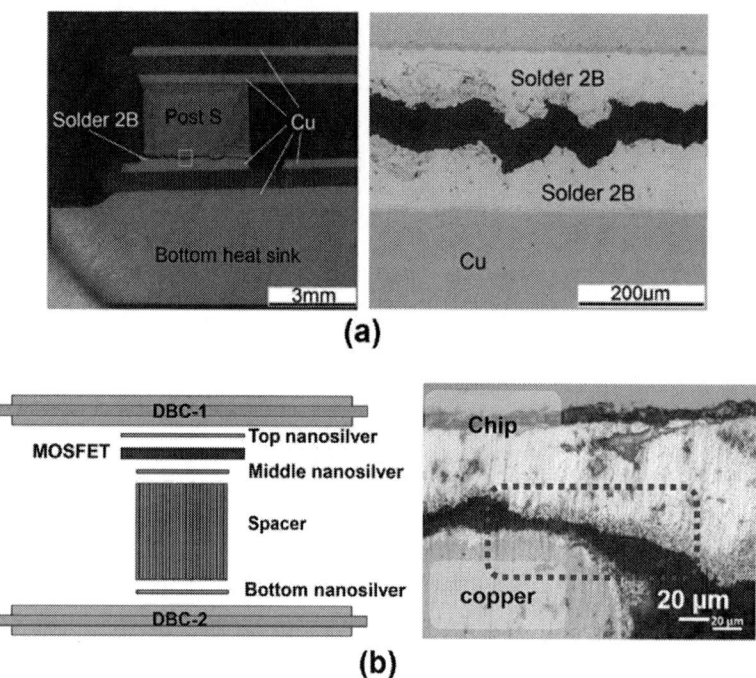

(a)

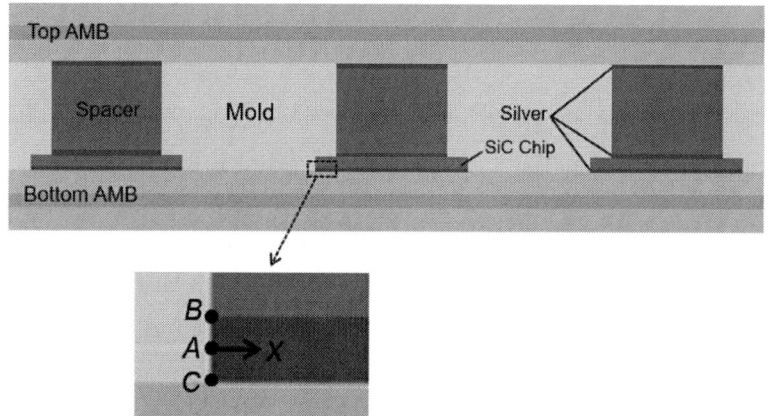

(b)

Fig. 1. Examples of crack propagation in a) solder, and b) silver layers between the spacer and the chip in double sided modules after temperature cycling [1-2].

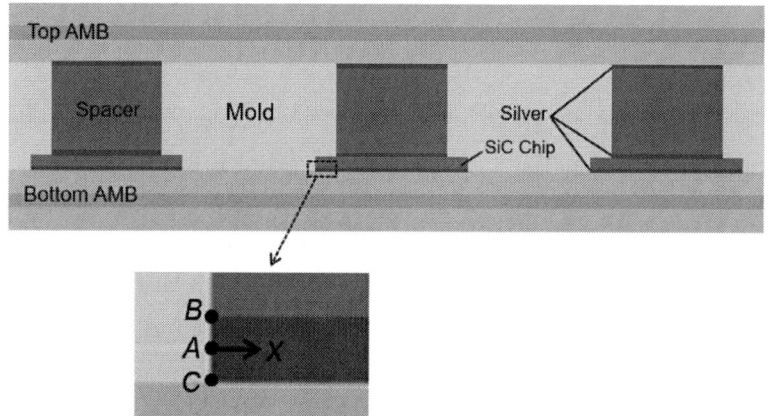

Fig. 2. Schematic representation of a double side module with three power dies. In the finite element model, damage was extracted along the midplane of each silver layer, e.g. X-axis starting at Point A. This is to avoid singularities and stress concnetration at Points B and C. Six elements were considered along the thickness of the silver layers. X-axis is defind from the left edge of each silver layer, and in this study is generally related to the middle silver layer formed between spacer and die, unless otherwise mentioned.

Only three power dies together with the bonded spacers and the joining layers were simulated in order to capture the coupling effects and interaction of the neighboring joints and spacers (Fig. 2).

In the original model, the dimensions of the dies were 5 mm × 5 mm × 0.2 mm. The distance between the dies was 3 mm. The thicknesses of the silver layers and the spacers were 30 μm and 1.7

2096

mm, respectively. For a parametric study, these values were changed subsequently to evaluate their effect on damage.

In single-sided modules, there is only one layer of die attach, connecting the chip to the ceramic substate. However, in double-sided modules, there are three sintered silver layers related to each die, named as follows:

 a) *Top Joint*: The silver layer between the spacer and the top ceramic AMB substrate.

 b) *Middle Joint*: The silver layer between the die and the spacer.

 c) *Bottom Joint*: The silver layer between the die and the bottom AMB substrate.

Thermal and mechanical properties of different layers are reported in Table 1. The epoxy mold compound was modeled using a viscoelastic model [6].

Table 1. Material properties of different layers of a double sided module [7-9].

Property	SiC	Silver	Si$_3$N$_4$	Mold	Copper	CuMo (CPC300)
k (W/m.k)	380	200	120	5	410	300
c (J/kg.k)	750	236	800	800	385	360
ρ (kg/k)	3210	10490	3170	1000	8960	9160
E (GPa)	420	20	320	7	124	155
α (ppm/°C)	6.8	15	3.2	45	16	12.1
ν	0.3	0.3	0.3	0.3	0.34	0.32

Inelastic behavior of the silver sintered bonding layers was described using Anand model [5]. Accumulated equivalent inelastic strain (ε_{in}) was used as damage parameter for relative comparison of different layouts. For simplicity, ε_{in} accumulated in each power cycle is referred to as *damage* in this work. Based on Anand model, damage in the sintered layers is mostly governed by temperature and stress:

$$\dot{\varepsilon}_{in} = A \exp\left(\frac{-Q}{RT}\right)\left[\sinh\left(\frac{\xi\sigma}{s}\right)\right]^{\frac{1}{m}} \quad (1)$$

The constants of Eq. (1) were extracted from [5]. To simulate power cycling, three cycles were modelled with $T_{j,min} = 65°C$ and $T_{j,max} = 175°C$ ($t_{on} = 5$ s, $t_{off} = 10$ s).

3. Results

As explained in previous section, the inelastic strain accumulated per cycle was used as *damage* to compare different module layouts. To show the importance of power cycling reliability in double sided modules, damage distribution in the silver layer between the SiC die and the ceramic substrate in single and double sided modules was calculated and plotted in Fig. 3. It shows that the damage is mostly concentrated in the joining layer edges, and is significantly larger in double sided module. Higher damage in double sided modules compared with single sided modules is a result of the constraint applied on the silver layers from the spacer and the top substrate.

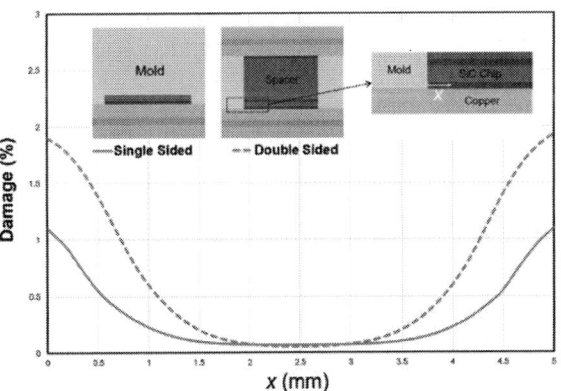

Fig. 3. Comparison of damage distribution in the silver layer betwee the chip and the ceramic substrate in single module as well as double sided module with spacer covering the entire chip surface.

As shown in Fig. 4, in double sided modules, the spacer does not cover the entire die surface, but only the source pads of power dies. Therefore, in a more accurate analysis, the spacer area is smaller than that of the chip. To investigate this in more detail, in Fig. 5 three different cases of single and double sided modules were considered. Case A is a single sided module, while Cases B and C are double sided modules. In Case B, the spacer is 5 mm long, and covers the entire die surface, while in Case C, the die is only 3.5 mm long, and partly caovers the die surface. For each case, damage distribution was obtained at silver layer betwee die and substrate.

Fig. 4. The layout of spacers covering source pads of SiC power dies in a double sided module [10].

It is seen in Fig. 5 that in Case C the maximum damage is lower than that in Case B, and is located below the spacer edge. Then damage remains constant to the end of silver layer. Since Case C better represents spacer layout in a power module, in the next results only this layout is considered.

In single sided modules, there is only one bonding layer for each power die, connecting the power die to the ceramic substrate. However, introduction of spacers and a second ceramic substrate in double sided modules creates two more bonding layers: one in between the spacer and the power die (middle joint), and the other between the spacer and the top ceramic substrate (top joint). These two additonal silver layers can be potential locations for failure as well.

Fig. 6 compares damage in all three joining layers of a double sided module with copper spacers. For a better accuracy, it was assumed the spacer does not cover the entire die surface, but the source connection pads. Fig. 6 shows that the maximum damage is induced in the middle joint (the silver sintered layer between the spacer and the power die). This is consistent with the crack propagation reported in previous studies (Fig. 1).

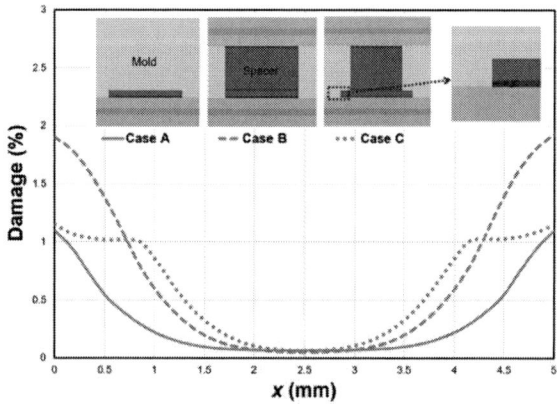

Fig. 5. Comparison of damage distribution in silver layer betwee the chip and the ceramic substrate in single and double sided modules.

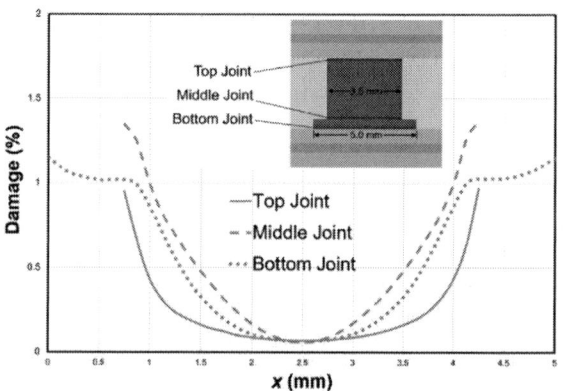

Fig. 6. Comparison of damage distriction in three different silver layers of a double sided module.

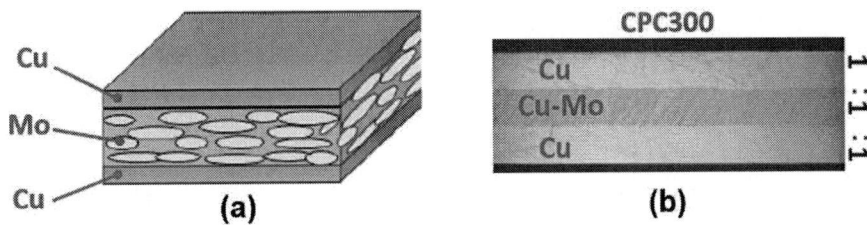

Fig. 7. a) Layered structure of Copper-Mollybdenum (CuMo) spacers. The middle layer is a composite structure consisting of Mollybdenum particles suspended in Copper. b) Relative thickness of the three layers specifies the thermal and mechanical properties of the spacer. For example, in CPC300, all the three layers have the same thickness [8].

For simplicity, it was assumed in Fig. 6 that middle joint is continuous, but it is not, as shown in Fig. 4, and only covers the source pads. This even makes damage growth in middle joint more plausible. Compared with the middle joint, the bottom joint is less critical, becase this joint is larger, and it takes a longer time for crack to propagate along this path. The top joint in double sided modules with copper spacers is the least critical silver joint with minimum damage, because it is formed between two copper layers (copper spacer and copper layer of top substrate).

Copper (Cu) has a significantly larger CTE than the SiC chip. This CTE mismatch creates significant stresses in the silver layers, especially the middle joint, as confirmed by Fig. 6. To address this issue, novel spacer materials, e.g. Molybdenum (Mo), has been used. An example is CPC (Cu, Cu-Mo, Cu) composite structures developed by Allied Material [8]. In CPC spacers, a Cu-Mo composite layer is sandwiched between Cu layers, as shown in Fig. 7. CTE and thermal conductivity of the CPC spacers can be adjusted by changing composition and lamination ratio of Cu-Mo. Table 2 shows thermal and mechanical properties of CPC spacers with different weight frac-

tions of Molybdenum. With increase of Molybdenum weight fraction, The CTE mismatch between the spacer and the SiC chip decreases. On the other hand, the CTE mismatch between the spacer and the top substrate increases.

Table 2. Effect of Molybdenum weight fraction on thermal and mechanical properties of CPC spacers, as demonstrated in Fig. 7.

Property	Copper	CPC300	CPC111	CPC141	Molybdenum
Mo wt%	0	16	23	47	100
k (W/m.k)	410	300	260	200	138
c (J/kg.k)	385	360	354	322	250
ρ (kg/k)	8960	9160	9245	9543	10200
E (GPa)	124	155	169	216	320
α (ppm/°C)	16	12.1	9.8	7.6	5.2
v	0.34	0.32	0.32	0.32	0.31

Fig. 8 compares damage in middle silver joint of double sided modules with Cu and CuMo (CPC300) spacers. It shows that the introduction of low CTE CPC300 spacers can significantly lower damage in middle joints, and possibly prevent failure modes observed in previous studies (Fig. 1).

PCIM Europe 2024, 11– 13 June 2024, Nuremberg   DOI: 10.30420/566262296

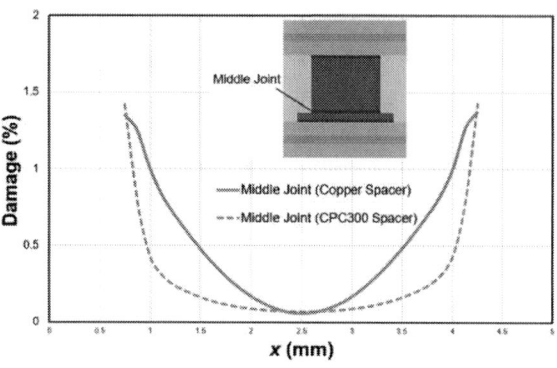

Fig. 8. Effect of spacer material (Cu vs. CuMo) on damage distribution in the middle joint (the silver layer betwee the spacer and the chip).

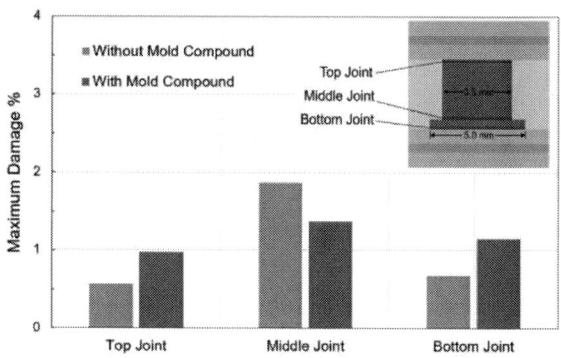

Fig. 9. Importance of modelling mold compound to capture damage values correctly in silver layers. The mold can change the damage values up to 40%.

Some simulation studies do not consider the epoxy mold compound in thermomechanical simulations [11-12]. The question may arise whether it is important to model epoxy mold in reliability prediction. To investigate the effect of mold compound on damage distribution in interconnection layers, another model was considered without mold comound. The results were then compared with the original model, containing mold comound, and plotted in Fig. 9. It shows that the presence of mold compound increases damage in the top and bottom joints, and lowers it in the middle joint. This means that if the failure happens in the middle joining layer (examples shown in Fig. 1), then using a stiffer mold compound, e.g. with higher ratio of filler particles, can reduce damge in the middle layer.

The reason is probably that the mold compound limits deformation of the spacer, and the die in the horizontal directin. As a result, less damage will be induced in the middle silver joint, sandwiched between the spacer and the die. The top and the bottom ceramic substrates are already fixed by the power terminals, and are not much affected by the mold compound.

The parameteric finite element mode developed in this work enables the study of effect of variables such as spacer and die thicknesses on damage in silver sintered layers. For this purpose, the maximum damage developed at the edge of the middle silver layer was caculated versus these parameters.

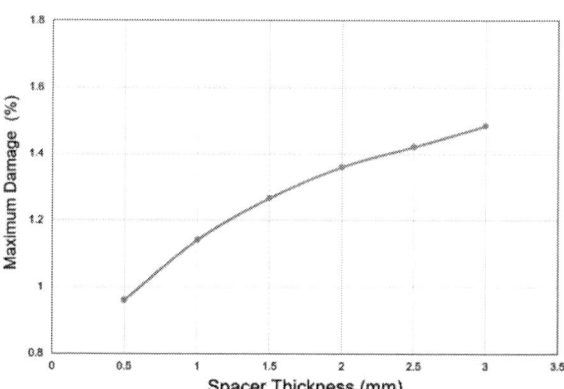

Fig. 10. Effect of copper spacer thickness on maximum damage in middle joint.

Fig. 10 shows that the maxiumum damage increases with spacer thickness. Similarly, Fig. 11 shows that the use of a thicker die can increase damage. These results show that an interconnection layer sandwiched between rigid thicker substrates is subject to more degraditation.

2100

On the contrary, the use of a thicker silver layer can decrease the maximum damage (Fig. 12).

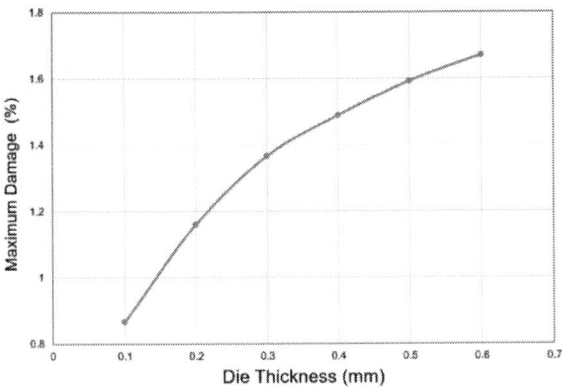

Fig. 11. Effect of die thickness on maximum damage.

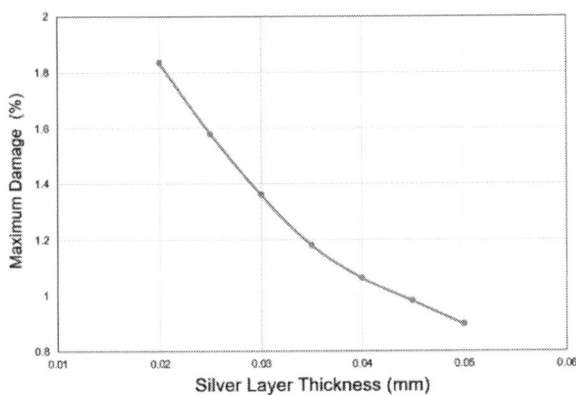

Fig. 12. Effect of silver layer thickness on maximum damage.

It is alreadyd known that there could be thermal coupling between power dies in a power module, meaning heat generation in one chip can incease temperature in neighboring chips. In addition, we may face structural coupling in double sided modules, which can influence damage distribution and failure of a silver joint due to mechanical constraints applied by neighboring joints and spacers. In order to investigate this further, the effect of adjacent dies and spacers on maximum damage in silver joints was studied by comparing two different cases demonstrated in Fig. 13. In Case A, the chip and the spacer are surrounded by adjacent dies and spacers on both sides, while in Case B there are no neighboring dies and spacers. The maximum damage was calculated for all three silver layers: top joint, middle joint, and bottom joint. The plotted data in Fig. 13 shows that the presense of neighboring joints can increase damage in all three joints. Many previous studies only model a single die and spacer for simplicity [13-15], hence underestimate damage values.

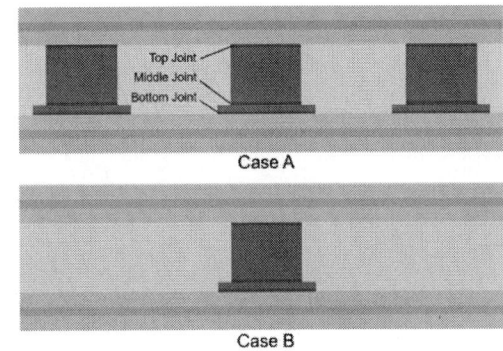

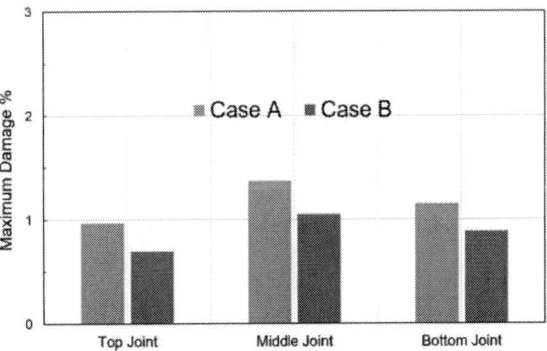

Fig. 13. Effect of structural coupling on damage in silver layers. Case A includes three spacers with dies to investigate the influence of neighboring spacers on damage in different silver layers. Case B includes only one spacer and one die.

4. Conclusions

Using a parametric model, we simulated damage distribution in different silver layers of double sided modules with SiC chips. It was found that in the modules with copper spacers, damage is maximum in the middle silver layer, defined between the spacer and the die due to CTE mismatch between copper and SiC. It was shown that the use of low CTE Copper-Molybdenum spacers can lower damage in the critical silver layer. In addition, mold compouned can decrease damage in middle layers, and increase it in the other two joints. The effect of adjacent dies and spacers on damage distribution was also demonstrated. Moreover, it was shown that thicker spacers and dies can incraese damage in ctritical silver layers.

Acknowledgement

This work was supported by Volvo Car Corporation.

References

[1] Ma Y, Li J, Dong F, Yu J. Power cycling failure analysis of double side cooled IGBT modules for automotive applications. Microelectronics Reliability. 2021 Sep 1;124:114282.

[2] Liu S, Mei YH, Li J, Li X, Lu GQ. Copper-Wire Stress Buffers for Extending Lifetime of Double-Sided Bidirectional SiC Modules. IEEE Transactions on Power Electronics. 2023 Mar 3;38(6):7118-27.

[3] Huesgen T. Printed circuit board embedded power semiconductors: A technology review. Power Electronic Devices and Components. 2022 Oct 1;3:100017.

[4] Durand C, Klingler M, Coutellier D, Naceur H. Power cycling reliability of power module: A survey. IEEE

Transactions on Device and Materials Reliability. 2016 Jan 8;16(1):80-97.

[5] Forndran F, Heilmann J, Metzler M, Leicht M, Wunderle B. Determination of Rate-and Temperature Dependent Inelastic Material Data for Sintered Silver Die Attach and Simulative Implementation. In2022 23rd International Conference on Thermal, Mechanical and Multi-Physics Simulation and Experiments in Microelectronics and Microsystems (EuroSimE) 2022 Apr 25 (pp. 1-6). IEEE.

[6] Durand C, Klingler M, Coutellier D, Naceur H. Study of fatigue failure in Al-chip-metallization during power cycling. Engineering Fracture Mechanics. 2015 Apr 1;138:127-45.

[7] Grams A, Jaeschke J, Wittler O, Fabian B, Thomas S, Schneider-Ramelow M. FEM-based combined degradation model of wire bond and die-attach for lifetime estimation of power electronics. Microelectronics Reliability. 2020 Aug 1;111:113683.

[8] https://www.allied-material.co.jp/en/research-development/heatspreader.html

[9] Heilmann DI. Lebensdauermodellierung für gesinterte Silberschichten in der leistungselektronischen Aufbau-und Verbindungstechnik durch isotherme Biegeversuche als beschleunigte Ermüdungstests.

[10] Yole Report, Automotive Power Module Comparison 2022.

[11] Chen C, Choe C, Kim D, Suganuma K. Lifetime prediction of a SiC power module by micron/submicron Ag sinter joining based on fatigue, creep and thermal properties from room temperature to high temperature. Journal of Electronic Materials. 2021 Mar;50:687-98.

[12] Hu B, Gonzalez JO, Ran L, Ren H, Zeng Z, Lai W, Gao B, Alatise O, Lu H, Bailey C, Mawby P. Failure and reliability analysis of a SiC power module based on stress comparison to a Si device. IEEE

Transactions on device and materials reliability. 2017 Oct 26;17(4):727-37.

[13] Thoben M, Sauerland F, Mainka K, Edenharter S, Beaurenaut L. Lifetime modeling and simulation of power modules for hybrid electrical/electrical vehicles. Microelectronics Reliability. 2014 Sep 1;54(9-10):1806-12.

[14] Hu B, Gonzalez JO, Ran L, Ren H, Zeng Z, Lai W, Gao B, Alatise O, Lu H, Bailey C, Mawby P. Failure and reliability analysis of a SiC power module based on stress comparison to a Si device. IEEE Transactions on device and materials reliability. 2017 Oct 26;17(4):727-37.

PCIM Europe 2024, 11– 13 June 2024, Nuremberg DOI: 10.30420/566262297

Tristate Modified Boost Converter

Felix A. Himmelstoss, Johannes V. Gragger

University of Applied Sciences Technikum Wien, Austria

Corresponding author: Felix Himmelstoss, felix.himmelstoss@technikum-wien.at
Speaker: Johannes Gragger, johannes.gragger@technikum-wien.at

Abstract

When the position of the capacitor in a Boost converter is changed from the output to bridge the positive input connector with the positive output connector, two interesting features occur: the voltage stress of the capacitor is reduced and the inrush current is avoided, when the converter is applied to a stable DC supply. Using the tristate concept, the active switch of the converter is replaced by a combination of two electronic switches and a diode. The two electronic switches are connected in series and an additional diode is connected to the connection point of the switches. Now, further interesting and helpful features occur. The voltage transformation ratio is linearized and the converter becomes a phase-minimum system, when the duty cycle of the second switch is held constant. The stress of the active switches is also reduced.

1 Introduction

The normal Boost converter, explained in all the textbooks of Power Electronics (e.g. [1], [2], [3]), has a large inrush current when connected to a stable input voltage source. Assuming constant values of the inductor and the capacitor, the inrush current is described by

$$i = U_1 \sqrt{\frac{C}{L}} \sin\left(\sqrt{\frac{1}{LC}}t\right) \qquad (1)$$

and the voltage across the output capacitor follows

$$u_C = U_1 \left[-\cos\left(\sqrt{\frac{1}{LC}}t + 1\right)\right]. \qquad (2)$$

The amplitude of the inrush current is even higher, because the coil saturates (except an air coil is used) and therefore reduces its value. The voltage across the capacitor reaches double the input voltage when no load is applied. When the load is connected to the output, the ringing is damped and the output voltage reaches a lower value and decreases by an exponential function to a value equal to the input voltage. To avoid the inrush-current, the position of the capacitor must be changed (or a pre-stage like Buck converter must be connect to the input terminals).

1.1 Modified Boost Converter

The modified Boost converter (Fig. 1) differs from the normal Boost converter by the position of the capacitor. The capacitor is connected between the positive output and the positive input connectors instead of between the output connectors. The output voltage is now the sum of the input voltage plus the voltage across the capacitor.

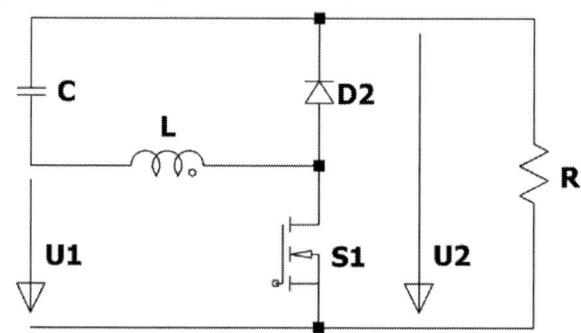

Fig. 1 Modified Boost converter.

The modified Boost converter has two interesting features. The voltage across the capacitor is reduced and an inrush current, when applied to a stable DC supply (e.g. car batteries or a battery buffered DC micro-grid) which stresses the components, is avoided. More details about this kind of modification can be found in [4], [5].

1.2 Modified Tristate Boost Converter

The tristate concept is shown and treated for the normal Boost converter in [6], [7]. The application to other converter topologies is treated in [8]. Here this concept is applied to the modified Boost converter. The circuit diagram is depicted in Fig. 2.

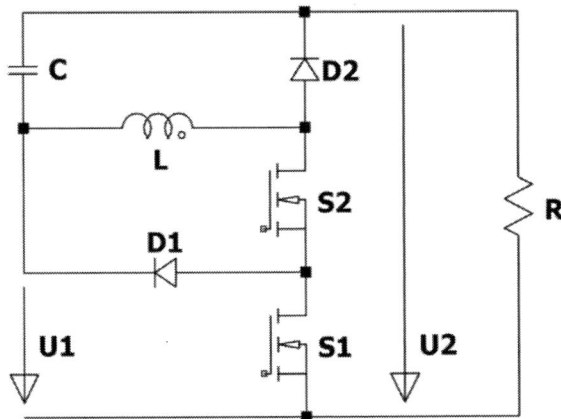

Fig. 2   Modified tristate Boost converter.

The converter has three modes in the continuous inductor current mode (CICM). In the first mode M1 both electronic switches are on and the input voltage lies across the inductor. The current through it increases. When the first switch S1 is turned off, the current of the coil commutates into the diode D1. The voltage across the coil is now nearly zero (only the on-resistor of S2 and the forward voltage produce a small negative voltage across the inductor) and the current stays nearly constant. When the first switch S1 is turned off, too, the free-wheeling diode D2 turns on. The voltage across the coil is now the negative voltage across the capacitor (or the difference between the input and the output voltages). The current through the coil decreases. The voltage across an inductor must be zero in the steady-state. Hence, the voltage-time balance of the duty cycles (the on-time of the switch referred to the switching period) for switch S1 d_1 and for switch S2 d_2 is therefore

$$U_1 d_1 = |U_1 - U_2|(1 - d_2) \tag{3}$$

which leads to the voltage transformation ratio of

$$M = \frac{U_2}{U_1} = \frac{1 + d_1 - d_2}{1 - d_2}. \tag{4}$$

The duty cycle of the second switch is greater than the one for switch S1. When both duty cycles are equal, the voltage transformation ratio is equal to the normal Boost converter. Fig. 3 shows the voltage transformation ratio with the duty cycle d_1 as variable and the duty cycle d_2 as parameter.

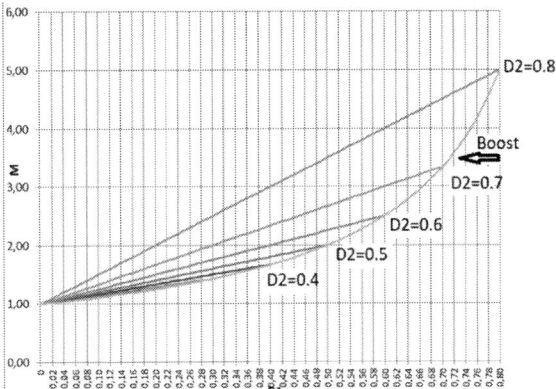

Fig. 3   Voltage transfer ratio M for the tristate

Boost converter: D1 variable and D2 as parameter.

The nonlinear curve shows the transformation ratio of the normal Boost converter. Using now a constant duty cycle for S2, the transformation ratio becomes linear. Later we will see that the transfer function of the converter between the output voltage and the duty cycle d1 is a phase-minimum one. This is helpful for the control.

It is also possible to hold the duty cycle of S1 constant and vary the duty cycle of S2 (Fig. 4).

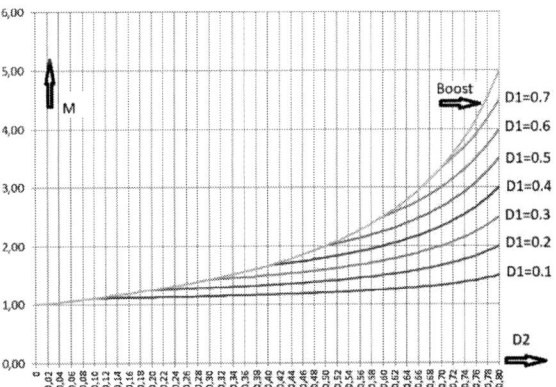

Fig. 4   Voltage transfer ratio M for the tristate

Boost converter: D2 variable and D1 as parameter.

In this case one gets a lower and less steep voltage transformation ratio compared to the Boost converter.

2 Description of the Function of the Tristate Boost Converter

The converter is now described in detail with the help of suggestive drawings. The duty cycle of the first switch is chosen to 25 % and for the second switch to 50 % (Fig. 5).

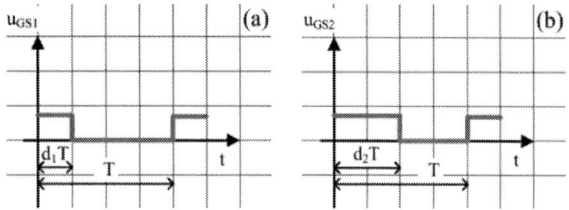

Fig. 5   (a) control signal 1, (b) control signal 2.

The voltage across the coil must be zero in the mean. With an input voltage of two divisions one gets Fig. 6.c. Now one can draw immediately the voltage across the capacitor (so that the voltage across the coil is zero in the mean) and by adding the input and the capacitor voltages one gets the output voltage (Fig. 6.d).

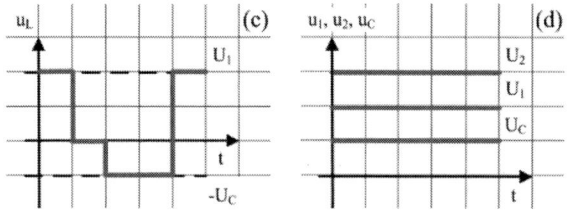

Fig. 6   (c) voltage across L, (d) input voltage, output voltage, and voltage across C.

Now one can draw the voltages across the semiconductors.

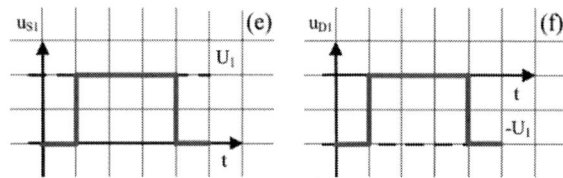

Fig. 7   (e) voltage across S1, (f) voltage across D1.

During M1, S1 is on and the voltage across it is zero. During M2, this voltage increases till U1 (the diode D1 is on during M2). In M3 the diode D1 turns off, because S2 is also off and the voltage will remain U1 across S1 (Fig. 7.e). Due to the capacities of the semiconductor devices, it can be a little different.

The voltage across D1 equals minus the input voltage during M1, zero during M2 and will stay near zero during M3 (Fig. 7.f).

The voltage across S2 is zero during the modes M1 and M2 and goes up to the difference between the input and the output voltages, which is also equal to the voltage across the capacitor (Fig. 8.g).

The voltage across D2 is the negative output voltage during M1, the difference between the input and the output voltages during M2, and zero during M3 (Fig. 8.h).

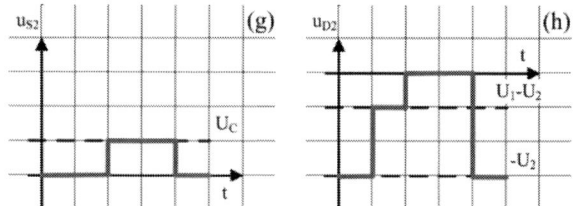

Fig. 8   (g) voltage across S2, (h) voltage across D2.

For the current through the coil we start arbitrarily with one division and let it rise by one division during the first mode. In M2 the voltage across the coil is short-circuited and therefore the current stays constant, and during M3 the voltage across the coil is negative, and the current decreases to the starting point in the steady state (Fig. 9.i). The current through S1 is equal to the current through the coil during M1 and zero during the rest of the period (Fig. 9.j).

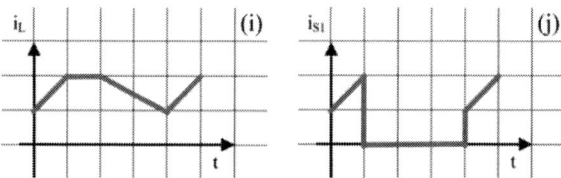

Fig. 9   (i) current through L, (j) current through S1.

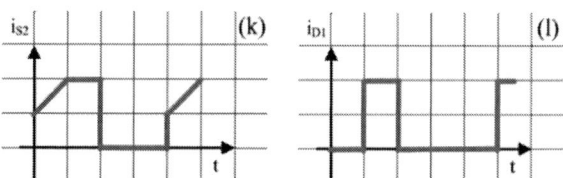

Fig. 10   (k) current through S2, (l) current through D1.

The current through the second active switch S2 is equal to the current through the coil during the first two modes and zero in the third mode (Fig. 10.k). Diode D1 conducts the current through the coil during M2, and during M1 and M3 no current is flowing through it (Fig. 10.l). The current through the coil free-wheels during M3 through D2 (Fig. 11).

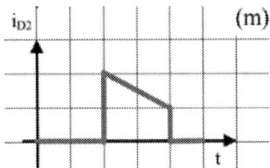

Fig. 11   (m) current through D2.

To get the connection between the currents, one has to inspect the current through the capacitor. It must be zero in the mean. During the complete period the negative load current flows through the capacitor and only during M3 a positive current flows through it.

The charge balance can now be written according to

$$I_{LOAD} \cdot d_2 = \left(\bar{I}_{LM3} - I_{LOAD} \right) \cdot (1 - d_2).$$ (5)

The value $\bar{I}_{LM3}$ is the mean of the current through the coil during M3, referred to the duration of the mode M3. This leads to

$$\bar{I}_{LM3} = \frac{I_{LOAD}}{1 - d_2}.$$ (6)

Now one can draw the currents through the load and through the capacitor (Fig. 12.n). Finally, the input current is sketched (Fig. 12.o)

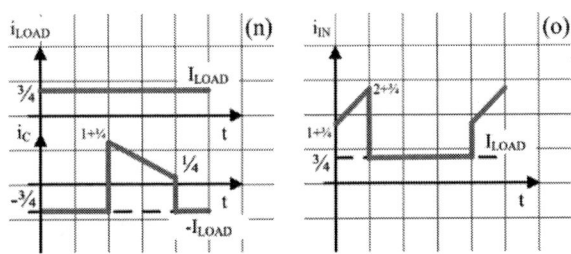

Fig. 12   (n) load current and current through C, (o) input current.

The input current is pulsating.

3   Modelling of the Tristate Boost Converter

3.1   Description of the Modes

During mode M1 (both electronic switches are on, both diodes are off) the state equations are

$$\frac{di_L}{dt} = \frac{u_1}{L}, \quad \frac{du_C}{dt} = \frac{-(u_1 + u_C)/R}{C},$$ (7)

or combined to the matrix description

$$\frac{d}{dt}\begin{pmatrix} i_L \\ u_C \end{pmatrix} = \begin{bmatrix} 0 & 0 \\ 0 & -\dfrac{1}{CR} \end{bmatrix}\begin{pmatrix} i_L \\ u_C \end{pmatrix} + \begin{bmatrix} \dfrac{1}{L} \\ -\dfrac{1}{CR} \end{bmatrix}(u_1).$$ (8)

During M2 (the first electronic switch is turned off, D1 turns on) one gets

$$\frac{di_L}{dt} = \frac{0}{L}, \quad \frac{du_C}{dt} = \frac{-(u_1 + u_C)/R}{C}$$ (9)

or again combined to the matrix description

$$\frac{d}{dt}\begin{pmatrix} i_L \\ u_C \end{pmatrix} = \begin{bmatrix} 0 & 0 \\ 0 & -\dfrac{1}{CR} \end{bmatrix}\begin{pmatrix} i_L \\ u_C \end{pmatrix} + \begin{bmatrix} 0 \\ -\dfrac{1}{CR} \end{bmatrix}(u_1).$$ (10)

During mode M3 (the second electronic switch is also turned off, the current through the coil now free-wheels over D2) the description is

$$\frac{di_L}{dt} = \frac{-u_C}{L}, \quad \frac{du_C}{dt} = \frac{i_L - (u_1 + u_C)/R}{C}$$ (11)

or again combined to the matrix description

$$\frac{d}{dt}\begin{pmatrix} i_L \\ u_C \end{pmatrix} = \begin{bmatrix} 0 & -\dfrac{1}{L} \\ \dfrac{1}{C} & -\dfrac{1}{CR} \end{bmatrix}\begin{pmatrix} i_L \\ u_C \end{pmatrix} + \begin{bmatrix} 0 \\ -\dfrac{1}{CR} \end{bmatrix}(u_1).$$ (12)

3.2   Large Signal Model

Weighting the three systems with the duty cycles during which they are valid (M1 by d_1, M2 by d_2-d_1, and M3 by 1-d_2) leads to the large signal system according to

$$\frac{d}{dt}\begin{pmatrix} i_L \\ u_C \end{pmatrix} = \begin{bmatrix} 0 & \dfrac{d_2 - 1}{L} \\ \dfrac{1 - d_2}{C} & -\dfrac{1}{CR} \end{bmatrix}\begin{pmatrix} i_L \\ u_C \end{pmatrix} + \begin{pmatrix} \dfrac{d_1}{L} \\ -\dfrac{1}{CR} \end{pmatrix}(u_1).$$ (13)

This description is a nonlinear one. To use the linear control theory and to get transfer functions and to construct Bode plots one has to linearize it.

3.3 Small Signal Model

With the perturbation ansatz, all variables are written by the operation point values (capital letters with a zero at the index) plus the perturbation of the operating point (small letters with a roof on top) one gets for the state variables and for the input variable u_1

$$i_L = I_{L0} + \hat{i}_L, \quad u_C = U_{C0} + \hat{u}_C, \quad u_1 = U_{10} + \hat{u}_1 \quad (14)$$

and for the two new input variables

$$d_1 = D_{10} + \hat{d}_1, \quad d_2 = D_{20} + \hat{d}_2 \quad . \quad (15)$$

One can now write for the first equation

$$\frac{d}{dt}\left(I_{L0} + \hat{i}_L\right) = \frac{1}{L}\left\{ \begin{array}{l} \left(D_{20} + \hat{d}_2 - 1\right)\left(U_{C0} + \hat{u}_C\right) + \\ \left(D_{10} + \hat{d}_1\right)\left(U_{10} + \hat{u}_1\right) \end{array} \right\} \quad (16)$$

Deleting the terms which are multiplications between two variables (perturbations) and doing the same for the second state equation of (13) leads to the linearization around the operating point

$$\frac{d}{dt}\begin{pmatrix} \hat{i}_L \\ \hat{u}_C \end{pmatrix} = \begin{bmatrix} 0 & \dfrac{D_{20}-1}{L} \\ \dfrac{1-D_{20}}{C} & -\dfrac{1}{CR} \end{bmatrix}\begin{pmatrix} \hat{i}_L \\ \hat{u}_C \end{pmatrix} +$$

$$\begin{bmatrix} \dfrac{D_{10}}{L} & \dfrac{U_{10}}{L} & \dfrac{U_{C0}}{L} \\ -\dfrac{1}{CR} & 0 & -\dfrac{I_{L0}}{C} \end{bmatrix}\begin{pmatrix} \hat{u}_1 \\ \hat{d}_1 \\ \hat{d}_2 \end{pmatrix} \quad . \quad (17)$$

This is the linearized model or small signal model. To obtain the transfer functions one has to use the Laplace transformation.

3.4 Connection Between the Operation Point Values

The constant terms on the right side of (16) must be zero and this leads to the same results for the operating point which we found by the graphical inspection

$$U_{C0} = \frac{D_{10}}{1 - D_{20}} U_{10} . \quad (18)$$

And for the output voltage one gets

$$U_{20} = \frac{1 + D_{10} - D_{20}}{1 - D_{20}} U_{10} . \quad (19)$$

For the currents one gets

$$I_{L0} = \frac{I_{LOAD}}{1 - D_{20}} . \quad (20)$$

One has to mention here that this current is equal to $\bar{I}_{LM3}$ in (6). The real mean value is a little bit larger

$$\bar{I}_L = \bar{I}_{LM3} \cdot d_1 + \bar{I}_{LM3} \cdot \left(1 - d_2\right) + \left(\bar{I}_{LM3} + \frac{\Delta I}{2}\right) \cdot \left(d_2 - d_1\right) \quad . \quad (21)$$

The current ripple corresponds with the change of the current through M1

$$\Delta I = \frac{U_1}{L} \cdot \frac{d_1}{f} . \quad (22)$$

(21) and (22) are valid always in the steady case. For the operational point values only a zero has to be supplemented and the small symbols for the duty cycle should be changed to capital letters.

3.5 Transfer Functions

Laplace transformation of (17) leads to

$$\begin{bmatrix} s & \dfrac{1-D_{20}}{L} \\ \dfrac{D_{20}-1}{C} & s + \dfrac{1}{CR} \end{bmatrix}\begin{pmatrix} I_L(s) \\ U_C(s) \end{pmatrix} =$$

$$= \begin{bmatrix} \dfrac{D_{10}}{L} & \dfrac{U_{10}}{L} & \dfrac{U_{C0}}{L} \\ -\dfrac{1}{CR} & 0 & -\dfrac{I_{L0}}{C} \end{bmatrix}\begin{pmatrix} U_1(s) \\ D_1(s) \\ D_2(s) \end{pmatrix} \quad . \quad (23)$$

One can now easily calculate the transfer functions with the help of Crammer's rule. With the abbreviations of the elements of the state matrix A and of the input matrix B

$$\begin{bmatrix} s & -A_{12} \\ -A_{21} & s - A_{22} \end{bmatrix}\begin{pmatrix} I_L(s) \\ U_C(s) \end{pmatrix} =$$

$$= \begin{bmatrix} B_{11} & B_{12} & B_{13} \\ B_{21} & 0 & B_{23} \end{bmatrix}\begin{pmatrix} U_1(s) \\ D_1(s) \\ D_2(s) \end{pmatrix} \quad . \quad (24)$$

one obtains the transfer functions between the capacitor voltage and the duty cycles

$$\frac{U_C(s)}{D_1(s)} = \frac{\dfrac{(1-D_{20})}{CL}U_{10}}{s^2 + s\dfrac{1}{CR} + \dfrac{(1-D_{20})^2}{CL}}, \qquad (25)$$

and

$$\frac{U_C(s)}{D_2(s)} = \frac{-s\dfrac{I_{L0}}{C} + \dfrac{(1-D_{20})}{CL}U_{C0}}{s^2 + s\dfrac{1}{CR} + \dfrac{(1-D_{20})^2}{CL}}. \qquad (26)$$

Here we see an interesting feature of the converter. It is a step-up converter, but when the duty cycle d_2 is held constant the converter can be controlled by changing the duty cycle d_1, and this system is a phase-minimum system. When one holds d_1 constant and controls the converter by d_2, the system is described by a non-phase-minimum transfer function (26). The zero is on the right side of the complex plane. A designed control loop will be slower than for the phase-minimum system. Therefore, one controls the converter by changing d_1 and not by changing d_2.

The influence of the disturbances caused by changes of the input voltage is given by

$$\frac{U_C(s)}{U_1(s)} = \frac{-s\dfrac{1}{CR} - \dfrac{(1-D_{20})D_{10}}{CL}}{s^2 + s\dfrac{1}{CR} + \dfrac{(1-D_{20})^2}{CL}}. \qquad (27)$$

The output voltage is the voltage across the capacitor plus the input voltage.

4   Dimensioning hints

During the on-time of the second active switch the load is supplied by the capacitor and the voltage across it decreases by

$$\Delta u_C = \frac{1}{C_2}\int_0^{d_2 T} I_{LOAD}\,dt. \qquad (28)$$

The capacitor is so large that the output voltage does not change noticeably, therefore one can use a constant load current. This leads to a capacitor value of

$$C = \frac{I_{LOAD}\cdot d_2}{\Delta u_C \cdot f}. \qquad (29)$$

In practice a larger capacitor is needed to compensate the influence of the parasitic series resistor of the capacitor which adds an additional voltage drop of

$$u_{RC} = R_C \cdot I_{LOAD}. \qquad (30)$$

With a chosen current ripple ΔI the inductor becomes

$$L = \frac{U_1}{\Delta I}\cdot\frac{d_1}{f}. \qquad (31)$$

The load current is equal to the mean value of the current through the second diode. When the converter is in the critical conduction mode (boundary mode: the current through L reaches zero at the end of the switching period), one can connect the minimal load current with the current ripple according to

$$\Delta I = \frac{2I_{LOAD,\min}}{(1-d_2)}. \qquad (32)$$

Now the converter is in the CICM for all load currents greater than $I_{LOAD,\min}$.

It is also possible to convert the circuit into a bidirectional one by replacing the diodes by current bidirectional switches (in Fig. 2 MOSFETs are drawn which have an intrinsic diode in parallel to the channel and are therefore current bidirectional switches). Fig. 13 shows the extension to a bidirectional converter. The diodes are included in the circuit diagram. The switches SD2 and SD1 are used to control the energy flow from the right side to the left side, and in this case S1 and S2 are blocked. One can use the switches also to reduce the losses by shunting the conducting diodes.

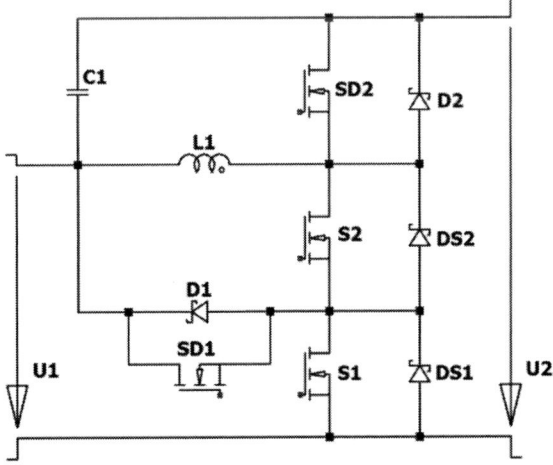

Fig. 13 Bidirectional tristate modified Boost converter.

5 Simulations

Fig. 14 shows the used simulation circuit. With the help of the fast comparators U1 and U2 the pulse width modulation of the switches is produced. The comparators are supplied by plus/minus 5 V. To avoid different time delays, the same driver stage (a voltage-controlled voltage-source, which also gains the pwm signal sigma) is used to drive the MOSFETs. With the piece-wise-linear voltage-sources V2 and V6 the duty cycle signals are produced.

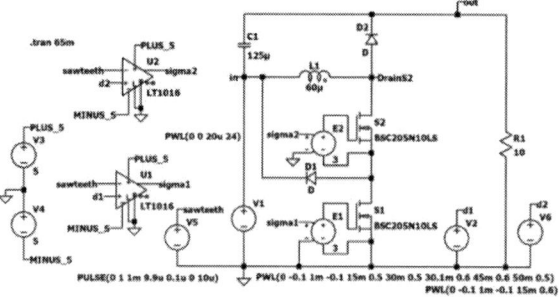

Fig. 14 Simulation circuit.

To achieve a soft-start the duty cycles are increased linearly as shown in Fig. 15. In this figure also some changes of the duty cycle of the first switch take place. The signals seen here are (up to down): the duty cycle of S2, the duty cycle of S1, the current through the coil, the current through the load, the output voltage, and the input voltage. The soft-start leads only to a small overshot and ringing. A step-change of the duty cycle d_1, however, leads to pronounced ringing and a large overshot. Therefore, the duty cycle should not be changed by steps but by ramps, as is also shown in this figure.

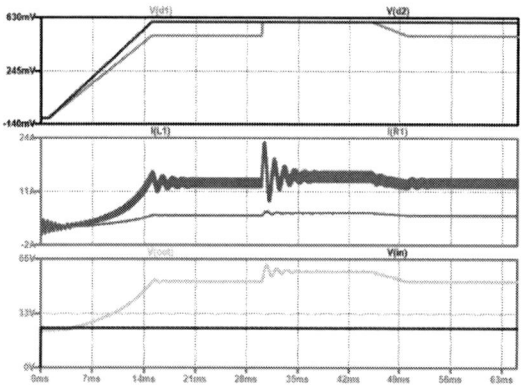

Fig. 15 Start-up and changes of the duty cycle of the first switch, up to down: duty cycle of S2 (black), duty cycle of S1 (turquoise); current through the coil (red), current through the load (brown); output voltage (green), input voltage (blue).

Fig. 16 shows the steady state. The signals seen here are (up to down): the voltage across the coil, the current through the coil, the current through the load; the output voltage, the input voltage, the control signal of S2, control signal of S1.

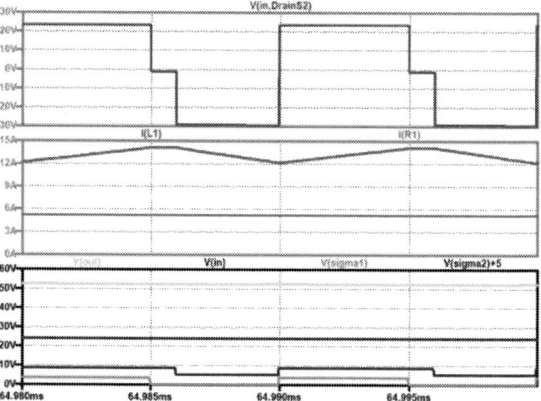

Fig. 16 Steady state, up to down: voltage across the coil (dark green); current through the coil (red), current through the load (brown); output voltage (green), input voltage (blue); control signal of S2 (black, shifted), control signal of S1 (turquoise).

In Fig. 17 the generation of pwm signals is shown. The signals seen here are (up to down): the control signal of S2, the control signal of S1, saw-teeth as the carrier of the modulation, the duty cycle of S2, the duty cycle of S1, the output voltage, the input voltage, the control signal of S2, the control signal of S1.

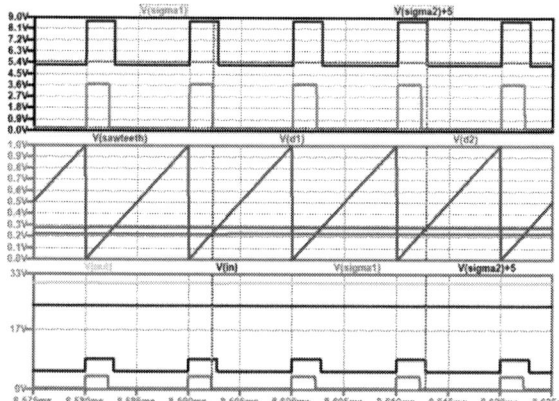

Fig. 17 Generation of the control signals, up to down: control signal of S2 (black, shifted), control signal of S1 (turquoise); saw-teeth (red), duty cycle of S2 (grey), duty cycle of S1 (violet); output voltage (green), input voltage (blue), control signal of S2 (black, shifted), control signal of S1 (turquoise).

Fig. 18 shows the Bode plot of the converter in this operation point.

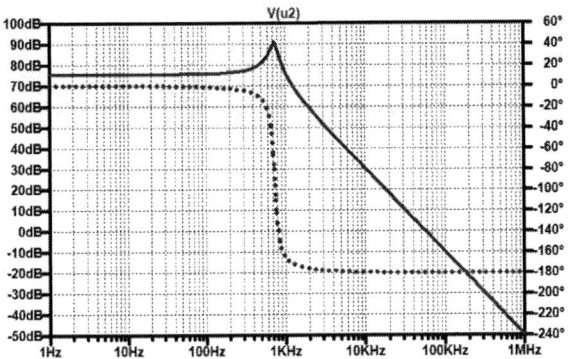

Fig. 18 Bode plot output voltage referenced to the duty cycle of switch S1, with constant duty cycle of switch S2 (solid line: gain response, dotted line: phase response).

The system is now a phase minimum system, because the phase tends to minus 180°. A non-phase minimum system would tend to minus 270°. The resonance is caused by the output capacitor and the coil. This resonance circuit is relatively low damped (keep in mind that the model was calculated for ideal devices and the damping is only caused by the load), which leads to a fast change of the phase from zero to minus 180°.

6 Conclusions

The tristate modified Boost converter has several interesting features:

- No inrush current when applied to a stable input source like batteries or a stable buffered DC micro-grid
- Reduced voltage stress across the capacitor
- Modified voltage transformation ratio
- When the duty cycle of the second switch is constant and the converter is controlled by the duty cycle of the first switch, the converter behaves as a phase-minimum system
- When the duty cycle of the second switch is constant, the voltage transformation ratio becomes linear
- When the duty cycle of the first switch is hold constant and the converter is controlled by the duty cycle of the second

switch, the converter behaves as a non-phase-minimum system and the voltage transformation ratio is lower and less steep as it is in the case of the classical Boost converter.

- Reduced voltage stress across the active switch
- A bidirectional version of the converter is possible

The converter can be used e.g. as switched mode power supply, battery charger or driver for lighting devices.

References

[1] F. Zach: Power Electronics, in German, Leistungselektronik, Frankfurt: Springer, 6th ed., 2022.

[2] N. Mohan, T. Undeland and W. Robbins, Power Electronics, Converters, Applications and Design, 3nd ed. New York: W. P. John Wiley & Sons, 2003.

[3] Y. Rozanov, S. Ryvkin, E. Chaplygin, P. Voronin, Power Electronics Basics, CRC Press, 2016.

[4] F. A. Himmelstoss, and K. H Edelmoser, "Modified Basic DC-DC Converters," in Power Conversion and Intelligent Motion PCIM 2018, Nuernberg, June 5-7, 2018, pp. 1076-1083, ISBN 978-3-8007-4646-0.

[5] F. A. Himmelstoss, and K. H. Edelmoser, "Modified Basic Converters in the Discontinuous Mode," in IEEE 19th International Power Electronics and Motion Control Conference (PEMC), 2021, pp. 118-123, Glivice, Poland, April 2021.

[6] K. Viswanathan, R. Oruganti and D. Srinivasan, "Dual-mode control of tri-state boost converter for improved performance," in IEEE Transactions on Power Electronics, vol. 20, no. 4, pp. 790-797, July 2005, doi: 10.1109/TPEL.2005.850907.

[7] K. Viswanathan, R. Oruganti and D. Srinivasan, "A novel tri-state boost converter with fast dynamics," in IEEE Transactions on Power Electronics, vol. 17, no. 5, pp. 677-683, Sept. 2002, doi: 10.1109/TPEL.2002.802197.

[8] F. A. Himmelstoss, "Tristate Converters," in WSEAS Transactions on Power Systems, Volume 18, 2023, pp. 259-269, E-ISSN: 2224-350X, DOI: 10.37394/232016.2023.18.27

PCIM Europe 2024, 11– 13 June 2024, Nuremberg   DOI: 10.30420/566262298

Comparative Evaluation of the Center Tapped Boost Converter Topology

Bryan Radix[1], Moheddin Shaik[2]

[1] Texas Instruments, Germany
[2] Texas Instruments, Germany

Corresponding author, speaker: Bryan Radix, b-radix@ti.com

Abstract

The scope of this paper is to evaluate the center tapped boost converter when comparing it to the boost converter and flyback converter topologies. The relationships of gain, efficiency and losses to the duty cycle are shown analytically and afterwards verified using a test setup built around a Texas Instruments LM51551-Q1 non-synchronous boost controller, researching if this integrated circuit is also capable of operating the center tapped boost converter topology.

1 Theory

1.1 Topology

The possible value of the center tapped boost converter is evaluated using a comparative approach, using analogous circuits with the boost converter and flyback converter topologies containing comparable components.

The equivalent circuit of the center tapped boost converter is shown below:

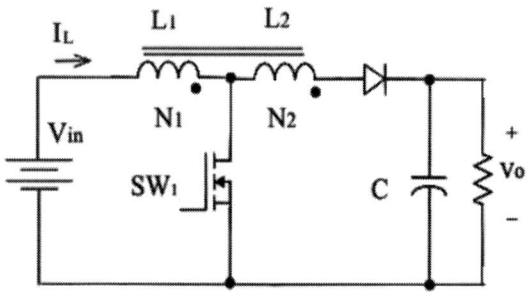

Fig. 1   Center tapped boost converter topology

To retain generality, the theoretical performance of the topology is evaluated with various turns ratios ($N = N_2 / N_1$) of the coupled inductor. Later, as the theoretical results are compared to the test setup, the calculations are limited to the special case of $N = 1$, in accordance to the center tapped inductor used on the test setup.

Additionally, one can say that the special case of $N = 0$ would be equivalent to the regular boost converter topology.

1.2 Gain

At first the lossless gain is determined, using the state equations that can be derived from the two equivalent circuits that arise from the active switch being either in an ON-state or OFF-state [1].
The lossless gain is solely dependent on the turns ratio and duty cycle (D), as shown in Eq. (1):

$$\frac{V_{out}}{V_{in}} = \frac{DN + 1}{1 - D} \tag{1}$$

Hence the lossless gain can be drawn for different turns ratios, dependent solely on the duty cycle, as shown in Fig. 2.
Due to the asymptomatic nature of the lossless gain equation, the drawn duty cycle is limited in its approach to 1. This is the reason of the graph limiting the range of the duty cycle between 0 and 0.9.

It becomes apparent that a higher gain can be achieved for a given duty cycle if the turns ratio is increased.

2112

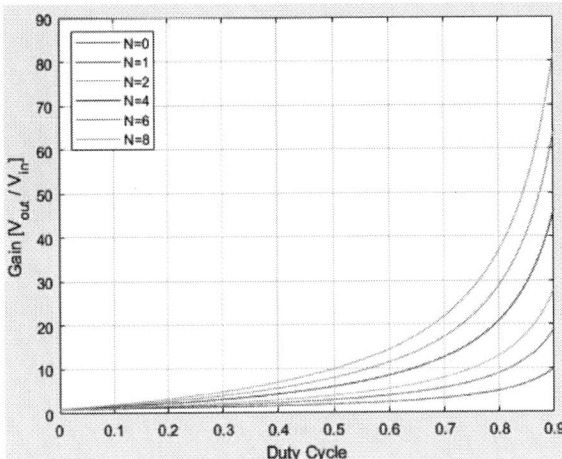

Fig. 2   Lossless gain

Equation (2) is synthesized to calculate the gain when taking (parasitic) resistive elements like the resistance of the primary and secondary copper windings R_{L1} and R_{L2}, the MOSFET drain-source resistance R_{DS}, the output resistance R_{OUT}, as well as the forward voltage V_F of the diode into account. With D' = 1 – D:

$$\frac{V_{out}}{V_{in}} = (D + 1)\left(1 - \frac{V_F D'}{V_{in}(D + 1)}\right) * $$
$$\left(\frac{D'R_{OUT}}{(3D + 1)R_{L1} + D'R_{L2} + 4DR_{DS} + (D')^2 R_{OUT}}\right) \quad (2)$$

The real gain has a critical point, its occurrence at which duty cycle being dependent on the turns ratio and the other component parameters. In order to remain in stable high gain operation, the relationship determining this maximum duty cycle can be found with the relevant parameters if Eq. (2) is differentiated. The duty cycle coupled with the maximum gain is then the solution of said differentiated function when it equals zero. It is advisable to limit the duty cycle to ensure a stable operating point.

1.3 Efficiency

The relationship between efficiency, duty cycle, the turns ratio and the real gain is derived in Eq. (3), determined in part by the chosen component parameters:

$$\eta = \frac{D'}{ND + 1}\left(\frac{V_{out}}{V_{in}}\right) \quad (3)$$

1.4 Compensation

An analysis of the small-signal behavior is made, to see if the usage of a center tapped boost converter changes the right-half-plane zero that is typically found in a boost converter topology. Thereafter the subsequent changes in the control loop are shown.

Assuming current mode control, the transfer function of the topology would be, with L_1 and C being the inductance of the primary winding and the output capacitance respectively:

$$G = \frac{V_{out}}{d} = \frac{\frac{(1 - D)R_{OUT}}{2}\left(1 - \frac{s(N + 1)L_1}{(1 - D)^2 R_{OUT}}\right)}{\frac{sR_{OUT}C}{2} + 1} \quad (4)$$

Comparing Eq. (4) with the standard transfer function for current mode control, as shown in Eq. (5), one can derive the right-half-plane zero ω_{RHP} and the first pole ω_p, which is dependent on the output capacitance and load resistance.

$$G = \frac{V_{out}}{d} = \frac{G_0\left(1 - \frac{s}{\omega_{RHP}}\right)}{\frac{s}{\omega_p} + 1} \quad (5)$$

$$\omega_{RHP} = \frac{(1 - D)^2 R_{OUT}}{(N + 1)L_1} = \frac{(1 - D)^2 R_{OUT}}{2L_1} \quad (6)$$

$$\omega_p = \frac{2}{R_{OUT}C} \quad (7)$$

If a type 2 transconductance amplifier is chosen to attain stability over the desired bandwidth of the system, the necessary additional passive components are shown in Fig. 3:

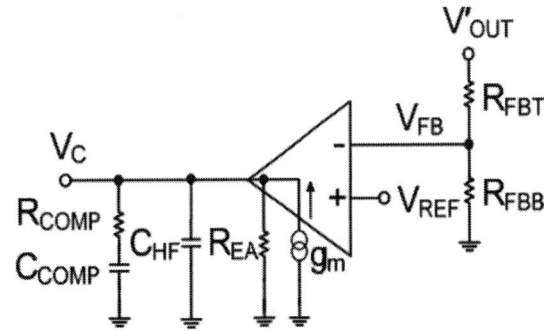

Fig. 3   Type 2 transconductance amplifier

For this type of compensation circuit, the transconductance error amplifier defines a pole at the origin, a zero ω_{ZEA} and a high-frequency pole ω_{HF} [4]. These can be set using Eq. (8) and Eq. (9), assuming $C_{COMP} \gg C_{HF}$ and $R_{EA} \gg R_{COMP}$:

$$\omega_{ZEA} = \frac{1}{R_{COMP}C_{COMP}} \qquad (8)$$

$$\omega_{HF} = \frac{1}{R_{COMP}C_{HF}} \qquad (9)$$

1.5 Comparison with boost converter

To compare the performance of the center tapped boost converter with the regular boost converter topology, various parameters are evaluated.

While the gain of the center tapped boost topology is higher at a given duty cycle than that of a boost converter, this comes at the cost of a lower efficiency in the higher duty cycle domain, which will be elaborated upon in section 2.2, where the simulations are described.

In terms of stability, the regular boost converter has the same first pole dependent on the output capacitance and load resistance as the center tapped boost converter and the following right-half plane zero:

$$\omega_{RHP} = \frac{(1-D)^2 R_{OUT}}{(N+1)L_1} = \frac{(1-D)^2 R_{OUT}}{L_1} \qquad (10)$$

When comparing Eq. (6) and Eq. (10), it is apparent that the right-half-plane zero of the center tapped inductor boost converter is located at half of the frequency of the right-half-plane zero of the regular boost converter.

1.6 Comparison with flyback converter

The comparison with a flyback converter is made, to analyze its attainable gain and transient response when compared to the center tapped boost converter.

To draw the comparison between the lossless gain of both discussed topologies, the lossless gain of the flyback converter with a turns ratio of $N = 1$ is given in Eq. (11):

$$\frac{V_{out}}{V_{in}} = \frac{DN}{1-D} = \frac{D}{1-D} \qquad (11)$$

The graph in Fig. 4 shows the consistently higher lossless gain when using the coupled boost converter in comparison to the flyback converter:

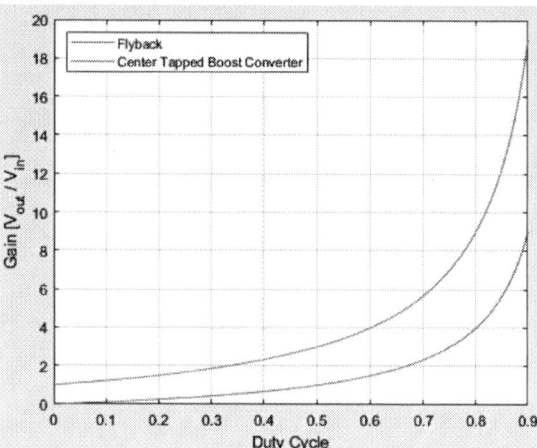

Fig. 4 Lossless gain of both topologies

For the current mode control flyback converter with a turns ratio of $N = 1$, the following right-half-plane zero can be expected:

$$\omega_{RHP} = \frac{(1-D)^2 R_{OUT}}{L_1 D}N^2 = \frac{(1-D)^2 R_{OUT}}{L_1 D} \qquad (12)$$

It is shown in Eq. (13) that the relationship between the right-half-plane zero of the flyback converter and the center tapped boost converter is:

$$\frac{\omega_{RHP,FLYBACK}}{\omega_{RHP,CENTER\_TAPPED\_L}} = \frac{2}{D} \qquad (13)$$

Which means that the frequency of the right-half-plane zero of the flyback converter is at least twice as large as the center tapped tapped boost converter's ω_{RHP} with a given duty cycle.
Since the bandwidth of the control loop is in large part set by the right-half-plane zero, the possibly smaller loop bandwidth of the center tapped boost converter would cause a slower transient response time [5].

The first pole, which is determined by the output capacitance and load resistance is determined in the following way:

$$\omega_p = \frac{1+D}{R_{OUT}C} \qquad (14)$$

When comparing the frequencies of the first pole for the two topologies, the following relationship is formulated:

$$\frac{\omega_{P,FLYBACK}}{\omega_{P,CENTER\_TAPPED\_L}} = \frac{1+D}{2} \qquad (15)$$

This shows that the frequency of the first pole of the flyback converter is maximally equal to that of the center tapped boost converter and minimally half of the frequency of the first pole of the newly evaluated topology.

The voltages on the switch nodes of the center tapped boost converter will be evaluated in the results section to give an indication whether this topology can reduce the voltage stress on the switching device and alleviate possible EMI issues by decreasing high frequency noise when compared to a flyback converter. The flyback converter has inherent overvoltages arising at the switch node once the primary side switch turns off, the causes of these overvoltages being either stored energy in the leakage inductance or reflected voltages from the secondary side arising from the change of energy source.

2 Test & Simulation

2.1 Test setup

To verify the theoretical findings, a printed circuit board is built, containing the TI LM51551-Q1 wide input range non-synchronous boost controller, to see if this integrated circuit which was developed with the classical boost converter topology in mind is capable of operating the center tapped boost converter topology. It is researched if changes in the compensation circuit are necessary because of changes in the control loop and if the theorized gain and efficiency at a given duty cycle can be achieved.

The circuit was optimized for an input voltage range between 11 V and 16 V, while a fixed output voltage was chosen of 90V, with a continuous output current and peak pulse current of respectively 350 mA and 625 mA. This would require an approximal load resistance of nominally 257.14 Ω and in the peak current case 144 Ω.

A coupled inductor of 10 µH ± 20% was chosen, with a maximum DCR of 36 mΩ, a high coupling coefficient of 0.98 and a self-resonant frequency of 20 MHz. Furthermore, the input capacitance consists of an electrolytic capacitor of 330 µF and five ceramic capacitors in parallel of 4.7 µF each. For the output capacitance, to limit the ESR, eight ceramic capacitors of 2.2 µF each are put in parallel.

Assuming a minimal duty cycle D_{MIN} = 0.782, occurring at V_{IN} = 11 V, the maximum inductance within the tolerance of the chosen coupled inductor L_{MAX} = 12µH and the maximum output current of 625 mA, we calculate the right-half-plane zero to be: ω_{RHP} = 285.1 kHz. The first pole, also evaluated at the operating point with maximum output current, i.e. R_{OUT} = 144 Ω, amounts to ω_p = 789 Hz.

The LM51551 uses peak current mode control and is programmed using an external resistor to support a switching frequency of 440 kHz [6]. In order to provide a slope of the compensation ramp that is greater than half of the sensed inductor current falling slope, to prevent subharmonic oscillations from happening at high duty cycle operation, a minimum amount of slope compensation is necessary. The internal circuitry of the LM51551 concerning slope compensation can be seen in Fig. 5:

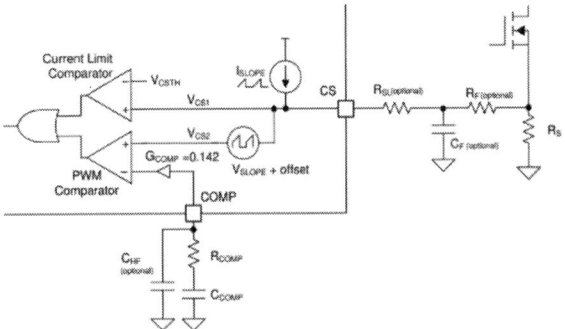

Fig. 5   Slope compensation circuit

An external slope resistor R_{SL} was added to the circuit to increase the slope of the compensation ramp in addition to the always present fixed slope compensation of the device.

When in the following sections of this paper switch node 1 (SW1) and switch node 2 (SW2) are mentioned, these describe the voltage nodes of the drain of the MOSFET and the anode of the diode respectively.

The internal transconductance error amplifier of the device has an output resistance of 10 MΩ and a bandwidth of 7 MHz. The output of said error amplifier is connected to the COMP pin of the device, which allows for a type 2 loop compensation network, containing R_{COMP}, C_{COMP} and optionally C_{HF}.

The following components of the compensation circuit set the gain and phase characteristic of the error amplifier were chosen to achieve a stable loop response:

Description	Value
R_{COMP}	6.98 kΩ
C_{COMP}	68 nF
C_{HF}	330 pF

Table 1   Compensation loop passive components

Therefore, the type 2 transconductance compensation circuit sets the following pole and zero:

Description	Value
ω_{ZEA}	2107 Hz
ω_{HF}	434.1 kHz

Table 2   Compensation loop pole and zero

2.2 Simulation

In order to accommodate a comparison between the simulations based on the theorical results and the measurements with the PCB, the following values are assumed, taken from the components chosen on the PCB:

Description	Value
V_F (diode forward voltage)	0.7 V
$R_{L1} = R_{L2}$	36 mΩ
R_{DS} (MOSFET drain-source resistance)	20 mΩ
R_{OUT} (output resistance)	257 Ω

Table 3   Parameters for simulation

The real gain for the cases N = 0 and N = 1, i.e. for the boost converter and center tapped boost converter respectively, is drawn in Fig. 6:

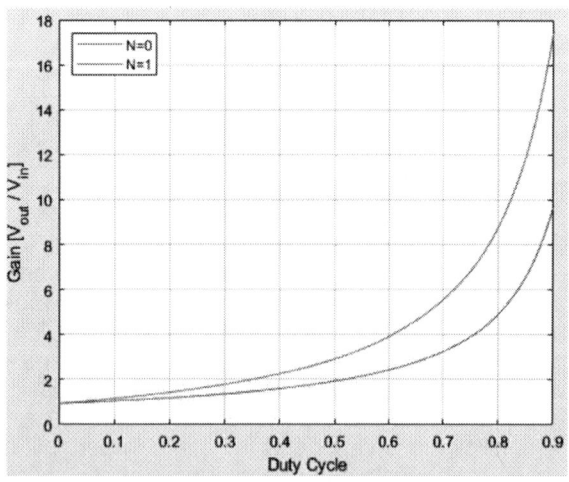

Fig. 6   Real gain of both topologies

Even with the considered losses, it is clear to see that the gain for a given duty cycle is increased in the case of the center tapped boost converter, especially in the higher duty cycle range. The theorized critical point after which the gain would drop, (thus determining an effective maximum duty cycle) is not visible, because of the low turns ratio.

The simulated efficiency is shown below in Fig. 7, again for the cases N = 0 and N = 1:

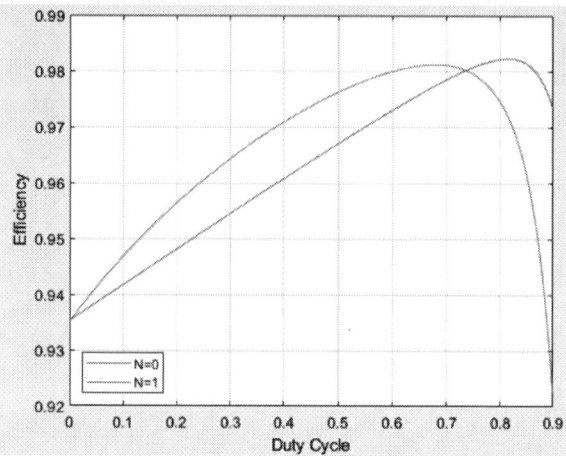

Fig. 7   Efficiency of both topologies

It's clearly visible that the improvement of the gain over the regular boost converter comes at the loss of some efficiency.

In order to determine the power losses per element over a progression of the duty cycle, expressions were derived which enable the visualization as seen in the Fig. 8:

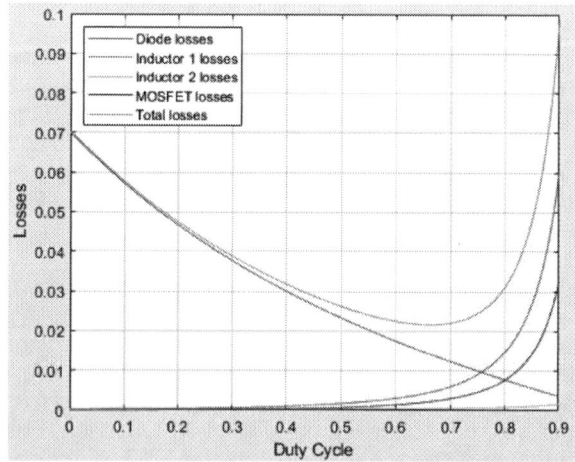

Fig. 8   Losses of center tapped boost converter

3 Results

Following are the measurements taken from the physical test setup with the LM51551-Q1 and the comparison of said test results with the performed simulations.

To give an insight into the performance of the test setup, the gain and steady state voltages are described in the following section.

3.1 Gain

The real gain is shown in Fig. 9 for the input voltage range of 11 V to 16 V. The graph includes the gain derived in Eq. (2) and the gain measured with the test setup:

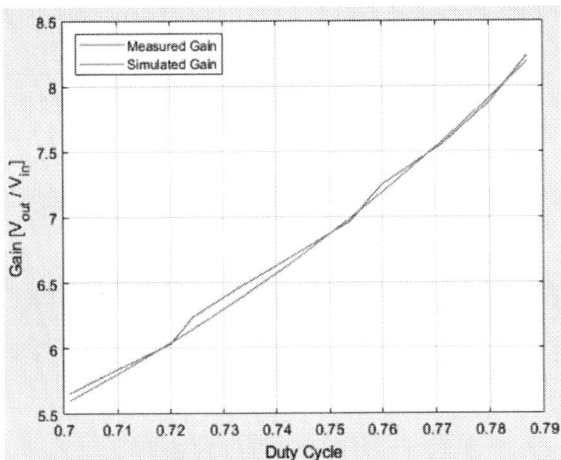

Fig. 9 Gain comparison

The deviation from the theoretical real gain lies between -0.4 % and 1.5 % across the input voltage range. It is therefore apparent that the derived gain equation is a good approximation of the behaviour of the circuit with input voltages from 11 V to 16 V.

3.2 Efficiency

For the evaluation of the efficiency, in Fig. 10 the comparison between the measured efficiency and the theoretically determined efficiency is made.

There is obviously an increased amount of losses in one of the already described elements or another element which has as of yet not been evaluated, since the measured efficiency is between 9.0 % and 13.8 % lower than the theoretically attainable efficiency. The overall shape of the efficiency profile is also deviant from the expected result, indicating there's either a need for improvement of the equations, or of the PCB design.

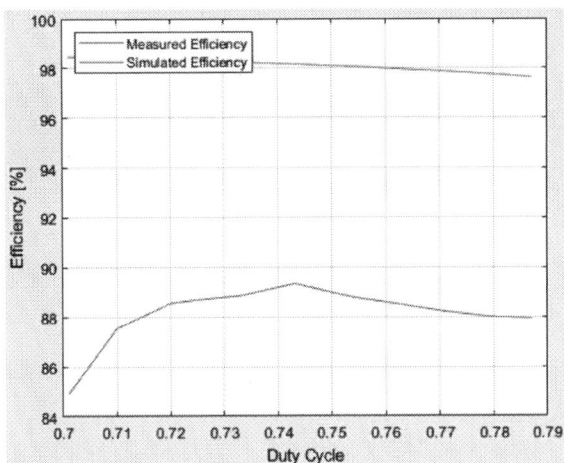

Fig. 10 Efficiency comparison

3.3 Steady State

For the evaluation of the steady state performance of the system Fig. 11 is shown, with channels 1, 2, 3 and 4 displaying the input voltage, the output voltage and both switch node voltages respectively. The output voltage is drawn AC coupled, to show the ripple on the output. For the graph, an input voltage of 13 V was chosen.

Channel	Value
Input voltage	10 V /div
Output voltage (ac-coupled)	50 mV / div
SW1 voltage	20 V / div
SW2 voltage	100 V /div

Table 4 Oscilloscope parameters

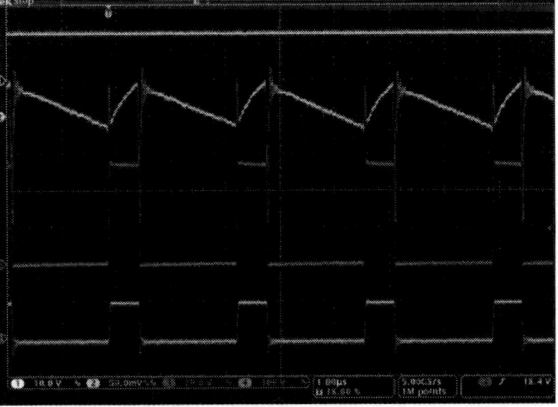

Fig. 11 Steady state operation, 1 μs / div

Both switch nodes voltages show that the system is operated with a switching frequency of 429 kHz.

The SW1 voltage shows a maximum of 56 V while the SW2 voltage has a maximum of approximately 100 V.

For the input voltage range between 11V and 16V, the output voltage accuracy is between 0.6% and 0.7%, considering the output voltage lies between 90.5 V and 90.6V.

The graphs in Fig. 12 shows the same system operating with the same parameters, but further enlarged on the time axis.

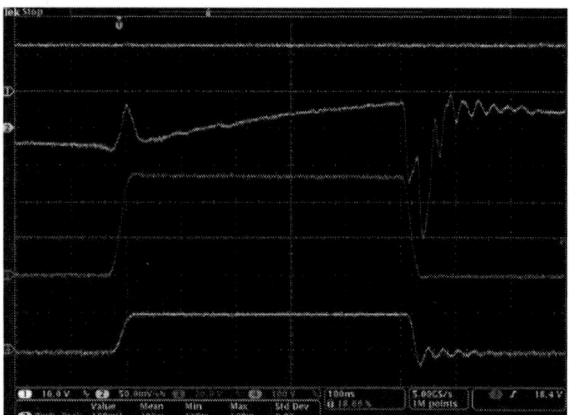

Fig. 12   Steady state operation, 100 ns / div

The output voltage has a peak-to-peak ripple of 198 mV. Considering the output voltage to be 90.5 V, this would mean the output ripple is 0.22 %.

Looking at the switch nodes, there are no major visible overvoltages. This would indicate a lower amount of voltage stress on the switching devices of the center tapped boost topology when comparing it to a flyback converter. In addition, the switching voltages are also apparently low in noise, possibly pointing towards a more optimized EMI performance.

4   Conclusion

The increased gain of the center tapped boost converter for a given duty cycle when compared to the regular boost converter can be reliably approximated and achieved using the test setup with the TI LM51551.

The printed circuit board used in the test setup showed much lower efficiency than expected, while the equations could be false, losses introduced by an as of yet unidentified element in the circuit board could very well be the cause of the

efficiency being suboptimal. The derived equations predict that the higher possible gain of the center tapped boost converter come at the price of a lower efficiency for a given duty cycle when compared to the regular boost converter.

The output voltage accuracy and output voltage ripple in steady state conditions are acceptable and there are no major overvoltages to be on the switch nodes. Because of the lack of secondary-side reflected voltages that would be present in the flyback converter topology, the voltage stress on the switching device is reduced and EMI issues can possibly be relieved by decreasing high frequency noise compared to a flyback converter.

The frequency of the right half plane zero of the flyback converter and regular boost converter is at least twice as large as the right-half-plane zero of the center tapped boost converter. This allows for a larger loop bandwidth to be set in the case of the flyback and boost converter topologies. Thus, the response time of the center tapped boost converter to a transient should be evaluated.

References

[1] Moheddin Shaik, Voltage boosted boost converter – Internship

[2] Robert Erickson and Dragan Maksimovic, Fundamentals of power electronics, 2020

[3] Brigitte Hauke, Basic calculation of a boost converter's power stage, 2022

[4] Robert Meehan and Louis Diana, Switch-mode power converter compensation made easy, 2016

[5] Lloyd H. Dixon, Jr., The right-half-plane zero — a simplified explanation, 1986

[6] Texas Instruments, LM5155x-Q1 2.2-MHz wide input nonsynchronous boost, SEPIC, flyback controller, 2023

PCIM Europe 2024, 11– 13 June 2024, Nuremberg   DOI: 10.30420/566262299

Comparison of Multi-level Topologies to Reduce the Components Voltage Stresses when Powered from Industrial DC Grids

Katharina Machtinger ⊙[1], Peter Jonke [1], Ulrich Boeke[2]

[1] Austrian Institute of Technology (AIT), Austria
[2] Signify Netherlands BV, Netherlands

Corresponding author: Katharina Machtinger, katharina.machtinger@ait.ac.at
Speaker: Katharina Machtinger, katharina.machtinger@ait.ac.at

Abstract

A promising approach for further CO_2 reduction is the use of industrial DC grids. Renewable energy sources typically have a DC output, and the DC grid eliminates the step of converting DC to AC (for the AC grid) and then back to DC. This paper evaluates the suitability of various multi-level topologies for use as LED drivers in a DC grid.

1   Introduction

One approach to further reducing CO_2 emissions is to improve the efficiency of power electronic devices and systems. There are various strategies for achieving increased efficiency e.g. using wide bandgap (WBG) semiconductors, zero voltage or current switching, interleaving converter, etc. . In most applications, such as motor drives or battery storage, the AC power of the grid has to be converted to DC, whilst renewable energy sources usually provide DC power anyway (which typically need to be converted to AC to be feed to the AC grid). Therefore, an approach is the use of an industrial DC grid, which avoids the conversion of DC to AC and AC back to DC. There are already industry pilot projects which utilizes an industrial DC grid [1].
For the DC grid are DC/DC converters essential components, whereby the stress of the semiconductors is increased by higher DC-link voltages. One approach to reduce stress on one semiconductor is a sophisticated connection of switching elements in series or using a multilevel converter topology. A multilevel approach brings the advantage of increased voltage high-power operation, reduced voltage stress of each individual semiconductor, reduced loss of each individual semiconductor, better power quality, and lower electromagnetic interference. But this advantages involves complex control and modulation, DC voltage balance, increased number of components and complex switching signal routing [2], [3]. Over the last decade, various multilevel converter topologies have been examined

and put to use not just in academic field but also within industrial applications. Multilevel converters are already widely applied in the fields of motor drivers, flexible AC transmission systems, converters for renewable energy sources, and high-voltage DC transmission systems.
Although this paper focus on the application of LED drivers, DC/DC converters for industrial 650V DC power systems are also used for photovoltaic maximum-power-point applications, battery and vehicle charging applications.

1.1   General Requirements

Figure 1 shows the configuration in which the topologies are analysed in this paper, where the different topologies are positioned in the blue highlighted box.

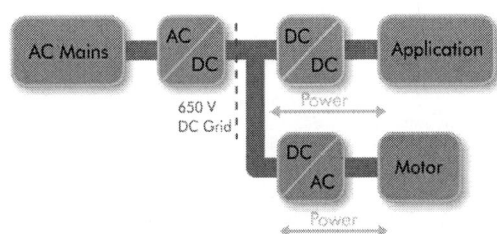

Fig. 1: Environment of the analysed topologies.

Different multilevel topologies are investigated, simulated and compared when fed from a 650 V_{DC} grid for LED lighting application. The focus in this work is on three level topologies, as they provide a good trade off of loss reduction and complexity of the topology. For the application as LED-driver, the output power of the converter is specified for a range of between 60 W to 150 W. The complete specification of the LED driver is listed in table 1.

Parameter	Value Min	Value Typ	Value Max	Unit
DC input voltage	620	650	750	V
DC input current	80		242	mA
DC output power	60		150	W
DC output voltage	100		350	V
DC output current			1.5	A
Semiconductor rated voltage			750	V
Switching frequency	20		2000	kHz

Tab. 1: Specification for LED driver application.

2 Converter Topologies

Some well-known multilevel converter topologies are neutral-point-clamped converter (NPC) initially introduced in a patent [4], cascaded H-bridge converter (CHB) introduced in [5], flying-capacitor converter (FC), T-type multilevel converter, modular multilevel converter (MMC), and active neutral-point-clamped converter (ANPC) [2], [3].

2.1 Two-level converter (reference)

The two level converter displayed in Fig. 2 is utilized as reference for the other three level topologies. Such an converter is widely used for low-voltage grid applications e.g. PV-converters, battery energy storage system (BESS). The switching states of the two-level converter are shown in table 2.

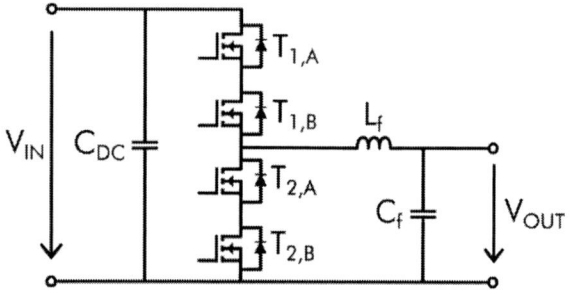

Fig. 2: Two level converter.

$T_{1,A}$	$T_{1,B}$	$T_{2,A}$	$T_{2,B}$	V_{OUT}
1	1	0	0	V_{IN}
0	0	1	1	0 V

Tab. 2: Switching states of two level converter.

2.2 Neutral point clamped (NPC)

Figure 3 shows the neutral point clamped topology. The advantage of this topology is that it has a simple design, no floating capacitor and good dynamic response [2], [6]. Furthermore, this topology is widely used commercially [3]. For a three level implementation, the design is compact and cost-effective [2]. For a higher number of levels, the number of clamping diodes and the complexity for balancing the DC link capacitors increases [6], [7]. Regardless of the number of levels, the NPC has the disadvantage of uneven utilisation of the semiconductors, as some components switch all the time and others only half the time [2], [3]. The DC link voltage must be balanced, which increases the complexity as the number of levels increases [3]. Depending on the number of NPC levels (n), the required clamping diodes (D) and capacitors (C) change. The change can be expressed as Eqs. (1) and (2)[6].

$$C = n - 1 \tag{1}$$
$$D = C \cdot (n - 2) \tag{2}$$

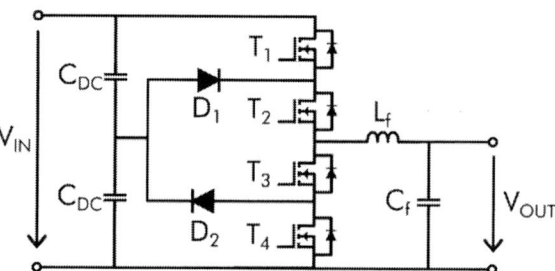

Fig. 3: Three level neutral point clamped (NPC).

The operating principle can be described as follows:

- T_1 and T_2 are closed: The phase output voltage results to $V_{OUT} = +V_{IN}$. Depending on the output current direction the current flows either through the power semiconductors T_1 and T_2 or their anti-parallel freewheeling diodes.
- T_2 and T_3 are closed: Depending on the output current direction the current flows either through the power semiconductors T_2 and the clamping diode D_1 or the power semiconductors T_3 and the clamping diode D_2. In any case the output voltage results in $V_{OUT} = +\frac{V_{IN}}{2}$.
- T_3 and T_4 are closed: The phase output voltage now is $V_{OUT} = 0$ V. Depending on the output current direction the current flows either

through the semiconductors T_3 and T_4 or the freewheeling diodes of T_3 and T_4.

The switching states of the three level neutral point clamped converter are summarized table 3.

T_1	T_2	T_3	T_4	V_{OUT}
1	1	0	0	V_{IN}
0	1	1	0	$\frac{V_{IN}}{2}$
0	0	1	1	0 V

Tab. 3: Switching states of the three level neutral point clamped (NPC) converter.

2.3 Active neutral point clamped (ANPC)

Replacing the diodes of the NPC with MOSFETs or IGBTs eliminates the drawback of uneven semiconductor loss distribution of the NPC. The resulting topology is called active neutral point clamped (ANPC) as shown in Fig. 4. The ANPC, just like

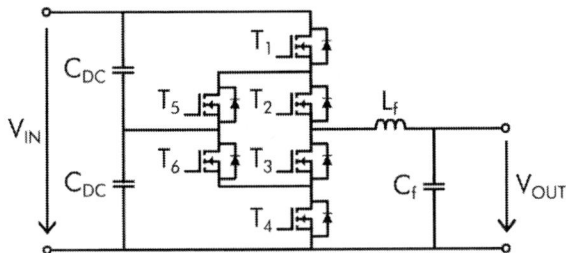

Fig. 4: Three level active neutral point clamped (ANPC).

the NPC, has the advantages of good dynamic behaviour and a simple design. With a three level design, this topology is compact and cost-effective and does not require a floating capacitor. Due to the replacement of clamping diodes, the ANPC has equal distributed switching losses. [2]. Compared to the NPC, the control method is more complex. When the ANPC is designed for more than three levels, floating capacitors are needed. For higher voltage levels the semiconductor have different voltage ratings.

The operating principle can be described as follows:

- T_1 and T_2 are closed: The phase output voltage results to $V_{OUT} = +V_{IN}$. Depending on the output current direction the current flows either through the power semiconductors T_1 and T_2 or their anti-parallel freewheeling diodes.
- T_2 and T_5 are closed: The output voltage results in $V_{OUT} = +\frac{V_{IN}}{2}$ and is used as complementary state when T_1 and T_2 are closed ($V_{OUT} = +V_{IN}$).

- T_3 and T_6 are closed: The output voltage results in $V_{OUT} = +\frac{V_{IN}}{2}$ and is used as complementary state when T_3 and T_4 are closed ($V_{OUT} = V$).
- T_3 and T_4 are closed: The phase output voltage now is $V_{OUT} = 0$ V. Depending on the output current direction the current flows either through the semiconductors T_3 and T_4 or the freewheeling diodes of T_3 and T_4.

The switching states of the three level active neutral point clamped converter are summarized in table 4.

T_1	T_2	T_3	T_4	T_5	T_6	V_{OUT}
1	1	0	0	0	0	V_{IN}
0	1	0	0	1	0	$\frac{V_{IN}}{2}$
0	0	1	0	0	1	$\frac{V_{IN}}{2}$
0	0	1	1	0	0	0 V

Tab. 4: Switching states of the three level active neutral point clamped (ANPC) converter.

2.4 Flying capacitor (FC)

Instead of clamping diodes, flying capacitors can be utilized to limit the device voltage to a capacitor voltage level. The resulting topology is called flying capacitor (FC) and shown in Fig. 5. The FC is common in the commercial sector[3]. The multi-carrier phase shifted PWM modulation method used for the FC is well known. It enables loss compensation and a negligible distortion factor in the voltage signals [3].

The control method of the FC is sophisticated and as the number of levels increases, the complexity increases further [2], [6]. Precharging of all the capacitors to same voltage level and startup are also elaborate [7]. The large number of capacitors are bulkier than clamping diodes of NPC [7].

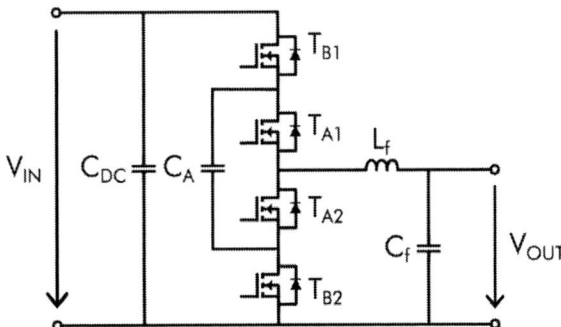

Fig. 5: Three level flying capacitor (FC).

The operation principle can be described as follows:

- T_{B1} and T_{A1} are closed: The phase output voltage results in $V_{OUT} = +V_{IN}$. Depending on the output current direction the current flows either through the power semiconductor T_{B1} or the freewheeling diode of T_{A1}.
- T_{B1} and T_{A2} are closed: Depending on the output current direction the current flows either through the power semiconductors T_{B1} and the freewheeling diode of T_{A2} (positive current) or the power semiconductors T_{A2} and the freewheeling diode of T_{B1} (negative current). For positive currents the voltage of the flying capacitor V_{CA} is increased and for negative currents the voltage of the flying capacitor V_{CA} is decreased. In any case the output voltage results in $V_{OUT} = \frac{V_{IN}}{2}$.
- T_{B2} and T_{A1} are closed: Depending on the output current direction the current flows either through the power semiconductors T_{B2} and the freewheeling diode of T_{A1} (positive current) or the power semiconductors T_{A1} and the freewheeling diode of T_{B2} (negative current). For positive currents the voltage of the flying capacitor V_{CA} is decreased and for negative currents the voltage of the flying capacitor V_{CA} is increased. In any case the output voltage results in $V_{OUT} = +\frac{V_{IN}}{2}$.
- T_{B2} and T_{A2} are closed: The output voltage $V_{OUT} = 0$ V , depending on the output current direction the current flows either through the power semiconductors T_{B2} and T_{A2} or the freewheeling diodes of T_{B2} and T_{A2}.

The switching states of the three level flying capacitor are summarized in table 5

T_{B1}	T_{A1}	T_{A2}	T_{B2}	V_{OUT}
1	1	0	0	V_{IN}
1	0	1	0	$\frac{V_{IN}}{2}$
0	1	0	1	$\frac{V_{IN}}{2}$
0	0	1	1	$0\,\mathrm{V}$

Tab. 5: Switching states of the three level flying capacitor (FC) converter.

2.5 Modular multilevel converter (MMC)

Figure 6 shows the modular multilevel converter (MMC), where the submodules are half-bridge modules. This topology has the advantage of its modular structure and good scalability [3], [6].
The MMC has three disadvantages. First, it has a large number of capacitors [2]. Second, the capaci-

tors need to be pre-charged. Third, the capacitors of each submodule (CX) have a high voltage ripple [6].

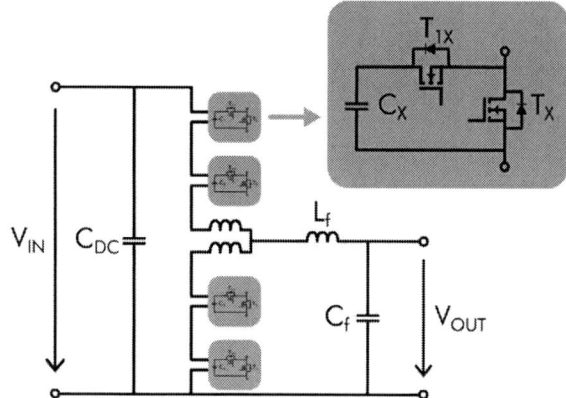

Fig. 6: Three level modular multilevel converter (MMC).

The operating principle can be described as follows:
- T_1 and T_2 are closed: The phase output voltage results to $V_{OUT} = +V_{IN}$. Depending on the output current direction the current flows either through the power semiconductors T_1 and T_2 or their anti-parallel freewheeling diodes.
- T_1, T_3, T_{12}, T_{14} or T_1, T_4, T_{12}, T_{13} or T_2, T_3, T_{11}, T_{14} or T_2, T_4, T_{11}, T_{13} are closed: The phase output voltage is $V_{OUT} = \frac{V_{IN}}{2}$, provided that the voltage at the capacitors $C_1 - C_4$ are balanced and the average voltage at the capacitor is $V_{C1} = V_{C2} = V_{C3} = V_{C4} = +\frac{V_{IN}}{2}$. The output voltage $V_{OUT} = \frac{V_{IN}}{2}$ shows redundancy states, which can be used for balancing the sub module capacitors [8], [9].
- T_3 and T_4 are closed: The phase output voltage is $V_{OUT} = 0V$. Depending on the output current direction the current flows either through the semiconductors T_3 and T_4 or the freewheeling diodes of T_3 and T_4.

The switching states of the three level modular multilevel converter are summarized in table 6.

T_1	T_2	T_3	T_4	T_{11}	T_{12}	T_{13}	T_{14}	V_{OUT}
1	1	0	0	0	0	0	1	V_{IN}
1	0	1	0	0	1	0	1	$\frac{V_{IN}}{2}$
1	0	0	1	0	1	1	0	$\frac{V_{IN}}{2}$
0	1	1	0	1	0	0	1	$\frac{V_{IN}}{2}$
0	1	0	1	1	0	1	0	$\frac{V_{IN}}{2}$
0	0	1	1	0	0	0	0	$0\,\mathrm{V}$

Tab. 6: Switching states of the three level modular multilevel converter (MMC) converter.

2.6 T-Type

The T-Type topology employs bidirectional switches to create a controllable path for the current between input and output, as shown in Fig. 7. There is no need for a floating capacitor or diode with the T-Type. Furthermore, it has the same simple modulation and control method as the NPC [2]. Compared to the NPC, this topology has fewer conduction losses. The outer semiconductor must block the entire DC link voltage, which is a limiting factor for applications in the high-voltage range and also leads to a high voltage stress on the switches [2], [3]. Due to the high switching losses the efficiency of this topology decreases with higher switching frequency, as this increases the switching losses [2].

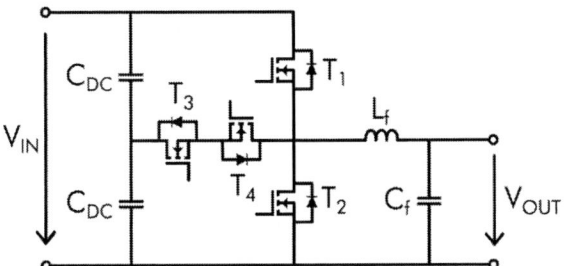

Fig. 7: Three level T-Type.

The operation principle can be described as follows:
- T_1 is closed: The phase output voltage results in $V_{OUT} = +V_{IN}$. Depending on the output current direction the current flows either through the power semiconductor T_1 or the freewheeling diode of T_1.
- T_2 and T_3 are closed: The phase output voltage now is $V_{OUT} = +\frac{V_{IN}}{2}$. Depending on the output current direction the current flows either through T_2 and the diode D_3 or T_3 and the diode D_2.
- T_4 is closed: The output voltage $V_{OUT} = 0$ V, depending on the output current direction the current flows either through the power semiconductor T_4 or the freewheeling diode of T_4.

The switching states of the three level T-Type converter are summarized in table 7.

2.7 Cascaded H-bridge (CHB)

The cascaded H-bridge (CHB) has several single-phase H-bridge converters connected in series, with each unit connected to a separate DC source. Figure 8 shows a three level H-bridge. The CHB is a well established topology and the phase-shifted

T_1	T_2	T_3	T_4	V_{OUT}
1	0	0	0	V_{IN}
0	0	1	1	$\frac{V_{IN}}{2}$
0	1	0	0	0 V

Tab. 7: Switching states of the three level T-Type converter.

PWM is also well known method [3]. The advantages are simple control method, modularity and no floating capacitor required [2]. Disadvantages are that each cell requires a separated DC source, and voltage unbalance between different phases of the topology [2], [3], [6].

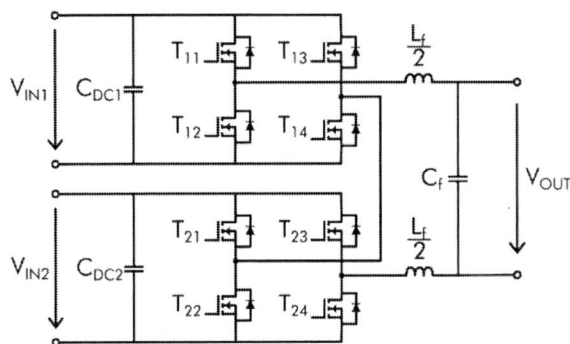

Fig. 8: Three level cascaded H-bridge (CHB).

The operating principle can be described as follows:
- T_{11} and T_{24} are closed: The phase output voltage results to $V_{OUT} = +V_{IN}$. Depending on the output current direction the current flows either through the power semiconductors T_{11} and T_{24} or their anti-parallel freewheeling diodes.
- T_{12} and T_{24} are closed: The phase output voltage is $V_{OUT} = +\frac{V_{IN}}{2}$. Depending on the output current direction the current flows either through the semiconductors T_{12} and T_{24} or the freewheeling diodes of T_{12} and T_{24}.
- T_{12}, T_{14}, T_{21}, and T_{24} are closed: The phase output voltage is $V_{OUT} = 0$ V.

The switching states of the three level cascaded H-bridge converter are summarized in table 8

T_{11}	T_{12}	T_{13}	T_{14}	T_{21}	T_{22}	T_{23}	T_{24}	V_{OUT}
1	0	0	0	0	0	0	1	V_{IN}
0	1	0	0	0	0	0	1	$\frac{V_{IN}}{2}$
0	1	0	1	1	0	1	0	0 V

Tab. 8: Switching states of the three level cascaded H-bridge (CHB) converter, $V_{IN1} = V_{IN2} = \frac{V_{IN}}{2}$.

2.8 Stacked Buck

Since only a few states of the cascaded H-bridge are used, a simplified topology can also be implemented [10]–[12]. One approach of such topology is called stacked buck and a version is shown in Fig. 9.

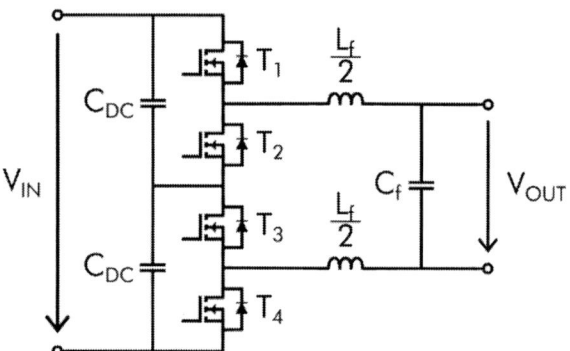

Fig. 9: Three level stacked buck converter.

The operating principle can be described as follows:
- T_1 and T_4 are closed: The phase output voltage results to $V_{OUT} = +V_{IN}$. Depending on the output current direction the current flows either through the power semiconductors T_{11} and T_{24} or their anti-parallel freewheeling diodes.
- T_1 and T_3 or T_2 and T_4 are closed: The phase output voltage is $V_{OUT} = +\frac{V_{IN}}{2}$. Depending on the output current direction the current flows either through the semiconductors T_2 and T_4 or the freewheeling diodes of T_1 and T_3.
- T_2 and T_3 are closed: The phase output voltage is $V_{OUT} = 0\text{ V}$.

The switching states of the three level stacked buck converter are summarized in table 9.

T_1	T_2	T_3	T_4	V_{OUT}
1	0	0	1	V_{IN}
1	0	1	0	$\frac{V_{IN}}{2}$
0	1	1	0	$\frac{V_{IN}}{2}$
0	1	1	0	0 V

Tab. 9: Switching states of the three level stacked buck converter.

2.8.1 Stacked buck for HIL systems

In [11] another three level stacked buck converter for HIL systems is presented, which is depicted in Figure 12. This converter topology allows three different voltage levels, which allows lower switching and conduction losses.

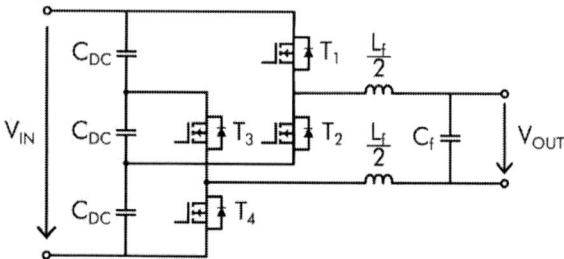

Fig. 10: Three level stacked buck converter for HIL systems.

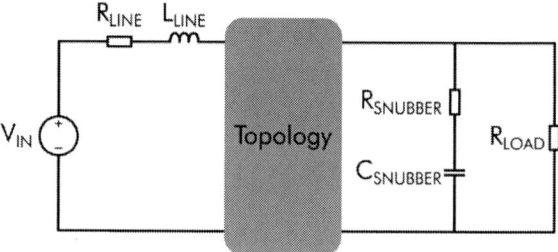

Fig. 11: Simulation setup for efficiency estimation.

The switching states of the three level stacked buck for HIL systems are summarized in table 10

T_1	T_2	T_3	T_4	V_{OUT}
1	0	0	1	V_{IN}
1	0	1	0	$\frac{V_{IN}}{3}$
0	1	0	1	$\frac{V_{IN}}{3}$
0	1	1	0	$-\frac{V_{IN}}{3}$

Tab. 10: Switching states of the three level stacked buck converter for HIL systems.

3 Comparison

3.1 Benchmarking setup

The benchmarking setup is based on a PLECS simulation environment, which contains a model of the DC-grid, the power electronic topology, a single stage LC-filter and a resistive load. In order to make the different simulations comparable, each converter simulation uses the same setup, which is depicted in Fig. 11. Therefore, only the converter topology needs to be exchanged. However, for the filter design, a distinction must be made between two level and three level topologies. For the two level converter (reference system) the output filter inductor calculates with Eq. (3) as well as the output filter capacitor is derived by Eq. (4).

$$L_F = \frac{V_{IN} \cdot (1 - \delta) \cdot \delta}{\Delta I_{pp} \cdot f_{SW}} \tag{3}$$

$$C_F = \frac{V_{IN} \cdot (1 - \delta) \cdot \delta}{8 \cdot L_F \cdot \Delta V_{PP} \cdot f_{SW}^2} \quad (4)$$

According to [13] the output filter inductor of a three level converters is given by Eq. (5) and the output filter capacitor computes with Eq. (6).

$$L_F = \frac{V_{IN} \cdot (1 - \delta) \cdot \delta}{2 \cdot \Delta I_{pp} \cdot f_{SW}} \quad (5)$$

$$C_F = \frac{V_{IN} \cdot (1 - \delta) \cdot \delta}{16 \cdot L_F \cdot \Delta V_{PP} \cdot f_{SW}^2} \quad (6)$$

The parameters for the simulation setup are summarized in table 11.

Parameter	Value
DC input voltage	$V_{IN} = 650$ V
DC output power	$P_{OUT} = 150$ W
DC output voltage	$V_{OUT} = 100$ V
DC output current	$I_{OUT} = 1.5$ A
Switching frequency	$f_{SW} = 50$ kHz
Filter inductor	$L_{F\_THREE\_LEVEL} = 2.8$ mH
	$L_{F\_TWO\_LEVEL} = 5.6$ mH
Filter capacitor	$C_F = 100$ nF
Snubber	$R_S = 158\ \Omega, C_S = 300$ nF
Load resistor	$R_L = 66\ \Omega$
Semiconductor temperature	25 °C
Si-IGBT	Infineon IKA08N65F5
GaN HEMT	Gan Systems GS-065-004-1-L

Tab. 11: Parameters for the efficiency estimation simulation setup.

3.2 Benchmarking results

In order to compare the topologies, a benchmark with various parameters is presented. On the one hand, the number of components, the control complexity as well as the reliability are analysed and on the other hand, a loss estimation based on a PLECS simulation provides information about the efficiency. Table 12 shows the benchmarking parameters of the proposed converter topologies.

The simulation results of the proposed converter topologies, an efficiency estimation and estimation of the energy in the filter inductor are summarized in table 13 and also shown in Figures 12 and 13. The maximum energy stored in the filter inductor is calculated with Eq. (7), which provides an indication of the volume of the required coil.

$$E_{L\_FILTER} = \frac{L_{FILTER} \cdot I_{L\_MAX}^2}{2} \quad (7)$$

The efficiency estimation is based on the semiconductor losses at 25 °C chip temperature, ohmic loss in L_{FILTER} and ohmic load. For the determination of the switching and conduction losses

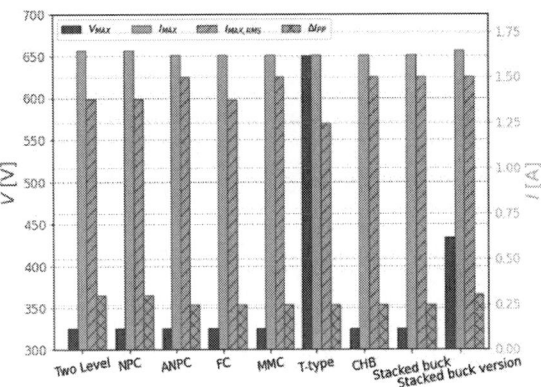

Fig. 12: Simulation result for current and voltages across the semiconductor of the different topologies.

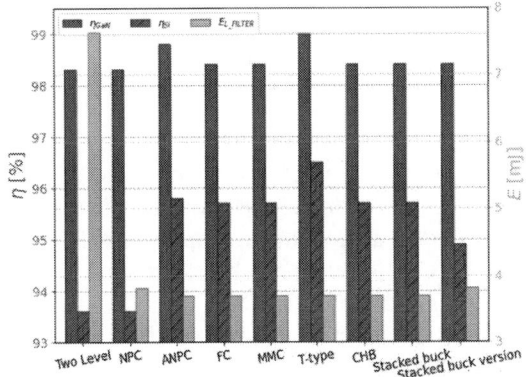

Fig. 13: Simulation result for the efficiency of the different topologies and estimation of the energy in the filter inductor.

and the resulting efficiency, the simulation was performed with both a GaN-HEMT model (Gan Systems GS-065-004-1-L) and a Si-IGBT model (Infineon IKA08N65F5), and with hard switching. Topology

The two level converter serves as a reference for benchmarking. In order to enable a comparison with a multilevel converter, two MOSFETs are connected in series on both the high-side and the low-side, so that the blocking voltage of each MOSFET can be reduced. Although the efficiency of the two level converter is comparable to a multilevel converter, the multilevel converter shows an advantage regarding the filter design. The NPC and the ANPC are both promising solutions regarding, number of components and control complexity. As expected,

Topology	Reliability	Control complexity	Number of active components	Number of passive components
Two Level	high	simple	4	1
NPC	high	medium	6 (2 Diodes)	2
ANPC	high	medium	6	2
FC	medium	medium-high	4	2
MMC	low	high	8	6
T-type	high	medium	4	2
CHB	low	high	8	2
Stacked buck	medium	medium	4	2
Stacked buck version	medium	medium	4	3

Tab. 12: Benchmarking parameters of the proposed converter topologies.

Topology	V_{MAX}	I_{RMS}	I_{MAX}	ΔI_{PP}	η_{GaN}	η_{Si}	$E_{L.FILTER}$
Two Level	325.0 V	1.381 A (T2)	1.651 A (T1)	0.302 A	98.3 %	93.6 %	7.630 mJ
NPC	325.0 V	1.381 A (T3)	1.651 A (T1)	0.302 A	98.3 %	93.6 %	3.815 mJ
ANPC	325.0 V	1.501 A (T3)	1.623 A (T6)	0.248 A	98.8 %	95.8 %	3.689 mJ
FC	325.1 V	1.379 A (TB2)	1.623 A (TA1)	0.248 A	98.4 %	95.7 %	3.689 mJ
MMC	325.0 V	1.501 A (T3)	1.623 A (T14)	0.248 A	98.4 %	95.7 %	3.689 mJ
T-type	650.0 V	1.247 A (T2)	1.623 A (T3)	0.248 A	99.0 %	96.5 %	3.689 mJ
CHB	325.0 V	1.501 A (T12)	1.623 A (T24)	0.248 A	98.4 %	95.7 %	3.689 mJ
Stacked buck	325.0 V	1.501 A (T2)	1.623 A (T4)	0.248 A	98.4 %	95.7 %	3.689 mJ
Stacked buck version	433.3 V	1.501 A (T3)	1.650 A (T1)	0.305 A	98.4 %	94.9 %	3.812 mJ

Tab. 13: Simulation results, voltage and current values on the semiconductors, energy in the filter inductance, and efficiency estimation.

the ANPC offers higher efficiency and is therefore the preferred converter topology. Also the flying capacitor (FC) converter shows similar efficiency results compared to the ANPC. The advantage of the flying capacitor is the reduced number of active components and and also a well-established balancing concept for flying capacitor. Although the T-Type converter shows the highest efficiency and is well established for industrial converters, this topology will not reduce the blocking voltage for the high-side and the low-side semiconductors. Therefore this topology will be not used for further consideration. For the modular multilevel converter (MMC) and the cascaded H-Bridge (CHB) converter a minimum of eight active components are used and also the balancing of the DC-link is challenging in for the DC/DC buck mode. Therefore both the MMC and CHB are rather inappropriate for this application. The stacked buck converter topologies can also be a promising solution, as long as no grounded output voltages are required. These kind of topologies are changing the reference potential of the output voltage during modulation, which as to be considered when these topologies are used.

4 Conclusion

The specification of DC/DC converter for LED driver applications and the comparison of different converter topologies are shown in this paper. The main goal, in addition to increasing efficiency and reduced volume, is the reduction of the blocking voltage of the used semiconductors. Therefore, several multilevel converter topologies have been analysed and compared to a two level converter system, which was used as reference system. Each topology was evaluated with respect to various parameters (number of components, control complexity etc.) and an efficiency analysis based on an offline simulation was also performed. It can be concluded that for the benchmarking of multilevel topologies, the ANPC converter and the FC converter are the most promising solutions. Therefore, these topologies are used for further investigations and prototype development, which will be part of future project steps.

5 Acknowledgment

The authors acknowledge the public funding of the open innovation project PowerizeD within the Key Digital Technologies Joint Undertaking (KDT JU) under grant agreement number 101096387 [14].

References

[1] "Dc industrie, open innovation project." (), [Online]. Available: https://dc-industrie.zvei.org/en/ (visited on 09/21/2023).

[2] A. Poorfakhraei, M. Narimani, and A. Emadi, "A review of multilevel inverter topologies in electric vehicles: Current status and future trends," *IEEE Open Journal of Power Electronics*, vol. 2, pp. 155–170, 2021. DOI: 10.1109/OJPEL.2021. 3063550.

[3] J. I. Leon, S. Vazquez, and L. G. Franquelo, "Multilevel converters: Control and modulation techniques for their operation and industrial applications," *Proceedings of the IEEE*, vol. 105, no. 11, pp. 2066–2081, Nov. 2017. DOI: 10.1109/JPROC. 2017.2726583.

[4] R. H. Baker, "Bridge converter circuit," U.S. Patent 4270163A, May 26, 1981.

[5] W. Mcmurray, "Fast response stepped-wave switching power converter circuit," U.S. Patent 3581212A, May 25, 1971.

[6] N. S. Hasan, N. Rosmin, D. A. A. Osman, and A. H. Musta, "Reviews on multilevel converter and modulation techniques," *Renewable and Sustainable Energy Reviews*, vol. 80, pp. 163–174, Dec. 2017. DOI: 10.1016/j.rser.2017.05.163.

[7] R. P. Vishvakarma, S. P. Singh, and T. N. Shukla, "Multilevel inverters and its control strategies: A comprehensive review," en, in *2012 2nd International Conference on Power, Control and Embedded Systems*, Allahabad, Uttar Pradesh, India: IEEE, Dec. 2012, 1–9. DOI: 10.1109/ICPCES. 2012.6508077.

[8] M. M. Harin, V. Vanitha, and M. Jayakumar, "Comparison of pwm techniques for a three level modular multilevel inverter," *Energy Procedia*, vol. 117, pp. 666–673, 2017, "First International Conference on Power Engineering Computing and CONtrol (PECCON-2017) 2nd -4th March .2017." Or-

ganized by School of Electrical Engineering, VIT University, Chennai, Tamil Nadu, India. DOI: https: //doi.org/10.1016/j.egypro.2017.05.180.

[9] M. Hagiwara and H. Akagi, "Control and experiment of pulsewidth-modulated modular multilevel converters," *IEEE Transactions on Power Electronics*, vol. 24, no. 7, pp. 1737–1746, 2009. DOI: 10.1109/TPEL.2009.2014236.

[10] Y. Du, X. Zhou, S. Bai, S. Lukic, and A. Huang, "Review of non-isolated bi-directional dc-dc converters for plug-in hybrid electric vehicle charge station application at municipal parking decks," in *2010 Twenty-Fifth Annual IEEE Applied Power Electronics Conference and Exposition (APEC)*, 2010, pp. 1145–1151. DOI: 10.1109/APEC.2010. 5433359.

[11] C. Carstensen and J. Biela, "Novel 3 level bidirectional buck converter with wide operating range for hardware-in-the-loop test systems," in *2012 15th International Power Electronics and Motion Control Conference (EPE/PEMC)*, 2012, DS2b.2–1–DS2b.2–8. DOI: 10.1109/EPEPEMC.2012. 6397271.

[12] C. Wang, P. Xing, L. Zhang, W. Kui, and Y. Li, "A modular cascaded multilevel buck converter based on gan devices designed for high power envelope elimination and restoration applications," Sep. 2018, pp. 4851–4857. DOI: 10.1109/ECCE. 2018.8558052.

[13] D. Boillat, T. Friedli, J. Muhlethaler, J. Kolar, and W. Hribernik, "Analysis of the design space of single-stage and two-stage LC output filters of switched-mode AC power sources," in *2012 IEEE Power and Energy Conference at Illinois (PECI)*, Feb. 2012, pp. 1–8. DOI: 10.1109/PECI.2012. 6184594.

[14] "Powerized, open innovation project." (), [Online]. Available: https://powerized.eu/ (visited on 09/21/2023).

PCIM Europe 2024, 11– 13 June 2024, Nuremberg   DOI:10.30420/566262300

Hard-Switching High-Frequency GaN-based DC-DC Converters with Concomitant Data Transmission Functionality

Abdelmoumin Allioua [1], Gerd Griepentrog [1]

[1] Technical University of Darmstadt, Germany

Corresponding author and speaker:
Abdelmoumin Allioua, abdelmoumin.allioua@lea.tu-darmstadt.de

Abstract

As the trend toward smaller, more compact, and downsized solutions intensifies, Power and Signal Dual Modulation (PSDM) within the field of Power Line Communication (PLC) has emerged as a cutting-edge methodology. PSDM not only integrates data transmission functionality within power converters; it additionally offers a safeguard against the inherent challenges of conventional PLC, including sensitivity to noise on Power-Lines (PLs) and fluctuating PLC channel characteristics. Despite these advancements, PSDM converters encounter a primary challenge with data transmission speed being limited to a few kbit/s, due to their reliance on low Switching Frequencies (SFs) within the kHz range.

This paper explores a new concept of utilizing tunable switching noise from Switched-Mode Power Supplies (SMPS) as signal carriers for data transmission on DC PLs, targeting up to 1 Mbit/s. With the increasing availability of Wide Bandgap (WBG) semiconductors such as Silicon Carbide (SiC) and Gallium Nitride (GaN), this approach can conceivably be feasible in the MHz-range. Yet, a key question persists: is high bit rate via hard-switching SMPSs achievable? Our research confirms this feasibility by encoding and analyzing Controller Area Network (CAN) messages, demonstrating direct and transparent propagation over PLs, in contrast to time delays faced when encoding and decoding using conventional PLC modems.

1 Introduction

PLC serves as an important communication technology within a broad range of fields. Its main benefit is cost, weight and complexity reduction by using a single cable for both power and communication signals. Thus, PLC has to deal with the conducted Electromagnetic Interference (EMI), particularly from SMPSs, where it is regulated by norms, namely the EN-50065-1 for conducted perturbations from mains communicating equipment. On the other hand, Power Electronic (PE) converters have to limit their emissions to meet the Electromagnetic Compatibility (EMC) standards [1], such as the DO-160G for airborne equipment and the CISPR25 in automotive sector.

The coexistence and "mutual acceptance" of PE and PLC represents a long-lasting conflict of interest in electrical engineering. Yet, the challenge remains, necessitating continuous research, and cross-disciplinary collaboration to find an optimal balance between efficient power conversion, low EMI emissions and undisturbed data transmission.

This research proposes embedding PLC in PE-SMPSs, employing SMPSs switching noise as a beneficial data carrier. It explores the technical aspects of using the switching noise for data transmission and its influence on hardware design. Aiming for low-latency and high bit rate demodulation with immediate data availability at the receiver, the study successfully integrates 1 Mbit/s CAN messages into the DC-DC converter, showcasing the feasibility of this innovative method in practical applications.

2 Merging PLC and SMPS

Incorporating PLC functionalities into PE designs is becoming increasingly popular, notably through an approach commonly termed as PSDM [2][3]. In PSDM, the data signal is inherently produced by the SMPSs alongside power. This method simplifies traditional PLC hardware by eliminating the neces-

2128

sity for coupling unit to merge data onto the PL, and it is not constrained by PLC channel transfer function non-flatness and multi-path effect. Comprehensive details on PLC channel characterization and coupling units are provided in [4], [5].

Although, several attempts [2], [3], [6]–[15] have explored embedding data in SMPS operations, aiming for simultaneous power and data transmission, bit rates have generally remained below 1 Mbit/s.

For instance, in [6] , a DC-DC buck converter was used to transmit Random Binary Data (RBD) by inducing a slight perturbation in the duty cycle at half the SF, serving as a carrier for data signals with minimal output voltage impact. This method achieved a maximum bit rate of 3.8 kbit/s at SFs up to 300 kHz. Conversely, in [2] and [7], the authors did not utilize switching noise as signal carriers. Instead, information was embedded within the DC-DC converter's switching ripple using a two-carrier strategy for modulating binary '1' and '0', achieving transmission bit rates of 2 kbit/s and 2.5 kbit/s, respectively, by varying the carrier frequencies.

The "talkative power" concept named in [8], utilized buck/boost converters and Binary-Frequency-Shift Keying (BFSK) modulation to achieve 2.78 kbit/s with SFs of 100 kHz. This technique was extended to Light Emitting Diode (LED) lighting systems for dual communication via PLs and visible light, employing GaN-based converters. In this setup, employing Frequency Hopping-Differential Phase Shift Keying (FH-DPSK) modulation allowed for a bit rate of 200 kbit/s at a SF of 1 MHz. This concept was further improved in [9] to reduce output voltage disturbances. In LED-based lighting system, the authors in [10] achieved 80 kbit/s by encoding data within the output voltage ripple through BFSK modulation, utilizing SFs between 0.9 and 1.1 MHz, and securing a Bit-Error Rate (BER) of 1×10^{-5}.

The study conducted in [11], demonstrated a method for altering the output voltage ripple by adjusting the output capacitor of a buck converter, adapted to either transmission or reception states, achieving a data rate of 10 kbit/s with SFs up to 240 kHz. Meanwhile, the use of PSDM in cascaded H-bridge converters was confirmed in [12], where switch drivers transmitted data following a Time-Division Multiple Access (TDMA) strategy. Similarly, [3] explored embedding data signals with Phase-Shift Keying (PSK) modulation over the power reg-

ulation signal in bridge topologies, reaching 5 kbit/s bit rate with SF of 100 kHz. The boost topology's potential for data transmission was investigated in [16], employing BFSK to modulate SFs up to 30 kHz in a two-carrier strategy and achieving 2 kbit/s. Similar communication was done in [13] on several buck converters, with SFs reaching 100 kHz.

This summary highlights diverse approaches and challenges in embedding data transmission within SMPS systems, emphasizing the need for further research to enhance data bit rates and efficiency in PEs. A recent study detailed in [17], have shown advancements in PSDM converters discipline, demonstrating the feasibility of encoding data at bit rates up to 1 Mbit/s, using Quasi-Resonant Zero-Voltage-Switching (QR-ZVS) DC-DC converter topologies, expanding the potential for PSDM technology in high-speed data communication.

These advances are detailed in the comparative analysis presented in Table 1, which contrasts previous PSDM attempts with the current study, underscoring our contributions to the field, which stands out due to several distinctions, such as targeting higher SFs up to 6 MHz with the use of WBG semiconductors. Also the enhancement of the transmission bit rate to 1 Mbit/s with reduced latency. Furthermore, the switching noise peaks is used as carriers instead of the switching ripple. Additionally, the feasibility of integrating the CAN bus communication with PSDM converters will be investigated, utilizing a hard-switching topology, going beyond the modest transmission of RBD.

3 Modulation Methods

PLC systems employ a variety of modulation schemes. Among them, the Orthogonal Frequency Division Multiplexing method (OFDM), based on multiple carrier frequencies modulation, stands out for its complexity and capability to transmit at higher bit rates. Additionally, the Code-Division Multiple Access (CDMA) technique is noted for its enhanced resilience to interference and noise. However, PLC networks also utilize simpler, single carrier modulation schemes such as Amplitude Shift Keying (ASK), FSK, and PSK. These methods achieve a compromise between transmission speed and communication reliability. Figure 1 shows their varying effectiveness against AWGN noise. ASK offers simplicity, yet has moderate noise susceptibility, lim-

Tab. 1: Comparative analysis of previous PSDM attemps along with the current study

Ref.[a]	Config.[b]	Mod.[c]	Bit rate	Data[d]	SF	Tech.[e]	Topo.[f]
[2]	PD	BFSK	2 kbit/s	-	83.3-100 kHz	-	Boost
[3]	PSE	BPSK	5 kbit/s	-	50 kHz	Si	Full Bridge
[6]	PD	BFSK	3.8 kbit/s	-	200 kHz	-	Buck
[7]	PD	BFSK	5 kbit/s	-	360 kHz	-	Buck
[8][*]	PD	BFSK	2.78 kbit/s	-	83.3-100 kHz	Si	Buck-Boost
[8][*]	PD	FH-DPSK	200 kbit/s	-	1 MHz	GaN	Buck-Boost
[9]	PSE/PD	FH-DPSK	16.7 kbit/s	-	83.3-100 kHz	-	Buck
[10]	PD	BFSK	80 kbit/s	-	0.9-1.1 MHz	-	-
[11]	PSE	BFSK	-	-	240 kHz	-	Forward
[12]	PSE	TDMA	-	-	40 kHz	-	Half Bridge
[17]	PSE/PD	BFSK	1 Mbit/s	-	2-5 MHz	GaN	QR-Buck
This research	PSE/PD	BFSK	1 Mbit/s	CAN	4-6 MHz	GaN	Buck

[a]Cited Reference, [b]PSDM Configuration, [c]Modulation Method, [d]Transmitted Data Type, [e]Switch Technology, [f]Converter Topology, [*]Indicates multiple results from the same reference.

iting its use in noisy environments. FSK improves noise immunity through frequency separation but at a bandwidth cost, and it's slightly less effective than PSK. PSK emerges as the most noise-resistant and bandwidth-efficient option, though it requires complex receiver design. These theoretical analyses highlight the trade-offs between modulation schemes, however real-world performance can differ based on transceiver performance, channel conditions, and noise type.

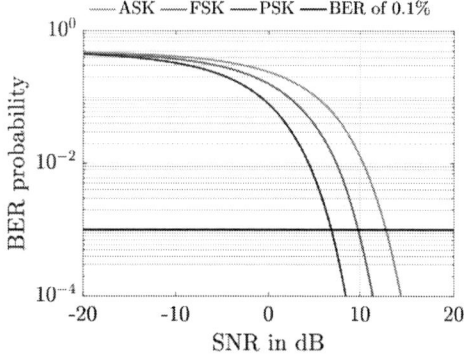

Fig. 1: Theoretical BER as a function of Signal-to-noise ratio (SNR) for different modulation schemes.

In this work, the BFSK modulation technique, depicted in Fig. 2, is selected for its straightforward and efficient method of transmitting binary data through variations in the carrier frequency. This technique encodes binary '0' and '1' as two distinct frequencies, simplifying the modulation and demodulation processes and enhancing signal robustness in noisy or fading conditions.

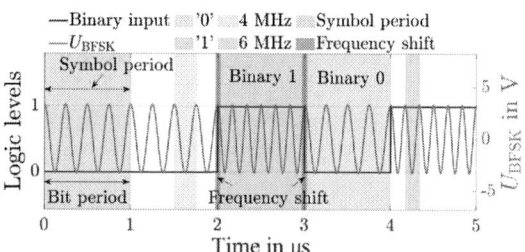

Fig. 2: Binary bit encoding over a 1 µs bit-period using BFSK modulation with 4 MHz space (f_0) and 6 MHz mark (f_1) frequencies.

Yet, careful selection of the carrier frequencies is necessary to minimize the BER, by ensuring phase continuity and enabling continuous symbol transitions. This continuity maintains signal coherence, supporting the receiver in accurately differentiating the binary states with reduced error rate. Moreover, to minimize the Inter-Symbol Interference (ISI), selecting the frequencies f_0 and f_1, to maintain orthogonality over the bit-period T_b is essential.

4 Switching Noise of SMPS

SMPS are commonly implemented using Constant Switching Frequency (CSF) strategies, such as conventional Pulse Wide Modulation (PWM). Despite their ubiquity in electronic systems – attributed to their efficiency, compact size, and adaptability to various power levels – SMPS with a periodic switching pattern produce a periodic EMI noise, which is harmful due its repetitive nature, and can be manifested as Differential Mode (DM) or Common Mode (CM) conducted noise. Furthermore, High-Frequency (HF) EMI can radiate from certain circuit components, primarily due to voltage drops from CM current across the system's parasitics [14].

The input and output voltages of SMPSs usually contain noise [15], referred to as switching noise and switching ripple. Primarily, the switching noise is sensitive to circuit design and switching components placement [2]. This turn-on commutation noise depends on the resonance between parasitic inductance and capacitance excited by the dv/dt variation. The total parasitic inductance comprises the capacitor's Equivalent Series Inductance (ESL), the switch bonding inductance, and stray inductance from interconnections. Meanwhile, the total parasitic capacitance is dominated by the switch parasitic output capacitance C_{oss}. The switching ripple can be approximately divided into two parts: the ripple $U_{C_{\text{out-ripple}}}$ caused by C_{out} (e.g., buck converter), and the ripple $U_{C_{\text{out-ripple}}}$ caused by $R_{\text{ESR}_{C_{\text{out}}}}$ its Equivalent Series Resistance (ESR), which can be calculated as follow [15]:

$$U_{C_{\text{out-ripple}}} = \frac{I_{\text{pp}}}{8 \cdot C_{\text{out}} \cdot f_{\text{s}}} \tag{1}$$

$$U_{\text{ESR}_{C_{\text{out-ripple}}}} = I_{\text{pp}} \cdot R_{\text{ESR}_{C_{\text{out}}}} \tag{2}$$

Where I_{pp} is the peak-to-peak inductor current ripple, f_{s} is the SF. Similarly, the input voltage ripple can be represented.

As an illustration, the output voltage AC content of a DC-DC converter was analyzed. In Fig. 3a, both the switching noise and the switching ripple are shown, while Fig. 3b displays the frequency content, highlighting the 6 MHz SF and its associated harmonics (e.g., at 12 MHz and 18 MHz).

5 Data Encoding Strategy

SMPSs generate HF content, manifested as voltage ripples and switching noise, either at their input or output, usable as potential data signal carriers. This method encodes the data frames as a sequence of switching pulses. The core idea is to

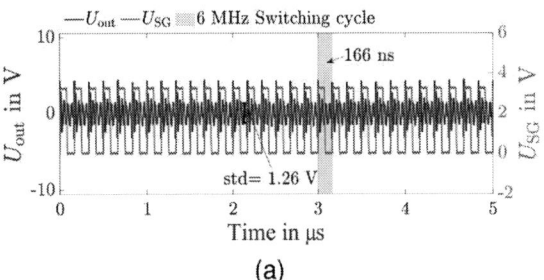

(a)

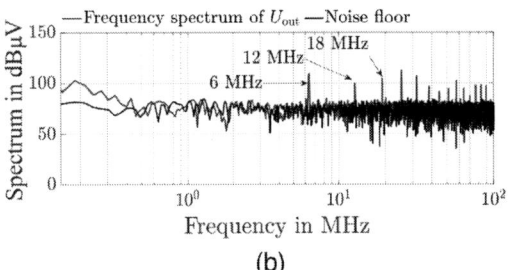

(b)

Fig. 3: SMPS AC content analysis showing: (a) time-domain and (b) frequency spectrum analysis, of a 6 MHz SF with constant pattern.

effectively encode the data into the switching noise pulses, where a predefined consecutive number of pulses with 4 MHz (f_0) represent binary "0", and a different set of pulses at 6 MHz (f_1), denote binary "1". The converter's idle mode (i.e., supplying power only) can be merged to the developed mechanism, by extending one of the initial two SFs beyond the set pulse count, and triggering new transmissions can be enabled with a predefined header.

5.1 PSDM Converter Operation

PSDM converters exhibit a notably flexible operation, enabling functionality across multiple configurations to meet diverse requirements of communication networks. These converters can function in Power Source Equipment (PSE) mode, wherein they utilize their output voltage to facilitate communication over the output DC bus. Conversely, in Powered Device (PD) mode, where communication is reliant upon their input voltage. This inherent adaptability allows PSDM converters to coherently transition between PSE and PD modes or concurrently operate in both, adapted to the specific demands of the network. Furthermore, they can act as a bridge, connecting the input and output DC buses, thus enhancing the connectivity and integration within the network system. Figure 4 illustrates the PSDM converter operating in both PSE and PD modes, showcasing its ability to adapt within the network architecture.

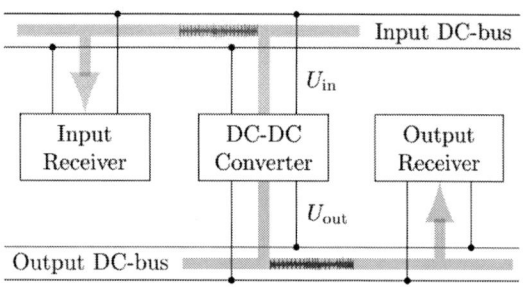

Fig. 4: Block diagram of a DC-DC converter functioning as PD and PSE through its input/output voltage.

5.2 Converter Specifications

The investigation is focused on a 60 W 48/24 V buck converter, evaluated under an open control loop while encoding binary bits using BFSK-PWM technique as illustrated in Fig. 5a and Fig. 5b.

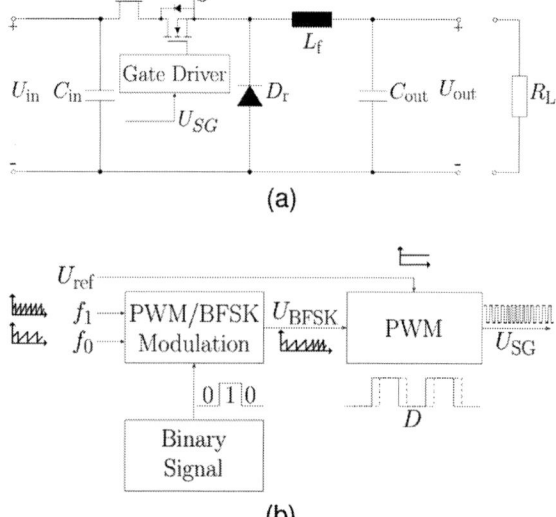

Fig. 5: PSDM converter: (a) equivalent circuit, and (b) BFSK-PWM modulation for bit encoding.

The specifications of the prototype, including key design parameters and considerations, are presented in Table 2, providing a comprehensive overview of the converter's capabilities.

The primary drawback of operating at High Switching Frequencies (HSF) is the increased overall losses, which adversely affects efficiency. To mitigate this, a GaN-HEMT is selected for its superior

Tab. 2: Specifications of the Converter Prototype

Parameter	Variable	Value/Model
Rated power	P (W)	60
Input voltage	U_{in} (V)	48
Output voltage	U_{out} (V)	24
Load	R_L (Ω)	12
Inductance	L_f (μH)	10
Input capacitor	C_{in} (μF)	3.2
Output capacitor	C_{out} (nF)	110
GaN-HEMT	S	TP65H300G4LSG
Rectifier diode	D_r	STPS20150C

performance at HFs, characterized by fast switching capabilities, lower on-resistance, and reduced output capacitance. As Fig. 6a and Fig. 6b illustrate, GaN notably exhibits reduced turn-off losses and less turn-off ringing than its turn-on phase, without relying on additional clamping or snubber techniques for ringing mitigation.

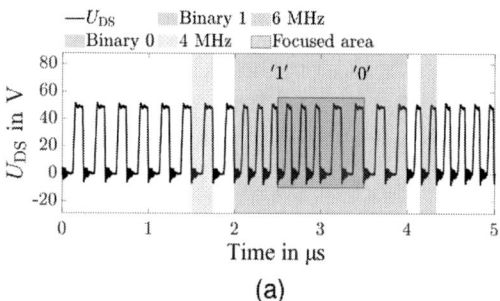

(a)

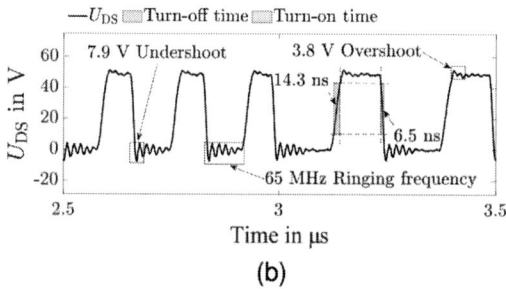

(b)

Fig. 6: Analysis of U_{DS}: (a) U_{DS} with binary signal modulation, (b) detailed waveform in zoomed view.

The elevated switching losses at HSFs necessitate an advanced design to maintain system reliability and performance. Further details on HSF converter design are elaborated in [18].

5.3 Optimal Switching Frequencies

Regarding PLC functionality, the selection of SFs is a critical decision influenced by multiple factors to ensure efficient and reliable data transmission. The primary considerations include:

- Achieving the desired bit rate by guaranteeing at least one switching cycle within the targeted bit period ($T_b = 1\ \mu s$).

- Avoiding data collisions via careful selection of SFs and their harmonics to prevent overlap.

- Ensuring orthogonality of SFs over T_b to eliminate ISI and minimize voltage distortions on the converter's input and output.

Considering these factors, the SMPS is designed to function similarly to PLC modems, equipped with communication capabilities. As shown in Fig. 7, the bits are encoded with BFSK/PWM using a frequency combination that fulfills the orthogonality criteria, ISI avoidance, and phase continuity. The figure shows how each binary bit is represented by distinct frequency, demonstrating the practical application of these design principles.

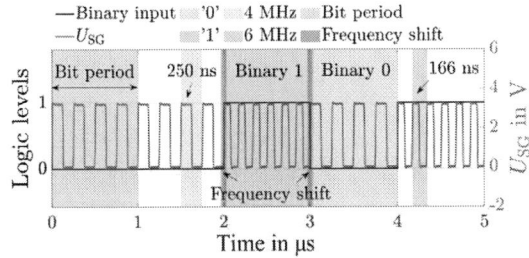

Fig. 7: Bit-period ($T_b = 1\ \mu s$) and tunable BFSK-PWM SF gate signals (e.g., 4 MHz space (f_0) and 6 MHz mark (f_1)).

The essence of this design is the strategic bit encoding over tunable SFs, customized to meet the specific requirements of the data communication protocol while maintaining signal integrity and preventing transmission errors.

5.4 BFSK-PWM Encoding Logic

The PSDM converter topology is developed to facilitate instantaneous data presence at the receiver end, for demodulation in the form of switching pulses. This design ensures low latency and a high bit rate encoding of 1 Mbit/s, as demonstrated

in Fig. 8. Such a capability significantly broadens the potential applications of these converters, particularly in enhancing PLC usage.

Figure 8 showcases the measured time-domain data frames, where binary data is encoded using a tunable BFSK-PWM switching pattern U_{SG} across the PL. Figure 8a illustrates the input voltage waveform after DC content filtering, presenting the binary bits as distinct voltage levels corresponding to their respective frequencies. Similarly, Fig. 8b displays the output voltage waveform, confirming the presence of binary-encoded information.

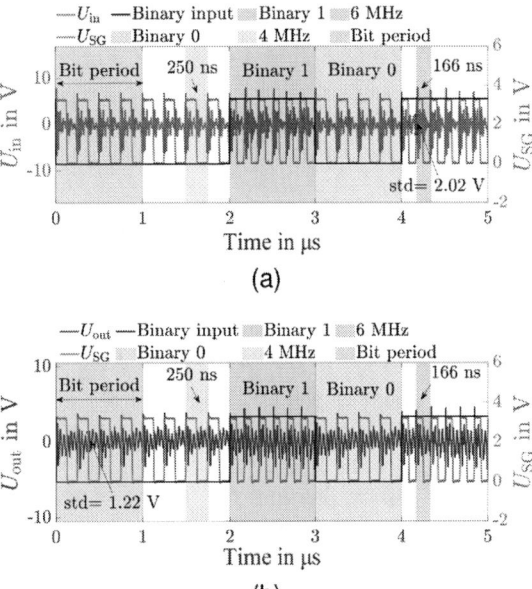

Fig. 8: Time-domain BFSK-PWM encoded data frames on PLs: (a) input and (b) output voltage.

The capability to deliver data with minimal delay and high bit rate are crucial for applications requiring real-time data transfer, indicating a significant step forward in the field of integrated power and data systems. By enabling more efficient power and data transfer, PSDM converters with FSK-PWM encoding logic hold the promise of advancing PLC technology, offering improved performance for both existing and emerging applications.

5.5 Dual Verification: Data and Power

A critical aspect of this dual-functionality converter is its ability to encode binary data without compromising the stability of the output voltage. To validate this, we examine the converter's ability to maintain voltage stability during data transmission.

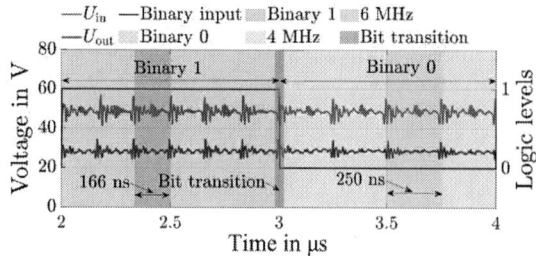

Fig. 9: Time-Domain waveforms of input and output voltages with highlighted bit transition.

Figure 9 presents a detailed view on the transition between binary bits '1' and '0', demonstrating the PSDM converter's proficient encoding of binary data onto the input and output voltages in the form of switching pulses. The figure provides a clear comparison between stable voltage levels during both high ('1') and low ('0') data states, and smooth transition between them, underscoring the converter's stable power delivery alongside data encoding. This stability marks an improvement over the findings in [17], where the output voltage demonstrated slight fluctuations during bit transitions, potentially causing minor power disturbances.

The adaptability of the PSDM converter is further illustrated by its performance across a range of operational frequencies, a factor that significantly affects efficiency. Insightful efficiency analysis at different SFs is detailed in Table 3.

Tab. 3: Performance Parameters for 4 and 6 MHz SFs Operation

	4 MHz	6 MHz	4 and 6 MHz
$\bar{U}_{\text{in}}$ (V)	47.37	47.41	47.42
$\bar{U}_{\text{out}}$ (V)	24.86	23.07	23.92
$\bar{I}_{\text{in}}$ (A)	1.20	1.13	1.16
$\bar{I}_{\text{out}}$ (A)	1.98	1.83	1.90
$\bar{P}_{\text{in}}$ (W)	56.85	53.70	55.01
$\bar{P}_{\text{out}}$ (W)	49.23	42.33	45.64
D_{cycle} (%)	47.00	43.50	-
$\bar{\eta}$ (%)	86.6	79.5	83

The highest efficiency is observed at 4 MHz with 86.6%, while operation at 6 MHz shows a reduced efficiency of 79.5%. Interestingly, an intermediate efficiency of 83% is achieved when combining both frequencies. This pattern underscores the impact of operational frequency on conversion efficiency.

The efficiency of the converter at 4 MHz under a 12

Ω load was notably higher than the 72.5% efficiency reported with a 6 Ω load in [18]. This improvement can be attributed to the light load condition, which leads to a decrease in conduction losses due to the reduced current flow. Lighter loads naturally result in lower energy dissipation within the converter, enhancing overall efficiency. This finding confirms that optimizing load conditions is crucial for minimizing losses and achieving higher efficiency in power conversion systems.

For operations combining 4 and 6 MHz frequencies, the duty cycle adjusts between 47% and 43.5%, aligning with the requirements of each specific frequency. This strategy of duty cycle adjustment ensures the system's efficiency across different operational modes, facilitating optimal data transmission and power conversion by matching the duty cycle to the frequency in use.

6 Evaluating CAN Integration with PLC and PSDM

6.1 CAN-PLC Setup

Prior efforts, as referenced in [19] and [20], have explored the potential of merging CAN and PLC.

In CAN-PLC setups, CAN messages are transmitted over PLs through signal modulation facilitated by PLC modems and the use of coupling/decoupling units as shown in Fig. 10a. The process begins with the reception of CAN messages by the PLC modem, which then prepares the data for transmission. This preparation involves dividing the data into packets, adding stuffing bits, and incorporating preamble bits to ensure synchronization and error detection.

The encoding of CAN messages into PLC frames significantly increases their length and transmission time. For instance, if a 1 Mbit/s CAN message (i.e., CAN 2.0A with Data Length Code (DLC) of 6 bytes) duration is 92 µs, the embedding process into a PLC frame introduces additional preamble and arbitration phases. This preliminary phase, combined with the message encoding, results in a total transmission time of approximately 339 µs as detailed in Fig. 11. This substantial latency, pose challenges for latency-critical applications where immediate data availability is crucial.

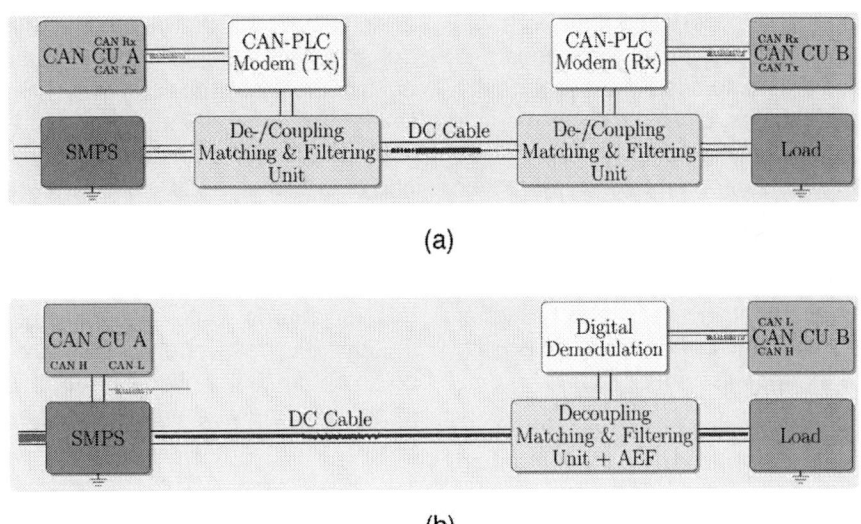

(a)

(b)

Fig. 10: llustration of a two-node communication infrastructure for (a) CAN-PLC, and (b) CAN-PSDM setups.

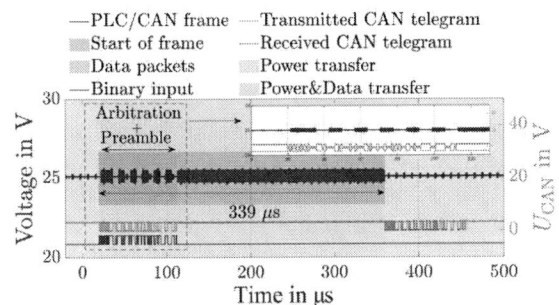

Fig. 11: Latency analysis of 1 Mbit/s CAN message transfer over PLs: transmitted/received CAN telegrams, and modulated PLC signal.

Fig. 12: Bit encoding and latency visualization in DC-DC converter output/input voltage, for direct CAN message observation on the PLs.

6.2 CAN-PSDM Setup

This approach allows the data to be transmitted over PLs without the need for traditional PLC modems, reducing the encoding and transmission delays as shown in Fig. 10b. The switching noise, which typically considered undesirable, is utilized here as a medium for data transmission.

The primary benefit of the CAN-PSDM approach lies in its ability to significantly reduce transmission delays. Unlike the CAN-PLC setup, where the encoding process introduces substantial latency, the CAN-PSDM method ensures that CAN messages are directly and immediately visible at the receiver end, as shown in Fig. 12. This instantaneity is crucial for applications requiring real-time data exchange, making the CAN-PSDM setup a more viable option for latency-sensitive environments.

The previous tests revealed that the CAN-PLC setup shown in Fig. 10a and Fig. 11, while unique in its integration of data communication over PLs, suffers from significant latency issues due to the encoding and modulation processes involved. The total transmission time for a single CAN message, including preamble, arbitration, and coding, could extend up to 339 μs, a substantial delay for critical applications. In contrast, the CAN-PSDM setup shown in Fig. 10b and Fig. 12 demonstrated a marked improvement in reducing transmission delays, with CAN messages becoming instantly available at the receiver end, closely mimicking the performance of traditional CAN bus systems.

7 Safety and EMI Compliance

In assessing the safety of the PSDM converters, a detailed examination of CM noise was undertaken as shown in Fig. 13, to understand their potential for EMI. CM current measurements at the input and output, alongside the noise floor, shed light on the interference these devices might introduce in practical scenarios. Noteworthy are the significant noise peaks observed at SFs of 4 and 6 MHz. Moreover, a distinct noise spike at 24 MHz was detected, arising from the overlap of the 4th harmonic of the 6 MHz frequency with the 6th harmonic of the 4 MHz frequency. This cumulative effect lead to higher noise levels at this specific frequency, imposing stricter EMI filtering requirements.

The EMI analysis was carried out without referencing specific EMC standards, and the intended applications for the converters. The aim was to construct a thorough profile of the PSDM converters' EMI behaviors to inform their deployment in real-world contexts. Adherence to PE EMC norms is undoubtedly necessary, and with advanced EMI filtering, compliance is achievable. However, the categorization of this data transmission technique within the PLC EMC standards remains ambiguous. It is not yet clear whether the use of PSDM converters should be classified under intentional signaling or as non-intentional emissions. This ambiguity highlights a gap that needs to be addressed by standardization bodies within the PLC community to ensure that such innovative approaches can be clearly understood and appropriately regulated.

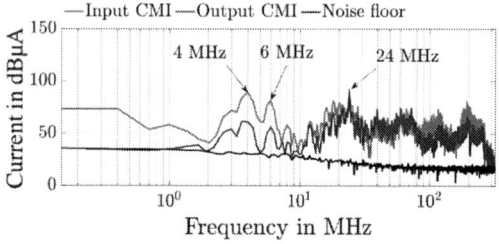

Fig. 13: Frequency spectrum of the PSDM converter input and output CM noise.

8 Conclusions and Future Work

This research explores merging PE and PLC, targeting HSFs and enhanced bit rate through hard-switching GaN-based DC-DC converters. The SMPS undesirable switching noise peaks are used as signal carriers, highlighting the potential to achieve high bit rate up to 1 Mbit/s over DC PLs. Despite our approach's promise, notable challenges

arise with HF hard-switching techniques, necessitating focus on the converter reliability, particularly against thermal stress and EMI. The efficiency data reveal the converter's capability to maintain relatively high efficiency across varying frequencies and operational modes, with an evident preference for lighter loads. The alternating duty cycle strategy for combined frequency operation illustrates the system's flexibility in handling diverse data transmission needs while managing power efficiency. Optimizing these operational parameters further could enhance the overall performance, especially in applications where both data transmission and power conversion efficiency are critical.

The comparative analysis between the CAN-PLC and CAN-PSDM setups showed that while the CAN-PLC approach offers a distinctive way to use PLs for data communication, its latency issues limit its applicability for time-sensitive applications. On the other hand, the CAN-PSDM setup provides a promising alternative, ensuring rapid data availability and lower latency, aligning closely with the real-time communication needs of CAN bus systems. Future applications may benefit from adopting the CAN-PSDM approach, especially in environments where latency is a critical factor.

References

[1] W. E. Sayed, P. Lezynski, R. Smolenski, N. Moonen, P. Crovetti, and D. W. Thomas, "The effect of EMI generated from spread-spectrum-modulated SiC-based buck converter on the G3-PLC channel," *Electronics*, vol. 10, no. 12, p. 1416, 2021.

[2] J. Wu, J. Du, Z. Lin, Y. Hu, C. Zhao, and X. He, "Power conversion and signal transmission integration method based on dual modulation of DC–DC converters," *IEEE Transactions on Industrial Electronics*, vol. 62, no. 2, pp. 1291–1300, 2014.

[3] J. Du, J. Wu, R. Wang, Z. Lin, and X. He, "DC power-line communication based on power/signal dual modulation in phase shift full-bridge converters," *IEEE Trans. Power Electron*, vol. 32, no. 1, pp. 693–702, 2016.

[4] A. Allioua, G. Griepentrog, M. Vögel, J. Eitler, and N. Mahdavi, "Design of PCB-based planar coil inductive coupler," in *2021*

IEEE Southern Power Electronics Conference (SPEC), IEEE, 2021, pp. 1–8.

[5] A. Allioua, G. Griepentrog, M. Vögel, J. Eitler, and N. Mahdavi, "Aircraft DC networks characterization and adaptive stage design for PLC use," in *2021 IEEE/AIAA 40th Digital Avionics Systems Conference (DASC)*, IEEE, 2021, pp. 1–8.

[6] W. Stefanutti, S. Saggini, P. Mattavelli, and M. Ghioni, "Power line communication in digitally controlled DC–DC converters using switching frequency modulation," *IEEE Transactions on Industrial Electronics*, vol. 55, no. 4, pp. 1509–1518, 2008.

[7] T. Kohama, S. Hasebe, and S. Tsuji, "Simple bidirectional power line communication with switching converters in dc power distribution network," in *2019 IEEE International Conference on Industrial Technology (ICIT)*, IEEE, 2019, pp. 539–543.

[8] X. He, R. Wang, J. Wu, and W. Li, "Nature of power electronics and integration of power conversion with communication for talkative power," *Nature communications*, vol. 11, no. 1, p. 2479, 2020.

[9] J. Chen, J. Wu, K. Liu, R. Wang, W. Li, and X. He, "Improved switching ripple modulation strategy for simultaneous power conversion and data communication in DC–DC converters," *IEEE Trans. Power Electron*, vol. 37, no. 8, pp. 9275–9284, 2022.

[10] N. Bertoni, S. Bocchi, M. Mangia, F. Pareschi, R. Rovatti, and G. Setti, "Ripple-based power-line communication in switching DC-DC converters exploiting switching frequency modulation," in *2015 IEEE International Symposium on Circuits and Systems (ISCAS)*, IEEE, 2015, pp. 209–212.

[11] A. Katsuki, K. Morita, K. Masutomo, and S. Maeyama, "Signal transmission by high-ripple DC-DC converter in a new wire communication system," in *2014 IEEE 36th International Telecommunications Energy Conference (INTELEC)*, IEEE, 2014, pp. 1–7.

[12] I. Mandourarakis, E. Koutroulis, and G. N. Karystinos, "Power line communication method for the simultaneous transmission of power and digital data by cascaded H-

bridge converters," *IEEE Trans. Power Electron*, vol. 37, no. 10, pp. 12 793–12 804, 2022.

[13] R. Wang, J. Du, S. Hu, J. Wu, and X. He, "An embedded power line communication technique for DC-DC distributed power system based on the switching ripple," in *2014 International Power Electronics and Application Conference and Exposition*, IEEE, 2014, pp. 811–815.

[14] J. Yao, Y. Lai, Z. Ma, and S. Wang, "Advances in modeling and reduction of conducted and radiated EMI in non-isolated power converters," in *2021 IEEE Applied Power Electronics Conference and Exposition (APEC)*, IEEE, 2021, pp. 2305–2312.

[15] R. Wang, Z. Lin, J. Du, J. Wu, and X. He, "Direct sequence spread spectrum-based PWM strategy for harmonic reduction and communication," *IEEE Transactions on Power Electronics*, vol. 32, no. 6, pp. 4455–4465, 2016.

[16] Z. Lin, J. Du, J. Wu, and X. He, "Novel communication method between power converters for DC micro-grid applications," in *2015 IEEE First International Conference on DC Microgrids (ICDCM)*, IEEE, 2015, pp. 92–96.

[17] A. Allioua and G. Griepentrog, "Power and Signal Dual Modulation with QR-ZVS DC/DC Converters using GaN-HEMTs," in *2024 IEEE Applied Power Electronics Conference and Exposition (APEC)*, IEEE, 2024, pp. 1–8.

[18] A. Allioua, D. Krause, A. Zingariello, and G. Griepentrog, "Reduction of DC/DC Converters EMI Emission Using Bi-and Unidirectional QR-ZVS Topologies," in *2023 IEEE 10th Workshop on Wide Bandgap Power Devices & Applications (WiPDA)*, IEEE, 2023, pp. 1–6.

[19] A. Allioua, G. Griepentrog, M. Vögel, J. Eitler, and N. Mahdavi, "Feasibility Analysis of a Commercial CAN-PLC Communication Interface over Aircraft DC Network," *The 14th Workshop for Power Line Communications*, pp. 1–4, 2023.

[20] F. Grassi, S. A. Pignari, and J. Wolf, "Assessment of CAN performance for Powerline Communications in dc differential buses," in *2009 IEEE International Conference on Microwaves, Communications, Antennas and Electronics Systems*, IEEE, 2009, pp. 1–6.

PCIM Europe 2024, 11– 13 June 2024, Nuremberg   DOI: 10.30420/566262301

Efficient Design of High-Current, Low-Output Voltage DC-DC Converters Using Artificial Intelligence-Based Topology Selection and Optimization

Thomas Harmand[1,3], Raphael Filipe[1], Patrick Dubus[2], Denis Labrousse[3,4]

[1] 3D PLUS, France
[2] Powerlogy, France
[3] Université Paris-Saclay, ENS Paris-Saclay, CNRS, SATIE, F-91190, Gif-sur-Yvette, France
[4] SATIE, Le Cnam, CNRS, F-75003 Paris, France, HESAM Université

Corresponding author: Thomas Harmand, tharmand@3d-plus.com
Speaker: Thomas Harmand, tharmand@3d-plus.com

Abstract

This paper introduces a new method for choosing and optimizing the best topology for a DC-DC converter in a specific application. To demonstrate the benefits of this approach, it has been implemented in the design of a non-isolated, low-voltage(0.6 to $1.5V$), high-current ($\geq 40A$) output converter. Finding the best design that meets specific application requirements while minimising development time can be challenging. A three-step approach is proposed using an artificial intelligence-based algorithm to address this challenge. This method selects the optimal topology for the application and optimizes the magnetic component(s) choice and output capacitor values. Firstly, the topology has been selected and optimized using the proposed method. Then, the converter was experimentally implemented to evaluate its performance.

1   Introduction

As the industry continues to trend toward consuming more power, the significance of an efficient and reliable power delivery architecture is increasing [1]. Also, the introduction of GaN transistors has significantly advanced power density and energy delivery capabilities, making them particularly valuable in low-voltage, high-current converters due to their high switching frequency capability [2], [3]. However, the growing power demand for computing requiring high currents ($\geq 40A$) represents a significant challenge. Tab. 1 summarizes the desired converter specifications. The main objective is to achieve 90% efficiency with minimal surface area while providing a $40A$ output current. Additionally, a constraint requires that the maximum voltage output drop should not exceed 10% under 0 to $30A$ with a $2A/us$ current transient. The decision to use an input voltage range of 5-12V instead of the standard 24-48V is based on the converter's specific requirements for space applications. However, the same approach would be applicable if the converter

were developed for a more conventional application, such as a datacenter power supply.
In response to these challenges, a four-step approach has been developed:

1. Appropriate topologies are selected to match the intended application.

2. A loss map is generated for GaN transistor, GaN drivers, and Schottky diode by conducting a SPICE simulation that considers current ripple, input voltage and output current.

3. A genetic algorithm is used to optimize the inductor(s) selection and the output capacitor value of each selected topology.

4. A Pareto front of efficiency functions of the power cell area is plotted for each topology.

It is now possible to compare different optimized topologies for a desired application, making the comparison relevant and allowing for a pre-optimized selection. The proposed method is used for a specific application but can be applied to other specifications.

2138

Tab. 1: PPOL40A Converter summary specifications

Symbol	Name	Values
V_{IN}	Input Voltage	$[5-12]$V
V_{OUT}	Output Voltage	$[0.6-1]$V
$V_{OUT_{ACC}}$	Output Voltage accuracy	$\pm 1\%$
$V_{OUT_{TRS}}$	Transient (30A, 2A/us)	$\pm 10\%$
I_{OUT}	Output Current	$[0-40]$A
η	Full Load Efficiency	$>90\%$
Area	Total converter surface	$<10\text{cm}^2$

The paper is organized into three sections. Section II describes the optimization process, which includes topology preselection, generation of loss maps, and integration of a genetic algorithm coupled with the Coilcraft database to optimize the selection of inductors and output capacitor values. Section III details the experimental implementation, covering power cell routing and performance evaluation. Finally, the advancements are discussed, and avenues for future enhancements are proposed.

2 Optimization Process

2.1 Topology Pre-selection

It is crucial to preselect the most suitable topologies for the target application to ensure relevant optimization and time-saving in the final power cell design. Topology preselection should be based on a literature review, incorporating the target application-specific constraints. Numerous studies have already been conducted on low voltage, high output current DC-DC converter [4]. The first topology chosen is the Multiphase Buck converter [5]. Multiphase topology is commonly employed for high-current, low-voltage, non-isolated Voltage Regulator Module [VRM] applications. Several variants of the Multiphase Buck converter have been proposed in the literature, such as the Double-Step Down Converter [6], which involves inserting a series capacitor between the two phases. This addition offers two advantages: halving the voltage perceived by the High-Side transients and distributing currents among phases without regulating them via control [7]. Consequently, this topology is the second selection.

The third topology studied is the switched-capacitor topology. Several variants have already been proposed in the literature [8], [9]. However, this topology is limited due to the maximum duty cycle,

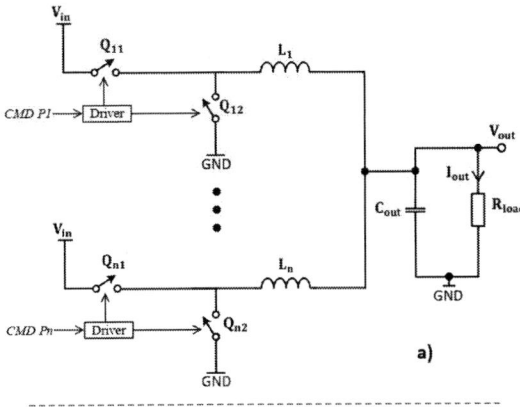

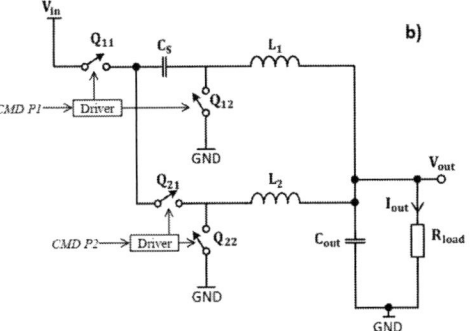

Fig. 1: Preselected Topologies: a) Multi-phase Buck (n phases), b) Double Step-Down Buck Converter

capped at 0.125. To achieve a 1V output voltage, a minimum input voltage of 8V is required, which does not align with the targeted specifications ($V_{in-min} = 5V$). Despite its promising performance, this topology is therefore excluded.

Resonant converters like LLCs are excluded for the following reasons:

- The converter must operate at a frequency close to resonance, preventing the use of non-linear control. The response during load transients and the converter stability are consequently degraded.

- Selecting the resonant inductance is particularly complex for applications involving high currents since the entire current will flow through the resonant inductor. Obtaining a sufficiently high value can be complicated to prevent excessive resonant current while limiting its DCR.

- Resonance helps to avoid switching losses, but these represent only a portion of total losses (less than 30% in the targeted application).

Therefore, the efficiency lost due to resonance is not necessarily recovered by soft switching of transistors.

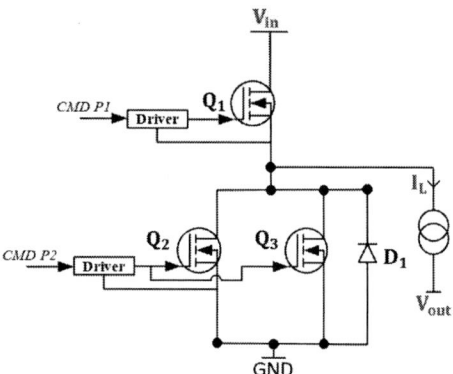

Fig. 2: Simulation Model of a Half-Bridge for Generating GaN Transistor Loss Map Based on Phase Number, Level, and Current Ripple

2.2 Mapping GaN Transistor Losses

The optimization of inductance selection and output capacitance value can only be accurately achieved by accounting losses in GaN transistors, Schottky diode and GaN Drivers. Since it has been chosen to operate in hard switching mode for reasons outlined in the introduction, the current ripple in the inductors(s) directly affects the switching losses of the transistors. Additionally, the number of phases and levels, which vary depending on the topology, influence the AC and DC losses of the inductors(s) and the switching and conduction losses of the GaN transistors. Analytical calculation of GaN transistor switching losses is challenging [10], [3], [11], [12]. One method to obtain satisfactory results is to use SPICE simulation utilizing models provided by the manufacturer (EPC in the studied case). A simplified model comprising a single arm (one phase) where a triangular current is injected with a current source is proposed to overcome the drawback of time-consuming SPICE simulations. After this improvement, a simulation typically lasts around 30 seconds on an average-powered machine. Consequently, generating a loss map within a few hours is possible, considering the perceived voltage (level(s) number), average current (phase(s) number), and current ripple (inductor(s) value).

The losses attributed to GaN are significantly impacted by the choice of transistor references used and the number of paralleled transistors. Although this could be an additional parameter considered in the loss map, internal constraints and the realization of a preliminary prototype have already facilitated the selection of high-side transistors (EPC2015C), low-side transistors (EPC2066), and the number of paralleled transistors (one on the high side and two in parallel on the low side).

2.3 Artificial Intelligence-Based Optimization Algorithm

The genetic algorithm is the core of this work. It is programmed in Python and relies on the Pymoo library [13], [14], [15]. Genetic algorithms belong to a class of optimization algorithms inspired by Darwin's theory of evolution. They are employed to solve challenging optimization problems by mimicking the process of natural selection. Pymoo is a Python library designed for solving multi-objective optimization problems. It implements the NSGA-II (Non-dominated Sorting Genetic Algorithm II) algorithm. NSGA-II is an efficient genetic algorithm for solving multi-objective optimization problems with constraints. The operation of this type of algorithm involves defining multiple objectives and constraints. In the studied case, the objectives are to minimize the total occupied area and the loss density. At the same time, the constraints that must be respected include the output voltage ripple in both static and dynamic states and the maximum loss value corresponding to a 90% efficiency at the operating point where $I_{out} = I_{max} = 40A$.

This work is centred around optimizing the selection of the power cell inductance(s) and output capacitance(s) values. The frequency value is considered fixed at 600kHz. A previous prototype determined this frequency, demonstrating the benefits of operating at this frequency with GaN transistors. However, including this parameter in the optimization process would be worthwhile.

The operation of the proposed algorithm is as follows: At each iteration, the algorithm selects a value for the inductance and output capacitance. The current ripple is calculated based on the input voltage (number of levels) and the inductor value. Using the output current values and current ripple, the FBI algorithm (Find Best Inductor) uses the Power Analyzer Tools available on the Coilcraft website to determine the inductance that meets the total loss criteria (loss generated by the inductors(s) $(AC + DC)$ and losses generated by the GaN transistors, diodes and GaN drivers at the operating point). If multiple inductances meet this loss criterion, the algorithm selects the one with the

PCIM Europe 2024, 11– 13 June 2024, Nuremberg   DOI: 10.30420/566262301

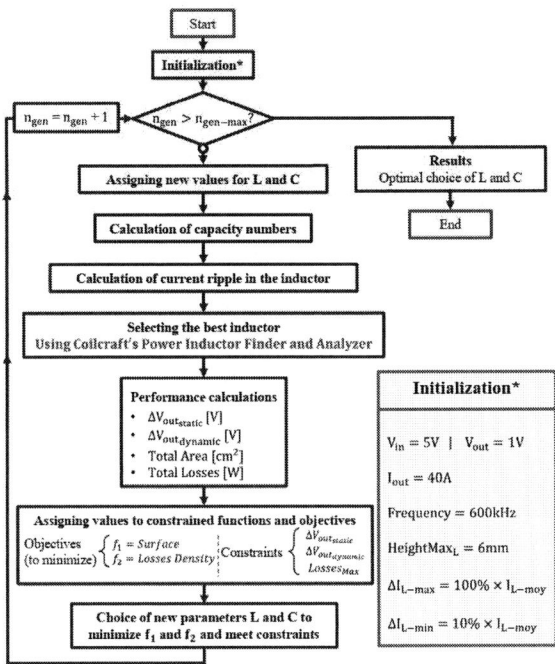

Fig. 3: Flowchart of the Genetic Algorithm for Optimizing Inductor Selection and Output Capacitor Value

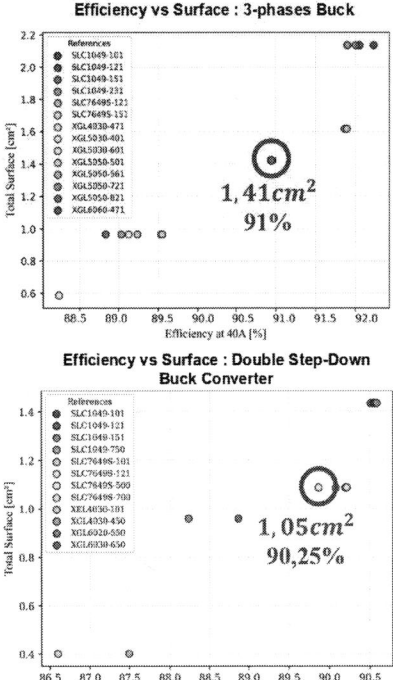

Fig. 4: Genetic Algorithm Results for Three-Phase Buck Topology and DSDBC: Pareto Frontiers Mapping Occupied Area versus Power Cell Efficiency at Full Load ($I_{out} = 40A$)"

smallest area. Subsequently, the area occupied by the output capacitors is added to that occupied by the inductor(s) (in multi-phase case). The various parameters calculated and determined using the tool provided by CoilCraft are then fed into the genetic algorithm to propose a better L/C configuration in the next iteration. After 1000 iterations, the algorithm converges towards a specific inductance and output capacitance. This optimization considers not only a real inductor(s) but also the influence of current ripple on the losses of the rest of the cell (GaN driver, GaN transistors, diode).

One of the challenges associated with using a genetic algorithm is visualizing and determining the "optimal" result. A common strategy is plotting a diagonal line x=y, implying that both minimized functions carry equal weight in the optimization process. The optimal pair (L, C) will be determined based on the criteria outlined in the introduction, which entails:

– Plotting the Pareto front: Total surface area versus efficiency.

– Selecting the result that offers the smallest surface area for a configuration that achieves efficiency greater than 90%.

– Isolating the results that validate this criterion. The optimal inductance for a given number of phases is then obtained.

After selecting the optimal inductance and output capacitance, SIMetrix simulations were conducted to determine the efficiency of two topologies, Buck 3-phase and DSDBC, based on their output current. Results indicated that the DSDBC topology is more efficient for currents below $25A$ due to its ability to reduce voltage stress on the High-Side transistors, thus decreasing their switching losses. Despite this, the three-phase Buck topology emerged as the least lossy at full load, reducing losses by 10% compared to DSDBC (4.2 to $3.8W$, or 91% to 90.25% efficiency). This reduction comes with a 25% increase in surface area. Both topologies met the initially chosen thermal dissipation constraints. Therefore, the Double Step-Down Converter is the chosen topology since it respects loss constraints while offering a surface area 25% smaller than the 3-phase Buck.

2141

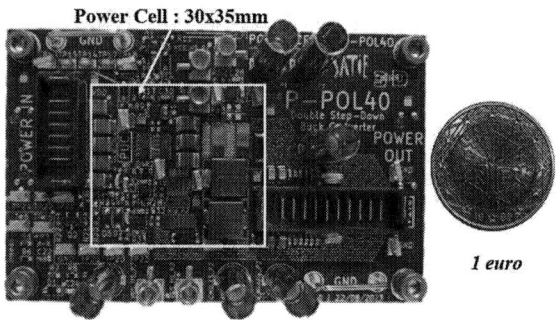

Fig. 5: Photograph of the experimental prototype on which measurements were conducted. $[5; 12]V \rightarrow [0.6; 1]V - 40A$.

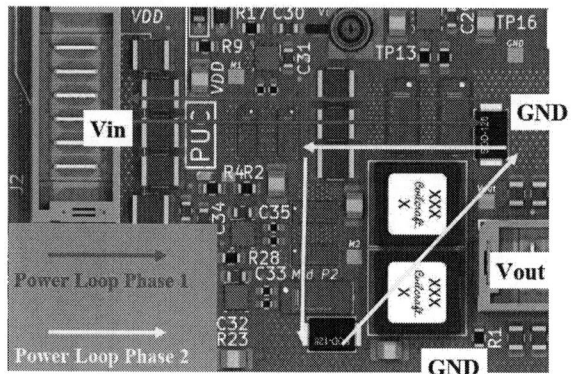

Fig. 6: 3D Visualization of PPOL40A Power Cell Routing on KiCad PCB Layout: Red Arrows Represent Phase One Current Paths, Yellow Arrows Represent Phase Two

3 Experimental Implementation

An experimental prototype (Fig.5) has been developed to validate the operation of the power module and evaluate the converter's performance regarding power dissipation and load transient response. The prototype consists of two printed circuit boards (PCBs), one for the power cell and the other for the control. The control part is non-linear and built using analog components, but it can also be accomplished with a microcontroller. However, only the PCB containing the power cell will be examined. The power cell occupies an area of $30x35mm$, which could be reduced using GaN drivers integrated with GaN transistors. The input and series capacitors used are $X7R$ from KEMET (low ESR). The two inductances were determined using a genetic algorithm presented in section II. The maximum height is defined by the sum of the PCB and inductor thicknesses, which in this case is $6.5mm$.

3.1 Power cell routing

Power cell routing significantly impacts module performance. To ensure optimal routing, three essential criteria need consideration:

- Maintain low parasitic inductance (less than $1.5nH$ per phase).

- Achieve low track resistance (less than $3m\Omega$ per loop).

- Limit power cell surface area.

A common technique to reduce parasitic inductance is to overlap the front and rear paths of the power loop while keeping the distance to a minimum. This is possible through superposition, as it is crucial to reduce parasitic inductance, even if it means increasing parasitic capacitance, especially when dealing with low voltages.. Solutions to solve PCB track resistance issues include maintaining a substantial copper thickness ($70um$ here) and providing adequate track width.

Discrete-level shifters for the GaN drivers are used. However, using GaN drivers integrated into the GaN transistor would make it possible to reduce the parasitic inductances of the gate-source loops between the GaN driver and the gate of the GaN transistor [16], [17]. Fig.6 shows the power loop path (phases one and two); red arrows represent phase one's current flow, and yellow arrows represent phase two's current flow.

3.2 Converter Performance

In the following measurements, the term "power cell" encompasses level shifters, GaN drivers, GaN transistors, inductors, and input and output capacitors. The power consumption of the control board, ranging from $120mW$ ($V_{in} = 5V$) to $250mW$ ($V_{in} = 9V$), is excluded from the efficiency calculations. Output current regulation and measurement (I_{out}) are facilitated by an electronic load (Keysight's EL34243A). Input voltage is measured using a voltmeter (Keysight's 34461A) across the input capacitors, while output voltage is measured across the output capacitors, also utilizing a voltmeter (Keysight's 34461A). The power cell is designed to operate at a frequency of $600kHz$. Yet, the converter frequency varies between $550kHz$ and $650kHz$ depending on the duty cycle, which fluctuates with input and output voltage and current. The transistors ' dead times and gate resistances have been optimized ($t_{DT} = 10ns$). Thermal inspection

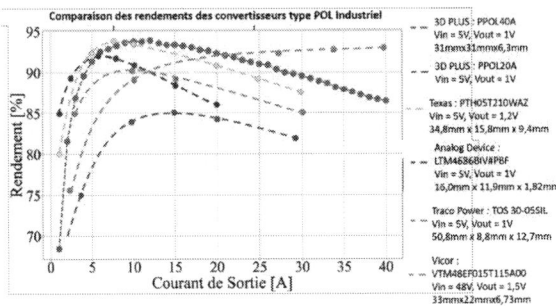

Fig. 7: Comparative Analysis of PPOL40A Efficiency versus industrial POL Converters at $V_{in} = 5V$ and $V_{out} = 1V$ Operating Conditions

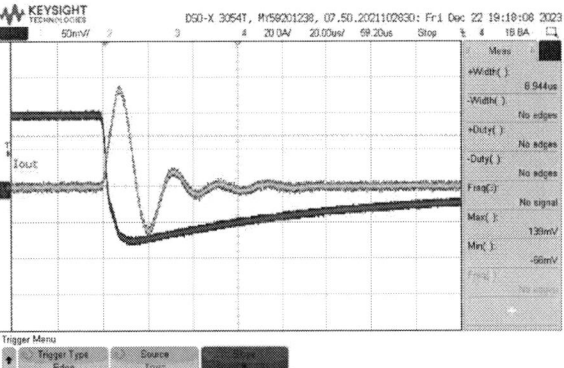

Fig. 8: AC Output Voltage Response (V_{out-AC}) to Transient Output Current Variation from 40 to $0A$ at a Rate of $15A/us$. V_{out} is the yellow curve (measured in AC), while the red curve is the output current I_{out}

using a thermal camera reveals uniform heating, indicating adequate converter sizing. The maximum converter temperature for $V_{in} = 5V$ increases by $15°C$ at $I_{out} = 20A$ ($T_{(final|I_{out}=20A} = 35°C)$, by $45°C$ at $I_{out} = 30A$ ($T_{(final|I_{out}=30A} = 70°C)$, and by $85°C$ at $I_{out} = 40A$ ($T_{(final|I_{out}=40A} = 110°C)$. These measurements were taken at an ambient temperature of $T_{ambient} = 25°C$, without external cooling (fan, PCB plating against a heat sink, water cooling, etc.). Therefore, the PCB cools solely through natural convection from its surface area $10.85cm^2$ at ambient temperature. Figure 7 compares the efficiency at $V_{in} = 5V$ and $V_{out} = 1V$ for the PPOL40A converter (this work) to other industrial Point Of Load [POL] converters with known efficiency.

As observed, the efficiency measured at full load is lower than that calculated using SPICE simulation after optimizing passive components (90% vs 86%). This can be explained by the fact that the loss map produced using SPICE simulation does not consider the effects of parasitic elements of the PCB. However, these parasitic elements, such as inductances and resistors, contribute significantly to the losses. A study is underway to precisely determine the nature of the various losses in the power cell. One way to improve the system is to consider the influence of these losses on the power cell. For instance, we could estimate the inductance and parasitic resistance of the tracks using finite-element simulation (2D or 3D), but this would require assumptions to be made about each preselected topology.

A transient load test is conducted to validate the converter's dynamic performance. Figure 8 depicts a current step from 40 to $0A$ at $15A/us$. The output voltage increases by $130mV$ (13%). This can be improved by adding more output capacitors ($600uF$ on the actual prototype). When the output current changes from 0 to $40A$ (at $2A/us$), the output voltage decreases by $75mV$, representing 7.5% of V_{out}. The PPOL40A is a serious competitor to other industrial POL converters. It is competitive regarding the surface area occupied and the current output capacity. Moreover, it offers enjoyable dynamic performance despite the low value of the output capacitance included in the prototype.

4 Conclusion

This study proposes an innovative method to optimize the passive components of various power cells by optimizing the choice and value of inductors and output capacitors. To demonstrate the performance of this method, a POL-type converter (PPOL40A) has been developed to validate the performance obtained in the simulation. Experimental validation showcased exceptional performance, achieving a peak efficiency of 94% and a full load efficiency of 86% under natural convection cooling conditions.

While the study demonstrates promising results, there exist avenues for further improvements. The discrepancy between the simulated and measured full-load efficiencies, amounting to 4%, can be partly attributed to the omission of the PCB's parasitic elements, including parasitic resistance and inductance, in the simulation. It is imperative to incorporate these parasitic elements into the loss map generated through SPICE simulations to achieve more accurate results. Exploration of coupled inductance and planar magnetics could enhance the

PPOL40A's compactness and efficiency.

In summary, optimizing the passive elements of a power cell presents numerous advantages, including optimizing the actual magnetic components, considering both AC and DC losses, and accounting for the impact of current ripple on losses in GaN transistors. Although this work has demonstrated good functionality, further exploration is warranted by incorporating additional parameters for optimization, such as operating frequency and transistor type, and considering the potential utilization of coupled inductance.

5 Bibliography

References

[1] J. W. Kwak and D. Brian Ma, "Comparative topology and power loss analysis on 48v-to-1v direct step-down non-isolated DC-DC switched-mode power converters," in *2020 IEEE Energy Conversion Congress and Exposition (ECCE)*, Detroit, MI, USA: IEEE, Oct. 11, 2020, pp. 943–949. DOI: 10.1109/ECCE44975.2020.9235451.

[2] Yi-Feng Wu, D. Kapolnek, J. Ibbetson, P. Parikh, B. Keller, and U. Mishra, "Very-high power density AlGaN/GaN HEMTs," *IEEE Transactions on Electron Devices*, vol. 48, no. 3, pp. 586–590, Mar. 2001. DOI: 10.1109/16.906455.

[3] A. Lidow, J. Strydom, M. D. Rooij, and D. Reusch, "GaN transistors for efficient power conversion,"

[4] M. Chen, S. Jiang, J. A. Cobos, and B. Lehman, "Design considerations for 48-v VRM: Architecture, magnetics, and performance tradeoffs," in *2023 Fourth International Symposium on 3D Power Electronics Integration and Manufacturing (3D-PEIM)*, Miami, FL, USA: IEEE, Feb. 1, 2023, pp. 1–9. DOI: 10.1109/3D-PEIM55914.2023. 10052608.

[5] J. Baek, Y. Elasser, K. Radhakrishnan, H. Gan, J. P. Douglas, *et al.*, "Vertical stacked LEGO-PoL CPU voltage regulator," *IEEE Transactions on Power Electronics*, vol. 37, no. 6, pp. 6305–6322, Jun. 2022. DOI: 10.1109/TPEL.2021.3135386.

[6] K. Nishijima, K. Harada, T. Nakano, T. Nabeshima, and T. Sato, "A double step-down two-phase buck converter for VRM," in *2005 European Conference on Power Electronics and Applications*, Dresden, Germany: IEEE, 2005, 8 pp.–P.8. DOI: 10.1109/ EPE.2005.219347.

[7] P. S. Shenoy, O. Lazaro, M. Amaro, R. Ramani, W. Wiktor, *et al.*, "Automatic current sharing mechanism in the series capacitor buck converter," in *2015 IEEE Energy Conversion Congress and Exposition (ECCE)*, Montreal, QC, Canada: IEEE,

Sep. 2015, pp. 2003–2009. DOI: 10.1109/ECCE. 2015.7309943.

[8] R. Das and H.-P. Le, "A regulated 48v-to-1v/100a 90.9%-efficient hybrid converter for POL applications in data centers and telecommunication systems," in *2019 IEEE Applied Power Electronics Conference and Exposition (APEC)*, Anaheim, CA, USA: IEEE, Mar. 2019, pp. 1997–2001. DOI: 10.1109/APEC.2019.8722246.

[9] M. Halamicek, T. McRae, and A. Prodic, "Cross-coupled series-capacitor quadruple step-down buck converter," in *2020 IEEE Applied Power Electronics Conference and Exposition (APEC)*, New Orleans, LA, USA: IEEE, Mar. 2020, pp. 1–6. DOI: 10.1109/APEC39645.2020.9124412.

[10] M. Koszel, P. Grzejszczak, B. Nowatkiewicz, and K. Wolski, "Design, analysis and comparison of si- and GaN-based DC-DC wide-input-voltage-RangeBuck-boost converters," *International Journal of Electronics and Telecommunications*, pp. 337–343, Dec. 1, 2020. DOI: 10.24425/ ijet.2021.135986.

[11] R. A. Mdanat, S. Saeed, R. Georgious, and J. Garcia, "Analytical power loss model for GaN transistors," in *2021 IEEE Vehicle Power and Propulsion Conference (VPPC)*, Gijon, Spain: IEEE, Oct. 2021, pp. 1–6. DOI: 10.1109/VPPC53923.2021. 9699229.

[12] K. Peng, S. Eskandari, and E. Santi, "Characterization and modeling of a gallium nitride power HEMT," *IEEE Transactions on Industry Applications*, vol. 52, no. 6, pp. 4965–4975, Nov. 2016. DOI: 10.1109/TIA.2016.2587766.

[13] A. Dahiya and S. Sangwan, "Literature review on genetic algorithm," vol. 05, no. 16, 2018.

[14] K. De Jong, "Learning with genetic algorithms: An overview," *Machine Learning*, vol. 3, no. 2, pp. 121–138, Oct. 1988. DOI: 10.1007/ BF00113894.

[15] J. Blank and K. Deb, "Pymoo: Multi-objective optimization in python," *IEEE Access*, vol. 8, pp. 89 497–89 509, 2020. DOI: 10.1109/ACCESS. 2020.2990567.

[16] K. Wang, L. Wang, X. Yang, X. Zeng, W. Chen, and H. Li, "A multiloop method for minimization of parasitic inductance in GaN-based high-frequency DC–DC converter," *IEEE Transactions on Power Electronics*, vol. 32, no. 6, pp. 4728–4740, Jun. 2017. DOI: 10.1109/TPEL.2016. 2597183.

[17] L. Pace, N. Idir, T. Duquesne, and J.-C. De Jaeger, "Parasitic loop inductances reduction in the PCB layout in GaN-based power converters using s-parameters and EM simulations," *Energies*, vol. 14, no. 5, p. 1495, Mar. 9, 2021. DOI: 10.3390/en14051495.

PCIM Europe 2024, 11– 13 June 2024, Nuremberg   DOI: 10.30420/566262303

A SiC-Based 60kW LLC Converter with Novel Transformer Design for Improving Voltage Balance

Chen Wei, Zongzeng Hu, Jianlong Chen, Fulin Zhang
Wolfspeed, China

Corresponding author: Chen Wei, Frank.Wei@Wolfspeed.com
Speaker: Chen Wei, Frank.Wei@Wolfspeed.com

Abstract

This paper introduces a novel transformer design to address the voltage imbalance issue for an LLC converter with multiple transformers in series to support a wide output voltage range. The proposed transformer design is implemented on a digital controlled SiC-based 60kW prototype with a switching frequency of 125kHz-250kHz. Identical voltage balance is demonstrated in a three phase LLC converter with 200Vdc-1000Vdc wide output voltage range and exceeding 98.5% in peak efficiency for EV fast charger.

1 Introduction

With continuously improving batteries and consumers demanding quicker turnaround on charging, the required power rating of a single charging pile is increased from 100kW to 500kW or even higher. The design trend of the DC fast charging module is towards high power rating, high efficiency, wide output voltage range and wide constant power operation range. The typical output voltage range is from 200Vdc to 1000Vdc to cover different EVs.

In each of the EV charging power module, there is a PFC stage followed by an isolated DC/DC stage. LLC converter is very attractive for the isolated DC/DC stage due to its ZVS operation and high efficiency [1]-[2]. However, as the power rating increases, the current ripple on the output and input capacitors is so large. The size and cost of the input and output capacitors become impractical. With the natural current sharing and the input and output ripple current cancellation, the three-phase resonant LLC converter was proposed and studied in [3]-[5]. It is the best fit for high power DC/DC converter. On the other hand, the output voltage range of LLC converter is limited. 2 level DC/DC solution was studied for wide output voltage range in EV chargers [6]. The constant current operation is achieved for a wide output voltage range with the flexible control scheme. But it is not able to support the constant power in a wide output voltage range. To support the constant power operation with wide voltage range, the outputs of two independent LLC converters can be connected by three switches. The outputs of the converters can be connected in series for high output voltage or in parallel to support the larger output current with lower output voltage. Compared to a single converter, the number of the required PWM outputs, the gate drivers, the resonant tanks and the current sensors have to be doubled. The cost is higher.

To simplify the control and reduce the cost, based on 1200V rated SiC MOSFETs, single converter solution with two transformers was studied in [7]. The current sharing is good with primary-series–secondary-parallel for the converter. However, the voltage difference can be large with the conventional transformer design in the primary-series–secondary-series mode. This has a significant impact on the selection of the output rectifiers, output filtering capacitors and the design of the power transformers.

This work will systematically investigate factors affecting the voltage balance. A SiC MOSFET-based digital controlled 60kW 2 level three-phase interleaved LLC resonant converter is designed. A novel transformer design is proposed and studied to achieve wide output voltage range. The experimental results for the converter manifest both high efficiency, high power density and good voltage balance at wide output voltage range.

2 The Specifications and Architecture of the DC/DC

2.1 Specifications

2146

The specifications are listed in Table 1. A resonant frequency of 180kHz was selected as a trade-off between efficiency and power density. The output voltage range is from 200V to 1000V while the DC link voltage range is from 650V to 870V. The converter supports wide voltage range from 300V to 1000V in constant power mode. The maximum output power is 60kW. The max output current rating is 200A. The target peak DCDC efficiency is above 98%. And the target efficiency is above 97% for full load. Forced air cooling is applied to the design.

DC Input Voltage	650Vdc-870Vdc
Battery side Voltage	200Vdc-1000Vdc
Rated Power	60kW Vout>=300Vdc; I_out_Max=200A
Peak Efficiency	> 98% at half load and > 97% at full load @300Vdc

Table 1: Specifications of the 60kW DC/DC converter

2.2 Topology

Figure.1 shows the block diagram of the 60kW three-phase interleaved LLC DC/DC converter. The input is isolated from the output through three high frequency transformers. Each transformer has one primary winding and two secondary windings. At primary side, there are three half-bridges based on 1200V SiC MOSFETs and three sets of resonant tanks. At secondary side, there are full-bridge rectifiers connected to the secondary windings of the transformers. The two outputs can be connected in series or in parallel through the configuration of S1~S3.

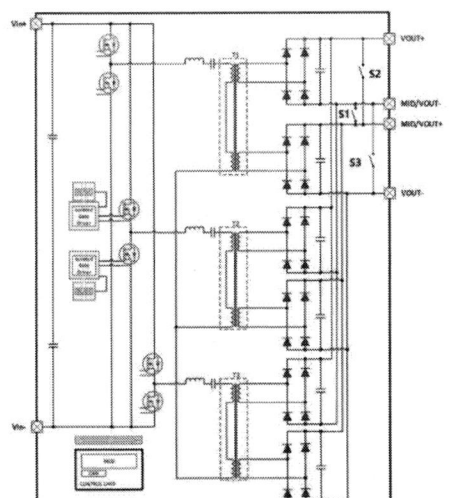

Fig. 1: Block diagrams of the 60kW DC/DC converter

2.3 Power Semiconductors and Resonant Frequency Selection

The maximum DC-link voltage is 870Vdc. The battery voltage is up to 1000Vdc. SiC MOSFET C3M0040120K 1200V 40mohm in TO-247-4L package is selected for the half-bridges at the primary side. SiC Schottky Diodes C6D20065D 650V 20A is selected for full-bridge rectifiers at the secondary side. The total usage is 12pcs for C3M0040120K. And it is 24pcs for C6D20065D in the design.

To get high efficiency and high power density, it is a trade-off to select the resonant frequency. 180kHz is selected for the resonant frequency.

Key design parameters are shown in Table 2.

Resonant frequency	180 kHz	Resonant choke primary	7 uH
Minimum switching frequency	125 kHz	Resonant cap primary	108 nF
Maximum switching frequency	250 kHz	Magnetizing inductance	30 uH

Table 2: Key design parameters.

3 The Output Voltage Balance Issue, the Factors Affecting the Voltage Balance and the Impact

3.1 The Output Voltage Balance

For the 60kW three-phase LLC converter, as shown in Figure 1, the primary windings are connected in series and then connected to the resonant tank. In this way, when the secondary side is connected in parallel, the secondary windings can naturally achieve current sharing. However, when the secondary side is connected in series, these two transformers may encounter a voltage imbalance issue. At primary side, the two outputs share the same primary power switches and the resonant tank. There is no difference. At output, the voltage drop on the output rectifiers can be different due to the device to device difference. But the voltage drop difference about ±0.3V. It is very small compared to the output voltage. The parasitic inductance of the output commutation loop is also part of the equivalent resonant circuit. It can also impact the voltage gain and output voltage. But with proper PCB design, the parasitic inductance can be managed very well. So the voltage balance mainly depends on the symmetry of the relevant parameters of the two transformers for each phase, including the coupling of the windings and the magnetizing inductance.

3.2 The Simulation Results and the Test Results of the Conventional Design

Considering the yield rate and the cost of the power transformers based on litz wire windings, the allowed coupling between primary and secondary windings can vary from 0.97 to 0.99. Refer to the simulation result in Fig. 2. With the coupling of 0.99 for the high side and 0.97 for the low side, the voltage difference between the two outputs is about 34V at 800V output.

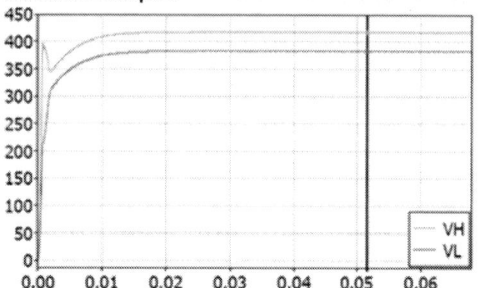

Fig. 2: Simulation results of the output voltage difference for the coupling mismatch.

Considering the yield rate and the cost of the power transformers, the achievable tolerance of the magnetizing inductance is about ±7%. Refer to the simulation result in Fig. 3, set the 15uH magnetizing inductance at upper limit for the high side and at the lower limit for the low side, the voltage difference between the two outputs is about 50V at 800V output.

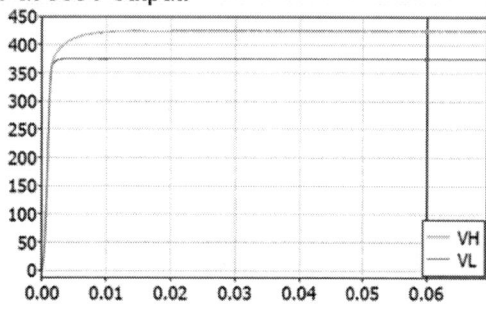

Fig. 3: Simulation results of the output voltage difference for the magnetizing inductance mismatch

Combine the coupling and magnetizing inductance mismatch in the simulation, the voltage difference

between the high side and low side is about 78V at full load as shown in Fig. 4.

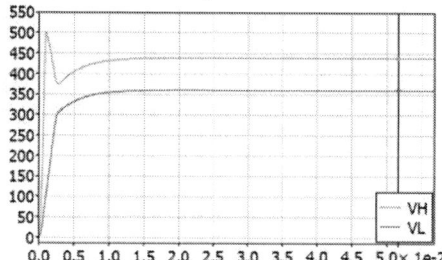

Fig. 4: Simulation results of the output voltage difference with the coupling and magnetizing inductance mismatch.

3.3 The Test Results and the Impact

As shown in Fig. 5, a digital controlled 60kW DC/DC prototype including the DC/DC power board, the control board, and the auxiliary power board is built to verify the simulation.

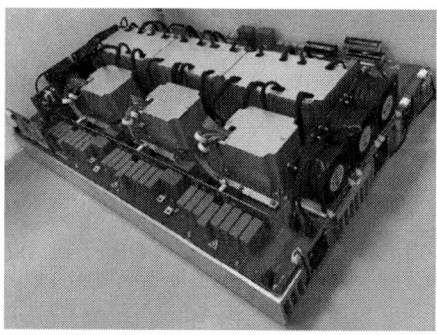

Fig. 5: Photo of the 60kW DC/DC prototype

As shown in Table. 3, the measured voltage difference between high and low side outputs is up to 71 V for 800V output in series mode.

Test result			
Output	Load	VH	VL
500V	20%	240V	262V
	50%	228.6V	271.9V
	100%	222.3V	278.6V
800V	20%	383.6V	416.9V
	50%	377V	422.7V
	100%	364.7V	435.3V

Table. 3: Test results of output voltage difference with the coupling and magnetizing inductance mismatch.

We didn't perform the test at 1000V output. Because the output voltage of each output is up to 500V while the 650V output rectifiers were selected in the design. It is ok on the voltage stress but without too much margin if the voltage sharing between the two outputs is good. With the voltage difference, they can be over voltage stress. If we change the output rectifiers to 1200V devices, the system level performance will be impacted since the voltage drop of the 1200V rectifiers is higher and the parasitic capacitor is larger. It will also impact the overall cost. In addition, with the voltage difference, the power loss on the transformers are different. It will impact the thermal of the transformer and the efficiency of the converter.

4 The Proposed Solution and the Results

4.1 The Proposal and the Simulation Results

To get a better coupling between the two secondary windings of each transformer, as shown in Fig.6, a special winding coupling method is proposed to reduce the impact of the parameter tolerances of the two transformers on the voltage balance. First, for each transformer, the output winding can be split into two windings with good coupling. Two wires in parallel for the two windings. Then for each phase, the two transformers have the output windings cross coupled with each other. In this way, even with the tolerance on the transformer, with the coupling of 0.99 for the high side and 0.97 for the low side, magnetizing inductance 15uH at upper limit for the high side and at the lower limit for the low side, refer to the simulation results as shown in Fig.7, the two outputs match with each other very well.

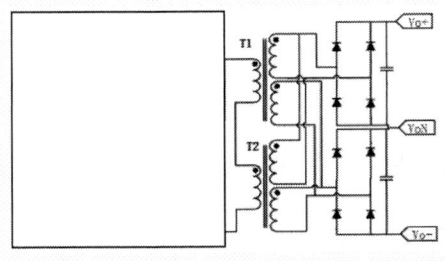

Fig. 6: The proposed solution for one of the three phases.

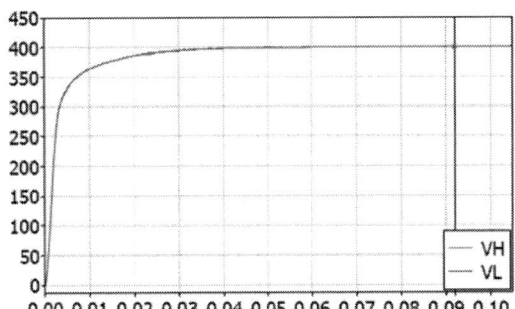

Fig. 7: Simulation results of the proposed solution with the coupling and magnetizing inductance mismatch.

4.2 Test Results

With the proposed transformer design, the testing waveform for 500Vdc output, 800Vdc output and 1000Vdc output are shown in Fig. 8, Fig. 9 and Fig. 10. The two outputs match with each other very well.

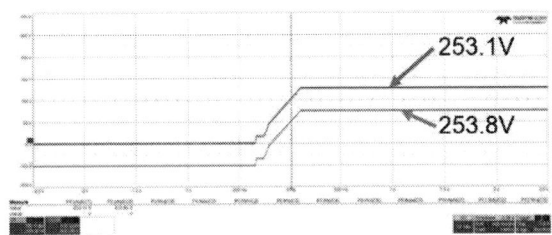

Fig. 8: Output waveforms of the high low side during start-up 500V output.

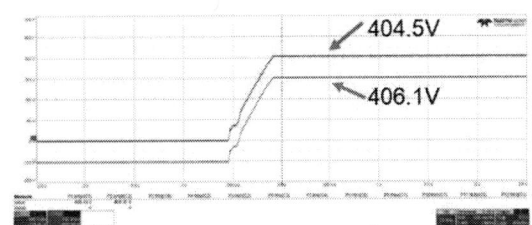

Fig. 9: Output waveforms of the high low side during start-up 800V output.

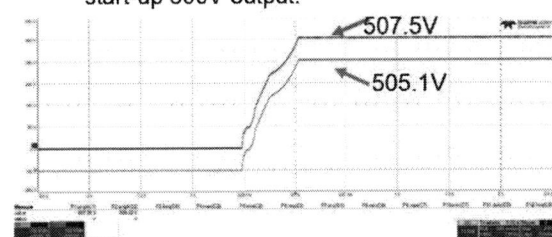

Fig. 10: Output waveforms of the high low side during start-up 1000V output.

As shown in Table. 4, the measured output voltage of the high and low side outputs also match with each other very well.

Test result			
Output	Load	VH	VL
500V	20%	250V	249.8V
	50%	250V	250V
	100%	253.1V	253.8V
800V	20%	401.7V	400.4V
	50%	403.2V	402.1V
	100%	404.5V	406.1V
1000V	20%	506.8V	505.7V
	50%	506.3V	505.5V
	100%	507.5V	505.1V

Table. 4: Test results of output voltage with the proposed transformer design.

The testing waveform for 300Vdc output voltage and 720Vdc input is shown in Fig.11. ZVS is achieved for all three half-bridges.

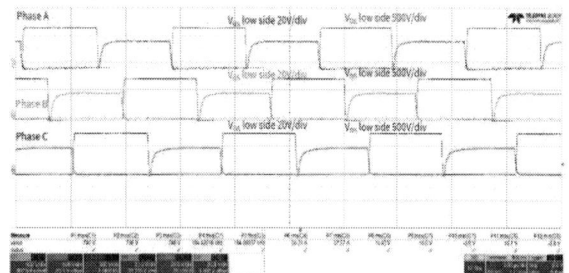

Fig. 11: Vgs and Vds waveforms

As shown in Fig.12, The current sharing can be achieved among the three resonant tanks.

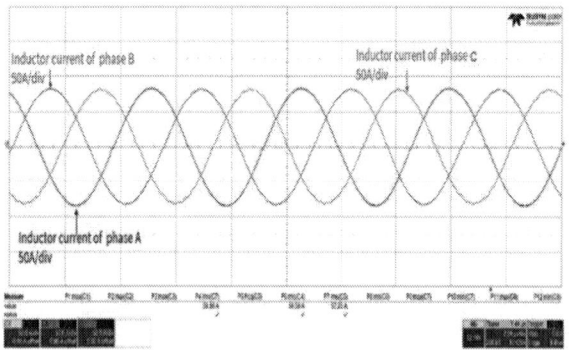

Fig. 12: Current waveform of resonant tanks

4.3 Efficiency Test Result

The efficiency of the 60kW DC/DC converter at different test conditions is shown in Fig.13. Input voltage and output voltage are marked in the curve. With the proposed flexible control scheme, for 300V output and above, above 98% peak efficiency and above 97% full load efficiency is demonstrated by the prototype with a switching frequency range 125-250kHz.

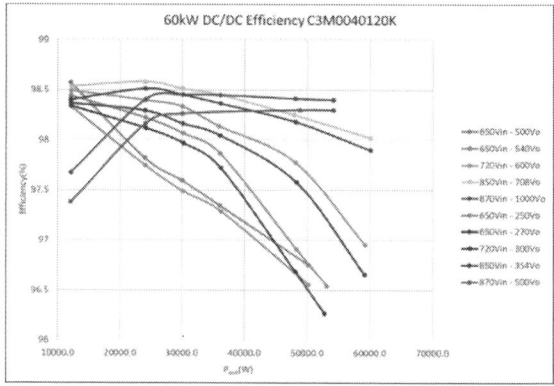

Fig. 13: Efficiency Curve of the 60KW DC/DC Converter

4.4 Thermal Test Result

In the thermal test of the prototype, forced air cooling was applied to the board using the attached heatsinks. T-type thermal couplers and KEYSIGHT 34972A acquisition unit are used to measure the case temperature of components.

Description	Rth j-c (c/w)	Calculated Power loss(watts)	Measured Case Temp. (°C)	Calculated Junction Temp. (°C)
Input: 730Vdc, Output: 300Vdc 60KW 200A				
MOSFET	0.46	33.2	96.6	111.9
Input: 650Vdc, Output: 200Vdc 130A				
MOSFET	0.46	47.5	101.6	123.4
Input: 870Vdc, Output: 500Vdc 60KW				
MOSFET	0.46	31	95.8	110.1

Table 5: Thermal Test Results.

The thermal test results are shown in Table 5. The junction temperature is calculated based on the measured case temperature, thermal resistance of the MOSFET and calculated component power loss. The maximum junction temperature of C3M0040120K is 175°C in the datasheet.

Referring to the test result, the max junction temperature of the SiC MOSFET is 123.4°C in the application. We conclude that the SiC MOSFETs meet the thermal de-rating requirement in the design.

5 Summary

In this paper, A novel transformer design is proposed and verified on a digital controlled SiC-based 60kW prototype to address the voltage imbalance issue for LLC converter with multiple transformers. Identical voltage balance is achieved. And the SiC-based prototype with a switching frequency of 125kHz-250kHz is demonstrated with 200Vdc-1000Vdc output voltage range and exceeding 98.5% in peak efficiency. It is very useful for the high-power LLC converter with wide voltage range such as EV charger applications.

6 References:

[1] B. Yang, F. C. Lee, A. J. Zhang, and G. Huang, "LLC resonant converter for front end DC/DC conversion," in Proc. Appl. Power Electron. Conf. and Expo.(APEC '02), 2002, pp. 1108-1112 vol.2.

[2] B. Lu, W. Liu, Y. Liang, F. C. Lee, and J. D. van Wyk, "Optimal design methodology for LLC resonant converter," in Proc. Appl. Power Electron. Conf. and Expo.(APEC '06), 2006, p. 6 pp.

[3] E. Orietti, P. Mattavelli, G. Spiazzi, C. Adragna and G. Gattavari, Current sharing in three-phase LLC interleaved resonant converter, IEEE Energy Conversion Congressand Exposition, SanJose, CA,2009.

[4] Yusuke Nakakohara, Hirotaka Otake, Tristan M. Evans, Tomohiko Yoshida, Mamoru Tsuruya, and KenNakahara, Three-Phase LLC Series Resonant DC/DC Converter Using SiC MOSFETs to Realize HighVoltage and High-Frequency Operation, IEEE Trans. on Industrial Electronics, vol. 63, no. 4, pp. 2103 - 2110, April 2016.

[5] Ho-Sung Kim, Ju-Won Baek, Myung-Hyu Ryu, Jong-Hyun Kim, JeeHoon Jung, The High-Efficiency Isolated ACDC Converter Using the Three-Phase Interleaved LLC Resonant Converter Employing the Y Connected Rectifier, IEEE Trans. on Power Electronics, vol. 29, no. 8, pp. 4017 - 4028, August 2014.

[6] Chen Wei, Dongfeng Zhu, Haitao Xie, Ying Liu, Jianwen Shao, "A SiC-Based 22kW Bi-directional CLLC Resonant Converter with Flexible Voltage Gain Control Scheme for EV On-Board Charger," in Proc. PCIM, 2020

[7] Chen Wei, Zongzeng Hu, Jianlong Chen, Fulin Zhang, Haiming Zhan, Anuj Narain, "A SiC Based 60kW Three Phases Interleaved LLC Converter for EV Fast Charger," in Proc. PCIM Europe, 2023

PCIM Europe 2024, 11–13 June 2024, Nuremberg   DOI: 10.30420/566262304

Analysis of Inverter Operation Modes of an IGBT-Based ZCS LLC Converter for a 2 kW Automotive On-Board DC-DC

Daniel Urbaneck, Jan Wiegard, Frank Schafmeister

Paderborn University, Paderborn, Germany

Corresponding author: Daniel Urbaneck, urbaneck@lea.uni-paderborn.de

Abstract

State-of-the-art LLC resonant converters use MOSFETs in their inverter stage, allowing high switching frequencies and thus the use of compact magnetic components. The large parasitic output capacitance and the poor reverse-recovery behaviour of the inherent body diode of high-voltage (600 V) silicon MOSFETs require soft switching, i.e. zero-voltage switching (ZVS) in half- and full-bridge configurations. Otherwise, the high turn-on switching losses would lead to excessive heating and the destruction of the switch. Therefore, MOSFET-based LLC converters are operated in the so-called *inductive* region only, which enables ZVS. The use of robust and cost-effective IGBTs instead of MOSFETs is particularly advantageous for automotive applications, since in addition to high reliability low costs are an important objective here. Since IGBTs are characterized by dominant turn-off losses and generally higher switching losses compared to MOSFETs, the aim is to operate them with zero-current switching (ZCS) and at lower overall switching frequencies. In the *capacitive* operation region of the LLC converter, both requirements are fulfilled. In this region also, the voltage transfer characteristic is steeper, which qualifies for applications with a strongly varying input-to-output voltage ratio, such as given for automotive on-board DC-DC converters connecting the high-voltage traction battery with the 12 V auxiliary battery. In this paper, a stress value analysis based on a switched-model simulation is used to design a ZCS LLC converter and take advantage of the mentioned benefits of IGBTs as well as of the steeper voltage transfer characteristic. A full-bridge inverter stage is employed, which can be operated in full-bridge, half-bridge or phase-shifted full-bridge mode. The efficiency of MOSFET-based ZVS LLC converter can be increased for applications with strongly varying voltage transfer ratios (as described above), by applying the mentioned inverter operation modes. However, the analysis of loss mechanisms of the IGBT-based ZCS LLC shows that the simple and robust full-bridge operation is globally the most efficient operating mode here. The main cause are the higher switching losses of the inverter stage in half-bridge and phase-shifted full-bridge operation mode.

1 Introduction

Due to their high efficiency, LLC resonant converters **(Fig. 2)** are used in many applications where high power-density is required, e.g. automotive on-board DC-DC converters. State-of-the-art LLC resonant converters employ 600 V power MOSFETs in their primary-side inverter stage. In half-bridge configurations, as required for the inverter stage, those MOSFETs have dominant turn-on losses due to their large parasitic output capacitance combined with poor reverse-recovery behaviour

of their intrinsic body diodes under hard-switching conditions [1]. Therefore, continuous operation with hard (turn-on) commutation would lead to excessive heating of the MOSFET and finally to its destruction. For this reason, power MOSFETs in such inverter stages are operated in zero-voltage-switching mode (ZVS), in which the voltage across the drain-source path is already about zero at the turn-on process. In order to enable ZVS, the LLC converter is operated in the so-called *inductive* region only [2],[3], cf. **Fig. 1**. Compared to MOSFETs, IGBTs have advantages such as a higher robustness against overload and lower costs at the same power rating [4]. On the other hand, IGBTs show dominant turn-off losses due to their large current tail during the turn-off process [5]. At ZCS, the current through the IGBT is already zero when it is forced to turn off. Therefore, ZCS is the preferred operation mode for this type of power switch. In order to take advantage of IGBTs in an LLC converter, the converter has to be operated in the *capacitive* region, cf. **Fig. 1**. This results in a remarkable characteristic of an ZCS-operated LLC: The steeper transfer characteristic at light up to medium loads leads to a significantly smaller variation of the switching frequency as well as a generally lower switching frequency [6]. Despite the more lossy switching behaviour of IGBTs compared to MOSFETs, the ZCS LLC achieves almost identical efficiency levels than a ZVS LLC for applications with a strongly varying voltage transfer ratio by utilizing the steep transfer characteristic [7]. While a MOSFET-based benchmark converter system employs advanced combined modulation strategies [12] the IGBT-based LLC is simply frequency-modulated only. This work analyzes the distribution of the power losses across the inverter stage, resonant circuit, transformer and rectification stage of a ZCS LLC converter. The influence of three different inverter operation modes (full-bridge, half-bridge and phase-shifted full-bridge) is then investigated and the loss-minimizing mode is presented. A converter is designed using a switched-model simulation. Finally, the simulative loss-modelling results are verified based on a 2 kW prototype of an on-board DC-DC converter for electric vehicles.

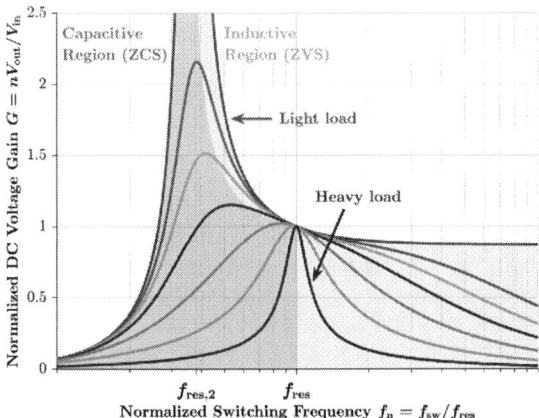

Fig. 1: Voltage transfer characteristic of the LLC's resonant circuit based on the fundamental harmonic approximation (FHA).

2   System Analysis

Electric vehicles typically employ two batteries, a traction and an auxiliary battery with a nominal voltage of about 400 V and 14 V, respectively. To charge the auxiliary battery out of the traction battery, an on-board DC-DC converter is required. Due to its high efficiency and power density, the LLC resonant converter is a suitable candidate for this task (cf. **Fig. 2**).

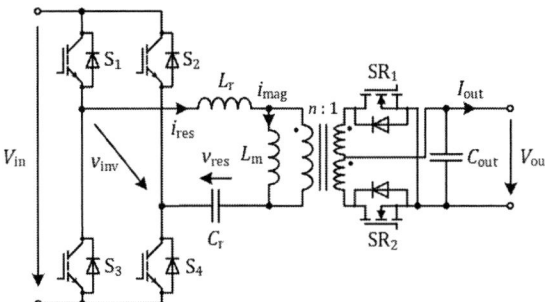

Fig. 2: LLC resonant converter with IGBT-based inverter stage, center-tapped transformer and synchronous rectifier.

Table 1: Characteristic operation points of the on-board DC-DC converter.

Operation Point	V_{in}/V	V_{out}/V	I_{out}/A
OP_{Gmin}	420	8	140
OP_{Gmax}	200	16	140
OP_{Gmid1}	420	11	140
OP_{Gmid2}	420	11	70

The major challenge of the described application is the strongly varying DC voltage transfer ratio between input and output voltage. Both voltages vary depending on the states of charge of the two batteries. Several approaches are presented in [8] to achieve high efficiencies over a wide voltage transfer ratio by utilizing morphing techniques. However, due to the steep gain characteristic (**Fig. 1**), a ZCS LLC can also achieve a wide variation of the transfer ratio by standard frequency modulation only. The minimum voltages of the traction battery and the auxiliary battery are 200 V and 8 V, respectively. The end-of-charge voltages are 420 V and 16 V, respectively. The maximum charging current is 140 A. Based on this data, the corner operation points can be defined according to **Table 1**. The converter is designed based on the minimum (OP_{Gmin}), maximum (OP_{Gmax}) and average (OP_{Gmid1}) voltage transfer ratio at full load as well as on a light-load operation point (OP_{Gmid2}).

2.1 Fundamental Harmonic Approximation vs. Simulation

The fundamental harmonic approximation (FHA) is often used to design the LLC resonant converter for inductive operation mode. The method approximates the voltages and currents of the resonant circuit assuming only a sinusoidal voltage v_{inv} excitation by the inverter stage. The analysis of the converter can then be done with the complex AC circuit analysis. By applying the FHA, the following equations are obtained [9]:

$$\lambda = \frac{L_r}{L_m}, \qquad Z = \sqrt{\frac{L_r}{C_r}}, \qquad f_{\mathrm{res}} = \frac{1}{2\pi\sqrt{L_r C_r}},$$
$$Q = \frac{Z}{R_{\mathrm{AC}}}, \qquad R_{\mathrm{AC}} = \frac{8}{\pi^2} n^2 R_0, \qquad R_0 = \frac{V_{\mathrm{out}}}{I_{\mathrm{out}}}. \qquad (1)$$

The gain of the resonant tank results as

$$G(\lambda, Q, f_n) = \frac{1}{\sqrt{\left(1 + \lambda - \frac{\lambda}{f_n^2}\right)^2 + Q^2\left(f_n - \frac{1}{f_n}\right)^2}} \qquad (2)$$

and is shown in **Fig. 1** for different values of Q. The voltage transfer characteristic of the LLC converter is commonly separated into a capacitive (i_{res} leads v_{inv}) and an inductive region (i_{res} lags behind v_{inv}). If the inverter voltage v_{inv} and the resonant current i_{res} are sinusoidal, capacitive behaviour enables zero-current switching (ZCS) and inductive behaviour enables zero-voltage (ZVS) of the inverter stage. The further the switching frequency f_{sw} deviates from the resonant frequency f_{res}, the more the resonant current distorts, so that capacitive behaviour no longer necessarily leads to ZCS.

$$f_{\mathrm{res}} = \frac{1}{2\pi\sqrt{L_r C_r}} > f_{\mathrm{res,2}} = \frac{1}{2\pi\sqrt{(L_r + L_m)C_r}} \qquad (3)$$

Since it is not the behaviour of the resonant circuit but the conditions at the switches of the inverter that are decisive for the design of the IGBT-based converter, the characterization as capacitive or inductive is unsuitable. For this reason, in contrast to the usual characterization of *capacitive* and *inductive* operation, as it is common with FHA, ZCS and ZVS are used in the following, as these terms describe the switching behaviour of the inverter more accurately. In [6],[7] it was shown that the FHA is not a suitable method for designing the LLC converter for ZCS operation. A simulation of a switched model of

the LLC converter ($n = 16$, $\lambda = 0.75$, $Z = 23$, $f_{\mathrm{res}} = 200$ kHz) shows that the voltage transfer characteristic for rising switching frequencies below $f_{\mathrm{res,2}}$ is not a monotonically increasing function as expected from the FHA. In fact, this area shows several local minima and maxima (**Fig. 3**). Based on the simulation data, it can be shown that within each section with positive slope ZCS is possible. Similarly, ZVS is possible in each section with negative slope. Therefore, the in the FHA as *capacitive* referred region actually consists of several ZCS and ZVS areas.

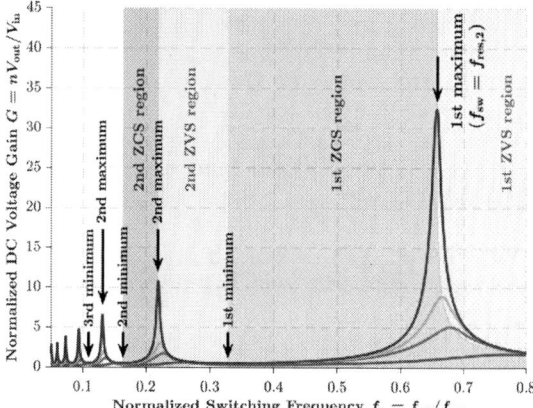

Fig. 3: LLC voltage transfer characteristic of a switched-model simulation with different loads. Several peaks below $f_{\mathrm{res,2}}$ are visible, which do not appear in the FHA.

2.2 Simulation-Based Converter Design

The comparison between the results from the FHA and the switched-model simulation show significant differences. The modelling of the sub-resonant operation region as a monotonically increasing function by the FHA has been shown to be wrong [6]. Therefore, a simulation-based design method with a switched model is presented. According to Eq. (1), the components of the resonant circuit C_r, L_r and L_m can be obtained by choosing suitable values for λ, Z and f_{res}. The stress values of the converter components (e.g. inverter turn-on current levels, resonant-capacitor voltages, etc.) are independent of the resonant frequency f_{res} [11]. Thus, the component stress can be analysed using only two variables (λ and Z) and visualized in a 3D-representation such as a contour plot [12]. An appropriate value for f_{res} can be then determined in a later design step. For the following design procedure, a data set of $\lambda = 0.2 \ldots 0.8$, $Z = 10 \ldots 30\ \Omega$ and $n = 12 \ldots 18$ is chosen.

2.2.1 Transformer Turns Ratio

The first design step is to determine an appropriate turns ratio n of the transformer. The influence of n is shown in **Fig. 4** and **Fig. 5**. For the operation point $\mathrm{OP_{Gmin}}$ and $n = 13$ a steep gradient between two areas of switching frequencies is visible (transition from orange to blue).

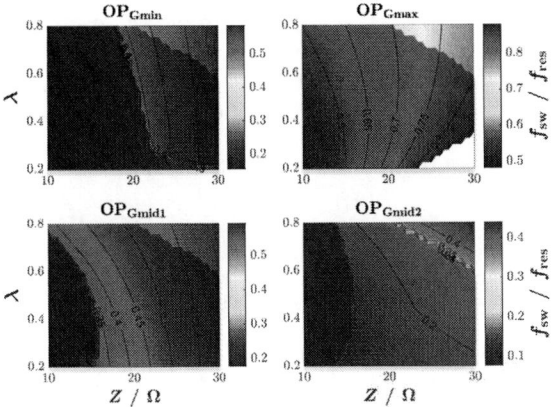

Fig. 4: Simulation-based normalized switching frequency $f_\mathrm{n} = f_\mathrm{sw}/f_\mathrm{res}$ of each operation point for the selected λ/Z-set and $\boldsymbol{n = 13}$. $\mathrm{OP_{Gmax}}$ is achievable in the first ZCS region only (no discontinuous changes of f_n). All other operation points are located in several ZCS regions depending on the λ/Z-set. Grey area: Operation point not enabled by λ/Z-set in first ZCS region.

Fig. 5: Simulation-based normalized switching frequency $f_\mathrm{n} = f_\mathrm{sw}/f_\mathrm{res}$ of each operation point for the selected λ/Z-set and $\boldsymbol{n = 16}$. Grey area: λ/Z-set does not enable operation in the first ZCS region. Compared to $n = 13$ the usable λ/Z-set is clearly enlarged. Blue star mark: chosen design for ZCS-operated LLC converter.

At the border of these areas, the frequency f_n changes from 0.25 to 0.45 only by a small variation of λ or Z. A similar behavior can be seen in $\mathrm{OP_{Gmid1}}$. Depending on the λ/Z-combination these operation points are located in the first or the second ZCS region (cf. **Fig. 3**). $\mathrm{OP_{Gmid2}}$ shows two gradients of discontinuous changes of the switching frequency meaning that the operation point is located at three different ZCS regions. In contrast, at $\mathrm{OP_{Gmax}}$ the switching frequency is continuous over λ and Z and therefore the operation point is located in the first ZCS region for all λ/Z-combinations. If the turns ratio is increased to $n = 16$, the results are similar, but the λ/Z-set with which all four operation points are located within the first ZCS region increases. White areas in **Fig. 4** and **Fig. 5** mean that the operation point is not reachable at all. The influence of the turns ratio to the ZCS region in which an operation point is reachable is shown in detail in **Fig. 6**.

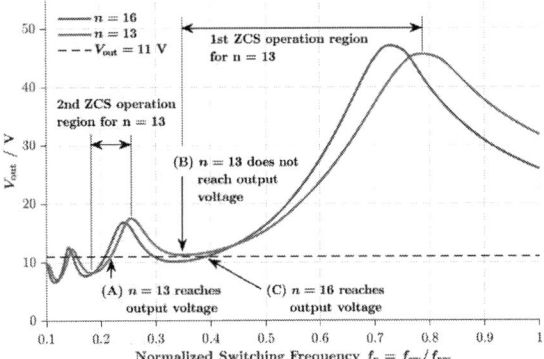

Fig. 6: Influence of the turns ratio n to the gain characteristic: By decreasing the value of n the minimum achievable output voltage of the first ZCS operation region increases.

For $n = 16$, an exemplary operation point with $V_\mathrm{out} = 11\,\mathrm{V}$ is reached at a normalized switching frequency of around $f_\mathrm{n} = 0.4$ and located in the first ZCS operation region (cf. **Fig. 6: (C)**). To reach the same operation point with $n = 13$, the second ZCS region has to be utilized (cf. **Fig. 6 (A)**) because within the first ZCS region the required output voltage of $11\,\mathrm{V}$ is not achievable (cf. **Fig. 6 (B)**). The converter parameters should be chosen such a way that all operation points can be reached in the same ZCS operation region to avoid discontinuous changes of the switching frequency. For $n = 13$ only $\mathrm{OP_{Gmid2}}$ defines the border of the λ/Z-area which enables the operation in the first ZCS region (cf. **Fig. 4**: non-grey area). To be more flexible in choosing values for λ and Z, this area can be enlarged by increasing n. **Fig. 5** visualizes the enlargement of the λ/Z-area by increasing n to 16. On the other hand, this increase of n reduces the usable λ/Z-area of OP_{Gmax} (cf. **Fig. 5**: white area in which OP_{Gmax} is not achievable expands). Therefore, a turns ratio which leads to a large λ/Z-set, valid for the utilization of the first ZCS operation region for *all* operation points, is given with $n = 16$.

2.2.2 Resonant Tank

To design the resonant circuit (C_{r}, L_{r}, L_{m}), a parameter set has to be selected for λ, Z and f_{res}. The most important component-stress values of the LLC converter are shown in **Fig. 7**. The turn-on current $I_{\text{sw,on}}$ as well as the conduction-loss causing current $I_{\text{sw,RMS}}$ of the IGBTs are significant for the inverter stage losses. The maximum resonant voltage $V_{\text{res,max}}$ determines the voltage rating of the resonant capacitors. On the secondary side, the current through the MOSFETs $I_{\text{rect,RMS}}$ is determining the synchronous-rectifier losses.

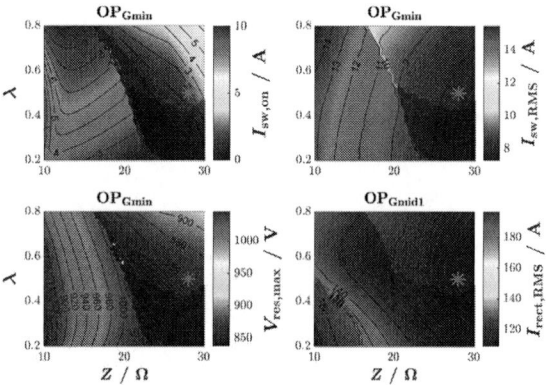

Fig. 7: Characteristic stress values of the LLC circuit components for $n = 16$: Low stress in the first ZCS region for large values of Z and small values of λ. Grey area: Operation points not reachable within the first ZCS region.

According to **Fig. 7**, λ should be chosen small to reduce switching and conduction losses of the IGBTs and the voltage stress of the resonant capacitors. Z should be chosen large to reduce rectifier losses. On the other hand, the values of λ and Z are restricted by the requirement that all operation points should belong to the first ZCS region (cf. **Fig. 3** and non-grey area in **Fig. 5**). To ensure operation exclusively in the first ZCS

region (including a safety margin to compensate for component deviations) and moreover to obtain low component stress values and losses, $\lambda = 0.5$ and $Z = 28\ \Omega$ are chosen as a suitable compromise (cf. star marking in **Fig. 5** and **Fig. 7**). The switching frequency f_{sw} in all operation points should be limited to about 70 to 80 kHz to keep the switching losses of the IGBTs at a moderate level. With the selected λ/Z-set the maximum normalized switching frequency is $f_{\text{n}} = 0.72$ (cf. **Fig. 5**: OP_{Gmax}). So, choosing a resonant frequency of $f_{\text{res}} = 100$ kHz results in a maximum switching frequency of $f_{\text{sw}} = 72$ kHz. Applying Eq. (1) the resonant components can be calculated as:

$$C_{\text{r}} = 56.8\ \text{nF},\ L_{\text{r}} = 44.6\ \mu\text{H},\ L_{\text{m}} = 89.1\ \mu\text{H}. \quad (4)$$

3 Inverter Operation Modes

The LLC resonant converter can be realized either with a half-bridge or a full-bridge inverter stage. Regardless of the inverter type, the output voltage is controlled via the switching frequency only while the duty cycle of the individual half-bridges is constant at $D = 0.5$. The design of the presented ZCS LLC is based on a full-bridge inverter stage as depicted in **Fig. 2**.

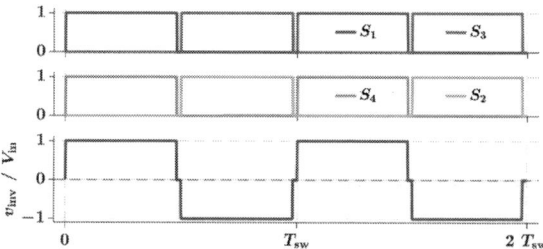

Fig. 8: Full-bridge inverter in full-bridge operation mode: Bipolar output voltage v_{inv} symmetrically around 0 V.

In (complementary) *full-bridge* operation mode, both half-bridges of the inverter are switched at the same time. This results in an inverter output voltage v_{inv} with both positive and negative amplitude equal to the input voltage V_{in} (**Fig. 8**). To avoid a bridge short-circuit, a switch is only switched on after a delay time (inter-locking time) after the switching partner has been switched off. This time is particularly important for slowly switching IGBTs. During the inter-locking time, v_{inv} is zero, when no inductively impressed inverter output current is assumed. While v_{inv} is non-zero, energy can be transmitted from the input via the resonant tank to the output of the converter. However, the full-bridge inverter can also be operated in *half-bridge* mode by disabling one of the two half-bridges, e.g. one switch is permanently turned on and the other one is turned off. The duty cycle of the actively pulsating half-bridge is still $D = 0.5$. The output voltage v_{inv} is unipolar and zero during half of the switching period (**Fig. 9**). During the zero-intervals, no energy is transmitted from the input to the output of the LLC. Therefore, the voltage gain is half the value of full-bridge operation.

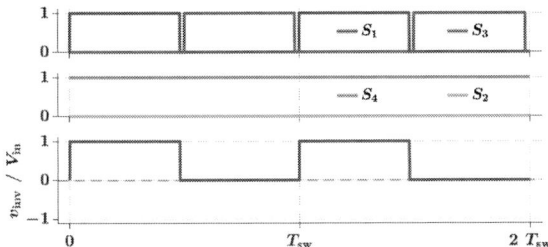

Fig. 9: Full-bridge inverter in half-bridge operation mode: Unipolar output voltage v_{inv} with only positive amplitude.

In full-bridge operation, both half-bridges can also be switched with a steered time delay to each other, resulting in *phase-shifted full-bridge* operation. The interval during which the output voltage of the inverter is zero (due to the interlocking time) can be extended by the phase shift time T_{PS}. The phase-shift angle can be defined as

$$\varphi = \frac{T_{PS}}{T_{sw}} \cdot 360°. \tag{5}$$

An angle of $\varphi = 0°$ corresponds to complementary full-bridge operation, while at $\varphi = 180°$ the output voltage v_{inv} is permanently $0\,V$ (**Fig. 10**). The phase-shift therefore

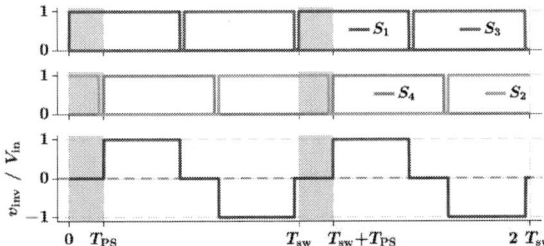

Fig. 10: Full-bridge inverter in phase-shifted operation mode: Intervals in which no energy is transmitted ($v_{inv} = 0\,V$) can be modulated by phase shift (red areas).

represents a further method to control the amount of energy injected into the resonant circuit. Thus, in contrast to half-bridge and complementary full-bridge operation, two degrees of freedom for controlling the output voltage V_{out} are enabled: switching frequency f_{sw} and phase-shift angle φ.

4 Loss Modeling

The influence of the inverter operation modes on the total LLC converter efficiency and loss distribution across the converter's components will be analysed in the following. The characteristic operation points (cf. **Table 1**) are simulated in a switched-model simulation using the *Plexim PLECS* software. The simulation data (currents, voltages, frequencies) are then analysed in a subsequent step with respect to the losses in the LLC's major stages inverter, resonant circuit, transformer and rectifier.

4.1 Inverter Stage

The inverter stage is built with 650 V automotive-qualified IGBTs with co-packed diodes Infineon AIKW40N65DF5. The gate voltage to control the IGBTs is 15 V and 0 V respectively. To determine the conduction losses, the output characteristic (IGBT) and forward characteristic (co-packed diode) are taken from the data sheet, given for a junction temperature of $T_j = 25\,°C$ and $T_j = 150\,°C$. Conduction losses for temperatures between these values are interpolated linearly. The switching losses of power semiconductors depend not only on current and voltage but also on the driver circuitry and commutation loop.

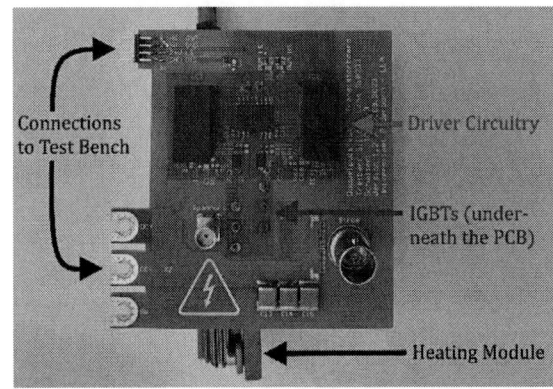

Fig. 11: Test board for characterizing the employed IGBT by double-pulse tests (DPT).

For this reason, a printed circuit board was designed to perform double-pulse tests (DPT) for characterization of the employed IGBT and its co-packed diode under real operation conditions (cf. **Fig. 11**). Several tests were performed at currents from 0.5 A to 20 A and voltages from 200 V to 420 V. In order to also model the influence of the temperature on the switching behaviour, the tests were performed at temperatures between 20 °C and 140 °C. Switching losses above 140 °C are extrapolated linearly.

4.2 Resonant Circuit

The resonant circuit consists of the resonant capacitance C_r, the resonant inductance L_r and the magnetizing inductance L_m. The resonant capacitance C_r (cf. **Fig. 2**) is realized with 12 1200 V C0G capacitors (KEMET CKC21C562JEGACAUTO) connected in parallel. The losses of this capacitor battery are modelled by the equivalent series resistance (ESR). The resonant inductor consists of an ETD 54/28/19 core made of N97 with several air gaps in the center lag in order to reduce fringing losses. The winding uses a 400 x 0.1 mm stranded wire with $n_r = 19$ turns. An ER54/28/38 core of the material DMR95N is used for the transformer. The primary winding consists of 600 x 0.07 mm stranded wire with $n = 16$ turns and the secondary windings consist each of four single-turned 11 mm x 1.5 mm copper sheets connected in parallel. The winding as well as the core losses are calculated using an open source 2D-Finite-Element-Method (FEM) simulation [13].

4.3 Rectifier Stage

The maximum output current of the LLC converter is 140 A (cf. **Table 1**). In order to keep the losses of the rectification stage low, a center-tapped transformer in combination with a synchronous rectifier (SR) is used. Three low-voltage MOSFETs Infineon IPT014N08NM5 are connected in parallel per rail resulting in a total $R_{DS,on} = 0.5$ mΩ at $T_j = 100$ °C. Synchronous rectification of a ZCS LLC resonant converter is challenging. In each SR rail, several current pulses occur within a single switching period, which also merge together depending on the operation point. To ensure proper rectification,

the concept presented in [10] is implemented, which proposes an active steering of the SR-MOSFET's gate signal during critical commutation states of the rectification process. Switching losses can therefore be neglected and only conduction losses caused by the MOSFET channel resistance $R_{DS,on}$ are considered. Based on the data sheet, a modelling for $R_{DS,on}$ as a function of the temperature T_j at a gate voltage of $u_{GS} = 12$ V is derived. Due to the high output current, also the conduction losses in the copper layers and connector terminals of the PCB as well as in the bus bar are considered.

5 Loss Analysis

A switched model of the ZCS LLC converter is used to simulate the characteristic operation points (cf. **Table 1**) in full-bridge, half-bridge and phase-shifted full-bridge operation for $\varphi = 5°; 10°; 15°; 20°$. The losses of the converter components are then calculated using the presented loss models and are shown in **Fig. 12**. In addition to full-bridge mode, all operation points can also be achieved in half-bridge as well as in phase-shifted full-bridge mode. One exception is OP_{Gmax}, which can not be achieved in half-bridge mode. This operation point is already close to the border to the first ZVS region during full-bridge operation (cf. **Fig. 3**).

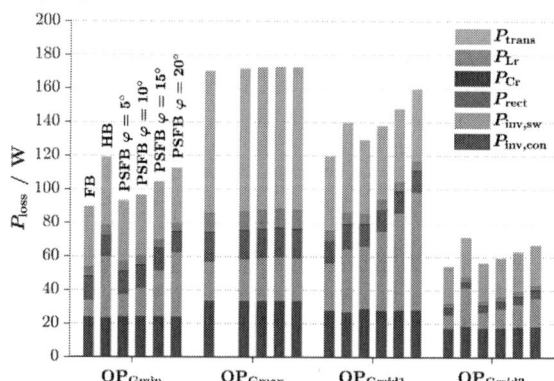

Fig. 12: Model-based power losses of the ZCS LLC converter in full-bridge (FB), half-bridge (HB) and phase-shifted full-bridge operation (PSFB). OP_{Gmax} cannot be achieved in half-bridge operation due to insufficient gain of the chosen design, which is optimized in order to reach OP_{Gmax} in full-bridge operation mode only.

In half-bridge mode, the voltage gain drops to half the value of full-bridge operation, resulting in a maximum output voltage of $V_{\mathrm{out}} \approx 6$ V. For all operation points, full-bridge operation results in lowest total losses and therefore highest efficiency. Half-bridge operation shows significantly higher switching losses in the inverter stage as well as higher losses in the transformer P_{trans}, while the remaining losses are at a similar level. The same applies in phase-shifted full-bridge operation: As the phase shift increases, also the switch-on currents of the IGBTs increase, especially those of the lagging half-bridge. Simultaneously, the switching frequency f_{sw} has to be increased to maintain the output voltage constant, as the intervals in which no energy is transferred from the converter input to the output is proportional to the phase-shift (cf. **Fig. 10**). The smaller switching losses $P_{\mathrm{inv,sw}}$ in full-bridge operation increase and eventually dominate the conduction losses $P_{\mathrm{inv,con}}$. Phase shifts of $\varphi > 20°$ were also analysed, but resulted in higher losses than for full-bridge operation as well. A 2 kW prototype of the ZCS-operated LLC converter (**Fig. 13**) was built up to evaluate the simulation-based loss analysis. An electronic load (EA ELR9980 − 340) serves as output load. The output current is measured via a precision shunt resistor (Burster 1282 − 0,001).

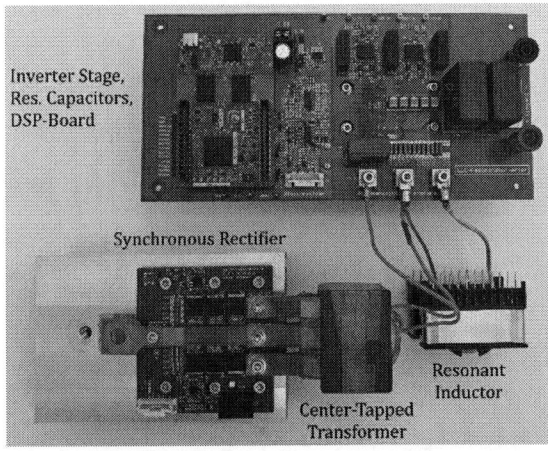

Fig. 13: Developed 2 kW LLC prototype: Mainboard with IGBT-based inverter (underneath PCB), resonant capacitors and attached DSP (digital signal processor) control board. Inductive components and SR-rectifier stage are separated from the mainboard.

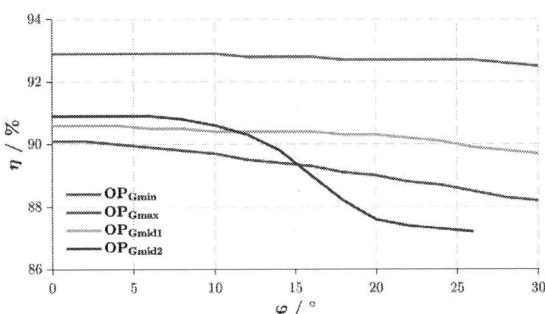

Fig. 14: Influence of the phase-shift angle φ on the total efficiency η of the developed prototype (measurements). OP_{Gmid2}: Due to distortion of the resonant current caused by the phase shift, the synchronous rectification operates properly only until $\varphi \leq 26°$.

Table 2: Comparison of the inverter operating modes of the developed prototype (measurements). Grey field: OP_{Gmax} not reachable in half-bridge mode.

Operation Point	η_{FB} %	η_{HB} %	$\eta_{\mathrm{PSFB,10°}}$ %	$\eta_{\mathrm{PSFB,20°}}$ %
OP_{Gmin}	90.1	84.3	89.7	89.0
OP_{Gmax}	92.9	-	92.9	92.7
OP_{Gmid1}	90.6	85.9	90.4	90.3
OP_{Gmid2}	90.9	89.4	90.6	87.6

Input and output power are determined via a power analyser (ZES LMG640). The results are shown in **Fig. 14** and **Table 2** and indicate that (complementary) full-bridge operation achieves the highest efficiency. The influence of the phase-shift angle φ is shown in detail in **Fig. 14**. The measurements confirm a decreasing efficiency with increasing phase shift, as shown by the simulative loss modelling. In addition to decreasing efficiency, another effect can be observed: The phase shift results in increased intervals of $v_{\mathrm{inv}} = 0$ V in which no energy is transferred to the resonant circuit (cf. **Fig. 10**). Particularly at low output loads (cf. OP_{Gmid2}), this leads to a distortion of the output current, which further intensifies the challenge with synchronous rectification (SR) of the ZCS LLC converter. Even the SR concept presented in [10] reaches its limits under these conditions.

6 Conclusion

This paper principally proves that the zero-current-switching LLC resonant converter can be operated in full-bridge, half-bridge and phase-shifted full-bridge operation. To ensure that all operation points can be achieved with the mentioned inverter modes, the resonant circuit has to be designed accordingly. In contrast to the ZVS/MOSFET-based LLC converter, which achieves improvements in efficiency through the use of half-bridge and phase-shifted full-bridge operation mode-transitions, the ZCS LLC shows a decrease in efficiency, mainly caused by increasing switching-losses of the inverter stage. Also, synchronous rectification becomes more complex during phase-shift mode. For these reasons, a globally applied complementary full-bridge operation is the most appropriate modulation mode for the ZCS LLC. Moreover, this robust modulation also simplifies the control of the converter at its different stages.

7 References

[1] Z. Yang *et al.*, "Investigations of inhomogeneous reverse recovery behavior of the body diode in superjunction MOSFET," *2017 29th International Symposium on Power Semiconductor Devices and ICs (ISPSD)*, Sapporo, Japan, 2017, pp. 155-158, doi: 10.23919/ISPSD.2017.7988934.

[2] Bo Yang, *Topology Investigation for Front End DC/DC Power Conversion for Distributed Power System*. Manuscript University Dissertation. Blacksburg, Virginia, 2003.

[3] J. Xiang, X. Ren, Y. Wang and Y. Zhang, "Investigation of cascode stucture GaN devices in ZCS region of LLC resonant converter," *2017 IEEE Energy Conversion Congress and Exposition (ECCE)*, Cincinnati, OH, USA, 2017, pp. 1374-1378, doi: 10.1109/ECCE.2017.8095950.

[4] C. Ma, K. Yoshida and K. Honda, "Si-IGBT versus SiC-MOSFET — An isolated bidirectional resonant LLC DC-DC converter for distributed power systems," *2015 54th Annual Conference of the Society of Instrument and Control Engineers of Japan (SICE)*, Hangzhou, China, 2015, pp. 894-899, doi: 10.1109/SICE.2015.7285440.

[5] K. Chen and T. A. Stuart, "A study of IGBT turn-off behavior and switching losses for zero-voltage and zero-current switching," *[Proceedings] APEC '92 Seventh Annual Applied Power Electronics Conference and Exposition*, Boston, MA, USA, 1992, pp. 411-418, doi: 10.1109/APEC.1992.228381.

[6] D. Urbaneck, P. Rehlaender, F. Schafmeister and J. Böcker, "LLC Converter Design in Capacitive Operation utilizes ZCS for IGBTs – a Concept Study for a 2.2 kW Automotive DC-DC Stage," *PCIM Europe digital days 2020; International Exhibition and Conference for Power Electronics, Intelligent Motion, Renewable Energy and Energy Management*, Germany, 2020, pp. 1-8.

[7] D. Urbaneck, P. Rehlaender, J. Böcker and F. Schafmeister, "LLC Converter in Capacitive Operation Utilizing ZCS for IGBTs – Theory, Concept and Verification of a 2 kW DC-DC Converter for EVs," *2021 IEEE Applied Power Electronics Conference and Exposition (APEC)*, Phoenix, AZ, USA, 2021, pp. 2753-2760, doi: 10.1109/APEC42165.2021.9487109.

[8] Q. Cao, Z. Li and H. Wang, "Wide Voltage Gain Range LLC DC/DC Topologies: State-of-the-Art," *2018 International Power Electronics Conference (IPEC-Niigata 2018 -ECCE Asia)*, 2018, pp. 100-107, doi: 10.23919/IPEC.2018.8507899.

[9] Silvio De Simone, "LLC resonant half-bridge converter design guideline," STMicroelectronics, Application note, 2014.

[10] D. Urbaneck, F. Schafmeister and J. Böcker, "Advanced Synchronous Rectification for an IGBT-Based ZCS LLC Converter with High Output Currents for a 2 kW Automotive DC-DC Stage," *PCIM Europe 2023; International Exhibition and Conference for Power Electronics, Intelligent Motion, Renewable Energy and Energy Management*, Nuremberg, Germany, 2023, pp. 1-9, doi: 10.30420/566091239.

[11] L. Keuck, P. Hosemann, B. Strothmann and J. Böcker, "A Comparative Study on Si-SJ-MOSFETs vs. GaN-HEMTs Used for LLC-Single-Stage Battery Charger," *PCIM Europe 2017; International Exhibition and Conference for Power Electronics, Intelligent Motion, Renewable Energy and Energy Management*, Nuremberg, Germany, 2017, pp. 1-8.

[12] P. Rehlaender, T. Grote, S. Tikhonov, M. Schröder, F. Schafmeister and J. Böcker, "A 3,6 kW Single-Stage LLC Converter Operating in Half-Bridge, Full-Bridge and Phase-Shift Mode for Automotive Onboard DC-DC Conversion," *PCIM Europe digital days 2020; International Exhibition and Conference for Power Electronics, Intelligent Motion, Renewable Energy and Energy Management*, Germany, 2020, pp. 1-8.

[13] N. Förster, J. Hölscher, T. Piepenbrock, P. Rehlaender, O. Wallscheid, F. Schafmeister, J. Böcker, "An Open-Source FEM Magnetic Toolbox for Calculating Electric and Thermal Behavior of Power Electronic Magnetic Components", *2022 24th European Conference on Power Electronics and Applications (EPE'22 ECCE Europe)*, Hanover, Germany, 2022, pp. P.1-P.9.

PCIM Europe 2024, 11– 13 June 2024, Nuremberg   DOI: 10.30420/566262305

Dual Output Hybrid Converter for 48 V Data Centers: M-HS

Roberto Rizzolatti [1], Mario Ursino [1], Erik Medeossi[1], Stefano Saggini [2]

[1] Infineon Technologies AG, Austria
[2] University of Udine, Italy

Corresponding author: Roberto Rizzolatti, roberto.rizzolatti@infineon.com
Speaker: Simone Mazzer, simone.mazzer@infineon.com

Abstract

In the quest for energy-efficient and dependable power distribution solutions for 48 V data centers, novel converter topologies have emerged in recent years, both within the Industry and Academia. The two-stage approach remains the most prevalent architecture for stepping down the input 48 V in two stages: initially, an intermediate voltage rail is obtained by employing an IBC (Intermediate Bus Converter), and then a VRM (Voltage Regulator Module) is utilized to regulate the ASIC core voltage. In this study, a dual-output IBC provides multiple voltage outputs from a single conversion stage within a 48 V architecture. The proposed conversion is an 8:1 + 4:1 conversion, where the 8:1 rail is designated as a high-performance VRM voltage supply, for instance, while the 4:1 rail is utilized to provide power to other loads such as PCIe, FANs, and legacy systems, which are nearly always present on the board. The suggested dual-output IBC is a Hybrid Switched Capacitor converter (abbreviated as M-HSC), which is a fully-soft-switched resonant converter based on consolidated previous topologies. Experimental results for a 500 W, 1.17 kW/in^3 TDP discrete solution are included, showing the efficiency and power density of M-HSC.

1 Introduction

The increasing electricity consumption in modern data centers requires continuous improvements at both the system and converter levels[1]. The 48 V rack power distribution is now a widespread solution in Data Centers [2]. This typically involves a two-stage approach: a resonant converter [3], such as LLC or Switched-Tank (STC) converters, at the first stage, and a multi-phase buck converter with a 12 V input at the second stage. For non-isolated setups, switched-capacitor converters can be used [4] to achieve high power density, as demonstrated in [5] with a novel load-independent zero-voltage switching (ZVS) solution. Recently, Hybrid Switched Capacitor (HSC) unregulated converters have become one of the most efficient and dense solutions to implement a non-isolated IBC [6]. In many cases, unregulated dual-output versions can be desirable [7]. Following this trend, a dual-output version of an HSC is here proposed, called M-HSC (Multirail-HSC), addressing a combined 8:1 + 4:1 conversion, as shown in 1. To achieve this conversion ratio, N1=2 and N2=1 is chosen for the multi-tapped autotransformer (MTA). The low voltage 8:1 rail

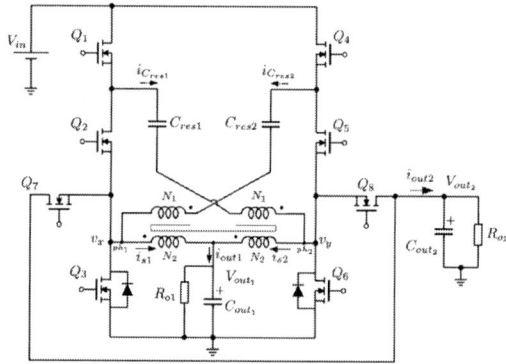

Fig. 1: M-HSC topology overview.

(6 V nominal) is intended as a high-performance VRM voltage supply, addressing compact and high-frequency VRM designs. Meanwhile, a 4:1 rail (12 V) can be used to provide a peripheral supply to other loads such as PCIe, FANs, and legacy systems, which are almost always present on the board. A single M-HSC can be used with the addition of only two MOSFETs with respect to [6]. The M-HSC is a fully-soft-switched resonant converter based on consolidated previous topologies. Experimental results for a 500 W, 1.17 kW/in^3 TDP discrete solution are included, demonstrating the

2162

efficiency and power density of this novel solution.

2 M-HSC operation theory

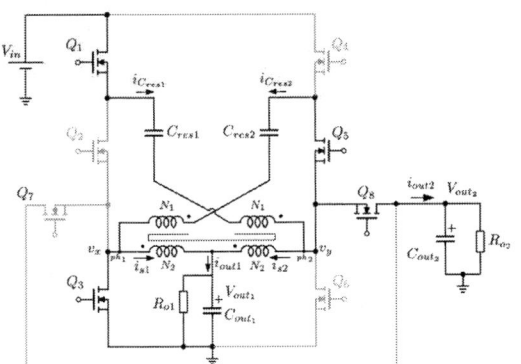

Fig. 2: M-HSC configuration in the interval $t_0 \rightarrow t_1$.

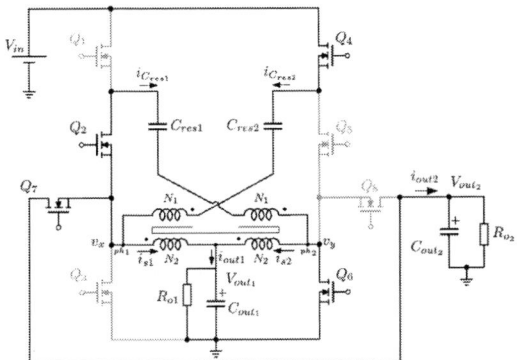

Fig. 3: M-HSC configuration in the interval $t_2 \rightarrow t_3$.

M-HSC operates in the two sub-intervals shown in Figure 2 and Figure 3, interleaved by a dead-time. During each sub-interval, a rectified sine current is delivered to the outputs by resonance between capacitors C_{res1}, C_{res2} and the leakage inductance of the autotransformer, composed by primary windings N_1 and secondary windings N_1. During the dead-time, the magnetizing inductance of the auto-transformer (not shown) allows for MOSFETs soft-switching (or ZVS, Zero-Voltage Switching). A single sub-interval is actually the combination of two resonances: the first one delivers power to V_{out1}, exactly as in [6], while the second one delivers power to V_{out2}. The two resonances cannot match, but the low leakage energy of this hybrid converters enables high efficiency even when non-zero turn-off currents are used, as shown in Section 3. For the sake of explanation, it is important to note that:

$$\frac{V_{out1}}{V_{in}} = \frac{1}{4 + 2\frac{N_1}{N_2}},\qquad(1)$$

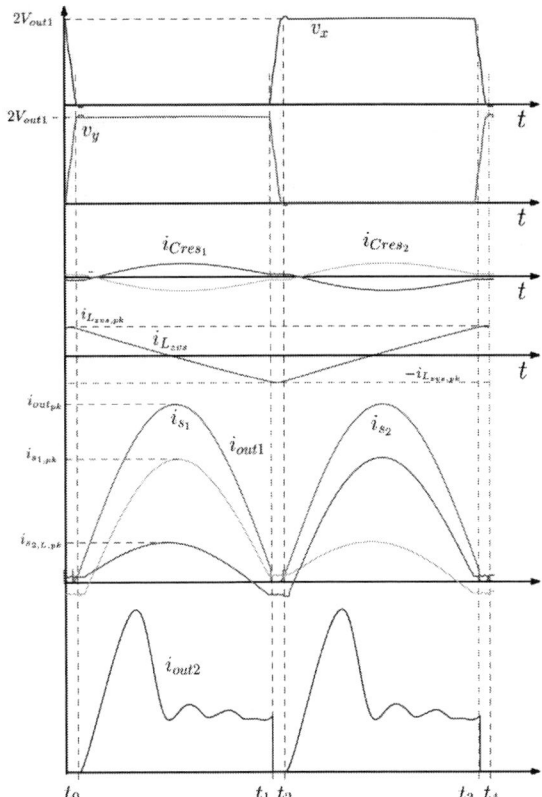

Fig. 4: M-HSC topology overview.

as derived in [6].

The additional rail V_{out2} is always connected to v_x and v_y nodes with MOSFETs Q_7 and Q_8, alternatively. The connection is established when v_x or v_y has a voltage of $2V_{out1}$, inducing a resonant current restoring the lost charge on C_{out2}. This happens with a soft-charging mechanism that depends on the leakage inductance of the MTA, C_{out2} and the resonant capacitors. As shown in Figure 4, the turn-off current related to instants t_1 and t_3 is in general not zero. As a consequence, Q_7 and Q_8 operate in ZVS but farther away from ZCS when compared to the other devices. Associated switching losses are proportionally lower, as $i_{out2} = 1/2 \cdot i_{out1}$ for a given power. From these considerations, it follows that:

$$V_{out2} = 2V_{out1}.\qquad(2)$$

To better understand the converter operation, each sub-interval is here described in detail. Each sub-interval can be better understood with the support of Figure 4, while the detailed equations in each sub-interval can be found in [6].

$t_0 \rightarrow t_1$: switches Q_1, Q_3, Q_5 and Q_8 are turned on in ZVS. An equivalent resonant tank is composed by C_{res1}, C_{res2} and the leakage current of the MTA, therefore a sinusoidal current flows in each primary winding. In this phase C_{res1} is soft-charged from the input voltage source V_{in} and C_{res2} is soft-discharged through a primary winding of the MTA. In this phase, one secondary winding is delivers to V_{out1} the two primary-windings currents (i_{s2}), while i_{s1} delivers a higher current magnitude depending on the turns ratio. C_{out1} and C_{out2} charges are restored as stated in (2) and (1), respectively.

$t_1 \rightarrow t_2$: all MOSFETs are turned OFF and the magnetizing current $i_{L_{zvs}}$ begins to discharge the C_{oss} of Q_2, Q_4, Q_6 and Q_7, which reach 0 V at t_2, while charging the C_{oss} of the other MOSFETs.

$t_2 \rightarrow t_3$: switches Q_2, Q_4, Q_6 and Q_7 are turned on in ZVS. The resonant capacitors currents change direction. In this phase C_{res2} is soft-charged from the input voltage source V_{in} and C_{res1} is soft-discharged through a primary winding of the MTA. In this phase, one secondary winding is delivers to V_{out1} the two primary-windings currents (i_{s1}), while i_{s2} delivers a higher current magnitude depending on the turns ratio. C_{out1} and C_{out2} charges are restored as stated in (2) and (1), respectively. This interval is the symmetrical version of $t_1 \rightarrow t_2$.

$t_3 \rightarrow t_4$: all MOSFETs are turned OFF and the magnetizing current $i_{L_{zvs}}$ begins to discharge the C_{oss} of Q_1, Q_3, Q_5 and Q_8, which reach 0 V at t_4, while charging the C_{oss} of the other MOSFETs.

3 Experimental results

Tab. 1: M-HSC 500 W prototype specification.

Size	26 x 36 mm
Max. thickness	8 mm
Baseboard x-section	8 layers, 2 oz copper
TX x-section	12 layers, 4 oz copper
Turns ratio	N1 = 2, N2 = 1
Q_1, Q_2, Q_4, Q_5	4 × IQE046N08LN5SC 80 V, 4.6 mΩ
Q_3, Q_6, Q_7, Q_8	4 × IQE008N03LM5 30 V, 800 µΩ
Drivers	4 × 1EDN7550B 2x2EDN7534G
f_{sw}	270 kHz

A 500 W Thermal-Design Power (TDP) discrete prototype is used to demonstrate the solution effectiveness under different loading scenarios. Table 1 summarizes its characteristics: 80 V MOSFETs are used as input half-bridges to enable startup without eFuse, reducing converter area occupation. The synchronous rectifiers and the *pass-transistors* are 30 V, best-in-class Infineon MOSFETs. Figure 5, shows both sides of the converter. The top side includes the power semiconductors (every FET shown in Figure 1), the SMD autotransformer, resonant capacitors C_{res1} and C_{res2}, and the output capacitors for the two voltage rails. The bottom side includes the driving system, which is based on a two-step bootstrap, the controller and its auxiliary circuits. The MTA is implemented as an SMD component with copper pins: this device constitutes the resonant tank, with its leakage inductance, and stores ZVS energy in its magnetizing inductance.

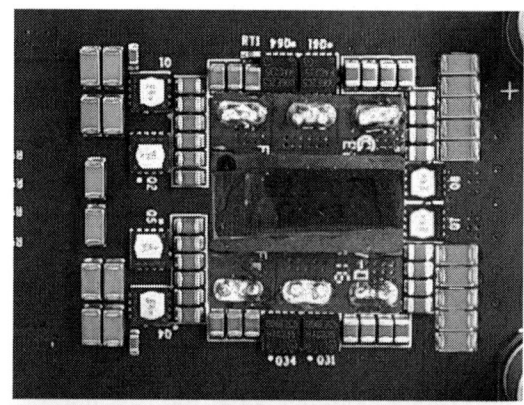
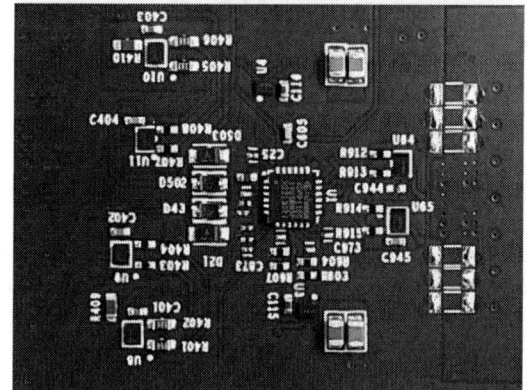

Fig. 5: 500 W M-HSC discrete prototype overview. This figure shows the top and bottom view of the converter, which occupy the same position on the motherboard.

Figure 6 shows the overall converter efficiency when different I_{out2} are applied. The total output

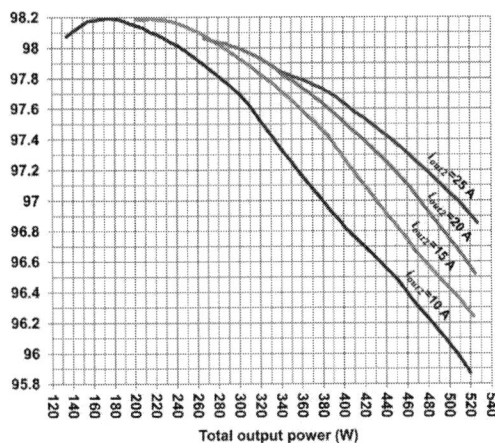

Fig. 6: Total converter efficiency (%) at different load currents i_{out2}. V_{in} = 54 V. V_{in} = 54 V, V_{out1} = 5.4 V (nominal), V_{out2} = 10.8 V (nominal).

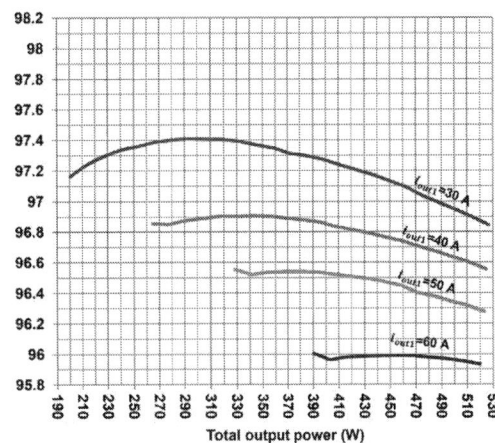

Fig. 7: Total converter efficiency (%) at different load currents i_{out1}. V_{in} = 54 V, V_{out1} = 5.4 V (nominal), V_{out2} = 10.8 V (nominal).

power includes the output power of both rails. The same measurement is shown in Figure 7 for different I_{out1} loads. From this graphs, the effect of I^2R losses is clear: higher efficiency is achieved when using the higher-voltage rail, as per every fixed-ratio converter.

Q_3 and Q_6 drain voltages V_x and V_y together with the respective V_{gs} are shown in Figure 8 for different I_{out1} currents. The leakage energy can be observed to help the transition to the final value with the increasing load, while the dead-time is fixed to ensure ZVS in the whole load and input voltage range. At zero load, ZVS is enforced by the magnetizing current (as shown by the V_x waveform reaching 1.5 V at time $t = 200$ ns, corresponding to no-load operation): this means that the worst-case ZVS transition is solely determined by the autotransformer gap (determining the magnetizing inductance), and does not depend on the resonant tank natural frequency. As a consequence, the resonant capacitor mismatch and/or temperature variation do not affect ZVS behavior, allowing to use standard MLCC capacitors.

The average *droop* resistance for the two rails is measured as follows:

$$R_{droop}^{Vout_1} = -\frac{\Delta V_{out1}}{\Delta I_{out1}} = 5.3 \text{ m}\Omega \quad (3)$$

$$R_{droop}^{Vout_2} = -\frac{\Delta V_{out2}}{\Delta I_{out2}} = 8.7 \text{ m}\Omega \quad (4)$$

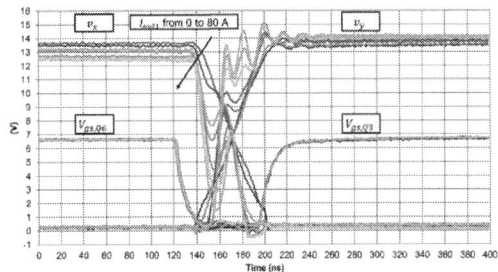

Fig. 8: Main converter waveforms during a ZVS transition. $V_{in} = 54$ V.

4 Conclusions

This paper presents a new multi-rail, hybrid switched-capacitor converter called M-HSC. High-and intermediate- step-down capability is achieved for the 48 V bus in a single converter. This novel concept relies on a single magnetic autotransformer than embeds the resonant inductance and the ZVS inductance. The low-voltage rail (6 V nominal) is meant to power a high-performance VRM, while the high-voltage rail (12 V nominal) can be used for peripherals powering, if the input range variation allow for it. A 500 W discrete prototype is shown to demonstrate converter operation and efficiency. The prototype has a 98.2% peak efficiency and a 1.17 kW/in^3 power density at the thermal design point.

References

[1] Yole Intelligence, *Computing and ai technologies for data center 2022, market and technology report*, 2022.

[2] S. Oliver, *From 48 v direct to intel vr12.0: Saving 'big data' $500,000 per data center, per year*, Jul. 2012.

[3] G. Deng, Y. Sun, G. Xu, X. Chen, S. Xie, *et al.*, "Zvs analysis of half bridge llc-dcx converter considering the influence of resonant parameters and loads," in *2020 IEEE Energy Conversion Congress and Exposition (ECCE)*, 2020, pp. 1186–1190. DOI: 10.1109/ECCE44975.2020.9235371.

[4] S. Jiang, S. Saggini, C. Nan, X. Li, C. Chung, and M. Yazdani, "Switched tank converters," *IEEE Transactions on Power Electronics*, vol. 34, no. 6, pp. 5048–5062, 2019. DOI: 10.1109/TPEL.2018.2868447.

[5] C. Rainer, R. Rizzolatti, and D. Varajao, "High density cascaded zvs switched capacitor converter for 48-v data-center application," in *PCIM Europe digital days 2020; International Exhibition and Conference for Power Electronics, Intelligent Motion, Renewable Energy and Energy Management*, 2020, pp. 1–8.

[6] R. Rizzolatti, C. Rainer, S. Saggini, and M. Ursino, "High density hybrid switched capacitor converter for data-center application," in *2021 IEEE Applied Power Electronics Conference and Exposition (APEC)*, 2021, pp. 1288–1293. DOI: 10.1109/APEC42165.2021.9487136.

[7] M. Ursino, S. Saggini, and O. Zambetti, "Configurable dual output non-isolated resonant converter for 48 v applications," in *2020 IEEE Applied Power Electronics Conference and Exposition (APEC)*, 2020, pp. 2199–2205. DOI: 10.1109/APEC39645.2020.9124575.

PCIM Europe 2024, 11– 13 June 2024, Nuremberg   DOI: 10.30420/566262306

3.6 kW High Efficiency SiC-based HV/LV DC/DC Converter for EVs

Veera Bharath, Gandluru ®[1], Yuequan, Hu[2]

[1] Wolfspeed, USA
[2] Wolfspeed, USA

Corresponding author: Veera Bharath, Gandluru, veera.gandluru@wolfspeed.com
Speaker: Veera Bharath, Gandluru, veera.gandluru@wolfspeed.com

Abstract

Electrical Vehicles (EVs) employ a DC/DC converter to power auxiliary loads with lower voltages from the high-voltage battery. In this work, a SiC based $3.6kW$ high-voltage (HV) to low-voltage (LV) auxiliary power supply (APS) converter is designed based on silicon carbide (SiC) technology to achieve high efficiency. SiC devices enable higher switching frequencies, leading to the development of power-dense converters. A phase-shifted full-bridge (PSFB) topology is chosen for this application. The design details a PSFB-based DC/DC converter for EVs with $800V$ DC bus voltage, employing state-of-the-art, automotive grade $1200V$ SiC MOSFETs to demonstrate the advantages of SiC for superior efficiency and power density. A comprehensive design procedure is provided to support the chosen topology and components. Finally, a thorough simulation and analytical analysis are performed across the entire input and output voltage range.

1   Introduction

Electric vehicles (EVs) are experiencing a surge in popularity, driven by environmental concerns. This rapid growth has fueled innovation in power conversion circuits, with a focus on cost effectiveness, efficiency, and power density [1, 2]. A reliable and efficient auxiliary power supply (APS) is crucial for powering essential low-voltage components such as controllers, sensors, and pumps in EVs [3, 4]. Traditionally, silicon (Si)-based DC/DC converters have served this purpose. However, the growing emphasis on extended driving range demands advancements in high-efficiency solutions.

Silicon Carbide (SiC) technology has emerged as a transformative option for EV DC/DC converters. SiC offers superior material properties, including a wider bandgap, higher thermal conductivity, and significantly lower switching losses compared to Si. These advantages enable SiC-based converters to achieve industry-leading efficiency and a more compact design, making them a perfect fit for next-generation EVs.

The EV industry is shifting from conventional Si-based solutions to SiC-based APS converters due to several key benefits. Higher efficiency in SiC converters minimizes energy losses during power conversion, directly contributing to extended driving range. Additionally, the compact size and

lightweight nature of SiC converters allow for better space utilization and reduced overall vehicle weight. Finally, superior thermal conductivity of SiC enables efficient heat dissipation for reliable operation under demanding conditions.

In contrast to other industrial applications with standardized input/output ratings, the automotive industry is experiencing rapid changes in DC-bus voltages, power requirements, and the number of auxiliary loads [1]. EV manufacturers are increasingly advocating for higher DC-bus voltages, faster charging speeds, and more features to enhance the customer experience. While converter ratings in the EV APS industry typically range from $3-5kW$, the advent of features like autonomous driving demands higher power and efficiency.

This paper proposes a specifically designed $3.6kW$, high-efficiency SiC-based high-voltage to low-

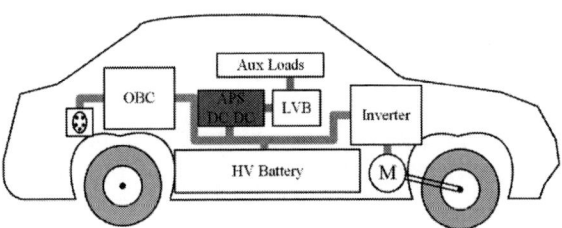

Fig. 1: Electric vehicle's power architecture.

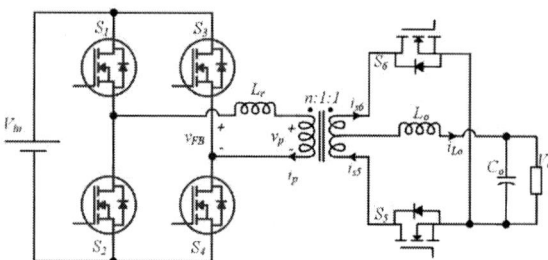

Fig. 2: Schematic of PSFB converter.

voltage DC/DC converter for EV auxiliary power supply applications. By harnessing the benefits of SiC technology, the proposed converter aims to achieve industry-leading efficiency, making it a compelling solution for the future of electric vehicles.

2 APS Requirements and Topology Selection

In this work, a DC/DC converter is designed and developed for EVs with $800V$ High Voltage (HV) bus and $12V$ Low Voltage (LV) DC bus. As noted in [2], an $800V$ battery can operate within a wide voltage range of $400V$-$900V$. Similarly, the low-voltage side bus can vary between $9V$-$16V$ [3]. Most manufacturers in the market target a power rating of $3 - 4kW$. In a typical configuration, the car chassis serves as the return path for the LV bus. Hence, it is crucial to galvanically isolate the LV bus from HV bus (ISO 6469-3). Additionally, OEMs are seeking reverse power (LV→HV) delivery capability during start-ups to precharge the HV bus capacitors, and as an active filter during charging [5] to reduce the passive storage components of On-Board chargers (OBCs).

Power conversion efficiency and power density are two key requirements in designing an APS as they directly impact the range of an EV [6]. Achieving higher power density requires switching at higher frequencies. Wide-bandgap devices enable engineers to work at higher switching frequencies with lower switching losses. Among WBG devices, SiC stands out due to its higher thermal conductivity, making it the preferred choice for engineers as it facilitates easier management of system thermals. Several popular isolated DC/DC topologies are available in the literature [3, 4]. Moreover, Dual-Active- Bridge (DAB), and resonant converter topologies have been favourites among designers for other EV applications like OBCs. In this work, all these topologies are investigated to identify the most suitable one that best meets the requirements of the EV APS application.

2.1 DAB for APS Application

Considering the aforementioned requirements, a DAB converter is designed for APS. In the DAB converter, the selection of transformer turns-ratio(n_d) and leakage inductance(L_l) is crucial. The power transferred from HV to LV side in a DAB converter operated using Single pulse width modulation scheme can be expressed as,

$$P = \frac{n_d V_{in} V_o}{8 f_{sw} L_r} sin(\theta) \tag{1}$$

where, f_{sw} is the switching frequency and θ is the phase difference between the primary and secondary side full-bridges. From control stability θ would practically be limited to $60°$. Considering a switching frequency of $100kHz$, substituting other design values in (1) gives the design condition,

$$\frac{L_l}{N_d} \leq 1.44 \times 10^{-6} \tag{2}$$

To have ZVS capability in a wide load range, n_d=60, and $L_l = 75\mu H$ are the optimum design values. In a DAB converter, transformer secondary side peak-peak current can be expressed as,

$$I_{s(pp)} = \left| \frac{n_d(V_{in} - n_d V_o)}{4L_l f_{sw}} \left[1 + \left[1 - \frac{8 f_{sw} L_l P}{n_d V_{in} V_o} \right]^{0.5} \right] \right| + \left| \frac{n_d(V_{in} + n_d V_o)}{4L_l f_{sw}} \left[1 - \left[1 - \frac{8 f_{sw} L_l P}{n_d V_{in} V_o} \right]^{0.5} \right] \right| \tag{3}$$

This current is plotted in Fig.3 with respect to turns-ratio at various input and output voltages. From these plots, it can be observed that the ripple current surpasses $2.5kA$ under certain operating conditions. Additionally, the magnitude of the first harmonic in this case is observed to be over $900A$. Although this may not significantly affect core losses, the copper losses in the transformer and the secondary-side devices would become too high to handle. Considering these findings, it is evident that the Dual-Active-Bridge (DAB) topology is not well-suited for APS applications.

2.2 Resonant Converters for APS

The CLLLC converter is investigated for its suitability in APS application. To understand the impact of the wide voltage gain requirement, a typical CLLLC transformer turns-ratio(N_l) value of $800/12$ is considered which corresponds to the nominal input and output voltages. Based on this turns-ratio, the maximum and minimum resonant tank voltage gain requirements for the APS can be calculated as,

$$G_{max} = \frac{N_l V_{o(max)}}{V_{in(min)}} = \frac{800}{12} \times \frac{16}{400} = 2.67$$

$$G_{min} = \frac{N_l V_{o(min)}}{V_{in(max)}} = \frac{800}{12} \times \frac{9}{900} = 0.67 \tag{4}$$

To assess the necessary variation in switching frequency, the resonant tank voltage gain with respect to normalised frequency is plotted in Fig. 4 for various quality factor (Q) values. Magnetizing to resonant inductance ratio of three is used to generate these voltage gain plots. These plots reveal that achieving the desired voltage gain range of 2.67 to 0.67 requires a substantial variation in switching frequency, spanning from below $0.6 f_{res}$ to above $2.5 f_{res}$. Building a transformer optimized to operate effectively across such a wide frequency range would pose significant challenges and could result in higher losses under extreme operating conditions.

Furthermore, resonant converters suffer from higher peak-peak transformer secondary currents due to the absence of output side filter inductance, similar to the issue observed with DAB. This would further contribute to increased losses within the system. Therefore, due to these limitations, particularly the challenges associated with transformer design and potential for higher losses, resonant topologies are considered not suitable for the APS application.

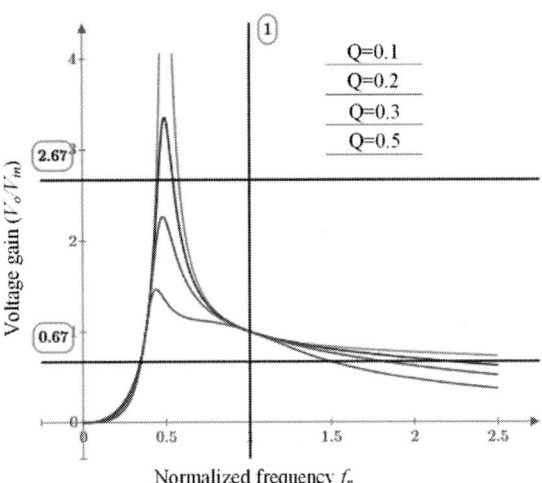

Fig. 4: CLLLC resonant tank voltage gain (vs) normalised frequency (f_n).

APS applications due to several key advantages. Unlike DAB and CLLLC converters, the PSFB utilizes an output inductor to effectively reduce output current ripple, eliminating the need for bulky capacitor banks. This allows for a potentially more compact and lighter converter design.

Furthermore, the PSFB converter ensures ZVS operation in at least two of the high-voltage (HV) side devices regardless of the load conditions. This combination of features significantly improves the converter's efficiency across a wide input/output voltage range, making it ideal for the APS application where high efficiency is crucial.

3 Design and Working Principle

This section delves into the design and operating principle of the PSFB converter for APS application. Firstly, the equations governing PSFB operation are derived. A conventional PSFB converter, as shown in Fig.2, employs four switches in a high-voltage (HV) side full-bridge configuration. These switches operate with a fixed 50% duty cycle, with the controllable parameter being the phase difference between the two legs of the bridge. The leg with S_1, and S_2 switches is considered the leading leg, while the one with S_3, and S_4 switches is the lagging leg.

The full-bridge voltage (V_{FB}) appears as a quasi-square wave with a width dependent on the controllable phase-shift (D_a) between the leading and lagging legs. However, the voltage across the primary side of the transformer (V_p) will have a slightly narrower width compared to V_{FB} due to phase-

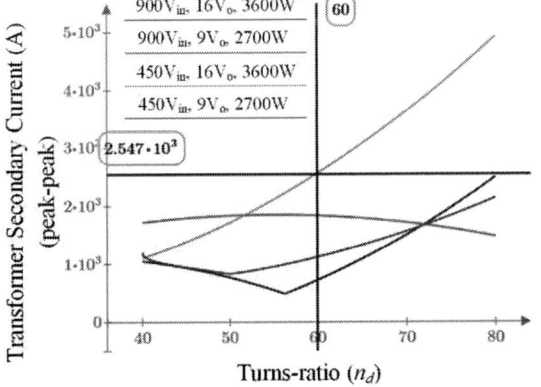

Fig. 3: Transformer secondary side peak-peak current magnitude (vs) turns-ratio.

2.3 PSFB for APS

The Phase-Shifted Full-Bridge (PSFB) converter stands out as the most suitable topology for the

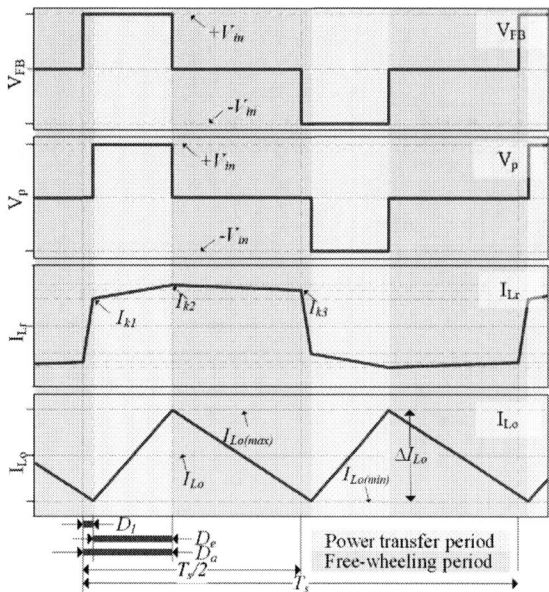

Fig. 5: Various waveforms in PSFB converter.

shift loss (D_l), as observed from the various PSFB waveforms depicted in Fig.5.

3.1 PSFB Voltage Gain

The relationship between a PSFB converter's output voltage and its input voltage (voltage-gain) in terms of transformer turns-ratio (n), and the effective phase-shift $(D_e = D_a - D_l)$ across the transformer primary side can be derived using volt-seconds balance in the output inductor, expressed as follows,

$$\left(\frac{V_{in}}{n} - V_o\right) D_e T_s + (-V_o)(0.5 - D_e)T_s = 0$$

$$\Rightarrow \frac{V_o}{V_{in}} = \frac{2D_e}{n} \tag{5}$$

where, T_s is the switching time period.

3.2 Phase-loss in PSFB

In PSFB topologies, phase-loss is a known phenomenon. During this period, the current in the transformer primary side changes polarity. Therefore, the greater the leakage inductance, the greater the phase-loss which gives less window to transfer the power from primary to secondary side. This can be inferred from the expression derived for the phase-loss in the equation (6),

$$D_l = (2I_o - \Delta I_o) \cdot \left(\frac{nV_{in}}{L_r f_{sw}} - \frac{\Delta I_o}{0.5 - D_e}\right)^{-1} \tag{6}$$

$$\approx 2I_o L_r f_{sw}/nV_{in}$$

where, I_o is the average output current, ΔI_o is the output peak-to-peak current ripple magnitude, L_r is

the primary side leakage inductance, and f_{sw} is the converter switching frequency. A detailed derivation of this equation is given in [7]. This equation shows that phase-loss (D_l) increases with higher leakage inductance (L_r) and switching frequency (f_{sw}), and decreases with a higher turns ratio (n) and input voltage (V_{in}).

3.3 Selecting Magnetic Components in PSFB Design

The first crucial step in designing a PSFB converter involves selecting the magnetic component parameters: transformer turns-ratio (n), transformer leakage inductance(L_r), and output filter inductance (L_o) value.

Transformer turns-ratio (n) needs to be selected considering the wide input and output voltages of the APS application. A lower turns ratio (n) can lead to higher phase-loss (6), but also requires higher voltage ratings for the secondary devices, increasing cost and conduction losses. Conversely, a higher turns ratio reduces phase-loss but limits the achievable output voltage.

Leakage inductance (L_r) value determines the ZVS in the leading leg devices. During the turn-on of a MOSFET, sufficient current in its body-diode is required to completely discharge the device output capacitance through charge re-circulation without losing energy. The peak value of the body-diode current at the beginning of the dead-time period determines the ZVS feasibility at that switching cycle. In PSFB, this current equals to the sum of magnetizing current and the transformer secondary-side current reflected to the primary side which can be expressed as,

$$I_{Lk(peak)} = I_{Lm} + \frac{I_{Lo(peak)}}{n}$$

$$= \frac{nV_o}{4L_m f_{sw}} + \frac{1}{2n}(I_o + \Delta I_o) \tag{7}$$

The dead-time period also affects the ZVS feasibility. If the body-diode current falls to zero before the gate pulse is applied (before the dead-time ends), the device output capacitance recharges to half the bus voltage, discharging in the channel at the turn-on, resulting in E_{oss} losses and thus only a partial ZVS. Therefore, the maximum usable dead-time value to achieve complete ZVS can be computed using,

$$T_d \leq \frac{L_k \cdot I_{Lk(peak)}}{V_{in}} \tag{8}$$

From (7), and (8), the minimum possible value for L_k can be computed.

Output filter inductance (L_o) value determines the ripple current in the secondary side devices, transformer, and even in the HV side SiC device turn-off currents. The ripple in the filter inductor can be expressed as,

$$\Delta I_{Lo} = \frac{(1-2D_e)V_o}{2L_o f_{sw}} \quad (9)$$

The output inductance value can be computed for the allowed ΔI_{Lo} value in the design.

3.4 Design Parameters of PSFB

To determine the design parameters, the following design specs are considered: $V_{in(min)}$=400V, $V_{in(max)}$=16V, $D_{a(max)}$=0.45, f_{sw}=100kHz, $\Delta I_{Lo(max)}$=30% of I_{Lo}. Additionally, ZVS is required in all HV side devices for loads $\geq 50\%$ of full-load at nominal voltages.

From the design specification of (D_a), an inequality expression can be formulated from (5), and (6) which can be expressed as,

$$D_a = D_e + D_l \leq 0.45$$
$$\Rightarrow \frac{nV_o}{2V_{in}} + \frac{2I_oL_rf_{sw}}{nV_{in}} \leq 0.45 \quad (10)$$

by solving this inequality expression, the maximum possible transformer turns-ratio can be obtained. Considering a leakage inductance L_r=10μH, and magnetizing inductance L_m=600μH as starting points, solving the inequality expression derived in (10) yields a maximum allowed n value of 20. From (6), lower n value result in higher phase-loss and require higher voltage rating for the secondary side devices. Thus, the maximum possible n value of 20 is chosen for this design, requiring the secondary side switch blocking voltage to be greater than 90V.

From the output inductor current ripple limit condition, the minimum inductance value required can be computed by substituting (5) into (9),

$$L_o \geq \frac{(V_{in}-nV_o)V_o}{0.6\ I_o f_{sw}} \quad (11)$$

Substituting appropriate parameter values in (11 results in the necessary inductance value being greater than 0.76μH.

The maximum allowed dead-time needs to be determined to ensure ZVS for loads > 50% of full-load.

The $I_{Lk(peak)}$ value is computed using (7), considering L_r=10μH, resulting in a value of 10.01A. Substituting this into (8) gives a maximum allowed dead-time of 125ns. This dead-time value is practical, so the initially considered $L_r = 10\mu H$ is selected as the final design value.

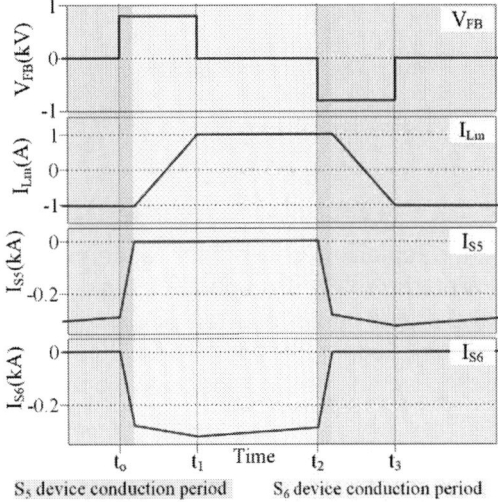

Fig. 6: Understanding current sharing in the secondary side MOSFETs during freewheeling period.

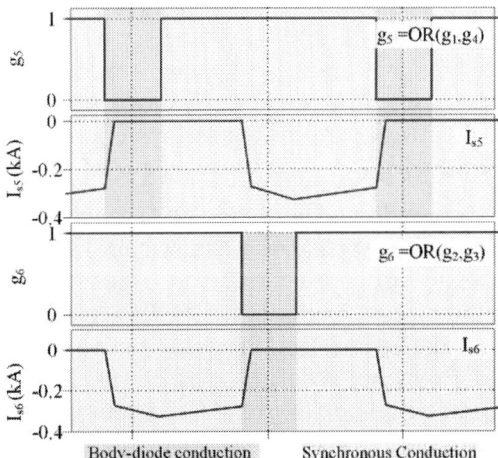

Fig. 7: PWM scheme used for synchronous rectification.

3.5 Freewheeling Current Sharing in the Secondary Side Switches

In PSFB topology, unlike in a full-bridge DC/DC converter, the two switches on the LV side do not share the current during the freewheeling mode. This is explained using the waveforms presented in Fig.6 under operating conditions of $800V_{in}$, $12V_o$, and $3.6kW$. Power transfer occurs from the HV side to LV side during intervals $t_0 - t_1$ and $t_2 - t_3$.

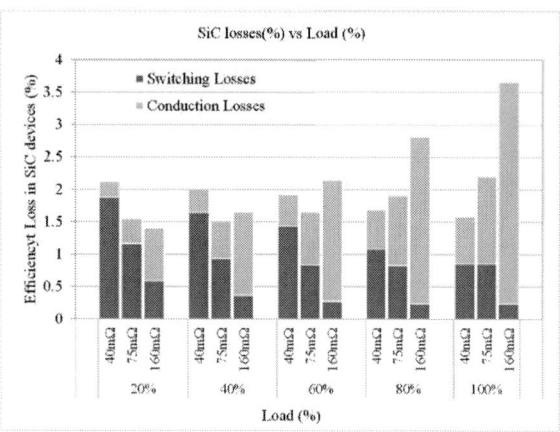

Fig. 8: Loss of % Efficiency in SiC devices vs Load (%).

During the remaining periods, the leakage inductor current freewheels through the top-side MOSFET of the leading-leg and top-side body-diode of the lagging-leg, or the bottom-side MOSFET of the leading-leg and bottom-side body-diode of the lagging-leg. As depicted in Fig. 6, during this freewheeling time, the current circulates through one of the two switches on the secondary side. This phenomenon can be explained as follows: when the primary side current circulates in S_1 and S_3, the transformer terminal with dot polarity exhibits positive voltage, causing a reverse voltage bias across the S_5 MOSFET body-diode, directing all current to flow through S_6. Similarly, when S_2 and S_4 are conducting, the S_6 MOSFET body-diode is reverse biased, and the entire current flows through S_5.

3.6 Synchronous Rectification Scheme

On the secondary side of the PSFB transformer, two diodes can replace the two switch positions. However, to reduce the conduction losses, and improve the system efficiency, MOSFETS are used as synchronous rectifiers. Precisely turning-on the MOSFET channel while its body-diode is conducting requires zero-crossing detecting circuits, which could be challenging due to the very high current in these switches. To eliminate the complexity, a simple synchronous rectification technique is adopted which is demonstrated in Fig.7. Although simple, when this technique is used, body-diode conducts for a brief duration which can also be observed in Fig.7, resulting in slightly higher conduction losses.

3.7 SiC Device Selection and Loss Estimation

The selection of a suitable SiC device for a given application is crucial for striking a balance between price and performance. Devices with larger die

sizes offers less $R_{ds(on)}$, significantly reducing conduction losses. However larger dies also means higher costs. Moreover, the larger die size correlates with higher output capacitance, leading to elevated switching losses.

In this work, devices with different $R_{ds(on)}$ values are compared for $3.6kW$ APS application to identify the most suitable device for optimizing benefits for EV customers. A comparative analysis is conducted using PLECS thermal models for Wolfspeed SiC devices with $R_{ds(on)}$ values $40m\Omega$, $75m\Omega$, and $160m\Omega$. The efficiency losses in different SiC devices at various loads are depicted in Fig. 8. These plots reveal that switching losses predominate in low $R_{ds(on)}$ devices, whereas conduction losses dominate in higher $R_{ds(on)}$ devices. Based on this analysis, the $75m\Omega$ device emerges as the most suitable choice from both price and performance perspectives.

4 Simulation Analysis and Hardware Design

The design and simulation parameters used are summarised in Table.1. The maximum current on the LV side is limited to $300A$ in this design. Using PLECS software tool, the designed PSFB converter is simulated under various operating conditions, and the resulting waveforms are presented in Fig.9. Using the loss thermal models of the witching devices, losses are estimated, while the losses in the magnetic components are estimated using data collected from manufacturers.

For the HV side, Wolfspeed's 1200V, $75m\Omega$ device is selected. Taking advantage of its low-profile and ease of mounting, surface mount package TO-263-7XL, automotive grade SiC device $E3M0075120J2$ is chosen. For the LV side, $150V$ rated, LittleFuse's $IXTK400N15X4$ Si MOSFET in a TO-247-3 package is selected. Four devices are used for each secondary-side switch position to handle high secondary side currents. Custom-made HF transformers and output filter inductors from Sunlord are used in this design. Two $1.54\mu H$ inductors are paralleled to handle $300A$ of output current, resulting in an effective output inductance of $0.77\mu H$.

4.1 Simulation Analysis

Losses and efficiency at various operating points are summarised in Table.2. From the loss analysis table, the following observations and conclusions can be drawn:(i) Switching losses of the SiC devices in case-1 are almost three times those in

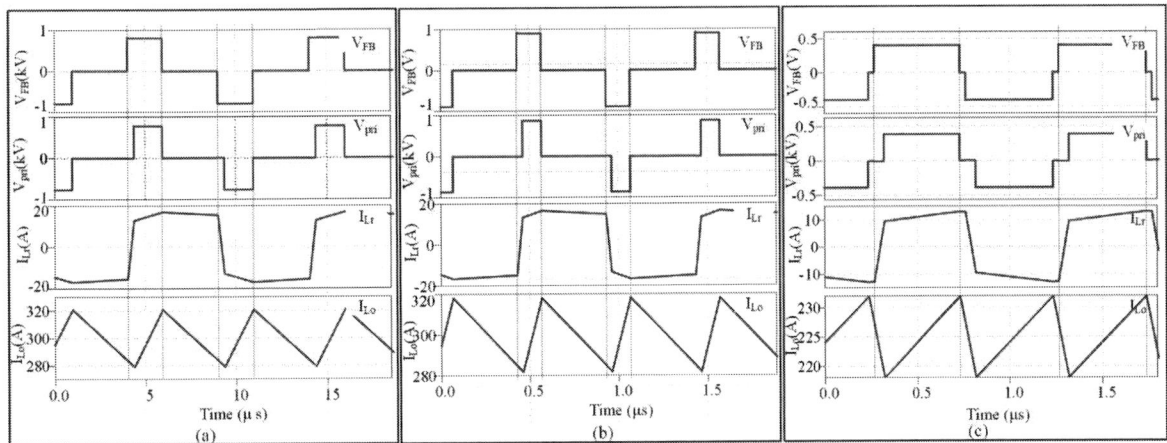

Fig. 9: Simulation waveforms corresponding to (a) case-1: V_{in}=800V, V_o=12V, and P_o=3.6kW, (b) case-2: V_{in}=900V, V_o=9V, and P_o=2.7kW, (c) case-3: V_{in}=400V, V_o=16V, and P_o=3.6kW.

Par	Value	Par	Value
V_{in}	$400-900V$	L_o	$0.77\mu H$
V_o	$9-16V$	C_o	$19.89mF$
$P_o/I_o(max)$	$3.6kW/300A$	$R_{Th(ch)}$	$2.5K/W$
f_{sw}	$100kHz$	$R_{g(on)}$	2.5Ω
n	20	$R_{g(off)}$	2.5Ω
L_r	$10\mu H$	T_d	$100ns$
L_m	$600\mu H$	T_{Amb}	$65°C$

Tab. 1: Design and simulation parameters.

case-3, despite both cases having same output power. At these operating conditions, only E_{off} losses occur in the HV devices, depending on V_{ds}, and turn-off current magnitude. This difference can be attributed to the 55% lower V_{ds}, 25% lower I_o (higher V_o), and lower ΔI_{Lo} in case-3 compared to case-1. (ii) Conduction losses of the SiC device in case-1 are more than twice those in case-3. This is despite of lower V_{in} in case-3, which should ideally draw higher current from the DC bus for the same operating power. The reason lies in the power-transfer period, which is almost three times longer in case-3, because of which the current peak has to be lower than case-1, leading to higher device rms currents in case-1. This can be observed from Fig.9. (iii) Output inductor and secondary-side device conduction losses are half in case-3 compared to case-1, explained by the higher V_o and thus lower current magnitudes.

At full load, the maximum efficiency of the designed PSFB converter is 97.29%. Even considering the worst-case scenario of $2.5K/W$ thermal resistance between the device case and the heatsink, SiC junction temperatures remain well below $175°C$, the manufacturer's recommended maximum value.

A modular construction approach is employed for the converter design. The four SiC devices on the primary side are positioned on the daughter board, while the rest of the circuitry is integrated into the main board. This modular approach facilitates the experimental validation of loss comparisons among various $R_{ds(on)}$ valued SiC devices and packages with convenience. The hardware prototype of both the daughter board, and complete converter assembly are shown in Fig.10.

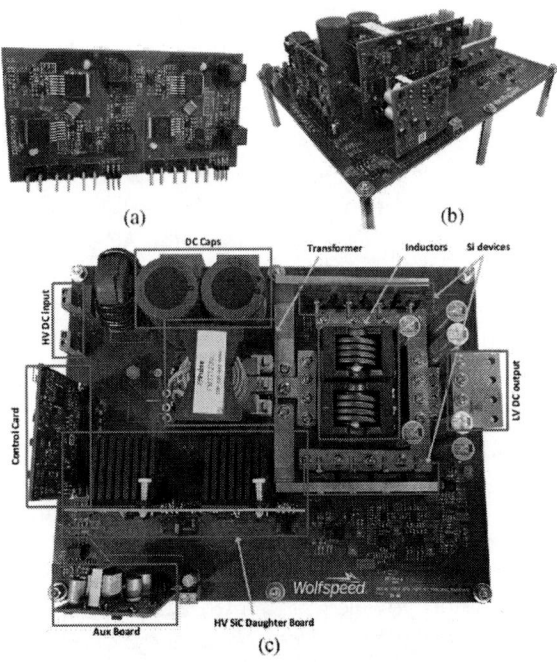

Fig. 10: Hardware prototype (a) daughter board with surface mount SiC devices. (b) whole prototype assembly, (c) component details.

Parameter	$800V/12V/3600W$	$900V/9V/2700W$	$400V/16V/3600W$
SiC devices Switching losses	$21.07W$	$23.81W$	$6.92W$
SiC devices Conduction losses	$56.70W$	$56.32W$	$26.12W$
Si devices Switching losses	$0W$	$0W$	$0W$
Si devices Conduction losses	$77.77W$	$80.13W$	$40.19W$
Xfrmr Primary Copper losses	$4.73W$	$4.69W$	$2.56W$
Xfrmr Secondary Copper losses	$20.04W$	$20.08W$	$8.54W$
Xfrmr Core losses	$1.48W$	$0.66W$	$3.31W$
Output inductor Copper losses	$15.94W$	$16.36W$	$9.45W$
Output inductor core losses	$0.04W$	$0.12W$	$0.16W$
Output Capacitor Losses	$1.44W$	$1.09W$	$0.15W$
Total Losses	$199.21W$	$203.26W$	$97.29W$
Efficiency	94.47%	92.47%	97.29%
SiC device Junction temperature	$129.56°C$	$131.71°C$	$92.80°C$
Si device junction temperature	$90.22°C$	$90.26°C$	$78.10°C$

Tab. 2: Power losses summary at various operating points.

5 Conclusion

This work presents a detailed design for a 3.6kW auxiliary power supply (APS) intended for use in Electric Vehicles (EVs) with 800V high-voltage (HV) DC-bus and 12V low-voltage (LV) DC-bus. The Phase-Shifted Full-Bridge (PSFB) topology is identified as the most suitable for this application. A comprehensive discussion comparing it with other popular isolated DC/DC converter topologies is provided. Detailed derivations for optimizing various design parameters are presented. An analysis is conducted to select the most appropriate $R_{ds(on)}$ valued SiC device for this application, optimizing both price and performance. Simulation results and loss analysis at multiple operating points are presented and thoroughly discussed to understand the nuances of the PSFB converter in the APS application. A hardware prototype is constructed using the selected design parameters. The modular design approach of this prototype allows for the experimental testing of various $R_{ds(on)}$ rated devices for APS application.

References

[1] J. A. Baxter, D. A. Merced, D. J. Costinett, L. M. Tolbert, and B. Ozpineci, "Review of electrical architectures and power requirements for automated vehicles," in *2018 IEEE Transportation Electrification Conference and Expo (ITEC)*, 2018, pp. 944–949.

[2] A. Emadi, Y. J. Lee, and K. Rajashekara, "Power electronics and motor drives in electric, hybrid electric, and plug-in hybrid electric vehicles," *IEEE Transactions on Industrial Electronics*, vol. 55, no. 6, pp. 2237–2245, 2008.

[3] C. Wang, P. Zheng, and J. Bauman, "A review of electric vehicle auxiliary power modules: Challenges, topologies, and future trends," *IEEE Transactions on Power Electronics*, pp. 1–12, 2023.

[4] S. M. N. Hasan, M. N. Anwar, M. Teimorzadeh, and D. P. Tasky, "Features and challenges for auxiliary power module (apm) design for hybrid/electric vehicle applications," in *2011 IEEE Vehicle Power and Propulsion Conference*, 2011, pp. 1–6.

[5] G. Vitale, "Dc/dc converter for hevs and resonant active clamping technique," in *AEIT Annual Conference 2013*, 2013, pp. 1–6.

[6] R. Muhammad, S. Kim, C. Suk, S. Choi, B. Yu, and S. Park, "Integrated planar transformer design of 3-kw auxiliary power module for electric vehicles," in *2020 IEEE Energy Conversion Congress and Exposition (ECCE)*, 2020, pp. 1239–1243.

[7] V. B. Gandluru, K. Autkar, and Y. Hu, "Identifying suitable psfb topology for hvlv auxiliary power supply (aps) application in evs," in *2024 IEEE Applied Power Electronics Conference (APEC)*, 2024.

PCIM Europe 2024, 11– 13 June 2024, Nuremberg   DOI: 10.30420/566262307

Bidirectional DC-DC Topologies Comparison for 800 V Automotive Applications Integrating 650 V GaN-on-Si Devices

Ilias Chorfi[1], Julio Brandelero[1], Corinne Alonso[2], Romain Monthéard[3], Thierry Sutto[1]

[1] STMicroelectronics, Automotive & Discrete Group (ADG). Labège, France
[2] LAAS-CNRS, Université de Toulouse, CNRS, UPS. Toulouse, France
[3] Commissariat à l'énergie atomique et aux énergies alternatives (CEA). Labège, France.

Corresponding author: Ilias Chorfi, ilias.chorfi@st.com
Speaker: Ilias Chorfi, ilias.chorfi@st.com

Abstract

Wide band gap (WBG) devices and particularly Gallium Nitride (GaN) switches are attractive solutions to increase the compactness of power converters in automotive applications, thanks to their improved switching characteristics and lower on-resistance. Nevertheless, their 650 V voltage rating limits their use in 800 V on-board chargers (OBC). In this paper, a side-by-side comparison between two isolated 800 V dc-dc topologies is proposed, backed by consistent experimental data. These topologies are based on 650 V GaN-on-Si devices and derived from the dual active bridge (DAB): the series-input series-output dual active half bridge (SISO DAHB) topology and the three-level active neutral point clamped DAHB (3L-ANPC DAHB) topology. This work compares the trade-offs between these two power circuits and evaluates the impact of the switching frequency and transformer design on the GaN switches losses, the overall conversion efficiency, as well as thermal dissipation. Experiments conducted on the realized prototypes show very good efficiencies even at high switching frequencies. Slightly lower losses are observed in SISO DAHB compared to 3L-ANPC DAHB, that has a greater switch count but fewer passive components.

1   Introduction

Upgrading the battery voltage from 400 V to 800 V reduces the wiring harness and decreases the charging time, allowing improvement of the overall performance of electrical vehicles (EV) [1][2]. The dual active bridge isolated dc-dc topology is used in various applications, but it is particularly interesting for EV charging applications due to its high efficiency, ease of zero voltage switching (ZVS), reliability and bidirectional power flow [3][4]. The DAB topology could benefit from using GaN devices with their improved switching characteristics and lower on-resistance ($R_{DS,on}$) [5]. Furthermore, GaN devices are particularly interesting in ZVS dc-dc topologies such as DAB topology, because of their very low turn-off energy which is 10 times lower than that of Silicon Carbide (SiC) MOSFET with equivalent ratings as can be seen in Fig. 1. Thus, the switching frequency could be increased to reduce the size of the passives

while maintaining good efficiency compared to SiC as shown in Fig. 2. Furthermore, thanks to the very low parasitic capacitance of the GaN devices, ZVS operation is achieved even at light loads. Nevertheless, the most commonly available and mature GaN transistors are qualified at 650 V [6].

This paper proposes a comparison between SISO DAB and multilevel DAB topologies in half-bridge configuration, used to overcome the voltage rating limitation of GaN devices in 7.4 kW/800 V battery charging systems. In section 2, the operational principle of each topology is explained, and the theoretical waveforms are provided, as well as the comparison between the two topologies. In section 3, the high frequency magnetic design is discussed. Finally, the experimental realization and results of both topologies are shown in section 4.

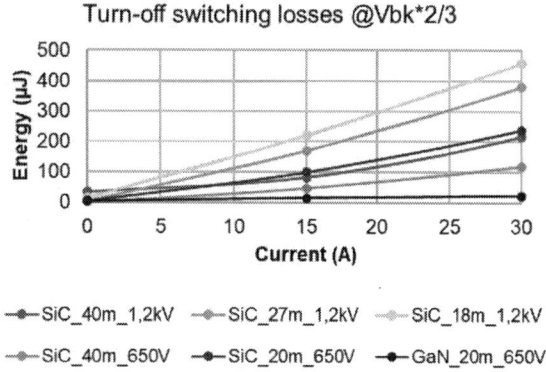

Fig. 1: Turn-off switching energies of different GaN and SiC power devices

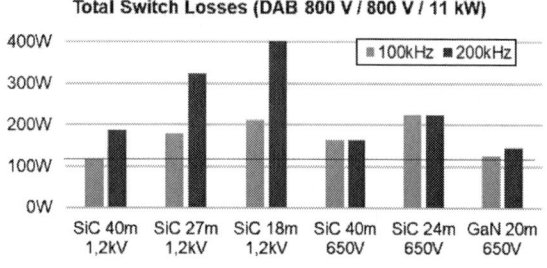

Fig. 2: Total losses for a conventional DAB different GaN and SiC power devices

2 Operation principle of the proposed topologies

The DAB topology can be implemented in different configurations while maintaining the same principle of operation [7], which consists of introducing a phase shift between the control signals of the primary and the secondary to create a power flow from the leading bridge to the lagging bridge. DAB topology is inherently bidirectional. In a full bridge configuration, any slight difference between the duty cycles of the devices can cause a DC bias current in the transformer which leads to core saturations and additional losses. This phenomenon is naturally avoided when using the dual active half bridge (DAHB) by using a capacitive voltage divider that blocks the DC current bias that leads to smaller magnetics, as they do not have to account for the DC current. Figure 3 represents the two proposed 800 V DAHB topologies using 650 V GaN devices. In Fig. 3a, two two-level DAHB converters are stacked in series to sustain 800 V, while in Fig. 3b, the half bridges of the DAHB are

implemented using three-level Active Neutral Point Clamped (3L-ANPC) topology to support 800 V.

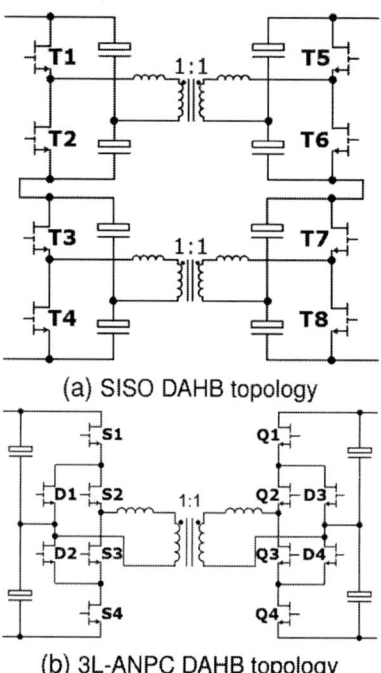

(a) SISO DAHB topology

(b) 3L-ANPC DAHB topology

Fig. 3: The proposed topologies for 800 V dc-dc converter using 650 V GaN devices

Table 1 shows the technical parameters of both topologies, the same technical specifications is fixed for both topologies to have a meaningful comparison.

Parameter	Value
Input voltage (V_{in})	800 V
Output voltage (V_{out})	800 V
Output power (P_{out})	7.4 kW
Switching frequency (f_{sw})	250 kHz

Tab. 1: Technical specifications of the proposed topologies

Phase-Shifted Pulse Width Modulation (PS-PWM) is applied to the primary and secondary ANPC half-bridges in order to generate the three-level square voltage. Fig. 4 shows the gate drive signals of the different devices in the proposed converter, all devices switch at 50% duty cycle with a phase shift applied between the outer switches (S_1–S_4 and S_5–S_8) and inner switches (S_2–S_3 and S_6–S_7). The phase shift d_1 is applied to the primary half-bridge, while d_2 is applied to the secondary half-bridge. To simplify the analysis and the design of the proposed converter, d_1 and d_2 are considered equal and are referred to as d_1 in the following. PS-PWM modulation is selected for its multiple advantages such as natural voltage balancing of the different devices, even losses distribution and ease of implementation. The natural voltage balancing is highly desirable as it eliminates the need of using sophisticated and resource-heavy active voltage balancing algorithms. The transferred power from the pri-

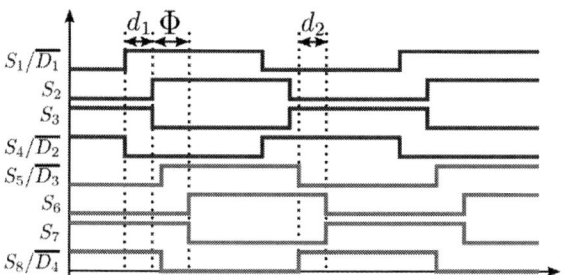

Fig. 4: PS-PWM gate drive signals applied to the three-level DAHB

mary side to the secondary side is described by (1), where P_{out} is the average output power, ω is the switching angular frequency, and n is the turn ratio of the AC-link transformer. This equation is valid only in the case of $d_1 < \Phi < \frac{\pi}{2}$. The value and the direction (direct or reversed) of the transferred power can be controlled by choosing the appropriate combination of d_1 and Φ. If $d_1 = 0°$ in (1), the equation is reduced to the SISO DAHB converter as can be seen in equation (2).

$$P_{out3L} = \frac{n \cdot V_p \cdot V_s}{\omega \cdot L_{lk3L}} \cdot (\Phi - \frac{\Phi^2}{\pi} - \frac{d_1^2}{\pi}) \qquad (1)$$

$$P_{out2L} = \frac{n \cdot V_p \cdot V_s}{\omega \cdot L_{lk2L}} \cdot (\Phi - \frac{\Phi^2}{\pi}) \qquad (2)$$

Figure 5 shows the different theoretical waveforms of the three-level ANPC DAHB under the condition of $d_1 < \Phi < \frac{\pi}{2}$. This condition can be considered the generalized case [8]. The near-sinusoidal shape of the transformer current contains very few harmonics (THDi) which results in more efficient and smaller magnetics, eventually leading to greater compactness.

2.1 Comparison of the two topologies

Table 2 shows the required components and their voltage/current stress for each topology, assuming a 1:1 transformer turn ratio. SISO DAHB requires more passives which impacts the power density,

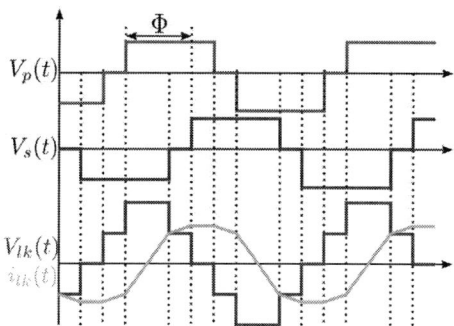

Fig. 5: Three-level DAHB converter waveforms

and may result in reliability issues compared to the 3L-ANPC topology. Furthermore, the capacitors of the SISO DAHB need an active voltage balancing algorithm in order for the converter to operate correctly, while the capacitors voltages are balanced naturally in the case of 3L-ANPC DAHB due to the use of phase-shift PWM modulation [9]. The major advantage of the SISO DAHB topology is the ability to operate at a wide voltage range, as well as its lower switch count compared to 3L-ANPC DAHB. On the other hand, 3L-ANPC DAHB requires four more GaN devices which results in more losses, however it should be noted that the D1, D2, D3 and D4 devices are clamping devices, and should theoretically present near zero losses. In practice they present low non-negligible losses since they operate at ZVS and ZCS. The major advantage of the 3L-ANPC DAHB topology is the additional voltage level in each bridge, leading to less harmonics, thus increasing the efficiency of the magnetics and reducing dv/dt and di/dt (EMI). Furthermore, the additional voltage level extends the ZVS range of the topology, especially at light loads [10], although it adds complexity to the control algorithm.

3 Transformer design

Table 3 shows the technical specifications of the transformer. In order to improve further the power density of the converter, the series power transfer inductor will be integrated in the transformer. Furthermore, the stray inter-winding capacitance is limited to 300 pF to limit the oscillations due to the fast transients of the GaN devices. The same transformer will be used for both topologies, in the case of SISO-DAHB topology, each DAHB is implemented using a 3.7 kW transformer to have 7.4 kW in total, as for the 3L-ANPC DAHB, two transformers connected in series will be used to support ±400 V and 7.4 kW operation.

	Quantity		Current stress		Voltage stress	
	SISO DAHB	ANPC DAHB	SISO DAHB	ANPC DAHB	SISO DAHB	ANPC DAHB
GaN devices	8	12	$2I_{out}$	$2I_{out}$	$V_{bus}/2$	$V_{bus}/2$
Transformer	2	1	$2I_{out}$	$2I_{out}$	$V_{bus}/4$	$V_{bus}/2$
Capacitor	8	4	N/A	N/A	$V_{bus}/4$	$V_{bus}/2$
Inductor	4	2	$2I_{out}$	$2I_{out}$	N/A	N/A

Tab. 2: Comparison of different active and passive elements of SISO and 3L-ANPC DAHB topology

Parameter	Value
Turn ratio	1:1
f_{sw}	250 kHz
Input/output voltage	± 200 V
Peak current	22 A_{pk}
RMS current	20 A_{RMS}
Nominal power	3.7 kW
Magnetizing inductance	100 µH
Total leakage	4.2 µH
Stray inter-winding capacitance	≤ 300 pF

Tab. 3: Technical specifications of the high frequency transformer

The transformer is constructed using two PQ60/52 cores in KF9A material for the transformer, and one PQ60/52 core for the inductor. The inductor core is fitted on the top of the transformer cores to improve magnetic coupling and to better control the value of the leakage inductor. In addition, the winding are realized using six 660x0.05mm Litz wire in parallel to support the high RMS current at the high switching frequency. Figure 6 shows the realized 3.7 kW/250 kHz transformer.

Fig. 6: The experimental realization of the high frequency transformer

4 Experimental results

The two converters are constructed using a modular approach. Each pair of devices is implemented in a daughterboard using 25 mΩ top-cooled GaN HEMTs in a PFLAT8x8 package, with a bus bar motherboard PCB ensuring the appropriate connections for each topology. The GaN devices are assembled on the bottom of the board to take advantage of the top side cooling, as seen in Fig. 7, where the heatsink is a 60 °C cold plate.

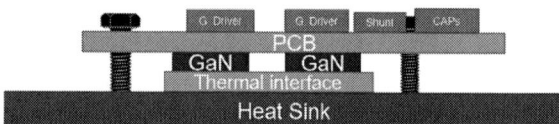

Fig. 7: Mechanical assembly of the GaN daughterboards

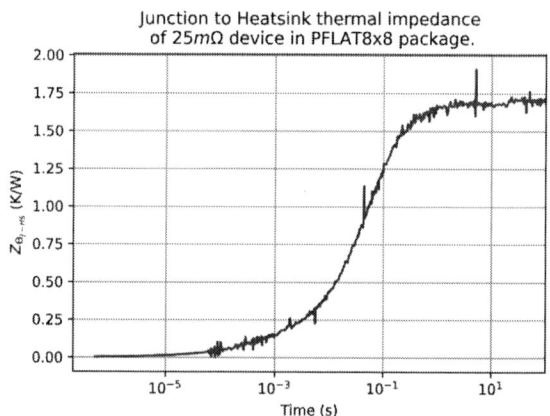

Fig. 8: Junction-to-heatsink thermal impedance of the 25 mΩ GaN device in PFLAT8x8 using GP5000S35 TIM

Figure 9 shows the experimental prototypes of the proposed topologies, the transformer is not assembled in the pictures. The experimental waveforms of both topologies are presented in Fig. 10, the current shape of the 3L-ANPC DAHB topology is nearly sinusoidal which result in lower THDi and better EMI performances. Furthermore, the additionnal voltage level reduce the dv/dts. The wave-forms in both topologies have minimal oscillations and overshoot due to the optimized layout of the GaN daughterboards, and the small inter-winding stray capacitance of the transformer.

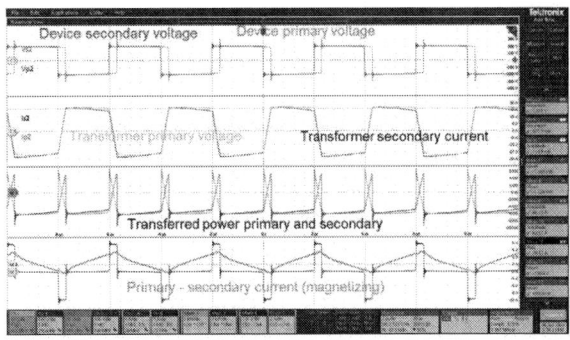

(a) SISO DAHB experimental waveforms

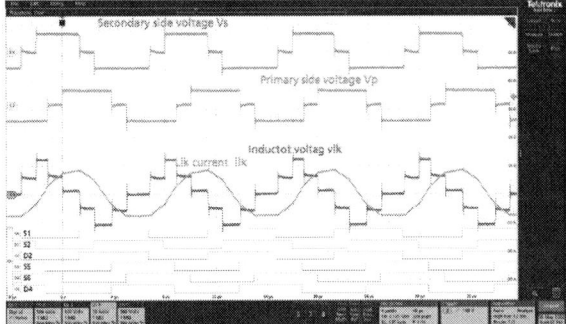

(b) 3L-ANPC DAHB experimental waveforms

Fig. 10: Experimental waveforms of the proposed topologies

The efficiency of both realized converters is measured at $V_{in}=_{out}=800$ V and for cold plate temperature of 20 °C and 60 °C as can be seen in Fig. 11. The temperature of the cold plate has very little effect on the efficiency of both converter up to the rated power of 7.4 kW. However, when the output power is higher than the rated power, the 20 °C cold plate allows the operation of both converters upto to 10 kW while maintaining a good efficiency of about 97%. While the 60 °C cold plate limit the operation to 8 kW.

The efficiency of the 3L-ANPC DAHB converter is slightly less than that of the SISO DAHB converter, due to the additionnal clamping devices in the 3L-ANPC DAHB converter, which they add about 1.5 W of losses per devices on the entire power range. The clamping devices operates at ZVS and ZCS and zero RMS current, therefore, the losses of these devices are due to other mechanisms such as deadtime and associated third-quandrant losses.

When the output power is less than 1 kW,

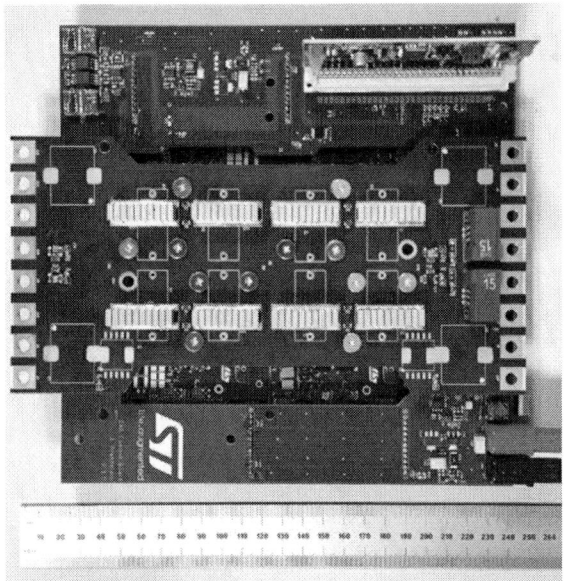

(a) SISO DAHB experimental prototype

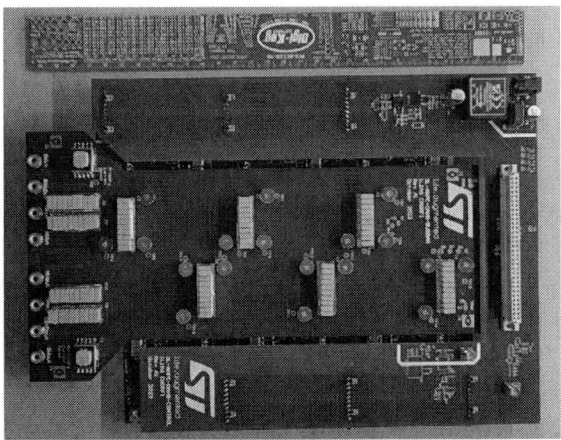

(b) 3L-ANPC DAHB experimental prototype

Fig. 9: Experimental prototypes of the proposed topologies

all the devices operates at hard-switching in both converters which explain the low efficiency. However, the moment the devices operates in ZVS, the efficiency is quite flat on the remaining power range due to the very low turn-off energy of GaN devices.

Figure 12 shows the thermal imagery of the realized transformer at the rated power of 3.7 kW, the hot spot is measured at 94 °C with natural cooling, which is reasonable for automotive applications. The thermal imagery shows a well balanced temperature gradient between the copper and the core. Furthermore, the transformer will be capable of passing more power if is cooled by the same cold plate used for the GaN devices.

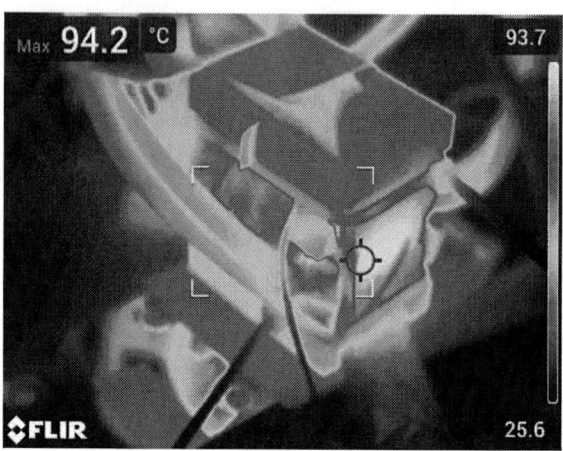

Fig. 12: Thermal image of the transformer at the maximum power

5 Conclusion

With a focus on providing side-by-side GaN-based experimental data, this article proposed a fair comparison of the SISO DAHB and the 3L-ANPC DAHB, which are derived from the well-known two-level DAB topology. Both can be proposed as a solution to enable the use of 650 V GaN devices in 800 V/11 kW battery charging systems for EVs. The topologies take full advantage of the low GaN turn-off energy to achieve a very good efficiency, even when operating at 1 MHz switching frequency. The SISO topology shows slightly better efficiency due to its lower switch count. The 3L-ANPC DAHB has fewer magnetics and capacitors and can therefore be expected to achieve greater power density and reliability

References

[1] C. Jung, "Power up with 800-v systems: The benefits of upgrading voltage power for battery-electric passenger vehicles," *IEEE Electrification Magazine*, vol. 5, no. 1, pp. 53–58, 2017. DOI: 10.1109/ MELE.2016.2644560.

[2] M. Safayatullah, M. T. Elrais, S. Ghosh, R. Rezaii, and I. Batarseh, "A comprehensive review of power converter topologies and control methods for electric vehicle fast charging applications," *IEEE Access*, vol. 10, pp. 40753–40793, 2022. DOI: 10.1109/ACCESS.2022.3166935.

[3] B. Zhao, Q. Song, W. Liu, and Y. Sun, "Overview of dual-active-bridge isolated bidirectional dc–dc converter for high-frequency-link power-conversion system," *IEEE Transactions on Power Electronics*, vol. 29, no. 8, pp. 4091–4106, 2014. DOI: 10.1109/TPEL.2013.2289913.

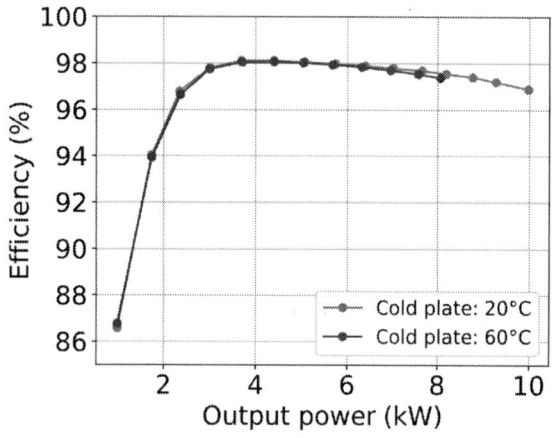

(a) Efficiency of SISO-DAHB at 800 V

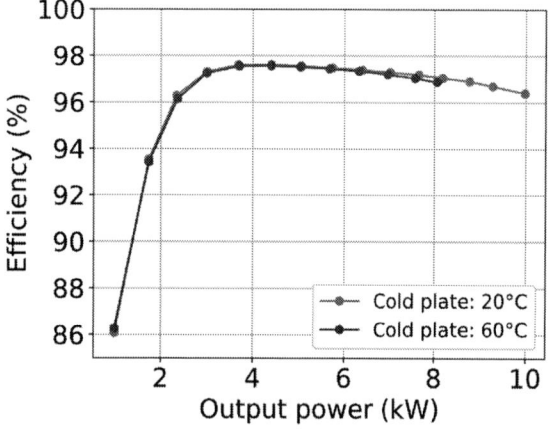

(b) Efficiency of 3L-ANPC DAHB at 800 V

Fig. 11: Efficiency of the realized converters at 250 kHz

[4] P. He and A. Khaligh, "Comprehensive analyses and comparison of 1 kw isolated dc–dc converters for bidirectional ev charging systems," *IEEE Transactions on Transportation Electrification*, vol. 3, no. 1, pp. 147–156, 2017. DOI: 10.1109/TTE. 2016.2630927.

[5] J. Millán, P. Godignon, X. Perpiñà, A. Pérez-Tomás, and J. Rebollo, "A survey of wide bandgap power semiconductor devices," *IEEE Transactions on Power Electronics*, vol. 29, no. 5, pp. 2155–2163, 2014. DOI: 10.1109/TPEL.2013.2268900.

[6] S. Chowdhury, Y. Wu, L. Shen, P. Smith, J. Gritters, *et al.*, "650 v highly reliable gan hemts on si substrates over multiple generations: Expanding usage of a mature 150 mm si foundry," in *2019 30th Annual SEMI Advanced Semiconductor Manufacturing Conference (ASMC)*, 2019, pp. 1–5. DOI: 10.1109/ASMC.2019.8791814.

[7] H. Higa, S. Takuma, K. Orikawa, and J.-i. Itoh, "Dual active bridge dc-dc converter using both full and half bridge topologies to achieve high efficiency for wide load," in *2015 IEEE Energy Conversion Congress and Exposition (ECCE)*, 2015, pp. 6344–6351. DOI: 10.1109/ECCE.2015. 7310549.

[8] M. A. Moonem, C. L. Pechacek, R. Hernandez, and H. Krishnaswami, "Analysis of a multilevel dual active bridge (ml-dab) dc-dc converter using symmetric modulation," *Electronics*, vol. 4, no. 2, pp. 239–260, 2015. DOI: 10.3390/ electronics4020239.

[9] Y. Li, H. Tian, and Y. W. Li, "Generalized phase-shift pwm for active-neutral-point-clamped multilevel converter," *IEEE Transactions on Industrial Electronics*, vol. 67, no. 11, pp. 9048–9058, 2020. DOI: 10.1109/TIE.2019.2956372.

[10] Y. Ikai and N. Hoshi, "Expanding zvs range for dual active bridge dc-dc converter using three-level neutral-point-clamped inverter topology," in *2016 IEEE International Conference on Renewable Energy Research and Applications (ICRERA)*, 2016, pp. 472–477. DOI: 10.1109/ICRERA. 2016.7884382.

Analysis of Phase Shielding Method Based on Δ-Cr-Y Three-Phase Interleaved LLC Converter

Jin Wen[1], Jiajia Guan[1], Zongheng Wu[1], Wenzhe Xu[1], Cai Chen[1], Yong Kang[1].

[1]Huazhong University of Science and Technology, China

Corresponding author: Cai Chen, caichen@hust.edu.cn
Speaker: Jin Wen, wenjin@hust.edu.cn

G02 DC-DC Hard- and Soft-Switched Converters
Preferred presentation form: poster presentation

Abstract

Based on the Δ-Cr-Y three-phase interleaved LLC converter, a series of possible phase shielding methods are proposed, and one of them is analyzed in detail. After the phase shielding operation, the resonant frequency of the converter does not change, which means smooth switching in the field of DC transformers. After loss analysis and comparison, the efficiency can still be improved at light load due to the reduction in volt-second product, even if the cores cannot be removed from the circuit.

1 Introduction

Three-phase interleaved LLC has soft switching characteristics, voltage regulation capabilities and EMI advantages like LLC converters with more power. There are currently many studies on three-phase interleaved LLC converters, including the impact of different resonant tank connection methods on circuit performance, the improvement of converter performance by different modulation methods, and light load efficiency optimization strategies [1]-[4]. The light load optimization strategy is mainly realized through phase shielding technology. However, the existing phase shielding technologies are all based on Y-Y connected three-phase LLC converters [3]. For this structure, it is easy to remove the resonant components and transformers from the circuit through control switch. However, no phase shielding technology has been proposed for the three-phase LLC converter with a Δ connection structure which has less transformer losses [4].

This article proposes phase shielding technologies for a delta-type three-phase LLC converter. Circuit analysis and loss comparisons will be demonstrated based on one of the phase shielding methods.

2 Design and Optimization of the Converter

2.1 Phase Shielding Methods

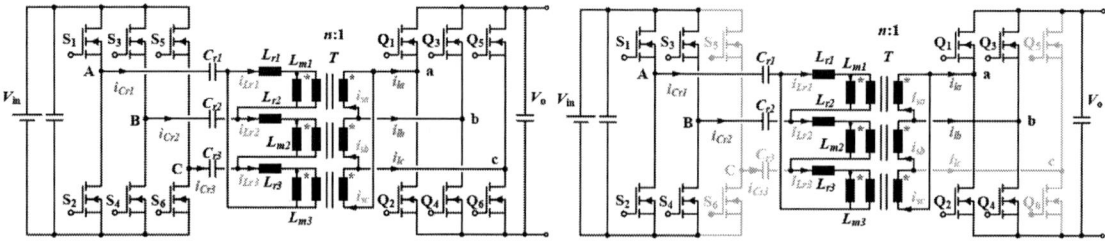

Fig. 1: Δ- Cr -Y three-phase LLC converter   **Fig. 2:** ABX mode of the converter

This paper will focus on the discussion of phase shielding technology and will not introduce the operating mode of the circuit in detail. The possible phase shielding methods and classifications are shown in Table I, where 0 means that the corresponding half-bridge is operating in the grounded state and X means that the corresponding half-bridge is completely disconnected and does not operate.

TABLE I

POSSIBLE OPERATING MODES

Normal operating condition	Phase shielding condition				
Three-phase interleave	Full bridge	Full bridge	Parallel half bridge	Half bridge	Parallel half bridge
ABC	ABX	AB0	AB0-SYN	A0X	A00

According to this rule, ABC represents normal three-phase interleaved operation as shown in Fig. 1, while ABX represents the mode as shown in Fig. 2.

Half-bridge phase shielding causes the gain to be cut in half and can generally be used in battery and energy storage applications with wide voltage variations. To simplify the analysis, only the full-bridge mode with similar voltage gains will be discussed here. Taking ABC and ABX modes as an example, the gain and loss analysis of the circuit will be given.

2.2 Gain Characterization

The circuit is analyzed using the first harmonic approximation method, where it is assumed that the circuit is completely symmetrical. The gain shown in (1) is the same as the gain form of the traditional LLC converter, but the characteristic parameters are different. The gain curves are shown in Fig. 3.

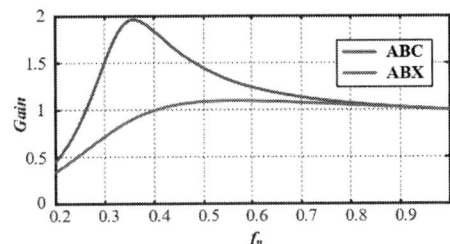

Fig. 3: Gain curves under the two modes

$$M = \frac{1}{\sqrt{\left(1 + \frac{1}{k} - \frac{1}{kf_n^2}\right)^2 + Q^2\left(f_n - \frac{1}{f_n}\right)^2}} \tag{1}$$

$$f_n = \frac{f_s}{f_r}, f_r = \frac{1}{\sqrt{L_r C_r / 3}}, k = \frac{L_m}{L_r} . \text{ For ABC: } Q = \frac{\sqrt{L_r / 3 / C_r}}{R_{eq}}, R_{eq} = \frac{18n^2}{\pi^2} R . \text{ For ABX: } Q = \frac{2\sqrt{L_r / 3 / C_r}}{R_{eq}}, R_{eq} = \frac{8n^2}{\pi^2} R$$

2.3 Evaluation of Main Losses

The following will take ABC mode and ABX mode as examples to analyze the main losses under light load including winding losses $P_{winding}$, Core losses of transformer P_{T-core}, Core losses of resonant inductor $P_{Lr-core}$, Switching losses P_{sw}, gate-driver losses P_{dr}.

ABC mode:

$$P_{winding} = \left(\frac{5}{108}\left(\frac{nV_oT}{L_m}\right)^2 + \frac{1}{18}\left(\frac{\pi I_o}{n}\right)^2\right)R_{pri} + \frac{1}{18}\left(\pi I_o\right)^2 R_{sec} \tag{2}$$

$$P_{T-core} = 3V_0 \frac{k_i}{T}\left(\frac{nV_oT}{3A_eN}\right)^{\beta-\alpha}\int_0^{T/3} 2\left|\frac{nV_o}{NA_e}\right|^\alpha dt \tag{3}$$

$$P_{Lr-core} = 3V_{Lr}\frac{K}{T^\alpha}\left(\frac{L_r}{N_{Lr}A_{Lr}}\sqrt{\frac{5}{162}\left(\frac{nV_oT}{L_m}\right)^2 + \frac{1}{27}\left(\frac{\pi I_o}{n}\right)^2}\right)^\beta \tag{4}$$

$$P_{cond} = \left(\frac{5}{36}\left(\frac{nV_oT}{L_m}\right)^2 + \frac{1}{6}\left(\frac{\pi I_o}{n}\right)^2\right)R_{dson,pri} + \frac{1}{6}\left(\pi I_o\right)^2 R_{dson,sec} \tag{5}$$

ABX mode:

$$P_{winding} = \left(\frac{1}{32}\left(\frac{nV_oT}{L_m}\right)^2 + \frac{1}{12}\left(\frac{\pi I_o}{n}\right)^2\right)R_{pri} + \frac{1}{12}\left(\pi I_o\right)^2 R_{sec} \tag{6}$$

$$P_{T-core} = V_0\frac{k_i}{T}\left((\frac{nV_oT}{2A_eN})^{\beta-\alpha}\int_0^{T/2} 2\left|\frac{nV_o}{NA_e}\right|^\alpha dt + 2(\frac{nV_oT}{4A_eN})^{\beta-\alpha}\int_0^{T/2} 2\left|\frac{nV_o}{2NA_e}\right|^\alpha dt\right) \tag{7}$$

$$P_{T-core} = V_{Lr}\frac{K}{T^\alpha}\left(\left(\frac{L_r}{N_{Lr}A_{Lr}}\sqrt{\frac{1}{24}\left(\frac{nV_oT}{L_m}\right)^2 + \frac{1}{9}\left(\frac{\pi I_o}{n}\right)^2}\right)^\beta + 2\left(\frac{L_r}{N_{Lr}A_{Lr}}\sqrt{\frac{1}{96}\left(\frac{nV_oT}{L_m}\right)^2 + \frac{1}{72}\left(\frac{\pi I_o}{n}\right)^2}\right)^\beta\right) \tag{8}$$

$$P_{cond} = \left(\frac{3}{32} \left(\frac{nV_oT}{L_m} \right)^2 + \frac{1}{4} \left(\frac{\pi I_o}{n} \right)^2 \right) R_{dson,pri} + \frac{1}{4} \left(\pi I_o \right)^2 R_{dson,sec} \tag{9}$$

Switching losses:
$$P_{sw} = m_1 f_s \frac{V_{in}I_{off}}{V_s I_s} \left(E_{off} - E_{oss} \right) \tag{10}$$

Gate-driver losses:
$$P_{dr} = m_2 Q_g f_s V_g \tag{11}$$

The loss and efficiency curves in the two modes are shown in Figure 4. Transformer core losses can be significantly reduced under phase shielding operation.

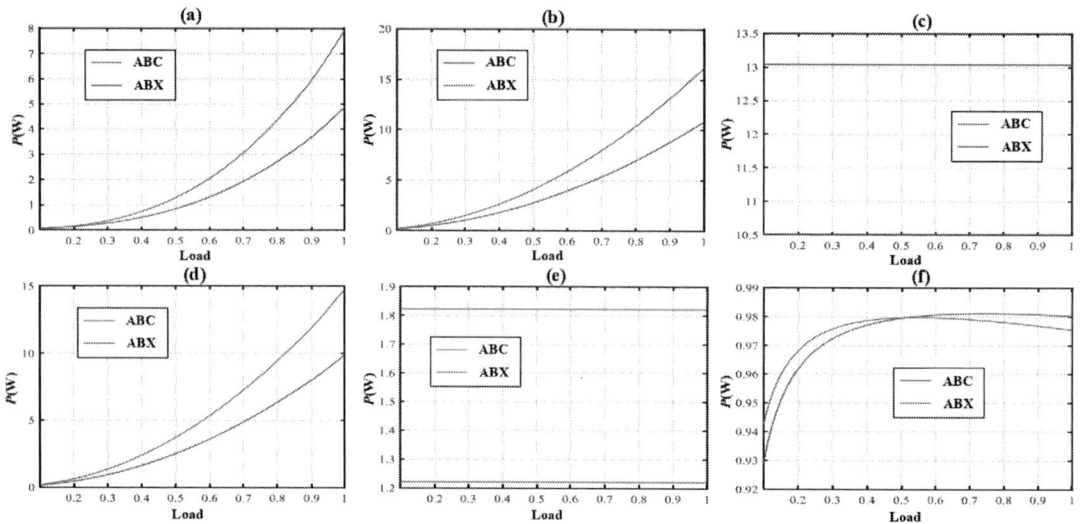

Fig. 4: Comparison of losses between ABC and ABX. (a) inductors' core losses; (b) winding losses; (c) transformers' core losses; (d) conduction losses; (e) switching and driver losses; (f) efficiency.

3 Conclusion

This article proposes phase shielding technologies for a delta-type three-phase LLC converter, and provides a specific analysis based on one of them. Under this phase shielding operation, although the removal of magnetic components is not achieved, the efficiency improvement can still be obtained at light load due to the reduction of the number of components and the volt-second product of the magnetic core. According to theoretical analysis, light load can achieve a 1% efficiency improvement. Figure 5 on the right is a picture of the prototype, and relevant experiments will be conducted in the future for verification.

References

[1] A. de Juan, D. Serrano, P. Alou, J. -N. Mamousse, R. Denéieport and M. Vasic, "Analytical Modelling of Single-Phase and Three-Phase DC/DC LLC Converters," *2022 IEEE Applied Power Electronics Conference and Exposition (APEC)*, Houston, TX, USA, 2022, pp. 2106-2113.

[2] Z. Shi, Y. Tang, Y. Zhang, Y. Guo, H. Sun and L. Jiang, "A Secondary-Side Semiactive 3-Phase Interleaved Resonant Converter Employing Multimode Modulation Scheme for Fast EV Charger Applications," in *IEEE Transactions on Power Electronics*, vol. 37, no. 11, pp. 13385-13397, Nov. 2022.

[3] S. A. Arshadi, M. Ordonez, M. Mohammadi and W. Eberle, "Efficiency improvement of three-phase LLC resonant converter using phase shedding," *2017 IEEE Energy Conversion Congress and Exposition (ECCE)*, Cincinnati, OH, USA, 2017, pp. 3771-3775.

[4] J. Guan et al., "A High Efficiency Δ-Lr-Y Type Three-Phase Interleaved LLC Converter with Less Transformer Loss," in *IEEE Transactions on Power Electronics*, vol. 38, no. 9, pp. 11152-11168, Sept. 2023.

PCIM Europe 2024, 11– 13 June 2024, Nuremberg   DOI: 10.30420/566262309

22 kW IMS-based Bidirectional DC-DC Converter Using Surface Mount SiC MOSFETs for OBCs

Hamlin Wang[1], Yuequan Hu[2], Zongzeng Hu[1]

[1] Wolfspeed, China
[2] Wolfspeed, USA

Corresponding author: Hamlin Wang, hamlin.wong@wolfspeed.com
Speaker: Hamlin Wang, hamlin.wong@wolfspeed.com

Abstract

Recent developments in the electric vehicle automotive market have fueled the demand for higher power on-board chargers (OBCs), which are necessary to support the power levels and densities required for longer range and reduced charging time. Using a discrete SiC MOSFET-based 22 kW bidirectional OBC DC-DC CLLC resonant converter, this paper investigates the impact of power semiconductor device selection, power circuit board (PCB) design, thermal interface material (TIM) selection, and hardware implementation on thermal management and peak efficiency. Experimental results show a 98.6% peak efficiency and 9.4 kW/L power density can be achieved for both charging and discharging modes.

1   Introduction

Investment in electric vehicles (EVs) is accelerating in many locations around the world, but range anxiety and charging time remain significant barriers to widespread adoption. Recently, many battery electric vehicle (BEV) original equipment manufacturers (OEMs) have started to integrate larger batteries with higher voltage and output power to improve range (reaching up to over 600 km on a single charge) and reduce charging time [1]. Presently, BEV makers are moving towards on-board chargers (OBCs) with more than 6.6 kW output power, with many new models focusing on 11 kW and 22 kW systems. These systems must maintain high efficiency at high switching frequency, making thermal management a significant challenge due to the considerable amount of heat dissipation required during power conversion. Therefore, this paper investigates the impact of power semiconductor device selection, power circuit board (PCB) design, thermal interface material (TIM) selection, and hardware implementation for the peak efficiency and power density of high-power bidirectional OBCs.

2   Device Selection and Design Specifications

Previous studies have shown that a SiC MOSFET-based 22 kW bidirectional OBC DC-DC converter with standard through-hole (TH) TO-247-4 package devices can meet automotive requirements for high-power OBCs, including high power density, compact size, and lightweight design [2]. Building on these findings, the present study utilizes bottom-side cooled surface mount technology (SMT) SiC MOSFETs in a standard TO-263-7XL package. SMT packages enable power systems to benefit from a smaller footprint, lower parasitic inductance, and automated assembly [3]. These devices help maximize power density, efficiency, and thermal performance within demanding automotive applications.

As shown in Fig. 1, this study utilizes Wolfspeed's SiC MOSFET (E3M0032120J2, 1200 V / 32 mΩ) in a bottom-side cooled TO-263-7XL package for both the primary and secondary full bridges of the CLLC resonant converter topology.

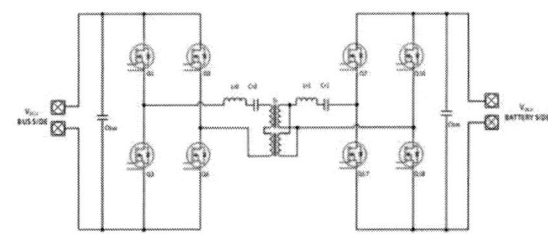

Figure 1   Block diagram of 22 kW DC-DC converter

Tab. 1 shows the major design specifications of the 22 kW DC-DC converter. The DC-link voltage is designed to range from 650 V - 900 V during charging mode to achieve high efficiency, regulate the CLLC converter, and achieve a wide output voltage range of 200 V to 800 V.

The DC-link voltage is designed to range from 360 V - 750 V during discharging mode to achieve high power density and efficiency. 200 kHz is selected for the CLLC converter resonant frequency, and 135 kHz - 250 kHz for the switching frequency range. As these specifications demonstrate, there is a trade-off between power density, efficiency, and thermal performance [2].

Description	Charging Mode	Discharging Mode
Input Voltage (V)	650 - 900	300 - 800
Output Voltage (V)	200 - 800	360 - 750
Output Current (A)	36	19
Rated Power (kW)	22	6.6
Operating Temperature (°C)	-40 - 65 at Heatsink	
Switching Frequency (kHz)	135 - 250	
Peak Efficiency (%)	> 98.6	

Table 1   Design specifications of 22 kW DC-DC converter

3   Thermal Resistance Analysis

In addition to power switching device selection, PCB design is an important consideration for high power OBCs. A significant amount of power (33 W) is dissipated in a small area due to increased power demand, increased power density, and reduced chip size with the introduction of WBG devices. For greater heat dissipation, most OBCs utilize either FR4 PCBs or insulated metal substrate (IMS) PCBs. Although FR4 PCBs with thermal vias offer the possibility of full flexibility and multilayer routing, the thermal conductivity across the PCB is poor. Therefore, a single-layer IMS PCB can be utilized in a high power OBC to improve the thermal performance of SMT cooling.

Figure 2 shows the cross-section of the major heat dissipation path of bottom-side cooled devices with IMS PCB in this study. A liquid cooling baseplate is commonly used in OBCs for EVs, and the temperature of cooling plate can be well controlled at about 65°C.

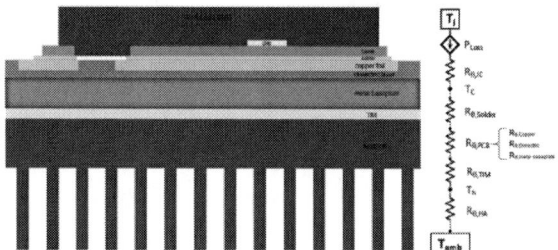

Figure 2   Thermal path of bottom-side cooled devices with IMS PCB

The total thermal resistance from junction to ambient can be estimated as in Eq. (1).

$$R_{\theta,JA} = R_{\theta,JC} + R_{\theta,solder} + R_{\theta,PCB} + R_{\theta,TIM} + R_{\theta,HA} \tag{1}$$

The junction temperature of the device can be estimated as in Eq. (2).

$$T_j = P_{LOSS} \times R_{\theta,JA} + T_{AMB} \tag{2}$$

Where:

$R_{\theta,JA}$ is the thermal resistance from junction to ambient (°C/W),

$R_{\theta,JC}$ is the thermal resistance from junction to case (°C/W),

$R_{\theta,solder}$ is the thermal resistance of the solder (°C/W),

$R_{\theta,PCB}$ is the thermal resistance of the PCB (°C/W),

$R_{\theta,TIM}$ is the thermal resistance of the TIM (°C/W),

$R_{\theta,HA}$ is the thermal resistance from heatsink to ambient (°C/W),

P_{LOSS} is the total power dissipation (W),

T_{AMB} is the ambient temperature (°C).

From Tab. 2 we can conclude that $R_{\theta,PCB}$ and $R_{\theta,TIM}$ contribute the major thermal resistance for a certain selected MOSFET, so it is very important to select and specify the most appropriate PCB solution within an OBC design.

Thermal Resistance	Typical Value	Units
$R_{\theta,JC}$	0.44	°C/W
$R_{\theta,solder}$	0.015	°C/W
$R_{\theta,PCB}$	1.1	°C/W
$R_{\theta,TIM}$	0.38	°C/W

$R_{\theta,HA}$	0.02	°C/W	
$R_{\theta,JA}$	1.955	°C/W	
T_{AMB}	65	°C	
P_{LOSS}	33	W	
Calculated T_j	129.5	°C	

Table 2 Lists of the major thermal resistance in the heat dissipation path

Note 1: The drain pad area of MOSFET is 7.78 mm × 8.0 mm, and package landing pad is about 9.4 mm × 11.8 mm.

Note 2: PCB size is 50 mm × 56 mm × 1.77 mm, including 1.57 mm aluminum (5052H32), 127 μm dielectric (HT-07006) and 70 μm Cu.

Note 3: TIM is THF 5000UT, 254 μm.

Note 4: Thermal resistance is calculated value, and horizontal heat dissipation effect is not considered.

4 Thermal Interface Material

Thermal interface material (TIM) also plays a critical role in the efficiency of heat transfer as $R_{\theta,TIM}$. Often, TIM selection is based on cost, availability, assembly complexity, insulation, and other factors [4][5]. Although most phase change materials have a lower thermal conductivity than other TIMs, the available thickness is thinner than gap pad, gap filler, and gel in the same application. However, phase change materials enable easier handling and reworking, and their thermal resistance does not change significantly with pressure, making them a good choice of TIM for this design. Here we use BERGQUIST HI FLOW THF 5000UT with a thickness of 10 mil and thermal conductivity of 5.3 W/(m-K).

Type	Part Number	Thermal Conductivity (W/m-K)	Available Thickness (mil)
SIL pad	TSP 3500	3.5	10-20
Gap pad	TGP HC5000	5.0	20-125

Gap filler	TGF 4500	4.5	custom
Phase change material	THF 5000UT	5.3	8,10,12,16
Gel	LIQUI-FORM 3500	3.5	Custom
Grease	LOC-TITE TG 100	3.4	Custom

Table 3 High performance TIM materials from Hankel/BERGQUIST

5 IMS PCB Design

Thermal management of SMT devices is different from TH devices. There are many types of mounting options that can be used to mount SMT power devices. While thermal vias is the traditional solution, insulated metal substrate (IMS) PCB is another popular SMT solution used in high-power level and high-power-density applications. Compared to the former, IMS can provide much lower thermal resistance at an appropriate cost.

As shown in Fig. 3, the structure of a standard single layer IMS PCB includes a metal base plate, thin dielectric layer, and copper layer. This configuration offers excellent heat transfer from power devices. The metal baseplate is often aluminum due to low cost and density, but copper is offered as an alternative in certain applications. The copper foil layer is generally a standard foil as used in regular copper-clad PCB laminates. The dielectric layer is resin-based and serves to bond to the metal layers as well as provide electrical insulation between the copper foil and the aluminum plate. The dielectric thickness should be as small as possible to provide the shortest thermal path, while maintaining acceptable electrical insulation and maximizing the thermal conductivity of the dielectric [6].

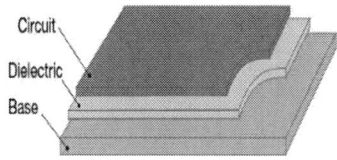

Figure 3 Cross-section view of standard single layer IMS board

6 Hardware Implementation

A digitally controlled 22 kW bidirectional OBC DC-DC prototype was built and tested to verify the design. Figure 4 shows a photo of the hardware, including main board, IMS daughter card, control board, transformer, resonant inductor, baseplate, fans, cover, and so on.

Figure 4 Prototype of 22kW BI-OBC DC-DC converter

In this paper, a half bridge configuration IMS daughter card assembly is used as shown in Fig. 5 with a size of 50 mm (W) × 56 mm (D). A driver card is mounted on the top of IMS board, and a heatsink is mounted on the bottom of IMS board with a phase change TIM. Also screws and nuts are used for securing all the boards together.

Figure 5 Photo of half bridge IMS daughter card with surface mount SiC MOSFETs

7 Experimental Results

The efficiency plots of the prototype are shown in Fig. 6 and Fig. 7, respectively. The peak efficiency is above 98.6% in both charging mode and discharging mode.

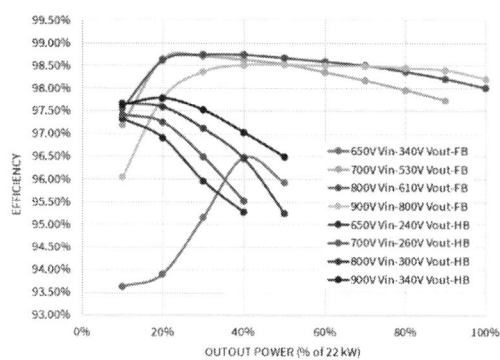

Figure 6 Plot of efficiency vs. output power in charging mode

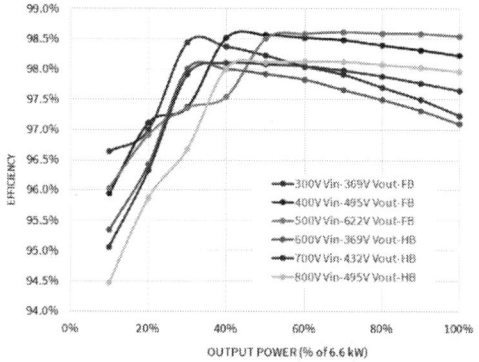

Figure 7 Plot of efficiency vs. output power in discharging mode

In the thermal test of this prototype, forced air cooling is applied, and the temperature of heat sink is kept as 65°C to emulate the liquid cooling system in an automotive application. To simulate the real application, air flow over power semiconductors is blocked, so there is no air flow to the power MOSFETs. The test results are shown in Tab. 4. The junction temperature of all SiC MOSFETs is lower than 105°C, which meets the thermal derating requirement for reliability.

Description	R_{TH_J-C} (°C/W)	Power Loss (W)	Measured Case Temp. (°C)	Calculated Junction Temp. (°C)
Charging Mode, temperature of heatsink: 65°C, input: 650 V, output: 340 V × 36 A.				
High Side MOSFET of DC-link Side	0.44	33.6	88.6	103.4
Low Side MOSFET of DC-link Side	0.44	33.6	89	104.0
High Side MOSFET of Battery Side	0.44	32.9	87.6	102.1
Low Side MOSFET of Battery Side	0.44	32.9	89.7	104.2

Table 4 Thermal test results of MOSFETs

8 Summary

Semiconductor device selection, power circuit board (PCB) design, thermal interface material (TIM), and hardware implementation all impact the efficiency, power density, and reliability of power converters. This paper discusses how to achieve a high peak efficiency (above 98.6%) and high power density (9.4 kW/L) for EV OBC systems using a SiC-based 22 kW CLLC resonant converter.

For bottom-side cooled devices, PCB design and TIMs selection play a critical role in thermal resistance. In this work, daughter card IMS PCBs and phase change TIMs are used to handle the huge amount of heat dissipation. Experimental results show that IMS is an excellent choice in high-power, high-density applications with forced air or liquid cooling system.

References

[1] A. Khaligh, and M. D'Antonio, "Global Trends in High-Power On-Board Chargers for Electric Vehicles," IEEE Trans. Vehicular Technology, Vol. 68, no. 4, pp. 3306-3324, 2019.

[2] C. Wei, D. Zhu, and H. Xie, "A SiC-Based 22kW Bi-directional CLLC Resonant Converter with Flexible Voltage Gain Control Scheme for EV On-Board Charger," PCIM Europe Digital Days 2020; Renewable Energy and Energy Management, Germany, 2020, pp. 1-7.

[3] C. Wei, J. Shao, and B. Agrawal, "New Surface Mount SiC MOSFETs Enable High Efficiency High Power Density Bi-directional On-Board Charger with Flexible DC-link Voltage," APEC 2019, Conference Paper, pp. 1904-1909.

[4] M. T. Demko, J. Yourey, and A. Wong, "Thermal and mechanical properties of electrically insulating thermal interface materials," 2017 16th IEEE ITherm, USA, 2017, pp. 237-242.

[5] S. Chen, and N. Lee, "High Performance Thermal Interface Materials with Enhanced Reliability," 2012 28th Annual IEEE SEMI-THERM Symposium, USA, 2012, pp. 348-353.

[6] D. Mauve, and I. Mayoh, Thermal Management with Insulated Metal Substrates, BR Publishing Inc., Salem, 2018.

Comparative Analysis of DC-DC Converters for Electrolyzers using Geometric Programming.

Tim McRae [*1], Ramkrishnan Maheshwari [2], Thomas Ebel [2]

[1] University of Southern Denmark, Digital and High Frequency Electronics, Denmark
[2] University of Southern Denmark, Centre for Industrial Electronics, Denmark

Corresponding author: Tim McRae, mcrae@sdu.dk
Speaker: Tim McRae, mcrae@sdu.dk

Abstract

This paper uses a multi-objective optimization algorithm to compare different DC-DC converter topologies in a $6kV - 500V$ electrolyzer interface in a Power-to-X (P2X) framework. The algorithm models the volumes and losses of different candidate DC-DC converters and sets constraints to construct a geometric programming problem. Optimal solutions for each converter are generated by sweeping the relative weight of loss to volume to construct a Pareto front, which are then compared. Results indicate that a modular design using Dual Active Bridge converters offers a significantly smaller and more efficient solution compared to the full-bridge and half-bridge converters.

1 Synopsis

Power-to-X (P2X) is a powerful energy framework that allows the reconversion and redistribution of surplus energy generated in off-peak hours to be used in a variety of sectors. Electrolyzers are of particular interest in Denmark, Germany, and Europe in general for P2X. The application here is to connect a $500V$, $50kW$ electrolyzer to a general-use $6kV$ DC grid. This power and voltage present significant challenges in the design of a highly efficient medium-voltage isolated converter. The goal of this paper is to recommend a suitable DC-DC converter in an otherwise open design space.

Typically in these situations, design variables are often selected at the beginning of the design phase to allow simple equations to be developed. Making important design decisions at the very beginning of the design process unnecessarily reduces the design space and potentially limits possible solutions. Exhaustive or brute-force solutions which make thousands to millions of designs and compare results have been used [1], but require intensive computation. Genetic algorithms have also been proposed to generate optimal solutions across dif-

ferent objective, but are also quite computationally expensive [2]. Recently, authors have used geometric programming (GP) - a sub problem within convex optimization - to quickly generate a set of optimal solution in a multi-objective design space [3], [4]. It has been shown that it is possible to make objective decisions about optimality in this framework without relying on fixing certain design variables beforehand. A Pareto front is generated by varying the relative importance of losses and volume over several optimization cycles.

In this work, the method is applied to propose a suitable isolated DC-DC converter for a $6kV - 500V$ electrolyzer interface. A flow diagram of the optimization algorithm can be seen in Fig. 1.A set of converters are compared. Each converter topology is organized in a serial-input, parallel-output configuration so that the input voltage and output current are divided among multiple converters. The set of converter topologies and module organization can be seen in Fig. 2. CVX, a convex optimization tool, is used to solve these optimization problems [5], [6].

This paper 1) develops posynomial models for loss and volume of the four converters, 2) develops posynomial fits for non-convex equations, 3) develops monomial constraints for size, number, and geometry of components in the converters of interest, and 4) constructs Pareto fronts by varying

---

*The authors gratefully acknowledge Erhvervsstyrelsen - the Danish Business Authority, EU's Regionalfond, and Energy Cluster, Denmark for supporting this project.

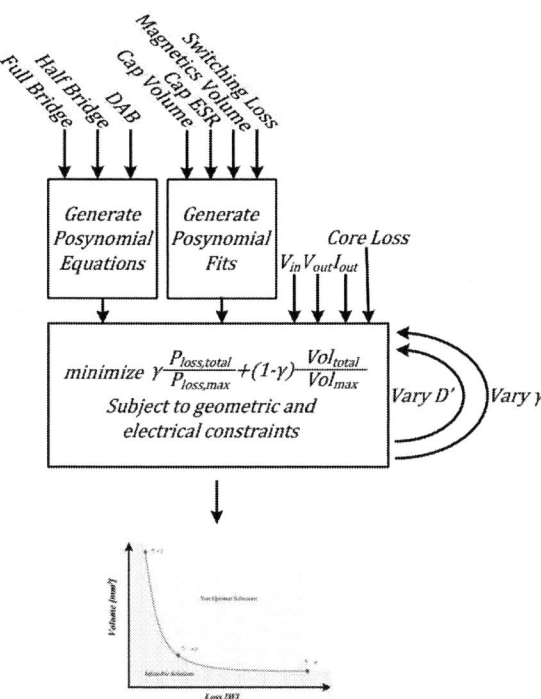

Fig. 1: Top level flow diagram of the optimization algorithm

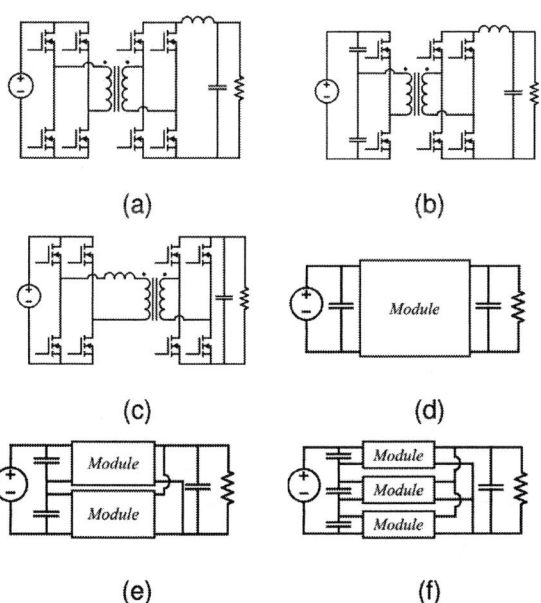

Fig. 2: a)Full Bridge Topology, b) Half Bridge Topology, c) DAB Topology, d) 1 Module Configuration, e) 2 Module Configuration, and f) 3 Module Configuration.

the weight of normalized loss and volume. Once the Pareto fronts are constructed, comparisons between the converters can be made objectively in the Loss-Volume design space.

2 Principle of Operation

This optimization algorithm is similar to the work presented in [7] in that it uses geometric programming to produce sets of Pareto optimal solutions for different converter topologies which can be compared in a multi-objective design space. This comparison allows all system variables to be considered simultaneously without the necessity for pre-selection of variables.

For this application, the full-bridge, the half-bridge, and the dual-active bridge - are compared. Each converter topology can be configured in a serial-input, parallel-output configuration so that the input voltage and output current are divided among multiple converters. Equations for the current of the DAB converter are non-posynomial when left completely unconstrained. To overcome this, two different design choices for the DAB were made. The first forces the turns ratio to be equal to the ratio of the input voltage to the output voltage. In other words this design should generate entirely trapezoidal waveforms. The second is that a fixed phase-shift is set between the primary side bridge and the secondary side bridge. These two choices collapse the equations into posynomials and allows gemeotric programming to be used. This results in two version of the DAB, fixed duty ratio and fixed turns ratio.

Referring to Fig. 1, different converter topologies are input into the algorithm. Each converter is analyzed and posynomial models are generated for the currents and voltages across each component where possible. In cases where a posynomial cannot be constructed, a posynomial equation is used to approximate the formula. Next, component data is input into the algorithm. In this process, data is collected from manufacturers and posynomial models are fit to the data. External parameters are then also included into the algorithm, including input and output voltage, output current, and data regarding core loss. Equations for constraints are also developed for each converter, ensuring that the converter operates as expected given the optimizer output.

These models are then used as inputs to a geometric program, forming the objective function of both loss and volume and the constraint functions. For each converter topology and number of modules, a single-objective optimization for loss and volume respectively is completed to generate the maximum loss and maximum volume solutions in the design space. These values of maximum loss and maximum volume are used to normalize the loss and volume in the multi-objective optimization problem for each converter. This avoids adding loss and volume together, which have different units and their direct sum has no physical meaning. The relative importance of loss and volume can be varied by varying the dummy variable γ. $\gamma = 0$ implies that only volume is prioritized and a minimal volume solution will be produced. $\gamma = 1$ means that only loss is prioritized and a minimal loss solution will be produced. $\gamma = 0.5$ implies an equal trade-off between the two objectives and a solution somewhere between the two extremes of $\gamma = 0, 1$ will be provided.

The set of optimization variables used in this project can be seen in Table 2. Constants were used to account for environmental conditions, materials used, and assumptions about the underlying operation of the converters.

Because the serial-input and parallel-output of modules changes the blocking voltage across the primary side switches, different FETs were selected for the application depending the required blocking voltage. A table of the selected components for the comparison can be seen in Tab. 1. The one-module configurations use the 10 kV SiC Module, the two-module configuration uses the 3300 V FET, and the three-module configuration uses the 2000 V FET. The secondary side uses the 2000 V FET for the full bridge and half brige, and the 650 V FET for the DAB converters. Using the different FETs allows the comparison to take into account the improved figure of merit associated with lower voltage rated switches.

The set of optimal solutions forms a Pareto front, which is seen at the bottom of the flow diagram. The solutions for each converter are simulated to confirm the validity of the solution in terms of conduction loss, switching loss, and operation. In this case PLECS was used to simulate the converters. PLECS intentionally avoids simulation of switching events, so switching loss was estimated using thermal models in the simulation software. Switching loss information from the switch data sheets was used as input to the thermal models of the switches

Tab. 1: Transistors

Manufacturer	Name and Parameters
Alborg University	10 kV SiC Module [8] 10 kV 0.5 Ω
GeneSiC	G2R50MT33k 3300 V 50 $m\Omega$
Infineon	IMYH200R024M1H 2000 V 24 $m\Omega$
ONSemi	NTBG025N065SC1 650 V 19 $m\Omega$

in PLECS. This can be used to confirm the correct voltages and currents were flowing through the device at the switching instant.

By comparing the Pareto fronts of all the converters in all configurations, designers can select the best solution objectively as each Pareto front represents the set of all non-dominated solutions. In the particular case of loss and volume, Pareto fronts which lie closer to the origin, i.e. smallest loss and volume, are objectively better than others.

Some models cannot be made convex or fit within the geometric programming framework with enough precision to provide a meaningful result. This presents a significant challenge to the use of this optimization tool. This paper proposes a method of using the iterative optimization process to overcome the non-convex models, trading off optimization time for modelling time.

3 Models and Constraints

To develop the optimization program, convex equations are needed for all losses and volumes as well as constraints. All equations are posynomials (positive polynomials) or monomials (a single term of a posynomial). An overview of the breakdown can be seen in Table 3, and more detailed descriptions are provided in the following subsections.

3.1 Primary and Secondary Side Switch Conduction Loss

The conduction loss is calculated by developing an equation for the RMS current flowing through the switch and multiplying the square of this term by the on-resistance of the switch. The on-resistance is a variable in the optimizationn problem and cannot be

Tab. 2: Variables and Constants

Variable	Variable Name
Switching Frequency	f_{sw} [Hz]
Inductance	L [H]
L Turns	n_L
Current Ripple	ΔI [A]
L Core Area	$A_{core,L}$ [m^2]
Tx Core Area	$A_{core,trans}$ [m^2]
L Path Length	l_{eff} [m]
Output Capacitor	C_{out} [F]
Duty Ratio	D
Primary Turns	n_p
Secondary Turns	n_s
Primary $R_{DS,on}$	$R_{on,p}$ [Ω]
Secondary $R_{DS,on}$	$R_{on,s}$ [Ω]
Avg Temperature	T_{avg} [$^\circ C$]
Pri FET Temp	$T_{j,p}$ [$^\circ C$]
Sec FET Temp	$T_{j,s}$ [$^\circ C$]

Constant	Constant Name
Amb Temp	25 $^\circ C$
FET Therm Res	0.2-0.38 $^\circ C/W$
Heat-Sink	0.1 $^\circ C/W$
Copper Temp Coeff	3.9m $1/^\circ C$
Cap Leakage Res	1 MΩ
FET Volume	0.05 L
Res per turn	1-5 $m\Omega$

Tab. 3: Loss and Volume Breakdown

Component
Inductor Conduction Loss
Inductor Core Loss
Transformer Conduction Loss
Transformer Core Loss
Primary Side FET Conduction Loss
Primary Side FET Switching Loss
Secondary Side FET Conduction Loss
Secondary Side FET Switching Loss
Input Capacitor Loss
Output Capacitor Loss
Inductor Volume
Transformer Volume
Input Capacitor Volume
Output Capacitor Volume
Active Component Volume
Heatsink

larger than the resistance of a single FET module. The RMS current equations are functions of D and current ripple for the full bridge and half bridge converter. For a general DAB converter, the equation for the RMS current is not a posynomial. However, if the phase shift between the primary and side bridges is fixed (fixed D) or if the turns ratio is fixed to $\frac{V_{out}}{V_{in}}$, then the equations can be simplified. Both types of DAB were optimized. Here the on-resistance is a function of temperature, where the temperature coefficient is extracted from the FET datasheet.

$$P_{p,cond,hb} = (I_o \frac{n_s}{n_p})^2 D (1 + \frac{(\Delta I_o/I_o)^2}{3}) R_{on,pri}(T)$$

$$P_{p,cond,fb} = (I_o \frac{n_s}{n_p})^2 D (1 + \frac{(\Delta I_o/I_o)^2}{3}) R_{on,pri}(T)$$

$$P_{p,cond,dab_{fd}} = \frac{V_g^2 + (V_o \frac{n_p}{n_s})^2}{96 f_{sw}^2 L^2} R_{on,pri}(T)$$

$$P_{p,cond,dab_{fn}} = \Delta I^2 (\frac{D}{3} + D') R_{on,pri}(T)$$

$$P_{s,cond,hb} = I_o^2 D (1 + \frac{(\Delta I_o/I_o)^2}{3}) R_{on,sec}(T)$$

$$P_{s,cond,fb} = I_o^2 D (1 + \frac{(\Delta I_o/I_o)^2}{3}) R_{on,sec}(T)$$

$$P_{s,cond,dab_{fd}} = \frac{V_g^2 + (V_o \frac{n_p}{n_s})^2}{96 f_{sw}^2 L^2} R_{on,sec}(T)$$

$$P_{s,cond,dab_{fn}} = (\Delta I \frac{n_p}{n_s})^2 (\frac{D}{3} + D') R_{on,sec}(T)$$

3.2 Switching Loss

Switching loss is calculated with the assumption that all switches can be turned either on or off (depending on the situation) with soft-switching, while the other edge is hard switched. For each FET, the switching energy is taken from the data sheet and is assumed to be linear with the applied voltage and drain current. For the SiC module from Aalborg university, a model for switching loss was generated from data from the publication [8].

$$P_{sw} = (E_{sw,off} + V_{gatedrive} Q_{gate}) f_{sw}$$

3.3 Inductor and Transformer Loss

The inductor loss is simplified so that it contains two components: conduction loss and core loss. The conduction is calculated similarly to the conduction loss of the switches, where an equation for the RMS current flowing through the inductor is derived, squared and then multiplied by an equivalent resistance of the inductor. It is assumed that AC resistance is not significant at the switching frequencies of interest, so an average resistance is assumed for each turn on the inductor. This resistance is then adjusted based on the estimate temperature of the copper conductor.

The core loss is estimated using the square wave core loss estimates for magnetic materials shown by Sullivan [9]. This previous work provides some models derived from curve fitting for several materials. In this work, the core material selected was 3C81. This curve fit model is a max-function of monomials. The basic loss model for the transformer is the same in that both conduction and core loss are accounted for.

$$P_{L,cond,hb} = (I_o \frac{n_s}{n_p})^2 (1 + \frac{(\Delta I_o/I_o)^2}{3}) R_{ind}(T)$$

$$P_{L,cond,fb} = (I_o \frac{n_s}{n_p})^2 (1 + \frac{(\Delta I_o/I_o)^2}{3}) R_{ind}(T)$$

$$P_{L,cond,dab_{fd}} = \frac{V_g^2 + (V_o \frac{n_p}{n_s})^2}{48 f_{sw}^2 L^2} R_{ind}(T)$$

$$P_{L,cond,dab_{fn}} = 2\Delta I^2 (\frac{D}{3} + D') R_{ind}(T)$$

$$P_{tr,cond,hb} = I_o^2 (1 + \frac{(\Delta I_o/I_o)^2}{3})$$
$$(R_{pri}(\frac{n_s}{n_p})^2 + R_{sec})(1 + \alpha_{cu}\Delta T)$$

$$P_{tr,cond,hb} = I_o^2 (1 + \frac{(\Delta I_o/I_o)^2}{3})$$
$$(R_{pri}(\frac{n_s}{n_p})^2 + R_{sec})(1 + \alpha_{cu}\Delta T)$$

$$P_{tr,cond,dab_{fd}} = \frac{V_g^2 + (V_o \frac{n_p}{n_s})^2}{48 f_{sw}^2 L^2}$$
$$(R_{pri} + R_{sec}(\frac{n_p}{n_s})^2)(1 + \alpha_{cu}\Delta T)$$

$$P_{tr,cond,dab_{fn}} = 2\Delta I^2 (\frac{D}{3} + D')$$
$$(R_{pri} + R_{sec}(\frac{n_p}{n_s})^2)(1 + \alpha_{cu}\Delta T)$$

$$P_{L,core}, P_{tr,core} = max(K_1(\frac{f_{sw}}{100kHz})^{\alpha_1}(\frac{B}{0.1T})^{\beta_1},$$
$$K_2(\frac{f_{sw}}{100kHz})^{\alpha_2}(\frac{B}{0.1T})^{\beta_2})$$

3.4 Input and Output Capacitor Loss

The final loss component included in this optimization problem is input and output capacitor loss. The two components of loss for the capacitors is leakage and ESR loss. Leakage was assumed to be case dependent and leakage resistance was estimated to be $1M\Omega$. A posynomial model for the current flowing through either the input or output capacitor is derived based on the topology. An equation for the ESR was found by curve fitting to data points taken from data sheets. A single monomial fit was used [10], relating ESR to the capacitance of the capacitor and the required blocking voltage. This provided an R-squared value of approximately 0.75. Although this is not ideal, it provides a reasonable estimate for overall capacitor loss. The equations are provided below.

$$P_{C_{in},hb} = \frac{V_{in}^2}{R_{leak}} + \frac{\Delta I^2}{3} R_{esr}$$

$$P_{C_{in},fb} = \frac{V_{in}^2}{R_{leak}} + \frac{\Delta I^2}{3} R_{esr}$$

$$P_{C_{in},dab_{fd}} = \frac{V_{in}^2}{R_{leak}} +$$
$$(\frac{(V_{in}^2)}{f_{sw}^2 L^2 192})^2 + \frac{(\frac{n_p}{n_s})^2 V_o^2}{f_{sw}^2 L^2 48})^2) R_{esr}$$

$$P_{C_{in},dab_{fn}} = \frac{V_{in}^2}{R_{leak}} +$$
$$\Delta I^2 (\frac{2D(1+2D')^2}{3} + 8D'D^2) R_{esr}$$

$$P_{C_{out},hb} = \frac{V_o^2}{R_{leak}} + \frac{(\Delta I \frac{n_p}{n_s})^2}{3} R_{esr}$$

$$P_{C_{out},fb} = \frac{V_o^2}{R_{leak}} + \frac{\Delta I \frac{n_p}{n_s})^2}{3} R_{esr}$$

$$P_{C_{out},dab_{fd}} = \frac{V_o^2}{R_{leak}} +$$
$$(\frac{(\frac{n_p}{n_s})^2 V_{in}^2}{f_{sw}^2 L^2 192})^2 + \frac{(\frac{n_p}{n_s})^4 V_o^2}{f_{sw}^2 L^2 48})^2) R_{esr}$$

$$P_{C_{out},dab_{fn}} = \frac{V_o^2}{R_{leak}} +$$
$$(\Delta I \frac{n_p}{n_s})^2 (\frac{2D(1+2D')^2}{3} + 8D'D^2) R_{esr}$$

$$R_{esr} = e^{\alpha_{esr,1}} C^{\alpha_{esr,2}} V^{\alpha_{esr,3}}$$

3.5 Input and Output Capacitor Volume

The capacitor volume was modelled using a monomial fit based on data gathered from data sheets of capacitors suitable for this voltage range. The model results in an R-squared value of 0.95, indicating that for these capacitors, a scaled energy-based equation predicts the final volume quite well.

$$Vol_C = k_{c_{out/in}} CV^2$$

3.6 Inductor and Transformer Volume

The volume of the magnetics was estimated with a very simple model to avoid the complexity of inductor and transformer optimization, which can be seen in [4]. Instead, the volume of the components is based around a scaled version of the product of the core area of component, the number of turns required, and a factor relating the number of turns to the required window area. It is assumed that the overall volume will be linearly related to the number of turns.

$$Vol_L = A_{core,L} n_L l_{turn}$$
$$Vol_{trans} = A_{core,trans}(n_p + n_s)l_{turn}$$

3.7 Active Volume

The volume of the active components is the volume of a single switch module multiplied by the number of modules used. In the optimization, the on-resistance of each FET is a continuous variable, so instead of counting the number of modules used, we scale the volume of each switch by a factor $\frac{R_{ds,on,single}}{R_{on}}$, which assumes that if the required on-resistance is halved, then the volume of that FET is doubled.

$$Vol_{active} = 4Vol_{module,pri}\frac{R_{ds,on,pri}}{R_{on,pri}} +$$
$$4Vol_{module,sec}\frac{R_{ds,on,sec}}{R_{on,sec}}$$

3.8 Heatsink Volume

The heat sink volume is estimated by determining the total loss and assuming the total required heatsink volume is proportional to that loss.

$$Vol_{hs} = k_{vol/W} P_{loss,total}$$

3.9 Maximum and Minimum Constraints

The multi-objective optimization problem is constrained with both equality and inequality constraints to ensure the final output of the optimizer is able to be made. Some maximum and minimum constraints were placed on geometry, temperature, turns, and component values. This is to ensure that the optimal values can be implemented. The maximums and minimums are shown in Table 4.

Tab. 4: Maximums and Minimums

Variable	Min-Max
f_{sw}	100 Hz-500 kHz
L	1 nH-100 mH
C_{out}	1 μF-10 mF
D	0.05-0.95
n_p	1-100
n_s	1-100
n_L	1-1000
Turns Ratio	N/A-100
R_{on}	0.1 mΩ - N/A
ΔI	1 μA-100 A
$A_{core,L}$	0.001 m^2-1 m^2
$A_{core,trans}$	0.001 m^2-1 m^2
l_{eff}	1 cm - 1 m
B_L FB, HB	N/A - 0.002 T
B_{trans} FB, HB	N/A - 0.5 T
B_L DAB	N/A - 0.3 T
B_{trans} DAB	N/A - 0.3 T
T_{avg}	T_{amb}-120 $^\circ C$
$T_{j,FET}$	T_{amb}-120 $^\circ C$

3.10 Thermal Constraints

Inequality constraints are placed on the temperature rise of overall converter and the FETs. This is calculated using the ambient temperature and adding the overall loss multiplied by the heatsink thermal resistance, or the thermal resistance of the FET. The thermal resistance of the heatsink was estimated and the thermal resistances of the FET were taken from their datasheets.

$$T_{amb} + P_{loss,total}R_{th,hs} \le T_{avg,max}$$
$$T_{amb} + (P_{pri,sw} + P_{pri,cond})R_{th,pri} \le T_{j,FET,pri}$$
$$T_{amb} + (P_{sec,sw} + P_{sec,cond})R_{th,sec} \le T_{j,FET,sec}$$

3.11 Inductor Constraints and Maximum Flux Density

An equality constraint is placed on the inductance to relate the number of turns, the core area, the effective path length, and the core material. For simplicity, the inductor is assumed to be a solenoid. Because constraint is an equality constraint, it must be a monomial function. The inductance for the DAB is assumed to be on the primary side.

$$L = \frac{n_L^2 A_{core,L}\mu}{l_{eff}}$$

There is also a constraint on the inductor current ripple, which varies based on the topology. This is also an equality constraint.

$$\Delta I_{hb} = \frac{V_o D'}{2 f_{sw} L}, \qquad \Delta I_{fb} = \frac{V_o D'}{2 f_{sw} L}$$

$$\Delta I_{dab,fd} = \frac{V_{in}}{4 f_{sw} L} \qquad \Delta I_{dab,fn} = \frac{I_o \frac{n_s}{n_p}}{2 D'}$$

Constraints on placed on the maximum flux density for the inductor and transformer to ensure that the components don't saturate when implemented. The applied voltage $V_{applied}$ depends on the topology and mode of operation.

$$\frac{V_{applied}}{A_{core,t} f_{sw} n_{turns}} \leq B_{max}$$

3.12 Duty Ratio and Phase Shift

A constraint is placed on the duty ratio or phase shift to ensure the correct output voltage is achieved for the given operating point. For the fixed phase shift DAB, a power equation constraint is used as a constraint.

$$D_{hb} = \frac{2 V_o \frac{n_p}{n_s}}{V_{in}}, \; D_{fb} = \frac{V_o \frac{n_p}{n_s}}{V_{in}}$$

$$D f_{sw,DAB,fd} = \frac{V_{in} \frac{n_p}{n_s}}{8 I_o}, \; D_{DAB,fn} = \frac{I_o L f_{sw} \frac{n_s}{n_p}}{2 D' V_{in}}$$

3.13 Output Capacitance

The output capacitor constraints are inequality constraints, allowing posynomial constraints.

$$C_{out,hb} = \frac{\Delta I}{\Delta V_o 8 f_{sw}}, \; C_{out,fb} = \frac{\Delta I}{\Delta V_o 8 f_{sw}}$$

$$C_{out,DAB,fd} = \frac{9 I_o}{\Delta V_o 32 f_{sw}}$$

$$C_{out,DAB,fn} = (D' + \frac{D^2 V_{in}}{8 V_o}) \frac{n_p}{n_s} \frac{V_{in} D^2}{\Delta V_o L f_{sw}^2}$$

3.14 The influence of D'

In conventional analysis of SMPS, the factor D' is used to indicate the period after DT_{sw}, and before the beginning of the next switching period. In fact, $D' = 1 - D$. However, this minus sign in the definition of D' prevents the use of D' in any posynomial equation. To solve this problem, the iterative optimization method shown in [4] is used for D'. First, D' is estimated, then the optimization is performed. After the optimization, the estimated D' and calculated D' are compared. If the estimated D' is closed to the calculated value, γ can be incremented. If not, then D' is recalculated based on the optimization and the system is reoptimized. The novelty here is the difference between the predicted and the calculated D' is scaled and added to the current D', implementing a simple integrator. This is saturated to ensure D' does not go beyond 0.95. The equation as implemented in MATLAB can be seen below.

$$D'_{new} = min(D'_{old} + k(1 - D_{opt} - D'_{old}), 0.95) \quad (1)$$

4 System Design and Simulation Results

Putting together all the previously discussed models and constraints, a full GP can be constructed for each converter and number of modules:

$$\text{minimize} \quad \gamma \frac{P_{loss,total}}{P_{loss,total,max}} + (1 - \gamma) \frac{Vol}{Vol_{max}}$$

$$\text{subject to} \quad T_{amb} + P_{loss,t/p/s} R_{th,hs/p/s} \leq T_{avg/fet,max}$$

$$f_{sw,min} \leq f_{sw} \leq f_{sw,max}$$

$$L_{min} \leq L \leq L_{max}$$

$$C_{o,min} \leq C_o \leq C_{o,max}$$

$$L = \frac{n_L^2 A_{core} \mu_0 \mu_e}{l_{eff}}$$

$$\mu_{e,min} \leq \mu_e \leq \mu_{e,max},$$

$$l_{eff,min} \leq l_{eff} \leq l_{eff,max},$$

$$\Delta I_{eff,min} = f(V_g, V_o, f_{sw}, D)$$

$$\Delta I_{eff,min} \leq \Delta I \leq \Delta I_{eff,max},$$

$$C_o = f(V_g, V_o, I_o, f_{sw}, D, \Delta V_o)$$

$$D = f(V_g, V_o, n_p, n_s)$$

$$D_{min} \leq D \leq \Delta D_{max},$$

$$B_{peak,T} \leq B_{max,T},$$

$$B_{peak,L} \leq B_{max,L},$$

$$n_{p/s/L,min} \leq n_{p/s/L} \leq n_{p/s/L,max}$$

$$\frac{n_p}{n_s} \leq TR_{max}$$

$$R_{on,p/s,min} \leq R_{on,p/s} \leq R_{on,p/s,max}$$

PCIM Europe 2024, 11– 13 June 2024, Nuremberg   DOI: 10.30420/566262310

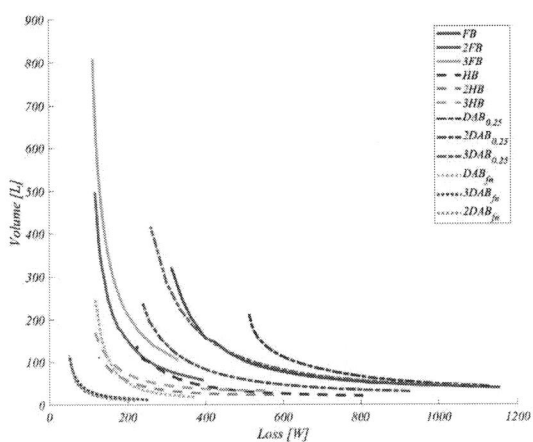

Fig. 3: Pareto fronts of the 1, 2, and 3 module versions of the Full Bridge, Half Bridge, fixed phase shift DAB, and fixed turns ratio DAB.

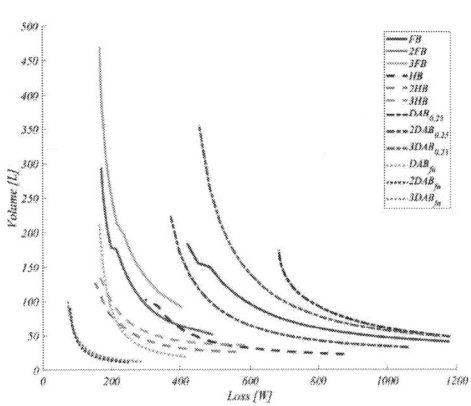

Fig. 4: Pareto fronts of the 1, 2, and 3 module versions of the Full Bridge, Half Bridge, fixed phase shift DAB, and fixed turns ratio DAB, assuming a larger resistance for each turn of the magnetic components.

The weighting variable γ is varied from 0.05 to 0.95 in steps of 0.05 to construct the Pareto front for each combination, resulting in a total of 12 separate Pareto Fronts which can be compared. This means that a minimum of 228 Optimizations are done for the comparison. Due to the iterated optimization to find D', additional optimizations are done.

The output Pareto fronts of the optimization can be seen in Fig. 3. Objectively better designs in this solution space are ones which are closer to the origin, i.e. with lower loss and lower volume. This is a comparison between topologies using models and approximations, giving a relative comparison between the converters and between different loss and volume components within a design.

The system was also optimized with a larger resistance per turn on the magnetics to determine if variation of the conduction loss in the solution has an impact on the final suggestion. The Pareto front can be seen in Fig. 4. Loss and volume breakdowns of the low resistance solutions can be seen in Fig. 5.

4.1 Simulation Results

To verify if the equations for loss were correct, the designs generated by the optimizer were simulated in PLECS. A simulation script was written to accept the generated designs, load them into the simulation, extract the conduction and switching loss values and compare them to the optimized results. Close matching indicates that the equations for loss have correctly assumed the current waveshape in

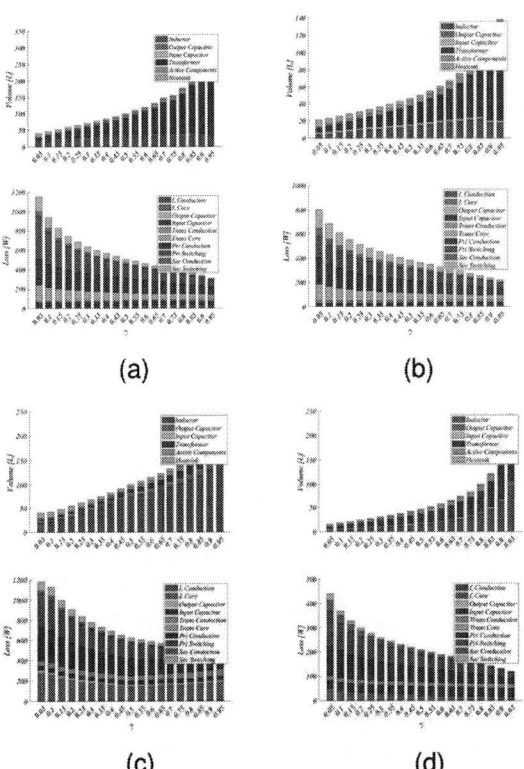

Fig. 5: Distribution of loss and volume for each compoents for the a)Full Bridge Topology, b) Half Bridge Topology, c) DAB Topology, fixed D, d) DAB fixed turns ratio all in the 1 module configuration.

2197

steady state as well as the voltages and currents applied to the switching elements. PLECS uses instantaneous switching actions to decrease simulation time, which unfortunately prevents switching loss prediction because the voltages and currents during the transition are simply not calculated. To overcome this, thermal models can be used in PLECS, which allow switching loss to be calculated using a look-up table. A look-up table for switching loss was generated for each FET used in the simulation (see Table 1) using the switching energy provided in the FET data sheet and extrapolating it to different voltages and currents linearly. This allows verification that the voltages and currents applied to FETs during switching transitions.

Core losses were not compared in this simulation as core losses are not commonly calculated with PLECS and a specific geometry of windings and core shape is not specified by the optimizer. Because of these two factors, conclusive results could not be obtained from this simulation comparison. An additional complexity that should be noted is the optimization model assumes that the conversion ratio is not affected by losses. This means that an ideal duty ratio or phase shift is calculated. In simulation, the losses will result in a lower output voltage than expected. To ensure the output voltage is the same for all cases (and thus all converters operating with the same external conditions), the duty or phase shift is adjusted.

Percent error between calculated and simulated results can be seen in the Fig. 6 for the main optimization. Some error is quite low, typically within a few percent, but there are others reaching up to 45% deviation. This error occurred for the ESR loss associated with the output capacitor for the two-module DAB with fixed turns ratio at the smallest γ, indicating a minimum volume design. At this extreme, the loss was calculated to be 5 W and simulated to be around 10 W. This means that for this particular design the current waveshape was significantly different from the one assumed in the optimization model. While simulation-aided optimization in this case is helpful, further confirmation using experimental setups is needed to verify the validity of the convex models.

5 Discussion and Conclusion

Based on the optimization and simulation results, the most appropriate converter for this $6kV$-$500V$ DC-DC converter electrolyzer application is a 2

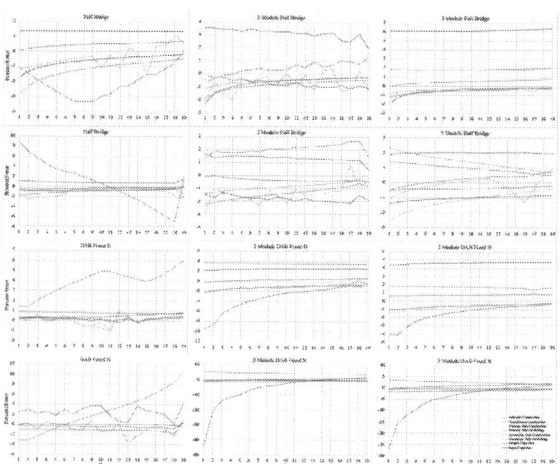

Fig. 6: Error between the calculated loss from the optimizer and the loss predicted through simulation using the component values from the optimizer. Simulations were done in PLECS.

module DAB with a turns ratio equal to the conversion ratio. This is determined simply by looking at the Pareto fronts and noticing that the two-module DAB Pareto front is closest to the origin, meaing that any of these solutions selected will be both smaller and more efficient than any of the other converters. It should be noted that this analysis was done assuming a fixed output voltage and output current. Typically, the DAB with a turns ratio tuned to the conversion ratio exhibits peak efficiency at that single operating point and deviates significantly as the conversion ratio deviates from the turns ratio. This means that in a practical situation, this solution may be not be best for all operating points as the output voltage varies based on output power.

This paper has demonstrated a comprehensive optimization framework which can be used to compare different converter topologies in a multi-objective design space. Models for loss, volume, and constraints for the full-bridge, the half-bridge, and the dual active bridge in two different modes of operation were determined, and they were placed in one, two, and three-module configurations. Each of these circuits was optimized as a geometric program using CVX [5], using an iterative process to generate a Pareto front in the loss-volume design space. During this iteration, a modified approach was used to overcome the non-convexity of D' in the design equations. Finally, the framework was aided by simulation in PLECS, used to verify if the designs generated by the optimizer produced feasible results.

Despite these limitations, this optimization process shows interesting results which could not be achieved by conventional design methods. Future work will be done to account for increasing the scope of the optimization by taking into account different operating points. Extending this simulation aided design to experimental design is the next logical step. Simulation indicates a good correlation, but due to the limitation of PLECS, switching loss and core loss were not directly compared.

References

[1] A. Stupar, T. Friedli, J. Minibock, M. Schweizer, and J. W. Kolar, "Towards a 99% efficient three-phase buck -type pfc rectifier for 400 v dc distribution systems," in *Proc. 26th Applied Power Electronics Conference and Exposition (APEC)*, Fort Worth, Texas, USA, 2011.

[2] G. Marsala and A. Ragusa, "A design method of zero ripple input current dc-dc converter assisted by constrained ga&pso algorithms to minimize power losses," in *2016 19th International Conference on Electrical Machines and Systems (ICEMS)*, 2016, pp. 1–6.

[3] A. Stupar, D. Flumian, B. Gouédard, and T. Meynard, "Generation and Derivation of Practical Optimization-Oriented Models of Inductors," in *COMPEL 2019*, Jun. 2019, pp. 1–8. DOI: 10.1109/COMPEL.2019.8769669.

[4] T. McRae, A. Stupar, and T. Meynard, "Multi-objective optimization of ee-core transformers using geometric programming," in *2022 IEEE 23rd*

Workshop on Control and Modeling for Power Electronics (COMPEL), 2022, pp. 1–8. DOI: 10.1109/COMPEL53829.2022.9829984.

[5] M. Grant and S. Boyd, *CVX: Matlab Software for Disciplined Convex Programming, version 2.1*, http://cvxr.com/cvx, Mar. 2014.

[6] M. Grant and S. Boyd, "Graph implementations for nonsmooth convex programs," in *Recent Advances in Learning and Control*, ser. Lecture Notes in Control and Information Sciences, V. Blondel, S. Boyd, and H. Kimura, Eds., http://stanford.edu/~boyd/graph_dcp.html, Springer-Verlag Limited, 2008, pp. 95–110.

[7] A. Stupar, T. McRae, N. Vukadinović, A. Prodić, and J. A. Taylor, "Multi-objective optimization and comparison of multi-level dc-dc converters using convex optimization methods," in *2016 18th European Conference on Power Electronics and Applications (EPE'16 ECCE Europe)*, 2016, pp. 1–10.

[8] D. N. Dalal, H. Zhao, J. K. Jørgensen, N. Christensen, A. B. Jørgensen, *et al.*, "Demonstration of a 10 kv sic mosfet based medium voltage power stack," in *2020 IEEE Applied Power Electronics Conference and Exposition (APEC)*, 2020, pp. 2751–2757. DOI: 10.1109/APEC39645.2020.9124441.

[9] C. R. Sullivan and J. H. Harris, "Testing core loss for rectangular waveforms, phase ii final report," *The Power Sources Manufacturers Association*, 2011.

[10] A. Furlan, A. Morentin, G. Fontes, G. Delamare, M. Heldwein, and T. Meynard, "Homothetic Method to Compute Winding Losses in the Design of Power Inductors," Dec. 2019, pp. 1–6. DOI: 10.1109/COBEP/SPEC44138.2019.9065497.

PCIM Europe 2024, 11– 13 June 2024, Nuremberg   DOI: 10.30420/566262311

Design Consideration of Bi-Directional CLLLC Resonant Converter in Energy Storage Systems

Guangzhi Cui[1], Sheng-Yang Yu[2]

[1]Texas Instruments, China,

[2]Texas Instruments, USA

Corresponding author: Guangzhi Cui, guangzhi-cui@ti.com
Speaker: Sheng-Yang Yu, seanyu@ti.com

Abstract

The bidirectional DC-DC converter plays a key role in order to realize power distribution in systems including batteries such as energy storage systems (ESS) and automotive on-board chargers. The capacitor-inductor-inductor-inductor-capacitor (CLLLC) resonant converter topology with its symmetric resonant tank, soft switching characteristics, which helps the system achieve higher efficiency and works convenient in bidirectional power flow. In this abstract, key CLLLC resonant converter design considerations in ESS are introduced. A CLLLC resonant converter prototype with rated power 3.6kW, and battery voltage 40~60 V is built to evaluate the performance which achieves 97.4% peak efficiency.

1   Introduction

A commonly used residential ESS system is illustrated in Fig.1, a high voltage DC bus is interfacing electric grid and the battery system. It is notable that 48V battery system is widely applied in residential ESS applications because of safety considerations. Therefore, isolated bidirectional topologies should be used here to realize voltage conversion from high voltage DC bus to low voltage battery system. Instead of using two stage isolated structure, such as buck/boost converter + LLC resonant converter, a single stage isolated converter can save system cost and improve power density. Among these single stage topologies, converters with soft-switching capabilities like CLLLC resonant and Dual Active Bridge (DAB) converters would be a better choice to improve the efficiency. The performance and key features have been compared in [1], and it is found that the efficiency of a DAB converter is generally lower than a CLLLC resonant converter under the same specification. Moreover, DAB converter may lose zero voltage switching (ZVS) on the active switches at light load. For CLLLC resonant converter, it could achieve soft-switching over full load range without an additional circuit and the symmetric resonant tank ensure the similar gain and resonance characteristics in both forward and reverse operation, which helps improve the efficiency and reduce the complexity of converter design. However, the gain curve of CLLLC resonant converter is flatter when the switching frequency is higher than series resonant frequency, especially at light load condition. Therefore, design the CLLLC resonant converter to cover a wide battery voltage range while maintain high efficiency is the key challenge for CLLLC resonant converter in ESS applications. Based on its high efficiency and ZVS features, CLLLC resonant converter is a popular study topic in recent years, [2] introduce the design method of CLLLC resonant converter parameters, but it does not introduce the control scheme and experiment is verified in high battery voltage system, [3] introduces the design and control method, but it also verifies in high voltage system and do not consider the light load condition method. In this paper, the parameters design method, control algorithm of the CLLLC resonant converter and key design considerations in ESS will be discussed and analyzed. The proposed design methodology and digital control algorithms are verified using a 3.6kW prototype converter.

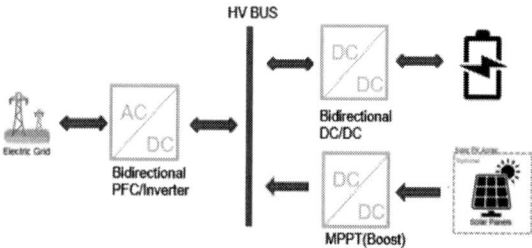

Fig. 1   Block diagram of residential ESS

2200

2 Design methodology of bidirectional CLLLC resonant converter

2.1 Design considerations of power stage

Fig. 2 shows the circuit topology of the full-bridge CLLLC resonant converter. This topology consists of a symmetric resonant tank and full-bridge structure. Among the resonant tanks, the resonant inductor L_{r1} and L_{r2} could be merged with the transformer T_1 to improve the power density.

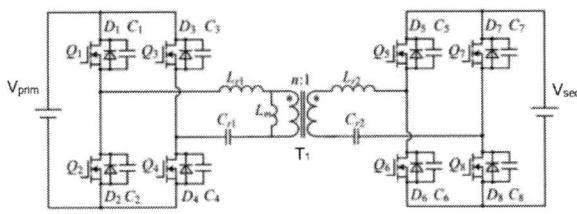

Fig.2 Schematic of CLLLC resonant converter

Similar to the LLC resonant converter, FHA (First Harmonic Analysis) model could be used to analyze the CLLLC resonant converter gain characteristic. The detailed operating principles and gain analysis using FHA model has been extensively researched in [2]. Based on the FHA approach, the steady-state models are obtained as shown in Fig. 3, which refer to all the resonant components from the load side to the source side. And the charging mode and discharging mode gain equation can be written as (1) and (2) using KCL and KVL.

$$\frac{V_{prim}}{V_{sec}} = \frac{n * [Z_m||(Z_{r1} + R_{L1})] * R_{L1}}{(Z_{r2} + [Z_m||(Z_{r1} + R_{L1})]) * (R_{L1} + Z_{r1})} \quad (1)$$

$$\frac{V_{sec}}{V_{prim}} = \frac{R_L' * [Z_m||(Z_{r2} + R_L')]}{n * (Z_{r1} + [Z_m||(Z_{r2} + R_L')]) * (R_L' + Z_{r2})} \quad (2)$$

Note here the effective R_{Li} is account as

$$R_{Li} = 8 * Rdc/\pi^2 \quad (3)$$

Where R_{dc} is the DC resistive load at the output. V_{prim} is the voltage at primary side and V_{sec} is the voltage at secondary side as shown in Fig.3. V_{sec}' is equal to $n*V_{sec}$, Z_m is equal to $s*L_m$, Z_{r1} is equal to $s*L_{r1}+1/s*C_{r1}$, Z_{r2} is equal to $s*L_{r2}+1/s*C_{r2}$.

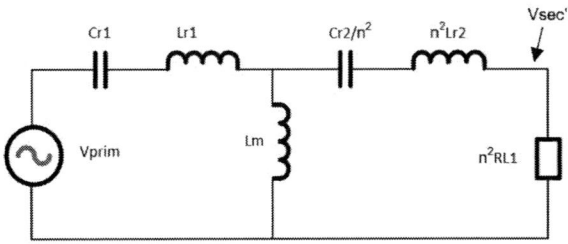

(a) Battery charging mode

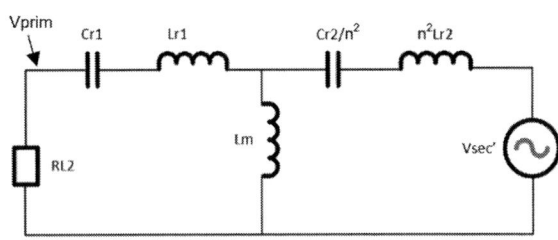

(b) Battery discharging mode

Fig. 3 FHA models of CLLLC resonant converter

2.1.1 Design considerations of Resonant Network

Due to a wide range of battery voltage, normally ranging from 40V to 60V, the CLLLC converter should work in a wide gain range. As the CLLLC converter has its efficiency optimized when the switching frequency (f_{sw}) is close to the series resonant frequency (f_r), when C_{r1} & L_{r1} or C_{r2} & L_{r2} are resonant, we will have to set $f_{sw}\sim f_r$ at the operating point where maximum power is delivered. And the turn ratio of the transformer is also calculated using the same maximum power delivery point. Design of magnetizing inductor (L_m), resonant inductor (Lr), and resonant capacitor (C_r) should consider the gain range, acceptable maximum f_{sw}, soft switching conditions, and efficiency optimization. Some definitions and assumptions of the resonant tank are given by

$$k = L_m/L_{r1} \quad (4)$$

$$a = n^2 * Lr2/Lr1, \qquad b = Cr2/(n^2 * Cr1) \quad (5)$$

k is the inductance ratio, a is the inductance ratio and b is the capacitance ratio. Based on the definitions, assuming a and b is equal to 1, which means the resonant tank is totally symmetrical. The gain curve with k varying is shown in Fig.4, where y-axis is the voltage gain and x-axis is the frequency.

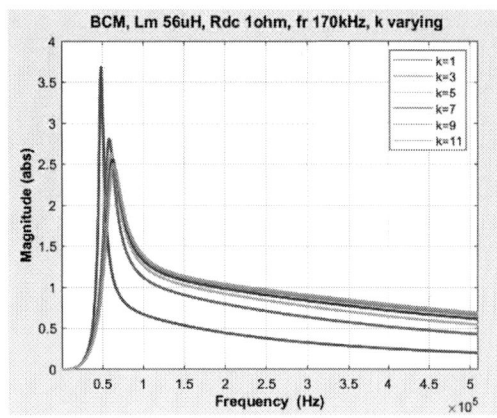

Fig. 4 Gain curve with different k value (1ohm Rdc)

In order to cover a wide gain range, the inductance ratio *k* has to be low, because the peak gain is reduced with high *k* ratio and the gain curve will be flatter at high frequency. That is, the converter needs to operate in a wide frequency range to meet required gain. In addition, as shown in Fig.5, there will be part of non-monotonic range when frequency is lower than series resonant frequency with load current increase, which will also limit the peak gain in this situation.

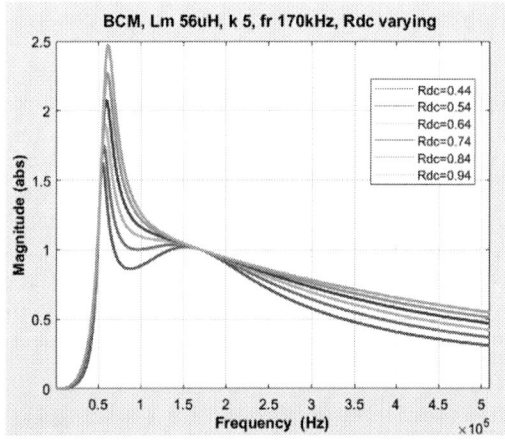

Fig.5 Gain curve with different R_{dc} value

Basically, high *k* means a larger L_m helps reduce circulating current to improve the efficiency, but it needs to consider if such a larger L_m could help achieve ZVS over full load range. In a resonant converter, the energy stored in magnetizing inductor Lm during the dead time is used to charge and discharge the MOSFETs output capacitance (C_{oss}) to achieve ZVS. Assuming an ideal transformer and ideal switches on the output full bridge switches (no parasitic capacitance). In order to

achieve ZVS on the input full bridge, the energy stored in L_m needs to be larger than the energy stored in all Coss. Hence, the following equation must be met:

$$0.5 * Lm * I_{Lm}^2 \geq 0.5 * (4 * Coss) * Vin^2 \qquad (6)$$

$$I_{Lm} = n * Vout/(4 * Lm * fswmax) \qquad (7)$$

Actually, the transformer winding capacitance and C_{oss} capacitors of output full bridge side are non-negligible. To achieve ZVS, these parasitic capacitances should be minimized and a smaller L_m is chosen in real cases. It is recommended to verify if ZVS is achievable in the simulation model with these parasitic parameters. Besides, these parasitic capacitors will also cause non-linear gain at high frequency at light load conditions, which will be discussed in the next section.

2.1.1 Effect of resonant network's tolerance

The gain characteristic is analyzed above based on the condition a=b=1, the effect will be analyzed assuming a≠b≠1 but a*b=1 with a certain k and Rdc. It means the series resonant frequency is the same in both sides but the resonant inductor and capacitor has the tolerance. Taking the different values of a and b that are listed in below table 1, the waveform in charging and discharging mode is shown in Fig.6 and Fig.7.

a	0.4	0.7	1	1.3	1.6
b	2.5	1.42	1	0.77	0.625

Table 1   value of a and b

In charging mode, if a>1, the gain curve is non-monotonic when frequency is lower than series resonant frequency, it is hard to regulate the output voltage by adjusting the frequency. There is no problem when a<1 in charging mode, but it will cause the obvious gain difference when it operates in discharging mode, the gain is flat around the series resonant frequency, which will lead to the converter operating in a wide frequency range to get high gain.

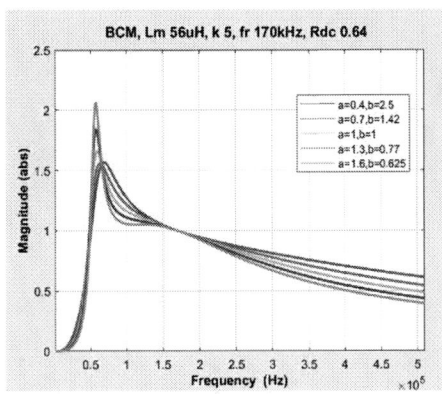

Fig. 6 Charging mode gain curve when a*b=1

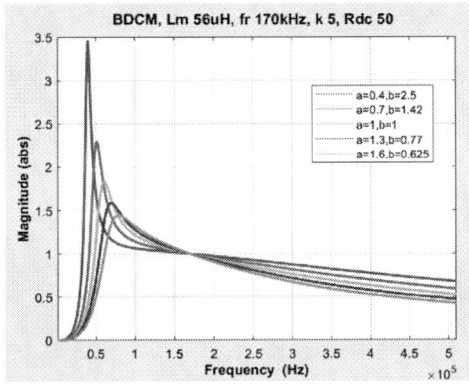

Fig. 7 Discharging mode gain curve when a*b=1

It is noticeable that the effect of resonant capacitor and inductor on the gain curve is different. The gain curve with constant b=1 and variable a is shown in Fig.8. For charging mode, when a<1, the tendency of the curve is similar as the one when a=1, but the slope increases sharply when a>1 at the left part of series resonant frequency.

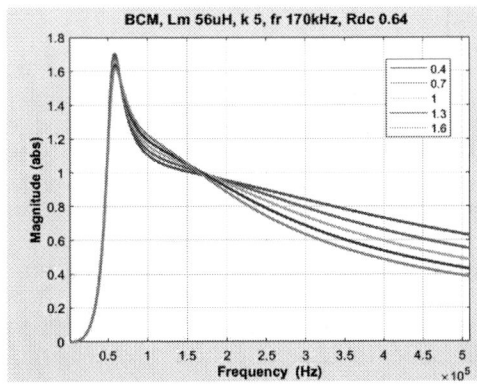

Fig. 8 Charging mode gain curve with different a

Similarly, the gain curve with constant a=1 and variable b could be drawn in Fig.9. According to the waveform, the tendency is similar to the one when b=1, but when the b<1, the gain curve is non-linear with the frequency decrease.

Compare Fig.8 with Fig.9, it should point out that the resonant capacitor mainly influences the part which is lower than series resonant frequency, resonant inductor mainly influences the part which is higher than series resonant frequency. Besides, the resonant capacitor has a worse effect on the gain curve, which will make the gain regulation difficult with frequency modulation below the series resonant frequency.

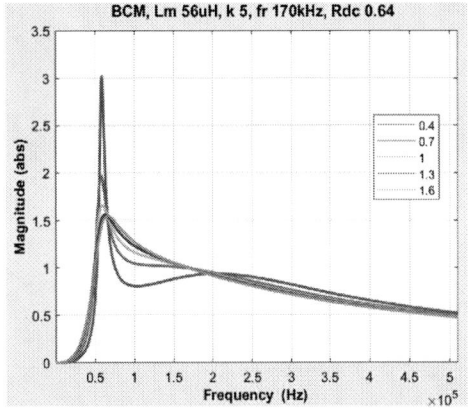

Fig. 9 Charging mode gain curve with different b

For different values of a and b, when both of them are close to 1, the gain curve in charging mode and discharging mode will be closer, the symmetry and uniformity of the converter will be better. However, it is hard to make the resonant parameters totally symmetrical in real design. Considering 20% margin of components tolerance and design error, the gain curve in different situation could be drawn in Fig.10 to Fig.13. It is found that when the tolerance of resonant parameters is within 20%, the gain curve could maintain the monotonicity in the whole frequency range. From Fig.10, the resonant inductor has a weaker effect on the gain curve, especially when the frequency is lower than series resonant frequency. From Fig.11, three gain curves coincide when the converter operates above series resonant frequency, however, it is noticeable that when the converter operates under series resonant frequency, a non-monotonic point may occur if the load continually increases. From Fig.12, when b=0.8 in charging mode, there will be inflection point when frequency is under resonant point, and this non-monotonic will be serious with b value decrease. In Fig.13, the gain curve seems no non-linear part but the gain curve will have ob-

2203

vious drifting when both a and b has 20% tolerance, which will lead to gain character difference, especially at series resonant frequency. Therefore, when design the resonant parameter with a*b=1, which means the series resonant frequency in both sides is same, if the tolerance is within 20%, the difference is acceptable and it will not have serious effect on gain curve. Besides, it is better to control the lower end value of b within 20% to avoid the gain curve non-monotonic.

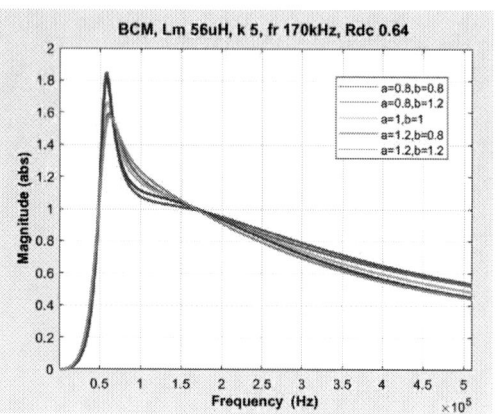

Fig. 12 Charging mode gain curve when both

a and b have 20% tolerance

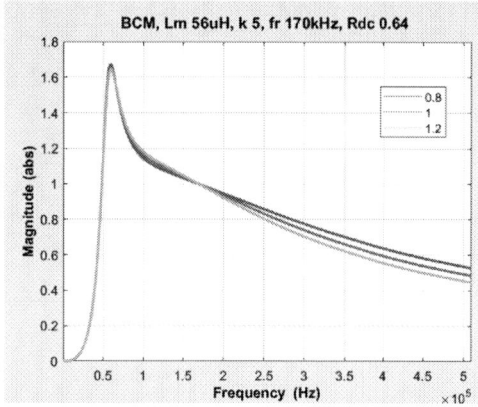

Fig. 10 Charging mode gain curve with 20%

tolerance of a

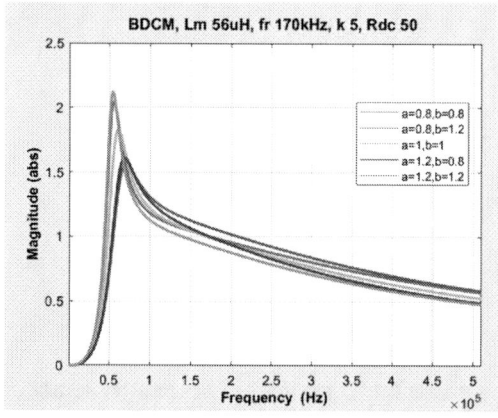

Fig. 13 Discharging mode gain curve when both

a and b have 20% tolerance

2.1.2 Design methodology of resonant network

During the design procedure, the voltage gain curves should keep monotonic within the acceptable range of operating frequency and meet the voltage gain requirements at full load range in both stages. Besides, the ZVS conditions should be achieved.

1) Design of the turns ratio of transformer (n)

Because the converter operates in a wide voltage range, it is better to design the turn ratio at its typical output to achieve the maximum efficiency.

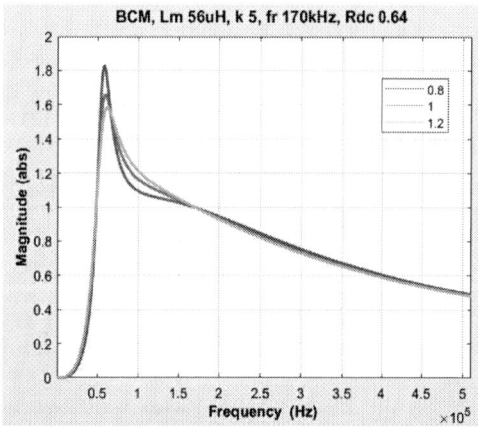

Fig. 11 Charging mode gain curve with 20%

tolerance of b

$$n = Vbus/Vout\_typ \qquad (8)$$

2) Calculating the maximum and minimum gain

It should calculate the maximum and minimum required gain on both sides and make sure the gain curve could cover this range. Gain equation in charging mode is shown below:

$$Gmax = n * Vout\_max/Vin\_min \qquad (9)$$

$$Gmin = n * Vout\_min/Vin\_max \qquad (10)$$

3) Design of the magnetizing inductor (Lm)

To ensure the output capacitor could be fully charged or discharged for ZVS, maximum L_m is calculated in equation (5) and (6). Basically, half of the calculated value is used because of FETs and transformer parasitic parameter effect.

4) Design of the inductor ratio (L_m/L_{r1})

Normally, a high k value helps reduce the RMS current, but the gain will decrease with k increase as shown in Fig.4, which will require wider operation frequency and be difficult to compact magnetic components design. So, select a higher k value to optimize the efficiency and make sure the gain curve could cover the calculated gain range in the specified frequency range.

5) Calculation of the resonant components

When k is defined, L_{r1} is known and C_{r1} could be calculated based on series resonant frequency. According to the analysis above, 20% tolerance of resonance is acceptable in the design but it is necessary to make sure the resonance on both sides are same (a*b=1). Besides, the resonant capacitor cannot be too large, as it will decrease the impedance of the resonant tank which will weaken the effect of startup current limitation at high frequency. Once the parameters are confirmed, it is recommended to do the simulation to verify the gain curve in different working conditions. Repeating the last 3 steps to find the available result if the simulation result cannot meet the specification.

2.2 Design considerations of control stage

Variable frequency control (VFC) is generally used as the CLLLC resonant converter control method. However, the gain curve is flat when $f_{sw}>f_r$, and considering the parasitic parameters of transformer intra-winding capacitance and C_{OSS}, the non-monotonic gain curve at high frequency is happening at light load. For the low voltage side, it is necessary to parallel more FETs to reduce the FET's on resistance in high power applications. In this case, the effect of C_{OSS} is serious, we can't rely just on VFC for output voltage regulation. Hiccup mode is a popular method to deal with CLLLC resonant converter non-monotonic features, but this method is not suitable in battery applications because the converter needs to deliver high current when battery voltage is low. PWM control and phase-shift control could solve this problem, but PWM control will make FETs work at hard switching state, which decreases the efficiency and limits the operation frequency. Therefore, the phase shift control is a better choice to cover this situation. The area of operations of a battery charger with variable frequency and phase-shift mixed control is proposed in Fig.14.

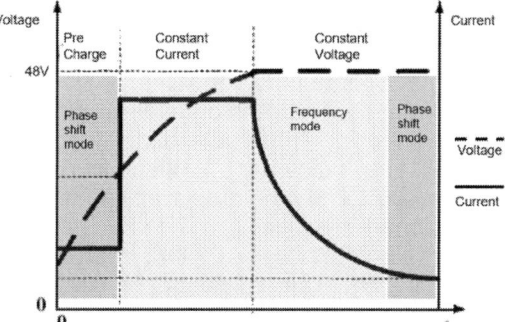

Fig. 14 Control scheme in different charge stage

During the startup stage, the output voltage is low, soft-start function is necessary to limit the high current spike if the pre-charge circuit is not used in the system. It is a limited effect to soft start from high frequency if the resonant inductor value or frequency is not high enough, so in this design the phase shift control is used as the soft start strategy. The converter will enter into frequency modulation when soft start is over. When the battery is almost fully charged, it will trickle charge with a small current and maintain a constant voltage. At this light load, output voltage tends to rise due to parasitic capacitance and could eventually go out of regulation, so the phase shift control is also needed to regulate the output voltage in this region. Whether the converter enters phase shift mode or not is decided by the controller's calculation result, Fig.15 shows the modulation switch between frequency and phase shift. When the load decrease, the frequency will increase to regulate the output voltage, if the calculated maximum frequency is higher than the setting value, the converter will enter into phase shift modulation; then when the load increase, the phase shift angle will decrease to regulate the output voltage, the converter will enter frequency mode again when phase shift angles decreases to zero.

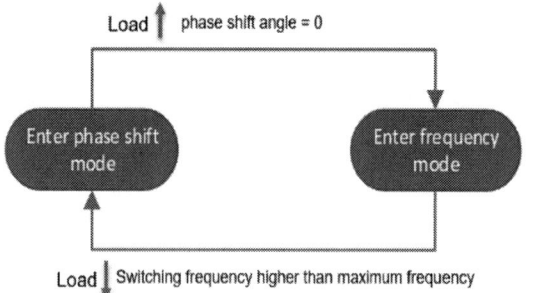

Fig. 15 Modulation switch between frequency and phase shift modes

3 Experimental results

A prototype [5] is built to verify the circuit design and performance. The transformer turns ratio $N=9$, with L_m=56µH, L_{r1}=10µH, C_{r1}=88nF, L_{r2}=200nH, C_{r2}=4.4µF. The typical input voltage and output voltage is 400V and 48V, output power is 3.6kW, series resonant frequency in 170kHz, peak efficiency in charging mode is 97.4%, in discharging mode is 95.1%. The efficiency in discharging mode is lower than charging mode, because the tank current is much higher in discharging mode when converting the primary tank current to secondary side with transformer turns ratio, which means higher condition loss in transistors and wire loss in transformer. In addition, Si FETs were used in the secondary side, compared with the GaNFETs used in the primary side, it has higher turn off loss and the increment of conduction resistance is larger with the junction temperature increasing. The efficiency test result is shown in Fig.16, soft start waveform is shown in Fig.17 and frequency/phase shift modulation switch test is shown in Fig.18. From the test waveforms, the startup current is limited within 28A with 750W output power. The converter could change the modulation smoothly in different working conditions.

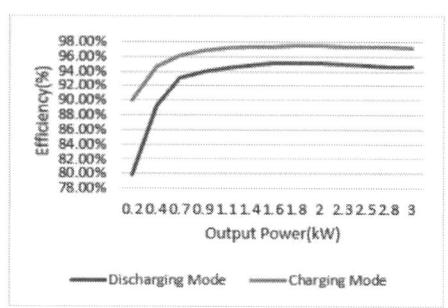

Fig. 16 Efficiency test result

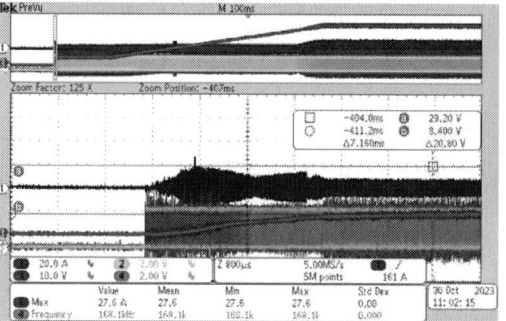

Fig. 17 Phase shift soft start with 750W output power

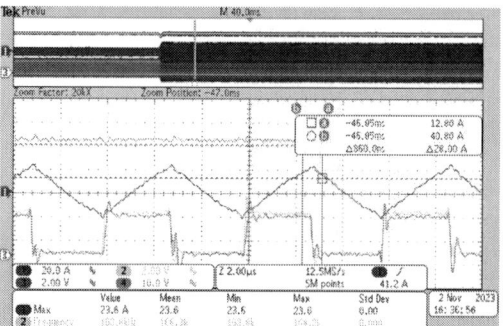

(a) Working at frequency mode with 5A load

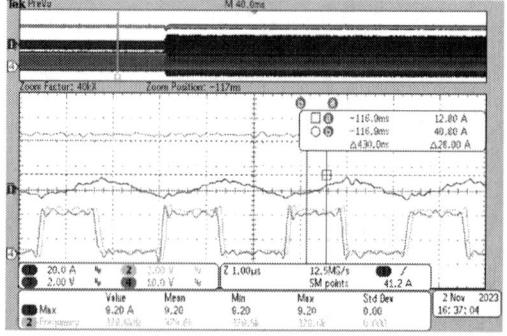

(b) Working at phase shift mode with 1A load

Fig. 18 Phase shift and frequency modulation switch

4   Summary

The bidirectional CLLLC resonant converter can achieve ZVS under full load range, which reduces the switching loss and improves the system efficiency. Also, the operation principle is similar in both directions and the parameters are easy to design because of its symmetrical structure. In this paper, the design considerations of the power stage and control stage are analyzed and the design steps are also discussed. Experimental data using 3.6kW prototype verify the validity of proposed methodology and control algorithm. The maximum efficiency is 97.4% in charging mode and 95.1% in discharging mode.

References

[1] S. -Y. Yu, C. Hsiao and J. Weng, "A High Frequency CLLLC Bi-directional Series Resonant Converter DAB Using an Integrated PCB Winding Transformer," *2020 IEEE Applied Power Electronics Conference and Exposition (APEC)*, New Orleans, LA, USA, 2020, pp.1074-1080.

[2] H. -T. Chang, T. -J. Liang and W. -C. Yang, "Design and Implementation of Bidirectional DC-DC CLLLC Resonant Converter," *2018 IEEE Energy Conversion Congress and Exposition (ECCE)*, Portland, OR, USA, 2018, pp. 2712-2719.

[3] J. -H. Jung, H. -S. Kim, M. -H. Ryu and J. -W. Baek, "Design Methodology of Bidirectional CLLC Resonant Converter for High-Frequency Isolation of DC Distribution Systems," in *IEEE Transactions on Power Electronics*, vol. 28, no. 4, pp. 1741- 1755, 2013.

[4] L. Qu, X. Wang, D. Zhang, Z. Bai and Y. Liu, "A High Efficiency and Low Shutdown Current Bidirectional DC-DC CLLLC Resonant Converter," *2019 22nd International Conference on Electrical Machines and Systems (ICEMS)*, Harbin, China, 2019, pp. 1-6.

[5] Guangzhi Cui, "Bidirectional CLLLC Resonant Converter Reference Design for Energy Storage System," *Texas Instruments Reference Design PMP41042*, 2024

PCIM Europe 2024, 11– 13 June 2024, Nuremberg   DOI: 10.30420/566262312

Adaptive Fast Charging System with Second Life Batteries - an Overview of a Research Project

Lukas Böhning[1], Mathias Herget[1], Alexander Menzel[1], Patrick Stock[1], Raphael Kress[1], Nils Kasseckert[1], Ulf Schwalbe[1]

[1] Hochschule Fulda, Germany

Corresponding author: Lukas Böhning, lukas.boehning@et.hs-fulda.de
Speaker: Lukas Böhning, lukas.boehning@et.hs-fulda.de

Abstract

An adaptive fast charging system offers a solution for sustainable electromobility. By reusing used vehicle batteries, a system is created, consisting of two energy storage units (each with 88 kW, 90 kWh) and a charging infrastructure with a total power of 172 kW. The prototype is designed to boost the grid connection power by a factor of 3.5 - 5. This makes it possible to install a fast charging infrastructure at grid connections with low power, independent of the location. This paper will present the development of the prototype and the current status of the project as well as first practical measurements and improvement potentials.

1   Introduction

The acceptance of electric mobility is a key element in the transition to a more sustainable solution for the use of renewable energy through an intelligent charging infrastructure [1]. One of the main barriers to this acceptance is the availability of charging stations [2]. This paper provides an insight into the development of the research project „Intelligent Reuse of Used e-Mobility Batteries for a Sustainable Transformation into an Environmentally Friendly and Flexible Charging Infrastructure", known by the german acronym „iWEnT". The system is hereafter referred to and named as the charge buffer unit (CBU), shown in Fig. 1.
It consists of the following components two battery storage units, each with an output power of 88 kW and a capacity of 90 kWh, connected to a low voltage grid connection with a total output power of 22 to 50 kW. The charging infrastructure with a total power of 172 kW, including a direct current (DC) charging station with a power of 150 kW and an alternating current (AC) charging station with a power of 22 kW, is connected to the same low voltage distribution. The battery storage systems are used to cover peak loads higher than the possible grid connection power and are recharged when the charging infrastructure needs less power than the

Fig. 1: Prototype in Fulda [3]

possible grid connection power. All components except for the AC charging station have been integrated into a ten foot shipping container to ensure maximum flexibility.

This paper first discusses the theory behind the CBU, followed by a description of the design of the prototype and the operation of the system. A practical measurement of all power and temperature curves is shown, as well as an efficiency study and the impact on the electrical grid. Current statistics from one year of operation and a comparison of the economics of the system are also given. Finally, potential improvements for future projects are highlighted, further project work is outlined and a summary of this work is provided.

1.1 Fast charging of electric vehicles

For the successful transformation to a sustainable mobility, the availability of charging infrastructure for battery electric vehicles must be increased or made quickly available [4]. In Germany, the number of public charging points amounts to 78,918 charging stations as of July 1, 2023, of which 23.5% are represented by DC-based fast charging [5]. Fast charging of electric vehicles in particular represents a major challenge for the electrical distribution grid [6].

One challenge for the use of fast charging stations is that they cause high peak loads compared to the amount of energy used. This results in high network charges, which in Germany are divided into fixed costs, costs per output and costs per energy [7]. The advantages of electromobility can only be exploited if it is powered from renewable sources [8]. These challenges prevent the development of a fast charging infrastructure for electric vehicles [9]. One possible solution to this problem is the use of stationary battery storage, which is described in the following sections [9].

1.2 Second life battery storage systems

Second life batteries from electric vehiceles have a capacity between 85% and 96% depending on the stress within the first life [6]. The use of battery storage systems in the field of peak load provision enables a solution to the problem from chapter 2. This could enable the use of fast charging stations at grid connection points with a low capacity to be made possible [10]. One challenge here is the aging monitoring of the second life cells to ensure safe operation [11][12]. The use of new batteries would further strain supply chains and due to the fact that used vehicle batteries could be used beyond the vehicle life, the use of second life batteries for peakshaving of fast charging stations results as an interesting application possibility [6].

2 Design of the prototype

In order to cover high peak loads through a charging infrastructure for electric vehicles by using second life batteries, the CBU is presented in the following section, which enables the simultaneous charging of four vehicles (two DC fast charging and two AC charging) with a total output of 172 kW on a grid connection with a minimum of 22 to 50 kW.

2.1 Concept and components in Detail

The following concept, shown in Fig.2, has been developed for the purpose of fast charging vehicles at low power grid connections.

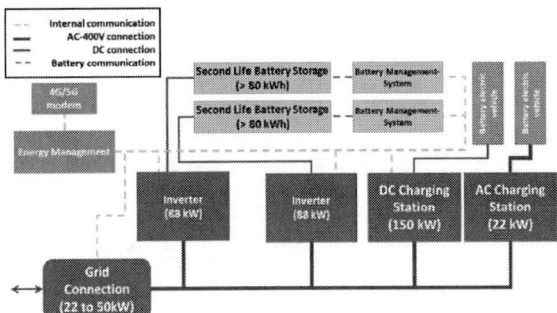

Fig. 2: Prototype concept. Interconnection of the battery storage, inverter, charging stations at a low-voltage distribution.

The grid connection is made to a 400 V distribution system. Two „REFUstor 88k 400V" battery storage converters are also connected to this distribution system. It has a 63 A (80 A maximum) input connection, two 125 A output connections for the inverters and a 224 A output connection for the DC fast charging station. Also several 32 A output connections for the controllers and the AC charging station are included. In Fig. 3 all components can be seen in a 3D sketch.

The „Alpitronic Hypercharger HYC 150" fast charging station with a total DC Power of 150 kW or AC input power of 160 kW is used in the system. The cars can be connected with the Combined Charging System 2 (CCS2) plug. The charging station provides a maximum current of 500 A DC and a maximum of 150 kW to one connected car. It also supplies 250 A direct current and a maximum power of 75 kW each for two vehicles. The DC voltage is provided in a range of 150 to 1,000 V, the full power can be provided at a voltage of 300 V [13]. The „Alfen eve double pro-line 22 kW" charging station are connected to the same low voltage distribution. The input current of the station can be statically regulated between 6 A 230 V AC 1 phase up to 32 A 400 V AC 3 phases. Cars can be linked with a IEC 62196 type 2 connector. Both charging stations are connected to the accounting backend of a regional energy supplier via OCPP1.6 [14]. As can be seen in the illustration, both charging stations are installed outdoors, which means that the heat generated is released into the ambient air.

Fig. 3: Inside view of the container. Fast charging station in blue, battery inverter in red, battery modules in gray-green. The low voltage distribution is located behind the Fast charging station.

Battery modules from a southern German car manufacturer are used to buffer the electrical energy. A battery module consists of 48 cells with an internal connection of two cells in parallel and 24 in series (2p24s). Eight of these modules form one of the two strings, resulting in a total configuration per string of two cells in parallel and 192 cells in series (2p192s). The voltage range of the string is therefore between 633.6 V and 806.4 V. A universal battery management system is used, which enables the balancing of the cells. Balancing is activated for 4 seconds from a cell voltage difference of absolut 20 mV. The maximum charging current of the balancing is 4 A, the maximum discharging current of the balancing is 15 A. In addition, balancing is only active above a cell voltage of 3.4 V exclusively during charging. Two „REFUstore 88k 400V" are used as converters between the two battery strings and the AC grid. The DC voltage of the inverter is in the range 585 V to 900 V. The maximum apparent power is 88 kVA. The inverters are cooled using an active air cooling system that feeds into the inside of the container. A display is also provided for the user of the system, which shows the current state of charge (SOC) of both battery strings, the current charging power and the current grid power.

All components inside the CBU (battery strings, inverter, low voltage distribution, display) are conditioned by an air-to-air heat pump from the manufacturer Samsung with the model code „09 TX-EAAWKN" with a maximum cooling power of 3.5 kW and a maximum heating power of 5 kW. The setpoint for heating is 14 °C and the setpoint for cooling is 28°C. The air-to-air heat pump itself operates with a temperature hysteresis of 2 K.

2.2 Energy management system

The energy management system (EMS), which is described in more detail in the following section, provides the control function to ensure that the grid connection of the CBU is constantly loaded with a specified setpoint value. The first step is to check whether the setpoint value does not exceed the physical load capacity of the CBU batteries. The EMS also limits the maximum charging and discharging currents of the individual strings as shown in Fig. 4.

It is important to note that a characteristic curve independent of the battery cell manufacturer was selected. These are the maximum possible cell currents determined from various tests, so that the control requirements can be kept low and the battery cells can also be passively cooled by an air-to-air heat pump installed in the system. As shown in Fig. 4 , the maximum cell currents are maintained at a level between a voltage of approximately 3.55 V and 3.8 V during charging, which also enables the CBU to be charged with an output of up to approx. 110 kW.

It is important to highlight that the energy management system does not refer to the SOC of the system, rather to the maximum individual cell voltage in a string when charging and to the minimum cell voltage in a string when discharging. In the current application, charging currents are additionally limited by the maximum available power at the grid connection. The power target of the batteries is calculated using the current characteristic curve and the current phase voltage.

In the upper cell voltage range, the charging current is reduced to 0 with increasing cell voltage. At a voltage of 4.08 V, a hysteresis limit is exceeded, at which point the system is put into standby mode so that constant recharging of the system is prevented. Only when the voltage drops below the limit of 3.925 V, the system continues to charge.

The discharge of the system is similar to charging, but the maximum current load is obtained in the upper cell voltage range. A maximum output of approx. 150 kW is achieved from the specified values. The discharge currents will be significantly reduced from a cell voltage of 3.6 V, although up to a cell volt-

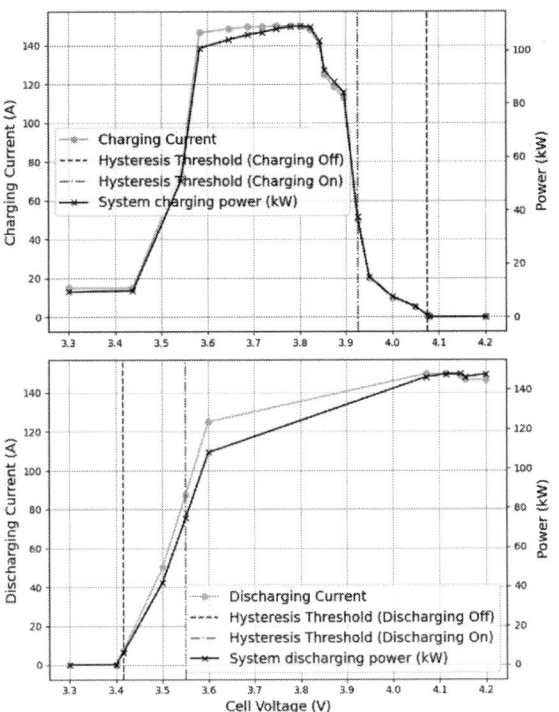

Fig. 4: Battery currents (orange) as a function of the voltage of two parallel cells (2p1s) and the system power (black) resulting from the total system voltage of a 2p192s connection. Hysteresis limit for stopping the discharging or charging of the batteries (blue). Hysteresis limit for restarting the discharging or charging of the batteries (red).

age of 3.4 V this is reduced to 0 A. From a voltage of 3.425 V, discharging of the system is stopped. At this point, the connected charging infrastructure only has an output power corresponding to the grid connection power. The system can only start discharging again once the second hysteresis limit has been exceeded. The hysteresis limits are summarized in Tab 1.

Description	Unit	Value
Upper limit of charging hysteresis	V	4.080
Lower limit of charging hysteresis	V	3.925
Upper limit of discharging hysteresis	V	3.550
Lower limit of discharging hysteresis	V	3.415

Tab. 1: Hysteresis limits for charging and discharging the battery storage system

The charging power available for the charging infrastructure will then be calculated from these two given values, the available and usable grid connection power added to the available power from both battery storage strings. If the charging stations are used at the same time, the AC charging station is prioritized in this case. The energy management system ensures that the fast-charging station receives the available power with a ramp of 5 kW per second. A total value for all additional consumers, the losses of the DC fast charging station and a power factor of 0.8 were also taken into account to ensure that there is still a sufficient buffer to cope with any disturbance quantity (e. g. the power consumption of the heat pump or other flexible power consumers).

Finally, a PI controller is used to maintain the power at the grid connection at a specified setpoint. The controller is limited by the available discharging and charging power of the two battery storage strings. The specified power for the DC fast-charging station is also fed to a disturbance variable so that the regulation of the system is improved.

The setpoint for the local energy management system is set depending on time in consultation with the grid connection owner. In this case, the output of the system is throttled to 15 kW between 10 a.m. and 2 p.m. on Mondays to Fridays. In the other time ranges, the setpoint is 28 kW. The physical maximum of the grid connection is 50 kVA, whereby only a maximum of 40 kVA is currently available for reasons of selective protection.

3 Operation and practical measurements

The function of the CBU when charging a vehicle with a large battery can be described as follows. The power curves of an example CBU operation day can be seen in Fig. 5.

The user starts charging at the DC fast charging station with any payment method (except cash, debit and credit card), the charging station immediately starts preparing for charging. If the stationary battery is fully charged, the power of the charging station is then increased until the maximum power of the batteries or the vehicle is reached. With decreasing state of charge of the CBU, the power is reduced due to the limitation of the battery currents, due to a high SOC of the vehicle battery or too high a temperature of the battery cells, until the vehicle is fully charged or the battery storage of the CBU is empty. In the second case, only the available grid connection power is available to the DC fast charging station.

Once a vehicle is charged, the batteries are

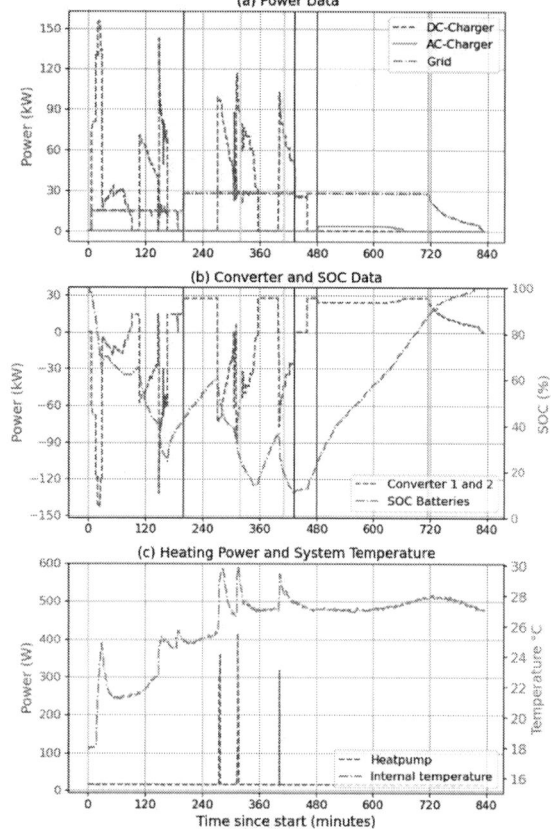

Fig. 5: (a) Powercurve of the charging stations and grid-connection, (b) state of charge und powercurve of the inverters and (c) temperature curves of the CBU during a complete discharge and recharge. Grid setpoint change (red), throttling the charging station (gold), discharge hysteresis (purple), starting AC charging (gray) and reducing charging power (turquoise).

recharged from the grid connection. The charging power of the batteries is increased until the power corresponds to the maximum power level of the batteries or the power of the setpoint value for the grid connection is reached. As the SOC of the CBU increases, the recharging is continued up to a state of charge of the individual cell of 4.08 V and then stopped. Hysteresis limits have been introduced in the upper and lower range of the cell voltage level to enable standby operation when the CBU is full or empty.

The curves shown represent a period in which a large number of vehicles have been charged on the system. In the diagrams a complete operating cycle of the system can be seen, i.e. the complete discharge of the system including the subsequent

recharging. In total, seven vehicles were charged at the DC fast charging station and one vehicle at the AC charging station during this observed period. As described in the previous chapter, the setpoint is regulated to a value of 15 kW up to minute 200 (red). Subsequently, the regulation is adjusted to a setpoint value of 28 kW. From minute 432 (purple), the hysteresis of the storage system for discharging is activated, after which only the available grid connection power is available for the entire charging infrastructure. In both time ranges (gold), the temperature setpoint is exceeded and the heat pump is switched to cooling mode. From minute 320 (gold) and from minute 410 (gold), the output of the DC fast charging station was significantly reduced due to the low state of charge. From minute 480 (gray), a vehicle was charged at the AC charging station with an output of approx. 3.6 kW. From minute 720 (turquoise), the storage system charging power is throttled due to the high state of charge.

3.1 Efficiency determination

In this section, losses and the efficiency of the system are determined. For this reason, the power flows and the resulting energy amounts are broken down below. During the period in focus, an amount of energy of 309 kWh was drawn from the grid connection, while an amount of energy of 273 kWh was converted on the AC measurement at the DC fast charging station during the same period. 10.4 kWh was transferred at the AC charging station. Over the same period, 0.242 kWh was consumed by the air conditioning unit. During the system's standby times, the auxiliary consumers were identified with an output of 0.3 kW, resulting in an energy quantity of 4.2 kWh for this period. Likewise, energy of 87.7 kWh was drawn from battery string one and an energy of 95.7 kWh was charged during this period. In addition, an amount of energy of 88.6 kWh was stored out from battery string two and 95.9 kWh was stored within this period.

To determine the efficiency of the battery storage system $\eta_{Storage}$, the stored energy $E_{StorageDischarged}$ is taken in relation to the stored energy of a battery string $E_{StorageCharged}$ and can bee seen in Eq. (1).

$$\eta_{Storage} = \frac{E_{StorageDischarged}}{E_{StorageCharged}} \tag{1}$$

The efficiency of 91.6 % for string one and 92.3 % for string two refers to a complete cycle, i.e. the discharge and subsequent full charging of the battery storage system. The calculation Eq. (2) is used to determine the efficiency of the overall system.

$$\eta_{CBU} = \frac{E_{DCCharger} + E_{ACCharger}}{E_{Grid}} \quad (2)$$

The energy $E_{DCCharger}$ at the DC fast charging station added to the energy at the AC charging station $E_{ACCharger}$ divided by the energy E_{Grid} drawn from the grid connection results in the overall system efficiency η_{CBU}. In relation to the time range under consideration, the efficiency is 91.71 %. However, it is important to note that the losses at the DC fast charging station were not taken into account. The manufacture specifies a maximum efficiency at full load of >94 % [13].

3.2 Power quality and effects on the grid

This section discusses the effects of the system on the electrical grid. In this regard, a power quality study was carried out over a period of 28 days using a Fluke 1777 [15] measuring device. In this context, the measurement of the normalized fundamental harmonics and the observation of the harmonic voltages, which are shown in Fig. 6.

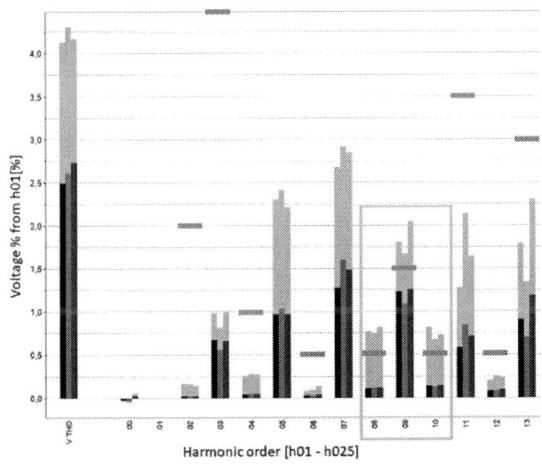

Fig. 6: Illustration of the mean voltage harmonics per phase related to the nominal voltage of 400V AC. Phase 1 (black), phase 2 (red), phase 3 (blue), limit values according to EN50160:2010+A1,2,3 (green) and problematic harmonics (orange) are shown. The maximum values of the individual harmonics are shown in pink. The fundamental is not shown in this illustration.

According to EN50160:2010+A1,2,3, the limit value for the maximum harmonic h08 and h10 is 0.5 % and the harmonic h07 is 1.5 % of the fundamental harmonic [16]. The maximum value within the measurement period, however, exceeds these limit values in relation to the grid connection of the CBU. This could be caused by the injection of harmonic currents, which are shown in Fig. 7 are shown.

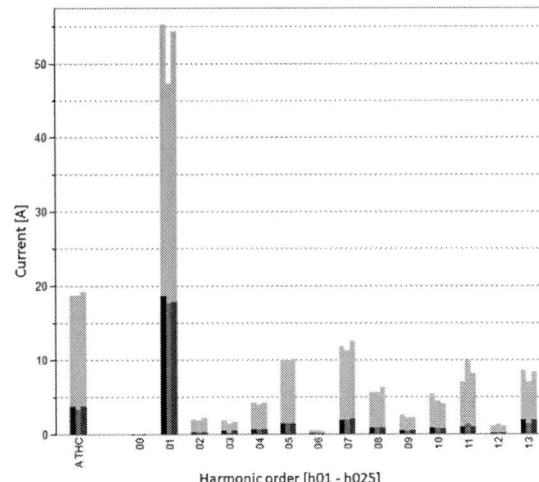

Fig. 7: Illustration of the absolut current harmonics per phase (1 black, 2 red, 3 blue) The maximum values of the individual harmonics are shown in pink. The reference value is 63 A.

The sum of the maximum current harmonic injection reaches up to 30.5 % (19.2 A) of the nominal current of 63 A of the grid connection. The individual components are each dimensioned for a sufficiently sized grid connection with corresponding grid impedance. In this case, however, a grid connection with a lower grid connection power is used. Therefore, for a commercial operation, measures must be taken to reduce the emission of current harmonics.

4 Statistics

The following section describes the use of the system over the last 12 months. The system was installed in January 2023, with the first tests and public use starting in 03/2023. Figure 8 shows the energy amounts broken down by the two charging stations.

The amount of energy sold has increased since September 2023, as shown in the figure, and has stabilized at around 4,500 kWh per month since December 2023. The AC charging station accounts for approx. 1,289 kWh to 2,070 kWh per month and the DC fast charging station for approx. 2,684 kWh to 3,207 kWh. Figure 9 also shows the number of charging processes, also divided into charging

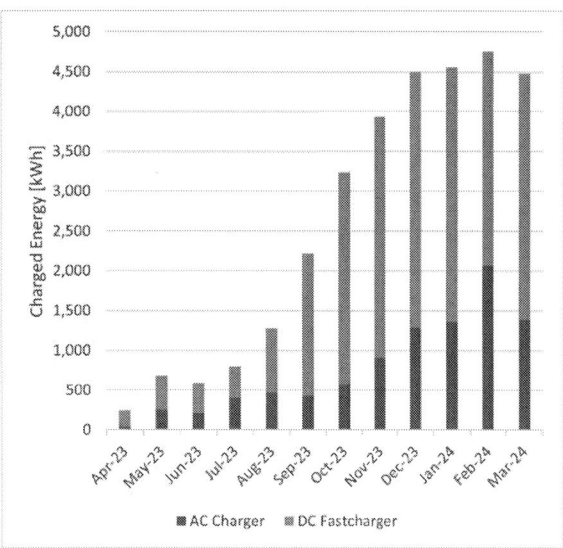

Fig. 8: Charged energy per month divided into AC charging station and DC fast charging station.

processes of the DC fast charging station and the AC charging station.

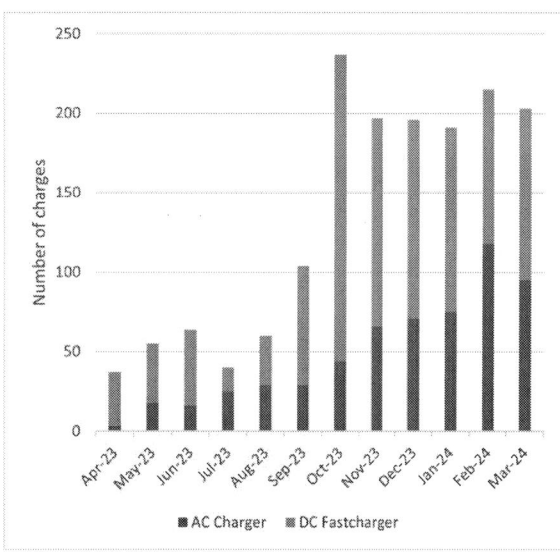

Fig. 9: Number of charges per month divided into AC charging station and DC fast charging station.

As from October 2023, there will be an increasing demand for charging at this location, especially at the DC fast charging station. This can be explained by the advertisement in various electric car charging portals. Since October 2023, the number of charges has increased to around 200 per month. Since December 2023, 71 to 118 charges per month have been completed at the AC charging station and 97 to 125 charges at the DC fast

charging station.

5 Economics

In this section, a cost comparison is made between a standard charging infrastructure and a charging infrastructure with the charging buffer presented in an extended expansion stage. As shown in chapter 3.2, it cannot be assumed that a significantly smaller grid connection can be operated with the described topology due to the high influence on the electrical grid. For this reason, the operation of a 220 kVA AC (200 kW DC) charging station is compared with a typical grid connection via a 20 kV transformer infrastructure and a charging infrastructure with 220 kVA at the same grid connection, with the power reduced to 50 kW. Although this does not reduce the investment costs for the grid connection, it does reduce the running costs via grid charges. The grid fees are based on the costs of the grid operator Osthessennetz for the year 2024, which for a medium voltage connection with a usage duration of more than 2,500 h per year has a performance price of 140.1 € per kW and a energy price of 0.0205 € per kWh. For a usage duration of less than 2,500 h per year, the performance price is 31.94 € per kW and a energy price of 0.0205 € per kWh [7]. All prices can be found in Tab. 2.

Voltage level	Usage duration	Unit	Value
Medium voltage peak load price	>2500 hours	€/kW	140.1
Medium voltage peak load price	<2500 hours	€/kW	31.94
Medium voltage energy price	>2500 hours	€/kWh	2.05
Medium voltage energy price	<2500 hours	€/kWh	6.37

Tab. 2: Grid fees according to [7].

Due to the additional efficiency loss of the CBU of 91.7 %, there are additional energy losses, which in turn increase the consumption of the system. A value of 0.17 € per kWh is assumed as the working price excluding grid fees. Therefore, in Figure 11, the running costs of the two options are determined as a function of the energy consumption.
The cost advantage of the system decreases with increasing energy turnover. It amounts to up to 23,628 € for very low energy volumes and at least 16,679 € per year for very high turnover.

6 Conclusion

The presented concept allows the operation of fast charging stations with a power of up to 150 kW by using second life vehicle batteries. The disadvantages of the concept are the running costs for the

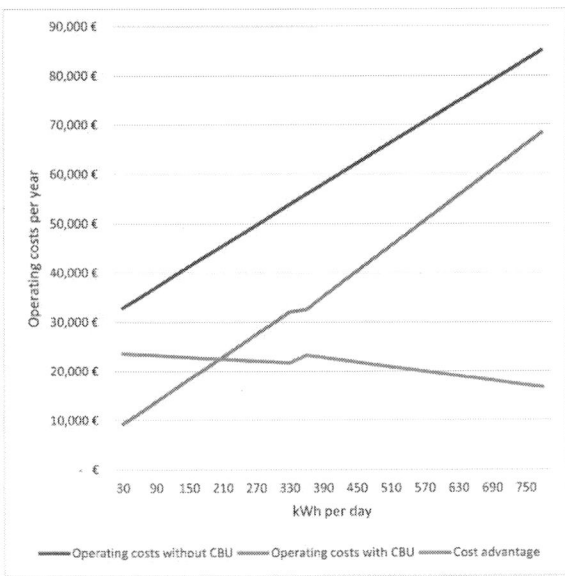

Fig. 10: Comparison of operating costs including electricity costs between a 200 kW charging infrastructure with and without CBU.

operation of the system (air conditioning and losses of the storage system). The advantage is the reduction of the maximum power at the grid connection and the resulting lower grid fees and avoided grid expansion costs. Furthermore, the concept creates a use case in which used vehicle batteries are given a second life. The statistics have also shown that the system has been well accepted by the public during the first year of operation.

However, the concept also has a weakness. As the components are connected to each other by means of low voltage distribution, problems arise due to the topology, which causes a high impact on the electrical grid in relation to the nominal power of the grid connection. For this reason, it is recommended that the topology of the system be modified for future work. The use of a fast charging station supplied directly from the DC bus of the battery storage could be a highly promising method to drastically reduce the impact on the electrical grid. An additional power supply unit could recharge the system's battery storage system using only the nominal power of the grid connection.

The economic impact of the system enables a significant reduction in running costs, especially in the case of low utilization of the system. This could have a significant influence on the decision to build a charging infrastructure.

In the next project steps, a detailed ageing analysis

of the batteries will be carried out by monitoring the individual cell voltages. The power consumption of the electrical grid will be dynamized in order to address additional use cases, e. g. demand charge reduction. Preparations are also currently being made to feed the power back into the system's grid connection so that it can be used as self-consumption storage for commercial purposes or for grid services.

This project (HA project no.: 1073/21-75) is funded by the funding program Electromobility in Hesse. Dieses Projekt (HA-Projekt-Nr.:1073/21-75) wird aus Mitteln des Förderprogramms Elektromobilität in Hessen gefördert.

References

[1] A. D. Gorbunova and I. A. Anisimov, "Assessment of the use of renewable energy sources for the charging infrastructure of electric vehicles," *Emerging Science Journal*, vol. 4, no. 6, pp. 539–550, 2020. DOI: 10.28991/esj-2020-01251.

[2] D. Efthymiou, K. Chrysostomou, M. Morfoulaki, and G. Aifantopoulou, "Electric vehicles charging infrastructure location: A genetic algorithm approach," *European Transport Research Review*, vol. 9, no. 2, 2017. DOI: 10.1007/s12544-017-0239-7.

[3] Steffen Bötcher, *Fotografie erneuerbare energien hochschule fulda.*

[4] P. Chakraborty, R. Parker, T. Hoque, J. Cruz, L. Du, *et al.*, "Addressing the range anxiety of battery electric vehicles with charging en route," *Scientific reports*, vol. 12, no. 1, p. 5588, 2022. DOI: 10.1038/s41598-022-08942-2.

[5] Bundesnetzagentur, *Elektromobilität: Öffentliche ladeinfrastruktur.*

[6] K. Richa, C. W. Babbitt, G. Gaustad, and X. Wang, "A future perspective on lithium-ion battery waste flows from electric vehicles," *Resources, Conservation and Recycling*, vol. 83, pp. 63–76, 2014. DOI: 10.1016/j.resconrec.2013.11.008.

[7] OsthessenNetz GmbH, *Netzzugangsentgelte strom preisblatt für den netzzugang strom 2024.*

[8] P. Nunes, T. Farias, and M. C. Brito, "Day charging electric vehicles with excess solar electricity for a sustainable energy system," *Energy*, vol. 80, pp. 263–274, 2015. DOI: 10.1016/j.energy.2014.11.069.

[9] Y. Sehimi, K. Almaksour, E. Suomalainen, and B. Robyns, "Mitigating the impact of fast charging on distribution grids using vehicle–to–vehicle power transfer: A paris city case study," *IET Electrical Systems in Transportation*, vol. 13, no. 1, 2023. DOI: 10.1049/els2.12051.

[10] S. Bae and A. Kwasinski, "Spatial and temporal model of electric vehicle charging demand," *IEEE Transactions on Smart Grid*, vol. 3, no. 1, pp. 394–403, 2012. DOI: 10.1109/TSG.2011.2159278.

[11] M. F. Börner, M. H. Frieges, B. Späth, K. Spütz, H. H. Heimes, *et al.*, "Challenges of second-life concepts for retired electric vehicle batteries," *Cell Reports Physical Science*, vol. 3, no. 10, p. 101 095, 2022. DOI: 10.1016/j.xcrp.2022.101095.

[12] M. Najeeb, U. Schwalbe, and M. Herget, "Improved approach for online monitoring of second life lithium-ion batteries to optimize the performance in stationary storage systems," in *2023 14th International Renewable Energy Congress (IREC)*, IEEE, 2023, pp. 1–6. DOI: 10.1109/IREC59750.2023.10389283.

[13] alpitronic GmbH, *Product data shee hyc 150: 75 kw / 150 kw rapid charging point for electric vehicles*, Bolzano Italy, 2021.

[14] Alfen N.V., *Eve double pro-line: Die eve double pro-line: Alfens flaggschiff für intelligente lösungen*, 2024.

[15] Fluke Corporation, *Dreiphasige netzqualitäts-analysatoren fluke serie 1770*.

[16] Deutsches Institut für Normung e. V., *Din en 50160:2020-11, merkmale der spannung in öffentlichen elektrizitätsversorgungsnetzen; deutsche fassung en_50160:2010_+ cor.:2010_+ a1:2015_+ a2:2019_+ a3:2019*, Berlin. DOI: 10.31030/3187943.

PCIM Europe 2024, 11– 13 June 2024, Nuremberg   DOI:10.30420/566262313

Parallel Operation and Synchronization of Microgrids by Using the Thevenin theorem

Marius Block[1], Stefanie Orlik[1], Wilfried Holzke[1] Holger Raffel[2]

[1] Institute for Electrical Drives, Power Electronics and Devices (IALB), University of Bremen, Germany
[2] Bremen Center of Mechatronics (BCM), University of Bremen, Germany

Corresponding author: Marius Block, mblock@ialb.uni-bremen.de

Abstract

In the future, microgrids will play a major role in the stability of the electrical grid and the integration of renewable energies. Unlike conventional power plants, renewable energy sources are often connected to the grid through converters. With the concept of the virtual synchronous machine, renewable energy sources contribute to grid services and are expected to provide stable grid operation. By using the Thevenin theorem, the modelling of parallel operation and synchronization of renewable energy sources can be considered, especially for the black start simulation scenario. For this purpose, a model consisting of two microgrids is set up. Each microgrid consists of two batteries, two renewable energy sources and a load.

1 Introduction

The complexity of the grid will increase as each conventional power plant is replaced by a large number of renewable energy plants [1]. The microgrid concept offers a solution by dividing the grid into sub grids. These microgrids can, either be operated independently or together [1].

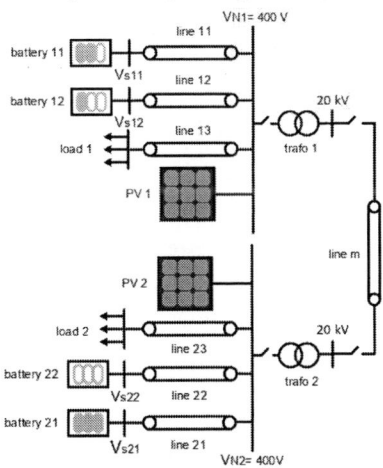

Fig. 1: Coupling of two microgrids

Figure 1 shows the structure of two microgrids that are connected at medium voltage via a line. The two microgrids each consist of a load, a photovoltaic system (PV) and two battery storage systems. According to [2], batteries are referred to as Distribution Generation (DG) units.

Renewable energy sources such as photovoltaic power plants and also battery storage systems are connected to the grid by power converters. By using the virtual synchronous machine (VSG) as control concept for the power converters, renewable energy sources can increase the grid stability [3]. The advantage of the virtual synchronous generator compared to a normal converter control is the virtual inertia. This allows the rate of change of the grid frequency to be reduced [4]. Wind power plants can improve grid stability using a virtual synchronous machine. This is made possible by transferring the control strategy from conventional power plants to the wind power plants [5].

The coupling of microgrids can be expressed as parallel operation of synchronous machines, as shown in Figure 1, if the virtual synchronous machine is employed as control concept. This control concept is defined for the batteries in Figure 1, as batteries according to [6] are suitable for grid restoration. In this paper the synchronization and parallel operation of two microgrids are examined.

2 Modelling

2.1 Generator model

The model of the synchronous generator from [7] simulates the generator behavior. This is an electrically excited synchronous machine with a full-pole rotor.

2217

Due to the full-pole rotor, no pronounced poles are created, which simplifies the mathematical description [7]. This generator model considers no grooving effects, no end effects, no saturation and no iron losses. In addition, there are symmetrical three-phase windings in the stator and rotor. This means that the separation of the field and damper windings in the rotor is not considered. The rotor is supplied with a single-phase DC voltage.

As shown in Figure 2, the model uses the field voltage V_f, the grid voltage V_N, the stator resistance R_s, the stator reactance X_d, the mechanical torque m_m and the transmission line resistance R_L as input variables. The factor a describes the relation between transmission line reactance X_L and stator reactance X_d. The output variable is the stator voltage V_s and the speed Ω.

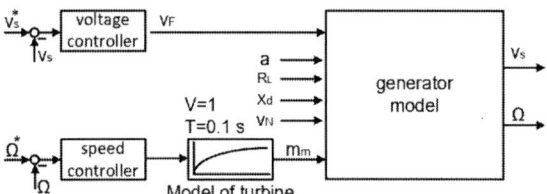

Fig. 2: Generator model with voltage and speed controller [7]
The speed and the stator voltages are subject to fluctuations, which are caused by load changes. In order to compensate these fluctuations, a speed and voltage control is needed [7].

The grid voltage V_N of the generator model is rigid according to frequency and voltage [7]. This means that the grid voltage can be described as a voltage source. As can be seen from Figure 1, the grid voltage $V_{N1,2}$ is subject to fluctuations in parallel operation caused by changes in load and feed-in. This means, the grid voltage is no longer rigid and the requirement for the generator model according to [7] is no longer met. With the help of the Thevenin theorem, the grid voltage $V_{N1,2}$ is described by an equivalent voltage source V_{sub} and an equivalent impedance Z_{sub}. By describing the grid voltage $V_{N1,2}$ using a voltage source V_{sub} and an impedance Z_{sub}, the requirement for the generator model according to [7] is once again fulfilled. This will be described in detail in the next section.

2.2 Parallel operation by using the Thevenin theorem

The Thevenin theorem states that a complex electrical circuit, consisting of multiple energy sources and impedances, can be described by a single voltage source with series-connected impedance [8].

In order to apply the Thevenin theorem, the two microgrids from Figure 1 have to be converted into an electrical circuit (Fig 3). The lines R_{Lij}, X_{Lij} of the two microgrids are specified by an ohmic inductive impedance [9]. The loads R_{Load}, X_{Load} can also be described by an ohmic inductive impedance [10]. The load R_{Load}, X_{Load} and the associated line R_{Li3}, X_{Li3} are combined into a resistance R_{ld} and a reactance X_{ld}.

PV systems are connected to the grid via an inverter. When using grid-following control for the PV inverter, the PV inverter can be represented as a current source. Virtual generators are used as a control concept for the battery inverters. Accordingly, the battery inverters can be defined as a voltage source [4]. The transformers are simply represented by a reactance X_{ki} [9]. These two reactances X_{ki}, together with the line impedance R_{LM}, X_{LM}, are combined to form a resistance R_{lM} and a reactance X_{lM}.

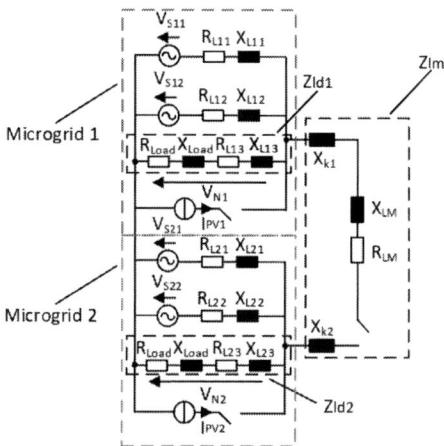

Fig. 3: Equivalent circuit of both microgrids

The application of the Thevenin theorem is first described for the island grid operation of microgrids and then expanded for parallel operation.

2.2.1 Stand-alone operation

The two microgrids are decoupled from each other for island grid operation. Accordingly, the two transformers and the line Z_{LM} between them are not considered. The functionality of the Thevenin theorem is explained on microgrid 1 (see Figure 3).

The Thevenin theorem is applied to model the grid voltage V_{N1} of microgrid 1 by calculating an equivalent circuit. Therefore, the grid voltage V_{N1} has to be viewed from the perspective of each individual Distributed Generation (DG) unit, and summarized to an equivalent circuit.

Figure 4 shows, how the Thevenin theorem is employed if the grid voltage V_{N1} is viewed from the perspective of DG 1. Employing the Thevenin theorem, DG 2 and the impedance Z_{Ld1} can now be added to a series connection of a voltage source V_{sub} and an impedance Z_{sub}, if the current source is not connected. If the current source is switched on, it has to be considered when determining the voltage source V_{sub}.

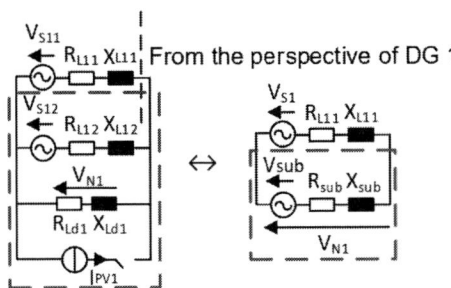

Fig. 4: Method of the Thevenin theorem for island grid operation and for microgrid 1

The voltage source V_{sub} is now the input of the generator model from Figure 2 and gets calculated by equation (1), if the current source is not connected. This voltage depends on the virtual stator voltage V_{s12} of the second Distributed Generation unit. The input variables a and R_L are calculated for parallel operation using the equations (2) and (3).

$$V_N = V_{sub} = \frac{\underline{Z}_{ld1}}{\underline{Z}_{ld1} + \underline{Z}_{L12}} \cdot V_{s12} \qquad (1)$$

$$a = \frac{X_{L11} + \mathrm{Im}\{Z_{sub}\}}{X_d} \qquad (2)$$

$$R_L = R_{L11} + \mathrm{Re}\{Z_{sub}\} \qquad (3)$$

$$Z_{sub} = \frac{\underline{Z}_{ld1} \cdot \underline{Z}_{L12}}{\underline{Z}_{ld1} + \underline{Z}_{L12}} \qquad (4)$$

$$V_{sub,PV} = \frac{\underline{Z}_{ld1} \cdot \underline{Z}_{L12}}{\underline{Z}_{ld1} + \underline{Z}_{L12}} \cdot I_{PV1} \qquad (5)$$

The impedance Z_{sub} is calculated with equation (4). If the current source is switched on, the voltage source V_{sub} results from the superposition of equations (1) and (5).

Additionally, the same has to be done from the perspective of DG 2. A circular process arises. The voltage source V_{sub}, the resistance R_L and the factor a are calculated applying the Thevenin theorem. The virtual stator voltage V_s is calculated using these variables according to Figure 2. The virtual stator voltage V_s is the input variable for the, Thevenin theorem.

2.2.2 Grid-connected operation

Figure 5 shows how to use the Thevenin theorem for the grid connected operation. This will be explained in detail for microgrid 1. The same steps can be done for microgrid 2. Here, microgrid 2 is first combined into a voltage source $V_{sub,m}$ and an impedance $Z_{sub,m}$, from the perspective of microgrid 1.

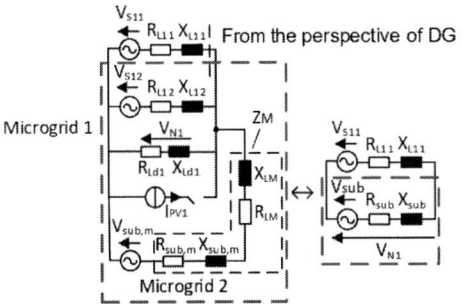

Fig. 5: Method of the Thevenin theorem for grid connected operation for microgrid 1

The impedance $Z_{sub,m}$ is described by equation (6). The voltage source $V_{sub,m}$ results from the superposition theorem according to equation (7). The voltage component $V_{sub,m1}$ arises from virtual stator voltage V_{s21} and results from equation (8). Equation (9) shows the voltage component $V_{sub,m2}$, which results from the virtual stator voltage V_{s22}. The voltage component $V_{sub,m3}$ only arises, when the current source is switched on (e.g. (10)).

$$Z_{sub,m} = \frac{Z_{L21} \cdot Z_{L22} \cdot Z_{Ld2}}{Z_{L22} \cdot Z_{Ld2} + Z_{L21} \cdot Z_{Ld2} + Z_{L21} \cdot Z_{L22}} \qquad (6)$$

$$V_{sub,m} = V_{sub,m1} + V_{sub,m2} + V_{sub,m3} \qquad (7)$$

The equivalent impedance $Z_{sub,m}$ of the microgrid 2 is combined with the impedance Z_{LM} to form an equivalent impedance Z_M.

$$V_{sub,m1} = \frac{Z_{L22} \cdot Z_{Ld2} \cdot V_{s21}}{Z_{L22} \cdot Z_{Ld2} + Z_{L21} \cdot Z_{Ld2} + Z_{L21} \cdot Z_{L22}} \qquad (8)$$

$$V_{sub,m2} = \frac{Z_{L21} \cdot Z_{Ld2} \cdot V_{s22}}{Z_{L22} \cdot Z_{Ld2} + Z_{L21} \cdot Z_{Ld2} + Z_{L21} \cdot Z_{L22}} \qquad (9)$$

$$V_{\text{sub,m3}} = \frac{Z_{\text{L21}} \cdot Z_{\text{L22}} \cdot Z_{\text{Ld2}} \cdot I_{\text{PV2}}}{Z_{\text{L22}} \cdot Z_{\text{Ld2}} + Z_{\text{L21}} \cdot Z_{\text{Ld2}} + Z_{\text{L21}} \cdot Z_{\text{L22}}} \quad (10)$$

As already described in Section 2.2.1, the grid voltage V_{N1} of microgrid 1 has to be replaced by an equivalent circuit, in order to be able to use the generator model according to [7]. However, this equivalent circuit also depends on the equivalent circuit of microgrid 2 (see Fig. 5). The equivalent impedance Z_{sub} is calculated according to equation (11).

$$Z_{\text{sub}} = \frac{Z_{\text{L12}} \cdot Z_{\text{Ld1}} \cdot Z_{\text{M}}}{Z_{\text{L12}} \cdot Z_{\text{Ld1}} + Z_{\text{Ld1}} \cdot Z_{\text{M}} \cdot Z_{\text{L12}} \cdot Z_{\text{M}}} \quad (11)$$

The input variables a and R_{L} are again calculated using equations (2) and (3). However, the equivalent impedance Z_{sub} now results from equation (11). The equivalent voltage V_{sub} for grid connecting operation can be derived from the superposition law (see Eq. (12)).

$$V_{\text{N}} = V_{\text{sub}} = V_{\text{sub,1}} + V_{\text{sub,2}} + V_{\text{sub,3}} \quad (12)$$

$$V_{\text{sub,1}} = \frac{Z_{\text{M}} \cdot Z_{\text{Ld1}} \cdot V_{\text{s12}}}{Z_{\text{L12}} \cdot Z_{\text{Ld1}} + Z_{\text{Ld1}} \cdot Z_{\text{M}} \cdot Z_{\text{L12}} \cdot Z_{\text{M}}} \quad (13)$$

$$V_{\text{sub,2}} = \frac{Z_{\text{L12}} \cdot Z_{\text{Ld1}} \cdot V_{\text{sub,m}}}{Z_{\text{L12}} \cdot Z_{\text{Ld1}} + Z_{\text{Ld1}} \cdot Z_{\text{M}} \cdot Z_{\text{L12}} \cdot Z_{\text{M}}} \quad (14)$$

$$V_{\text{sub,3}} = \frac{Z_{\text{L12}} \cdot Z_{\text{Ld1}} \cdot Z_{\text{M}} \cdot I_{\text{PV1}}}{Z_{\text{L12}} \cdot Z_{\text{Ld1}} + Z_{\text{Ld1}} \cdot Z_{\text{M}} \cdot Z_{\text{L12}} \cdot Z_{\text{M}}} \quad (15)$$

The voltage component $V_{\text{sub,1}}$ arises from virtual stator voltage V_{s12} and results from equation (13). Equation (14) describes the voltage component $V_{\text{sub,2}}$, which results from the equivalent voltage source $V_{\text{sub,m}}$ from microgrid 2. The voltage component $V_{\text{sub,m3}}$ only arises, when the current source from microgrid 1 is switched on (eg. (15)).

The procedure just described has to be carried out for DG2, as well. A circular process arises again. Only this time, the input variables V_{sub}, a and R_{L} are influenced by the second microgrid, too.

2.3 Design of load, transmission line and transformer

According to Section 2.2, the lines, transformers and loads can be defined by ohmic-inductive impedances. This section explains the design of the impedances.

In the first step, the nominal apparent power of the load has to be determined. According to [2], microgrids always refer to low-voltage grids.

In [11] the low-voltage grids are divided into clusters. The high-rise cluster is selected for microgrid 1 and the urban apartment building cluster for microgrid 2. These two clusters have a typical nominal apparent power of 400 kVA [11]. This refers to the local grid transformer. In this paper, this typical nominal apparent power is equated with the nominal apparent power of the load. The lines of the two microgrids have to be designed for the nominal load apparent power. For this purpose, the nominal current flowing through the cables is calculated. With a nominal apparent power of 400 kVA, was results in 580 A by a line to line voltage of 0,4 kV. The cable that can transport this rated current is selected from the table for cable parameters according to [9]. This leads to a resistance coating of 0.1188 $\frac{\Omega}{\text{km}}$ and a reactance coating of 0.08 $\frac{\Omega}{\text{km}}$ for this nominal current [9]. The resistance R_{Lij} and reactance X_{Lij} of the lines can then be calculated using equations (16) and (17) [9].

$$R_{\text{Lij}} = 0.1188 \, \frac{\Omega}{\text{km}} \cdot l_{\text{ij}} \quad (16)$$

$$X_{\text{Lij}} = 0.08 \, \frac{\Omega}{\text{km}} \cdot l_{\text{ij}} \quad (17)$$

The line lengths required can be found in Table 1. The line lengths $l_{3\text{j}}$ are the grid lengths of the high-rise cluster and urban apartment building cluster [11].

l_{ij}		i		
		1	2	3
j	1	0,3 km	0,3 km	1,6 km
	2	0,3 km	0,3 km	1 km

Table 1: Line lengths of the microgrids

The transformers are designed according to equation (18). The transformer is a medium-voltage transformer. The two microgrids are located within a city due to the chosen clusters. This means the distance is short, so that medium voltage is sufficient for transmission. The variable S_{rT} represents the nominal apparent power of the transformers, which is 400 kVA. The top side rated voltage U_{r1} of the transformer is 20 kV.

The nominal voltage U_{r2} on the bottom side of the transformers is, however, 0.4 kV [11]. The relative short-circuit voltage u_k is 5% for local grid transformers [9].

$$X_{ki} = \frac{u_k \cdot U_{r1}^2}{S_{rT} \cdot \left(\frac{U_{r1}}{U_{r2}}\right)^2} \tag{18}$$

The design of the line between the microgrids is based on a guideline for medium voltage lines. The resistance coating is 0.3 $\frac{\Omega}{km}$ and the reactance coating is 0.34 $\frac{\Omega}{km}$. The line resistance R_{LM} and line reactance X_{LM} are calculated according to equations (18) and (19) [9].

$$R_{LM} = 0.3 \ \frac{\Omega}{km} \cdot l_M \tag{18}$$

$$X_{LM} = 0.34 \ \frac{\Omega}{km} \cdot l_M \tag{19}$$

The length l_M of the medium voltage line is chosen to be 0.3 km since both microgrids are located within one city.

The following section deals with the calculation of the low voltage load. According to [10] the load is described by a resistance R_{load} and a reactance X_{load}. Equations (20) and (21) show how the load resistance R_{load} and load reactance X_{load} are calculated. The nominal load voltage $|V_{load}|$ is chosen to be 400 V.

$$R_{load} = \frac{|V_{load}|^2 \cdot P_{load}}{P_{load}^2 + Q_{load}^2} \tag{20}$$

$$X_{load} = \frac{|V_{load}|^2 \cdot Q_{load}}{P_{load}^2 + Q_{load}^2} \tag{21}$$

As can be seen from equations (20) and (21), the current load active P_{load} and reactive power Q_{load} are required to calculate the load resistance R_{load} and load reactance X_{load}. The current active power P_{load} is calculated from the current apparent load power S_{load} and the power factor $\cos(\varphi)$ (see equation 22). According to [9] the power factor is 0.9. The current reactive power Q_{load} is calculated using equation (23).

$$P_{load} = |S_{load}| \cdot \cos(\varphi) \tag{22}$$

$$Q_{load} = \sqrt{|S_{load}|^2 - P_{load}^2} \tag{23}$$

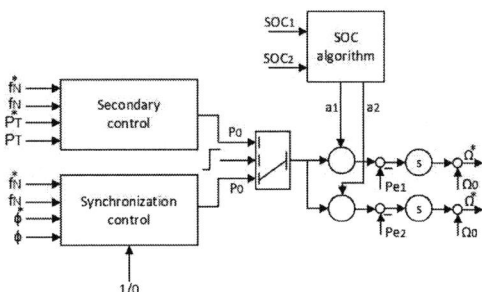

Fig. 6: Determination of the speed setpoints Ω^*

3 Black start strategies

3.1 Methods

According to [12] there are several black start strategies for microgrids:

1) Bottom-up Black start
2) Minimum-Resource Black start
3) Top-down Black start

The first one is considered in this paper. In the first step, a converter in each microgrid performs a black start. However, the microgrids are still decoupled from each other. In the next step, the other converter units synchronize with the respective microgrid and share the load through droop control. In the last step, the microgrids synchronize with each other [12].

3.2 Synchronization and control

This section deals with the synchronization and control of the microgrids. The behavior of the microgrid is influenced by the adjustment of the speed setpoint Ω^* and the voltage setpoint v_s^* of the virtual generators (see Figures 6 and 7).

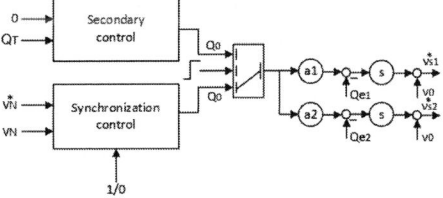

Fig. 7: Determination of the voltage setpoints v_s^*

The speed setpoints Ω^* and voltage setpoints v_s^* are determined using droop control. The active power setpoint P_0 and the reactive power setpoint Q_0 for the droop control are determined via the synchronization control or a secondary control.

The factors a_1 and a_2 define how the target power P_0 and Q_0 are distributed among the batteries. The secondary control is only activated when the two microgrids are connected to each other.

The two microgrids are synchronized when the frequency, phase and amplitude of the grid voltages match [1]. It will always synchronize to the microgrid that is closest to 50 Hz. Figure 8 shows how the synchronization control of a microgrid works. The phase difference is added to the frequency setpoint f_N^* using a droop control.

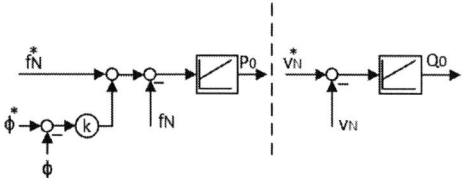

Fig. 8: Synchronization control of microgrids

This forms the actual frequency setpoint, which is compared with the frequency f_N of the synchronizing grid voltage. A PI controller then specifies the active power setpoint P_0. The desired voltage amplitude v_N^* is also set using a PI controller.

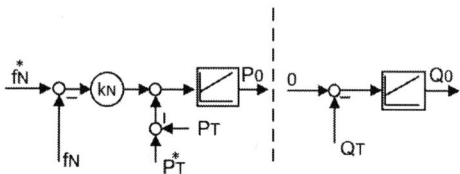

Fig. 9: Secondary control of microgrid [13]

Figure 9 shows the structure of the secondary control. Each microgrid has its own secondary control [1]. The task of the secondary control is to regulate the frequency f_N and the active power P_T exchanged between the microgrids. The chosen control concept, for this task, is the grid characteristic method. With the grid characteristic method, the secondary control power is only provided in the disrupted microgrid [13]. This is explained in the following section. The stationary deviation of the exchanged active power ΔP_{Ti} between the microgrids is described using equation (24). This stationary deviation occurs when the secondary control is not activated [13].

$$\Delta P_{Ti} = -k_{Ni} \cdot \Delta f + \Delta P_{Load,i} \tag{24}$$

The variable $\Delta P_{Load,i}$ describes the change in active power caused by a change in load or feed-in. Δf specifies the stationary frequency deviation, which is identical for both microgrids. The variable k_{Ni} describes the statics of the respective microgrid. If there is no disruption in a microgrid, the stationary deviation of the exchanged active power ΔP_{Ti} results only from the stationary frequency deviation.

By weighting the grid frequency error with the constant k_N (equal to the statics), the secondary control power P_0 is only made available in the disrupted microgrid, as the control error in the undisturbed microgrid adds up to zero (see Eq. 24) [13]. In addition, the reactive power Q_T that is, exchanged between the microgrids, is controlled. The target value for the exchanged reactive and active power is zero.

An algorithm determines the factors a_1 and a_2 depending on the current state of charge SOC_i, according to equations (25) and (26).

$$a_1 = 0.5 \pm a(SOC_1 - \overline{SOC}) \tag{25}$$

$$a_2 = 0.5 \pm a(SOC_2 - \overline{SOC}) \tag{26}$$

The variable SOC_i describes the current state of charge of the respective battery, while $\overline{SOC}$ represents the average of the state of charge of all batteries in a microgrid. If the current state of charge SOC_i is greater than the average state of charge $\overline{SOC}$, the factor a_i is greater than 0.5, with a positive secondary control power P_0. This will drain the battery more. With a negative secondary control power P_0, the factor a_i is smaller than 0.5, if the current state of charge SOC is higher than the average state of charge $\overline{SOC}$. The constant a decides the strength of the increase or decrease. This is chosen to $0.001 \frac{1}{\%}$.

4 Simulation

The grid reconstruction is now simulated for the previously described model, consisting of two microgrids. The grid reconstruction is divided into three steps:

1) Black start of the two microgrids in island operation
2) Synchronization of the microgrids
3) Parallel operation

In the first step of grid reconstruction one batterie of each microgrid performs a black start. The second battery then synchronizes with the first. In the second step of grid reconstruction, the microgrids synchronize with each other. The synchronization control from Figure 8 is used for this. In the third and final step of the grid reconstruction, the grid frequency of the microgrids will be brought back to 50 Hz. The secondary control from Figure 9 is used for this purpose. The grid reconstruction is then completed. During the grid reconstruction phase, the load is divided equally between the batteries.

Load and feed-in changes in grid operation are distributed independently of the state of charge and once depending on the state of charge. In section 4.1 the division occurs independently of the charging status. In contrast to section 4.1, the load distribution in section 4.2 takes place depending on the current charge status. Figure 10 shows the load requirements and the feed-in of the PV systems during grid reconstruction and grid operation.

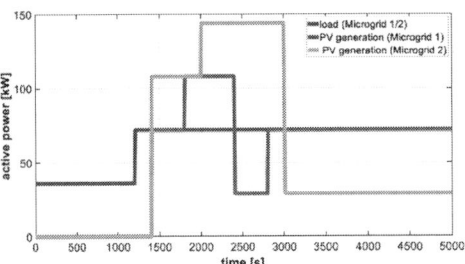

Fig. 10: Load and feed-in profile of the two microgrids

During the grid reconstruction, the active power requirement is 36 kW and the PV generation of both microgrids are zero. After the microgrids are interconnected, the active power requirement increases to 72 kW. The PV generation is different in both microgrids as there is a spatial distance between the microgrids. This means that there may be slightly different weather conditions. After 2400 s, the PV generation from microgrid 1 briefly stops as there is a cloud train. In microgrid 2, the PV production drops permanently after 3000 s since it is very cloudy in this area.

Table 1 shows the parameters of the two microgrids that are used in this study. The parameters SOC_{01} and SOC_{02} describe the state of charge of the two batteries, that exists at the beginning of the grid reconstruction. According to [1], the grid reconstruction is carried out with a minimum load, which was selected to be 10 % of the nominal load

Parameter	Microgrid 1	Microgrid 2
SOC_{01}	80 %	80%
SOC_{02}	30 %	30%
S_{load}	10 % $S_{load,N}$= 40 kVA	10 % $S_{load,N}$= 40 kVA

Table 2: Parameters of the two microgrids

4.1 Without charging status algorithm

This section describes grid reconstruction and grid operation when the state of charge algorithm is not activated. This means, the active power is distributed among the batteries independently of the state of charge.

The batteries each have the same rated power, so the active power distribution is identical. In the first step of grid reconstruction, both microgrids are in island grid operation. In this phase, a battery in each microgrid performs the black start. The other battery then synchronizes with the grid.

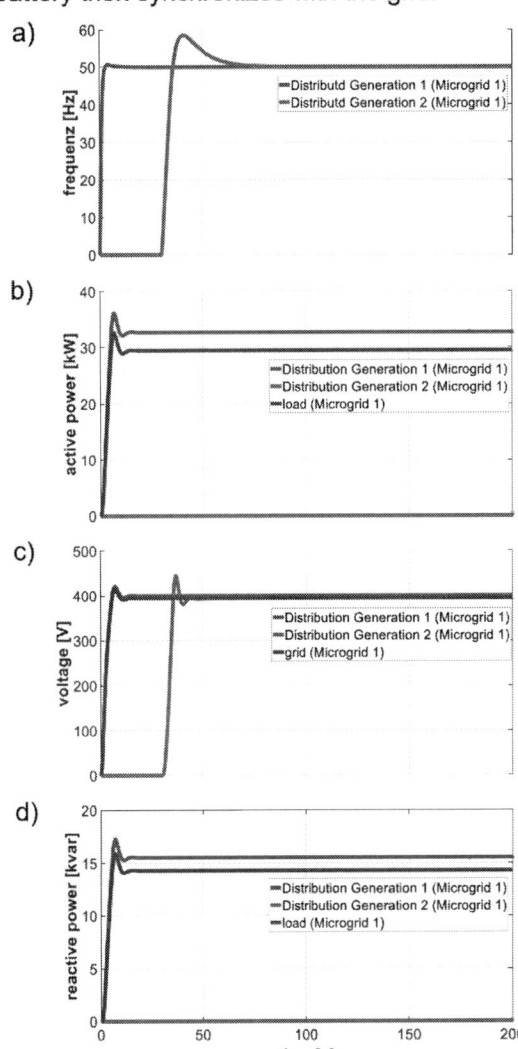

Fig. 11: Black start and synchronization of the DGs

The battery with the highest initial state of charge carries out the black start as it can supply the load for the longest time. Figure 11 shows the black start of microgrid 1. This is identical to microgrid 2 because of the identical load. DG 1 increases its frequency and voltage to the nominal values (Fig. 11 a). This increases the active and reactive power requirements of the load. DG 1 supplies the load at all times (Fig. 11 b, d). The grid voltage does not correspond to the nominal value of 400 V because there are losses across the line (Fig. 11 c).

After 30 s, DG2 begins to synchronize with the microgrid. DG 2 increases the frequency to 50 Hz (Fig. 11 a). To ensure that no compensation process occurs, the voltage of DG 2 have to be equal to the grid voltage (Fig. 11 c). DG 2 is not yet connected to the microgrid and therefore no active and reactive power transmission takes place (Fig. 11 b, d). After 200 s, DG 2 is switched on in both microgrids and this is shown in Figure 12. The two DGs share the load equally. This reduces the losses across the line which in turns increases the grid voltage. Through this the load requirement increases. Due to statics, the frequency and voltage of the DGs deviate from the nominal values. The increase in frequency and voltage can be explained by load sharing.

In the second step of rebuilding the grid, the two microgrids synchronize with each other. The first microgrid synchronizes with the second microgrid because the frequency of the second microgrid is closer to 50 Hz. The synchronization process is shown in Figure 12. The frequency of the DGs in microgrid 1 adjusts to the frequency of the DGs in microgrid 2 (Fig. 12 a). The frequency increases briefly since the phase angles also have to match. This is within the permissible range as the dynamic change in frequency is not greater than 800 mHz [13]. In addition, the load voltage of microgrid 1 has to be synchronized with the grid voltage of microgrid 2. This is done by varying the voltages of the DGs (Fig. 12 c). After 900 s, the synchronization process is completed and the microgrids are connected to each other.

In the third step, the frequency of both microgrids is brought back to 50 Hz (Fig. 12 a). This is done via a secondary control. In order to bring the frequency back to 50 Hz, the active power is reduced because the active power feed-in is too high (Fig. 12 b). This causes the grid voltage in both microgrids to increase due to the lower voltage losses (Fig. 12 c). The grid reconstruction is complete.

After 1200 s, a load increase occurs in both microgrids (Fig. 10). The frequency in both microgrids drops (Fig. 12 a). The active power feed of the DGs in both microgrids is increased and the frequency of both microgrids is returned to 50 Hz (Fig. 12 b). In addition, the reactive power requirement of the load on both microgrids has also increased. Accordingly, the DGs deliver more reactive power (Fig. 12 d). The grid voltage of both microgrids drops due to the increased load requirement (Fig. 12 c).

After 1400 s, the PV feed-in is increased in both microgrids (Fig. 10). This causes the frequency of both microgrids to increase (Fig. 12 a). In order to bring the frequency back to 50 Hz, the active power of the DGs in microgrid 1 is reduced and the DGs in microgrid 2 even absorb active power (Fig. 12 a, b). This is due to the fact that the active power requirement in microgrid 2 is lower than the PV feed-in. Due to the PV feed-in, the grid voltage of both microgrids increases (Fig. 12 c).

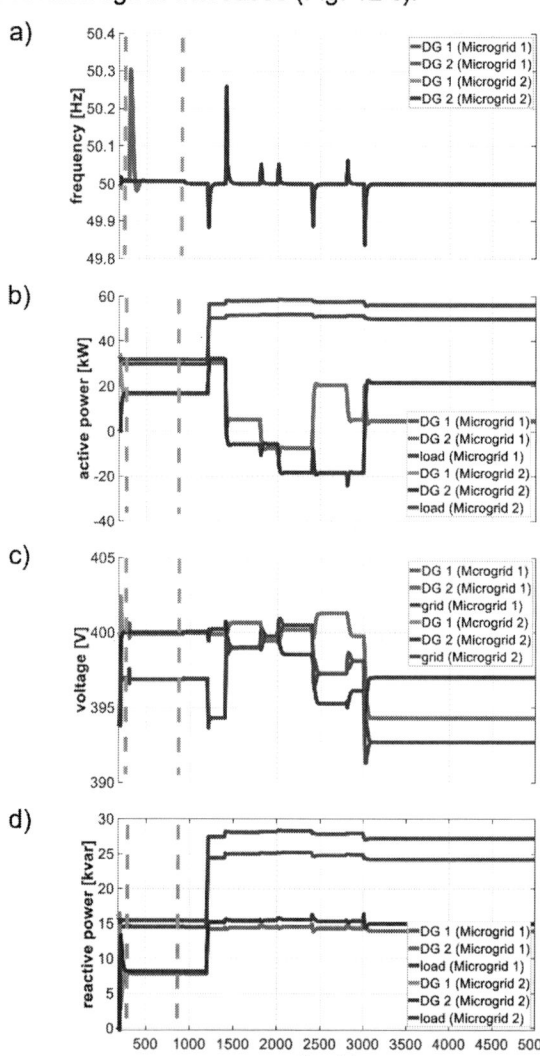

Fig. 12: Synchronization and parallel operation of microgrids

The grid voltage in microgrid 2 is higher than in microgrid 1 because the PV feed-in is higher. That's why reactive power flows from microgrid 2 into microgrid 1.

The aim of the secondary control is that no reactive power exchange takes place between the microgrids. To ensure that no reactive power exchange occurs, the grid voltages have to be equal.

Accordingly, the grid voltage of microgrid 1 increases and the grid voltage of microgrid 2 is reduced. This is achieved by increasing the voltages of the DGs of microgrid 1 and decreasing them of microgrid 2. The PV feed into the microgrid increases after 1800 s (Fig. 10). This causes the frequency of both microgrids to increase (Fig. 12 a). The disruption of the active power balance is only in microgrid 1, accordingly the active power delivery of the DGs only changes permanently in microgrid 1. Microgrid 2 also takes part in the disruption for a short time. DGs of microgrid 1 consume active power because the PV feed-in is higher than the current load requirement (Fig. 12 b). The grid voltage of microgrid 1 increases due to the PV feed-in, while the grid voltage of microgrid 2 remains the same. In order to stop the exchange of reactive power between the microgrids, the grid voltages have to be equalized. To achieve this, the grid voltage of microgrid 1 decreases and microgrid 2 increases. This is achieved by reducing or increasing the voltages of the DGs (Fig. 12 c). These processes just described are repeated when there are feed-in changes. During grid operation, all voltages are above 90% of the nominal voltage and below 110% of the nominal voltage. These are within the permissible range for low-voltage grids. [9].

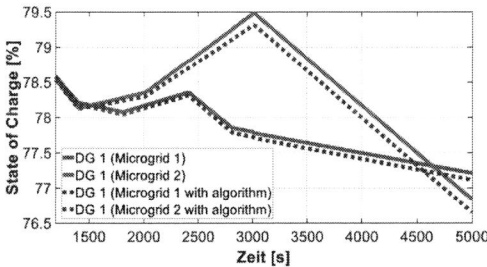

Fig. 14: State of charge of DG 2 for both microgrids

Figures 13 and 14 show the charge status of the DGs of both microgrids during grid reconstruction and parallel operation. During the grid reconstruction, the charge level of all DGs decreases as active power is supplied. During parallel operation, the state of charge increases or decreases depending on whether active power is supplied or consumed. The gradient is determined by the level of active power. The higher the active power is, the greater the gradient.

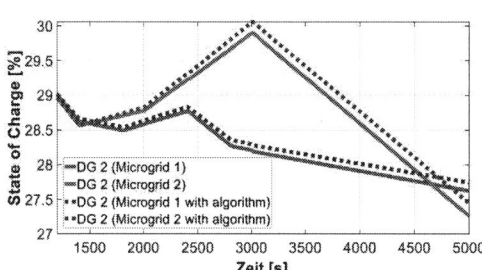

Fig. 14: State of charge of DG 2 for both microgrids

4.2 With charging status algorithm

In the following section, the algorithm described in 3.2 is activated, which divides the active power depending on the state of charge. The grid reconstruction is identical to section 4.1. The effects of the algorithm on active power, reactive power and voltage are shown on microgrid 1.

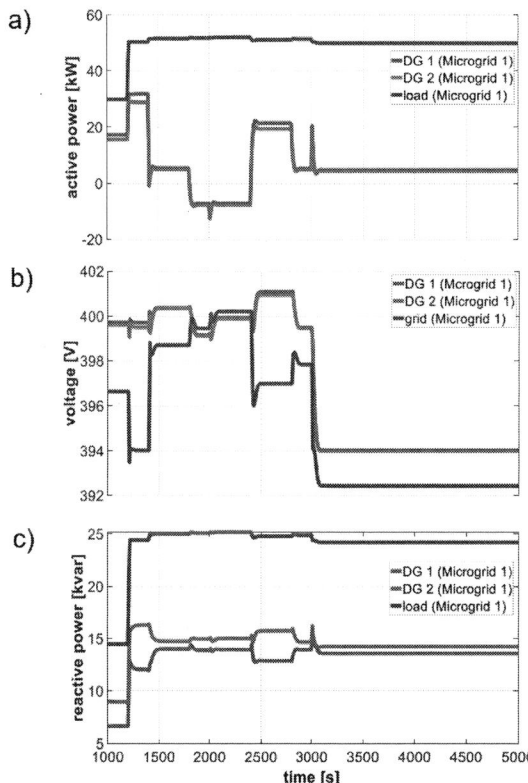

Fig. 15: Parallel operation of microgrids for activated state of charge algorithm

The effect on the frequency remains the same as the load or feed profile does not change. Figure 15 a show the active power distribution of the DGs of microgrid 1. When delivering active power, DG 1 delivers more active power since it has a higher state of charge (Fig. 13).

If active power is consumed, DG 2 consumes more power because it has a lower state of charge (Fig. 14). By distributing the active power to the DGs depending on the state of charge, the battery with a higher state of charge is discharged more than the battery with a lower state of charge. This is reversed when charging. Figure 15 b shows the voltage. It is noticeable here that DG 1 has a higher voltage than DG 2. The voltage of DG 1 has to be greater to compensate the higher voltage loss. As a result, due to the voltage droop control, DG 2 supplies more reactive power (Fig. 15 c).

5 Conclusion

The grid reconstruction and the parallel operation, for different loads and PV feeds, of two microgrids were examined. To rebuild a microgrid, batteries with a fictitious synchronous generator as a control method have to be used. The parallel operation of batteries is therefore considered to be the parallel operation of synchronous generators. In order to study the parallel operation of generators, the generator model had to be extended using the Thevenin theorem. Changes in the load or PV feed lead to frequency changes. These are compensated by a secondary control. Through the grid characteristic curve method, the secondary control power was only made available in the faulty grid. The secondary control allowed the grid frequency of the respective microgrid to be returned to 50 Hz. In order to return the grid frequency to 50 Hz, the batteries absorb or release active power. This leads to an increase or decrease in the charge level. By activating the algorithm, the secondary control power is distributed to the batteries depending on the current charge status. This made it possible to improve the charging status. It is important to note that the battery with greater active power output has a higher voltage.

6 Acknowledgement

This work has been funded by the German Federal Ministry for Economic Affairs and Climate Action (BMWK) as part of the project "STROM" under grant no. 03EI6084A.

7 References

[1] Bernhard Hammer: "Netzseitige Umrichterregelung in Microgrids und Mircrgrid-Verbundnetzen" Dissertation. Technische Universität Darmstadt 2021.

[2] Magdi S. Mahmoud, Fouad M. Al-Sunni: „Control and Optimization of Distributed Generation Systems ", Springer International, 2015.

[3] Florian Redmann, Alexander Ernst, Bernd Orlik: "Black Start Capability and Islanded Operation of Power Converters with Virtual Synchronous Generator Control" PCIMEurope 2021, Nürnberg.

[4] Florian Mahr, Stefan Henninger, Martin Biller, Johann Jäger: „Elektrische Energiesysteme Wissensvernetzung von Stromrichter, Netzbetrieb und Netzschutz ", Springer Vieweg, 2021.

[5] David Matthies, Alexander Ernst, Henning Sauerland, Rene Reimann, Wilfried Holzke, Bernd Orlik: "Provision of Power Plant Equal Ancillary Services by Wind Turbines: From Maximum to Grid-demanded Power Point Tracking" PCIM Europe 2022, Nürnberg.

[6] Leonard Wilkening: „Netzorientierter Betrieb von Batteriespeichersystemen in Verteilnetzen" Dissertation. TU Hamburg 2021.

[7] Werner Leonhard: "Regelung in der elektrischen Energieversorgung", B. G. Teubner Stuttgart, 1980.

[8] Thomas Harriehausen, Dieter Schwarzenau: „Moeller Grundlagen der Elektrotechnik" Springer Vieweg, 2019.

[9] Klaus Heuck, Klaus-Dieter Dettmann, Detlef Schulz: „Elektrische Energieversorgung", Springer Vieweg, 2013

[10] Wilfried Holzke, Florian Redmann, Matthias Joost, Bernd Orlik: "Separation of Models for the Distributed Simulation of Electric Grids" PCIM Europe 2021.

[11] Andreas Wieß, Elisabeth Springmann, Maximilian Hecker: „Steckbriefe der modellierten Typnetze", FfE GmbH, 2023.

[12] Elliott Fix, Abhishek Banerjee, Ulrich Muenz, Gab-Su Seo: „Investigating Multi-Microgrid Black Start Methods Using Grid Forming Inverters" ISGT NA 2023, Washington.

[13] Lutz Hofmann: „Elektrische Energieversorgung Band 3: Systemverhalten und Berechnung von Drehstromnetzen", Walter de Gruyter GmbH, 2019.

PCIM Europe 2024, 11– 13 June 2024, Nuremberg   DOI: 10.30420/566262315

21 kA Solid State DC Breaker for Supergrid Institute's High Power Test Facility

Christophe Conilh[1], Mohammad Kabalo[1], Christophe Creusot[2] , Guillaume Amodeo[2]

[1] GE Vernova Power Conversion, France
[2] Supergrid Institute, France

Corresponding authors: Christophe Conilh, christophe.conilh@ge.com
                        Christophe Creusot, christophe.creusot@supergrid-institute.com

Speaker: Christophe Conilh, christophe.conilh@ge.com

Abstract

SuperGrid Institute commissionned its High Power Source (HPS) test platform on June 2022. This platform is a new tool for the industry to develop future high-voltage direct and alternating current equipment for the massive integration of renewable energy. The HPS can produce the real high voltage direct current that devices will encounter in the field. This capacity is made possible by the platform's current rectifier, which converts the current generated by an alternator into a high-voltage direct current. With this platform, SuperGrid Institute can perform both DC (200 kV DC 40kA) and AC short circuit tests at various frequencies from 10 Hz to 60 Hz, and up to 80 kA, on circuit breakers, fault current limiters, etc. The short-circuit generator (SC Gen) is rated 3.45 GVA and is equipped with two different excitation systems: a "conventional excitation" system connected to the distribution grid, that supplies the SC Gen rotor through a 24-pulse controlled rectifier, and a so-called "super-excitation" system, that excites the SC Gen rotor with a current pulse generated from a supercapacitor bank and controlled by a solid-state DC breaker. For this specific switching function, a medium voltage 21 kA static DC current breaker has been developed. Based on sub-assemblies using press-pack IGBT and press-pack diodes that are connected in parallel and in series, this solid-state breaker has been manufactured, tested, installed, and commissioned for a nominal direct current of 21 kA and a nom. voltage of 2.85 kV DC. After a description of the SC Gen test facility with super excitation system, the technical requirements for solid-state breaker and the developed solutions are presented, including measurement results that were taken during breaker qualification testing and commissioning of the new HPS test platform.

1 Introduction

SuperGrid Institute is a R&D center dedicated to the energy transition focusing on developing technologies on High Voltage Direct Current (HVDC) as well as on Medium Voltage Direct Current (MVDC) to prepare the future of the electrical grid. One key development field for the deployment of such networks is protection devices in case of short circuit. Contrary to the MVAC and HVAC sectors, the protection technologies for DC current are currently emerging and need to be experimentally evaluated and then qualified. For this reason, SuperGrid Institute has invested in a High Power Source (HPS) that enables to test the performance and qualify HVDC and MVDC protection devices.

This article describes first the functionalities and performance of the High Power Source. Furthermore, the short circuit generator and its auxiliaries are detailed, with a particular focus on a solid-state DC breaker, developed for the super-excitation function of the short circuit generator.

2 HPS description

The HPS installation was commissioned in June 2022 and handed over to CERDA Testing Laboratory to be operated.

HPS consists of a 3450 MVA short circuit generator connected to a large test hall, either to perform AC short circuit tests or DC short circuit tests. This testing hall is itself designed such that all power sources of the site can be parallelized so to enhance the total available short circuit power for AC tests.

Fig. 1 Short circuit generator

2.1 HPS architecture

As shown on Fig. 2, the circulation of the current between the generator and the test hall follows a common path to the short circuit test transformers: through the generator protection circuit breaker, the making switch, the short circuit current setting reactors and the circuit switches.

These devices are necessary to protect and transmit the maximum of power that the generator can offer in bi-phase or three-phase configuration.

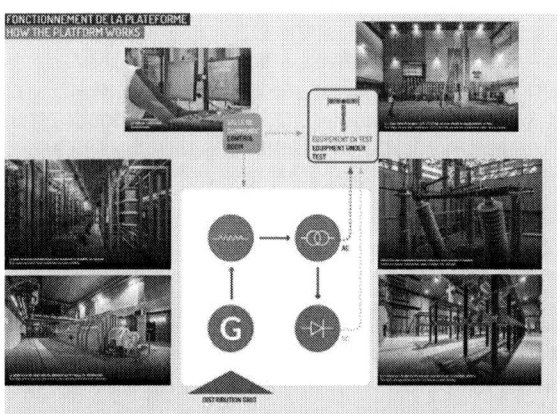

Fig. 2 Simplified architecture of HPS

From the short circuit transformers, two main options are available: the first one is to feed a six-pulse diode rectifier, able to deliver DC short circuit current up to 200 kVdc and 40 kAp; the second one is to go directly to the test hall either to perform bi-phase or three-phase AC short circuit tests for MVAC or HVAC applications. In AC, HPS can perform short-circuit tests at frequencies from 10 Hz to 60 Hz and at test voltage up to 190 kVrms and test current up to 80 kArms.

2.2 Generator electrical auxiliaries

Mainly for layout reasons, the generator is equipped with a 18 pulse Static Frequency Converter (SFC). This soft-starter is designed to fulfill the specific requirements on operating modes, among others, the possibility to parallelize the generator with other sources of CERDA Testing Laboratory.

The excitation function is designed considering the generator driving requirements and the combination of the short circuit current requirements, generator characteristics and available auxiliary power supply. Indeed, one of the main constraints of the project is to cope with the existing connection to the public distribution grid. According to the generator and the power circuit external reactance characteristics from generator to test hall, the existing connection allows to manage the excitation for short circuit tests only to half of the maximum rated generator short circuit current. To build a new connection with a higher power supply capability was excluded for reasons of cost and planning. The way to solve this issue was to design and install a supercapacitor bank. This bank is charged by the 24-pulse excitation converter prior to performing a short circuit test, and the super-excitation current is transferred to the rotor at the right instant by triggering a solid-state breaker. After a test sequence, the excitation converter is used again, this time to discharge the residual energy of the supercapacitor bank on the distribution grid. The specificities and challenges of the super-excitation solid-state breaker is the object of the following sections.

3 Solid-state breaker for super-excitation system of HPS test platform

3.1 Super-excitation system overview

As mentioned above, the SC Gen is equipped with two different excitation systems: a "conventional excitation system based on a 24-pulse controlled rectifier connected to the distribution grid and a so-called "super-excitation" system where the SC Gen is excited by a current pulse generated by the closing of a solid-state breaker upon a pre-charged supercapacitors bank. The choice between the conventional excitation or the super-excitation system depends on the test performed with the SC Gen. When the SC Gen rotor needs an excitation current lower than 10 kA, the conventional system is used. Beyond this value, the super-excitation is activated.

Figure 3 shows a diagram of this super-excitation system.

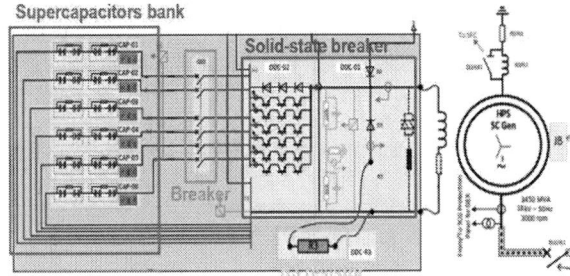

Fig. 3 Overview of the super-excitation system

The super-excitation system is composed of:

- A supercapacitors bank to store the necessary energy for the SC Gen rotor super-excitation.
- A solid-state breaker to connect/disconnect the supercapacitors bank to/from the SC Gen rotor.
- DC breakers to protect the super-excitation system.
- A resistor (R3) to guarantee the recovery voltage across the equipment under test after the disconnection of the super-excitation system.

The supercapacitors bank has the following characteristics:

- Rated voltage: 2856 V.
- Capacitance: 41.62 F.
- Stored energy: 170MJ.

It is realized by connecting in parallel and in series elementary supercapacitors with a rated voltage of 102 V and a rated capacitance of 95.6 F.

The selected structure of the solid-state breaker is a chopper structure, i.e., using an igbt to connect/disconnect the supercapacitors bank to/from the rotor and a freewheeling diode with a series resistor (R3) to dissipate the stored inductive energy of the rotor. In the next paragraphs, the igbt part of the switching cell "igbt + freewheeling diode" will be called igbt sub-assembly and the diode part will be called diode sub-assembly.

3.2 Solid-state breaker overview

3.2.1 Main requirements and constraints

Among all the tests duties requiring the super-excitation system, the most severe one is an Open –

Close - Open cycle called "O-0.3s-CO". This testing cycle requires the Solid-state breaker to follow the same switching cycle. As illustrated by Fig 4, during this test, the rotor is powered from the supercapacitors bank by the switching-on of the solid-state breaker. The duration of each cycle is 150 ms with 300 ms time interval. The values of the rotor current are:

- Around 1550 A at the beginning of each shot corresponding to the rotor no-load current.
- Around 21000 A at the end of each shot.

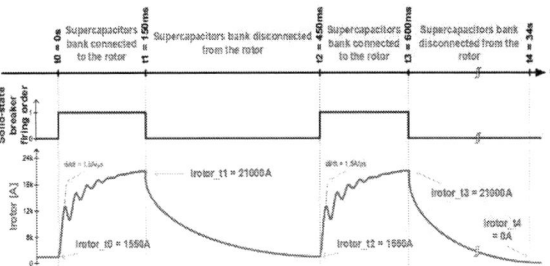

Fig. 4 "O-0.3s-CO" Solid-state breaker switching cycle

Such a cycle has the following constraints on the solid-state breaker:

- Driving two times a current of 21000 A during 150 ms.
- Turning off a current of 21000 A under a switching voltage of 2850 V.
- Turning on with a 1550 A initial current under a switching voltage of 2850 V.
- Requirement to repeat such switching cycle as often as needed for the short-circuit tests, i.e., multiple 21 kA switching operations per day and thousands of switching operations during the lifetime of the DC breaker.

Another constraint is related to the parasitic inductance that is involved in the switching cell. This parasitic inductance is due to the large length of the cables used to connect the supercapacitors bank or the resistance R3 to the solid-state breaker. Indeed, due to the high nominal current, the links between the supercapacitors bank and the solid-state breaker are realized with several cables in the go and return path (a consequence of the mechanical arrangement of the supercapacitors bank). The total estimated length of this link is between 16 m and 24 m (wire section = 150 mm²). The link between the R3 resistor and the solid-sate breaker is realized with one cable (section = 150 mm²). The total length is estimated at 40 m. A special care was paid to the way in which all these links were installed to minimize the value of the parasitic inductance of the switching cell. The measurements made at site show that the parasitic inductance is around 15 to 20 μH. This comparably big value makes the realization of a such solid-state breaker very challenging.

3.2.2 The IGBT sub-assembly of the solid-state breaker

Figure 5 shows a picture of the IGBT sub-assembly. The size of the cabinet is approximately: L = 3.1 m x P = 1.5 m x H = 2 m.

Fig. 5   Picture of the IGBT sub-assembly

Due to the high value of the current to drive/turn off and, depending on the current rating of the IGBTs available on the market, the IGBT sub-assembly is realized by six legs connected in parallel. To ensure a good sharing of the current in each leg, the arrangement is mechanically symmetrical [1]. A particular attention is paid to the wiring of the legs to manage a uniform length of the copper bars. In this way, each leg turns off or drives a current equal to 3500 A.

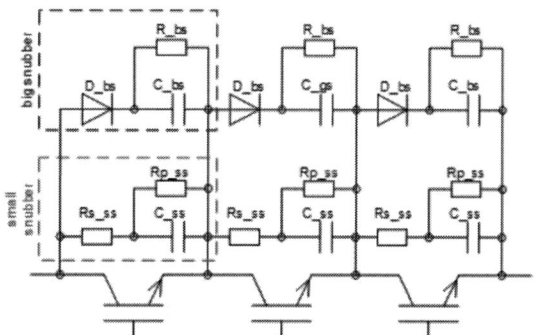

Fig. 6   Electrical schematic of one leg

Each leg is composed of three 4500V press-pack IGBTs connected in series (see Fig 6). The need to use three IGBTs connected in series comes

from the presence of the R3 resistor in series with the diode sub-assembly of the solid-state breaker. Indeed, when the SC Gen is in free-wheeling mode at the end of each cycle (free-wheeling phase of the solid-state breaker), a voltage drop across the R3 resistor appears. The voltage seen by the IGBT sub-assembly is then equal to the sum of the voltage of the supercapacitors bank and the voltage drop across the R3 resistor. Each IGBT is equipped with a big snubber and a small one. The big snubber is an RCD snubber and is used to protect the IGBTs against overvoltage during their turn-off due to the total parasitic inductance of the switching cell [2]. The small snubber is an RC snubber and is used to protect the IGBTs against overvoltage during their turn-off due to the parasitic inductance of the big RC snubber circuit. To guarantee a good current sharing between the six legs, the eighteen IGBTs, which compose the IGBT sub-assembly, are selected according to their threshold voltage $V_{GE(off)}$ and their saturation voltage $V_{CE(sat)}$ (manufacturer recommendation).

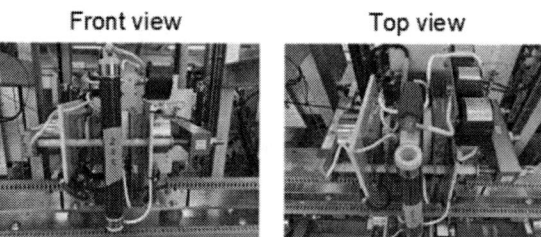

Fig. 7   Picture of one of the three IGBTs and their snubbers in each leg

From a mechanical point of view, the realization of each leg is done with a modular approach as illustrated by Fig 7. Each IGBT of the leg is stacked with the diode of its associated big snubber. The heatsinks used are only copper pieces and the igbts are cooled by natural convection. The thermal calculations have shown that, with this cooling method, the IGBT maximum junction temperature is not exceeded during all the "O-0.3s-CO" switching cycles. All the other elements of the snubbers are fixed on the IGBT stack.

3.2.3 The diode sub-assembly of the solid-state breaker

At the end of the "O-0.3s-CO" switching cycle, the diode sub-assembly must withstand an i²t value equal to 148 MA²s. In accordance with the rating of devices available on the market, this high value of i²t implied to connect two diode legs in parallel. Each leg is composed of two 4500V press-pack diodes connected in series. The need to use two diodes connected in series comes from the fact that the overvoltage across the diode must not exceed 3600 V during its turn-off (beginning of each 150ms shot). However, this voltage limit can be reached and thus seen by the diode sub-assembly due to the di/dt value of the current during the recovery phase of the diode (manufacturer data) and due to the high value of the stray inductance.

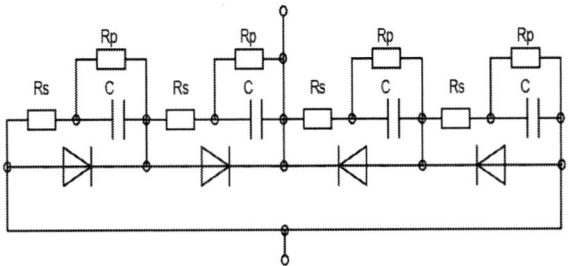

Fig. 8   Electrical sketch of the diode sub-assembly

To ensure a good sharing of the current in each leg, the two legs are installed mechanically symmetrical. The diodes are selected according to their voltage forward V_f criteria. The voltage balance between the diode connected in series is guaranteed by using an RC snubber connected in parallel to each diode. The schematic of this RC snubber is the same as the small RC snubber used for the igbts of the igbt sub-module. Forced-air cooling is used for the diode sub-assembly.

3.3   Experimental results

The solid-state breaker was installed at site in October/November 2020. Its commissioning was done in January 2021. During this phase, several tests and measurements have been performed to verify its operation. Among all the tests performed, specific tests and measurements have been done:

- On the IGBT sub-assembly to verify the balance of the current between the six legs during a shot of a duration equal to 150 ms, and the behavior of the IGBTs when they turn off a current around 3500 A each.

- On the diode sub-assembly to verify the current balance between the two legs during a shot of a duration equal to 150 ms, and the behavior of the diodes when they turn-off a current of around 1000 A.

3.3.1   The IGBT sub-assembly results

Figure 9 presents a measurement of the current in each leg for a shot duration equal to 150 ms and a total current value equal to 21 kA. This result illustrates the fact that the current is well shared between the six legs. In this case, the highest value of the current seen by a leg is 3603 A and the lowest is 3430 A. More generally, all the measurements show that the current unbalance between the legs is less than 5 %. This result validates the mechanical arrangement of the legs and the selection of the IGBTs. The capability of the solid-state breaker to drive a 21kA current is also demonstrated.

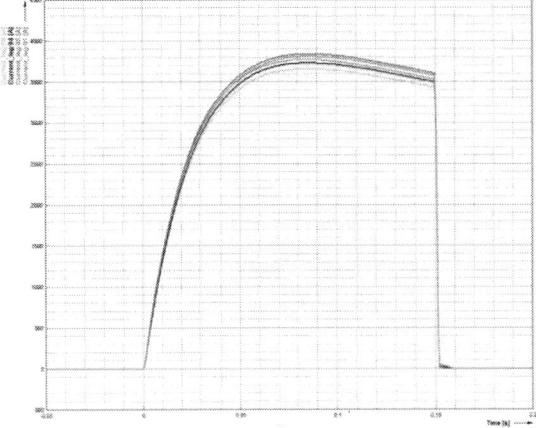

Fig. 9   IGBT sub-assembly – Leg current waveforms

Figure 10 presents the collector-emitter voltage Vce and the collector current Ic of an IGBT of one leg, when it turns off a current equal to 3500A (t = 150 ms in Fig. 9) for the nominal voltage applied by the supercapacitors bank on the load. Figure 11 shows the obtained Ic versus Vce curves where the red curve represents the reverse bias safety operating area (RBSOA) given by the manufacturer. These measurements show that the Ic versus Vce curve is inside the RBSOA. This demonstrates the capability of the solid-state breaker to turn off 21 kA current without any risk.

Fig. 10   IGBT sub-assembly – IGBT turn-off waveforms

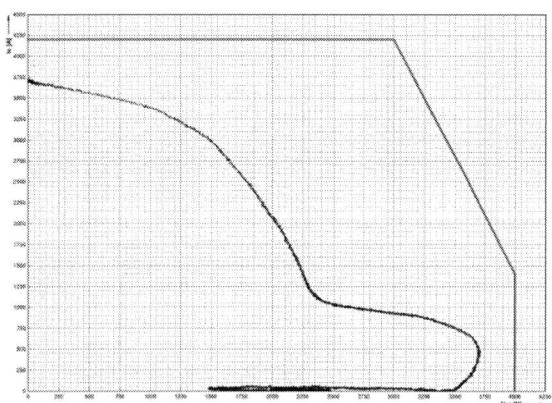

Fig. 11   IGBT sub-assembly – Ic versus Vce igbt turn-off curve at 2850 V/ 3500 A

3.3.2   The diode sub-assembly results

Figure 12 presents the kathode-anode voltage Vka and the kathode-anode current Ika of the diode of one leg when it turns off a current equal to 2500 A for the nominal voltage applied by the supercapacitors bank on the load. These measurements show that during the recovery phase, the Vka voltage of each diode reaches a value around 1800/1900 V. This result validates the use of two diodes connected in series (to follow the diode's manufacturer recommendation). The voltage during the switching is well shared between the diodes: The voltage unbalance is lower than 50 V. These two results demonstrate the capability of the solid-state breaker to turn-on 2.5 kA current without any risk.

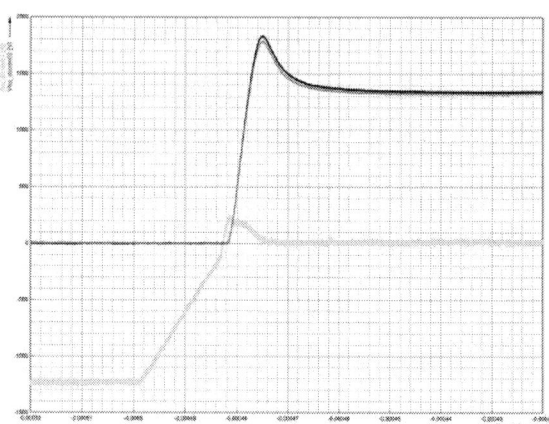

Fig. 12   Diode sub-assembly – Diode turn-off waveforms of one leg

All the tests and measurements performed to verify the current balance between the two legs give similar results as the one presented in Fig. 13. In this case, the free-wheeling phase starts at t = 0 and finishes at t = 7.6ms. The current is equal to 16500 A at the beginning of the phase. During all these tests, the measured current difference between the two legs was below 10%.

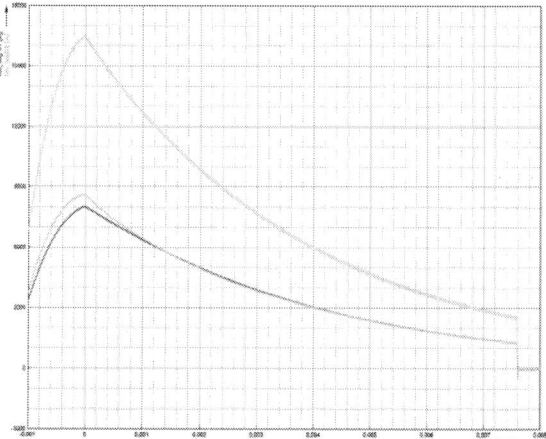

Fig. 13   Diode sub-assembly – Current waveforms in the two legs and total current during the free-wheeling phase.

4   Conclusion

The new High Power Source test platform of the Supergrid Institute, in operation since June 2022, has been presented. In this platform, the short-circuit generator can be excited by a super-excitation system based on a supercapacitors bank. The focus of this paper is the DC solid-state breaker, which allows to connect/disconnect the supercapacitors bank to/from the SC Gen rotor. It is based on several press-pack IGBTs and diodes connected in series and parallel, to drive/turn off a DC current equal to 21 kA for a switching voltage equal to 2850V. The tests performed and the measurements obtained during the breaker qualification and commissioning confirm the described design and manufacturing decisions, and that all switching operations meet the test platform operational requirements. This work has demonstrated that such solid-state DC breaker design could also be used to address other research or industrial medium voltage applications that require multiple reliable DC current switching operations with similar or other current and voltage ratings.

References

[1] Ahmed Majed Saif, Karsten Fink, Hao Wang. "Hardware Optimization of Current Distribution for Parallel-Connected, High-Power, Press-Pack IGBTs." PCIM Asia 2022; International Exhibition and Conference for Power Electronics, Intelligent Motion, Renewable Energy and Energy Management. VDE, 2022.

[2] Philip C. Todd. "Snubber Circuits: Theory, Design and Application." https://www.thierry-lequeu.fr/data/SLUP100.pdf.

PCIM Europe 2024, 11– 13 June 2024, Nuremberg   DOI: 10.30420/566262316

Design and Analysis of a 50kW SiC-based Active-Front-End with a very small line choke for DC-Grids

Raphael Otte[1], Jan-Niklas Koch[1], Tim Stuckmann[1], Prof. Dr.-Ing. Holger Borcherding[1]

[1] University of Applied Sciences and Arts Lemgo, Germany

Corresponding author: Raphael Otte, raphael.otte@th-owl.de
Speaker: Raphael Otte, raphael.otte@th-owl.de

Abstract

As DC grids become more widespread in industry, the demand for DC industry-compliant power supplies is increasing. This paper describes the design and analysis of a SiC-based Active-Front-End with a switching frequency of 100 kHz and an output power of 50 kW. The focus of this Active-Front-End is on the small input inductance (u_k=0,6%), which has been deliberately reduced in order to reduce package size. For a small inductance a high switching frequency and a fast and precise current measurement are basic requirements. The increase of the switching frequency reduces the size of the whole filter.

1   Industrial DC-Grids

In industrial production, the use of renewable energy sources has gained importance in recent years due to rising energy prices. The integration of these is particularly efficient in an open industrial DC grid as described in DC-INDUSTRIE and DC-INDUSTRIE2 [1]. An exemplary system architecture is shown in Figure 1.

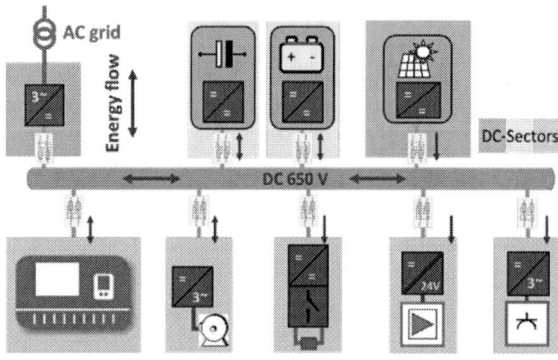

Fig. 1   System concept DC-INDUSTRIE [1]

The participants in the DC grid are divided into DC sectors and are fused with smart DC breakers [7]. Typical applications in industrial production such as drives for machines, passive loads and auxiliary supplies are comparable to AC grids, except that they are directly coupled via the DC bus. Due to

this it is advisable to provide bidirectional energy flow in those participants. Energy is supplied by infeed converters from the AC grid, which can be controlled or uncontrolled rectifiers, as well as by renewable energy sources. As additional supplement energy storage units can be used for peak load reduction, save shut down of facilities during a power failure or short time bridging of power failures.

The integration of, for example, photovoltaic systems into the DC grid provides an energy surplus at low load in the DC grid, which can be fed back into the AC grid if it cannot be stored. This circumstance makes a feed-back capable device topology for the AC/DC converters of the DC grid indispensable. Existing Active-Front-End devices at the current state of the art often have a system-related filter unit in the AC network which requires almost the same installation space as the power electronic components of the device. These filters are necessary to reduce the electromagnetic interference emissions of the active infeed converter. They consist largely of copper, iron and aluminum, which are expensive and require a very high energy input in production. Thus, reducing the filter also minimizes component costs.

2   Circuit topology and dimensioning

Existing active front-end power supplies based on silicon semiconductors typically work with a switching frequency of 8 or 16 kHz, as otherwise the losses, especially the switching losses, in the semiconductors would be too high. For these power supplies, very large inductance values and designs are required in order to maintain acceptable current ripple values of approx. 20 % of the rated current. The mains chokes are often so large that they have to be installed outside the power supply unit in the switch cabinet, as they require the same installation space as the power supply unit itself.

With SiC semiconductor-based power supply units, the mains chokes can be reduced to such an extent that the mains choke and the filter capacitance can be integrated back into the housing without significantly increasing the size of such devices. This is possible due to the high switching frequencies and the reduced switching losses in the semiconductors.

2.1 Circuit topology

In most cases, the filter topology of a silicon based Active-Front-End consists of an LCL-filter composed of a line choke in the direction of the Active-Front-End (L_{in}) and a filter choke to the AC supply grid (L_{grid}). For an optimal filter L_{in} equals L_{grid}. In most cases, L_{in} is two times bigger than L_{grid}.

For an Active-Front-End with a switching frequency of 100 kHz, it is possible to reduce the filter to an LC-filter. This leads to enormous savings in installation space and resources and optimizes the efficiency of the power supply.

The functional architecture of the Active-Front-End comprises an LC-filter, three SiC-MOSFET half bridges, and the DC link. Figure 3 shows the schematic structure.

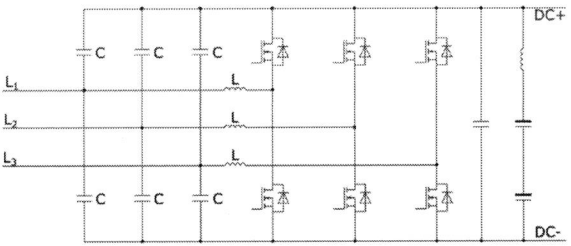

Fig. 2 Concept representation of the Active-Front-End

The filter capacitor is directly coupled to the DC link as it is typical for Active-Front-End converters. By reducing the size of the filter choke and capacitor and integrating it into the case of the power supply the feedback path for the EMI is significantly reduced.

2.2 Dimensioning

For the design of the filter components, the focus was on reducing the size and deliberately trying to reach the functional limits. In advance, promising functional designs were determined using simulation models.

2.2.1 LC-Filter

In order to keep the main choke of the Active-Front-End as small as possible, it is necessary to know the minimum inductance required to maintain the filter effect and the current ripple.

According to [4], the minimum inductance for the line choke with a current ripple of 15 % is as follows

$$L_{min} = x \cdot \frac{U_{zk}}{f_{sw} \cdot \Delta I_L} \qquad (1)$$

where U_{zk} = 650 V, ΔI_L = 15 % of I_N, f_{sw} = 100 kHz and x = 0,1137. The variable x depends on the converter topology, the modulation method and the modulation factor. This results in a minimum inductance of 48 µH at I_N. Table 1 shows the other parameters of the functional model.

Parameter	Variable	Value
DC-Grid voltage	U_{DC}	620-720 V
AC-Grid voltage	U_{AC}	400 V
Switching frequency	f_{SW}	100 kHz
Output power @50 kHz	P_{out50}	50 kW
Output power @100kHz	P_{out100}	40 kW
Inverter-side inductor @ I_N	L_{in}	60 µH
Capacitor	C	11 µF

Table 1 Parameter of the functional model

2.2.2 Comparison of the LC-Filter

To evaluate the reduction of the input filter, the physical dimensions and the weight of the individual components were compared with a similar Active-Front-End based on silicon semiconductors and a switching frequency of 8 kHz. The comparison is shown in table 2.

Parameter	SiC-AFE		IGBT-AFE	
Component	L	C	L	C
Value	60 µH	11 µF	716 µH	8 µF
Volume [cm³]	512	58	7020	129
Weight [kg]	1,3	0,072	4	0,14
Amplitude reduction @f_{sw}	-24,16 dB @100 kHz		-14,62 dB @8 kHz	
cutoff frequency f_{g3}	6195 Hz		1487 Hz	

Table 2 Parameters of the mains choke of the SiC-based Active-Front-End compared to a state-of-the-art Active-Front-End

The volume of the mains choke could be reduced by 93 % and the weight by 67 %.
The volume of the filter capacitor has been reduced by 55 % and the weight was reduced by 48 %. The reduction in weight and volume of the capacitor is possible due to a reduction in the insulation voltage.

2.3 Voltage and Current control

The current value of the inductor current is measured by using a shunt and sigma-delta converter. This measuring method enables simple galvanic isolation and very high precision. The disadvantage of using this measurement method in a current control loop is the dead time of the sigma-delta modulation. The dead time of this measurement method is significantly greater than that of other analog-to-digital converters. The dead time can be calculated according to [6]

$$T_{tot} = y \cdot \frac{OSR}{f_{SD}} \tag{2}$$

where OSR = 125, f_{SD} = 12,5 MHz and y = 3. The variable y represents the order of the sigma-delta-filter. The dead time for the used settings in the functional model is 30µs. This dead time must be considered in the controller and limits the dynamics of the control loop.
Instead of a regular voltage control loop typically used in cascade control, the voltage controller is replaced by a voltage droop curve [8]. This essentially represents a proportional element with different gain sections. For the droop curve shown in

Figure 4, Equation 3 describes the relation of the current I_D and the DC bus voltage U_{DC} in the range from 620V to 720V and $|I_D|$ > 20 A. In this range I_D = I_{set}.

$$I_D = -I_{DE} - \left(\Delta I_{DE} \cdot \frac{U_{DC} - U_{DEmax}}{\Delta U_{DE}} \right) \tag{3}$$

Due to stability problems of the control loop at currents below 20 A, a quadratic behavior was implemented. For the range $|I_D|$ < 20 A, the equation 4 shows the relation between I_D and I_{set}.

$$I_{set} = I_D \cdot \frac{|I_D|}{20A} \tag{4}$$

The exact function of the droop curve is described in the DC-INDUSTRIE system concept [1] and in [8]. The complete control structure with voltage and current control is shown in Figure 3.

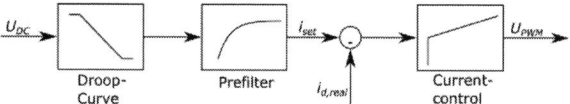

Fig. 3 Droop-Control loop for voltage control of a supply module in the DC-grid [1]

For the structure described above, the droop curve shown in Figure 4 was used.

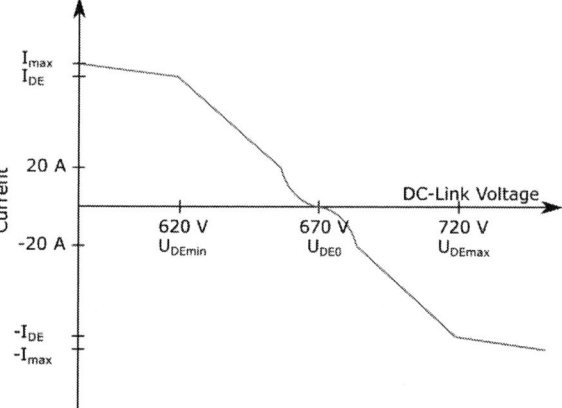

Fig. 4 Droop curve for voltage control of a supply module in the DC-grid [1]

The Droop-Curve was modified for the power measurements so that the maximum possible current of the test environment (Is = 60 A) is reached at 680 V. To achieve this, the left corner of the droop curve was moved from 620 V to 675 V. This point is shown in the measurement results in Chapter 3.

2236

2.4 Functional model

Figure 5 shows the assembled functional model including LC filter in the open housing. The external dimensions of the model are 290x450x250 mm and thus results in a volume of 32 L including the filter. The arrangement of the components in the housing was not volume-optimized and therefore there is still potential for optimizing the power density.

A comparable Active-Front-End based on Si-semiconductors including filters achieves a volume of 63 L.

Fig. 5 SiC-based functional model of the Active-Front-End

3 Measurements

The measurements at the Active-Front-End were adjusted so that the voltage at an output current of 60 A is 680 V. This was necessary to enable the maximum possible output power in the test environment. Due to the existing loads, only 40 kW output power was possible. Figure 6 shows the measurement points in the schematic of the Active-Front-End.

The junction temperatures shown in Figure 11 were measured directly on the SiC-Mosfet using an optical measurement method.

For all Measurements a three-phase pulse-width modulation was used as the modulation method. This method is not optimal regarding switching losses.

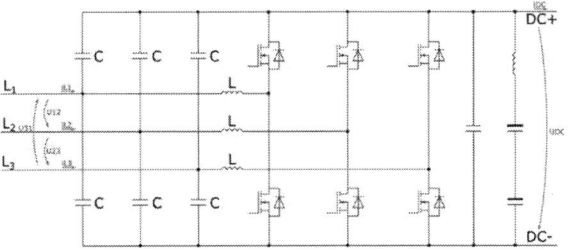

Fig. 6 Measuring points of Voltage and Currents

3.1 Voltage and Current of the Active-Front-End

The following Figures 7 and 8 show the currents and voltages at a power of 40 kW. The line currents are almost sinusoidal. This results in a power factor of 99.6% measured with an LMG500 power meter.

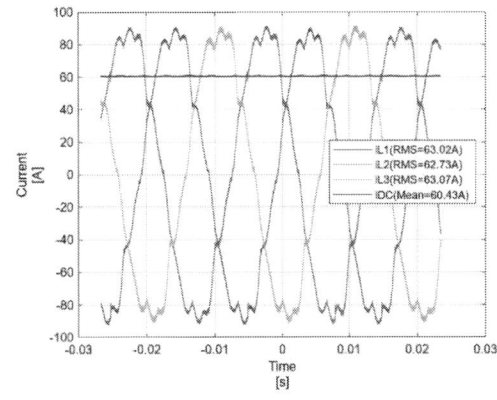

Fig. 7 AC current of the Active-Front-End with switching frequency of 50kHz and 100kHz

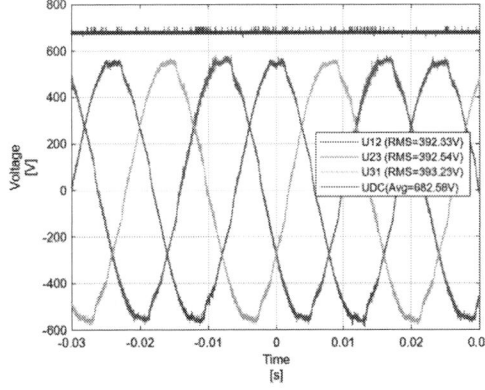

Fig. 8 AC voltages of the Active-Front-End with switching frequency of 50 kHz and 100 kHz

3.2 Efficiency and Losses

Figures 9 and 10 show the efficiency and combined losses of the Active-Front-End converter including the filter at 50 and 100 kHz switching frequency. At 50 kHz a peak efficiency of 97.3% was achieved, while at 100 kHz the peak efficiency was at 96.6%. This results in losses of 1.07 kW and 1.42 kW respectively.

At both switching frequencies a three-phase pulse-width modulation was implemented as the modulation method.

The used modulation method is not ideally suited to minimize switching losses. Therefore, other modulation methods will be tested in the future to reduce the switching losses and reach higher efficiency.

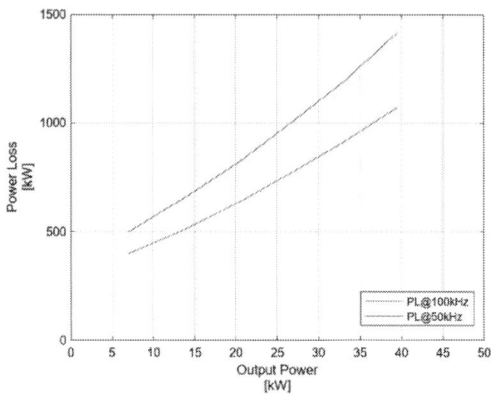

Fig. 10 Losses of the Active-Front-End with switching frequency of 50 kHz and 100 kHz

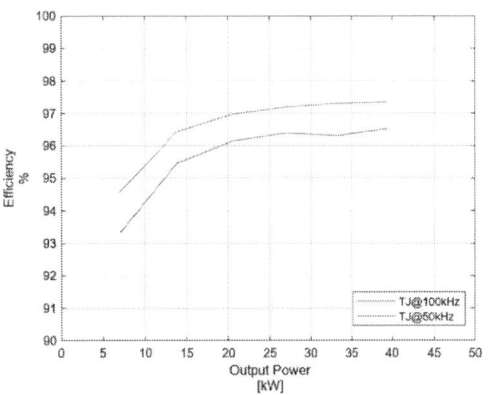

Fig. 9 Efficiency of the Active-Front-End with Switching frequency of 50 kHz and 100 kHz

The losses mainly result in a temperature increase in the semiconductors and the line choke. Figures 11 and 12 show the junction temperature of the SiC-semicondcutor and the choke temperature during the efficiency measurements. All measurements were taken after a settling time of 10 minutes for the thermal processes to reach a steady state.

Notably the switching losses at 50 kHz are significantly reduced compared to 100 kHz, which can be seen in the temperature difference in Figure 11. The temperature at approximately 40 kW is almost 50 K lower. In contrast to this the losses in the line choke are much higher at 50 kHz due to the increase in ripple current.

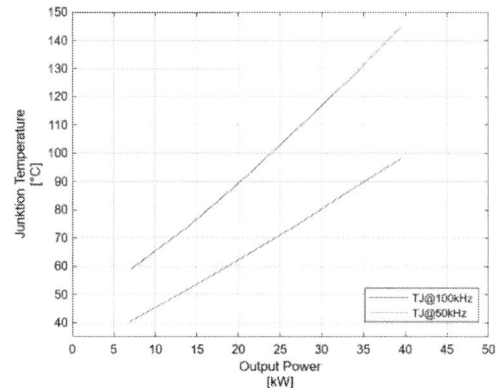

Fig. 11 SiC-Mosfet-Junction-Temperature of the Active-Front-End with switching frequency of 50 kHz and 100 kHz

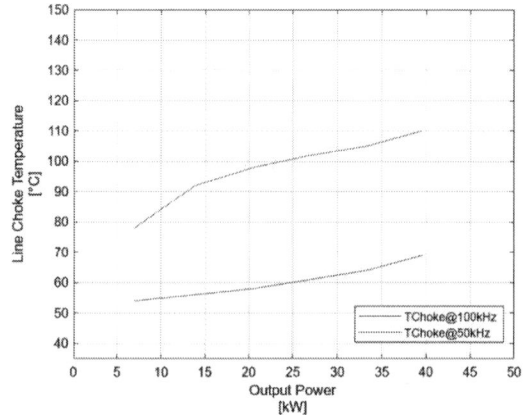

Fig. 12 Temperature of the line chokes at 50 kHz and 100 kHz

4 Conclusion

This paper shows the successful design of an Active-Front-End converter with SiC-semiconductors and a very small line choke for the use in an open industrial DC grid.

Due to the high switching frequencies it was possible to minimize the size of filter unit compared to state-of-the-art IGBT-based Active-Front-End, while still maintaining a high efficiency.

The volume of the line choke was reduced by 93 % and the weight by 67 %. While the functional model is not size optimized, it is still significantly smaller than its IGBT counterpart.

For future improvements it is planned to implement different pulse-width modulation strategies to lower the switching losses, such as space vector modulation. Furthermore, the windings of the line chokes will be done with high frequency stranded wire to lower the impact of the skin effect at higher switching frequencies and thus increase the efficiency of the converter.

Additionally, the temperature measurements strongly suggest that the losses in the semiconductors and in the line chokes are opposed to each other. Therefore, it should be possible to find an optimal switching frequency to minimize the combined losses of the converter.

Acknowledgement

This paper is funded by the German Federal Ministry of Economic Affairs and Climate Action (BMWK) pursuant to a decision of the German Parliament in the project STIM (Smart Transformers as Power Supply for the Future Mechanical Engineering Industry). Funding number: 03EN2010F and DC-Schiene (Highly Efficient, Resource Saving DC Conductor Systems for Production and Manufacturing). Funding Number: 03EN4045E

References

[1] ZVEI & consortium DC-INDUSTRIE2: System Concept DC-INDUSTRIE2 (Version 2, 29.04.2022). URL: https://dc-industrie.zvei.org/en/publications/system-concept-for-dc-industrie2, 16.10.2023

[2] H. Borcherding; J. Austermann; T. Kuhlmann; B. Weis; and A. Leonide: "Concepts for a DC network in industrial production", in 2017 IEEE Second International Conference on DC Microgrids (ICDCM), 2017, pp. 227–234.

[3] J.-N. Koch; Otte, R; Borcherding, H: "Investigation of Multi Feed In of Active Infeed Converters for Industrial DC Micro Grids with Voltage Droop Control", NEIS 2021 Conference on Sustainable Energy Supply and Energy Storage Systems, 2021

[4] D. Kampen; L. Fräger; N. Badenhop; A. Mambetow: "Difference in the design process of LCL filters for grid connected VSI when using SiC/GaN instead of Si semiconductors", in EPE 2022 ECCE Europe

[5] Y. Liu et al., "LCL Filter Design of a 50-kW 60-kHz SiC Inverter with Size and Thermal Considerations for Aerospace Applications," in IEEE Transactions on Industrial Electronics, vol. 64, no. 10, pp. 8321-8333, Oct. 2017, doi: 10.1109/TIE.2017

[6] TMS320F2837xS Real-Time Microcontrollers Technical Reference Manual. URL: https://www.ti.com/lit/ug/spruhx5h/spruhx5h.pdf, August 2014-Revised October 2023

[7] K. Askan; M. Bartonek; K. Weichselbaum: "Power Module for Low Voltage DC Hybrid Circuit Breaker", IEEE Third International Conference on DC Microgrids, Matsue, 2019

[8] L. Ott et al.: "An advanced voltage droop control concept for grid-tied and autonomous DC microgrids", 2015 IEEE International Telecommunications Energy Conference (INTELEC), 2015

Investigation of Load Transitions Between Loaded and Load Free Conductor Segments in Industrial Conductor Systems

Jan-Niklas Koch[1], Raphael Otte[1], Tim Stuckmann[1], Slavi Warkentin[1], Holger Borcherding[1]

[1] University of Applied Sciences and Arts Lemgo, Germany

Corresponding author: Jan-Niklas Koch, jan-niklas.koch@th-owl.de
Speaker: Jan-Niklas Koch, jan-niklas.koch@th-owl.de

Abstract

Due to the increasingly widespread use of efficient power electronics, DC grids offer more and more advantages over AC grids. In the field of energy supply the use of DC technology is also growing and is state of the art in offshore, high-voltage, vehicle and data center applications. The spread of industrial open DC grids is currently starting and is completely different due to the requirements and the specifics of certain industrial applications such as conductor systems, e.g. monorail conveyer or rack feeder systems. This paper investigates multi feed in operation in industrial DC conductor systems, especially for transitions between loaded and load free conductor segments. For this purpose, a test bench consisting of two conductor segments is developed. Each segment is individually fed by an infeed converter. The transitions between segments will be analyzed regarding control behavior and stability.

1 Industrial DC Grids

In industrial production energy efficiency and the use of renewable energy became even more important within the last few years due to the energy crisis. For companies to stay competitive they must optimize production processes and reduce energy consumption significantly. To achieve this goal, the introduction of open industrial DC grids is a game changer due to high energy saving potential for highly dynamic processes and an easy way to integrate photovoltaic with a minimum of conversion stages. Recent research has provided a variety of advantages of open industrial DC grids compared to AC grids [1] and the research projects DC-INDUSTRIE and DC-INDSTURIE 2 have provided an extensive system concept for building up such a grid [2]. In Fig. 1, the structure of an open industrial DC grid can be seen.

The participants in the DC grid are divided into DC sectors and are fused with smart DC breakers [3]. Typical applications in industrial production such as drives for machines, passive loads and auxiliary supplies are comparable to AC grids, except that they are directly coupled via the DC bus. Due to this it is advisable to provide bidirectional energy flow in those participants. Energy is supplied by infeed converters from the AC grid, which can be controlled or uncontrolled rectifiers, as well as by renewable energy sources. As additional supplement energy storage units can be used for peak load reduction, save shut down of facilities during a power failure or short time bridging of power failures.

The DC grid described in [2] implements voltage droop control, which is a control scheme that requires all infeed converters in the grid to work as current controlled voltage sources. This means

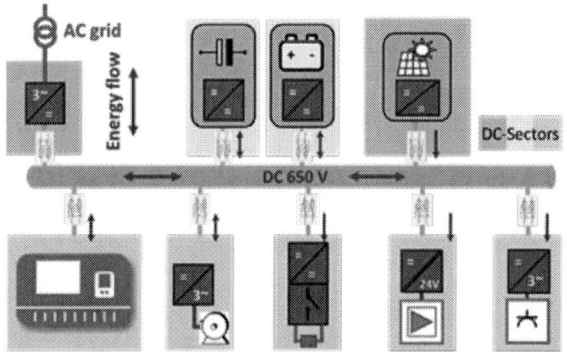

Fig. 1: Concept for an Open Industrial DC grid with several Devices located in DC Sectors

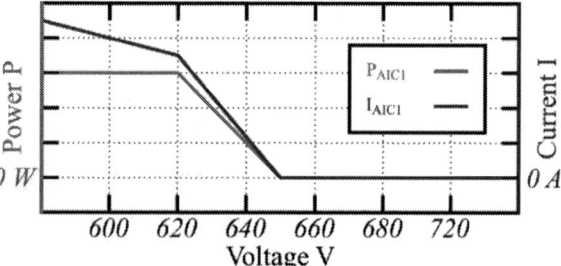

Fig. 2: Exemplary I-U and P-U voltage droop curve of a unidirectional infeed converter

that the voltage on the common DC bus is load dependent. Higher loads result in a lower DC bus voltage. A benefit of this control strategy is, that no further communication between infeed converters regarding the control is needed. All the information can be transmitted via the DC bus voltage. Fig. 2 shows an exemplary I-U and P-U voltage droop curve of a unidirectional infeed converter.

2 DC Conductor Systems

Conductor systems are an integral part of many industrial facilities. They are used in cranes, sliding stages and elevators, rack storage and monorail conveyor systems. Therefore, industrial production heavily relies on the safe and reliable operation of conductor systems. In addition to the energy saving potential of industrial DC grids in DC conductor systems the usage of copper can be reduced significantly compared to state-of-the-art AC systems. The reduction from three to only two active conductors combined with the elimination of the reactive power in the DC grid can lead to a reduction in copper usage of 50% and more.

Still, some challenges must be met to implement safe and reliable DC conductor systems in industrial production. Typically, state-of-the-art conductor systems are equipped with separating points, because of track switches, multiple infeed locations and different infeed grids for the reduction of voltage drop and safety reasons. The possibility to include those separating points in DC grids is mandatory to operate a DC conductor system. Fig. 3 shows an example of a possible setup of a maintenance area in a DC grid. The conductor system of the maintenance section and the conductor system of the production line are each separately fed by an individual infeed converter. This way the maintenance area can be disconnected from the grid without interfering with the production line, ensuring save maintenance work when a vehicle needs to be checked.

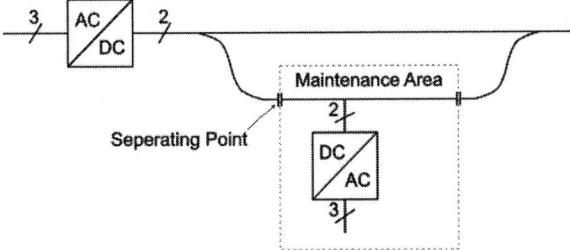

Fig. 3: Example of a separated maintenance area

However, using separating points in a droop controlled industrial DC grid comes with a cost. Using the example above, while the production line is al-

ways burdened by several drives, the maintenance area will be load free in most circumstances. Due to the droop control this leads to a voltage difference between the two areas. While a load free segment is operated at 650V, a loaded segment is operated in the range between 580 to 650V without power de-rating (see Fig. 2). When a load transitions between a loaded and load free segment, it connects the two segments for a short period of time. This is extremely challenging for the control of the infeed converters due to the short moment during transition when both converters are working in parallel. This is mainly because they must eliminate the voltage difference, but also because the converters work in parallel for a short amount of time and must divide the load accordingly. In [4] it is shown that there is the possibility of circulating currents when interconnecting infeed converters for multi feed in operation. Those circulating currents must be avoided, because they lead to power reduction. Usually, DC output filters are used to avoid circulating currents.

Therefore, the crossing of separating points in DC grids must be investigated. A modular test bench for this purpose is set up.

3 Circulating Currents

Connecting infeed converters in parallel without dedicated measures to suppress circulating currents leads to undesirable effects such as over currents, power de-rating and life-time reduction of components. Fig. 4 shows the parallel connection of two three-phase Boost PFC converters [5]. The following measurements were made without additional filtering to suppress circulating currents in the two converters.

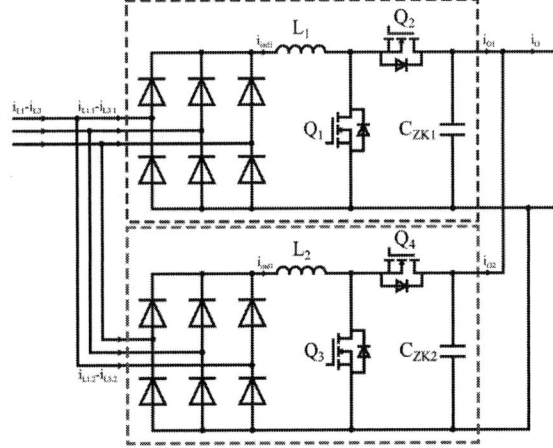

Fig. 4: Parallel connection of three-phase Boost PFC converters

Fig. 5 shows the combined input currents i_{L1}, i_{L2} and i_{L3} (top) and the distorted currents of a single infeed converter $i_{L1.1}$, $i_{L2.1}$ and $i_{L3.1}$ (bottom). While the combined currents are as expected, the currents of a single inverter are heavily influenced by circulating currents. The same can be seen in Fig. 6 for the output currents. The output currents of each of the single inverters are influenced by the circulating currents, while the combined output current is as expected.

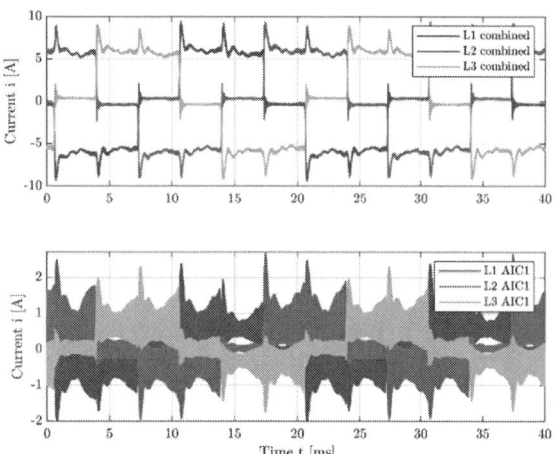

Fig. 5: Combined input currents (top) and input current of AIC1 (bottom)

The increase of the output current due to the circulating currents results in temporary currents that are 2.5 times the average output current, which may lead to unwanted heating of the converter, over-current turn-off and power de-rating of the device. In [4] a more detailed analysis of the circulating currents is shown.

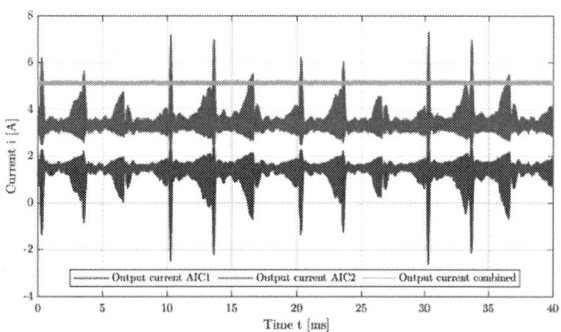

Fig. 6: Combined output current and output currents of AIC1 and AIC2 with circulating currents

4 Modular Test Bench

Due to the parasitic characteristics of a conductor system, namely the parasitic resistance, inductance and capacitance, and the voltage difference between conductor segments with different load condition, the transient behavior of the voltage droop control must be examined when a load shifts between two conductor segments. Therefore, a modular test bench was designed. The basic principle is shown in Fig. 7.

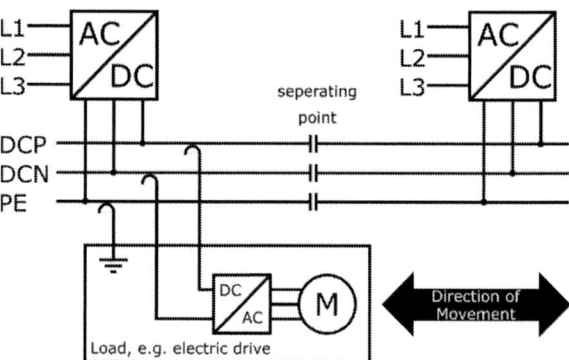

Fig. 7: Concept for the setup of the test bench

The test bench consists of two conductor segments that are separated by a separating point. Each segment can be fed by an AC/DC infeed converter. For the purposes of this paper the used infeed converters are equipped for parallel operation. A current collector connected to a load will be used to transfer the load from one segment to the other. In the current state of the test bench the load must be moved manually.

4.1 Test Scenarios

The focus of the test bench and the topic of this paper is the investigation of transitions between conductor segments with different voltage levels. For this purpose, the test bench can be adjusted in different ways to illustrate typical challenges of industrial conductor systems.

Improper Installation:
The test bench is designed in a way to display all kinds of influences of an industrial environment. Therefore, the separating point can be shifted horizontally and vertically to investigate the influences of improper installation. Additionally, the contact pressure of the current collector can be adjusted.

Influence of conductor length:
Furthermore, it is possible to include additional conductor segments to increase the distance between the separating point and the infeed converters. Due to the parasitic impedance of a conductor

segment the transient behavior can be influenced when adding more distance between the infeed converter and the separating point. The distance can be adjusted in segments of four meters to a maximum of 32 meters between the separating point and each infeed converter or to a maximum of 64 meters between the separating point and a single infeed converter. Fig. 8 is a schematic representation of the test bench. The additional conductor between the separating point and the infeed converter is represented by its length in meters multiplied by the specific inductance and specific resistance. The influence of the specific capacitance is neglected, because in conductor systems it is relatively small.

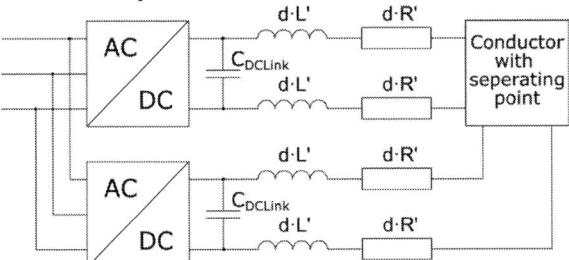

Fig. 8: Schematic representation of the test bench with simplified equivalent circuit of a conductor

The addition of distance between infeed converter and separating point resembles the circumstances in industrial production facilities. It is necessary to evaluate the influence of the parasitic properties on the stability of the DC link voltage when load transients occur or separating points are passed.

Transition speed:
In industrial production there are many different applications of conductor systems. Therefore, the vehicles operate at different speed. For example, storage and retrieval machines have high dynamics and operate at a relatively high speed. In comparison to this port cranes or cranes in heavy industry usually are quite slow. To evaluate the transient behavior in the test bench for different transition speeds, it will be equipped with a linear drive axis with a speed controlled drive.
For the measurements of this paper the load will be shifted manually across the separating point.

Load type:
In industrial facilities the majority of loads are electrical machines. Therefore, a combination of several 1.5 and 3 kW drives [6] can be used as a load for the test bench. The drives can be interconnected in a way to represent loads measured in real industrial plants. Ohmic load can be applied as well and will be used for the measurements of this paper.

4.2 Test Bench Realization

The test bench consists of two three pole conductor systems (DC+, DC- and PE). Both conductor systems feature a separating point. One of the separating points is fixed and perfectly aligned, while the other separating point can be adjusted horizontally and vertically. This way the effect of the transition can be compared between a proper and an improper installation.
The current collector is fixed to a linear sled, which can be adjusted in height. This way the contact pressure of the current collector can be changed. The two conductor systems with separating points, the linear sled and the current collector can be seen in Fig. 10.

Fig. 9: Test bench with separating point and current collector

Additional conductor segments with a length of four meters each are mounted on a rack. They can be added to the test bench to increase the distance between the separating point and the infeed converter.
The addition of a controlled drive for the horizontal movement of the current collector to achieve repeatable results for variable transition speed and the use of a modular experimental plant for industrial DC grids described in [6] is planned.

5 Controller Implementation

Voltage droop control has been used successfully in both AC and DC power distribution for many years. It offers an easy way of load distribution between power sources which does not need additional communication. In [7] the fundamentals for the design of voltage droop curves of LVDC grids with several different power sources and consumers is explained. In [5] a droop control for a three-phase PFC converter with a PI current controller is described. The dynamics of the controller are mainly determined by the slope of the droop curve, which basically is a proportional voltage controller, and the integral part of the current controller. The dynamics of the of the current controller are limited by the hardware it is designed for, because the inductor, describing the controlled system, determines the minimum time constant of the controller. Additional measurement filters and computation time further reduce the dynamics of the current controller.

Fig. 10 shows the control architecture with droop curve and PI current controller. The droop curve consists of a number of voltage to current set points and is interpolated linearly between those. The input for the droop curve is the measured DC bus voltage. The output is the set point for the current controller. The current controller consists of a preliminary filter and PI controller. The input of the PI controller is the difference between filtered set point value and measured current. The output of the PI controller is the set point DC bus voltage.

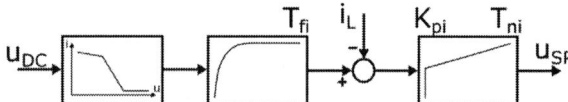

Fig. 10: Droop curve and PI current controller

For the following measurements both infeed converters use almost the same droop curve and controller settings.

6   Measurements

There is a variety of options in which the test bench can be configured for different test scenarios and measurements. Installation, conductor length, transition speed and load type are all modifications that can be applied to the test bench. This enables a broad range of tests and measurements that will result in a better understanding of transitions of separating points in DC fed industrial conductor systems.

In the following section the connecting and releasing of two individually fed conductor segments with different ohmic loads will be shown. Both infeed converters are installed close to the separating point. In Fig. 11 a burdened current collector with (a) 2 kW, (b) 2.5 kW and (c) 3.5 kW connects the two individually fed conductor segments when crossing the separating point. The inductor current (left) and the output voltage (right) of each converter is measured. At first converter B supplies the load alone. The output voltage of converter B is lower than the output voltage of converter A, according to their droop curves. At approximately 1 ms the current collector connects the two conductor segments, which leads to a parallel connection of the two infeed converters. The load is divided between both converters and the output voltages are identical. It can be seen that the transient be-

havior slightly differs for the different load scenarios, but always reaches a stable state. This shows that the droop control is not prone to oscillation, if both infeed converters are installed close to the separating point.

Fig. 12 shows the releasing of a loaded conductor segment with the same loads of (a) 2 kW, (b) 2.5 kW and (c) 3.5 kW. Both converters divide the load according to the droop settings and the output voltages are identical. At 1 ms the current collector disconnects from converter A and converter A is load free again. The transient behavior of converter B, which is now carrying the whole load again, is similar to the behavior shown in Fig. 11. The load shedding on the other hand causes the output voltage of converter A to rise to approximately 670 V before slowly decreasing to the no-load voltage of 650 V. The reason for this is the limitation of the negative set point current limit in the droop curve. Therefore, a non-loaded DC link can only be discharged slowly. Still, both converters enter a steady state and the system remains stable.

Similar investigations must be done for other test conditions to evaluate the transient behavior and stability in all possible scenarios.

7   Conclusion

The investigation of the crossing of separating points is an integral part for the success of industrial DC conductor systems.

The proposed test bench can be used to investigate the challenges of transitions across a separating point in an industrial DC conductor system. The main challenges of industrial applications such as installation errors, influence of parasitic components, varying transition speed and loads can be replicated and analyzed.

For the example of ohmic loads and infeed points close to the separating point, it was shown that the transition does not cause system instability and the chosen infeed converters reach a steady state and divide the load as expected. The control meets the requirements and does not need to be adjusted for this test. Nevertheless, the control has to be reconsidered if other scenarios cause instability or excessive oscillation.

For future measurements the test bench will be extended by a speed controlled linear axis to move the sled to ensure reproducibility and multiple drives will be included to investigate dynamic loads instead of steady ohmic loads.

Furthermore, measurements with additional conductor segments will be made to investigate the behavior of parasitic impedance on the transient behavior of infeed converters.

PCIM Europe 2024, 11– 13 June 2024, Nuremberg   DOI: 10.30420/566262317

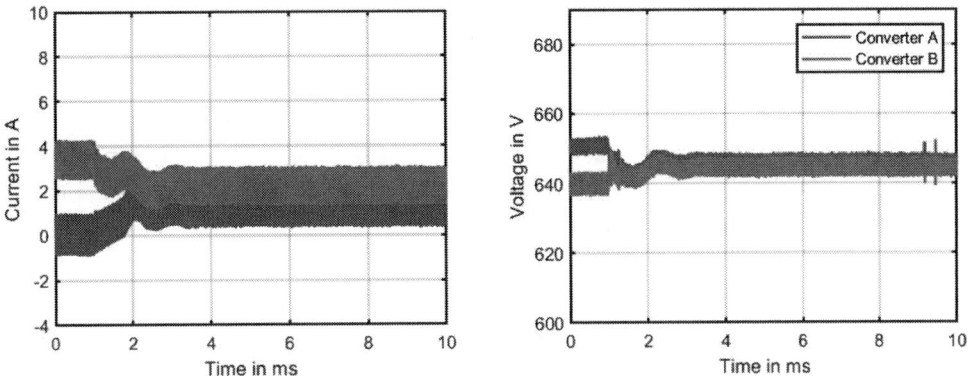

a) Infeed converter A starts with no load and infeed converter B starts with a load of 2 kW; at approximately 1 ms the current collector bridges the separating point and connects both infeed converters

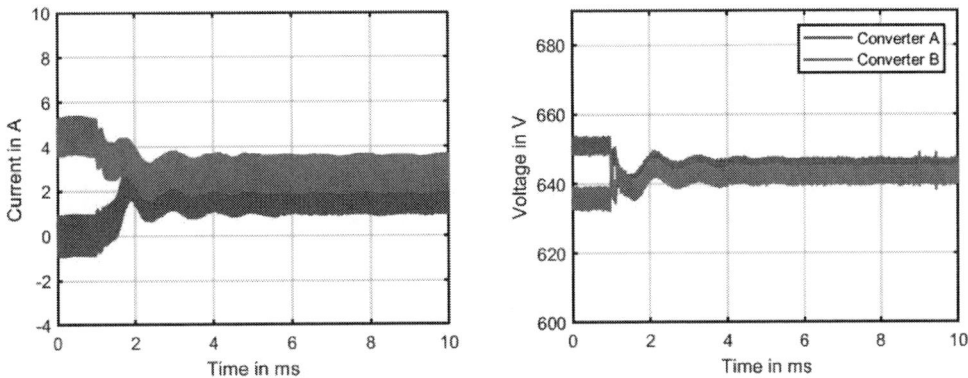

b) Infeed converter A starts with no load and infeed converter B starts with a load of 2.5 kW; at approximately 1 ms the current collector bridges the separating point and connects both infeed converters

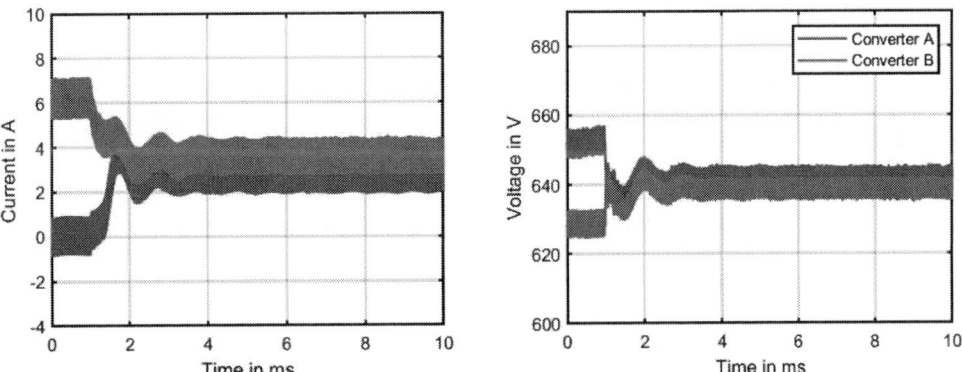

c) Infeed converter A starts with no load and infeed converter B starts with a load of 3.5 kW; at approximately 1 ms the current collector bridges the separating point and connects both infeed converters

Fig. 11: Loaded current collector (ohmic load) connecting two separately fed conductor segments; DC output voltage (left) and inductor current (right) waveforms of the two infeed converters

2245

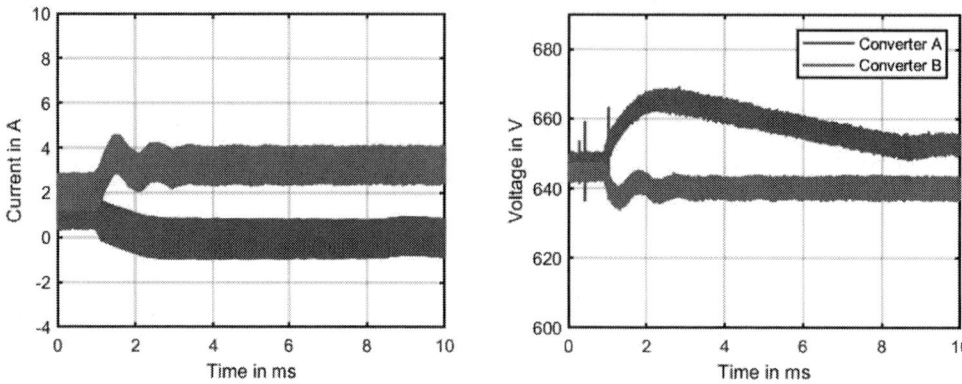

a) Infeed converter A and B divide a load of 2 kW according to their droop settings; at approximately 1 ms the current collector leaves the separating point and disconnects the infeed converters

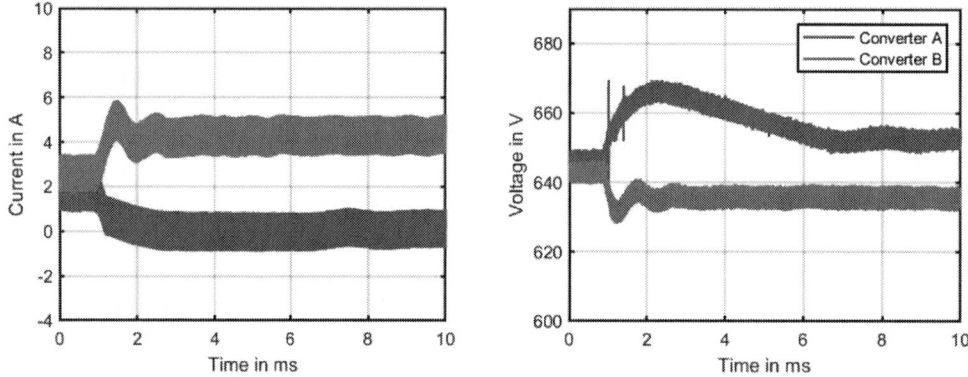

b) Infeed converter A and B divide a load of 2.5 kW according to their droop settings; at approximately 1 ms the current collector leaves the separating point and disconnects the infeed converters

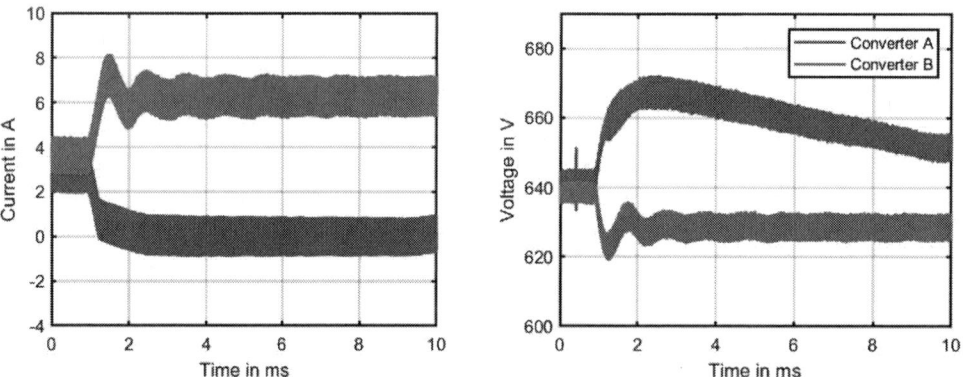

c) Infeed converter A and B divide a load of 3.5 kW according to their droop settings; at approximately 1 ms the current collector leaves the separating point and disconnects the infeed converters

Fig. 12: Loaded current collector (ohmic load) releasing separately fed conductor segments; DC output voltage (left) and inductor current (right) waveforms of the two infeed converters

8 Acknowledgements

This paper is funded by the German Federal Ministry of Economic Affairs and Climate Action (BMWK) pursuant to a decision of the German Parliament in the project DC-Schiene (Highly Efficient, Resource Saving DC Conductor Systems for Production and Manufacturing). Funding Number: 03EN4045E

References

[1] H. Borcherding; J. Austermann; T. Kuhlmann; B. Weis; and A. Leonide: "Concepts for a DC network in industrial production", in 2017 IEEE Second International Conference on DC Microgrids (ICDCM), 2017, pp. 227–234.

[2] ZVEI & consortium DC-INDUSTRIE2: System Concept DC-INDUSTRIE2 (Version 2, 29.04.2022). URL: https://dc-industrie.zvei.org/en/publications/system-concept-for-dc-industrie2, 16.10.2023

[3] K. Askan; M. Bartonek; K. Weichselbaum: "Power Module for Low Voltage DC Hybrid Circuit Breaker", IEEE Third Internatinal Conference on DC Microgrids, Matsue, 2019

[4] J.-N. Koch; Otte, R; Borcherding, H: "Investigation of Multi Feed In of Active Infeed Converters for Industrial DC Micro Grids with Voltage Droop Control", NEIS 2021 Conference on Sustainable Energy Supply and Energy Storage Systems, 2021

[5] J.-N. Koch; Otte, R; Borcherding, H: "Highly Efficient SiC-based Active Infeed Converter for Industrial DC Conductor Systems", PCIM Europe 2021, 2021

[6] S. Warkentin; J. Austermann; and H. Borcherding: "Modular Experimental Plant for Industrial DC Grid with Controlled Electrical Drives", NEIS 2020 Conference on Sustainable Energy Supply and Energy Storage Systems, 2020

[7] L. Ott et al.: "An advanced voltage droop control concept for grid-tied and autonomous DC microgrids", 2015 IEEE International Telecommunications Energy Conference (INTELEC), 2015

PCIM Europe 2024, 11– 13 June 2024, Nuremberg   DOI: 10.30420/566262318

A Method to Control Voltage And Power Flow in a DC Grid

P.J. van Duijsen[1], D.C. Zuidervliet[1]

[1] THUAS, DC-Lab, Delft, The Netherlands

Corresponding author: Peter van Duijsen, p.j.vanduijsen@hhs.nl
Speaker: Peter van Duijsen, p.j.vanduijsen@hhs.nl

Abstract

Control of power flow in a DC grid is done by power electronic converters. The requirement on their functionality and two typical topologies are presented. Voltage based droop control, short-circuit and earth leakage protection, as well as start-up and islanding properties are discussed. Start-up functionality of the converters, and protection in the DC grid is integrated in the converters and discussed. Droop control and their parameters are presented.

1   Introduction

A typical converter can control the output voltage and has a limitation on the maximum current it can supply [1].

In a DC grid there is voltage based droop control, bidirectional controlled power flow, short circuit and earth leakage protection, and soft-start to prevent inrush currents [1]. The power flow is regulated purely based on the level of the voltage in the DC Grid [2]–[4], These various requirements have to be controlled via the DC converters in the DC grid [5]. Protection is mainly based on detecting a sharp increase of the outgoing current in the DC grid [6], [7].

In this paper the control methods and converters required to achieve this functionality and structure, see section 2, will be outlined. The voltage levels are detailed in section 3. In section 4, two typical types of converters are discussed, being the interlink and the grid manager.

In section 5, the required functionality will be outlined, such as voltage control, power flow control and soft-start. Short-circuit detection, earth leakage detection, islanding detection and impedance measurement are outlined in section 6.

In section 7, the regulation of the operation and droop control are outlined. Here a flowchart shows the basic algorithm for the overall control. Subjects like soft-start, short-circuit and earth leakage protection are shown in the control algorithm, whose main task is the voltage based power control, the so called droop control.

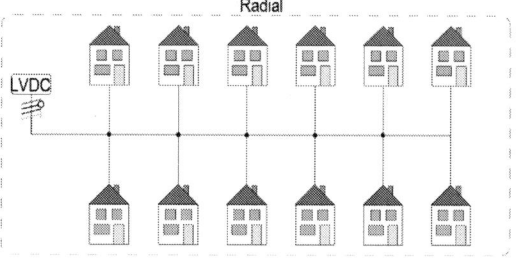

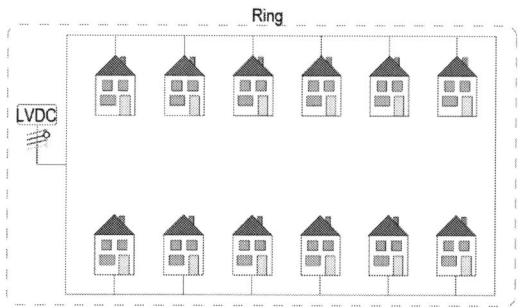

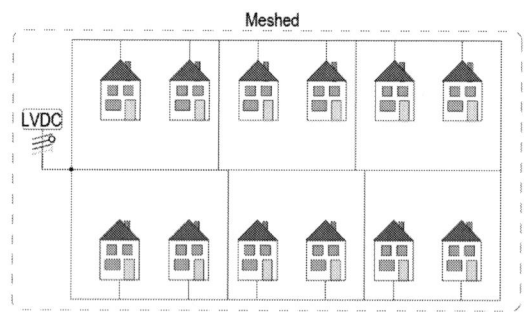

Fig. 1: DC Grid structures. From top to bottom: Radial, Ring and Meshed grid.

2248

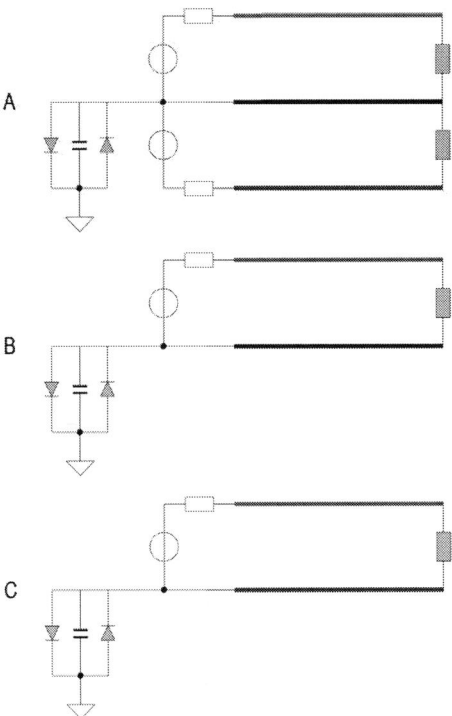

Fig. 2: Bipolar and unipolar grid structure with wiring color, A: Bipolar grid, B: Unipolar grid with black neutral wire, C: Unipolar grid with double voltage level and blue neutral wire.

The Interlink [8] and Grid Manager [9] are presented as the building blocks for the DC Grid. They contain all the functionality and protection as outlined in the next sections.

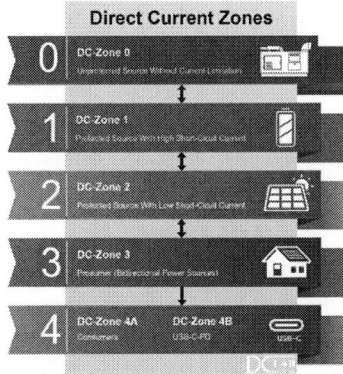

Fig. 3: Different zones and their protection level.

2 DC Grid structure

Different grid structures, like Radial, Ring or Mesh, are possible, see Fig. 1. In contrary to AC grids, there is no need for phase synchronization, when connecting two different grids.

The DC network can consist of a bipolar network or a unipolar network. The difference between the bipolar and unipolar grid structure is shown in Fig. 2 With a bipolar net you are dealing with a positive and negative voltage. With a unipolar net you are dealing with a positive and a zero voltage. The advantage of the bipolar network is that with longer lines, the current through the middle line becomes zero and thus the losses in the line are reduced. Depending on the power level, different zones are identified, where protection is required, see Fig. 3

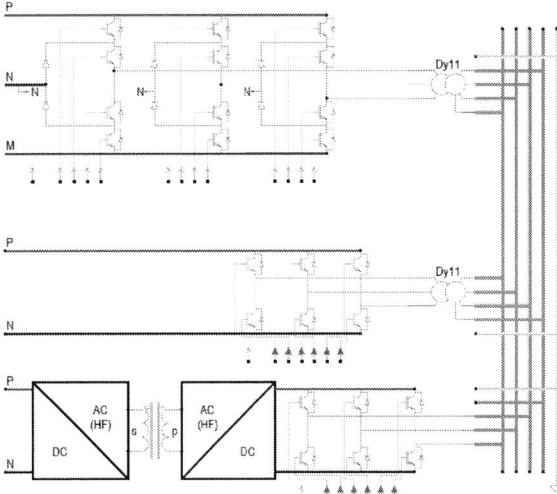

Fig. 4: Galvanic isolation between the DC and AC grid. From top to bottom, three-level inverter with $50Hz$ AC transformer, two-level inverter with $50Hz$ AC transformer, two-level inverter with high-frequency galvanic isolation on the DC grid side.

Galvanic isolation between the AC and DC grid is required, because the connection to the grounding earth might be different. For example, in Fig. 4 the galvanic isolation is either on the AC side, or inside an isolating DCDC converter. Reduction of size and weight for high-frequency coupled inductors, compared to a 50Hz transformer, is the main reason why the isolation is mostly performed in the DC grid, instead of the AC grid.

3 Voltage Level

There are various initiatives to standardize DC grids. In any case, DC networks must meet the standards according to the IEC and national standards, such as NEN1010 in the Netherlands. Within these standards, there are sufficient options for setting up the DC grid. In particular, the way in which protection must take place and which standards the voltage levels must meet are in principle already included in these national an international standards. The only thing that needs to be changed is to translate these standards towards DC. This is also what is happening now, if you look at the voltage levels proposed for DC networks. The current voltage levels in DC networks fall neatly within this standard. However, it is necessary to add some nuance here.

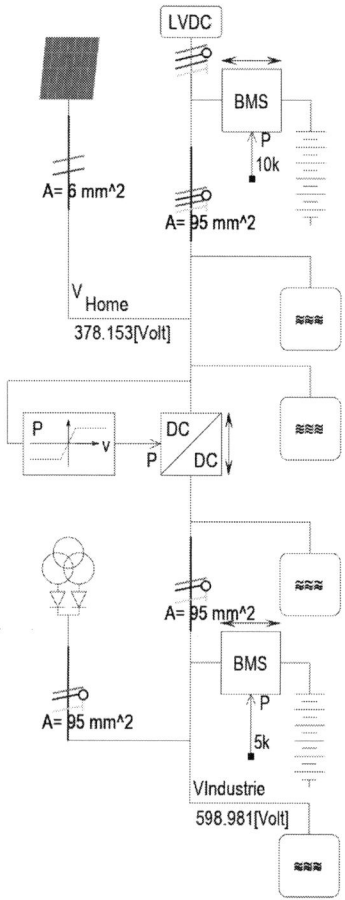

Fig. 5: Interlink with droop-control between the upper household DC grid of $V_{Home} = 380v$, to an industrial of $V_{Industry} = 600v$

In particular, the difference between low voltage and medium voltage and high voltage is an important item that, although described in the standards, must be worked out more specifically for DC networks. Low voltage in particular is a very broad area where the voltage levels for the DC grid must be redefined. It should be noted that on the one hand safety is important, but on the other hand it is important that the current equipment available for AC networks also remains applicable within DC

networks.

There are already different voltage ranges for the low voltage level, within the current standards. Particularly in Automotive, the voltage levels are already clearly defined. It is also clearly visible that the voltage levels of 12, 24 and 48 volts are already used in current Automotive applications. These voltage levels also remain within the low voltage standard in the IEC standards.

It is clear that this lower voltage is not suitable for somewhat higher powers, due to impermissible large currents through the cabling. This low voltage would cause such high currents in the cabling that the risk of fire or spark formation would become too great. It is therefore logical to choose a higher voltage for the higher powers than the lower SELV voltages. Various proposals have been made for a voltage level between 200 and 400 volts.

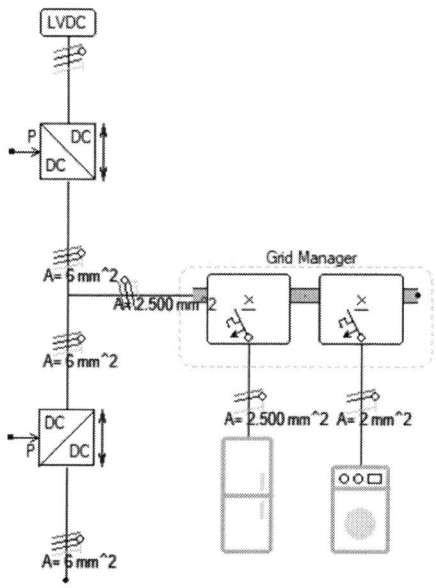

Fig. 6: Interlink and Grid Manager

There is something to be said for each of these voltage levels. Various proposals have been made, which show that, for example, a lower voltage around 200 volts is in line with the current RMS voltage of 230 volts in Europe. But there are also many proposals around 350 and 375 volts, which are more in line with industrial applications, see Fig. 5. This voltage level can be doubled by using a bipolar network. This is especially important for the higher powers. A voltage level of 700 or 750 volts is suitable for powers between 3 and 20 kilowatts. This is also in analogy with the current three phase 400 volt AC network in Europe. However, when we look at industrial applications, such as inverters for asynchronous machines and permanent magnet machines, we must also look at the current voltage level, which is used internally in these inverters. Here a voltage level between 500 and 600 volts is a common level. This has to do with the voltage level obtained after rectifying a 400 volt AC 3 phase voltage using rectification. However, if we look at smaller household applications, a lower voltage level, below 230 volts, is more likely to be applicable. This is due to some passive loads, such as irons, kettles and other household applications, which pose a resistive load on the current AC network. These resistance loads assume a certain power that is passively extracted from the AC grid. Applying these applications directly to a DC grid therefore implies that a matching DC voltage level must be present to keep the power at the same level when switching from an AC grid to a DC grid. This would argue in favor of using a lower voltage level in a household DC grid, for example 200 volts. A completely different theme that comes up when we talk about voltage levels is the issue of how we will connect the applications to the DC grid. By combining the combination of switching method and protection in one piece of power electronics, something can also be done about the different voltage levels. In particular, the Interlink and the Grid Manager are discussed here, which can be used for the different voltage levels.

In the electrical energy transition, a transition from an AC grid to a DC grid will not take place all at once. There will be a gradual transition, with applications from the ac grid being transferred to a DC grid. It is therefore important to see whether the voltage levels from the AC grid can somehow be maintained in a DC grid. Particularly in industrial applications, major benefits can be achieved if current inverters can be directly connected to a DC grid. But this would argue in favor of setting the voltage level in the DC grid equal to the current voltage levels in the DC link in the current inverters. However, this voltage level would be too high for passive resistive loads in domestic applications. This dilemma remains, even if a middle path is chosen, for example a voltage level between 350 and 400 volts. A solution to this is to allow multiple voltage levels that can be connected to each other using power electronics. This is the function of the

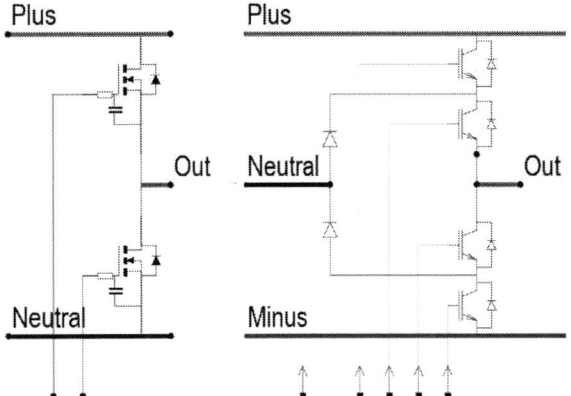

Fig. 7: Single leg of a Two-level(left) and Three-level(right) converter, for interfacing to the AC grid.

Interlink, see Fig. 5, where a $380v$ household grid is connected to a $600v$ industry grid. Creating voltages equivalent to the 230 volt RMS voltage level of the AC grid is a function of the Grid Manager. The task of this is to provide a voltage level to an outgoing group and also to protect this connection.

4 Converters

There are two types of converters, the Interlink and the Grid Manager, see Fig. 6.

The Interlink converters are the main routers between two DC grids. Their task is to control the power flow between grids and to guard the voltage levels of the grids as well as to provide short circuit and earth leakage detection. In case of a fault in on of the sections of a DC grid, their task is to isolate that fault and to keep the remaining sections up and running. The Grid Manager provides the connections to the consumers and prosumers. The Grid Manager is connected to the DC grid and its output terminals provide voltage control and bidirectional power flow control. Short circuit and earth leakage protection is included in each of the output terminals. Impedance measurement and soft-start is provided during turn-on of the output terminal.

4.1 Interlink

Depending on the requirement for isolation, a Dual Active Bridge [DAB] or Dual Active Half Bridge [DAHB] can be applied. The latter has no galvanic isolation [10].

4.2 Grid Manager

The Grid Manager basically provides a terminal per single prosumer. The output voltage is generally lower than the DC grid voltage, but in case of low voltage grids, voltage levels might be close or the output voltage can be higher that the grid voltage. Therefore there are two types of Grid Manager topologies, the synchronous buck for lower output voltages and the DAHB topology for equal voltage levels.

In this paper, most converter schematics are presented showing a two-level converter. However, multilevel converters can also be applied. The advantage of the multilevel converter is, that it can directly be applied in a bipolar grid. For example, a three phase AC inverter connects directly to the plus, minus and neutral nodes in the DC grid, see Fig. 4 and Fig. 7. In an bipolar DC grid, the voltage level is double the voltage level in a unipolar DC grid. The Three-level converter allows these higher voltages, since two semiconductors are connected in series. The maximum voltage across a single semiconductor is therefore half of that of a semiconductor in a Two-level inverter [11].

can be obtained. Only if the impedance is above a minimum requirement, the converter can build a voltage on the output terminal. In case of islanding, this procedure is required to prevent feeding in a too large grid. The converter can only start if the connected prosumer/grid has a high enough impedance that can be fed from the converter. In case of islanding, the converters that link the grids, should switch off and thereby create small islands, to isolate fault conditions in the grids. To prevent inrush currents, a soft-start is included to slowly increase the voltage and thereby load any capacitance at the connected prosumer or grid. Short circuit detection is included using the principle of Rate of Change or Current [RoCoC] [2], [10], see Fig. 8. Earth leakage detection is included by measuring the difference between the outgoing and returning current, using a current sensor.

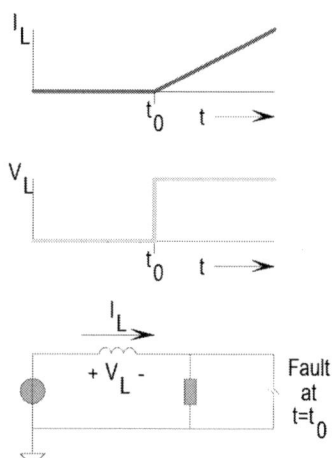

Fig. 8: Fault at $t = t_0$ causes a voltage V_L across the sense inductor.

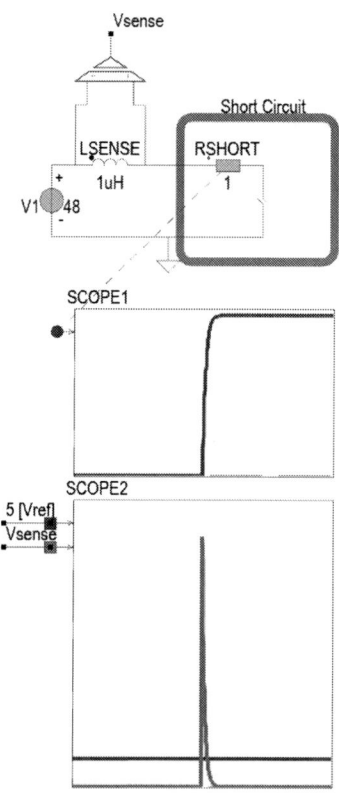

Fig. 9: Sense voltage V_{Sense} across the short circuit detection inductor L_{Sense}, exceeds the reference voltage $V_{ref} = 5v$ at the moment of short circuit.

5 Functionality

Each output terminal requires voltage control. Also the maximum bidirectional power flow is controlled, using a maximum allowable current in the output terminal. At start-up of the converter output terminal, the converter has to detect if it is safe and possible to increase the output terminal voltage.

First the impedance of the grid or prosumer connected to the output terminal, is measured. This is done using short low voltage pulses, and by measuring the current, an indication on the impedance

6 Protection

The voltage level in this network is built up by power electronics. However, it is not that easy to directly energize the DC grid. The biggest problem is the inrush current, which arises when a device is connected. Related to this is the short circuit current, which is measured by looking at a change in current increase, see Fig. 8. In particular, suddenly switching on an application causes an increase in current, which could be seen as a short circuit by the protection. This shows that both the energizing of the DC grid and the short-circuit protection, must be investigated together. A common solution must be provided for energizing the DC grid and checking the short-circuit current.

First we look at the different types of protection that are necessary in the DC Grid. These are short-circuit protection and Earth-Leakage protection.

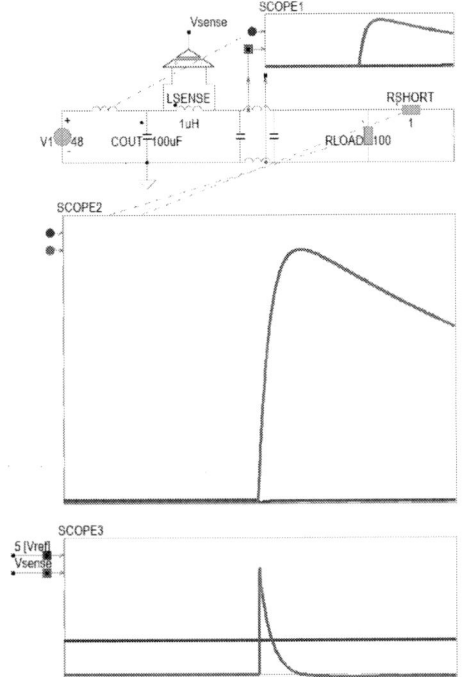

Fig. 10: Simulation in Caspoc [12], of a short circuit inside the appliance(R_{Short}), causes the sense voltage V_{Sense} to exceed $V_{ref} > 5$ volt.

6.1 Short Circuit Detection

The short-circuit protection consists of detecting an increase in current. An increase with a certain slope can indicate a short circuit in the network, for example if a short circuit occurs in the cabling, or if an accidental short circuit occurs in an application. In most cases, in the event of a short circuit, the current will increase rapidly in a short time [2], [6], [7]. By measuring this increase in current and comparing it with the maximum permitted increase,

a statement can be made about a possible short circuit in the DC Grid.

Measuring this increase in current is best done using a small inductance of the order of 1 to $10\mu H$ over which a voltage is measured. Any change in current will result in a voltage across this measuring inductance. In normal operation with a constant DC current, only a constant DC current will flow through the inductance and will not cause significant losses in the inductance, see Fig. 8.

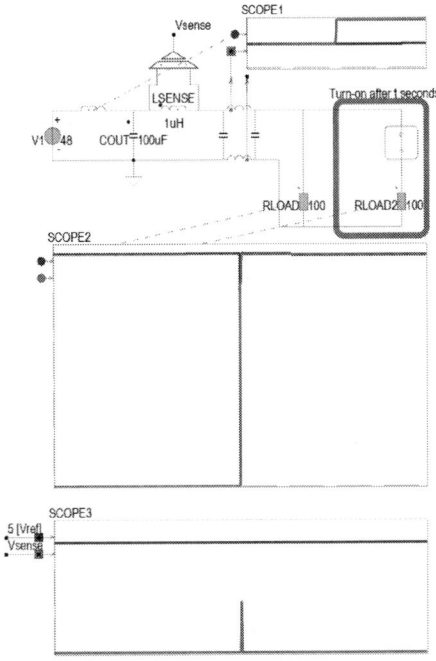

Fig. 11: Simulation in Caspoc [12], showing a limited V_{Sense} after adding a second load.

Fig. 9 shows the simulation of a short circuit detection. At $t = t_0$, the connection on the right side is closed, creating a short circuit through the short circuit resistance $R_{short} = 1\Omega$. At $t = t_0$, the current increases rapidly, causing a voltage over the inductance L_{sens}. The voltage exceeds the reference voltage $V_{ref} = 5v$.

Fig. 10 shows the simulation of a short circuit, when the fault occurs inside an appliance. The appliance is connected via a cable of 10 meter. The inductance of the cable will limit the current rise $\Delta i/\Delta t$. At the moment of short circuit, the voltage V_{sens} exceeds the reference voltage level $V_{ref} = 5v$, see Scope3 in Fig. 10.

In Fig. 11, a second resistive load $R_{Load2} = 100\omega$ is connected in parallel to the existing load of $R_{Load} = 100\omega$. The steepness of the rise of the current is limited, and therefore V_{sense} remains below the reference voltage level of $V_{ref} = 5v$. In case of a capacitive input, an inrush current will occur, as further detailed in section 6.3.

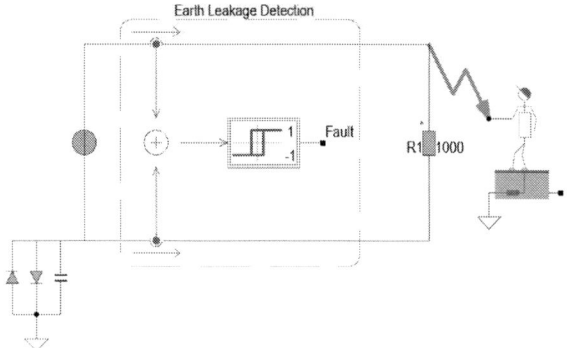

Fig. 12

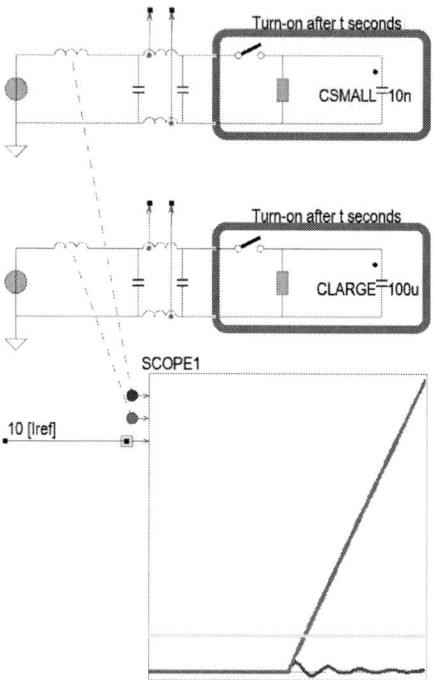

Fig. 13: Inrush current exceeding the maximum current of $10A$.

6.2 Earth Leakage Detection

Measuring a ground fault is done in the same way as in an AC Grid. The current through the forward conductor and return conductor are measured together and must be equal to zero, see Fig. 12. Any deviation from zero amperes means that a leakage current flows from the DC Grid to ground. The detection method is therefore analogous to the earth leakage measurement in an AC Grid. Instead of an electromechanical detection, an electronic detection can also be used to turn off the semiconductor switches in the inverters. Because electronic detection is faster and more accurate than electromechanical detection, an earth leak can be detected more quickly. A smaller leakage current can also be detected more accurately, because this leakage current does not have to supply an electromagnetic relay. This makes a DC Grid inherently safer than an existing AC grid with electromagnetic protection.

6.3 Inrush

A practical value for short-circuit detection is approximately one ampere per microsecond.

$$\frac{\partial i}{\partial t}^{max} = \frac{\Delta i}{\Delta t} < 10^6 [A/s] \qquad (1)$$

being equal to an increase of current of maximum 1 Ampere per μs. The reason for this value follows from the fact that the current can increase very quickly in the event of a short circuit, due to the high power density of capacitors in DC applications and the low impedance of the cabling between these capacitors and the location of the short circuit.

Connecting an application that has a filter with an input capacitance will always have a higher current increase than this one ampere per microsecond. See Fig. 13 for the inrush current of a capacitive input of $100\mu F$ and a connection cable of 10 meter. For this reason, applications that are connected to the DC network using a plug must have inrush protection. This inrush protection must be able to regulate a slow increase in current at the input. This can be achieved with power electronics, but requires an inductive input on the side of the power electronic converter.

Suitable power electronic converters include the Boost converter, the Four-Switch Buck/Boost converter and the Sepic converter.

6.4 Pre-Charging

To energize a DC grid that is de-energized, it is not easy to supply voltage directly to the output using a power converter. To limit the current increase in the DC Grid, the voltage must be built up slowly. This so-called precharging is a functionality that is built into both the Interlink and the Grid manager to supply all applications that are directly connected with power in such a way that the power increase remains below the limit. Controlling the voltage at the output of a power electronics converter is a basic functionality of the control in the converter. The limited voltage increase is to limit the current increase, generally several hundred milliseconds for the current to increase. Before the converter can apply voltage to the output, it must first measure whether the output impedance of the converter is high enough so that no short-circuit current can flow when switched on. The converter must perform this measurement internally using an impedance measurement. The measurement is performed by applying a signal to the output and measuring the associated current. Naturally, the converter must contain a current measurement at the output in order to perform this measurement. Only if the output impedance is high enough can the inverter start pre-charging. The value of the measured impedance must of course be greater than the maximum connection impedance suitable for the inverter. A lower connection impedance will always result in an overcurrent at the output of the inverter and is therefore not permitted. In that case, the inverter will not start pre-charging, but will give an error message that the output impedance is too low.

6.5 Overcurrent protection

Each appliance that can source power tot the DC grid requires and over-current protection. Overcurrent can arise after a short-circuit fault in the DC Grid, which in turn causes disconnection of other power sources in the DC grid.

6.6 Restart after fault

Every application that can source power to the DC grid, is probably able to sustain the current for the short circuit. Therefore, all of these applications should have short-circuit detection at their connection terminals. In the event of a short-circuit in the DC grid, these applications have to turn off and disconnect from the DC grid. Restoring the connection tot he DC grid is only possible, once the fault in

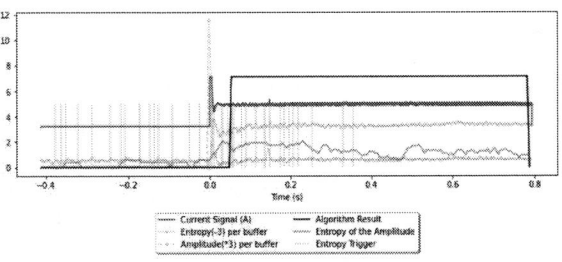

Fig. 14: Arc detection by calculating the entropy of the current.

the DC grid is removed. Each application can test this by carrying out an impedance measurement at their connection terminals and only restart, when they are able to provide enough power to maintain the power request from the DC grid.

6.7 Arc Detection

Detection of arcs and arc flashes is important, to prevent hazardous fires. Arc detection can be implemented inside the Grid Manager or Interlink, as there is already a provision for current measurement. Arcs do have a certain specific frequency spectrum footprint, that can be detected via the entropy of the measured current [13]. In Fig. 14, the entropy(orange trace), of the current signal(blue trace) is created and used for detection. As soon as the entropy increases, arcing is detected, see the black trace in Fig. 14.

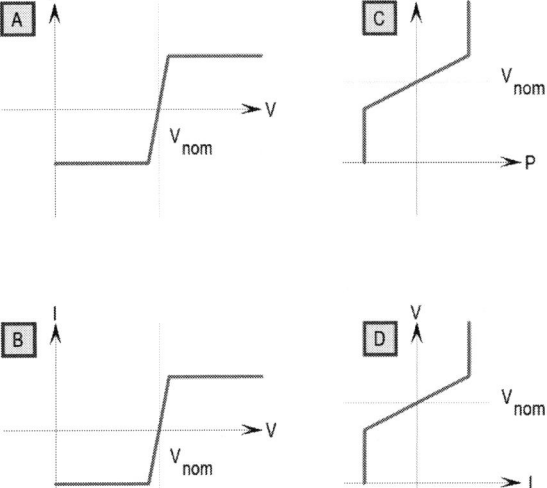

Fig. 15: Typical droop characteristics. A: Power as function of voltage, B: Current as function of voltage, C: Voltage as function of Power, D: Voltage as function of current.

7 Droop Control

Droop control [3]–[5] is implemented in each converter. The Interlink converters have droop control on both output terminals, see Fig. 15. The voltage level on either side of the Interlink defines the direction and size of the power flow, see Fig 16. The Grid Manager uses the input voltage level to decide the power flow of each connected prosumer.

The control of the power in the DC grid is done using a droop control. Power is available depending

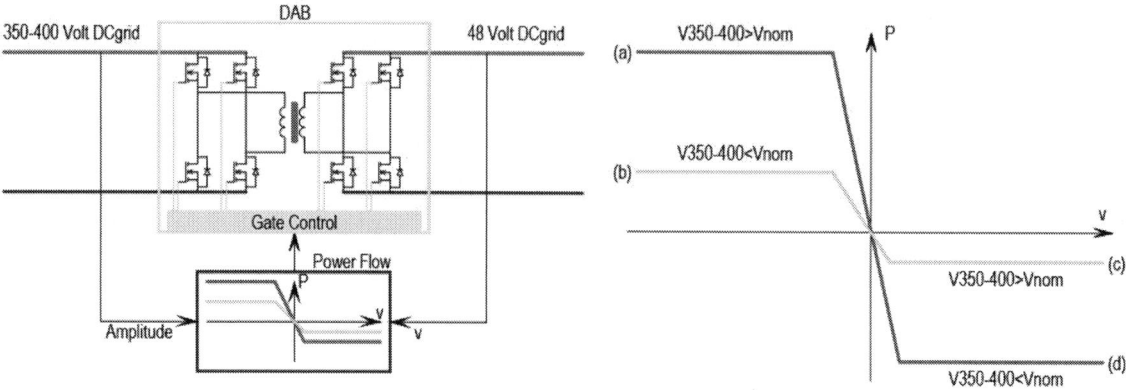

Fig. 16: Droop characteristics for a DAB

on the voltage level in the DC network. The more power available, the higher the voltage. This allows each application to immediately read to what extent it can draw power from the DC grid or put it into it. In particular, the Interlink uses the voltages on both sides for power regulation, and the Grid manager uses the input voltage as a measure for the power regulation of its outputs.

The droop control characteristic is usually displayed

as power over voltage a shown in Fig.15A, but other visualizations are also possible, although not common. The advantage of showing the voltage on the X axis, is that is complies with the characteristics of semiconductors. To indicate the power flow by power P, instead of current, is given by the fact that the amount of power remains constant by varying voltage, in contrary to a constant currrent value. The parameters of the droop control are shown in Fig. 17

7.1 Bidirectional power control

Fig. 16 shows the droop control in an Interlink converter. The direction and size of the power flow is determined by the voltage levels on either side of the Interlink converter, as indicated by the 4 curves a, b,c, and d. The interlink has a bidirectional power flow and in Fig. 16 is indicated the power flow is positive (from left to right) if the voltage of the $48v$ grid is below the nominal voltage of 48 volt. The power is negative if the voltage is above the nominal voltage of $48v$, as indicated by the vertical line in the figure. Depending on the voltage of the $350-400v$ grid, the amount of power flow is controlled. The amplitude of the characteristic is defined by the voltage level, and reduces from the red trace down to the green trace.

7.2 Power congestion control

Typical droop characteristics are shown in Fig. 18. Depending on the voltage level in the DC grid, each of the four prosumers can sink or source power from the DC grid, as indicated by the characteristics. The priority is defined by the distance from the nominal voltage level, indicated by the dashed grey line in all four characteristics.

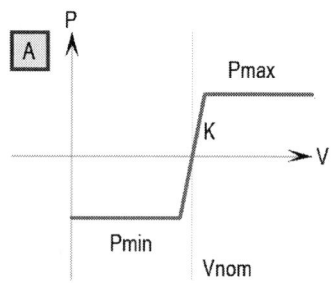

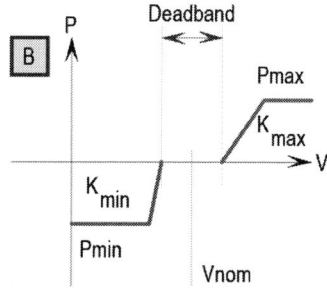

Fig. 17: Parameters of the droop characteristic centered at V_{nom} with limits P_{min} and P_{max}. A: Continuous characteristic with gain K. B: Discontinuous characteristic with centered blanking time around V_{nom}, with gains K_{min} and K_{max}.

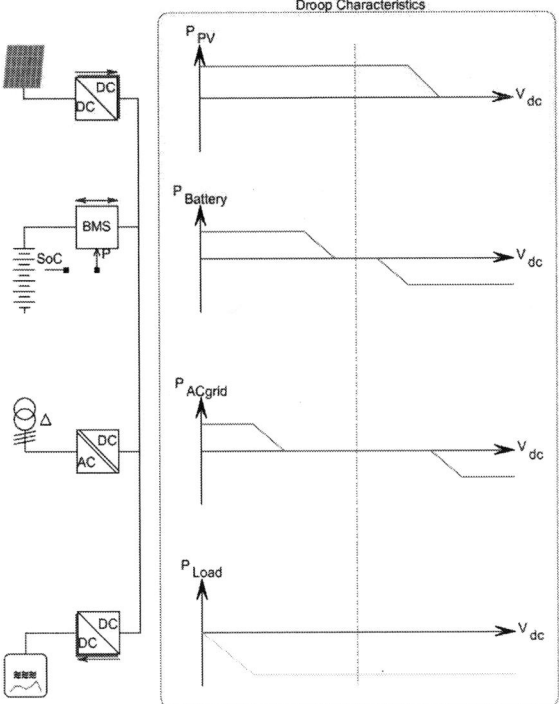

Fig. 18: Droop characteristics for prosumers

Conclusion

An overview of methods for controlling voltage levels and power flow in a DC grid is presented. Two converters, Interlink and Grid Manager, are presented. The Interlink can couple DC grids of different voltage level. The Grid Manager controls the power flow to an from connected appliances. Protection is build inside the Grid Manager and the Interlink, being short circuit(RoCoC: $\Delta i/\Delta t < 1A/\mu s$) and earth leakage detection. Inrush limitation and pre-charging as well as arc detection are also added functionality. Droop control regulates the power flow.

References

[1] N. Kondrath, "Bidirectional dc-dc converter topologies and control strategies for interfacing energy storage systems in microgrids: An overview," in *2017 IEEE International Conference on Smart Energy Grid Engineering (SEGE)*, 2017, pp. 341–345. DOI: 10.1109/SEGE.2017.8052822.

[2] A. Meghwani, S. C. Srivastava, and S. Chakrabarti, "A non-unit protection scheme for dc microgrid based on local measurements," *IEEE Transactions on Power Delivery*, vol. 32, no. 1, pp. 172–181, 2017. DOI: 10.1109/TPWRD.2016.2555844.

[3] J. M. Guerrero, J. C. Vasquez, J. Matas, L. G. de Vicuna, and M. Castilla, "Hierarchical control of droop-controlled ac and dc microgrids—a general approach toward standardization," *IEEE Transactions on Industrial Electronics*, vol. 58, no. 1, pp. 158–172, 2011. DOI: 10.1109/TIE.2010.2066534.

[4] B. Wunder, L. Ott, J. Kaiser, K. Gosses, M. Schulz, *et al.*, "Droop controlled cognitive power electronics for dc microgrids," in *IEEE International Telecommunications Energy Conference (INTELEC)*, 2017, pp. 335–342. DOI: 10.1109/INTLEC.2017.8214158.

[5] T. Dragičević, J. M. Guerrero, J. C. Vasquez, and D. Škrlec, "Supervisory control of an adaptive-droop regulated dc microgrid with battery management capability," *IEEE Transactions on Power Electronics*, vol. 29, no. 2, pp. 695–706, 2014. DOI: 10.1109/TPEL.2013.2257857.

[6] C. Li, P. Rakhra, P. Norman, P. Niewczas, G. Burt, and P. Clarkson, "Practical computation of di/dt for high-speed protection of dc microgrids," in *IEEE Second International Conference on DC Microgrids (ICDCM)*, 2017, pp. 153–159. DOI: 10.1109/ICDCM.2017.8001037.

[7] J. Yang, J. E. Fletcher, and J. O'Reilly, "Short-circuit and ground fault analyses and location in vsc-based dc network cables," *IEEE Transactions on Industrial Electronics*, vol. 59, no. 10, pp. 3827–3837, 2012. DOI: 10.1109/TIE.2011.2162712.

[8] P. Duijsen and D. Zuidervliet, "Control of bidirectional power flow in railway catenary overhead lines," in *PCIM Europe Conference*, 2024.

[9] S. Koning, D. Zuidervliet, and P. Duijsen, "Multifunctional grid manager topology with configurable output," in *PCIM Europe Conference*, 2024.

[10] P. J. van Duijsen and D. C. Zuidervliet, "Structuring, controlling and protecting the dc grid," in *2020 International Symposium on Electronics and Telecommunications (ISETC)*, 2020, pp. 1–4. DOI: 10.1109/ISETC50328.2020.9301065.

[11] B. Wu and M. Narimani, *High-Power Converters and AC Drives* (IEEE Press Series on Power Engineering). Wiley-IEEE Press, 2017.

[12] Simulation-Research. "Caspoc: Simulation and animation." (2024), [Online]. Available: https://www.caspoc.com.

[13] J. El Bhwas, "Dc series fault arc detection in photo-voltaic panels using current entropy and current amplitude changes," in *Bachelor-thesis, DC-Power-Lab.com*, vol. 1, 2022. DOI: dc-lab.org.

PCIM Europe 2024, 11– 13 June 2024, Nuremberg   DOI: 10.30420/566262319

Considerations on a High-Cell-Count Converter-Based Battery Storage System with Reduced Communication Effort

Paul Aspalter[1], Markus Vogelsberger[2], Hans Ertl[1]

[1] TU Wien, Inst. of Energy Systems & Electrical Drives, Gusshausstr. 27, A-1040 Vienna, Austria
[2] Alstom Austria GmbH, Hermann Gebauer Strasse 5, A-1220 Vienna, Austria

Corresponding author: Paul Aspalter, paul.aspalter@tuwien.ac.at
Speaker: Paul Aspalter, paul.aspalter@tuwien.ac.at

Abstract

This paper presents a concept study on a very high-cell-count modular multilevel converter battery storage system. With its high number of cells, the system inherits known advantages like improved voltage/current ripple, low filtering effort, reduced semiconductor voltages, redundancy etc. In case of very high cell counts like $n \geq 100$, which would be applicable for medium voltage applications, a drawback of such a topology however, is the high effort for the communication from the central controller to the n individual isolated switching cells. Addressing this challenge, a control/communication approach is presented which to be finally implemented in a truly wireless manner. Furthermore, the applicability of this method for second life batteries, as well as the effect on the control loop are discussed.

1 Introduction

Facing ever growing demand for and deployment of renewable energy systems (RES) the need for energy storage systems (ESS) in our electrical grid also increases. Due to the dependence of these systems on environmental factors, which do not necessarily correlate with consumption needs, ESS need to fill the gap [1]. A composition of different types of ESS is necessary to fulfill the diverse needs for energy storage. For example, peak shaving or the provisioning of primary control reserve have different requirements than a system that stores the energy for daily or seasonal provisioning. Furthermore, the reduction of fossil fuel based power plants coincides with a reduction of synchronous generators. These, however, make up the majority of the inertia of the grid, which is necessary for grid stability.

The objectives of ESS can be divided into two major groups. Firstly, energy management in the grid like balancing and congestion management (e.g. peak shaving). Secondly, ancillary services, such as frequency regulation, voltage support and black start capabilities [2]. Different kind of ESS (mechanical, electrical, thermal, chemical) have different characteristics considering response time,

cost per kWh, round-trip efficiency or lifetime. For the ancillary services, like synthetic inertia and voltage support, a response time in the ms range is necessary [2], [3]. The comparison in [2] *Table 33 to 35* shows that, within ESS technologies that have a response time in the ms range, Li-Ion based ESS show a sweet spot regarding cost, maturity and efficiency. This makes this technology ideal to fulfill the aforementioned task. One of the main downsides of Li-Ion batteries, the environmental impact of the source materials, could be lessened by the use of second life batteries. Due to the growing electric car market [4], these will be available in large quantities in the near future. This topic will be addressed in more detail later.

The considerations from above have led to choosing a Li-Ion based Battery Energy Storage System (BESS) as the base technology for a ESS concept in this paper.

2 State of the art BESS solutions

The ac interconnect voltage is seen as the first ac voltage with the base frequency of the grid, that is created from the DC battery voltage through power electronics. Currently deployed BESS solutions interconnect at the low voltage level (<1000 V). In [11] the ac interconnect voltage is in the range

Name	AC Interconnect Voltage [V]	Power [kVA]	Capacity [kWh]	DC Voltage Level [V]
Nidec ES1000i [5]	320-380	1935	-	575-1000
TRICERA Flex100e Indoor [6]	400	92	96	640-840
TRICERA Flex400e Outdoor [7]	400	368	384	640-840
Tesla Megapack [8]	380-550	1573	2965	-
Xelectric Power Box Unlimited M20 [9]	400	150	480	585-744
BatterStabil [10]	550	2500	2200	588-823

Tab. 1: Comparison off some currently deployed or commercially available BESS. "-" means that the value was not available to the public.

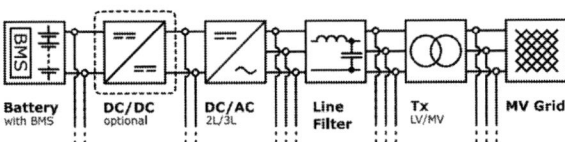

Fig. 1: Schematic of a typical BESS. The dotted vertical interconnections show the options for parallelization.

of 300 V to 690 V and Tab. 1 shows some currently available BESS products with a similar voltage range. Figure 1 shows the functional blocks of these BESS, starting from the battery and ending at the medium voltage grid connection. As high capacity/high power batteries are necessary for gird stabilization purposes, multiple of the products from Tab. 1 need to combined in parallel. This is possible at the points in between the blocks, as indicated in Fig. 1. However, interconnections before the transformer result in high currents and therefore high conduction losses, in all blocks after the parallelization. To reduce this effect one could combine the BESS at the medium voltage level. However, this increases the overall system size, as well as the price due to the additional transformers, filters and power electronics being needed.

2.1 Medium Voltage BESS concepts

Multiple concepts and papers of BESS that connect to the medium voltage grid without a transformer have been published. They often utilize a multi-cell/mutli-level topology to achieve the necessary output voltage [12], [13]. Due to the before mentioned high currents with systems that connect to the low voltage grid, these BESS have higher losses in comparison to multilevel topologies [11], [14]. These two sources also show that multilevel converters have a lower total harmonic distortion (THD). This results in less filtering efforts which further reduces cost, complexity and losses. Figure 2 and Figure 3 show two options for a multilevel medium voltage battery.

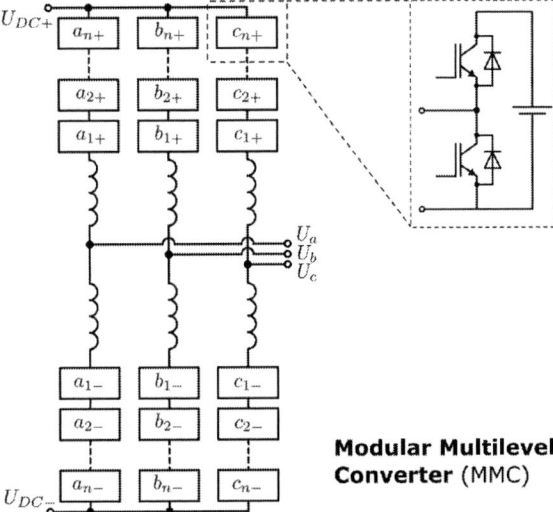

Modular Multilevel Converter (MMC)

Fig. 2: Topology of a modular multilevel converter (MMC).

The modular multilevel converter (MMC) topology (Fig. 2) is typical for high voltage direct current systems (HVDC) [15], but could also be used for a BESS [16], [17]. In comparison to the cascaded H-bridge converter (CHB) (Fig. 3) the MMC has the advantage of also having a DC bus available which would make it possible to interface with other batteries on the DC side or a DC grid.

A CHB-based BESS on the other hand, only needs half of the cells for the same AC output voltage as a MMC, if the CHB is configured as a star. Otherwise the reduction in cells is only a factor of $\sqrt{3}/2 = 0.87$. However, the delta-configuration is better suited for unbalanced grids [18]. Both topologies have the advantage of being highly modular with a base cell design being used to build most of the BESS. This also enables the possibility for $n+j$ reliability meaning that up to j modules per arm can fail and the system is still operational. Furthermore, with cells being designed to be hot-swappable modules, meaning that modules

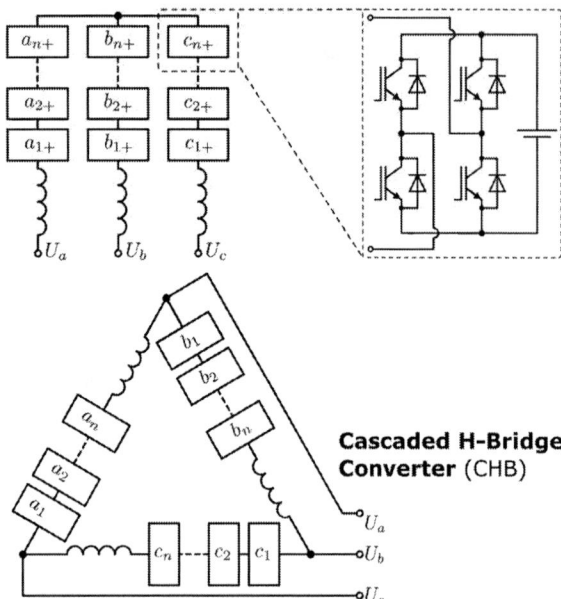

Fig. 3: Topology of a cascaded H-bridge converter in the variants: star-configuration and delta-configuration.

can be changed while the system is running, the system is ideally suited for ancillary services applications. For these applications continuous operation of the systems can be a requirement. The topic of reliability and hot-swappable modules are discussed later in this paper.

2.2 Battery cell balancing

Almost all BESS need more than one battery cell. Consequently multiple cells are connected in parallel and in series. All battery cells need to be kept inside their operating voltage range in order not to irreversible damage them. This is especially important for Lithium-Ion cells where a violation of these limits can lead to fire or even explosion of the cell [19]. Since it is impossible to manufacture all cells perfectly equal and the operating conditions of the cells (e.g. temperature) are not equal, cell capacity, self discharge rate and other parameters will vary within one BESS of multiple batteries. Therefore, it is not only important to monitor cell voltage and state of charge (SoC) of each voltage level, but to also manage it. This is done by the battery management system (BMS) of the BESS. It monitors the SoC and tries to equalize it by the different means of balancing, which can be divided into three different groups[20], [21]. The first is called passive discharge balance: a bleed resistor can be connected to cells that are charged too high compared to the rest. Second group can be

described as active balance or energy reallocation: the energy of cells that are charged too high is redistributed to a battery with lower SoC, a stack of batteries or even the whole BESS. The third group is characterized by indirect balancing/load balancing: the load itself is distributed amongst the batteries. The passive discharge circuits are simple and cheap but the energy is lost and may even make additional cooling necessary. If only a small percentage of the cells have a too high SoC, these cells will be loaded with the resistors and discharged to the pack average. However, if it is the other way round, almost all cells need to be discharged for the pack to equalize. Furthermore, if the inbalance is occurring because some cells have a lower capacity than the average (e.g. due to aging), this wasteful method of balancing needs to occur every charge cycle. Active methods are desired since it conserves the energy within the pack, disregarding conversion losses. This, however, requires (bi-directional) isolated DCDCs and complex circuits [22].

Balancing by load distribution is only possible by having each cell or stack one wants to balance interact with the rest of the BESS through a converter [21]. This would be possible in multi-cell topologies like in Fig. 3. Additionally, this method also makes the usage of second life batteries easier, since good capacity matching of the cells/stacks is not necessary. If a cell has 10% lower capacity than the average in the BESS, which could be due to aging, it has drastic effects on the overall BESS performance depending on the BMS solution. The *TRICERA Flex 400e* [7] from Tab. 1 is made up of 200 cells stacks (multiple battery cells in parallel) connected in series, which all need to be SoC controlled by a BMS. With a total BESS capacity of 384 kWh this means that each stack has a capacity of 1.92 kWh. Would one stack have a 10% lower capacity, a passive discharge BMS would lead to the BESS losing 10% capacity overall, which would be 38.4 kWh in that case. This is because, either the BESS has to stop discharging because the low capacity stack is at its lower voltage level, or it has to stop charging because the low one is already at its maximum voltage level.

An active balancer can redistribute energy, which theoretically would make it possible to use the full 382.27 kWh of the BESS. However, if the BESS operates at it's nominal power point, the output power could be reduced because of one stack. A simple active balancer, e.g. switched capacitor, can only operate well in the lower or upper voltage region of the cell [22]. This is due to the necessary voltage difference to transfer the energy, but battery chemistries like Nickel–metal hydride (NMH) or Li-Ion have a really flat voltage curve over most of their SoC [19]. Therefore, an active balancer like a flyback, that can operate over the whole voltage range, is necessary. Besides the before mentioned downsides of these systems, it would also need a balancing power of 185 W, considering no losses. At 3.7 V nominal cell voltage, this would be a balancing current of nearly 50 A. This number can be calculated considering the *Flex400e Outdoor* the following way: At full power the BESS is discharged from 100% SoC to 0% in $t_{dis} = E_{BESS}/P_{BESS} = 384kWh/368kW = 1.04h$. The missing 10% of energy of the bad stack needs to be added in t_{dis} to this stack which determines the balancer power requirement $P_{Bal} = 0.1 \cdot E_{BESS}/n_{stacks}/t_{dis} = 184.6W$.

Many BESS can be operated with downtime, where this equalization time is not a problem. But for a system that should run 24/7 to recoup cost faster or because it provides ancillary services, this downtime would be problematic. Load balancing can solve this problem and enable solutions with second life batteries without high matching efforts!

3 Downsides of high-cell-count CHB converters

So far, a CHB BESS that directly interfaces with the medium voltage grid would be a favorable design. It has a high dynamic in its power output making it suitable for ancillary grid services, the higher output voltage reduces conduction losses, the many voltage levels improve the THD of the output waveform and, therefore, reduce filter efforts. Additionally, the module based design makes load balancing possible and opens up the possibility for hot-swappable modules. This, in combination with a further amount of j modules being installed than the n necessary modules to reach the desired output voltage, would make the BESS redundant and serviceable without interrupting the operation of the BESS. However, the high converter cell count raises the question of reliability of the system. In principle, the reliability of semiconductors can be seen as a function of the junction temperature T_j and the utilization of the blocking voltage V_{CE} or V_{DS} [23]. This means that by increasing the cooling capacity or improving efficiency and reducing ratio of operating voltage to the blocking voltage, the reliability per semiconductor goes up. However, the reduction in operating voltage per transistor necessitates a higher cell count and therefore a higher semiconductor count. Consequently, this also reduces the overall reliability. The way to address this issue is to make the system redundant via additional cells and hot-swappable/repairable without interrupting operation. According to [23], this can increase the reliability, described as mean time between failure by a factor of 80 (comparison of 2 cells with no spare cells to 13 cells with one spare cell, both systems have the same power output).

3.1 Communication

With the increase in cells the communication effort increases drastically. Either each cell is addressed individually by the master controller [24] which necessitates a high bandwidth bus communication or lots of parallel connections. This issue can be reduced to a third by using sub masters that control each arm of a CHB or MMC system [25]. However, the wiring is still a major cost, installation and maintenance problem in HVDC systems. Research is therefore being carried out to make this system wireless [15].

4 CHB BESS solution with minimal communication effort

Current CHB BESS prototypes address each converter cell individually [12] [13]. Nevertheless, to build an easily scaleable CHB based BESS the communication needs to be wireless at best or at least with one single bus to reduce the wiring effort. Furthermore, the required communication bandwidth needs to be kept as low as possible. A high cell count converter ($n < 100$) is aspirational to reach a high output voltage of the BESS to enable new heights in storage capacity and power capability at high system efficiency. Otherwise, this high count could be used to reduce the number of batteries in series per cell and therefore, improve the load balancing ability of the BESS. Section 4 presents a method to enable these options.

4.1 Global addressing local balancing

The convert cells of the CHB function as the actor in the control loop. Therefore, the most important messages is the desired output voltage or modulation index of the cells. The modulation index represents the cell output voltage in relation to the cell battery voltage ($m = u_{out}/u_{batt}$). To solve the problem of needing to transmit it to each cell individually, the central controller only sends the general desired output voltage. For example, there are n cells, therefore, the controller broadcasts the desired cell voltage of U^*/n to each cell.

However, as discussed in section 2.2, this would only be possible if each battery and converter cell are perfectly identical. In reality, this is not possible due to the aforementioned balancing issues. Furthermore, this system should be able to handle cells with different capacities, e.g. second life batteries, where active balancing is a must. To solve this, not only the needed output voltage is transmitted, but also a statistical value about the SoC distribution of the cells. This way, the cells itself can handle the balancing. To achieve this, a local balancing function such as Eq. (1), where U^* describes the sum voltage of the converter arm and u_i^* the output voltage of the individual converter cells, is used. This function will be called exponential balancing function.

$$u_i^* = U^* \left(\frac{SoC_i}{\overline{SoC_\alpha}} \right)^\alpha, \quad \overline{SoC_\alpha} = \sqrt[\alpha]{\sum_{i=1}^{n} SoC_i^\alpha} \quad (1)$$

This function performs a higher load to cells that have a higher SoC and vice versa, resulting in an equalization of the overall SoC. With the parameter α, the distribution of the load can be adjusted. The sum of all cell voltages u_i^* still has to be U^* to operate in the desired working point. Equation (2) shows that this condition is satisfied, as the sum term in the second line represents $\overline{SoC_\alpha}^\alpha$, as can be seen in Eq. (1).

$$\sum_{i=1}^{n} u_i^* = \sum_{i=1}^{n} U^* \left(\frac{SoC_i}{\overline{SoC_\alpha}} \right)^\alpha =$$
$$= \frac{U^*}{\overline{SoC_\alpha}^\alpha} \sum_{i=1}^{n} SoC_i^\alpha = \quad (2)$$
$$= U^*$$

The underlying principle can be done with any function where the sum over all voltages stays the same. A more simple solution is a linear distribution function as in Eq. (3) with the tuning parameter $\lambda > 0$ and $\overline{SoC}$ being the arithmetic mean of all SoC_i. This function will be called linear balancing algorithm.

$$u_i^* = \frac{U^*}{n} \left[1 + \lambda \left(\frac{SoC_i}{\overline{SoC}} - 1 \right) \right] \quad (3)$$

Furthermore, in Eq. (1) and Eq. (3) SoC can be replaced with any state (e.g. battery cell voltage, stored charge) of the cell and U^* can be replaced with any other value one wants distributed between the cells. To analyze the system further, at a given time instant the distribution of the voltage U^* can be seen as a distribution of load P^* since the current through each cell is equal. Disregarding losses and combining this with the fact that the change in stored energy can be described by cell output power as in $dSoC_i/dt = -P_i$, means that this represents a system of coupled non linear differential equations. Therefore, the effects of the balancing algorithm will be analyzed analytically by starting at a given SoC and numeric integration.

Figure 4 shows the result of the introduced linear respectively exponential balancing algorithm applied to a 10-cell-system of equal capacity. This system is alternately charged and discharge at nominal power of the BESS (discharge current of 1 C). The SoC at the beginning is selected with the margin values 0.5 and 0.8 and in between eighth random values are selected (full list SoC=[0.50, 0.74, 0.59, 0.66, 0.55, 0.68, 0.58, 0.70, 0.71, 0.80]). Vertical lines in Fig. 4 show when each balancing function has reached the point, where the spread of the SoCs is below 0.1%. It is clearly visible that with $\alpha = 4$ a well balanced state is reached in roughly half the time compared to $\lambda = 2$ or $\alpha = 2$. However, this aggressive balancing comes at the cost of saturation in the modulation index. As seen in Fig. 4 in the bottom plot, the modulation index of cell-10 is saturated at 1 for approximately the first 600 s. This non linear saturation effect has to be handled by the controller of the output voltage. It has to be noted that the saturation in this case is due to the unrealistic initial span of the SoCs of 30%. This is chosen to better show the effectiveness of the algorithm while the BESS operates at full output power.

PCIM Europe 2024, 11– 13 June 2024, Nuremberg DOI: 10.30420/566262319

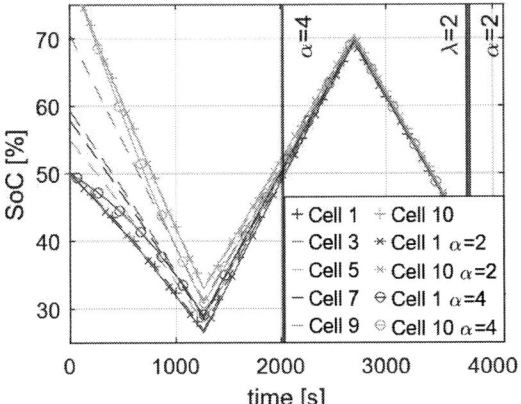

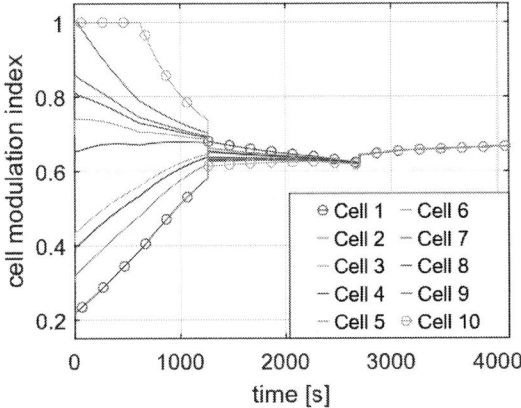

Fig. 4: Convergence of the cell SoCs (top): the dashed lines use the linear algorithm with $\lambda = 2$, for the exponential algorithm with $\alpha = 2$ resp. 4 only cell-1 and cell-10 are displayed. The corresponding modulation indices of the cells for the exponential algorithm with $\alpha = 4$ are shown in the bottom.

Furthermore, the bottom plot in Fig. 4 shows a step change of the modulation indices around t=1250 s where the BESS is switched from discharge to charge. This step is due to the fact that during charging, cells with lower SoC shall have a higher output voltage, and therefore higher charging power, than those with higher SoC. This is achieved by replacing SoC in Eq. (1) with the State of Discharge ($SoD = 1 - SoC$). Determined by the sign of the output power of the BESS, either SoC or SoD is used.

4.2 Combination of different capacities

The application of the algorithm of Fig. 4 shows it's usability for a BESS made up of cells with identical capacities. However, a goal of the proposed BESS

is to be able to handle cells of different capacities. To test the capabilities of the balancing algorithm in that regard, it was tested using 8 identical cells with a capacity of Q_0 and cell-1 being $0.7\,Q_0$ and cell-10 being $1.3\,Q_0$. All cells started with a relative SoC of 70% and the total system repeatedly is charged and discharged at 1 C, meaning that it will discharge from completely full to empty in one hour.

As noted, for Eq. (1) and Eq. (3) different attributes of the battery can be used for distributing the load throughout the cells. To balance the cells, an attribute linked to the stored charge, is necessary. Besides the relative SoC, the absolute stored charge Q_i could be used alternatively. However, in this case, the algorithm will try to converge the parameter used for the distribution, meaning that it will try to equalize the absolute stored charge. Figure 5 demonstrates that this is a problematic approach because of multiple reasons:

(1) At the beginning of the discharge cycle, the load on the cell with lower capacity (cell-1) is too small (their stored charge is below the average). As the stored charge equalizes, cell-1 will be overloaded. At the point of equal charge, the absolute value of the load on the cells would be the same, however, cells should be loaded relative to their capacity. Otherwise, they are overloaded which accelerates the aging process or may even destroy the battery [26].

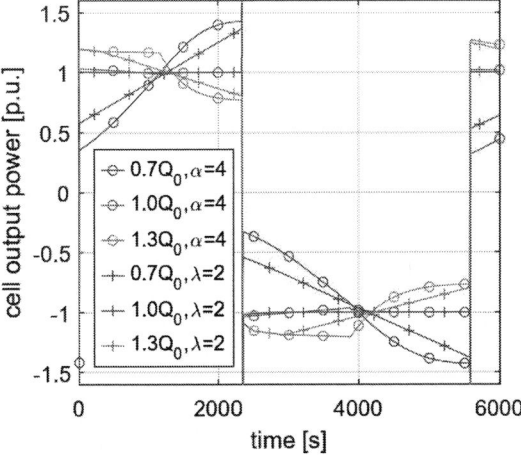

Fig. 5: Balancing with the absolute SoC used as distribution parameter. The output power of the cells is normalized with the respective capacities.

2263

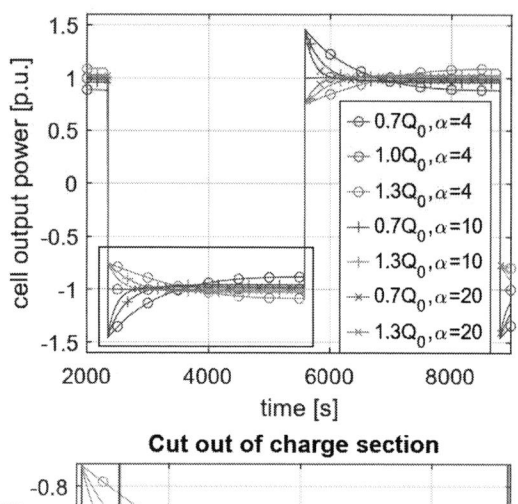

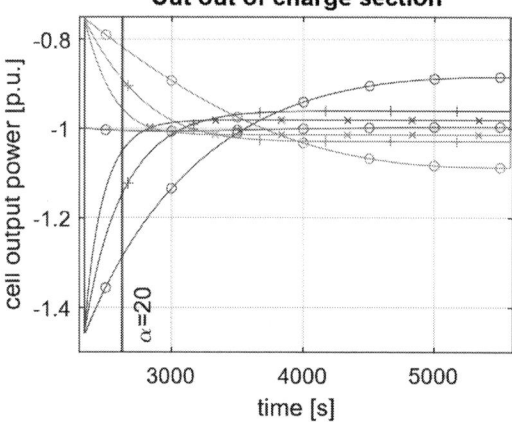

Fig. 6: Balancing with the exponential method and the relative SoC used as distribution parameter. The output power of the cells is normalized with the respective capacities.

(2) A second problem is that with the stored charge being held equal, during the charging process, cell-1 will be full first. It needs to be disconnected from the BESS, which does not necessarily stop the operatbility of the system, but it increases the load on the remaining cells. If the BESS operates at a high load, this would overload the remaining cells. Despite being possible to control the multi-cell BESS in this way, balancing based on the stored charge is not an option for the longevity of the battery cells.

Figure 6 shows the balancing algorithm applied to an identical setup as before but now the relative SoC is used as distribution parameter. Multiple α parameters (cf. Eq. (1)) are compared. The linear algorithm (Eq. (3)) produces similar trajectories and therefore only the exponential algorithm is shown. Cell-1 (cell with 70% of Q_0) being loaded

above 1p.u. and cell-10 being loaded below 1p.u. (cell with 130% of Q_0), at the beginning of a charge or discharge cycle, is determined by the capacity values of the cells itself. The tuning parameters α and λ do not influence it as can be seen in the bottom plot of Fig. 6. They, however, influence how fast the relative load on each cell convergences towards 1p.u. and the steady state error. This is highlighted in the bottom plot which just shows the charging cycle. With $\alpha = 20$ the relative load on cell-1 deviates only for the first 9% of the charging cycle more than 5% from a nominal load of 1. This point is shown by the black vertical line in the bottom plot of Fig. 6.

However, the longevity of the cells needs to be considered, especially regarding the overloading of the cells. It has to be kept in mind that in this paper an overload is considered as a load of more than 1p.u. which is a relative current of 1 C, meaning that the battery would be discharged in 1 h from full to empty. The peak discharge of cell-1 is 1.44 C, however, the average value is just 1.0014 C, for $\alpha = 20$. Batteries usually can be loaded with higher relative currents than 1 C at the cost of lifetime [26]. Investigations into the effect of this specific curve on the lifetime of the batteries would need to be performed. However, Li-Ion cells show a degradation in capacity of 9.5% after 300 full cycles at 1 C. At 2 C and 300 full cycles it decreases by 13.2% [26].Therefore, with the observed low average value a significant impact is not to be expected.

4.3 Effect of SoC estimation error

The proposed balancing methods rely on an accurate determination of the stored charge in the battery. Coulomb counting is a widely applied method that, as the name suggests, counts the charge leaving and entering the battery by integrating the battery current [27]. The discrete implementation of this is shown in Eq. (4) with C_{batt} being the storage capacity in Ah, which batteries are typically characterized.

$$SoC\,(k) = SoC\,(k-1) + \frac{\Delta_k i(k)}{3600 C_{batt}} \qquad (4)$$

This method has multiple sources of error and from [27] a normal distributed error with $\sigma = 0.03$ is chosen as a realistic value. With this error applied to the SoC, the used exponential balancing algorithm, with $\alpha = 20$, was tested with the same setup as

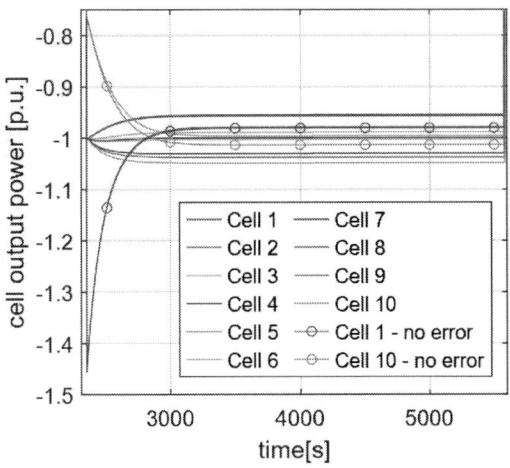

Fig. 7: The exponential balancing algorithm with an error in the SoC measurement is shown. The power is normalized with the true capacity values.

Message [Bits]	Conventional communication	Proposed communication
Communication overhead	3	3
Power Sign	N.A.	1
Average SoC	N.A.	10
Cell-to-Master	20	20
Master-to-Cell	20	20
Cell output voltage	10x200	10
Sum [Bits]	**2043**	**64**

Tab. 2: Comparison of the communication package size that is needed to control one converter arm with 200 cells.

before. Figure 7 shows that this error leads to a worse load distribution than before. However, the continuous overload is still below 5% which should not lead to faster degradation of the cells [26]. Furthermore, the biggest contributor to the error in SoC is C_{batt} in Eq. (4) [27]. This value is difficult to determine online during operation of the BESS. The proposed BESS in this paper opens up the possibility for the individual cells to be removed at fixed intervals and measured externally with high accuracy. Due to the cells being hot-swappable the operation of the BESS would not be interrupted by such an identification procedure.

4.4 Communication effort

A goal of the proposed BESS is to achieve simple scalability and enable cell counts in the three digits and above. The authors see wireless communication as a key technology for such a concept. For this broadcast messages are transmitted to all cells at a relatively high sample rate of the digital control loop (e.g. 5 kHz). The proposed balancing method reduces the required communication bandwidth of the system. An exemplary comparison for a BESS with 200 cells per arm is shown in Tab. 2. The *Cell-to-Master* and *Master-to-Cell* messages represent low speed messages, like cell SoC or temperature. Within each message only one specific cell is addressed for these low priority messages. Resulting in an effective data rate of 5khz/200 in this case. If the digital control loop of a conventional BESS output voltage runs at a sample rate of 5kHz and the central controller needs to address each cell with

an individual message, this would require a communication speed of 2043 Bit·5 kHz=10.43 MBit/s per arm. With the proposed balancing algorithm this speed decreases to 64 Bit·5 kHz=320 kBit/s per arm. For the conventional system the requirement for speed scales with the number of cells n while the proposed algorithm has a fixed requirement. In [15] a wireless communication for a HVDC MMC system is tested and achieves a data speed of 18 MBit/s. This however, would not be sufficient to communicate to all cells of the three phase system from one master.

4.5 Effect of faults on dq-control

The task of the main control of the BESS is to provide the required active and reactive power to the grid. This done by a dq-current-control loop. For the sake of brevity, this section only contains the simulation result and does not go into detail. To show the resilience of the system to a random cell-disconnection disturbance of the output voltage, a transient simulation was implemented. A 330-cell CHB BESS (110 per phase arm, c.f. Fig. 3) was simulated with a control loop that has a delay of 100 µs to incorporate a realistic scenario of a wireless communication. The connection impedance of the BESS to the three-phase grid is modeled by a simple RL-element. Figure 8 shows the system's step response from 0 to 1p.u. for the d current at t=0.5 s and for q current at t=0.7 s. At t=0.88 s a single cell disconnects. Due to the implementation of the aforementioned cell-to-master communication, the time until the central controller reacts to the missing cell depends on the subsequent arbitrary delay until this cell would send information to the master. Consequently, the delay is between 1 (150 µs) and 330 (16.6ms) sample instances. Fig-

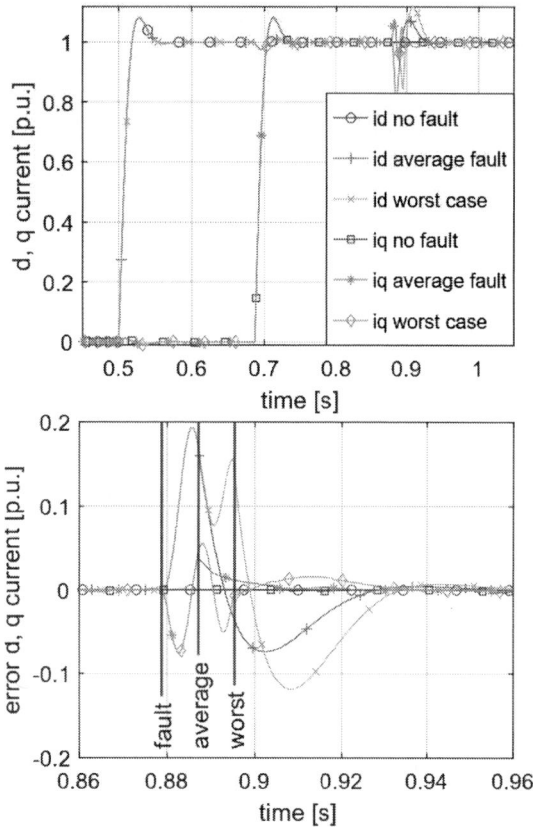

Fig. 8: Effect of one cell disconnecting from the stack at t=0.88 s on the d and q current of the system. Top shows the normed d and q current traces, and bottom shows the normed error during the fault.

ure 8 bottom indicates that even for highest possible delay the error is limited and the control loop operates in a stable manner. Here d-current error is almost 4 times higher than the q-current due to the fact that the fault occurring takes place in a time instant when the rotating voltage phasor $U_{grid,\alpha\beta}$ is aligned with the d-axis of the rotating reference frame.

5 Conclusion

With the continuous trend of an increasing share of renewable energy systems in the grid, the requirement for storage systems coincides. This paper established that for high capacity, high power storage systems that can provide ancillary services for the gird, a battery-based multilevel multi-cell topology, which directly interfaces to the medium voltage grid, would be a very attractive approach. The multi-cell concept is the basis for using second life batteries, with different capacities. Furthermore, the multi-cell

design enables 24/7 operation due to redundancy and hot-swap-capability. It can also achieve higher overall BESS efficiency compared to low voltage solutions. A major drawback of multi-cell designs, the communication effort, has been drastically reduced, by the proposed method and decoupled from the number of cells. The effectiveness of balancing, load distribution, and tolerance to errors in the SoC estimation, of has also been shown. Finally, a short outlook on the effect of the balancing algorithm in combination with the dq-current control loop of a grid connected BESS and its handling in case of a cell fault is demonstrated. Future work will focus on the concrete implementation of the balancing algorithm in combination with the control method and wireless communication with a prototype.

References

[1] S. Sahoo and P. Timmann, "Energy storage technologies for modern power systems: A detailed analysis of functionalities, potentials, and impacts," *IEEE Access*, vol. 11, pp. 49 689–49 729, 2023. DOI: 10.1109/ACCESS.2023.3274504.

[2] G. Coppez, S. Chowdhury, and S. Chowdhury, "The importance of energy storage in renewable power generation: A review," in *45th International Universities Power Engineering Conference UPEC2010*, 2010, pp. 1–5.

[3] Y. Chen, R. Hesse, D. Turschner, and H.-P. Beck, "Improving the grid power quality using virtual synchronous machines," in *2011 International Conference on Power Engineering, Energy and Electrical Drives*, 2011, pp. 1–6. DOI: 10.1109/PowerEng. 2011.6036498.

[4] S. Micari, S. Foti, A. Testa, S. Caro, F. Sergi, *et al.*, "Reliability assessment and lifetime prediction of li-ion batteries for electric vehicles," *Electrical Engineering*, vol. 104, pp. 1–13, Feb. 2022. DOI: 10.1007/s00202-021-01288-4.

[5] Nidec Industrial Solutions. "Es1000i 1000vdc inverter battery energy storage." (2024), [Online]. Available: https://www.nidec-industrial.com/wp-content/uploads/2017/09/TSD2017.09.01.00EN-US_ES1000.pdf.

[6] TRICERA energy GmbH. "Tricera energy - flex100e indoor." (2024), [Online]. Available: https://tricera.energy/industriespeichersystem-indoor/.

[7] TRICERA energy GmbH. "Tricera energy - flex400e outdoor." (2024), [Online]. Available: https://tricera.energy/industriespeichersystem-outdoor/.

[8] Tesla, Inc. "Order megapack — tesla." (2024), [Online]. Available: https://www.tesla.com/megapack/design.

[9] xelectrix Power GmbH. "Tricera - power box unlimited m20." (2024), [Online]. Available: https://www.xelectrix-power.com/en/products/unlimited-m20/.

[10] W. Vitovecs, W. Gawlik, J. Marchgraber, C. Alács, and P. Jonke, "Endbericht: Batteriestabil," Netz Niederösterreich GmbH, EVN Platz, 2344 Maria Enzersdorf, Tech. Rep., Nov. 2019.

[11] G. Wang, G. Konstantinou, C. Townsend, J. Pou, S. Vazquez, *et al.*, "A review of power electronics for grid connection of utility-scale battery energy storage systems," *IEEE Transactions on Sustainable Energy*, vol. 7, pp. 1–1, Jul. 2016. DOI: 10.1109/TSTE.2016.2586941.

[12] N. Kawakami, S. Ota, H. Kon, S. Konno, H. Akagi, *et al.*, "Development of a 500-kw modular multilevel cascade converter for battery energy storage systems," *IEEE Transactions on Industry Applications*, vol. 50, no. 6, pp. 3902–3910, 2014. DOI: 10.1109/TIA.2014.2313657.

[13] L. Maharjan, S. Inoue, H. Akagi, and J. Asakura, "A transformerless battery energy storage system based on a multilevel cascade pwm converter," in *2008 IEEE Power Electronics Specialists Conference*, 2008, pp. 4798–4804. DOI: 10.1109/PESC.2008.4592732.

[14] L. Xavier, W. Amorim, A. Cupertino, V. Mendes, W. Boaventura, and H. Pereira, "Power converters for battery energy storage systems connected to medium voltage systems: A comprehensive review," *BMC Energy*, vol. 1, Jul. 2019. DOI: 10.1186/s42500-019-0006-5.

[15] B. Çiftçi, S. Schiessl, J. Gross, L. Harnefors, S. Norrga, and H.-P. Nee, "Wireless control of modular multilevel converter submodules," *IEEE Transactions on Power Electronics*, vol. 36, no. 7, pp. 8439–8453, 2021. DOI: 10.1109/TPEL.2020.3045557.

[16] M. Schroeder, S. Henninger, J. Jaeger, A. Raš, H. Rubenbauer, and H. Leu, "Integration of batteries into a modular multilevel converter," in *2013 15th European Conference on Power Electronics and Applications (EPE)*, 2013, pp. 1–12. DOI: 10.1109/EPE.2013.6634328.

[17] I. Trintis, S. Munk-Nielsen, and R. Teodorescu, "A new modular multilevel converter with integrated energy storage," in *IECON 2011 - 37th Annual Conference of the IEEE Industrial Electronics Society*, 2011, pp. 1075–1080. DOI: 10.1109/IECON.2011.6119457.

[18] P. Sochor and H. Akagi, "Theoretical comparison in energy-balancing capability between star- and delta-configured modular multilevel cascade inverters for utility-scale photovoltaic systems," *IEEE Transactions on Power Electronics*, vol. 31, no. 3, pp. 1980–1992, 2016. DOI: 10.1109/TPEL.2015.2442261.

[19] V. Pop, H. Bergveld, P. Notten, and P. Regtien, "State-of-the-art of battery state-of-charge determination," *Meas. Sci. Technol. Prof. Holstlaan*, vol. 16, Dec. 2005. DOI: 10.1088/0957-0233/16/12/R01.

[20] L. Zheng, J. Zhu, and G. Wang, "A comparative study of battery balancing strategies for different battery operation processes," in *2016 IEEE Transportation Electrification Conference and Expo (ITEC)*, 2016, pp. 1–5. DOI: 10.1109/ITEC.2016.7520204.

[21] D. F. Frost and D. A. Howey, "Completely decentralized active balancing battery management system," *IEEE Transactions on Power Electronics*, vol. 33, no. 1, pp. 729–738, 2018. DOI: 10.1109/TPEL.2017.2664922.

[22] G.-H. Min and J.-I. Ha, "Active cell balancing algorithm for serially connected li-ion batteries based on power to energy ratio," in *2017 IEEE Energy Conversion Congress and Exposition (ECCE)*, 2017, pp. 2748–2753. DOI: 10.1109/ECCE.2017.8096514.

[23] J. E. Huber and J. W. Kolar, "Optimum number of cascaded cells for high-power medium-voltage multilevel converters," in *2013 IEEE Energy Conversion Congress and Exposition*, 2013, pp. 359–366. DOI: 10.1109/ECCE.2013.6646723.

[24] A. Lesnicar and R. Marquardt, "An innovative modular multilevel converter topology suitable for a wide power range," in *2003 IEEE Bologna Power Tech Conference Proceedings,*, vol. 3, 2003, 6 pp. Vol.3-. DOI: 10.1109/PTC.2003.1304403.

[25] B. Xia, Y. Li, Z. Li, G. Konstantinou, F. Xu, *et al.*, "Decentralized control method for modular multilevel converters," *IEEE Transactions on Power Electronics*, vol. 34, no. 6, pp. 5117–5130, 2019. DOI: 10.1109/TPEL.2018.2866258.

[26] G. Ning, B. Haran, and B. N. Popov, "Capacity fade study of lithium-ion batteries cycled at high discharge rates," *Journal of Power Sources*, vol. 117, no. 1, pp. 160–169, 2003. DOI: https://doi.org/10.1016/S0378-7753(03)00029-6.

[27] K. Movassagh, A. Raihan, B. Balasingam, and K. Pattipati, "A critical look at coulomb counting approach for state of charge estimation in batteries," *Energies*, vol. 14, no. 14, 2021. DOI: 10.3390/en14144074.

PCIM Europe 2024, 11– 13 June 2024, Nuremberg   DOI: 10.30420/566262320

Studying Convertors for Voltage Equalization in Energy Storage System with Active BMS

Dimitar Arnaudov[1], Krasimir Kishkin[1], Vladimir Dimitrov[1]

[1]Department of Power Electronics, Technical University of Sofia, Bulgaria

Corresponding author: Dimitar Arnaudov, dda@tu-sofia.bg

Abstract

The paper presents study for converters, designed for voltage equalization in energy storage system, made out of connected in series lithium ion sells. The concept of charging the battery uses resonant converters. The obtained results are verified ones by computer simulations and second by testing on a laboratory stand. A comparison is made between different voltage equalization methods by using resonant inverters, operating in ZVS and ZCS modes. The characteristics of the presented algorithms are compared.

1 Introduction

The methods for voltage equalization can be divided in three main groups: *battery selection method* where the battery pack is built by selecting cells with equal characteristics; *passive balancing method* where no active control is used for controlling the excess energy; *active balancing method* where actively controlled external circuits (DC/DC converters) are used for transferring the excess energy and thus, balancing the battery pack [1] [2].

Passive methods for balancing, uncontrollably remove the excess energy from the fully charged cell(s) by constantly dissipating it as a heat. This is usually done by resistors which are permanently connected in parallel to individual battery cells, i.e. there is no active (external) control of the charge flow. The amount of current drawn by the resistive element is proportional to the cell voltage, meaning that when the cell voltage increases, more current passes through the resistor. In addition, this current is not regulated which leads to not fully regulated cell voltages [3] [4].

In order to achieve control over the charge flow, it has to be used some kind of active circuitry. As such, a single transistor can be used. The transistor can be used directly, by replacing the passive resistive element or it can be connected in series with the resistive element. By actively controlling ON and OFF states of the transistor, the amount of the dissipated energy can be controlled more efficiently.

This is the basic approach to active balancing technique, even though the excess energy is still dissipated as a heat. That's why, this equalization methods are also called dissipative methods [5].

Dissipative methods for equalization are applicable for low power applications and low current charge/discharge rates.

For applications like modern electric vehicles, that require high current rates for their charge or discharge modes that also have to happen for quite short period of time, energy losses have to be reduced.

Minimizing the energy losses during equalization and thus improving the efficiency of the system requires active balancing techniques where the excess energy can be transferred among the cells of the energy storage system instead of wasting it as a heat.

This task can be relatively easy done by using different kinds of topologies according to what element is used for storing the excess energy. As a storage element capacitive or inductive component can be used as well as DC/DC converters like flyback, buck, boost, buck-boost and so on [6] [7] [8] [9] [10].

Active cell balancing can be divided into five different topologies according to energy distribution among the cells. These are *cell to cell* where the excess energy is transferred from higher charged cell to the lower charged one; *cell to pack* where the excess energy is transferred from the most charged cell to the whole battery pack or particular string; *pack to cell* technique transfer energy from the whole battery pack to the least charged cell; *cell to pack to cell* that combines the previous two techniques; *cell bypass* where cell currents are bypassed or

2268

redirected when the cell voltage reaches the highest value [11] [12].

Choosing the topology, depends on the application of the energy storage system.

As we know, for ensuring safe and reliable operation of the battery, a battery management system (BMS) is needed. This need comes from the fact that in a battery pack, each cell has its own characteristics that need to be checked and controlled in order to extend the battery life.

BMS is simply an electronic control circuit that has the task to monitor and regulate charge and discharge processes in the battery. To do that job, multiple of parameters have to be monitored. For instance voltage, temperature, capacity, state of charge, power consumption, charging cycles and so on. More detailed review on BMS can be found in [13] [14] [15] [16].

There are four basic topologies that are used in the design of BMS as shown in figure 1 [17].

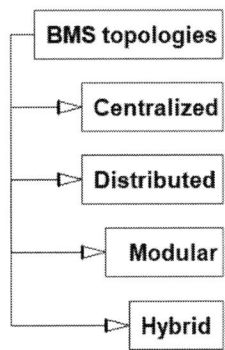

Fig. 1. BMS topologies

In the centralized topology, all cell are connected to a central master control unit. In the distributed topology, each single cell has its own control unit. In the modular topology, the battery pack is separated into smaller modules, each equipped with its own BMS. Usually, these modules are independent of each other. The hybrid topology combines elements of centralized, distributed and modular topologies.

2 Converter for active BMS

2.1. Description of the converter

The basic circuit of the studied converter is shown in Fig. 2.

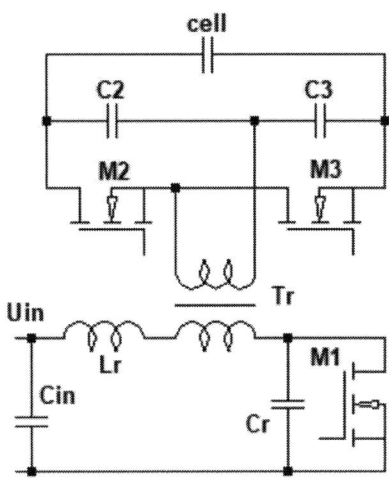

Fig. 2 DC/DC resonant converter

The circuit is single switch resonant DC to DC converter.

2.2. Block diagram of BMS system

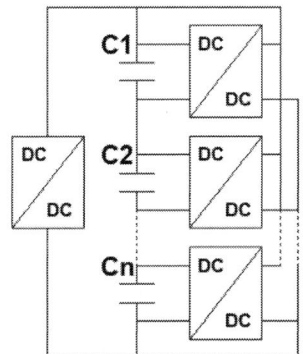

Fig. 3. Block diagram of BMS system

3 Simulation model

3.1. Single cell description

A model of the resonant converter was developed in PLECS software with the ability to set the converter output current through the control system (not shown in the figure). Part of the model is shown in Fig. 4. It was used as a subcircuit in the complex ESS model. Fig. 5 shows typical current and voltage timing diagrams for the resonant inverter, demonstrating ZVS soft-switched operation.

3.2. Equalization model description

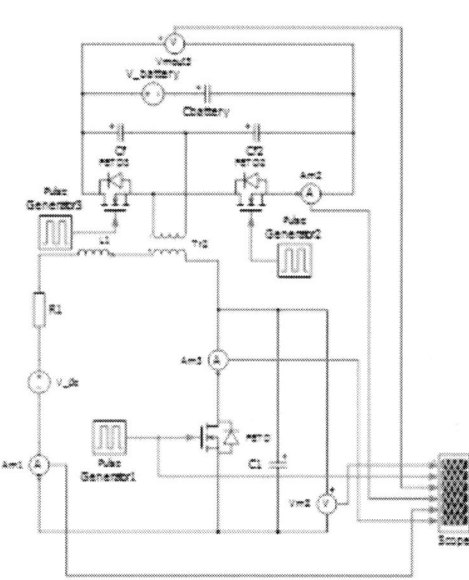

Fig. 4. PLECS model of ASC circuit

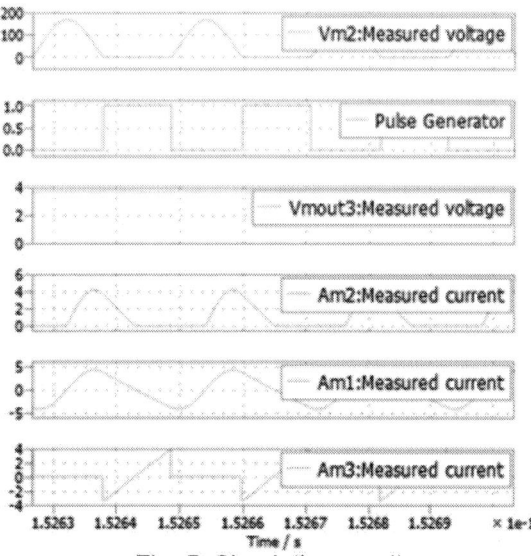

Fig. 5. Simulation results.

4 Experiments

4.1. Description of scenario

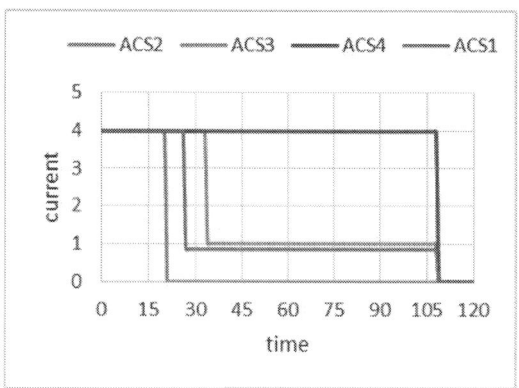

Fig. 6. Current change scenarios through the cells

Figure 6 shows an example scenario of the change in current values from the different additional sources. Such a scenario has been applied to a storage system based on supercapacitors.

5 Conclusion

An energy storage system model with active balancing blocks is developed. The additional charging sources are controlled by different algorithms to achieve voltage balancing. The losses in the converters are measured and these algorithms are compared in terms of their applicability with resonant converters as sources of additional charging currents. Results about the losses are achieved infrared thermography measurements. The algorithms and the concept of operation of the circuit are verified with ESS based on Lithium Ion cell type 18650.

ACKNOWLEDGEMENT

The carried out research is realized in the frames of the project "Optimal design and management of electrical energy storage systems", КП-06-Н37/25/18.12.2019, Bulgarian National Scientific Fund.

References

[1] Gallardo-Lozano, J., Romero-Cadaval, E., Milanes-Montero, M. I., & Guerrero-Martinez, M. A. (2014). Battery equalization active methods. Journal of Power Sources, 246, 934–949. doi:10.1016/j.jpowsour.2013.08.026.

[2] Uzair, M.; Abbas, G.; Hosain, S. Characteristics of Battery Management Systems of Electric Vehicles with Consideration of the Active and Passive Cell Balancing Process. World Electr. Veh. J. 2021, 12, 120. https://doi.org/10.3390/wevj12030120

[3] Kutkut, N. H., & Divan, D. M. (n.d.). Dynamic equalization techniques for series battery stacks. Proceedings of Intelec'96 - International Telecommunications Energy Conference. doi:10.1109/intlec.1996.573384

[4] Qi, J., & Dah-Chuan Lu, D. (2014). Review of battery cell balancing techniques. 2014 Australasian Universities Power Engineering Conference (AUPEC). doi:10.1109/aupec.2014.6966514

[5] Maneesut, K., & Supatti, U. (2017). Reviews of supercapacitor cell voltage equalizer topologies for EVs. 2017 14th International Conference on Electrical Engineering/Electronics, Computer, Telecommunications and Information Technology (ECTI-CON). doi:10.1109/ecticon.2017.8096311

[6] Daowd, M., Omar, N., Van Den Bossche, P., & Van Mierlo, J. (2011). Passive and active battery balancing comparison based on MATLAB simulation. 2011 IEEE Vehicle Power and Propulsion Conference. doi:10.1109/vppc.2011.6043010

[7] Samaddar, N., Senthil Kumar, N., & Jayapragash, R. (2020). Passive Cell Balancing of Li-Ion batteries used for Automotive Applications. Journal of Physics: Conference Series, 1716(1), 012005. https://doi.org/10.1088/1742-6596/1716/1/012005

[8] Qu, Y., Zhu, J., Hu, J., & Holliday, B. (2013). Overview of supercapacitor cell voltage balancing methods for an electric vehicle. 2013 IEEE ECCE Asia Downunder. doi:10.1109/ecce-asia.2013.6579196

[9] Daowd, M., Omar, N., Bossche, P., & Van Mierlo, J. (2012). Capacitor Based Battery Balancing System. World Electric Vehicle Journal, 5(2), 385–393. doi:10.3390/wevj502038

[10] Farzan Moghaddam, A., & Van den Bossche, A. (2019). Flyback Converter Balancing Technique for Lithium Based Batteries. 2019 8th International Conference on Modern Circuits and Systems Technologies (MOCAST). doi:10.1109/mocast.2019.8741893

[11] Kurpiel, W.; Deja, P.; Polnik, B.; Skóra, M.; Miedziński, B.; Habrych, M.; Debita, G.; Zamłyńska, M.; Falkowski-Gilski, P. Performance of Passive and Active Balancing Systems of Lithium Batteries in Onerous Mine Environment. Energies 2021, 14, 7624. https://doi.org/10.3390/en14227624

[12] Gallardo-Lozano, J., Romero-Cadaval, E., Milanés-Montero, M. I., & Guerrero-Martinez, M. A. (2014). Battery Equalization Control Based on the Shunt Transistor Method. Electrical, Control and Communication Engineering, 7(1), 20–27. doi:10.1515/ecce-2014-0019

[13] Balasingam, B.; Ahmed, M.; Pattipati, K. Battery Management Systems—Challenges and Some Solutions. Energies 2020, 13, 2825. https://doi.org/10.3390/en13112825.

[14] Lelie, M.; Braun, T.; Knips, M.; Nordmann, H.; Ringbeck, F.; Zappen, H.; Sauer, D.U. Battery Management System Hardware Concepts: An Overview. Appl. Sci. 2018, 8, 534. https://doi.org/10.3390/app8040534.

[15] Ali, M.U.; Zafar, A.; Nengroo, S.H.; Hussain, S.; Junaid Alvi, M.; Kim, H.-J. Towards a Smarter Battery Management System for Electric Vehicle Applications: A Critical Review of Lithium-Ion Battery State of Charge Estimation. Energies 2019, 12, 446. https://doi.org/10.3390/en12030446.

[16] Liu, K., Li, K., Peng, Q. et al. A brief review on key technologies in the battery management system of electric vehicles. Front. Mech. Eng. 14, 47–64 (2019). https://doi.org/10.1007/s11465-018-0516-8.

[17] Barreras, J.V.; de Castro, R.; Wan, Y.; Dragicevic, T. A Consensus Algorithm for Multi-Objective Battery Balancing. Energies 2021, 14, 4279. https://doi.org/10.3390/en14144279

[18] Fadlaoui, Elmahdi & Ismail, Lagrat & Masaif, Noureddine. (2021). Fitting the OCV-SOC relationship of a battery lithium-ion using genetic algorithm method. E3S Web of Conferences. 234. 00097. 10.1051/e3sconf/202123400097.

[19] Ahmed, M. S., Raihan, S. A., & Balasingam, B. (2020). A scaling approach for improved state of charge representation in rechargeable batteries. Applied Energy, 267, 114880. doi:10.1016/j.apenergy.2020.114880

[20] B. Pattipati, B. Balasingam, G.V. Avvari, K.R. Pattipati, Y. Bar-Shalom, Open circuit voltage characterization of lithium-ion batteries, Journal of Power Sources, 2014, Pages 317-333, ISSN 0378-7753,

https://doi.org/10.1016/j.jpowsour.2014.06.152.

Challenges of High Side Gate Driver and Disconnect MOSFET for Battery Protection Unit during Start-up, Turn-off and Over Current Events

Niranjan Reddy Suravarapu[1], Hrach Amirkhanian[2], Jianfu Fu[3],
[1] Infineon Technologies, Austria
[2] Infineon Technologies, USA
[3] Infineon Technologies, USA

Corresponding author: Niranjan Reddy Suravarapu, NiranjanReddy.Suravarapu@infineon.com
Speaker: Niranjan Reddy Suravarapu, NiranjanReddy.Suravarapu@infineon.com

Abstract

This paper will provide an overview of challenges encountered in high side battery protection applications including high side gate driver as well as high side disconnect switch (7TH MOSFET). The objective of this study is to make users better understand and design battery protection solutions that allow right design-in, ensure safe battery power operation, and prevent battery damage and catastrophic failures. The conditions that are encountered during system turn-on and turn-off, specifically for a low-cost open loop system are discussed in detail.

1 Introduction

Today, driven by more and more safety regulations, it is a major trend to have a disconnect switch system between the main battery pack and the downstream load system. Failing to disconnect the main battery system can lead to thermal runaway of battery cells and even fires. Therefore, there is a great interest in high side gate driver and disconnect switch applications (7th FET) to provide a cheaper safety solution for battery power tool application such as power tools, gardening tools, drones, e-bikes as well. The overall solution can provide a more compact bill of material, smaller footprint of system protection, integrated short circuit protection, adjustable timer delay, turn on inrush current control and safely turn-off control. With such protection mechanisms, the unit can be used to isolate the battery modules causing the occurrence of a fault from the main rail. Furthermore, such disconnect switch mechanisms helps prevent a load inverter side short-circuit from affecting the battery which can lead to a fatal error or even catastrophic battery destruction.

2 High Side MOSFET Gate drive

2.1 Turn-off performance and trade-off

It is always critical to perform a timely turn-off of the high side disconnect switch due to Over Current (OC), Short Circuit (SC) event using high side gate driver with passive gate resistance network. Whenever a short circuit or over current event occurs at the load side, it always requires the high side disconnect switch to be shut down fast enough to ensure no avalanche occurs and SOA (Safe Operating Area) operation of the 7th high side disconnect switch is maintained. The peak voltage occurs due to the

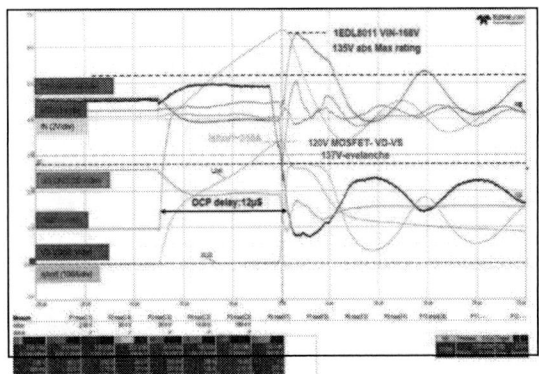

Fig. 1: SC event with 1EDL8011& small Rg_off @80V.

For example, as can be seen in Fig. 1, during short circuit event, the high side gate driver's VIN pin reaches 168V which exceeds 135V Abs Max rating of device. It also shows that the 7th MOSFET is exposed to Avalanche with Drain to Source voltage of 137V which is above the breakdown voltage of the 120V MOSFET IPT030N12N being used. Therefore, the existing parameter of Rg_off=470Ohms for turnoff to a short circuit event is not suitable for both 1EDL8011 and 7th FET. Increasing Rg_off to slowdown turn-off or potential use of two MOFETs in parallel is required.

When Rg_off is increased to 2kOhm, Fig. 2 shows improved gate driver VIN voltage under Abs Max condition and MOSFET VD-VS voltage within breakdown voltage for a short circuit condition.

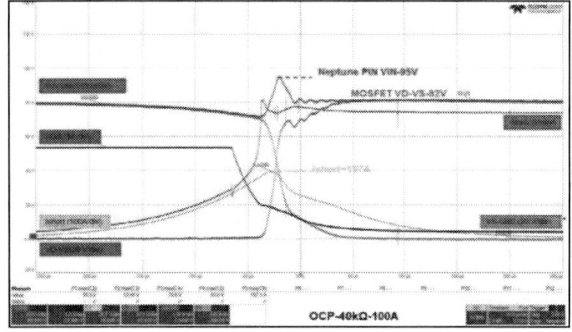

Fig. 2: SC event with 1EDL8011& large Rg_off @80V

Furthermore, based on the SOA region of the MOSFET, the energy dissipated during short circuit event must be evaluated to ensure the safe operation of the device. For example, as shown in Fig. 3. During a short circuit event, the maximum power can be obtained from the measurement waveforms.

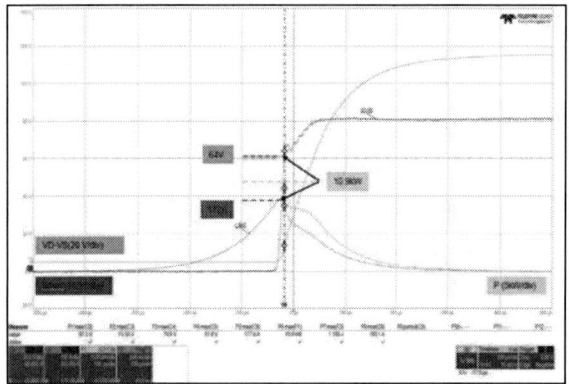

Fig. 3: Energy and power dissipation of MOSFET

The corresponding operating voltage and current can be obtained as well. Therefore, based on the pulse time duration, energy and power, the operating point with voltage, current and pulse time can be drawn on the SOA curve in the MOSFET datasheet shown in Fig.4.

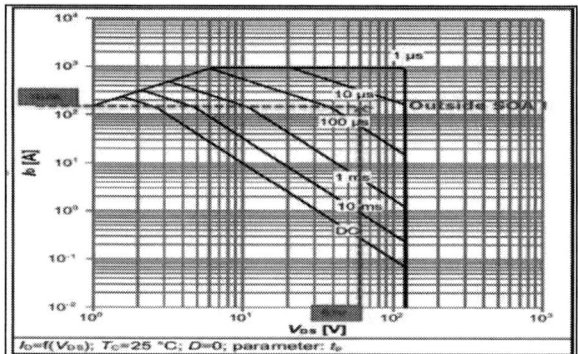

Fig. 4: SOA of Infineon MOSFET: IPT030N12N

This is the conservative way to check if such operating condition is within the SOA region of the MOSFET. If it is outside SOA region, the alternative way to ensure that the MOSFET is within SOA is to reduce the OCP threshold setting or add a capacitor between MOSFET drain and ground or add another MOSFET in parallel [3]. During a fast transient event, the snubber takes most of the energy of the peak pulse.

2.2 Turn-on performance with inrush current control

The presence of inrush currents happens when the battery system is connected for the first time to the downstream inverter and DC capacitor bank system. If the MOSFET does not have a controlled startup behavior, the battery system sees a low impedance path to the fully discharged bus capacitors. Such inrush current can increase to dangerous levels and even destroy semiconductor devices due to violating the SOA operation of 7TH high side disconnect switch. Therefore, in the absence of a pre-charge circuit, a slowed down turn-on event using the external Rg_on network can mitigate the inrush current level to operate safely. As an illustrated in Fig. 5, for a 2xIPT015N10N MOSFET in a 40V battery system during start up, the inrush current can reach up to 180A with high side gate driver using Rg_on of 0 Ω. If this current is referred to the SOA curve in Fig 4, the safe limit for a 10ms turn on event with a steady reduction of V_{ds} could vary between couple of amps to few tens of amps.

PCIM Europe 2024, 11– 13 June 2024, Nuremberg   DOI: 10.30420/566262321

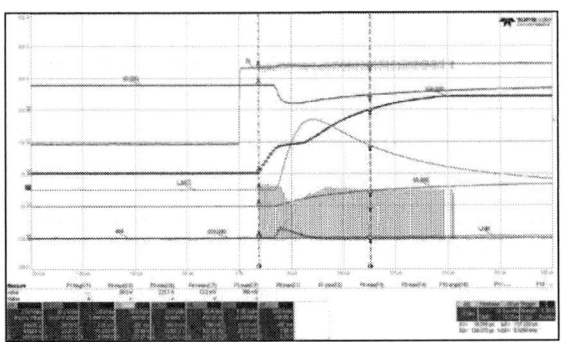

Fig. 5: High inrush current with no gate resistor Rg_on

For the same system, when the Rg_on was increased to 100 kΩ, the inrush current came down to 32A peak. The consequence would be an increased startup time as seen in Fig 6 which is acceptable for most systems. To have an even better control, an alternative system with a regulated current pullup and pulldown drive could be implemented. This needs a slightly more complex system with current sense circuitry to regulate the load current which is often expensive.

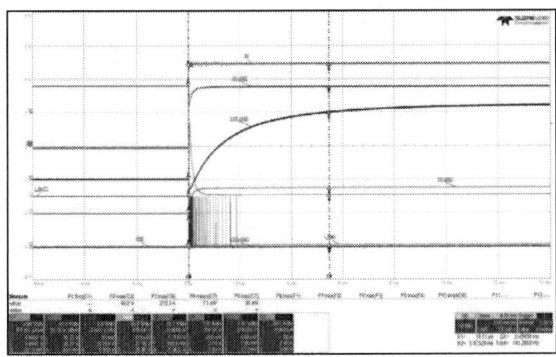

Fig. 6: Lower inrush current with 100 kΩ R$_{g\_on}$

3   Parasitic Gate Oscillation

3.1   Issue

A conventional gate driver for a switching application consists of large boot capacitor for the high side driver circuitry for a N type MOSFET. The charge necessary to drive the MOSFET and the associated voltage drop during the pull up activation of the high side driver defines the value of the boot capacitor needed. A typical high side static application like a simple disconnect switch needs a separate strategy. A charge pump is an absolute necessity in such cases. The drive capability of the

charge pump is defined by the value of the fly capacitor. There are two options now to use this chargepump output drive. Either drive the MOSFET directly from the chargepump output or store the charge onto the reserve tank capacitor and switch the pull up drive when needed. The second option needs a separate pin exclusively for storage phase. This can be an extra cost for low pin count ICs. Focus for the current study is only on low-cost solution. The considerations now for the low-cost solution are discrete charge dumps onto the gate of the MOSFET. To have a smoother voltage step, a tank capacitor is used to reduce the voltage steps after each charge pump cycle.

It is observed that during the start-up event, parasitic inductance and capacitance of PCB trace and cables in the gate circuit loop, external Ctank and Rg network between high side gate driver and MOSFET self-capacitance [1] can form oscillations and create ringing. Such oscillations and ringing can cause false switching in the devices and even lead to destruction of the7th MOSETs. The original arrangement of Rg-Ctank network comprises of a first-order RC filter network including the gate driver internal impedance and MOSFET internal capacitance.

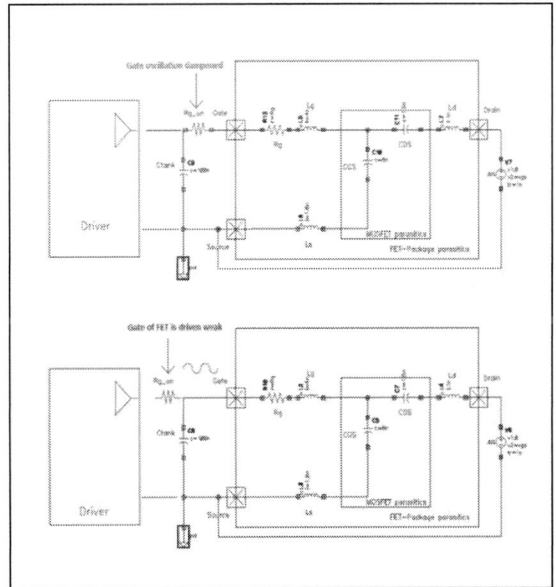

Fig. 7: RLC oscillation model of FET and Rg_on drive.

The issue can be modelled [2] with a simple first order linear system with a basic inductor, capacitor, and resistor circuit as show in Fig 7. The consequence of placing the Rg_on resistor between the driver and the Ctank results in a weak drive

2275

seen at the Gate terminal of the MOSFET. This results in an underdamped RLC circuit of the MOSFET parasitics. The simulation results of the entire system shown in Fig 8, with the actual gate drive and the MOSFET system for turn-on event shows that there is a significant lack of damping for oscillations that results in peak inrush currents.

The consequences of such an event are often catastrophic for the MOSFET if there is no voltage clamp across the gate and voltage. Besides, the overvoltage, there is a modulation effect on the load current. High di/dt can result in a lot of noise on the ground system of the load. In some cases, during our evaluation, the enable signal for the driver experienced large voltage swings. This can be a real time scenario if the enable signal arrives from a controller which might have long routing distance combined with a long ground return path. The voltages swings can be wide enough to cause an unwanted turn on-turnoff events.

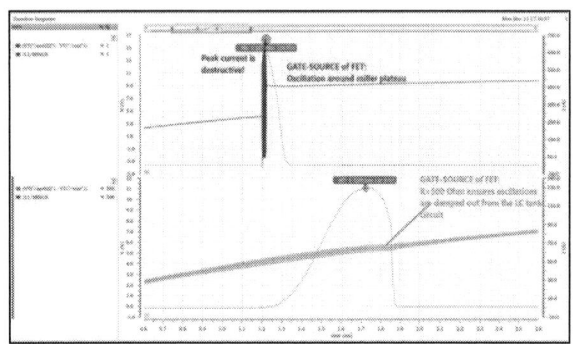

Fig. 8: System simulation with FET and Rg_on(0, 500 Ohm) drive.

3.2 Potential solution

The root cause of gate source oscillation can now be traced to the damping element's position in the drive circuit. In the case study of placing a Ctank directly at the gate of the MOSFET, there is a clear oscillation observed as soon the Gate source voltage of the MOSFET reaches the threshold voltage region as shown in Fig 9. An important observation is this oscillation occurs around the miller plateau of the MOSFET. During the first oscillation event, the enable signal (IN pin) sees a series of toggling events. This results in the gate-source voltage never exceeding the threshold voltage to get over the oscillatory state. At a later stage, because the voltage on the load side was precharged during the previous phase, despite the oscillations, the Vgs gets over the miller plateau. Nevertheless, this is still not an ideal situation.

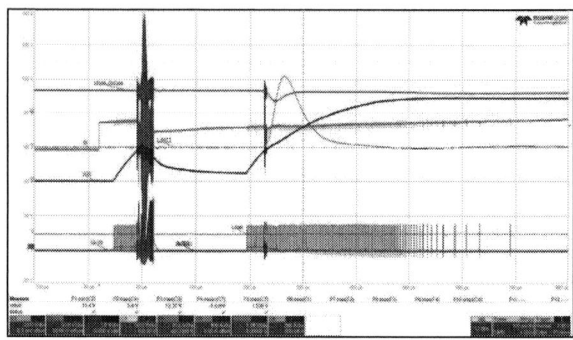

Fig. 9: Parasitic oscillation during startup with C_{tank} connected through 0 Ω R_{g_on} to MOSFET.

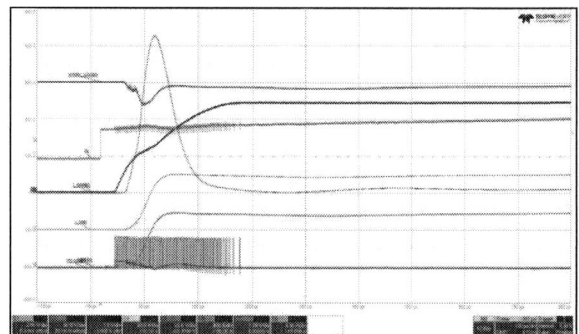

Fig. 10: Parasitic oscillation during startup with C_{tank} connected through 500 Ω R_{g_on} to MOSFET.

Proposed solution is a rearrangement of Ctank-Rg_on network. If Ctank is placed close to the IC and Rg_on placed between the MOSFET and IC-that, a second-order RC filter network can be realized with the IC's internal resistance, external Ctank-external capacitance, Rg_on and the MOSFET's internal capacitance Ciss. This addition of Rg_on acts as a dampener for the LC circuit despite the presence of a weak current drive from the gate driver. Test waveform in Fig.8 shows that oscillations can be eliminated, and the system can operate safely without any ringing.

4 Conclusion

Special consideration needs to be taken for Gate Drives which involves parallel tank capacitor to the gate source of the FET. The parasitics components of the MOSFET package could form an oscillatory circuit result in unwanted or even damaging consequences. Startup and power down of the switch should also need to be controlled to avoid potential issues like inrush currents, large voltage spikes etc.

References

[1] Application note for Induced turn on and shoot through in MOSFET half bridges: https://www.infineon.com/dgdl/Infineon-De-signing_with_power_MOSFETs-Application-Notes-v01_02-EN.pdf?fileId=8ac78c8c7ddc01d7017e6c619a490f47

[2] Application note for MOSFET oscillations: https://toshiba.semicon-storage.com/info/ap-plica-tion_note_en_20180726_AKX00066.pdf?did=59456

[3] Application note for RC snubber network: https://www.infineon.com/dgdl/Infineon-Ap-pNote_Introduction-To-Power-PROFET-Ap-plicationNotes-v02_00-EN.pdf?fileId=8ac78c8c869190210186a80f2b985278

PCIM Europe 2024, 11– 13 June 2024, Nuremberg   DOI: 10.30420/566262322

Electric Insulation Coordination to Prevent Electric Arcs in Lithium-ion Batteries

Daniel Chatroux[1], Julien Chauvin[1,2]

[1] Univ. Grenoble Alpes, CEA, Liten, DEHT, 38000 Grenoble, France
[2] French Environment and Energy Management Agency 49004 Angers France

Corresponding author: Daniel Chatroux, daniel.chatroux@cea.fr
Speaker: Daniel Chatroux, daniel.chatroux@cea.fr

Abstract

Battery systems, particularly those for electric vehicles, uses a simple electrical architecture with few components. The battery management system (BMS) is often considered as the key component for safety. Literature and our own experience tend to prove that multiple insulation defects for example due to coolant leakage is often the initial cause of batteries fires. Batteries, such as lithium-ion technology ones, for example, have specific risks. It is necessary to design electric architecture and electric insulation coordination in accordance.

1   Introduction

1.1   Batteries fire due to multiple insulation defects

According to the literature and our experience, electric arc due to multiple insulation defects is an important cause of Li-ion battery fires [1, 2].

The consequence is short-circuit of a part or the totality of the battery without the action of the classical system-wide protections (Battery Management System (BMS) and fuses). In this case, unlike studies [3] relating to the propagation, or not, of thermal runaway [4] from a single cell to others, several cells can go into thermal runaway simultaneously.

The risk is the thermal runaway of all the accumulators in the short-circuit loop simultaneously, a very fast initiation of the fire, huge quantity of combustible gas generation and of energy release.

A part of our research work is the characterization of the maximum interrupting power of the internal protections of the accumulators [5]. This work has demonstrated the inability of built-in battery protection to interrupt the current in such situations. For this reason, an effective insulation strategy must be implemented in all situations. In this paper, we examine the various concepts that need to be taken into account to create correctly isolated battery systems.

2   Electrical architecture and battery management system

2.1   Battery standard electric architecture in vehicle application

The Figure 1 describes the classical electrical architecture of a battery pack in an electric vehicle.

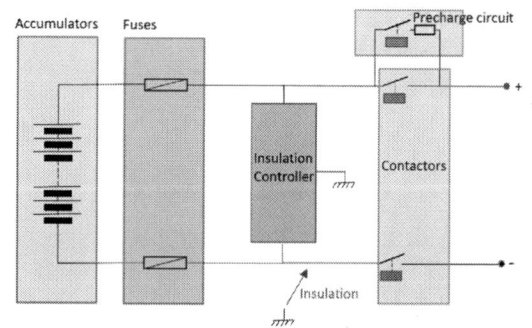

Figure 1: lithium-ion battery and vehicle electric architecture

On this figure, we can firstly observe that all the electrical architecture of the battery and the vehicle are isolated from the ground (IT system) and from the metallic body of the vehicle. Due to IT architecture, an initial insulation defect as no impact. The system can continue to operate with a single insulation fault. Driving is still possible.
An insulation controller is mandatory to inform the driver of the first insulation defect, and the need to

do a maintenance before the occurrence of a second fault.

We can also mention the presence of two contactors on both pole of the battery. Their purpose is to disconnect the battery after use. Contactors are for functional insulation to disconnect the battery after use, there are not safety disconnections. Safety disconnection to operate on the vehicle is necessary by the battery plug or by an additional safety operating device. It's mandatory to have no current before opening the disconnection device because this component is unable to switch-off the current.

At the starting of the vehicle, a precharge circuit is mandatory for the charge of all the capacitors of the power electronics of the vehicle before closing the main contactor. The pre-charge phase is the closing of the contactor in the minus pole and the pre-charge one. The capacitors of the power electronics to supply charges by the current throw the pre-charge resistor.

When the voltage difference is only a few tens of volts, the positive main contactor can switch on with a controlled overcurrent. During this pre-charge phase, all power electronics must remain in the off state to prevent not to limit the voltage reached. Because of the high energy dissipated in the resistor body in a short time, the precharge resistor has to be a high energy one, for example vitrified or ceramic wound resistance. As mentioned, it is crucial to ensure that all power electronics remain turned off throughout the entire precharge duration to prevent voltage limitation and the welding of the precharge relay contacts.

To provide the energy in the starting phase an auxiliary 12 V battery is kept. Its second function is to provide energy in case of accident for car lighting and warning.

2.2 Two contactors and two fuses are mandatory in the electrical architecture

Due to the insulating architecture, two contactors are mandatory to disconnect completely the battery from the rest of the system, including all the power electronics. This ensures effective disconnection, even in the event of a battery insulation defect. Likewise, because of the insulating architecture, a minimum of two fuses (one in the positive and one in the negative terminals) is required to protect the two poles of the battery.

A classic mistake is to have only one fuse in the positive pole, which can result in a short-circuit

loop through the negative pole without protection in the event of insulation defects in both the battery and the supplied circuit.

In case of an insulation defect in the battery and another one in the supplied circuit, a short-circuit loop by the minus pole without protection.

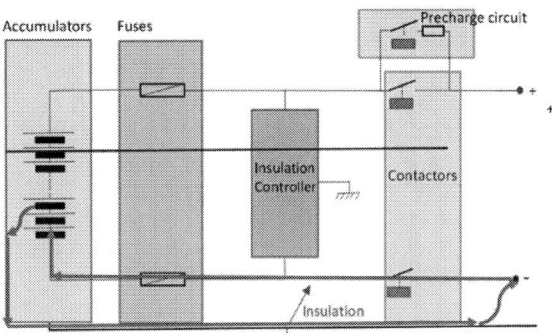

Figure 2 Example of short-circuit loop due to battery insulation defect and another one in the supplied circuit involving the minus pole.

Figure 2 is an example of short-circuit current loop in the case of a double insulation defect, one in the battery, another in the minus of the supplied circuit. For this kind of defect, a fuse in the minus is mandatory.

The fuses are sized to interrupt the short-circuit current under the worst-case scenario conditions, which include factors such as the lowest internal resistance, maximum battery voltage, and highest temperature of the accumulators.

The described electric architecture is a minimum, but we observed that this minimum can be not respected, with for example, no fuse in the negative pole, because of lack of knowledge on insulated electric network and/or cost of each component.

Other battery suppliers or automakers with higher experience minimize the risk by a higher number of fuse, for example a third fuse in the middle of the battery string.

2.3 Role of the battery management system (BMS) in the safety chain

Associated to the presented electrical architecture, we often think that a battery management system (BMS) will provide a major or total safety to the battery.

In fact, BMS has only two roles linked to safety [6]. A BMS measures all the voltages of the stages of accumulators in series to prevent overcharge (thermal runaway risk [7]) and over-discharge (accumulator degradation by copper diffusion of the negative electrode and risk of thermal runaway at the recharge [8]). Additionally, a BMS measures some temperatures of the pack to inform of local over-temperature. In the event that the BMS detects a fault, its only means of action is the opening of the contactors, generally located in the vehicle's powerbox, outside the battery pack.

Firstly, BMS cannot protect for electrical risks, upstream the main contactors.

On the other hand, contactors don't have the breaking capacity to interrupt the short-circuit current of a battery pack, which can reach for example around 10,000 A at 400 V for a conventional automotive battery pack. The contactors are absolutely not designed to switch off a short circuit current in place of an missing fuse.

3 Insulation and dielectric strength of air

3.1 Insulation requirements for electrical objects

The insulation requirements are described in a huge number of regulations. The number of regulations is increasing continuously, but the central one is the EN 60664-1.

This regulation deals with electric insulation coordination for low voltage (<1000 V AC or 1500V DC) with rated frequencies up to 30 kHz.

It specifies the requirements for clearances, creepage distances and solid insulation, and the coherence between these requirements, the insulation coordination.

The regulation includes methods of electric testing, by pulse generator or dielectric tester according to the altitude of the test bench. The physical basis of this regulation is the physic of air breakdown voltage, which is described by the Paschen's law.

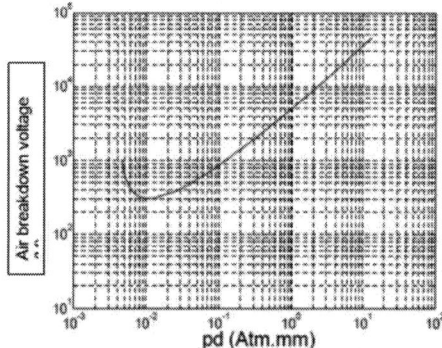

Figure 3: Diagram of the Paschen's curve for air

The figure 3 represents the breakdown voltage between homogenous electrode in Volts depending of the product distance between the electrodes (in mm) and pressure of air (in atmospheres).

The two axes, representing the product of pressure and distance in the air and the breakdown voltage of the air, are logarithmic axes. In the right part of the curve the free electron in air are accelerated by the electric field up to the collision with an atom or molecule. If the collision energy is sufficient, ionization of the atom and secondary electrons occur.

If the electric field is to low or the density of gas is too high, the electron energy is too low to provide a disruptive avalanche.

To prevent arcing phenomenon, the distance between electric parts at different voltage, or the surrounding gas pressure, should be increased.

In the left part of the curve, the density of the gas is low and the electron may be mainly ballistic up to the positive electrode without a sufficient probability to ionize an atom or a molecule, so the disruptive arc cannot occur [10].

To prevent arc, the distance between electric parts or the pressure should be lower.

There is a minimum voltage of around 300 V for air, known as the Paschen minimum, below which this voltage, sparks and arcing phenomenon cannot occur.

3.2 Presentation of the EN 60664-1

EN60664-1 is designed in accordance with the maximum transient overvoltage for electric equipment. Transient overvoltage can be caused by various factors related to grid connection, including atmospheric disturbances such as indirect lightning strikes through magnetic induction in the lines, as well as switch-on and off events in the grid distribution. Additionally, transient overvoltage can also be generated internally within the equipment itself.

If there is a connection to the grid, the maximum transient overvoltage is probably conducted by the grid.

Four overvoltage categories are defined I, II, III IV depending on equipment position in the distribution network:.

- Overvoltage category IV is for equipment at the origin of the installation as energy meters or breakers.

- Overvoltage category III is for equipment connected directly to the grid, fixed equipment and equipment whose reliability or availability is important.
- Overvoltage category II is for non-critical energy-consuming equipment, by example appliance connected by plug.
- Overvoltage category I should not be connected directly. Additional overvoltage protections are mandatory.

For a 230/400V grid, the withstand transient voltage are respectively:
- Overvoltage category IV: 6 kV
- Overvoltage category III: 4 kV
- Overvoltage category II: 2.5 kV
- Overvoltage category I: 800V

Based on these values, we can deduce that voltage surges on the electrical network are of the order of a few kilovolts.

In the case of electrical architecture that is not connected to the grid, as in the case of a vehicle battery for example, any surges that may occur in the system are estimated on a case-by-case basis.

In the case of a system with no long external cable runs, a first guideline for the minimum insulation voltage (U_{min} in V) is represented by Eq.1 :

$$U_{min} = 2*U_n + 1000V. \qquad (1)$$

U_n is the maximum voltage of the system between the two poles.

The factor two correspond to a possible resonance between stray inductance and stray capacitance. And the 1000 V is a safety margin.

U_n represents the potential difference between the two poles, but care must be taken in the case of negative polarity to consider the difference and not the maximum value.

For example, if the positive pole is + 1500 V and the minus pole is - 1500 V, U_n = 3 kV and 2 U_n + 1000 V= 7 kV.

This basic rule should be adjusted, and the minimum insulation voltage increased, in the following examples:

If the circuit has an inductive switch-off, the transient voltage may be higher, and has to be calculated.

If the circuit has a breaker or a fuse in series with an inductive circuit, the transient voltage is due to the arc voltage of the opening of the breaker or the fuse.

If the wiring is long, atmospheric disturbance are important and generate transient voltage. This situation is common, for example, in the case of photovoltaic installations, and further information on this subject can be found in the UTE-C 15-712 standard.

The clearances (the distances in air between pieces at different voltage) and solid insulations must be designed to respect this transient voltage requirement.

In accordance with Paschen's law, the maximum electric field strength for air at ambient pressure is approximately 3 kV/mm, assuming a homogeneous field with specific electrode configurations. However, in practical scenarios, the electric field is non-homogeneous, particularly near tips and angles, resulting in field reinforcement known as the peak effect. As a result, the EN 60664 standard utilizes a conservative estimate of around 1 kV/mm to account for these non-uniform field conditions.

In the case of solid insulation, the creepage distance, at the surface of the insulators materials is higher than clearance because of the risk of tracking. Due to moisture or pollution spark can occur at the surface of the insulator and burn the insulator material locally. This phenomenon progress regularly spark after spark, so the vocabulary tracking is used. Due to this physical phenomenon, creepage distance depends on the voltage (and not the transient voltage) and of the class of material (resistance to tracking).

3.3 Insulation of lithium-ion batteries

Lithium-ion batteries are known for their numerous advantages, but they also come with specific risks that must be carefully managed. These risks include the potential for water condensation due to thermal and pressure variations, particularly at varying altitudes. Additionally, the batteries contain a vast number of cell poles, each functioning as live conductors, increasing the risk of electrical hazards. Moreover, if the battery management system is centralized, the system may involve a significant number of wires, adding complexity and potential points of failure. Furthermore, there is a risk of leakage from the cooling circuit, which contains a conductive dielectric fluid such as water and glycol, as well as from the electrolyte of the accumulators, which is conductive due to its ionic conduction properties. In the event of a thermal runaway within one or several accumulators, there is a generation of hot gas and conductive material projection, posing both thermal and electrical hazards. To address battery fires, flooding of the battery may be necessary, introducing additional safety considerations and risks.

Due to all this risks, the electric insulation has to be designed not only in accordance with the EN60664 requirements, but with additional margin or additional solid insulation and/or fire resistant materials.

Now, because of their electrochemistry, the batteries are electrical components but there are not specified as electrical components. All the power supplies, laptop or smartphone chargers are specified with dielectric test as 4 kV AC for example, but classically, not the batteries

3.4 Flooding of Li-ion batteries

The flooding of the battery is now the only solution to stop a battery fire because positive material is mainly an oxide, which generate oxygen. To cut the air access by CO_2, bicarbonate powder or foam can be not sufficient to extinguish the fire. Flooding the battery with water to cool the accumulators is the only efficient proven solution for lithium-ion batteries fires extinguishing.

The flooding of the battery has to be considered in the conception phase, as Renault do for its batteries. The conception of dielectric insulations in accordance with flooding phase seems to be an interesting improvement for safety.

3.5 Partial discharge risk

Between conductors separated by some layers of insulators, the voltage repartition depends on the resistance of the layers for DC voltage and the parasitic capacitance for AC voltages.

In insulating materials comprising layers of air and solid insulators, distinct characteristics govern the distribution of electrical fields. In the case of direct current (DC) fields, the primary field is concentrated within for example, the polymer layer due to its exceptionally high resistivity. This phenomenon occurs as resistances in series channel the flow of current through the insulating material. Conversely, for alternating current (AC) fields, the predominant field is situated within the air layer. This is attributed to the permittivity of the

material, particularly the polymer, which results in capacitances in series influencing the distribution of the AC electrical field. Therefore, understanding the interplay between resistivity and permittivity is essential for comprehending the behavior of electrical fields across layers of air and solid insulators. When the field voltage is too high, a spark can occur to discharge the capacitance of the air. The arc is not between the electrodes, but only in air, so the vocabulary is partial discharge. The spark of partial discharge degrade gradually the insulation materials.

If the AC voltage is below the Paschen's minimum, there is no risk of arc due to the physical principle.

In that case, there is no risk of partial discharge for power supplies connected to the grid and low voltage outputs. For transformers or coupled inductors vacuum impregnation or molding are not mandatory. The case of partial discharge occurs when these power supplies are in AC dielectric test, and partial discharge are audible during the test. Due to the degradation of the insulation, the duration test is classically limited to 1 minute.

For high voltages, power supplies, classically around 1 kV, partial discharge are an issue and vacuum impregnation or molding oil or oil substitute immersion is mandatory.

So there is a difference between 400V batteries and higher ones (800V, 1000V, 1500V). For higher ones, the AC common mode voltage, due to grid connections or power electronic switching, has to be known to estimate the risk of partial discharge. As indicated, partial discharge is the discharge in air layers or air bubbles in a solid insulation, because the electric field is higher in air than in insulator due to the permittivity of the material. For 400V and less batteries, there is no risk of partial discharge in solid insulation. For higher voltage batteries, the AC component of the voltage must be controlled to avoid partial discharge or insulation must be designed to avoid air bubbles or air layers in solid insulators.

4 Electric arcs

4.1 From dielectric insulation to establish arc

Traditionally, insulation defects have been considered less problematic due to several factors. Firstly, electrical appliances are typically operated in dry environments, minimizing the likelihood of moisture-related issues that could compromise insulation. Additionally, appliances are often designed as class II devices, featuring double or reinforced insulation in compliance with specific appliance or generic electric regulations. For class I appliances, which require an earth connection, insulation defects are mitigated by the presence of this grounding, which serves as a protective measure. In the event of an insulation defect, electrical protection mechanisms are in place to automatically switch off the current, preventing potential hazards. Moreover, appliances are typically used in conjunction with short-circuit and earth current electrical connections, further enhancing safety measures and minimizing the risks associated with insulation defects.

For lithium-ion batteries, all the accumulators are electrical generators classically without individual electrical protections.

Battery has a huge quantity of live poles to insulate from each other and from the ground. For a 400 V battery, with 96 accumulators in series, there is 97 live poles with voltage.

Due to all the risks listed above, insulation defect as contact between live conductors or conductive pollution between some live conductors and metallic enclosure can initiate electric arcs.

As explain before, air is a very good dielectric, but when the electric arc occur, the physic changes totally. Ionized air is a very good conductor.

In the arc column, there is two voltage drops, one on the cathode side, another one on the anode side. Depending on the metals, the voltage drop near the electrodes is twelve to thirty volt only.

In the plasma region, where electron and ion exist, the conduction in very good with a voltage drop of some volts per millimeter, for example 3V/mm.

The higher the current, the higher the plasma temperature, the higher the conductivity.

There is a minimum voltage to the arc existence which is the sum of the anode and cathode drop for electron and ion generation.

There is a minimum of current to the arc existence which is the current to maintain the temperature of the plasma. A current of 1 A can be sufficient for arc initiation.

When the two conditions are respected the arc can stay establish as long as a sufficient level of current is delivered to keep the plasma temperature.

Low current AC arc are easy to extinguish because of the zero current crossing. The zero current crossing can allow plasma cooling and arc extinction. So, it is possible removing a plug from a socket for a 10 A appliance without maintained arc.

For DC current, the electric arc stay maintained. DC arc are described in photovoltaic applications where the low level of current (a few tens of amps) limits the power of the arc. Due to this limited power, there is some time to react and due to this fact and the electric architecture detection and protection is possible [10].

For batteries, because the short circuit currents is in the kA range, this strategy is very challenging perhaps impossible due to the power involved, so the very short delay and the level of current to switch off.

Due to the level of current in case of battery short circuit (some kA), the arc voltage is very low even if the arc length is long (around 3V/mm in the plasma region), and a huge energy is dissipated in a very short time.

4.2 Illustration of DC arc on a relay switching

To illustrate the change of physic between air insulation and conduction with an electric arc, we conduct test on automotive relays with a distance between contacts of 0.5mm.

The specification of the relays is 10A for 250V for AC current, but only for 12V for DC one.

We do first test on dielectric strength.

The dielectric performances at atmospheric pressure is 2.4 kV DC and goes up to 6 kV for 2.8 bars.

As Paschen curve indicates, pressure allows linear dielectric improvement.

Secondly, we do some switch-off of DC current.

For the arc at the opening, the performances at switch-off in DC are poor.

The pressure has no favorable impact. There is a small increase of the arc duration in correlation with the pressure increase.

On the Figure 4, we see the degradation of the contact due to the arc.

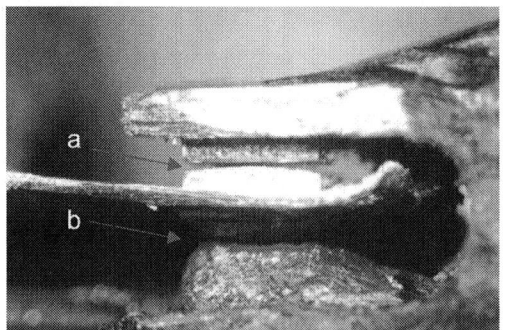

Figure 4: photograph of the surface condition of a relay without arc (a) and after an arc of 270 ms (b).

On the figure 5, we see a first phase where the arc is maintained at 26 V. This voltage is mainly the sum of the anode and cathode drop of the electrodes. After 270 ms, the current is switch-off, but high degradation of the surfaces.

The specification of this relay is coherent:

- For AC current, after a maximum time of 10 ms, the zero crossing can interrupt the AC current
- For DC current, 12V is a to low voltage, lower than the sum of the anode and cathode voltage drop, so the arc can't be maintained. The relay is non-adapted for higher voltage.

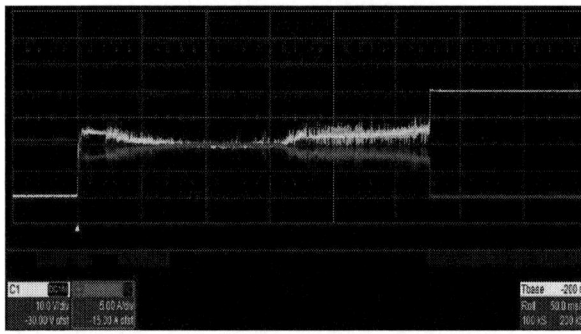

Figure 5: oscillogram of arc current and voltage of the relay

Conclusion

Initially, lithium-ion battery technology was marketed in 1991 for camscope, then for mobile electronics and handtools. Today, due to the possibility to provide high energy and high continuous power this technology is the reference one for autonomous transportation with electric energy.

According to the literature and our experience, electric arcing due to multiple insulation defects is an important cause of battery fires. In this case, several or all the accumulators of the battery pack can enter into thermal runaway simultaneously, with the risk of very fast initiation of the fire, huge quantity of combustible gas generation and energy release.

In terms of electrical architecture, due to the global insulation, the two pole of the battery must be open after use. Two fuses or two protected poles breaker are mandatory. These components are often located in a power box.

The battery management system (BMS) has a safety protection role, but only to prevent over-charge or over-discharge and to monitor some localized temperature.

Upstream of the two fuses, unless other additional fuses, a double insulation defect will create a very dangerous short-circuit loop.

Insulation is well defined in the regulation for electrical products. A reference is EN-60664-1. This regulation is based on the physic of dielectric rigidity of air as a function of the pressure, and therefore altitude. The physic reference curve is Paschen's law. The input data for this regulation is the peak voltage in use, the transient overvoltage. For grid connection devices, centuries-old feedback makes it possible to indicate figures in the regulation. For unconnected or specific devices, the transient overvoltage must be studied. On the surface of solid insulators the creepage distance is greater than the clearance one, in air, because of the risk of tracking.

Due to the specific risks of lithium-ion batteries with a large amount of active poles, different voltages and environmental conditions that can change during the operation time, compliance with electric regulation such as EN-60664-1 is a minimum, but additional protection as a preference for solid insulation or higher distances is suitable. Even flooding of the battery, in case of thermal runaway, by filling with water or immersing the battery should be taken in account.

For 800, 1000 V and higher voltages systems, the alternative component of the common mode voltage has to be limited, to avoid partial discharge in the solid insulations.

Air is very good dielectric, but if an electric arc occurs due to an overvoltage, conductive liquid or solid pollution, the physic completely changes and high default currents can be establish into the electric arc. An air plasma exhibits exceptional conductivity, with its conductivity increasing proportionally with higher currents. For alternating current, zero crossing can allow the arc to disappear. With direct current, the electric arc can be maintained for a long time. A minimum voltage of twenty to thirty volts is necessary for anode and cathode voltage loss and maintain a DC electric arc. To illustrate the transition between insulation behavior in air and the arcing behavior, some test on a standard relay proves a 2.8 kV insulation at ambient pressure, higher insulation for higher pressure, but only the ability to open 12 amps under 40 volts after an arc is established. Due to the long arcing time (280 ms) before current interrupting, the contacts are damaged.

These tests underscore the complexity of ensuring adequate breaking capacity, particularly in light of the high continuous fault currents that Li-on battery systems can generate. Moreover, the recent and anticipated future increases in system voltage, reaching up to 1200 V, highlight the ongoing need for the community to address and refine these aspects. Continued research and development efforts are crucial to meet the evolving demands of high-voltage systems and ensure their safety and reliability.

References

[1] A. Blum, « Victorian Big Battery Fire: July 30, 2021 », Fisher Engineering, Inc.; Energy Safety Response Group, janv. 2022. [Online].Available at: https://victorianbigbattery.com.au/wp-content/uploads/2023/10/VBB-Fire-Independent-Report-of-Technical-Findings.pdf

[2] Rosevear, J. (2023, august 14). Nikola shares fall after EV maker recalls all of its battery-electric semitrucks following a fire. CNBC. https://www.cnbc.com/2023/08/14/nikola-recalls-all-battery-electric-trucks-following-a-fire.html

[3] Wen, J., Yu, Y., & Chen, C. (2012). A Review on Lithium-Ion Batteries Safety Issues: Existing Problems and Possible Solutions. Materials Express, 2(3), 197-212. https://doi.org/10.1166/mex.2012.1075

[4] Feng, X., Ouyang, M., Liu, X., Lu, L., Xia, Y., & He, X. (2018). Thermal runaway mechanism of lithium ion battery for electric vehicles: A review. Energy Storage Materials, 10, 246-267. https://doi.org/10.1016/j.ensm.2017.05.013

[5] Chauvin J. Key Points Regarding Electrical Safety in Small Cylindrical Li-Ion Cell Assemblies During Overcharge or Partial Short-Circuit. PCIM Europe 2023, p. 399-406 DOI:10.30420/566091053 ISBN 978-3-8007-6091-6

[6] Xing, Y., Ma, E. W. M., Tsui, K. L., & Pecht, M. (2011). Battery Management Systems in Electric and Hybrid Vehicles. Energies, 1840-1857. https://doi.org/10.3390/en4111840

[7] Ohsaki, T., Kishi, T., Kuboki, T., Takami, N., Shimura, N., Sato, Y., Sekino, M., & Satoh, A. (2005). Overcharge reaction of lithium-ion batteries. Journal of Power Sources, 146(1-2), 97-100. https://doi.org/10.1016/j.jpowsour.2005.03.105

[8] Guo, R., Lu, L., Ouyang, M., & Feng, X. (2016). Mechanism of the entire overdischarge process and overdischarge-induced internal short circuit in lithium-ion batteries. Scientific Reports, 6(1), 30248. https://doi.org/10.1038/srep30248

[9] Yuri P. Raizer. Physics of gas discharge, 1987.

[10] W. Xu et al. A comprehensive review of DC arc faults and their mechanisms, detection, early warning strategies, and protection in battery systems. Renewable and Sustainable Energy Reviews 186 (2023) 113674

PCIM Europe 2024, 11– 13 June 2024, Nuremberg   DOI: 10.30420/566262324

Battery Charger with Impedance Spectroscopy Capability for Li-Ion Cells

Christian Brañas[1], Juan C. Viera[2], Francisco J. Azcondo[1], Alberto Pigazo[1], Rosario Casanueva[1], E. Valdés[2], Francisco J. Diaz[1], Paula Lamo[3]

[1] Universidad de Cantabria, Spain

[2] Universidad de Oviedo, Spain

[3] Universidad Internacional de la Rioja, Spain

Corresponding author: Christian Branas, branasc@unican.es

Abstract

This paper presents the design and modeling of a charger for LiFePO$_4$ lithium-ion cells based on a multiphase resonant converter. The combined phase-shift control together with frequency control of the resonant converter as well as its wide bandwidth enables the built-in implementation of the Electrochemical Impedance Spectroscopy (EIS) technique for carrying out the battery cell characterization. The resonant converter is designed as a voltage controlled current source. In this way, the maximum current is limited inherently, while the maximum voltage is set by a control loop. The EIS study is carried out in galvanostatic mode. The high-capacity ANR26650M1 LiFePO$_4$ is chosen as experimental battery cell to validate the proposal.

1 Introduction

Lithium iron phosphate (LiFePO$_4$) batteries exhibit a very low internal resistance and offer a high current rating [1]. Their cycle life is significantly longer compared to other technologies [2]. The typical applications of LiFePO$_4$ batteries are storage systems in renewable energy facilities, powering electric vehicles and uninterruptible power supplies (UPS). Nowadays, some additional functionality in battery chargers are of interest in order to optimize the battery operation and reducing costs. This work proposes the design of a battery charger with capabilities for carrying out EIS characterization [3]. EIS is a widely used technique for investigating the dynamic behavior of electrochemical systems like batteries and supercapacitors (SC). In batteries, the internal impedance, z_{Bat}, can be directly obtained through the application of EIS. Others battery parameters like state of charge (SoC) and state of health (SoH) can be also inferred from the EIS study [3]. The EIS study is mostly implemented in potentiostatic mode, which means that the battery voltage is controlled and the battery current perturbation is measured. However, in this work, the galvanostatic mode is adopted for the EIS study in order to avoid large current variations considering the low internal impedance of the high-current capacity LiFePO$_4$ cells.

2 LC$_p$C$_s$ resonant converter

The proposed battery charger is shown in Fig. 1 and consists of a two-phase LC$_p$C$_s$ resonant converter [4].

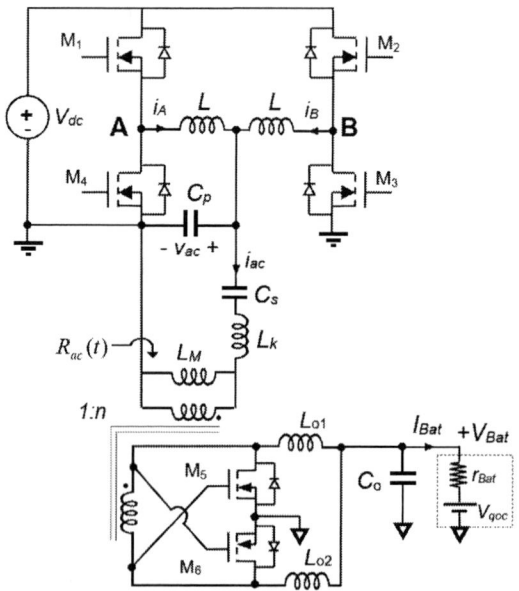

Fig. 1 Charger based on the two-phase LC$_p$C$_s$ resonant converter with synchronous current-doubler rectifier.

The battery is represented in steady-state by its internal impedance, r_{Bat}, in series with the quasi-open-circuit voltage at a given SoC, $V_{qoc}(SoC)$ [5]. The current-doubler rectifier is chosen as output stage considering its high performance for high current applications [6]. The resonant converter is designed as current source, where the phase displacement between the midpoint voltages v_A and v_B achieves an effective current control at constant switching frequency.

2.1 Resonant inverter stage

The circuit analysis is carried out assuming the fundamental approximation (FHA) [7]. The square midpoint voltages (nodes marked as A and B in Fig. 1) are replaced by its first-harmonic components v_A and v_B. The current-doubler rectifier is represented by its equivalent impedance, R_{ac}, reflected in the primary side of the transformer. The simplified circuit is depicted in Fig. 2.

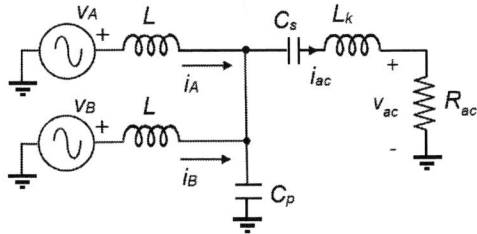

Fig. 2   Simplified LC_pC_s resonant inverter stage.

The exponential forms of v_A and v_B, are given in (1), where Ψ is the angle of phase displacement between v_A and v_B.

$$V_{A,B} = \frac{2V_{dc}}{\pi} e^{\pm j(\Psi/2)} \qquad (1)$$

The circuit in Fig. 2 exhibits a symmetrical structure. Carrying out the decomposition of input voltages V_A and V_B in (1) into their orthogonal components, it is observed that the imaginary parts of phasors V_A and V_B have equal amplitude and 180° phase displacement, therefore they cancel each other,

$$V_{A,B} = \frac{2V_{dc}}{\pi} [cos(\Psi/2) \pm j\, sin(\Psi/2)] \qquad (2)$$

In this way, the resonant inverter is reduced to the circuit, shown in Fig. 3, where the input voltage, V_{AB}, is the sum of the real parts of V_A and V_B.

$$V_{AB} = \frac{4V_{dc}}{\pi} cos(\Psi/2) \qquad (3)$$

Further simplifications are performed considering that the impedance of the capacitor C_s and the leakage inductance L_k of the transformer cancelled out each other at the switching frequency. The parallel parameters of the resonant inverter are defined in Table 1.

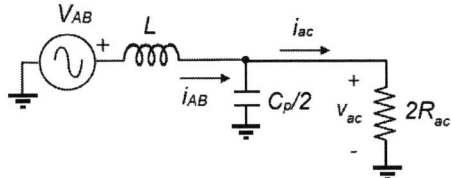

Fig. 3   Reduced circuit to perform the analysis.

Table 1. Parameters of the LC_pC_s resonant inverter.

Parallel resonant frequency	Parallel characteristic impedance	Parallel Quality factor
$\omega_p = \dfrac{1}{\sqrt{\dfrac{LC_p}{2}}}$	$Z_p = \omega_p L$ $= \dfrac{2}{\omega_p C_p}$	$Q_p = \dfrac{2R_{ac}}{Z_p}$

The EIS technique in galvanostatic mode is carried out by superimposing an ac current to the nominal dc battery charging current to obtain impedance measurements. In this work, in order to implement the EIS technique in galvanostatic mode, the resonant inverter stage is designed as a voltage-dependent-current-source, so that the circuit presents an inherent maximum current limitation. The current source behavior of the converter is achieved working at the parallel resonant frequency, ω_p, given in Table 1. The phasor of the primary side current as a function of the input DC voltage, V_{dc}, parallel resonant tank impedance, Z_p, and the control parameter, Ψ, results in (4)

$$I_{ac} = -j\frac{4V_{dc}}{\pi Z_p} cos(\Psi/2) \qquad (4)$$

The switching losses within the resonant converters are minimized by ensuring the zero-voltage switch (ZVS) of the transistors [10]. The effect of the leakage inductance referred to the primary side of the transformer, L_k, could compromise the ZVS mode. Making the most of the series disposition of L_k and C_s, it is possible to achieve the cancelation of the L_k effect by calculating C_s according to (5).

$$C_s = \frac{L}{2L_k} C_p \qquad (5)$$

Cancelling out the effect of L_k, the value of the power factor angle, ϕ_i, depends essentially on the value of the quality factor, Q_p. Some reactive energy must be accepted for ensuring the ZVS mode of all transistors. The minimum value of power factor angle for achieving ZVS, φ_{zvs}, is given in (6) and depends on the dead time, t_d, of the transistors' driver and ω_p [8].

$$\varphi_{zvs} = \frac{t_d \omega_p}{2\pi} \cdot 360^0 \quad (6)$$

As design criteria, a value of power factor angle $\phi_i = 2\varphi_{zvs}$ is assumed at nominal conditions for achieving a reliable operation of the converter, which also defines the value of Q_p.

$$Q_p = \frac{1}{\tan 2\,\varphi_{zvs}} \quad (7)$$

The transformer's turn ratio, n, is calculated according to (8).

$$n = \frac{\pi^2 V_{Bat(Max)} \tan 2\,\varphi_{zvs}}{2V_{dc}} \quad (8)$$

2.2 Current-doubler rectifier stage

The current-doubler rectifier is shown in Fig 4, where the current path for positive and negative cycles of the voltage in the secondary side of the transformer, nv_{ac}, is indicated.

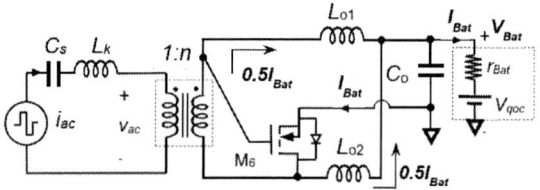

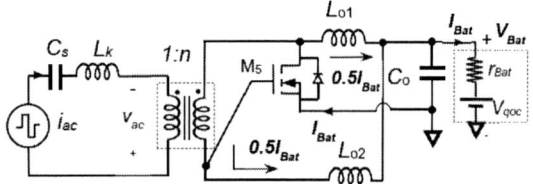

Fig. 4 Current-doubler rectifier with synchronous rectification. Current path for positive and negative cycles of the AC current.

The low voltage of the battery cell imposes the use of synchronous rectifiers, which are directly driven by the voltage across the secondary. In this way, the transistors M_5 and M_6 turn on alternatively according the positive or negative cycle of the quasi-sinusoidal voltage v_{ac}. Considering that the output filter removes the high-frequency ripple, the low ripple approximation [9] is used to obtain the

equivalent impedance, R_{ac}, of the rectifier in steady-state. Using the first harmonic of the square waveform of the currents in the transformer windings and neglecting the magnetizing current, the relation between the amplitude of the current in the primary winding and the DC output current is given in (9).

$$\hat{I}_{ac} = \frac{2n}{\pi} I_{Bat} \quad (9)$$

The amplitude of the voltage across the primary side of the transformer is obtained from the transformer power balance.

$$\hat{V}_{ac} = \frac{\pi}{n} V_{Bat} \quad (10)$$

From (9) and (10), the battery is modeled in the AC side by,

$$\hat{V}_{ac} = \frac{\pi^2}{2n^2} r_{Bat} \hat{I}_{ac} + \frac{\pi}{n} V_{qoc}(SoC) \quad (11)$$

The rectifier stage is reflected into the AC side as the equivalent resistance R_{ac} in (12).

$$R_{ac} = \frac{\pi^2}{2n^2} R_{Bat} = \frac{\pi^2}{2n^2} \left(r_{Bat} + \frac{V_{qoc}(SoC)}{I_{Bat}} \right) \quad (12)$$

The filter components are calculated according to the volt-seconds balance in the inductor's voltage waveforms.

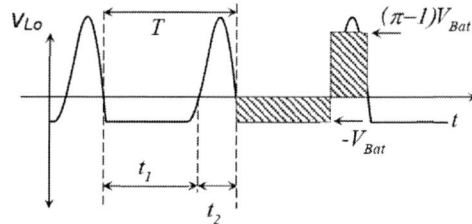

Fig. 5 Theoretical voltage waveform in the inductors filter L_{o1} and L_{o2}.

The conduction time, t_1, of diodes D_1 and D_2 is obtained assuming the approximate areas shown in Fig. 5.

$$t_1 = \frac{\pi - 1}{\pi} T \quad (13)$$

The average current through each inductor, $L_{o1,2}$, is equal to a half of the charging current I_{Bat}. The amplitude of the current ripple in each inductor is determined by,

$$\Delta i_L = \frac{(\pi - 1)V_{Bat(Max)}}{\omega_p L_o} \quad (14)$$

The total ripple current through the filter capacitor C_o is calculated considering the ripple cancellation effect due to the 180° phase displacement between the current through each inductor [10] thus,

$$\Delta i_C = \frac{(\pi - 1)V_{Bat(Max)}}{2\omega_p L_o} \quad (15)$$

The output voltage ripple is,

$$\Delta v_{Bat} = \frac{\pi(\pi - 1)V_{Bat(Max)}}{8\omega_p^2 L_o C_o} \quad (16)$$

From (16), the ripple of the charging current is a function of the switching frequency, output filter components and battery parameters.

$$\Delta i_{Bat} = \frac{\pi(\pi - 1)V_{Bat(Max)}}{8 r_{Bat}\omega_p^2 L_o C_o} \quad (17)$$

From (17), the value of the output capacitor C_o is obtained. To achieve the proper filtering of the current ripple, the equivalent series resistance (*ESR*) of C_o must be much lower than the internal impedance of the battery, $r_c << r_{Bat}$. In fact, C_o is implemented as an array in order to minimize its *ESR*.

3 Design of the resonant converter

The converter is supplied from a DC bus, V_{dc} = 48 V. The main characteristics of the ANR26650M1 LiFePO$_4$ cell are: 3.3 V nominal voltage, 2.3 Ah nominal capacity [11]. The maximum battery voltage, during charge, is $V_{Bat(Max)}$ = 3.6 V. The ANR26650M1 cell tolerates fast charging protocols so, the maximum current is set at 2C thus, $I_{Bat(max)}$ = 5 A. The switching frequency is set at ω_p = 2π(125kHz). The dead time of the drive signals is t_d = 650 ns for a φ_{zvs} = 29.25°. Upon substitution at (7-8), the quality factor is Q_p = 0.6128 and the transformer's turn ratio is n = 0.603. The maximum charging current, I_{Bat} = 5 A, is achieved at Ψ_o = 22.5°. Working with (9) and (4), the parallel characteristic impedance is obtained, Z_p = 31 Ω. The parameters of the resonant circuit are obtained from table I: L = 40 µH and C_p = 82 nF.

The filter inductors are Vishay IHLP-8787MZ with L_o = 100 µH and r_{LF} = 37 mΩ at 25° C. From (14), the amplitude of the current ripple in each inductor is Δi_L = 100 mA. The r_{Bat}, in a first approximation, is estimated at 20 mΩ. The limitation of the output current ripple, Δi_{Bat}, is mandatory to avoid errors during the EIS test as well as to prevent the battery degradation. The output capacitor, C_o, is calculated to achieve a maximum current ripple of 0.1% of the charging current, Δi_{Bat} = 5 mA. From (17), C_o = 470 µF. The transformer has been built with an ETD34 core of material N87. The leakage inductance of the transformer was L_k = 1.2 µH. Once L_k is known, the series capacitor C_s is calculated with (5) to cancel out the effect of L_k on the resonant circuit, C_s = 1.3 µF.

4 Modelling the battery-charger system

The battery cell is modeled by the electrical parameters-based model [5], shown in Fig. 7.

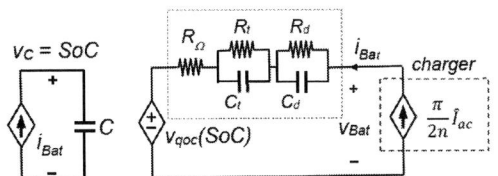

Fig. 6 Electrical parameters-based model of the battery.

The model calculates the *SoC* by integrating the current-dependent-current-source, i_{bat}, which charges / discharges the capacitor C. The voltage-controlled voltage source, $v_{qoc}(SoC)$, represents the quasi-open-circuit battery voltage. The v_{qoc} as a function of the *SOC* was experimentally obtained by charging and discharging the battery at a very low current rate, C/50. The result is shown in Fig. 7. From Fig. 7, the hysteresis voltage has a maximum value of about ~40 mV within the 30% *SoC* region, and averages ~20 mV within 40% to 80% S_oC region.

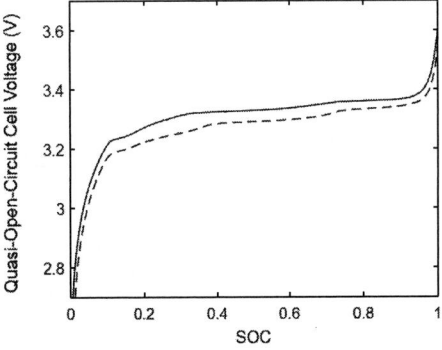

Fig. 7 Quasi-open-circuit voltage of the cell as a function of the SOC obtained at C/50 for a complete charge / discharge cycle. Solid line: Charging trajectory. Dashed line: Discharging trajectory.

The experimental results show a cell capacity of C = 2.048 Ah, which is represented in the model by the capacitor C = 7372.8 F. The electrolyte and electrode resistance are modeled by R_Ω. The time constants $R_t C_t$ and $R_d C_d$ are associated to transport and charge diffusion phenomena into the electrolytic volume. The model is tuned by using curve fitting [12]. The resulting parameters are: R_Ω = 10 mΩ and $R_t = R_d = 7$ mΩ for a total internal resistance of the battery, r_{Bat} = 24 mΩ. This result is in good agreement with experimental measurements carried out in [13]. The transport time constant is $R_t C_t = 5$ s, thus C_t = 714.3 F. The diffusion time constant is $R_d C_d = 35$ s then C_d = 5000 F.

4.1 Envelope model of the resonant converter

The cycle-by-cycle simulation of the charger-battery system is not practical due to the high frequency events of the inverter stage. In order to make achievable the simulation of the charger-battery system, the envelope model of the resonant inverter stage is obtained [14]. Given that the control of the converter is performed at constant switching frequency, the variation of the control parameter $\Psi(t)$ results in the amplitude modulation of currents and voltages according (18).

$$x(t) = Re[A(t)e^{j\omega t}] \qquad (18)$$

Upon substitution of (18) into equations of the inductor voltage and capacitor current, the envelope model of the resonant circuit components is obtained in (19) and (20).

$$L\frac{dI(t)}{dt} + j\omega_p LI(t) = V(t) \qquad (19)$$

$$C\frac{dV(t)}{dt} + j\omega_p CV(t) = I(t) \qquad (20)$$

According to (19), (20) and the circuit shown in Fig. 3, the envelope model of the resonant inverter stage is shown in Fig. 8, where the "imaginary resistor" [14] represents the steady-state impedance of the corresponding reactive element at the switching frequency.

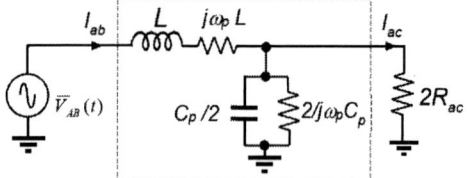

Fig. 8 Envelope model of the resonant inverter stage.

The complex large signal model in Fig. 8 requires its decomposition into its real and imaginary parts [15] as it is shown in Fig. 9 (a) and (b). The envelope of the input voltage, V_{ab}, is defined by (3) where the control parameter, $\Psi(t)$, is the modulator variable. The phase of V_{ab} is adopted as the reference for the model, so that $V_{1ab} = V_{ab}$ and V_{2ab} is null.

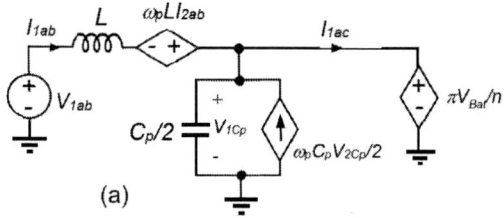

(a)

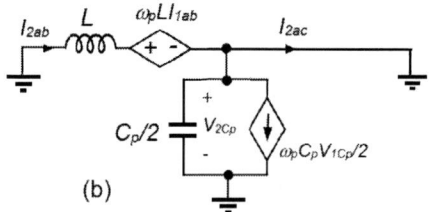

(b)

Fig. 9 Spice-compatible (a) real and (b) imaginary sub circuits of the envelope model of the resonant inverter stage.

The amplitude of the primary side current $\hat{I}_{ac}$ is calculated from its orthogonal components I_{1ac} and I_{2ac} according (21).

$$\hat{I}_{ac} = \frac{n\pi}{2}\sqrt{I_{1ac}^2 + I_{2ac}^2} \qquad (21)$$

Sub circuits in Fig. 9 and (21) are given in their general form. However, working at ω_p, where the inverter behaves as a current source, the real part of the envelope, I_{1ac}, is null thus, $\hat{I}_{ac} = |I_{2ac}|$.

4.1.1 Battery impedance characterization

Carrying out the IES study in galvanostatic mode, the battery is biased by a constant charging current I_{Bat} [5]. By perturbing the control parameter, $\Psi(t)$, in a frequency range from 1mHz to 10 kHz, an AC component Δi_{Bat} is superimposed to I_{Bat}. The resulting battery voltage variation, Δv_{Bat}, must be limited to 10 mV in order to preserve the electrochemical equilibrium of the battery. The internal battery impedance z_{Bat} is obtained, $z_{Bat} = \Delta v_{Bat} / \Delta i_{Bat}$. The joint simulation, as it shown in Fig.10, of the sub circuits of Fig. 9 with the battery model yields the variation of the battery voltage, Δv_{Bat}, for a sinusoidal perturbation of the battery current, ΔI_{Bat}. The bias current to carry out the EIS study can be adjusted by varying Ψ_o.

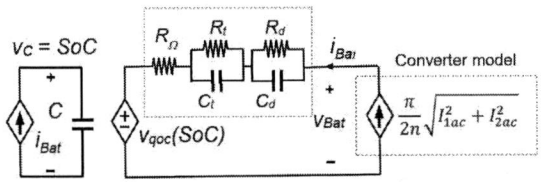

Fig. 10 Envelope model of the converter loaded by the battery model.

The result, obtained for a sinusoidal perturbation, ΔI_{Bat}, at a frequency of 1 kHz, it is shown in Fig. 11, where the battery impedance is purely resistive [13]. The bias current was I_{Bat} = 3.52 A. The measured Δv_{Bat} was 10 mV achieved with Δi_{Bat} = 0.445 A. The corresponding battery impedance is therefore r_{Bat} = 24 mΩ.

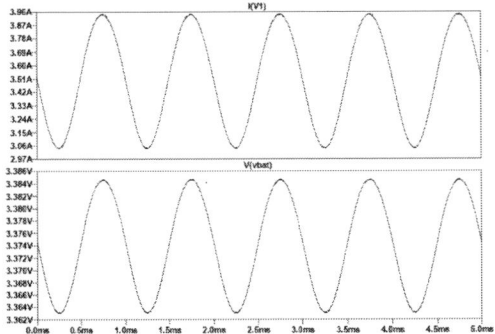

Fig. 11 Envelope model of the converter loaded by the battery model.

4.1.2 Frequency response of the charger – battery system

The linearization of the envelope model, shown in Fig. 8, allows the evaluation the frequency response of the resonant converter. In order to obtain the bandwidth, the control parameter, Ψ, has to be perturbed [16], which results in the perturbation of the input voltage. Then, the small-signal perturbation, ϕ, is added to Ψ:

$$\Psi(t) = \Psi_o + \phi \qquad (22)$$

Substituting (22) into (3) and extracting the amplitude of the small-signal perturbation,

$$\hat{v}_{ab} = -\phi \cdot \frac{2V_{dc}}{\pi} \cdot sin\left(\frac{\Psi_o}{2}\right) \qquad (23)$$

The small-signal envelope model is shown in Fig. 12 where the reflected impedance of the output filter on the primary side is neglected. For a given *SoC*, the battery is represented by its r_{Bat}. In order to obtain $\hat{i}_{ac}$, the model in Fig. 12 was implemented in LT-Spice, which requires of the model decomposition into real and imaginary sub circuits [15].

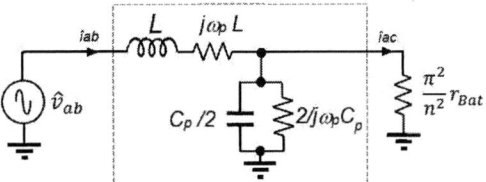

Fig. 12 Small-signal envelope model of the resonant inverter stage.

The amplitude of the small-signal perturbation of the primary current i_{ac} is obtained by linearizing (21).

$$\hat{i}_{ac} = \frac{I_{1ac} \cdot \hat{i}_{1ac} + I_{2ac} \cdot \hat{i}_{2ac}}{\sqrt{I_{1ac}^2 + I_{2ac}^2}} \qquad (24)$$

In (24) $\hat{i}_{1ac}$ and $\hat{i}_{2ac}$ are the small-signal components of I_{1ac} and I_{2ac}. At the frequency ω_p, where the inverter behaves as a current source, the phasor of the steady-state primary side current (4) is purely imaginary, i.e. I_{1ac} = 0 so that,

$$\hat{i}_{ac} = \hat{i}_{2ac} \qquad (25)$$

From (25), the small-signal response of the current in the primary side is obtained. Once, $\hat{i}_{ac}$ is obtained, the converter can be approximated to a first order system as it is shown in Fig. 13.

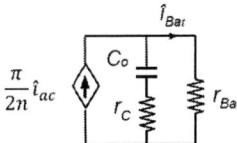

Fig. 13 First order envelope model of the resonant converter.

From the model in Fig. 13, and using (4), (9) and (23), the DC gain of the system is obtained in (26).

$$\frac{\hat{i}_{Bat}}{\phi} = \frac{V_{dc}}{nZ_p} \cdot sin\left(\frac{\Psi_o}{2}\right) \qquad (26)$$

In this case, the control angle is adjusted at Ψ_o = 93° for setting the bias current at I_{Bat} = 3.52 A. Upon substitution in (26), the DC gain is 5.4 dB. The system bandwidth, ω_B, is given in (27) and strongly depends on the filter capacitor C_o, its ESR r_c, and the battery impedance r_{Bat}. From (27), it is observed that the calculation of the output filter capacitor C_o requires a trade-off between the ripple restriction and the desired bandwidth.

$$\omega_B = \frac{1}{(r_c + r_{Bat})C_o} \qquad (27)$$

Considering a minimum value of r_c = 5 mΩ, r_{Bat} = 24 mΩ and C_o = 470 μF, the resulting bandwidth is ω_B = 2π(11.6 kHz), which is enough to implement the EIS study of the battery. The Bode diagram showing the frequency response of the converter is shown in Fig. 14.

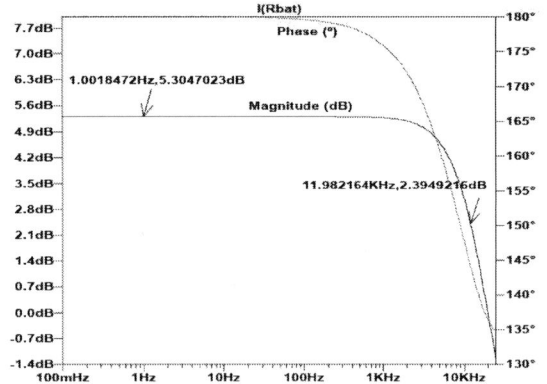

Fig. 14 Bode diagram of the charging current-to-control angle transfer function

The Bode diagram is in good agreement with the expected values of DC gain and bandwidth obtained from the theoretical analysis.

5 Conclusions

The design of a charger with EIS analysis capability for a LiFePO$_4$ battery cell have been presented. The proposal is based on a two-phase LC_pC_s resonant converter with a synchronous current-doubler rectifier as output stage. The converter is designed as a current source with current limitation inherent to the circuit. The envelope modelling demonstrates to be a useful tool which enables the co-simulation of both, the resonant converter and the battery model. The small-signal response of the converter loaded by the battery can be approximated to the response of a first order system. The bandwidth of the converter is wide enough to implement the EIS study of the battery up to 10 kHz.

Acknowledgment

This work was supported in part by the EU Regional Development Fund (FEDER) and the Spanish Ministry of Science and Innovation under the research project PID2021-128941OB-I00 "Efficient Energy Transformation in Industrial Environments" (TRENTI) and by the Regional Government of Cantabria, Spain, and EU FEDER under the project 2023-TCN-008 "Ultra-Efficient Technologies for AI-based Anomaly Detection" (UETAI).

6 References

[1]- A. Khaligh and Z. Li, "Battery, ultracapacitor, fuel cell, and hybrid energy storage systems for electric, hybrid electric, fuel cell, and plug-in hybrid electric vehicles: State of the art," IEEE Trans. Veh. Technol., vol. 59, no. 6, pp. 2806–2814, Jul. 2010.

[2]- P. Keil and A. Jossen, "Charging protocols for lithium-ion batteries and their impact on cycle life—An experimental study with different 18650 high-power cells," J. Energy Storage, vol. 6, pp. 125–141, May 2016.

[3]- S. M. Rakiul Islam, Sung-Yeul Park, Balakumar Balasingam, "Circuit parameters extraction algorithm for a lithium-ion battery charging system incorporated with electrochemical impedance spectroscopy", 2018 IEEE Applied Power Electronics Conference and Exposition (APEC), 2018.

[4]- C. Branas, F. J. Azcondo, R. Casanueva, "A Generalize Study of Multiphase Parallel Resonant Inverters for High-Power Applications", Circuits and Systems I: Regular Papers, IEEE transactions on, Vol. 55, No. 7, pp. 2128-2138, Aug. 2008.

[5]- M. Chen and G. A. Rincón-Mora, "Accurate Electrical Battery Model Capable of Predicting Runtime and I-V Performance," IEEE Trans. on Energy Conversion, vol. 21, pp. 504-511, June 2006.

[6]- Nasser H. Kutkut; Deepakraj M. Divan; Randal W. Gascoigne, "An Improved Full-Bridge Zero-Voltage Switching PWM Converter Using a Two-Inductor Rectifier". IEEE Transactions on Industry Applications Vol: 31, Issue: 1, Jan-Feb 1995, pp. 119-126.

[7]- Marian K. Kazimierczuk, Dariusz Czarkowski, Resonant Power Converters, 2nd Ed., Wiley, Nov. 2012.

[8]- Lopez, V.M.; Navarro-Crespin, A.; Schnell, R. W.; Branas, C.; Azcondo, F. J.; Zane, R. "Current Phase Surveillance in Resonant Converters for Electric Discharge Applications to Assure Operation in Zero-Voltage-Switching Mode", Power Electronics, IEEE Trans. on, Vol. 27, 2012, pp. 2925 – 2935.

[9]- R. W. Erickson, D. Maksimovic, "Fundamentals of Power Electronics", 2nd Ed. Kluwer Academic Publishers, 2001.

[10]- Sang-Min Park, Dong-Hee kim, Dong-Myoung Joo, Min-Jung kim, and Byoung-Kuk Lee, "Design of Output Filter in LLC Resonant Converters for Ripple Current Reduc-

tion in Battery Charging Applications", 9th International Conference on Power Electronics-ECCE Asia, June 1 - 5, 2015. Convention Center, Seoul, Korea.

[11]- ANR26650M1-B Power Cell Datasheet. https://store-cvt2jlgam8.mybigcommerce.com/content/Lithium-Werks%20ANR26650M1-B%20Power%20Cell%20%28030921%29%20Data%20Sheet.pdf, Accessed on April 8, 2024.

[12]- G. Plett, "Extended Kalman Filtering for Battery Management Systems of LiPb-Based HEV Battery Packs—Part 2: Modeling and Identification," Journal of Power Sources, Vol. 134, No. 2, August 2004, pp. 262–276.

[13] -A. Jossen, "Fundamentals of battery dynamics," J. Power Sources, vol. 154, no. 2, pp. 530–538, Mar. 2006.

[14]- C.T. Rim, G.H. Cho, "Phasor Transformation and its Application to the dc/ac Analyzes of Frequency Phase-Controlled Series Resonant Converters (SRC)", IEEE Trans. on Power Electron. Vol.5, pp. 201-211, April 1990.

[15]- S. Lineykin and S. Ben-Yaakov, "Unified SPICE Compatible Model for Large and Small-Signal Envelope Simulation of Linear Circuits Excited by Modulated Signals", IEEE Trans. on Ind. Electron., Vol. 53, No.3, pp. 745-751, June 2006.

[16]- Y. Yin, R. Zane, J. Glaser, R.W. Erickson, "Small-Signal Analysis of Frequency-Controlled Electronic Ballast", IEEE Trans. on Circuit and Systems-I: Fund. Theory and Appl., Vol. 50, No.8, August 2003, pp. 1103-1110.

PCIM Europe 2024, 11– 13 June 2024, Nuremberg   DOI: 10.30420/566262325

Efficiency, Volume and CO_2 Emissions Comparison in a PFC Converter with an Active Filter Solution for OBC Application

Kelly Ribeiro de Faria[1], Jean-Raphael Capounda[1], Vineel Rajagopal[1], Pascal Menegazzi[1], Benjamin Paul[1], Nabil Kamil[1], Soleiman Galeshi[1], Norbert Messi[1]

[1] Valeo Powertrain Electrified Mobility, France

Corresponding author: Kelly Ribeiro, kelly.ribeiro@valeo.com
Speaker: Kelly Ribeiro, kelly.ribeiro@valeo.com

Abstract

This paper presents the impact of the carbon footprint in the design of a bridgeless totem-pole PFC converter present in a two-stage on-board charge. The conventional design of the PFC is compared with an active filter solution operating as a pulsating power absorber proposed to reduce the volume of the bulk capacitors necessary in the converter. The carbon footprints of these two designs are evaluated for an application up to 7kW. The evaluation of different capacitor technologies and the efficiency comparison gives an idea about the environmental impact of the active filter solution and the guidelines to improve the future designs for on-board chargers.

1   Introduction

In order to contribute to the sustainable development of our society, we must reduce our carbon footprint. In this context, electric vehicles become one of the mobility alternatives to help in the reduction of CO_2, and thereby, many efforts are being made in the latest years to make the electric vehicles more attractive and popular in terms of cost and power density [1].

To contribute in this scenario, Valeo targets to achieve zero emission by the end of 2050 by making efforts in three main scopes: Scopes 1 and 2 refer to the CO_2 emissions inside the company, such as the equivalent emissions from product manufacturing and our energy consumptions related to the other work activities inside the company. In these two scopes, we can improve the energy efficiency and use low carbon energy sources. The Scope 3 considers the emissions outside the company: 3U (upstream) related to our supply chain in terms of material extraction, and components manufacturing process that we use inside the products; and 3D (downstream), related to the use of the products during its lifetime. In this last scope, the design of the power converters with improved efficiency and higher power density are good indicators to contribute with a low carbon emission.

In order to achieve higher efficiency and power density, the increase of the switching frequency associated with soft-switching techniques are usually implemented in the design of on-board

chargers (OBC) and the DC-DC isolated converters for the low-voltage auxiliary battery (LDC). With the same purpose, the magnetic integration of these two converters, OBC and LDC, into a single topology is one of the ways to achieve higher power density [2].

Generally, for the OBC design, a two-stage conversion is preferred to perform a power factor correction in the first AC/DC stage and optimize the design of the high-frequency transformer in the second DC/DC stage. In the first conversion stage, power factor correction (PFC) topologies able to perform the charge in single and three-phase systems [3] are also desirable, but in most of the cases, the converter topology will be optimized for the local grid conditions. In these applications, the single-phase mains will require the higher capacitance in the output of the AC/DC PFC converter stage for the same power level, which will lead to the use of aluminum electrolytic capacitors between the two conversion stages. These capacitors will not only decrease the power density of the converter, but it is also well known by its low lifetime. One of the solutions to overcome this drawback is the use of active filter solutions, also known as pulsating power absorbers or buffers [4], very explored a few years ago during the Google Little Box Challenge. This solution consists of deviating the second harmonic ripple from the DC bus to auxiliary passive components by using an active compensation. Because these passive components are not directly connected to the DC bus,

2294

their size can be reduced leading to a volume reduction of the AC/DC converter [5]-[6].

However, despite the overall decrease of the converter volume, additional components such as semiconductors and drivers are also necessary in that solution and an evaluation of the CO_2 impact becomes necessary in the actual context.

Therefore, in this paper, we propose a comparison of volume, efficiency and CO_2 emissions between a PFC converter with the bulk capacitors and the same converter topology with an active filter solution to decrease the total volume of the DC bus capacitors. Using the carbon footprint analysis of both designs, the combination of different capacitor technologies can also be proposed to improve the solution for the OBC applications.

The design will focus on the AC/DC PFC stage for single-phase OBC application in Europe, with maximum phase current of 32 A_{RMS} [7]. In this application, the totem-pole AC/DC bridgeless converter is selected because of the small number of components and reduced power loss in the conductive path when compared to other topologies [8].

In the first part of the paper, the design of the AC/DC converter stage is presented, and the active filter solution or pulsed power absorber is also included to deviate the low-frequency ripple from the DC bus and therefore decrease the total capacitance required in the design.

The second section presents the selection of the additional components necessary for the active filter solution highlighting the capacitor types, as well the equivalent volumes and efficiency comparison between the PFC conventional solution and the PFC with the active filter solution.

The third section presents the methodology used to estimate the CO_2 emissions and the results considering the different components used in both designs.

Finally, the last section presents the experimental results to validate the design and further proposals to improve the design and the results.

2 The active filter solution

The interleaved bridgeless totem-pole converter in Fig.1 is associated with the active filter, a Buck converter connected in parallel to the DC bus. To perform 7 kW, two high-frequency switched legs are interleaved by 180° (S1-S4), as presented in Fig. 1. The switches (S5-S6) are switched at the electrical grid frequency and are bidirectional to allow the reverse operation mode of the charger.

The minimum inductance is designed to allow continuous conduction operation mode and all the transistors are switched under hard-switching conditions.

The active filter operating as a pulsating power absorber will allow reducing the value of the capacitance connected to the DC bus (C_1) because part of the low-frequency voltage ripple can be disassociated from the DC bus by using the auxiliary converter performed by the switches (S7-S8). This auxiliary converter performs the low-frequency ripple compensation by transferring this pulsating energy to the inductor (L_{aux}) and the capacitor (C_2) connected in its output. As presented in [4], different topologies can be used for this purpose, but the buck-type is recommended in order to obtain the minimum capacitance requirement and to limit the voltage in C_2 to the maximum DC bus voltage, an important condition for the selection of the new capacitor bank.

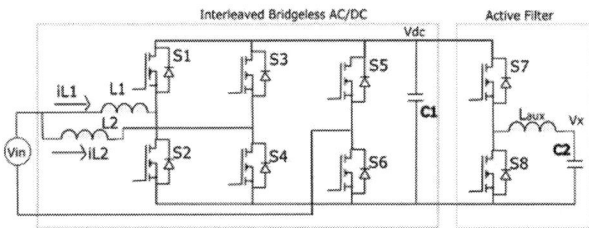

Fig. 1. The interleaved bridgeless totem-pole converter with an auxiliary active filter.

V_{in}	V_{dc}	P_{out}/ I_{RMS}	L (µH)	Voltage ripple
200-240 V_{ac}	340-500Vdc	7 kW/ 32 A	100	6 %

Table 1. Design parameters of the PFC converter

The design parameters of the PFC converter are presented in Table 1. All the semiconductors considered in this study are SiC type, and both converters are operated under hard-switching conditions.

As well explained in [5], the design of the active filter consists in selecting the final capacitance C_2 for the maximum allowed voltage variation. If C_2 can be fully charged and discharged between zero and the maximum DC bus voltage, then the capacitance C_2 can be much smaller compared to its initial value. Nevertheless, the equivalent RMS current will increase.

The minimum required capacitance can be calculated according to the ripple power that depends on the boost inductors L_1 and L_2, the maximum power and the maximum allowed voltage on the auxiliary capacitors, as following [5]:

$$C_r = \frac{I_s \ \sqrt{V_s^2 + (\omega L_1 \ I_s)^2} \ (K+1)}{\omega \ V_{C_{r_{max}}}^2} \qquad (1)$$

$$K \ is \ a \ constant > 1$$

Fig. 2 shows the minimum required capacitance according to the maximum allowed voltage in the auxiliary circuit.

Fig.3 gives the equivalent current for voltage level in the auxiliary capacitors of 450V and 500V to obtain the minimum required capacitance.

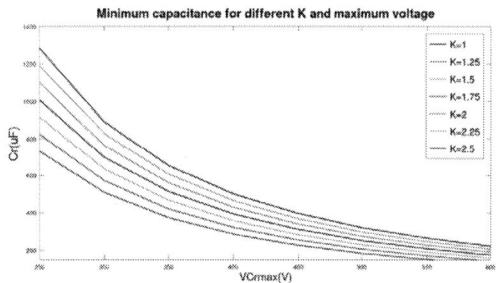

Fig. 2. The minimum required capacitance for different voltage ranges in the auxiliary capacitors.

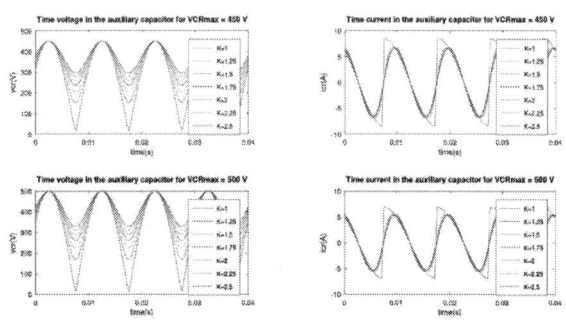

Fig. 3. The equivalent current in the auxiliary converter for voltage levels of 450/500 V for different K values.

It should be noted that as the value of K increases, the voltage of the auxiliary capacitors increases, and, in contrast, the current in the inductor decreases. Therefore, the value of K must be chosen carefully to avoid exceeding the design limits.

The minimum K value gives the minimum capacitance, but to avoid saturation in the control of the active filter and to keep a margin between V_{dc} and the maximum voltage in the capacitors, K must be > 1. For this specific design, we have selected C_2 = 430 µF. Compared with the initial capacitance required on the DC bus, 1.6 mF, related to the voltage ripple presented in Table 1, the overall decrease in the equivalent capacitance is higher than 3 times.

The auxiliary inductor is designed to have a minimum value of 70 µH and nominal current of 10 A_{RMS}.

In addition to the two transistors and the inductor magnetic component, auxiliary circuitry is also required for the control of the power absorber converter, such as drivers and resistors for the voltage acquisition. Table 2 presents the list of components for the PFC converter in standalone mode and with the active filter solution. Both converters are switched at fixed frequency and the design has been optimized by considering the volume of the passive elements and the losses in the AC/DC converter.

Components	AC/DC PFC only	AC/DC PFC + Active filter
L_1, L_2	100 µH/16 A_{RMS}	100 µH/16 A_{RMS}
C_1	1.6 mF / 450V_{dc}	30 µF / 450V_{dc}
L_{aux}	-	70 µH/10A_{RMS}
C_2	-	460 µF / 450V_{dc}
N° of Transistors	6 (SiC 650 V)	8 (SiC 650 V)
N° of Drivers (16-lead SOIC)	3 drivers (HS/LS)	4 drivers (HS/LS)
Voltage meas.	1 (V_{dc})	2 (V_{dc} and V_x)
Current meas. (8-lead SOIC)	2 current sensors	2 current sensors

Table 2. List of the components required for the PFC bridgeless converter and the PFC with the buck active filter solution.

For the inductors design, iron powder core material with copper wire has been selected for L_1, L_2 and L_{aux}.

For the capacitor C_1, in the conventional design, 3 aluminum electrolytic capacitors of 560 µF/450V are connected in parallel. For all the semiconductors, S1-S8, SiC 650 V devices are considered in this study.

Table 3 summarizes the volume of different combinations for the output capacitors. In order to compare the volume between the AC/DC PFC conventional solution and the active pulsating power absorber, five proposals are considered for the new capacitor bank: The two first proposals consist in only using Ceralink® capacitors from TDK. This capacitor type supports high current and frequency and therefore has a better performance for medium-high frequency switching converters. The third solution consists in the use of polypropylene film capacitors. Although this capacitor solution is not good in terms of power density, it has better lifetime and in the future, if we also take into consideration the recycling of components, it could represent a good option for the implementation. The fourth solution is a combination between aluminum electrolytic and Ceralink® capacitors and the last one, between MLCC and electrolytic capacitors. Here, the electrolytic capacitors are included again because of its high power density and relatively low-cost.

Reference	Quant.	Ref.	Vol.
AC/DC only	3	Aluminum type: 560 µF/450 V_{dc}	106 cm³
AC/DC+ AF 1	20	Ceralink (Solder pin) 20 µF/500 V_{dc}	167cm³
AC/DC+ AF 2	40	Ceralink type: 10 µF/500 V_{dc}	81cm³
AC/DC+ AF 3	4	Film PP type: 100 µF/450 V_{dc}	403cm³
AC/DC+ AF 4	1	Aluminum type: 380 µF/450 V_{dc}	34cm³
	2	Ceralink type: 10 µF/500 V_{dc}	4 cm³
AC/DC+ AF 5	1	Aluminum type: 380 µF/450 V_{dc}	34cm³
	10	MLCC type: 1.5 µF/450 V_{dc}	1.5 cm³
L_{aux}	1	70 µH/10A_{RMS}	7 cm³ /55 turns

Table 3. Comparison of capacitors and equivalent volume of different capacitors type in the output of the active pulsating filter.

The volume of the additional inductor connected in the active pulsating power absorber also needs to be included.

Despite the gain in the volume for some proposals, the total equivalent PCB area also needs to be evaluated because it represents carbon footprint impact on the global warming potential. We consider a PCB with 6 layers/105µm. Table 4 summarizes the total PCB area occupied for each one of the configurations, including as well the auxiliary circuitry required in the active filter solution and the auxiliary inductor L_{aux}.

Reference	Equivalent PCB area (cm²)
AC/DC only	278 cm²
AC/DC+ AF 1	420 cm²
AC/DC+ AF 2	364 cm²
AC/DC+ AF 3	336 cm²
AC/DC+ AF 4	287 cm²
AC/DC+ AF 5	285 cm²

Table 4. Comparison of the total area required for the PCB.

3 Efficiency comparison

As expected, the power losses in the additional transistors and in the inductor of the auxiliary converter decrease the total conversion efficiency. The expected efficiency drop in the AC-DC converter is presented in Fig. 4. This first result does not consider the power losses in the auxiliary capacitors, but as shown in [9], despite the difference between the capacitor technologies, the power losses generated in these components do not represent an important amount of the total losses. Simulation results are shown in Fig. 5 to present the voltage ripple decreasing on the DC bus when the active filter is initiated.

The additional losses will also have an impact on the CO_2 emissions during the 15 years of the converter lifetime. In the case of a 100% electrified vehicle, the emission associated with the charging scenario depends on the carbon footprint of the available electricity, on the resource type and it can also change with the time in a day for a certain region or country. As an example, consider a scenario for a battery of 50 kWh, where the charger is activated during 7 hours, 220 days in a year. In fact, the power depends on the SOC (state of charge) and it is not constant. If the vehicle is always in a region where the electricity can be classified as a medium-low carbon type [15], the equivalent $kgCO_2$ by day is represented in Fig. 6. After 15 years, the PFC standalone converter will be responsible for 282 $kgCO_2$ against 377 $kgCO_2$ of the PFC + active filter.

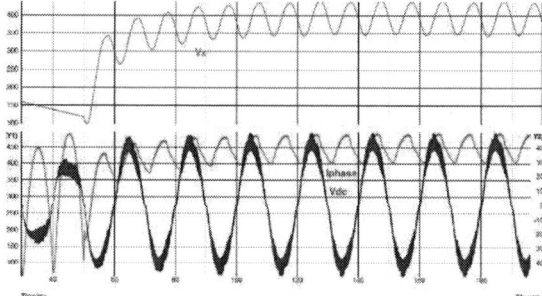

Fig. 5. Simulation results with the AC/DC PFC converter and the active filter: Vx and the Vdc voltage results.

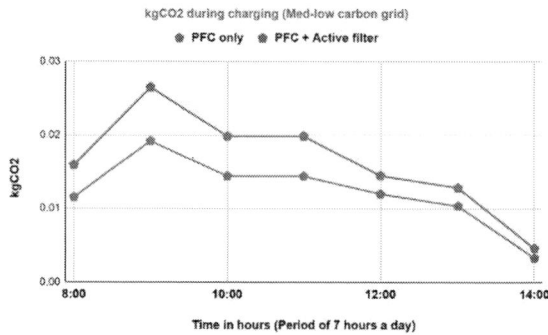

Fig. 6. Example of emission in $kgCO_2$ associated to the charging of the vehicle considering a med-low carbon grid.

4 CO₂ footprint of the solutions

The CO_{2eq}[1] footprint analysis focuses on the differences in the BOMs (Bill of Materials) of each proposed solution. The auxiliary 'active filter' allows, in some of the proposals, to decrease the overall equivalent volume of the DC-link capacitors, but because of the additional required circuitry, other components, as listed in Table 2, are required. In addition, when looking at the different available capacitors on the market, despite the short lifetime of the aluminum electrolytic capacitors, this technology remains a good option concerning the required area on the PCB.

Therefore, each identified element in Table 2 has its own materials with their equivalent weights.

In theory, if we know all the materials present in a certain component with its respective %, by check-

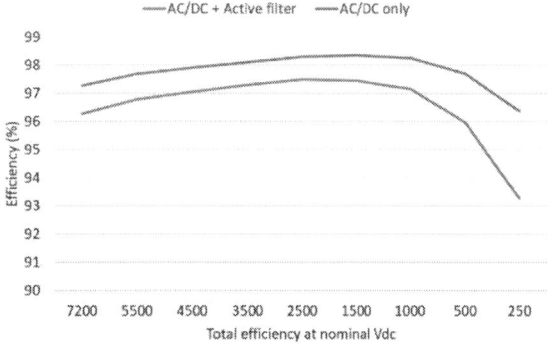

Fig. 4. Efficiency comparison between the AC/DC PFC conventional solution and the active filter to minimize the DC bus capacitance.

---

[1] CO_{2eq}: Carbon dioxide equivalent, metric based on the global warming potential (GWP) to consider different greenhouse gases.

ing on the material datasheet provided by the suppliers, CO_{2eq} emission factors in $kgCO_{2eq}$/kg or $kgCO_{2eq}$/part attached to these substances can be [2]verified in literature and international databases, such as LCA (Life Cycle Assessment) for Experts (GaBi) [10] and Ecoinvent [11].

These databases give the emission factor in $kgCO_{2eq}$/kg for the different materials that multiplied by the mass in kg gives the emission in $kgCO_{2eq}$ of the material under analysis. Then, the contribution of each material gives the equivalent CO_{2eq} emission of an electronic component.

In practice, however, sometimes it is not possible to find the Full Material Declaration of the component materials or more information about the manufacturing process that also has influence on the component carbon footprint. This is the case, for example, for the SiC transistor used in our application.

In Ecoinvent, for example, there is the emission factor associated to Si MOSFETs depending on the chip size. As has been proposed in [12] and [13], it is possible to associate the emission factors related to the SiC and Si substrate manufacturing processes. This relation can then be used to scale the emission factor associated with the SiC MOSFET. In this paper, we use the emission factor related to Silicon MOSFET available in [10]. Although the package does not have an important impact, it has also been considered.

Regarding the inductors, there is an emission factor in GaBi database [10] that uses an estimation for the equivalent ring core coil based on ferrite materials. As already mentioned in [12] there is a lack of information concerning iron powder core materials.

For the PCB, we have used the available data in GaBi, to estimate the CO_{2eq} of a 6 layers-PCB of 105 µm depending on the equivalent area.

For the capacitors, the same database in [10] has been used to obtain the emission factor for ceramic, film and aluminum capacitors. For Ceralink® capacitors, since their composition is about 85% of ceramic material [14], the emission factor associated with this capacitor is considered the same as the one of the ceramic MLCC capacitors.

In the OBC application, all the components are integrated inside an aluminum casing, and therefore we have considered the volume given by the board, magnetic components and different capacitors presented in Table 3 to estimate their equivalent

CO_{2eq} impact. The emission factor for the die-casting aluminum, including the material and the manufacturing process in China is estimated at 9.1 $kgCO_{2eq}$/kg using the database [10].

Fig. 7 shows an estimation of the contribution of each different part of the AC/DC PFC traditional converter, designed for 7 kW. Notice that together, the power transistors, the PCB and the aluminum casing are responsible for the major global warming impact.

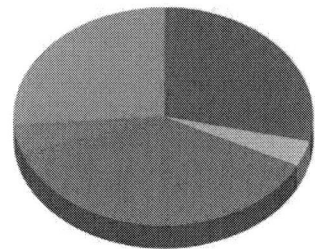

Fig. 7. Distribution of the CO_{2eq} emissions by component part in the PFC converter.

Fig. 8 presents the comparison in $kgCO_{2eq}$ between the proposed configurations in Table 2. The estimations are based on the emission factors considering the available databases. Due to the large variation in the emission factors, the results can be different when considering other databases or more details about the materials. Nevertheless, the obtained results can be used to give an idea about the influence of the different components, but it should not be interpreted as an absolute result.

Despite the reduction in the total equivalent capacitance, the gain in the aluminum casing is not enough to compensate for the additional footprint of new transistors, magnetic component and PCB. In this case, the use of a green Aluminum or secondary Aluminum could lead to an overall reduction in the final CO_{2eq} impact for all the configurations but without changing the final comparison result.

The two last solutions present almost the same CO_{2eq} footprint of the initial PFC standalone converter.

The use of a different structure rather than the buck for the pulsating power absorber at this power level, can also require transistors with smaller chip area and also reduce the amount of current in the new capacitor's bank, thus reducing the number of capacitors assembled in parallel.

In addition, we also noticed that the use of film capacitors in the solution AF 3 results in the increase of the equivalent CO_{2eq} footprint because despite its high current capability, the capacitance per volume unit is low.

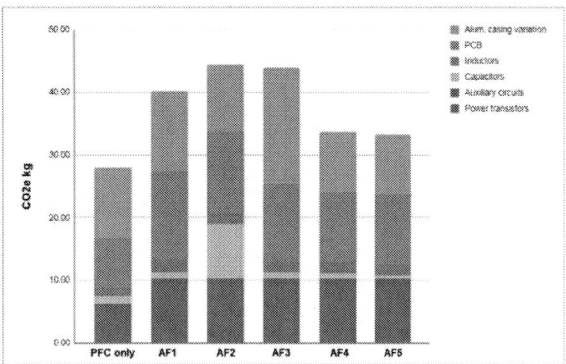

Fig. 8. Distribution of the CO_{2eq} emissions by components part in the initial PFC standalone converter and with the active filters (PFC converter + Active filter).

5   Experimental results

In order to validate the design, some tests were performed in an OBC prototype with nominal DC bus capacitance. At first, the tests were performed using the simulation with the hardware in the loop to see possible instabilities in the system when both converters are operated together. Fig. 9 presents the current in both inductors on the AC side, iL_1, iL_2, the voltage in the DC bus, V_{dc}, and the voltage in the auxiliary capacitors, V_x.

During the experimental tests, at first, we performed the measurement of the initial efficiency of the AC/DC PFC converter with its nominal capacitance on the DC bus and without the active filter.

After this step, the nominal capacitance on the DC bus, C_1, was decreased step by step, to verify the PFC controller stability. Even after the activation of the auxiliary filter to compensate the low-frequency ripple, it was not possible to increase the output power with a very low capacitance at the DC bus. Therefore, Fig. 10 presents the measured efficiency for output power up to 4.5 kW of the PFC converter with the auxiliary filter with C_1 equal to 100 µF instead of 30 µF as proposed in the previous section.

Fig. 11 and Fig. 12 present the voltage on the DC link, the current on the AC side, the voltage and current in the auxiliary capacitors.

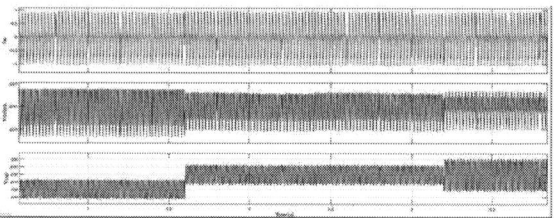

Fig. 9.   Results obtained using the simulation hardware in the loop: AC side current, DC bus voltage V_{dc}, and voltage in the auxiliary capacitors, V_x.

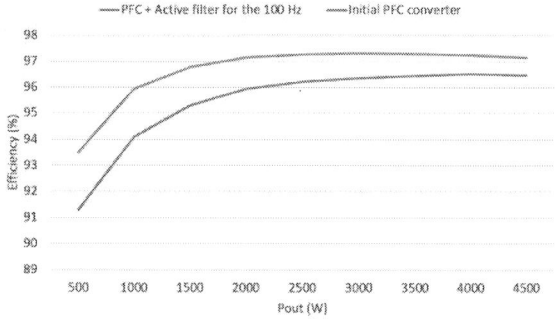

Fig. 10. Measured efficiency comparison between the PFC with the bulk capacitors and the PFC with the Active filter (V_{dc} = 500 V).

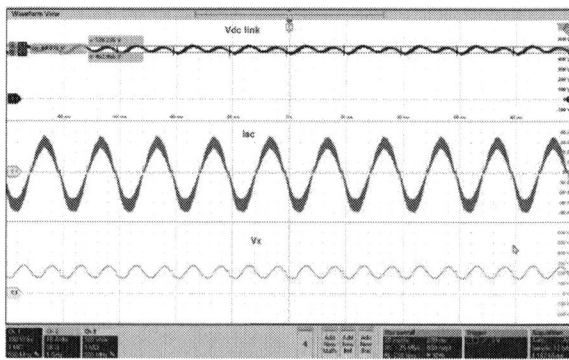

Fig. 11. DC bus voltage, V_{dc}, AC side current, and voltage in the auxiliary capacitors, V_x. P_{out} = 4.5 kW.

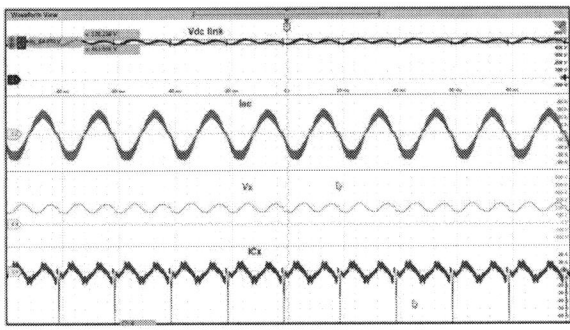

Fig. 12. DC bus voltage, V_{dc}, AC side current, voltage and current in the auxiliary capacitors, V_x and I_{Cx}. P_{out} = 4.5 kW.

6 Conclusion

The active filter solution seems to be promising in terms of power density, but the additional required PCB area related to the new capacitor's arrangement to support the high current, the transistors and the auxiliary components associated with the lower obtained efficiency has an impact on the CO_{2eq} of the converter.

These obtained results are an example to evaluate different designs considering the environment aspects, but it should not be considered as final conclusion due to the variation of the emission factors associated with the components. In fact, the variation of the emission factors of the components used in the converter design needs to be reduced to allow a more accurate estimation of the equivalent CO_{2eq} footprint. In addition, the use of other converter topologies for the pulsating power absorber can also be part of a next study regarding the equivalent CO_{2eq}, considering that we can reduce the chip size of the transistors and the number of capacitors connected in parallel to support the high current. The use of GaN also can lead to a better solution by increasing the switching frequency, however more information provided by manufacturers is necessary concerning not only this semiconductor, but also to allow more accuracy on the emission factors of each component.

Acknowledgment

The author would like to thank Saad Balbard, Lucian Cassand, Abou Diomande and Mohamed Abasse for the help in providing the software, materials, building the mockup versions and for all the support and help during the experimental tests.

References

[1] S. Habib et al., "Contemporary trends in power electronics converters for charging solutions of electric vehicles," in CSEE Journal of Power and Energy Systems, vol. 6, no. 4, pp. 911-929, Dec. 2020, doi: 10.17775/CSEE-JPES.2019.02700.

[2] H. Ma, Y. Tan, L. Du, X. Han, J. Ji: An integrated design of power converters for electric vehicles, Proc. IEEE 26th International Symposium on Industrial Electronics 2017, 2163-5145

[3] P. Papamanolis, F. Krismer and J. W. Kolar, "22 kW EV Battery Charger Allowing Full Power Delivery in 3-Phase as well as I-Phase Operation," 2019 10th International Conference on Power Electronics and ECCE Asia (ICPE 2019 - ECCE Asia), Busan, Korea (South), 2019, pp. 1-8, doi: 10.23919/ICPE2019-EC-CEAsia42246.2019.8797063.

[4] D. Bortis, D. Neumayr, and J. W. Kolar, "ηρ - Pareto optimization and comparative evaluation of inverter concepts considered for the Google Little Box Challenge," in Proc. of 17th IEEE workshop on Control Model. Power Electron. (COMPEL) [Submitted for Publication], 2016

[5] R. Wang et al., "A High Power Density Single-Phase PWM Rectifier With Active Ripple Energy Storage," in IEEE Transactions on Power Electronics, vol. 26, no. 5, pp. 1430-1443, May 2011, doi: 10.1109/TPEL.2010.2090670.

[6] R. Wang, F. Wang, D. Boroyevich and P. Ning, "A high power density single phase PWM rectifier with active ripple energy storage," 2010 Twenty-Fifth Annual IEEE Applied Power Electronics Conference and Exposition (APEC), Palm Springs, CA, USA, 2010, pp. 1378-1383,doi:10.1109/APEC.2010.5433409.

[7] Plugs, socket-outlets, vehicle connectors and vehicle inlets – conductive charging of electric vehicles," IEC 62196.

[8] X. Gong, G. Wang and M. Bhardwaj, "6.6kW Three-Phase Interleaved Totem Pole PFC Design with 98.9% Peak Efficiency for HEV/EV Onboard Charger," 2019 IEEE Applied Power Electronics Conference and Exposition (APEC), Anaheim, CA, USA, 2019, pp. 2029-2034, doi: 10.1109/APEC.2019.8722110.

[9] 2016 IEEE Proceedings of the 8th International Power Electronics and Motion Control Conference (IPEMC 2016-ECCE Asia), Hefei, China,

May 22-25, 2016 Ultra Compact Power Pulsation Buffer for Single-Phase DC/AC Converter Systems D. Neumayr, D. Bortis, J. W. Kolar.

[10] https://sphera.com/product-sustainability-gabi-data-search/

[11] https://support.ecoinvent.org/impact-assessment

[12] F. Musil, C. Harringer, A. Hiesmayr and D. Schoenmayr, "How Life Cycle Analyses are Influencing Power Electronics Converter Design," PCIM Europe 2023; International Exhibition and Conference for Power Electronics, Intelligent Motion, Renewable Energy and Energy Management, Nuremberg, Germany, 2023, pp. 1-9, doi: 10.30420/566091368.

[13] L. Imperiali, D. Menzi, J. W. Kolar, J. Huber, Multi-Objective Minimization of Life-Cycle Environmental Impacts of Three-Phase AC-DC Converter Building Blocks, Proceedings of the 39th Applied Power Electronics Conference and Exposition (APEC 2024), Long Beach, CA, USA February 25-29, 2024.

[14] https://product.tdk.com/en/products/capacitor/ceramic/ceralink/index.html

[15] Emissions Associated with Electric Vehicle Charging: Impact of Electricity Generation Mix, Charging Infrastructure Availability, and Vehicle Type. Joyce McLaren, John Miller, Eric O'Shaughnessy, Eric Wood, and Evan Shapiro. National Renewable Energy Laboratory.

PCIM Europe 2024, 11– 13 June 2024, Nuremberg   DOI: 10.30420/566262326

Analytical and Experimental Validation Common Mode Feedback Loop for a Three-Phase/Level Vienna Rectifier

Daniel San Laureano Igartuburu[1], Gonzalo Moreno Huerta[1], Diego Ochoa Moreno[2], Antonio Lázaro Blanco[3]

[1] Indra Sistemas S.A, Spain

[2] Power Smart Control S.L, Spain

[3] Universidad Carlos III de Madrid, Spain

Corresponding author: Daniel San Laureano Igartuburu, dsan@indra.es

Abstract

The EMI filter can be separated into common mode filter (CM) and differential mode filter (DM). In Vienna rectifier, it is very frequent to place a capacitor that connects the star-point formed by the DM capacitors with the midpoint of the DC-link. This capacitor is known as the feedback capacitor. In this paper the effect of the feedback capacitor on the common mode current is studied when the feedback capacitor is connected to different points of the rectifier. As a result of the study, a mathematical model is obtained, and it is validated on a 1 kW Vienna rectifier prototype.

1   Introduction

From the point of view of all the sensors required for control, high noise emissions in the systems have drawbacks [1]. To analyse the problems, the first step is separated the input current of the three-phase Vienna rectifier into differential mode and common mode, essentially to analyse the contribution level of each mode. To obtain an equivalent common-mode noise model, it is crucial to specify the heat sink connection to establish the common-mode equivalent circuit, thereby causing variations in noise distribution. The possibilities are described in [2]. In this work, the option where the heat sink is connected to the chassis, and consequently to ground, has been considered.

Fig. 1. shows the three-phase Vienna rectifier with parasitic capacitances, and Fig. 2 shows the noise equivalent circuit in DM and CM.

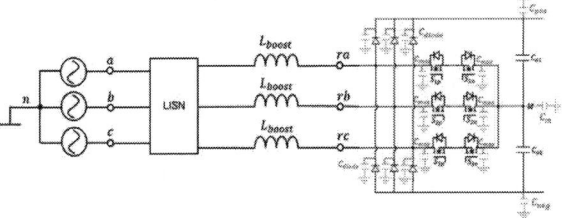

Fig. 1   Three phase Vienna Rectifier with parasitic capacitances.

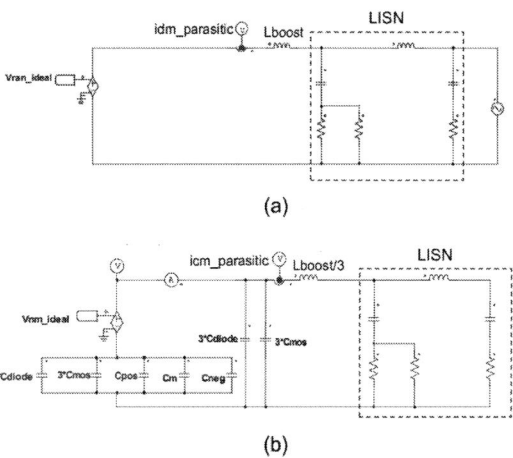

(a)

(b)

Fig. 2   a) One phase equivalent DM circuit. b) Equivalent CM circuit.

The current through the input inductor, L_{boost}, is analysed in detail. As explained in [3] [4], it should be noted that the current flowing through each phase is composed of the differential mode of its phase and can be approximated to one third of the total common mode, in the case of a balanced three-phase system. Applying the Fast Fourier Transform (FFT) to the input current, it is observed that the maximum common mode peak occurs at the switching frequency, in this case 100 kHz. The result in the frequency domain is shown in Fig. 3.

2303

At the switching frequency the differential mode noise is 89 dBµA, while the common mode is 102 dBµA. Applying (1), this translates into 28 mA and 125 mA respectively.

$$10^{(db\mu A - 60)/20} = X \ (mA) \qquad (1)$$

The CM Noise currents should not be circulated through the chassis because they can be introduced into the sensing circuits and make the converter unstable.

To avoid such problems, all components in the circuit should have a high Common Mode Rejection Ratio (CMRR). In addition to these problems, circulating such current through the chassis also exposes the load connected to the active rectifier, so as much current as possible should be avoided. This work presents a technique to reduce this common-mode noise by recirculating some of the common-mode noise through the output capacitors to the EMI filter connected to the three-phase input.

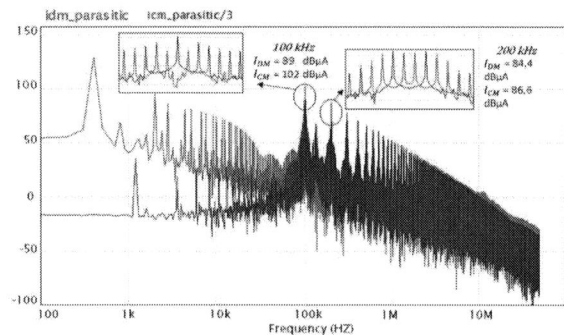

Fig. 3   DM and CM input current in frequency domain.

2 Equivalent Circuit for CM Feedback Loop

Several ways to implement the CM feedback loop, C_{FB}, can be found in the state of the art [1] [2] [5] [6]. Mainly, the designers implement the CM feedback loop from the point M, or it can be separated into P and N [5] [6], which leads to the division of the capacitance into two capacitors of half the capacitance. In this work, two new cases are also analysed: performing the CM feedback loop from P and from N individually. Fig. 4 show all the cases. The main advantage of recirculating through P and N at the same time is that a capacitor of half the capacity can be chosen and will therefore be smaller in size. On the other hand, two capacitors will have to be used.

The choice of one option or the other will depend on the design requirements and constraints. [1]

The CM feedback loop is achieved by adding a capacitor from the midpoint of the output DC-LINK capacitors to the differential mode capacitors of the EMI filter. This modifies the common mode circuit of the system to hinder the common mode current propagation, thereby preventing issues that could arise for both the load connected to the rectifier and ensuring compliance with EMC regulations for the input three-phase. Therefore, the new common mode circuit is shown in Fig. 5.

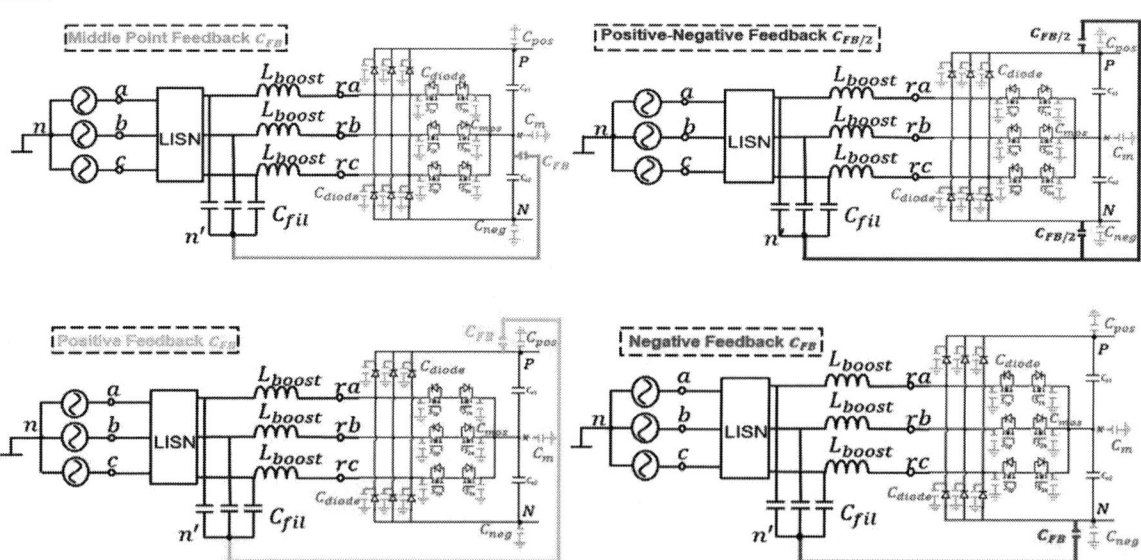

Fig. 4.   Three phase Vienna rectifier with different CM feedback loop

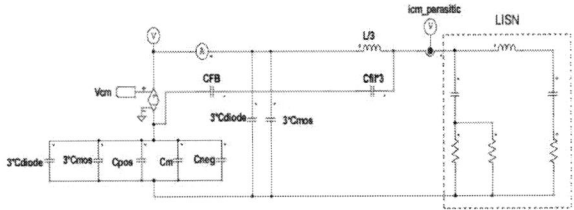

Fig. 5   Equivalent CM circuit with feedback loop.

When choosing the CM capacitors, it is essential to consider safety regulations that restrict the leakage earth current to several milliamperes. This limitation significantly impacts the total capacitance connected to the earth and influences the fundamental aspects of CM filter design [4]. The main advantage of this method is that, in any of the analysed cases, the recirculation capacitor C_{FB} is not connected to ground, similar to the C_{DM} capacitors, and is thus unrestricted in capacitance according to equipment safety regulations.

The selection of the C_{FB} value will depend on the required attenuation. In other works where a common-mode choke is incorporated in series with L_{boost}, it is also utilized to prevent choke saturation. [2] [7]

The equivalent circuit of the CM feedback loop is identical for the four cases studied, given that, at the frequency of common mode noise, the DC-LINK capacitors act as a short circuit. The addition of the differential mode filter to the CM feedback loop introduces resonance, which is a significant aspect involving L_{boost}, the filter, and the feedback capacitors.

To determine the frequency at which this resonance occurs, it is necessary to analyse the circuit depicted in Fig. 5. Additionally, an improvement that can be implemented involves introducing a series resistance with the feedback capacitor to provide damping, as explained in [7]. Both cases will be analysed.

Applying Kirchhoff's second law to the circuit shown in Fig. 6 and using the impedances defined in (2), (3), (4), and (5), equations (6), (7), and (8) are obtained. Note that R_{LISN} represents the Line Impedance Stabilization Network (LISN) in the CM circuit.

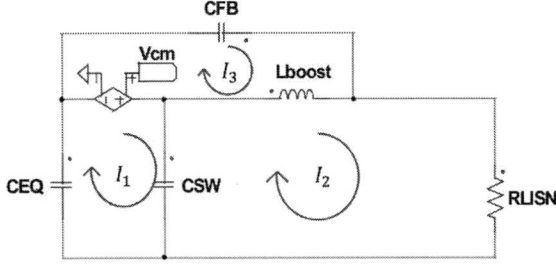

Fig. 6   Equivalent CM simplified circuit with feedback loop.

$$Z_{C_{eq}} = \frac{1}{s \cdot (C_m + C_{pos} + C_{neg} + (3 \cdot C_{diode}) + (3 \cdot C_{mos}))} \quad (2)$$

$$Z_{C_{sw}} = \frac{1}{s \cdot ((3 \cdot C_{diode}) + (3 \cdot C_{mos}))} \quad (3)$$

$$Z_L = \frac{1}{s \cdot L_{boost}} \quad (4)$$

$$Z_{C_{FB}} = \frac{1}{s \cdot ((3 \cdot C_{fil}) + C_{FB})} + R_{FB} \quad (5)$$

$$I_1 = \frac{1}{(Z_{Ceq} + Z_{Csw})} \cdot (V_{cm} + Z_{Csw} \cdot I_2) \quad (6)$$

$$I_3 = \frac{Z_{Csw} + Z_L + R_{lisn}}{Z_L} \cdot I_2 - \frac{Z_{Csw}}{Z_L} \cdot I_1 \quad (7)$$

$$I_2 = \frac{(Z_L + Z_{C_{Fb}})}{Z_L} \cdot I_3 + \frac{1}{Z_L} \cdot V_{cm} \quad (8)$$

Solving equations (6), (7) and (8) gives the transfer function of I_2 / V_{CM}. Note that I_2 is equal to I_{CM} as it is the current flowing into the LISN. The transfer function is shown in (9).

$$\frac{I_2}{v_{cm}} = \frac{\left(\frac{1}{Z_L} - \frac{Z_{Csw} \cdot (Z_L + Z_{C_{Fb}})}{Z_L^2 \cdot (Z_{Ceq} + Z_{Csw})} \right)}{\left[1 - \frac{(Z_L + Z_{C_{Fb}})}{Z_L} \left(\frac{Z_{Csw} + Z_L + R_{lisn}}{Z_L} - \frac{Z_{Csw}^2}{Z_L \cdot (Z_{Ceq} + Z_{Csw})} \right) \right]} \quad (9)$$

The bode plot of the transfer function of (9) is shown in Fig. 7. Consideration has been given to the case of using (5), with or without R_{FB}, to mitigate the amplification of harmonics at the resonant frequency [8]. The analysis includes variations using resistors of one and ten ohms. Including this resistor, as shown in Fig. 7, reduces the resonance peak and prevents the amplification of harmonics that would otherwise occur at this frequency.

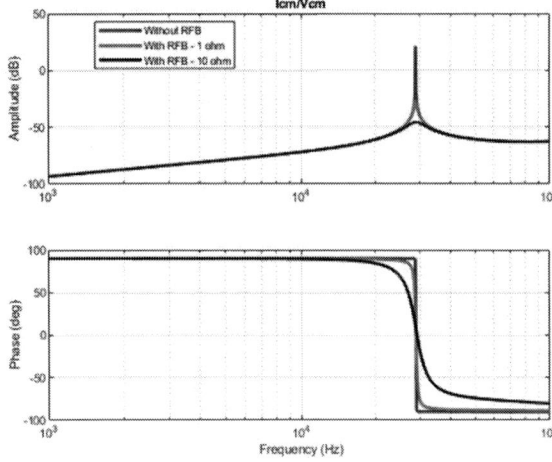

Fig. 7 Bode diagram of the transfer function I_{CM} / V_{CM}.

3 Simulation and Experimental Results

As mentioned in section 2, all equivalent circuits in CM are the same, regardless of the case.

To validate the accuracy of the equivalent circuit, the currents through the equivalent circuits are obtained at the simulation level and compared with the actual currents obtained from the laboratory prototype in all three phases. Validation has been performed using the current of all cases presented in Fig. 4. In this case, the results obtained by recirculating the midpoint are shown in Fig. 8, as all cases obtained identical results.

The FFT of both currents up to 1 MHz has also been performed to verify that the harmonic content is similar in both cases, as shown in Fig. 9. Based on the experimental results obtained, the proposed equivalent circuit is validated.

Texas Instrument prototype of the Vienna rectifier controlled by a TMS320F28370D DSP has been used for experimental validation [9]. The prototype parameters are listed in Table 1.

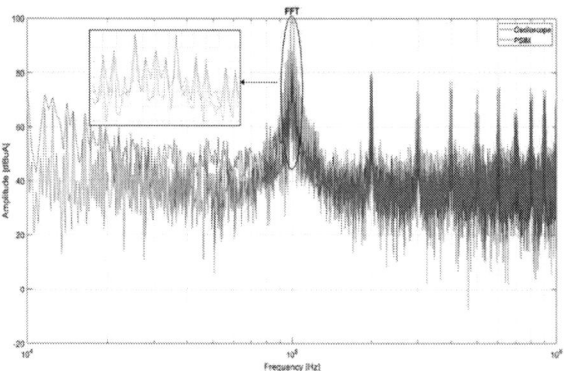

Fig. 9 Comparison of real and simulated current FFT.

To complete the analysis, the FFT is applied to the experimental currents to identify the differences obtained at the switching frequency. The results are shown in Fig. 10 and listed in Table 2. This table shows that in each of the four cases studied, the CM current noise is reduced compared to the case where the feedback loop is not included, but there is a distribution of energy in the wide bands.

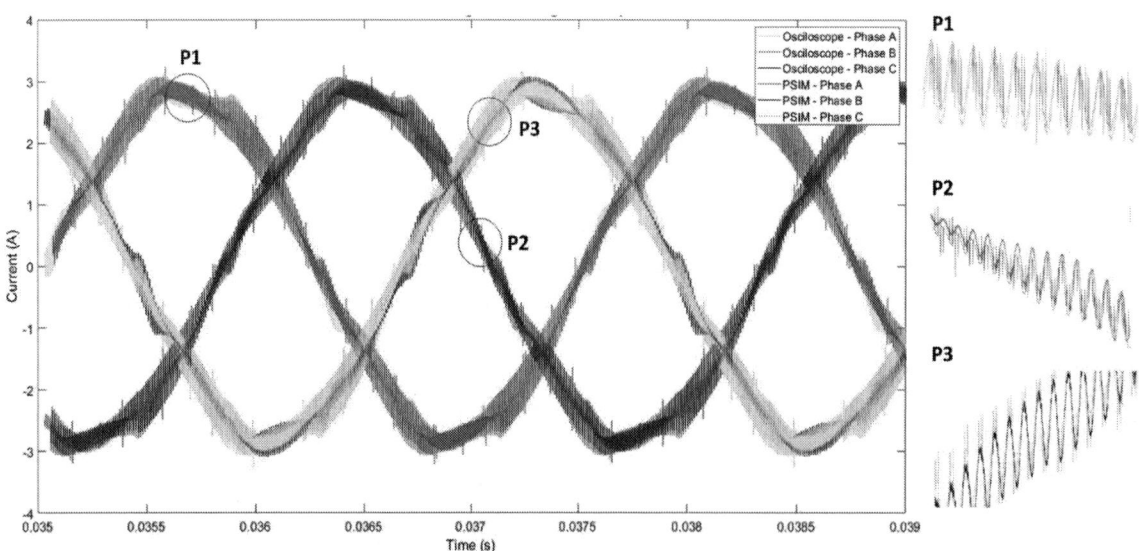

Fig. 8 Experimental result comparing with mathematical model.

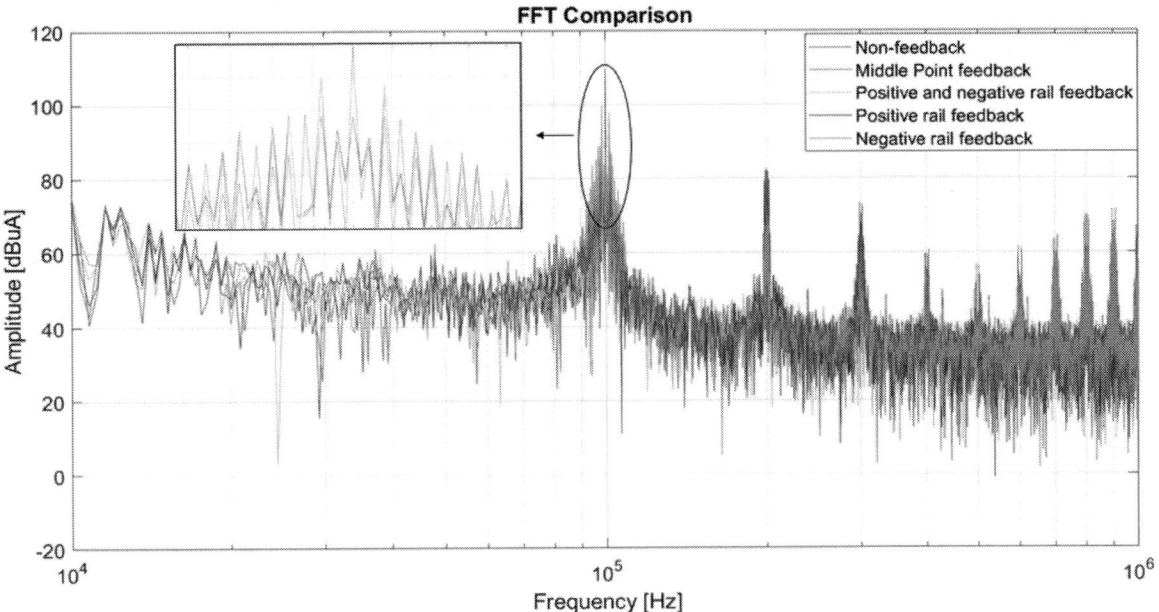

Fig. 10 FFT comparison of the different paths for the CM feedback loop.

Converter parameters	Value
Inductance (L_{boost})	1 mH
Filter capacitors (C_{fil})	330 nF
Feedback capacitor (C_{FB})	94 nF
Switching frequency	100 kHz

Table 1. Values used in the 1 kW prototype.

	Switching frequency (dbμA)	Wide Bands (dbμA)
Non-feedback	109	87
Middle Point Feedback	88	97
Positive and Negative rail Feedback	88	97
Positive rail Feedback	88	97
Negative rail Feedback	88	97

Table 2. Maximum peaks of the FFT

As shown in Table II, the same result is obtained at the switching frequency in all four feedback cases. The peak at the switching frequency is attenuated but amplified in the wide bands. Nevertheless, Fig. 10 shows that the noise in the converter input current is still attenuated at multiples of the switching frequency. At the frequency studied in this work, the result is equivalent for all feedback cases.

4 Conclusions

This paper presents a study of the common mode noise in the Vienna three-phase rectifier. It compares several ways of reducing this noise by connecting the DM capacitors to the input at different points of the rectifier output.

Furthermore, the equivalent circuit with this new CM feedback loop is shown and experimentally validated in the laboratory. Four feedback cases have been studied and it has been observed that at the frequency of study, all obtain similar results. The implementation of one case or another will depend on the layout and available space. The advantage of recirculating through P and N at the same time is that capacitors of half the capacity are used, whereas it is necessary to implement two instead of one.

In all cases, a noise reduction is observed at the switching frequency and at a higher frequency with similar results. In addition, the transfer function to damp the resonance produced by implementing this feedback capacitor has been obtained.

References

[1] J. W. Kolar, U. Drofenik, J. Minibock and H. Ertl, "A new concept for minimizing high-frequency common-mode EMI of three-phase PWM rectifier systems keeping high utilization of the output voltage," APEC 2000. Fifteenth Annual IEEE Applied Power Electronics Conference and Exposition (Cat. No.00CH37058), New Orleans, LA, USA, 2000, pp. 519-527 vol.1, doi: 10.1109/APEC.2000.826153.

[2] M. Hartmann, H. Ertl and J. W. Kolar, "EMI Filter Design for a 1 MHz, 10 kW Three-Phase/Level PWM Rectifier," in IEEE Transactions on Power Electronics, vol. 26, no. 4, pp. 1192-1204, April 2011, doi: 10.1109/TPEL.2010.2070520

[3] G. Moreno Huerta, D. San Laureano Igartuburu, D. O. Moreno and A. Lazaro Blanco, "Experimental Validation of Differential and Common Mode Equivalent Circuit of a Three-Phase/Level Vienna Rectifier," PCIM Europe 2023; International Exhibition and Conference for Power Electronics, Intelligent Motion, Renewable Energy and Energy Management, Nuremberg, Germany, 2023, pp. 1-5, doi: 10.30420/566091337.

[4] Dey S, Mallik A, "A Comprehensive Review of EMI Filter Network Architectures: Synthesis, Optimization and Comparison," *Electronics.* 2021; 10(16):1919.

[5] T. Nussbaumer, M. L. Heldwein and J. W. Kolar, "Common mode EMC input filter design for a three-phase buck-type PWM rectifier system," Twenty-First Annual IEEE Applied Power Electronics Conference and Exposition, 2006. APEC '06., Dallas, TX, USA, 2006, pp. 7 pp.-, doi: 10.1109/APEC.2006.1620757.

[6] M. L. Heldwein, "EMC Filtering of Three-Phase PWM Converters," Ph.D. dissertation, Swiss Federal Institute of Technology (ETH Zurich), 2008.

[7] S. Chen, W. Yu and D. Meyer, "Design and Implementation of Forced Air-cooled, 140kHz, 20kW SiC MOSFET based Vienna PFC," 2019 IEEE Applied Power Electronics Conference and Exposition (APEC), Anaheim, CA, USA, 2019, pp. 1196-1203, doi: 10.1109/APEC.2019.8721979

[8] K. Jayaraman and M. Kumar, "Design of Passive Common-Mode Attenuation Methods for Inverter-Fed Induction Motor Drive With Reduced Common-Mode Voltage PWM Technique," in IEEE Transactions on Power Electronics, vol. 35, no. 3, pp. 2861-2870, March 2020, doi: 10.1109/TPEL.2019.2930825.

[9] S. Texas Instruments April 2020. Design Guide:TIDM-1000 Vienna Rectifier-Based, Three-Phase Power Factor Correction (PFC) Reference Design Using C2000™ MCU. Accessed: March. 2024 [Online]. Available: https://www.ti.com/tool/TIEVM-VIENNARECT

Robustness of Frequency-Domain Terminal Modeling of Electromagnetic Interferences in Static Converters

Mehyeddine SINGER, Arnaud VIDET[ID], Nadir IDIR[ID]

Univ. Lille, Arts et Metiers Institute of Technology, Centrale Lille, Junia
ULR 2697 - L2EP Lille F-59000, Lille, France

Corresponding author and speaker: Mehyeddine Singer, mehyeddine.singer.etu@univ-lille.fr

Abstract

Power electronics converters generate conducted electromagnetic interference (EMI) over a wide frequency range. In order to prevent pollution of the electrical network and thus comply with electromagnetic compatibility (EMC) standards, it is necessary to install filters at the input side of converters. The optimal design of these filters requires modeling the entire energy conversion system and using the proposed models in circuit simulations (SPICE, etc.). Depending on whether the conversion system is already implemented or in the design stage, there are several modeling methods. When the system is already implemented, the method that treats the converter with its load as a black-box model, called "Terminal Modeling" is particularly suitable in this case. In this paper, we highlight its limitations and propose a gray-box model to enhance its accuracy and robustness up to 100 MHz.

1 Introduction

The EMC standards require conducted EMI levels to be below specified limit values. To comply with these standards, an EMC filter placed upstream of the converter reduces conducted EMI returned to the power grid. Typically, such a filter is expensive, heavy, and bulky. To optimize the sizing of this filter, it is necessary to perform simulations using circuit simulation software such as SPICE. If the converter is in the design phase or if it has already been implemented, there are several modeling methods available.

For a converter in the design phase, the first step is to characterize the active and passive components through measurements or using modeling tools (finite element method, etc.). The components thus characterized will be modeled using equivalent electrical circuits. The next step is to simulate the entire conversion chain in the time domain. This simulation requires long computation times associated with convergence issues. For this reason, frequency-domain simulation is of interest using the impedances $Z(f)$ of components associated with current or voltage generators. The accuracy of frequency-domain models depends on the identification method and the type of final application.

In [1], the converter is modeled by fixed impedances and equivalent generators, which makes the model very simple and thus reduces computation time. However, the non-linearity of the semiconductor state change in the switching cell, and consequently the change in propagation paths, are not considered. Furthermore, this model separates the propagation modes of conducted EMI, common mode (CM) and differential mode (DM), by neglecting mode transformation, which affects the model accuracy. In [2], [3], the "Multi-Topology Equivalent Sources" (MTES) method considers the non-linearity of the switching cell and mode transformation, and provides acceptable accuracy up to $50MHz$. However, this method becomes more complex when the number of switching cells is large.

For an already implemented converter, modeling in the frequency domain is directly performed based on measurements on the conversion system without going through the characterization phase of each component. In [4], the converter without its load is considered as a black box. It is modeled only in CM, neglecting DM perturbations and mode transformation. The results obtained show that this model is not sufficiently accurate. In [5], [6], [7], [8], the "Terminal Modeling" (TM) method offers significant simplification by considering the converter and its load as a single black-box model, without separating propagation modes. This method yields satisfactory results when the converter operates near the identification point of model parameters. However, this model has limitations that we highlight in this paper. Therefore, we propose to evaluate and improve the robustness and accuracy of this method. For this study, we consider that the conversion system has already been implemented. The test setup consists of a Buck converter using SiC components feeding into an $R - L$ load as shown in figure 1.a. The results of conducted perturbation currents obtained using the "TM" method are compared to circuit-type simulation in the time domain. This allows us to fully understand the propagation paths of the system. In the following sections, we will present the drawbacks of the "TM" method, and then we will detail the proposed gray-box model.

2 Identification of the "TM" model of the converter

The "TM" model of the considered conversion system consists of two current sources and three impedances as shown in figure 1.b. To identify these five parameters, a system of five equations to solve is necessary. For this reason, three tests are carried out in time-domain simulation (using SIMetrix software), by inserting a shunt impedance Z_{shunt} consisting of a $2\mu F$ capacitor in series with a 1Ω resistor in three positions as depicted in figure 1.b. For each test, the voltages V_{pg} and V_{ng} are measured using an oscilloscope [5]. The measured voltages, from one test to another, differ due to the change in propagation paths induced by the position of Z_{shunt}. Calculating the spectra (FFT) of these voltages allows us to solve the equations system and identify the parameters of the "TM" model. These tests are conducted for an operating point of the Buck converter corresponding to a DC bus voltage $E = 300V$, a load current $I_c = 6A$, and a duty cycle $\alpha = 0.3$.

The first test is conducted without the insertion of Z_{shunt}, yielding equations (1) and (2). The second test is carried out with the insertion of Z_{shunt} between terminals (p, g), resulting in equations (3) and (4). The third test is performed with the insertion of two identical impedances Z_{shunt} between terminals (p, g) and (n, g), providing equations (5) and (6). For these three tests, we

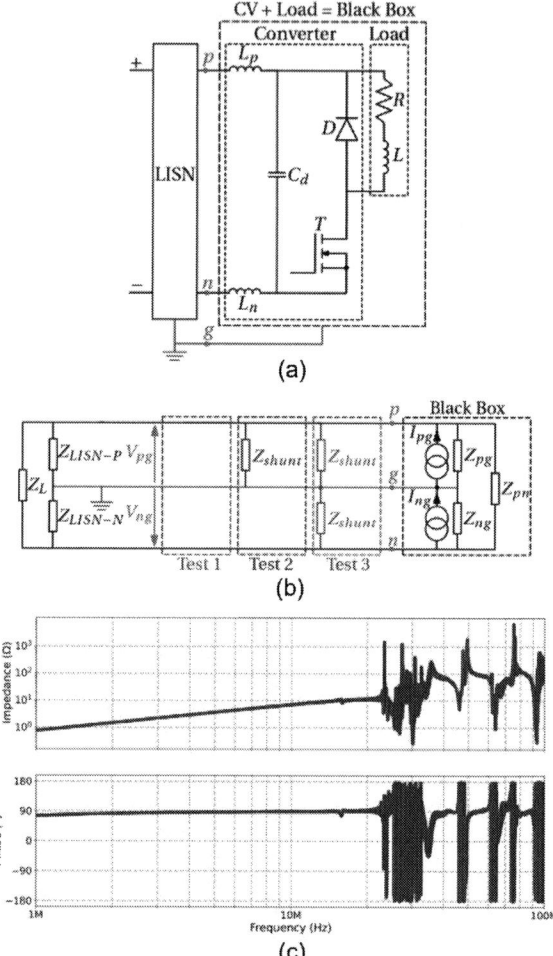

Fig.1 (a) DC/DC conversion chain. (b) "TM" model and test configurations for its identification. (c) Evolution of impedance Z_{pn}.

obtain six different voltages (two voltages for each test). Therefore, among the six equations obtained, five are chosen, corresponding to equations (1-4) and (6). Solving them allows the determination of complex values for each frequency of the two current generators and the three impedances of the black-box model shown in figure 1b.

Significant variations in the amplitudes and phases of the black-box model impedances Z_{pg}, Z_{ng}, and Z_{pn} are observed, especially beyond $20MHz$ (Fig. 1.c), indicating a non-physical behavior of these impedances. In Section 3, we will verify the validity of this method when applying the model in a new configuration, in our case, after adding a cable at the input of the converter.

$$V_{pg1} = \left(I_{pg} - \frac{V_{pg1} - V_{ng1}}{Z_{pn} \| Z_L} \right) . Z_{pg} \| Z_{LISN-P} \qquad (1)$$

$$V_{ng1} = \left(-I_{ng} + \frac{V_{pg1} - V_{ng1}}{Z_{pn} \| Z_L} \right) . Z_{ng} \| Z_{LISN-N} \qquad (2)$$

$$V_{pg2} = \left(I_{pg} - \frac{V_{pg2} - V_{ng2}}{Z_{pn} \| Z_L} \right) . Z_{pg} \| Z_{LISN-P} \| Z_{shunt} \qquad (3)$$

$$V_{ng2} = \left(-I_{ng} + \frac{V_{pg2} - V_{ng2}}{Z_{pn} \| Z_L} \right) . Z_{ng} \| Z_{LISN-N} \qquad (4)$$

$$V_{pg3} = \left(I_{pg} - \frac{V_{pg3} - V_{ng3}}{Z_{pn} \| Z_L} \right) . Z_{pg} \| Z_{LISN-P} \| Z_{shunt} \qquad (5)$$

$$V_{ng3} = \left(-I_{ng} + \frac{V_{pg3} - V_{ng3}}{Z_{pn} \| Z_L} \right) . Z_{ng} \| Z_{LISN-N} \| Z_{shunt} \qquad (6)$$

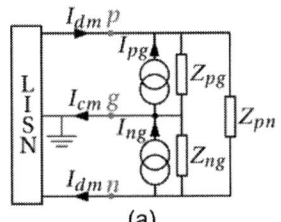

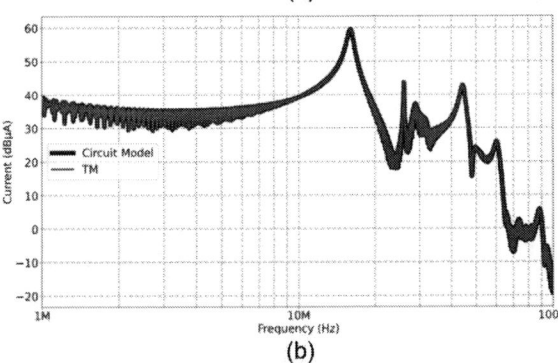

Fig.2 (a) Application of the "TM". (b) Spectral envelopes of the DM currents obtained with the circuit model and "TM".

Initially, we compare the simulation results obtained from the "TM" model presented in figure 2.a with those obtained using the circuit model with SIMetrix software. These results demonstrate, as expected, that the "TM" model perfectly reproduces the spectra of the circuit model with high precision for the same identification conditions of test 1 (Fig. 2.b).

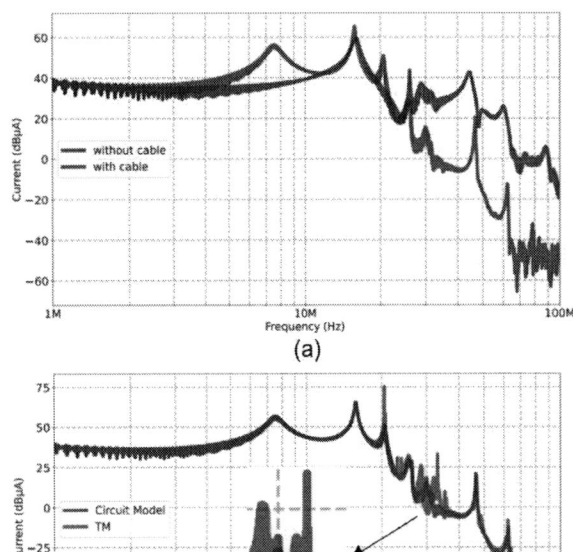

Fig.3 Spectral envelopes of DM currents: (a) obtained with the circuit model with and without inserted cable. (b) obtained with the circuit model and "TM" with the inserted cable.

3 Application of the "TM" model in a new configuration

To study the robustness of the "TM" model, we apply this model in a new configuration by inserting a cable between the LISN and the conversion system. This significantly modifies the propagation paths of conducted perturbations. A time-domain simulation (with the circuit model) including the cable shows a considerable difference in the spectra of DM and CM currents, as depicted in figure 3.a.

To demonstrate the robustness of the "TM" model obtained in the presence of a cable at the system input, we combine the "TM" model with the cable model. The simulation results obtained are compared to those obtained using circuit-type simulation. This comparison shows that the conducted perturbations obtained with the "TM" model generally follow the spectra of the circuit model. However, some large variations are detected (beyond $20MHz$), especially the amplitude deviation at $33MHz$ reaching $36dB\mu A$ in DM (Fig. 3.b).

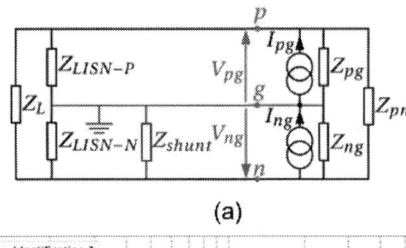

(a)

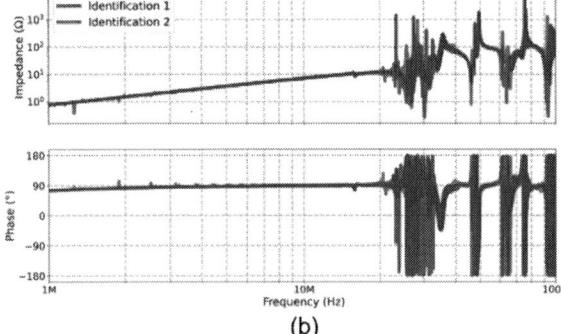

(b)

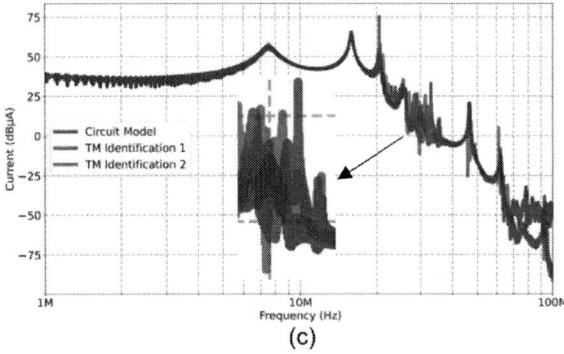

(c)

Fig.4 (a) Configuration of Z_{shunt} insertion between terminals (n, g). (b) Z_{pn} impedances of "TM" identifications 1 and 2. (c) Spectral envelopes of DM currents obtained with "TM" of identifications 1 and 2, and the circuit model with the inserted cable.

This could be attributed to significant impedance variations at the system input compared to the Z_{LISN} impedance, and especially to the resonance frequencies of the input cable.

4 The influence of identification conditions on the "TM" model accuracy

4.1 Study of the sensitivity of the Z_{shunt} position

To study the influence of the Z_{shunt} position on the "TM" model accuracy, the third test is conducted (without the cable) by inserting not two Z_{shunt} as in figure 1b, but only one between terminals (n, g),

$$V_{pg4} = \left(I_{pg} - \frac{V_{pg4} - V_{ng4}}{Z_{pn}||Z_L}\right).Z_{pg}||Z_{LISN-P} \quad (7)$$

$$V_{ng4} = \left(-I_{ng} + \frac{V_{pg4} - V_{ng4}}{Z_{pn}||Z_L}\right).Z_{ng}||Z_{LISN-N}||Z_{shunt} \quad (8)$$

giving us equations (1-4) and (8). This procedure is subsequently referred to as identification 2 (Fig. 4.a). Comparison of the results shows a modification of the model obtained with this configuration (Fig. 4.b). Thus, we clearly highlight the low robustness of the "TM" model as its parameters and accuracy depend on the Z_{shunt} position during the identification phase. Applying the same approach in the presence of a cable also shows a variation in peak amplitudes (Fig. 4.c).

4.2 Voltage measurement constraints

The voltages V_{pg} and V_{ng} are measured using an oscilloscope, enabling us to obtain the amplitude and phase of the spectra. However, these measurements come with practical constraints such as the jitter phenomenon in the transistor gate drive control chain and the propagation phenomenon in the passive voltage probes, which affects the synchronization of voltages during the measurements of the three tests.

A signal temporally shifted relative to an initial signal presents the same spectrum amplitude but with a different phase. To investigate the influence of such a shift on the "TM" model, in simulation, the voltage measured between the terminals (n, g) of the second test V_{ng2} has been shifted by $10ns$, and a new "TM" model with the shifted V_{ng2} has been identified. The results show a change in the model response (Fig. 5a). Applying this model with a cable also shows a lack of precision (Fig. 5.b).

5 Enhaced "TM" : gray-box model

We have just demonstrated the weaknesses of the "TM" model by considering the converter with its load as a black-box model. In this section, we propose to improve the "TM" model by replacing the black-box model with a gray-box model (Fig. 6.a) through a detailed modeling of the switching cell. The structure of the converter and the duty cycle are two essential parameters for identifying the gray-box model. Figure 6.b shows the converter with the switching cell associated with its various parasitic elements of the power loops. In this section, the duty cycle is set to $\alpha = 0.3$.

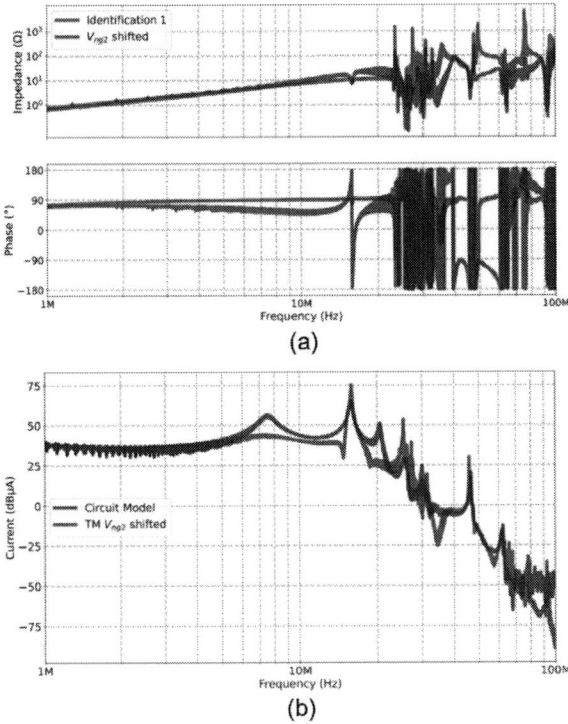

Fig.5 (a) Impedances Z_{pn} of "TM" with and without shifted V_{ng2}. (b) Spectral envelopes of DM currents obtained with "TM" with V_{ng2} shifted and circuit model with inserted cable.

The identification method involves measuring the impedances of the system under voltage, between the terminals (p, g), (n, g), and (p, n), based on the state of the power components, which allows for the consideration of the variation in semiconductor impedance values depending on their ON or OFF state. When the MOSFET is conducting, V_g is equal to $15V$ (the diode is blocked). When the MOSFET is blocked, V_g is equal to $0V$ (the diode is conducting). In the case of the MOSFET being blocked, a DC current source of $6A$ is connected in parallel to the load, which serves to turn on the diode. This current value corresponds to the average value of the current passing through the load for $\alpha = 0.3$.

To measure an impedance between two terminals in simulation, a voltage generator with an amplitude of $1V$ is inserted between these two terminals, while varying the frequency from 1 to $100MHz$. Thus, the measured impedance $Z(f)$ is the ratio between the voltage and the current generated by the source. It should be noted that to measure the impedance between the terminals (p, n), a capacitor is placed in series with the

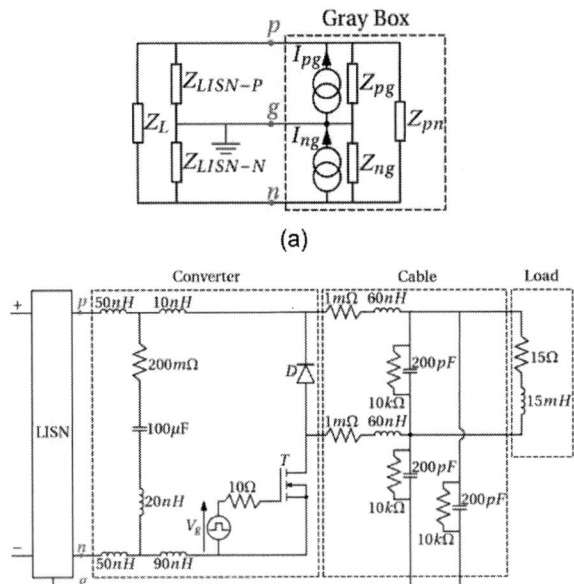

Fig.6 (a) "TM" gray-box model. (b) Buck converter in a conversion system.

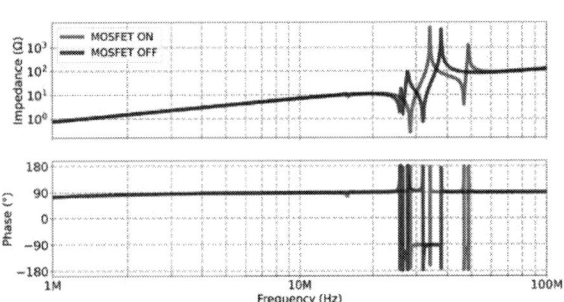

Fig.7 Impedance Z_{pn} of the "TM" model for a stable state of the switching cell.

sinusoidal voltage generator to block the DC component from the power source.

From these measured impedances, we deduce the individual node-to-node impedances in figure 6a. These are calculated, and individually, they may exhibit non-physical behavior due to phases exceeding $\pm 90°$ (Fig. 7). However, the apparent impedances of the model between any two terminals, so called "input impedances", remain passive, as will be shown later. Every input impedance is calculated as a function of the individual node-to-node impedances.

To obtain the impedances of the gray-box model based on the states of the semiconductors over a switching period, a linear interpolation is performed, based on the duty cycle α, of the two

PCIM Europe 2024, 11– 13 June 2024, Nuremberg   DOI: 10.30420/566262327

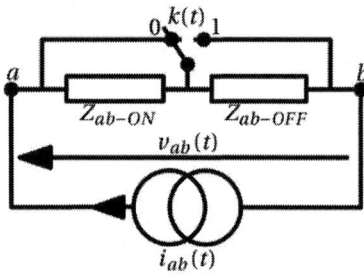

Fig.8 Input impedance seen between two terminals of the model for each frequency over a switching period.

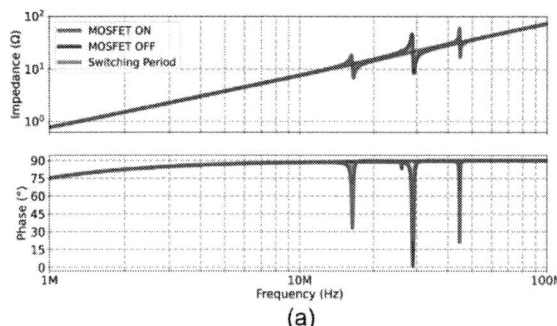

(a)

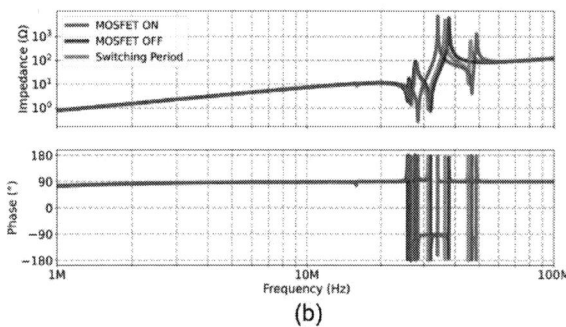

(b)

Fig.9 (a) Input Impedances of the "TM" model between terminals (p, n). (b) Impedance Z_{pn} of the "TM" model.

input impedances of the gray-box model, between the same two terminals, each corresponding to a state of the MOSFET.

The linear interpolation of input impedances between any two terminals (a, b) of the model is justified by the figure 8 which shows that for each frequency, the sinusoidal current $i_{ab}(t)$ is imposed by the current source of the model. Therefore, as seen from these two terminals, a change in the state of the switch $k(t)$ according to the duty cycle α results in a change in the input impedances of the model with Z_{ab-ON} and Z_{ab-OFF} according to the state of the switching cell, and consequently causes a change in the voltage $v_{ab}(t)$ across the terminals (a, b).

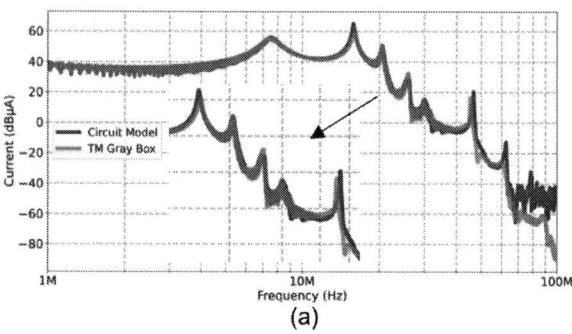

(a)

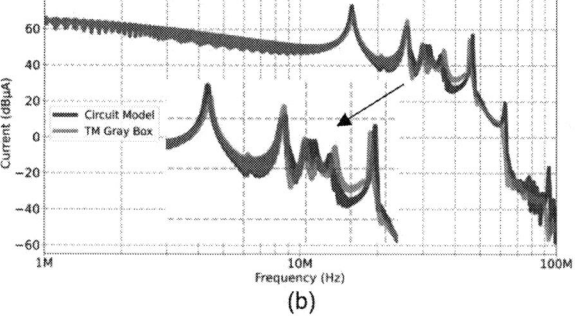

(b)

Fig.10 Envelope of the current spectra obtained with the circuit model and gray-box "TM" model: (a) DM, (b) CM.

The results show that the input impedances of the model, as seen from the terminals, when the MOSFET is conducting or blocked, remain physical impedances. Therefore, the interpolation yields a physical impedance that reproduces the resonance frequencies due to the MOSFET ON and OFF states, thus accounting for the distinct paths associated with the different semiconductor states (Fig 9.a). Therefore, the input impedances of the gray-box model exhibit continuous amplitudes and phases and reflect a physical impedance system, as seen from the terminals, while the individual impedances of the model may exhibit a non-physical nature in certain frequency bands as previously discussed (Fig 9.b).

After identifying the impedances of the gray-box model, the current sources of the model are identified by solving a simple equation system based on voltage measurements between the terminals (p, g) and (n, g).

To test the robustness of the model, it is applied with the insertion of the cable at the input. The simulation results show high accuracy of the gray-box model up to $60MHz$ and a significant improvement compared to the black-box model (Fig. 10). The spectra obtained by the gray-box

2314

model precisely follow the spectra obtained by the circuit model, correctly reproducing the peaks.

6 Conclusion

In this study, we have highlighted the weaknesses of the "TM" method based on the black-box model. The drawbacks of this method mainly concern the influence of the external impedance Z_{shunt} position, during the identification phase, on the accuracy and robustness of the obtained model, as well as the limited validity of the latter around the characterization point. In order to improve the "TM" model, we have proposed a gray-box model that more finely models the switching cell.

The advantage of the proposed model is that the impedances effectively represent the impedances of the system over a switching period, and the identification of current sources is achieved through a single test without the insertion of external impedances, which has a major influence on the accuracy and robustness of the obtained model. The results of the gray-box model show good accuracy up to $60MHz$ in an application with the presence of an external cable, demonstrating a notable improvement in the accuracy and robustness of this model.

In future work, the gray-box method will be applied to an experimental bench that includes a Buck converter, where the system impedances under voltage can be extracted using the two-current-probe method [9] or three-current-probe method [10].

References

[1] S. Zhang, Q. Lin, Y. Noge, M. Shoyama, E. Takegami et G. M. Dousoky, "Developed Common Mode Noise Modeling Approach for DC-DC Flyback Converters", IEEE Letters on Electromagnetic Compatibility Practice and Applications, vol. 2, pp. 147-151, 2020.

[2] C. Marlier, A. Videt, N. Idir, H. Moussa et R. Meuret, "Modeling of switching transients for frequency-domain EMC analysis of power converters", 2012 15th International Power Electronics and Motion Control Conference (EPE/PEMC), 2012.

[3] C. Marlier, A. Videt, N. Idir, H. Moussa et R. Meuret, "Hybrid time-frequency EMI noise sources modeling method", 2013 15th European Conference on Power Electronics and Applications (EPE), 2013.

[4] B. Sun, R. Burgos et D. Boroyevich, "Common-Mode EMI Unterminated Behavioral Model of Wide-Bandgap-Based Power Converters Operating at High Switching Frequency", IEEE Journal of Emerging and Selected Topics in Power Electronics, vol. 7, pp. 2561-2570, 2019.

[5] H. Bishnoi, A. C. Baisden, P. Mattavelli et D. Boroyevich, "Analysis of EMI Terminal Modeling of Switched Power Converters", IEEE Transactions on Power Electronics, vol. 27, pp. 3924-3933, 2012.

[6] M. Foissac, J.-L. Schanen et C. Vollaire, ""Black box" EMC model for power electronics converter", 2009 IEEE Energy Conversion Congress and Exposition, 2009.

[7] L. Wan, A. Beshir, X. Wu, X. Liu, F. Grassi, G. Spadacini et S. Pignari, "Assessment of Validity Conditions for Black-Box EMI Modelling of DC/DC Converters", 2021 IEEE International Joint EMC/SI/PI and EMC Europe Symposium, 2021.

[8] B. Kerrouche, M. Bensetti et A. Zaoui, "New EMI Model With the Same Input Impedances as Converter", IEEE Transactions on Electromagnetic Compatibility, vol. 61, pp. 1072-1081, 2019.

[9] V. Tarateeraseth, B. Hu, K. Y. See et F. G. Canavero, "Accurate Extraction of Noise Source Impedance of an SMPS Under Operating Conditions", IEEE Transactions on Power Electronics, vol. 25, pp. 111-117, 2010.

[10] K. Li, A. Videt et N. Idir, "Multiprobe Measurement Method for Voltage-Dependent Capacitances of Power Semiconductor Devices in High Voltage", IEEE Transactions on Power Electronics, vol. 28, pp. 5414-5422, 2013.

PCIM Europe 2024, 11– 13 June 2024, Nuremberg   DOI: 10.30420/566262328

Study of EMI Behavior of a 2-Level GaN-Inverter – Simulation and Measurement

Benedikt Kohlhepp[1] , Yassin Fal[2], Julian Dobusch[2], Daniel Kübrich[2], Thomas Dürbaum[2]

[1] Department of Power Electronics, Technical University of Berlin, Berlin, Germany
[2] Institute of Optoelectronics, Friedrich-Alexander Universität Erlangen-Nürnberg, Erlangen, Germany

Corresponding author: Benedikt Kohlhepp, benedikt.kohlhepp@fau.de
Speaker: Benedikt Kohlhepp, benedikt.kohlhepp@fau.de

Abstract

Wide-bandgap semiconductors are more and more applied in power electronics converters, as these components feature lower parasitics and as a consequence, lower losses compared to their silicon counterparts. The advantage from the power electronics perspective mainly results from the higher du/dt and di/dt during switching operation and the lower on-state resistance. Unfortunately, faster switching transients may result in increased problems regarding the electromagnetic interference (EMI). This fact makes studies regarding the EMI inevitable for power electronic circuits applying e.g. GaN power semiconductors. One way is to study the EMI behavior of converters with a prototype at the workbench. As this method is very time consuming and costly, carrying out EMI simulations during the development phase of the converter is more efficient. Thus, this paper studies a 2-level GaN-inverter regarding conducted emissions. The frequency domain simulations, which are carried out for a simple DC/DC-converter as well as for an inverter with load are validated using measurements with a standardized test setup for conducted emission. The simulations match the results gained by measurement, which proves the validity of the simulation approach.

1 Introduction

Nowadays, 2-level inverters are widely used in industry and household. The inverters generate sinusoidal output voltages from a DC-link or vice versa. Recently, wide-bandgap semiconductors, e.g. GaN-HEMTs, replace conventional ones in these applications [1] [2]. As those fast switching devices lead to lower switching losses, they are attractive candidates for high efficient power converters. Unfortunately, high di/dt and du/dt resulting from these new semiconductor devices may also be a problem regarding the electromagnetic interference (EMI) [3] [4] [5] [6]. Especially the higher du/dt causes higher common mode (CM) currents, which worsens EMI behavior of the converter and thus, require higher filter effort. Furthermore, higher power density requires higher switching frequencies in order to shrink the size of passives [7]. Both, the increased switching speed (du/dt and di/dt) and higher switching frequencies call for proper studies and improved EMI

measures before applying these modern semiconductors widely.

Thus, the importance of simulating the EMI behavior of power electronic systems to predict the high frequency noise generated by the converters rises. Simulating the EMI characteristics delivers the generated noise already during the design phase, saving time and cost during development of power converters regarding EMI measures (e.g. filter design) and final EMI tests by avoiding design iterations due to EMI trouble. Modeling approaches can be divided into two categories [8] [9]. Behavioral modeling bases its results on measurements of a complete (or bigger part of a) setup [10], while detailed modeling considers many components of the circuit under study [11] [12] [13]. This paper follows the second approach in order to be flexible enough to examine the influence of circuit changes. The presented model extends the approach of [11] in order to incorporate the influence of the output filter while being more general then [14].

2316

Thus, this paper studies a 2-level inverter with load that uses GaN-HEMTs as power switches and operates at high frequency. Investigating a single phase inverter with load is sufficient for the purpose of this paper. A custom simulation model implemented in Julia [15], which uses linear subsystems and also models the nonlinear behavior of the half-bridge, is exploited to predict the conducted emissions from 9 kHz to 30 MHz. Measurements using a standardized test environment with line impedance stabilization network (LISN) and EMI receiver gives measurement results. A comparison to the simulated results allows to verify the model.

Section 2 introduces the circuit topology and modulation schemes, which underlies all the measurements conducted. After that, the section 3 and 4 describe the EMI modeling approach as well as the EMI measurement setup. The first EMI study, which compares simulated results with the measurement, uses a DC/DC-converter to start with a simple circuit. Afterwards, a single phase inverter with load is studied.

2 Circuit Topology and Modulation Scheme

This section describes the circuit topology and the modulation scheme which are used for the EMI studies. First, this paper studies a DC/DC-converter without load in order to have a simple setup to evaluate the EMI behavior. In a second step, the sinusoidal-triangular modulation is used for driving the GaN-inverter, which uses the same printed circuit board as the DC/DC-converter. Furthermore, for the study of the inverter, a load, which represents an electrical machine, is introduced and the EMI characteristics of this system, comprising inverter and load, are studied.

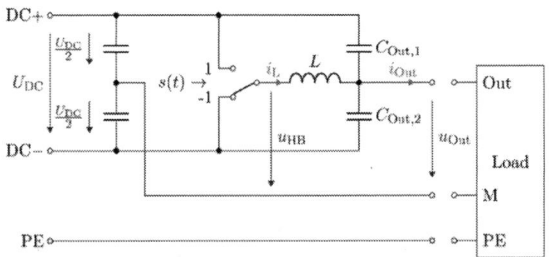

Fig. 1: Circuit topology of the DC/DC-converter and the DC/AC-inverter.

2.1 DC/DC-Converter

The schematic of the DC/DC-converter is shown in Fig. 1. It consists of a DC-link, a half-bridge ($s(t)$ and the switch), and an LC-filter (L, $C_{Out,1}$ and

$C_{Out,2}$) at the output. The half-bridge, which is built by two GaN-HEMTs (100 V-class, EPC2045 [16]) operates with a fixed switching frequency of 300 kHz and a duty-cycle of 0.5. The output filter uses a custom made inductor with 2.4 µH and ceramic capacitors of type X7R with 3x4.7 µF for the upper and lower capacitor ($C_{Out,1}$ and $C_{Out,2}$). A laboratory power supply feeds the input voltage of 48 V to the converter. First, the DC/DC-converter operates without a load connected, resulting in zero output power of the converter. The power supply only feeds the losses to the circuit. From this operation mode, the waveforms in steady-state shown in Fig. 2 result. The half-bridge voltage u_{HB} has a rectangular shape periodic with the switching frequency. From the half-bridge switching action, a triangular inductor current i_L arises, which allows for zero voltage switching of both half-bridge switches [1]. This explains the smooth switching transitions (zoomed diagram in Fig. 2, rising edge of the half-bridge voltage) with a rise and fall time below 4 ns. The component values and operating parameters are summarized in Tab. 1.

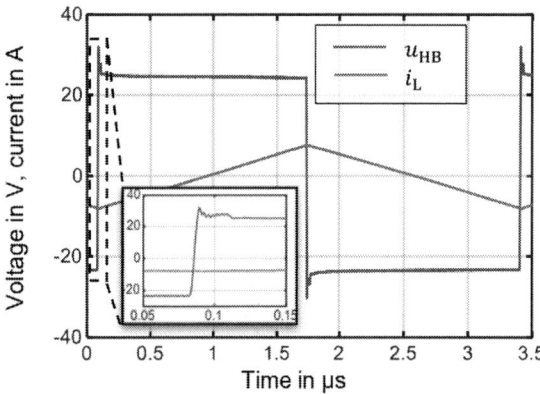

Fig. 2: Waveforms of the DC/DC-converter in steady state and zoomed diagram of the rising edge of the half-bridge voltage.

Tab. 1: Overview of the component values and operating conditions.

Parameter	Value	Parameter	Value
L	2.4 µH	U_{In}	48 V
$C_{Out,1}, C_{Out,2}$	3x4.7 µF	$\hat{u}_{Out}$ (Inverter)	17 V
L_F	33 µH	$\hat{\imath}_{Out}$ (Inverter)	11 A
C_F	680 µF	f_{Out}	100 Hz

2.2 Inverter with Load

After studying the DC/DC-converter, the converter operating in DC/AC mode (GaN-inverter) with load shall be considered. Now, the half-bridge operates using sinusoidal-triangular modulation with a switching frequency of 300 kHz. The duty-cycle varies due to the intended sinusoidal output voltage. Fig. 3 shows the carrier signal ($T(t)$, black), the sinusoidal control signal ($m(t)$, blue) as well as the resulting PWM signal ($s(t)$, red) required to operate the inverter.

In conjunction with the load, the waveforms for a complete sinusoidal period shown in Fig. 4 result. The output waveforms feature sinusoidal shape and show only a small ripple (voltage as well as current). The fundamental frequency f_{Out} (frequency of the output current i_{Out}) is 100 Hz. Studying a single phase inverter is sufficient in order to obtain the relevant disturbance. With the same approach, studying a three phase inverter with load is possible requiring much more effort while providing only a little more insights.

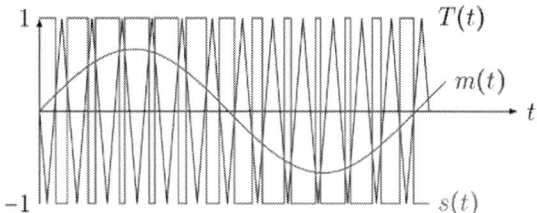

Fig. 3: Modulation scheme of the inverter: Carrier T, sinusoidal control signal m, and PWM signal s.

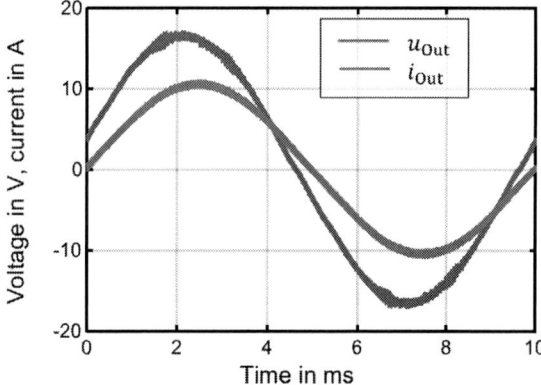

Fig. 4: Output waveforms of the inverter in steady state.

3 EMI Modeling Approach

A proper model of the inverter provides several benefits. First, it allows to analyze the circuit for the source and propagation path of a disturbance. The second application is the prediction of the influence of changes in design and component values. A third advantage results from the comparison between measurement and simulation. This is especially useful during development of a circuit

when a mental model of the system and deep understanding can lead to improved designs.

Simulating EMI disturbances exhibits some unique challenges. The results should be as accurate as possible for a wide frequency range. While at some frequencies, e.g. the frequency of the output current, the signal values are quite high, in the range of several amps and volts, the standards demand millivolts and less for other frequencies. This requires a high dynamic range from the simulation model (up to 100 dB). Additionally, parasitic characteristics of the components shows big influence especially at high frequencies. Another important property of the model is the simulation speed which should be very high especially when the model is used in parallel to the development process. This allows studying the influence of model parameters.

The standards virtually exclusively require measurements with an EMI receiver and evaluation in the frequency domain. Also, the frequency dependent behavior of components is often hard to model exactly using only linear components in the time domain. Thus, the simulation model operates in the frequency domain where frequency dependent impedances are easily implemented. This however poses a challenge for the switching cell which exhibits time dependent behavior. This time dependency is difficult to translate into the frequency domain. Therefore, the simulation model applies an approach similar to Harmonic Balance [17] where different parts are "stitched together".

In order to apply the simulation approach, the inverter has to be split into three parts. The linear components left and right of the switching cell form one part each and are described in the frequency domain easily. The networks of passive components to the left and the right are summarized with one frequency dependent impedance $\boldsymbol{Z}_{le}$ or admittance matrix $\boldsymbol{Y}_{ri}$, respectively. The switching cell itself requires a different approach. It needs to be replaced by a piecewise linear transfer function. In the time domain, most of the time one of the half-bridge switches conducts, resulting in the simplified switching cell of Fig. 1 ($s(t)$ and the switch itself). Thus, the voltage drop of one switch is either identical to the DC-link voltage or zero. A similar relation is true for the current. A square wave shaped function $j_{HB}(t)$ (very similar to $s(t)$) multiplied with the left side voltage describes this behavior. An additional rise and fall time can be added. In case of the inverter, this function becomes more complicated but is still a PWM-signal. A transformation into the frequency domain results in a convolution of the voltage with the Fourier series (or DFT) of the square wave $\mathcal{F}(j_{HB}(t))$. Finding voltages and current that satisfy all requirements at the two connections of the three parts represents the end of the simulation.

A mathematic description helps to shed more light on the simulation approach. Fig. 5 serves as an example. The model uses a reference potential, in this case PE, which is present at each element in the simulation. The figure shows potentials instead of voltages related to the reference potential. At the left and right side of the simulation source, which includes the DC voltage source and the switches, the voltages and currents are collected in

$$v_{\text{le}}, v_{\text{ri}}, i_{\text{le}} \text{ and } i_{\text{ri}} \tag{1}$$

respectively. Each vector includes elements for every connection (e.g. DC+, DC-, and M on the left side) and all considered frequencies (DC, fs, and multiples of fs up to the upper limit in case of the DC/DC-converter). The relation of these is given by

$$\begin{bmatrix} v_{\text{le}} \\ i_{\text{ri}} \end{bmatrix} = \begin{bmatrix} Z_{\text{le}} & 0 \\ 0 & Y_{\text{ri}} \end{bmatrix} \begin{bmatrix} i_{\text{le}} \\ v_{\text{ri}} \end{bmatrix} \tag{2}$$

for the left and right part and

$$\begin{bmatrix} i_{\text{le}} \\ v_{\text{ri}} \end{bmatrix} = \begin{bmatrix} 0 & J_{1,2} \\ J_{2,1} & 0 \end{bmatrix} \begin{bmatrix} v_{\text{le}} \\ i_{\text{ri}} \end{bmatrix} + \begin{bmatrix} 0 \\ v_{\text{sour}} \end{bmatrix} \tag{3}$$

for the center part with the simulation source. v_{sour} includes voltages for the three connections to the right of the simulation source: $U_{\text{DC}}/2$ for DC of $v_{\text{ri,DC+}}$, $-U_{\text{DC}}/2$ for DC of $v_{\text{ri,DC-}}$ and for $v_{\text{ri,HB}}$ the Fourier series of square wave with $U_{\text{DC}}/2$ as high

and $-U_{\text{DC}}/2$ as low potential. It has to be noted that the direct influence of U_{DC} to the left side is neglected as it has no relevance for the EMC standards.

Equation (3) features $J_{1,2}$ and $J_{2,1}$ for the function of the simulation source assuming the idealized switching described above. In this case, $J_{2,1}$ includes all components of the convolution of v_{le} with the Fourier series of the switches in the frequency domain and $J_{1,2}$ the same for the currents. E.g. for the mth frequency of the HB voltage, the convolution is given by

$$v_{\text{ri,HB},m} = \sum_{n=-M_{\text{max}}}^{M_{\text{max}}} \mathcal{F}(j_{\text{HB}}(t))_{m-n}(v_{\text{le,DC+},n} - v_{\text{le,DC+},n}) \tag{4}$$

with M_{max} as the relation of the highest considered frequency to the fundamental. The influence of higher frequencies is considered negligible. To use this for $J_{2,1}$, the negative frequencies (m and n) of (5) have to be combined with the positive. Other connections are much simpler, e.g.

$$v_{\text{ri,DC+}} = v_{\text{le,DC+}} \tag{5}$$

leads to a diagonal in $J_{2,1}$.

The last step is to solve this system of equations. This can be performed by transforming (2) and (3) into

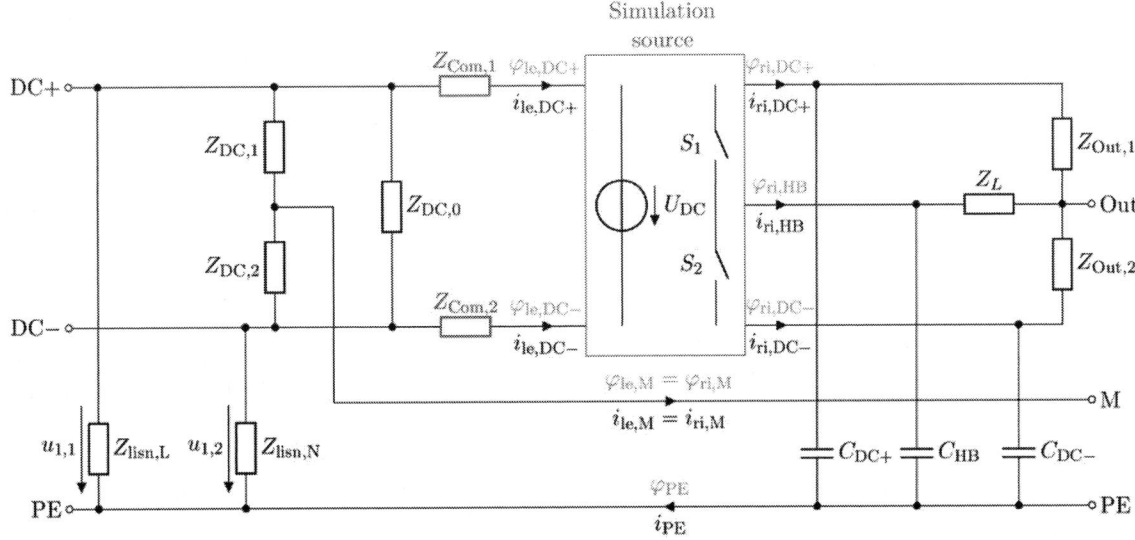

Fig. 5: Simulation model comprising the DC-link, half-bridge (with DC-source), LC-output filter, and parasitic elements.

$$\left(I - J_{1,2}Y_{ri}J_{2,1}Z_{le}\right)i_{le} = J_{1,2}Y_{ri}v_{sour} \qquad (6)$$

where I is the identity matrix. This allows to obtain the left side current i_{le} from v_{sour}. The other voltages and currents follow with

$$v_{le} = Z_{le}i_{le}, \quad v_{ri} = J_{2,1}v_{le} + v_{sour} \quad \text{and}$$
$$i_{ri} = Y_{ri}v_{ri}. \qquad (7)$$

When studying an inverter application, the fundamental frequency f_{Out} is much lower than the switching frequency leading to a higher number of frequencies to be evaluated. Further optimizations allow for a fast calculation of the model even for the high amount of frequencies. However, these are beyond the scope of this paper.

4 EMI Measurement Setup

A validation of the simulation requires the measurement setup (Fig. 7) explained hereinafter. It is a standard test environment to evaluate conducted emissions and is described in [18]. Instead of connecting the grid voltage to the test setup, a laboratory power supply (PS 1) feeds the power through a low pass filter to the LISN. The low pass filter prevents disturbances from PS 1 supply to reach the measurement setup. This ensures that all noise in the setup originates from the inverter and thus, prevents misinterpretations of the noise of PS 1, which may be unintentionally present in the system. A preliminary test (inverter not operating) shows, that no noise (e.g. from the laboratory supply) reaches the LISN, as only the noise floor is visible at the receiver. The inverter with load is

connected at the output of the LISN (or the DC/DC-converter in the first step). Furthermore, the LISN separates [19] the noise generated by the inverter from the DC-component and transfers it to the EMI receiver. A modification of the LISN allows to separate the disturbances into differential mode (DM) and common mode noise, which simplifies filter design (in a second step) by knowing the dominant noise component [19]. An auxiliary power supply (PS 2) is required to power the microcontroller and gate driver of the inverter. It is also decoupled by a low pass filter.

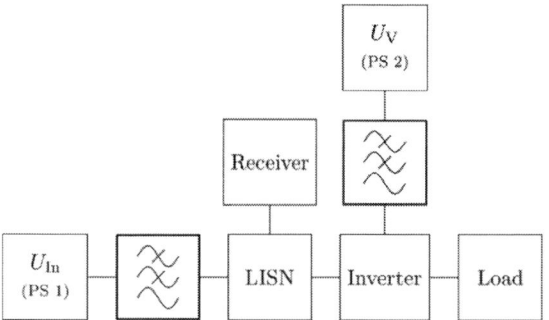

Fig. 7: Measurement environment: DC-supply, LISN, EUT, load, auxiliary supply and low pass filters.

5 EMI Study of DC/DC-Converter

This test setup operates as a DC/DC-converter without load (see section 2.1) for the first results.

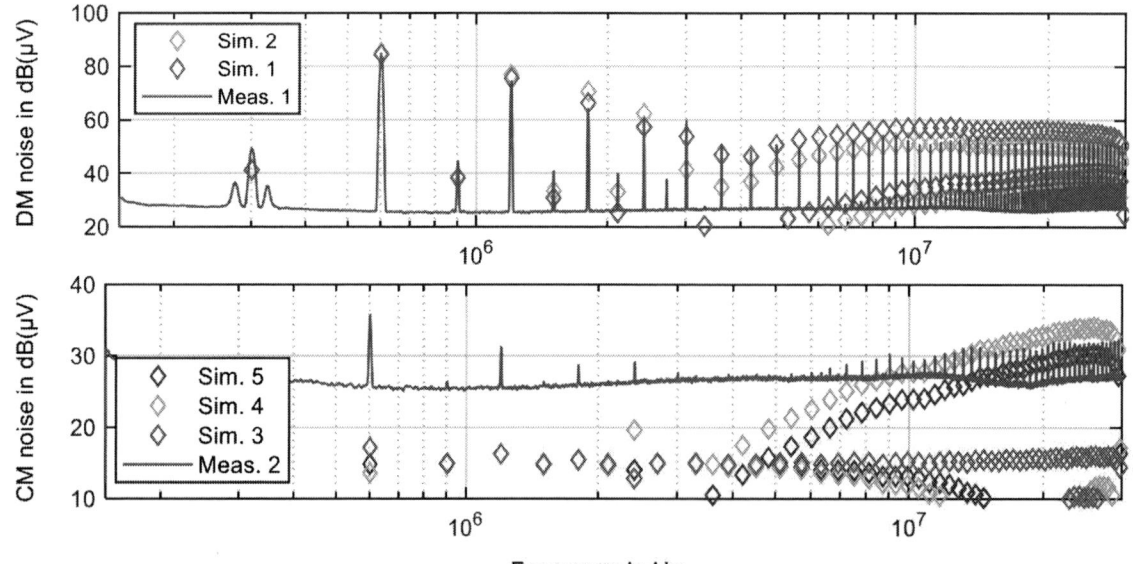

Fig. 6: DM noise of the DC/DC-converter (upper) and CM noise of the DC/DC-converter (lower).

Fig. 6 (upper) shows the DM noise as 'Meas. 1'. All measurements and simulations are summarized in Tab. 2. The peak at 300 kHz, the switching frequency, is significantly (>30 dB) lower than the one at 600 kHz, which is due to the symmetrical waveforms (duty-cycle) and circuit topology (filter capacitor $C_{Out,1}$ and $C_{Out,2}$ connected to $DC+$ and $DC-$) within the converter. This resulting ripple at the DC-link is dominated by twice the switching frequency. Up to 3 MHz the odd harmonics of the signal are significantly lower than the even ones. For higher frequencies the measurement shows noise slightly below 60 dB(µV). Due to slight variations of the switching frequency, not covered by the simulation, the side bands at 300 kHz (and harmonics of the switching frequency) result.

Tab. 2: Overview of the simulations and measurements.

Symbol	Comments
Meas. 1	DM, DC/DC-converter
Meas. 2	CM, DC/DC-converter
Meas. 3	DM, inverter
Meas. 4	CM, inverter
Meas. 5	DM, inverter with filter
Meas. 6	CM, inverter with filter
Sim. 1	DM, DC/DC-converter
Sim. 2	DM, DC/DC-converter with $Z_{Com,1} = Z_{Com}$ and $Z_{Com,2} = 0$
Sim. 3	CM, DC/DC-converter with $Z_{Com,1} = Z_{Com}$ and $Z_{Com,2} = 0$
Sim. 4	CM, DC/DC-converter with $Z_{Com,1} = 0$ and $Z_{Com,2} = Z_{Com}$
Sim. 5	CM, DC/DC-converter with 33 % to $Z_{Com,1}$ and 67 % to $Z_{Com,2}$
Sim. 6	DM, inverter with $Z_{Com,1} = Z_{Com,2} = 0$
Sim. 7	DM, inverter with 33 % to $Z_{Com,1}$ and 67 % to $Z_{Com,2}$
Sim. 8	CM, inverter with 33 % to $Z_{Com,1}$ and 67 % to $Z_{Com,2}$
Sim. 9	CM, inverter with Z_{Com}, $L_{PE} = 150$ nH
Sim. 10	DM, inverter with Z_{Com}, with filter
Sim. 11	CM, inverter with Z_{Com}, $L_{PE} = 150$ nH, with filter

For the DM noise, two simulations have been carried out. Both use the simulation model depicted in Fig. 5. From left to right, it comprises the line impedance of the LISN ($Z_{lisn,L}$ and $Z_{lisn,N}$), the impedance of the electrolytic DC-link capacitors ($Z_{DC,1}$ and $Z_{DC,2}$) as well as the ceramic capacitors $Z_{DC,0}$ (four in parallel), which form the high frequency commutation loop. The impedance of the commutation loop represent $Z_{Com,1}$ and $Z_{Com,2}$. These stem from a 3D FEM simulation of the PCB [20] [21] [22]. The simulation source comprises the modulation as well as the DC-link voltage source

U_{DC}. Z_L, $Z_{Out,1}$, and $Z_{Out,2}$ are the impedances of the filter components from Tab. 1. A proper model of the common mode current paths requires parasitic capacitances from the PCB's copper planes to ground (PE). The capacitors C_{DC+}, C_{DC-} and C_{HB} represent this capacitive coupling to ground. Those parasitic capacitances are also derived from a 3D numerical simulations and are in the range of a few pF ($C_{DC+} = 1.07$ pF, $C_{DC-} = 5.32$ pF, and $C_{HB} = 5.5$ fF) [23]. For the study of the DC/DC-converter, the terminals 'Out', 'M', and 'PE' are left open.

The first simulation, denoted 'Sim. 1', neglects the commutation loop impedance ($Z_{Com,1} = Z_{Com,2} = 0$). For frequencies below 4 MHz, the main peaks (even harmonics) the measurement fits to the simulated results ('Sim. 1' vs. 'Meas. 1'). But, for higher frequencies, neglecting the commutation loop impedance leads to higher DM noise in the simulation than in the measurement. Therefore, 'Sim. 2' takes the commutation loop impedance into account ($Z_{Com,1}$ equals the complete impedance and $Z_{Com,2} = 0$, but the distribution of to the two components does not affect the DM noise) and leads to a better match of the simulation and measurement at high frequencies. The minimum at ≈4 MHz results from the impedance of the ceramic capacitors, which are used as DC-link ($Z_{DC,0}$). These components exhibit their self-resonance at this frequency.

Next, the CM noise (Fig. 6 (lower)) is measured ('Meas. 2') as well as simulated ('Sim. 3'-'Sim. 5') resulting in noise levels significantly lower than the DM noise. Here, the simulations differ in the distribution of the commutation loop impedance to $Z_{Com,1}$ and $Z_{Com,2}$. 'Sim. 3' sets the complete impedance to $Z_{Com,1}$ and $Z_{Com,2} = 0$, for 'Sim. 4' the arrangement is vice versa, and 'Sim. 5' distributes it to both according to the PCB track length (33 % to $Z_{Com,1}$ and 67 % to $Z_{Com,2}$). This study reveals, that the commutation loop impedance affects the results gained by simulation mainly at high frequencies. All simulations differ from the measurement. This can be explained by the significantly higher DM noise (compared to CM) and the limited DM suppression of the DM/CM separator within the LISN [19]. This means, that the separator outputs noise levels at the CM output, which are contributed by DM noise, as the DM suppression is not high enough.

As the CM noise level is only slightly above the noise floor, it is not of relevance for EMI limits. Thus, the EMI study can progress to the DC/AC-inverter.

6 EMI Study of Inverter with Load

Before the inverter with load can be studied regarding conducted emissions, all components of the system must be accurately modeled. As the PCB of the inverter is the same as for the DC/DC-converter, the model of Fig. 5 can be used for the inverter. Thus, only the load needs to be modeled. The load is an electrical machine, but for the single phase inverter, an artificial load, which includes the EMI behavior (high frequency impedance) of the real machine shall be applied. It serves as a load for the EMI measurements as well as the simulations. First, two small signal impedance measurements versus frequency give the small signal impedance for DM and CM, which are depicted in Fig. 8 ('Machine DM') and Fig. 9 ('Machine CM') according to [24] [25] [11]. For the EMI model of the load, only the frequency range between the two dashed lines in Fig. 8 and Fig. 9 is relevant. The first measurement equals the impedance between the terminals 'Out' and 'M', which equals the DM impedance. The CM impedance results, when 'Out' is shorted to 'M' (first port) and 'PE' forms the second port. These requirements allow to derive the artificial load. This electrical network must be designed and parameterized to satisfy both measured impedances (DM and CM) versus frequency.

For DM, the machine impedance (Fig. 8) exhibits inductive behavior up to 1 MHz and capacitive behavior for higher frequency. From the CM configuration (Fig. 9, 'Machine CM'), capacitive behavior is visible in a wide frequency range. The resonance at around 2 MHz should be modeled as well. From the two measured impedances of the electrical machine, the schematic with lumped elements for the artificial load in Fig. 10, which models the DM as well as CM impedances of the electrical machine, can be derived. The CM impedance of the artificial load fits very well to the machine impedance (see Fig. 9). Especially at low frequency, the DM impedance of the artificial load differs from the electrical machine, as a single phase equivalent circuit cannot fit both CM and DM impedances of a three phase electrical machine easily. However, the difference of the artificial load to the electrical machine for DM is not of relevance as the impedance of the parallel path (capacitors within the inverter C_{Out}, 'Parallel path' in Fig. 8) is significantly lower. Thus, the parallel path dominates the impedance for DM, which renders better modeling of the load unnecessary. The main components of the artificial load (Fig. 10) are a wire wound inductor (Z_{Lmot}) as well as a resistor R_{mot}, which dissipates the output power of the inverter. Those components give the ohmic-inductive behavior known from electrical machines. Further elements are placed to fit both, the CM and DM impedance of the electrical machine.

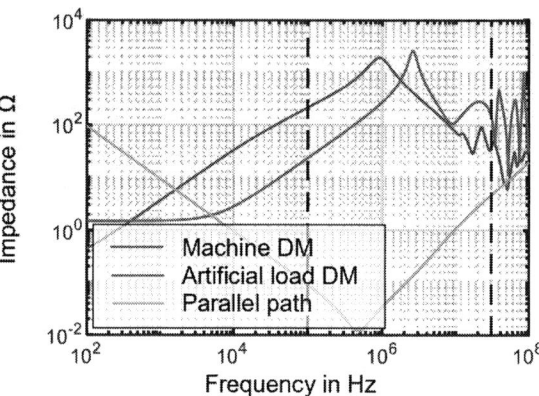

Fig. 8: DM impedance of the electrical machine and the load model.

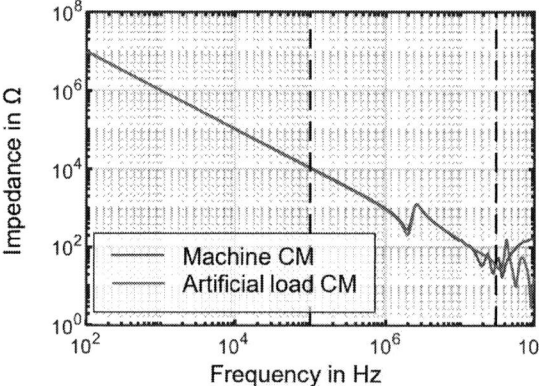

Fig. 9: CM impedance of the electrical machine and the load model.

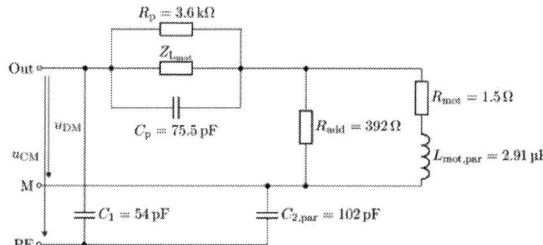

Fig. 10: Artificial load representing an electrical machine.

When using the artificial load (Fig. 10) in conjunction with the model of the inverter's PCB (Fig. 5) connected at the clamps 'Out', 'M', and 'PE', the measurement as well as the simulation of the complete system can be carried out. The modulation scheme of Fig. 3 applied to the circuit gives the output waveforms shown in Fig. 4.

Fig. 11 compares the DM noise generated by the inverter with artificial load. Besides the measurement ('Meas. 3'), again, two simulation results are depicted in Fig. 11. 'Sim. 6' neglects the commutation loop impedance ($Z_{\mathrm{Com,1}} = Z_{\mathrm{Com,2}} = 0$) and gives similar behavior as for the DC/DC-converter (simulation results slightly higher than measurement for high frequencies). Taking the commutation loop impedance into consideration ('Sim. 7') minimizes the difference at high frequency between simulation and measurement. Now, when using the inverter with PWM, the even as well as the odd harmonics of the switching frequency with side bands are present with significant amplitude. These originate from the varying duty-cycle, which is not 0.5 anymore. The simulation fits to the measured results with sufficient accuracy.

The CM noise is studied next. Fig. 12 compares the simulated noise with the measured one. Again, CM noise is significantly lower than the DM noise (compare Fig. 11). Fig. 12 shows two results gained by simulation. The first one uses the simulation model depicted in Fig. 5 with the artificial load of Fig. 10. As the measured CM noise ('Meas. 4') drops at 20 MHz, but the simulation ('Sim. 8') does not, further investigation is necessary. As the CM current is propagating through the parasitic capacitances $C_{\mathrm{DC+}}, C_{\mathrm{DC-}}$ and C_{HB}, reaches the LISN by the PE conductor and returning to the DC-link, a significant current loop is formed. In order to account for this, an inductance of 150 nH is introduced to the PE-conductor (between Z_{lisn} and the parasitic capacitances). With this enhanced model, 'Sim. 9' results. Now, the drop in CM noise is also visible in the simulation. This shows, that mainly parasitic elements need to be taken into consideration to accurately model high frequency EMI noise.

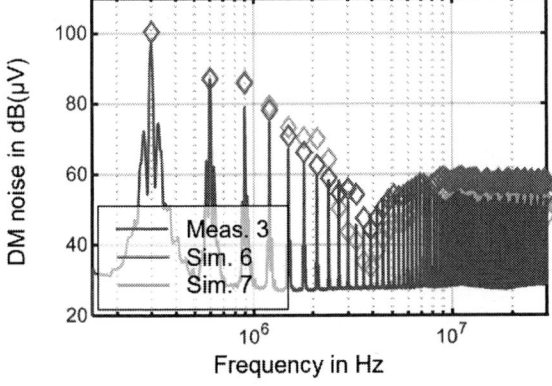

Fig. 11:DM noise of the inverter with load.

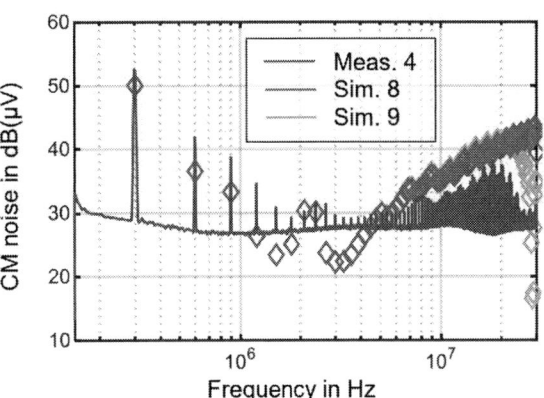

Fig. 12:CM noise of the inverter with load.

7 Filter Design for the Inverter

As visible from the simulations and measurements depicted in Fig. 11 (DM) and Fig. 12 (CM), the noise is dominated by DM. In order to satisfy common EMI limits regarding conducted emissions, the emitted DM noise must be reduced. Therefore, a filter should be placed at the input side of the inverter to prevent noise to reach the LISN. To circumvent conversion from DM to CM, a symmetrical filter is placed between inverter and LISN. The schematic is shown in Fig. 13. It uses two inductors and one electrolytic capacitor (parameters can be found in Tab. 1) and filters predominantly DM noise. By introducing this filter into the test setup, the DM noise is drastically reduced (from Meas. 3 to Meas. 5 and Sim. 6 to Sim. 10 in Fig. 11 vs. Fig. 14). For frequencies higher than 1.5 MHz, the emitted noise disappears in the noise floor of the test environment turning it irrelevant. Fig. 14 and Fig. 15 display the limit line for 'EN55014QP' in green [26]. The relative difference between simulation and measurement for the setup with filter is significantly higher than before, but both are well below limits, and thus accuracy is not that relevant anymore.

The CM-noise with DM filter is depicted in Fig. 15. It shows a deviation between simulation and measurement of less than 6 dB, which is well within the desired tolerance for EMI prediction. Above 6 MHz, the measured peaks vanish in the noise floor of the test environment and thus are negligible.

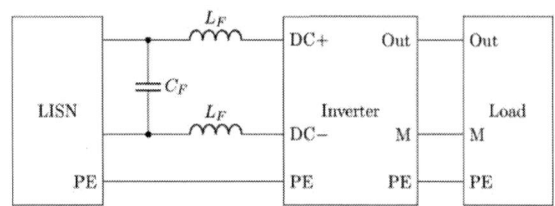

Fig. 13: DM filter between LISN and inverter.

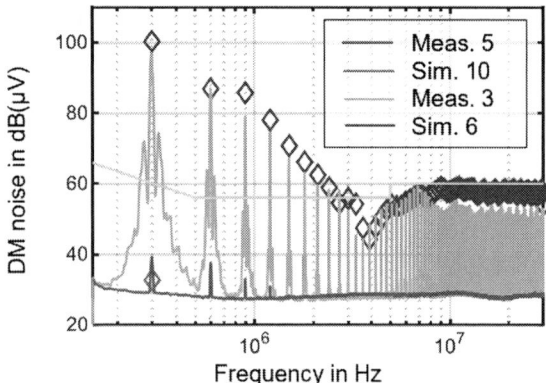

Fig. 14: DM noise of the inverter with load and filter.

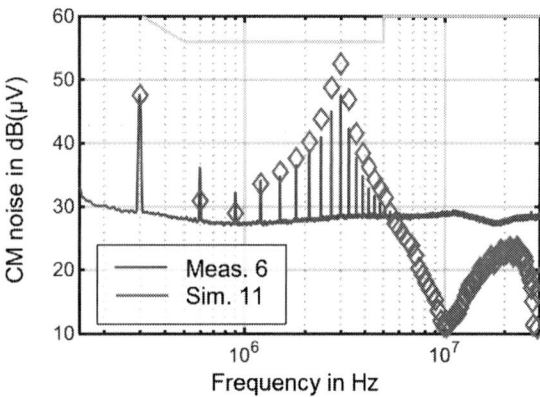

Fig. 15: CM noise of the inverter with load and filter.

8 Conclusion

Due to the faster switching transitions and higher switching frequencies, the application of wide-bandgap semiconductors in inverters makes EMI studies of these systems inevitable. Thus, this paper studies a 2-level inverter with load featuring GaN-HEMTs. The investigation of one phase of a three phase inverter suffices for the desired outcomes.

The presented approach provides many insights into the inverters behavior regarding EMC. Sufficient knowledge of the parasitic behavior of the circuit is crucial in order to obtain useful results. Given this, the presented simulation approach shows good agreement with the measurements. The investigations reveal that a DM filter suffices in order to comply with the standard.

References

[1] B. Kohlhepp, J. Heubeck and T. Duerbaum, "Non-ideal Behavior of ZVS Inverter Comprising Variable and Fixed Frequency Operation: Analysis, Compensation, and Verification," *Power Electronic Devices and Components*, p. 100007, Apr. 2022, doi: 10.1016/j.pedc.2022.100007.

[2] B. Kohlhepp, D. Kübrich, M. Tannhäuser and T. Dürbaum, "Adaptive dead time in high frequency GaN-Inverters with LC output filter," in *The 10th International Conference on Power Electronics, Machines and Drives (PEMD 2020)*, Online Conference, 2020, pp. 372–377, doi: 10.1049/icp.2021.0977.

[3] B. Zhang and S. Wang, "A Survey of EMI Research in Power Electronics Systems With Wide-Bandgap Semiconductor Devices," *IEEE J. Emerg. Sel. Topics Power Electron.*, vol. 8, no. 1, pp. 626–643, Mar. 2020, doi: 10.1109/JESTPE.2019.2953730.

[4] B. Kohlhepp, J. Dobusch and D. Kübrich, "Radio Disturbance Evaluation of Si- and GaN-Inverters," DE, Jun. 2023, Accessed: Sep. 22, 2023. [Online]. Available: https://doi.org/10.30420/566091338.

[5] J. Dobusch, D. Kuebrich, T. Duerbaum and F. Diepold, "Evaluating EMI of Unshielded Cables in High Frequency GaN Inverter Application," in *2021 IEEE International Joint EMC/SI/PI and EMC Europe Symposium*, Raleigh, NC, USA, Jul. 2021, pp. 401–406, doi: 10.1109/EMC/SI/PI/EMCEurope52599.2021.9559259.

[6] B. Kohlhepp, T. Dürbaum and D. Kübrich, "Common Mode of Inverters: Survey and Study of Filter Placement on Grid and Load Side," presented at the PCIM Europe 2022; International Exhibition and Conference for Power Electronics, Intelligent Motion, Renewable Energy and Energy Management, 2022, Accessed: Sep. 10, 2022. [Online]. Available: https://doi.org/10.30420/565822059.

[7] B. Kohlhepp, V. Zeller and T. Dürbaum, "Full GaN Asymmetrical Half-Bridge PWM Converter with Synchronous Rectifier for a High Efficient 90W Laptop Charger," in *PCIM Europe digital days 2021; International Exhibition and Conference for Power Electronics, Intelligent Motion, Renewable Energy and Energy Management*, 2021, pp. 1–7.

[8] Q. Liu, W. Shen, F. Wang, D. Borojevich and V. Stefanovic, "On discussion of motor drive conducted EMI issues," in *The 4th International*

Power Electronics and Motion Control Conference, 2004. IPEMC 2004., 2004, vol. 3, pp. 1515-1520 Vol.3.

[9] F. A. Kharanaq, A. Emadi and B. Bilgin, "Modeling of Conducted Emissions for EMI Analysis of Power Converters: State-of-the-Art Review," *IEEE Access*, vol. 8, pp. 189313–189325, 2020, doi: 10.1109/ACCESS.2020.3031693.

[10] Q. Liu, F. Wang and D. Boroyevich, "Conducted-EMI Prediction for AC Converter Systems Using an Equivalent Modular–Terminal–Behavioral (MTB) Source Model," *IEEE Trans. on Ind. Applicat.*, vol. 43, no. 5, pp. 1360–1370, 2007, doi: 10.1109/TIA.2007.904435.

[11] B. Revol, J. Roudet, J.-L. Schanen and P. Loizelet, "EMI Study of Three-Phase Inverter-Fed Motor Drives," *IEEE Trans. on Ind. Applicat.*, vol. 47, no. 1, pp. 223–231, Jan. 2011, doi: 10.1109/TIA.2010.2091193.

[12] F. Costa, C. Gautier, E. Labouré and B. Revol, "Electromagnetic Compatibility in Power Electronics," 1st ed. Wiley, 2013.

[13] J. Dobusch and T. Duerbaum, "EMC Focused SiC Half-Bridge Modeling in the Frequency Domain: Procedure, Advantages and Limitations," in *PCIM Europe digital days 2020; International Exhibition and Conference for Power Electronics, Intelligent Motion, Renewable Energy and Energy Management*, 2020, pp. 1–8.

[14] W. Zhou, X. Pei, Y. Xiang and Y. Kang, "A New EMI Modeling Method for Mixed-Mode Noise Analysis in Three-Phase Inverter System," *IEEE Access*, vol. 8, pp. 71535–71547, 2020, doi: 10.1109/ACCESS.2020.2983084.

[15] J. Bezanson, A. Edelman, S. Karpinski and V. B. Shah, "Julia: A Fresh Approach to Numerical Computing," *SIAM Rev.*, vol. 59, no. 1, pp. 65–98, Jan. 2017, doi: 10.1137/141000671.

[16] Efficient Power Conversion Corporation, "Datasheet EPC2045 – Enhancement Mode Power Transistor." El Segundo, 2018.

[17] M. Nakhla and J. Vlach, "A piecewise harmonic balance technique for determination of periodic response of nonlinear systems," *IEEE Trans. Circuits Syst.*, vol. 23, no. 2, pp. 85–91, Feb. 1976, doi: 10.1109/TCS.1976.1084181.

[18] Adjustable speed electrical power drive systems - Part 3: EMC requirements and specific test methods (IEC 61800-3:2017).

[19] J. Stahl, D. Kuebrich and T. Duerbaum, "Modification and characterization of a standard LISN for effective EMI noise separation," in *2010 International Conference on Electromagnetics in Advanced Applications*, Sydney, Australia, Sep. 2010, pp. 39–42, doi: 10.1109/ICEAA.2010.5652246.

[20] B. Kohlhepp, S. Faber, J. Kaiser, D. Kübrich and T. Dürbaum, "Extraction of Parasitic Elements of a Printed Circuit Board applied to a GaN Half-Bridge," presented at the PCIM Europe 2022; International Exhibition and Conference for Power Electronics, Intelligent Motion, Renewable Energy and Energy Management, 2022, Accessed: Sep. 10, 2022. [Online]. Available: https://doi.org/10.30420/565822168.

[21] B. Kohlhepp, S. Faber, J. Kaiser and T. Duerbaum, "Study on Commutation Loop Inductance and Current Distribution to DC-link Capacitors in a GaN Half-bridge," in *2022 24th European Conference on Power Electronics and Applications (EPE'22 ECCE Europe)*, 2022, pp. 1–8.

[22] B. Kohlhepp, S. Faber, D. Kübrich and T. Dürbaum, "PCB Layout Parasitics Extraction of a GaN Half-Bridge: Simulation and Experimental Validation," in *2023 25th European Conference on Power Electronics and Applications (EPE'23 ECCE Europe)*, Aalborg, Denmark, Sep. 2023, pp. 1–9, doi: 10.23919/EPE23ECCEEurope58414.2023.10264516.

[23] S. Faber, B. Kohlhepp, J. Dobusch, J. Kaiser and T. Dürbaum, "In-Depth Study of the Parasitic Capacitances of a Half-bridge Circuit," in *2023 25th European Conference on Power Electronics and Applications (EPE'23 ECCE Europe)*, Aalborg, Denmark, Sep. 2023, pp. 1–8, doi: 10.23919/EPE23ECCEEurope58414.2023.10264660.

[24] M. Schinkel, S. Weber, S. Guttowski, W. John and H. Reichl, "Efficient HF Modeling and Model Parameterization of Induction Machines for Time and Frequency Domain Simulations," in *Twenty-First Annual IEEE Applied Power Electronics Conference and Exposition, 2006. APEC '06.*, USA, 2006, pp. 1181–1186, doi: 10.1109/APEC.2006.1620689.

[25] J. Dobusch, D. Kuebrich, T. Duerbaum and F. Diepold, "Implementation of Current Based Three-Phase CM/DM Noise Separation on the Drive Side," in *2018 International Symposium on Electromagnetic Compatibility (EMC EUROPE)*, Amsterdam, Aug. 2018, pp. 220–225, doi: 10.1109/EMCEurope.2018.8485048.

[26] Electromagnetic compatibility - Requirements for household appliances, electric tools and similar apparatus - Part 1: Emission (CISPR 14-1:2005 + A1:2008 + Cor. :2009 + A2:2011); German version EN 55014-1:2006 + A1:2009 + A2:2011.

PCIM Europe 2024, 11– 13 June 2024, Nuremberg   DOI: 10.30420/566262328

Common Mode Currents in Resonant Circuits Generated with a Delta-Sigma Modulated Voltage Source Inverter

Tobias Haas[1], Theo Zeihsel[1], Henning Kasten[1], Abbas Mehraban[2], Michael Terörde[2]

[1] Technische Hochschule Würzburg-Schweinfurt, Germany
[2] Technische Universität Braunschweig, Germany

Corresponding author: Tobias Haas, tobias.haas@thws.de
Speaker: Tobias Haas, tobias.haas@thws.de

Abstract

According to parasitic effects, unwanted resonant circuits occur in electric applications. In power electronic systems, resonances might be excited. Modulation methods are decisive for the frequency dependent voltage excitation generated by power electronic components. In this paper a delta-sigma modulated voltage source inverter is considered with the aim of reducing common mode disturbances. Simulative and experimental results are discussed in order to state out advantages and disadvantages in comparison to a pulse width modulation. Especially significant peaks in voltage spectra can be reduced with a delta-sigma modulated inverter.

1 Introduction

Since the development of new semiconductor materials like silicon-carbide and gallium-nitride enable faster switching processes, advantages like a better waveform in effective signals with lower ripple can be achieved. Looking at electric drives, this means lower losses and an increased efficiency [1]. On the other hand, high frequency effects, like common mode (CM) and differential mode (DM) disturbances have to be considered in system designs [2].

Voltage source inverters are prevalent in electric drive systems. Especially in industrial applications, the power electronic is commonly used at velocity controlled electric drives. In compliance with DC-link voltage and switching frequency, an inverter allows generating customized voltages at the fundamental frequency. Besides the desired behaviour, CM and DM disturbances are generated through switching operations. Hence, parasitic capacitances are polarised, e.g. between a transmission line and its shielding mesh or between an electric drive's windings and the stator housing. As a result, the parasitic capacitances and inductances generate series resonant circuits. Driven common mode currents depend on the system's impedance and on the voltage excitation. Inverters with high DC link voltage and fast

switching processes are able to generate strong disturbances. The resulting excitation of parasitic paths may violate regulations by disturbing or damaging other components in the grid. For example a drive's bearing can be damaged [3] or ageing effects in inductive components might be accelerated. Even control units or measurement equipment might be influenced through electromagnetic interference (EMI).

Thereby a reduction of disturbance currents is desirable. This goal can be achieved for example by designing and implementing appropriate filters [4]. Another way for reduction is to choose a favourable modulation method in power electronic systems.

Well known procedures are pulse width or space vector modulations. One advantage is the simple implementation and handling of these techniques. On the other hand, high peaks occur in the output voltage's spectrum [5]. In case that a series resonant circuit is excited through those peaks, high currents are driven. Several possibilities are given to reduce this effect, for example using a varying switching frequency smooths the spectral peaks. Hysteresis current controllers operate in general with a variable switching frequency, since they switch voltages in dependency of the momentary current. Instead of certain peaks in the spectrum, hills can be determined in region of the averaged switching frequency. It is even possible to reduce CM disturbances by shifting the pulse patterns in combination with a hysteresis current controller [6].

In comparison to these approaches, the delta-sigma modulation shows different spectral behaviour with vales [7] instead of peaks or hills. Hence, the current caused by a series resonant circuit can be reduced under certain circumstances, which gives an effective alternative. Therefore the vale has to be placed in the resonance frequency. It has to be emphasised, that the spectral behaviour between those vales shows a higher excitation compared to the previous mentioned methods.

This paper investigates a useful case for implementing a delta-sigma modulation with the aim to reduce common mode currents. The procedure is compared to a pulse width modulation. Several components of a test setup are analysed with small signal analysis and modelled in order to design accurate common mode models and compare simulative to experimental results.

2 Modulation Methods

Previously mentioned approaches for modulating a two-level voltage source inverters are explained in this section. CM currents in a system depend on the switching patterns and the system's frequency dependent behaviour. Each pattern can create a different voltage spectrum and associated current, but they are all created to generate a desired reference signal at fundamental frequency.

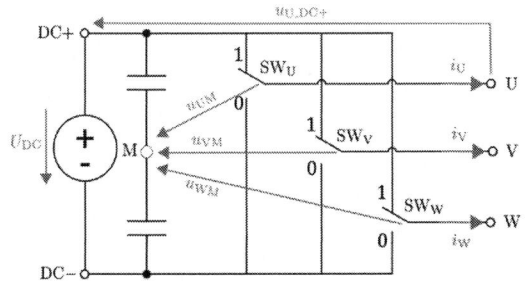

Fig. 1: Two-level inverter structure

Since there is a variety of possibilities, only fundamental methods are shown, starting with conventional approaches like pulse width modulation (PWM), space vector pulse width modulation (SVPWM) and further the delta-sigma space vector modulation (DSSVM). Each method generates different switching states, which in this paper are all

described for a two-level voltage source inverter as shown in Fig. 1. According to the structure, u_{UM}, u_{VM} and u_{WM} can be switched between $\pm\frac{U_{\mathrm{DC}}}{2}$. Since the common mode currents analysed in sec. 4, 5 flow between AC and DC side, $u_{\mathrm{U\_DC+}}$ is used to analyse the CM voltage excitation.

2.1 Pulse Width Modulation

The most commonly known approach for modelling a desired voltage with a two-level inverter is PWM. Based on a comparison between carrier u_{car} and reference signal u_{ref}, which are normalized to $\frac{U_{\mathrm{DC}}}{2}$, switching states are determined as shown in Fig. 2 exemplary for $\mathrm{SW_U}$. The carrier signal's frequency matches the PWM period and the half bridge's switching frequency $f_{\mathrm{SW\_HB}}$, which determines the voltage's spectral behaviour. An important parameter to discuss different approaches is the modulation degree (Eq. 1). It is determined by the fundamental frequency's amplitude $\hat{u}_{\mathrm{set}}$ of desired modulated voltage in relation to $\frac{U_{\mathrm{DC}}}{2}$.

$$M = \frac{\hat{u}_{\mathrm{set}}}{\frac{1}{2}U_{\mathrm{DC}}} \tag{1}$$

By injecting a third harmonic to the reference signal with an amplitude of $\frac{1}{6}$ (PWM3), several improvements can be achieved, like a higher modulation degree without operating in block mode. Another advantage is that the behaviour of total harmonic distortion is closer to the space vector pulse width modulation [8]. For this reason, PWM3 is used in sec. 4 for comparison to DSSVM.

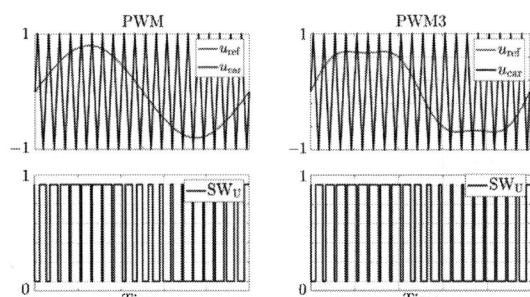

Fig. 2: Pulse generation in PWM and PWM3

2.2 Space Vector Pulse Width Modulation

Transforming a three phase system's voltages into an alpha-beta system, leads to a space vector description, where the switching states of a two-level inverter can be represented as vectors. Any vector within the circle in Fig. 3 can be modulated by using neighboured space vectors and zero vectors.

The length of the needed vectors in relation to the space vector determines how long the corresponding spacevector has to be switched over the time of one PWM period.

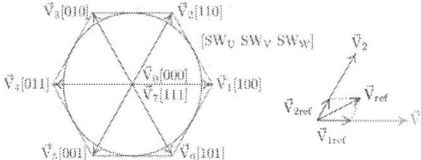

Fig. 3: Space vector diagram and SVPWM

2.3 Delta-Sigma Space Vector Modulation

In the last decades, a variety of researches investigated variants of the delta-sigma modulation (DSM) in combination with power electronics systems [9], [7], [10]. The technique can be classified as a pulse frequency modulation. Commonly used procedures like SVPWM and PWM work with a fixed pulse frequency and varying duty cycle.

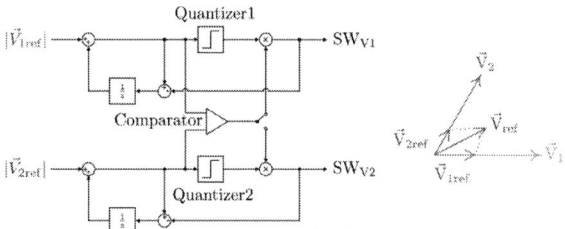

Fig. 4: Delta-sigma Modulation in combination with space vector modulation

In this paper a DSSVM scheme, first presented in [7] is used. Figure 4 shows the operating principle. On the right side, the split of a specific reference voltage vector V_{ref} in space dimensions is depicted. Instead of determining the switching duration over one pulse period with the split reference vectors $V_{1\mathrm{ref}}$ and $V_{2\mathrm{ref}}$, as it is the case in conventional space vector pulse width modulation, the vectors' magnitudes are used as input quantity for a delta-sigma modulation. Hence, the values of one reference vector is summed up until the threshold value of the quantizer is exceeded. An additional comparator determines which space vector will be set to the inverter. It has to be stated out, that for the spectral behaviour, the maximum switching frequency of space vectors $f_{\mathrm{SW\_SV}}$ is decisive in this structure. It is determined by sample time of the unit delays in Fig. 4.

3 Common Mode Impedances

The CM current is defined as a flow in the same direction at all connections of one component. Hence, multiple conductors can behave as a single transmission line. Such effects are often caused by parasitic couplings for example to the ground system. In this paper, CM currents from AC to DC side of a two-level voltage source inverter are considered (Fig. 5). The switching operations drive currents through these paths in dependency of the system's impedances.

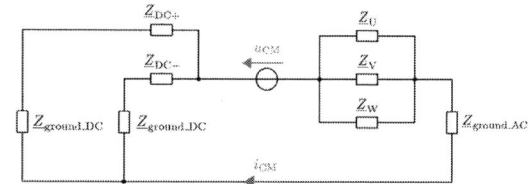

Fig. 5: Significant CM impedances for the proposed setup

Figure 5 depicts the considered, significant impedances for the proposed system in order to describe CM behaviour. The inverter is described as a single voltage disturbance source u_{CM}, while i_{CM} is driven in the grounding system between AC- and DC-side. The assignment of impedances and components from experimental setup is explained in sec. 3.2.

3.1 Experimental Setup

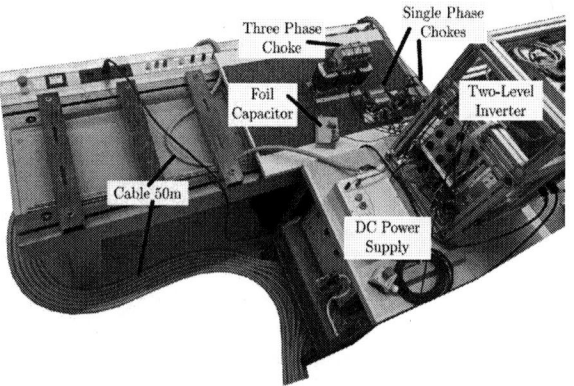

Fig. 6: Experimental setup

For investigation on driven common mode currents, a DC-grid is considered (Fig. 6). An inverter is electrically placed close to the load, a three-phase choke with a single core in star point connection. A ground path from AC-side to the inverter's housing

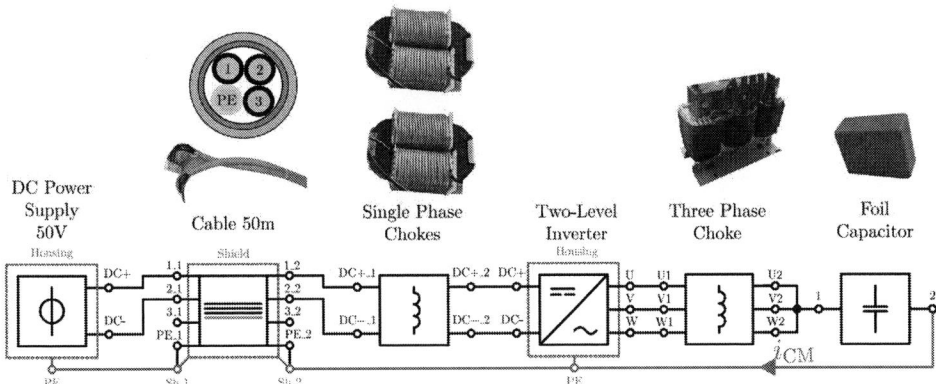

Fig. 7: Block diagram of experimental setup

is applied with an additional star point connected foil capacitor. This leads to a well defined ground system, where the system's resonance frequency can be changed through varied capacitances.

The DC-power supply is placed electrically far away and is connected via a 50 m shielded cable (shield is applied on both sides). Additional single phase chokes are used as an inductive filter component on DC-side. The setup is designed to enable CM currents through the star point connection on AC-side to the ground system, flowing back to the inverter through parasitic capacitances, generated by 50 m cable on DC-side.

Having a closer look to the proposed ground system in Fig. 7, parasitic capacitances to inverter's and DC-power supply's housings occur in the system as well. These values are neglected in modelling process, because the dominant capacitances in the system are determined by foil capacitor and 50m cable. The considered, restricted frequency range allows a simplification (sec. 3.2). Although investigating higher frequencies, these effects have to be considered. The system is designed to show a CM resonance frequency at 72 kHz and no other in the regarded range, to create a well defined system.

3.2 Component Modelling

The approached simulation models are specified to be valid up to a frequency of 200 kHz. This allows to have a look at lower frequency impacts, but neglecting high frequency regions with electro dynamic effects. Figure 8 depicts the component's measured and modelled common mode impedances.

Three components are modelled with high accuracy in the system: Single phase chokes (filter), three-phase choke (load) and 50 m cable. Frequency dependent effects like skin- and proximity-effect shall be designed sufficiently accurate and are determined from impedance measurement results with small signal analysis. It has to be stated out, that the components are modelled with mathematical approaches. This means, that not every electric circuit component in the model corresponds to a physical effect.

The chokes (Fig. 8 (a) and (c)) are both designed to have a frequency dependent resistance, what can be achieved by using multiple RL parallel circuits in series connection [11]. The single phase chokes are considered to have no magnetic coupling since each consists of its own core. In contrast, the three-phase choke's coupling cannot be neglected and is modelled according to [12]. The parallel resonance, seen in both chokes over 500 kHz comes from parasitic winding capacitances, but will be neglected for the presented model. Up to this frequency region, the models show high accuracy to small signal analysis.

A four-wire cable with an additional shield is implemented on DC-side. As shown in Fig. 7, only two wires are connected to guide DC-currents. Shield and PE-wire are short circuited on both sides and connected to DC-power supply's and inverter's housings. The presented model in Fig. 8 (d) consists of magnetic and capacitive couplings between wires and shield/PE-wire. The symmetric design enables CM currents to flow from both sides of the cable through equal impedances. Ongoing researches will focus on DC-grids with several disturbance sources where this model can be reused. The model's frequency dependent behaviour fits the measured values from small signal analysis very well. The implemented foil capacitor is modelled

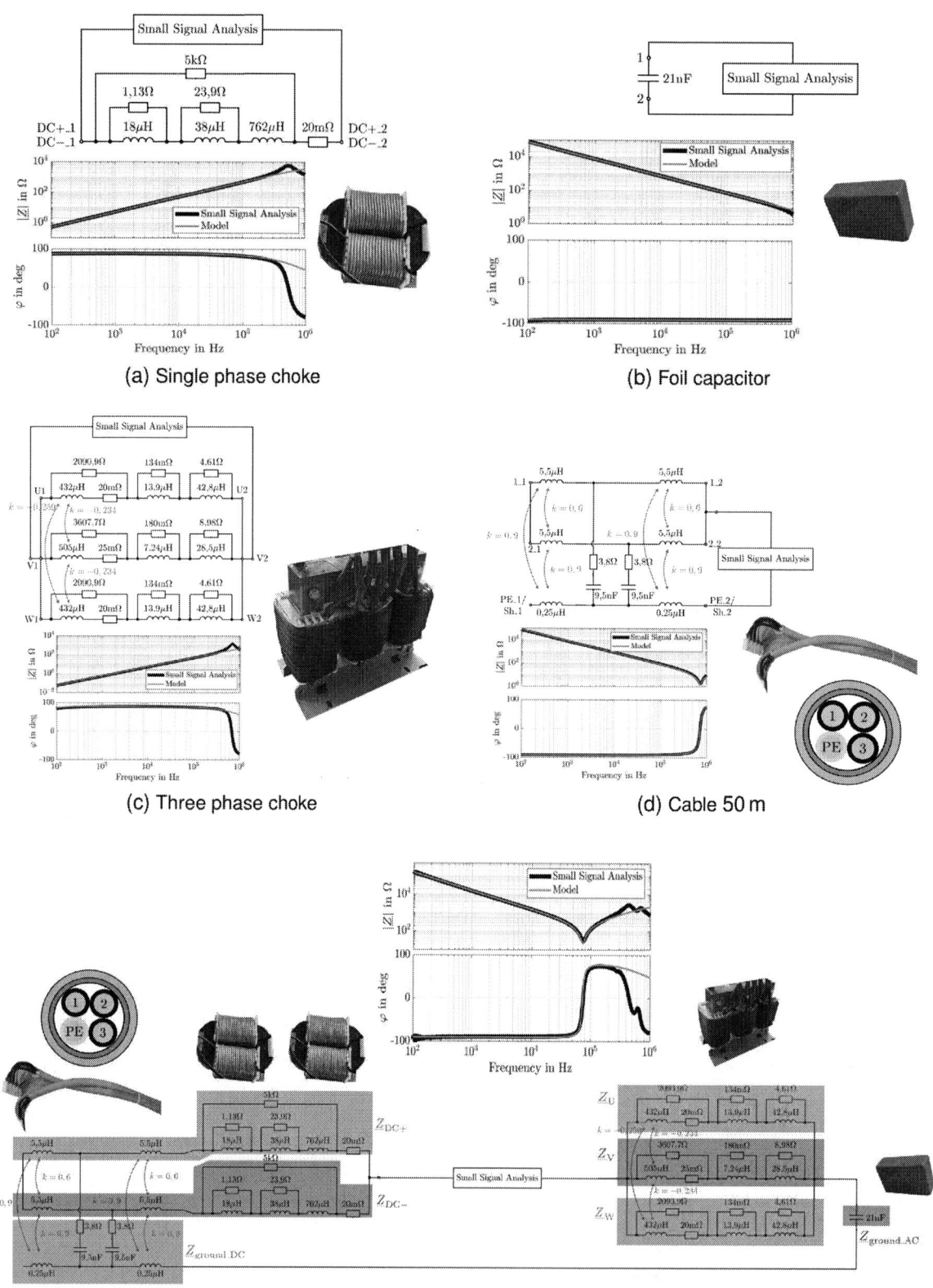

Fig. 8: Frequency dependent models of single components and overall structure in CM

without frequency dependent effects with a single capacitance (Fig. 8 (b)) and shows satisfactory behaviour.

Figure 8 depicts the overall CM structure of the system, composed from all single components. As mentioned before, the total resonance frequency can be determined at 72 kHz. The model and small signal analysis show high compliance up to 200 kHz. Over this frequency, the before mentioned neglected effects come into play. Comparing the structure to Fig. 5, all shown impedances can be determined with the single components. For example $\underline{Z}_U$, $\underline{Z}_V$, $\underline{Z}_W$ and $\underline{Z}_{\text{ground\_AC}}$ are directly assigned with the three-phase choke and foil capacitor. $\underline{Z}_{DC+}$ and $\underline{Z}_{DC-}$ are formed from single-phase chokes and the cable's connected wires, while $\underline{Z}_{\text{ground\_DC}}$ compounds of the parasitic capacitances in the cable and shield's/PE's inductive behaviour.

4 Simulative Results

In order to compare PWM3 and DSSVM with each other, u_{U_DC+} (Fig. 1) is used to analyse the CM voltage excitation. Several modulation degrees are simulated and the signal is transformed into frequency domain, shown in Fig. 9 to discuss the frequency dependent amplitudes. The THD according to Eq. 2 and the peak values are compared for every modulation degree. In Eq. 2, U is meant to be the signal's RMS (AC) value in total, while U_1 stands for the fundamental frequency's RMS value.

$$\text{THD} = \frac{\sqrt{U^2 - U_1^2}}{U_1} \qquad (2)$$

Spectral analysis is performed with a fast fourier transform (FFT), without any window function. A maximum hold voltage spectrum over several fundamental periods is performed and depicted. The DC-link voltage is set to 50 V.

The main sufficient difference between both approaches is the excitation in switching frequency region. Peaks occur in PWM3 spectrum around $f_{\text{SW\_HB}}$ (half bridge switching frequency) and its multiples. For the proposed setup, $f_{\text{SW\_HB}}$ is set to 24 kHz, so that the third excited peak lies in the system's CM resonance frequency at 72 kHz. This leads to a strongly driven CM current.

As mentioned before, in DSSVM vales are found instead of peaks in spectrum. The variable switching frequency behaves advantageous and generates a vale around $f_{\text{SW\_SV}}$ (maximum space vector

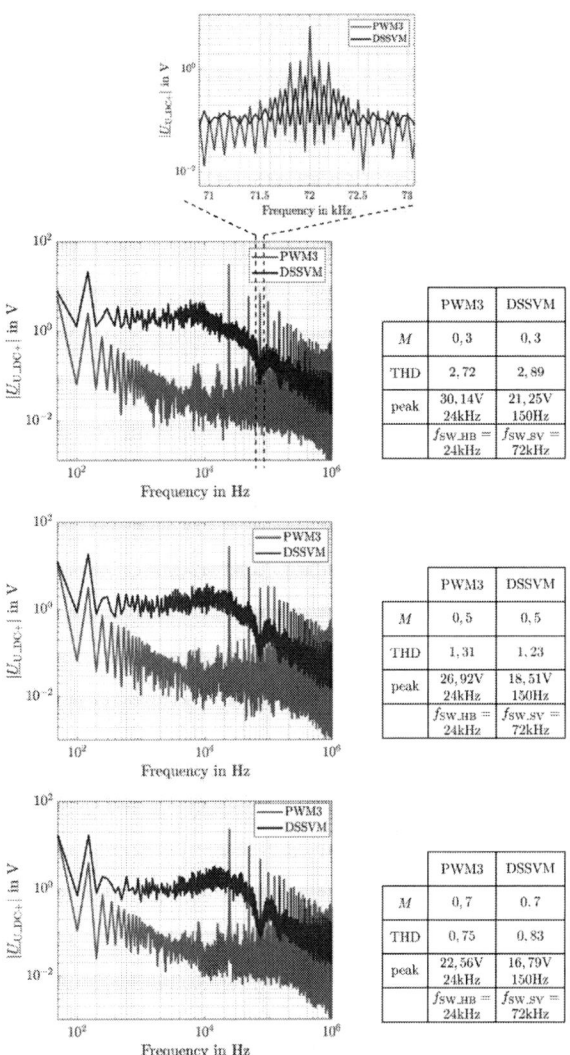

Fig. 9: CM voltage in comparison for PWM3 and DSSVM

switching frequency). With the aim of reducing CM currents, $f_{\text{SW\_SV}}$ is set to 72 kHz.

Comparing both methods with each other, it is recognizable, that DSSVM generates a higher voltage excitation in low frequency regions. Especially the third harmonic shows a high amplitude and gives the spectral peak. The spectrum shows consequently larger values below 24 kHz compared to PWM3 for all modulation degrees. In PWM3, the excitation becomes more significant in higher frequency regions. After all it can be stated that DSSVM brings a higher THD and stronger excitation in lower frequency regions and seems to be an unfavourable approach.

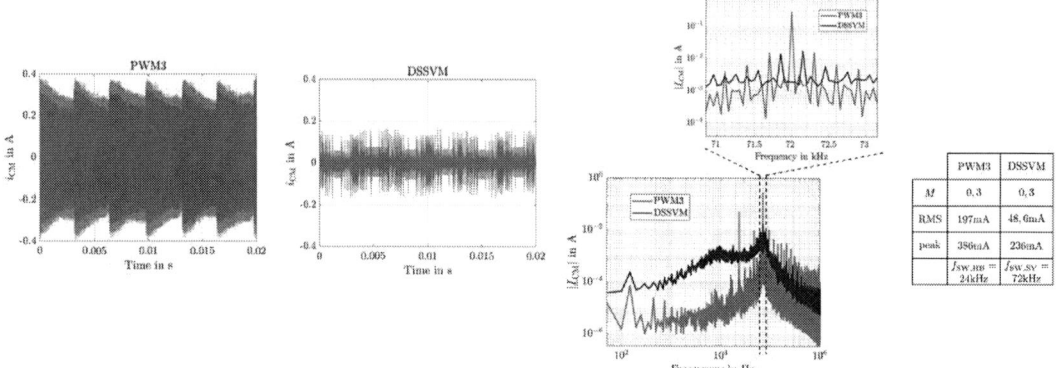

Fig. 10: Driven CM currents in comparison for PWM3 and DSSVM

Although the peaks generated through PWM3 could be obviously decreased. Especially in case of hurting CM voltage regulations with those peaks, a DSSVM might be useful, since regulations are stricter for higher frequencies.

In order to decrease the voltage excitation at a resonance, $f_{\mathrm{SW\_HB}}$ could be decreased. Thereby high frequency excitation can be lowered in total. Anyway, since new semiconductor materials appear on market, the trend is moving towards higher switching frequencies. Hence, new use cases might appear for DSSVM. In any case, advantages and disadvantages of possible modulation approaches should be weighed up.

Figure 10 depicts the simulated CM currents driven by PWM3 and DSSVM. Clearly recognizable from time domain, peak and RMS values can be decreased with DSSVM. The frequency domain shows compliance with the analysis of CM voltage. In lower frequency regions, PWM3 shows lower amplitudes until 24 kHz. Especially at the resonance frequency (72 kHz), an advantage for DSSVM can be stated out. Hence, PWM3 generates a high amplitude according to the voltage peak at this frequency.

5 Experimental Results

With the aim of validating the shown models from sec 3, experimental results are directly compared to simulative results in Fig. 11. A comparison is performed for DSSVM at $M = 0,3$. Having a look at i_{U} shows, that the models have high accuracy for DM values and the same load current is driven. The experimental results are measured with a data recorder with an implemented line filter (cut-off

frequency 200 kHz). As performed for simulative analysis, a FFT with maximum hold over several fundamental periods is used to depict spectral behaviour for experimental results. The depicted spectrum of i_{U} shows, that simulated values are highly accurate up to 200 kHz.

Considering i_{CM} for a comparison, the validation of simulation model is confirmed. The differences in time and frequency domain are negligible. Although it has to be emphasised, that the resonance frequency in experimental setup lies at 62 kHz instead of 72 kHz as predetermined. This may come from non-linear effects due to the inductive components. Core materials can show increased inductance values, because of magnetisation during operation. The models are derived from small signal analysis and cannot include this effect.

6 Conclusion

This paper presented a comparison between PWM3 and DSSVM, regarding CM effects. It is recognizable from simulative results, that DSSVM brings certain advantages and disadvantages. Although exciting lower frequency regions with higher amplitudes in CM voltage, no peaks occur in the spectrum. This can lead to reduced CM currents, if $f_{\mathrm{SW\_SV}}$ is set correctly. Especially regarding CM voltage regulations, DSSVM can bring an advantage in the case of hurt regulations with PWM3. In any case, the advantages and disadvantages of possible modulation methods have to be compared. Furthermore, CM models of several components were generated from small signal analysis. Although the resonance frequency in experimental setup differs to the presented model, high

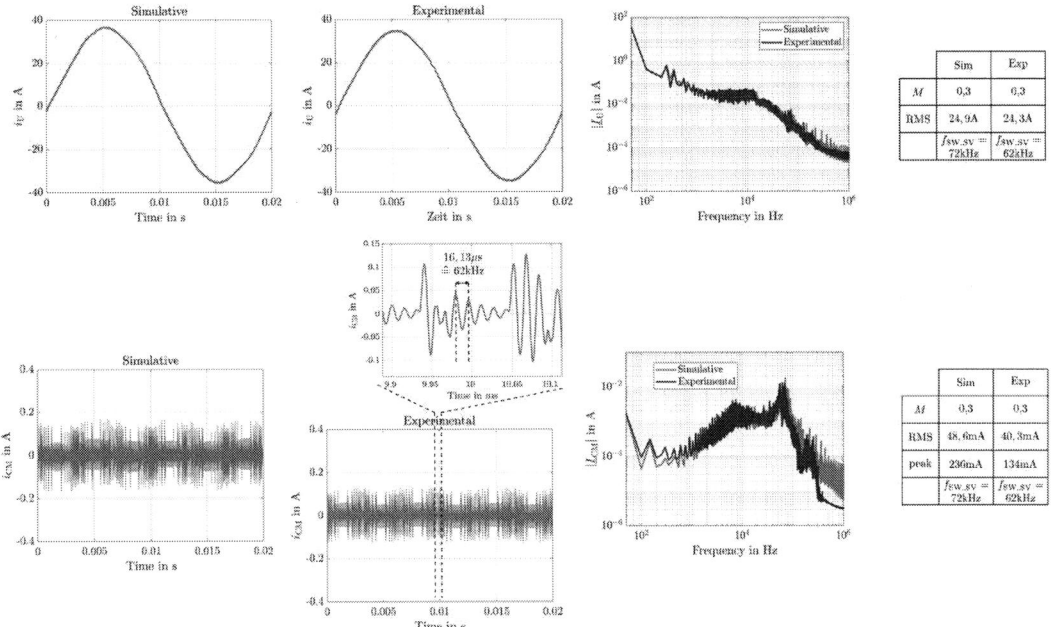

Fig. 11: Driven CM currents in comparison for simulative and experimental results

accuracy between simulative and experimental results is shown.

This work lays the groundwork for further investigation into the influence of CM currents on complex power electronics systems, particularly regarding EMI challenges in emerging applications like electrified aircraft. Furthermore, the potential of CM currents as an index for DC power quality, encompassing both low and high-frequency disturbances, warrants exploration. This would necessitate the development of standardized measurement techniques and the investigation of correlations between CM currents and existing power quality metrics.

References

[1] J. P. Kozak, R. Zhang, M. Porter, Q. Son, J. Liu, *et al.*, "Stability, Reliability, and Robustness of GaN Power Devices: A Review," IEEE Transactions on Power Electronics, 2023.

[2] A. Hirota, B. Saha, S.-P. Mun, and M. Nakaoka, "Common Mode EMI Analysis in Power Electronics Enabled Power System," 2021 IEEE Energy Conversion Congress and Exposition, 2021.

[3] A. Muetze, "Bearing Currents in Inverter-Fed AC-Motors," en, Dissertation, Technische Universität Darmstadt, Darmstadt, 2003.

[4] I. Manushy, "Design and Optimization of EMI Filters for Power Electronics Systems," en, Dissertation, Technische Universität Darmstadt, Darmstadt, 2019.

[5] S. Bernet, *Selbsgeführte Stromrichter am Gleichspannungszwischenkreis*, de. Springer, 2012.

[6] M. Schmitt and A. Ackva, "Active damping of common-mode oscillations in electric drive systems using direct current control," 19th European Conference onPower Electronics and Applications, 2017.

[7] A. Hirota, B. Saha, S.-P. Mun, and M. Nakaoka, "An Advanced Simple Configuration Delta-Sigma Modulation Three-Phase Inverter Implementing Space Voltage Vector Approach," IEEE Power Electronics Specialists Conference, 2007.

[8] G. Holmes and T. Lipo, *Pulse Width Modulation for Power Converters*, en. John Wiley & Sons, 2003.

[9] A. Mertens and V. Ganesan, "Three-phase sigma-delta modulation using zero-sequence components," IEEE European Conference on Power Electronics and Applications, 2005.

[10] B. Jacob and M. Baiju, "Space Vector based pulse Density Modulation scheme for two level voltage source inverter," IEEE Conference on Industrial Electronics and Applications, 2011.

[11] L. Heinemann, R. Schulze, P. Wallmeier, and H. Grotstollen, "Modeling of high frequency inductors," Proceedings of 1994 Power Electronics Specialist Conference - PESC'94, 1994.

[12] A. Wist, T. Haas, B. Dressel, and A. Ackva, "High Frequency Model with Ohmic Behave of a Three-Phase Coil," PCIM Europe 2023, 2023.

PCIM Europe 2024, 11– 13 June 2024, Nuremberg   DOI: 10.30420/566262329

Analysis of Common-Mode Noise Generated due to Fast-Switching GaN Devices in Totem-Pole PFCs

Ali Tausif [1], Serkan Dusmez [2]

[1] Yildiz Technical University, Istanbul, Turkey
[2] Huawei Technologies Duesseldorf GmbH, Nuremberg Research Center, Germany

Corresponding author: Serkan Dusmez, serkan.dusmez@huawei.com

Abstract

The adoption of GaN devices has led to an increase in the switching frequency of high-power single-phase power-factor-correction converters, rising from 20-50 kHz to 65-135 kHz. This shift is primarily attributed to the rapid switch-node transitions and the absence of reverse recovery losses in GaN devices. However, the higher switching frequency and switch-node transitions have implications for common-mode (CM) noise filtration. This paper proposes mathematical models to estimate CM noise generated by GaN device switching and analyzes the influence of various factors such as switching frequency, parasitic elements, dv/dt, and voltage ringing on CM noise. Additionally, the study suggests that unlike Si-based converters, CM noise peaks resulting from switch-node voltage ringing appear in the radiated emission range for GaN-based converters.

1   Introduction

As the demand for higher efficiency and miniaturization intensifies, high switching frequency converters utilizing fast GaN devices have become a prominent choice. These converters offer the advantage of utilizing smaller passives, contributing to space-saving benefits [1, 2]. However, this transition to higher frequencies comes with a trade-off in the form of increased CM noise [3–5]. In contrast to typical Si-based power-factor-correction (PFC) converters, GaN-based hard-switched totem-pole (TP) converters often operate at hundreds of kHz of the switching frequency, the switch-node voltage transitions are quite fast, and the loop inductances are low [6, 7]. Considering the poor attenuation performance of EMI filters at high frequencies due to self-parasitics or mutual couplings, these otherwise beneficial features of GaN devices introduce additional challenges in terms of EMI compatibility [8]. Estimating the CM noise is highly challenging, as it depends on various converter parasitics, resulting from PCB layouts, heatsinks, and trace inductances [9]. Accurate CM noise modeling is a cumbersome task, particularly for PFC converters, as it involves several intricate aspects such as varying duty cycle, different turn on/off transitions and zero-crossing distortion. Modeling parasitics in high-frequency converters is not straightforward and often requires estimation and physical measurements of the prototype under test. However, once these measured parasitics are accurately obtained, predicting common-mode (CM) noise can be done accurately, relying on appropriate CM noise models.

Therefore, this paper presents a comprehensive mathematical model for CM noise in TP PFC converters. Using these developed models, the CM noise resulting from the use of fast GaN devices is analyzed, and its impact on the EMI spectrum is evaluated. The main contribution of this paper lies in determining crucial parameters in GaN-based TP PFC converters concerning CM noise evaluation. These findings not only aid in achieving optimal design and performance of EMI filters but also help in achieving the optimal design of the converter.

2   Common mode noise modeling and estimation

A TP PFC topology with parasitics impacting the CM noise is shown in Fig. 1. Here, the key parasitic capacitances and loop inductances between the earth and (1) the switch-node of GaN half-bridge, (2) power ground, (3) switch-node of rectifier half-bridge are labeled. There are two sources of CM noise in a TP PFC based on switching; (a) noise

2334

PCIM Europe 2024, 11– 13 June 2024, Nuremberg   DOI: 10.30420/566262329

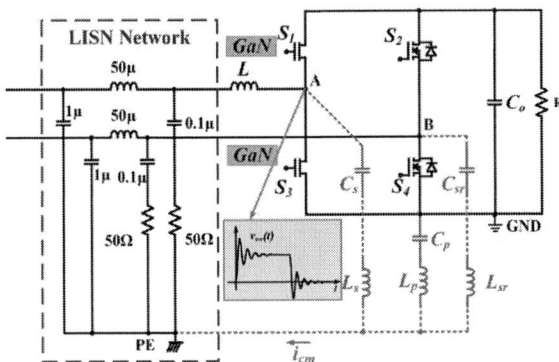

Fig. 1: CM noise causing parasitics in a TP PFC.

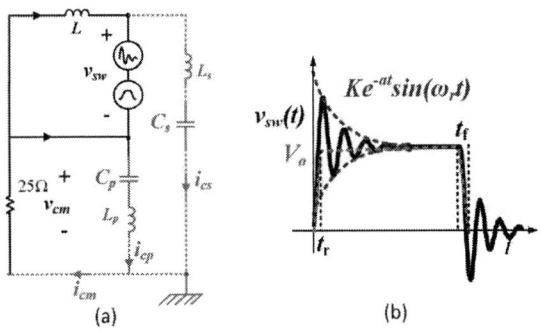

Fig. 2: Noise modeling during high frequency switching: (a) equivalent CM noise model, (b) voltage ringing model.

due to high frequency switching, (b) noise due to line frequency switching. In this paper, only the high frequency switching noise is considered.

TP PFC converter exhibits hard-switching behavior when operated in CCM. It is essential to take into account the overshoot and ringing caused by the stray loop inductance (L_{loop}) of the DC power rail and the output capacitance (C_{oss}) of half-bridge during the on/off transition. The equivalent CM noise model is given in Fig. 2(a). In this paper, the voltage ringing component is modeled within the source of CM noise. Therefore, the CM noise source resulting from high-frequency switching includes the ringing voltage component along with the trapezoidal-shaped voltage waveform characterized by rise-time (t_r) and fall-time (t_f), as illustrated in Fig. 2(b).

To model the ringing part, the CM equivalent circuit incorporates an RLC circuit formed by L_{loop}, C_{oss}, and some stray resistance R_{stray} associated with DC link and PCB traces. The general expression approximating this ringing phenomenon can be given as; $v_{\text{ring}} = Ke^{-\alpha t}\sin(\omega_d t)$, where $\omega_d = \sqrt{\omega_r^2 - \alpha^2}$ and $K = V_o e^{-\pi\zeta/\sqrt{1-\zeta^2}}$. Here, K is the maximum amplitude of oscillation, α is the attenuation constant and is equal to $R_{\text{stray}}/(2L_{\text{loop}})$, and ζ is the damping coefficient equal to $\alpha\omega_r$.

2.1 CM noise source modeling

Considering Fig. 2(b), the noise source comprises a ringing component and a trapezoidal waveform characterized by rise-time (t_r) and fall-time (t_f).

2.1.1 CM noise source modeling without ringing part

Since CM noise is highly dependent on dv/dt, it is important to take the effects of t_r and t_f into account. For simplicity in analysis, t_r can be assumed

equal to t_f for now. However, later, an intuitive approach will be used to explain scenarios where t_r and t_f are not equal.

The Fourier expression for trapezoidal wave part is given as:

$$
c_{n,\text{trap}}(n, d(t)) = V_o d(t) \left(\frac{\sin\left(\frac{n\omega d(t)T_s}{2}\right)}{\frac{n\omega d(t)T_s}{2}} \right) \times
$$
$$
\left(\frac{\sin\left(\frac{n\omega t_r}{2}\right)}{\frac{n\omega t_r}{2}} \right) e^{\frac{-jn\omega(t_r - d(t)T_s)}{2}} \tag{1}
$$

Where n = harmonic number given as $\lceil 150k/f_s \rceil$.

Impact of duty cycle on CM noise source spectrum: According to the expression in Eq. (1), it is quite evident that a change in the duty cycle affects both the magnitude of low-frequency components and the point of the first cutoff frequency in the EMI spectrum of the switch node voltage, as shown in Fig. 3. It is important to understand the impact of the duty cycle on the overall spectrum. Fig. 3 shows how duty cycle variation affects the upper bound of the EMI spectrum for a non-ringing square wave. It can be observed that the duty ratio affects only the low-frequency spectral envelope of the waveform.

The expression mentioned above also elucidates how the envelope of the CM noise source behaves when taking into account the effects of rise and fall times. Fig. 4 illustrates their impact on harmonic spectrum asymptotes. The first cutoff frequency correlates with the switch voltage duty ratio, resulting in an overall 20 dB/dec reduction in slope beyond that point. Similarly, the second cutoff frequency relates to the rise/fall time, introducing another 20 dB/dec reduction.

2335

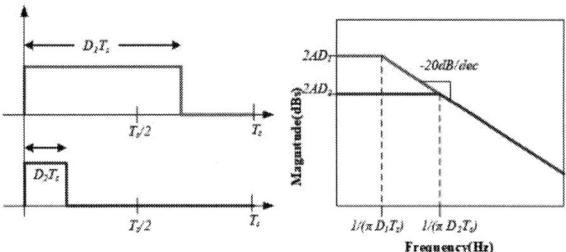

Fig. 3: Effect of duty cycle on CM noise source spectrum.

Impact of rise-time/fall-time on CM noise source spectrum: To understand the effects of diverse rise and fall times on the switch voltage's EMI spectrum, a sequence of impulses are considered in a time derivative of finite order as discussed in [10] and shown in Fig. 4. Fourier series expansion coefficients of a waveform are expressed as,

$$c_n = \frac{1}{T_s(jn\omega_o)^k} \sum_{i=1}^{p} M_i e^{-jn\omega_o \tau_i}, n \neq 0. \quad (2)$$

M_i and τ_i represent the (signed) amplitudes and times of the p impulses occurring in the kth derivative of a waveform with a period T_s and angular frequency $\omega_o = \frac{2\pi}{T_s}$. Although inspection of this expression does not directly determine the overall spectral bound of the waveform, it provides information about the amplitude and gradient of its terminal high-frequency roll-off, given by:

$$\lim_{n \to \infty} c_n = \frac{1}{T_s(jn\omega_o)^k} \sum_{i=1}^{p} |M_i| \quad (3)$$

For a waveform with equal rise and fall times, the impulses associated with each transition contribute half of the total magnitude M_i. If the duration of one transition is increased, such that $t_r < t_f$, the total magnitude of the impulses will be reduced. As t_f approaches infinity, the value of M_i will be halved because the impulses associated with the falling edge of the waveform will have zero amplitude. This results in a 6 dB reduction of the amplitude of the high-frequency spectral content of the waveform as illustrated in bottom part of Fig. 4. Therefore, in the case of asymmetrical transition times, the faster transition dominates the spectral characteristics of the waveform. If the duration of one transition is increased, the high-frequency spectral amplitude of the waveform will be suppressed by a maximum of 6 dB relative to the original symmetrical waveform.

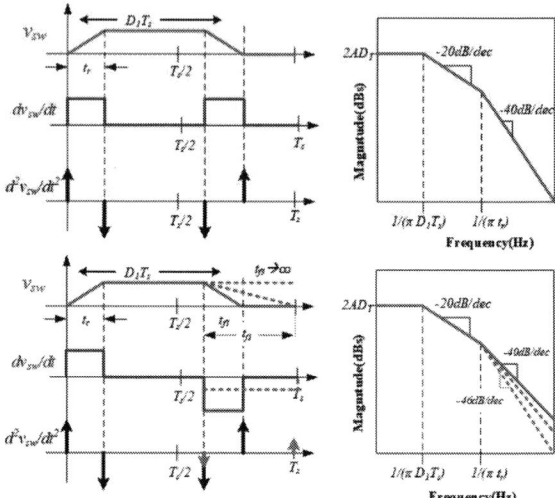

Fig. 4: Illustration of the impact of rise/fall time on CM noise source spectrum.

2.1.2 CM noise source modeling with ringing part

The common mode noise spectrum including the noise due to switch node voltage ringing can be given by the expression below:

$$c_{n,\text{sw}}(n, t) = c_{n,\text{trap}}(n, d(t)) + c_{n,\text{ring}}(t) \quad (4)$$

Here $c_{n,\text{ring}}$ can be given as the ringing voltage occurring at turn on and turn-off instances. For the ease of analysis we assume that oscillations at turn on and turn-off instances are identical, so the expression for ringing spectrum can be given using Fourier analysis,

$$c_{n,\text{ring}}(n, d(t)) = \frac{1}{T_\text{s}} \int_0^{T_s} K e^{-\alpha t} \sin(\omega_r t) e^{-jn\omega t} dt -$$
$$e^{\frac{-(jn\omega d T_s)}{2}} \left[\frac{1}{T_\text{s}} \int_0^{T_s} K e^{-\alpha t} \sin(\omega_r t) e^{-jn\omega t} dt \right] \quad (5)$$

Putting Eq. (1) and Eq. (5 in Eq. (4) and taking the absolute of it yields,

$$|c_{n,\text{sw}}(n, d(t))| = V_o d(t) \left(\frac{\sin\left(\frac{n\omega d(t)T_s}{2}\right)}{\frac{n\omega d(t)T_s}{2}} \right) \left(\frac{\sin\left(\frac{n\omega t_r}{2}\right)}{\frac{n\omega t_r}{2}} \right) \times$$
$$\left| \frac{(-j(k\omega)^2 - (2\alpha + (\frac{K}{V_o})\omega_r)(k\omega)^2 + (\alpha^2 + \omega_r^2)jk\omega)}{((jk\omega^2 + 2\alpha jk\omega + (\alpha^2 + \omega_r^2))} \right| \quad (6)$$

2.1.3 Propagated CM noise across LISN

CM noise appearing at the LISN network terminals can be determined by considering all the circuit

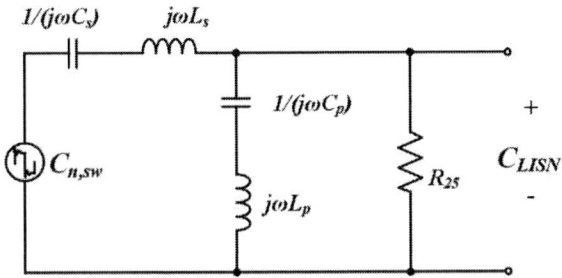

Fig. 5: Equivalent noise model of TP PFC considering parasitics in high frequency switching.

parasitics of the circuit which includes C_s, L_s, C_p and L_p, as shown in Fig. 5. The expression for CM noise spectrum can be given by finding the transfer function of the equivalent circuit in Fig. 5. It can be seen that from Fig. 5, a 2nd order RLC resonance circuit is formed, whose resonance frequencies due to pole and zero are given in Eq. (7) and Eq. (8) respectively,

$$\omega_p = \frac{1}{\sqrt{(L_p + L_s)(C_p \| C_s)}} \quad (7)$$

$$\omega_z = \frac{1}{\sqrt{L_p C_p}} \quad (8)$$

The transfer function of the Fig. 5 is represented as

$$\frac{c_{\mathsf{LISN}}(n, d(t))}{c_{n,\mathsf{sw}}(n, d(t))} = \frac{jn\omega R_{25}C_s(A-1)}{(A+B) - A \cdot B - 1 + jn\omega R_{25}(BC_p + AC_s) - jn\omega R_{25}C} \quad (9)$$

where, $A = n^2\omega^2 C_p L_p$, $B = n^2\omega^2 C_s L_s$ and $C = C_p + C_s$. The final expression for n^{th} harmonic of the propagated CM noise is given as

$$c_{\mathsf{LISN}}(n, d(t)) = \frac{jn\omega R_{25}C_s(A-1) \cdot c_{n,\mathsf{sw}}(n, d(t))}{(A+B) - A \cdot B - 1 + jn\omega R_{25}(BC_p + AC_s) - jn\omega R_{25}C} \quad (10)$$

The magnitude of CM noise appearing across the LISN is given by averaging over whole line cycle and taking absolute value of it.

$$|v_{\mathsf{cm}}| = \left| \frac{1}{T_g} \int_0^{T_g} c_{\mathsf{LISN}}(n, t)) dt \right| \quad (11)$$

Finally, CM noise in dBμV can be obtained as,

$$v_{\mathsf{cm}}[\mathsf{dB\mu V}] = 20 log_{10}\left(\frac{|v_{\mathsf{cm}}|}{1\mu V}\right) \quad (12)$$

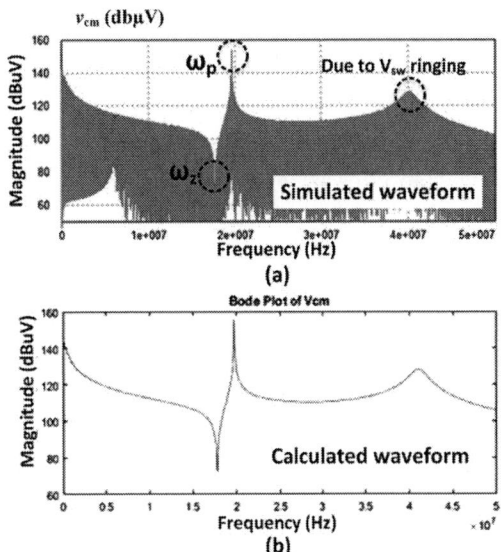

Fig. 6: Validation and comparison of; (a) simulation based CM noise spectrum, (b) calculation based CM noise spectrum.

2.2 CM noise model validation

The simulated CM noise envelope is shown in Fig. 6(a) along with the calculated one as shown in Fig. 6(b). It can be seen that the CM noise model results closely match the simulation results obtained with the same parasitic elements. The simulation parameters of the design example are provided in Tab. 1.

According to the results in Fig. 6, two resonance peaks occur inside the CM noise spectrum. The first peak occurs at around 19.4 MHz, which is due to w_p, and the second peak occurs at 40 MHz, attributed to the voltage ringing in the CM noise source. It should also be noted that there is a notch at w_z due to the zero present in Eq. (9).

3 Impact of fast switching GaN devices on CM noise

CM noise at the switching frequency and some of its lower harmonics can be effectively filtered using a CM filter comprising CM chokes and Y-capacitors. However, due to the parasitics involved in circuit elements and loop paths, the behavior of CM noise changes, especially at higher frequencies, as these parasitics typically have low values in GaN based systems. These parasitics cause multiple resonances within the EMI spectrum, where noise peaks often occur. These resonances can lead to deterioration in converter performance in

Tab. 1: Simulation parameters

Input voltage (V_g)	220 V_{rms}
Output voltage (V_o)	400 V
FET's output capacitance (C_{oss})	150 pF
Power loop inductance (L_{loop})	50 nH
Switch-node to earth Capacitance (C_s)	3 nF
Ground to earth capacitance (C_p)	5 nF
Switch-node to earth loop inductance (L_s)	18 nH
Ground to earth loop inductance (L_p)	18 nH

terms of EMI and pose challenges in the optimum design of an EMI filter. Additionally, the filter insertion loss does not remain consistent throughout the frequency range and deteriorates at higher frequencies due to self-parasitics and mutual couplings within the filter components. Therefore, it is crucial to consider all these parameters while designing a converter using fast GaNs.

In the following sections, the impacts of different parameters such as switching frequency, parasitic elements (i.e., C_s, C_p, L_s, and L_p), and dv/dt on GaN-based TP PFC are discussed, and critical factors influencing the design of a GaN-based TP PFC converter in terms of CM noise are highlighted.

3.1 Impact of switching-frequency on CM noise spectrum

The CM noise spectrum, portrayed as a bode plot, is derived using the proposed model across a broad frequency range, spanning from 150 kHz to 200 MHz. Notably, an increase in the switching frequency (f_s) correlates with an escalation in the magnitude of the CM spectrum at lower frequencies, assuming all other parameters remain constant. As depicted in Fig. 7(a), when f_s is set at 55 kHz, the filter's design frequency stands at 165 kHz with a magnitude of 96 dB. Elevating f_s to 100 kHz shifts the filter design frequency to 200 kHz, concurrently raising the magnitude to 103 dB. Similarly, at $f_s = 135$ kHz, the magnitude peaks at 106 dB, aligning with the filter's design frequency of 270 kHz.

3.2 Impact of parasitics on CM noise

The presence of parasitics in the switch-node to earth and ground to earth loops is critical in determining the CM noise spectrum, especially at high-frequency regions, as they directly influence the propagation of CM noise, as shown in Fig. 7. These parasitics typically result in resonance peaks and notches in the CM noise spectrum.

It can be demonstrated from Fig. (7) that C_s not only contributes to resonance at high frequencies when combined with L_s but also affects the amplitude of CM noise at low frequencies. Figure 7(b) illustrates the impact of C_s on the CM noise spectrum while keeping L_s constant, highlighting that CM noise is highly sensitive to C_s, impacting it even at low frequencies. Lower C_s values result in reduced CM noise levels. Additionally, the design of the CM noise filter is significantly dependent on C_s, as it plays a crucial role in determining the magnitude of CM noise at the specified frequency where the CM filter design is implemented. On the other hand, C_p does not have as significant an impact on either low frequencies or higher frequencies, except for changing the notch frequency, as illustrated in Fig. 7(c).

3.3 Impact of dv/dt and voltage ringing on CM noise spectrum

Similarly, Fig. 7(d) illustrates the impact of increasing dv/dt. At low frequencies, the influence of dv/dt on CM noise is negligible; however, it becomes more pronounced at the resonance frequency of the voltage ringing. These frequencies coincide with the onset of filter couplings impacting the filter response [8]. Consequently, any enhancement in this frequency range contributes significantly to achieving EMI compliance.

In order to comprehensively assess the influence of dv/dt in GaN based TP PFCs, it is essential to draw a comparison with Si-based PFCs. This comparative analysis holds significant importance due to the substantial differences in voltage ringing, L_{loop}, and C_{oss} between Si and GaN based converters. In Si-based half-bridges, the switch-node voltage ringing usually remains within the conducted EMI range as illustrated in Fig. 8(a). At these frequencies, the filter response deteriorates due to self-parasitics and mutual couplings of the filter components. For instance, attenuation of a two-stage EMI filter for CM noise has been illustrated in Fig. 8(b). In the frequency range of 5-30 MHz, the filter's attenuation experiences a significant decline.

In the realm of Si-based converters, engineers commonly adopt the practice of reducing dv/dt in fast switches to successfully pass conducted EMI tests. This approach proves effective in mitigating CM

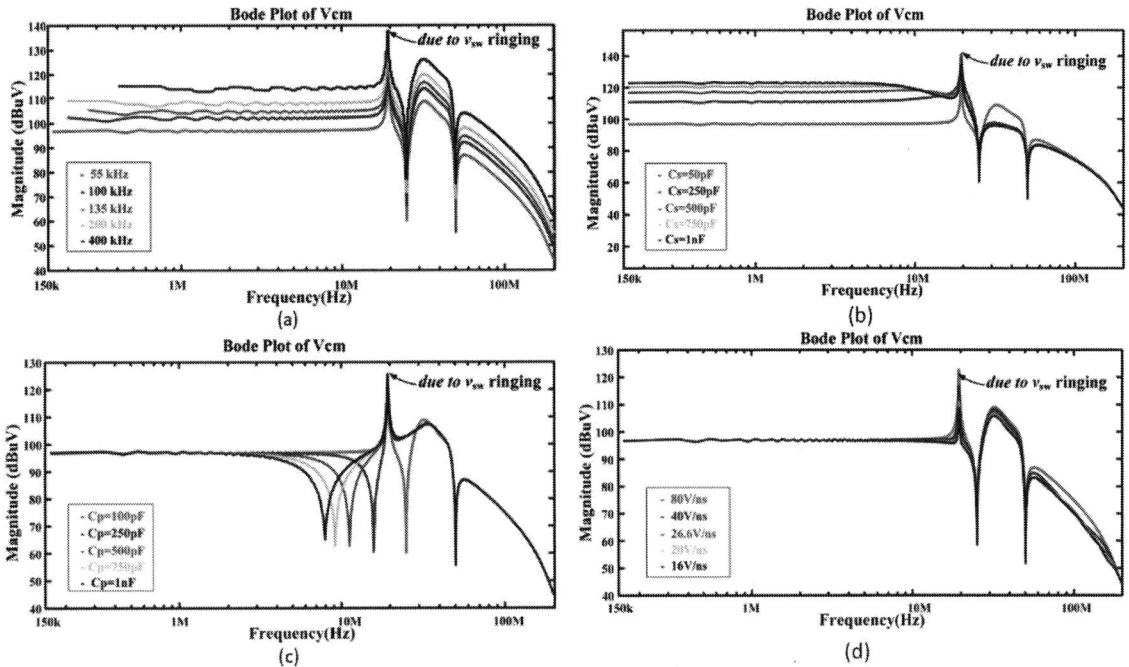

Fig. 7: CM noise spectrum for different; (a) switching frequencies, (b) switch-node to earth parasitic capacitance (C_s), (c) Ground to earth parasitic capacitance (C_p), (d) dv/dt.

noise caused by ringing within the conducted EMI range as seen from Fig. 8(a).

However, the scenario shifts significantly when dealing with GaN-based converters. This is attributed to the relatively smaller parasitics involved, specifically the L_{loop} and C_{oss} of GaN FETs. In GaN-based systems, noise resulting from ringing manifests within the radiated emission range as shown in Fig. 8(c). Consequently, it becomes imperative to consider factors such as loop area and energy stored in the loop.

3.4 dv/dt adjustment for reduced CM noise and its impact on switching losses

After developing a comprehensive CM noise model and understanding the crucial parameters that impact the converter's performance in terms of CM noise, it is valuable to apply this model practically to estimate and predict the performance of fast GaN-based converters with high dv/dt. It is a known fact that in GaN TP PFCs, high dv/dt results in low switching losses, especially IV overlap losses. However, it also leads to increased CM noise at high-frequency regions where ringing occurs. This is due to the rapid transitions in the presence of

L_{loop} and C_{oss}, generating voltage ringing in the switch-node voltage with high overshoot. This overshoot appears as a peak in the CM noise spectrum and can sometimes exceed the maximum switch voltage limit. Therefore, the careful selection of dv/dt is critical not only for minimizing switching losses but also for managing CM noise levels and preventing device failure due to voltage overshoot beyond the maximum limit.

In this section, the impact of changing dv/dt on switching losses and CM noise is presented and discussed. As an example, the 650V GaN FET (GS66508B) is considered for implementing a single-leg TP PFC converter operating at $f_s = 67$ kHz where the C_{oss} and L_{loop} is taken as 65 pF and 5.6 nH respectively. The SPICE model of GS66508B is utilized to determine the dv/dt and voltage ringing on the V_{DS} of the GaN FET. Additionally, switching loss models presented in [11] and [12] are applied to estimate the half-bridge switching losses, and the results are provided in Tab. 2. The table results indicate that the CM noise at the ringing frequency is highest at 71 dB, with a voltage overshoot of around 18%, which is within the allowable limit when considering 650 V GaN. The switching losses are minimized at 3.75 W due

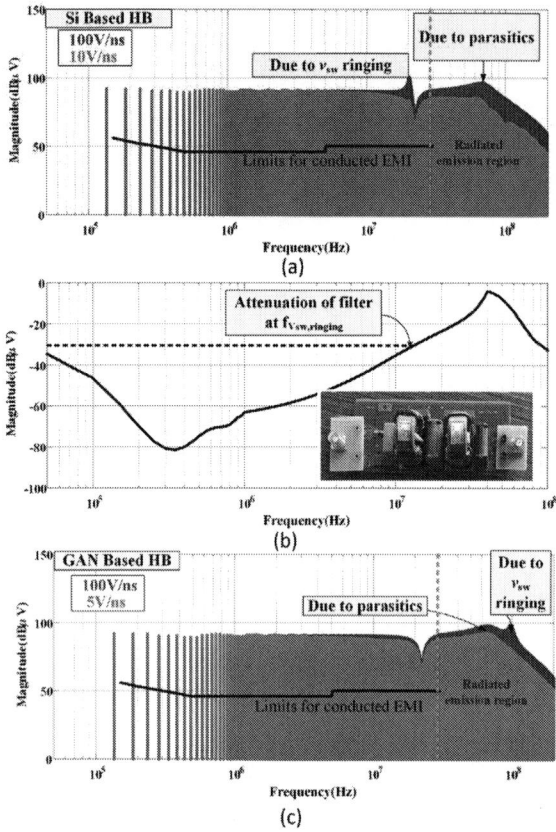

Fig. 8: Noise spectrums and filter responses; (a) CM noise spectrum for Si based PFC, (b) insertion loss of a typical 2-stage EMI filter, (c) CM noise spectrum for GaN based PFC.

to small IV overlap losses. This is achieved with a high dv/dt of 118 V/ns and R_g=21 Ω. To reduce CM noise caused by ringing, the dv/dt is decreased to 104 V/ns by increasing R_g to 24 Ω. This adjustment lowers the CM noise value to 67.1 dB but introduces 3.9 W of switch power loss. The lowest dv/dt at 39 V/ns requires 65 Ω of R_g, resulting in a CM noise value of 49 dB. However, switching losses in this scenario are highest at 5.9 W.

Consequently, a large L_{loop} also increases the voltage ringing amplitude and, consequently, CM noise. In such scenarios, decreasing dv/dt can be beneficial. According to optimum design guidelines, after designing a filter and identifying the peak points that are violating the EMI limits, reducing dv/dt can effectively decrease high-frequency CM noise to meet EMI standards at the expense of increasing switching losses.

Tab. 2: CM noise and switching losses at different switch-node voltage slew rates.

R_g (Ω)	dv/dt (V/ns)	CM noise at ringing freq. (dBμV)	Switching losses (W)
21	118	71	3.75
24	104	67.1	3.9
29	88	62	4.13
37	68	54	4.56
65	39	49	5.97

4 Conclusion

This paper studied the mathematical background and impact of various parasitics and parameters on generated common-mode noise, including parasitic capacitance and inductances, switching frequency, and switch node voltage slew rate (dv/dt). Among these, switch-node to earth parasitic capacitance and switching frequency are shown as the major contributing factors to generated CM at the switching frequency and its low-order multiples, whereas the switch-node voltage transition only impacts the high-frequency noise, particularly, the peak occurring at the resonance frequency of $2C_{oss} - L_{loop}$.

Minimizing switching losses becomes achievable by driving GaN devices rapidly, resulting in high dv/dt transitions. Fast switching, however, necessitates low loop inductance to keep ringing amplitude low. In instances where conducted EMI tests reveal noise spikes at the $2C_{oss} - L_{loop}$ resonance, it is prudent to consider slowing down the dv/dt, which results in increased switching losses.

For GaN-based converters, reducing dv/dt may not always effectively mitigate CM noise, especially in the conducted EMI range. This is because the noise ringing frequency can extend into the radiated noise region. However, detailed investigations into the impact of different dv/dt values on radiated CM noise characteristics and switching losses will provide valuable insights that can aid in optimizing GaN-based converter designs for improved EMI performance.

References

[1] X. Huang, Z. Liu, F. C. Lee, and Q. Li, "Characterization and enhancement of high-voltage cascode GaN devices," IEEE Trans. Electron. Devices, vol. 62, no. 2, pp. 270–277, Feb. 2015.

[2] J. Sun, J. Li, D. J. Costinett, and L. M. Tolbert, "A GaN-based CRM totem-pole PFC converter with fast dynamic response and noise immunity for a multi-receiver WPT system," in Proc. IEEE Energy Convers. Congr. Expo. 2020, pp. 2555–2562.

[3] Bingyao Sun and R. Burgos, "Assessment of switching frequency impact on the prediction capability of common-mode EMI emissions of sic power converters using unterminated behavioral models," 2015 IEEE Applied Power Electronics Conference and Exposition (APEC), Charlotte, NC, USA, 2015, pp. 1153-1160.

[4] K. Shi, M. Shoyama and S. Tomioka, "Common mode noise reduction in totem-pole bridgeless PFC converter," 2014 International Power Electronics and Application Conference and Exposition, Shanghai, China, 2014, pp. 705-709, doi: 10.1109/PEAC.2014.7037943.

[5] J. Yao, Y. Li, S. Wang, X. Huang, and X. Lyu, "Modeling and reduction of radiated EMI in a GaN IC-based active clamp flyback adapter," IEEE Trans. Power Electron., vol. 36, no. 5, pp. 5440–5449, May 2021

[6] M. Sasaki, J. Imaoka and M. Yamamoto, "An Investigation on the Relationship between CM Noise and Distribution of Parasitic Capacitance," 2022 International Power Electronics Conference (IPEC-Himeji 2022- ECCE Asia), Himeji, Japan, 2022, pp. 753-758.

[7] S. Wang and F. C. Lee, "Common Mode Noise Reduction for Power Converters with Parasitic Capacitance Cancellation," APEC 07 - Twenty-Second Annual IEEE Applied Power Electronics Conference and Exposition, Anaheim, CA, USA, 2007, pp. 923-928, doi: 10.1109/APEX.2007.357625.

[8] H. K. Comert, G. Odabas and S. Dusmez, "A 3-D Single-stage Differential Mode Filter Modeling with Mutual Couplings and Layout Optimizations," ACEMP & OPTIM, Brasov, Romania, 2021, pp. 265-273.

[9] Haoyi Ye, Zhihui Yang, Jingya Dai, Chao Yan, Xiaoni Xin and Jianping Ying, "Common mode noise modeling and analysis of dual boost PFC circuit," INTELEC 2004. 26th Annual International Telecommunications Energy Conference, Chicago, IL, USA, 2004, pp. 575-582, doi: 10.1109/INTLEC.2004.1401526.

[10] C. R. Paul, "Signal Spectra—the Relationship between the time domain and the frequency domain," Introduction to Electromagnetic Compatibility, 2nd ed. Hoboken, NJ, USA: Wiley, 2006, pp. 91–175.

[11] E. B. Bulut, M. O. Gulbahce, D. A. Kocabas and S. Dusmez, "Simplified Method to Analyze Drive Strengths for GaN Power Devices," PCIM Europe digital days 2021; International Exhibition and Conference for Power Electronics, Intelligent Motion, Renewable Energy and Energy Management, Online, 2021, pp. 1-8.

[12] S. Satpathy, P. P. Das and S. Bhattacharya, "Study of Switching Transients based on dv/dt and di/dt for a GaN-based Two-Level Pole," 2021 IEEE 12th Energy Conversion Congress & Exposition - Asia (ECCE-Asia), Singapore, Singapore, 2021, pp. 19-25, doi: 10.1109/ECCE-Asia49820.2021.9479426.

PCIM Europe 2024, 11– 13 June 2024, Nuremberg   DOI: 10.30420/566262330

Conducted EMI from GaN-based 48 V to 12 V DC-DC Converters for Automotive Applications

Marita Wendt[1], Erik Kampert[1], Jost Wendt[2], Klaus F. Hoffmann[1], Stefan Dickmann[1]

[1] Faculty of Electrical Engineering, Helmut-Schmidt-University, Germany
[2] Senior Consultant, Germany

Corresponding author: Marita Wendt, wendtm@hsu-hh.de
Speaker: Erik Kampert, erik.kampert@hsu-hh.de

Abstract

Gallium Nitride (GaN)-based converters have emerged as a promising technology to increase the efficiency and power density of power electronic systems in electric vehicles (EVs). However, the integration of GaN semiconductor devices introduces challenges regarding electromagnetic interference (EMI). This work focusses on EMI measurements and on exploring the relationship between defined parameters such as output current or switching frequency and conducted emissions (CE) in converters with GaN switches. Moreover, the relationship between the switching transients of the GaN-FETs and the interference frequency range is examined and the challenges of using GaN-based converters in EVs related to EMI are discussed.

1 Introduction

1.1 Preface

Electric vehicles (EVs), as opposed to vehicles with internal combustion engines, feature a significantly higher number of electronic modules and components. These pose additional risks to critical and non-critical sections and functions of the car due to the high frequency noise being generated. Such noise producing modules are largely found in the areas of power conversion such as the main variable speed drive, several smaller motor drives and, particularly, in the numerous step-up or step-down voltage converters. Especially, if a vehicle or a segment thereof is operating in an autonomous mode, all such risk factors need to be eliminated.

Autonomous driving combined with electromobility is crucial for the competitiveness of the automotive industry. Both trends require fundamental progress in ensuring electromagnetic compatibility (EMC), in particular, immunity to system-intrinsic and external electromagnetic interference (EMI). Hence, the susceptibility of autonomous systems to malfunction or failure is to be systematically examined [1].

All the data of individual sections and modules of such a vehicle in combination are then evaluated regarding the existing risks. The vast data set from

the experimental results is entered into a model utilising machine learning which will finally ease the prediction of risk factors and methods of their mitigation.

1.2 Structure and content

This paper looks specifically at a voltage conversion section situated between an intermediate battery section often used with a nominal DC-voltage of 48 V and the 12 V battery circuit for most control and interface functions. The main drive converter and the on-board charger with nominal voltages from 400 V to 800 V are not part of this examination. One part of the research project examines the actual high frequency emissions produced and conducted by a significant section of an implicit EV. It also determines the origin of the emissions and gives recommendations both regarding any possible degradation of functionality of the vehicle as well as their mitigation.

2 Selection of semiconductors and topology

2.1 Challenges related to EMI when using GaN-based converters in EVs

In order to minimise the footprint and weight of converter circuits, it has become common practice to increase switching frequencies, thus minimising inductive components. Wide bandgap (WBG) devices, such as those based on Gallium Nitride (GaN) or Silicon Carbide (SiC), are particularly suitable for this task. They are able to achieve very high switching speeds due to their high carrier mobility, reduced device capacitances and the lack of reverse recovery losses (GaN) [2].

Using high switching frequencies (up to 500 kHz in case of hard switching) has a positive effect on the efficiency. It increases the power density - one of the crucial factors - which leads to compact and light-weight power electronic systems in EVs.

In this paper, EPC® eGaN-HFETs are used, which are normally-off switches with a lateral device structure and a two-dimensional electron gas (2DEG) channel, which provides the high charge carrier mobility. These devices have low input/output capacitances in combination with low parasitic inductance and no reverse recovery charge. This results in fewer losses when charging or discharging the device capacitances and, therefore, in lower switching losses.

Nevertheless, despite the excellent properties in terms of efficiency and power density of eGaN-HFETs, the high turn-on and turn-off transition characteristics may produce considerable noise levels and could therefore lead to high EMI frequencies [3]. Both the conducted emissions (CE) and radiated emissions (RE) generated by GaN-based converters are potentially able to impact the overall EMC within EVs. All these points have been considered during the selection of the components and topology for the work presented here. Bi-directional (buck-boost) converters featuring two half-bridges operating interleaved are used for the conducted EMI measurements, and their different eGaN-HFETs packages are compared: in passivated die-form with solder bars and as quad flat no-leads (QFN).

2.2 eGaN-HFETs in on-board DC-DC converters

In this paper, different types of GaN transistors from EPC® are examined, all of which function as the power switches in a buck-boost converter with two conversion sections acting in interleaved mode. To gauge the risk stemming from the generated interference posed to other car functions, the latter can be separated into two major categories – actuators and sensors. It then needs to be further divided into the 48 V DC section and the 12 V DC section. Apart from mild hybrid vehicles which often include starter-generator systems, the 48 V section in a full hybrid EV might contain voltage and current sensors and some high-powered actuators, such as servo steering and pump motors. Most importantly, the servo motors (linear or rotational) are each fed from an electronic controller. These modules are susceptible to EMI to a certain degree.

The 12 V section will power most of the near-user actuators such as window motors, wiper motors, seat heating, car media centre etc. Most importantly, the electronic vehicle controllers are supplied from the 12 V rail. These controllers in turn generate the various voltages for all the sensors such as for temperatures, pressures, voltages and currents. It is thus primarily the controllers and only indirectly the sensors that need attention regarding EMI immunity. Very common and absolutely crucial failures or malfunctions can be found in engine control units, navigation systems, anti-lock braking systems, air-bag controls, car alarms, audio systems, collision warnings and avoidance control systems [4].

3 Measurement methods for conducted EMI

3.1 Basic considerations

The first section of this paper focusses on the CE of both the 48 V and the 12 V lines. Initially, the converter is configured in its buck function, so the flow of energy is from 48 V to 12 V. The test set-up is depicted in Fig. 1.

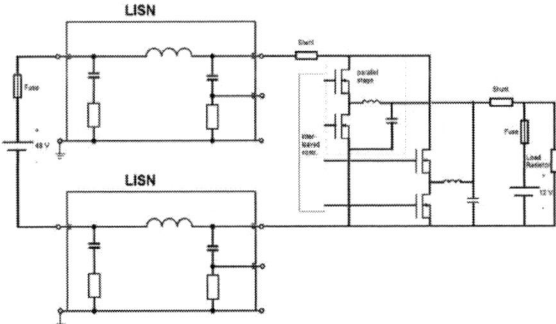

Fig. 1: Test set-up featuring 48 V supplying the converter via two LISNs.

The supply is fed via two line impedance stabilization networks (LISNs), as the standard CISPR 25 requires [5]. A current transducer is used to measure both the common mode (CM) and differential mode (DM) CE. Line currents are verified by inserting shunt resistors.

3.2 Experimental set-up

3.2.1 Converter selection

Central to the test set-up is a buck-boost converter from EPC® which is available in various configurations [6], [7]. The main variables are the GaN switches themselves as well as the main switching frequency. The two frequencies 250 kHz and 500 kHz are used here. Each converter operates in an interleaved mode, hence featuring two separate conversion sections (switches and inductors). Control of all main parameters is achieved via a programmable interface PCB by Microchip®. The output voltage is adjustable within a narrow window (12 V to 14 V). The maximum power is limited to 1.2 kW. With the option of adding a second converter PCB, a total power of 2.4 kW can be converted, as depicted in Fig. 2.

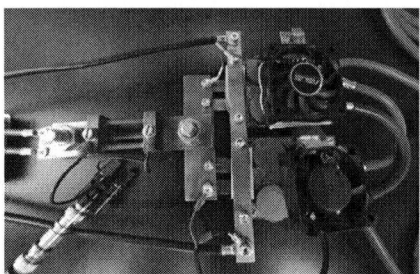

Fig. 2: 2.4 kW EPC® buck converter with power shunts.

3.2.2 Load selection

The load consists primarily of a bank of switchable power resistors (Fig. 3). In a later testing stage, a 12 V battery is added to the load circuit.

Fig. 3: Switchable resistor load banks.

3.2.3 Measurement equipment

The positive and return rails for the DC-DC converters under test (EPC9137 and EPC9165) are provided with a 48 V input voltage using a Delta Elektronika SM70-CP-450 power supply through a Schwarzbeck NNBM 8125 LISN in each line. The induced CM and DM currents are detected with an FCC F-52B current transformer and recorded with either a Keysight N9048B PXE EMI test re-

ceiver in spectrum analyser and max-hold mode or a Keysight MXR608A oscilloscope. The converter's DC input and output currents are measured with appropriate power shunts. A Keysight N2873A passive probe is used for the measuring the turn-on and turn-off voltages across the GaN transistors. It needs to be pointed out, that the two types of buck converters EPC9137 [6] and EPC9165 [7] used in these tests do not feature any EMI filters or intentional EMI mitigation measures.

3.2.4 Including car batteries as power source or power sink

In order to adapt the test conditions to the real state as much as possible, the next step is to set up the measurements using car batteries: a 48 V battery on the input side as a power source and a 12 V on the output side as an additional variable load or power source for the load (Fig. 4).

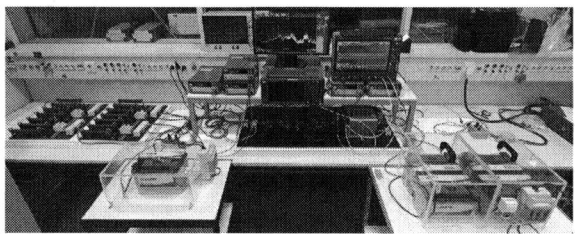

Fig. 4: Experimental set-up including car batteries.

4 Results and analysis of the CE measurements

4.1 Conducted interference currents for different modes and operating points

CE in EVs exhibit large spectral content across a frequency band beginning with the fundamental switching frequency. The different types of emitted interference currents will have different effects. CM currents flowing through the vehicle chassis (as a supply return path) most prominently contribute to radiated EMI, which is not part of this publication. Nevertheless, CM currents are measured and will be documented and depicted below, in order to provide an overview of the disturbance to be expected. Also, DM CE emanating from an inverter propagate through both DC power bus paths to the battery and all connected users. It then interacts with electronic circuits, causing crosstalk and interfering with other signals at the point of common coupling (PCC).

With the test results selected below it is possible to gauge later compliance with the individual standards applying to EVs [8]. As the origin of the EMI is primarily found at the 48 V input (due to the fact that the hard switching occurs directly on the positive supply), a number of different operating points and conditions are examined. Most important is the spectrum from 150 kHz to 30 MHz. The 12 V output is of most interest where it couples into all low voltage devices in the vehicle. Thus, the test results will focus on the absolute interference current (regardless if CM or DM) at the +12 V bus. To a certain degree this is superimposed by the noise at the input. The spectral analysis of the converter output will conclude the chapter.

Due to the fact that the set-up does not comprise an AC main or carrier signal (e.g. 50 Hz), but consists purely of a DC component, the usual considerations of noise signals need to be somewhat varied. Any and every AC current or voltage, both on the supply lines and on the load side, is to be considered as EMI. In addition, measurements on the load side will show a superimposed waveform of the interference present at the supply side.

At no-load (only 160 mA drawn by the 2" cooling fan) the converter works at the lowest operating point in discontinuous mode. At 100 A and above this has changed completely to the continuous mode. Results of the noise signals show that the DM EMI amplitude is heavily dependent on the load current, whereas the CM noise is, to a large degree, independent of the transferred power.

Several high-level peaks exist in the range from 50 MHz to 250 MHz, the origin of which will be explained in the following section. However, it should be noted that, although present as conducted EMI, their potential to lead to degradation of purpose or in fact to the malfunctioning of any modules or components will be most prevalent as radiated interference [9]. This type of EMI is not part of the segment of tests presented here.

4.2 Correlation between CE and rise or fall times

Since some of the high-power portions of the spectrum emanate – amongst others – from the switching transitions of the power transistors (i.e., the $\mathrm{d}v/\mathrm{d}t$ at turn-on and turn-off), this has been evaluated in more detail by direct measurements of the voltages across the transistors, as shown in Figs. 5 and 6.

The rise and fall times of the evaluated eGaN-HFETs during turn-on and respective turn-off have been determined, with selected results shown in Figs. 7 and 8. In general, the turn-on is slower

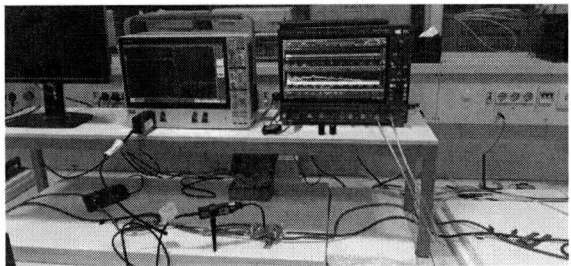

Fig. 5: Experimental set-up to determine $\mathrm{d}v/\mathrm{d}t$ of the switches.

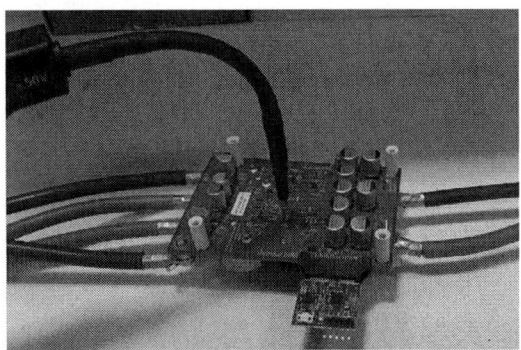

Fig. 6: Measurement of $\mathrm{d}v/\mathrm{d}t$ with differential voltage probe.

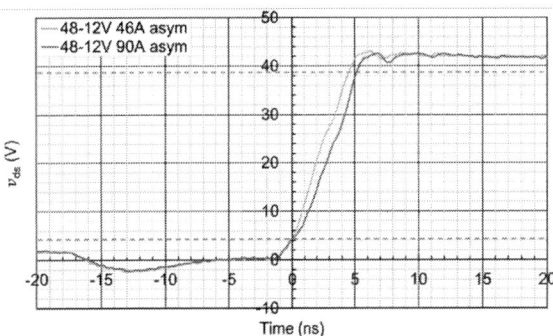

Fig. 7: v_{ds} turn-on performance of the upper switch eGaN-HFET (here referenced to the +48 V rail) depending on load current.

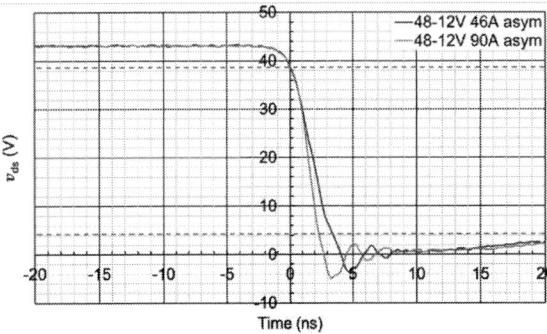

Fig. 8: v_{ds} turn-off performance of the upper switch eGaN-HFET (here referenced to the +48 V rail) depending on load current.

with average slopes in the range from 7 V/ns to 8 V/ns, whereas turn-off ranges from 10 V/ns to 20 V/ns, where absolute values also depend on the current and transistor type in operation.

The method of angular approximation assists in forecasting certain high-power sections within the interference spectrum. Essentially, this model relies on the observation that the switching times (10 % to 90 % of v_{ds}) approximate one third of a period of the generated interference frequency [10]. CE results of 40 MHz to 60 MHz can be predicted for the lower slope range, whereas the turn-off slope range is likely to produce signals at 70 MHz to 120 MHz. These ranges roughly match the experimental observations as will be detailed below, exemplarily shown in Fig. 9. In addition, the spectrum shows a prevalent peak at 200 MHz, which originates from oscillations generated at the end of a switching process.

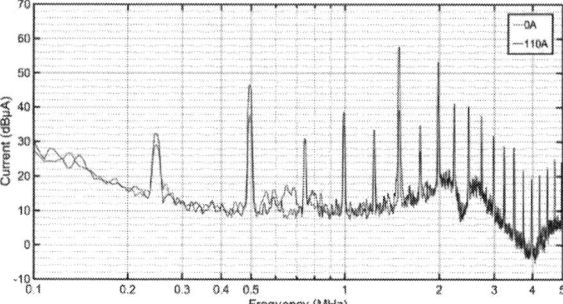

Fig. 10: CM spectrum to 5 MHz, EPC9137 input at no-load and near full load.

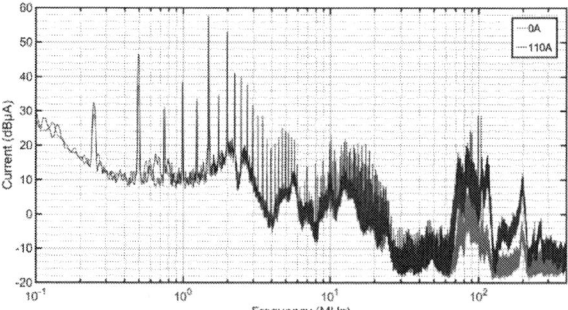

Fig. 11: CM spectrum to 400 MHz, EPC9137 input at no-load and near full load.

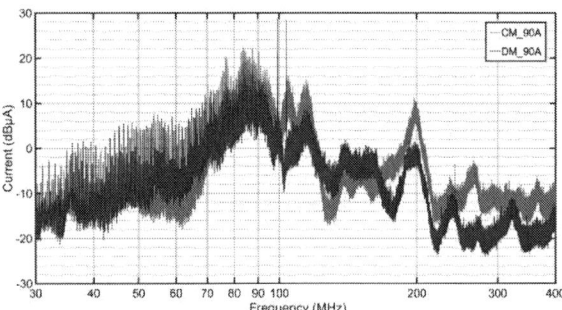

Fig. 9: Example CE emission spectrum resulting from switching dv/dt.

4.3 Results and analyses for various converters under selected conditions

4.3.1 Input emissions from single EPC9137

The converter operates at a switching frequency of 250 kHz. Due to the interleaved structure, the main spectral components will be mirrored in the second phase. The lower spectrum detailed in Fig. 10 shows the main switching frequency and the various harmonics. Very prevalent are the 1.5 MHz and 2 MHz peaks, where the two-phase currents coincide with the third and fourth harmonics. The load current also shows its strongest influence here.

The spectrum up to 400 MHz in Fig. 11 reveals not only the upper harmonic content of the main switching frequency, but also the usual multitude of noise-producing edges of a converter. Whereas the no-load condition is almost converging with full load

between 2 MHz and around 60 MHz (with the exception of some individual harmonics), the section above differs significantly.

DM currents are of more interest in this part of the examination, as they contribute less to RE, but instead have great influence on the PCC. Since this is the 48 V side, there will likely be components connected which are less susceptible to this kind of disturbance. Distinct is the difference in current amplitude with the changed load condition (Fig. 12). The upper part of the DM spectrum depicted in Fig. 13 is similar to the CM spectrum in Fig. 11. Again, upwards of around 60 MHz the no-load condition shows a significant change to full load. As noise currents are comparatively low in the high frequency part of the spectrum, there will be little influence on other users connected to the PCC.

4.3.2 Input emissions from two semi-parallel EPC9137

One goal of the research project is to test a converter capable of a maximum power of 3 kW. This can be achieved with a second EPC9137 converter, with both PCBs controlled from the same source. Therefore, a two-phase interleaved converter with four separate synchronous buck sections is being

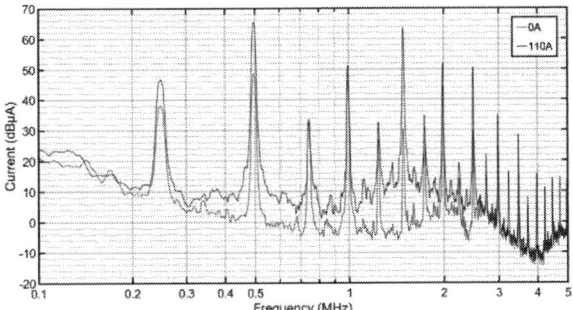

Fig. 12: DM spectrum to 5 MHz, EPC9137 input at no-load and near full load.

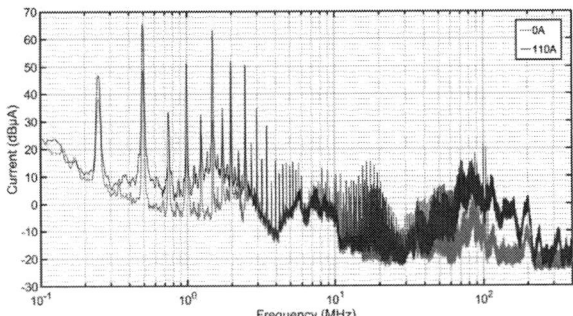

Fig. 13: DM spectrum to 400 MHz, EPC9137 input at no-load and near full load.

operated (Fig. 2). In the lower range of the resulting spectrum (Fig. 14), where the most influential noise appears, the currents for two PCBs have between double and quadruple peak values compared to a single PCB, as expected. Between 2 MHz and 20 MHz the semi-parallel operation appears beneficial compared to the single board. It would need to be examined whether this is purely a cancellation effect.

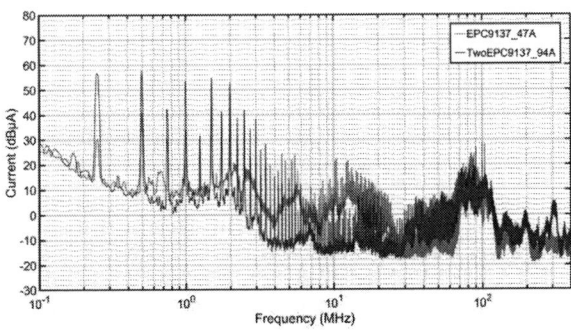

Fig. 14: CM spectrum to 400 MHz, comparison of a single EPC9137 input at half load and two EPC9137 semi-parallel both at half load.

4.3.3 Input emissions from single EPC9165

To portray differences occurring at a higher switching frequency, an almost identical converter as the EPC9137 is employed. The EPC9165 utilises the same control board and features two separate converter sections operating in two-phase interleaved mode. However, it switches at 500 kHz and for this it employs different eGaN-HFETs.

Clearly visible in Fig. 15 is the fundamental of a single converter section and, at 1 MHz and 2 MHz, the addition of the second phase with its nearest harmonic. Over the entire spectrum the no-load noise currents are significantly below the full load amplitude.

As with its counterpart EPC9137, the EPC9165 yields a DM spectrum (Fig. 16) which details a much stronger change between no-load and full load in the fundamental and its nearest harmonics. Here too, upwards of around 60 MHz the no-load condition drops significantly below full load. The low-power noise currents in and above this region will have little influence on other users connected to the 48 V PCC.

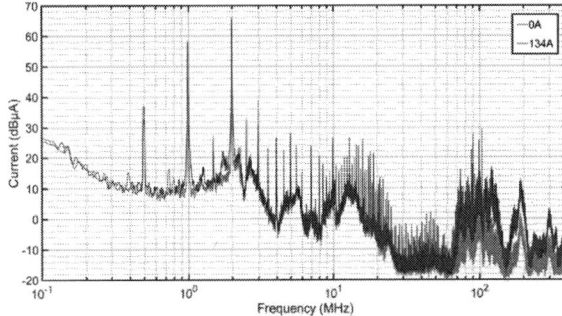

Fig. 15: CM spectrum to 400 MHz, EPC9165 input at no-load and full load.

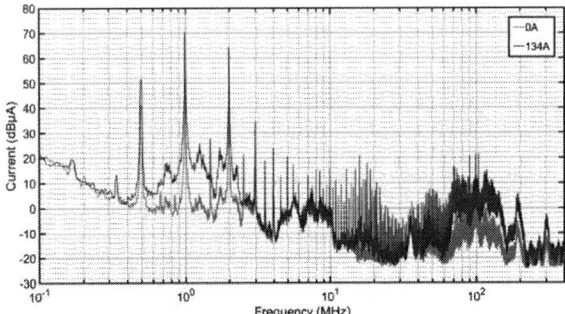

Fig. 16: DM spectrum to 400 MHz, EPC9165 input at no-load and full load.

4.3.4 Comparison of input emissions from single EPC9137 and EPC9165

To give an overview of the two structurally equivalent converters, Fig. 17 shows their CM spectra near full load. Apart from the obvious differences around their fundamentals and direct harmonics, there is a great similarity in the noise spectra.

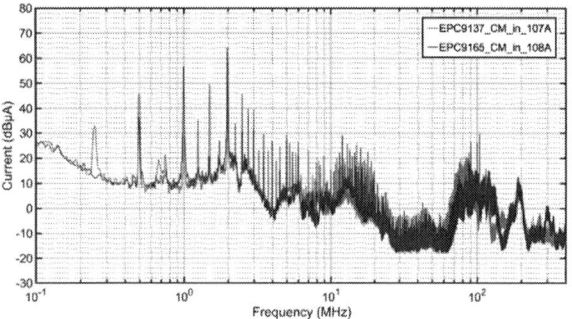

Fig. 17: CM spectrum to 400 MHz, comparing EPC9137 and EPC9165 input near full load.

4.3.5 Output emissions from single EPC9137 and EPC9165

As mentioned in section 4.1, the 12 V output is of interest where it couples into the other low voltage devices in the vehicle. Examined is the absolute EMI current (regardless if CM or DM) at the +12 V bus. To a certain degree this is superimposed by the noise at the input. The EMI currents at the 12 V output of the converter are, up to ≈2 MHz, heavily dependent on the load current. The frequency band above is dominated by capacitive components, which erases the load differences almost completely, as can be seen in Figs. 18 and 19. Generally, the depicted spectra replicate those of the DM at the input. This is well-founded by the experimental set-up, in which the current transducer incloses the positive output conductor only, which

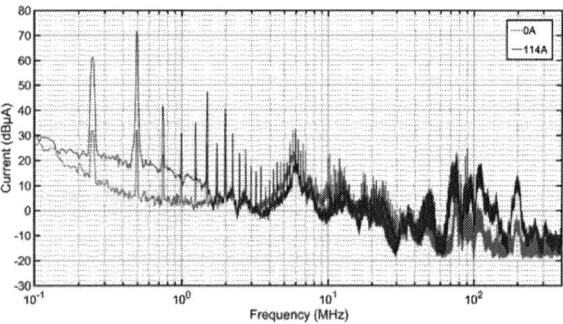

Fig. 18: Spectrum to 400 MHz, EPC9137 output at no-load and near full load.

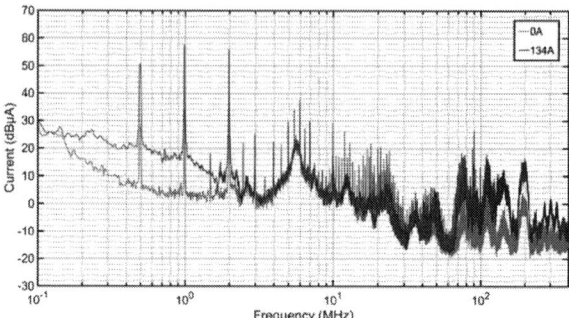

Fig. 19: Spectrum to 400 MHz, EPC9165 output at no-load and near full load.

emphasises the differential noise portion.

5 Conclusions

As mentioned in section 1, one aim of this research project is to point out factors of EMI generation by a buck converter without any measures of noise mitigation. These data will serve to construct a model of EMI sources and targets within EVs.

In the actual implementation in an EV the following points need particular attention. Both the input and output of the converter require EMI filters, in accordance with any mitigation measures with respect to switching frequency and switching speed. Avoiding a significant performance reduction of the converter or increasing its footprint, the guideline should be the appropriate standard [5], [8] in order to stay well below the permissible EMI levels for any actuators and sensors employed both at the 48 V and 12 V side. To mitigate the resulting EMI spectrum, a switching speed reduction could be appropriate. In order to avoid the resulting increase of switching losses, changes to the switching pattern concept (SCM or ZVS) are a possible solution [11].

Existing testing will be further developed using numerical modelling and simulation as well as methods including artificial intelligence [12]. New virtual and real testing strategies are designed and evaluated on demonstrators. In the future, they will enable a development time and cost reduction for operationally reliable autonomous systems.

Acknowledgements

This paper is funded by dtec.bw – Digitalization and Technology Research Center of the Bundeswehr, which we gratefully acknowledge [project ESAS – Elektromagnetische Störfestigkeit autonomer Systeme]. dtec.bw is funded by the European Union – NextGenerationEU.

References

[1] U. Aizpurua, T. Brandt, M. Hagel, E. Kampert, M. Wendt, *et al.*, "Avoiding Electromagnetic Interference Induced Risks for Autonomous Driving," in *dtec.bw-Beiträge der Helmut-Schmidt-Universität*, vol. 1, 2022, pp. 174–180, Band 1. DOI: 10.24405/14548.

[2] M. Buffolo, D. Favero, A. Marcuzzi, C. De Santi, G. Meneghesso, *et al.*, "Review and Outlook on GaN and SiC Power Devices: Industrial State-of-the-Art, Applications, and Perspectives," *IEEE Transactions on Electron Devices*, vol. 71, no. 3, pp. 1344–1355, 2024. DOI: 10.1109/TED.2023.3346369.

[3] A. S. Abdelrahman, Z. Erdem, Y. Attia, and M. Z. Youssef, "Wide Bandgap Devices in Electric Vehicle Converters: A Performance Survey," *Canadian Journal of Electrical and Computer Engineering*, vol. 41, no. 1, pp. 45–54, 2018. DOI: 10.1109/CJECE.2018.2807780.

[4] B. Deutschmann, G. Winkler, and P. Kastner, "Impact of electromagnetic interference on the functional safety of smart power devices for automotive applications," *e & i Elektrotechnik und Informationstechnik*, vol. 135, no. 4-5, pp. 352–359, 2018. DOI: 10.1007/s00502-018-0633-4.

[5] "Vehicles, Boats and Internal Combustion Engines – Radio Disturbance Characteristics – Limits and Methods of Measurement for the Protection of On-Board Receivers," CISPR 25:2021, Geneva, Switzerland, Dec. 2021.

[6] "EPC9137 Quick Start Guide," EPC®, Rev. 4.0, 2021, [Online].

[7] "EPC9165 Quick Start Guide," EPC®, Rev. 1.0, 2022, [Online].

[8] "Regulation No 10 of the Economic Commission for Europe of the United Nations (UN/ECE) — Uniform provisions concerning the approval of vehicles with regard to electromagnetic compatibility," UN ECE R10, Geneva, Switzerland, Sep. 2012.

[9] B. Kohlhepp, J. Dobusch, D. Kuebrich, D. Kuebrich, and T. Duerbaum, "Radio Disturbance Evaluation of Si- and GaN-Inverters," in *PCIM Europe 2023; International Exhibition and Conference for Power Electronics, Intelligent Motion, Renewable Energy and Energy Management*, 2023, pp. 1–7. DOI: 10.30420/566091338.

[10] C. R. Paul, *Introduction to Electromagnetic Compatibility*. New York: Wiley, 1992.

[11] D. Han, S. Li, W. Lee, W. Choi, and B. Sarlioglu, "Trade-off between switching loss and common mode EMI generation of GaN devices-analysis and solution," in *2017 IEEE Applied Power Electronics Conference and Exposition (APEC)*, 2017, pp. 843–847. DOI: 10.1109/APEC.2017.7930794.

[12] M. Stiemer, M. Hagel, U. Aizpurua, M. El-Sayed, and I. Cahani, "Machine Learning Based Data Analysis for Electromagnetic Reverberation Chambers," in *2023 Kleinheubach Conference*, 2023, pp. 1–4.

PCIM Europe 2024, 11– 13 June 2024, Nuremberg   DOI: 10.30420/566262331

Applied Design Automation for Finding Feasible Designs for High-Frequency Planar Transformers

Rando Raßmann[1], Knud Gripp[2], Ulf Schümann[1], Victor Golev[1]

[1] University of Applied Sciences Kiel, Institute for Power Electronics and Drives, Germany
[2] University of Applied Sciences Kiel, Institute for Mechatronics, Germany

Corresponding author: Rando Raßmann, rando.rassmann@fh-kiel.de

Abstract

The development of planar transformers for high frequency applications is a multi-physics design problem. Various parasitic effects and thermal limitations must be considered. Parasitic elements behave in opposite ways, so a trade-off must be found to meet the design requirements. During the design process, the designs are optimized iteratively. For this purpose, several designs are manually generated and analyzed by electric field simulation tools. However, this optimization is a time-consuming procedure with repetitive tasks. Design automation methods can be used to automate and, thus, to accelerate this optimization process. This paper applies design automation to identify feasible designs for high frequency planar transformers. Automated 2D electric field simulations in conjunction with an evolutionary algorithm are utilized to optimize the designs. The implemented algorithm can perform complex operations to make substantial changes in the transformer layout in order to consider design improvement techniques during the optimization process. Selected prototypes are assembled, electrical characterized and compared with the simulation results to validate the algorithm applied.

1   Introduction

Power electronic converters play a central role in the energy transition [1, 2]. The need to develop efficient and cost-effective power electronic systems is constantly increasing. Since the introduction of wide bandgap semiconductors, the switching frequency of power converters is increased up to the MHz range to achieve higher power densities [1, 2].

For DC-DC converters which require a galvanic insulation, the transformer is an important subcomponent. Planar transformers are preferred over wire-wound transformers for higher switching frequencies, because of their advantages in terms of thermal properties, predictable parasitic elements (resistances, leakage inductance and stray capacities), and simple manufacturing [2–4].

Developing a planar transformer is a challenging and time-consuming process, where different design requirements and constraints must be considered. At higher switching frequencies effects caused by the parasitic elements have a more crucial impact on the electrical characteristics of the transformer and the whole converter [1, 5, 6]. In order to meet the requirements, potential combinations of design parameters and their impact on both the parasitic elements and the thermal behavior must be investigated. However, there is a wide range of design parameters, that can be changed to improve the electrical and thermal characteristics. Consequently, the design of a high frequency planar transformer is a multi-objective optimization problem.

Design automation in power electronics has established itself as a key topic for reducing the development effort and time [7–10]. Workflows for design automation allow for a coaction with various simulations or tools and use the co-generated data for holistic design improvements to solve multi-objective design problems [7–9]. In order to reduce the workload for the generation and optimization of planar transformer models, existing software tools and workflows can be used [11]. However, these tools cannot perform all necessary steps automatically and there are still process steps that have to be carried out manually by the designer. For this reason, there is a clear demand for workflows which can automatically perform all required tasks.

To address the problems of the high development effort and to cope with shorter development times, in this work a workflow for design automation for planar transformer is presented. This is achieved by coupling automated electric field simulations with an evolutionary algorithm. With respect to both the computational costs and the simulation time, 2D simulations are used. This allows to analyze a transformer design in less than one minute, enabling the exploration of larger search spaces compared to 3D simulations.

Selected designs are assembled, and impedance measurements are used to electrical characterize the assembled transformers. The measurements are compared with the simulation results to validate the implemented simulations and evaluation of the designs.

The rest of the paper is organized as follows: Section 2 gives a short overview about the design considerations and the optimization of power transformers. Then in Section 3 the applied optimization algorithm which is used to identify feasible designs for a planar transformer is explained. The measurement results of the assembled prototypes and the comparison with the simulations are presented in Section 4. Section 5 discusses the results and Section 6 concludes the study.

2 Planar transformer optimization

2.1 Design considerations

The voltage V of a transformer winding can be calculated with (1), where K_v is a factor depending on the waveform, f_s is the switching frequency of the converter, N represents the number of turns of the winding, ΔB is the peak ac flux density and A_{Core} is the cross section of the transformer core.

$$V = K_v \cdot f_s \cdot N \cdot \Delta B \cdot A_{Core} \qquad (1)$$

The core losses P_{Core} can be calculated by the Steinmetz equation (2). V_C is the effective volume of the transformer core and K_c, α, and β are the Steinmetz parameters for the transformer core material.

$$P_{Core} = V_c \cdot K_c \cdot f_s^{\alpha} \cdot \Delta B^{\beta} \qquad (2)$$

The minimum number of turns for the primary winding N_p for a given value for ΔB and A_{Core} can be calculated with (3).

$$N_p \geq \frac{V_p}{K_v \cdot \Delta B \cdot f_s \cdot A_{Core}} \qquad (3)$$

After N_p has been determined, the number of turns of the secondary winding (N_s) can be calculated.

If the winding geometry is known, the dc resistance for the windings and thus the dc copper losses can be calculated. At higher frequencies, additional loss mechanisms such as the skin effect and the proximity effect occur and cannot be neglected [4–6]. The structure of the windings influence the intensity of these losses, which is why the arrangement of the windings must also be considered during the optimization [4–6].

The copper losses P_{Cu} of the primary and secondary winding can be calculated with (4), where I is the current in the respective winding and $R_{p,ac,i}$ as well as $R_{s,ac,i}$ are the ac resistances of the single layers of the windings considering high frequency effects.

$$P_{Cu} = \sum_{i=1}^{N_p} R_{p,ac,i} \cdot I_{p,rms}^2 + \sum_{i=1}^{N_s} R_{s,ac,i} \cdot I_{s,rms}^2 \qquad (4)$$

Using (1) and (2) it is evident, that P_{Core} decrease if N_p increase because ΔB decreases. P_{Cu} behave in an opposite way and increase with the number of turns. A trade-off between P_{Cu} and P_{Core} can minimizes the total loss. Fig. 1 illustrates this problem.

After the calculation of P_{Cu} and P_{Core}, the temperature rise ΔT of the transformer can be determined with (5), where A_t is the surface area of the core and h_t is the heat transfer coefficient.

$$\Delta T = \frac{P_{Cu} + P_{Core}}{h_t \cdot A_t} \qquad (5)$$

Besides the thermal behavior, the electrical characteristics of the transformer must be considered during the design process. Fig. 2 presents the lumped electrical equivalent circuit of a two-winding transformer. The magnetization inductance L_m of a transformer can be calculated with (6), where A_L is the inductance factor of the respective transformer core.

$$L_m = A_L \cdot N_{pri}^2 \qquad (6)$$

Fig. 3 illustrates effects caused by parasitic elements of a transformer which adversely affect the power converter during operation. The leakage inductance L_{lk} leads to a trapezoidal current which increase the switching losses (Fig. 3a) [3–5]. Moreover, it causes peaks in the voltage waveform which can damage the semiconductor switches.

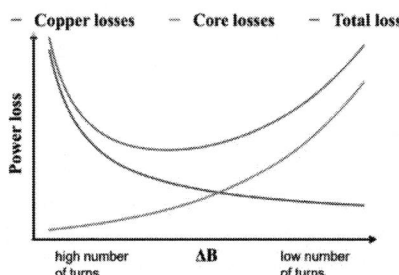

Fig. 1: Trade-off between core and copper losses to minimize the total losses.

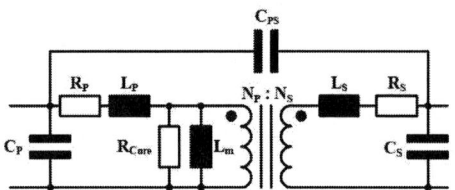

Fig. 2: Electrical equivalent circuit of a two-winding transformer.

The capacity of a transformer can be divided into two types as follows: The first one is the inter-winding capacity C_{PS} formed between the primary and the secondary winding. It is charged and discharged at each switching cycle that results in a high current peak during turn-on of the switches [5]. Fig. 3b shows the influence of C_{PS} on the current waveform at low and high load. The current peak is independent from the load and can causes EMI problems [4, 5]. Therefore, C_{PS} should be kept low. The second capacity is the intra-winding capacity of each winding (C_P and C_S). These capacities are formed between the turns within a winding. Together with L_m, C_P and C_S determine the first self-resonance frequency of the transformer, which can be calculated with (7).

$$f_{res} = \frac{1}{2\pi\sqrt{L_m \cdot C_\sigma}} \quad with\ C_\sigma = C_p + C_s \quad (7)$$

2.2 Design improvement techniques

To improve the electrical characteristics, design techniques can be used to manipulate a particular parasitic element by changing design parameters. TABLE I lists some basic design techniques to influence a particular parasitic element. However, the reduction of one parasitic element generally leads to an increase of another parasitic element. For identifying a feasible designs various design parameter combinations have to be investigated and trade-offs must be defined which depend on the application and the constraints.

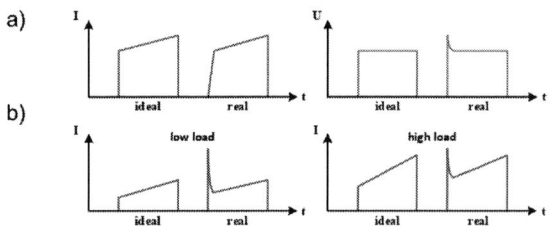

Fig. 3: Effects caused by parasitic elements: a) Impact of the leakage inductance on the current and voltage waveforms of the power switches. b) Current peaks caused by the inter-winding capacity during turn on of the power switches [5].

TABLE I: General design techniques to manipulate particular parasitic elements for planar transformers.

Reducing the intra-winding capacity:
• reducing the number of turns
• reducing the width of the winding traces
• decrease the overlap area between the turns
• interleaving of the windings
Reducing the inter-winding capacity:
• reducing the number of turns
• reducing the overlap area of the primary and the secondary winding
• increasing the insulation thickness
• do not interleave the windings
Reducing the leakage inductance:
• decrease the insulation thickness between the primary and secondary winding
• interleaving of the windings
• overlap of primary and secondary winding tracks

There are also more complex techniques for improving the electrical characteristics of a design, which are described, for example, in [12, 13]. However, these are not considered in the later implemented algorithm and are therefore not described further.

2.3 Transformer optimization tools

Optimization algorithms based on analytical equations for calculating transformer designs can be used to investigate thousands of design variants in a short time, considering different design trade-offs. [13–16]. However, analytical equations have limitations that can affect the accuracy of the calculations [2, 11]. For this reason, there is still a demand to build up selected designs as 2D or 3D models and analyze them with electric field simulation software to validate the analytical results. Within the simulation software only linear parameter variations are possible which limits the range of design variations. Complex operations such as changing the excitation of an object, as required for studying multiple interleaving strategies, or adding new objects are not possible. This means that different designs and their sub-variants must be generated manually. In addition, the corresponding simulation settings and excitations have to be configured and all results must be evaluated. As a result, the conventional optimization process for high-frequency transformers based on electric field simulations leads to a time-consuming procedure with repetitive tasks.

There are design tools for transformer and inductors available which simplify the development and optimization process (e. g. PExpert from Ansys) [11, 17, 18]. These tools can be used to identify initial designs or automate the model generation as well as the setup of the simulations. However, the search space is limited, and these tools cannot work automatically or cannot apply general design techniques to improve the electrical characteristics. Design changes and the evaluation of all generated data must still be carried out by the developer.

3 Planar transformer optimization process

In the following, a workflow for design automation for planar transformer is presented. It is exemplarily applied to identify feasible designs for a planar transformer for a gallium nitride based dual active bridge dc-dc converter. The design specifications for the transformer are listed in TABLE II. Based on the design specifications, the search space to be explored can be defined which is presented in TABLE III.

3.1 Applied methodology for the optimization process

To perform the simulations the Ansys Electronics Desktop (AEDT) 2022 R2 software is used because it provides the Python PyAEDT [19] library. This library allows directly accessing to the AEDT API. By using the provided class and method structures, all required steps of a conventional design process can be scripted and, thus, the entire simulation process can be automated. This allows to develop functions that can represent transformer design improvement techniques. Moreover, since all variables in Python are treated as objects and PyAEDT offers a corresponding interface, data can be exchanged in a bidirectional manner. This allows to take results from different simulations into account during the optimization or use them in subsequent simulations.

TABLE II: Transformer design specifications.

Symbol	Description	Value
V_{in}	Input voltage	400 V
V_{out}	Output voltage	400 V
$I_{out,rms}$	Output current	5 A
P_{out}	Output power	2000 W
f_s	Switching frequency	400 kHz
T_a	Ambient temperature	40 °C
ΔT	Maximum temperature rise	80 °C
N	Turns ratio	1:1
h_t	Heat transfer coefficient	15 W/m²K
e_r	Permittivity of the insulation material (FR-4)	4.2

TABLE III: Defined search space to identify feasible designs.

Parameter	Values
Insulation thicknesses (μm)	75, 100, 125, 150, 200, 225, 250, 300, 350, 375, 400, 450
Copper thicknesses (μm)	35, 70, 105, 200, 300
Transformer cores	EI/43/10/28, EE/43/10/28, EI/58/11/38, EE/58/11/38, EI/64/10/50, EE/64/10/50, EER/64/13/51, EER/51/10/38, EER/41/7.6/32
Number of turns	min: 4; max: 16
Core material	Ferroxcube 3F36
Winding width	min: 10%; max: 100%
Interleaving strategy	All possible winding arrangements from non-interleaved to full interleaved

Considering the computational costs and in order to keep the simulation times low, 2D simulations are used. With the eddy current solver, the winding resistances and the leakage inductance of the designs are determined. For calculating the winding capacities, the electrostatic solver is used. For the generation of the transformer cores, a database is created where all relevant information about material properties and geometrical parameters of the different cores are stored.

A cylindrical coordinate system with a symmetrical field distribution around the z-axis is assumed. In order to be able to consider different core shapes, the methodology presented in [20] is applied to modify the geometric description of the 2D transformer designs.

For the optimization of the planar transformer designs, an evolutionary algorithm used. An evolutionary algorithm is a stochastic optimization method based on the mechanisms of natural evolution, such as selection, mutation, and recombination [16]. It can be used to solve complex optimization problems. The algorithm is implemented in a Python script which can call the developed functions for the automated electric field simulations. Additional to the simulations, analytical calculations are used to evaluate the designs.

3.2 Optimization algorithm

Fig. 4 presents the schematic structure of the applied algorithm. It works as follows: Before the optimization process starts, the search space and the design specifications have to be defined. An initial population is created by selecting random values for the design parameters within the defined search space. Subsequently, analytical equations are used to calculate the core loss, the expected peak flux density, and the height of the windings. At this step, the transformers are checked for design errors. A possible design error could be, for example, if the windings do not fit into the core window or the calculated peak flux density is higher than the maximum allowed value. A design error caused by a too wide winding is not possible as the winding width is specified as a percentage of the core window. Designs with a design error are not further analyzed with simulations.

After the simulations have been performed, the results are extracted from each respective simulation and are scored by a fitness function. In this work a weighted optimization function $f(\vec{x})$ with the objective to minimize the inter-winding capacity and to maximize the power density ρ and the resonance frequency is applied (8). The vector $\vec{x}$ represents the design parameters which can be variated during the optimization.

$$f(\vec{x}) = \max\left\{W_1 \cdot f_{res} + W_2 \cdot \rho + W_3 \cdot \frac{1}{C_{PS}}\right\} \quad (8)$$

Furthermore, constraints can be imposed. If a result parameter (e. g. temperature rise) of a design violates a constraint, the fitness score can be penalized. The fitness score determines which designs are taken into the next epoch.

At the end of each epoch, all design information and results are saved in a csv-file. Thus, the information is available after the optimization process and can be used for further evaluations. A new population for the subsequent epoch is created by selecting the best designs of the current epoch and applying genetic operators such as crossover and mutation to generate new designs. The designs are, therefore, continuously adapted so that the designs are successively optimized with respect to the implemented objective function.

The algorithm can vary the following design parameters (compare Fig. 5):

- core shape
- trace thickness
- trace width
- insulation thickness
- number of turns
- interleaving of layers
- overlap of both primary and secondary winding

The values for the trace thickness and trace width can have different values for the primary and secondary windings. The thickness of the insulation is the same for all layers. A more detailed description of the implemented algorithm and simulations can be found in [10].

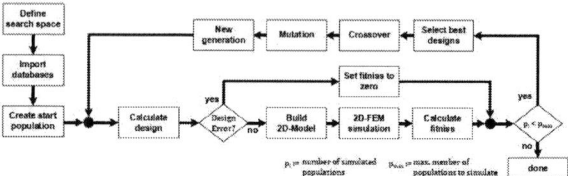

Fig. 4: Schematic structure of the implemented EA.

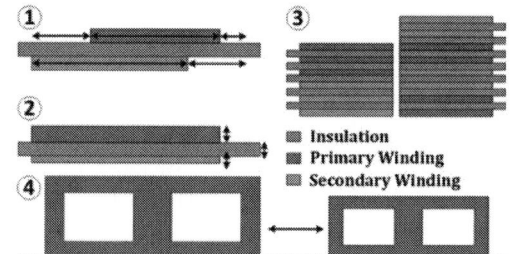

Fig. 5: Variable design parameters: Variation of (1) winding widths and overlapping, (2) thicknesses of windings and insulation, (3) number of turns and interleaving and (4) core size.

3.3 Optimization process

In this study, the number of epochs and the population size are set to 40 and 20, respectively. The algorithm is executed on a computing system with an AMD Ryzen 7 5700U 1.80 GHz 8-Core processor and the simulations are running sequentially.

The algorithm analyzes 800 designs, of which 91 show a design error and 709 designs are investigated. In average, the algorithm needs about 33 s to analyze a single design.

Fig. 6a presents the calculated and simulated electrical parameters of the best transformer design within a population over the epochs. In Fig. 6b the losses and the resonance frequency of the best design within a population over the epochs are shown. It is evident that over the epochs the resonance frequency increases while the inter-winding capacity remains low.

TABLE IV presents three selected designs which achieve the highest fitness score within their population at different epochs. $Design_1$, $Design_2$ and $Design_3$ are selected designs from epochs at the beginning, the middle and the end of the optimization process. From the comparison of the design parameters, it is evident, that during the optimization the insulation thickness increases to reduce the inter-winding capacity and the width of the windings are decreasing to reduce the intra-winding capacities. The leakage inductance is reduced by interleaving the windings which also reduces to copper losses. In addition, by varying the core shape, core size and the number of turns the algorithms tries to reduce the total loss.

a)

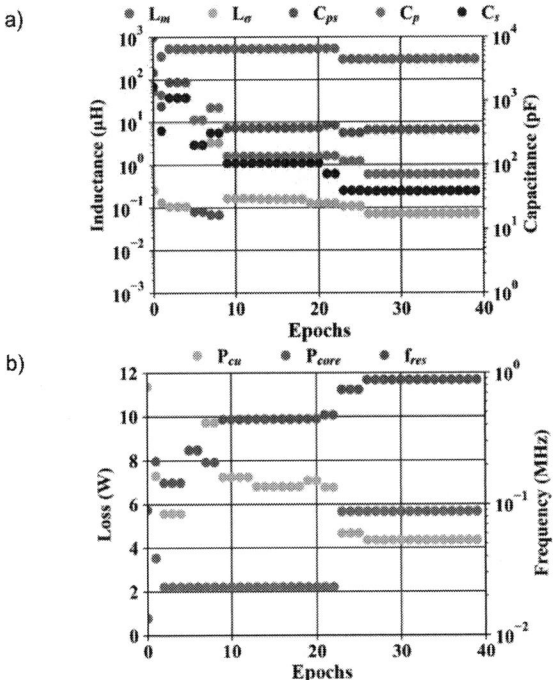

b)

Fig. 6: Results of the optimization process: a) Simulated values for the parasitic elements. b) Calculated values for the losses and the resonance frequency.

TABLE IV: Design parameters of the selected transformer designs at different epochs during the optimization.

	Design₁	Design₂	Design₃
Epoch	0	20	39
Isolation thickness (µm)	225	400	400
Number of turns	6	8	6
Layer order *⁾	PSSPSP PSSPSP	PSPSSPPS PSSPSSPP	SPSPSP PSPSSP
Transformer core	EE 43/10/28	EER 64/13/51	EER 64/13/51
Trace thickness pri. / sec. winding (µm)	105 / 35	105 / 35	105 / 300
Power density (kW/l)	62.0	29.6	29.6
Width pri. / sec. (mm)	2.54 / 3.81	5.380 / 4.035	4.035 / 4.035
Shift pri / sec. (mm)	8.89 / 7.62	0 / 0	0 / 0

*⁾ P: Layer of the primary winding; S: Layer of the secondary winding

4 Experimental validation

In the next step, the calculated and simulated values of the applied optimization algorithm are validated. For this purpose, selected designs are manually manufactured as physical prototypes and further investigated. To simplify the prototype assembly process, easily manufacturable designs

are selected. The thickness of the winding tracks is set to 300 µm and for the insulation a PVC foil (e_r = 2.9) is used. The parameters of the selected designs are presented in TABLE V.

4.1 Simulation results

Due to the adjustment of the insulation material the value for e_r is changed from 4.2 to 2.9. For this reason, the 2D simulations are repeated for the selected designs. Furthermore, the designs are also investigated with 3D simulations. TABLE VI compares the results of the 2D and 3D simulations.

4.2 Prototype assembly

For the transformer windings, multiple parts of traces are cut out of a copper foil and laser-welded to form a winding. In the next, step the windings are interleaved and the PVC foil for the insulation is placed between the copper layers. In the last step the windings are pressed together and set into the transformer core. Fig. 7a shows how the windings are interleaved and insulated from each other and Fig. 7b shows the three assembled transformer prototypes.

TABLE V: Design parameters of the selected transformers that are assembled as prototypes.

	T1	T2	T3
Isolation thickness (µm)	250	250	375
Number of turns	7	5	6
Layer order	PSPSPSP SPSPSPS	PSPSP SPSPS	PSPSPS PSPSPS
Core shape	EER 64/13/51	EE 64/10/50	EER 64/13/51
Trace thickness pri. / sec. winding (µm)	300 / 300	300 / 300	300 / 300
Core Material	Ferroxcube 3F36	Ferroxcube 3F36	Ferroxcube 3F36
Width pri. / sec. (mm)	12.105 / 12.105	12.720 / 12.720	5.38 / 4.035
Shift pri / sec. (mm)	1.345 / 1.345	4.24 / 4.24	0 / 0

TABLE VI: Comparison of 2D and 3D simulation results of the prototypes.

	T1		T2		T3	
	2D	3D	2D	3D	2D	3D
L_m (µH)	406.7		220.0		298.8	
L_σ (nH)	60.00	57.01	58.57	55.57	127.8	129.6
C_{PS} (nF)	2.121	2.008	2.282	2.345	0.319	0.322
C_σ (pF)	9.14	9.50	10.48	11.92	16.39	17.1
f_{res} (MHz)	2.61	2.56	3.31	3.11	2.27	2.22

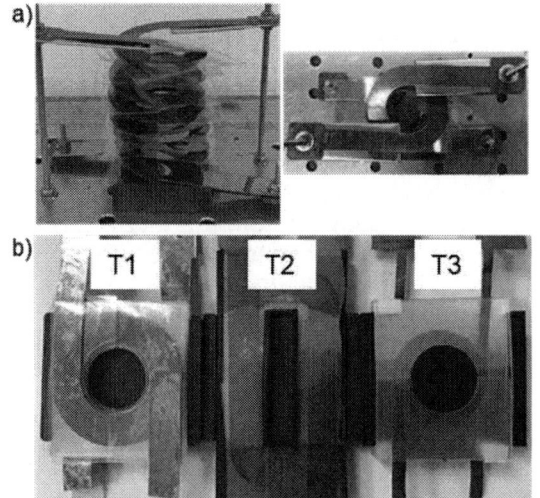

Fig. 7: Assembly of the prototypes: a) Interleaving and insulation of the windings. b) Assembled prototypes.

4.3 Electrical characterization

The transformer prototypes are electrical characterized using an impedance analyzer. Three different measurement configurations are used to determine the electrical parameters.

With the first configuration, L_m is extracted by connecting the primary winding of a prototype to the impedance analyzer and the secondary winding is in open-circuit (Fig. 8a). Moreover, by determine the resonance frequency f_{res}, C_σ can be calculated using (9).

$$C_\sigma = \frac{1}{(2\pi \cdot f_{res})^2 \cdot L_m} \qquad (9)$$

The value of L_{lk} is determined by using the second configuration. Here, the primary winding is connected to the impedance analyzer and the secondary winding is in short-circuit (Fig. 8b). In the last configuration, the value for C_{PS} is determined by shorting the primary and secondary winding and connecting both to the impedance analyzer (Fig. 8c). TABLE VII contains the extracted values from the impedance measurements and the corresponding impedance curves are shown in Fig. 9.

5 Discussion

From TABLE VI it is evident, that the 2D and 3D electric field simulation matches well. The differences between the capacities for the prototype T2 can be explained by the differences of the surface areas between the winding layers for the z-axis symmetrical 2D model and 3D model explained.

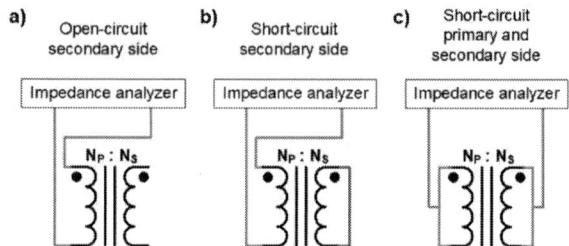

Fig. 8: Impedance measurement configurations for determining the values for the parasitic elements: a) Secondary winding in open circuit for determining L_m. b) Secondary winding in short circuit for determining L_{lk}. c) Primary and secondary winding in short circuit for determining C_{PS}.

TABLE VII: Extracted values of the assembled prototypes by measurements with an impedance analyzer.

	L_m (μH)	L_σ (nH)	C_{PS} (nF)	C_σ (pF)	f_{res} (MHz)
T1	448.4	72.82	1.312	15.32	1.92
T2	233.7	66.12	1.501	26.56	2.02
T3	295.2	180.2	0.271	16.22	2.30

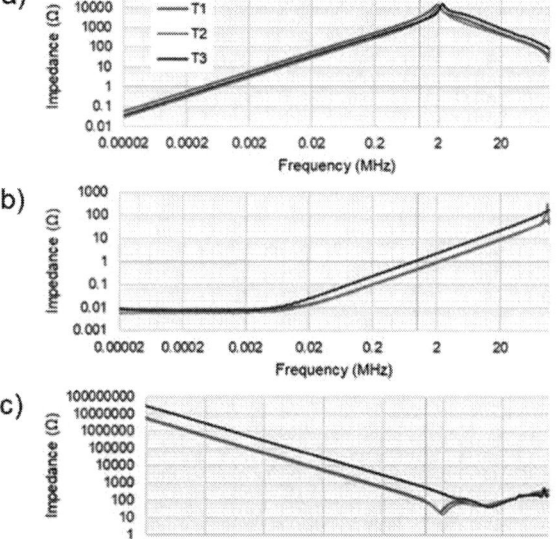

Fig. 9: Measured impedance curves: a) Secondary winding in open-circuit. b) Secondary winding in short-circuit. c) Primary and secondary winding in short-circuit.

TABLE VIII compares the electric field simulations of the 2D designs with the experimental measurements. In general, the measurement results agree well with the simulated values. There are some differences between the simulated and the measured values, which can be explained by inaccuracies due to the manual manufacturing process and the tolerances of the transformer cores.

Furthermore, the substantial differences of the capacities can be explained because the windings are not sufficiently pressed together to achieve the assumed thickness of the insulation layers. The measurements for C_{PS} are therefore repeated without the core so that the windings could be mechanically pressed together. TABLE IX shows the measured values for C_{PS} when the transformer core is excluded and the windings are compressed. It is evident, that the capacity increases.

In the presented of the algorithm the temperature rise of the transformer is determined based on the surface area of the transformer core and the total loss. By implementing a temperature model or an additional simulation for the thermal behavior of the planar transformers, a more accurate prediction of the thermal behavior would be possible. This would allow to identify feasible designs with higher power densities and lower parasitic elements. Moreover, extending the functionality of design improvement techniques would improve the algorithm.

6 Conclusion

In this paper a workflow for design automation to improve the development process for high frequency planar transformers was presented. Automated 2D electric field simulations in conjunction with an evolutionary algorithm are used to identify feasible designs. All required steps from model creation over setting up the electric field simulations and extracting the simulation results are implemented in a tool chain using the Python library PyAEDT. With an average time of 33 s to evaluate a design, the algorithm is able to investigate thousands of designs with electric field simulations. This allows to explore large search spaces in an acceptable time.

In addition, selected designs are assembled and electrically characterized in order to validate the implemented algorithm. The measured values for the electrical parameters agree with the simulated values. Consequently, the developed algorithm can be applied for the optimization of planar transformers. Moreover, this works demonstrates that the implementation of design automation can improve the development process by effectively linking multiple simulation tools and optimization algorithms. Thus, the need of manual optimization steps by a designer can be avoided.

TABLE VIII: Comparison between the 2D simulation results and the measured values of the assembled transformer prototypes.

	T1		T2		T3	
	2D	Measured	2D	Measured	2D	Measured
L_m (µH)	406.7	448.4	220.0	233.7	298.8	295.2
L_σ (nH)	60.00	72.82	58.57	66.12	127.8	180.2
C_{PS} (nF)	2.121	1.312	2.282	1.501	0.319	0.271
C_σ (pF)	9.14	15.32	10.48	26.56	16.39	16.22
f_{res} (MHz)	2.61	1.92	3.31	2.02	2.27	2.30

TABLE IX: Extracted values for the inter-winding capacities without the transformer core and compressed windings.

	T1	T2	T3
C_{PS} (nF)	1.776	1.870	0.292

Such workflows enable a holistic optimization and can help to face the design challenges for power electronic components in terms of achieving a higher power density and increased efficiency while reducing development times.

7 References

[1] Y. Wang, O. Lucia, Z. Zhang, S. Gao, Y. Guan, and D. Xu, "A Review of High Frequency Power Converters and Related Technologies," *IEEE Open J. Ind. Electron. Soc.*, vol. 1, pp. 247–260, 2020, doi: 10.1109/OJIES.2020.3023691.

[2] Z. Ouyang and M. A. E. Andersen, "Overview of Planar Magnetic Technology— Fundamental Properties," *IEEE Trans. Power Electron.*, vol. 29, no. 9, pp. 4888–4900, 2014, doi: 10.1109/TPEL.2013.2283263.

[3] Z. Ouyang, O. C. Thomsen, and M. A. E. Andersen, "Optimal design and tradeoffs analysis for planar transformer in high power DC-DC converters," in *2010 International Power Electronics Conference: IPEC-Sapporo 2010 - [ECCE Asia] ; Sapporo, Japan, 21 - 24 June 2010*, Sapporo, Japan, 2010, pp. 3166–3173.

[4] Z. Ouyang, O. C. Thomsen, and M. A. E. Andersen, "Optimal Design and Tradeoff Analysis of Planar Transformer in High-Power DC–DC Converters," *IEEE Trans. Ind. Electron.*, vol. 59, no. 7, pp. 2800–2810, 2012, doi: 10.1109/TIE.2010.2046005.

[5] C. W. T. McLyman, *Transformer and inductor design handbook*, 4th ed. Boca Raton, London, New York: CRC Pr, 2011.

[6] W. G. Hurley and W. H. Wölfle, *Transformers and inductors for power electronics: Theory, design and applications*. Chichester: Wiley, 2014.

[7] A. Bindra and A. Mantooth, "Modern Tool Limitations in Design Automation: Advancing Automation in Design Tools is Gathering Momentum," *IEEE Power Electron. Mag.*, vol. 6, no. 1, pp. 28–33, 2019, doi: 10.1109/MPEL.2018.2888653.

[8] K. Hermanns, Y. Peng, and A. Mantooth, "The Increasing Role of Design Automation in Power Electronics: Gathering What Is Needed," *IEEE Power Electron. Mag.*, vol. 7, no. 1, pp. 46–50, 2020, doi: 10.1109/MPEL.2019.2959706.

[9] A. J. Marques Cardoso, "Power Electronics Design Methods and Automation in the Digital Era: Evolution of Design Automation Tools," *IEEE Power Electron. Mag.*, vol. 7, no. 2, pp. 36–40, 2020, doi: 10.1109/MPEL.2020.2988077.

[10] R. Rasmann, V. Golev, U. Schumann, and J. Schnack, "Automated Design and Optimization of Planar Transformers for High-Frequency Applications," in *2023 IEEE Design Methodologies Conference (DMC)*, Miami, FL, USA, Sep. 2023 - Sep. 2023, pp. 1–6.

[11] S. Vaisambhayana, C. Dincan, C. Shuyu, A. Tripathi, T. Haonan, and B. R. Karthikeya, "State of art survey for design of medium frequency high power transformer," in *Asian Conference on Energy, Power and Transportation Electrification (ACEPT 2016): 25-27 October 2016, Sands Expo and Convention Centre, Marina Bay Sands, Singapore*, Singapore, Singapore, 2016, pp. 1–9.

[12] M. Pahlevaninezhad, D. Hamza, and P. K. Jain, "An Improved Layout Strategy for Common-Mode EMI Suppression Applicable to High-Frequency Planar Transformers in High-Power DC/DC Converters Used for Electric Vehicles," *IEEE Trans. Power Electron.*, vol. 29, no. 3, pp. 1211–1228, 2014, doi: 10.1109/TPEL.2013.2260176.

[13] M. I. Hassan, N. Keshmiri, A. D. Callegaro, M. F. Cruz, M. Narimani, and A. Emadi, "Design Optimization Methodology for Planar Transformers for More Electric Aircraft," *IEEE Open J. Ind. Electron. Soc.*, vol. 2, pp. 568–583, 2021, doi: 10.1109/OJIES.2021.3124732.

[14] A. Garcia-Bediaga, I. Villar, A. Rujas, L. Mir, and A. Rufer, "Multiobjective Optimization of Medium-Frequency Transformers for Isolated Soft-Switching Converters Using a Genetic Algorithm," *IEEE Trans. Power Electron.*, vol. 32, no. 4, pp. 2995–3006, 2017, doi: 10.1109/TPEL.2016.2574499.

[15] A. Fouineau, M. Guillet, B. Lefebvre, M.-A. Raulet, and F. Sixdenier, "A Medium Frequency Transformer Design Tool with Methodologies Adapted to Various Structures," in *2020 Fifteenth International Conference on Ecological Vehicles and Renewable Energies (EVER)*, Monte-Carlo, Monaco, 2020, p. 1.

[16] S. D. Sudhoff, *Power magnetic devices: A multi-objective design approach*. Hoboken, NJ: IEEE Press Wiley, 2022. [Online]. Available: https://ieeexplore.ieee.org/servlet/opac?bknumber=9622344

[17] *Ansys Maxwell | Electromechanical Device Analysis Software*. [Online]. Available: https://www.ansys.com/products/electronics/ansys-maxwell (accessed: Mar. 19 2024).

[18] *Ansys Electronics Transformer ACT*. [Online]. Available: https://blog.ozeninc.com/resources/ansys-electronics-transformer-act (accessed: Mar. 19 2024).

[19] *API reference — PyAEDT*. [Online]. Available: https://aedt.docs.pyansys.com/version/stable/API/index.html (accessed: Mar. 19 2024).

[20] R. Prieto, L. Ostergaard, J. A. Cobos, and J. Uceda, "Axisymmetric modeling of 3D magnetic components," in *Conference proceedings / APEC '99, Fourteenth Annual Applied Power Electronics Conference and Exposition, 14 - 18 March, 1999, Adam's Mark Hotel - Dallas, Dallas, Texas, USA*, Dallas, TX, USA, 1999, 213-219 vol.1.

Frequency Dependent Area Product Method

Alfonso Martínez[1]
[1] Würth Elektronik, Spain

Corresponding author: Alfonso Martínez, alfonso.martinez@we-online.com
Speaker: Alfonso Martínez, alfonso.martinez@we-online.com

Abstract

The selection of the magnetic core is the most critical decision when designing magnetic components for Power Converters, as it shapes its dimensions and constrains the rest of parts, like the insulation or the wires. Current selection methods are inherited from decades ago, designed for lower frequencies, failing to take eddy currents and core losses effects into account, and thus increasing the number of iterations needed to design a working component. A method is introduced that expands the current Area Product core selection method to the medium-high frequencies used in modern converters..

1   Introduction

One of the dreams of the Power Electronics Engineers is the automatic synthesis of the magnetic components needed in the converters, the design of an optimal inductive component tailored to each individual application, or at least its simplification to some simple process that can be followed by Engineers. One of the steps of such a process is the selection of the magnetic core, where several methods have been developed over the years that aim to simplify the initial selection of the suitable core for a given operating point. Examples of those methods are the Kg [1] or Kge [7], although the most extensively used is the Area Product (AP), which can be directly applied from the parameters offered by the core manufacturers in their datasheets.

Several forms of the AP method exist in the literature. Although they are designed for low frequency, neglecting the effects produced by medium or high switching frequency, such as eddy current effects in the wires or an increase of core losses at equal magnetic flux density amplitude. Of these methods, most of them are for inductor design [1, 2, 4, 5], while a few [3, 6] take into account transformers (or multi winding magnetics in general).

In this paper we will use one of the existing methods (described in [3], as it is the most general one) and we will develop two improvements that allow us to take into account

the effects of any frequency into the selection of the optimal core for a given application.

2 The Base Area Product Method

The Area Product method states that the suitability of a core depends on two geometrical areas from its shape: The area of the magnetic core, where the induced magnetic flux has to go through, and the area of the winding window, where all the turns, the wires carrying all the current of the component, must go through. These two areas reflect the magnetic and electrical behaviors of our system.

The method calculates the product of these two physical areas for a given core, and compares it against a needed electrical value calculated from the parameters of our converter. If the core AP, the product of the two areas, is greater than the needed value (in m^4), then our core is valid for that application.

That concept can be expanded to a list of available cores, and the AP value will allow us to discard any core that has an AP lower than our needed value, and will even provide a ranking, as cores with a much greater AP than the needed value will be overdimensioned.

Section 4.A.1 of [3] provides a formula to calculate the needed value for a given converter:

$$AP = \frac{P_{in}}{K_t K_u K_p J \, \Delta B \, 2 f} \; m^4 \quad \text{(Eq.1)}$$

Where P_{in} is the input power, K_t is the ratio of the DC input current to the maximum primary current, K_u is the ratio of the area of window occupied by copper to the total available window area, K_p is the ratio of the winding area provided for the primary to the total window area, J is the maximum current density we want, ΔB is the magnetic flux density swing, and f is switching frequency.

The formula provided in [3] is a step forward from the traditional one, but still lacks the necessary frequency dependency for medium-high frequency magnetics.

From the point of view of this author, the two problems with it arise from the terms K_u and ΔB:

1. The term K_u is typically chosen as 0.4-0.3, assuming round wires and bobbin, but at higher frequencies a lower value must be used, as wire diameter goes down to keep skin effect under control. To take this into account, a formula to calculate K_u for a given core is proposed in this paper.

2. The ΔB is commonly chosen as either a proportion of the lowest saturation point given by the manufacturer (typically 50-60% of the saturation value at 100°C), or a fixed constant value than the designer's experience dictates (typically 170-200 mT). The problem with this approach is that as the frequency increases, the limit of the magnetic field is dictated by the losses heat it can dissipate, not by ΔB. To take this effect into account a frequency dependent ΔB is proposed in this paper.

2.1 First improvement, the Window Utilization Factor, Ku

The Window Utilization Factor is composed of two parts, the conducting-to-insulating material ratio and the Bobbin Filling Factor (K_b)

In order to calculate the latter, a database with all the bobbin defined in IEC-62317 was created, storing the width and height of the winding window of the core, with and without bobbin. The data for width and height can be seen in Fig.1 and Fig.2

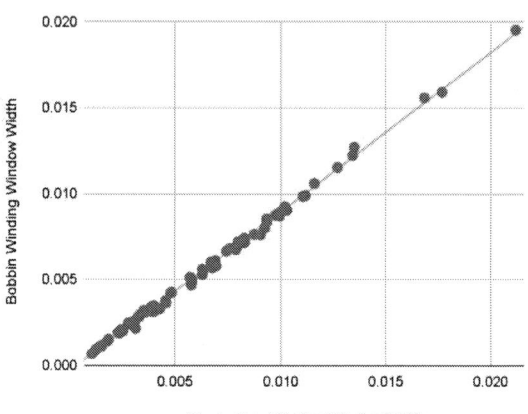

Fig. 1 Core Winding Window Width

Fig. 1 Bobbin winding window width versus Core winding window width

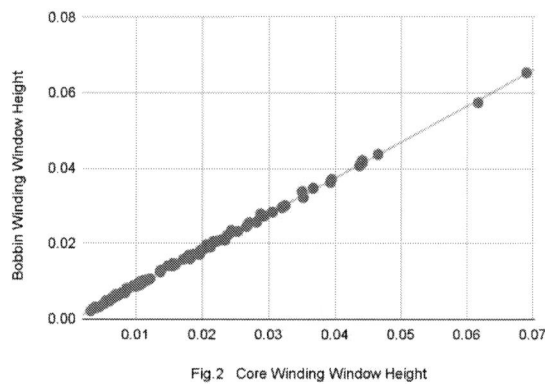

Fig.2 Core Winding Window Height

Fig. 2 Bobbin winding window height versus Core winding window height

With this information an empirical equation was fitted:

For bobbin winding window width:

$$B_{www} = 0.927 \cdot C_{www} - 0.000332 \quad \text{(Eq. 2)}$$

For bobbin winding window height:

$$B_{wwh} = 0.953 \cdot C_{wwh} - 0.000734 \quad \text{(Eq. 3)}$$

Where C_{www} and C_{wwh} are the winding window width and height of the core, respectively.

With that we can easily find the Bobbin Filling factor as:

$$K_b = \frac{\left(B_{www} \cdot B_{wwh}\right)}{\left(C_{www} \cdot C_{wwh}\right)} \qquad \text{(Eq. 4)}$$

The same process has been done for the wires, following standard IEC 60317, choosing for this case wires of grade 1, as that way we can cover the case of strands inside a Litz wire. The data is presented in Fig. 3.

Which gives is the following equation for the outer diameter of a round wire:

$$\phi_{outer} = -118.11\phi_{cond}^2 + 1.12\phi_{cond} + 9.96 \cdot 10^{-} \qquad \text{(Eq. 5)}$$

The conducting-to-insulating material ratio then can be calculated as:

$$K_{c2i} = \frac{\phi_{cond}^2}{\phi_{outer}^2} \qquad \text{(Eq. 6)}$$

In order to apply Eq.6, we need a conducting diameter, which we obtain from designing the wire to minimize the skin effect losses, meaning that diameter of the wire must be two times the skin depth.

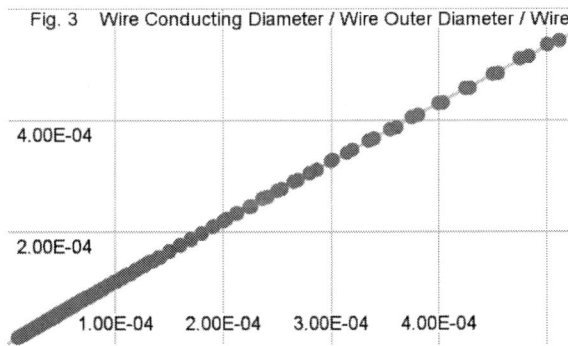

Fig. 3 Wire Conducting Diameter versus Wire Outer Diameter

To calculate this skin depth we use the effective frequency presented by Sullivan in [8], which will provide us with an accurate estimation in the case of non-sinusoidal waveforms, or when a DC bias is present.

$$f_{eff} = \sqrt{\frac{\sum\limits_{j=0}^{\infty} I_j^2 f_j^2}{\sum\limits_{j=0}^{\infty} I_j^2}} \qquad \text{(Eq. 7)}$$

$$\sigma_{eff} = \sqrt{\frac{\rho}{\mu_0 \mu_r \pi f_{eff}}} \qquad \text{(Eq. 8)}$$

$$\phi_{cond} = 2\sigma_{eff} \qquad \text{(Eq. 9)}$$

Where ρ and μ_r are the resistivity and the permeability of the conducting material of the wire, μ_o the permeability of free space, σ_{eff} is the skin depth at the effective frequency f_{eff}.

Finally the frequency dependent Window Utilization Factor is calculated:

$$K_{ufd} = K_b \cdot K_{c2i} \qquad \text{(Eq.10)}$$

2.2 Second improvement, the Frequency-dependent Magnetic Flux Density Swing, ΔB

As was mentioned, the same ΔB at different frequencies can produce totally different core losses, and can be the difference between a working design or a burning one. In order to control the core losses, *ΔB* must be scaled up or down inversely proportional to the switching frequency. This section presents one method to achieve this by keeping constant the core losses.

The process consists on the following steps:

1. Choosing a reference ΔB$_{ref}$ at a f$_{ref}$ that we are familiar with. We recommend 0.2 T at 100 kHz.
2. We calculate the core losses of that core at ΔB$_{ref}$ and f$_{ref}$.
3. We calculate the ΔB at the frequency of our converter that has the same losses as the reference. This is the ΔB$_{fd}$ that we will use in the Area Product method.

Step 3 can be easily done with the graphical data provided by the manufacturers, as shown in Fig.4, with the following substeps:

1. Find the magnetic flux density that we want to use as reference in the x axis.
2. Going vertically to the frequency that we want to use a reference, and on the intersection going left we obtain the volumetric losses for the reference point

3. From those reference volumetric losses we go horizontally to the right until we cross the target frequency, the one for our operating point.
4. We move down vertically to obtain the final magnetic flux density that we will use as the maximum in our Area Product method.

With these steps we obtain a ΔB_{fd} of 71 mT for 400 kHz, which produces the same core losses as ΔB_{ref} of 0.2 T and f_{ref} of 100kHz, for Ferroxcube 3C98.

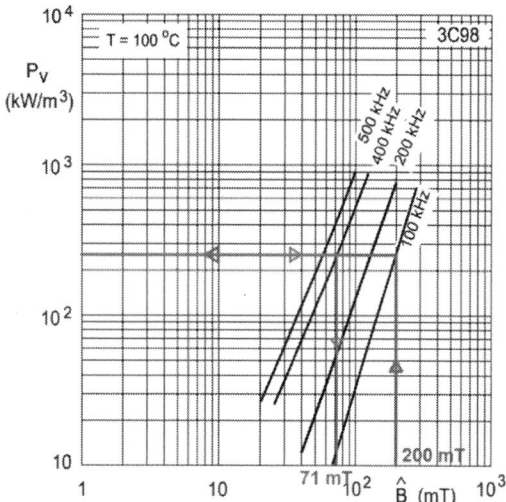

Fig. 4 Proposed process represented for Ferroxcube 3C98.

This process can be easily automated in software by using any of the models available, although the classical Steinmetz equation is the easiest choice, as it allows us to calculate ΔB_{fd} as:

$$\Delta B_{fd} = 2 \left(\frac{k f_{conv}^{\alpha}}{\left(k B_{ref}^{\beta} f_{ref}^{\alpha} \right)} \right)^{\frac{1}{\beta}} \quad \text{(Eq.11)}$$

The method for calculating the volumetric core losses is not important in itself, as the objective of this calculation is applying the formula and the inverse operation, so any advantages of advanced methods will be canceled.

2.3 The Improved AP

Finally, using Eq. 10 and Eq.11 in Eq.1, we can obtain an AP formula that scales with the frequency and harmonic content of our excitation.

$$AP_{fd} = \frac{P_{in}}{K_t K_{ufd} K_p J \Delta B_{fd} 2 f} \quad m^4 \quad \text{(Eq.12)}$$

3 Conclusion

Two improvements have been presented for expanding the existing Area Product method for selecting cores for magnetic components in order to make it work better at medium and high switching frequencies.

The first improvement consists of taking into account that at these frequencies Litz wires are commonly used, which will decrease the utilization factor as the frequency goes up.

The second improvement consists of scaling the maximum magnetic flux density used in the Area product method in order to keep the core losses constant. This improvement works in both directions, allowing a larger magnetic flux density for low frequency, and a smaller one for higher frequencies.

References.

[1] Colonel Wm. T. McLyman, Transformer and Inductor Design Handbook, Fourth Edition. CRC Press, 2011.

[2] Marian K. Kazimierczuk, High-Frequency Magnetic Components. Second Edition.

[3] Pressman, Abraham, Billings, Keith, and Taylor Morey. 2011. Switchmode Power Supply Handbook. 3rd ed. New York: McGraw-Hill Education.

[4] P. Wallmeier, "Pre-optimization of linear and nonlinear inductors using area-product formulation," Conference Record of the 2002 IEEE Industry Applications Conference. 37th IAS Annual Meeting (Cat. No.02CH37344), Pittsburgh, PA, USA, 2002, pp. 2445-2450 vol.4, doi: 10.1109/IAS.2002.1042788.

[5] S. Janghorban, D. G. Holmes, B. P. McGrath, W. G. Hurley and X. Yu, "Selecting magnetic cores for higher power inductors," 2016 IEEE 7th International Symposium on Power Electronics for Distributed Generation Systems (PEDG), Vancouver, BC, Canada, 2016, pp. 1-6, doi: 10.1109/PEDG.2016.7527060.

[6] N. Ekekwe, J. E. Ndubah, K. White and O. Ben, "Practical process in high frequency distribution transformer design," Proceedings: Electrical Insulation Conference and Electrical Manufacturing and Coil Winding Technology Conference (Cat. No.03CH37480), Indianapolis, IN, USA, 2003, pp. 121-128, doi: 10.1109/EICEMC.2003.1247867.

[7] S. Barqi, "Optimum Design Approach of High Frequency Transformer: Including the Effects of Eddy Currents," 2018 15th International Multi-Conference on Systems, Signals & Devices (SSD), Yasmine Hammamet, Tunisia, 2018, pp. 298-303, doi: 10.1109/SSD.2018.8570390.

[8] C. R. Sullivan, "Optimal choice for number of strands in a litz-wire transformer winding," in IEEE Transactions on Power Electronics, vol. 14, no. 2, pp. 283-291, March 1999, doi: 10.1109/63.750181.

High Resolution Mixed-Signal Pulse Width Modulator for High-Frequency DC-DC Converters

Tim McRae [*][1], Kasper Paasch [2], Thomas Ebel [2]

[1] University of Southern Denmark, Digital and High Frequency Electronics, Denmark
[2] University of Southern Denmark, Centre for Industrial Electronics, Denmark

Corresponding author: Tim McRae, mcrae@sdu.dk
Speaker: Tim McRae, mcrae@sdu.dk

Abstract

This paper introduces a mixed-signal pulse-width modulator which uses a low clock-frequency digital pulse-width modulator and a binary-weighted resistive divider, ramp, and comparator to generate switching signals with sub-clock-cycle accuracy. A voltage reference is set prior to a desired switching event and a compared against a ramp, generating an asynchronous gate turn-off signal. The effective resolution of the PWM is increased shifting from a time-resolution required to a voltage resolution requirement. A discrete experimental prototype is built to verify the operation of the circuit. 29 ps time resolution is achieved on FPGA without the need for a PLL.

1 Synopsis

Both digital and analog controllers have been used in industry for quite some time [1]. Often analog controllers are described as being efficient in terms of energy consumption during steady-state operation [2]–[4], component count, and resolution, while also being somewhat limited in their complexity. Digital control methods on the other hand are seen as more flexible with the possibility for more complex control methods [3], but are more energy intensive and have limitations in terms of resolution or clock speed in fully integrated solutions. Mixed-signal control design has been a popular method of combining these two distinct controller types to gain the advantages of both while avoiding their drawbacks [5], [6].

One of the main challenges with digital control is resolution, for both the output voltage sensing and the digital pulse-width modulator (DPWM). Without proper consideration, mismatches in resolution between these two components can lead to limit-cycle oscillations in controlled variables [4]. One way to avoid this is to ensure the resolution of the DPWM is high enough such that the output voltage is able

---

*This work is part of the PE-Region Platform Project and the authors gratefully acknowledge the support of Interreg Deutschland-Danmark and the European Regional Development Fund, Denmark.

to fall within the zero error bin of the ADC [5], [7]. This can be achieved by using a high frequency internal clock for a counter-based DPWM, but this leads to higher energy consumption in steady-state and is limited by the maximum bandwidth of a PLL. Other methods to overcome this challenge include delay-line based PWMs, which are able to achieve very high resolution [8]. Some of these delay-line based PWMs also include an analog controlled supply voltage to reach even higher resolutions [9]. Access to a flexible, high-resolution DPWM should be available which can interface with a fast prototyping system, while also being viable for integration. This paper proposes a novel mixed-signal PWM which can be implemented both discretely and fully integrated with different structures. By combining both a counter-based DPWM with a variable reference ramp, a simple hardware implementation for a high resolution mixed-signal pulse-width modulator (MSPWM) is achieved. Experimental results indicate an average time resolution of $29ps$, equivalent to a conventional counter-based DPWM with a clock frequency of $34GHz$.

The paper is organized as follows: Section 2 describes the principle of operation of the MSPWM and discusses design considerations, Section 3 provides the experimental results to verify the operation of the proposed MSPWM, and Section 4 discuss the results and concludes the paper.

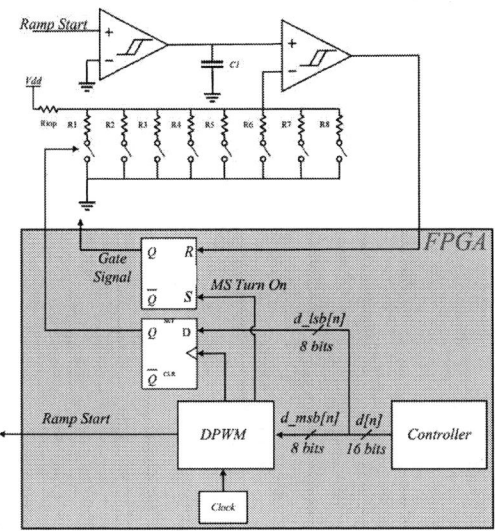

Fig. 1: System level diagram of the proposed pulse width modulator.

2 Principle of Operation

A diagram of the proposed MSPWM can be seen in Fig. 1. The MSPWM is comprised of a low frequency counter-based DPWM, a binary-weighted resistive divider, a ramp-generator, a comparator, and an SR-latch. This system is able to generate sub-clock cycle resolution within a selected clock cycle within the switching period. A timing diagram of the sequence can be seen in Fig. 2. The following describes the generation of a single high resolution gate signal.

The duty-cycle provided to the MSPWM is divided into two components, d_{MSB}, the most significant bits sent to the DPWM, and d_{LSB}, the least significant bits sent to the resistive divider. The switch gate signal starts by setting SR latch at the beginning of the switching cycle. The ramp reference is set to a new value, and the ramp begins by setting the comparator input high, the ramp enters the ramp reference voltage range at the beginning of the selected switching cycle. The ramp is compared to the ramp reference and generates a reset signal which forces the switch gate signal low, completing the generation of a high resolution gate-signal in the overall switching cycle. The blocks of the system are described below.

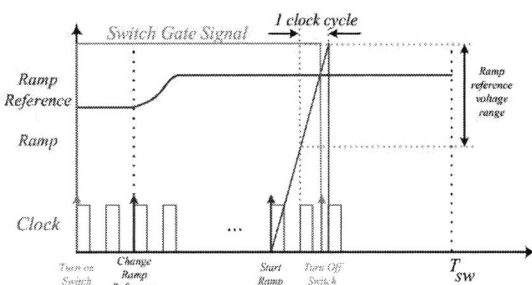

Fig. 2: Timing diagram to implement sub-clock-cycle resolution

2.1 Low Frequency Counter-Based DPWM

The DPWM in the system is the backbone of the digital or mixed signal controller, providing clocking signals to all blocks and providing the overall timing framework for the switching cycle of the SMPS. By avoiding the use of a PLL, all blocks in the system can be run at a much lower frequency, thus saving energy. The switching frequency of the SMPS is determined by selecting how many cycles of the internal clock are counted, and can be seen in equation 2.1

$$f_{sw} = \frac{f_{clk}}{N_{counter}} \tag{1}$$

The duty cycle generated by the controller is divided into two components, $d_{controller} = [d_{MSB}, d_{LSB}]$. The n_{MSB} most-significant bits output from the duty cycle generated by the controller, corresponding to d_{MSB} are sent to the low frequency DPWM. m cycles before time $\frac{d_{MSB}}{f_{clk}}$, the n_{LSB} least significant bits, corresponding to d_{LSB} are sent to the resistive divider. At $\frac{d_{MSB}-1}{f_{clk}}$, a signal is sent from the DPWM to trigger the ramp. A set signal is sent from the DPWM to the SR-latch and turn on the gate signal, waiting for the reset signal from the resistive divider.

2.2 Binary-Weighted Resistive Divider

The resistive divider emulates a simple digital-to-analog converter and sets the reference for the comparator. The n_{LSB} least significant bits are sent to the resistive divider, with each bit controlling one switch, adjusting the resistance of the lower resistor of the resistive divider. This variation of resistance is not linear. A diagram of the change of the reference over the input codes can be seen in Fig. 2.2. The reference generated by the resistive

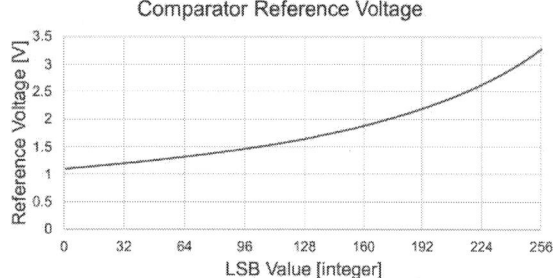

Fig. 3: Nonlinear variation of voltage reference due to resistive divider variation

divider is triggered m cycles before the switching event occurs so the reference voltage has time to settle before the comparison. The settling is related to the total resistance of the divider and the effective capacitance seen at the divider node. The time constant can be seen in equation 2.2. This time constant varies based on the state of the resistive divider; fastest when the R_{bottom} is minimized and slowest when it is maximized, corresponding to the smallest and largest value of d_{LSB} respectively.

$$\tau = \frac{R_{top}R_{bottom}}{R_{top} + R_{bottom}}C_{eq} \qquad (2)$$

2.3 Ramp Generator

The ramp generator controls the sub-clock-signal timing of the MSDPWM. Effectively the ramp generator, the resistive divider and the comparator form an analog PWM whose reference is controlled digitally. The generator in this case is implemented with a capacitively loaded comparator to approximate a ramp. Other implementations are possible, and a fully integrated solution would perform better with a fixed reference and a binary weighted current source or binary weighted capacitance to generate a variable ramp. All that is required of this block is that a sawtooth signal can be formed with a sufficient slope.

2.4 Comparator and SR-latch

The comparator compares the digitally controlled reference to the ramp generated by the capacitively-loaded comparator. The output of the comparator is sent to an SR-latch, which generates the gating signal of the desired switch.

2.5 Effective Time Resolution

The addition of the analog portion of the PWM results in an effective time resolution of

$$\Delta t = \frac{1}{f_{clk}2^{n_{LSB}}} \qquad (3)$$

where Δt is the minimum time resolution, f_{clk} is the digital clock frequency, and n_{LSB} is the number of least significant bits sent from the duty ratio $d[n]$ to the resistive divider if the digital voltage references are able to lie entirely evenly within a single low frequency clock cycle and all variations of reference are linear. If the range of voltage reference causes a reset signal to the sent before or after the completion a low frequency cycle, then the effective resolution is not increased by this amount as some values of the LSBs become mapped onto the values which could be reproduced by the low frequency clock. Additionally, not all changes in reference are linear due to the resistive divider implementation. The effective time resolution can alternatively be calculated by considering the slope of the ramp and the minimum change in the voltage reference.

$$\Delta t = \frac{\Delta V}{slope} \qquad (4)$$

where

$$\Delta V = V_{ref}\frac{R_{bottom}}{R_{top} + R_{bottom}}\frac{R_{LSB}}{R_{top} + R_{bottom} + R_{LSB}} \qquad (5)$$

As can be seen from the above equation, ΔV varies based on the state of the resistive divider, meaning that the time resolution will vary. The time resolution is smallest when the bottom resistance is smallest and increase as the bottom resistance increases. This non-linearity is inherent to the implementation. Other implementations should be considered if this non-linearity is unacceptable for the application.

2.6 Extension to Multiple High Resolution Switching Edges

As mentioned previously, the system as described here only generates one high-resolution switching edge in a single switching cycle. This means that the turn-on edge of a single switch lies on the low frequency DPWM timing grid and the turn-off edge lies on the high-resolution grid.

The only requirement to extend this to multiple switches is to 1) ensure the ramp and comparator periods of different switches do not overlap and 2) the voltage reference for the comparison reaches its final value before the comparison takes place.

In the current implementation, the minimum ramp and comparator period is 2 cycles of f_{clk}. The time-constant for the settling time is seen in Eq. 2.2, and if m cycles are used to ensure the final value is reached, then a total of $m + 2$ f_{clk} cycles are needed for a high-resolution edge to take place. In this case m was selected to be 2. With this considered, multiple high resolution edges can be placed within a single switching cycle with a single ramp-comparator-reference circuit simply by ensuring these time regions do not overlap. Typically, even with multiple switches, most circuits do no require multiple switches to switch within a short time of one another, except for dead-time.

Enabling a fully controllable dead-time for this system requires a separate ramp-comparator-reference circuit to be run in parallel, because the time required to prepare the turn-on for the complementary switch would coincide precisely with the turn-off of the main switch. Such a trade-off depends on the space and cost requirements of the system.

3 Experimental Results

An experimental prototype was made to test the effective time resolution of the proposed PWM. Oscilloscopes with a high enough bandwidth or a fast enough sample rate were not available to verify or characterize the performance of the circuit due to the extremely small time resolution. Instead, the resolution is tested by operating a phase-shifted full bridge DC-DC converter and measuring the change in steady-state output voltage, ΔV_{LSB}, across the full range of the analog portion of the MSPWM. The converter is run in open loop with an e-load. A table of the components used for the full-bridge can be seen in Table 1. Operation of this converter is demonstrated in [10].

A table of ideal and practical resistances for the resistive divider are shown in Table 2. A fixed resistance is placed in parallel with the binary weighted resistors to tuned the reference voltage so that the comparison point occurs at least one switching cycle after the ramp start signal is sent by the FPGA. Results can be seen in Fig. 3 demonstrating an average time resolution of $29ps$ across the 8-bit range of the LSBs of the duty ratio. These results only demonstrate a single high resolution edge in the operation of the full bridge DC-DC converter.

It can be seen that the variation in duty ratio is not completely linear, as we expect from the analysis

Tab. 1: Full Bridge DC-DC Converter Experimental Setup

Component	Value
Switches	C3M0045065K Cree SiC Fet
Transformer	Custom tape wound, E32/16/9-N87 TDK, Turns Ratio 10:5 $229\mu H$ Magnetizing $93nH$ Leakage
Inductor	Custom wirewound, Toroidal core 77617A7 Magnetics, $297\mu H$
Output Capacitor	$120\mu F$
Switching Frequency	$200kHz$
Clock Frequency	$50MHz$
$N_{counter}$	250
D_{MSB}	109
Input Voltage	$48V$
Output Voltage	$5.2V$
Multimeters	Keysight 34465A
Power Supply	Elektro-Automatik EA-PS 9500 20
Electronic Load	Zentro Elektrik EL 1000/800/20
FPGA	DE0 Nano
Oscilloscope	Teledyne Lecroy HDO4054-MS

Tab. 2: Resistive Divider

Binary Weighted	Practical
$2k\Omega$	$2k\Omega$
$4k\Omega$	$2x2k\Omega$
$8k\Omega$	$8.2k\Omega$
$16k\Omega$	$16k\Omega$
$32k\Omega$	$2x16k\Omega$
$64k\Omega$	$68k\Omega$
$128k\Omega$	$2x68k\Omega$
$256k\Omega$	$261k\Omega$

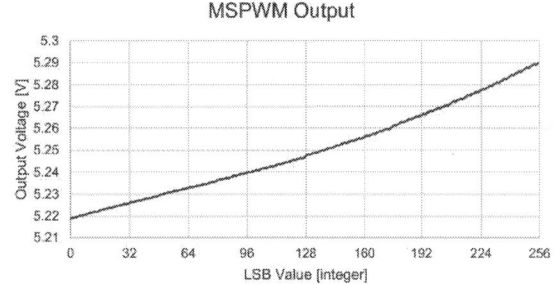

Fig. 4: Experimental results showing the change in the output voltage a phase-shifted full-bridge converter corresponding the change in LSBs of the duty ratio.

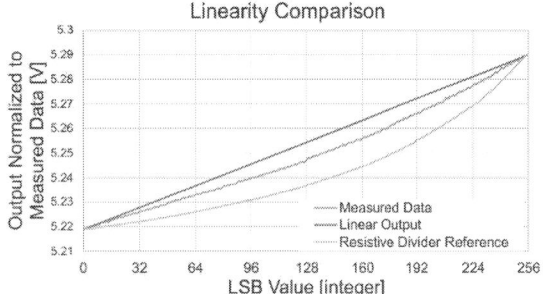

Fig. 5: Comparison to demonstrate the linearity of the output of the LSB component of the MSPWM. The linear output and the calculated output for the resistive divider are normalized to the measured output.

of the variation of reference voltage with respect to the least significant bit. Fortunately, for closed loop systems, this type of non-linearity typically does not impede the controller from achieving steady-state operation. Additionally, the variation of output voltage is not smooth with variation in LSB. This is likely due to variation in operating point of the converter during testing.

A comparison between the measured output, a linear output and the reference voltage generated by the binary weighted resistive divider using practical resistor values can be seen in Fig. 3.

Due to the small variation over the range, absolute errors are quite small, while it is still clear that non-linearity is introduced. The results show that the output of the MSPWM lie between a linear PWM and the reference generated by the resistive divider. This indicates that there is additional non-linearity or error in the measurement setup which partially compensates for the non-linearity introduced by the resistive divider. This could arise from significant variations in resistance values or from a non-linear slope generated by the ramp generator. Both these components have a large impact on the overall precision and linearity of the system.

4 Conclusion

A high resolution mixed-signal pulse width modulator (MSPWM) has designed and experimentally verified. It has been shown that sub nanosecond time resolutions can be easily achieved using conventional off-the-shelf FPGAs and the design of a simple circuit without the need for extremely high internal clock frequencies or delay line based implementations.

The system is able to produce such high resolution by removing the requirement for precision at every point in the switching cycle. Instead, the system prepares a single sub-switching period to have precise time resolution within the overall switching cycle. The number of such moments in the switching period is limited by the time-constant of the variable voltage reference for the ramp comparison. Including equally high time resolution to enable tune-able dead-time using this system can also be done, but requires a secondary ramp comparator to be implemented.

Implementation on chip presents promise for further development along with precise characterization of the introduced non-linearity and its impacts on the performance of closed-loop switch mode power supplies.

References

[1] F. G. R. Ramos, T. C. Pimenta, and L. H. Ferreira, "A mixed-signal pulse width modulator for portable smps applications," *Integration*, vol. 55, pp. 265–273, 2016. DOI: https://doi.org/10.1016/j.vlsi.2016.07.005.

[2] B. Patella, A. Prodic, A. Zirger, and D. Maksimovic, "High-frequency digital pwm controller ic for dc-dc converters," *IEEE Transactions on Power Electronics*, vol. 18, no. 1, pp. 438–446, 2003. DOI: 10.1109/TPEL.2002.807121.

[3] Z. Lukic, K. Wang, and A. Prodic, "High-frequency digital controller for dc-dc converters based on multi-bit /spl sigma/-/spl delta/ pulse-width modulation," in *Twentieth Annual IEEE Applied Power Electronics Conference and Exposition, 2005. APEC 2005.*, vol. 1, 2005, 35–40 Vol. 1. DOI: 10.1109/APEC.2005.1452883.

[4] M. D. Hagen and V. Yousefzadeh. "Applying digital technology to pwm control-loop designs." (2008), [Online]. Available: https://api.semanticscholar.org/CorpusID:17097730.

[5] A. Prodic and D. Maksimovic, "Mixed-signal simulation of digitally controlled switching converters," in *2002 IEEE Workshop on Computers in Power Electronics, 2002. Proceedings.*, 2002, pp. 100–105. DOI: 10.1109/CIPE.2002.1196722.

[6] O. Trescases, Z. Lukic, W. T. Ng, and A. Prodic, "A low-power mixed-signal current-mode dc-dc converter using a one-bit /spl delta//spl sigma/ dac," in *Twenty-First Annual IEEE Applied Power Electronics Conference and Exposition, 2006. APEC '06.*, 2006, 5 pp.-. DOI: 10.1109/APEC.2006.1620615.

[7] A. Syed, E. Ahmed, and D. Maksimovic, "Digital pwm controller with feed-forward compensation," in *Nineteenth Annual IEEE Applied Power Electronics Conference and Exposition, 2004. APEC '04.*, vol. 1, 2004, 60–66 Vol.1. DOI: 10.1109/APEC.2004.1295788.

[8] T. Carosa, R. Zane, and D. Maksimovic, "Implementation of a 16 phase digital modulator in a 0.35 /spl mu/m process," in *2006 IEEE Workshops on Computers in Power Electronics*, 2006, pp. 159–165. DOI: 10.1109/COMPEL.2006.305669.

[9] D. Costinett, M. Rodriguez, and D. Maksimovic, "Simple digital pulse width modulator under 100 ps resolution using general-purpose fpgas," *IEEE Transactions on Power Electronics*, vol. 28, no. 10, pp. 4466–4472, 2013. DOI: 10.1109/TPEL.2012.2233218.

[10] C. S. Kjeldsen and C. Østergaard, "Experimental demonstration of the transformer interwinding capacitance voltage waveform in an isolated full-bridge forward converter," in *2023 25th European Conference on Power Electronics and Applications (EPE'23 ECCE Europe)*, 2023, pp. 1–8. DOI: 10.23919/EPE23ECCEEurope58414.2023.10264368.

PCIM Europe 2024, 11–13 June 2024, Nuremberg   DOI: 10.30420/566262333

Designing a Control Library for Grid-following and Grid-forming Power Converters

Lars Lindner [1], Matthias Klee[1], Daniel Stracke [1], Jonas Steffen [1], Axel Seibel [1], Marco Jung [1,2]

[1] Fraunhofer Institute for Energy Economics and Energy System Technology, Germany
[2] Bonn-Rhein-Sieg University of Applied Sciences, Germany

Corresponding author: Lars Lindner, lars.lindner@iee.fraunhofer.de
Speaker: Lars Lindner, lars.lindner@iee.fraunhofer.de

Abstract

As a result of the energy transition, more and more power converters will be required in the future, which will lead to distributed energy generation across all voltage levels. Amongst others, these power converters take on tasks, such forming a grid or being capable of black starting. In order to be able to answer new research tasks, rapid control prototyping (RCP) systems are required, which must be flexibly programmable and configurable as laboratory devices. The RCP system offers great potential for fast and targeted laboratory validation of control algorithms and supports this with a user-friendly development platform for algorithm development. In order to simplify the programming of these RCP systems for the user, a software library is needed, which combines common building blocks for grid-following and grid-forming control algorithms. This software library enables the rapid development of new control algorithms for power converters, providing the user with tools for modeling, simulation and deployment of these control algorithms on a RCP system. This paper presents the development and verification of this software library, introduces the RICOSO system and presents aspects of the soft- and firmware design.

1 Introduction

One important aspect of the energy system transformation is the transition from large centralized to smaller and locally distributed renewable energy sources, such as wind turbines and photovoltaic systems [1], [2]. In order to connect these energy sources to the public power grid, power converters with varying tasks and characteristics will be required in the future. Amongst others, these power converters must operate as current or voltage sources, i.e. either as grid-following or grid-forming converters [3]. Thereby, grid-following and grid-forming control algorithms must be implemented on these power converters, which are made up of common building blocks. Furthermore, flexible and programmable power converters as laboratory devices are therefore needed to research new algorithms for grid-following or grid-forming control [4]. To enable the user to program these power converters as laboratory devices, a *control library* was designed, which makes the common

building blocks of grid-following and grid-forming control available to the user in an organized manner.

Furthermore, there will be a necessity to deploy these new control algorithms on real-time hardware, like *Rapid Control Prototyping* (RCP) systems. To meet this necessity, a flexible and easy-to-use RCP system was developed at Fraunhofer IEE [5]–[7]. The *Rapid Inverter Control Prototyping Solutions* (RICOSO) provides a RCP system that allows developers and researchers to test ideas for power converter control in practice [8].

The purpose of this article is to introduce the above-mentioned control library, present the RICOSO platform, give an overview of the underlying software and firmware design, present an application example of this control library and thus to provide a comprehensive development ecosystem for power converter control systems.

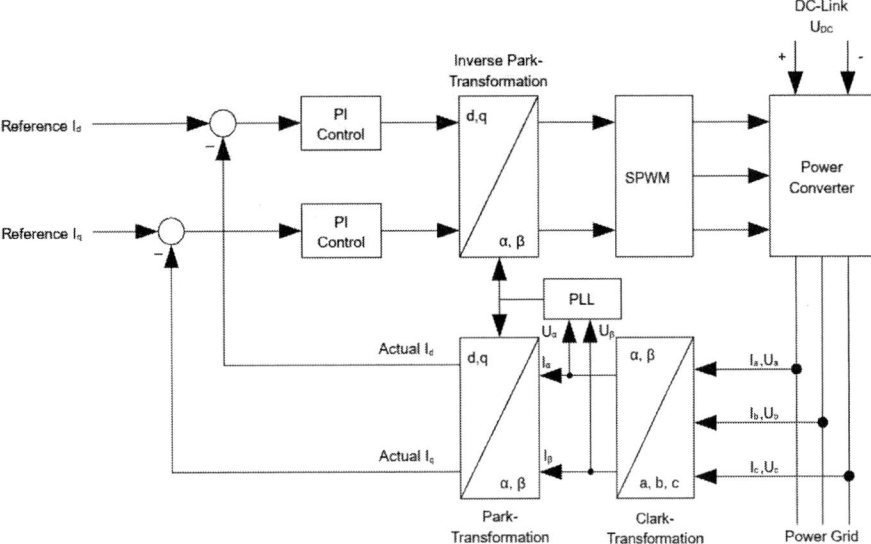

Fig. 1: Overview of the grid-following control algorithm

2 Basic control of grid-following power converters

In order to convert the direct current generated by a photovoltaic system, for example, into alternating current for the power grid, power converters are required. In addition to this main task of power conversion, power converters nowadays perform numerous other tasks, like efficiency improvement, network and system monitoring, as well as safety-critical functions to protect users and the overall system. The control concepts of these power converters can be classified roughly according to the method of determining the reference phase angle [9]. Today, the relevant technical literature distinguishes between the two concepts of grid-following and grid-forming control. In grid-forming control, the reference phase angle is generated by the control loop itself and in grid-following control it is generated externally [9].

One way to define a grid-following algorithm for power converters is to derive it from the method of *Vector Control* (field-oriented control) used in asynchronous motors [10], [11]. The basic control loop of the field-oriented control is shown in Fig.1. Using the field-oriented control for defining the grid-following algorithm, the power converter operates as a current source for feeding into the three phase power grid. Thereby, the three phase currents fed into the power grid must be controlled in closed loop configuration. Since the three grid phase current represent alternating variables, they

can be transformed into a stationary coordinate system (d/q) which rotates with the power grid fundamental frequency, using the Clarke and subsequent Park transformation [12]. In the transformed d/q coordinate system, the three phase currents represent stationary variables, which are easier to regulate, than alternating variables. To utilize the Park transformation, precise knowledge of the current phase angle is necessary, which can be achieved using a *Phase-locked Loop* (PLL) [13]. The PLL locks on the current angle of the power grid, which is required for the subsequent coordinate transformations. For controlling the I_d and I_q currents, PI controllers are suitable for achieving a residual control deviation of zero. The controller output variables are then transformed back to alternating variables using the inverse Clarke and Park transformation, which represent the reference voltages for the subsequent modulation scheme, which generates the switching signals for the semiconductor components of the power converter half-bridges [12].

3 Basic components of the control library for grid-following converters

The control library for developing new algorithms for power converters was designed, which allows users to quickly simulate control algorithms and test them on embedded systems. This control library represents a proprietary custom library, developed using the software tool Matlab/Simulink. For this purpose, existing and specially developed control models were divided into separate entities. Each entity was defined within a single model with in- and outputs. All these entities were logically grouped into different classes. Each class was defined using a sub-library in the overall library. This library was then integrated in the Simulink browser. For usability, this library must address the following general key aspects, summarized in following table 1.

Tab. 1: Control library key aspects

Different control or modulation algorithms
Tools for modeling and simulation of power converters
Enabling the implementation of real-time control algorithms on embedded systems
Must contain signal processing and filtering functions for sensor outputs
Should support different communication interfaces
Must implement safety features and fault detection mechanisms
Should offer data logging and analysis
Should provide a well-designed graphical user interface (GUI)
Must include comprehensive documentation and user tutorials

Different control or modulation algorithms are necessary, since they are integral part of the power converters control concepts. Modeling and simulation is required to create a mathematical model of the power converter algorithms and to use this model to numerically determine the time characteristics of internal signals. Also, the developed power converter algorithms should be able to be tested in a real-time hardware environment, hence the library must provide functionality to deploy these algorithms on embedded systems. Signal processing and filtering functions for sensor outputs are required, to feed back actual power converter measurements into the control system running on the targets hardware. The control library should support different communication interfaces and protocols in order to be able to exchange process data with external hardware. It must also contain safety functions and error detection mechanisms to protect the user and the devices from unpredictable errors. Furthermore, the control library should offer options for implementing data acquisition so that real-time data can be recorded for later analysis.

Finally, the library should provide a well-designed graphical user interface (GUI) to interact with and configure control algorithms.

From the grid-following control method, the control library for developing new control algorithms was derived. Therefore, the control library must contain the following basic building blocks: Measuring and filtering of power converter output signals, like e.g. the actual DC-link voltage, the actual grid voltage or the actual converter voltage; different controller blocks, like e.g. the power controller or the DC balance controller; different configuration blocks like e.g. for the current-controller, the PWM or the harmonic controller; different PLL blocks for power grid synchronization; different transformation blocks and a parameter block to define global parameters like e.g. controller parameters. The control library was defined as a project in Matlab/Simulink. When the project is opened, an initialization script is executed which adds the control library to the Simulink browser. This initialization script also installs the toolchain for code generation, defines constants and parameters required for code generation and simulation of models.

4 Overview of the development environment

The control library was developed as part of a RCP system designed to operate on converter hardware. The development environment depicted in Fig.2 combines modular hardware with the control library to form a system called RICOSO [14]. The system utilizes two bidirectional high-performance power converters as *Devices Under Test* (DUTs) for the control library, supporting the model-based approach for designing, testing and tuning control algorithms. RICOSO features a highly flexible FPGA and processor, allowing certain parts of the simulated models to be implemented on the targets hardware and executed in real-time. Overall, RICOSO presents a dynamic and versatile converter system that can be programmed via optical fiber cable or Ethernet connection.

The power converters are capable of operating in four quadrants, offering versatility in various applications [14]. They exhibit the ability to handle unbalanced and nonlinear loads, making them suitable for a broad spectrum of scenarios including island grids and fault scenarios. Utilizing the active-neutral-point-clamped (ANPC) three level semiconductor topology with SiC semiconductor devices,

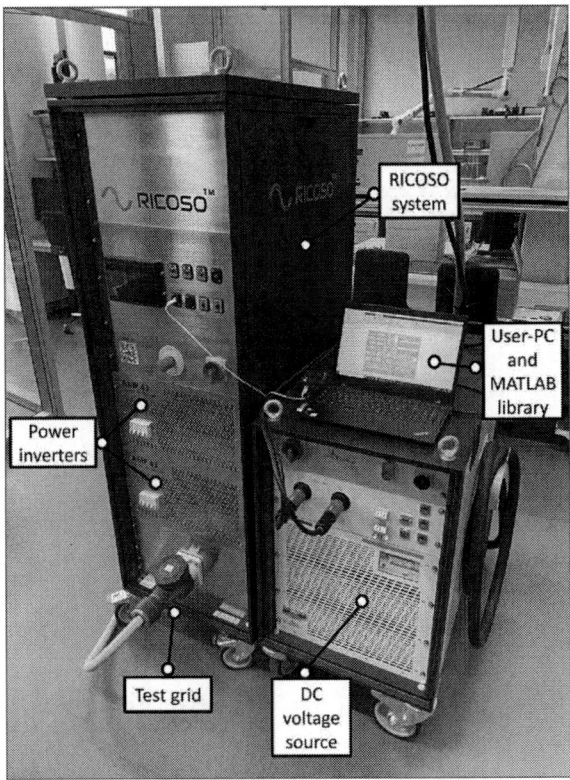

Fig. 2: RICOSO system development environment

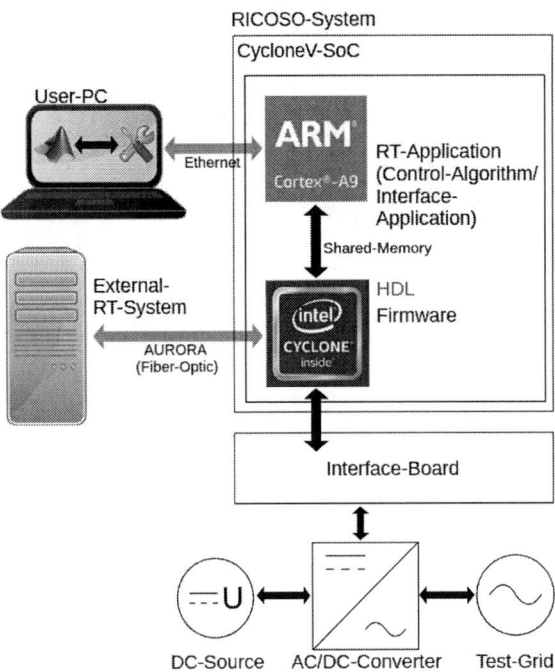

Fig. 3: RICOSO control system overview

the system's hardware is compact compared to systems employing silicon devices and come in 19-inch slide-in units. The compactness is further driven by the switching frequency of 70 kHz.

5 Design of soft- and firmware

The RICOSO RCP uses a System-on-Chip (SoC) as its control system. The SoC, an Intel Cyclone V, consists of an ARM-Cortex-A9 processor and a Field-Programmable-Gate-Array (FPGA). During the following section *firmware* will refer to the design implemented on the FPGA portion of the SoC, while *software* refers to the designs portion implemented on the ARM-processor. Fig. 3 displays an overview of the various control systems, as well as the soft- and firmware on each platform.

The firmware implements all parts of the control that either rely on high bandwidth or strict deterministic execution. Examples would be the actual pulse generation of the PWM, processing of sensor data, as well as all safety features. Additionally, a simple current controller is implemented (referred as *Embedded-current-controller*).

The software running on the ARM-processor consists of a multitude of different small programs. The starting point of the system is a minimalist Linux operating system, which includes a Real-Time (RT) patch-set, making it a Real-Time-Operating-System (RT-OS) suitable for control tasks. A set of services running in the background of the RT-OS take care of measurement calibration, error-logging, as well as providing an application programming interface for GUI applications. The application containing the actual control will be dynamically generated using the mentioned control library in combination with Matlab/Simulink's embedded coder toolbox and a custom toolchain, that directly integrates into Matlab/Simulink (referred as *ARM-based-control*).

Aside from the two already mentioned approaches to control the RICOSO system (Embedded-current-control and ARM-based-control), recently a third mode of operation was added. It allows the RICOSO system to be controlled using an external RT-device, connected using a fiber-optic connection.

As previously described, the system supports three different modes of control: Embedded-current-controller, ARM-based-control and external-control. Table 2 summarizes the different control modes. Each mode differs by a set of variables, such as bandwidth, latency, flexibility and ease-of-use. With the bandwidth mainly depending on the computing power of the processor system. Latency is particu-

Tab. 2: Control-mode comparison

Control Property	Embedded-Current-Control	ARM-based-Control	External-Control
Bandwidth	+	-	+
Latency	+	O	O
Flexibility	-	O	+
Ease-of-use	+	+	O

larly important for external controllers, as communication and possible clock drifts of the processor systems are added here. Flexibility describes the degree of freedom each mode provides to change the structure of the controller. In a similar fashion, ease-of-use describes the amount of work needed to change a controllers structure or parameters.

The embedded-current-controller is designed for quick and easy deployment. As it is part of the firmware, it provides the maximum bandwidth and lowest latency. It can be controlled by either an external GUI or using the control library. The current controller is implemented as a classical PI-controller, with optional parallel resonant controllers to damp current harmonics. The control library and GUI both allow to modify the controller gains and the setpoints of the PI-controller. A drawback of the control-mode is its lack of flexibility, while it is possible to tune various parameters, the underlying structure is fixed and can not be changed.

Using the presented control library and toolchain it is possible to develop control algorithms in Matlab/Simulink and directly deploy them to the converter system. Due to the control library which provides access to all internal measurements and the possibility to use almost all standard Matlab/Simulink blocks, controller design in this control-mode is simple and flexible. The biggest disadvantage of this approach is the bandwidth. Due to the limited computing power of the ARM processors, the controllers implemented here cannot fully utilize the bandwidth of the converter. The control library provides a set of tools that can be used to partly overcome this limitation. One approach would be to combine the ARM-based controller with embedded-current-control, using a cascaded controller design. Here the outer controller is used to provide the set points of the embedded-current-controller.

The external-control-mode was developed to fully overcome the limitations given by the ARM-based-controller design. Here, an external Real-Time Processing System (RPS) is used to control the converter. When combined with a sufficient powerful RPS, the converter can be controlled at switching frequency, allowing it to fully utilize its bandwidth. However, this approach also faces a couple of drawbacks, aside from the general need for an external RPS, which increases costs, one would be the additional delay added by the communication between the converter and the external RPS. Flexibility and ease-of-use mainly depend on the chosen RPS system. Table 2 summarizes the individual benefits of each approach.

6 Example of using the control library

A power control algorithm using grid-following is presented as an application example of the control library. This power control algorithm contains a cascaded control, featuring a super-imposed power and a under-laid current controller. With the former being implemented in the ARM-processor-system and the latter using the embedded-current-controller. Additionally, the model features multiple resonant controllers (optional feature of the embedded-current-controller) which are used to damp current harmonics. All blocks used in the design are provided by the control library and can be directly deployed to the system using a custom toolchain, which is also provided with the control library. The toolchain uses Matlab/Simulinks Embedded-Coder and ARM-Cortex-A Board-Support-Package to generate code from the model itself. This code will then be compiled using the GNU Compiler Collection based cross-compiler-toolchain and automatically deployed to the ARM-processor-system. Due to the code being based on the ARM-Cortex-A Board-Support-Package, controls designed this way allow for live monitoring using Matlab/Simulinks *external mode* setting.

This power control algorithm uses setpoints for the active and reactive power and generates the setpoints for the subordinate current controller using

PI controllers. As current control loop the grid-following control algorithm (Fig. 1) is used. Using the power controller, the step response of the RICOSO power converter at nominal apparent power ($S = 43.47\,kVA$) was measured and is shown in following Fig. 4.

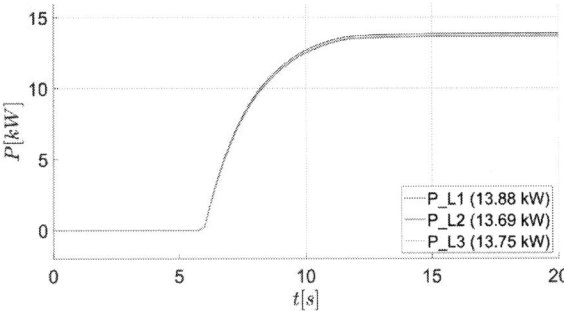

Fig. 4: Active power step response of grid following power control

This figure shows the step responses of the active power for each of the three phases L1-L3, as well as the steady-state values. Since a setpoint ramp was used internally for the reference voltage of the sinusoidal pulse width modulation (SPWM) modulation (Fig. 1), the step responses requires approx. $5s$ to rise to its final value. Also, the sum of the 3 phases active power is not exactly $43.47\,kVA$, as part of the apparent power is converted into reactive power for the harmonic (LCL) and the electromagnetic interference (EMI) filter.

7 Conclusion and further work

This publication presents a control library and a development ecosystem, which enables the rapid development of control algorithms for power converters and which provides the user a set of standard building blocks of grid-following and grid-forming control. The control library represents a proprietary custom library, which was designed using the software tool Matlab/Simulink. All necessary components of the library and their functions were presented.

The paper also gives a short overview of the soft- and firmware design of the development environment, as well as an application example of using this control library. This example presents the model of a grid-following power control algorithm with a subordinate current controller and describes how this model is compiled for the target and deployed on it. It also shows measurement results of the step responses of active power at the output of

the converter when feeding into the grid.

The advantage of this control library is that it enables the user to flexibly program and configure an RCP system. This control library combines common building blocks, which are typical needed to design a controller for grid-following and grid-forming power converters. Another advantage of this control library together with the RCP is that the developed models can be deployed directly on a hardware development environment. The Rapid Control Prototyping system RICOSO was introduced, which enables developers and researchers to efficiently design and test control algorithms for power converters on an embedded system. Using the control library and the RCP system, new control algorithms for power converters can be designed and tested on this real-time system. Using the control library, further use cases for the RCP system can be defined and tested, such as amplifier for Power Hardware-in-the-Loop, grid code tests, emulator for power electronic devices (e.g., battery, photovoltaic system, charging station) and training purposes. Amongst others, further work involves testing the control library by end users of the RICOSO development environment.

Acknowledgment

The authors acknowledge the support of the presented work by the *Forschungszentrum Jülich GmbH* within the project *F-HiL-Reloaded* (FKZ 03EI6104B). Only the authors are responsible for the content of this publication.

References

[1] Bundesnetzagentur für Elektrizität, Gas, Telekommunikation, Post und Eisenbahnen, "Netzentwicklungsplan strom," zweiter Entwurf 2037 / 2045 (2023), p. 157.

[2] C. Pape and D. Geiger, "Regionalisierung des Ausbaus der Erneuerbaren Energien," 2023, Publisher: Fraunhofer IEE.

[3] A. Lunardi, L. F. Normandia Lourenço, E. Munkhchuluun, L. Meegahapola, and A. J. Sguarezi Filho, "Grid-connected power converters: An overview of control strategies for renewable energy," *Energies*, vol. 15, no. 11, p. 4151, Jun. 5, 2022. DOI: 10.3390/en15114151.

[4] S. Anttila, J. S. Döhler, J. G. Oliveira, and C. Boström, "Grid forming inverters: A review of the state of the art of key elements for microgrid operation," *Energies*, vol. 15, no. 15, p. 5517, Jul. 29, 2022. DOI: 10.3390/en15155517.

[5] D. Stracke, F. Schnabel, S. Sprunck, and M. Jung, "Efficiency comparison of three-phase four-wire inverter topologies for unbalanced and nonlinear loads," in *PCIM Europe digital days 2021; International Exhibition and Conference for Power Electronics, Intelligent Motion, Renewable Energy and Energy Management*, May 2021, pp. 1–7.

[6] T. Gühna, D. Stracke, M. Klee, F. Schnabel, A. Seibel, and M. Jung, "Hardware and software concept for distributed grid-forming inverters in microgrids," in *2021 23rd European Conference on Power Electronics and Applications (EPE'21 ECCE Europe)*, Ghent, Belgium: IEEE, Sep. 6, 2021, P.1–P.10. DOI: 10.23919/EPE21ECCEEurope50061.2021.9570688.

[7] D. Stracke, F. Schnabel, S. Sprunck, and M. Jung, "Comparison of three-level grid-forming inverter topologies for unbalanced and nonlinear load conditions in microgrids," in *PCIM Europe 2022; International Exhibition and Conference for Power Electronics, Intelligent Motion, Renewable Energy and Energy Management*, May 2022, pp. 1–8. DOI: 10.30420/565822193.

[8] "Ricoso- rapid inverter control prototyping solutions." (), [Online]. Available: https://www.iee.fraunhofer.de/de/geschaeftsfelder/leistungselektronik-und-elektrische-antriebssysteme/hil-systeme-und-digitale-zwillinge/ricoso.html (visited on 04/15/2024).

[9] P. Hackl, Z. Zhang, and R. Schürhuber, "Vergleich von regelkonzepten von umrichtern für eine 100 % erneuerbare energieerzeugung: 17. symposium energieinnovation : Future of energy - innovationen für eine klimaneutrale zukunft," Feb. 2022.

[10] F. Jenni and D. Wüest, *Steuerverfahren für selbstgeführte Stromrichter*. vdf Hochschulverlag an der ETH Zürich etc., 1995, Accepted: 2017-06-13T01:38:10Z. DOI: 10.3929/ethz-a-001427314.

[11] M. Kazmierkowski and L. Malesani, "Current control techniques for three-phase voltage-source PWM converters: A survey," *IEEE Transactions on Industrial Electronics*, vol. 45, no. 5, pp. 691–703, Oct. 1998, Conference Name: IEEE Transactions on Industrial Electronics. DOI: 10.1109/41.720325.

[12] D. Schröder and R. Marquardt, Eds., *Leistungselektronische Schaltungen: Funktion, Auslegung und Anwendung*, Berlin, Heidelberg: Springer, 2019. DOI: 10.1007/978-3-662-55325-1.

[13] B. Sahan, *Wechselrichtersysteme mit Stromzwischenkreis zur Netzanbindung von Photovoltaik-Generatoren* (Elektrische Energiesysteme). Kassel University Press, vol. 1.

[14] T. Gühna, A. Seibel, D. Stracke, F. Schnabel, and M. Jung, "Regelungsstrategie und Multisimulationstool für dezentral verteilte netzbildende Wechselrichter," in *PV-Symposium*, Bad Staffelstein, 2020.

PCIM Europe 2024, 11– 13 June 2024, Nuremberg   DOI: 10.30420/566262334

Intelligent Optimisation of a Wind Turbine Digital Twin Model

René Reimann[1], Steffen Menzel[1], Wilfried Holzke[1], Holger Raffel[2], Bernd Orlik[1]

[1] Institute for Electrical Drives, Power Electronics and Devices (IALB), University of Bremen, Germany
[2] Bremen Center of Mechatronics (BCM), University of Bremen, Germany

Corresponding author: René Reimann, rreimann@ialb.uni-bremen.de
Speaker: René Reimann, rreimann@ialb.uni-bremen.de

Abstract

In this work a digital twin framework is proposed, to improve the detection and analysis of anomalies within wind turbine operations. By integrating a model of a wind turbine with real-time data from an operational turbine, this framework allows a comparison between the expected and actual operational behaviour. To ensure that the model behaves correctly, the parameters of the model are optimised based on the measurement data and by using intelligent optimisation algorithms to match the behaviour of the real wind turbine. For the optimisation process surrogate optimisation, particle swarm optimisation and a genetic optimisation algorithm are used, and the results are compared to each other.

1   Introduction

The increasing global demand for sustainable energy solutions has put the wind energy sector into a key position in the search for clean and renewable power sources. Wind turbines have made significant advances in design, operation, and maintenance over the past few decades. In this context, the development of digital twins for wind turbines has emerged as a transformative approach for optimising their performance, enhancing their reliability, and reducing the operational costs [1].

A digital twin can be defined as a virtual replica of a physical system or process, faithfully mirroring its behaviour and dynamics in real-time. Digital twins in the wind energy domain provide intricate simulations of their components, operational scenarios, and environmental interactions of wind turbines. By seamlessly integrating real-time data from sensors and historical performance data with advanced computational models, a digital twin offers an immensely powerful tool for monitoring, analysing, and optimising the operation of wind turbines. As a result, the turbine efficiency can be significantly improved due to the anticipation maintenance requirements and fast responds to environmental changes [2].

In this work, the digital twin of a wind turbine is constructed through a developed model, as detailed in a previous paper [3]. This model serves as the foundational framework for the digital twin, to replicate and simulate the behaviour of a physical wind turbine in a virtual environment. To ensure the precision of the digital twin, real measurement data collected directly from a research wind turbine shown in figure 1, is fed into the model. The used distributed measurement system was shown in previous papers [4] [5].

Furthermore, an extended Kalman filter is used to calculate non-measurable values from the real measured data. The signals generated by the digital twin are then compared to the measured signals as well as estimated values from the extended Kalman filter. This comparative analysis is important to evaluate the accuracy of the simulated digital twin. In this way, the accuracy of the model in replicating the real behaviour of the wind turbine can be assessed. However, achieving a seamless alignment between the behaviour of the physical wind turbine and the digital twin performance is a complex task due to the dynamic non-linear behaviour of the model and the required parameters, that are not always precisely available [6]. To address this challenge, intelligent optimisation algorithms are used. This algorithm systematically adjusts selected parameters of the model in order to bring it into closer alignment with the real-world behaviour of the wind turbine. Through iterations and fine-tuning, the model parameters are optimised until an acceptable level of agreement is achieved between the digital twin and the observed performance of the physical wind turbine.

The structure of the digital twin and the implemented model of the wind turbine are outlined in section two. In addition, the possible application of the digital twin to detect anomalies in the operating of the wind turbine is described there. In section three the used optimisation algorithm is introduced. The methodology of the optimisation including the measurement and the pre-processing of the data is shown. Section five provides the results of the optimisation process. Finally, the main conclusions are discussed.

2 Digital Twin

The literature presents diverse interpretations of the digital twin concept, with a comprehensive definition provided in [7] highlighting the essential link between a physical system and its digital counterpart. This relationship is distinguished not only by the fidelity of the virtual image to the real system but also by its real-time responsiveness, customised to corresponding applications. Moreover, the digital twin enhances the operational value of the physical system by enabling monitoring and efficiency improvements. It is possible to use the digital twin to monitor the real plant in real time and analyse the internal mechanism and to react to results instantaneously by changing the control parameters. Additionally, emerging defects and errors can be detected in advance resulting in a better scheduling of maintenance [8].

The proposed digital twin structure for a wind turbine is shown in figure 3. The real wind turbine shown in figure 1 is equipped with a distributed measurement system. This measurement system records the wind speed v_w, the rotation speed ω_m, the generator and grid voltages v_s and v_g, the generator and grid currents i_s and i_g, the DC link voltage V_{dc} as well as the excitation voltage V_e of the synchronous machine. Additionally, the set values for current of the grid-side converter $i_{s,ref}$ and the pitch angle β are recorded [4]. These values serve as inputs for the digital twin. Thus, the digital twin has the identical input variables found in the real system and the

Fig. 1 Picture of the research wind turbine

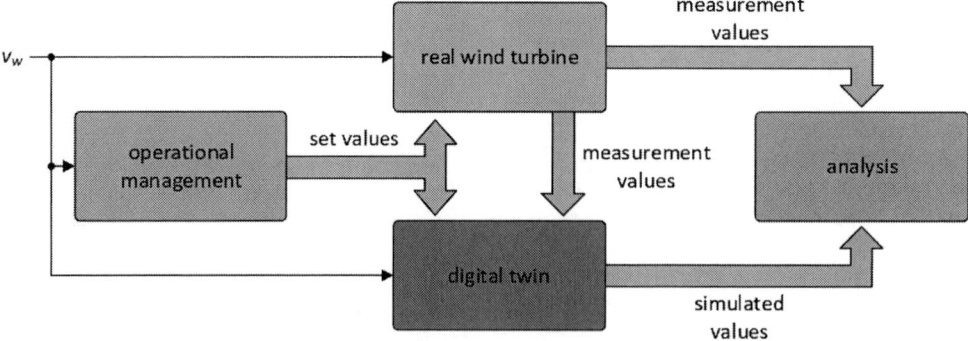

Fig. 2 Overview of the proposed structure of a digital twin for a wind turbine

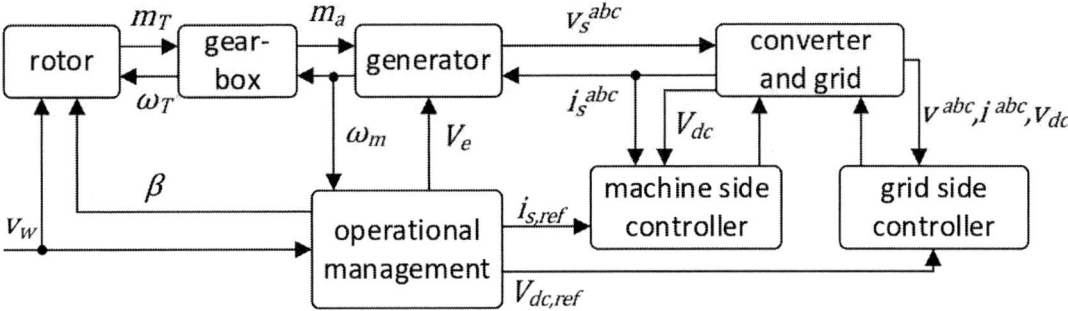

Fig. 3 Overview of the model of the wind turbine

behaviour of the operation control does not have to be reproduced within the digital twin. Incorporating partial models and observer structures for the wind turbine components, the digital twin becomes a virtual replication of the real wind turbine.

The model of the wind turbine, consisting of partial models for the different components, is shown in figure 2. The model was implemented using MATLAB/Simulink and PLECS for the electrical components. The model of the rotor uses the dynamic, unsteady Blade Element Momentum (BEM) method to compute the driving torque of the rotor. Therefore, the blades of the rotor are divided into different elements and forces and velocities are calculated for each element individually. The resulting torque is then derived by the integral over all blade elements. The generator model is based on the description of the electrically excited synchronous generator shown in [9]. It calculates the rotation speed according to the driving torque and the current. The model of the IGBT based converter is implemented in PLECS. The controller structures for the generator side converter and the grid side converter are based on a cascaded PI controller. As described above, the set values are also recorded by the measurement system, but for standalone simulations of the wind turbine model the operational management is also modelled to calculate the set values, according to the current operating state of the wind turbine.

The output of the digital twin are simulated values for the rotation speed $\hat{\omega}_m$, the torque of the rotor $\hat{m}_a$, the generator voltages $\hat{v}_s^{abc}$ and the generator currents $\hat{\imath}_s^{abc}$. These values are compared to the corresponding signals from the real wind turbine in the analysis block. For the non-measurable value – the torque of the rotor – an extended Kalman filter is used to derive the torque from the other measurement values [3]. The goal is to find a discrepancy between the simulated and the real values to detect an anomaly in the operation of the real wind turbine. For this anomaly detection a robust random cut forest algorithm (RRCF) can be used [10]. The RRCF algorithm uses an ensemble of decision trees to model the normal behaviour of the input data. An anomaly is identified by evaluating the degree of isolation of data points within these trees. RRCF is particularly well-suited for real-time anomaly detection due to its incremental learning capability. It can process data streams efficiently, adapting to new patterns as they emerge, without the need to retrain the model from scratch. Furthermore, the implementation of the algorithm ensures low computational complexity [11]. Thus, RRCF can be used in real time and for high-dimensional data which makes it well suited for the anomaly detection in the wind turbine digital twin. The anomalies can be used to detect and identify failures in different parts of the wind turbine [12].

For this application of the digital twin the accuracy of the simulated values is important. The accuracy depends directly on the model of the wind turbine. The two main reasons for inaccuracy are simplifications by the model and parameter uncertainties. Where the errors caused by model simplifications can be resolved by more detailed models, the correction of errors due to parameter uncertainty is more difficult. The reason for this is that some of the parameters are not fully known, especially for older wind turbines. In addition, the parameters differ from one another due to tolerances, even for identical turbines, meaning that there are also deviations to be found there. To solve this problem, in this work the intelligent optimisation of the model of the wind turbine is shown. The intelligent optimisation algorithm shown in the next chapter uses the data from the real wind turbine to optimise the parameter of the model to get a better alignment between the model and the real wind turbine.

3 Intelligent Optimisation

In the field of computer science, intelligent optimisation algorithms are a pivotal tool for solving a variety of complex and multidimensional problems. These algorithms emulate natural processes and cognitive strategies to find optimal or near-optimal solutions in scenarios where traditional analytical approaches fall short. The two intelligent optimisation algorithms used in this work are surrogate optimisation and particle swarm optimisation. Both algorithms are explained in more detail in the following.

3.1 Surrogate Optimisation

Surrogate optimisation (SO) is an optimisation technique designed to efficiently solve optimisation problems where the objective function evaluations are extremely costly regarding the computing time. It finds the solution by constructing an approximate model of the objective function based on previously evaluated solutions called surrogate model. The surrogate model is then used to predict the behaviour of the objective function in unevaluated regions of the search space, guiding the optimisation process towards areas with potential improvement [13].

$f(x)$ is the objective function where $x \in \mathbb{R}^n$ are the parameters. The goal of SO is to minimise $f(x)$ in dependence of $x \in \Omega$, where Ω is the feasible search space for the parameters. Therefore, a surrogate model $g(x)$ is constructed, which approximates the objective function based on the available sampled data. The optimisation process can then be formulated as [13]:

$$x[k] = \arg \min_{x \in \Omega} g(x[k-1]) \qquad (1)$$

where $x\{k\}$ is the point selected for evaluation with the true objective function f. The decision to select $x[k]$ is influenced not only by the value of $g(x)$ but also by a strategy that balances exploration – searching unexplored regions – and exploitation, which means to refine the search around known good solutions. The update of the surrogate model after each true function evaluation incorporates the new information. This process iterates, with the surrogate model becoming an increasingly accurate representation of the true objective function, thereby guiding the search towards the optimum. The ability of the SO to approximate and predict the behaviour of the objective function with a limited number of evaluations makes it a valuable tool in the optimisation of the wind turbine model [13].

3.2 Particle Swarm Optimisation

The particle swarm optimisation (PSO) is an evolutionary computation technique that simulates the social behaviour observed in flocks of birds or insects swarming. The algorithm is esteemed for its effectiveness in navigating complex, multidimensional search spaces to locate optimal solutions. PSO operates through a population of candidate solutions, referred to as particles, which move through the search space according to simple mathematical rules. These movements are influenced by the particles own best-known positions and the best-known positions of their neighbours. This mechanism allows the swarm to explore and exploit the search space, converging towards optimal solutions over time [14].

Let x_i represent the position of particle i in the search space, and v_i denote the velocity of this particle. Each particle keeps track of its coordinates in the solution space, which are associated with the best solution b it has achieved so far. The particles update their velocity at each iteration with following equation [14]:

$$v_i[k+1] = wv_i[k] + c_1 r_1 (b - x_i[k]) \\ + c_2 r_2 (g - x_i[k]) \qquad (2)$$

w is the inertia weight, balancing the global and local exploration abilities of the swarm, c_1 and c_2 are cognitive and social parameters, respectively, indicating the particle's tendency to return to its own best position, the swarm's best position, r_1 and r_2 are random numbers between 0 and 1, introduced to maintain diversity in the swarm's movement and g the best value obtained by any particle in the population. After the updated velocity is calculated the new position can be derived [14]:

$$x_i[k+1] = x_i[k] + v_i[k+1] \qquad (3)$$

At first the population of particles is initialised with random positions and velocities in the search space. The objective function is than evaluated for each particle and then the positions and velocities are updated. This process is done iteratively until a stopping criterion is met [14].

4 Methodology

The process of the optimisation of the model can be divided into three steps: the data acquisition, the simulation and data pre-processing, and the actual optimisation. The data acquisition is only done one time, where the other two steps are carried out for each iteration of the optimisation algorithm.

4.1 Data Acquisition

An Overview of the measurement system used for the data acquisition is shown in figure 4. The system consists of three measurement units placed at different locations inside the wind turbine. These measurement units record the voltages and currents on the machine side and the grid side, the temperatures and humidity as well as the rotation speed of the generator. The data is transmitted over the EtherCAT field bus system to an industry PC (IPC). Additionally, the PLC of the wind turbine is also connected over EtherCAT to record the set points of the converter controllers. The voltages, currents and the rotation speed are sampled with 50 kHz, where the data from the PLC is sampled with 100 Hz and the temperatures and humidities with 1 Hz. The recorded data is than saved on the IPC. But it is also possible to transmit the recorded data to a cloud platform. This can be used to conduct the real time anomaly detection in the cloud where more computing resources are available than on the IPC. For the purpose of the optimisation of the model the data is saved on the IPC and then used on another PC to run the simulations of the model and the optimisation algorithm.

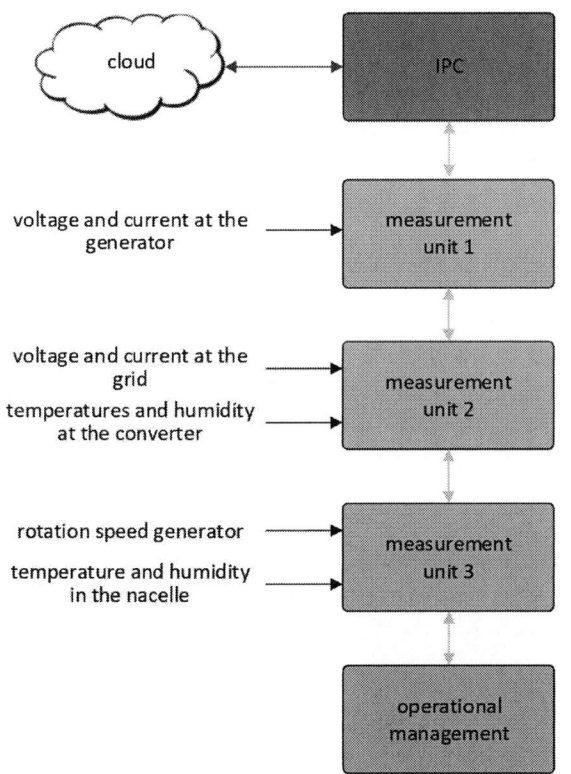

Fig. 4 Overview of the distributed measurement system

4.2 Simulation and Data Pre-Processing

The next step of the process is the simulation of the model of the wind turbine and the pre-processing of the data from the model and the measurement data of the real wind turbine. At first the simulation of the model in figure 3 is carried out with the measured wind speed and pitch angle from the real wind turbine as inputs. After that the measured data and the data from the simulation is pre-processed as shown in figure 5. One part of the pre-processing is the calculation of the root mean square (RMS) values of the voltages and currents. This is important for the comparability of the data from the model and the real wind turbine. Therefore, the frequency f of the signals are needed. On the generator side the rotation speed n can be used to calculate the frequency:

$$f = \frac{n \cdot p}{60} \quad (4)$$

p is the number of pole pairs of the generator. For the grid side a phase locked loop (PLL) is used to get the frequency of the grid voltage. With the frequency the number of samples n_s needed to calculate the RMS values over a full period of the signals can be determined:

$$n_s = \frac{f_s \cdot m}{f} \quad (5)$$

f_s is the sample rate of the measured signal and m is the factor that can be used to calculate the RMS value over more full periods and not only for one period. In this work $m = 10$. The RMS value x_{rms} of the signal x can then be calculated:

$$x_{rms} = \sqrt{\frac{1}{n_s} \sum_{k=1}^{n_s} x^2[k]} \quad (6)$$

For the optimisation of the model and later for the anomaly detection additional values which cannot be measured directly like the torque of the rotor and the magnetic flux in the generator are needed.

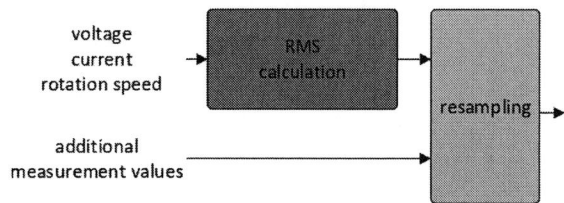

Fig. 5 Overview of the data pre-processing

Table 1 Parameter values of the baseline model and the optimised models

Parameter	Baseline Model	Surrogate Optimisation	Particle Swarm Optimisation
moment of inertia J	18.23 kgm^2	17.02 kgm^2	17.61 kgm^2
stator resistance R_s	0.0287 Ω	0.0315 Ω	0.0402 Ω
reactance of the longitudinal axis x_d	2.8000 Ω	1.8341 Ω	3.8248 Ω
reactance of the transverse axis x_q	1.5500 Ω	1.6306 Ω	1.2123 Ω
transient reactance of the longitudinal axis x_d'	0.2250 Ω	0.3094 Ω	0.2459 Ω
sub transient reactance of the longitudinal axis x_d''	0.1110 Ω	0.1003 Ω	0.0829 Ω
sub transient reactance of the transverse axis x_q''	0.1330 Ω	0.1530 Ω	0.1655 Ω

These values are determined by using an observer structure with an extended Kalman filter [3].

The last step of the pre-processing is the resampling of the data. This is necessary because not all data for the optimisation is sampled with the same sample rate and for the optimisation a lower sample rate is adequate. The sampling rate for the resampling is 100 Hz. The whole process of the pre-processing is implemented in MATLAB and Simulink.

4.3 Implementation of the Intelligent Optimisation

The intelligent optimisation process is shown in figure 6. The data from the measurements, the model and the pre-processing, hence the digital twin is the input to the objective function. The value of the objective function is a metric on how precisely the model behaves like the real wind turbine. The first part of the objective function is the calculation of the mean absolute error (MAE) for each pair of signals from the model and the real wind turbine. In this case the voltages and currents of the generator, the rotation speed and the torque of the rotor are used for the calculation. The MEA for each signal e_i is calculated as follows [15]:

$$e_i = \frac{1}{N} \sum_{k=1}^{N} |x_i[k] - \hat{x}_i[k]| \tag{7}$$

Here x_i is the measured value of the signal and $\hat{x}_i$ the value from the simulation at sample k. N is the number of samples used for the calculation. Hence the duration of the time frame. Here 50 s is used for the optimisation. For the calculation of the objective function only normalised values are used. The value of the objective function f_e is the sum of the MAE of each signal:

$$f_e = \sum_{i=1}^{P} e_i \tag{8}$$

This value is the input of the optimisation algorithm, which uses this value to determine a new set of parameters as described in section 3. The new set of parameters are than used again to run a simulation of the model. This process is done iteratively until the maximum of iterations of the optimisation algorithm is reached.

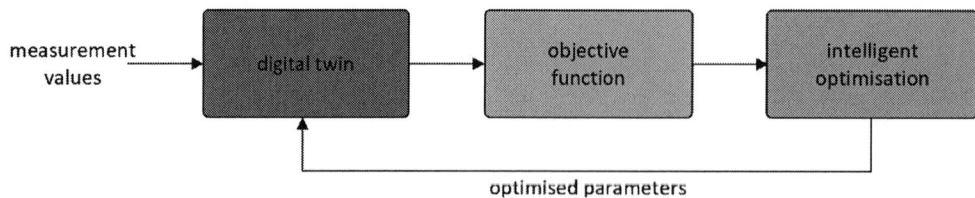

Fig. 6 Overview of the optimisation process

The parameter used by the algorithms to optimise the model are shown in table 1. These parameters are chosen because their values are not precisely known from the documentation of the wind turbine and are therefore good candidates for the optimisation of the model. For the optimisation algorithms the integrated MATLAB functions are used. The objective function is also implemented in MATLAB.

5 Results

In this section the results of the optimisation processes are shown. The values for the parameters of the baseline model and values after the optimisation processes are shown in table 1.

5.1 Surrogate Optimisation

The settings for the surrogate optimisation are shown in table 2. The algorithm should run for 100 iterations and the range in that the algorithm can change the parameter values is $\pm40\%$ from the initial values. The progression of the objective function value over the iterations can be seen in figure 7. As can be seen that the value decreases

Table 2 Settings for the surrogate optimisation algorithm

Parameter	Value
max. evaluations	100
min. sample distance	10^{-6}
min. surrogate points	20
lower parameter bound	60 % baseline parameters
upper parameter bound	140 % baseline parameters

significantly over the first 10 iterations. After that the value is converging towards a minimal value. Consequently, the surrogate optimisation is able to optimise the parameters in order to make the model behave more like the real wind turbine.

5.2 Particle Swarm Optimisation

For the particle swarm optimisation, the settings are shown in table 3. The settings are chosen similar to the settings for the surrogate optimisation algorithm with the maximum iterations of 100 and the upper and lower bounds of $\pm40\%$ of the parameters. In figure 8 the progression of the objective function value over the iterations is shown. The progression of the objective function value is similar to the progression of the surrogate optimisation. Over the first iterations the value decreases significantly and is than converging to the minimal value. The reached minimal objective function value by the particle swarm optimisation is lower than by the surrogate optimisation.

Table 3 Settings for the surrogate optimisation algorithm

Parameter	Value
max. iterations	100
swarm size	10
initial swarm span	200
max. stall iterations	20
min. neighbors fractions	0.25
lower parameter bound	60 % baseline parameters
upper parameter bound	140 % baseline parameters

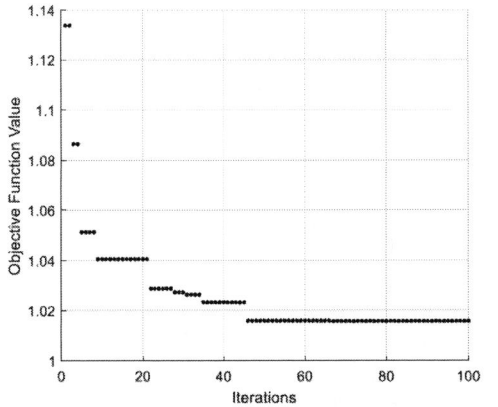

Fig. 7 Progression of the fitness value during the surrogate optimisation

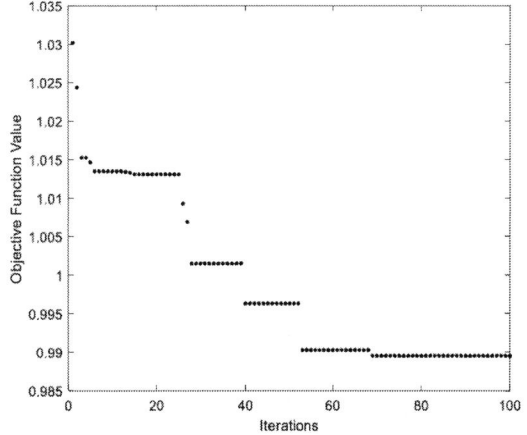

Fig. 8 Progression of the fitness value during the particle swarm optimisation

5.3 Discussion

In the section before, the results for the optimising with the surrogate optimisation and the particle swarm optimisation were shown. The best objective function values for the baseline model and the optimised models are listed in table 4. It can be seen, that the optimisation resulted in a significant improvement of alignment in the behaviour between the real wind turbine and the model. Nevertheless, there are still deviations between the values, which may be due to simplifications in the model, as described in section 2, or from errors in other parameters.

Table 4 Best fitness value of the baseline model and the optimised models

	Best Objective Function Value
Baseline Model	1.15891
Surrogate Optimisation	1.01568
Particle Swarm Optimisation	0.95064

6 Conclusion

In this work, the development of a digital twin for a wind turbine is shown. A digital twin framework was defined, which uses the model of the wind turbine and measurement data from the real wind turbine to detect anomalies in the operational behaviour of the wind turbine. The anomaly detection can be done by analysing the deviations between the behaviour of the real turbine and the digital twin. Therefore, the parameters of the model were optimised to match the behaviour of the real wind turbine. For this optimisation process the performance of the surrogate optimisation algorithm and the particle swarm optimisation were analysed.

7 Acknowledgement

This work has been funded by the German Federal Ministry for Economic Affairs and Climate Action (BMWK) as part of the project "Wind IO" under grant no. 03EE2015A.

on the basis of a decision
by the German Bundestag

8 Reference

[1] R. Issa, M. S.Hamad, and M. Abdel-Geliel, 'Digital Twin of Wind Turbine Based on Microsoft® Azure IoT Platform', in *2023 IEEE Conference on Power Electronics and Renewable Energy (CPERE)*, Feb. 2023, pp. 1–8. doi: 10.1109/CPERE56564.2023.10119576.

[2] O. O. Olatunji, P. A. Adedeji, N. Madushele, and T.-C. Jen, 'Overview of Digital Twin Technology in Wind Turbine Fault Diagnosis and Condition Monitoring', in *2021 IEEE 12th International Conference on Mechanical and Intelligent Manufacturing Technologies (IC-MIMT)*, Cape Town, South Africa: IEEE, May 2021, pp. 201–207. doi: 10.1109/IC-MIMT52186.2021.9476186.

[3] R. Reimann, S. Menzel, W. Holzke, H. Raffel, and B. Orlik, 'Development and Evaluation of a Model for the Implementation of a Digital Twin for a Wind Turbine', in *PCIM Europe 2023; International Exhibition and Conference for Power Electronics, Intelligent Motion, Renewable Energy and Energy Management*, May 2023, pp. 1–9. doi: 10.30420/566091280.

[4] R. Reimann *et al.*, 'Development of a Distributed Measurement System for the Digitalisation of a Wind Turbine', in *PCIM Europe 2022; International Exhibition and Conference for Power Electronics, Intelligent Motion, Renewable Energy and Energy Management*, May 2022, pp. 1–8. doi: 10.30420/565822248.

[5] R. Reimann, W. Holzke, S. Menzel, and B. Orlik, 'Synchronisation of a Distributed Measurement System', in *PCIM Europe 2019; International Exhibition and Conference for Power Electronics, Intelligent Motion, Renewable Energy and Energy Management*, May 2019, pp. 1–8.

[6] A. Ebrahimi, 'Challenges of developing a digital twin model of renewable energy generators', in *2019 IEEE 28th International Symposium on Industrial Electronics (ISIE)*, Jun. 2019, pp. 1059–1066. doi: 10.1109/ISIE.2019.8781529.

[7] S. Mihai *et al.*, 'Digital Twins: A Survey on Enabling Technologies, Challenges, Trends and Future Prospects', *IEEE Commun. Surv. Tutor.*, pp. 1–1, 2022, doi: 10.1109/COMST.2022.3208773.

[8] A. Rasheed, O. San, and T. Kvamsdal, 'Digital Twin: Values, Challenges and Enablers From a Modeling Perspective', *IEEE Access*,

vol. 8, pp. 21980–22012, 2020, doi: 10.1109/ACCESS.2020.2970143.

[9] A. Ernst, D. Matthies, W. Holzke, and B. Orlik, 'Validation of a Generator-Side Boost Converter with Load by a Fictitious Synchronous Machine', in *PCIM Europe digital days 2021; International Exhibition and Conference for Power Electronics, Intelligent Motion, Renewable Energy and Energy Management*, May 2021, pp. 1–7.

[10] S. Guha, N. Mishra, G. Roy, and O. Schrijvers, 'Robust Random Cut Forest Based Anomaly Detection On Streams', in *Proceedings of the 33 rd International Conference on Machine Learning*, New York, NY, 2016.

[11] M. Bartos, A. Mullapudi, and S. Troutman, 'rrcf: Implementation of the Robust Random Cut Forest algorithm for anomaly detection on streams', *J. Open Source Softw.*, vol. 4, no. 35, p. 1336, Mar. 2019, doi: 10.21105/joss.01336.

[12] M. Schlechtingen and I. F. Santos, 'Wind turbine condition monitoring based on SCADA data using normal behavior models. Part 2: Application examples', *Appl. Soft Comput.*, vol. 14, pp. 447–460, Jan. 2014, doi: 10.1016/j.asoc.2013.09.016.

[13] A. Bhosekar and M. Ierapetritou, 'Advances in surrogate based modeling, feasibility analysis, and optimization: A review', *Comput. Chem. Eng.*, vol. 108, pp. 250–267, Jan. 2018, doi: 10.1016/j.compchemeng.2017.09.017.

[14] D. Wang, D. Tan, and L. Liu, 'Particle swarm optimization algorithm: an overview', *Soft Comput.*, vol. 22, no. 2, pp. 387–408, Jan. 2018, doi: 10.1007/s00500-016-2474-6.

[15] M. Z. Naser and A. Alavi, 'Insights into Performance Fitness and Error Metrics for Machine Learning', *Archit. Struct. Constr.*, Nov. 2021, doi: 10.1007/s44150-021-00015-8.

PCIM Europe 2024, 11– 13 June 2024, Nuremberg   DOI: 10.30420/566262335

Thermal Transient Digital Twin Modelling for Power Converters

Xianghao Mo [*] [©1], Daniel Ríos Linares [©1], Regina Ramos [©1], Miroslav Vasić [©1]

[1] Centro de Electrónica Industrial (CEI), Universidad Politécnica de Madrid (UPM), Spain

Corresponding author: Xianghao Mo, xianghao.mo@upm.com
Speaker: Xianghao Mo, xianghao.mo@upm.com

Abstract

Over recent years, digital twins have gained attention in monitoring power converters propelled by advancements in data-driven model. To achieve higher precision in power converter temperature monitoring, in this paper a voxel-based approach for real-time thermal modelling of transient behaviour in power converters is introduced. It is based on analytical model and conceptualized as a 3-D matrix, which integrates heat transfer equations and thermodynamic principles. Aiming to build up a digital twin in embedded system to operate in parallel with physical prototype, this work details its implementation on hardware, highlighting system requirements and computational needs, and finally compares its results with FEM simulation and experimental data.

1   Introduction

In recent years, power converter design has prioritized thermal management due to rising power densities. Temperature is the key factor in determining the health of power converters, accounting for over 55% of failures [1]. Semiconductors and capacitors are particularly failure-prone, while the degradation of PCBs and magnetics also crucially affects system performance [2]. These insights emphasize the need of an advanced technique for real-time monitoring [3]. Concurrently, the rise of Digital Twins(DTs), which is a virtually replica of real-world systems, allows the real-time monitoring of the converters physical behaviour. The key part of every DT in power electronics is the prediction of the electric [4], electromagnetic [4], and thermal behaviour [5]. In the literature, thermal models applied to components of power converter are thermal network for power modules [6], analytical model for printed circuit board (PCB) [7], fast model for planar

PCB magnetics [8], ANN-based model for transistor [9] and magnetics [10]. However, the model based on thermal network is more feasible at device-level and certain trade-off needs to be taken for presenting the transient behaviour in real-time. The analytical model focuses more on the design stage due to the need to know the boundary condition. The limitation of application of ANN is mainly due to the lack of generalization, wrong prediction due to abstract learning of physical laws and the trade-offs between accuracy and computational cost. Hence, for predicting the transient thermal behaviour of power converters in real-time these models do not meet the the requirement [11].

This paper, in order to provide a more comprehensive representation of transient thermal behaviour for power converters, introduces a novel approach based on the existing voxelization of the geometry of interest which, through analytical equations, is solved numerically for consequent time steps, leading to a transient model for the power converter. Following this approach, the dynamic thermal evolution is predicted accurately employing an analytical model rather than a prior model trained by database. This makes the geometry more scalable without the need of training, obtaining the complete temperature profile of the converter for temperature monitoring.

---

[*]This work was possible thanks to the ALL2GAN-SP, PCI2023-143375, funded by MICIU/AEI/10.13039/501100011033 and co-funded by the European Union.

The ALL2GaN Project(Grant Agreement No 101111890) is supported by the Chips Joint Undertaking and its members including the top-up funding by Austria, Belgium, Czech Republic, Denmark, Germany, Greece, Netherlands, Norway, Slovakia, Spain, Sweden, and Switzerland. Views and opinions expressed are however those of author(s) only and do not necessarily reflect those of the European Union or the national granting authorities. Neither the European Union nor the national granting authorities can be held responsible for them.

PCIM Europe 2024, 11– 13 June 2024, Nuremberg   DOI: 10.30420/566262335

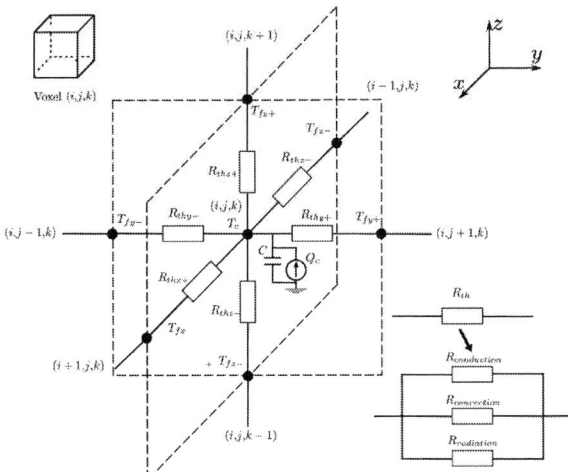

Fig. 1: Presentation of a voxel with its thermal properties

age, and the Fig.2b shows how this technique is applied to its cooling layer. When the number of intersections is odd, such as 1 and 3, the points are located inside the solid. Conversely, points with an even number of intersections, like 2 and 4, reside outside the solid.

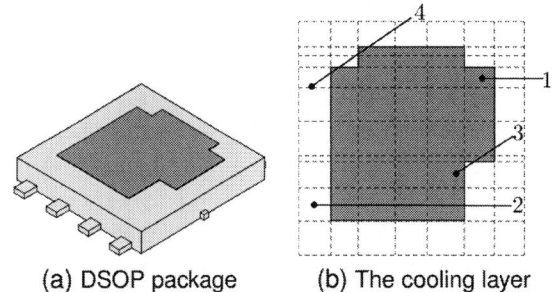

(a) DSOP package   (b) The cooling layer

Fig. 2: The Möller–Trumbore ray-triangle intersection

2 Power Converter Voxelization

Considering the variety of geometries among components of the power converter, this article introduces a Finite-Difference Time-Domain (FD-TD) approach to solve the analytical equations in real time. This method converts the three-dimensional heat problem into a voxel-based format, employing both spatial and temporal discretization. As a result, it provides a transient numerical solution to the heat equation. To be able to do that, the power converter needs to be divided into granular cells. As presented in the Fig.1, each cell called voxel, presenting a portion of the converter, contains the physical properties, the centre temperature T_c, self heat generation Q_c and its heat capacity C, the face temperature T_{fx+}, T_{fx-}, T_{fy+}, T_{fy-}, T_{fz+}, T_{fz-}, conduction resistance $R_{conduction}$, convection resistance $R_{convection}$ and radiation resistance $R_{radiation}$.

To realize spatial discretization of the power converter layout, the Möller–Trumbore ray-triangle intersection algorithm based on Jordan curve theorem, which efficiently determines a ray's intersection with a geometry's triangular facets, is adopted. This technique calculates the intersection points of a ray with a solid's triangular facets. Rays are projected from points outside of the solid in space to find where they meet the triangles, determining whether a point lies within the solid. An example of the power transistor package Illustrated in Fig.2, a point is deemed internal if an odd number of intersections with the solid's surface occurs. The Fig.2a is the 3D model of a dual-direction cooling pack-

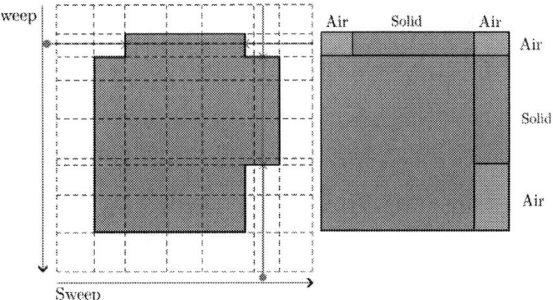

Fig. 3: Accelerated approach for ray triangle intersection

This method can be accelerated by scanning the polygon mesh along three spatial axes shown in the Fig3. The region between two consecutive intersection points is classified as solid and all other regions are identified as air. In this manner, each voxel serves as a unit for detailed heat transfer analysis, offering the flexibility of thermal modelling and the transient details.

3 Voxel-Based Thermal Modelling

While analytical solutions to heat equations are very precise, their effectiveness is restricted to simplified scenarios. Recognizing the complexity of power converters, which consist of components with diverse geometries, the FD-TD method is applied to obtain the numerical solution. This approach converts the three-dimensional issue into a voxel-based model via spatial and temporal discretization, thereby providing the transient thermal behaviour of power converter.

2387

The heat is transferred by three different modes: conduction, convection, and radiation and their mechanisms are given as following:

$$\dot{Q}_{\text{cond}} = kA\frac{dT}{dx} \tag{1}$$

$$\dot{Q}_{\text{conv}} = hA_s(T_s - T_\infty) \tag{2}$$

$$\dot{Q}_{\text{rad}} = \varepsilon\sigma A_s(T_s^4 - T_\infty^4) \tag{3}$$

where the $\dot{Q}_{\text{cond}}$, $\dot{Q}_{\text{conv}}$ and $\dot{Q}_{\text{rad}}$ are the heat transferred by conduction, convection, and radiation respectively, k is the thermal conductivity of the material, A is the conduction transfer area which is always normal to the heat transfer direction, h is the convection heat transfer coefficient, $\frac{dT}{dx}$ is the temperature gradient, ε (0-1) is the emissivity of the surface, σ is the Stefan-Boltzmann constant, A_s is the surface area, T_s is the surface temperature and T_∞ is the temperature of the fluid sufficiently far from the surface, typically the ambient temperature.

To solve the heat equation numerically, discretization across space and time are applied. By dividing the 3D object into voxels, the heat equation is expressed inside each voxel´s bounds:

$$\rho \cdot c \cdot \Delta x \cdot \Delta y \cdot \Delta z \cdot \frac{\partial T_c}{\partial t} = \sum_l \dot{Q}_l \tag{4}$$

$$\dot{Q}_{l,\text{cond}} = \frac{k_l}{\Delta x_i l/2} A_l(T_c - T_l) \tag{5}$$

$$\dot{Q}_{l,\text{conv}} = h_l A_l(T_c - T_l) \tag{6}$$

$$\dot{Q}_{l,\text{rad}} = \varepsilon_l\sigma A_l(T_c^4 - T_l^4) \tag{7}$$

where ρ is the density of the material, c is the specific heat, T_c is the average temperature of the cell and $\dot{Q}_l$ is the heat flow from the neighbour voxel and self heat generation. The index l represents the space direction of voxel deciding the heat flux path. Δx_l is the conduction heat path length of the voxel, and T_l is the surface temperature at the corresponding heat path direction.

Subsequently, the Euler method is employed to discretize the time domain of the heat equation into an explicit form,

$$\rho c \Delta x \Delta y \Delta z \frac{T_c^{(n+1)} - T_c^{(n)}}{\Delta t} = \sum_i \dot{Q}_l^{(n)} \tag{8}$$

$$T_c^{(n+1)} = T_c^{(n)} + \Delta T_c^{(n)} \tag{9}$$

where $Q_l^{(n)}$ is calculated with equation (5,6,7) for $T_c^{(n)}$ and $T_l^{(n)}$ by forcing equal overall heat transfer into the of the voxel and out of it. This interaction is trivial for the radiationless case. When radiation heat is added, and to simplify the calculation, equation (7) can be linearized when $\Delta T_c^{(n)}$ is considered small.

It is important to note, as mentioned in [12], that the application of the explicit method is constrained by stability considerations, which impose a limit on the time step Δt. The stability criterion, derived from the analytical solution, applies to a first-order system. This criterion becomes relevant when analyzing the interaction between two voxels: one maintaining a constant temperature and the other experiencing self-generated heat. This interaction can be discretized as follows:

$$T_c^{(n+1)} = T_c^{(n)} + \Delta T_c^{(n)} \tag{10}$$

$$T_c^{(n+1)} = T_c^{(n)} + \frac{\Delta t}{\tau}\left[T_c^{(\infty)} - T_c^{(n)}\right] \tag{11}$$

The stability condition is that $T_c^{(n)}$ will not overpass $T_c^{(\infty)}$ in Δt, this is that $\Delta t \ll \tau$, where τ is given by,

$$\tau = \frac{\rho \cdot c \cdot \Delta x \cdot \Delta y \cdot \Delta z}{\left(\frac{k}{l_i}\right)A_i} \tag{12}$$

$$\tau = \frac{\rho \cdot c \cdot \Delta x \cdot \Delta y \cdot \Delta z}{\left(\frac{k}{2l_i} + h\right)A_i} \tag{13}$$

(12) is the conductive case and (13) is the convective case and where l_i and A_i depend on the chosen voxel.

The thermal solver calculates the solution of the heat equation at each time step, providing it as transient behaviour. As the variation in heat flux diminishes, the heat equation reaches convergence, indicating that a steady-state condition has been attained. The entire transient behavior is depicted through the progression from the initial condition to the steady-state.

The complete iteration process of solving the heat equation is given in Fig4. Each iteration begins with calculating the voxel surface temperature, derived from the centers of the neighboring voxels, which then requires an adjustment in the voxel's thermal impedance as a response to temperature variations. Following this, the heat flux is calculated based on

PCIM Europe 2024, 11– 13 June 2024, Nuremberg   DOI: 10.30420/566262335

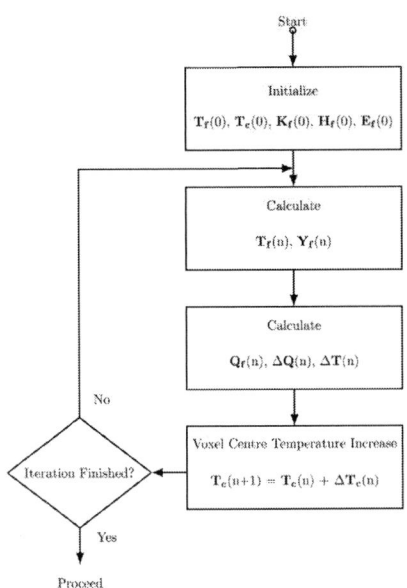

Fig. 4: Thermal solver iteration workflow

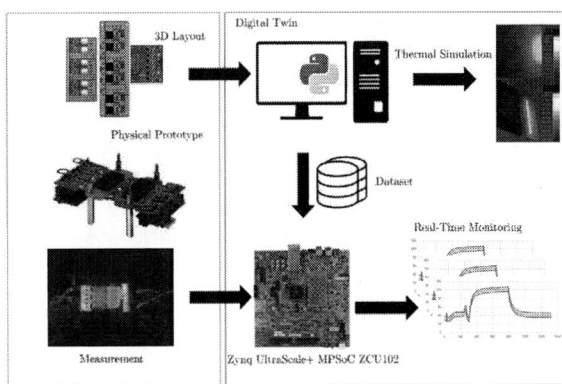

Fig. 5: Conception of proposed thermal DT for temperature monitoring

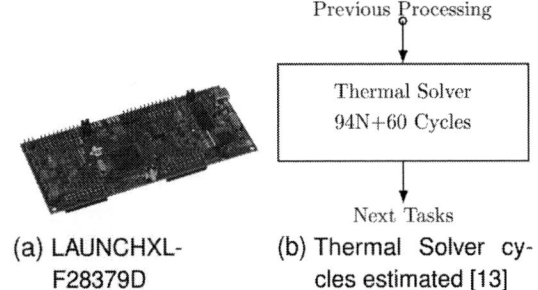

(a) LAUNCHXL-F28379D

(b) Thermal Solver cycles estimated [13]

Fig. 6: The launch pad and the estimation of thermal solver on LAUNCHXL-F28379D

the temperature difference between the voxel and its adjacent neighbor. The total heat for each voxel is computed as the sum of the heat flows in the six directions and its own heat generation. Considering the time step and the heat capacity of the voxel material, the rise in the voxel's central temperature is then determined.

4 Digital Twinning

DTs are tasked with data sampling, processing, and transfer, necessitating a platform capable of supporting these functionalities effectively. The real-time operation of DTs alongside physical prototypes benefits prognostics and health management. It requires timely and precise temperature data, which permits ongoing system evaluation and the anticipation of maintenance needs.

As illustrated in Fig.5, the Digital Twin (DT) mirrors the thermal behavior of the physical prototype, enabling real-time temperature monitoring. In this work, the simulation is developed using Python and encompasses four stages: power converter geometry translation, voxelization, thermal modelling, and thermal simulation. The Python-based simulation software constitutes the foundational and central component of the digital twin, where the heat problem of power converter is built up, which translates the layout of the power converter into interactive data for thermal modelling and simulation. This process facilitates early prediction of temperature profiles during the design phase. The thermal sim-

ulation offers insights into the thermal performance of power converters at the design stage. Additionally, it acts as the central hub for data management, where thermal data undergoes both preprocessing and postprocessing. The voxelized model is ported to a dedicated embedded system platform, where real-time monitoring is conducted. This embedded system prioritizes data processing and interacts seamlessly with real-world measurements. Operating concurrently with the prototype, the DT dynamically responds to changes, ensuring accurate and timely temperature monitoring.

Voxels define the heat problem in discretization, and solving the proposed discretized equation with a large number of them is time-consuming. In the purpose of choosing the appropriate platform, a study has been made comparing the LAUNCHXL-F28379D (Fig. 6a) and Zynq UltraScale+ MPSoC ZCU102 (Fig. 5). The LAUNCHXL-F28379D board, based on the TMS320F28379D microcontroller, features dual C2000 series 32-bit cores, multi-channel high-resolution ADCs suitable for various control applications, while the ZCU102 includes ARM Cortex-

PCIM Europe 2024, 11– 13 June 2024, Nuremberg   DOI: 10.30420/566262335

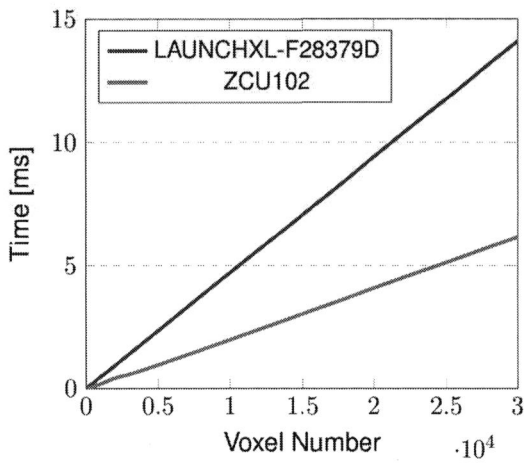

Fig. 7: The comparison of one thermal iteration between LAUNCHXL-F28379D (Clock 200MHz) and ZCU102 (Clock 50MHz) in relation with the number of voxels

Voxels	BRAM(%)	DSP(%)	FF(%)	LUT(%)
100	11	16	3	16
1000	15	16	3	16
5000	49	17	3	16
10000	75	17	3	16
30000	173	17	3	16

Tab. 1: Resource usage estimation of ZCU102 in relation with the number of voxels

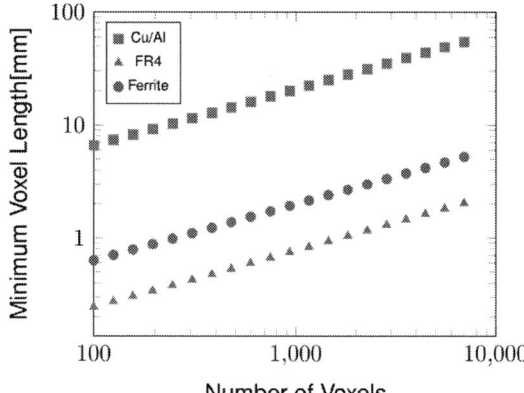

Fig. 8: Logarithmic Scale of Number of Voxels vs. Minimum Voxel Length

A53 and Cortex-R5 processors for complex and real-time tasks, FPGA logic units for highly configurable logic resources, supports high-speed DDR4 memory and multiple storage interfaces, and advanced high-speed interfaces for high-speed data transmission needs.

For the LAUNCHXL-F28379D, the computation cycles of realizing one thermal iteration in relation with the number of voxels is estimated in Fig. 6. And the comparison with ZCU102 is shown in the Fig.7 where for the thermal iteration, the relationship between the time consumed and the complexity of the grid is linear. The clock of ZCU102 is 50MHz as the time margin in each cycle is reserved for other DT operations. The clock of LAUNCHXL-F28379D is 200MHz and the estimation is made theoretically as the memory limits the operation of voxel thermal data. The thermal solver executed by ZCU102 presents advantage in time comparing to the LAUNCHXL-F28379D. The resources usage of ZCU102 is shown in the Tab.1. As the increase the number of voxels, the memory burden is also increasing and when it reaches 30000 voxels, the memory requirement exceeds the capacity of the DSP.

Additionally, considering the low latency, parallelism, and connectivity of the FPGA platform, it presents a unique advantage for developing thermal DTs for power converters, enabling processing and fast response to thermal changes. The platform, central to this evolution, enhances DT construction by employing Programmable Logic for complex thermal calculations and Processing System for operational management, optimizing real-time temperature monitoring and diagnostics. This approach accelerated by Direct Memory Access, ensures efficient data flow and updates, making the ZCU102 the optimal choice for implementing a comprehensive DT platform that translates thermal behaviors into actionable insights for real-time management.

From another perspective on digital twinning, due to its significant memory and computational demands, voxelisation of a solid is limited. It is essential to estimate the cycles needed to perform the iterative process faster than the reality, which represents the thermal transient. This value depends on the material and geometry whose dependency is given in Fig. 8, constraining the hardware implementation and thus, composing the 3-D grid. This time step is then limited by the hardware ability to make the calculations in real time. For being thermally conductive material, copper and aluminium limit the minimum size of the grid cell, hence they will require bigger voxels to satisfy the stability condition according to equation (13), leading to an inaccurate

PCIM Europe 2024, 11– 13 June 2024, Nuremberg   DOI: 10.30420/566262335

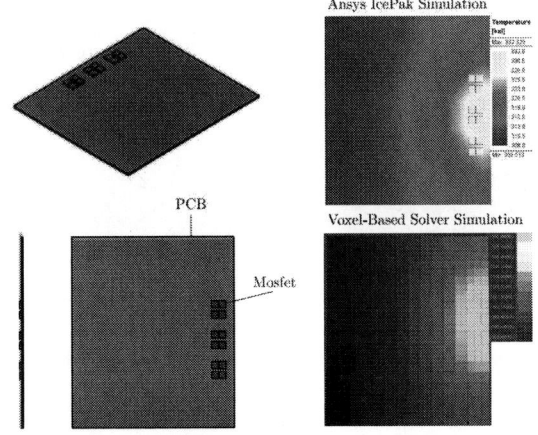

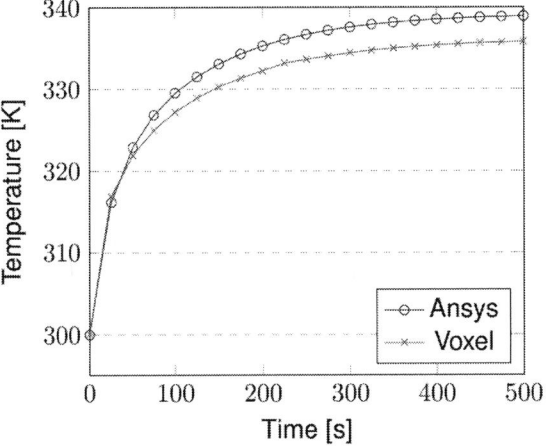

(a) 3D views of case A and its temperature distribution in IcePak and Voxel-Based Solver

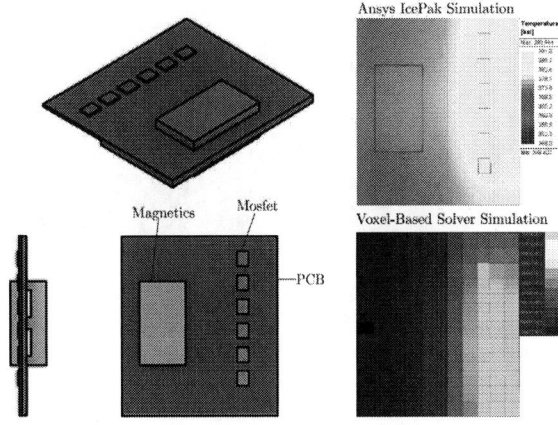

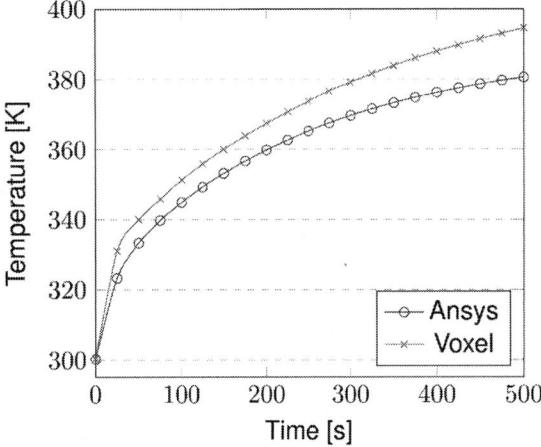

(a) 3D views of case B and its temperature distribution in IcePak and Voxel-Based Solvere

(b) Simulation results in 500 seconds of case A

(b) Simulation results in 500 seconds of case B

Fig. 9: Study case A: A PCB with 12 transistor each dissipating 0.24 W

Fig. 10: Study case B: A PCB with 6 transistor with 1 W on each of them and a 4224 m³ planar magnetic with 500 kW m⁻³

transient calculation with a coarser grid. To model circuit boards, as they present preferred conduction axis depending on the PCB layout (composition of FR-4 and copper), a homogenising of the thermal conductivity in each conduction axis has to be performed. The result shows that the critical length is reduced by an order of magnitude compared with the copper, leading to finer grid requirement.

5   Results and Discussion

To validate the numerical solution given by the thermal solver, the comparisons of two study case have been done between the transient thermal simulation between Ansys IcePak and proposed voxel thermal solver.

Shown in the Fig.9, the simulation of 500 seconds is made for a PCB with 12 transistors on it. Each of the transistors has 0.24 W power losses. In the Fig.9a its 3D presentation and the temperature distribution in steady-state of two simulation is illustrated. In Fig.9b the transient behaviour during the 500 seconds is shown. In Fig.10, another example, case B is shown. The model consists of a PCB, six transistors, and a planar magnetic. Fig.10a illustrates the 3D representation and the steady-state temperature distribution from two simulations. Each transistor has 1 W power losses and the magnetic has 500 kW m⁻³ distributed in its volume of 4224 m³ and their transient behaviours are shown

2391

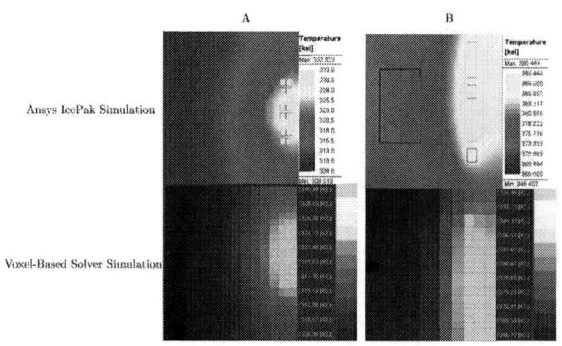

Fig. 11: Detailed comparison of temperature distribution between IcePak and Voxel-Based solver

in the Fig.10b.

The detailed comparison of two simulator is shown in Fig.11. As presented in Fig.9b that difference of temperature occurs since 75 seconds and there is about three kelvins of difference in the steady-state. This is because for the current version of the voxel-based solver, the heat transfer coefficient of convection is considered as constant while in the reality, it increases as the rise of temperature [12]. Combing the temperature difference shown in Fig.11 and Fig.9b highlights the significance of modelling the heat capacity of the power converter's materials as the complexity of the converter's construction grows. For the case B, the materials of each component are treated as homogeneous to enhance simulation speed. However, achieving higher precision necessitates more accurate definition of the materials, which requires smaller voxels and, consequently, increases computational demand.

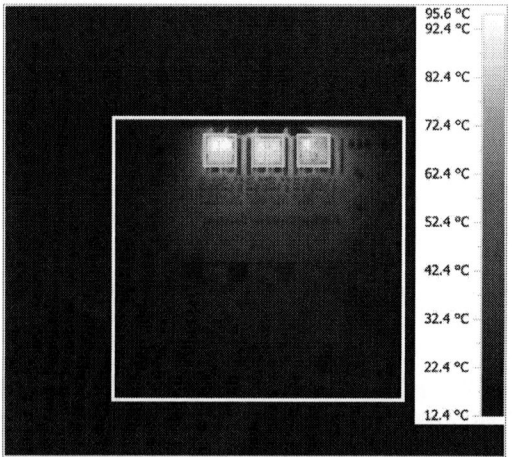

Fig. 12: Image taken by the thermal camera PY-ROVIEW 640L of DIAS Infared Systems

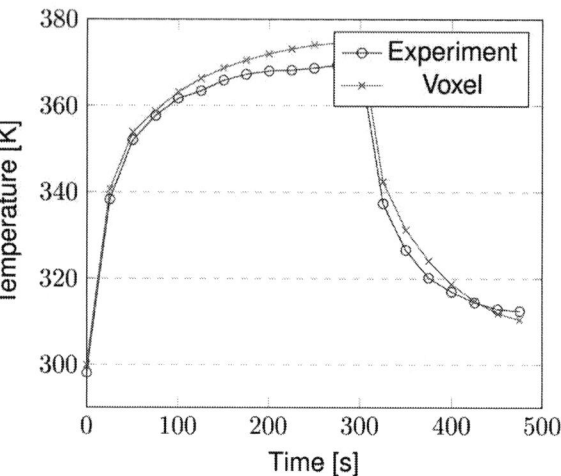

Fig. 13: A comparison of temperature change over time between experiment and voxel-based solver simulation.

An experiment was conducted to validate the simulation results. A three-phase rectifier PCB, equipped with twelve transistors, was tested to natural convection testing. DC voltage was applied, causing the transistors to act as diodes and generate consistent power losses, thereby heating the package and transferring heat to the PCB. As shown in Fig.12, the transistors are highlighted within a green rectangle and PCB is highlighted within a white rectangle. The experiment's transient behavior and the corresponding results from the proposed solver simulation are presented in Fig.13. During the initial 300 seconds, the rectifier is heated, followed by a cooling phase in air over the next 200 seconds, as illustrated by the curve in Fig.13.

6 Conclusion

From the perspective of monitoring a power converter's condition, temperature is a important factor. The interplay of heat among components calls for a proper technique to accurately capture thermal transient behavior. The accuracy of the voxel-based thermal model highly depends on the grid cell's size and the specificity of material definitions. Among these factors, the grid cell's volume and the solver's time step are critical in determining the feasibility of real-time monitoring by the hardware. And once the voxel size goes small, it brings more computational burden. Reducing the voxel size increases computational demands. Future efforts will focus on refining the thermal model's adaptive mesh to facilitate parallel processing, thereby reducing computational loads. Additionally, development will progress on implementing the Thermal DT system on the ZCU102 platform.

Acknowledgements

This work was possible thanks to the ALL2GAN-SP, PCI2023-143375, funded by MICIU/AEI /10.13039/501100011033 and co-funded by the European Union.

The ALL2GaN Project(Grant Agreement No 101111890) is supported by the Chips Joint Undertaking and its members including the top-up funding by Austria, Belgium, Czech Republic, Denmark, Germany, Greece, Netherlands, Norway, Slovakia, Spain, Sweden and Switzerland. View and opinions expressed are however those of author(s) only and do not necessarily reflect those of the European Union or the national granting authorities. Neither the European Union nor the national granting authorities can be held responsible for them.

References

[1] A. Sundaram and R. Velraj, "Thermal management of electronics: A review of literature," *Thermal Science - THERM SCI*, vol. 12, pp. 5–26, Jan. 2008. DOI: 10.2298/TSCI0802005A.

[2] S. Kalker, L. A. Ruppert, C. H. van der Broeck, J. Kuprat, M. Andresen, *et al.*, "Reviewing thermal-monitoring techniques for smart power modules," *IEEE Journal of Emerging and Selected Topics in Power Electronics*, vol. 10, no. 2, pp. 1326–1341, 2022. DOI: 10.1109/JESTPE.2021.3063305.

[3] O. Olanrewaju, Z. Yang, N. Evans, A. Fayyaz, T. Lagier, and A. Castellazzi, "Investigation of temperature distribution in sic power module prototype in transient conditions," in *2019 20th International Symposium on Power Electronics (Ee)*, 2019, pp. 1–5. DOI: 10.1109/PEE.2019.8923270.

[4] M. Milton, C. D. L. O, H. L. Ginn, and A. Benigni, "Controller-embeddable probabilistic real-time digital twins for power electronic converter diagnostics," *IEEE Transactions on Power Electronics*, vol. 35, no. 9, pp. 9850–9864, 2020. DOI: 10.1109/TPEL.2020.2971775.

[5] J. Kuprat, K. Debbadi, J. Schaumburg, M. Liserre, and M. Langwasser, "Thermal digital twin of power electronics modules for online thermal parameter identification," *IEEE Journal of Emerging and Selected Topics in Power Electronics*, vol. 12, no. 1, pp. 1020–1029, 2024. DOI: 10.1109/JESTPE. 2023.3328219.

[6] M. Musallam and C. M. Johnson, "Real-time compact thermal models for health management of power electronics," *IEEE Transactions on Power Electronics*, vol. 25, no. 6, pp. 1416–1425, 2010. DOI: 10.1109/TPEL.2010.2040634.

[7] Y. Shen, H. Wang, F. Blaabjerg, H. Zhao, and T. Long, "Thermal modeling and design optimization of pcb vias and pads," *IEEE Transactions on Power Electronics*, vol. 35, no. 1, pp. 882–900, 2020. DOI: 10.1109/TPEL.2019.2915029.

[8] L. C. Ordonez, A. D. Exposito, P. A. Cervera, M. Bakic, and T. Wijekoon, "Fast and accurate analytical thermal modeling for planar pcb magnetic components," *IEEE Transactions on Power Electronics*, vol. 38, no. 6, pp. 7480–7491, 2023. DOI: 10.1109/TPEL.2023.3259064.

[9] Z. Shuai, S. He, Y. Xue, Y. Zheng, J. Gai, *et al.*, "Junction temperature estimation of a sic mosfet module for 800v high-voltage application in electric vehicles," *eTransportation*, vol. 16, p. 100 241, 2023. DOI: https://doi.org/10.1016/j.etran.2023. 100241.

[10] D. Santamargarita, D. Molinero, E. Bueno, M. Marrón, and M. Vasić, "On-line monitoring of maximum temperature and loss distribution of a medium frequency transformer using artificial neural networks," *IEEE Transactions on Power Electronics*, vol. 38, no. 12, pp. 15 818–15 828, 2023. DOI: 10.1109/TPEL.2023.3308613.

[11] M. Musallam and C. M. Johnson, "Real-time compact thermal models for health management of power electronics," *IEEE Transactions on Power Electronics*, vol. 25, no. 6, pp. 1416–1425, 2010. DOI: 10.1109/TPEL.2010.2040634.

[12] Y. A. Cengel, *Heat Transfer: A Practical Approach*, 3rd ed. New York: McGraw-Hill, 2002.

[13] *FPU DSP Software Library, USER'S GUIDE*, Accessed: 2021-01-01, Texas Instruments, 2020.

PCIM Europe 2024, 11– 13 June 2024, Nuremberg   DOI: 10.30420/566262336

A Digital Twin Approach Toward Lifetime Analysis and Predictive Maintenance of Power Semiconductors for Railway Application

Emmanuel Batista[1], Michel Piton[1], Nicolas Alferez[1], Damien Tisné-Grimaud[1], Vincent Escrouzailles[1].

[1] Alstom SA, France

Corresponding author: Emmanuel Batista, emmanuel.batista@alstomgroup.com
Speaker: Emmanuel Batista, emmanuel.batista@alstomgroup.com

Abstract

This paper proposes an approach to electrical and thermal modeling of a high-voltage power semiconductor module with a view to set up a digital twin representation for assessing degradation and remaining life of component in railway applications. Based on experimental results obtained at test bench level where optic fibers and Negative Temperature Coefficient (NTC) thermistors have been used to monitor internal temperatures of the power semiconductor module, this paper will focus on an example of fitting methodology between digital twin and real application data from tests. An application of optimization algorithm, more specifically a genetic algorithm, will be described. Results and limitation will be then discussed.

1   Semiconductor use context

Widely used nowadays, converters based on silicon Insulated Gate Bipolar Transistors (Si IGBT) or silicon carbide Metal Oxide Semiconductor Field Effect Transistor (SiC MOSFET) multichip modules are key elements also very critical for reliability and lifetime of a railway traction system. New High Power Half-Bridge Modules referred as Low-Voltage Module (LVM) in this paper packaging is becoming the standard solution for electric power applications [1] [2] [3]. Prognostics and health management of such component are directly linked to its internal thermal layer interfaces degradation, see figure 1, to the cooling performance of the converter and to the mission profiles applied during its life [3].

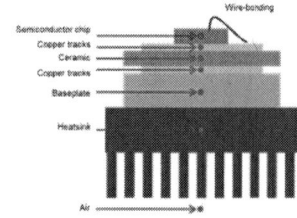

Fig. 1   Simplified cross-section view of a power semiconductor device.

2   Type of degradation of interest in this study

There are several types of failures in power semiconductor devices, an exhaustive synthesis is proposed in [4]. Among the stressors, the semiconductor chip internal temperature e.g. the junction temperature is usually considered as the main one. As the constituent materials of each layers have different Coefficient of Thermal Expansion (CTE), the generated thermomechanical constraints at the materials interfaces lead to wear-out failures as wire bonding lift-off, cracks, solder delamination, see figure 2.

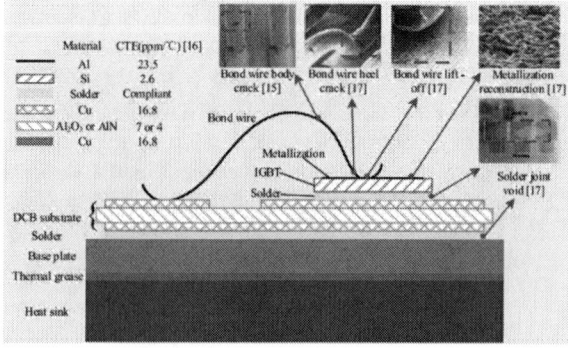

Fig. 2   Example of wear-out and CTE[ppm/C] values for material layers [2]

2394

	Values		Component impedance variation [%]			
	Min	Max	DC inductance	DC resistance	AC inductance	AC resistance
Number of cutted Wire-bondings per switch	1	11	0,5	5	0,3	2
Chip solder delamination surface ratio [%]	5	100	0,007	0,05	0,1	0,3
High-voltage terminals delamination surface ratio [%]	5	100	<0,001	<0,01	0,2	0,3

Table 1 Sensitivity analysis of electrical contact quality on overall impedance of the component This is the title of table

It is interesting to note that other factors are also influent on the degradation evolution such as duration of heat increase (t on), absolute junction temperature value (Tj) and current value [4].

3 Setting up a digital twin for electrothermal analysis

In order to validate our digital twin, a power module have been instrumented with optic fibers for junction temperature of chips (IGBTs and Diodes) [5], thermocouples for case temperature and the internal NTC thermistors have been monitored. This instrumented power module is illustrated on the following figure.

Several PWM strategy have been applied on a 3 leg inverter topology and all temperatures are measured up to steady-state regime.

a)

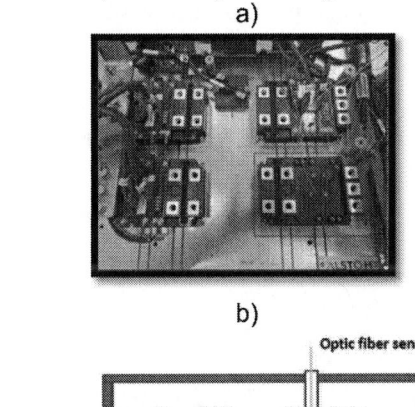

b)

Fig. 3 Power module implantation with instrumented LVM power components a) and illustration of optic fiber sensor insertion method in the packaging b) [5]

In this paper, the digital twin methodology will be based on 3D electrical and thermal models where the 3D geometry can be changed in order to estimate the impact of the performance degradation.

For the electrical degradation analysis, the following degradation have been taken into account:

Wire-bonding lift-off, chip solder delamination, high-voltage terminals solder delamination.

In order to evaluate the range of the resistance and inductance parasitic variation, a sensitivity analysis is presented in the table 1 with DC and AC@1MHz calculations.

Regarding the on-state resistance (Ron), the wire-bonding lift-off has the main impact (5% in DC). The other degradation modes show a variation lower than 0.3% on resistance and inductance, raising the impossibility of measuring such a small variation in real component during operation.

In the following steps, we will focus on the impact of wire bonding lift-off on the DC parasitic resistance of the component. A detailed parametric study is realized to compute the component DC resistance value by successively deleting wire-bonding in the model. Results are detailed on figure 4.

a)

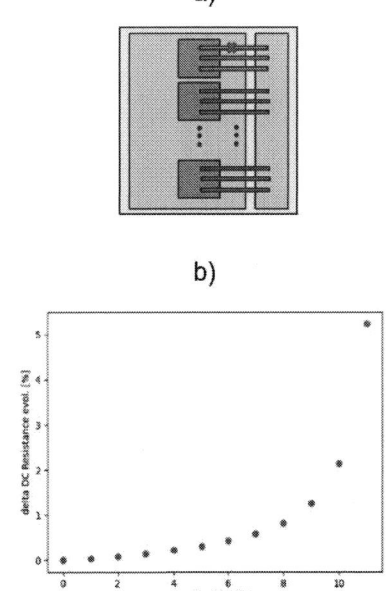

b)

Fig. 4 a) Representation of solder degradation in the 3D model b) Evolution of NTC temperature in function of % of heatsink solder delamination

This DC resistance value evolution is added in series to the component Ron value (ESR bonding resistor in the electrical scheme illustrated in figure 5c) and has an impact on the conduction losses with the following equation (1).

$$Pcond = V0.Imean + (Ron\_init+Rdc\_component(number\ of\ wire)).Irms^2 \quad (1)$$

With a predictive maintenance approach to monitor this parameter, the On-state resistance of the component needs to be taken into account and since this parameter is thermo-dependent, the junction temperature is needed to have an accurate estimation. A parametric simulation of this drain-voltage value at low current including the resistance evolution is described in figure 5. By monitoring periodically this parameter over time, prediction of electrical degradation can be made. An application to wire-bonding damage prediction is detailed in [6].

a)

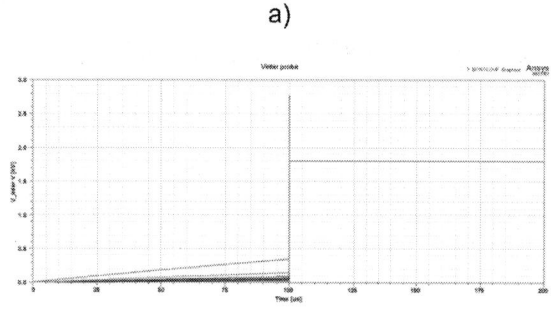

b)

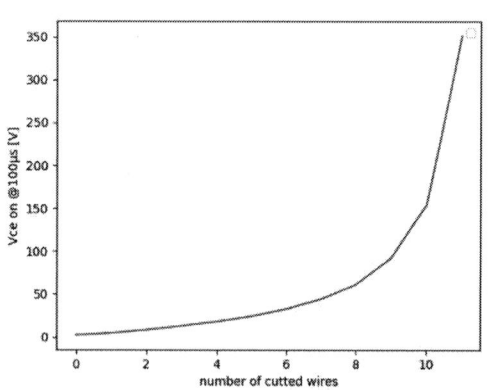

c)

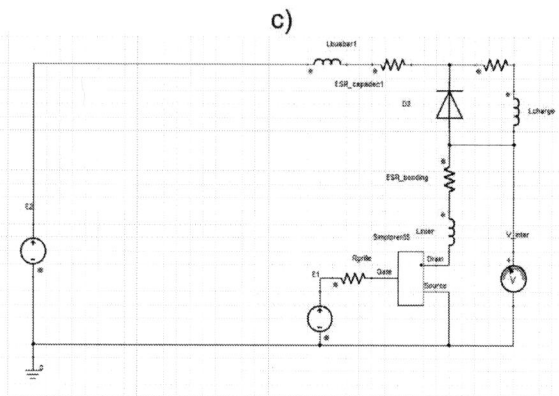

Fig. 5 a) Simulation of Vce voltage with impact of wire-bonding lift-off on Vce_on voltage during double pulse procedure simulation b) extraction of the Vce_on value at 100µs @Tj=125°C and c) the corresponding electrical scheme with ESR bonding resistor

For the thermal modelling part, two methodologies are addressed in this paper:
- 3D FEM with parametric 3D CAD file
- Electrical equivalent circuit approach fitted to real data with optimization algorithm

Regarding the 3D FEM models, the shape and size of the solders in the model have been changed and their impact on the temperature is analyzed. The example of the ceramic substrate solder delamination on the internal NTC steady-state temperature is presented on the figure 6b.

a)

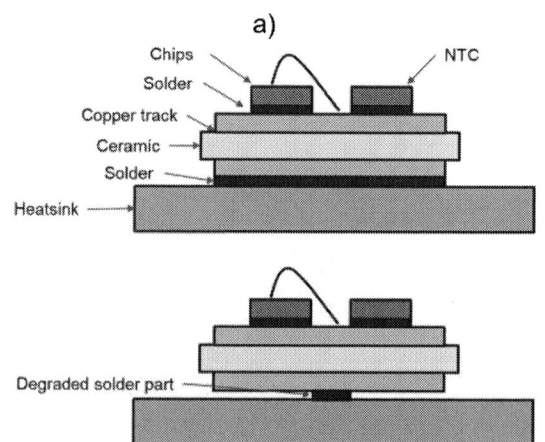

b)

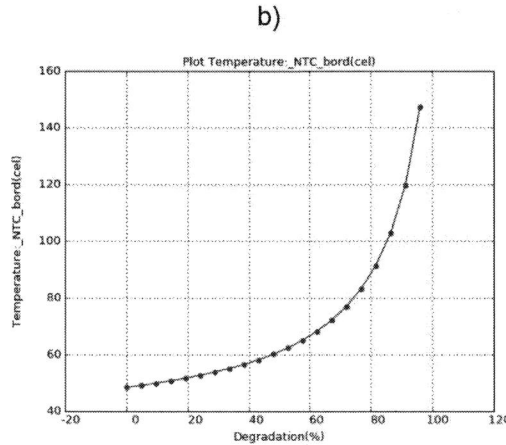

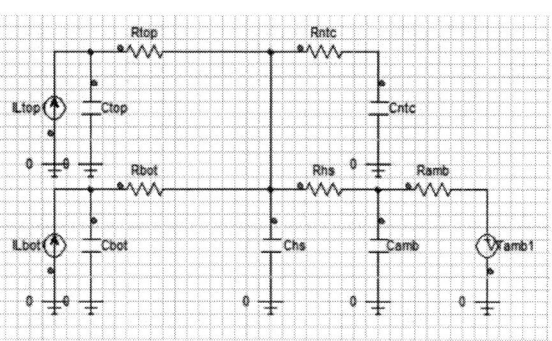

Fig. 7 Electric equivalent circuit for Thermal analysis

Fig. 6 a) Representation of solder degradation in the 3D model b) Evolution of NTC temperature in function of % of heatsink solder delamination

For ageing evaluation and in order to accelerate the calculation time, an electrical equivalent model can be set-up including some exponential laws for Rth and Cth value as function of each layer degradation. As an use case, a simple electrical equivalent thermal model of the half-bridge component will be used and the scheme is presented on the next figure.

An example is given in equation 2 for a Rth function.

$$Rth(deg)=Rth\_init.e(\alpha.deg) \qquad (2)$$

With the following parameters:
- Rth_init = the initial value of the thermal resistance
- deg = percentage of degradation of the thermal interface
- α = the constant to be fitted on the thermal 3D modelling results

Where :
- Rtop, Ctop are respectively the thermal resistance and capacitance of the Top switch
- Rbot, Cbot are respectively the thermal resistance and capacitance of the Bottom switch
- Rntc, Cntc are respectively the thermal resistance and capacitance of the internal NTC thermistor
- Rhs, Chs are respectively the thermal resistance and capacitance of the heatsink
- Ramb, Camb are respectively the thermal resistance and capacitance linked to the ambient temperature Tamb
- Ltop and Lbot are respectively the losses of the top and bottom switches

By reducing the error between the outputs of this thermal model and the corresponding experimental temperature measurements obtained at test bench level (and described in section Setting up a digital twin for electrothermal analysis) an optimization algorithm methodology can help to determine the evolution of each thermal parameters. An example of fitting this equivalent model to real measurements value is presented in the next figure by using a genetic optimization algorithm NSGA-II [7].

PCIM Europe 2024, 11– 13 June 2024, Nuremberg   DOI: 10.30420/566262336

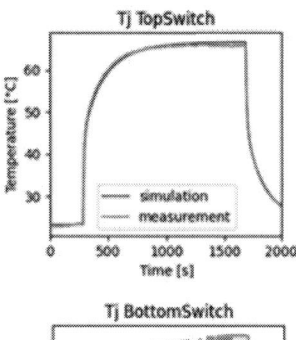

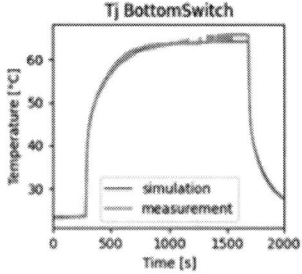

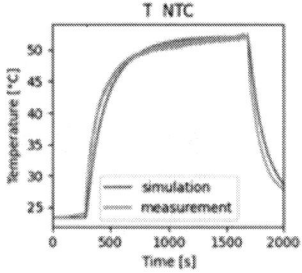

Fig. 8 Comparison of the optimized thermal model with experimental data. Respectively, junction temperature of Top and Bottom switch and internal NTC thermistor

The convergence criteria of the NSGAII algorithm is based on the following equation and the overall convergence is plotted on figure 9. The best solution is obtained around the 2000th iteration with a complete calculation time of 4 minutes.

$$convergence\ criteria = \sqrt{\sum (Tsimulated(t) - Tmeasured(t))^2}$$

(3)

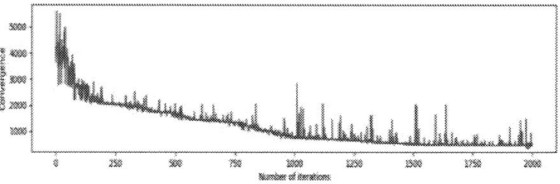

Fig. 9 NSGAII Optimisation algorithm convergence

The fitted R and C values of the thermal model will be considered as reproducing the behavior of the LVM as initial condition without degradation.

4   Evaluating degradation with the digital twin

In order to evaluate the degradation of a power component installed on real power converter application, the output of the fitted model can be compared to real signals coming from the field with an instrumented power module.

To validate this approach, a power component of the converter used in the test bench described in this paper have been changed by a degraded power component where the top switch have been aged with power cycling methodology.

The comparison of the measured junction temperature of the top switch (aged) and the bottom switch (not aged) with the calculated one from the digital twin is proposed in the next figure.

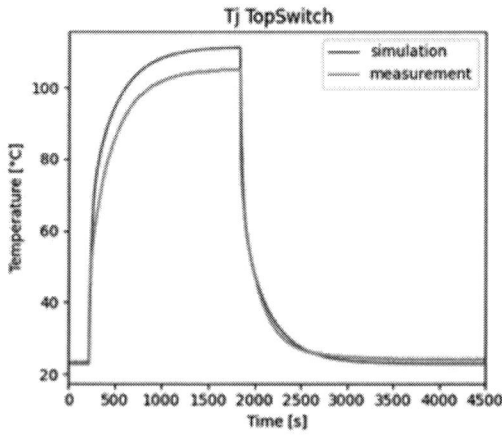

2398

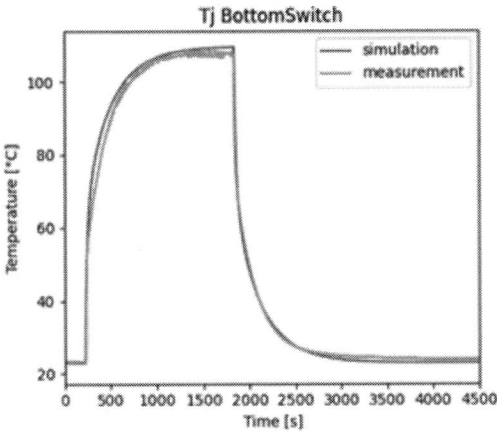

Fig. 10 Comparison of the model with real data where top switch is degraded.

During the first power step (t<2000s), a gap of 7°C between the measure and the simulation from the digital twin is observed for the top switch and indicates the degradation of the component where the bottom switch (non-aged) is presenting a gap lower than 1°C and indicates no degradation.

A simple way to use the digital twin is to manage the deviation between the expected temperature (from the digital twin) and the real measurement in function of the time and by checking if a specified threshold value is reached. Another way to use the digital twin is to re-run the optimization phase developed in section "Setting up a digital twin for electrothermal analysis" in order to fit the resistance, capacitance and power values of the thermal model to the real measurement. This optimization step could give some indication of the localization of the degradation such as losses increase or thermal interface deterioration.

As a perspective, a periodic use of this methodology can give an image of the degradation evolution over time. It is interesting to note that prediction can be make based on this evolution tracking [8].

5 Conclusion

In this article, a digital twin approach have been developed to estimate electrical and thermal degradation i.e. wire bonding lift-off and thermal impedance degradation of power semiconductors in LVM (half-bridge) multi-chips module. An application of optimization algorithm was presented based on thermal experimental results at test bench level showing a good comparison with models.

References

[1] Common specification of new generation power semiconductors for railway traction applications – Roll2Rail - R2R-T1.1-T-BTS042-01 – November 2016.

[2] R. Tsuda, S. Iura, E. Thal, T. Negishi, N. Soltau and E. Wiesner, "LV100 High Voltage Dual Package in Paralleling Operation," PCIM Europe 2018; International Exhibition and Conference for Power Electronics, Intelligent Motion, Renewable Energy and Energy Management, Nuremberg, Germany, 2018, pp. 1-6.

[3] M. Piton, E. Batista and V. Escrouzailles, "Improved Stress Distribution in Railway Traction Converters Using New High Power Half-Bridge Modules," CIPS 2020; 11th International Conference on Integrated Power Electronics Systems, Berlin, Germany, 2020, pp. 1-6.

[4] K. Hu, Z. Liu, Y. Yang, F. Iannuzzo and F. Blaabjerg, "Ensuring a Reliable Operation of Two-Level IGBT-Based Power Converters: A Review of Monitoring and Fault-Tolerant Approaches," in IEEE Access, vol. 8, pp. 89988-90022, 2020, doi: 10.1109/ACCESS.2020.2994368

[5] M.Piton, B.Chauchat, JF.Serviere, "Implementation of Direct Chip Junction Temperature Measurement in High Power IGBT Module in Operation - Railway Traction Converter", 29th European Symposium on Reliability of Electron Devices, Failure Physics and Analysis, ESREF 2018, Aalborg, Denmark, 2018.

[6] F. Qin, X. Bie, T. An, J. Dai, Y. Dai and P. Chen, "A Lifetime Prediction Method for IGBT Modules Considering the Self-Accelerating Effect of Bond Wire Damage," in IEEE Journal of Emerging and Selected Topics in Power Electronics, vol. 9, no. 2, pp. 2271-2284, April 2021, doi: 10.1109/JESTPE.2020.2992311.

[7] K. Deb, A. Pratap, S. Agarwal and T. Meyarivan, "A fast and elitist multiobjective genetic algorithm: NSGA-II," in IEEE Transactions on Evolutionary Computation, vol. 6, no. 2, pp. 182-197, April 2002, doi: 10.1109/4235.996017.

[8] Celaya, Jose & Saxena, Abhinav & Saha, Sankalita & Goebel, Kai. (2011). Prognostics of power MOSFETs under thermal stress accelerated aging using data-driven and model-based methodologies. Proceedings of International Conference on Prognostics and Health Management, Montreal. 2.

PCIM Europe 2024, 11– 13 June 2024, Nuremberg   DOI: 10.30420/566262337

Saturable Ferrite Core Inductors in LCL Filters of Three-Phase Voltage Source Inverters

Marius Kaufmann-Bühler[1], Hannah Riepe[1], Sibylle Dieckerhoff[1]
[1] Technische Universität Berlin, Germany

Corresponding author and speaker: Marius Kaufmann-Bühler, kaufmann-buehler@tu-berlin.de

Abstract

This article presents a filter inductance design methodology for saturable ferrite inductors with stepped air gaps in three-phase LCL filters. The analysis shows that allowing saturation of the filter inductor enables the choice of smaller cores or reduces the inductors winding resistance. This results in higher power density or efficiency. The paper contains a design methodology for stepped air-gap inductors based on a reluctance model including fringing flux effects. The findings are verified by means of field simulation and time domain simulation.

1   Introduction

Filter inductors contribute significantly to the weight, volume and overall efficiency of power electronic systems. Recent research shows that applying saturating filter inductors in three phase LCL-filters of voltage source inverters (VSI) allows for a significant reduction of the overall filter inductance which can result in smaller inductors with less winding resistance [1–3]. The authors of the stated publications use iron powder cores, which exhibit soft saturation characteristics for their filter inductors.

Ferrite cores are well suitable in filter applications for medium or high frequency range, due to their high resistivity and low cost [4, 5]. In contrast to iron powder cores, they exhibit an abrupt saturation at high flux density levels. Ferrite-core inductors often contain discrete air gaps in order to limit the flux density. It is possible to manipulate the saturation characteristics of gapped ferrite cores by changing the shape of the air gap. There are various publications that concern the use of saturable inductors to optimize the system efficiency and power density, reduce component stress or to improve electromagnetic compatibility, controllability or system dynamics [6]. A design approach for stepped air gaps based on a simple reluctance model has been presented in [7].

The aim of the present work is to propose a design methodology for saturable, stepped air-gap ferrite cores in three phase LCL filters and to evaluate the resulting designs in terms of inductor volume and winding resistances. A reluctance model to deter-

mine the inductance characteristic of stepped-air-gap inductors is presented in section 2. Section 3 explains the dependence of the required filter inductance to limit the inverter output current ripple on the desired core saturation. The presentation of the design methodology consists of the core selection, the determination of the relevant geometry parameters and the design verification using field simulation and time domain simulation, see section 4. Evaluating the presented design methodology leads to the insight, that saturable ferrite inductors can contribute to reduced winding resistances and more compact inductors, cf. section 5.

2   Inductance of a Stepped Air-Gap Inductor

The examined arrangement consists of two E-shaped cores and N turns of copper winding. The

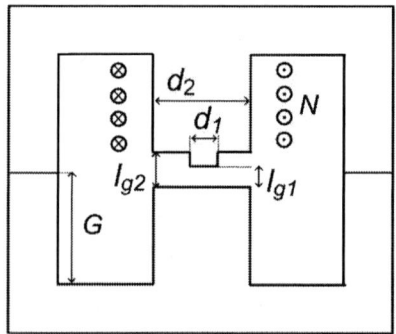

Fig. 1   Geometry of a stepped air-gap EE core

air gap has two regions with different air gap lengths l_{g1} and l_{g2}, cf. Fig. 1. The width d_1 further characterizes the air-gap geometry. The center leg width d_2 and the arrangement's depth are determined by the selected core.

The reluctance model proposed in [7], extended by a fringing flux estimation, is used to derive the inductance of the arrangement in the following. In unsaturated operation, the inductor is modelled by two reluctances in parallel. R_{g1} in Fig. 2 represents the air gap region below the step with the air gap length l_{g1} and the width d_1, (1). R_{g2} describes the air gap region, which surrounds R_{g1}, with the air gap length l_{g2} (2). A_c in (1)-(2) denotes the core's cross section area, and $F_{F1,2}$ are fringing flux factors. Section 4 contains details on the fringing flux factors.

$$R_{g1} = \frac{l_{g1}}{\mu_0 A_c \cdot \frac{d_1}{d_2} \cdot F_{F1}} \qquad (1)$$

$$R_{g2} = \frac{l_{g2}}{\mu_0 A_c \cdot \frac{d_2 - d_1}{d_2} \cdot F_{F2}} \qquad (2)$$

The relative magnetic permeability of the core material depends on the magnetic field strength and is constant in two sections: $\mu_r(B) = \mu_i$ for $|B| < B_{sat}$ and $\mu_r(B) = 1$ otherwise. μ_i and B_{sat} are the specified relative permeability and saturation flux density.

Due to the reduced cross-section area and air gap length of R_{g1}, the small core region above R_{g1} will saturate at lower load current than the rest of the arrangement. Thus, the overall inductance has three operating regions: the unsaturated inductance L_0, the partly saturated inductance L_S and the fully saturated inductance. Operation in the latter region must be avoided. The corresponding current boundaries are i_S and i_{max}, see Fig. 3.

The flux density is lower than the saturation flux density throughout the core in unsaturated region. In the partly saturated region, only the part of the center leg with reduced width d_1 saturates. In this state, the stepped air-gap model simplifies to one discrete air gap with the length l_{g2}, the width d_2 and the reluctance $R_{g,s}$ (3). F_{FS} denotes the fringing flux factor of the air gap reluctance in partly saturated state.

$$R_{g,S} = \frac{l_{g2}}{\mu_0 A_c F_{FS}} \qquad (3)$$

The corresponding inductances L_0 and L_S follow from dividing the squared number of turns by the parallel connection of (1) and (2) or by (3) in the latter case, see (4)-(5).

$$L_0 = \frac{N^2 \mu_0 A_c}{d_2} \left(\frac{d_1 F_{F1}}{l_{g1}} + \frac{(d_2 - d_1) F_{F2}}{l_{g2}} \right) \qquad (4)$$

$$L_S = \frac{N^2 \mu_0 A_c F_{FS}}{l_{g2}} \qquad (5)$$

Combining the reluctance (1) and the given geometry, we can derive an expression for the magnetic flux (6), which is the product of the field density and the cross-section area if homogenous field distribution is assumed. This assumption will be discussed in section 4. The saturation current i_S can thus be derived as in (7).

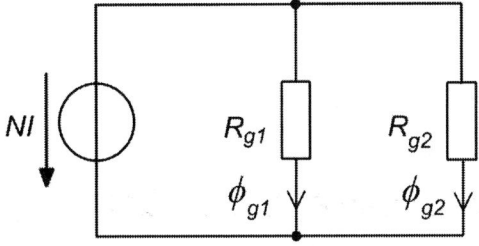

Fig. 2   Simplified reluctance model of the unsaturated stepped air gap inductor.

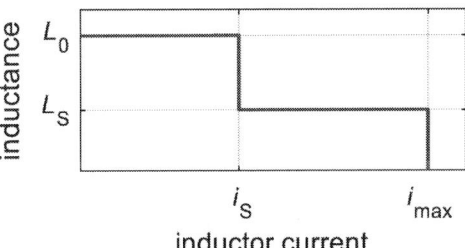

Fig. 3   Current-depending inductance of the stepped air-gap inductor based on the simplified reluctance model.

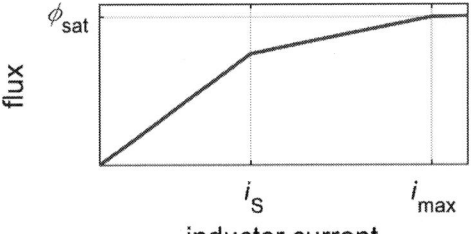

Fig. 4   Magnetic flux depending on the inductor current

For currents higher than i_S, the slope in the flux-current-plot is decreased due to the reduction of the inductance in the partly saturated region, see Fig. 4. At maximum current, the flux reaches the value ϕ_{sat}, which can be derived as in (8).

$$\phi_{g1} = B_{g1}A_c\frac{d_1}{d_2} = \frac{Ni}{R_{g1}} \quad (6)$$

$$i_S = \frac{B_{\mathrm{sat}}l_{g1}}{N\,\mu_0} \quad (7)$$

$$N\phi_{\mathrm{sat}} = NB_{\mathrm{sat}}A_c = L_0 i_S + L_S(i_{\max} - i_S) \quad (8)$$

3   Filter Inductance Calculation

The maximum current ripple of the output current of a three-phase, two level VSI occurs either during the zero crossing (index "zc") or during the peak value (index "pk") of the fundamental current, if PF = 1 is assumed, [8–10]. In three-phase systems with saturable inductors, the output current ripple depends on the instantaneous inductances in all three phases and the parameters m (modulation index), V_{dc} (dc-link voltage) and f_{sw} (switching frequency). The derivation of (9) and (10) was presented extensively in [3] and therefore won't be repeated here.

$$\Delta i_{\mathrm{zc}} = \frac{\sqrt{3}mV_{\mathrm{dc}}}{8f_{\mathrm{sw}}} \frac{1}{2L(i=0) + L\left(i=\frac{\hat{i}\sqrt{3}}{2}\right)} \quad (9)$$

$$\Delta i_{\mathrm{pk}} = \frac{3mV_{\mathrm{dc}}}{8f_{\mathrm{sw}}} \cdot \frac{1 - 3m/4}{2L(i=\hat{i}) + L\left(i=\frac{\hat{i}}{2}\right)} \quad (10)$$

Applying the derived current-dependent inductance of a stepped air-gap inductor (Fig. 3) allows to replace the inductances in (9) and (10) with L_0 and L_S. The saturation current should be in the range $\hat{i}/2 < i_S < \hat{i}\sqrt{3}/2$. Smaller i_S would significantly increase the current ripple during peak current (10). For higher i_S, Δi_{zc} would be unaffected by saturation effects. The required inductances to limit the current ripple to Δi during fundamental current zero crossing and peak value follow as (11) and (12).

$$2L_{0,\mathrm{zc}} + L_{S,\mathrm{zc}} = \frac{\sqrt{3}mV_{\mathrm{dc}}}{8f_{\mathrm{sw}}\Delta i} \quad (11)$$

$$L_{0,\mathrm{pk}} + 2L_{S,\mathrm{pk}} = \frac{3mV_{\mathrm{dc}}}{8f_{\mathrm{sw}}\Delta i_{\mathrm{pk}}} \cdot \left(1 - \frac{3m}{4}\right) \quad (12)$$

The ratio between L_S and L_0 will be referred to as saturation factor k_{sat} in the following. The inductance saturates to $L_S = k_{\mathrm{sat}}L_0$, whenever the fundamental frequency current exceeds the saturation current i_S. Lower k_{sat} leads to smaller L_S and larger L_0. Evaluating (11) and (12) for variable k_{sat} shows that during fundamental current zero crossing, a higher saturated inductance L_S is required to limit the current ripple to Δi, when compared to the fundamental current peak value, see Fig. 5.

4   Stepped Air-Gap Design

4.1   Application Parameters

The application scenario is a three-phase and two-level VSI with a rated power of 10 kVA, which is connected to the German low-voltage grid (230 V, 50 Hz). It is built of silicon-carbide MOSFETs. Further application parameters are listed in Table 1. The inductor under investigation is the inverter-side inductor in an LCL filter.

4.2   Inductor design methodology

The aim of the design methodology is to find a suitable core and to determine the parameters l_{g1}, l_{g2}, d_1 and N. In this work, we focus on E-shaped cores because of their general availability, especially in

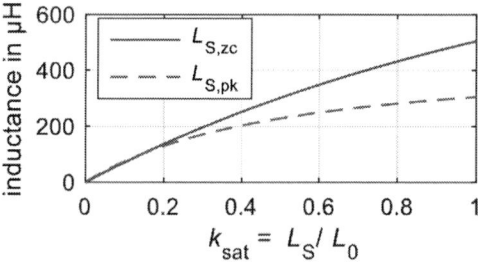

Fig. 5 L_S depending on the occurring saturation factor. Comparison between fundamental current zero crossing (blue) and peak value (red).

Fundamental current	$I_{\mathrm{rms}} = 14.5\,\mathrm{A}$
Max. current ripple	$\Delta i_{\mathrm{m}} = 2.9\,\mathrm{A}$
Switching frequency	$f_{\mathrm{sw}} = 32\,\mathrm{kHz}$
DC-Link voltage	$V_{\mathrm{dc}} = 750\,\mathrm{V}$

Table 1   Selected inverter parameters

various sizes and prefabricated air-gap lengths and easy assembly [11].

The input values of the process are the given system parameters (Table 1), the saturation and maximum current i_S and i_{max}, the desired saturation k_{sat} and some further inductor parameters, see Table 2. Throughout this paper, the max. current is defined as the sum of the fundamental current amplitude, the max. current ripple and 10 % reserve. The saturation current is the sum of half the amplitude and the max. current ripple.

4.2.1 Saturation and Core Selection

The core selection follows the well-established area-product method [12, 13]. The area-product A_p is a composition of the inductance L, the current i_{max}, the wire area A_w, the maximum flux density B_m and window utilization factor K_u, see (13). On the other hand, it is the product of the core's cross section and window area A_c and W_a, describing the core geometry. The saturated inductance L_{sat} depends on the choice of the saturation depth k_{sat}, Fig. 5. Inserting the reduced L_{sat} for lower k_{sat} in (13) results in a decreased area product, which directly affects the core selection. Note that Fig. 6 contains the calculated area product (solid line) as well as the area product of available E-cores (dashed line) [11].

$$A_p = \frac{A_w i_{max} L}{K_u B_m} = A_c \cdot W_a \tag{13}$$

4.2.2 Air Gap Geometry

Once saturation and the core are selected, equation (8) only contains known design parameters and N. Manipulating (8) leads to (14). The gap length l_{g2} follows from (5) as (15). The saturation current i_S only depends on the air-gap length l_{g1}, material parameters and N, leading to (16). Finally, d_1 results from rearranging (4) as (17).

$$N = \frac{L_0 i_S + L_S (i_{max} - i_S)}{B_{sat} A_c} \tag{14}$$

Max. current	$i_{max} = 25.75$ A
Saturation current	$i_S = 13.15$ A
Core Material	N27
Saturation flux density	$B_{sat} = 410$ mT
Wire Area	$A_w = \pi$ mm^2
Window utilization factor	$K_u = 0.3$

Table 2 Selected inductor parameters.

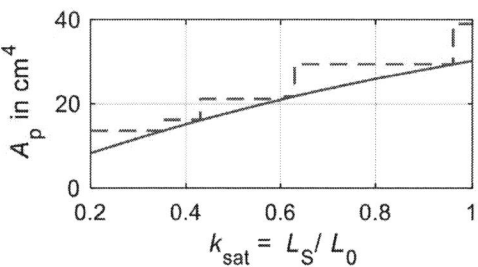

Fig. 6 Area product depending on the occurring saturation factor. Calculated value (solid) and area product of available cores (dashed).

$$l_{g2} = \frac{N^2 \mu_0 A_c F_{FS}}{L_S} \tag{15}$$

$$l_{g1} = \frac{F_{F1} i_S N \mu_0}{B_{sat}} \tag{16}$$

$$d_1 = \left(\frac{d_2 L_0}{N^2 \mu_0 A_c} - \frac{d_2 F_{F2}}{l_{g2}} \right) \cdot \left(\frac{F_{F1}}{l_{g1}} - \frac{F_{F2}}{l_{g2}} \right)^{-1} \tag{17}$$

Equations (15)-(17) contain different fringing flux factors, which contribute decisively to the design parameters. Each of the factors F_{F1}, F_{F2} and F_{FS} is related to one of the described reluctances (1)-(3). There are different approaches to estimate the fringing flux' effect on the inductance of gapped ferrite inductors [4, 13]. The energy, which is stored in the magnetic field close to the air gap, affects the inductance as well as the saturation current. A fringing flux factor greater than one increases the inductance and decreases the saturation current of the inductor.

We use equation (18) to estimate the fringing flux effect in the present work. l_g and A_c are the air gap length and cross-section area of the respective reluctance (1)-(3). G is the height of the winding window, cf. Fig. 1, and $k_{F1/2/S}$ a correction factor for each respective reluctance (1)-(3), which is 1 in [13]. These corrections are necessary, as the mentioned fringing flux estimations apply for air gaps without steps. Similarly to the process described in [12], l_g and F_F are calculated iteratively. $F_F = 1.3$ is a realistic starting value.

$$F_F = 1 + k_{F1/2/S} \cdot \frac{l_g}{\sqrt{A_c}} \ln \left(\frac{2G}{l_g} \right) \tag{18}$$

Prior to further investigations, it should be checked, if the number of turns with the chosen wire area result in a window fill factor close to the desired value.

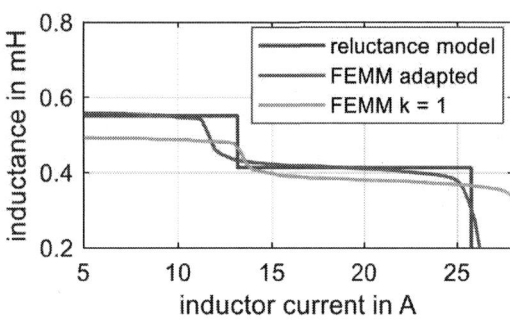

Fig. 7   Current-depending inductance of the stepped air-gap inductor. Blue: reluctance model, red: FEMM simulation with $k_{F2} = k_{FS} = 0.7$ and $k_{F1} = 0.6$, yellow: FEMM simulation with $k_{F1} = k_{F2} = k_{FS} = 1$.

	$k_{F1/2/S} = 1$	$k_{F2} = k_{FS} = 0.7$ $k_{F1} = 0.6$
N	43	43
l_{g2}	5,69 mm	5,1 mm
l_{g1}	2,76 mm	2,27 mm
d_1	5,15 mm	5,29 mm

Table 3   Design parameters for an EE 70/33/32 inductor with $k_{sat} = 0.75$ and two different sets of factors $k_{F1/2/S}$

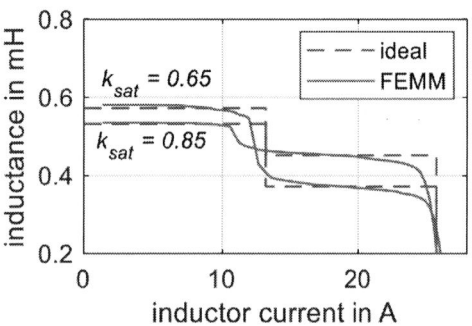

Fig. 8   Inductance of stepped E70/33/32 inductors with $k_{sat} = 0.85$ and $k_{sat} = 0.65$

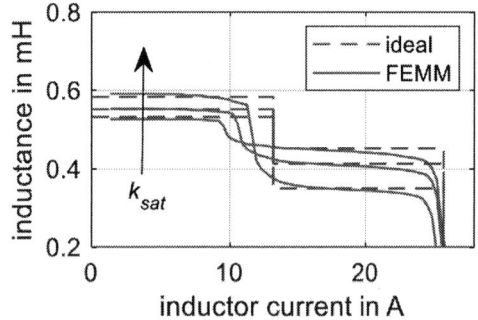

Fig. 9   Inductance of stepped E65/32/27 inductors with $k_{sat} = [0.60, 0.75, 0.85]$

4.2.3 Design Verification

Every design consisting of parameters l_{g1}, l_{g2}, d_1 and N is verified with the field simulation software FEMM [14] to evaluate the accuracy regarding the desired inductance depending on the current. Fig. 7 depicts the simulated current-depending inductance of a stepped air gap inductor consisting of two E70/33/32 cores with $k_{sat} = 0.75$ and the further parameters listed in Table 1 and Table 2. The blue line is the desired inductance. The red and yellow lines are the results of FEMM simulations based on two different sets of $k_{F1/2/S}$. The design which is obtained with the original fringing flux factor $k_{F1/2/S} = 1$ clearly has too small inductances L_S and L_0. Besides, the maximum current is larger than specified.

A design with adapted parameters $k_{F2} = k_{FS} = 0.7$ and $k_{F1} = 0.6$ leads to less deviations between the simulated and desired inductance. The inductances L_S, L_0 and the maximum current show good agreement with the reluctance model. The saturation current i_S is too low, which is acceptable if it is in the range $\hat{\imath}/2 < i_S < \hat{\imath}\sqrt{3}/2$. This condition is met, as $\hat{\imath}/2 = 10.2$ A in our application. The corresponding geometry parameters are listed in Table 3.

The focus of the present paper is not to find optimal air-gap parameters which lead to minimum deviation between the ideal inductances of the reluctance model and the simulated inductances. It rather proves that it is possible to find design parameters which result in acceptable inductance characteristics for variable saturation and different cores. To underline this, Fig. 8 depicts the inductance characteristics of two stepped E70/33/32 designs with $k_{sat} = 0.85$ and $k_{sat} = 0.65$. Fig. 9 shows the inductances of two stepped E65/33/32 inductors with $k_{sat} = 0.85/0.75/0.60$. All designs were obtained with the adapted $k_{F1/2/S}$ parameter set used in Fig. 7 and Table 3.

The MATLAB script and FEMM models that generate Fig. 7 - Fig. 9 are available online [15].

4.3 Time Domain Simulation

A time domain simulation model in MATLAB/Simulink is built to prove the compliance with the maximum current ripple limit. The model consists of a three-phase, two level VSI, which is supplied by a DC voltage source and connected to the grid model with an LCL-filter, see Fig. 10. The grid is modelled as ideal three-phase voltage source without harmonic disturbances.

The applied voltage-oriented control (Fig. 11) has the measured inverter side current i_i, the filter capacitor voltage v_{Cf} and filter output voltage v_o

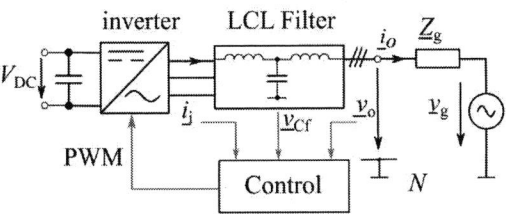

Fig. 10 Overview of the grid connected inverter

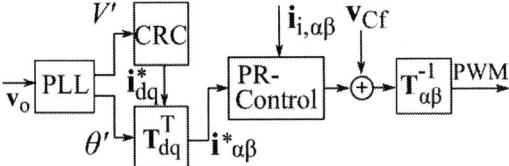

Fig. 11 Scheme of the voltage-oriented control

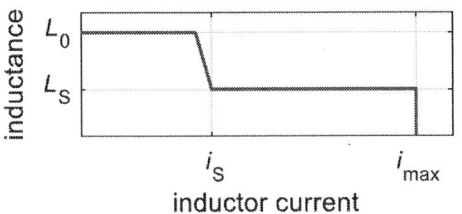

Fig. 12 Implementation of the saturable inductance in the time domain simulation

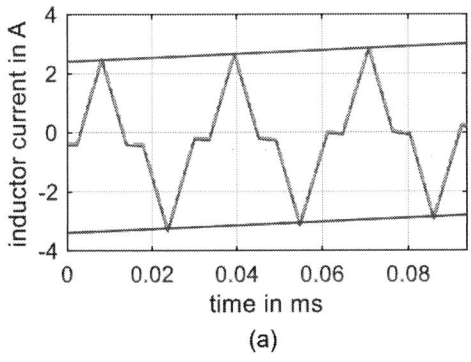

(a)

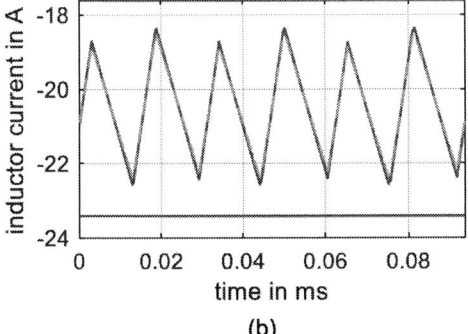

(b)

Fig. 13 Simulated inductor current and current ripple boundaries (red) during fundamental current (a) zero crossing and (b) amplitude for $k_{sat} = 0.65$ (blue), 0.75 (red) and 0.85 (yellow)

as inputs. A notch-filter phase-locked loop determines the grid voltage V' amplitude and phase angle θ' [16, 17]. The current reference calculator (CRC) provides the reference currents in synchronous reference frame. A proportional-resonant controller of the inverter-side current with filter capacitor voltage feed forward generates the voltage references for the space vector modulation [18, 19]. The dimensioning of the inverter-side filter inductance follows (11). A variable inductance represents the stepped inductor. The instantaneous inductor current is used to determine the corresponding inductance. A transition zone for $i_S - 1\,A < i(t) < i_S$ as depicted in Fig. 12 is introduced to avoid a step in the inductance.

The simulation was repeated for different k_{sat}. Figure 13 (a) shows that the resulting inductor current during fundamental current zero crossing complies with the specified maximum current ripple for all displayed k_{sat}. As expected, the current ripple during the fundamental current amplitude is smaller than during zero crossing, cf. Fig. 13 (b).

5 Evaluation for Variable Saturation

It is possible to integrate the desired occurring saturation in the inductor design with the presented method. Saturable ferrite inductors allow for the choice of smaller cores, if the area product method is applied, see Fig. 6. The specified system parameters (Table 1 and Table 2) and $k_{sat} = 1$ ($L_0 = L_S$) require a set of E 70/33/32 cores. Choosing $k_{sat} = 0.7$ leads to a core volume reduction of 13 %, as the smaller E 65/32/27 cores suffice to realize the inductor.

Further, a lower number of turns follows from reduced k_{sat}. This leads to smaller winding resistance, as the resistance is directly proportional to the length of the copper wire. Fig. 14 depicts the number of turns N and the corresponding DC winding resistance for varying k_{sat}. The resistances are obtained from FEMM simulations and normalized to the value for $k_{sat} = 1$. N follows from (14), where L_0 and L_S are functions of k_{sat}. Allowing $k_{sat} = 0.7$ reduces N and DC resistance to 93% when compared to $k_{sat} = 1$.

The winding resistance was also evaluated for the switching frequency $f_{sw} = 32\,kHz$, as the high frequency components of the inductor current cause significant parts of the winding losses. The resulting AC winding resistance decreases to 85 % for $k_{sat} = 0.7$ when compared to $k_{sat} = 1$. This is not only caused by the reduced wire length. Skin and proximity effects cause less ohmic losses for the arrangement with reduced N. To illustrate this, Fig.

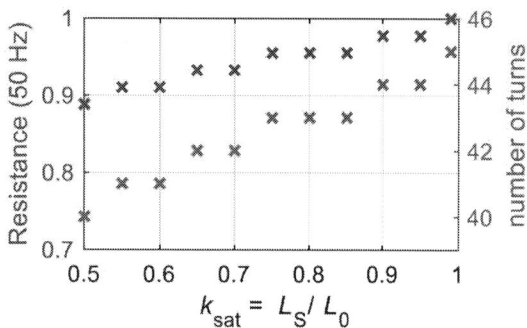

Fig. 14 Number of turns and dc resistance of an EE 70/33/32 inductor for varying k_{sat}.

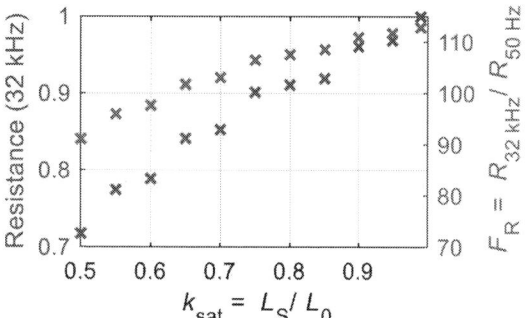

Fig. 15 AC winding resistance at 32 kHz and resistance ratio F_R of an EE 70/33/32 inductor for varying k_{sat}.

15 shows the simulated AC winding resistance as well as the resistance factor $F_R = (50\,\text{Hz})/R(32\,\text{kHz})$ for varying k_{sat}.

Different FEMM models of inductor designs that were used to determine the winding resistances in Fig. 14 and Fig. 15 are available online [15].

6 Conclusion

We propose a methodology to determine the filter inductance of saturable stepped air-gap ferrite core inductors to limit the max. output current ripple in three-phase VSIs and suggest a process to determine the air-gap geometry parameters to realize the required inductances. If saturation is specified in the inductors requirements, this will result in decreased core volume or in reduced dc and ac winding resistances because of the lower required number of turns. In the presented design example, a saturation of $k_{sat} = 0.7$ allows for a volume decrease of 13 % or a reduction of the dc and ac winding resistance by 7 % and 15 % respectively. Allowing more saturation will enhance these benefits.

Future investigations will be focused on experimental validation of the presented findings and improvements of the design methodology in terms of minimizing the deviation between the inductance characteristic of the ideal reluctance model and the FEMM simulation results or optimizing the winding resistances.

7 References

[1] Q. Wei, B. Liu, and S. Duan, "Current Ripple Analysis and Controller Design for Grid-Connected Converters Considering the Soft-Saturation Nature of the Powder Cores," *IEEE Trans. Power Electron.*, vol. 33, no. 10, pp. 8827–8837, 2018, doi: 10.1109/TPEL.2017.2777906.

[2] Q. Li, D. Jiang, and Y. Zhang, "Analysis and Calculation of Current Ripple Considering Inductance Saturation and Its Application to Variable Switching Frequency PWM," *IEEE Trans. Power Electron.*, vol. 34, no. 12, pp. 12262–12273, 2019, doi: 10.1109/TPEL.2019.2903884.

[3] M. Kaufmann-Bühler, H. Özeloglu, E. Eichstädt, and S. Dieckerhoff, "Benefits of saturated powder core inductors in LCL filters of three-phase voltage source inverters," in *2023 25th European Conference on Power Electronics and Applications (EPE'23 ECCE Europe)*.

[4] Kazimierczuk and M. K, *High-Frequency Magnetic Components*: John Wiley & Sons Ltd, 2014.

[5] M. S. Rylko, B. J. Lyons, J. G. Hayes, and M. G. Egan, "Revised Magnetics Performance Factors and Experimental Comparison of High-Flux Materials for High-Current DC–DC Inductors," *IEEE Trans. Power Electron.*, vol. 26, no. 8, pp. 2112–2126, 2011, doi: 10.1109/TPEL.2010.2103573.

[6] J. Kaiser and T. Dürbaum, "An Overview of Saturable Inductors: Applications to Power Supplies," *IEEE Trans. Power Electron.*, vol. 36, no. 9, pp. 10766–10775, 2021, doi: 10.1109/TPEL.2021.3063411.

[7] J. Kaiser and T. Dürbaum, "Design Approaches for Nonlinear Inductors with a Stepped Air-Gap," in *2023 25th European Conference on Power Electronics and Applications (EPE'23 ECCE Europe)*.

[8] Jelena Loncarski, "Peak-to-Peak Output Current Ripple Analysis in Multiphase and Multilevel Inverters," Ph.D. Thesis, University of Bologna, 2014.

[9] A. Kouchaki and M. Nymand, "Analytical Design of Passive LCL Filter for Three-Phase

Two-Level Power Factor Correction Rectifiers," *IEEE Trans. Power Electron.*, vol. 33, no. 4, pp. 3012–3022, 2018, doi: 10.1109/TPEL.2017.2705288.

[10] J. Mühlethaler, M. Schweizer, R. Blattmann, J. W. Kolar, and A. Ecklebe, "Optimal Design of LCL Harmonic Filters for Three-Phase PFC Rectifiers," *IEEE Trans. Power Electron.*, vol. 28, no. 7, pp. 3114–3125, 2013, doi: 10.1109/TPEL.2012.2225641.

[11] "Magnetics 2022 Ferrite Cores Catalog: Toroids, Shapes, Pot Cores," [Online]. Available: https://www.mag-inc.com/Media/Magnetics/File-Library/Product%20Literature/Ferrite%20Literature/Magnetics-2022-Ferrite-Catalog.pdf?ext=.pdf

[12] M. K. Kazimierczuk and H. Sekiya, "Design of AC resonant inductors using area product method," in *2009 IEEE Energy Conversion Congress and Exposition*, San Jose, CA, 2009, pp. 994–1001.

[13] Colonel Wm. T. McLyman, *Transformer and Inductor Design Handbook: Third Edition, Revised and Expanded*. New York: Marcel Dekker Ink, 2004.

[14] D. C. Meeker, *Finite Element Method Magnetics*. [Online]. Available: https://www.femm.info/

[15] Marius Kaufmann-Bühler, *Stepped Air Gap FEMM Models and Scripts*. [Online]. Available: https://git.tu-berlin.de/mrkb_011/stepped-air-gap

[16] H. Just, "Modeling and control of power converters in weak and unbalanced electric grids," Dissertation, Technische Universität Berlin, 2021.

[17] H. Just, H. Yang, M. Eggers, P. Teske, and S. Dieckerhoff, "Multi-Fidelity Model-based PLL Design for Enhanced Dynamics and Transient Stability during Fault Ride-Through," in *2020 IEEE 21st Workshop on Control and Modeling for Power Electronics (COMPEL)*, Aalborg, Denmark, 2020, pp. 1–7.

[18] M. Kaufmann-Buhler, M. Eggers, H. Just, and S. Dieckerhoff, "Can SiC MOSFETs improve the dynamics of grid-connected voltage source inverters?," in *2021 IEEE 12th International Symposium on Power Electronics for Distributed Generation Systems (PEDG)*, Chicago, IL, USA, 2021, pp. 1–7.

[19] A. Vidal *et al.,* "Assessment and Optimization of the Transient Response of Proportional-Resonant Current Controllers for Distributed Power Generation Systems," *IEEE Trans. Ind. Electron.*, vol. 60, no. 4, pp. 1367–1383, 2013, doi: 10.1109/TIE.2012.2188257.

PCIM Europe 2024, 11– 13 June 2024, Nuremberg   DOI: 10.30420/566262338

2D Copper Loss Analytical Model for Planar Inductor Combining High and Low Permeability Materials

Idriss Nachete[1,2] , Xavier Margueron[1] , Frédéric Gillon[1] , Guillaume Lefevre[2]

[1] Univ. Lille, Arts et Metiers Institute of Technology, France
[2] Mitsubishi Electric R&D Center Europe (MERCE), France

Corresponding author: Idriss Nachete, idriss.nachete@centralelille.fr
Speaker: Idriss Nachete, idriss.nachete@centralelille.fr

Abstract

Blocks of low permeability magnetic materials can be used to effectively replace lumped airgaps in planar inductors. With this solution, a significant reduction of High Frequency (HF) copper losses can be obtained compared to standard planar E and Plate (PLT) core-based inductors with airgap. In order to estimate HF copper losses in such gapless planar inductors, a 2D copper loss model based on analytical equations is developed in this paper. The results are compared to the well-known 1D Dowell model and finite element (FEA) simulations to show its benefits and limitations on Planar inductors.

1   Introduction

For power planar inductors, the addition of an airgap in the core is mandatory to avoid magnetic saturation but also to adjust the permeability and reach a targeted inductance value. However, the fringing field stemming from the airgaps induces additional copper losses, which significantly degrades the inductor's efficiency. In order to limit these fringing effects, the quasi-distributed airgap technology can be applied. It consists in dividing a lumped airgap into several smaller airgaps [1]. Despite the effectiveness of this solution when considering AC copper losses, the process of cutting ferrite adds complexity into the manufacturing of components. On the other hand, an alternative approach involves the use of low permeability materials [2], which are essential to replace airgap while achieving the desired inductor behavior. Thus, by employing Low Permeability (LP) materials, the magnetic core can be kept from saturating while mitigating the limitations associated with fringing field and manufacturability.

For planar inductors, such LP material can be used to replace planar E or plate (PLT) ferrite core. In the following, such planar inductors composed of high permeability (HP) and LP magnetic materials are referred as HLP planar inductors.

The analytical modeling of copper losses in planar inductors is quite complex because it requires

accounting the effect of the field expansion in the winding window, which limits the use of 1D approaches like conventionally adopted Dowell model [3]. Several researches later established 2D analytical models for the copper loss calculation in different planar inductor cases. For example, in [4], Gao et al. developed an approximate 2D model for lumped airgap inductors, which is able to capture the fringing effect for a wide range of core geometries and winding configurations. In [5], Ahmed et al. proposed a 2D analytical model for multilayered air-cored inductors with a single-turn per layer. It is based on an empirical equation, through extensive finite element (FEA) simulations, to determine H-field boundary values for modeling the foil edge effect. In this paper, a 2D copper loss model is specifically developed for HLP planar inductors with multi-layers and multi-turns per layer. This model, hereinafter called Simplified 2D model (S2DM), enables to deduce the correct H-field values created by the homogeneous distribution of the magnetic energy stored in the magnetic core.

The paper is organized as follows. In section 2 the issue of planar inductors using HLP magnetic materials is highlighted. Then, in section 3, the 2D copper loss model is developed based on the H-field distribution in the winding window. Results are compared, in section 4, to 1D approach and FEA simulations.

2 HLP Planar Inductors

In order to highlight the value of HLP planar inductors, a test case is studied, based on an E/PLT combination of standard planar E38/8/25 and PLT38/25/3.8 cores, with mixed materials (i.e., LP or HP magnetic materials). The ferrite magnetic material used in this test case is 3F3, with an initial relative permeability of $\mu_r(HP) = 2000$. The LP's relative permeability, $\mu_r(LP)$, is between 10 and 125. Printed Circuit Board (PCB) winding is made of 8 layers for a total thickness of 4 mm. The dimensions of the copper turns (width w = 9 mm, height h = 400 µm) are based on a stack-up achievable by PCB manufacturers.

FEA simulations are performed with FEMM to compare 3 HLP planar inductor configurations (Table 1) to a reference one with airgap, hereinafter called HPHP.

Configuration		Planar E	PLT
HPLP		HP	LP
LPHP		LP	HP
LPLP		LP	LP

Table 1 Planar inductor configurations.

The FEA simulations are carried out in 2D to estimate the values of the total magnetic energy W and the copper losses $Loss_{Cu}$. The current I in the inductor is assumed to be sinusoidal without DC Bias. The inductance L and the AC resistance R_{AC} are extracted from W and $Loss_{Cu}$, respectively, with Eq. (1) and Eq. (2). The core depth d is taken into account to model the third dimension of the inductor.

$$L = \frac{2 \times W}{I^2} \qquad (1)$$

$$R_{AC} = \frac{Loss_{Cu}}{I^2} = F_R \times R_{DC} \qquad (2)$$

The variation of the resistance factor F_R (2) as a function of frequency is shown, for the 3 configurations plus the reference case in Fig. 1. For each configuration, F_R is plotted for 3 different inductance values, 18 µH, 29 µH and 43 µH, obtained by setting the LP block to specific permeability values.

Figure 1 enables to highlight the impact of replacing airgap in planar inductor with HLP configurations. The airgap fringing field (HPHP config.) induces a higher resistance factor, and thus more copper losses. Other HLP planar inductor's configurations offer lower copper losses. Depending on the operating frequency, a better configuration can be determined. The various F_R variation profiles, can be explained with the H_x and H_y distribution in the winding window, as it will be shown in the next section.

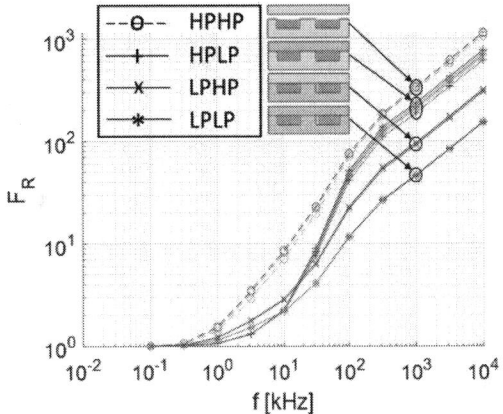

Fig. 1 Resistance factor as a function of frequency.

In Fig. 2, the R_{AC} is plotted as a function of the inductance at 250 kHz. One can note that the R_{AC} remains practically constant on the inductance value range. Furthermore, the reduction of R_{AC} compared to the airgap inductor solution (HPHP) for a same inductance value, is:

- between 20% to 40% for the HPLP config.
- around 70% for the LPHP config.
- up to 85% for the LPLP config.

This clearly demonstrates the benefits of HLP planar inductors for copper loss reduction.

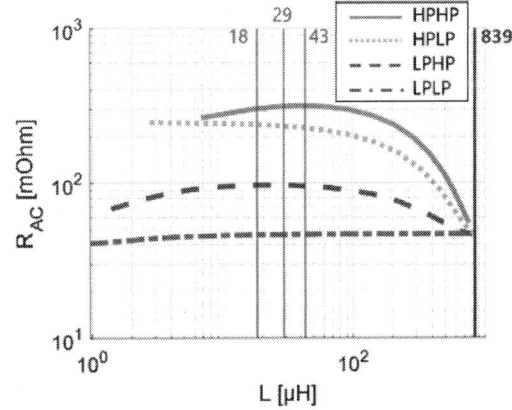

Fig. 2 AC resistance as a function of inductance at 250 kHz.

Nevertheless, this result must be mitigated. Indeed, the benefit of copper loss reduction should not be compensated by a significant increase in core losses due to the use of LP materials. It is known that such LP materials can present higher core losses than ferrite magnetic material. However, with a DC bias, LP core losses are almost invariant [6], while for ferrites, the latter tend to significantly increase. Then, a global evaluation must be stringently done.

The magnetic saturation is also an important factor to consider. Indeed, LP materials such as Kool Mμ® and XFlux® saturate at higher values, 1 T and 1.6 T respectively [7], while ferrites saturate at roughly 0.4 T. HP ferrite is then the most constraining element in a HLP configuration.

In order to size efficient HLP inductors, fast and accurate models are needed. In this paper, an analytical copper loss model is developed to estimate HLP planar inductor winding loss in a design stage.

3 HLP Copper Loss Modeling

In this section, a Simplified 2D analytical Model (S2DM) for HLP planar inductor copper loss estimation is developed. S2DM considers skin and proximity effects in the windings.

3.1 HLP inductor's assumptions: Window Corner Effect

S2DM is based on a major assumption for the H-field. The Fig. 3 presents the H_x distribution in the window for the LPLP configuration. As it can be seen window corner effects are important in the window corners. It is the same for the H_y component. Such effect is difficult to correctly model. In S2DM, this effect is neglected in order to simplify the estimation method of H-field values H_x and H_y.

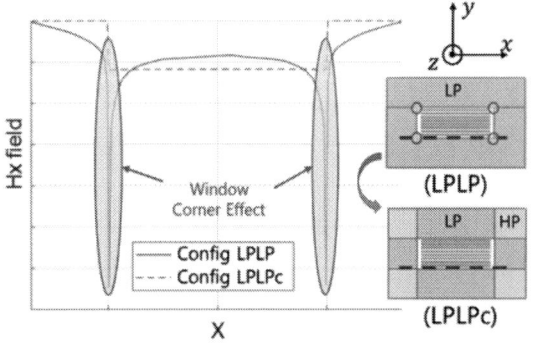

Fig. 3 Window corner effect on the H_x distribution in magnetostatics, illustrated for LPLP config.

Neglecting the window corner effect means that the magnetic energy stored in the corners is assumed to be zero, i.e., their core relative permeability is high (HP).

For each HLP planar inductor configuration, a corner-modified simplification can be considered as illustrated in Fig.4. These modified configurations are denoted HPLPc, LPHPc and LPLPc. The impact of such modification is shown in Fig.3, comparing LPLP and LPLPc.

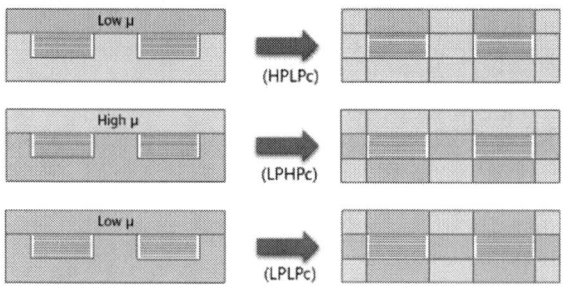

Fig. 4 The 3 configurations and their corner-modified simplification.

3.2 Skin and Proximity Effects

Inductor winding is considered as multi-layers with multi-turns per layer. While using the corner-modified configurations, it can be assumed that, inside the window, x-axis and y-axis components of the H-field only depend on y and x, respectively. Thus, only Hx(y) and Hy(x) must be calculated based on the Helmholtz equation applied to a single conductor:

$$\nabla^2 H = \gamma^2 H \tag{3}$$

with $\gamma = (1+j)/\delta$ the complex propagation constant and δ the skin depth

The general form solution of Eq. (3) is given by Eq. (4), where H_{x1}, H_{x2}, H_{y1} and H_{y2} are constant parameters related to H-field boundary condition (BC) of each conductor.

$$\begin{cases} H_x(y) = H_{x1}e^{\gamma y} + H_{x2}e^{-\gamma y} \\ H_y(x) = H_{y1}e^{\gamma x} + H_{y2}e^{-\gamma x} \end{cases} \tag{4}$$

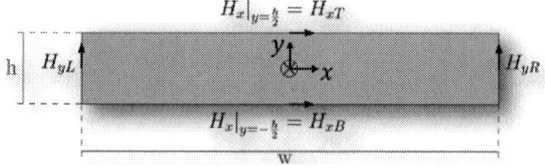

Fig. 5 H-field boundary condition of a conductor.

Considering the four H-field BC of each conductor H_{xT}, H_{xB}, H_{yR} and H_{yL} (Fig. 5), the H-field distribution in the conductor can be written as in Eq. (5):

$$\begin{cases} H_x(y) = \underbrace{\dfrac{H_{xB} + H_{xT}}{2}\dfrac{\cosh(\gamma y)}{\cosh\left(\gamma\frac{h}{2}\right)}}_{H_{Xprox}} - \underbrace{\dfrac{H_{xB} - H_{xT}}{2}\dfrac{\sinh(\gamma y)}{\sinh\left(\gamma\frac{h}{2}\right)}}_{H_{Xskin}} \\[4mm] H_y(x) = \underbrace{\dfrac{H_{yR} + H_{yL}}{2}\dfrac{\cosh(\gamma x)}{\cosh\left(\gamma\frac{w}{2}\right)}}_{H_{Yprox}} + \underbrace{\dfrac{H_{yR} - H_{yL}}{2}\dfrac{\sinh(\gamma x)}{\sinh\left(\gamma\frac{w}{2}\right)}}_{H_{Yskin}} \end{cases} \quad (5)$$

Skin fields are defined in the positive direction of the current along the z-axis while proximity fields are defined in the positive direction of its axis: x-axis for H_{Xprox} and y-axis for H_{Yprox}.

The current density (J) is derived from Maxwell-Ampere's Law with displacement current being ignored:

$$J_z(x,y) = \frac{\partial H_y(x)}{\partial x} - \frac{\partial H_x(y)}{\partial y} \quad (6)$$

Then, the copper losses are calculated from $J(x,y)$ and its conjugate $\bar{J}(x,y)$ using Eq. (7):

$$P_{loss} = \frac{d}{2\sigma}\int_{-\frac{w}{2}}^{\frac{w}{2}}\int_{-\frac{h}{2}}^{\frac{h}{2}}|J_z(x,y)|^2\, dy\, dx$$
$$= \frac{d}{2\sigma}\int_{-\frac{w}{2}}^{\frac{w}{2}}\int_{-\frac{h}{2}}^{\frac{h}{2}} J_z \times \bar{J}_z \, dy\, dx \quad (7)$$

Where σ is the conductor's electrical conductivity. By combining Eq. (5), Eq. (6) and Eq. (7), the copper loss is reformulated in Eq. (8). The obtained equation is divided into five terms: Each H-field component (H_x and H_y) results in two copper loss terms (skin and proximity), plus a fifth common term P_{XY} due to the non-linearity between H_x and H_y.

$$P_{loss} = P_{Xskin} + P_{Xprox} + P_{Yskin} + P_{Yprox} + P_{XY} \quad (8)$$

where each copper loss term is defined as:

$$\begin{cases} P_{Xskin} = R_{DC} \times I_{RMS_{Xskin}}^2 \times F_{RXskin} \\ P_{Xprox} = R_{DC} \times I_{RMS_{Xprox}}^2 \times F_{RXprox} \\ P_{Yskin} = R_{DC} \times I_{RMS_{Yskin}}^2 \times F_{RYskin} \\ P_{Yprox} = R_{DC} \times I_{RMS_{Yprox}}^2 \times F_{RYprox} \\ P_{XY} = 2R_{DC} \times I_{RMS_{Xskin}} \times I_{RMS_{Yskin}} \end{cases} \quad (9)$$

With $R_{DC} = d/(\sigma h w)$ and RMS current expressions and F_R equations given in Eq. (10) and Eq. (11), respectively.

$$\begin{cases} I_{RMS_{Xskin}} = \dfrac{1}{\sqrt{2}}w(H_{xB} - H_{xT}) = \sqrt{2}\, w \times H_{Xskin} \\[3mm] I_{RMS_{Xprox}} = \dfrac{1}{\sqrt{2}}w(H_{xB} + H_{xT}) = \sqrt{2}\, w \times H_{Xprox} \\[3mm] I_{RMS_{Yskin}} = \dfrac{1}{\sqrt{2}}h(H_{yR} - H_{yL}) = \sqrt{2}\, h \times H_{Yskin} \\[3mm] I_{RMS_{Yprox}} = \dfrac{1}{\sqrt{2}}h(H_{yR} + H_{yL}) = \sqrt{2}\, h \times H_{Yprox} \end{cases} \quad (10)$$

$$\begin{cases} F_{RXskin} = \dfrac{h}{2\delta}\dfrac{\sinh\left(\frac{h}{\delta}\right) + \sin\left(\frac{h}{\delta}\right)}{\cosh\left(\frac{h}{\delta}\right) - \cos\left(\frac{h}{\delta}\right)} \\[4mm] F_{RXprox} = \dfrac{h}{2\delta}\dfrac{\sinh\left(\frac{h}{\delta}\right) - \sin\left(\frac{h}{\delta}\right)}{\cosh\left(\frac{h}{\delta}\right) + \cos\left(\frac{h}{\delta}\right)} \\[4mm] F_{RYskin} = \dfrac{w}{2\delta}\dfrac{\sinh\left(\frac{w}{\delta}\right) + \sin\left(\frac{w}{\delta}\right)}{\cosh\left(\frac{w}{\delta}\right) - \cos\left(\frac{w}{\delta}\right)} \\[4mm] F_{RYprox} = \dfrac{w}{2\delta}\dfrac{\sinh\left(\frac{w}{\delta}\right) - \sin\left(\frac{w}{\delta}\right)}{\cosh\left(\frac{w}{\delta}\right) + \cos\left(\frac{w}{\delta}\right)} \end{cases} \quad (11)$$

Based on these equations, S2DM can enable to calculate each term of copper losses in an HLP planar inductor. However, before using S2DM formulations, the values of the H-field BC must be calculated for each conductor.

3.3 Boundary Condition

The distribution of the H-field in the window depends on the location of the LP material in the planar core. The Magnetomotive force (MMF) module distribution NI_X and NI_Y along the x and y axis is illustrated in Fig. 6 for HPLP LPHP and LPLP configurations. For all those configurations, the S2DM automatically considers their corner-modified simplification (Fig. 4). N_L layers and N_T number of turns per layer are also regarded in the winding.

For the HPLP configuration (Fig. 6a). MMF is concentrated only in the top edge of the window. The H-field distribution in the window is practically 1D, with only a H_x component. In this case, the S2DM equations become similar to a 1D model like the Dowell one [3]. H_x-field variation can be deduced from Ampère's circital law as shown with the red dashed lines in Fig. 6a.

The MMF distribution for LPHP config, in Fig. 6b, is different. The LP of the E core results in the homogeneity of the H-field intensity in the bottom, left and right edges of the window. In this case, the reluctance of the HP plate is negligible due to the high value of its relative permeability. Thus, the H-field in the top edge of the window is zero. H_x-field and H_y-field variation can be deduced from Ampère's circital law with the superposition theorem. Red dashed-lines illustrate how to calculate H_x-field while supposing H_y is zero. The blue dashed-lines shown the same approach for H_y-field while assuming H_x is zero.

Finally, in the last configuration LPLP (Fig. 6c), all the window edges have the same H-field absolute value. In this case, the H-field is null at the window center. The H-field distribution is deduced in the same way as the previous configuration (LPHP).

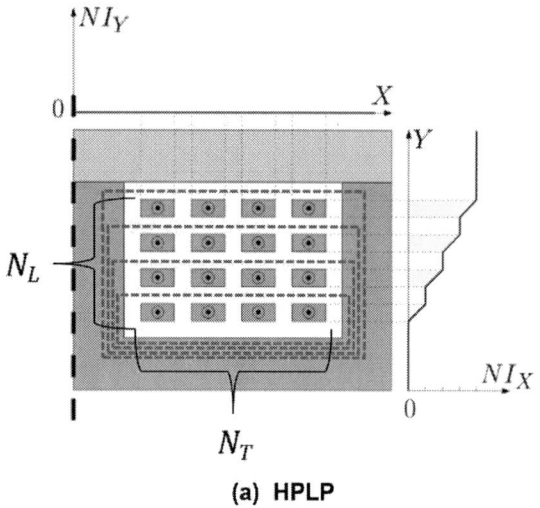

(a) HPLP

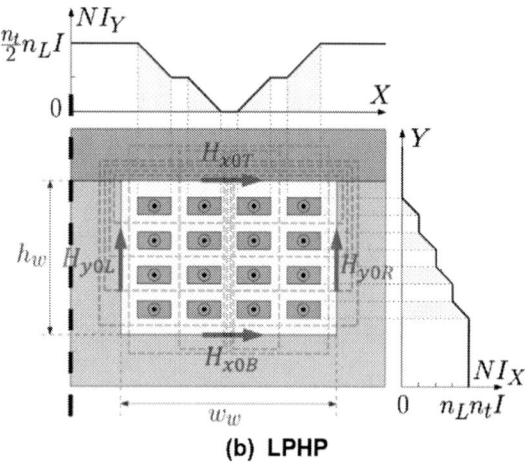

(b) LPHP

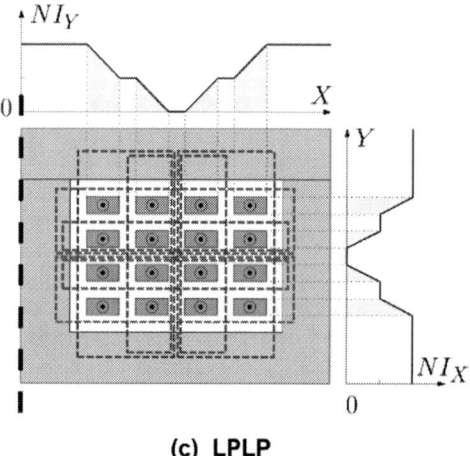

(c) LPLP

Fig. 6 MMF module distribution in x and y direction in S2DM: (a) HPLP, (b) LPHP, (c) LPLP.

The value of the H-field on the LP window edges obtained from Maxwell-Ampere, with I_{AC} being the current amplitude, is:

$$H_0(LP) = \frac{N_L N_T I_{AC}}{l_H} \tag{12}$$

Where l_H is the circulation length depending on the configurations. l_H values are reported in Table 2.

Configuration	l_H
HPLP	w_w
LPHP	$w_w + 2h_w$
LPLP	$2w_w + 2h_w$

Table 2 Circulation length for the field; the dimensions are those of the window.

The H_x-field values between the layers of conductors can be calculated using the Eq. (13). They are parameterized by the variable n_L which varies from 0 at the top of the window to N_L at the bottom.

$$H_x(n_L) = \frac{n_L}{N_L}(H_{x0B} - H_{x0T}) + H_{x0T} \tag{13}$$

Where H_{x0T} and H_{x0B} are the H-field values in the top and bottom window edge, respectively. They can be either 0 if the edge has HP, or calculated from Eq. (12) if the edge has LP.

Similarly, H_y-field values between the columns of conductors can be calculated using the Eq. (14). They are parameterized by the variable n_T which vary from 0 at left of the window to N_T on the right.

$$H_y(n_T) = \frac{n_T}{N_T}(H_{y0R} - H_{y0L}) + H_{y0L} \tag{14}$$

Where H_{y0R} and H_{y0L} are the H-field values in the right and left window edge respectively. They similarly can be either 0 if the edge has HP or calculated from Eq. (12) if the edge has LP.

3.4 Validation on a window without porosity

The H-field BC can be calculated based on the previous assumptions and equations. However, in a window with a low porosity factor η (Eq. 15), those H-field values are no longer correct as shown in the Fig. 7 for different η values. This occur because the values of H_x and H_y are no longer independent and start impacting each other. Indeed, in Fig. 7, the magnetostatic H-field inside the window varies linearly for $\eta = 1$. However, as porosity factor decreases, i.e., when the conductor's weight w decreases, the H_y-field

BC of each conductor significantly changes due to the presence of the H_x-field

$$\eta = \frac{N_T w}{w_w} \qquad (15)$$

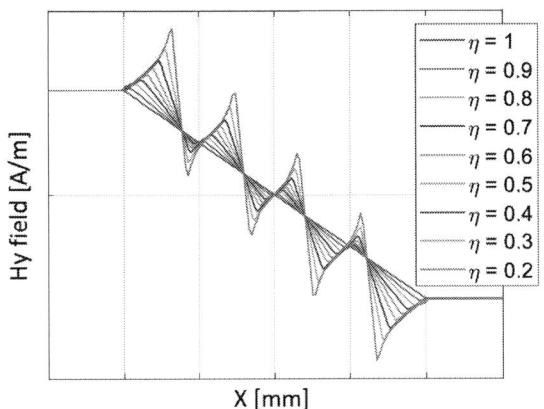

Fig. 7 Example of H_y-field in a window for different porosity factor values for N_T=4.

In order to validate the S2DM approach, a window without porosity is first considered. Figure 8 presents a H-field comparison between S2DM and FEMM numerical results, for the LPHPc configuration with porosity factor ($\eta = 1$). Both curves, for H_x and H_y are in good agreement. The analytical modeling provides an accurate field distribution and thus a good copper loss estimation with an error under 1%.

4 Application on HLP inductors

In this section, S2DM is improved for accounting the window's porosity factor. Then, S2DM is validated on HLP planar inductor configurations.

4.1 Window's porosity factor: an issue

Figure 9 presents a window with a low porosity factor. Ampère's circuital law, applied on $\mathcal{C}_1$ and $\mathcal{C}_2$, Eq. (16), show two contradictions, in Eq. (17), for the H-field BC.

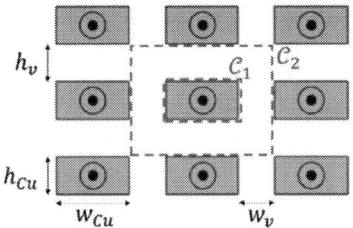

Fig. 9 Ampère's Law in a window with a low porosity factor.

$$\begin{cases} \begin{cases} -H_{xT}w_{Cu} + H_{xB}w_{Cu} = I_X \\ -H_{xT}(w_{Cu} + 2w_v) + H_{xB}(w_{Cu} + 2w_v) = I_X \end{cases} \\ \begin{cases} -H_{yL}h_{Cu} + H_{yR}h_{Cu} = I_Y \\ -H_{yL}(h_{Cu} + 2h_v) + H_{yR}(h_{Cu} + 2h_v) = I_Y \end{cases} \end{cases} \qquad (16)$$

$$\begin{cases} w_v = 0 \\ h_v = 0 \end{cases} \qquad (17)$$

This BC method is then valid only when there is no porosity in the winding window. Therefore, a transformation to consider the window's porosity is needed.

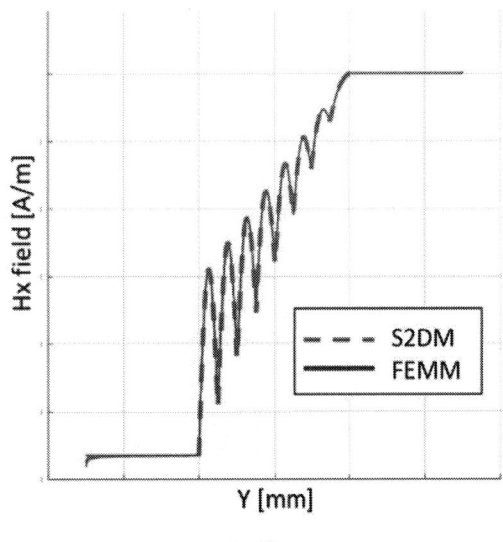

(a)

(b)

Fig. 8 Comparison of S2DM and FEA simulation (FEMM) results for the LPHPc configuration at 250 kHz: (a) H_y-field, (b) H_x-field.

4.2 Transformation Method

In S2DM, it is assumed that the H-field in the inter-conductor region is uniform, when moving in the y direction for H_x and in the x one for H_y. Then, a transformation method can be applied, like in Dowell method [3],[8], to introduce a double porosity factor transformation Eq. (18), one for each direction.

$$\begin{cases} \eta_X = \dfrac{N_T w}{w_w} \\ \eta_Y = \dfrac{N_L h}{h_w} \end{cases} \qquad (18)$$

The objective is to transform winding into a set of non-porous conductors of equivalent conductivity $\sigma\eta_X$ for the H_x-field and $\sigma\eta_Y$ for the H_y-field. The transformation process is illustrated in Fig. 10. The obtained porosity factor is to be included in S2DM as reported in Table 3.

$\sigma\eta_X$	σ	$\sigma\eta_Y$
w/η_X	w	w
h	h	h/η_Y

Table 3 Transformation coefficient

4.3 Validation

In this section, S2DM is compared to FEMM simulations results and Dowell 1D Model (D1DM). The resistance factor *Fr* is plotted for the 3 HLP configurations with an 8-layers PCB of 75µm copper thickness.

4.3.1 HPLP Configuration

Figure 11 presents the resistance factors for the HPLP configuration. The results from S2DM and D1DM are identical, due to the unidirectionality of the H-field in the HPLP configuration. It is possible to verified that the HPLPc configuration leads to a perfect condition for an 1D H-field. However, for an HPLP configuration, the distribution of the H-field in the window in not exactly unidimensional due to the window's corner effect. Thus, the impact of the H_y-field observed in mid frequencies (MF –

300Khz) is underestimated, while the impact of the H_x-field observed in high-frequency (HF – 10MHz) is overestimated compared to FEMM results.

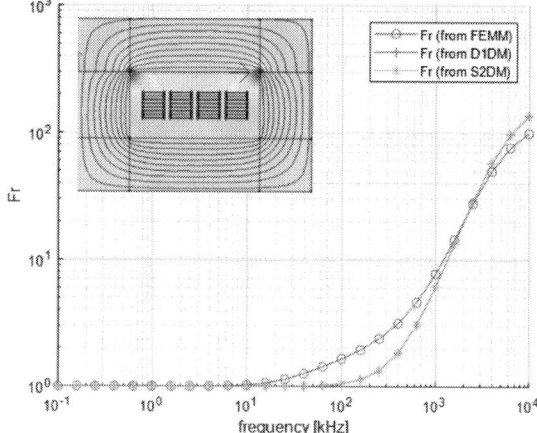

Fig. 11 Comparison of HPLP resistance factor obtained with FEMM, D1DM and S2DM.

4.3.2 LPHP Configuration

For the LPHP configuration (Fig. 12), S2DM gives a better estimation than D1DM. As it can be seen in Fig. 12, S2DM correctly models both impact of the H_y-field observed in MF and the impact of the H_x-field observed in HF. The S2DM is particularly efficient for this configuration.

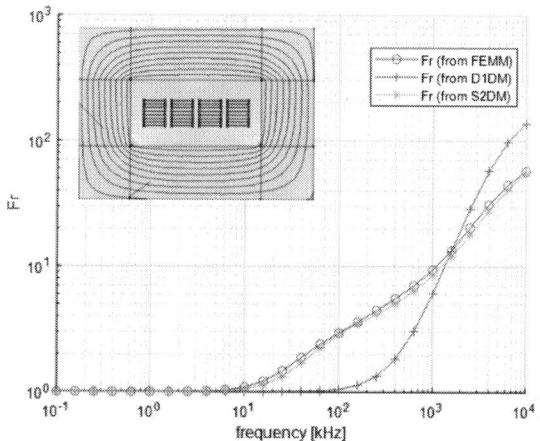

Fig. 12 Comparison of LPHP resistance factor obtained with FEMM, D1DM and S2DM.

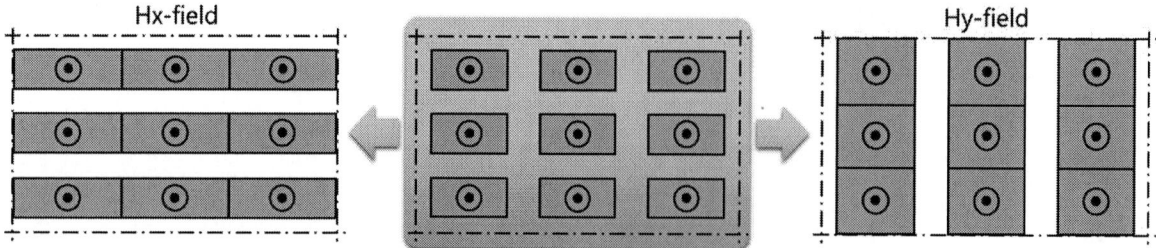

Fig. 10 Porosity transformation for S2DM application

4.3.3 LPLP Configuration

For the LPLP configuration (Fig. 13), S2DM gives an estimation slightly underestimated with an error below 45%. This is due to the corner effect assumption that modifies the H-field distribution in the window and so limits the precision. However, S2DM remains largely better than D1DM model, particularly in MF.

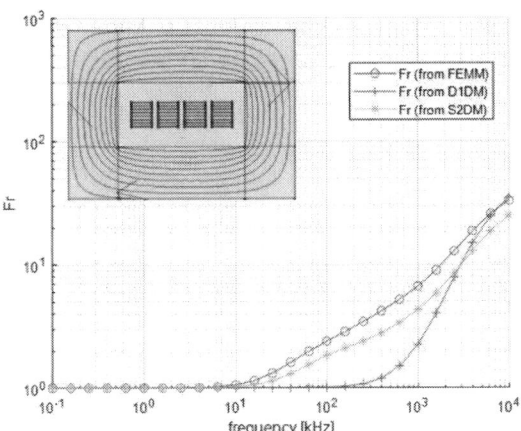

Fig. 13 Comparison of LPLP resistance factor obtained with FEMM, D1DM and S2DM.

4.3.4 Discussion

Results presented in the previous section enable to validate the S2DM approach. The 2D developed model considers both H_x-field and H_y-field as well as their corresponding porosity factor η_X and η_Y. However, the window corner effects are neglected to assume a constant H-field BC in each conductor edge. This assumption limits its accuracy and its application.

As it can be seen, S2DM gives a better estimation than D1DM, for LPHP and LPLP configurations. Indeed, those configurations generate a 2D H-field in the winding window which can be divided into two independent H-field components, H_x and H_y. For the HPLP configuration, results from S2DM and D1DM are the same, since the H-field in the winding window is supposed to have only an x-axis component, making the problem 1D.

5 Conclusion

HLP planar inductors are a promising solution to reduce the copper losses of conventional planar inductors. In order to offer a fast and accurate model suitable for optimization, a simplified 2D analytical model to compute rapidly the HLP planar inductor's copper losses is developed in the paper.

The 2D model, named S2DM for Simplified 2D model, is based on the simplifying hypothesis that neglect the window corner effect. The model is fully analytical and it considers two porosity factors η_X and η_Y to precisely estimate skin and proximity effect contribution to the copper losses.

S2DM has been validated for the 3 configurations of HLP planar inductors: HPLP, LPHP and LPLP. The results are on good agreement with FEA simulations results, especially in the case of LPHP and LPLP configurations, where S2DM offers a better estimation than a conventional 1D Dowell model.

References

[1] T. Nomura, C.-M. Wang, K. Seto, and S. W. Yoon, "Planar inductor with quasi-distributed gap core and busbar based planar windings," in *2013 IEEE Energy Conversion Congress and Exposition*, Sep. 2013, pp. 3706–3710.

[2] P. Ren, W. Chen, X. Huang, Y. Chen, Y. Wang, and X. Yang, "AC Copper Loss Reduction in Planar Inductors With Magnetic Building Blocks-Based Gapless Parallel Symmetrical Magnetoresistance Structure," *IEEE Journal of Emerging and Selected Topics in Power Electronics*, vol. 11, no. 4, pp. 4295–4312, Aug. 2023.

[3] P. L. Dowell, "Effects of eddy currents in transformer windings," *Proceedings of the Institution of Electrical Engineers*, vol. 113, no. 8, pp. 1387–1394, Aug. 1966.

[4] Y. Gao *et al.*, "Modeling and Design of High-Power, High-Current-Ripple Planar Inductors," *IEEE Transactions on Power Electronics*, vol. 37, no. 5, pp. 5816–5832, May 2022.

[5] D. Ahmed, L. Wang, M. Wu, L. Mao, and X. Wang, "Two-Dimensional Winding Loss Analytical Model for High-Frequency Multilayer Air-Core Planar Inductor," *IEEE Transactions on Industrial Electronics*, vol. 69, no. 7, pp. 6794–6804, Jul. 2022.

[6] J. Muhlethaler, J. Biela, J. W. Kolar, and A. Ecklebe, "Core Losses Under the DC Bias Condition Based on Steinmetz Parameters," *IEEE Transactions on Power Electronics*, vol. 27, no. 2, pp. 953–963, Feb. 2012.

[7] Magnetics, "Powder Core Catalog," *https://www.mag-inc.com*, 2020.

[8] F. Robert, "A theoretical discussion about the layer copper factor used in winding losses calculation," *IEEE Transactions on Magnetics*, vol. 38, no. 5, pp. 3177–3179, Sep. 2002.

PCIM Europe 2024, 11– 13 June 2024, Nuremberg   DOI: 10.30420/566262339

CNC-Manufactured Power Inductors with High Bandwidth for Multi-Megawatt Converters

Thomas Kreppel[1], Thomas Brückner[1], Rainer Marquardt[1], Rene Weick[2]

[1] Universität der Bundeswehr München, Germany
[2] innovatek OS GmbH, Germany

Corresponding author / Speaker: Thomas Kreppel, thomas.kreppel@unibw.de

Abstract

Inductors are essential components of power converters - used as energy storage devices or filter elements. The progress of power semiconductors and converters (SiC, MMC) towards higher switching speeds and power levels enforces new inductor designs in order to meet the requirements for high bandwidth, low loss and power density. In this paper, a new planar inductor design with shielding plates is presented and evaluated, which enables excellent high-frequency behaviour and compact size. Moreover a simple CNC-based manufacturing process can be applied for industrial production. The new design is analyzed through FEM simulations and a medium power range demonstrator. Finally, the concept is applied to an arm inductor for a containerized 40-MVA Modular Multilevel Converter (MMC).

1   Introduction

Inductors are core components of power electronic converters. Their purpose can be broadly divided into three categories: intermediate energy storage, limitation of di/dt and electromagnetic interference (EMI) filtering. Depending on the category and many requirements of the specific application, inductors have to be tailored individually, quite often. The focus of the following paper is mostly directed on arm inductors for high-power MMCs in the multimegawatt range, but this concept can be applied analogously to high-power inductors for other converter topologies as well. A suitable concept for this application area must permit direct conductor cooling and magnetic shielding of the stray fields.

Recent progress of MMC and their control systems [1]–[3] enables essential reduction of the required inductance values, while the requirements in the high frequency (HF) range have become more demanding. Conventional air-core inductors, which are often applied up to now, can fulfill these requirements at the expense of power density. Conventional iron-core inductors are suffering from high parasitic capacitances in general, limiting the achievable bandwidth. This leads to large current spikes and disturbing resonance currents, enforcing additional passive filters and damping elements.

The first (parasitic) parallel resonance frequency is a useful parameter for comparison of the high frequency performance of power inductors. For the main focus of applications, as discussed here, values in the MHz range should be targeted, in order to eliminate the need for additional HF and EMI filters.

2   Planar Design Concept for High-Power Inductors

Planar inductors have found extensive application in recent years as transfer elements in wireless power transfer (WPT) [4]–[8]. In WPT systems two planar inductor layers forming a transformer, are used. To concentrate the magnetic flux between the two inductors to maximize the transferred power, plates of magnetic material is usually positioned at either side of the transformer [9]. For air-core inductors the planar design offers excellent magnetic coupling between the turns within one layer as well as between two facing inductor planes [10].

The proposed inductor concept is built from multiple planar winding layers similar to transformers in WPT systems. This is done to facilitate the accessability of the external terminals of the inductor. Especially for medium- or high- voltage devices isolation distances pose a major challenge for the placement of the inductor terminals. All layers are identical, in the full inductor the alternating layers are rotated by 180° so that the direction of the magnetic flux stays the same for currents traveling from the outside to the center in one layer, or from the center to the outside respectively.

This concept offers an attractive design space to shift the electrical performance of the inductor between several mutual exclusive optimization goals through the variation of various design parameters. In particular, this approach enables the adjustment of the parasitic capacitive coupling in order to fine-tune the high-frequency behaviour of the inductor.

2.1 Inductor Outline

The standard inductor consists of two winding layers. The outline of the winding layers is freely selectable - it can be approximately circular, oval or rectangular, amongst others (see Fig. 1). The conductor comprises multiple turns arranged in a spiral, with a cross-section that is small against the area of the winding layer. Circular outlines have the best ratio of inductor area to conductor surfaces facing each other and therefore show a slight advantage for HF properties. However, this effect is very minor and the outline should rather be selected to optimally utilize the available space, as the concept permits arbitrary inductor outlines. This allows the adaption to unusual shapes to fit into uncommon installation volumes. When having a rectangular space of $l_{\text{total}} \times w_{\text{total}}$ as displayed in Fig. 1, a rectangular inductor outline is preferred to optimally utilize the space compared to a circular one.

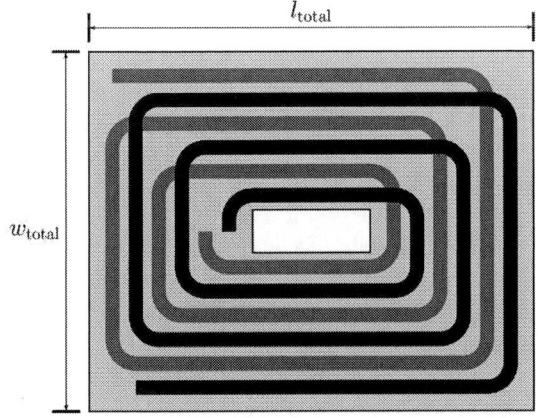

Fig. 1: Inductor and shielding plate shape, as seen from above. Different outlines for the windings are possible.

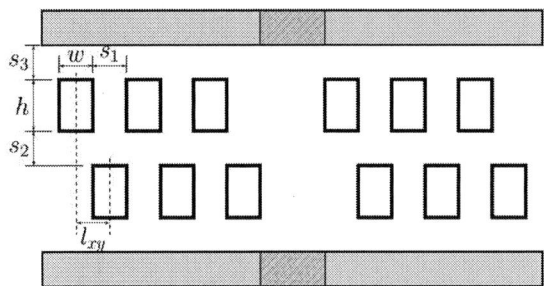

Fig. 2: Cross section of inductor, conductor geometry and arrangement with shielding plates at top and bottom. The available design parameters for performance optimization are highlighted.

2.2 Shielding of magnetic stray fields

To provide shielding of the stray field at either face of the inductor it is possible to arrange flat ferromagnetic disks on one or both sides of the conductor (see Fig. 2). This enables the installation in a space-saving manner in narrow installation spaces without interfering with neighboring components. The maximum induction in these disks is low, so that ferrite materials or powder composites with moderate permeability can be used. Due to the low conductivity, these materials are effective at suppressing eddy currents and therefore have very low losses at high frequencies - compared to iron cores. It is advantageous to allow these disks to protrude slightly beyond the outer edge of the conductor arrangement, the center section of the disk can be cut out to reduce weight, as the significant portion of flux is concentrated directly near the windings.

The shielding plates can be made from a cost-effective material based on recycled waste of ferrite production. This material is called "magnetic concrete" (Magment MC120) [11], and consists of ferrite fragments and powder mixed with cement as binding material to form concrete elements with magnetic properties. It enables reproducibility and efficient production of large quantities due to the mold-based production process. The outline is freely selectable and can be easily adapted in all three dimensions optimize the stray-field containment. As concrete possesses excellent compressive strength, these components could also be integrated as load-bearing elements of the converter structure.

2.3 Manufacturing concept

High-power inductors of this power range typically utilize hollow, directly liquid-cooled conductors. This requires proficiency in demanding manufacturing processes such as winding or bending of pipes. Special consideration has to be given to the risk of collapse that exists when bending hollow structures. Geometric restrictions due to these manufacturing processes, such as the minimal bending radius for the conductor, impose strict limits on the possible inductor shapes.

The proposed inductor concept enables an advantageous CNC-based process that can be fully automated. Larger quantities can therefore be produced with high precision and excellent reproducibility with tight tolerances. Arbitrary winding-layer outlines can be achieved with no increase in effort. In particular, the method also allows the production of cooling channels in the same production step. The entire contour and the cooling channels of one layer are milled in a single clamping, and a second clamping is then used only to remove the fastening structures that were left in place in the first pass. After machining, the cooling channels are closed by welding on a cover.

For the production of large quantities, efficiency can further be optimized by reducing waste and machining time by manufacturing the raw stock in the rough shape of the finished geometry using a low-pressure die-casting process.

2.4 Conductor arrangement

As the inductor shape is defined by the external volume constraints, the performance of the inductor can only be influenced by the conductor shape and the arrangement of the conductors within the winding layer, as well as spacing between the layers. The usable inductive bandwidth is limited by the parallel resonance between main inductance and the equivalent parallel stray capacitance [10], [12]–[15]. The influence of the geometry on the stray capacitance is therefore the focus of the following analysis. Figure 2 shows the cross section of the inductor, with available parameters for performance optimization highlighted.

The influence of variation of certain design parameters on the electrical properties and other design goals has been investigated through FEM simulations with Ansys.

Each winding is modeled separately to extract the inductance matrix $L_{i,j}$ and frequency corrected capacitance matrix $C_{i,j}^*$, containing the capacitive and inductive coupling between winding i and j for all windings. According to [15] the first parallel resonace can then be calculated as

$$\omega_0 \cong \frac{1}{\sqrt{\displaystyle\sum_{i=1}^{n}\sum_{l=1}^{n-1}\sum_{k=1}^{i}\sum_{m=i+1}^{n} \frac{L_{i,l} + L_{i,l+1}}{2} C_{k,m}^{\star}}}. \quad (1)$$

After selecting the number of turns and defining the outline, the remaining parameters (see Fig. 2) to influence the inductor design are the conductor shape (height h and width w), the YX offset between the layers l_{xy}, and the spacing between each winding s_1, between the layers s_2, as well as between layer and shielding plate s_3.

The optimization goals for an inductor design are typically its quality factor $Q = L/R$, the weight m, and the first resonance frequency f_0 limiting the bandwidth. Table 1 illustrates the effect of the variation of each geometrical parameter on the electrical characteristics, where the inductor losses are held constant.

Tab. 1: Effect of variation of geometrical parameters (see Fig. 2) on electrical properties

	L	$C_{\parallel}$	$C_=$	Q	m	f_0
$h \uparrow$	↓	↑	↓	↓	↓	↑
$w \uparrow$	↑	↓	↑	↑	↑	↓
$l_{xy} \uparrow$	↓	—	↓	↓	—	↑
$s_1 \uparrow$	↓	↓	↓	↓	↑	↑
$s_2 \uparrow$	↓	↓	↓	↓	—	↑
$s_3 \uparrow$	↓	↓	—	↓	—	↑

For a constant conductor cross-section, the height and width are reciprocal as the product of both is constant. The height determines the capacitance between the turns, and the width influences the capacitance between the layers.

The stray capacitance limits the bandwidth and is dominated by two main capacitive coupling mechanisms in this inductor concept. Firstly, the capacitive coupling within the winding layer ($C_{\parallel}$) depends mainly on two adjacent turns, and to a lesser extent on the coupling between all turns within the layer, due to the permittivity of shielding plate ($\epsilon_r >> 1$). For the reduction of stay capacitance within the layer, the conductor height can be reduced, or the spacing between the turns can be increased. The second effect is the capacitive coupling between

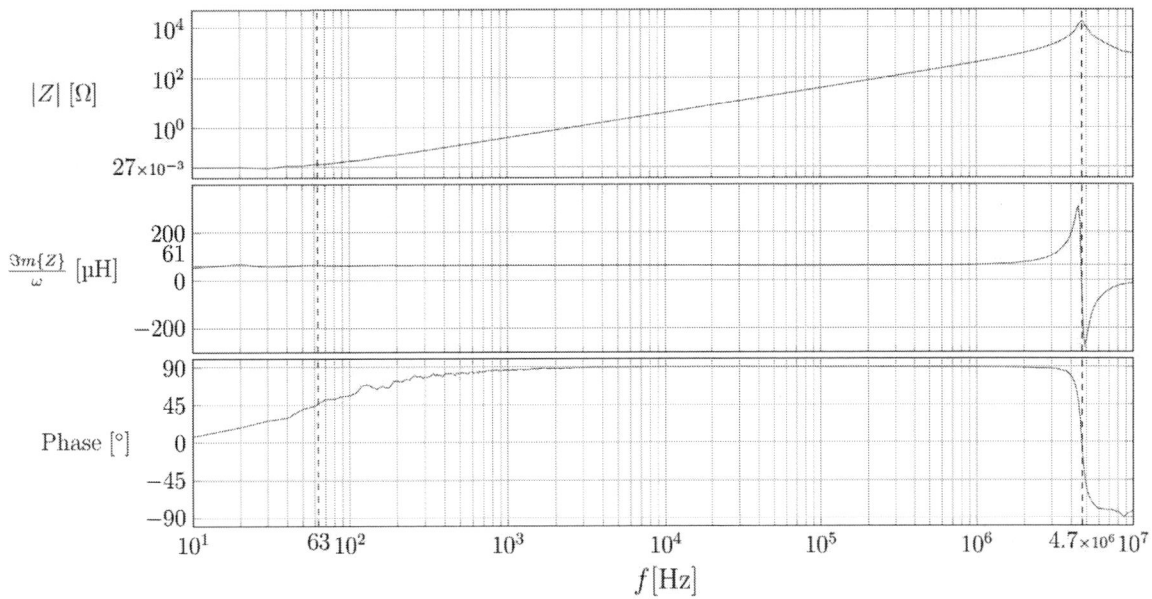

Fig. 3: Measurement of HF behaviour of prototype inductor - parasitic parallel resonance $f_0 \approx 5\,\text{MHz}$

two adjacent winding layers ($C_=$). To achieve a reduction of stray capacitance between the winding layers, the spacing can be increased and the windings can be shifted in the XY plane to further increase the distance between directly facing surfaces.

The weight of the inductor is mainly determined by the shielding plates. From the conductor parameters, only the conductor shape and the winding spacing has an influence on the total weight. Moreover, under the given constraints weight and bandwidth are opposing characteristics. The tradeoff between these two aspects has to be done according to the specific requirements for the application.

3 Medium-size demonstrator

To verify the simulation results, a demonstrator in the medium-power range (Multi-kW) with a continuous current rating of 250 A was manufactured (see Fig. 4). The mechanical dimensions are a length of 400 mm and a width of 250 mm. Table 2 displays a comparison between the expected parameters obtained through simulation and measured values. L_0 denotes the inductance of the windings without shielding plates, L_1 the inductance including the shielding plates.

Fig. 3 displays the frequency-dependent impedance of the inductor for the range of 10 Hz − 10 MHz. A nominal inductance of 61 μH and a first parallel resonance of almost 5 MHz

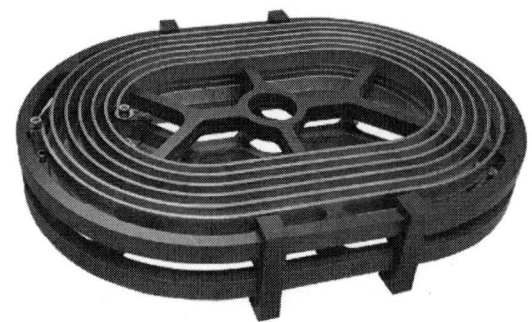

Fig. 4: Medium-size demonstrator, two winding layers without shielding plate, $w \times d \times h = 250 \times 400 \times 70$ mm

Tab. 2: Comparison of simulation and measurement of electrical parameters for demonstrator with a nominal current rating of 250 A

Parameter	Simulation	Measurement
L_0	38.66 μH	41.24 μH
L_1	60.11 μH	61.64 μH
R_{DC}	16.7 mΩ	17.8 mΩ
f_0	4.98 MHz	4.68 MHz

is achieved. The frequency response of the demonstrator exhibits constant inductive behaviour for a wide bandwidth up to well above 1 MHz, before showing capacitive behaviour above the resonance. At low frequencies the impedance converges against R_{DC}, which has been established

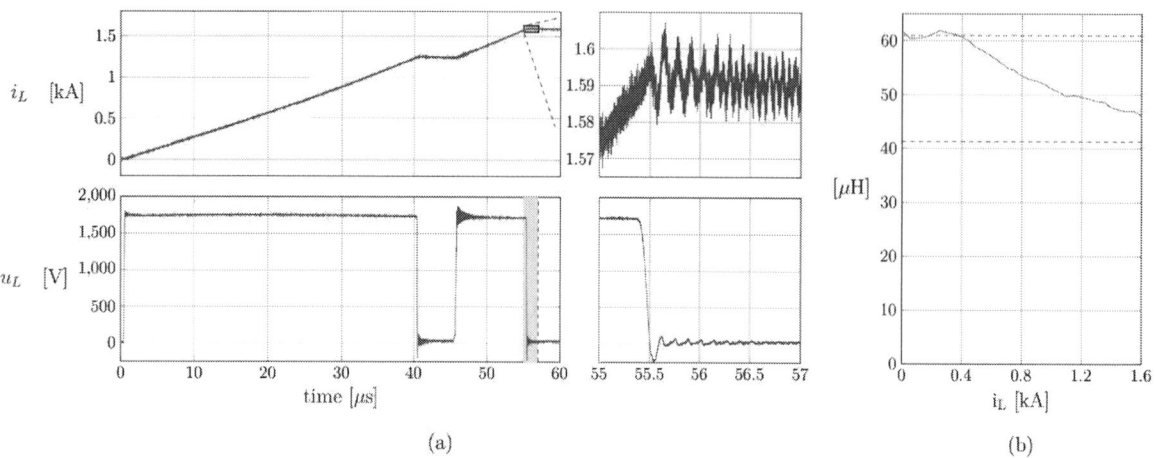

Fig. 5: Application as load inductor in double pulse test (a) and Inductance reduction at higher currents (b)

as 18 mΩ by measuring the voltage drop for DC currents up to 300 A. The phase crosses the 45° point at $\omega L = R_{DC}$ at around 63 Hz indicating a DC resistance of approximately 24 mΩ.

To explore the saturation of the shielding plate material, the inductor has been used as the load inductor in a double-pulse test for a 3.3 kV SiC-MOSFET, at a capacitor voltage of 1.8 kV and a peak turn-off current of 1.6 kA (see Fig. 5 (a)). The extremely high dv/dt of nearly 10 kV/µs shows nearly no capacitive current excitation, the visible oscillations exhibit a frequency of approximately 15 MHz which is well above the resonance of the inductor and originate from further parasitic elements of the measurement setup.

When determining the rate of current change of the current ramp and relating it to the instantaneous capacitor voltage value, the current dependency of the inductance value can be derived (as displayed in Fig. 5 (b)). At low currents of up to 400 A the shielding plates exhibit no saturation , whereas at high currents the inductance converges against 41 µH, the inductance value without the shielding plates.

4 Example application: requirements for MMC arm inductor

The proposed concept shall be applied to the design of an arm inductor for an MMC. Considerations for the electrical requirements are made in this section.

4.1 Selecting the inductance value L_e

To ensure proper and safe operation a certain inductance value is necessary. The permissible time

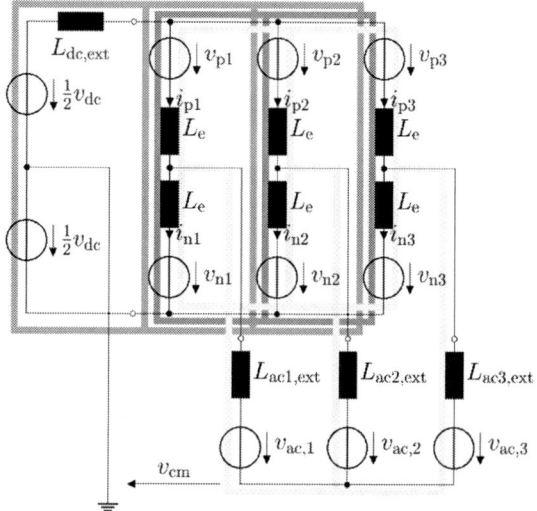

Fig. 6: Structure of the MMC with its current loops highlighted: DC (red), AC (yellow), circulating current (blue)

constants of the current dynamics are determined by the control bandwidth. By employing fast reacting control strategies with minimal dead times [2] a drastic reduction of the inductance can be achieved.

Figure 6 shows a simple representation for the internal structure of the MMC, as well as the immediate external components. The arm current of the MMC can be split into three independent parts, the external AC and DC currents, and the internal circulating currents (CC). The three AC currents flowing through the yellow loop are shared symmetrically between the upper and lower arms in each phase. The sum of all three AC currents in all three nodes is zero, resulting in an effective inductance of the external inductance plus two arm inductors in parallel,

$$L_{ac} = \frac{1}{2}L_e + L_{ac,ext}. \tag{2}$$

The DC current (red loop) flowing equally through all three phases, leading to an effective inductance of two arm inductors in series with three parallel paths,

$$L_{dc} = \frac{2}{3}L_e + L_{dc,ext}. \tag{3}$$

The internal circulating currents (blue loop) flow through one phase, returning through the other two phases in parallel. The effective inductance is therefore equal to three times the arm inductance,

$$L_{cc} = 3L_e. \tag{4}$$

From Eqs. (2) and (3) it can be seen, that the dynamics of DC and AC currents also depend on the external DC and AC inductance. Especially on the AC side, a grid connecting transformer is usually employed, therefore $L_{ac,ext} >> L_e$, so the internal arm inductance is not relevant to the AC and DC control performance.

A straightforward approach for selecting the arm inductance value L_e is used. When employing the MVC [2], all currents are guaranteed to be held within a defined tolerance band. The inductance value then only limits the maximum switching frequency during transients - limiting the di/dt and determining how fast the edge of the tolerance band can be reached. If through an internal error any of the arm voltages is differing from its intended reference value it will cause a deviation of $\Delta i_{cc,err}$ until the next sampling instance. For an assumed error voltage v_{err}, controller dead time Δt and the maximum circulating current deviation $\Delta i_{cc,err}$, it can be concluded, that the necessary inductance value L_e is

$$3L_e\frac{\Delta i_{cc,err}}{\Delta t} = v_{err} \implies L_e = \frac{v_{err} \cdot \Delta t}{3 \cdot \Delta i_{cc,err}}. \tag{5}$$

As displayed in Eqs. (2) to (4) the effective inductances are composed of multiple arm inductors. Therefore, in order to evaluate the saturation-dependent decrease of the effective inductances, the saturation effect of all individual arm inductors must be taken into account. Here, a characteristic property of the MMC is helpful: within a phase

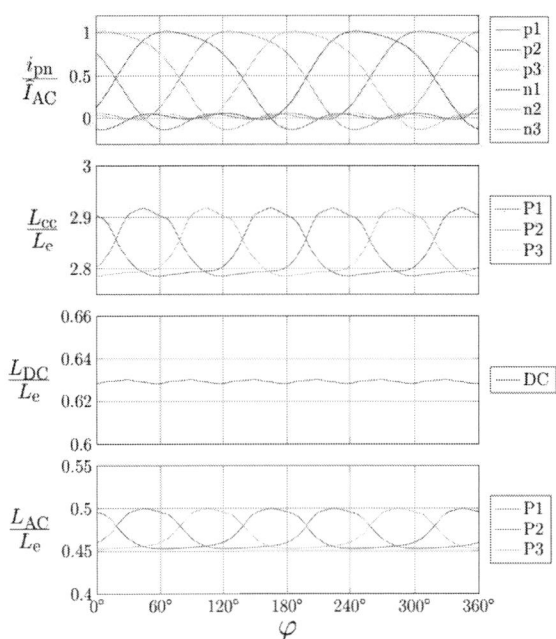

Fig. 7: Variation of the effective inductances due to saturation. External inductances are assumed to be zero. All inductances are normalized to the nominal arm inductance L_e.

only one arm at a time is conducting a substantial amount of current, the other arm of the same phase conducts almost no current. This effect also depends on the control algorithm used and can differ in the degree of intensity.

Figure 7 displays typical arm-current waveforms for an MMC with full-bridge submodules, an AC-DC voltage ratio of $k = 1.3$ and a power factor of $\cos\varphi = 0.966$. In this example, the individual arm inductance is assumed to decrease by 15% at peak current value. As can be seen in Fig. 7, the change in the effective inductances is significantly lower at 5%, 1.6%, and 9% for CC, DC, and AC, respectively.

4.2 HF-Reqirements

In contrast to many power converter topologies, which use medium-frequency transformers the spectral components of the arm current are essentially limited to f_{AC} and $2 \cdot f_{AC}$. An indcutor with wide bandwith is nevertheless required to limit capacitive currents triggered by switching transients, to prevent the excitation of oscillations withing the converter and avoid potential EMI problems.

Due to the recent spread of wide-bandgap power semiconductors, the switching speeds and therefore the steepness of the voltage edges has increased substantially. This entails increased requirements to the HF behaviour for power inductors, not just for MMC. If the use of additional HF filters is to be avoided, the parasitic equivalent-parallel capacitance of the inductor needs to be minimized to ensure a wide inductive operating range.

5 Inductor for containerized 40-MVA MVDC converter station

Tab. 3: Parameters of the 40-MVA arm inductor. Nominal RMS current is 1.1 kA

	Requirement	Simulation
L	100 µH	97.37 µH
R_{DC}	1.6 mΩ	1.46 mΩ
f_0	1 MHz	1.68 MHz
P_{loss}	2 kW	1.8 kW
m	350 kg	320 kg
w	800 mm	800 mm
d	1500 mm	1100 mm
h	300 mm	280 mm

The presented concept is finally applied to design an arm inductor for a 40-MVA, $20\,\text{kV}_{\text{DC}}$ MMC. To facilitate the installation of this converter in an urban scenario, the converter including the arm inductors shall be fully integrated into a 40-ft high-cube container. Table 3 lists the required electrical and mechanical parameters for this application. The required inductance is calculated according to Eq. (5), with the worst-case error voltage assumed to be two capacitor voltage steps of $v_c = 2000\,\text{V} \pm 25\%$, in total $v_{\text{err}} = 5000\,\text{V}$, controller dead time $\Delta t = 6\,\mu\text{s}$ and maximum circulating current deviation of $\Delta i_{\text{cc,err}} = 100\,\text{A}$, which yields an nominal arm inductance of $L_e = 100\,\mu\text{H}$. For the inductor losses $P_{\text{loss,L}} = 2\,\text{kW} < 10\% \cdot P_{\text{loss,HL}}$ at 40-MVA operation is targeted. The RMS arm current at this power is $1.1\,\text{kA}_{\text{RMS}}$, requiring a resistance of less than $1.6\,\text{m}\Omega$. The insulation requirements (not listed in table 3) are according to the operation at a 20 kV DC-cable grid. In typical DC grid fault scenarios, transient overvoltages of up to $2 \cdot v_{\text{DC}}$ can occur [3], which needs to be considered for terminal and winding insulation. As this fault voltage is applied over both inductors in each phase, the nominal isolation must therefore be rated for at least the DC grid voltage. The integration of the arm inductors into the converter container implies very demanding mechanical specifications. The form factor is heavily constrained due to the available installation space inside the container to

$w \times d \times h = 800 \times 1500 \times 300$ mm. Additionally, as the container should still be able to be transported by a 40-tons truck, the weight budget for the design is $m = 350\,\text{kg}$. The combination of these electrical and mechanical parameters could not be achieved by conventional designs.

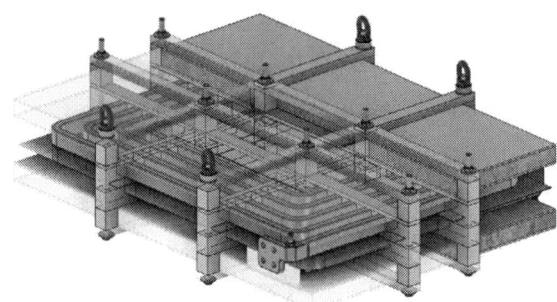

Fig. 8: 3D rendering of arm inductor design for 40 MVA-MMC $w \times d \times h = 800 \times 1100 \times 280$ mm

Figure 8 displays a rendering of the 3D model of the full-scale inductor design. It features two winding layers with 4 turns per layer. The layers are held in place by a glass-fiber reinforced plastic supporting structure. A PTFE sheet is placed between the winding layers to achieve the required surge voltage resistance between the external terminals. Figure 9 shows the section view, where the cooling

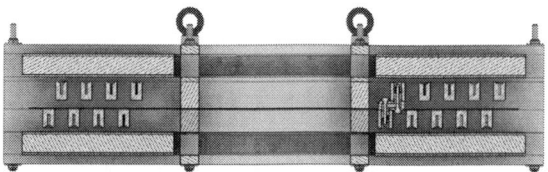

Fig. 9: Section view of arm inductor

channels can be seen. The design of the connection between the two winding layers was particularly challenging due to the XY-offset between the layers. The connection piece needs to achieve good electrical connection as well as ensuring no leakage in the cooling channel up to 5 bar water pressure.

The design was done and optimized with the help of the simulation tool, verified on the demonstrator (see section 3). The simulation results for the full-scale inductor, given in table 3, indicate that the presented concept is capable of delivering all the desired properties or even exceeding the requirements. Figure 10 shows the production of one winding layer, Fig. 11 shows the raw winding layer before welding and assembly.

Fig. 10: Machining of one layer of the 40-MVA inductor

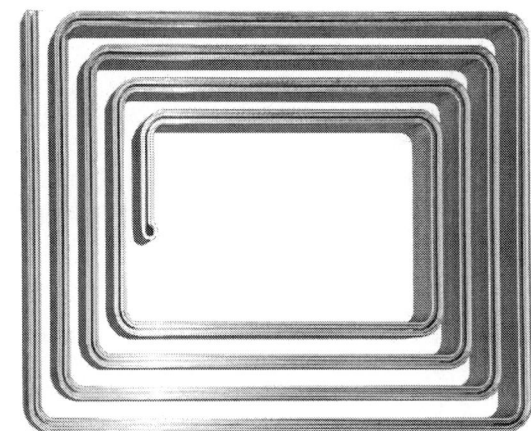

Fig. 11: Finished winging layer before welding of lid and terminals, $w \times d = 730 \times 1000$ mm

6 Conclusion

In this paper, a concept for compact high-power inductors with excellent bandwidth is presented. The proposed geometry allows the design of inductors with substantially reduced parasitic capacitances; the achievable resonant frequencies are up to about ten times higher than the state of the art - compared with reactors of the same inductance, current rating and size. The concept is analyzed and verified with simulations as well as measurements on a demonstrator. Finally, the concept is applied to create a compact high-power MMC arm inductor.

Further steps to validate the new inductor design are the electrical and thermal qualification tests on the full-scale prototype.

Acknowledgment

This research work has been carried out within the project DEFINE and is funded by dtec.bw - Digitalization and Technology Research Center of the Bundeswehr, which we gratefully acknowledge. dtec.bw is funded by the European Union - NextGenerationEU.

References

[1] R. Marquardt, "Modular Multilevel Converters: State of the Art and Future Progress," en, *IEEE Power Electronics Magazine*, vol. 5, no. 4, pp. 24–31, Dec. 2018. DOI: 10.1109/MPEL.2018.2873496.

[2] D. Dinkel, C. Hillermeier, and R. Marquardt, "Direct Multivariable Control for Modular Multilevel Converters," en, *IEEE Transactions on Power Electronics*, vol. 37, no. 7, pp. 7819–7833, Jul. 2022. DOI: 10.1109/TPEL.2022.3148578.

[3] S. Marquardt, C. Dahmen, and T. Brueckner, "Fault Management in Meshed MVDC Grids Enabling Uninterrupted Operation," en, in *PCIM Europe 2023; International Exhibition and Conference for Power Electronics, Intelligent Motion, Renewable Energy and Energy Management*, DE: VDE VERLAG GMBH, Jun. 2023, pp. 1–10. DOI: 10.30420/566091038.

[4] A. F. A. Aziz, M. F. Romlie, and Z. Baharudin, "Review of inductively coupled power transfer for electric vehicle charging," en, *IET Power Electronics*, vol. 12, no. 14, pp. 3611–3623, Nov. 2019. DOI: 10.1049/iet-pel.2018.6011.

[5] J. Acero, C. Carretero, I. Lope, R. Alonso, O. Lucia, and J. M. Burdio, "Analysis of the Mutual Inductance of Planar-Lumped Inductive Power Transfer Systems," en, *IEEE Transactions on Industrial Electronics*, vol. 60, no. 1, pp. 410–420, Jan. 2013. DOI: 10.1109/TIE.2011.2164772.

[6] Z. Li, X. He, and Z. Shu, "Design of coils on printed circuit board for inductive power transfer system," en, *IET Power Electronics*, vol. 11, no. 15, pp. 2515–2522, Dec. 2018. DOI: 10.1049/iet-pel.2018.5780.

[7] Y.-C. Hsieh, Z.-R. Lin, M.-C. Chen, H.-C. Hsieh, Y.-C. Liu, and H.-J. Chiu, "High-Efficiency Wireless Power Transfer System for Electric Vehicle Applications," en, *IEEE Transactions on Circuits and Systems II: Express Briefs*, vol. 64, no. 8, pp. 942–946, Aug. 2017. DOI: 10.1109/TCSII.2016.2624272.

[8] J. H. Kim, B. G. Choi, S. Y. Jeong, S. H. Han, H. R. Kim, *et al.*, "Plane-Type Receiving Coil With Minimum Number of Coils for Omnidirectional Wireless Power Transfer," en, *IEEE Transactions on Power Electronics*, vol. 35, no. 6, pp. 6165–6174, Jun. 2020. DOI: 10.1109/TPEL.2019.2952907.

[9] MAGMENT, *Why you need magnetic materials for WPT*, https://www.magment.co/wp-content/uploads/2023/10/WHY-YOU-NEED-MAGNETIC-CONSTRUCTION-MATERIALS-FOR-WPT.pdf, Accessed: 04/2024, Aug. 2023.

[10] P. Zacharias, *Magnetische Bauelemente: Grundlagen und Anwendungen*, de. Wiesbaden: Springer Fachmedien Wiesbaden, 2020. DOI: 10.1007/978-3-658-24742-3.

[11] MAGMENT, *MC120 Datasheet*, https://www.magment.co/wp-content/uploads/2023/09/MAGNETIZABLE-CONCRETE-MC120-DataSheet.pdf, Accessed: 04/2024, Aug. 2023.

[12] A. Massarini and M. Kazimierczuk, "Self-capacitance of inductors," en, *IEEE Transactions on Power Electronics*, vol. 12, no. 4, pp. 671–676, Jul. 1997. DOI: 10.1109/63.602562.

[13] A. Ayachit and M. K. Kazimierczuk, "Self-Capacitance of Single-Layer Inductors With Separation Between Conductor Turns," en, *IEEE Transactions on Electromagnetic Compatibility*, vol. 59, no. 5, pp. 1642–1645, Oct. 2017. DOI: 10.1109/TEMC.2017.2681578.

[14] M. Zdanowski, K. Kostov, J. Rabkowski, R. Barlik, and H.-P. Nee, "Design and Evaluation of Reduced Self-Capacitance Inductor in DC/DC Converters with Fast-Switching SiC Transistors," en, *IEEE Transactions on Power Electronics*, vol. 29, no. 5, pp. 2492–2499, May 2014. DOI: 10.1109/TPEL.2013.2281990.

[15] I. Lope, C. Carretero, and J. Acero, "First self-resonant frequency of power inductors based on approximated corrected stray capacitances," en, *IET Power Electronics*, vol. 14, no. 2, pp. 257–267, Feb. 2021. DOI: 10.1049/pel2.12030.

PCIM Europe 2024, 11– 13 June 2024, Nuremberg   DOI: 10.30420/566262340

Analytical Evaluation of Differential Model DC EMI Filter Inductors using Material Saturation Coefficient

Lukas Müller[1]

[1] Micrometals, Inc, U.S.A.

Corresponding author: Lukas Müller, lmueller@micrometals.com
Speaker: Lukas Müller, lmueller@micrometals.com

Abstract

This paper presents the use of a material grade's saturation coefficient with the core geometry method (K_g) to analytically determine the required size and achievable performance of differential mode EMI DC filter inductors. The traditional K_g method is only directly applicable to gapped ferrite designs, whereas the proposed method expands the use of this equations to gapped ferrites and powdered magnetic materials.

1   Introduction

Differential mode EMI filters form a crucial part of modern switched-mode power converters. They are generally required to meet regulatory limits on conducted emissions. An accurate metric on the performance and size of DC filter inductors is required to optimize the design of differential mode EMI filters in switch mode power converters. A modified version of the core geometry method using the magnetic material grade's saturation constant (Γ) is presented here. The saturation constant can be used to analytically evaluate the performance of a large variety of magnetic materials, including those featuring a distributed air gap. Based on the presented equations, the required volume for differential mode EMI filter inductors can be evaluated and optimized.

2   Classical DC Inductor Performance Evaluation

Traditionally, the area product (A_p) method or core geometry (K_g) method were used to design and evaluate DC inductors used in EMI filters analytically[1]. The area product method remains popular for inductor and transformer designs and optimization[2]–[6]. However, the area product has a number of significant drawbacks. The area product relies on a set current density values for the wiring. In thermally limited designs a fixed current density cannot be assumed[7].In addition, the area product method does not consider all aspects of the

core geometry in the design process, neglecting the impact of different aspect ratios of the component. This can lead to sub-optimal designs for a given volume, especially considering new types of core shapes, like planar cores, being introduced by magnetic core manufacturers.

The core geometry method and its derivatives addresses most of the shortcomings of the Area Product method when designing inductors or transformers[8], [9].The standard K_g expression for DC filter inductors is shown in Eq. (1).

$$K_g = \frac{A_{eff}^2 W_a}{MLPT} = \frac{\rho L_{eq}^2 I_{nom}^2}{k_{cu} R_{dc} B_{eq}^2} \quad (1)$$

with A_{eff} being the effective magnetic cross sectional area of the core used, W_a equaling the winding window area, MLPT being the mean length per turn, L_{eq} equaling the nominal differential mode inductance at the nominal load current I_{nom}, k_{cu} being the winding fill factor, R_{dc} being the effective DC resistance and B_{eq} being the equivalent saturation flux density of the material[1]. For gapped ferrites, the value of B_{eq} is traditionally set slightly below the material's saturation flux density.

The core geometries use of a materials' equivalent flux density only makes it directly applicable to designs using high permeability materials with one or multiple discrete air gaps. Powdered iron and alloy cores tend to be an ideal choice for DC inductors due to their high saturation flux density and soft saturation behavior[10]–[12]. Powdered materials, especially powdered iron, also tend to provide a more consistent complex permeability behavior over a wider frequency range than power ferrite[13]. Powdered materials cannot be evaluated accurately using the traditional K_g method. This is due to the fact that the K_g equation assumes a gapped core in which the effective permeability of the magnetic structure can be freely set equal to:

$$\mu_{eff} = \frac{B_{eq}}{\frac{0.4\pi N \times I_{nom}}{MPL}} \qquad (2)$$

with MPL representing the mean magnetic path length of the core.

Powdered magnetic materials feature a distributed air gap throughout the material. The distributed air gap results in a soft-saturation behavior, similar to using a slanted air gap in a ferrite core. The effective differential permeability of the material is a non linear function of applied magnetization force. Due to this fact, the underlining assumption in the K_g equation cannot be used for powdered materials and a different variable needs to be used.

3 Core Geometry Method using Saturation Coefficient

To address this shortcoming, the K_g equation for DC inductors can be modified to allow its use with both gapped high permeability materials and a large variety of powdered materials. The terms related to the relative core dimensions are left unchanged from the traditional K_g equation. This allows the use of already existing core size reference tables, like the ones found in [1], [8]. The equivalent flux density squared term in the original equation is replaced by the variable Γ ,which represents the relative saturation behavior of the magnetic material grade that is being considered. The modified K_g equation is then given as:

$$K_g = \frac{A_{eff}^2 W_a}{MLPT} = \frac{\rho L_{eq}^2 I_{nom}^2}{k_{cu} R_{dc} \Gamma} \qquad (3)$$

In EMI filter inductor designs, total loss is generally an important design consideration. The equation can be rearranged to solve for power loss and required energy storage directly:

$$K_g = \frac{A_{eff}^2 W_a}{MLPT} = \frac{4\rho E_{eq}^2}{k_{cu} P_{loss} \Gamma} \qquad (4)$$

where the equivalent energy storage is defined as:

$$E_{eq} = \frac{1}{2} L_{eq} I_{nom}^2 \qquad (5)$$

To use the equation effectively, the Γ coefficient of various magnetic materials has to be known.

Material	$\mu_{initial,max}$
Sendust	160
Molypermalloy	550
Permalloy	160
Silicon Steel	125
Hybrid Alloy	125
LF Iron	100

Fig. 1: Maximum initial permeability of various commercially available powderd magnetic material grades

4 Saturation Coefficient of Magnetic Materials

4.1 General Information

The saturation constant Γ has a large impact on the overall performance of the DC filter inductor. The achievable minimum losses for a given required energy storage E_{eq} and inductor volume V_t are inversely proportional to the saturation constant.

$$P_{loss} \propto \frac{E_{eq}^2}{\Gamma} \qquad (6)$$

Utilizing a material with twice the Γ value will half the DC losses in the EMI filter inductor. Choosing a material with the highest Γ value will therefore lead to the most efficient design per unit volume.

4.2 Saturation Coefficient for Powdered Materials

For powdered cores, the saturation constant Γ depends on a large number of material characteristics including the material's saturation flux density, its amplitude permeability, its differential permeability for a given magnetization force, and the geometry of the core. In turn, these factors are determined by the base powdered magnetic material, insulation, and binder used. Performing additional manufacturing processing steps or varying the compacting pressure of the powdered material during pressing will also effect Γ. Due to these complexities, there is no simple analytical equation to calculate Γ directly for a specific base magnetic material used in the production of powdered cores. Another complication is that Γ will only be valid for a certain range of magnetization forces as the starting permeability of powdered materials is generally limited depending on material types. A comparison between powdered magnetic material types and maximum commercially available starting permeability's in toroidal core shapes in shown in Fig. 1. A

Material-Type	Γ	$H_{min}(Oe)$
Sendust	≈ 0.08	35
Molypermalloy	≈ 0.08	18
Permalloy	$0.20 - 0.26$	38
Silicon Steel	≈ 0.17	60
Hybrid Alloy	$0.144 - 0.221$	65
LF Iron	$0.04 - 0.09$	35

Fig. 2: Approximate Γ factor for a variety of different magnetic material types

Material	Γ	$H_{min}(Oe)$	$\approx B_{sat}(T)$
MS	0.0784	35	0.9
MP	0.0812	18	0.9
HF	0.2025	40	1.5
HFP	0.2401	38	1.5
FS	0.1764	60	1.7
OD	0.2209	65	1.3
OE	0.1444	70	1.6
LF Fe	0.0640	35	1.7

Fig. 3: Γ factor for select Micrometals material grades

Core Shape	Γ	$H_{valid}(Oe)$	$\approx B_{sat}(T)$
T	0.0784	≥ 35	0.9
E	0.0676	$50 - 400$	0.9
PQ	0.0676	≥ 90	0.9

Fig. 4: Changes in MS materials' Γ factor depending on core geometry

similar issue was already present in the traditional K_g equation, where the equation would only be valid over a certain range of magnetization forces. The permeability of power ferrites is generally so high, that it was not a consideration in practical design. While the range of valid magnetization forces is smaller when using powdered magnetic materials, it will be demonstrated later that it is still wide enough to not be an issue in practical inductor designs and therefore not a draw back in the proposed design.

To find the Γ coefficient numerical methods have to be employed. To fit Γ, the differential permeability over a commonly used range of magnetization forces has to be fitted for the given powdered material type. Even if manufacturers use the same base magnetic material, there will be difference in the Γ coefficient for the materials due to differences in the manufacturing process. One can still determine approximate values of Γ for different powdered material types which are summarized in Fig. 2. The chart also shows the minimum magnetization force H_{min} for which the Γ value can be used in the K_g equation. The Γ terms for some of Micrometals' powdered materials in a toroid core shape are summarized in Fig. 3 for reference.

Comparing the Γ and B_{sat} values in Fig. 3 shows that the two are not necessarily proportional to one another. The most obvious example is the line

frequency powdered iron materials (LF Fe) which feature one of the highest saturation flux densities, while feature one of the lowest Γ values. The difference between the two values is due to the large variation in amplitude and differential permeability line frequency powdered iron materials have. The HF and HFP materials, which are variations of permalloy, also feature a higher saturation constant Γ than the FS material, which is a type of Silicon Steel, even though their saturation flux density is lower. This is due to the permalloy materials' more square saturation behavior. HF and HFP are both use permalloy as their base powder, but HFP has a higher Γ value due to the additional material processing compared to HF.

As mentioned previously, for some magnetic materials' Γ can change depending on the geometry of the core. A comparison between the Γ coefficient of Micrometals' MS material in different core shapes is shown in Fig. 4. For powdered materials, the more complex the core shape, the lower Γ will generally be.

4.3 Range of validity of Saturation Coefficient for Powdered Materials

The Γ value will only provide a good prediction of material performance in the range of magnetization forces shown. For a cost and space efficient DC inductor design, the magnetization force should always be chosen in such a away that the inductor's temperature rise is close to the maximum allowable one. Considering a maximum temperature rise of $40°C$ and natural convection, the range of optimum magnetization forces for standard toroid cores is in the range of approximately 30-215Oe as shown in Fig. 5. This makes the Γ values provided applicable for most practical designs using natural convection and all designs using forced convection cooling.

4.4 Saturation Coefficient for Ferrite

The expression below can be used to calculate the Γ factor for gapped ferrite materials:

$$\Gamma = \left(\frac{B_{sat}}{k_{core}k_{overload}} \right)^2 \qquad (7)$$

Toroid Core	K_g	$H_{opt,40C}$
T600	$876cm^5$	215
T300	$6.39cm^5$	171
T184	$2.1cm^5$	108
T106	$0.15cm^5$	85.2
T68	$0.11cm^5$	68.3
T37	$1171mm^5$	52.2
T16	$35.1mm^5$	31

Fig. 5: K_g and optimum magnetization force for operation with a $40°C$ temperature rise for a number of common powdered toroid sizes based on [12]

where B_{sat} is the saturation flux density of the ferrite material as a function of the maximum expected operating temperature, k_{core} is a derating factor for the core geometry and $k_{overload}$ describes the overload current the inductor has to be able to withstand while providing a meaningful inductance[14]. The equation assumes one or multiple discrete straight gaps in the magnetic structure. The Γ term will be different for slanted air gaps as this makes the saturation behavior non-linear[15]. A typical value for Γ for MnZn power ferrites operating at a peak temperature of $100°C$ is between 0.04-0.0544. This value assumes an overload current of 50% which is a common requirement in modern power supplies. Note that the inductance of the ferrite based inductor will remain constant, while that of the powdered core will swing.

5 Dimensional Analysis

For the optimization of differential mode EMI filters it is useful to have an accurate model relating inductor volume to energy storage. In this section both the theoretical and practical relationship between the two are evaluated.

5.1 Theoretical scaling

All dimensions terms in the K_g design equation can be related to the total wound volume of the inductor using:

$$A_{eff} = l_1 V_t^{\frac{2}{3}} \tag{8}$$

$$W_a = l_2 V_t^{\frac{2}{3}} \tag{9}$$

$$MLPT = l_3 V_t^{\frac{1}{3}} \tag{10}$$

where l_1, l_2 and l_3 are constant depending on the core geometry used.

The value of K_g can then be related to the total wound volume of the inductor using:

$$K_g = \frac{l_1 l_2}{l_3} V_t^{\frac{5}{3}} \tag{11}$$

The term $\frac{l_1 l_2}{l_3}$ can be combined into the term l_{Kg} which relates the total wound volume of a core shape to it's K_g value, which is easier to fit to a large set of commercially available core geometries.

The power loss for a DC inductor having to store some equivalent energy can then be approximated using:

$$P_{loss} = \frac{4\rho E_{eq}^2}{k_{cu} l_{kg} V_t^{\frac{5}{3}} \Gamma} \tag{12}$$

Based on this generalized equation, it can be seen that the power loss does not scale linearly with energy stored or volume assuming the same magnetic material is used.

$$P_{loss} \propto \frac{E_{eq}^2}{V_t^{\frac{5}{3}}} \tag{13}$$

This relationship can also be used to evaluate the impact of paralleling multiple inductors to form a large DC inductor. If x separate DC inductor are used to storage a total energy E_{total} in a total volume of V_t, then the total power losses will be proportional to:

$$P_{total} = x \times \frac{\left(\frac{E_{eq}}{x}\right)^2}{\left(\frac{V_t}{x}\right)^{\frac{5}{3}}} \tag{14}$$

which simplifies to:

$$P_{total} = \frac{E_{eq}}{V_t^{\frac{5}{3}}} x^{\frac{2}{3}} \tag{15}$$

which shows that the total power losses increase if the energy storage is split up over multiple separate inductors with the same volume compared to a single inductor. A similar relationship between number of separate inductors used and total losses versus volumes was also shown in [4]. This scaling of losses might have to be considered in multi-stage differential mode EMI filters where one differential model filter inductor is separated into multiple smaller inductors over a number of stages. While realizing a filter in multiple stages generally results in a lower energy storage requirements[16], having to use multiple separate components can lead to lower than expected loss or volume savings.

In many designs using natural convection, the maximum allowable temperature rise can be a more restrictive design parameter than the total losses. The minimum possible DC inductor volume will be dictated by the temperature rise in these cases. Modeling temperature rise accurately is challenging due to the large number of parameters involved. However, the following approximation can be used to relate power losses to temperature rise[11]:

$$P_{loss} = 0.1 \Delta T^{1.1} S_a \qquad (16)$$

with S_a equaling the surface area of the inductor core in m^2.

$$0.1 (\Delta T)^{1.1} \times S_a = \frac{4\rho E_{eq}^2}{k_{cu} l_{kg} V_t^{\frac{5}{3}} \Gamma} \qquad (17)$$

The surface area of the inductor can also be related to the total volume of the inductor using the following general expression:

$$S_a = l_4 V_t^{\frac{2}{3}} \qquad (18)$$

which results in:

$$0.1 (\Delta T)^{1.1} \times l_4 V_t^{\frac{2}{3}} = \frac{4\rho E_{eq}^2}{k_{cu} l_{kg} V_t^{\frac{5}{3}} \Gamma} \qquad (19)$$

which can be solved for:

$$l_{kg} l_4 V_t^{\frac{7}{3}} = \frac{40\rho E_{eq}^2}{k_{cu} (\Delta T)^{1.1} \Gamma} \qquad (20)$$

which shows the inductor volume required to maintain a certain maximum allowable temperature rise for a given material and energy storage requirement.

For a DC filter inductor with a certain maximum allowable temperature rise, the relation between volume and energy storage is not linear but:

$$E_{eq} = V_t^{\frac{7}{6}} \sqrt{\frac{k_{cu} (\Delta T)^{1.11} \Gamma}{40\rho} l_{kg} l_4} \qquad (21)$$

This is in contrast to the linear relationship between volume and energy storage capabilities of a differential mode EMI filter inductor reported in [6], [16]

5.2 Scaling of commercial core shapes

The generalized core scaling assumed ideal and constant relative dimensions in the magnetic core. However, standard magnetic core shapes do not scale all dimensions proportionally. For standard powdered toroid cores, the relationship between K_g and the total wound volume of the inductor can be fitted to the following general expression:

$$K_g \approx 4.2 \times 10^{-4} \times V_t^{1.64} \qquad (22)$$

Note that V_t is raised to the 1.64 instead of the $\frac{5}{3}$ power for commercially available cores.

For commercially available powdered toroid core sizes, the relationship between S_a and the total wound volume is given by the following expression:

$$S_a \approx \times 2.8307 V_t^{0.6481} \qquad (23)$$

Note that V_t is raised to the 0.681 instead of the $\frac{2}{3}$ power. Once again, this is due to variations in the relative proportions of the toroid as sizes increase. Assuming a fill factor of 0.6 and copper wiring at $100°C$ with resistivity of $23 \times 10^{-9} \Omega m$ and a temperature rise of $40°C$, the approximate relationship between energy storage in a DC filter inductor and its volume is equals to:

$$V_t \approx 0.00926769 \times E_{eq}^{0.874} \times \Gamma^{-0.437} \qquad (24)$$

which is also non-linear. The proposed equation can be use to approximate the required volume of a DC filter inductor for a given required energy storage and chosen magnetic material.

6 Summary

The K_g method is presented here using the material Γ coefficient which allows both gapped ferrite and powdered materials to be evaluated in an objective manner for DC inductor designs in EMI filters. The term Γ is used to describe the saturation characteristic of the material type. The Γ value can be used to directly compare the relative performance of different material grades to one another in DC inductor applications. Higher value of Γ indictate a better material performance.

The approximate relationships between material choice, inductor volume, temperature rise and equivalent energy stored were derived. An equation to approximate the relationship between required volume, energy storage and choice of material using commercially available toroid cores operating at a commonly specified operating point was presented as well. This relationship can be used to approximate the total volume of differential mode EMI filter designs.

References

[1] W. T. McLyman, *Transformer and Inductor Design Handbook*. Boca Raton, FL, USA: CRC Press, 2011.

[2] M. K. Kazimierczuk and H. Sekiya, "Design of ac resonant inductors using area product method," in *2009 IEEE Energy Conversion Congress and Exposition*, 2009, pp. 994–1001. DOI: 10.1109/ECCE.2009.5316501.

[3] J. Lavers and V. Bolborici, "Loss comparison in the design of high frequency inductors and transformers," *IEEE Transactions on Magnetics*, vol. 35, no. 5, pp. 3541–3543, 1999. DOI: 10.1109/20.800583.

[4] C. R. Sullivan, B. A. Reese, A. L. F. Stein, and P. A. Kyaw, "On size and magnetics: Why small efficient power inductors are rare," in *2016 International Symposium on 3D Power Electronics Integration and Manufacturing (3D-PEIM)*, 2016, pp. 1–23. DOI: 10.1109/3DPEIM.2016.7570571.

[5] W. J. Muldoon, "Analytical design optimization of electronic power transformers," in *1978 IEEE Power Electronics Specialists Conference*, 1978, pp. 216–225. DOI: 10.1109/PESC.1978.7072356.

[6] M. L. Heldwein and J. W. Kolar, "Design of minimum volume emc input filters for an ultra compact three-phase pwm rectifier," *Eletrônica de Potência*, vol. 14, no. 2, pp. 85–96, May 2009. DOI: 10.18618/REP.2009.2.085096.

[7] J. Watson, *Applications of Magnetism*. New York, NY, USA: John Wiley & Sons, 1980.

[8] R. W. Erickson and D. Maksimovic, *Fundamentals of Power Electronics*, 2ed. Springer, 2001.

[9] L. Müller and J. W. Kimball, "High frequency core coefficient for transformer size selection," in *2016 IEEE Energy Conversion Congress and Exposition (ECCE)*, 2016, pp. 1–6. DOI: 10.1109/ECCE.2016.7855151.

[10] D. O. Boillat, J. W. Kolar, and J. Mühlethaler, "Volume minimization of the main dm/cm emi filter stage of a bidirectional three-phase three-level pwm rectifier system," in *2013 IEEE Energy Conversion Congress and Exposition*, 2013, pp. 2008–2019. DOI: 10.1109/ECCE.2013.6646954.

[11] A. Van den Bossche and V. C. Valchev, *Inductors and Transformers for Power Electronics*. Boca Raton, FL, USA: CRC Press, 2005.

[12] J. Cox and D. Nicol, *Power Conversion & Line Filter Applications - Issue L*. Anaheim, CA, USA: Micrometals, 2007.

[13] M. Hartmann, H. Ertl, and J. W. Kolar, "Emi filter design for high switching frequency three-phase/level pwm rectifier systems," in *2010 Twenty-Fifth Annual IEEE Applied Power Electronics Conference and Exposition (APEC)*, 2010, pp. 986–993. DOI: 10.1109/APEC.2010.5433382.

[14] M. Esguerra, "Dc-bias specifications of gapped ferrites," *Power Electronics Technology*, Oct. 2003.

[15] W. Hurley and W. H. Wölfle, *Transformers and Inductors for Power Electronics - Theory, Design and Applications*. New York, NY, USA: John Wiley & Sons, 2013.

[16] K. Raggl, T. Nussbaumer, and J. W. Kolar, "Guideline for a simplified differential-mode emi filter design," *IEEE Transactions on Industrial Electronics*, vol. 57, no. 3, pp. 1031–1040, 2010. DOI: 10.1109/TIE.2009.2028293.

PCIM Europe 2024, 11– 13 June 2024, Nuremberg   DOI: 10.30420/566262341

Design and Performance Evaluation of Air Core Inductors for Very High Frequency Power Conversion

Florentin Salomez ©[1], Vincent Blanchon ©[2], Sébastion Carcouet[2], Jean-Luc Schanen ©[1], Ghislain Despesse ©[2], Yves Lembeye ©[1]

[1] Univ. Grenoble Alpes, CNRS, Grenoble INP, G2ELAB, 38000 Grenoble, France
[2] CEA-Leti, Univ. Grenoble Alpes, F-38000 Grenoble, France

Corresponding author: Florentin Salomez, florentin.salomez@grenoble-inp.fr
Speaker: Florentin Salomez, florentin.salomez@grenoble-inp.fr

Abstract

This paper shows a simple performance evaluation of rectangular cross section air core solenoids based on analytical approach validated on prototypes. The asymptotic behaviors of the quality factor as a function of geometrical parameters show the theoretical range of inductance and resistance of solenoids made out of planar substrate like Printed Circuit Board.

1 Introduction

The renewed interest in VHF converters [1] comes in part from the need to miniaturized the low power switch-mode power supply without sacrificing the efficiency [2], [3], and to improve wireless power transfer [4], [5]. The goal of this paper is to present a way to increase the mass power density of very high frequency (VHF) converters thanks to air core inductors without sacrificing the efficiency. In the literature one of the most used technology is the inductor integrated in the Printed Circuit Board (PCB) [6]–[10]. Numerical optimization is used to find the inductor with the best Q factor in [6], [7], [10], sometimes in conjunction with Finite Element software for solenoids [6], [8] and for ring cores [9]. None of the papers cited before has quantify the theoretical limit of Q factor one can achieve with PCB inductor technology. This paper addresses the theoretical limits of the Q factor of a rectangular cross-section solenoid for a given substrate thickness and area, which is very important to assess the potential of the air core technology for VHF power conversion with functionalised PCB [6], [7], [9], [10]. First the model of the Q factor of air core solenoids with a rectangular cross section is derived. Second, this model is used to study the theoretical limit of the Q factor and experiments are performed to validate the results. Third, experimental results of air core solenoids inside a VHF converter are presented.

Finally, some conclusions and perspectives are presented.

2 Modelling of the Solenoid

2.1 Geometrical Description

A solenoid with a rectangular cross-section and made with copper tape is considered hereafter and drawn in the Figure 1, with t the thickness of the substrate, w_t the width of the trace, w_{ti} the width of the trace at the edge, s_t the interturn space, θ the angle between the traces and the cross-section, w the width of the solenoid, l its length, $S_{top} = l \cdot w$ its footprint area, $k = l/w$ its form factor and N the number of turns. Some of the preceding variables

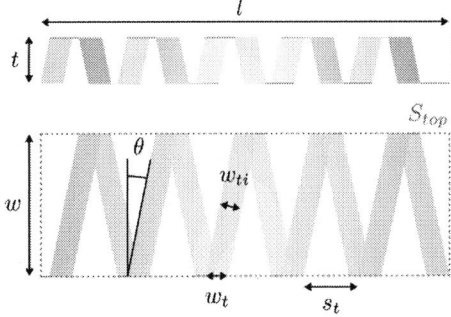

Fig. 1: Drawing of the rectangular cross-section solenoid with exaggerated inter-turns space s_t.

are linked by the following equations:

$$w_{ti} = \frac{w_t}{\cos(\theta)}, \tag{1}$$

2431

$$l = N \cdot (w_{ti} + s_t) + w_{ti}, \tag{2}$$

$$w_{ti} + s_t = 2 \cdot (h + w) \cdot \tan(\theta). \tag{3}$$

It is possible to express θ as a function of input parameters N, l, s_t such that

$$\theta = \arctan\left(\frac{\frac{l-s_t}{N+1} + s_t}{2\pi r}\right). \tag{4}$$

2.2 Inductance, Resistance and Q Factor

The parasitic capacitance of the solenoid is neglected. Its inductance is defined by

$$L = \mu_0 N^2 \frac{tw}{l} = \mu_0 N^2 \frac{t}{k}, \tag{5}$$

with μ_0 the permeability of the vacuum. The direct current (DC) resistance of the component is

$$R_{DC} = \rho \cdot \frac{2 \cdot N \cdot (t+w)}{t_{Cu} \cdot w_t \cdot \cos(\theta)}, \tag{6}$$

with ρ the copper resistivity and t_{Cu} its thickness. The alternating current (AC) resistance is approximated as the resistance of the skin depth $\delta(f)$ along the conductor (if the thickness of the conductor is greater than the skin depth at the considered frequency, proximity effects are neglected) and defined by

$$R_{AC} = \frac{t_{Cu}}{\delta(f)} R_{DC}. \tag{7}$$

The Q factor is defined as

$$Q = \frac{2\pi f L}{R_{AC}}. \tag{8}$$

3 Performance Evaluation

The evaluation of performance consists of studying the variation of the Q factor as a function of the geometrical parameters of the solenoids to assess the potential of the technology. The Q factor is frequency dependent. Since the ultimate goal of the paper is to design an inductance for a VHF converter, all computations and results are presented at the frequency 27.12 MHz of the second harmonic of the converter which is filtered by a parallel LC branch.

3.1 Asymptotic Behaviors

Assuming θ is small leads to $w_t \approx w_{ti}$ and $\cos(\theta) \approx 1$ and substituting w_{ti} thanks to Equation (2) leads to the approximation $Q \approx Q_a$ with

$$Q_a = \frac{\pi f \mu_0 \cdot \delta(f)}{\rho} \cdot \frac{N}{N+1} \cdot \frac{tw}{t+w} \cdot \frac{l - N \cdot s_t}{l}. \tag{9}$$

It is also possible to rewrite the previous equation with k and S_{top} such that

$$Q_a = \frac{\pi f \mu_0 \delta(f)}{\rho} \cdot \frac{N}{N+1} \cdot \frac{t}{t \cdot \sqrt{\frac{k}{S_{top}}} + 1} \\ \cdot \left(1 - \frac{N \cdot s_t}{\sqrt{k \cdot S_{top}}}\right) \tag{10}$$

The theoretical maximum Q factor for a given PCB thickness Q_{max} is defined thanks to the following assumptions: the number of turns is large enough such that

$$\frac{N}{N+1} \underset{N \gg 1}{\sim} 1, \tag{11}$$

the width w is greater than the substrate thickness t such that

$$\beta = \frac{tw}{t+w} \underset{w \gg t}{\sim} t, \tag{12}$$

the inter-tuns space s_t is small enough in comparison to l such that

$$\frac{l - N \cdot s_t}{l} \underset{N \cdot s_t \ll l}{\sim} 1, \tag{13}$$

and the skin depth is

$$\delta(f) = \sqrt{\frac{\rho}{\pi f \mu_0}}. \tag{14}$$

These asymptotic behaviors ($N \gg 1$, $w \gg t$, $N \cdot s_t \ll l$) lead to

$$Q_a \sim Q_{max} = \frac{t}{\delta(f)} = \sqrt{\frac{\pi f \mu_0}{\rho}} \cdot t, \tag{15}$$

meaning that **the thickness of the PCB t is the main parameter to increase the Q factor.** This is experimentally verified in the Figure 2 where two solenoids of same parameters apart thickness have been characterized. The solenoids are depicted in the Figure 3 and the details about their dimensions and electrical performances are given in the Table 1. An impedance analyzer E4990A equipped with the adapter 42942A and the socket 16092A has been used for Q measurement. The adapter has been calibrated with open, short, load and low loss standards and the socket has been compensated for open and short parasitic impedances to ensure the accuracy of the measurement of Q factor.

Tab. 1: Dimensions and electrical characteristics of the tested solenoids at $f = 27.12\,\text{MHz}$.

Id.	t (mm)	w (mm)	l (mm)	N	L (nH)	R (mΩ)	Q
$A1$	2	17	17	6	104	187	95
$A2$	2	21	21	6	113	214	89
$A3$	2	24	24	6	109	173	107
$B1$	2	24	22	4	58	93	107
$B2$	2	24	24	9	231	428	92
$T5$	5	24	24	6	222	224	168

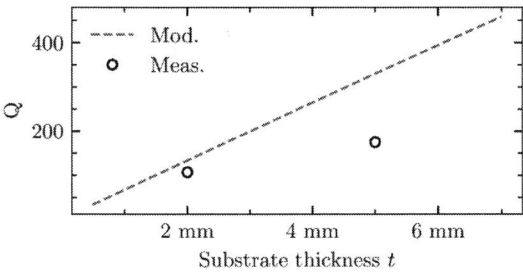

Fig. 2: Evolution of Q for two different substrate thickness t at $f = 27.12\,\text{MHz}$.

Fig. 3: Photograph of the solenoids characterised in the Figure 2, with ($l = w = 2.4\,\text{mm}$, $s_t = 0.5\,\text{mm}$, $N = 6$).

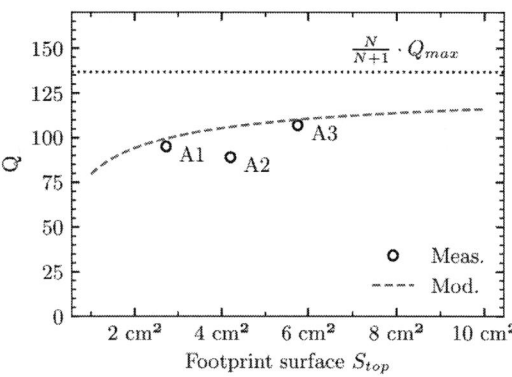

Fig. 4: Evolution of Q as a function of S_{top} for given frequency f, s_t, L and t.

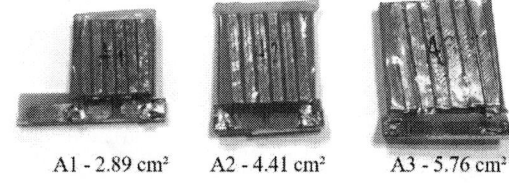

Fig. 5: Photograph of the solenoids A1, A2 and A3 characterized in Figure 4.

3.2 Evolution of Q as a Function of Footprint Area S_{top}

According to Equation (5) the inductance is only a function of the number of turns, the thickness and the form factor. Since L is independent of S_{top}, it is possible to study the evolution of Q as a function of S_{top} for a given inductance L. The Equation (10) shows that Q_a is increasing monotonically for an increasing footprint surface S_{top}. The theoretical evolution is plotted in the Figure 4 alongside three measurement points ($A1$, $A2$, and $A3$). The photographs of the characterized solenoids $A1$, $A2$, and $A3$ are presented in the Figure 5. The Figure 4 shows the asymptotic behavior of Q which reaches

$$\frac{N}{N+1} \cdot Q_{max}, \tag{16}$$

as S_{top} tends to $+\infty$. **The Q factor tends to a limit as S_{top} increases. This asymptotic behavior shows that increasing Q factor once the thickness has been fixed is possible but at the expense of a large footprint area.**

3.3 Evolution of Q as a Function of the Number of Turns N

Once the thickness is fixed, the maximum Q factor that one can achieve on a given footprint area S_{top} and form factor k is defined by the optimal number of turns N_{opt}. The optimal value is defined when

the derivative

$$\frac{dQ_a}{dN} = \frac{\beta}{\delta} \cdot \frac{-s_t \cdot N^2 - 2 \cdot s_t \cdot N + \sqrt{k \cdot S_{top}}}{\sqrt{k \cdot S_{top}} \cdot (N+1)^2}, \quad (17)$$

cancels. This is the case for

$$N_{opt} = \sqrt{\frac{\sqrt{k \cdot S_{top}} + s_t}{s_t}} - 1. \quad (18)$$

the evolution of Q as a function of the number of turn for given S_{top}, t, k is shown in the Figure 6. Three solenoids (B1, A3, B2) have been realized and characterized to validate experimentally the optimal number of turns. They are depicted in the Figure 7.

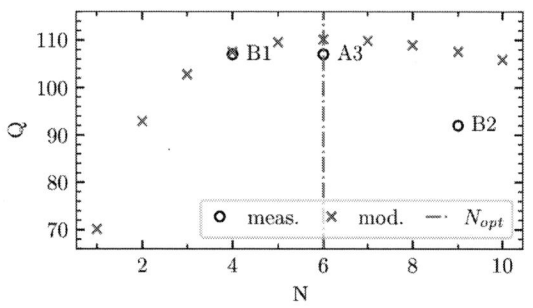

Fig. 6: Evolution of Q as a function of N for fixed thickness $t = 2\,\text{mm}$, surface $S_{top} = 5.76\,\text{cm}^2$, form factor $k = 1$ and frequency $f = 27.12\,\text{MHz}$.

Fig. 7: Photograph of the solenoids B1, A3 and B2 characterized in Figure 6, with $l = 2.4\,\text{cm}$, $w = 2.4\,\text{cm}$, $t = 2\,\text{mm}$, $s_t = 0.5\,\text{mm}$.

For this specific case the optimal number of turns is 6. Around this value the evolution of Q is flat which can explain why the measurements do not show clearly the optimal number of turns. The discrepancy between point B2 and model value (roughly 20 %) might be due to the irregularities on the width w_t of the copper tape cut by hand. Indeed, in this case, a 20 % error on the width of the tape (2 mm) is equal to 400 µm.

3.4 Optimal Q on a Given Footprint Area S_{top} and Thickness t

According to the previous results there is an optimal Q for given surface and thickness. This optimal Q is approximated by injecting Eq. (18) in Eq. (10) such that

$$Q_{a,opt} = \frac{\beta}{\delta} \cdot \frac{1}{\sqrt{k \cdot S_{top}}} \cdot \left(\sqrt{k \cdot S_{top}} + 2 \cdot s_t \right.$$
$$\left. - 2 \cdot \sqrt{s_t \cdot \left(\sqrt{k \cdot S_{top}} + s_t \right)} \right). \quad (19)$$

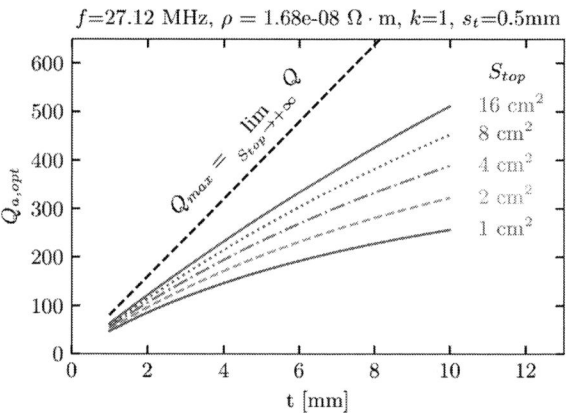

Fig. 8: Evolution of $Q_{a,opt}$ as a function of the thickness t and the footprint area S_{top}.

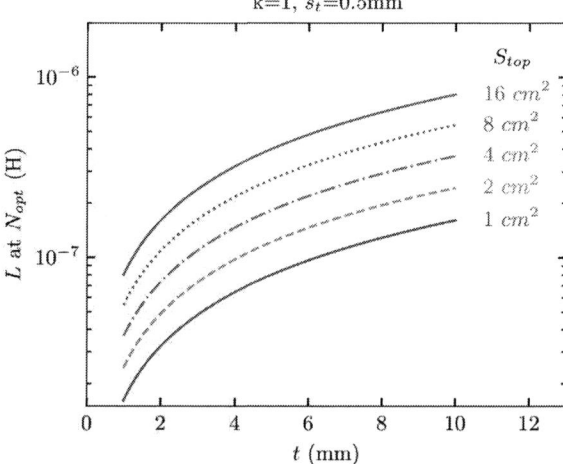

Fig. 9: Evolution of the inductance at N_{opt} as a function of the thickness t and the top surface S_{top}.

The evolution of $Q_{a,opt}$ as a function of the thickness t and the top surface S_{top} is plotted in Figure 8. According to Equation (10), under the assumption that s_t is small in comparison to $l = \sqrt{k \cdot S_{top}}$, the Q factor is greater for greater S_{top} and reaches Q_{max} when $S_{top} \to \infty$ as presented in the Figure 8. Then the corresponding inductance is computed and plotted in the Figure 9. These two figures show that the Q factor of planar solenoids reaches limit ($Q \approx 200$) for a size compatible with Printed Circuit Board (PCB) process. It shows also that the achievable range of inductance with good quality factor is quite short, roughly 10 nH to 2 µH.

4 Solenoids Tested Inside a VHF Converter

The converter under test is a $\Phi E2$ previously designed in [11]. The circuit diagram of the converter is presented in the Figure 10. The converter is made out of an inverter and a rectifier linked by a resonant series tank $L_s - C_s$. The inverter achieves naturally Zero Voltage Switching (ZVS) when sized properly [12]. The low voltage stress on the transistor is ensured thanks to the $L_{MR} - C_{MR}$ branch which resonates at two times the fundamental frequency of the converter. This filtering of the second harmonic ensures a maximum voltage of $2 \cdot V_{IN}$ on the transistor. The converter has been initially

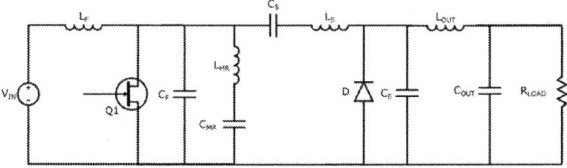

Fig. 10: Circuit diagram of the VHF converter.

designed with inductors made out of magnetic material (iron powder *Micrometal Mix-2*) as presented in Figure 11 a. Two inductors (L_{MR} and L_s) have been replaced by air core solenoids to show the performance and the limits of the air core technology. The air core solenoids have been designed with the following parameters and constraints: the thickness is fixed at $t = 5\,\text{mm}$, the length l is fixed by the available space on the existing layout. Then the width w and the number of turns N are chosen to obtain the desired inductance value with Eq. (5) while keeping a reasonable volume and a small R_{AC}. The designed solenoids are sub-optimized because of the existing constraint on the length. The measured inductance and Q values are shown in the Table 2. The L_{MR} inductance has been measured at 27.12 MHz because the MR branch resonates at the second harmonic. This air core

Tab. 2: Measured inductance L and Q factor of the inductors L_{MR} and L_s.

Inductor	f (MHz)	L (nH)		Q factor	
		core	air	core	air
L_{MR}	27.12	217	206	167	122
L_s	13.56	420	325	265	122

inductor shows similar performances to the core one with a relative difference on the inductance value of –5 % and relative difference on Q value of –27 %. The L_s inductor on the other hand works at the fundamental frequency 13.56 MHz and shows a relative difference on the inductance value of –22 % and relative difference of Q value of –54 %. The L_s case shows that it is harder to get a high Q with a lower frequency as expected by looking at Eq. (15). And at the same time the model for the inductance value is not accurate for solenoids with a greater width than length. This might be due to non uniformity of the magnetic field as already studied and well known on circular cross-section tape solenoids [13]–[15]. The adaptation of the formula to rectangular cross-section solenoids will be addressed in future work.

This drop of Q factor is mirrored on the global efficiency of the converter as shown in the table 3. The drop is due partly to a higher resistance value of the air core solenoid in comparison to the ring core with magnetic material and partly to the lower value of the inductance which causes the loss of the ZVS as depicted in the Figure 12 (higher magnitude oscillation around 0 in the case air).

Tab. 3: Measured efficiency of the converter with standard powder core components and air core solenoids.

Case	P_{in} (W)	P_{out} (W)	η (%)
core	6.53	4.99	76.4
air	7.9	5.02	63.5

While the electrical performance is smaller with air core solenoids the mass power density is better as described in the Table 4. The L_{MR} component is 3 times lighter than the powder core one. At the same time the overall volume is 1.72 smaller when considering the powder core without the winding (3.77 cm³) and 2.37 smaller when considering the overall volume (powder core and winding, $\approx 5.22\,\text{cm}^3$). The discrepancy of volume is due to

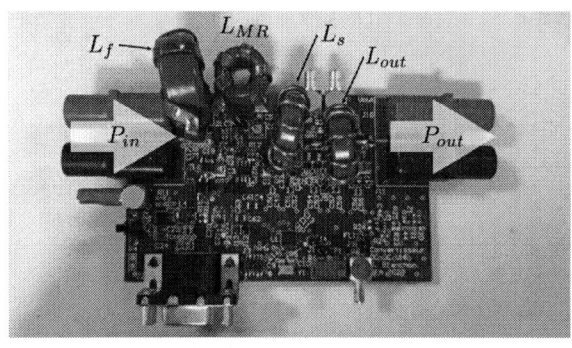 

a.   b.

Fig. 11: Photographs of the converter with a. the ring core inductors and b. two air core solenoids.

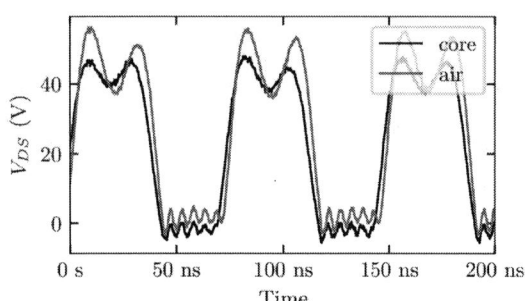

Fig. 12: Waveform of V_{DS} as a function of time.

the thickness and stiffness of the copper tape used for the powder core.

Tab. 4: Measured mass and volume of the inductance L_{MR}.

Case	Mass (g)	Core Volume (cm^3)
core	5.05	3.77
air	1.68	2.19

This comparison should be treated with caution, because the two types of components (the air core solenoids and the powder cores ones) have not been sized with exactly the same objectives and constraints due to the use of a pre-existing converter for the study case of one air core solenoids application. A more rigorous comparison will be applied in future study.

5 Conclusion

A simple analytical model has been checked against experiments and used to size two air core components for the replacement of standard powder core components. The design method is simple and fast, and the characteristics of the components obtained correspond to those expected from the modeling for cases with small thickness and width in comparison to the length of the solenoid. Thanks to this model, it has been shown that the Q factor of air core solenoids made out of planar substrate is mainly constrained by the substrate thickness. In addition the asymptotic behavior of the Q factor with the footprint area tends to limit the achievable increase in performance within a reasonable area range.

In comparison to standard powder core component, the air core solenoids use less raw material (in term of mass) and have a larger volume.

This work shows the difficulty to exploit the advantages of air core technology (less material use) for VHF converter while keeping the electrical performances the same. One way to take advantage of the air core technology for VHF converters is to increase the resonant frequency to decrease the requirement in inductance and to improve at the same time the Q factor.

In the future, the model will be refined to take into account the drop of inductance for solenoids with a width greater than the length. The performance of the solenoid will be compared to other shapes like ring cores in terms of quality factor, radiated magnetic field and thermal performances.

6 Acknowledgment

This work is supported by the French National Research Agency in the framework of the "Investissements d'avenir" program (ANR-15-IDEX-02), via the project CDP PowerAlps.

References

[1] J. Xu, Z. Tong, and J. Rivas-Davila, "1 kW MHz Wideband Class E Power Amplifier," *IEEE Open*

Journal of Power Electronics, vol. 3, pp. 84–92, 2022, Conference Name: IEEE Open Journal of Power Electronics. DOI: 10.1109/OJPEL.2022. 3146835.

[2] M. Madsen, A. Knott, and M. A. E. Andersen, "Low power very high frequency switch-mode power supply with 50 v input and 5 v output," *IEEE Transactions on Power Electronics*, vol. 29, no. 12, pp. 6569–6580, 2014. DOI: 10.1109/TPEL.2014. 2305738.

[3] L. Pace, M. Beley, M. E. Khattabi, and A. Bréard, "Design and modeling of a 100w 1mhz gan-based single-switch resonant converter for high power density inherent pfc led driver," in *2023 25th European Conference on Power Electronics and Applications (EPE'23 ECCE Europe)*, 2023, pp. 1–9. DOI: 10.23919/EPE23ECCEEurope58414.2023. 10264555.

[4] L. Gu and J. Rivas-Davila, "1.7 kW 6.78 MHz Wireless Power Transfer with Air-Core Coils at 95.7% DC-DC Efficiency," in *2021 IEEE Wireless Power Transfer Conference (WPTC)*, ISSN: 2573-7651, Jun. 2021, pp. 1–4. DOI: 10.1109/WPTC51349. 2021.9458037.

[5] M. Beley, M. El-Khattabi, L. Pace, and A. Bréard, "Analytical Design of a Finite Input Inductance 34.5 MHz Class E Inverter for Wireless Power Transfer," in *International Conference on Integrated Power Electronics Systems*, Power Engineering Society VDE ETG, Düsseldorf, Germany, Mar. 2024.

[6] M. Madsen, A. Knott, M. A. Andersen, and A. P. Mynster, "Printed circuit board embedded inductors for very high frequency Switch-Mode Power Supplies," in *2013 IEEE ECCE Asia Downunder*, Jun. 2013, pp. 1071–1078. DOI: 10.1109/ECCE-Asia.2013.6579241.

[7] J. D. Mønster, M. P. Madsen, J. A. Pedersen, and A. Knott, "Investigation, development and verification of printed circuit board embedded air-core solenoid transformers," in *2015 IEEE Applied Power Electronics Conference and Exposition (APEC)*, ISSN: 1048-2334, Mar. 2015, pp. 133–139. DOI: 10.1109/APEC.2015.7104343.

[8] M. Biglarbegian, N. Shah, I. Mazhari, and B. Parkhideh, "Design considerations for high power density/efficient PCB embedded inductor," in *2015 IEEE 3rd Workshop on Wide Bandgap Power Devices and Applications (WiPDA)*, Nov. 2015, pp. 247–252. DOI: 10.1109/WiPDA.2015. 7369285.

[9] G. Zulauf, W. Liang, and J. Rivas-Davila, "A unified model for high-power, air-core toroidal PCB inductors," in *2017 IEEE 18th Workshop on Control and Modeling for Power Electronics (COMPEL)*, Jul. 2017, pp. 1–8. DOI: 10.1109/COMPEL.2017. 8013401.

[10] Y. Wu and C. R. Sullivan, "Optimizations and Comparisons of Air-Core Inductors Based on a Semi-Analytical Calculation Toolkit," in *2021 IEEE 22nd Workshop on Control and Modelling of Power Electronics (COMPEL)*, ISSN: 1093-5142, Nov. 2021, pp. 1–7. DOI: 10.1109/COMPEL52922. 2021.9646075.

[11] V. Blanchon, S. Carcouet, X. Maynard, and G. Despesse, "A 13.56MHz DC-DC Converter with Innovative Output Voltage Regulation," in *13th International Conference on Power Electronics, Machines and Drives (PEMD 2024)*, Jun. 2024.

[12] J. M. Rivas, O. Leitermann, Y. Han, and D. J. Perreault, "A Very High Frequency DC–DC Converter Based on a Class Φ_2 Resonant Inverter," *IEEE Transactions on Power Electronics*, vol. 26, no. 10, pp. 2980–2992, 2011. DOI: 10.1109/TPEL.2011. 2108669.

[13] L. Lorenz, "Ueber die Fortpflanzung der Electricität," en, *Annalen der Physik*, vol. 243, no. 6, pp. 161–193, 1879. DOI: 10.1002/andp. 18792430602.

[14] E. B. Rosa and F. W. Grover. "Formulas and Tables for the Calculation of Mutual and Self-inductance." en. (1948), [Online]. Available: https: // nvlpubs.nist.gov / nistpubs / bulletin / 08 / nbsbulletinv8n1p1_A2b.pdf.

[15] D. W. Knight. "Solenoid Inductance Calculation." (Feb. 2016), [Online]. Available: https://g3ynh. info / zdocs / magnetics / part_1.html (visited on 04/02/2024).

PCIM Europe 2024, 11– 13 June 2024, Nuremberg   DOI: 10.30420/566262343

Improving Multi-Phase Ferrite Magnetics by Coupling for MV and UPS Converters

Michael Schmidhuber, Jonas Pfeiffer ⓘ, Philemon Wrensch, Manfred Wohlstreicher, Christian Blaum and Christoph Drexler

SUMIDA Components & Modules GmbH, Germany

Corresponding author: Michael Schmidhuber, mschmidhuber@eu.sumida.com
Speaker: Michael Schmidhuber, mschmidhuber@eu.sumida.com

Abstract

In the majority of power electronic systems, volume and weight are important design priorities, which can have a significant impact on other factors such as component costs. Magnetic components tend to be among the bulkiest and heaviest components in power electronics applications. Due to the high voltages and powers in medium voltage (MV) systems as well as in uninterruptible power supply (UPS) systems, the magnetic components used there are usually voluminous. Coupled inductors offer a good option for improving the design in this respect. In this paper, two coupled inductor designs for weight and volume reduction in MV systems are presented.

1   Introduction

Power electronics applications are subject to constant improvement and further development. In addition to efficiency, the main design priorities are volume and weight, as these have a significant impact on other design parameters such as costs or resource conservation. The rapid further development of semiconductor switches in particular has made it possible to increase switching frequencies or the power of converter systems.

In particular, the increased switching frequency has a positive effect on the volume reduction of the magnetic components, which are also part of numerous power electronics applications. Nevertheless, the magnetic components are generally among the heaviest and bulkiest components in a converter. This trend is further reinforced in medium voltage (MV) systems by increased demands on insulation due to the higher voltage level and cooling due to the increased power.

The use of coupled inductors is a promising option for reducing the volume or weight of the magnetic components and at the same time improving the converter in terms of the load on individual components, for example by reducing the phase current ripple. Coupling two or more inductors has been common practice for decades and has therefore been well studied. Basic analyses and design notes are given in [1,2]. However, it has to be mentioned that the design of coupled coils is by no means trivial.

For this reason, two designs of coupled inductors for MV systems are presented in this paper and compared with their conventional designs. Section 2 deals with a two-phase coupled design for an uninterruptible power supply (UPS) system in the form of an MV battery inverter system and section 3 with a three-phase coupled system in the form of an MV charge converter system.

2   Coupled inductors for medium voltage battery inverter systems

UPS systems fulfill an important role in grid-connected applications as they function as power backup in the event of a grid failure and protect the loads which are connected to the system from blackout and possible resulting damage.

2.1   System overview and specification

The investigated battery inverter system is shown in Fig. 1. A brief overview of the system is presented in [3].

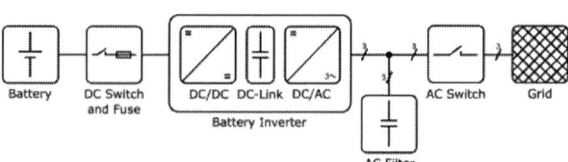

Fig. 1 MV battery inverter system [3]

Due to the specifications of the converter system, a Three-level Active Neutral-Point-Clamped 3Leg (3L-ANPC 3Leg) was chosen for the DC/AC stage. The topology is shown in Fig. 2. The investigated filter inductors are bordered in red.

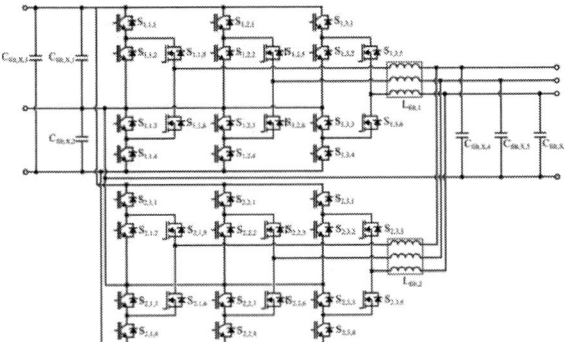

Fig. 2 Double interleaved Three-level Active Neutral-Point-Clamped 3Leg (3L-ANPC 3Leg). The filter inductors bordered in red are investigated. [3]

The standard non-interleaved version is analyzed in [4].

The specification of the conventional DC/AC filter inductor is shown in Table 1. The design of the conventional inductor is presented in detail in [3].

DC-Link voltage (V)	1200
Frequency (kHz)	70
RMS current (A)	113.5
Peak current (A)	190.1
Inductance (conventional inductor) (µH)	32
Maximum temperature (°C)	120

Table 1 Conventional DC/AC filter inductor specification of a MV battery inverter system [3]

2.2 Coupled inductor design

The coupled inductor was designed based on the conventional DC/AC filter inductor presented in [3]. The used mathematical model as well as simulation and measurement verifications are presented in the following subsections.

2.2.1 Mathematical optimization

In order to optimize the DC/AC filter inductor, the required inductance L as well as the coupling coefficient k have to be determined based on the given specifications (see Table 1). This is achieved using a mathematical model. The general inductance matrix for a two-phase case is as follows:

$$L_{coupled} = \begin{bmatrix} L & -k \cdot L \\ -k \cdot L & L \end{bmatrix} \quad (1)$$

The theoretical maximum value of the coupling coefficient in a two-phase coupled inductor is $k = 1$. In this case, the entire magnetic flux of one winding would flow through the other winding. This is not possible in practice. However, a proper design can achieve coupling coefficients very close to $k = 1$.

The greater the coupling coefficient, the lower the magnetic flux in the core material. The cross-section of the magnetic core can therefore be reduced so that the magnetic flux density remains approximately the same compared to the conventional inductor design. This saves core material and reduces the magnetic component's volume.

In most design routines for coupled inductors, the coupling coefficient is selected to be as large as possible in order to minimize the current ripple in the individual phases. However, it is often ignored that this simple optimization approach leads to shifts in the phase currents that increase the current ripple of the total current. However, an increased total current ripple would lead to a larger filter capacitance and thus to larger output capacitors, provided that the total current ripple at the output of the converter is to remain unchanged compared to the conventional case. As a result, the volume saved in the inductor would be partially or completely compensated by an increased volume of the capacitors.

Under the condition that the total current ripple of the coupled design should remain unchanged compared to the conventional case, so that no adjustment of the output capacitor is necessary, the inductance of the coupled inductor has to be increased accordingly. This can be done either by increasing the number of turns or by reducing the air gap while increasing the magnetic core's cross-section to avoid saturation effects. Both options lead to an increase in the volume of the magnetic component, which is contrary to the volume reduction described above. An optimum therefore has to be calculated.

The maximum value of the phase current ripple and total current ripple can be determined from the current increases of the phase currents. These are calculated as follows:

$$\frac{\Delta i_{ph}}{\Delta t} = L_{coupled}^{-1} \cdot v \quad (2)$$

Due to the high DC-Link voltage of 1200 V and if only a very small time section of one of the three phases within the positive half-wave is considered, the switching behavior of the 3L-ANPC 3Leg can be viewed in a very simplified form as a buck converter. This simplification results in the following relationship between the input and output voltage of the topology:

$$V_{out} = D \cdot \frac{V_{DClink}}{2} \qquad (3)$$

In order to be able to calculate the respective voltages applied on the two windings, the following two cases have to be distinguished depending on the duty cycle:

- $D < 0.5$
- $0.5 \leq D$

The differentiation between cases is based on whether there is a gap or non-gap resp. overlapping operation. The resulting time interval matrix below represents the first case ($D < 0.5$):

$$t_{c1} = T \cdot \left[D \quad \frac{1}{2} - D \quad D \quad \frac{1}{2} - D \right] \qquad (4)$$

Here, T is the period duration. The corresponding voltage matrix is:

$$v_{c1} = \frac{V_{DClink}}{2} \cdot \begin{bmatrix} 1-D & -D & -D & -D \\ -D & -D & 1-D & -D \end{bmatrix} \qquad (5)$$

If this is inserted into (2), the result is:

$$\frac{\Delta i_{ph,c1}}{\Delta t} = \frac{V_{DClink}}{2L} \begin{bmatrix} \frac{D \cdot k + (D-1)}{(k-1) \cdot (k+1)} & \frac{D}{k-1} & \frac{(D-1) \cdot k + D}{(k-1) \cdot (k+1)} & \frac{D}{k-1} \\ \frac{(D-1) \cdot k + D}{(k-1) \cdot (k+1)} & \frac{D}{k-1} & \frac{D \cdot k + (D-1)}{(k-1) \cdot (k+1)} & \frac{D}{k-1} \end{bmatrix} \qquad (6)$$

Since the curves of the individual phase currents can be viewed as a linear spline function, a section-by-section calculation is possible. To calculate this, the individual columns of the current rise matrix in (6) are multiplied by the individual columns of the time vector in (4). In the following, this multiplication was carried out for the first column of the matrix and the vector as an example:

$$\Delta i_{ph,c1,i1} = \frac{V_{DClink}}{2L} \cdot \begin{bmatrix} \dfrac{D \cdot k + (D-1)}{(k-1) \cdot (k+1)} \\ \dfrac{(D-1) \cdot k + D}{(k-1) \cdot (k+1)} \end{bmatrix} \cdot T \cdot D \qquad (7)$$

The four individual vectors as exemplarily calculated in (7) can be combined into a new phase current ripple matrix as follows:

$$\Delta i_{ph,c1} = \frac{V_{DClink} \cdot T}{2L} \begin{bmatrix} \frac{D \cdot (D \cdot k + (D-1))}{(k-1) \cdot (k+1)} & -\frac{D \cdot (2D-1)}{2 \cdot (k-1)} & \frac{D \cdot ((D-1) \cdot k + D)}{(k-1) \cdot (k+1)} & -\frac{D \cdot (2D-1)}{2 \cdot (k-1)} \\ \frac{D \cdot ((D-1) \cdot k + D)}{(k-1) \cdot (k+1)} & -\frac{D \cdot (2D-1)}{2 \cdot (k-1)} & \frac{D \cdot (D \cdot k + (D-1))}{(k-1) \cdot (k+1)} & -\frac{D \cdot (2D-1)}{2 \cdot (k-1)} \end{bmatrix} \qquad (8)$$

The two entries of every column of the phase current ripple matrix in (8) can be added to the total current ripple. In the case of the 3L-ANPC 3Leg topology with the assumed simplifications, the total current ripple oscillates between is maximum and minimum within a very small time interval, regardless of the duty cycle D. The evaluation of

the first column is therefore sufficient to calculate the maximum value of the total current ripple. The addition of the two entries in the first column in (8) results in the following:

$$\Delta i_{tot,c1} = \frac{V_{DClink} \cdot T}{2L} \cdot \frac{D \cdot (2D-1)}{k-1} \qquad (9)$$

Since both the coupling coefficient k and the total current ripple Δi_{tot} have an impact on the volume of the magnetic component, an optimum has to be found between the two variables. For this purpose, the total current ripple in (9) is set in relation to the coupling coefficient. Thus follows:

$$\frac{\Delta i_{tot,c1}}{k} = \frac{V_{DClink} \cdot T}{2L} \cdot \frac{D \cdot (2D-1)}{k \cdot (k-1)} \qquad (10)$$

To calculate the minimum of the ratio in (10), the zeros of its derivative are calculated. The result is a singular zero at $k = 0.5$. For the case of $D < 0.5$ the volume of the magnetic component is minimal with a minimal increase of the total current ripple for this coupling coefficient.

This mathematical optimization procedure now has to be repeated for the other case of $0.5 \leq D$. It leads to the same minimum $k = 0.5$. A graphical illustration of both investigated cases of duty cycle D is shown in Fig. 3.

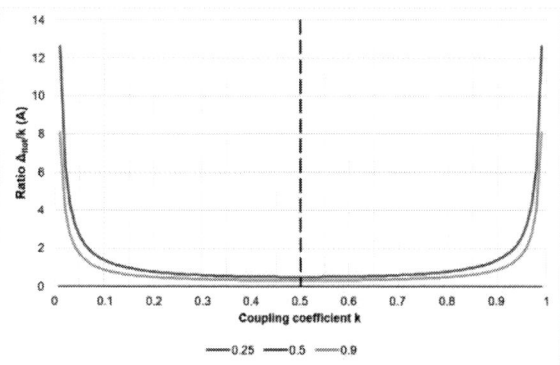

Fig. 3 Ratio of total current ripple to coupling coefficient over coupling coefficient for different duty cycles (0.25, 0.5 and 0.9)

The worst case of the total current ripple Δi_{tot} is $D = 0.25$ resp. $D = 0.75$. With a coupling coefficient of $k = 0.5$, the volume of the coupled design would be minimized. To achieve the same total current ripple as in the conventional design, the inductance value of the coupled design has to be increased. However, due to the specification defined for this application, the phase current ripple Δi_{ph} should also be minimized. For this purpose, the increase in phase current from (7) was set in relation to the coupling coefficient k:

$$\frac{\Delta i_{ph,c1,i1}}{k} = \frac{V_{DClink} \cdot T}{2L} \cdot \frac{D \cdot (D \cdot k + (D-1))}{k \cdot ((k-1) \cdot (k+1))} \quad (11)$$

A graphical illustration of (11) for the worst case of the total current ripple (D = 0.25) is shown in Fig. **4**. This worst case is used to prevent a massive increase in the total current ripple.

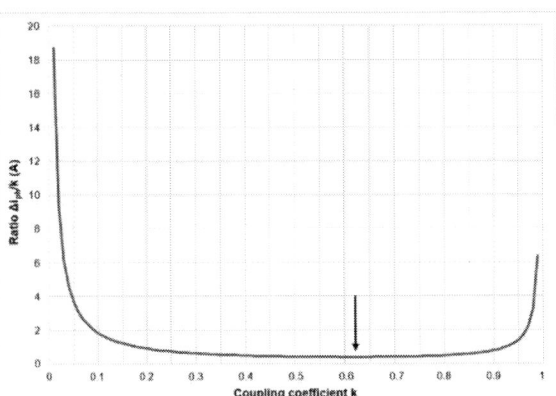

Fig. 4 Ratio of phase current ripple to coupling coefficient over coupling coefficient for D = 0.25

This results in a minimum at a coupling coefficient of k=0.62. At this value, the phase current ripple is minimized without massively increasing the total current ripple. This slight increase in the total current ripple can be compensated for by increasing the inductance. Although this means that the volume of the magnetic component is no longer minimal, it represents a good compromise between the required specifications.

For the selected coupling coefficient $k = 0.62$, the phase current ripple Δi_{ph} and the total current ripple Δi_{tot} are plotted in Fig. 5 for both investigated cases of the duty cycle D. The two maxima of the total current ripple at $D = 0.25$ and $D = 0.75$ are clearly visible. At a duty cycle of $D = 0.5$, both current ripples are minimal.

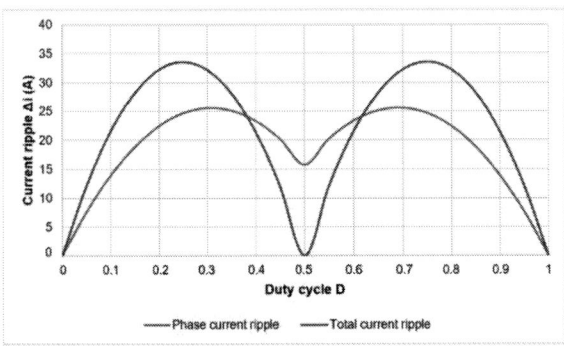

Fig. 5 Phase current ripple and total current ripple over duty cycle

The coupling coefficient k as well as the duty cycle D to be used were determined by mathematical optimization. The total current ripple Δi_{tot} is known from the conventional case. The required inductance value can therefore be calculated from (9). This results in $L = 84.2\,\mu H$.

2.2.2 Inductor design

The calculated values from the previous subsection were inserted into the inductance matrix in (1). This results in the following matrix:

$$L_{coupled,calc} = \begin{bmatrix} 84.2\,\mu H & -52.2\,\mu H \\ -52.2\,\mu H & 84.2\,\mu H \end{bmatrix}$$

To further improve the coupled design presented in [3], a customized core geometry made of the ferrite material Fi395 was designed. Fi395 is a power ferrite with low losses [5]. One half is shown in Fig. 6. The air gap in the center leg is used to adjust the coupling between both windings. The self-inductance can be adjusted by small air gaps in the outer legs.

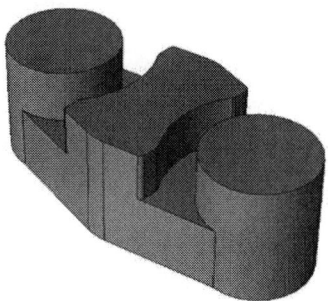

Fig. 6 Individual core shape of the improved coupled DC/AC filter inductor design

In contrast to the conventional inductor design, flat wire could be used in the improved coupled inductor design – even at a frequency of 70 kHz. This is possible due to the significant phase current ripple reduction from 67 A to 25 A (-62 %). In addition, the number of turns was adapted to the modified inductance value. The use of flat wire enables a smaller design with a higher copper filling factor as well as a cheaper manufacturing compared to litz wire.

2.2.3 Verification through Simulation

The improved coupled DC/AC filter inductor design was simulated using the electrical-magnetic 3D FEM simulations software JMAG. The simulation results confirmed the calculated design. A simulation of the magnetic flux distribution within the ferrite core is shown in Fig. 7.

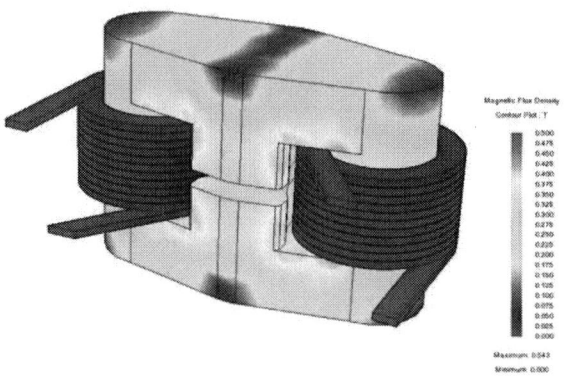

Fig. 7 Simulated flux density distribution of the improved coupled DC/AC filter inductor

The results of the electrical-magnetic 3D FEM simulation were combined with a thermal simulation. In order to adequately map the thermal behavior during subsequent installation, the simulated magnetic component is encapsulated in an aluminum pot. The pot is connected to the cooling system which provides a constant temperature of 65 °C. Figure 8 shows the simulated temperature distribution. Due to the higher winding losses, the windings become hotter than the ferrite core. The maximum is approx. 99 °C.

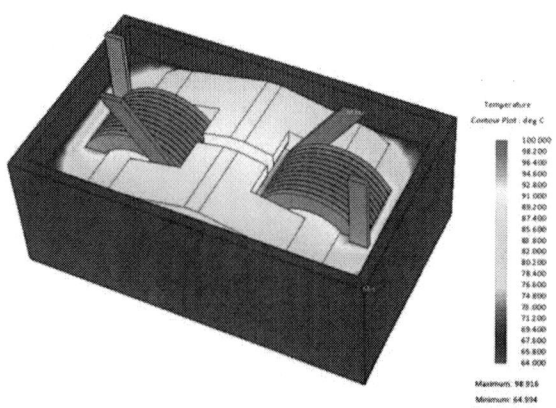

Fig. 8 Simulated temperature distribution of the improved coupled DC/AC filter inductor encapsulated in an aluminum pot with a connection to the cooling system

2.2.4 Verification through Measurement

The assembled component is shown in Fig. 9. The core was covered with tape to protect the insulation of the flat wire from damage due to repeated assembly and disassembly.

Fig. 9 Assembled improved coupled DC/AC filter inductor

The magnetic component was assembled and measured within the aluminum housing (without potting). Taking the aluminum housing into account in the measurements is very important because the housing changes the values of the inductance matrix. The measurement results are:

$$L_{coupled,meas} = \begin{bmatrix} 83.1 & -50.9 \\ -50.9 & 83.4 \end{bmatrix}$$

This results in the following coupling coefficient matrix:

$$k_{meas} = \begin{bmatrix} 1 & -0.61 \\ -0.61 & 1 \end{bmatrix}$$

2.3 Comparison of the designs

The conventional DC/AC filter inductor with its dimensions is shown in Fig. 10 a). A first design approach of a coupled inductor, which was also briefly presented in [3], is shown in Fig. 10 b), whereas in Fig. 10 c) shows the improved coupled DC/AC filter design with its dimensions. The weight values given in the caption refer to the components without housing and potting.

The weight reduction of the designs is shown in Fig. 11. The weight values presented there include housing and potting.

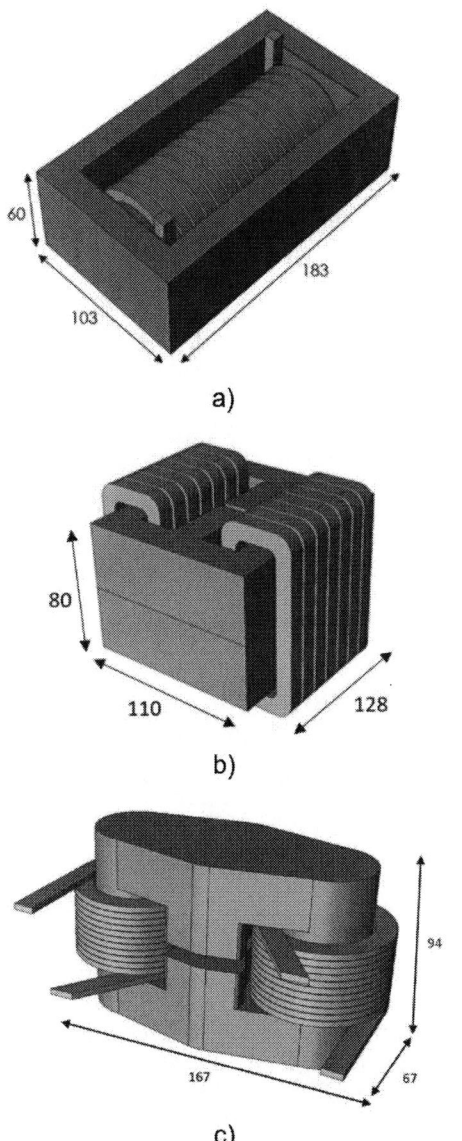

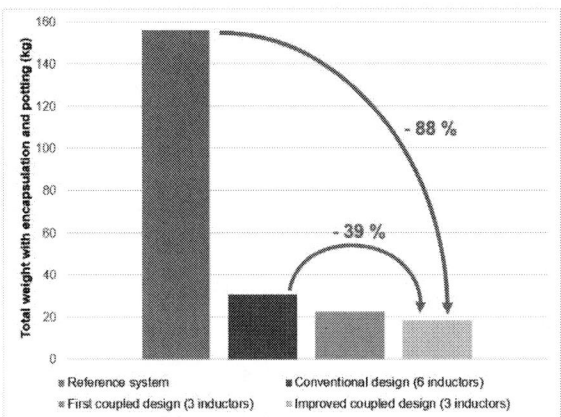

Fig. 11 Total weight (including housing and potting of the reference design (blue), the conventional design (red), the first coupled design (grey) and the improved coupled design (yellow)

The filter inductors of the reference system presented in [3] have a total weight of 156 kg. Including housing and potting, the improved coupled DC/AC filter inductor design achieves a weight of 18.3 kg per module. This results in a weight reduction of approx. 88 %. However, a large proportion of the weight reduction is due to the significant increase in the switching frequency from 1.6 kHz to 6 kHz in the reference system to 70 kHz in the current battery inverter system.

For this reason, a comparison with the conventional inductor design is more informative, as both were developed on the basis of 70 kHz. Here, the improved coupled inductor design leads to a weight reduction of approx. 39 %.

Fig. 10 Improvement of the DC/AC filter inductor (dimensions in mm)
a): Conventional DC/AC filter inductor with an overall weight of 30.3 kg (per module) [3];
b): First coupled DC/AC filter inductor with an overall weight of 22.4 kg (per module) [3];
c): Improved coupled DC/AC filter inductor with an overall weight of 17.3 kg (per module)

In the conventional case, six individual inductors are required. These can be replaced by three coupled inductors. This means that the improved coupled DC/AC filter inductor design can achieve a volume reduction of approx. 53 % (without housing and potting compared to the conventional inductor design.

The dimensions in Fig. 10 show that the volume of the coupled design could be reduced by approx. 53 % compared to two conventional inductors (without housing and potting).

Furthermore, the losses of the conventional DC/AC filter inductor design are compared with the improved coupled DC/AC filter inductor design. A distribution of the individual losses is shown in Table 2. The improved coupled design shows a reduction of approx. 42 % in losses compared to the conventional design.

Category	Conventional	Coupled
DC windings losses	117.5 W	58.7 W
AC winding losses	166.9 W	126.8 W
Core losses	113.9 W	47 W
Total losses	398.3 W	232.5 W
	Improvement	-42 %

Table 2 Loss distribution of the conventional inductor design (6 inductors) and the improved coupled inductor design (3 inductors)

3 Coupled inductors for medium voltage charging systems

DC charging systems for electric vehicles (EVs) are currently limited to the maximum transmitted power of the low voltage power grid. To increase the transmitted power and with that the number of EV which can be charged at the same time, a DC charging system connected to the medium voltage power grid is a promising approach.

3.1 System overview and specification

The approach investigated envisages that the MV power grid supplies an MV DC grid via a converter. Several converter systems are then connected to this MV DC grid, which charge the respective EVs. An overview of the overall converter system including its topologies, the used semiconductors as well as the magnetic components and measurement results is published in [6].

The investigated charging converter system is shown in Fig. 12.

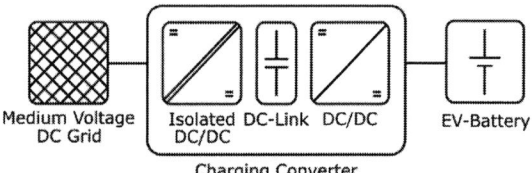

Fig. 12 MV charging converter system [7]

Due to the specifications of the converter system, a three-phase interleaved buck converter was chosen for the DC/DC stage. The topology is shown in Fig. 13.

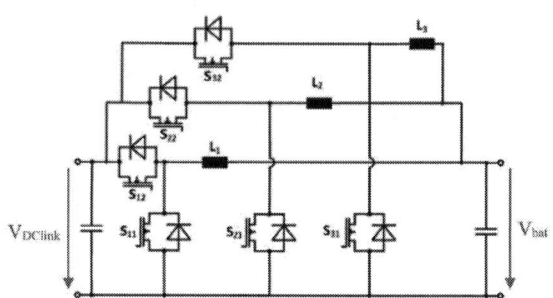

Fig. 13 Three-phase interleaved buck converter [7]

The topology is presented and analyzed for lower voltage levels in [8,9].

The design of the conventional inductor is presented in detail in [7]. The specification of the coupled inductor is shown in Table 3.

DC-Link voltage (V)	1200
Nominal battery voltage (V)	600
Frequency (kHz)	70
Maximum output current (A)	250
Maximum output current ripple (A_{pp})	9
Self-inductance (coupled inductor) (µH)	926
Maximum output power (kW)	175

Table 3 Coupled storage inductor specification of a MV charging converters system (acc. to [7])

3.2 Coupled inductor design

The design as well as the simulative and measurement verification of the coupled inductor is described in detail in [7]. It also briefly discusses the selection of materials for magnetic components in MV applications. A detailed design study regarding the choice of core material between metal powder and ferrite is presented in [10].

The design of the coupled inductor design was carried out and verified by simulation on the basis of a design study according to [11,12].

The assembled magnetic component with its housing (without potting) is shown in Fig. 14.

PCIM Europe 2024, 11– 13 June 2024, Nuremberg   DOI: 10.30420/566262343

Fig. 14 Assembled coupled inductor with housing

3.3 Comparison of the designs

Figure 15 shows the conventional as well as the coupled inductor design with their dimensions. The weight values given in the caption refer to the components without housing and potting.

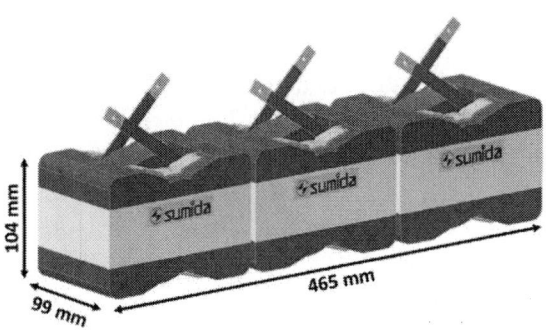

a)

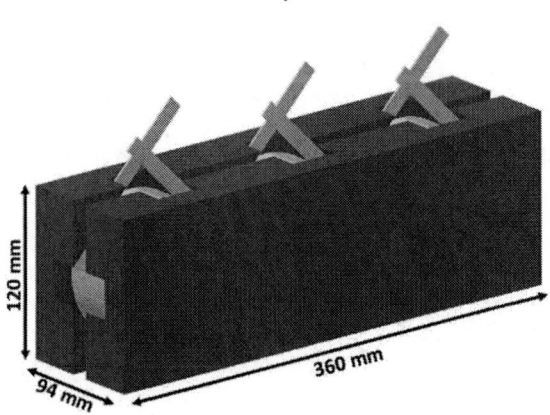

b)

Fig. 15 Improvement of the storage inductor (dimensions in mm)
a): Three Conventional storage inductors with an overall weight of 20.8 kg [7];
b): Coupled storage inductor with an overall weight of 17.7 kg [7]

The comparison of the two designs shown in Table **4** is presented in [7]. The values refer to the magnetic components as a module without housing and potting.

Category	Conventional inductors	Coupled inductor	Improvement
Enveloping volume	4.79 liters	4.06 liters	-15 %
Weight	20.82 kg	17.74 kg	-15 %
Total losses	262.8 W	233.1 W	-11 %

Table 4 Comparison between the conventional inductor module (3 inductors) and the coupled inductor (without housing and potting) [7]

Since the magnetic components are housed and potted for their installation in the DC/DC stage of the charging converter, it makes sense to compare both designs with housing and potting.

The comparison of both designs with regard to their total weight is shown in Fig. 16.

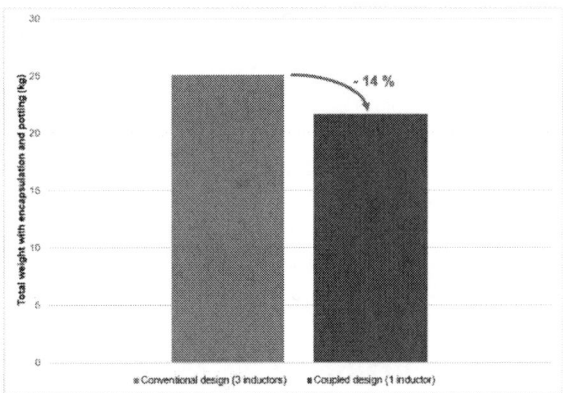

Fig. 16 Total weight (including housing and potting) of the conventional design (blue) and the coupled design (red)

The coupled inductor design has a total weight of 21.65 kg. Housing and potting have a similar weight in both designs. As a result, the weight reduction of the coupled design compared to the conventional case is slightly lower than in the comparison of the magnetic components in [7].

The reduction of the enveloping volume including housing and potting is shown in Fig. 17.

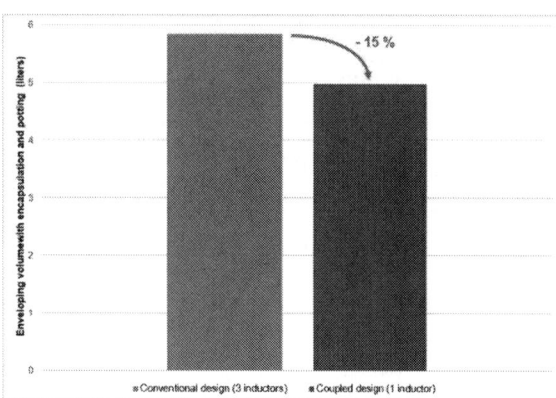

Fig. 17 Enveloping volume (including housing and potting) of the conventional design (blue) and the coupled design (red)

With regard to the enveloping volume, the improvement of the coupled design also decreases slightly in terms of volume reduction. However, this decrease due to housing and potting is insignificant.

It has to be mentioned that the reduction potential of this application is lower due to the given specifications. Furthermore the conventional inductor design is already highly optimized. For example, the conventional design already uses flat wire instead of litz wire, which greatly reduces the volume of the conventional inductor design and its winding losses. In addition, the PQ core geometry used in the conventional design already has a favorable ratio of cross-section to volume.

4 Conclusion

Magnetic components are subject to constant further development in terms of reducing their volume and weight. Particularly in medium-voltage (MV) applications, savings in terms of core and winding materials and the associated costs are very desirable.

The use of coupled inductors is a promising option for saving volume and weight while reducing the current ripple in the converter.

In this paper, two coupled inductors for MV applications were presented and compared with their conventional designs.

In the two-phase case for a battery inverter as part of an uninterruptible power supply (UPS) system, the weight of the coupled design could be reduced by approx. 39 % compared to the conventional design with a volume reduction of approx. 53 %. Both values are without housing and potting. The reduction in losses is approx. 42 %.

In the three-phase design for an electric vehicle (EV) charging converter, the reduction potential is smaller. Here, the reductions are 14 % in weight and 15 % in volume (including housing and potting) with a simultaneous loss reduction of 11 %. The reason for the lower reduction lies primarily in the already highly optimized design of the conventional inductor.

The use of coupled inductors can therefore massively improve the converter application and thus reduce costs and promote the sustainable use of resources.

5 Acknowledgements

Parts of the work presented in this paper have been supported by the German Federal Ministry of Economic Affairs and Climate Action (BMWK). The project funding reference numbers are 03EI6030F and 01MV21010B. Responsibility of the content of this publication lies with the authors.

References

[1] S. Cuk and Z. Zhang. *"Coupled-inductor anal sis and design"*. Conference publication. PESC. Vancouver, BC, Canada. June 1986. DOI: 10.1109/PESC.1986.7415621

[2] A. F. Witulski. *"Modeling and design of transformers and coupled inductors"*. Conference publication. APEC. San Diego, CA, USA. March 1993. DOI: 10.1109/APEC.1993.290725

[3] M. Wohlstreicher, J. Pfeiffer and M. Schmidhuber. *"Improved magnetic devices for battery inverter systems with a high power-to-weight ratio"*. Conference publication. PCIM Europe. Nuremberg, Germany. May 2023. DOI: 10.30420/566091182

[4] T. Brückner and S. Bemet; "Loss balancing in three-level voltage source inverters applying active NPC switches". IEEE Annual Power Electronics Specialists Conference (IEEE Cat. No.01CH37230), Vancouver (BC), Canada, 2001, DOI: 10.1109/PESC.2001.954272

[5] SUMIDA Components & Modules GmbH; "Magnetic material Fi395"; Datasheet; Obernzell, Germany; 2018.

[6] David Derix et al. *"High-voltage converter with 2 kV SiC MOSFETs for electric fast charging stations"*. Conference publication. ECCE Europe. Düsseldorf, Germany. September 2024. In press.

[7] Christian Blaum, Philemon Wrensch, Jonas Pfeiffer, Christoph Drexler, Michael Schmidhuber, David Derix and Andreas Hensel. *"A three-phase coupled inductor for a medium voltage high power charging system."* Conference publication. CIPS. Düsseldorf, Germany. March 2024.

[8] Praful Nandankar and Jyoti P. Rothe. *"Design and implementation of efficient three-phase interleaved DC-DC converter"*. Conference publication. ICEEOT. Chennai, India. March 2016. DOI: 10.1109/ICEEOT.2016.7754962

[9] Akriti Garg and Moumita Das. *"High efficiency three phase interleaved buck converter for fast charging of EV"*. Conference publication. ICPEE. Bhubanes-war, India, January 2021. DOI: 10.1109/ICPEE50452.2021.9358486

[10] Christoph Drexler, Jonas Pfeiffer, Michael Schmidhuber, David Derix, Michael Geiss and Juergen Thoma. *"Magnetic component design for medium voltage photovoltaic application"*. Conference publication. PCIM Europe. Nuremberg, Germany. May 2022. DOI: 10.30420/565822252

[11] Christoph Drexler, Manfred Wohlstreicher, Philemon Wrensch, Herbert Jungwirth and Michael Schmidhuber. *"Calculation and verification of high-frequency losses in power inductors for automotive application"*. Conference publication. PCIM Europe. Nuremberg, Germany, May 2021. ISBN: 978-3-8007-5515-8

[12] Christoph Drexler and Michael Schmidhuber. *"Towards a multiphysics FEM simulation model of an arbitrary inductive component for power electronic applications"*. Conference publication. PCIM Europe. Nuremberg, Germany, May 2023. DOI: 10.30420/566091059

PCIM Europe 2024, 11– 13 June 2024, Nuremberg   DOI:10.30420/566262344

22-kW Bidirectional Single-Stage Direct-Ac-Ac Power Conversion On-Board Charger with High-Power-Density Implementations

Héctor Sarnago[1], Ignacio Álvarez[1], and Oscar Lucía[1]

[1] Electronics Engineering and Communications Department. Universidad de Zaragoza, I3A, Spain.

Corresponding author: Óscar Lucía, olucia@unizar.es
Speaker: Óscar Lucía, olucia@unizar.es

Abstract

Modern electric vehicles rely on on-board chargers (OBC) to provide fast and efficient charging when connected to the grid. Classical OBC implementations use a dual-stage conversion featuring a PFC rectifier plus a dc-dc converter. Whereas effective, this implementation significantly affects power density, which is a key aim for automotive converters. In this context, this paper proposes a 22-kW bidirectional single-stage direct-ac-ac power conversion OBC. The proposed converter features a single-stage implementation with direct ac-ac power conversion, achieving higher power density due to the reduction in power, magnetic and capacitive components. The proposed converter has been experimentally verified by means of a 22-kW prototype, proving the feasibility of this proposal.

1 Introduction

Transition towards electric transportation is a key step forward to achieve the Paris agreement and some of the UN Sustainable Development Goals [1, 2]. To achieve such aims, transportation electrification is a key pillar for sustainability. Consequently, electric vehicle (EV) technology development and adoption has grown exponentially in recent years. Nowadays, EV includes a number of key power converters (Fig. 1) [3], including inverters [4], on-board chargers [5] and dc-dc converters [6-8] to manage the complete vehicle powertrain (0).

Nowadays, electric vehicles can perform ultrafast/fast battery charge (level 3) using fast chargers that provide the required dc voltage [9], or slower charge using the ac grid and relying on the OBC (level 1-2) to perform the ac-dc power conversion. In recent years, significant efforts have been paid to the development of higher efficiency, power density and performance OBCs to provide efficient implementations for EV manufacturers [3, 5, 10, 11].

Classical OBC converter implementation includes a dual conversion stage composed of a PFC rectifier plus a dc-dc converter [12]. The PFC rectifier utilizes different implementations depending on the required power density, single or multi-phase operation [13-15] and bidirectional capabilities. The dc-dc converter stage is usually based on a

resonant converter, LLC or CLLC [16], or a dual active bridge (DAB) [17] implementation. This allows to obtain a decoupled system with important benefits in terms of performance and control. However, this power conversion architecture leads to limited power density due to the required number of power devices, transformers/inductors and decoupling capacitors [18].

In order to overcome these limitations, this paper proposes a 22-kW bidirectional single-stage direct-ac-ac power conversion OBC. The proposed converter features a single-stage implementation with direct ac-ac power conversion, achieving higher power density due to the reduction in power, magnetic and capacitive components. The proposed converter has been experimentally verified by means of a 22-kW prototype, proving the feasibility of this proposal.

The remainder of this paper is organized as follows. Section II details the proposed power converter and Section III shows the implemented experimental prototype and the main experimental results. Finally, Section IV summarizes the conclusions of this paper.

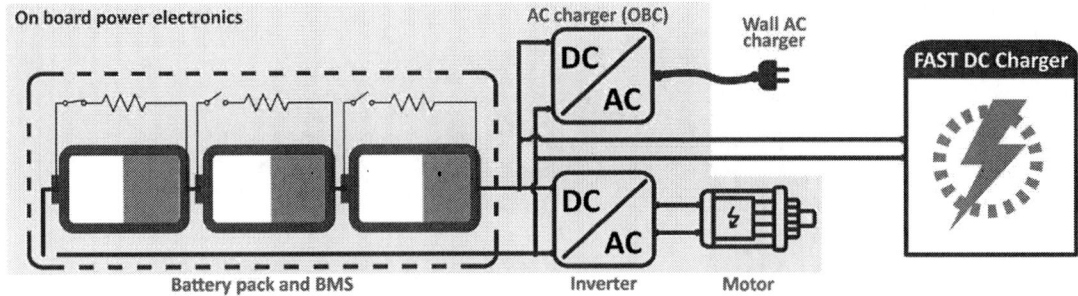

Fig. 1 EV power electronic architecture.

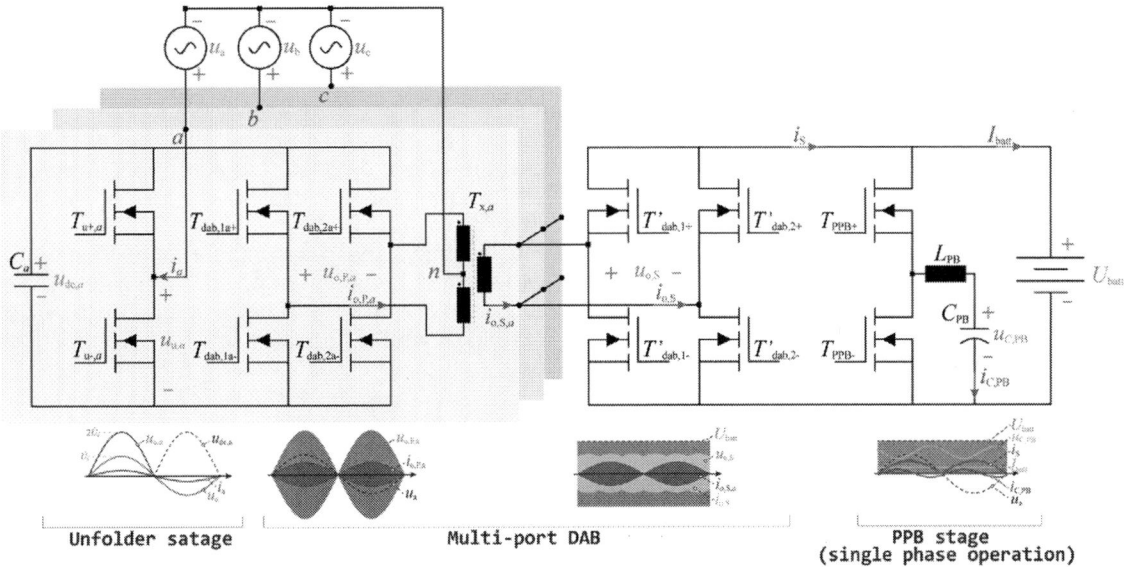

Fig. 2 Proposed converter and main waveforms.

2 Proposed power converter

The topology presented in this paper is illustrated in Fig. 2. It comprises three identical power rails on the mains side, one for each phase, and a single rail on the secondary side. Consequently, the former can operate at full power in both single and three-phase configurations, subjecting the power elements to similar stress.

The primary rails, where each line (i={a, b, c}) is connected, consist of an unfolder stage formed by the half-bridge ($T_{u+,i}$, $T_{u-,i}$) and a full-bridge converter ($T_{dab,1i+}$, $T_{dab,1i-}$, $T_{dab,2i+}$, $T_{dab,2i-}$) that supplies the primary side of each high-frequency (HF) transformer, $T_{x,i}$. The mains' neutral point is connected to the midpoint of the primary winding. By setting the full-bridge duty cycle to 50%, a voltage doubler configuration is achieved, reducing the re-

quired transformer currents and, consequently, increasing efficiency. This also leads to direct ac-ac conversion, resulting in enhanced efficiency by eliminating the traditional full-bridge rectifier stage and enabling full bidirectional operation. The heart of the converter's operation is a bidirectional multi-port DAB converter, featuring four ports: one for each mains phase and a fourth port for the secondary side, all controlled by the full-bridge ($T'_{dab,1+}$, $T'_{dab,1-}$, $T'_{dab,2+}$, $T'_{dab,2-}$). The power flow is regulated by the phase shift between the different primary power rails and the battery port, whose duty cycle is adjusted to accommodate battery voltage variations.

In the case of single-phase operation, a power pulsating buffer (PPB) structure is integrated into the converter to mitigate low-frequency ripple content to the battery. This structure is based on a buck configuration (T_{PPB+}, T_{PPB+}, L_{PB}, C_{PB}), enabling

2449

Fig. 3 Experimental prototype.

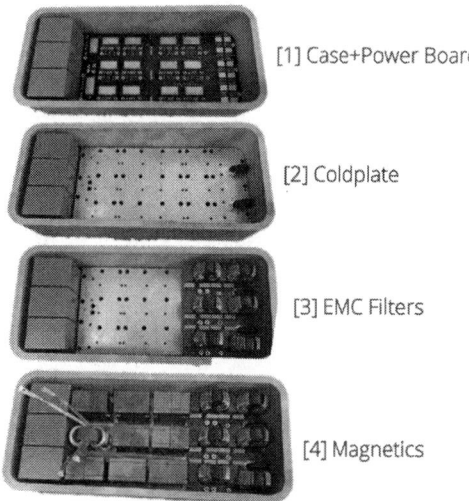

[1] Case+Power Board

[2] Coldplate

[3] EMC Filters

[4] Magnetics

Fig. 4 Detailed view of the experimental prototype architecture.

more efficient utilization of the energy storage capacitor. It can replace traditional electrolytic capacitors with film technology.

3 Implementation and experimental results

In order to test the proposed topology and control strategy a 800-V, 22-kW bidirectional OBC prototype has been designed and implemented. The implemented prototype is depicted in Fig 3 and Fig. 4. It features SiC power modules NVXK2TR80WDT (1200 V, 80 mOhm) from ON-SEMI, FILM capacitors for the PPB stage (3x210 µF), and PQ50 cores for the magnetics. A water-cooled cold plate is placed in the middle, serving for cooling both the magnetics (in the bottom side) and the power devices (top side). Finally, the converter is digitally controlled via FPGA, AMD-XILINX Artix 7-100 T.

Fig. 5 shows the main experimental results, key waveforms at maximum power for a single-rail operation (a), and 22 kW, 3ph operation (b). These results prove the feasibility of the proposed topology.

4 Conclusions

This paper has proposed proposes a 22-kW bidirectional single-stage direct-ac-ac power conversion OBC. The proposed converter features a single-stage implementation with direct ac-ac power conversion, achieving higher power density due to the reduction in power, magnetic and capacitive components. The proposed converter has been experimentally verified by means of a 22-kW prototype, proving the feasibility of this proposal.

Acknowledgment

This work was partly supported by Projects TED2021-129274B-I00, CNS2023-144980, and PDC2023-145837-I00 co-funded by MICIU/AEI/10.13039/501100011033, by "ERDF A way of making Europe", by the "European Union NextGenerationEU/PRTR", and by the DGA-FSE.

PCIM Europe 2024, 11– 13 June 2024, Nuremberg   DOI:10.30420/566262344

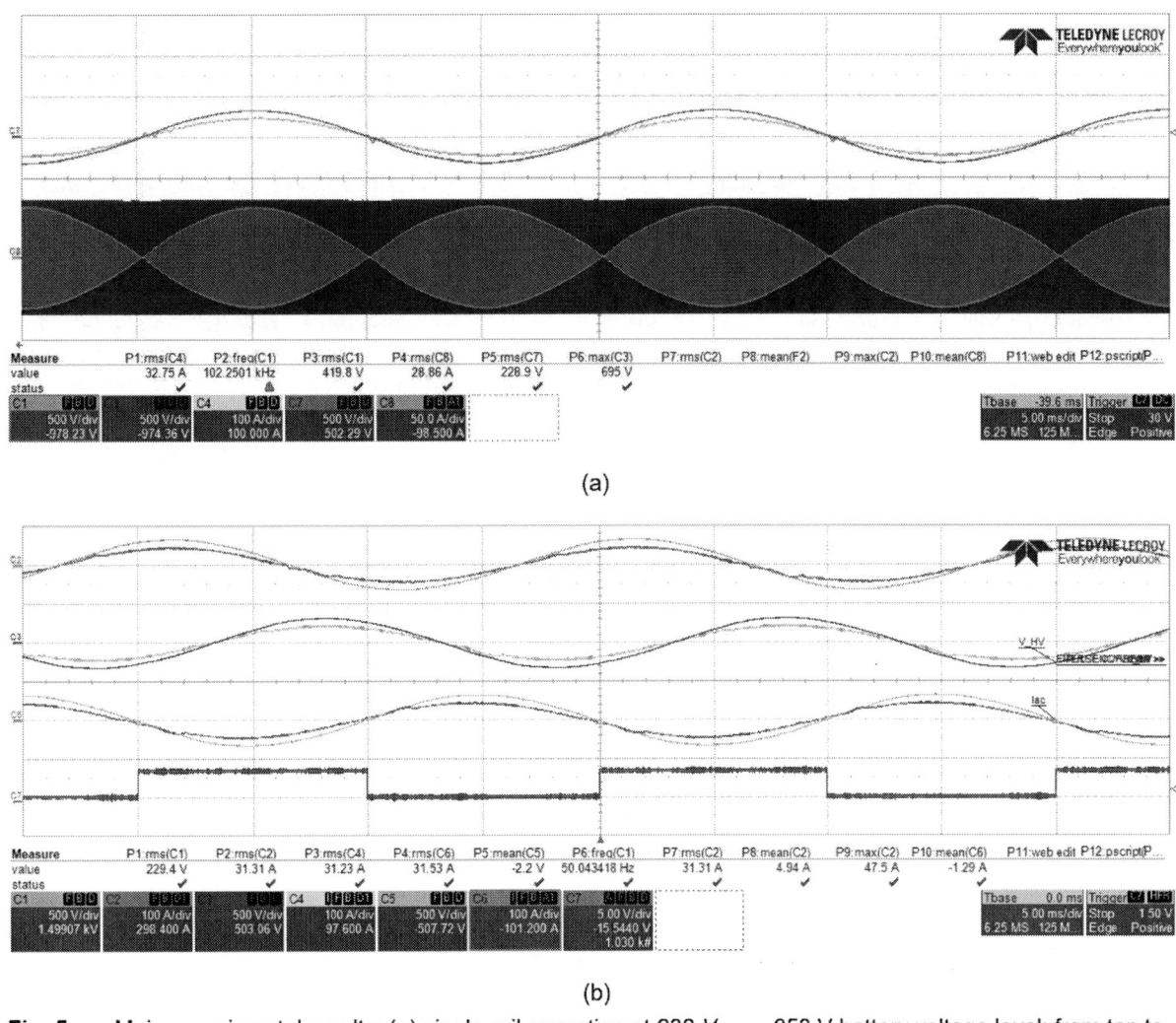

(a)

(b)

Fig. 5   Main experimental results: (a) single rail operation at 230 $V_{ac,rms}$, 650 V battery voltage level; from top to bottom: mains voltage and current and primary and secondary transformer voltages. (b) three-phase operation at 22 kW, 700 V battery voltage; from top to bottom: mains voltages and currents and digital synchro signal.

References

[1] WEC, "A vision for a sustainable battery value chain in 2030. Unlocking the full potential to power sustainable development and climate change mitigation.," World Economic Forum. Global Battery Alliance., 2019.

[2] *Transforming our world : the 2030 Agenda for Sustainable Development,* U. G. Assembly, 2015.

[3] M. Yilmaz, and P. T. Krein, "Review of Battery Charger Topologies, Charging Power Levels, and Infrastructure for Plug-In Electric and Hybrid Vehicles," *IEEE Transactions on Power Electronics,* vol. 28, no. 5, pp. 2151-2169, 2013, doi: 10.1109/TPEL.2012.2212917.

[4] S. S. Williamson, A. K. Rathore, and F. Musavi, "Industrial Electronics for Electric Transportation: Current State-of-the-Art and Future Challenges," *IEEE Transactions on Industrial Electronics,* vol. 62, no. 5, pp. 3021-3032, 2015, doi: 10.1109/TIE.2015.2409052.

[5] A. Khaligh, and M. D. Antonio, "Global Trends in High-Power On-Board Chargers for Electric Vehicles," *IEEE Transactions on Vehicular Technology,* vol. 68, no. 4, pp. 3306-3324, 2019, doi: 10.1109/TVT.2019.2897050.

[6] P. He, and A. Khaligh, "Comprehensive Analyses and Comparison of 1 kW Isolated DC–DC Converters for Bidirectional EV Charging Systems," *IEEE Transactions on*

2451

Transportation Electrification, vol. 3, no. 1, pp. 147-156, 2017, doi: 10.1109/TTE.2016.2630927.

[7]   H. Sarnago, and O. Lucía, "Bidirectional 400-12 V dc-dc Converter with Improved Dynamics and Integrated Transformer for EV Applications," in *IEEE Applied Power Electronics Conference and Exposition*, 2024, pp. 3081-3085.

[8]   H. Sarnago, I. Álvarez-Gariburo, and O. Lucía, "Bidirectional isolated 400-12V dc-dc converter with improved power density and full-range operation for EV applications," in *International Exhibition and Conference for Power Electronics, Intelligent Motion, Renewable Energy and Energy Management PCIM 2024*, 2024: VDE VERLAG Gmbh.

[9]   H. Sarnago, I. Alvarez-Gariburo, and O. Lucia, "High-performance bidirectional fast EV charger featuring full power/voltage range and cost-effective implementation," in *2023 IEEE 17th International Conference on Compatibility, Power Electronics and Power Engineering (CPE-POWERENG)*, Tallin, 2023, pp. 1-4, doi: 10.1109/CPE-POWERENG58103.2023.10227422.

[10] J. Yuan, L. Dorn-Gomba, A. D. Callegaro, J. Reimers, and A. Emadi, "A Review of Bidirectional On-Board Chargers for Electric Vehicles," *IEEE Access,* vol. 9, pp. 51501-51518, 2021, doi: 10.1109/ACCESS.2021.3069448.

[11] D. Cesiel, and C. Zhu, "A Closer Look at the On-Board Charger: The development of the second-generation module for the Chevrolet Volt," *IEEE Electrification Magazine,* vol. 5, no. 1, pp. 36-42, 2017, doi: 10.1109/MELE.2016.2644265.

[12] H. Sarnago, O. Lucía, R. Jiménez, and P. Gaona, "Differential-power-processing on-board-charger for 400/800-V battery architectures using 650-V super junction MOSFETs," in *IEEE Applied Power Electronics Conference and Exposition*, 2021, pp. 564-568.

[13] D. Menzi, J. W. Kolar, H. Sarnago, O. Lucia, and J. E. Huber, "New 600 V GaN single-stage isolated bidirectional 400 V input three-phase PFC rectifier," in *IEEE Energy Conversion Conference and Expo ECCE23*, Nashville, IEEE, Ed., 2023: IEEE.

[14] H. Sarnago, O. Lucía, S. Chhawchharia, D. Menzi, and J. W. Kolar, "Novel Bidirectional Universal 1-Phase/3-Phase-Input Unity Power Factor Differential AC/DC Converter," *Electronics Letters,* vol. 59, no. 13, pp. 1-4, 2023, doi: https://doi.org/10.1049/ell2.12857.

[15] P. Papamanolis, F. Krismer, and J. W. Kolar, "22 kW EV battery charger allowing full power delivery in 3-phase as well as I-phase operation," in *2019 10th International Conference on Power Electronics and ECCE Asia (ICPE 2019 - ECCE Asia)*, 27-30 May 2019 2019, pp. 1-8, doi: 10.23919/ICPE2019-ECCEAsia42246.2019.8797063.

[16] K. Siebke, T. Schobre, and R. Mallwitz, "Comparison of GaN based CLLC converters for EV chargers operating at different switching frequency ranges," in *2019 21st European Conference on Power Electronics and Applications (EPE '19 ECCE Europe)*, 3-5 Sept. 2019 2019, pp. P.1-P.9, doi: 10.23919/EPE.2019.8915565.

[17] H. Sarnago, and O. Lucía, "Optimized EV ON-Board Charging Power Converter Using Hybrid DCX-Dab Topology," in *IEEE Applied Power Electronics Conference and Exposition*, 2024, pp. 1305-1309.

[18] H. Kim, J. Park, S. Kim, R. M. Hakim, H. Belkamel, and S. Choi, "A Single-Stage Electrolytic Capacitor-less EV Charger with Single- and Three-Phase Compatibility," *IEEE Transactions on Power Electronics*, pp. 1-1, 2021, doi: 10.1109/TPEL.2021.3127010.

PCIM Europe 2024, 11– 13 June 2024, Nuremberg   DOI: 10.30420/566262345

Benchmarking DC Fast Chargers: A Comparative Analysis of Power Converter Structures for Wide Voltage Range

Sadik Cinik[1], Fangzhou Zhao[1], Xiongfei Wang[1,2], Giuseppe De Falco[3]

[1] Aalborg University, Denmark
[2] KTH Royal Institute of Technology, Sweden
[3] Infineon Technologies AG, Austria

Corresponding author: Xiongfei Wang, xwa@energy.aau.dk
Speaker: Sadik Cinik, sci@energy.aau.dk

Abstract

Advancements in DC fast chargers are pivotal for accelerating the adoption of Electric Vehicles (EVs), particularly by enhancing charger efficiency and cost-effectiveness within a wide output voltage range (150 V to 1000 V). This paper benchmarks three predominant DC/DC converter structures—single-stage, two-stage, and relay matrix—against consistent performance metrics to determine the most effective configuration for fast charging applications. Employing a standardized evaluation framework, we compare these systems based on their power efficiency and overall cost implications under wide output voltage range operation. Our findings indicate that the two-stage converter configuration provides the optimal balance, offering high efficiency across this broad voltage spectrum while maintaining cost-effectiveness. These insights are crucial for refining DC fast charger designs to meet future technological and commercial benchmarks.

1 Introduction

Sustainable transportation increasingly depends on the widespread adoption of EVs to decrease carbon emissions and fossil fuel reliance [1]. However, the acceptance of EVs is hindered by concerns such as range anxiety and long charging times, which critically influence consumer purchase decisions and, consequently, the broader impact of EVs on sustainability goals.

The prevalence of fast-charging stations with higher rated power is seen as a solution to reduce charging time and alleviate range anxiety among EV owners [2]. Achieving this is not solely about increasing charging power but also involves improving battery technologies to accommodate high-power charging demands. For instance, 800V batteries, which are utilized in vehicles like the Audi E-Tron GT, Hyundai Ioniq 6, and Porsche Taycan, can achieve quicker charge times compared to conventional 400 V batteries [3]. Such advancements necessitate compatible DC fast charging infrastructure that can handle a broad output voltage range (150 V to 1000 V), catering to both current and upcoming EV technologies.

Current DC fast chargers utilize a sophisticated multistage power conversion process to meet the rapid charging needs of modern EVs [4]. Initially, an AC/DC rectifier converts AC mains voltage to a stable DC output. This DC output is then precisely adjusted by a DC/DC converter stage to match the specific charging requirements of the EV battery, ranging from 150 V to 1000 V. Furthermore, isolation between the power grid and the EV battery during charging is essential for safety. This isolation is achieved through two principal architectures [4,5]: one that uses a conventional line-frequency transformer followed by the AC/DC rectifier and DC/DC stage, and another that connects the AC/DC stage directly to the grid, integrating isolation within the DC/DC converter stage. This latter approach, by operating at higher frequencies, allows for a smaller transformer size, reduced weight, and potential cost savings [5].

Therefore, existing literature predominantly focuses on isolated DC/DC converters, especially the Phase-Shift Full-Bridge (PSFB), Dual Active Bridge (DAB), and LLC resonant types, which are extensively explored across both commercial EV chargers and academic research [4-18]. These converters face a common significant challenge in maintaining high efficiency across the wide voltage range operation. To enhance efficiency within this range, various strategies have been investigated, including variable DC link [6,7], two-stage converter [8-12], and relay matrix [13,14]. Nevertheless, making direct comparisons of these methodologies to evaluate their impact on efficiency and cost—two critical factors for EV adoption—is notably complicated.

One reason for this complication is the literature's lack of consistency in reporting efficiency; some studies assess overall system efficiency which encompasses both AC/DC and DC/DC conversion stages, while others evaluate solely the DC/DC stage. Furthermore, these studies employ design considerations tailored for specific output powers, switching frequencies, and transformer turn ratios, none of which may directly correlate to the needs of a 350 kW system. Such systems require optimization procedure that specifically consider their high target power, which can significantly alter fundamental design parameters and the overall effectiveness of the converter. Additionally, the use of different semiconductor materials (such as Silicon (Si), Silicon Carbide (SiC), and Gallium Nitride (GaN)) further complicates direct comparisons. Lastly, while many studies focus on improving efficiency, the aspect of cost is often neglected. This omission is critical as cost-effectiveness is a key factor for the broader adoption and commercial viability of DC fast charging stations. Without integrating cost considerations, conclusions drawn about the efficiency of different approaches may not provide a realistic picture of their overall practicality and market potential.

Given these complexities, there is a need for a benchmarking study specifically designed for 350 kW DC fast chargers that evaluates the wide voltage range operation. This study aims to systematically benchmark three distinct approaches: the single-stage converter, the two-stage converter, and the relay matrix approaches. By employing standardized conditions across all approaches—such as using the same topology, modulation technique, power level, optimization methodology, and semiconductor technology—this study facilitates direct comparisons. This methodical approach determines which structure not only meets the high-efficiency standards required but also optimizes cost-effectiveness at the 350 kW level necessary for practical DC fast charging deployment.

2 Wide Voltage Range Management Approaches

To overcome the limitations of single-stage converter structure shown in Fig. 1.a in maintaining high efficiency across the wide output voltage range required for DC fast charging, several strategies have been adopted. These approaches are designed to optimize the power conversion process, ensuring fast and efficient charging capabilities under diverse conditions. This section details three key strategies: the variable DC link voltage approach, the two-stage converter approach, and the relay matrix approach. Each method enhances the conventional power conversion mechanisms, addressing specific challenges and operational inefficiencies. The following discussions provide insights into their functional principles and efficiency enhancements.

2.1 Variable DC Link Voltage Approach

The efficiency of the isolated DC/DC converters is optimal when the input-to-output voltage ratio matches the transformer's turns ratio [4]. Achieving this match is critical because it facilitates zero voltage switching (ZVS) and minimizes circulating currents which significantly enhance the overall efficiency of the converter.

To maintain this optimal condition for varying output voltage demands, the variable DC link approach is proposed which adjusts the converter's input voltage accordingly [6]. However, the capability of this approach is limited by the use of boost-type or buck-type rectifiers, which offer a restricted DC link voltage range. This limitation means they are inadequate for covering the full desired output voltage range from 150 V to 1000 V, thereby restricting efficiency improvements to certain segments of the output voltage range. Consequently, this strategy is often utilized as a complementary approach alongside other structures in the literature to ensure the system can achieve and maintain high efficiency across the entire range of required output voltages [9,15,16].

2.2 Two-Stage Converter Approach

To fully harness the efficiency benefits of matching the input-output voltage ratio with the transformer's turns ratio over the wide voltage range, a two-stage converter approach shown in Fig. 1.b is proposed in the literature [8,12]. This approach integrates a non-isolated DC/DC converter with the isolated DC/DC converter, positioning the non-isolated stage either prior to or following the isolated stage.

This configuration is designed to extend the voltage adjustment capabilities beyond what single-stage converters can achieve, thereby aligning the input-output voltage ratio more closely with the transformer's turns ratio across a broader range of voltages. By doing so, it enhances the efficiency of

PCIM Europe 2024, 11– 13 June 2024, Nuremberg   DOI: 10.30420/566262345

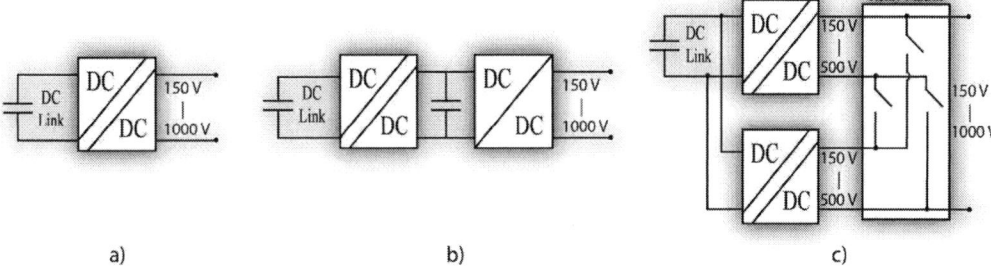

Fig. 1. Schematic representations of converter structures evaluated a) Single stage structure. b) Two-stage structure. c) Relay matrix structure.

the isolated DC/DC converter significantly, leveraging the advantages of both converter types to optimize system performance.

However, the introduction of an additional non-isolated stage comes with its own set of challenges. While it boosts the efficiency of the isolated DC/DC converter, it also introduces new losses that may offset some of the efficiency gains. Additionally, this added complexity can increase the overall system cost and may necessitate more sophisticated control strategies to ensure optimal operation.

2.3 Relay Matrix Approach

The relay matrix approach involves using relays to configure the connections of minimum two converter units in either series or parallel arrangements depending on the required output voltage and current demands of the charging process [13-15]. In this way, each isolated DC/DC converter needs to provide a fraction of the full 150 V to 1000 V range, which enhances the efficiency due to the reduced output voltage range per converter.

There are different ways of implementing the relay matrix approach. For example, more than two DC/DC converters can be used, or the input sides of the converters can be connected in series or parallel. Alternatively, instead of using multiple converter units, a single converter unit which has a common primary side, and multiple secondary sides can be employed [16-18]. However, this study considers the relay matrix structure shown in Fig. 1.c. This structure takes advantage of utilizing isolated DC/DC converters already optimized for 400V, the standard battery voltage in many EVs. In this way, the relay matrix approach leverages this mature technology to enhance efficiency at key charging voltages of 400V and 800V without extensive redesign, making it highly effective for EV charging applications.

However, implementing the relay matrix structure increases costs, primarily due to the need for additional converters to support both series and parallel configurations. Furthermore, incorporating relays introduces extra costs and conduction losses (due to the relay's resistance).

3 Benchmarking Methodology

The benchmarking analysis presented in this section aims to evaluate and compare the efficiency and cost of three distinct structures used in DC fast chargers: the single-stage, the two-stage, and the relay matrix configurations as depicted in Fig. 1. The process involves two critical steps: applying key assumptions to standardize conditions across all configurations, and executing a multi-objective optimization procedure for the isolated DC/DC converter in each structure. This approach ensures that each converter design is evaluated under consistent conditions, allowing for a fair and direct comparison based on efficiency and cost-effectiveness. The ultimate goal is to identify the optimal converter configuration that balances performance with economic viability.

3.1 Key Assumptions

We establish the following key assumptions to ensure a consistent and systematic evaluation across the three configurations examined:

- The analysis focuses exclusively on the performance of the DC/DC converter stage.

- The variable DC link voltage approach is applied as a complementary technique across all configurations to optimize each under identical operating conditions, assuming a boost type three-phase rectifier that can provide a DC link voltage ranging from 600 V to 850 V.

- The DAB converter, utilizing single-phase shift (SPS) modulation, is selected as the isolated DC/DC converter topology to maintain a consistent comparison basis.

- Only SiC MOSFETs are considered to maintain uniformity in semiconductor performance.

2455

- The evaluation of losses and costs is restricted to main converter components including semiconductors, cooling systems, and passive elements such as transformers, inductors, and capacitors.
- Each converter structure is designed and optimized for the 350 kW output capacity, following a detailed optimization routine that will be introduced in the subsequent section.

3.2 Optimization Procedure

Optimizing the DAB converter for each specific structure is crucial to accurately benchmark their performance under consistent and uniform criteria. To find the optimal DAB converter design, the systematic optimization procedure should be applied. This study expands the previously established optimization methodology [19], originally tailored for a single-stage DAB converter, to now encompass the relay matrix and two-stage converter structures.

The objective is to determine the optimal design of the DAB converter that achieves the highest efficiency and cost-effectiveness possible. Essential design parameters adjusted during this optimization include the switching frequency (f_{sw}), transformer turns ratio (n), and the number of converter modules (N). The process encompasses several critical steps: defining system requirements, exploring the design space, specifying converter configurations in detail, formulating a global optimization strategy, and modeling converter components.

Given the unique operating characteristics of the relay matrix and two-stage configurations, specific modifications to the optimization methodology are required. These modifications ensure that the methodology remains applicable and effective in identifying optimal design parameters under the varying operational conditions presented by each structure. Besides these adaptations, all other relevant data, models and parameters from the previous study [19] are utilized to maintain consistency.

3.2.1 Modifications for Two-Stage Converter

In this study, the two-stage converter structure combines a DAB converter with a synchronous buck converter (show in Fig. 2). The operational strategy is divided based on the output voltage requirements. For outputs ranging from 600 V to 1000 V, the DAB converter actively regulates its output utilizing the variable DC link voltage approach as a complementary method to enhance efficiency. During this phase, the synchronous

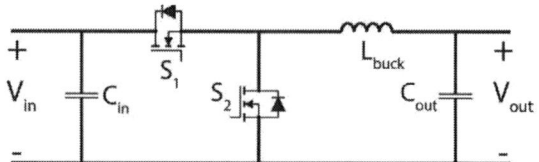

Fig. 2 Synchronous buck converter circuit diagram

buck converter is bypassed with switch S1 permanently closed, and thus does not participate in the conversion process. Conversely, for the 150 V to 600 V output range, the buck converter takes the lead in voltage conversion. In this operation, both the input and output voltages of the DAB converter are fixed at 600 V, simplifying the system's complexity and optimizing efficiency. Based on the operational principles described, the following modifications are applied

Reduced Output Voltage Range: The DAB converter in this approach is specifically optimized for the 600 V to 1000 V output voltage range.

Integration of Synchronous Buck Converter: The introduction of the synchronous buck converter optimization represents a new dimension not covered in [19]. This buck converter is specifically optimized for the 150 V to 600 V range operation, complementing the DAB converter's operation, and ensuring the system covers the full desired voltage range. The optimization process for the buck converter is similar to that of the DAB converter, focusing on determining the optimal switching frequency and the number of modules necessary to achieve a total power output of 350 kW. It utilizes the same semiconductor model, cooling system model, and capacitor model as established for the DAB converter in [19].

Additionally, an inductor optimization process is incorporated for the buck converter. This process uses similar principles for loss and cost calculation as the transformer optimization in the DAB converter, utilizing the same core structure (EE) and material type (N87). The calculation of the required inductance value and the capacitor value for the buck converter is guided by the formulas in [5].

3.2.2 Modifications for Relay Matrix

In this study, the relay matrix approach utilizes two identical isolated DC/DC converters, as illustrated in Figure 1.c. The operational strategy is segmented based on the output voltage requirements. For output voltages higher than 500 V, the two converters are configured in series, effectively doubling the voltage capacity to meet higher demands. Conversely, for output voltages lower than 500 V, the converters are configured in parallel, which allows the system to better manage lower voltage outputs by combining the capacity of both converters. Based on this operating principle, the following modifications are applied:

Reduced Output Voltage Range: The DAB converter in the relay matrix is specifically optimized for the 150 V to 500 V output voltage range.

Semiconductor Update: The narrower voltage range enables the use of 650 V SiC semiconductors, on the secondary side of the DAB converter, selected over 1.2 kV options for their better loss performance and lower cost. A range of semiconductors listed in Table I, each selected for different current ratings, were used to accommodate the output range of 10 kW to 70 kW for the DAB converter. The pricing of these components, which is vital for the cost analysis, was obtained from Mouser Electronics, showing the price for 100 units as of 06/04/2024.

Table I   Infineon 650 V SiC MOSFET List

Part Number	$R_{ds(on)}$	$I_d@$ 100 °C	$Cost_{100}$
IMBG65R007M2H	6.7 mΩ	167 A	$29.63
IMBG65R015M2H	14.5 mΩ	84 A	$16.81
IMBG65R020M2H	20 mΩ	64 A	$13.49
IMBG65R030M1H	30 mΩ	45 A	$10.3

Integration of Relays: The T9GV1L14-5 mechanical relay was selected due to its proven performance in similar wide voltage range applications, as noted in [18]. This relay can handle a maximum current of 30 A, and features a conduction loss resistance of 10 mΩ. In configurations requiring higher currents, a paralleling approach is employed, where multiple relays are used in parallel. The cost of the relay is $3.25 each, as of 06/04/2024, according to Mouser listings.

4   Benchmarking Results

The optimization outcomes for the DAB converters within the single-stage, two-stage, and relay matrix structures are comprehensively demonstrated in Fig. 3. Specifically, Fig. 3.a, 3.b, and 3.c showcase the Pareto front optimization results for each respective approach, highlighting the critical trade-offs between average efficiency and power-cost ratio. Unlike converging to a single optimal solution, the optimization presents a spectrum of potentially optimal configurations along the pareto front. This pareto front represents the trade-off curve where any further improvement in efficiency would come at the cost of higher system expense, or vice versa. To ease the selection of the optimal design and meet industry benchmarks, we introduced an additional criterion: selecting the design with the highest power/cost ratio that also ensures a minimum efficiency of 95% across the wide output voltage range at rated power.

Table II Optimal Configurations for Each Structure

Structure	N	P	fsw	n
Single Stage	7	50 kW	30 kHz	0.9
Two Stage - DAB	5	70 kW	40 kHz	1
Two Stage - Buck	7	50 kW	100 kHz	N/A
Relay Matrix	14	25 kW	50 kHz	0.5

Table II provides the selected design parameters for the DAB converters in each configuration optimized for the 350 kW DC fast charger. These parameters include the switching frequency (fsw), transformer turns ratio (n), and the number of converter modules (N) which were determined to achieve the best balance between cost and efficiency.

Table III Advantages and Limitations of Converter Structures: Single-Stage, Two-Stage, and Relay Matrix

Structure	Advantages	Limitations
Single-Stage	• Low cost • Less complex	• Low efficiency under the wide output voltage range and light loads • Limited ZVS • High circulating current • Bulkier, more expensive transformers
Two-Stage	• High efficiency • Full ZVS • Low circulating current	• More complex due to integration of additional converter stage • Moderate cost due to additional converter stage
Relay Matrix	• Moderate efficiency • Extended ZVS • Reduced circulating current • Reduced magnetic size	• Higher overall cost due to doubled component count and inclusion of relays • Efficiency drops at light loads

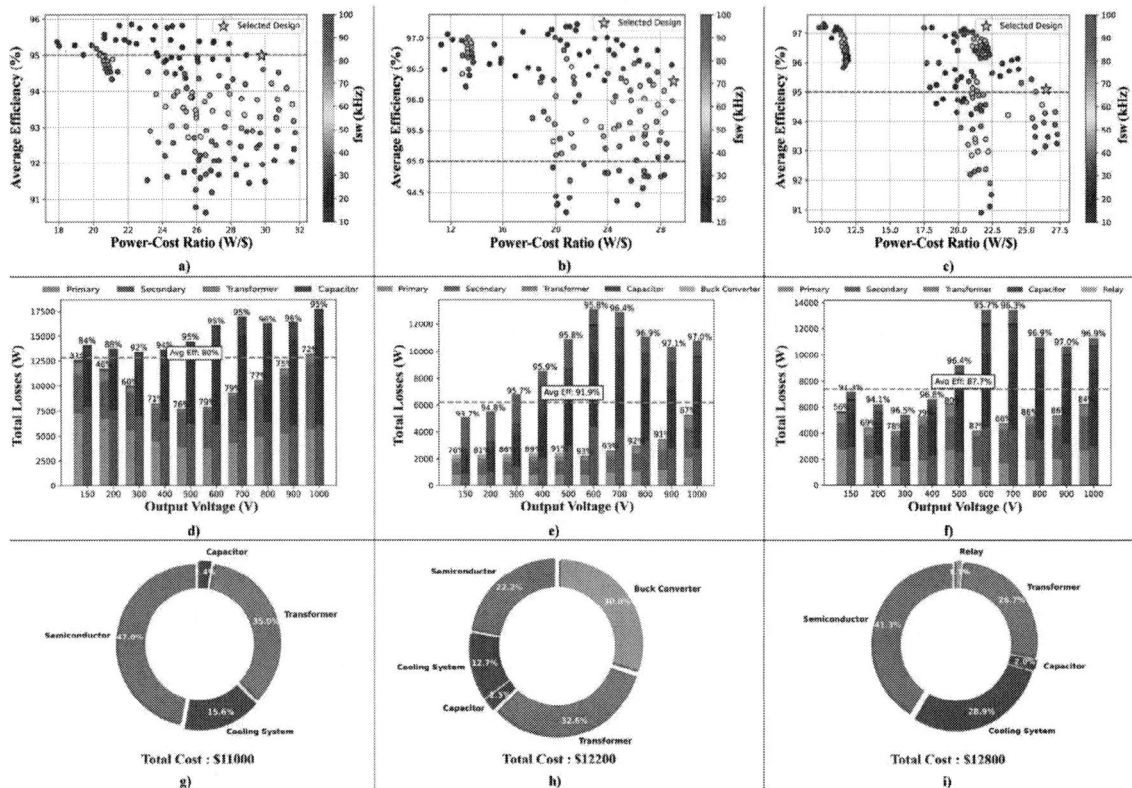

Fig. 3. A comparative analysis of three approaches: single-stage converter (a, d, g), two-stage converter (b,e,h), and relay matrix (c,f,i), across different performance and cost metrics. Sub-figures (a), (b), and (c) illustrate the Pareto front optimization results for each approach, highlighting the trade-off between average efficiency and power-cost ratio, with a color map reflecting the switching frequency. Sub-figures (d), (e), and (f) compare the efficiency of the three approaches at light load (10% of full power, denoted by lower color intensity) and full load (350 kW), across the output voltage range from 150 V to 1000 V. Sub-figures (g), (h), and (i) break down the main cost components of each converter, offering insights into the financial implications of their design and operational efficiencies.

To assess the performance of the selected designs, we conducted an analytical comparison over the voltage range from 150 V to 1000 V. Fig. 3.d, 3.e, and 3.f depict the total losses and efficiency across various voltages at rated power and at 10% of full rated power for each configuration, providing insights into their operational effectiveness under different load conditions.

Meanwhile, Fig. 3.g, 3.h, and 3.i focus on the total cost breakdown of the main components (semiconductors, transformers/inductors, cooling systems, capacitors, relays) for the full 350 kW charger. This detailed cost analysis helps in understanding the financial implications of each configuration and supports cost-effective decision-making.

Table III complements the analytical data illustrated in Fig. 3.d through 3.i by providing a structured comparison of the single-stage, relay matrix, and two-stage converter structures. It details the respective advantages and limitations of each

configuration, facilitating a straightforward assessment of their performance characteristics. The two-stage converter structure offers the most advantageous balance, achieving high efficiency across its operational range while maintaining a favourable power/cost ratio.

5 Conclusion

As electric vehicles become increasingly prominent, optimizing charger efficiency and cost within the 150 V to 1000 V range remains a challenge. While various wide voltage range approaches have emerged for isolated DC/DC converters, a direct comparison from the literature is complicated due to varying parameters. To address this gap, our study conducts a standardized benchmark analysis under uniform conditions, optimizing three approaches employing the DAB converter for the 350 kW DC fast charger. Efficiency and cost are then evaluated using the Pareto front optimization method to select the optimal design parameters. The benchmarking revealed the two-stage converter approach as notably effective, showing an optimal balance between efficiency and cost across the explored voltage range.

However, this study concentrated on single-phase shift modulation for the DAB converter across all examined approaches, acknowledging that adopting more advanced modulation methods could potentially unlock further efficiency gains for each structure. Moreover, our benchmarking is based primarily on analytical analysis, highlighting a need for future experimental verification to validate these findings in practical settings. Future research will extend beyond theoretical analysis, exploring experimental validations and the integration of advanced modulation techniques.

Acknowledgements

This work has received funding from the European Union's Horizon Europe research and innovation programme under the Marie Sklodowska-Curie Doctoral Networks grant agreement No 101072414 (E2GO).

References

[1] European Environment Agency, "Decarbonising Road Transport — The Role of Vehicles, Fuels and Transport Demand," European Environment Agency, 2021. [Online]. Available: https://www.eea.europa.eu/publications/transport-and-environment-report-2021 [Accessed: 21-Apr-2024].

[2] S. Rivera, S. Kouro, S. Vazquez, S. M. Goetz, R. Lizana and E. Romero-Cadaval, "Electric Vehicle Charging Infrastructure: From Grid to Battery," in IEEE Industrial Electronics Magazine, vol. 15, no. 2, pp. 37-51, June 2021.

[3] C. Jung, "Power Up with 800-V Systems: The benefits of upgrading voltage power for battery-electric passenger vehicles," in IEEE Electrification Magazine, vol. 5, no. 1, pp. 53-58, March 2017.

[4] H. Tu, H. Feng, S. Srdic and S. Lukic, "Extreme Fast Charging of Electric Vehicles: A Technology Overview," in IEEE Transactions on Transportation Electrification, vol. 5, no. 4, pp. 861-878, Dec. 2019.

[5] D. Aggeler, F. Canales, H. Zelaya-De La Parra, A. Coccia, N. Butcher and O. Apeldoorn, "Ultra-fast DC-charge infrastructures for EV-mobility and future smart grids," 2010

IEEE PES Innovative Smart Grid Technologies Conference Europe (ISGT Europe), Gothenburg, Sweden, 2010, pp. 1-8.

[6] H. Wang, S. Dusmez and A. Khaligh, "Maximum Efficiency Point Tracking Technique for LLC -Based PEV Chargers Through Variable DC Link Control," in IEEE Transactions on Industrial Electronics, vol. 61, no. 11, pp. 6041-6049, Nov. 2014.

[7] Y. Li et al., "Optimal Synergetic Operation and Experimental Evaluation of an Ultra-Compact GaN-Based Three-Phase 10 kW EV Charger," in IEEE Transactions on Transportation Electrification.

[8] B. O. Aarninkhof, D. Lyu, T. B. Soeiro and P. Bauer, "A Reconfigurable Two-stage 11kW DC-DC Resonant Converter for EV Charging with a 150-1000V Output Voltage Range," in IEEE Transactions on Transportation Electrification.

[9] J. Na et al., "Development of Bi-directional Charger with a Wide Voltage Range," The Transaction of Korean Institute of Power Electronics, vol. 27, no. 1, pp. 74-79, Feb. 2022.

[10] F. Jin, A. Nabih, C. Chen, X. Chen, Q. Li and F. C. Lee, "A High Efficiency High Density DC/DC Converter for Battery Charger Applications," 2021 IEEE Applied Power Electronics Conference and Exposition (APEC), Phoenix, AZ, USA, 2021, pp. 1767-1774.

[11] N. Zanatta, T. Caldognetto, D. Biadene, G. Spiazzi and P. Mattavelli, "Design and Implementation of a Two-Stage Resonant Converter for Wide Output Range Operation," in IEEE Transactions on Industry Applications, vol. 58, no. 6, pp. 7457-7468, Nov.-Dec. 2022.

[12] W. -S. Lee, J. -H. Kim, J. -Y. Lee and I. -O. Lee, "Design of an Isolated DC/DC Topology With High Efficiency of Over 97% for EV Fast Chargers," in IEEE Transactions on Vehicular Technology, vol. 68, no. 12, pp. 11725-11737, Dec. 2019.

[13] S. Götz and V. Reber, "Modular Power Electronics System for Charging an Electrically Operated Vehicle," US Patent 2018/0162229, Aug 14, 2019.

[14] M. Weisbach, T. Schneider, D. Maune, H. Fechtner, U. Spaeth, R. Wegener, S. Soter, and B. Schmuelling, "Intelligent Multi-Vehicle DC/DC Charging Station Powered by a Trolley Bus Catenary Grid," Energies, vol. 14, no. 24, Art. no. 8399, 2021.

[15] D. Cittanti, E. Vico, M. Gregorio, F. Mandrile and R. Bojoi, "Iterative Design of a 60 kW All-Si Modular LLC Converter for Electric Vehicle Ultra-Fast Charging," 2020 AEIT International

Conference of Electrical and Electronic Technologies for Automotive (AEIT AUTOMOTIVE), Turin, Italy, 2020, pp. 1-6.

[16] C. Wei, Z. Hu, J. Chen, F. Zhang, H. Zhan and A. Narain, "A SiC Based 60kW Three Phases Interleaved LLC Converter for EV Fast Charger," PCIM Europe 2023; International Exhibition and Conference for Power Electronics, Intelligent Motion, Renewable Energy and Energy Management, Nuremberg, Germany, 2023, pp. 1-6.

[17] O. Zayed, A. Elezab, A. Abuelnaga and M. Narimani, "A Dual-Active Bridge Converter With a Wide Output Voltage Range (200–1000 V) for Ultrafast DC-Connected EV Charging Stations," in IEEE Transactions on Transportation Electrification, vol. 9, no. 3, pp. 3731-3741, Sept. 2023.

[18] D. Lyu, T. B. Soeiro and P. Bauer, "Multiobjective Design and Benchmark of Wide Voltage Range Phase Shift Full-Bridge DC/DC Converters for EV Charging Application," in IEEE Transactions on Transportation Electrification, vol. 10, no. 1, pp. 288-304, March 2024.

[19] S. Cinik, F. Zhao, G. De Falco, and X. Wang, "Efficiency and Cost Optimization of Dual Active Bridge Converter for 350kW DC Fast Chargers," arXiv, 2024.

PCIM Europe 2024, 11– 13 June 2024, Nuremberg
DOI: 10.30420/566262346

Performance Optimization of Single-phase On-Board Chargers with Ripple Port

Davide Gottardo [1], Giorgio Valente [1]

[1] Hexagon Manufacturing Intelligence, UK

Corresponding author: Davide Gottardo, davide.gottardo@hexagon.com
Speaker: Davide Gottardo, davide.gottardo@hexagon.com

Abstract

This paper introduces a new modulation method for a ripple port integrated in an On-Board Charger (OBC), that includes an isolated resonant LLC DCDC stage. The ripple port switching frequency is synchronized with the variable frequency of the LLC, and is phase-shifted so that the combined DC link current ripple is reduced. The voltage offset applied to the ripple port capacitor is controlled to achieve ZVS operation of the ripple port switches across a broad range of switching frequencies, while limiting excessive current stress. Results show substantial reduction in the DC link capacitor current ripple.

1 Introduction

Increasing power density of OBCs is of growing interest, following the market trend of developing combo units, that include multiple converters in the same enclosure (such as OBC and LV/HV DCDC) [1], [2]. In single-phase OBCs, the necessity of including a large DC link capacitor to absorb the ripple power inherent in single-phase AC connections significantly reduces power density. The capability of the same OBC converter to operate both in Europe, where three-phase LV connections are common, and in US, where single-phase connections with high power ratings are available, is of commercial interest. To obtain this, a common way is to use three full-bridge or bridge-less boost stages that can be connected between each phase and neutral in a three-phase AC system, and connected to the same phase (interleaved) in case of single phase system, or, alternatively, an interleaved totem-pole topology.

While power transfer in a three-phase connection is continuous, power transfer in single-phase connections is inherently discontinuous, having a power ripple at twice AC grid frequency (120Hz if AC frequency is 60Hz), with an instantaneous peak to peak amplitude of twice the average power. Usually this ripple is partially absorbed by the internal DC link capacitor that is shared by the ACDC and DCDC section of the OBC while some ripple

residue is often present in the charging current of the battery. To limit ripple voltage on the DC link, a big capacitance is needed, negatively affecting the overall power density of the converter. A well known solution to solve this problem is the use of a ripple port (RP)[3]–[6]. The RP is an additional section of the converter that includes one or more components capable of energy storage (usually a capacitor), used to provide ripple power cancellation. This allows to substantially reduce the required DC link capacitance, however, this comes with the disadvantage of additional cost and complexity, and additional losses. Moreover, the required DC link capacitance can be drastically reduced, but the capacitor still has to be big enough to withstand the switching ripple current, that is, the current injected in the DC link by the switching devices, whose spectrum comprises the switching frequency of the active devices and its harmonics. In an OBC without ripple port, the DC link capacitor current stress is made up by the 120Hz component, the switching current of the ACDC stage, which is often operated at fixed frequency, and the switching current of the DCDC stage, which often is a resonant converter whose switching frequency changes according to the operating point. While the ripple port eliminates the 120Hz component, it introduces additional switching ripple current that adds up in the total current stress of the DC link capacitor.

In this paper a modulation strategy for the ripple port is proposed, that allows to significantly reduce the RMS current absorbed by the DC link capacitor. This is achieved by synchronizing the switching frequency of the ripple port and the LLC, and applying the optimum phase-shift as function of the working point.

Losses in the ripple port adversely affect the efficiency of the converter. A common practice to reduce losses, is to design the ripple port inductance and choose its modulation frequency to achieve partial soft switching (ZVS) on the active switches [7]. Synchronizing the RP modulation frequency with the LLC makes this difficult, because choosing the inductance value low enough to achieve ZVS for all working points results in unacceptably high switching ripple current when the LLC frequency is low. The solution proposed in this paper is to vary the DC voltage offset in the RP capacitor to achieve ZVS without excessive currents. The modulation strategy adopted is known as TCM (Triangular Current Mode) [8];

Practical implementation of the solution in a microcontroller is discussed. The proposed solution is based on look-up tables, where the optimum phase-shift of the RP modulator carrier and optimum RP capacitor DC voltage offset are chosen as function of the LLC working point.

The paper is organised as follows: in section 2 the main components of an on-board chargers are described, in section 3 the ripple port concept is examined, in section 4 the RP modulation strategy and control (which is the original contribution of the paper) is explained, and in section 5 the simulation results are shown and commented.

2 OBC converter topology

The On Board Charger (OBC) is a component in the subsystem of an electric vehicle, that allows to charge the HV battery from an AC connection. The OBC takes care of the physical conversion of the AC current to DC, and the control of the battery charging power, following the reference coming the BMS (Battery management system). In addition to that, the OBC provides galvanic isolation between the AC connection and the battery, which is required for safety reasons. Several topologies of OBC are available in the literature [9], [10], however, most of them have a similar structure, that can be divided in two sections: an ACDC section, connected to the AC port, and a DCDC section, where galvanic isolation is achieved via an high frequency transformer. The ACDC and DCDC section share

a common inner DC link, and thus a common DC link capacitor.

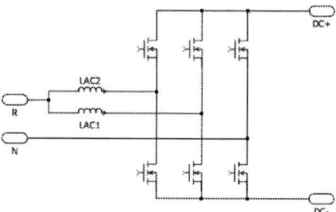

Fig. 1: ACDC stage topology

The ACDC stage considered in this paper is an interleaved totem-pole topology, shown in Fig. 1, while the DCDC stage, an LLC resonant converter, is shown in Fig. 2.

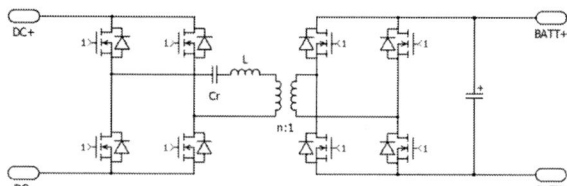

Fig. 2: DCDC stage: LLC

3 Ripple power in single phase connections, ripple port

3.1 Ripple port

As ripple port is a well-known topic, this paper does not address it in depth: the main concept is briefly introduced, with focus on the aspects that are relevant for the implementation of the proposed modulation strategy.

Various topologies of ripple port are available in the literature. In this paper, one of the simplest and more widely used topology is chosen. It comprises an half-bridge, an inductor, and a capacitor (see Fig. 3).

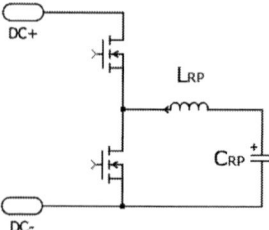

Fig. 3: Ripple port topology

In this topology, the capacitor is used as energy

buffer, and the inductor is needed to limit the switching frequency current ripple.

To provide the required power ripple cancellation, the voltage applied to the RP capacitor has to be the following:

$$V_C(t) = \sqrt{V_0^2 + \frac{P}{2\pi fC}\sin(4f\pi t)} \qquad (1)$$

where P is the active power exchanged trough the AC port, f_{AC} is the AC frequency, C is the RP capacitance, and V_0 is a DC voltage offset applied to the RP capacitor. The latter gives an additional degree of freedom, as the same ripple power can be absorbed by the capacitor with any value of V_0, as long as it satisfies the modulation constraints of the ripple port, that in this case are:

$$0 < V_C < V_{DC} \qquad (2)$$

this results in the following constraints:

$$\sqrt{\frac{P}{2\pi fC}} < V_0 < \sqrt{V_{DC}^2 - \frac{P}{2\pi fC}} \qquad (3)$$

These constraints should be met with a margin, to account for components tolerance, voltage distortion (e.g. due to the effect of dead-time), and the need for a control band.

The current flowing in the capacitor and the inductor of the RP can be split in two components: a switching "frequency ripple component", having a triangular waveform (assuming continuous conduction mode), and a low frequency component, associated to the charging and discharging of the RP capacitor, "pulsating power component". The latter is the following:

$$I_C(t) = \frac{P\cos(2\pi f_{AC}t)}{\sqrt{V_0^2 + \frac{P}{2\pi f_{AC}C}\sin(4\pi ft)}} \qquad (4)$$

The peak-to-peak amplitude of the switching ripple current applied to the output inductor of the RP half bridge is a function of the DC link voltage V_{DC}, inductance L_{RP}, switching frequency f_{RP} and duty cycle d, as in Eq. (5).

$$\Delta I_{RP,SW}(t) = \frac{V_{DC}}{L_{RP}f_{RP}}d(t)[1 - d(t)] \qquad (5)$$

where $d(t) = \frac{V_{DC}}{V_C(t)}$.

If the current ripple is high enough so that the current changes sign prior to switching, the RP switches operate in ZVS (zero voltage switching). This modulation is known as TCM (Triangular Current Mode).

The instantaneous current at the turn-on of the upper switch of the RP half bridge $I_{RP,ON+}$, and the turn-on current of the lower device $I_{RP,ON-}$ are the following:

$$I_{RP,ON+} = I_C(t) - \frac{1}{2}\Delta I_{RP,SW}(t) \qquad (6)$$

$$I_{RP,ON-} = I_C(t) + \frac{1}{2}\Delta I_{RP,SW}(t) \qquad (7)$$

To achieve TCM, the instantaneous current during switching have to be high enough to completely discharge the stray capacitance C_{oss} of the device prior to its turn on, during the deadtime. Assuming this current to be a constant I_{ZVS}, this translates in the following constraints that, if verified, result in ZVS operation:

$$I_{RP,ON+} < -I_{ZVS} \qquad (8)$$
$$I_{RP,ON-} > I_{ZVS} \qquad (9)$$

3.2 RP Effects on the DC link capacitor

In an OBC that does not include a ripple port, the ripple currents circulating through the DC link capacitor are the switching frequency components coming from the ACDC and DCDC stage, and the 120Hz component that results from the single phase AC connection. As these components have different frequency spectra, their total RMS is calculated as in Eq. (10):

$$I_{RMS,DClink} = \sqrt{I_{120Hz}^2 + I_{RMS,ACDC}^2 + I_{RMS,DCDC}^2} \qquad (10)$$

The presence of a ripple port eliminates the 120Hz power pulsation from the DC link capacitor; for this reason, the selection of a DC capacitor is not driven by the required capacitance, but by its capability to withstand the switching frequency current ripple. The capacitor has to absorb switching ripple currents coming from three sources: the ACDC and DCDC stages, and the ripple port.

4 Proposed RP modulation strategy

The ACDC stage is usually controlled at fixed frequency, to simplify the design of the AC line filter, and the contribution of its three sections to the DC link ripple current is significantly reduced by controlling the phase shift of the modulator carrier between each section (interleaving). Synchronization

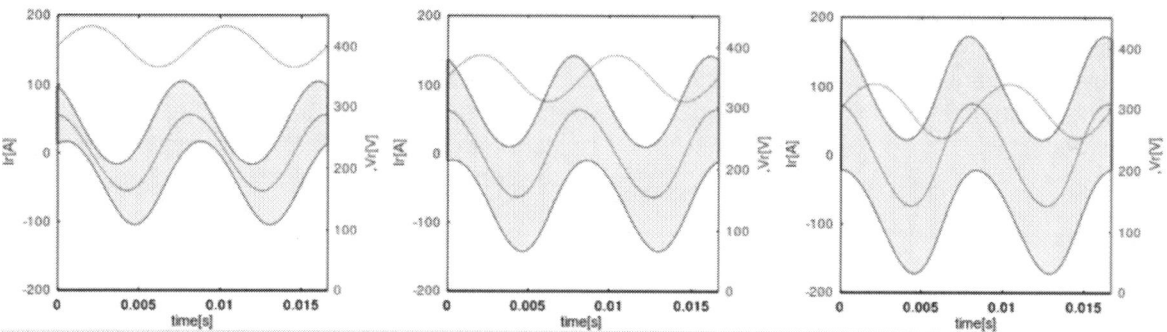

Fig. 4: Ripple port voltage and current waveforms, current ripple envelope

between the ACDC and DCDC stage, however, is impossible because LLC converters are controlled at variable frequency. The proposed modulation strategy consists in synchronizing the switching frequency of the ripple port with that of the LLC, to reduce the RMS of the combined ripple current. To do so, the switching frequency of the ripple port has to be 2 times the switching frequency of the LLC, since the chosen topology for the ripple port is an half bridge, while the primary side of the LLC is a full bridge. This comes at the cost of a higher complexity of the ripple port control (variable frequency control).

Param	Value	Unit
$P_{NOMINAL}$	11	kW
V_{DC}	400	V
C_{DC}	500	μF
V_{BATT}	250-425	V
f_{AC}	60	Hz
C_{RP}	1500	μF
L_{RP}	3	μH
$L_{AC1,2}$	50	μH
f_{ACDC}	50	kHz
f_{LLC}	70 - 160	kHz
L_R	18.5	μH
L_M	92.5	μH
C_R	135	nH
T_{RATIO}	4/3	

Tab. 1: Simulation parameters

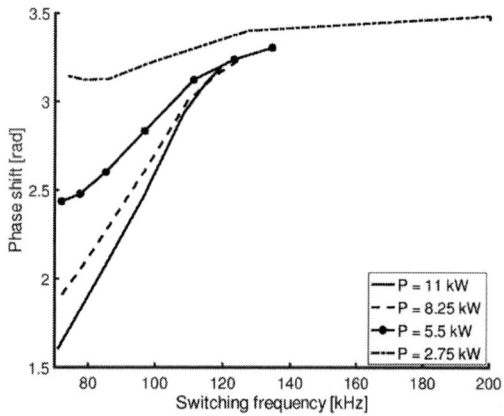

Fig. 5: Optimum phase shift value as function of DCDC switching frequency and power setpoint

4.1 Optimum RP capacitor DC offset calculation

As can be seen from the equations shown in section 3, the choice of V_0 impacts the amplitude of both I_C and $\Delta I_{RP,SW}$. Fig. 4 shows the voltage waveform in the RP capacitor, in red, and the current flowing in the RP inductor, in blue, including its switching frequency ripple envelope $\Delta I_{RP,SW}$. In

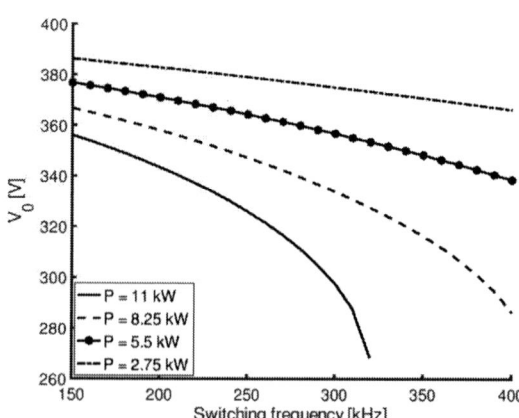

Fig. 6: Optimum V_0 value as function of f_{SW} and P

the three figures, the switching frequency, capacitance and inductance are the same, while the DC component in the RP capacitor voltage is changed. When the V_0 high, and the RP capacitor voltage is close to the DC link voltage, the amplitude of ripple current on the RP inductor $\Delta I_{RP,SW}$ is low, while when voltage offset is lower, and the capacitor voltage is closer to $V_{DC}/2$, the ripple current is higher. Similarly, the pulsating power component of the current $I_C(t)$, is low when V_0 is high. This is because the same ripple power can be exchanged by the capacitor with a lower current, when its DC voltage offset is higher.

For this reason, in general, choosing the highest possible value for V_0 is beneficial, as it allows to achieve the required ripple power with the minimum current stress in the RP components.

V_0 is initially chosen equal to the upper limit shown in Eq. (1), the currents I_C and $\Delta I_{RP,SW}$ are calculated and the conditions for ZVS verified. If the ZVS conditions are verified, the calculation stops returning the value of V_0, if not, V_0 is decremented by a fixed step and the analysis repeated. If V_0 reaches the lower limit shown in Eq. (1), the analysis is stopped. In the case, ZVS is not possible in this working point: to achieve ZVS, a smaller inductor has to be selected.

The values of V_0 resulting from this calculation are shown in Fig. 6. The minimum current required for ZVS I_{ZVS} assumed in this case is 5A. A RP inductor of $3\mu H$ allows to achieve ZVS in all the required operating points.

4.2 Optimum RP optimum phase calculation

To generate the optimum phase shift look-up table, the DCDC and RP stage are simulated individually, using PLECS. The circuit parameters used in the simulation are shown in table 1.

In the first simulation, only the DCDC stage is simulated, and the DC link is tied to an ideal voltage source. Several working points are simulated, across the battery side DC voltage range and across different power set-points. For each simulation, the steady-state switching frequency f_{LLC} of the DCDC (which is the output of the DCDC control loop) is saved, together with the DC link current. The Fourier analysis of the DC link current is then evaluated, and the phase φ_{LLC} of the fundamental component of the current is saved.

In the second simulation the RP alone is simulated, again connecting its DC link to an ideal voltage source. The same working points of the previous simulation are analysed: for each working point the RP modulator frequency is set to two times the DCDC modulator frequency, and the modulated voltage is calculated as function of the power, as in Eq. (1). The value of V_0 is chosen as the maximum that still achieves ZVS. This is obtained by an algorithm using a numerical approach. A Fourier analysis of the DC link current is again evaluated, and the phase φ_{RP} of the fundamental component of the current is saved.

At his point, all necessary data is available for generating the optimum phase look-up tables: the optimal phase shift, for each working point, is the following:

$$\Delta\varphi_{RP}^* = 180\deg + \varphi_{LLC} - \varphi_{RP} \qquad (11)$$

Applying this phase shift to the RP modulator carrier will put the fundamental component of the switching frequency current ripple of the RP in opposition of phase compared to the fundamental component of the switching frequency current ripple of the DCDC.

5 Simulation results

Fig. 7 shows a bar plot of the RMS of the current injected in the DC link by the three stages of the OBC (ACDC, DCDC and RP), and the RMS that appears on the DC link capacitor in three scenarios: when no RP is used, when there is a RP, but its switching frequency is not synchronized with the LLC, and when a RP is present and the proposed modulation is used. The RMS is split in the following spectral components:

- the DC component, associated with power transfer,

- the 120Hz component, due to the pulsating power of a single phase connection,

- the SW_{ACDC} component (switching frequency of the ACDC and its multiples),

- the SW_{DCDC} component (switching frequency of the LLC in this working point and multiples),

- the SW_{RP} component (switching frequency of the ripple port and its multiples), and

- the total RMS, that is the square root of the sum of the squares of each component.

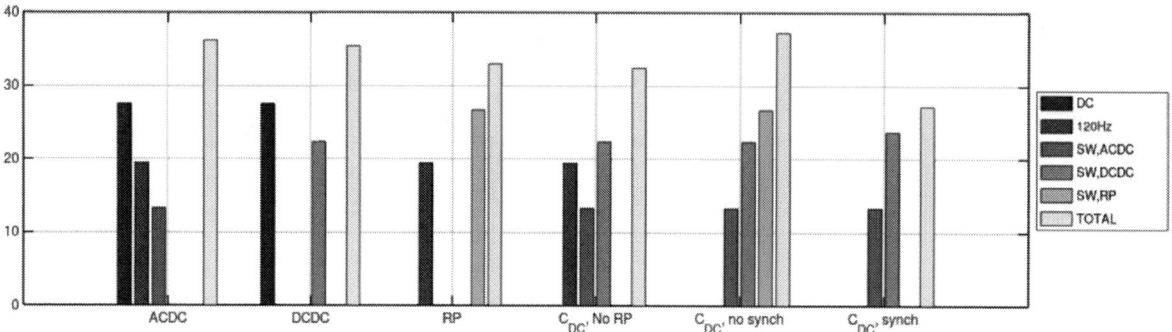

Fig. 7: DC link RMS current spectra ($P = 11kW$, $V_{BATT} = 400V$)

The RMS of DC link current of the ACDC stage is constituted of its DC component, 120Hz component and switching ripple component. The switching component is relatively small due to interleaving. The RMS of the DCDC stage has only the DC component and the DCDC switching component, and the RMS of the RP has the 120Hz component and the RP switching component.

If the ripple port is not used, the current RMS on the DC link capacitor will contain the ACDC and DCDC switching component, and the 120Hz component, if the RP is used, but switching at a different frequency of the ACDC and DCDC, the 120Hz does not appear, but is replaced by the RP switching component; finally, if the RP is used with the proposed modulation, the combined RP and DCDC switching frequency RMS is lower than the root of the sum of the squares of the two, due to the destructive interference between the fundamental component of the current, achieved by using the optimum phase shift.

While Fig. 7 focus on one working point, Figs. 8 to 11 span the entire frequency range of the LLC at different power set-points, to show the behaviour of the proposed modulation strategy across all the operating points of the OBC.

The results show that the proposed modulation strategy is particularly advantageous when the LLC switching frequency is low. This happens at high power, when the DC voltage on the LLC output (battery side) is high. In these operating points, the combined RMS of the switching ripple of the ripple port and LLC is even lower then the switching ripple of the LLC stage alone.

Conversely, when the LLC switching frequency is high, the DC voltage offset of the RP is reduced to achieve ZVS. This causes an increase of the current in the RP section, that has negative effects

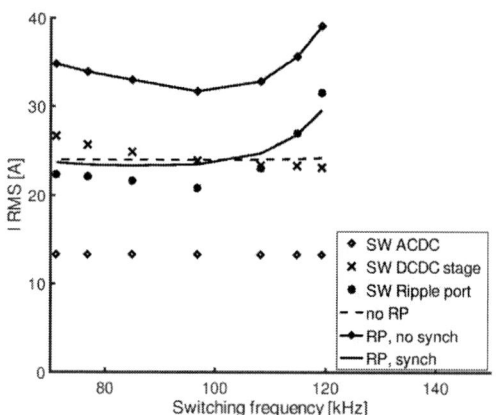

Fig. 8: DCDC and RP switching ripple RMS, P=11kW

on the converter efficiency. However, even in these working points, the total RMS appearing on the DC link capacitor is lower compared to not using the RP, or operating it out of synch.

6 Experimental results

At present, an 11 kW OBC prototype is in the process of being assembled; consequently, experimental results are unavailable as of the submission deadline for this paper.

7 Conclusions

The presence of a ripple port allows for a substantial increase of power density, simplifying the integration of the OBC, DCDC, and/or the traction inverter in the same combo unit.

A variable frequency control of the ripple port is proposed. Synchronization of the ripple port switching frequency with the LLC switching frequency allows a substantial reduction of ripple on the DC link capacitor, increasing its lifetime. Simulations

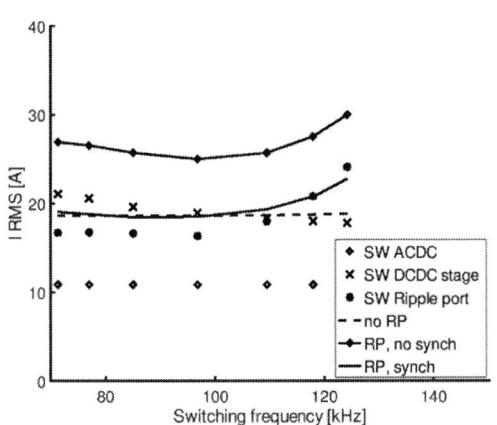

Fig. 9: DCDC and RP switching ripple RMS, P=8.25kW

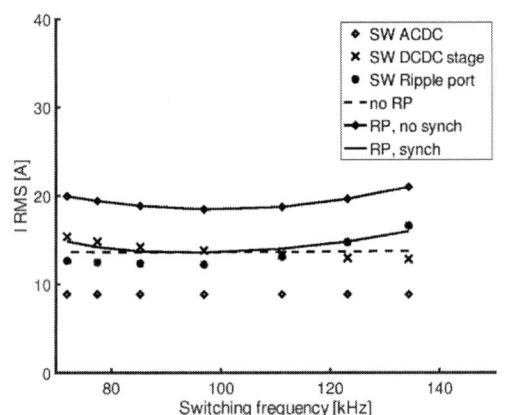

Fig. 10: DCDC and RP switching ripple RMS, P=5.5kW

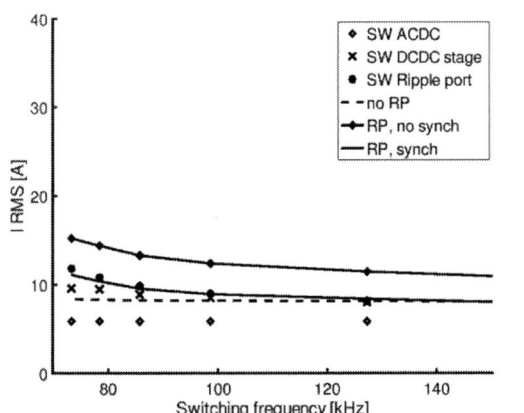

Fig. 11: DCDC and RP switching ripple RMS, P=2.75kW

results show a substantial impact on DC link capacitor RMS current stress, that is reduced by up to 27% compared to using the ripple port with no switching frequency synchronization, and by up to 16% if no ripple port is used. A method of deriving the optimum DC offset in the RP capacitor as function of the switching frequency is described, that allows soft switching of the RP active devices. ZVS of the RP switches is achieved in the entire operating frequency range.

References

[1] H. Wouters and W. Martinez, "Bidirectional on-board chargers for electric vehicles: State-of-the-art and future trends," *IEEE Transactions on Power Electronics*, vol. 39, no. 1, pp. 693–716, 2024. DOI: 10.1109/TPEL.2023.3319996.

[2] D.-H. Kim, M.-J. Kim, and B.-K. Lee, "An integrated battery charger with high power density and efficiency for electric vehicles," *IEEE Transactions on Power Electronics*, vol. 32, no. 6, pp. 4553–4565, 2017. DOI: 10.1109/TPEL.2016.2604404.

[3] P. T. Krein, R. S. Balog, and M. Mirjafari, "Minimum energy and capacitance requirements for single-phase inverters and rectifiers using a ripple port," *IEEE Transactions on Power Electronics*, vol. 27, no. 11, pp. 4690–4698, 2012. DOI: 10.1109/TPEL.2012.2186640.

[4] H. Sarnago and O. Lucía, "High power density on-board charger featuring power pulsating buffer," *IEEE Open Journal of Power Electronics*, vol. 5, pp. 162–170, 2024. DOI: 10.1109/OJPEL.2024.3359271.

[5] D. Gottardo, L. De Lillo, L. Empringham, and A. Costabeber, "A low capacitance single-phase ac-dc converter with inherent power ripple decoupling," in *IECON 2016 - 42nd Annual Conference of the IEEE Industrial Electronics Society*, 2016, pp. 3129–3134. DOI: 10.1109/IECON.2016.7793634.

[6] D. Gottardo, L. De Lillo, L. Empringham, and A. Costabeber, "Differential buck single phase grid connected ac-dc converter with active power decoupling using a flipping capacitor," in *2017 IEEE 8th International Symposium on Power Electronics for Distributed Generation Systems (PEDG)*, 2017, pp. 1–5. DOI: 10.1109/PEDG.2017.7972438.

[7] R. Wang, F. Wang, D. Boroyevich, R. Burgos, R. Lai, *et al.*, "A high power density single-phase pwm rectifier with active ripple energy storage," *IEEE Transactions on Power Electronics*, vol. 26, no. 5, pp. 1430–1443, 2011. DOI: 10.1109/TPEL.2010.2090670.

[8] C. Marxgut, J. Biela, and J. W. Kolar, "Interleaved triangular current mode (tcm) resonant transition, single phase pfc rectifier with high efficiency and high power density," in *The 2010 International Power Electronics Conference - ECCE ASIA -*, 2010, pp. 1725–1732. DOI: 10.1109/IPEC.2010.5542048.

[9] J. Yuan, L. Dorn-Gomba, A. D. Callegaro, J. Reimers, and A. Emadi, "A review of bidirectional on-board chargers for electric vehicles," *IEEE Access*, vol. 9, pp. 51 501–51 518, 2021. DOI: 10.1109/ACCESS.2021.3069448.

[10] S. A. Q. Mohammed and J.-W. Jung, "A comprehensive state-of-the-art review of wired/wireless charging technologies for battery electric vehicles: Classification/common topologies/future research issues," *IEEE Access*, vol. 9, pp. 19 572–19 585, 2021. DOI: 10.1109/ACCESS.2021.3055027.

PCIM Europe 2024, 11– 13 June 2024, Nuremberg   DOI: 10.30420/566262347

A Reduced-Sensor Modular Dual Active Bridge-Based Battery Charging System for Electric Vehicles Using an Improved Linear Extended State Observer

Armel A. Nkembi[1], Paolo Cova[1*], Iñigo Kortabarria[2], Emilio Sacchi[3], Nicola Delmonte[1], Marco Portesine[3]

[1] Department of Engineering and Architecture, University of Parma, Parma, Italy
[2] Department of Electronic Technology, University of the Basque Country, Bilbao, Spain
[3] Poseico S.p.A., 16153 Genova, Italy

Corresponding author: Paolo Cova, paolo.cova@unipr.it
Speaker: Armel A. Nkembi, armelasongu.nkembi@unipr.it

Abstract

Developing an off-board rapid battery charging station is quite challenging since it must handle numerous battery chemistries with varying power requirements while maintaining high power conversion efficiency and small size. Furthermore, numerous modular topologies are widely employed to meet the battery's high power requirements while decreasing voltage and current stress on power switches, permitting the use of low-power rating switches. In such modular architectures, power balance among modules must be ensured to avoid overvoltage/overcurrent. Furthermore, such modular systems require many sensors, raising system costs. The traditional observers commonly used to reduce the sensor number introduce large errors between the actual and estimated system states and possess poor transient performance. To address these shortcomings, this paper proposes an improved extended state observer for modular DAB-based battery charging of electric vehicles. Numerical simulations using MATLAB/Simulink demonstrate the proposed system's robustness against disturbances and show how the transient response of the system is greatly improved.

1   Introduction

With the growing popularity of electric vehicles (EVs), there's a critical need for fast and efficient charging stations. These stations need power converters that can handle the high power demands of fast DC charging for different EVs [1]. Modular DC-DC converters are a popular solution. These converters come in various designs, like input-series-output-parallel (ISOP), input-series-output-series (ISOS), input-parallel-output-series (IPOS), and input-parallel-output-parallel (IPOP). The key benefit of these modular designs is that they share the overall power requirement among several smaller modules. This reduces the current and voltage stress on individual components, allowing for the use of more affordable, lower-power switches.

Out of many battery charging methods studied [2], the constant current (CC) /constant voltage (CV) approach is the most popular. This is because it's both safe and keeps batteries healthy. There are two main ways typically adopted to implement CC-CV battery charging control [3]. One uses separate

loops for current and voltage that switch based on the charging stage. The other method puts the current control loop within the voltage control loop. This ensures the battery current is always regulated during both stages (CC and CV), preventing damage from overcurrent and maximizing battery life.

Out of all the different modular DC-DC converter architectures found in research papers for charging batteries, the one that has caught the attention of researchers is the modular dual active bridge (DAB). This is because DAB converters offer several benefits like: high power density capability, soft switching characteristics, high efficiency, galvanic isolation, and bidirectional ability (useful for features like vehicle-to-grid power transfer) [4]. Fig. 1 shows the IPOP modular DAB converter which will be used in this work.

DAB converters typically use single-phase shift (SPS) modulation, which creates misaligned 2-level square wave voltages on both full-bridges. This mismatch leads to high circulation current and limits the soft switching range. As a result, SPS suffers from high current stress, high switching

2469

losses, and losses in the magnetic components (inductors and transformer), ultimately hampering overall efficiency [5]. While more complex modulation techniques [6] exist to improve efficiency and reduce stress, this work will focus on the SPS method due to its simplicity and also because the goal is not efficiency improvement but rather a control method to accurately estimate battery current irrespective of the modulation method used.

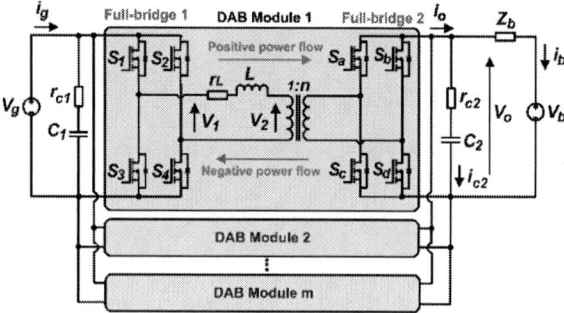

Fig. 1 Modular IPOP DAB converter.

Practically, parameter mismatches between DAB modules in the IPOP topology cause differences in each module's input and output currents, degrading the electro-thermal behavior of such systems, thus requiring the use of proper power balancing schemes, such as input current sharing (ICS), output current sharing (OCS), input voltage sharing (IVS), or output voltage sharing (OVS) [7], this is in addition to the CC-CV charging needs of the EV battery. All these requirements must be met by the controller, thus increasing the control effort.

Moreover, the above-mentioned power balancing schemes require many high-bandwidth sensors which increase costs, and noise, needing much effort to calibrate and filter out [8]. Adding to the complexity, the system also needs high-speed analog-to-digital conversion (ADC) to handle the measured data. For accurate measurements and better dynamic visualizations, the ADC sampling rate is typically a multiple of the switching frequency. However, achieving this can be difficult, especially when the switching frequency reaches hundreds of kHz, because high-speed ADCs are often more expensive and challenging to implement.

To reduce the number of sensors, observers based on active disturbance rejection control have proven to be very effective as implemented by authors [9] and [10]. However, these works focused only on basic converters and its application to more advanced modular topologies is still lacking. Moreover, in these works, the design of the load

current observer introduces huge errors between the actual and estimated values of load current, and the transient performance of the system wasn't considered.

Inspired by these previous works, this paper proposes an enhanced extended state observer for battery current estimation for use in fast charging of electric vehicles using a 3-module IPOP DAB topology. The proposed observer design removes a sensor from the system, while Kirchhoff's Current Law (KCL) redundancy allows for the elimination of an additional sensor. Furthermore, a novel feedforward scheme is proposed to improve the converter's dynamic response during the battery charging process.

The remainder of this work is structured as follows: Section 2 discusses the fundamental operation of the DAB converter employing single-phase shift (SPS) modulation and its accompanying waveforms, while Section 3 depicts the modeling of the system components, including the DAB and batteries. Section 4 provides the suggested enhanced extended state observer (ESO), as well as the OCS power balancing and CC-CV control algorithms. Section 5 presents numerical results from MATLAB/Simulink, and Section 6 concludes the work.

2 Operation Principle of the DAB Converter

The DAB converter's two symmetrical full bridges, seen in Fig. 1, convert the input and output DC voltages into two intermediate high-frequency square wave signals (V_1 and V_2). The power flow can be controlled by changing the phase shift angle between full-bridge 1 and full-bridge 2 while using SPS modulation on each H-bridge. As shown in Fig. 1, the power flow might be positive or negative depending on the sign of the phase shift angle. When the phase shift angle is positive, V_2 trails V_1, and power flows in the positive direction; when the phase shift angle is negative, V_1 trails V_2, and power flows in the opposite way.

Figure 2 displays typical DAB converter waveforms with positive power flow operating in both buck ($V_g > V_o$) and SPS modes. Each full-bridge operates at the same switching frequency and 50% duty cycle when using SPS modulation.

By neglecting the converter's resistive losses, the average transferred power depends on switching frequency (f_s) and can be represented as Eq. (1).

$$P_{av} = \frac{V_g V_o \varphi(\pi - |\varphi|)}{2\pi^2 nLf_s} \qquad (1)$$

According to (1), the magnitude of the power transferred in the positive direction increases as the phase shift angle (φ) increases up to a maximum at π/2, after which any increase in phase shift angle causes a decrease in power sent. A similar pattern is found in the reverse direction, with the highest reverse power obtained at -π/2.

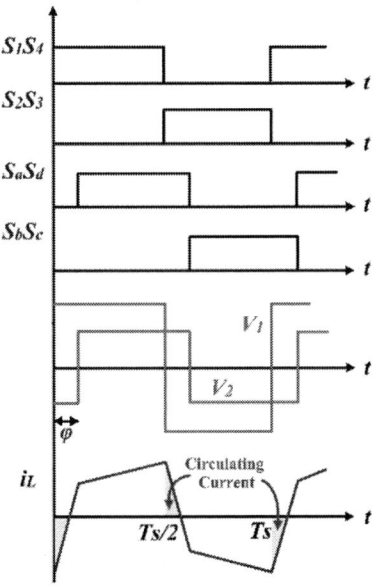

Fig. 2 Typical theoretical waveforms of DAB converter under SPS modulation.

3   System Modelling

3.1   Battery Modelling

Even though there are many different ways to model battery behavior [11], this study uses the second-order Thevenin equivalent model shown in Fig. 3. This model was chosen because of its simplicity while still providing good numerical results. This model doesn't account for the self-discharge of the battery. Additionally, while battery open circuit voltage, resistance, and capacitance can change depending on temperature and state of charge (SoC), this study considers only the SoC and ignores the temperature's influence.

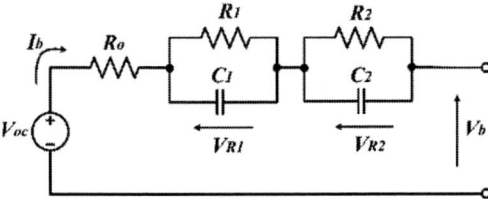

Fig. 3 Thevenin equivalent circuit battery model.

The terminal voltage in frequency domain is given by Eq. (2):

$$V_b(s) = V_{OC}(s) - R_o I_b(s) - V_{R_1}(s) - V_{R_2}(s) \quad (2)$$

where $V_{R1}(s)$ and $V_{R2}(s)$ are given by Eq. (3) and Eq. (4) respectively.

$$V_{R_1}(s) = \frac{R_1}{1 + sR_1C_1} I_b(s) \quad (3)$$

$$V_{R_2}(s) = \frac{R_2}{1 + sR_2C_2} I_b(s) \quad (4)$$

The SoC is computed based on the Coulomb counting technique. Here, the discharging current of the battery is measured and integrated over time to estimate the SoC [12]. This is given mathematically by Eq. (5):

$$SoC(t) = SoC(t_o) + \frac{\int_{t_o}^{t_o+\tau} I_b d\tau}{Q_{rated}} \quad (5)$$

where Q_{rated} is the rated or nominal capacity of the battery (Ah).

The resistance and capacitance values are derived experimentally at different SOCs and used to build a look-up table in MATLAB/Simulink from which the Thevenin equivalent model is then constructed using Eqs. (2)-(5).

3.2   OCS Control Strategy of Modular IPOP DAB Based on Enhanced GAM of Single DAB Converter

The generalized average modeling (GAM) method utilizes the complex exponential Fourier series expansion of periodic time-dependent variables to expand the state equations of a given converter circuit. This yields a differential equation governing the evolution of the complex magnitudes of successive harmonics for each state variable. While the choice of the number of harmonics for a specific state variable is discretionary, the accuracy of the approach is enhanced with an increased number of harmonics considered [13].

According to [14], the derivative of the kth coefficient for a state variable $x(t)$ is given by Eq. (6).

$$\frac{d\langle x \rangle_k(t)}{dt} = -jk\omega\langle x \rangle_k(t) + \langle\frac{dx}{dt}\rangle_k(t) \quad (6)$$

Furthermore, the kth coefficient of the product of two variables x and y is given by Eq. (7).

$$\langle xy \rangle_k(t) = \sum_i \langle x \rangle_{k-i}(t)\langle y \rangle_i(t) \quad (7)$$

Equations (6) and (7) are the basic GAM expressions which can be used to model the DAB converter. For this work, $k = 0$ for DC voltages while $k = \pm 1$ for the transformer AC currents and switching function components at the switching frequency ω_s. To perform GAM, we transform the DAB module 1 circuit in Fig. 1 into its equivalent shown in Fig. 4. In this circuit, $S_1(t)$ and $S_2(t)$ are the switching functions of the primary and secondary H-bridges, respectively, and they are given by Eq. (8) and Eq. (9):

$$S_1(t) = \begin{cases} 1, & 0 \le t < \dfrac{T_s}{2} \\ -1, & \dfrac{T_s}{2} \le t < T_s \end{cases} \quad (8)$$

$$S_2(t) = \begin{cases} 1, & \dfrac{dT_s}{2} \le t < \dfrac{dT_s}{2} + \dfrac{T_s}{2} \\ -1, & \dfrac{dT_s}{2} + \dfrac{T_s}{2} \le t < T_s \end{cases} \quad (9)$$

where d is the phase shift duty ratio and related to the phase shift angle by $\varphi = \pi d$.

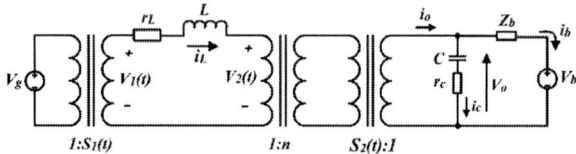

Fig. 4 Single DAB equivalent circuit for GAM.

From Fig. 4, the following circuit equations can be derived.

$$\frac{di_L}{dt} = \frac{v_g S_1}{L} - \frac{i_L r_L}{L} - \frac{(S_2)^2 i_L r_c}{n^2 L} + \frac{S_2 i_b r_c}{nL} - \frac{v_c S_2}{nL} \quad (10)$$

$$\frac{dv_c}{dt} = \frac{i_L S_2}{nC} - \frac{i_b}{C} \quad (11)$$

$$i_o = \frac{S_2 i_L}{n} \quad (12)$$

$$v_o = (i_o - i_b)r_c + v_c \quad (13)$$

Applying GAM Eq. (6) and Eq. (7) to equations (10)-(13) gives the DAB small signal model in Eq. (14) and Eq. (15):

$$\frac{d}{dt}\begin{bmatrix} \hat{i}_d \\ \hat{i}_q \\ \hat{v}_c \end{bmatrix} = A \begin{bmatrix} \hat{i}_d \\ \hat{i}_q \\ \hat{v}_c \end{bmatrix} + B \begin{bmatrix} \hat{v}_g \\ \hat{i}_b \\ \hat{\varphi} \end{bmatrix} \quad (14)$$

$$\begin{bmatrix} \hat{i}_o \\ \hat{v}_o \end{bmatrix} = G \begin{bmatrix} \hat{i}_d \\ \hat{i}_q \\ \hat{v}_c \end{bmatrix} + H \begin{bmatrix} \hat{v}_g \\ \hat{i}_b \\ \hat{\varphi} \end{bmatrix} \quad (15)$$

where $i_d = Re(\langle i_L \rangle_1)$, $i_q = Im(\langle i_L \rangle_1)$,

$$A = \begin{bmatrix} -\dfrac{1}{L}\left(r_L + \dfrac{r_c \cos(2\Phi)}{2n^2}\right) & \left(\omega - \dfrac{r_c \sin(2\Phi)}{2Ln^2}\right) & \dfrac{2\sin(\Phi)}{\pi nL} \\ \left(-\omega + \dfrac{r_c \sin(2\Phi)}{2n^2 L}\right) & -\dfrac{1}{L}\left(r_L + \dfrac{r_c \cos(2\Phi)}{2n^2}\right) & \dfrac{2\cos(\Phi)}{\pi nL} \\ -\dfrac{4}{\pi nC}\sin(\Phi) & -\dfrac{4}{\pi nC}\cos(\Phi) & 0 \end{bmatrix}$$

$$G = \begin{bmatrix} -\dfrac{4}{\pi n}\sin(\Phi) & -\dfrac{4}{\pi n}\cos(\Phi) & 0 \\ -\dfrac{4r_c}{\pi n}\sin(\Phi) & -\dfrac{4r_c}{\pi n}\cos(\Phi) & 1 \end{bmatrix}$$

$$H = \begin{bmatrix} 0 & 0 & \dfrac{4}{\pi n}[-X_1 \cos(\Phi) + X_2 \sin(\Phi)] \\ 0 & -r_c & \dfrac{4r_c}{\pi n}[-X_1 \cos(\Phi) + X_2 \sin(\Phi)] \end{bmatrix}$$

From Eq. (14) and Eq. (15), we can easily get the transfer function of the output current to phase shift in the form shown in Eq. (16).

$$\hat{i}_o = F(s)\hat{\varphi} \quad (16)$$

From Eq. (16), we can obtain the output current variation in any DAB module with reference to DAB module 1 as:

$$\hat{i}_{oj} = F(s)\left(\hat{\varphi}_j - \hat{\varphi}_1\right) + \hat{i}_{o1} \quad (17)$$

If the output currents are balanced, then $\sum_{j=1}^{3} \hat{i}_{oj} = \hat{i}_o = 0$. Using this, Eq. (17) becomes:

$$\hat{i}_{oj} = F(s)\left(\hat{\varphi}_j - \frac{1}{3}\sum_{j=1}^{3}\hat{\varphi}_j\right) \quad (18)$$

Equation (18) can be written in the form of a matrix as given in Eq. (19):

$$B = \begin{bmatrix} 0 & \left(-\dfrac{2r_c \sin(\Phi)}{\pi nL}\right) & \left[\dfrac{r_c}{Ln^2}\left(X_1 \sin(2\Phi) - X_2 \cos(2\Phi)\right) - \dfrac{2\cos(\Phi)}{\pi nL}\left(r_c I_b - V_c\right)\right] \\ -\dfrac{2}{\pi L} & \left(-\dfrac{2r_c \cos(\Phi)}{\pi nL}\right) & \left[\dfrac{r_c}{Ln^2}\left(X_2 \sin(2\Phi) + X_1 \cos(2\Phi)\right) + \dfrac{2\sin(\Phi)}{\pi nL}\left(r_c I_b - V_c\right)\right] \\ 0 & -\dfrac{1}{C} & -\dfrac{4}{\pi nC}\left(X_1 \cos(\Phi) - X_2 \sin(\Phi)\right) \end{bmatrix}$$

$$\begin{bmatrix} \hat{i}_{o1} \\ \hat{i}_{o2} \\ \hat{i}_{o3} \end{bmatrix} = F(s) \begin{bmatrix} \dfrac{2}{3} & -\dfrac{1}{3} & -\dfrac{1}{3} \\ -\dfrac{1}{3} & \dfrac{2}{3} & -\dfrac{1}{3} \\ -\dfrac{1}{3} & -\dfrac{1}{3} & \dfrac{2}{3} \end{bmatrix} \begin{bmatrix} \hat{\varphi}_1 \\ \hat{\varphi}_2 \\ \hat{\varphi}_3 \end{bmatrix} \qquad (19)$$

Equation (19) can then be used to perform the decoupled OCS for power balance control of the modular IPOP DAB converter.

4 Control System Design

The proposed control system which is used for the battery charging is shown in Fig. 5. It consists of two feedforward control loops (FF1 and FF2), a power balancing loop through output current sharing control, an enhanced extended state observer for reduced-error battery current estimation and a CC-CV control loop.

4.1 Feedforward Control

The feedforward paths bring about a smooth transition from CC charging mode to CV charging mode. This improves system stability and dynamic performance significantly even in the presence of input voltage disturbance.

To achieve feedforward compensation, we first consider that the converter average output power can be written as: $P_o = V_o I_b$ where I_b is the average battery current. Substituting this expression into

(1) and simplifying, an expression for the phase shift ratio used in FF2 is obtained as in Eq. (20).

$$\varphi^* = \pi \left(\frac{1}{2} - \sqrt{\frac{1}{4} - \frac{2n\,F_{sw}\,L\,I_b^*}{V_g}} \right) \qquad (20)$$

Furthermore, a seamless switch from CC to CV charging modes can be guaranteed if the ratio of the desired average battery current estimate to its instantaneous current is equivalent to the ratio of the desired average input current to its instantaneous current as given in Eq. (21).

$$\frac{I_{g\_est}^*}{I_g} = \frac{I_{b\_est}^*}{I_{b\_est}} \implies I_{b\_est}^* = \left(\frac{I_{g\_est}^*}{I_g} \right) I_{b\_est} \qquad (21)$$

$I_{b\_est}^*$ is then added to ΔI_b^* which is the output of the external voltage loop PI controller. Therefore, the effective battery reference current for the CV mode of charging is given by Eq. (22). ΔI_b^* helps in compensating for the difference between instantaneous output power and desired output power.

$$I_{b\_CV}^* = \Delta I_b^* + I_{b\_est}^* \qquad (22)$$

The desired input current is deduced considering input/output port power balance and this is given by Eq. 23.

$$I_{g\_est}^* = \left(\frac{I_{b\_est}}{V_g} \right) V_o^* \qquad (23)$$

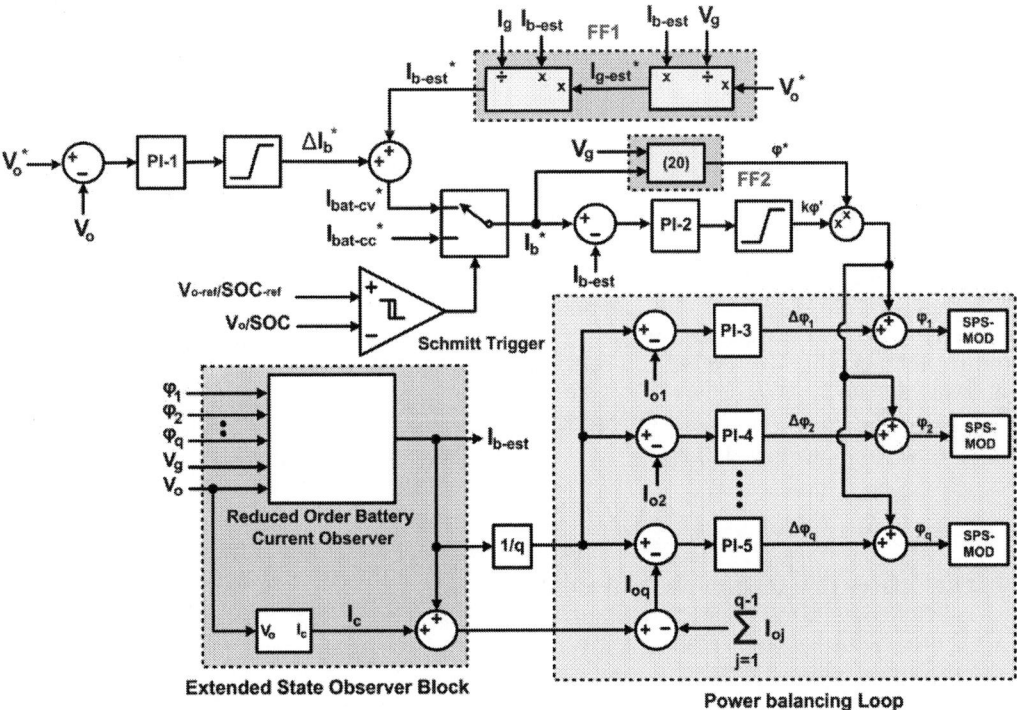

Fig. 5 Proposed control system for modular IPOP converter.

4.2 Extended State Observer (ESO)

The ESO begins with the dynamic equation of the output capacitor voltage derivative given by Eq. (24):

$$\frac{dV_c}{dt} = \frac{I_o}{C} - \frac{I_b}{C} \qquad (24)$$

$$= \frac{V_{in}}{2nF_{sw}C} \sum_{m=1}^{q} \frac{D_m(1 - D_m)}{L_m} + F \qquad (25)$$

where $D_m = \varphi_m/\pi$ and $F = -I_b/C$. Our study is based on a three module DAB system, so $q = 3$.

The small signal value of the capacitor voltage in Eq. (25) can be written in the form shown in Eq. (26):

$$\frac{d\hat{V}_c}{dt} = b\hat{u} + \hat{F} \qquad (26)$$

where $b = [b_1\ b_2\ b_3]$ and $\hat{u} = \begin{bmatrix} \hat{d}_1\ \hat{d}_2\ \hat{d}_3 \end{bmatrix}'$. $b_j\ \forall\ j \in [1\ 3]$ is given by

$$b_j = \frac{V_{in}D_j(1 - 2D_j)}{2nF_{sw}CL_j} \qquad (27)$$

In Eq. (26), $\hat{V}_c$ is taken as a state variable and expressed as $x_1 = \hat{V}_c$, while $\hat{F}$ is considered as an augmented state variable expressed as $x_2 = \hat{F}$. Furthermore, the time derivative of x_2 is signified h. From this analysis, the state-space model of Eq. (26) is given in Eq. (28) where g denotes a 1×3 zero vector.

$$\begin{bmatrix} \dot{x}_1 \\ \dot{x}_2 \end{bmatrix} = \begin{bmatrix} 0 & 1 \\ 0 & 0 \end{bmatrix} \begin{bmatrix} x_1 \\ x_2 \end{bmatrix} + \begin{bmatrix} b \\ g \end{bmatrix} \hat{u} + \begin{bmatrix} 0 \\ 1 \end{bmatrix} h \qquad (28)$$

Based on Eq. 28, the conventional ESO is constructed as follows:

$$\begin{bmatrix} \dot{z}_1 \\ \dot{z}_2 \end{bmatrix} = \begin{bmatrix} 0 & 1 \\ 0 & 0 \end{bmatrix} \begin{bmatrix} z_1 \\ z_2 \end{bmatrix} + \begin{bmatrix} b \\ g \end{bmatrix} u + \begin{bmatrix} \beta_1 \\ \beta_2 \end{bmatrix} [x_1 - z_1] \qquad (29)$$

where z_1 and z_2 are the estimated values of x_1 and x_2 respectively while $[\beta_1\ \beta_2]'$ is the Luenberger observer gain vector which is designed using pole placement method [15]. The battery current estimate is derived from z_2 as $I_{b\_est} = -Cz_2$.

By subtracting Eq. (29) from Eq. (28), the error state equation can be written as follows:

$$\begin{bmatrix} \dot{e}_1 \\ \dot{e}_2 \end{bmatrix} = \underbrace{\begin{bmatrix} -\beta_1 & 1 \\ -\beta_2 & 0 \end{bmatrix}}_{A} \begin{bmatrix} e_1 \\ e_2 \end{bmatrix} + \begin{bmatrix} 0 \\ 1 \end{bmatrix} h \qquad (30)$$

From control theory, the matrix A will be Hurwitz stable if all the roots of its characteristic polynomial, $\lambda(s)$, are in the left half plane. If the ESO poles are designed to be all located at $-\omega_o$ (ω_o is the observer bandwidth), then we obtain the following:

$$\lambda(s) = |sI - A| = s^2 + s\beta_1 + \beta_2 \equiv (s + \omega_o)^2 \qquad (31)$$

Therefore, we obtain the observer gains from Eq. (31) as $\beta_1 = 2\omega_o$ and $\beta_2 = \omega_o^2$.

The ESO's bandwidth is often chosen to be 5-15 times that of the controller in order to ensure that the estimated state dynamics have fast-tracking capabilities. However, the ESO's bandwidth cannot be too big because it reduces the system's noise immunity.

The conventional ESO assumes that the system disturbance is constant, but in fact it is time varying due to significant changes in the plant model, resulting in high estimation errors of the system states and disturbances. Furthermore, the time delay caused by signal transmission will result in additional estimation mistakes [16]. To estimate x_2 and hence battery current more correctly, an enhanced ESO is proposed as shown in Eq. 32, which introduces a new fictitious variable k_2 to provide a preliminary estimate of x_2 and subsequently z_2 to supplement k_2 and provide a more accurate estimate of x_2.

$$\begin{cases} \dot{z}_1 = k_2 + \beta_1(x_1 - z_1) + bu + M \\ M = \alpha_1\big(u(t) - u(t - h)\big) + \alpha_2\big(x_1(t) - x_1(t - h)\big) \\ \dot{k}_2 = \beta_2(x_1 - z_1) \\ z_2 = k_2 + \alpha_3 x_1 + \dfrac{d^\gamma(x_1 - z_1)}{dt^\gamma} + \beta_1(x_1 - z_1) \end{cases} \qquad (32)$$

In Eq. 32, $h > 0$ is a constant artificial delay, arbitrarily chosen to ensure that the error dynamics is exponentially stable. Adjusting the parameters α_1, α_2, and α_3 greatly improves observation performance. The gains β_1 and β_2 are same as the conventional ESO. Moreover, the proposed enhanced ESO introduces a fractional-order derivative (FOD) operator s^γ.

FOD implementation requires significantly more computational complexity than integer order transfer function models because of their irrational nature. As a result, FODs are typically approximated as higher-order rational transfer functions with a constant phase curve within a specific frequency band. Many frequency and time domain strategies for implementing FODs have been presented in literature [17]. In this work, Carlson's method has been chosen to implement the FOD found in the enhanced ESO and is discussed next.

Consider $G(s)$ to be a rational function and $H(s)$ to be a fractional order transfer function such that $H(s) = [G(s)]^r, r \in \mathbb{R}$. Then $H(s)$ can be recursively approximated as in Eq. 33.

$$\begin{cases} H_i(s) = H_{i-1}(s)\dfrac{G(s) + p[H_{i-1}(s)]^2}{pG(s) + [H_{i-1}(s)]^2} \\[2mm] where \ p = \dfrac{1-r}{1+r} \ and \ H_o(s) = 1 \end{cases} \quad (33)$$

For this work, first order derivative is considered in Eq. 32, so $\gamma = 1$. From this we implement s^1 using Eq. (33) by considering that $s^1 = s^{0.5}s^{0.5}$; so $r = 0.5$ and $G(s) = s$. The higher the number of iterations, the better is the performance as the ripples in the phase and magnitude responses are greatly reduced.

5 Simulation-Based Numerical Results

An IPOP modular DAB-based battery charging system is designed and implemented in MATLAB/Simulink based on the parameters in Table I. Furthermore, MATLAB's SISOTOOL is utilized to design the PI controllers, which are then tested against the control strategy shown in Fig. 5. For the current controller, K_{pi} = 0.002438, K_{ii} = 32.01, PM = 61° with a BW of 2 kHz, while for the voltage controller, K_{pv} = 0.4417, K_{iv} = 923.1948, PM = 58° and BW = 800 Hz. Each PI in the OCS loop has K_p = 0.094, K_i = 76.64, PM = 65° and BW = 700 Hz. Furthermore, the observer bandwidth, ω_o is chosen to be 6×10^4 rad/s.

The proposed ESO and feedforward control schemes presented in section 4 are tested in the presence of slight mismatches in inductor and transformer parameters while charging the battery at a CC of 200 A and then CV of 900 V. The results obtained with the conventional ESO and the proposed ESO are shown in Fig. 6 as well as the impact of the proposed feedforward system on the dynamic performance of the battery charging system.

Parameter	Value
Input voltage (V_{in})	750 V
Battery nominal voltage (V_o)	900 V
Rated power per module (P_{rated})	85 kW
Number of DAB modules	3
Switching frequency (F_{sw})	100 kHz
Leakage inductance per module (L_r)	7.8 µH
Parasitic resistance of inductor and transformer (r_L)	50 mΩ
Drain-source on resistance of MOSFET (R_{ds-on})	35 mΩ
Transformer turns ratio (n_p:n_s)	5:6
Output filter capacitor (C_{fo})	200 µF
Input filter capacitor (C_{fi})	400 µF
Filter capacitor ESR (r_c)	50 mΩ

Table 1 DAB converter specifications for single module.

Fig. 6a shows that the current estimation error during the CC charging stage with the conventional ESO is 5.5 A which reduces significantly to about 0.5 A when the proposed enhanced ESO is utilized. We also notice that during the CV stage, the

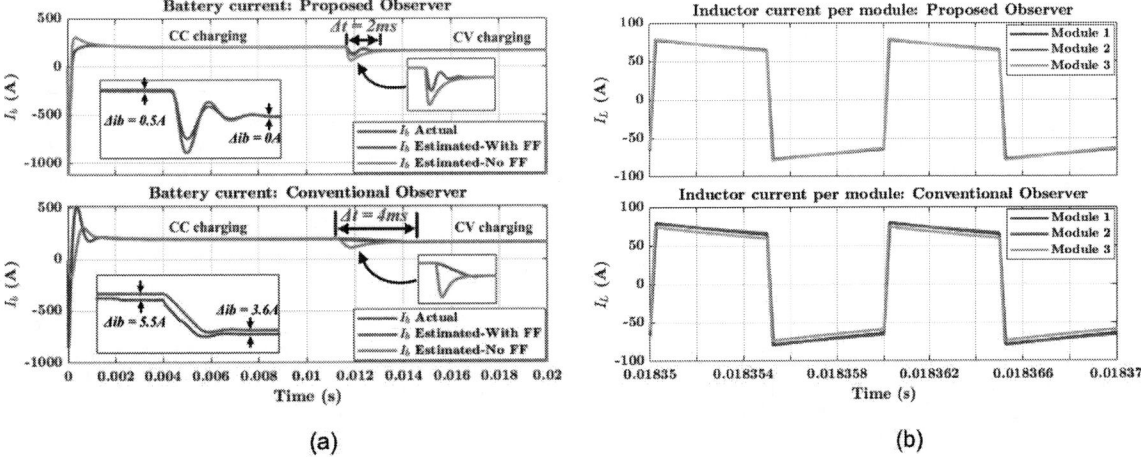

Fig. 6 (a) Battery current with proposed and conventional observer; (b) Inductor currents with proposed and conventional observer.

estimation error with the Luenberger-based conventional ESO is 3.6 A which reduces to 0 A with the proposed ESO. Also observable is the fact that the transition from CC to CV charge mode takes about 4 ms with the conventional observer while with the proposed ESO, this time reduces to 2 ms. We therefore see that implementing the proposed linear ESO (LESO) greatly reduces the current estimation error (Δi_b) and shortens the transition time (Δt) from CC to CV charging mode while ensuring inductor current balancing in each DAB module as seen in Fig. 6b.

Furthermore, we see that feedforward compensation independently reduces the current dips during the CC-CV transition for both the conventional and proposed observers.

6 Conclusion

A novel linear ESO is proposed for battery current estimation which improves observation performance by introducing a fractional order term and state delays to account for system disturbances. Consequently, the design of this LESO is more flexible such that the current estimation error is reduced in comparison to the typical LESO. The enhanced LESO also improves the dynamic response speed while the proposed feedforward control reduces the current undershoots. Overall, the enhanced ESO-based battery charging system has a good robustness to deal with system parameter variations and has a fast dynamic response for estimating the total disturbance. With the proposed control strategy, we can eliminate two sensors which save cost. As a next step in this work, the authors intend to apply the proposed control schemes on an actual hardware prototype to validate the numerical simulations obtained thus far.

7 Acknowledgement

This research was funded by the Italian "Ministero dell'Istruzione, dell'Universita e della Ricerca—Programma Operativo Nazionale 2014–2020 (PON): AZIONE IV.5-Dottorati su tematiche Green del PON R&I 2014–2020".

References

[1] S. S. Trivedi and A. V. Sant, "Comparative analysis of dual active bridge DC–DC converter employing si, SIC and gan mosfets for G2V and V2G operation," *Energy Reports*, vol. 8, no. 13, pp. 1011–1019, Nov. 2022. doi:10.1016/j.egyr.2022.08.100.

[2] N. Ghaeminezhad and M. Monfared, "Charging control strategies for lithium-ion battery packs: Review and recent developments," *IET Power Electron.*, vol. 15, no. 5, pp. 349–367, 2022, doi: 10.1049/pel2.12219.

[3] A. Urtasun, A. Berrueta, P. Sanchis, and L. Marroyo, "Parameter-Independent Control for Battery Chargers Based on Virtual Impedance Emulation," *IEEE Trans. Power Electron.*, vol. 33, no. 10, pp. 8848–8858, 2018, doi: 10.1109/TPEL.2017.2778041.

[4] O. M. Hebala, A. A. Aboushady, K. H. Ahmed, S. Burgess and R. Prabhu, "Generalized Small-Signal Modelling of Dual Active Bridge DC/DC Converter," 2018 7th International Conference on Renewable Energy Research and Applications (ICRERA), Paris, France, 2018, pp. 914-919, doi: 10.1109/ICRERA.2018.8567014.

[5] A. Rashwan, A. I. Ali, and T. Senjyu, "Current stress minimization for isolated dual active bridge DC–DC converter," *Scientific Reports*, vol. 12, no. 1, Oct. 2022. doi:10.1038/s41598-022-21359-1.

[6] F. Corti, V. Bertolini, A. Reatti, E. Cardelli and M. Giallongo, "Comparison of Control Strategies for Dual Active Bridge Converter," 2022 IEEE 21st Mediterranean Electrotechnical Conference (MELECON), Palermo, Italy, 2022, pp. 902-907, doi: 10.1109/MELECON53508.2022.9843012.

[7] X. Ruan, W. Chen, L. Cheng, C. K. Tse, H. Yan and T. Zhang, "Control Strategy for Input-Series–Output-Parallel Converters," in IEEE Transactions on Industrial Electronics, vol. 56, no. 4, pp. 1174-1185, April 2009, doi: 10.1109/TIE.2008.2007980.

[8] N. D. Dinh and G. Fujita, "Design of a reduced-order observer for Sensorless control of Dual-Active-Bridge converter," *2018 International Power Electronics Conference (IPEC-Niigata 2018 -ECCE Asia)*, Niigata, Japan, 2018, pp. 363-369, doi: 10.23919/IPEC.2018.8507888.

[9] J. Lu, M. Savaghebi, Y. Guan, J. C. Vasquez, A. M. Y. M. Ghias and J. M. Guerrero, "A Reduced-Order Enhanced State Observer Control of DC-DC Buck Converter," in *IEEE Access*, vol. 6, pp. 56184-56191, 2018, doi: 10.1109/ACCESS.2018.2872156.

[10] S. Ahmad and A. Ali, "Active disturbance rejection control of DC–DC boost converter: A review with modifications for improved performance," *IET Power Electronics*, vol. 12, no. 8, pp. 2095–2107, Jun. 2019. doi:10.1049/iet-pel.2018.5767.

[11] G. Saldaña, J. I. San Martín, I. Zamora, F. J. Asensio, and O. Oñederra, "Analysis of the current electric battery models for electric vehicle simulation," *Energies*, vol. 12, no. 14, p. 2750, 2019.

[12] S. Susanna, B. R. Dewangga, O. Wahyungoro, and A. I. Cahyadi, "Comparison of simple battery model and thevenin battery model for SOC estimation based on OCV method," in *2019 International Conference on Information and Communications Technology (ICOIACT)*, 2019, pp. 738–743.

[13] M. Rolak, M. Twardy, and C. Soból, "Generalized average modeling of a dual active bridge DC-DC converter with triple-phase-shift modulation," *Energies*, vol. 15, no. 16, p. 6092, Aug. 2022. doi:10.3390/en15166092.

[14] J. A. Mueller and J. W. Kimball, "Generalized average modeling of DC subsystem in solid state transformers," *2017 IEEE Energy Conversion Congress and Exposition (ECCE)*, Cincinnati, OH, USA, 2017, pp. 1659-1666, doi: 10.1109/ECCE.2017.8095992.

[15] J. Lu, S. Golestan, M. Savaghebi, J. C. Vasquez, J. M. Guerrero and A. Marzabal, "An Enhanced State Observer for DC-Link Voltage Control of Three-Phase AC/DC Converters," in *IEEE Transactions on Power Electronics*, vol. 33, no. 2, pp. 936-942, Feb. 2018, doi: 10.1109/TPEL.2017.2726110.

[16] F. Wang *et al.*, "Modified active disturbance rejection control scheme with sliding mode compensation for Airborne Star Tracker driven by permanent magnet synchronous motor," *Control Engineering Practice*, vol. 127, p. 105267, Oct. 2022. doi:10.1016/j.conengprac.2022.105267.

[17] F. N. Deniz, B. B. Alagoz, N. Tan, and M. Koseoglu, "Revisiting four approximation methods for fractional order transfer function implementations: Stability Preservation, Time and frequency response matching analyses," *Annual Reviews in Control*, vol. 49, pp. 239–257, 2020. doi:10.1016/j.arcontrol.2020.03.003.

PCIM Europe 2024, 11– 13 June 2024, Nuremberg   DOI: 10.30420/566262348

Bidirectional Non-Isolated Three-Phase Onboard Charger with a Low-Voltage Lower-Phase Operation Mode

Milad Khani[1], Steffen Frei[1], Gerd Griepentrog[1]

[1] Technische Universität Darmstadt, Germany

Corresponding author: Milad Khani, mldkhn23@gmx.de
Speaker: Steffen Frei, steffen.frei@lea.tu-darmstadt.de

Abstract

Compared to isolated design, non-isolated onboard chargers are more susceptible to common-mode issues but promise a higher power density by omitting bulky inductive elements. A single-stage, non-isolated design for an onboard charger is presented, which can be operated in three-phase or a lower-phase mode and offers measures against common-mode problems. The system voltage of the charger has been optimized for $800\,V$ traction battery packs and allows charging of the battery even below $650\,V$, despite a non-isolated, three-level design. Possible common-mode current problems which could trip residual current devices (RCDs) are investigated and drastically mitigated through a selection of suitable modulation schemes. Furthermore, a hardware prototype is presented, as well as measurement results of the single-phase operation, common-mode currents and efficiency.

1 Introduction

Onboard chargers (OBC) play a crucial role in plug-in hybrid (PHEV) or battery electric vehicles (BEV). OBCs convert the AC supplied power to DC and charge the traction battery of the vehicle. With the ongoing rapid adoption of electric vehicles (EV), advancements in the design of OBCs are a necessity. Space is a precious commodity in an EV; it is therefore essential to achieve high power densities for OBCs.

Higher power density enables faster charging or more compact designs. One of the main contributors to higher power density is the usage of new semiconductor materials such as Silicon Carbide (SiC) [1] and Gallium Nitrite (GaN) [2] and the successive increase of switching frequency, allowing for more compact designs by shrinking passives such as inductors, transformers and capacitors [3].

Newer developments also see a departure from viewing the OBCs as just a unidirectional charger [4]. Although national norms regarding bidirectional charging are still in the design phase, ISO 15118-20 provides a standard for bidirectional charging, detailing the communication between EV and the electric vehicle supply equipment (EVSE), better known as a wall box or charging station. Different car manufacturers already provide the ability to use OBCs to supply loads (V2L) or homes (V2H) with energy from the EVs traction battery. It is to be expected that bidirectional OBCs will become more widespread, especially after national norms are defined and adopted.

Most non-integrated OBCs employ a two-stage design with one of the stages implementing a galvanic isolation [4]–[8]. Some non-isolated designs for integrated OBCs exist [9], [10], but it is unknown if such designs are used in commercially acquirable vehicles. Further, most of these designs do not take common-mode issues into consideration. In order to meet rising power demands, simply increasing the switching frequency may not be a sufficient solution. Higher switching frequencies coupled with new topologies will be key to solving this problem.

This paper presents a non-integrated three-phase, non-isolated OBC design with the ability to operate in a lower-phase mode, consisting of a single stage, allowing the use of fewer passive components and therefore a higher power density.

PCIM Europe 2024, 11– 13 June 2024, Nuremberg   DOI: 10.30420/566262348

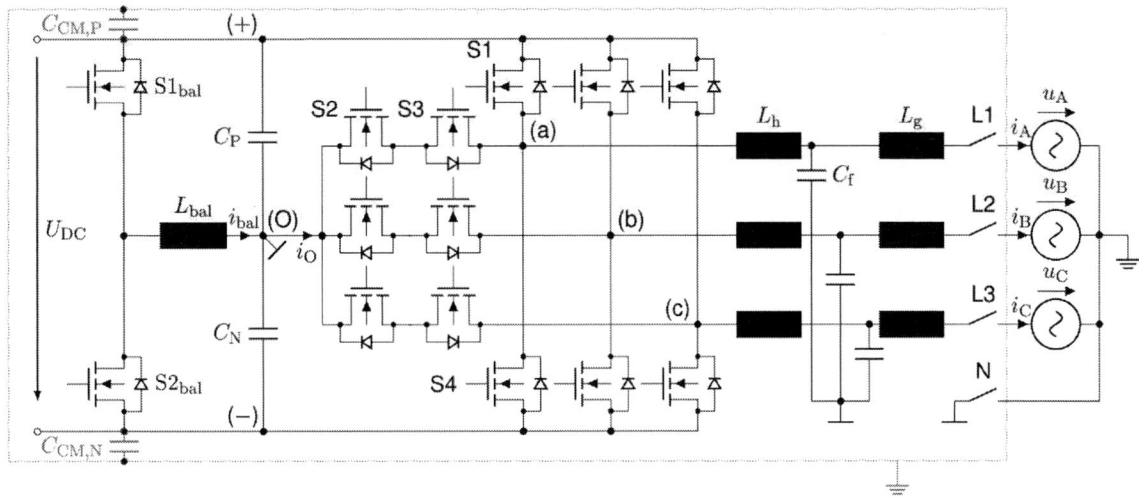

Fig. 1: Single stage non-isolated topology with parasitic CM-capacitors

2 Challenges and Goals

Non-isolated power electronics are more susceptible to common-mode problems. Due to many different factors, switching converters generate common-mode voltages (CMV) and subsequently common-mode currents (CMI) through parasitic or Y-capacitors which can lead to a multitude different issues. CMV are generated by switch states in the converter which change the voltage of the vehicle's high voltage DC bus in respect to ground i.e. protective earth (PE). These changes in the CMV generally do not have an impact on the performance of the converter. Unfortunately, the generation of CMVs lead to CMI through unwanted parasitic capacitances in the system. The housing of any device connected to the high-voltage bus is electrically connected to the chassis of the vehicle. The area and distance between the high voltage parts of the system and the chassis determine the magnitude of the parasitic capacitance in the system. Since OBCs are required to be connected to an EVSE, which are mandated to have an all-current sensitive type B residual current device (RCD) installed, and since CMIs travel through PE, these currents will be detected by the RCD and can lead to tripping of said RCD during normal operation. This is especially important for type B RCDs, as they also cover the range of common switching frequencies of converters. It is therefore vital to take special care in the design of the OBC to minimize any CMI in the system.

Another challenge that is a consequence of a non-galvanically isolated design is the operation voltage of the OBC. The absence of a transformer limits the lower end of battery voltages in which the charger will still be able to operate. In the case of a non-isolated three-phase inverter this voltage equals two times the peak voltage of the grid, if a CMV-free modulation is used. Taking a 10% tolerance (according to IEC 60038) into account and assuming a $230\,\mathrm{V}/400\,\mathrm{V}$ grid, this would equal a lower battery threshold voltage of $715\,\mathrm{V}$. $800\,\mathrm{V}$ traction battery systems, which are getting more popular, usually range from $600\,\mathrm{V}$ to $800\,\mathrm{V}$, depending on the state of charge (SOC) [11]. This means a solution is necessary to charge the battery from low SOCs which fall below the threshold voltage.

3 Topology

The chosen topology is depicted in Fig. 1. It consists of (from left to right) a balancer circuit, DC-link, three level T-type inverter, as well as the output sine-filter. On the left are the terminals to the traction battery and on the right is the three-phase grid. The symbol ⏚ is used for earth ground, i.e. PE, and the symbol ⊥ is used for circuit ground, i.e. the DC-link midpoint. The midpoint of the sine-filter is directly connected to the circuit ground. Through the relay "N" it is possible to connect circuit ground to earth ground. The dotted box represents the chassis of the car, which is connected to PE during the connection to an EVSE. The red capacitors represent the CM-capacitance of the EV.

The balancer ensures that both DC-link halves are properly balanced, i.e. the voltage of C_P and C_N

2479

are equal. The balancer can directly influence the potential at point (O). This plays a crucial part for the operation with lower DC-link voltages, which fall below the aforementioned threshold voltage for the operation of the inverter. The balancer is also used to minimize the necessary DC-link capacitance, by compensating for the fundamental frequency ripple generated by the operation of the inverter. The ripple in the DC-link can be derived with a fundamental frequency analysis as shown in [12]. The current i_O is mainly dependent on the state of the switches in each bridge. The averaged ripple current $i_{O,av}$ over one switching cycle can then be derived as follows:

$$
\begin{aligned}
i_{O,av}(t) &= -\frac{2}{U_{DC}} \cdot (|u_A(t)| \cdot i_A(t) \\
&\quad + |u_B(t)| \cdot i_B(t) + |u_c(t)| \cdot i_C(t)) \\
&= -\frac{2 \cdot m \cdot \hat{I}_N}{5 \cdot \pi} \cdot (5 \cdot \cos(3 \cdot \omega \cdot t - \varphi) \\
&\quad - \cos(3 \cdot \omega \cdot t + \varphi))
\end{aligned}
\tag{1}
$$

with

$$
\begin{bmatrix} |u_A(t)| \\ |u_B(t)| \\ |u_C(t)| \end{bmatrix} = \frac{2 \cdot \hat{U}_N}{\pi} \cdot \begin{bmatrix} 1 + 2 \cdot \sum_{n=1}^{\infty}(-1)^{n+1} \cdot \frac{\cos(2 \cdot n \cdot \omega \cdot t)}{4 \cdot n^2 - 1} \\ 1 + 2 \cdot \sum_{n=1}^{\infty}(-1)^{n+1} \cdot \frac{\cos(2 \cdot n \cdot \omega \cdot t + \frac{n \cdot 2 \cdot \pi}{3})}{4 \cdot n^2 - 1} \\ 1 + 2 \cdot \sum_{n=1}^{\infty}(-1)^{n+1} \cdot \frac{\cos(2 \cdot n \cdot \omega \cdot t + \frac{n \cdot 4 \cdot \pi}{3})}{4 \cdot n^2 - 1} \end{bmatrix}
$$

$$
\begin{bmatrix} i_A(t) \\ i_B(t) \\ i_C(t) \end{bmatrix} = \hat{I}_N \cdot \begin{bmatrix} \cos(\omega \cdot t - \varphi) \\ \cos(\omega \cdot t - \frac{2}{3} \cdot \pi - \varphi) \\ \cos(\omega \cdot t - \frac{4}{3} \cdot \pi - \varphi) \end{bmatrix}
$$

with $\hat{I}_N$ as the amplitude of the grid current, $\hat{U}_N$ as the amplitude of the grid voltage, m as the modulation index defined as $m = 2 \cdot \hat{U}_N / U_{DC}$ and φ as the grid current angle. It is now possible to decrease the necessary DC-link capacitance by compensating for the current in Eq. (1). By introducing a compensation factor k, it is possible to derive the voltage ripple in the DC-link as:

$$
\begin{aligned}
u_{O,av}(t) &= \frac{2 \cdot m \cdot \hat{I}_N}{15 \cdot \pi \cdot \omega \cdot C} \cdot ((1 - k) \cdot \sin(3 \cdot \omega \cdot t + \varphi) \\
&\quad - 5 \cdot (1 - k) \cdot \sin(3 \cdot \omega \cdot t - \varphi))
\end{aligned}
\tag{2}
$$

with $C = C_P = C_N$ as the capacitance of one DC-link half. Finally, the maximum ripple can be derived:

$$
\Delta u_{O,av} = \frac{4 \cdot \hat{I}_N \cdot m \cdot |k - 1|}{15 \cdot \pi \cdot \omega \cdot C} \cdot \sqrt{9 - 5 \cdot \cos^2(\varphi)}
\tag{3}
$$

With rising compensation factor k, the DC-link voltage ripple is reduced. This enables a significantly smaller DC-link.

4 Three-Phase Modulation

During three-phase operation, the relays "L1" to "L3" are closed. To gain the highest degree of freedom, instead of controlling the inverter with a sinusoidal pulse with modulation (SPWM), the space

vector PWM (SVPWM) is used. Out of the 21 available voltage vectors, six voltage vectors, as well as one zero voltage vector will not generate any CMV. By limiting the usage to these voltage vectors, it is possible to operate the inverter without generating any CMIs, with the exception of switching transients. One of the resulting modulations schemes, called 2 medium vectors 1 zero vector (2MV1Z) can be seen in Fig. 2. This modulation utilizes the two closest medium vectors and the zero vector and enables operation in the greyly shaded area [13].

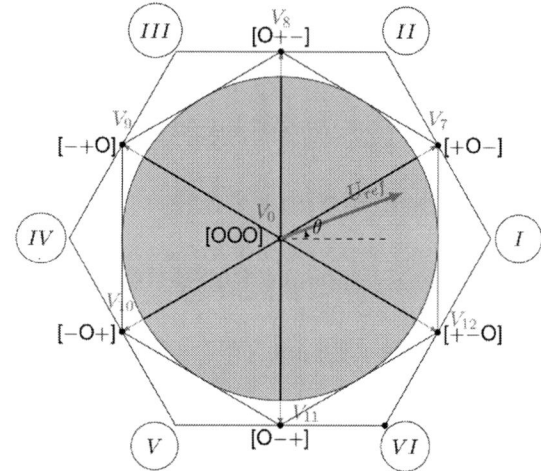

Fig. 2: 2MV1Z SVPWM with reference vector [13]

In theory it is possible to operate the inverter completely CM-free. In practice this is not possible for two reasons:

1. Imbalance of the DC-link will lead to shifts in the (O)-potential, which will introduce a CMV with three times the grid frequency. Most of the low frequency part of this imbalance can be compensated for with the balancer.

2. To avoid shoot-through events and subsequent destruction of the inverter, a dead-time is inserted into the PWM signal. This will lead to CMV pulses due to the generation of non-zero CMV space vectors. The result is a CMV with switching frequency. [14].

5 Lower-Phase Modulation

If the voltage falls below the lower voltage threshold needed to operate the inverter, it is possible to fall back to a lower-phase operation mode. Lower-phase meaning an operation with less than three phases. For this purpose, the relays "N" and one

2480

or two of the phase relays in Fig. 1 are closed. The neutral phase is directly connected to the midpoint of the DC-link. This enables lower-phase operation, but does not extend the voltage operation range of the inverter to lower voltages. The balancer now injects a current into the DC-link midpoint, creating an imbalance voltage defined as $u_{\mathrm{imb}} = u_{\mathrm{C,P}} - u_{\mathrm{C,N}}$. By carefully shaping the imbalance voltage depending on which half-wave the grid voltage is currently located in, it is possible to extend the useful voltage operation range of the system. This principle is shown for single-phase operation in Fig. 3. By generating a positive imbalance voltage during the positive half-wave, it is possible to extend the positive voltage vector by $u_{\mathrm{imb}}/2$. The same can be done in the negative half-wave with a negative imbalance voltage.

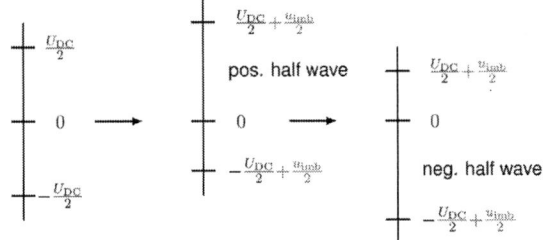

Fig. 3: Voltage vectors for single-phase operation

The resulting imbalance voltage is a sinusoidal wave with the same frequency as the grid voltage. For single-phase operation, the phase of that sinusoidal wave is identical to the grid voltage. The imbalance voltage for that case is defined as:

$$u_{\mathrm{imb,1ph}}(t) = 2 \cdot (1 + \alpha) \cdot \left(\hat{U}_{\mathrm{N}} - \frac{U_{\mathrm{DC}}}{2} \right) \cdot \cos(\omega \cdot t) \quad (4)$$

with α being a safety factor to account for voltage drops in the system. The same principle can be used for two-phase operation, the only difference being the amplitude and phase of the sinusoidal voltage:

$$u_{\mathrm{imb,2ph}}(t) = \left(\frac{1}{\sqrt{1 - \frac{3 \cdot \hat{U}_{\mathrm{N}}^2}{U_{\mathrm{DC}}^2}}} \cdot \left(\frac{3 \cdot \hat{U}_{\mathrm{N}}^2}{U_{\mathrm{DC}}} - U_{\mathrm{DC}} \right) + \hat{U}_{\mathrm{N}} \right)$$
$$\cdot (1 + \alpha) \cdot \cos\left(\omega \cdot t - \frac{\pi}{3} \right) \quad (5)$$

The corresponding necessary imbalance currents for both modes can be derived with

$$i_{\mathrm{imb}}(t) = C \cdot \frac{d u_{\mathrm{imb}}}{dt} \quad (6)$$

This shows why a smaller DC-link is beneficial for the operation in lower-phase mode. The imbalance current is directly proportial to the DC-link capacitance. Since the balancer current is the sum of the imbalance current and the fundamental frequency ripple current, ensuring a small DC-link will also ensure lower currents in the balancer choke.

The resulting currents and voltages in the system are shown in Figs. 4 and 5. The grid voltages need to stay between the DC-link voltage $u_{\mathrm{C,P}}$ and $-u_{\mathrm{C,N}}$ at all times. Otherwise, the diodes in the system will start conducting, taking away the ability to control the current, which may lead to the destruction of the inverter.

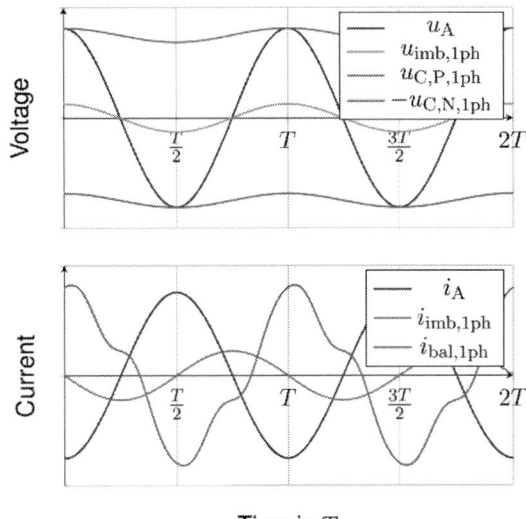

Fig. 4: Voltages and currents for single-phase operation with low DC-link voltage with $\varphi = \pi$

By examining i_{bal} it can be observed that the balancer needs to be able to provide the imbalance current i_{imb}, as well as compensate for the fundamental frequency ripple current caused by the lower-phase operation.

A lower-phase operation can also be achieved during operation with DC-link voltages higher than the lower voltage threshold for three-phase operation, in which case the imbalance voltage is set to zero. This would enable charging operation in locations where only a single-phase grid is present.

The CMI during lower-phase operation is expected to be worse compared to the three-phase operation.

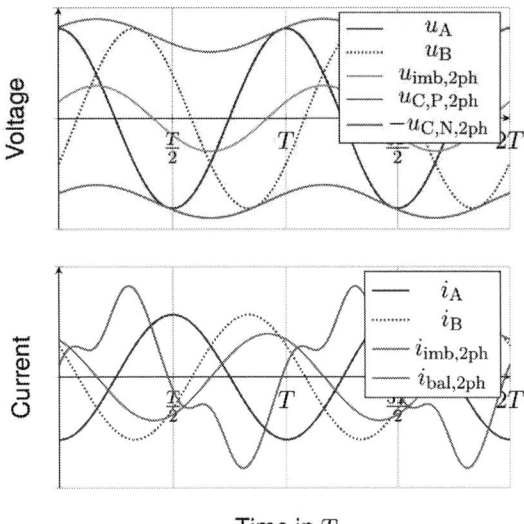

Fig. 5: Voltages and currents for two-phase operation with low DC-link voltage with $\varphi = \pi$

This is mainly due to two reasons:

1. The intentional imbalance voltage introduced to enable operation at lower DC-link voltages introduces a CMI with grid frequency, which cannot be compensated for.

2. The lower-phase grid inherently possesses a CMV. For example for a single-phase grid this CMV is defined as $u_{\mathrm{CM,N}} = u_A/2$. The CMV for single-phase inverters is defined as $u_{\mathrm{CM,1ph}} = u_{a0}/2$ [15]. The difference $u_{\mathrm{CM}} = u_A/2 - u_{a0}/2$ will generate a CMI. The inverter generates a voltage with the same frequency and amplitude as the grid voltage. Only the phase is different, dependent on the transferred power. Additionally, it is also generating a voltage with switching frequency. The fundamental frequency part will result in a very small CMV-difference, but the switching frequency parts result in a significant CMI peak at switching frequency. It is therefore not possible to operate the OBC CMV-free in single-phase operation.

6 Common-Mode Behaviour

To understand the mechanisms for the cause of CMI it is necessary to identify the CM paths in the system. These paths differ depending on the operation mode. For three-phase operation, the relay "N" in figure 2 stays open, i.e. the middle point of the LCL-filter is not connected to the grid.

The resulting CM equivalent circuit for this case is depicted in figure 6. The CMI travels through the CM-capacitance, LCL-filter ($L_{\mathrm{h}}, C_{\mathrm{f}}$ and L_{g}) and then into the grid (L_{N} and R_{N}). The transfer function can be derived with a two-port network analysis, with the result being a polynomial transfer function of 4th order:

$$H_{\mathrm{CM,3ph}}(s) = \frac{z_3 \cdot s^3 + z_2 \cdot s^2 + z_1 \cdot s}{n_4 \cdot s^4 + n_3 \cdot s^3 + n_2 \cdot s^2 + n_1 \cdot s + n_0} \tag{7}$$

with

$$z_3 = 3 \cdot C_{\mathrm{f}} \cdot L_{\mathrm{h}} \cdot C_{\mathrm{CM}}$$
$$z_2 = 3 \cdot C_{\mathrm{f}} \cdot R_{\mathrm{f}} \cdot C_{\mathrm{CM}}$$
$$z_1 = 3 \cdot C_{\mathrm{CM}}$$
$$n_4 = (L_{\mathrm{g}} + L_{\mathrm{N}}) \cdot L_{\mathrm{h}} \cdot C_{\mathrm{CM}} \cdot C_{\mathrm{f}}$$
$$n_3 = C_{\mathrm{CM}} \cdot C_{\mathrm{f}} \cdot ((L_{\mathrm{h}} + L_{\mathrm{g}} + L_{\mathrm{N}}) \cdot R_{\mathrm{f}} + L_{\mathrm{h}} \cdot R_{\mathrm{N}})$$
$$n_2 = C_{\mathrm{CM}} \cdot C_{\mathrm{f}} \cdot R_{\mathrm{N}} \cdot R_{\mathrm{f}} + (3 \cdot C_{\mathrm{f}} + C_{\mathrm{CM}}) \cdot L_{\mathrm{h}}$$
$$\qquad + C_{\mathrm{CM}} \cdot L_{\mathrm{N}}$$
$$n_1 = 3 \cdot C_{\mathrm{f}} \cdot R_{\mathrm{f}} + C_{\mathrm{CM}} \cdot R_{\mathrm{N}}$$
$$n_0 = 3$$
$$C_{\mathrm{CM}} = C_{\mathrm{CM,P}} + C_{\mathrm{CM,N}}$$

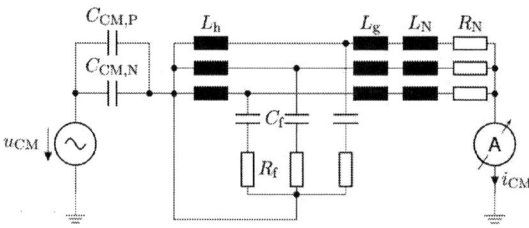

Fig. 6: CM equivalent circuit for three-phase operation

For lower-phase operation relay "N" is closed and therefore the midpoint of the LCL-filter is connected to the neutral phase. Also, one or two of the phases are not connected to the grid and can be neglected in the equivalent circuit. The resulting circuit is depicted in figure 7. The following evaluation is performed for three- and single-phase operation only. The transfer function for the single-phase and two-phase operation can be derived in the same manner, but yield polynomials of up to 5th and 6th order respectively.

The CM-behaviour was examined for charging operation of the OBC, i.e. current flowing into the DC-link. For this, a DC-source/sink was connected to the DC output of the OBC. This load is isolated from the grid, but still influences the CMI spectrum. The CM equivalent circuit of the DC load was also examined and can be seen in Fig. 8. The values for the output filter are $L_{\mathrm{CMC1}} = 320\,\mu\mathrm{H}$, $L_{\mathrm{CMC2}} = 700\,\mu\mathrm{H}$ and $C_{\mathrm{Y}} = 27.2\,\mathrm{nF}$.

2482

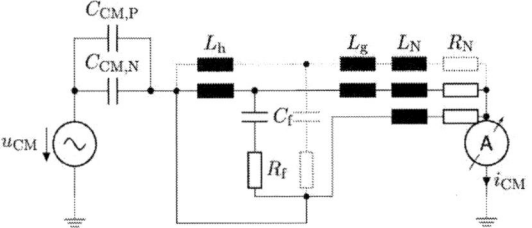

Fig. 7: CM equivalent circuit for lower-phase operation

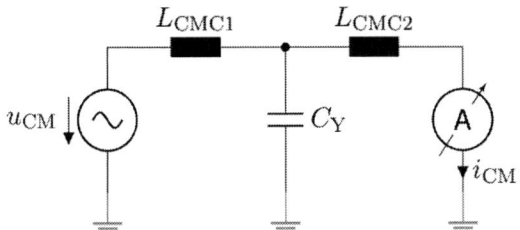

Fig. 8: CM equivalent circuit of DC load

The following analysis was performed for a strong grid connection, as later measurements were performed in a lab environment, which has its own substation transformer in the basement. According to [16], strong grid conditions in Germany equate to a grid inductance of $L_N = 12\,\mu\text{H}$ and grid resistance of $R_N = 1\,\text{m}\Omega$. The bode magnitude plot for these grid parameters, the circuit parameters given in table 1, and both operation modes are depicted in Fig. 9. Each operation mode possesses two resonance frequencies. The first resonance frequency is actually comprised of a resonance and anti-resonance point. The frequency difference between these two points is minimal, as well as damped by R_f and can therefore be evaluated as a single resonance point. The first resonance point for three-phase operation only depends on the LCL-filter values (Eq. (8)). For single-phase, the grid inductance, in addition to the LCL-filter values, always determines the frequency of the first resonance point (Eq. (9)). The formulas are defined as:

$$f_{1,\text{res,3ph}} = \frac{1}{2\pi \cdot \sqrt{L_h \cdot C_f}} \tag{8}$$

$$f_{1,\text{res,1ph}} = \frac{1}{2\pi \cdot \sqrt{\frac{L_h \cdot (2 \cdot L_N + L_g)}{L_h + 2 \cdot L_N + L_g} \cdot C_f}} \tag{9}$$

The second resonance point is only lightly damped by R_f and is dependent on the grid-side LCL-inductance, the grid inductance and the CM-

capacitance of the system:

$$f_{2,\text{res,3ph}} = \frac{1}{2\pi \cdot \sqrt{\frac{L_N + L_g}{3} \cdot C_{CM}}} \tag{10}$$

$$f_{2,\text{res,1ph}} = \frac{1}{2\pi \cdot \sqrt{\frac{L_N \cdot (L_N + L_g)}{2 \cdot L_N + L_g} \cdot C_{CM}}} \tag{11}$$

The output filter of the DC load forms a simple LCL-filter, whose resonance frequency can be calculated with:

$$f_{\text{res,DCload}} = \frac{1}{2\pi} \cdot \sqrt{\frac{L_{CMC1} + L_{CMC2}}{L_{CMC1} \cdot L_{CMC2} \cdot C_Y}} \tag{12}$$

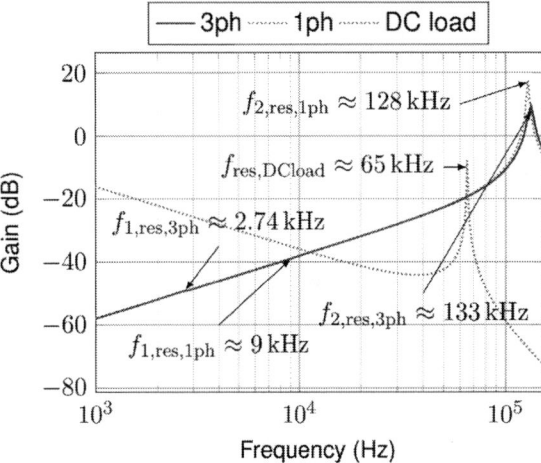

Fig. 9: Bode magnitude plot of CM equivalent diagrams for strong grid conditions

7 Hardware Prototype

A hardware prototype was designed and incorporated into a test setup to emulate the EV in addition to the EVSE. The parameters for the OBC prototype are given in table 1. The test setup depicted in Fig. 10 shows the prototype in the middle and the balancer as well as the LCL-filter chokes on the right side. The smaller PCB on the left side houses the L_{CMC2} CM-choke for the DC-side.

The DC-source/load consists of a bidirectional Regatron TC.GSS 1500V power supply. The prototype was connected to the three-phase grid through a Doepke DFS 4 016-4/0.03-B NK RCD type B. Grid voltages were measured with Testec TT-SI9001 differential probes. Grid currents were measured with

Tab. 1: OBC parameters

Switching frequency	f_{sw}	100 kHz
Main inductance	L_h	340 µH
Filter capacitance	C_f	10 µF
Damping resistance	R_d	1 Ω
Grid-side inductance	L_g	10 µH
DC-link capacitance	C_P, C_N	400 µF
Rated apparent power	S_N	10 kV A
Balancer inductance	L_{bal}	340 µH
Compensation factor	k	0.8
CM-capacitance	$C_{CM,P}, C_{CM,N}$	100 nF

Fig. 10: Hardware prototype setup

Keysight N7026A currents probes. DC-link voltages were measured using Testec TT-SI9110 differential probes. The balancer choke current was measured using a Fluke i30s current probe. The CM-spectrum was measured at the AC-output with a Pearson Current Monitor 6600. All probes were connected to a Keysight MXR058A oscilloscope. Efficiency was measured with a ZES LMG671 power analyser. Both three- and single-phase operation modes were implemented on the Texas Instruments C2000 TMS320F28379D microcontroller. Part of the PWM generation was implemented on a Xilinx XC2C256 CPLD.

8 Results

8.1 Single-Phase Operation Mode

The modulation scheme for single-phase operation with lower DC-link voltages was implemented on the hardware. Measurements were taken with a DC-link voltage of $U_{DC} = 620\,$V. The measure-

ments results are depicted in Fig. 11.

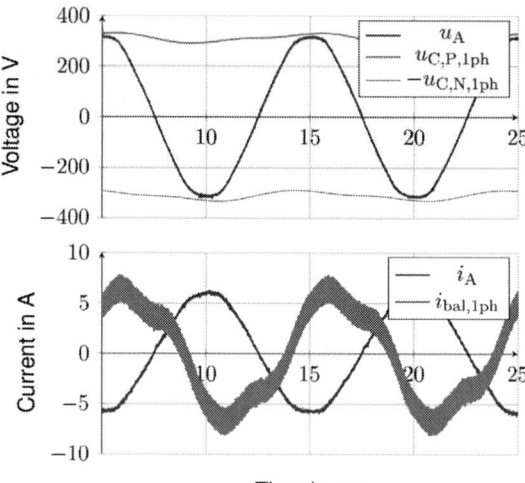

Fig. 11: Measurement of voltages and currents for single-phase operation with $U_{DC} = 620\,$V and $P = -1\,$kW

Comparing Fig. 4 and Fig. 11 shows that the model derived via fundamental frequency analysis is in good accordance with the measurements. The grid voltage u_A stays within the DC-link voltages $u_{C,P,1ph}$ and $-u_{C,N,1ph}$ at all times. The balancer current i_{bal} contains the current needed to inject the imbalance current into the midpoint, as well as compensate for the fundamental frequency ripple of the single-phase mode. In addition, the balancer current also contains the switching frequency ripple, which has been omitted in the fundamental frequency analysis. To summarize, the single-phase operation mode was successfully implemented and enables the operation of the charger with battery voltages down to $620\,$V.

8.2 Common-Mode Behaviour

A simulation model of the OBC, including the grid, common-mode paths, DC load as well as all operation modes was implemented in PLECS. Simulations and measurements were performed to investigate the CM-spectrum in the detection range of the employed RCD (up to $150\,$kHz). The first simulation and measurement was performed for the three-phase charging mode, using a DC-link voltage of $U_{DC} = 700\,$V, with a power of $P = -3\,$kW. The simulation result in Fig. 12 clearly depicts the influence of the CM transfer function shown in Fig. 9. The first peak at $f = 150\,$Hz is generated by DC-link voltage imbalances caused by the ripple during the operation of the OBC. The second peak

2484

PCIM Europe 2024, 11– 13 June 2024, Nuremberg   DOI: 10.30420/566262348

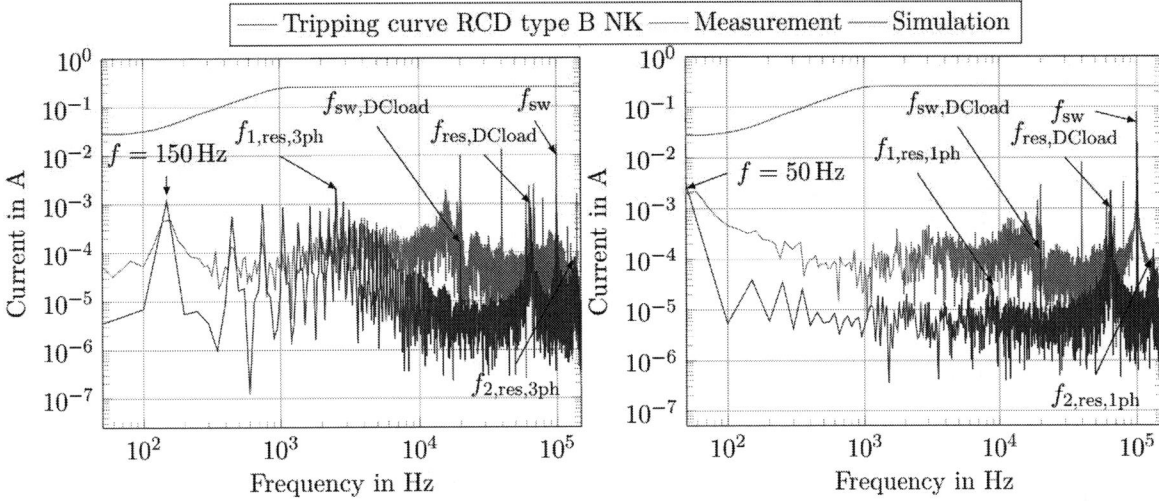

Fig. 12: Simulated and measured CM-spectrum for three-phase operation (left, $P = -3\,\mathrm{kW}, U_{\mathrm{DC}} = 700\,\mathrm{V}$) and single-phase operation (right, $P = -1\,\mathrm{kW}, U_{\mathrm{DC}} = 620\,\mathrm{V}$)

at $f = 2.7\,\mathrm{kHz}$ corresponds to the CM resonance $f_{1,\mathrm{res,3ph}}$ of the LCL-filter. The next two peaks are mainly generated by the DC-load, which has a switching frequency of $f_{\mathrm{sw,DCload}} = 20\,\mathrm{kHz}$, which also leads to the peak at $f_{\mathrm{res,DCload}}$. The second to last peak occurs at switching frequency, and is mainly generated by the dead-time. The last peak in the range occurs at $f = 133\,\mathrm{kHz}$, which marks the resonance point $f_{2,\mathrm{res,3ph}}$ of the CM capacitance and grid. This peak will shift to lower frequencies if the grid connection is weaker, i.e. the grid impedance rises.

The measured CM spectrum for three-phase operation show similar peaks compared to the simulation. The peak at $f_{1,\mathrm{res,3ph}}$ is not as pronounced, which suggests the LCL-filter of the prototype is more damped. The peaks at $f_{\mathrm{sw,DCload}}$, as well as $f_{\mathrm{res,DCload}}$ are also present in the measurement. The additional peaks at $40\,\mathrm{kHz}$ and $80\,\mathrm{kHz}$ are 2nd and 4th harmonics of the DC-load, which were not present in the simulation. $f_{2,\mathrm{res,3ph}}$ can not be identified in the measurement. It is unlikely that the peak at $15\,\mathrm{kHz}$ represents $f_{2,\mathrm{res,3ph}}$, as this is only achievable with unreasonably high grid impedances ($L_{\mathrm{N}} = 1.2\,\mathrm{mH}$). The grid impedance most likely has a value of around $L_{\mathrm{N}} = 20\,\mathrm{\mu H}$, which shifts $f_{2,\mathrm{res,3ph}}$ into the area of f_{sw}, masking the resonance point. To conclude the three-phase operation mode, the measurements show a margin of at least one order of a magnitude below the

tripping curve of the RCD type B for a total CM capacitance of $200\,\mathrm{nF}$. The OBC can be operated in combination with the RCD without any tripping events.

For the single-phase operation mode, the same peaks from the CM transfer function can be observed in the simulation. The first peak shifted to $f = 50\,\mathrm{Hz}$, which is caused by the injection of the imbalance voltage to enable operation at lower DC-link voltages. As expected, the CM performance for single-phase operation is worse. Due to the higher CMV, the peak at switching frequency is much more pronounced and less than one order of magnitude away from the tripping curve of the RCD. Small deviations in the system may lead to the tripping of the RCD. This can be easily mitigated with a simple and lightweight EMI-filter, which is designed to only filter high frequency interferences.

8.3 Efficiency

Finally, the efficiency of the OBC was measured using a power analyser. The results are shown in Fig. 13. Peak efficiency is reached at $98\,\%$ in discharging mode, i.e. power flow from the battery to the grid. The peak efficiency in charging mode is slightly lower with $97.7\,\%$. In both power flow directions, the efficiency is already above $90\,\%$ at $500\,\mathrm{W}$. In the current design the maximum power transfer during single-phase operation is limited to

2485

$2\,\mathrm{kW}$. This limitation is primarily caused by the maximum allowable peak current in the balancer choke and can be increased. Peak efficiency in this operation mode is $95.5\,\%$. The main reason for the lower efficiency is that the balancer is compensating for more ripple current caused by the single-phase operation, as well as that the injection of the imbalance current into the DC-link midpoint causes more current to flow through the balancer switches. The goal of this work was to present a proof-of-concept and not an efficiency-optimized design. Nevertheless, the efficiency in all operation modes is more than satisfactory and shows the benefit of a single-stage solution.

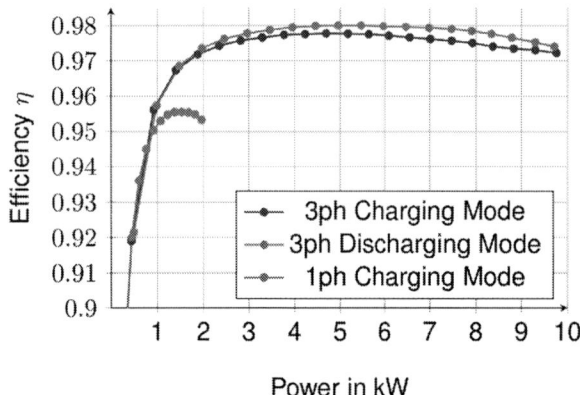

Fig. 13: OBC efficiency at $U_{\mathrm{DC}} = 700\,\mathrm{V}$ three-phase and $U_{\mathrm{DC}} = 620\,\mathrm{V}$ single-phase

9   Conclusion

A non-isolated, single-stage design for an OBC for $800\,\mathrm{V}$ battery packs was presented. Even though a non-isolated design was used, through the use of novel modulation techniques, it is possible to operate the OBC without any significant CM filter measures. Lower-phase operation is possible and enables operation of the OBC at low DC-link voltages, covering the entire battery voltage range of an $800\,\mathrm{V}$ battery pack. Simulation and measurement results show that an operation of the design with an RCD type B is possible without risking any tripping events. Improvements in the measurement setup are needed to further isolate the effect of the DC-supply to better understand the shape of the CM-spectrum. Also, a redesign of the hardware prototype is necessary, as further improvements, especially regarding the size of the DC-link, are possible. This would lead to further improvements

of the power density.

References

[1]   C. Wei, J. Shao, B. Agrawal, D. Zhu, and H. Xie, "New surface mount sic mosfets enable high efficiency high power density bi-directional on-board charger with flexible dc-link voltage," in *2019 IEEE Applied Power Electronics Conference and Exposition (APEC)*, 2019, pp. 1904–1909. DOI: 10.1109/APEC.2019.8721866.

[2]   A. D. Le Ta, N. D. Dao, and D.-C. Lee, "High-efficiency hybrid llc resonant converter for on-board chargers of plug-in electric vehicles," *IEEE Transactions on Power Electronics*, vol. 35, no. 8, pp. 8324–8334, 2020. DOI: 10.1109/TPEL.2020.2968084.

[3]   S. Li, S. Lu, and C. C. Mi, "Revolution of electric vehicle charging technologies accelerated by wide bandgap devices," *Proceedings of the IEEE*, vol. 109, no. 6, pp. 985–1003, 2021. DOI: 10.1109/JPROC.2021.3071977.

[4]   J. Yuan, L. Dorn-Gomba, A. D. Callegaro, J. Reimers, and A. Emadi, "A review of bidirectional on-board chargers for electric vehicles," *IEEE Access*, vol. 9, pp. 51 501–51 518, 2021. DOI: 10.1109/ACCESS.2021.3069448.

[5]   N. Kumar and M. P. R. Prasad, "A comparative analysis of different topologies of on-board charger (obc) with an approach of interfacing it with mcb," in *2021 6th International Conference for Convergence in Technology (I2CT)*, 2021, pp. 1–6. DOI: 10.1109/I2CT51068.2021.9417956.

[6]   A. Khaligh and M. D'Antonio, "Global trends in high-power on-board chargers for electric vehicles," *IEEE Transactions on Vehicular Technology*, vol. 68, no. 4, pp. 3306–3324, 2019. DOI: 10.1109/TVT.2019.2897050.

[7]   B. Li, F. C. Lee, Q. Li, and Z. Liu, "Bi-directional on-board charger architecture and control for achieving ultra-high efficiency with wide battery voltage range," in *2019 IEEE Applied Power Electronics Conference and Exposition (APEC)*, 2019, pp. 3688–3694. DOI: 10.1109/APEC.2017.7931228.

[8]   J. Lu, K. Bai, A. R. Taylor, G. Liu, A. Brown, *et al.*, "A modular-designed three-phase high-efficiency high-power-density ev battery charger using dual/triple-phase-shift control," *IEEE Transactions on Power Electronics*, vol. 33, no. 9, pp. 8091–8100, 2018. DOI: 10.1109/TPEL.2017.2769661.

[9] S. Jaman, S. Chakraborty, D.-D. Tran, T. Geury, M. El Baghdadi, and O. Hegazy, "Review on integrated on-board charger-traction systems: V2g topologies, control approaches, standards and power density state-of-the-art for electric vehicle," *Energies*, vol. 15, no. 15, p. 5376, 2022. DOI: 10.3390/en15155376.

[10] B. Li, M. Zhou, D. Jiang, and K. Jiang, "Integrated on-board battery charger based on t-type converter," in *2022 IEEE International Power Electronics and Application Conference and Exposition (PEAC)*, 2022, pp. 636–641. DOI: 10.1109/PEAC56338.2022.9959573.

[11] C. Jung, "Power up with 800-v systems: The benefits of upgrading voltage power for battery-electric passenger vehicles," *IEEE Electrification Magazine*, vol. 5, no. 1, pp. 53–58, 2017. DOI: 10.1109/MELE.2016.2644560.

[12] L. Guo, R. Liu, X. Li, and Y. Zhang, "Neutral point potential balancing method for three–level power converters in two–stage three–phase four–wire power conversion system," *IET Power Electronics*, vol. 13, no. 12, pp. 2618–2627, 2020. DOI: 10.1049/iet-pel.2019.1211.

[13] M. C. Cavalcanti, A. M. Farias, K. C. Oliveira, F. A. S. Neves, and J. L. Afonso, "Eliminating leakage currents in neutral point clamped inverters for photovoltaic systems," *IEEE Transactions on Industrial Electronics*, vol. 59, no. 1, pp. 435–443, 2012. DOI: 10.1109/TIE.2011.2138671.

[14] V. Karakasli, A. Allioua, and G. Griepentrog, "Common-mode emi noise modeling of three-level t-type inverter for adjustable speed drive systems," in *2022 24th European Conference on Power Electronics and Applications (EPE'22 ECCE Europe)*, 2022, pp. 1–8.

[15] T. K. S. Freddy, N. A. Rahim, W.-P. Hew, and H. S. Che, "Comparison and analysis of single-phase transformerless grid-connected pv inverters," *IEEE Transactions on Power Electronics*, vol. 29, no. 10, pp. 5358–5369, 2014. DOI: 10.1109/TPEL.2013.2294953.

[16] B. Valov, "Elektrotechnisches basiswissen zur netzintegration der erzeugungsanlagen," in *Handbuch Netzintegration Erneuerbarer Energien*, Springer Vieweg, Wiesbaden, 2020, pp. 39–127. DOI: 10.1007/978-3-658-28969-0_3.

PCIM Europe 2024, 11– 13 June 2024, Nuremberg   DOI: 10.30420/566262349

Control of a Three-Phase Inductive Power Transfer System Based on DD²Q Coil Topology

Nikola Mirković [1], Alberto Delgado [1], Pedro Alou [1], Miroslav Vasić [1]

[1] Universidad Politécnica de Madrid - Centro de Electrónica Industrial, España

Corresponding author: Nikola Mirković, n.mirkovic@upm.es
Speaker: Nikola Mirković, n.mirkovic@upm.es

Abstract

Problem of inter-phase coupling between the different phases of a poly-phase inductive power transfer (IPT) system is present in all of such systems when they operate under non-aligned conditions. This leads to uneven load sharing between the phases and to the increased reactive power circulating in the system. In this paper we propose a control strategy for a compact three-phase IPT charger based on the DD²Q magnetic coupler topology that ensures equal power distribution between the phases on the secondary side and equal current stress of the primary semiconductors, under misalignment conditions. The proposed control strategy has been tested on the laboratory prototype, for power of 6.83 kW, with 86.1% system efficiency, achieving power transfer increase of 20.5%, achieving and showing efficiency increase of 1.6% comparing to the case with no control strategy applied.

1   Introduction

Tendency towards poly-phase IPT systems in electric vehicle (EV) charging applications is notable in the last years [1,2]. Not only that such poly-phase systems offer the possibility of high-power charging, but they also reduces current and voltage stress on the coils of each of the phases individually as the total power is distributed between several phases. Out of all poly-phase IPT systems, three-phase ones are the most common [3-5]. These systems can be realized in two different ways: by separating them and having a complete magnetic structure for each phase [6], or, by compactly stacking the coils of different phases one on top of the other while having a single magnetic core [7,8]. One thing in common for all of the three-phase IPT systems is that they are based on a principle of zero-coupling between non-corresponding phases, i.e., each phase of the primary is coupled to only one phase of the secondary, thus achieving independent operation between the phases. In the applications such as EV charging, lateral misalignment between the receiver and transmitter is expected, as it is defined in the current standard proposal SAE J2954 [9]. Under misalignment conditions, in the case of having a compact three-phase IPT system, coupling between non-corresponding phases is inevitable, thus it is necessary to examine the influence of this coupling to the system functionality.

2   Three-phase IPT system based on DD²Q coil structure

A compact three-phase IPT system with a DD²Q coil structure is proposed in [8] and it is given in Fig. 1a. It is composed out of two mutually orthogonal DD coils and a Q coil, all stacked together, one on top of another. In Fig. 1b, magnetic fields generated by each of these coils are shown. Being that every two magnetic fields of all of the three coils are mutually orthogonal, there is no coupling between the coils of such a structure. Both transmitter and receiver side of the considered system are made using the DD²Q coil structure. System based on this structure would have several benefits over other compact three-phase IPT systems that are proposed in the current state-of-the-art:

1. It is composed out of simple coil structures comprised by the SAE J2954 [9].

2. It could be used both as three-phase charger in cases when there is a corresponding receiver, but also each of the phases can be used as a single-phase charger.

PCIM Europe 2024, 11– 13 June 2024, Nuremberg   DOI: 10.30420/566262349

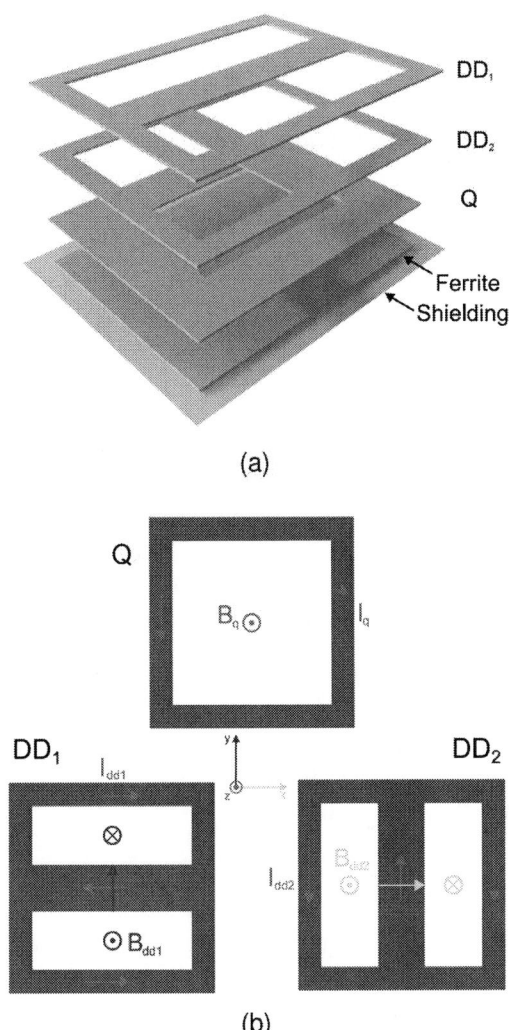

(a)

(b)

Fig. 1: DD²Q coil structure

3. Each of the coils utilizes the entire surface of the charging pad, thus maximizing the per-turn inductance of the windings

4. It has the benefits of high coupling coefficient of Q coil and low stray field of DD coils [8,10]

From the stated properties of the DD²Q structure, it is clear that, when transmitter and receiver structures are aligned, there is no coupling between the non-corresponding phases. However, in the case of lateral misalignment, the coupling between non-corresponding phases will arise, leading to the unbalanced power distribution across the phases of the system. The converter topology that will be used for testing the system in this work is given in Fig. 2. Each of the phases is compensated us-

ing the double-sided LCC compensation topology, that offers high efficiency even under misalignment conditions [11]. The selected topology is the one employing the minimum number of semiconductor devices while allowing for completely independent control of each of the phases. This is out of high importance, as it is the requirement of the power balancing control method that will be proposed.

3 Mathematical model of the system and proposed control method

To analyze the considered system, it is necessary to analyze the impedance matrix of the IPT link, that in this case is 6x6. The only input that the proposed control method requires is the impedance matrix that characterizes the inductive link. The impedance matrix of the system M is given with:

$$M = \begin{bmatrix} A & B \\ B^T & C \end{bmatrix} \tag{1}$$

where A is the inductance matrix of the transmitter, C is the inductance matrix of the receiver and B is the coupling inductance matrix between the transmitter and receiver.

$$A = \begin{bmatrix} L_Q & M_{QDD1} & M_{QDD2} \\ M_{QDD1} & L_{DD1} & M_{DD1DD2} \\ M_{QDD2} & M_{DD1DD2} & L_{DD2} \end{bmatrix} \tag{2}$$

$$B = \begin{bmatrix} M_{Qq} & M_{Qdd1} & M_{Qdd2} \\ M_{DD1q} & M_{DD1dd1} & M_{DD1dd2} \\ M_{DD2q} & M_{DD2dd1} & M_{DD2dd2} \end{bmatrix} \tag{3}$$

$$C = \begin{bmatrix} L_q & M_{qdd1} & M_{qdd2} \\ M_{qdd1} & L_{dd1} & M_{dd1dd2} \\ M_{qdd2} & M_{dd1dd2} & L_{dd2} \end{bmatrix} \tag{4}$$

When transmitter and receiver structures are manufactured correctly, A and C are equal to:

$$A = \begin{bmatrix} L_Q & 0 & 0 \\ 0 & L_{DD1} & 0 \\ 0 & 0 & L_{DD2} \end{bmatrix} \tag{5}$$

$$C = \begin{bmatrix} L_q & 0 & 0 \\ 0 & L_{dd1} & 0 \\ 0 & 0 & L_{dd2} \end{bmatrix} \tag{6}$$

If transmitter and receiver structures are aligned, the coupling inductance matrix B is diagonal:

$$B = \begin{bmatrix} M_{Qq} & 0 & 0 \\ 0 & M_{DD1dd1} & 0 \\ 0 & 0 & M_{DD2dd2} \end{bmatrix} \tag{7}$$

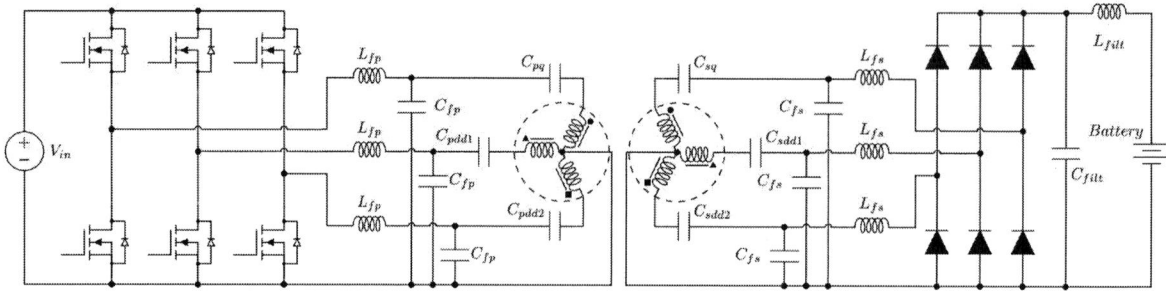

Fig. 2: Converter topology used for testing the DD²Q coils

In the case when the system is not aligned, matrix B is not diagonal. System is compensated using double-sided LCC compensation strategy. The system compensated in this way behaves as a triple cascaded gyrator, experiencing the current source behavior at the output, for voltage excitation at the input. Input voltage vector is defined with:

$$\underline{V}_p = \begin{bmatrix} \underline{V}_Q \\ \underline{V}_{DD1} \\ \underline{V}_{DD2} \end{bmatrix} = \begin{bmatrix} V_Q e^{j0} \\ V_{DD1}e^{j\varphi_{DD1}} \\ V_{DD2}e^{j\varphi_{DD2}} \end{bmatrix} \quad (8)$$

When the system is excited with this input vector, vector of the secondary currents can be calculated using the following equation:

$$\underline{I}_s = \begin{bmatrix} \underline{I}_q \\ \underline{I}_{dd1} \\ \underline{I}_{dd2} \end{bmatrix} = -j\omega^3 C_{fp}C_{fs}B\underline{V}_p \quad (9)$$

From here it is possible to deduce that in case that B is diagonal, the power will be exchanged only between corresponding phases. This means if coupling between non-corresponding phases exists, i.e., the receiver and transmitter are not aligned, each of the phases of the transmitter will send power to each of the phases of the receiver, if excited with symmetrical voltages at the input. This problem exists in all of the compact three-phase IPT systems. In this article we propose a control strategy devised to achieve equal power transfer per-phase on both primary and secondary sides of the considered system. As it was stated previously, to implement the proposed control, it is necessary to have determined the matrix M.

Once the secondary currents are determined, it is possible to determine the secondary voltages:

$$\underline{V}_s = \begin{bmatrix} \underline{V}_q \\ \underline{V}_{dd1} \\ \underline{V}_{dd2} \end{bmatrix} = f(\underline{I}_s) \quad (10)$$

Function f that determines the relation between the secondary voltages and currents depends on the parameters of the electrical circuit. In order to determine this relation precisely, due to the non-linear behavior of the diode, we recommend establishing it based on tests performed on the actual prototype.

Once secondary voltages are determined, it is possible to calculate primary currents:

$$\underline{I}_p = \begin{bmatrix} \underline{I}_Q \\ \underline{I}_{DD1} \\ \underline{I}_{DD2} \end{bmatrix} = -j\omega^3 C_{fp}C_{fs}B^T\underline{V}_s \quad (11)$$

Apparent powers at the primary and secondary sides are, respectively:

$$\underline{S}_p = \begin{bmatrix} \underline{S}_Q \\ \underline{S}_{DD1} \\ \underline{S}_{DD2} \end{bmatrix} = \begin{bmatrix} \underline{V}_Q \underline{I}_Q^* \\ \underline{V}_{DD1}\underline{I}_{DD1}^* \\ \underline{V}_{DD2}\underline{I}_{DD2}^* \end{bmatrix} \quad (12)$$

$$\underline{S}_s = \begin{bmatrix} \underline{S}_q \\ \underline{S}_{dd1} \\ \underline{S}_{dd2} \end{bmatrix} = \begin{bmatrix} \underline{V}_q \underline{I}_q^* \\ \underline{V}_{dd1}\underline{I}_{dd1}^* \\ \underline{V}_{dd2}\underline{I}_{dd2}^* \end{bmatrix} \quad (13)$$

Active powers are real part of the apparent powers:

$$\underline{P}_p = \begin{bmatrix} P_Q \\ P_{DD1} \\ P_{DD2} \end{bmatrix} = real(\underline{S}_p) \quad (14)$$

$$\underline{P}_s = \begin{bmatrix} P_q \\ P_{dd1} \\ P_{dd2} \end{bmatrix} = real(\underline{S}_s) \quad (15)$$

Now, that all of the values of interest of the system are determined, it is possible to start the process of determining the required excitation vector $\underline{V}_p$ to obtain desired system functionality. In the case of this work, we will seek $\underline{V}_p$ to achieve equal power distribution on both primary and secondary sides, i.e.:

$$\begin{aligned} P_Q &= P_{DD1} = P_{DD2} \\ P_q &= P_{dd1} = P_{dd2} \end{aligned} \quad (16)$$

Equation (8) has total of 5 variables - V_Q, V_{DD1}, V_{DD2}, φ_{DD1}, φ_{DD2} while the system in (16) has total of 4 equations. This means that there is 1 degree of freedom, that can be used to set the exact power that will be exchanged between the primary and secondary sides. Considering that the previous system is pretty complex and nonlinear due to equation (10), it is hard to solve it analytically. Rather, it is more convenient to apply false assumption method for solving, first by defining the upper and lower limits for each of the 5 previously mentioned variables, and dividing the ranges into n equally sized parts. After this, by trying all of the combinations by varying each of the variables between its n previously established discrete values, having total of n^5 iterations, it is possible to find certain number of combinations that will fulfill the requirement of equation (16). Since the domain was discretized, there will be certain tolerance, thus the previous algorithm will give solutions in vicinity of which the exact solutions are located. The unbalance in power distribution across the phases obtained by applying the discretized solutions will depend on the tolerance set on equation (16) during the execution of the process, and on the number of considered discrete values n, and it is realistic to expect this unbalance in range between [-5%,5%]. If one wants to obtain the exact solution, it is necessary to do the following:

1. Determine the starting point from the pool of obtained discrete solutions $[V_Q^0, V_{DD1}^0, V_{DD2}^0, \varphi_{DD1}^0, \varphi_{DD2}^0]$

2. For each of the per-phase powers in the given operating point P_Q^0, P_{DD1}^0, P_{DD2}^0, P_q^0, P_{dd1}^0, P_{dd2}^0 determine the values of partial derivatives:

    (a) $\frac{\partial P_x}{\partial V_Q}(V_Q^0, V_{DD1}^0, V_{DD2}^0, \varphi_{DD1}^0, \varphi_{DD2}^0)$

    (b) $\frac{\partial P_x}{\partial V_{DD1}}(V_Q^0, V_{DD1}^0, V_{DD2}^0, \varphi_{DD1}^0, \varphi_{DD2}^0)$

    (c) $\frac{\partial P_x}{\partial V_{DD2}}(V_Q^0, V_{DD1}^0, V_{DD2}^0, \varphi_{DD1}^0, \varphi_{DD2}^0)$

    (d) $\frac{\partial P_x}{\partial \varphi_{DD1}}(V_Q^0, V_{DD1}^0, V_{DD2}^0, \varphi_{DD1}^0, \varphi_{DD2}^0)$

    (e) $\frac{\partial P_x}{\partial \varphi_{DD2}}(V_Q^0, V_{DD1}^0, V_{DD2}^0, \varphi_{DD1}^0, \varphi_{DD2}^0)$.

for $x \in \{Q, DD1, DD2, q, dd1, dd2\}$. These derivative values are best to be taken by measuring the given values on the real-life setup, not by simulation.

3. It is necessary to find a delta vector (ΔV_Q, ΔV_{DD1}, ΔV_{DD2}, $\Delta \varphi_{dd1}$, $\Delta \varphi_{dd2}$) that is to be added to the vector of discrete solutions in order to perfectly equalize all of the powers across the phases. The system of equations that has to be solved is defined in equation (17). Solving this system of linear equations, one obtains the final values of the 5 variables V_Q, V_{DD1}, V_{DD2}, φ_{DD1}, φ_{DD2} that allow for equal power sharing between the phases of the system. As the given system has one more variable then the number of equations, it is possible to consider one of the delta values equal to 0.

In the previously described way, we have calculated the necessary excitation voltage vector that, as a result, leads to equal distribution of power levels across the phases of the considered system. As it was stated before, the discrete solution for achieving a certain power at a certain misalignment is not unique. This opens space for further improving of the presented control methodology, through analyzing the per-phase losses, and balancing them through applying different solutions that allow for the same power to be transferred, however that are producing different losses in different phases, thus balancing the system thermally over longer time spans. However, this will not be in the scope of this work.

4 Experimental results

To verify the proposed control method, a setup was built and it is given in Fig. 3. The exploded view of the DD^{2}Q coupler is given in Fig. 4. Coil sizes are in accordance with the SAE standard for Z3 class of chargers. The system was tested with

Fig. 3: Experimental setup

$$P_{AVG} = P_Q^0 + \frac{\partial P_Q}{\partial V_Q}\Delta V_Q + \frac{\partial P_Q}{\partial V_{DD1}}\Delta V_{DD1} + \frac{\partial P_Q}{\partial V_{DD2}}\Delta V_{DD2} + \frac{\partial P_Q}{\partial \varphi_{DD1}}\Delta \varphi_{DD1} + \frac{\partial P_Q}{\partial \varphi_{DD2}}\Delta \varphi_{DD2}$$

$$P_{AVG} = P_{DD1}^0 + \frac{\partial P_{DD1}}{\partial V_Q}\Delta V_Q + \frac{\partial P_{DD1}}{\partial V_{DD1}}\Delta V_{DD1} + \frac{\partial P_{DD1}}{\partial V_{DD2}}\Delta V_{DD2} + \frac{\partial P_{DD1}}{\partial \varphi_{DD1}}\Delta \varphi_{DD1} + \frac{\partial P_{DD1}}{\partial \varphi_{DD2}}\Delta \varphi_{DD2}$$

$$P_{AVG} = P_{DD2}^0 + \frac{\partial P_{DD2}}{\partial V_Q}\Delta V_Q + \frac{\partial P_{DD2}}{\partial V_{DD1}}\Delta V_{DD1} + \frac{\partial P_{DD2}}{\partial V_{DD2}}\Delta V_{DD2} + \frac{\partial P_{DD2}}{\partial \varphi_{DD1}}\Delta \varphi_{DD1} + \frac{\partial P_{DD2}}{\partial \varphi_{DD2}}\Delta \varphi_{DD2}$$

$$P_{avg} = P_q^0 + \frac{\partial P_q}{\partial V_Q}\Delta V_Q + \frac{\partial P_q}{\partial V_{DD1}}\Delta V_{DD1} + \frac{\partial P_q}{\partial V_{DD2}}\Delta V_{DD2} + \frac{\partial P_q}{\partial \varphi_{DD1}}\Delta \varphi_{DD1} + \frac{\partial P_q}{\partial \varphi_{DD2}}\Delta \varphi_{DD2} \qquad (17)$$

$$P_{avg} = P_{dd1}^0 + \frac{\partial P_{dd1}}{\partial V_Q}\Delta V_Q + \frac{\partial P_{dd1}}{\partial V_{DD1}}\Delta V_{DD1} + \frac{\partial P_{dd1}}{\partial V_{DD2}}\Delta V_{DD2} + \frac{\partial P_{dd1}}{\partial \varphi_{DD1}}\Delta \varphi_{DD1} + \frac{\partial P_{dd1}}{\partial \varphi_{DD2}}\Delta \varphi_{DD2}$$

$$P_{avg} = P_{dd2}^0 + \frac{\partial P_{dd2}}{\partial V_Q}\Delta V_Q + \frac{\partial P_{dd2}}{\partial V_{DD1}}\Delta V_{DD1} + \frac{\partial P_{dd2}}{\partial V_{DD2}}\Delta V_{DD2} + \frac{\partial P_{dd2}}{\partial \varphi_{DD1}}\Delta \varphi_{DD1} + \frac{\partial P_{dd2}}{\partial \varphi_{DD2}}\Delta \varphi_{DD2}$$

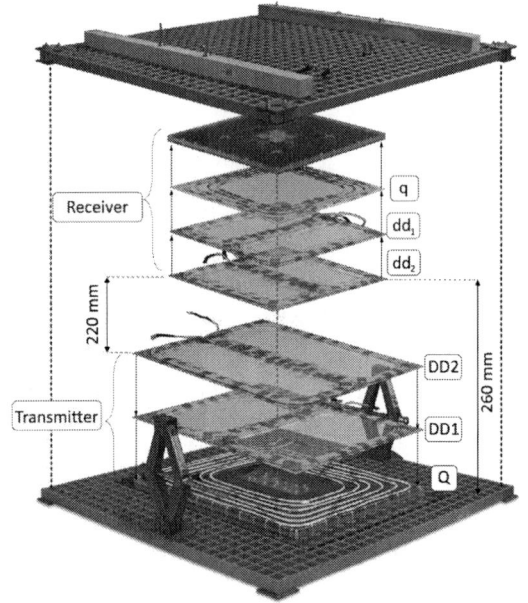

Fig. 4: Exploded view of the considered DD^{2}Q system

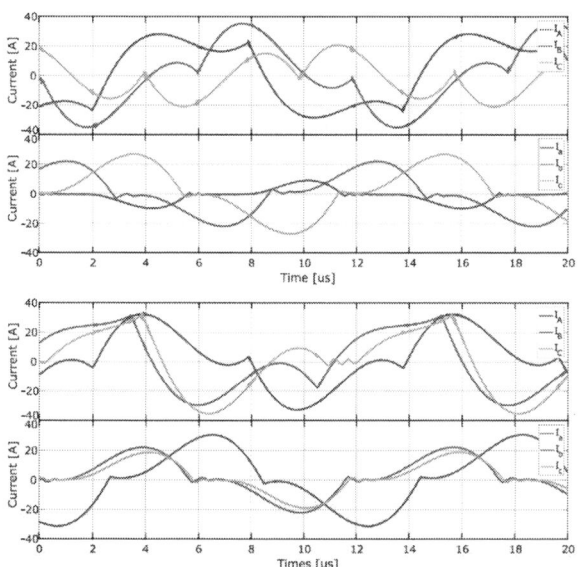

Fig. 5: Experimental waveforms without balancing and with balancing

the air gap between the two sides equal to 220 mm and with the misalignment of the secondary equal to 100 mm in the x direction, and 75 mm in the y direction. Measured impedance matrix for this case is given in Table 1. In the top of 2 graphs of Fig. 5, primary and secondary currents are given for the case when no control method is applied, i.e., all of the primary voltages have the same amplitude, and are phase shifted by 120°. In the bottom two graphs of Fig. 5 primary and secondary currents are given for the case where the amplitudes and phase-shifts are obtained using the proposed control method. Relative amplitudes and phase-shifts between the three phases are, respectively, 1, 0.95, 0.825, and

0°, 226° and 240°. Power that is transferred in the first case is equal to 5.67 kW, while in the latter it is equal to 6.83 kW. In the first case, efficiency of the system is equal to 84.5%, while in the latter one it is equal to 86.1%. Using the proposed control method we were able to obtain 20.5% increase in power transfer capability of the system, and system efficiency was increased for 1.6%, presenting a significant improvement. Certain unbalance remains due to the manufacturing tolerance of the coils, and slight deviation of matrices A and C from the values that are given in equations (5) and (6), as it can be seen in Table 1.

	Q	DD$_1$	DD$_2$	q	dd$_1$	dd$_2$
Q	32 µH	0 µH	−0.02 µH	2.42 µH	−0.83 µH	−1.24 µH
DD$_1$	0 µH	47 µH	−0.07 µH	1.82 µH	1.83 µH	−0.23 µH
DD$_2$	−0.02 µH	−0.07 µH	50.11 µH	1.48 µH	−0.16 µH	2.07 µH
q	2.43 µH	1.82 µH	1.48 µH	14.36 µH	0.09 µH	−0.4 µH
dd$_1$	−0.83 µH	1.83 µH	−0.16 µH	0.09 µH	20.04 µH	−0.05 µH
dd$_2$	−1.24 µH	−0.23 µH	2.07 µH	−0.4 µH	−0.05 µH	18.58 µH

Tab. 1: Measured impedance matrix of the considered IPT system at 220 mm air gap and 100 mm misalignment in x axis and 75 mm misalignment in y axis

5 Conclusions and Future work

In this work a three-phase IPT system for EV charging applications is considered, employing a DD^{2}Q coil topology for the inductive link. The control strategy for this system is proposed, allowing for equal load sharing between the secondary phases and equal current stress of the primary semiconductors, in the case of potential misalignment. The proposed control strategy was tested on the prototype, achieving power transfer of 5.67 kW and 6.83 kW respectively in the case without and with the proposed control method. Efficiency and power transfer capability of the system were increased, thus proving that, with a smart control strategy, use of three-phase IPT systems in EV charging applications is possible.

6 Acknowledgements

This work was possible thanks to the Grant PID2020-117582RB-I00 funded by MCIN/AEI/ 10.13039/501100011033 and, as appropriate, by "ERDF A way of making Europe", by the "European Union NextGenerationEU/PRTR".

References

[1] J. Pries, V. P. N. Galigekere, O. C. Onar and G. -J. Su, "A 50-kW Three-Phase Wireless Power Transfer System Using Bipolar Windings and Series Resonant Networks for Rotating Magnetic Fields," in IEEE Transactions on Power Electronics, vol. 35, no. 5, pp. 4500-4517, May 2020, doi: 10.1109/TPEL.2019.2942065.

[2] S. Chowdhury, M. T. B. Tarek and Y. Sozer, "Design of a 7.7 kW Three-Phase Wireless Charging System for Light Duty Vehicles based on Overlapping Windings," 2020 IEEE Energy Conversion Congress and Exposition (ECCE), Detroit, MI, USA, 2020, pp. 5169-5176, doi: 10.1109/ECCE44975.2020.9235697.

[3] T. Kurpat and L. Eckstein, "A Three-Phase Inductive Power Transfer Coil with SAE J2954 WPT3

Magnetic Interoperability," 2019 IEEE PELS Workshop on Emerging Technologies: Wireless Power Transfer (WoW), London, UK, 2019, pp. 150-155, doi: 10.1109/WoW45936.2019.9030687.

[4] J. -i. Itoh, K. Yamanokuchi, S. Takuma and K. Kusaka, "Three-phase Wireless Power Supply System Using Matrix Converter," 2019 21st European Conference on Power Electronics and Applications (EPE '19 ECCE Europe), Genova, Italy, 2019, pp. P.1-P.10, doi: 10.23919/EPE.2019.8914969.

[5] M. Aganti and C. Bharatiraja, "A New 3-Phase Wireless Power Transfer Circular Pad for Electric Vehicles Battery Charging Systems," 2022 Second International Conference on Power, Control and Computing Technologies (ICPC2T), Raipur, India, 2022, pp. 1-5, doi: 10.1109/ICPC2T53885.2022.9776832.

[6] G. -J. Su, O. C. Onar, J. Pries and V. P. Galigekere, "Variable Duty Control of Three-Phase Voltage Source Inverter for Wireless Power Transfer Systems," 2019 IEEE Energy Conversion Congress and Exposition (ECCE), Baltimore, MD, USA, 2019, pp. 2118-2124, doi: 10.1109/ECCE.2019.8912565.

[7] J. Pries, G. -J. Su, V. Galigekere and O. Onar, "Phase Shift Control of a Three-Phase Inverter for Balanced Secondary Currents in Misaligned Three-Phase Inductive Power Transfer Systems," 2020 IEEE PELS Workshop on Emerging Technologies: Wireless Power Transfer (WoW), Seoul, Korea (South), 2020, pp. 10-15, doi: 10.1109/WoW47795.2020.9291271.

[8] B. J. Varghese, A. Kamineni and R. A. Zane, "Investigation of a DD2Q Pad Structure for High Power Inductive Power Transfer," 2019 IEEE PELS Workshop on Emerging Technologies: Wireless Power Transfer (WoW), London, UK, 2019, pp. 129-133, doi: 10.1109/WoW45936.2019.9030644

[9] SAE International. "Wireless Power Transfer for Light-Duty Plug-In/Electric Vehicles and Alignment Methodology". In: SAE J2954 (Oct. 2022).

[10] J. Skorvaga and M. Pavelek, "Review on high power WPT coil system design," 2021 International Conference on Electrical Drives & Power Electronics (EDPE), Dubrovnik, Croatia, 2021, pp. 13-18, doi: 10.1109/EDPE53134.2021.9604108.

[11] S. Li, W. Li, J. Deng, T. D. Nguyen and C. C. Mi, "A Double-Sided LCC Compensation Network and Its Tuning Method for Wireless Power Transfer," in IEEE Transactions on Vehicular Technology, vol. 64, no. 6, pp. 2261-2273, June 2015, doi: 10.1109/TVT.2014.2347006.

PCIM Europe 2024, 11– 13 June 2024, Nuremberg   DOI: 10.30420/566262350

Comparison of Two Bidirectional 11KW 400V CLLC and CLLLC Resonant Converters for EV Applications

Hasan Mousavi Somarin[1], Norbert Messi[2], Farshid Sarrafin Ardebili[3], Luiz Braz[4]

[1,2,3,4] Valeo Systèmes De Contrôle Moteur, France

Corresponding author: Hasan Mousavi Somarin, hasan.mousavi-somarin@valeo.com
Speaker: Hasan Mousavi Somarin, hasan.mousavi-somarin@valeo.com

Abstract

Resonant converters are becoming increasingly popular due to their lower power loss, higher efficiency, improved electromagnetic interference (EMI) characteristics, and better bi-directional performance. To transfer power bi-directionally, it is necessary to have two resonant tanks on both the primary and secondary sides of the transformer. Two types of resonant topologies, namely, CLLC and CLLLC, are widely used in the EV sector. This paper aims to compare these two converters in terms of power loss, component count, operation, EMC behavior, volume, and efficiency. This way, the design process for two 11 kW 400 V CLLC and CLLLC converters is presented, and a comprehensive comparison is provided.

1 Introduction

Resonant converters have mostly emerged as prominent choices within the Electric Vehicle (EV) sector, particularly in the realm of onboard chargers. Renowned for their superior efficiency compared to other converter types, these DC/DC converters play a pivotal role in enhancing the charging infrastructure of EVs. While the design of resonant converters has evolved considerably over recent years, reaching a level of maturity, the pursuit of optimization remains a focal point of ongoing research. Despite significant advancements, there exists a continued interest in refining and fine-tuning these converters to maximize their performance in real-world EV applications.

In [1], a planar integrated transformer design for energy storage systems is introduced. However, the analysis lacks discussion on parasitic parameters and electromagnetic interference (EMI) behavior of the converter. Notably, factors such as complex manufacturing, limited power handling, higher cost, and sensitivity associated with planar transformers often favor the use of conventional transformers in many applications.

In [2], a Gallium Nitride (GaN) based CLLC resonant converter is developed, boasting an impressive efficiency of 97.02%. Additionally, [3] proposes a control strategy aimed at stabilizing the CLLLC resonant converter. However, there remains a lack of clarity regarding which topology, in terms of resonant capacitors and inductors, is optimal for specific applications.

A comprehensive survey on resonant tanks is conducted in [4], while [5] presents a comparison between two 3.5KW CLLC and CLLLC resonant converters, albeit with separately designed inductors and transformers in both cases.

In our current study, we focus on two 11KW 400V bidirectional CLLC and CLLLC converters, examining, designing, and comparing their performance. For the CLLC converter, we design the separated inductor and a transformer, whereas in the CLLLC converter, the leakage inductance of the transformer is utilized as the series resonant inductance [6-8]. In both cases, the magnetizing inductances serve as the parallel resonant inductance [9].

In typical electromagnetic designs, integrated transformer-inductors often feature separate winding types to achieve the required leakage inductance [10-12]. Consequently, high-frequency AC copper loss becomes unavoidable in such designs. Notably, the distinctive feature of our designed CLLLC transformer-inductor lies in its interleaved winding type, a rarity among integrated transformer-inductors. This design choice significantly reduces AC copper loss, thereby contributing to enhanced efficiency.

The subsequent sections of this paper delve into the modeling of the CLLC and CLLLC converters. Gain curves for both topologies are provided and compared, offering insight into their respective performances. Additionally, the transient behavior of both converters is analyzed and compared.

Furthermore, the paper details the design process of electromagnetic components, which are pivotal in resonant converters. A comparative study is conducted between two types of electromagnetic parts: separately designed transformers and inductors versus integrated transformer and inductor configurations.

Finally, the test results are presented, accompanied by an efficiency comparison between the two converter topologies.

2 Modelling

2.1 Schematic diagram

Figure 1 illustrates the schematic diagram of CLLLC and CLLC resonant converters. The gain function for these converters is obtained using the First Harmonic Approximation (FHA) method.

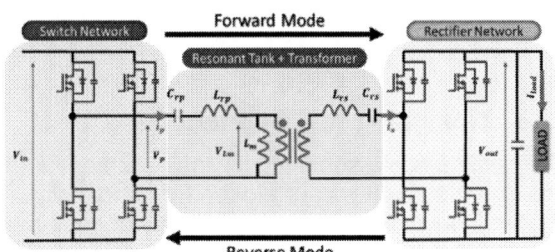

CLLLC resonant converter

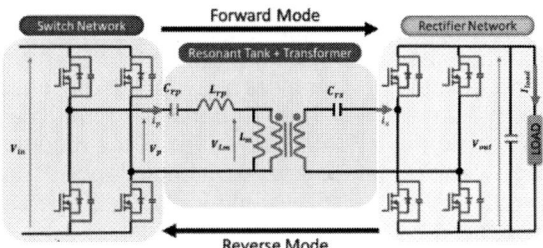

CLLC resonant converter

Figure 1. Schematic diagram of two 11KW 400V bidirectional resonant converter.

In the FHA approach, the primary side switch network, typically found in an inverter generating a quasi-sine voltage waveform, is replaced by its equivalent voltage source. For the primary side, the fundamental or first harmonic of the voltage at the resonant frequency equals $(4/\pi)$ times the maximum DC voltage, denoted as Vx.

On the secondary side, both the rectifier network and the DC load are substituted with their equivalent AC load. The equivalent AC load value is calculated as $R_{ac} = (8/\pi^2)n^2 R_{DC}$ where n represents the transformation ratio.

Figure 2 illustrates the FHA model of the CLLLC and CLLC resonant converters. Parameters labeled with "PRIM" signify elements transferred to the primary side of the transformer.

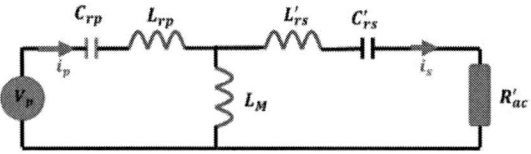

FHA model of CLLLC converter

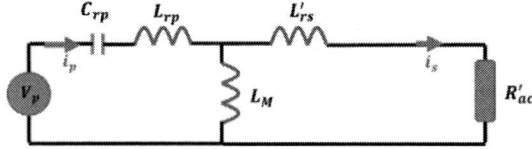

FHA model of CLLC converter

Figure 2. Simplified FHA model of two 11KW 400V bidirectional resonant converter

2.2 Mathematical model

The mathematical model of the converters can be found in [2] and [13]. Here, we present only the final gain functions.

The gain function of the CLLLC converter for the forward and reverse modes are as follows:

$$G_f(F_n) = \cfrac{1}{\sqrt{\begin{array}{l}(1+\frac{1}{\lambda}-\frac{1}{\lambda}\frac{1}{F_n^2})^2+Q^2(\frac{1}{\lambda}\frac{1}{g_c}\frac{1}{F_n^3}+F_n+\frac{g_L}{\lambda}\frac{F_n}{\lambda} \\ -\frac{1}{F_n}\frac{g_L}{\lambda}\frac{1}{F_n}-\frac{1}{\lambda}\frac{1}{g_c}\frac{1}{F_n}-\frac{1}{g_c}\frac{1}{F_n})^2\end{array}}} \quad (1)$$

$$G_r(F_n) = \cfrac{1}{\sqrt{\begin{array}{l}\left(1+\frac{1}{\lambda'}-\frac{1}{\lambda'}\frac{1}{F'^2_n}\right)^2+Q'^2(\frac{1}{\lambda'}\frac{1}{g_c'}\frac{1}{F'^3_n}+F'^3_n \\ +\frac{g_L'}{\lambda'}F'_n+g_L'F'_n-\frac{1}{F'_n}\frac{g_L'}{\lambda'}\frac{1}{F'_n} \\ -\frac{1}{\lambda'}\frac{1}{g_c'}\frac{1}{F'_n}-\frac{1}{g_c'}\frac{1}{F'_n})^2\end{array}}} \quad (2)$$

Where $F_{res-f} = \frac{1}{2\pi\sqrt{L_{rp}\,C_{rp}}}$ is the Resonant frequency in the forward mode, $F_{res-r} = \frac{1}{2\pi\sqrt{L_{rs}\,C_{rs}}}$ is the Resonant frequency in the reverse mode, $F_n = \frac{F_{switching}}{F_{res-f}}$ is the Normalized frequency in the forward mode, $F'_n = \frac{F_{switching}}{F_{res-r}}$ is the Normalized frequency in the reverse mode, $Q = \frac{\sqrt{\frac{L_{rp}}{C_{rp}}}}{R_{ac}}$ is the quality factor in the forward mode, $Q' = \frac{\sqrt{\frac{L_{rs}}{C_{rs}}}}{R_{ac}}$ is the quality factor in the reverse mode, $g_L = \frac{C'_{rs}}{C_{rp}}$ is

the capacitance ratio in the forward mode, $g_c' = \frac{C'_{rp}}{C_{rs}}$ is the capacitance ratio in the reverse mode,

$\lambda = \frac{L_M}{L_{rp}}$ is the inductance ratio in the forward mode,

and $\lambda = \frac{L_M}{L_{rs}}$ is the inductance ratio in the reverse mode

The gain function of the CLLC converter for the forward and reverse modes are

$$\|G_F\| = \|\frac{nV_o}{V_{F_{tank}}}\| = \| \ 1 + \frac{f^2-1}{L_n \ f^2} + $$
$$j\left\{\frac{Q_1[C_nL_nf^4-(C_nL_n+L_n+1)f^2+1]}{C_nL_nf^3}\right\}\|^{-1} \quad (3)$$

$$\|G_R\| = \|\frac{V_o}{nV_{R_{tank}}}\| = \|1 - \frac{1}{C_nL_n \ f^2} + $$
$$j\left\{\frac{Q_2[C_nL_nf^4-(C_nL_n+L_n+1)f^2+1]}{\sqrt{C_n}L_nf^3}\right\}\|^{-1} \quad (4)$$

Where

$Q_1 = \frac{\sqrt{L_{rp}/C_{rp}}}{R'_{F_{ac}}}$ is the quality factor in the forward

mode, $Q_2 = \frac{\sqrt{L_{rs}/C_{rs}}}{R'_{R_{ac}}}$ is the quality factor in the re-

verse mode, $C_n = \frac{C'_{rs}}{C_{rp}}$ is the capacitance ratio in

the forward mode, $L_n = \frac{L_m}{L_{rp}}$ is the inductance ratio

in the forward mode, $f_n = \frac{1}{2\pi\sqrt{L_{rp}C_{rp}}}$ is the basic

frequency, $f = \frac{f_s}{f_n}$ is the normalized operating fre-

quency, $f_{m1} = \frac{1}{2\pi\sqrt{(L_{rp}+L_m)C_{rp}}}$ is the Series in-

verter-side resonant frequency in the forward

mode, and $f_{m2} = \frac{1}{2\pi\sqrt{L_mC'_{rs}}}$ is the Series inverter-

side resonant frequency in the reverse mode

Unlike the CLLLC topology, the CLLC topology is not symmetrical; therefore, there will be two Quality Factors and two resonant frequencies in the forward and reverse modes of operation. Defining the $a = C_nL_n$ and $b = -(C_nL_n + L_n + 1)$, the normalized resonant frequencies for the forward mode f_{resF} and reverse mode f_{resR} will be

$$f_{resF} = \sqrt{\frac{-b-\sqrt{b^2-4a}}{2a}} \quad (5)$$

$$f_{resR} = \sqrt{\frac{-b+\sqrt{b^2-4a}}{2a}} \quad (6)$$

2.3 Designing the resonant tank

To design the resonant tank, gain curves of the resonant converters are plotted for various values of the Quality factor (Q) and inductance ratio (λ or Ln). Subsequently, from these curves, the gain curve that satisfies the minimum and maximum gain values is selected.

Once the Q and Ln values are determined from the plots, and with the resonant frequency (120 kHz) from the input data, three known parameters and three unknown parameters are identified. Consequently, a system of three Equations is formulated. By solving this system, the values of the resonant capacitances, series resonant inductances, and the parallel (magnetizing) inductance can be calculated.

2.4 Validation of the resonant tank size

The effectiveness of the selected parameters is validated through the AC simulation in the first step.

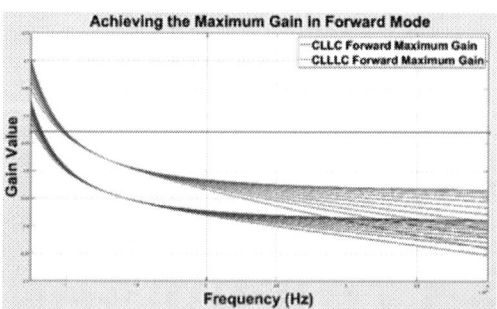

PCIM Europe 2024, 11– 13 June 2024, Nuremberg   DOI: 10.30420/566262350

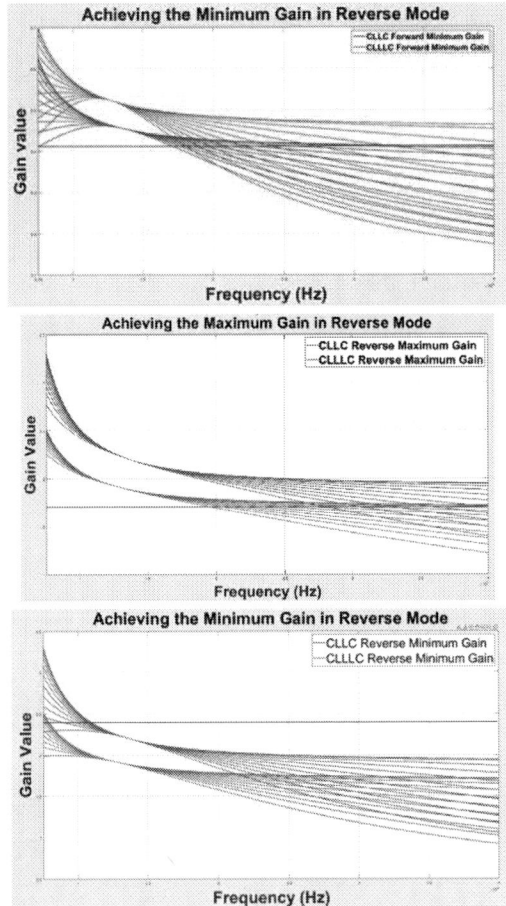

Figure 3. Achieving the minimum and maximum gain in forward and reverse modes of operation in the worst case condition. Red curves: CLLLC converter gain curves from 0.5KW to 10.5KW; blue curves: CLLC converter gain curves from 0.5KW to 10.5KW

Figure 3 depicts the gain curves of the CLLLC and CLLC converters in both forward and reverse modes of operation, spanning load values from 0.5 kW to 10.5 kW. Each plot includes a horizontal line indicating the critical value of the gain.

Observations reveal that in the forward mode, the CLLLC converter exhibits a higher gain compared to the CLLC converter. Conversely, in the reverse mode, this relationship is reversed. Notably, the resonant frequency in the CLLC converter is slightly shifted to the left.

Under forward operation, the CLLC converter can achieve slightly higher power output compared to the CLLLC converter for the same inductance value. However, when the output voltage decreases, the CLLLC converter demonstrates superior response. Despite both converters being

capable of reaching the minimum gain in the forward mode, the CLLC variant operates at lower frequencies, deviating significantly from the resonant frequency and resulting in higher harmonics. In the reverse mode, the CLLLC converter outperforms the CLLC converter in achieving both minimum and maximum voltages. Notably, the CLLLC converter exhibits less deviation when achieving maximum gain, leading to fewer harmonics, improved Zero Voltage Switching (ZVS), and increased efficiency.

2.5 Transient simulation

To assess the voltage and current stress on c omponents of the converters, transient simulati ons were conducted. The results of these sim ulations, reflecting the four modes of operatio under worst conditions, are presented in Table s 1 through 4, where V_Crp and V_Crs are the ap plied voltage to the primary and secondary reso nant capacitor networks respec tively, I_rp and I_rs are the current pass ing through the primary and secondary capacitor net works respectively, V_tr_p and V_tr_s are the ap plied voltage to the primary and secondary side wind ings, Fsw is the switching fre quency and I_mos_P and I_mos_S are the current pass ing through primary and secondary MOSFETs.

2.5.1 Achieving the minimum gain in forward mode

650V ->270V	5KW		8KW		10.5KW	
	CLLLC	CLLC	CLLLC	CLLC	CLLLC	CLLC
V_Crp(V)	85	95	160	164	224	220
V_Crs(V)	36	44	71	80	101	111
I_rp(A)	11	10	18	16	23	21
I_rs(A)	20	19	32	31	42	41
V_tr_p(V)	705	707	758	748	805	782
V_tr_s(V)	269	272	274	279	283	290
Fsw (KHz)	214	174	179	157	166	150
I_mos_P(A)	8	7	12	11	16	14
I_mos_S (A)	14	13	22	22	29	29

Table 1. Voltage and current stress of the components in CLLLC and CLLC converters in different operating point while achieving the minimum voltage in forward mode

2497

The comparison reveals that, in this operational point, the CLLLC converter experiences higher voltage and current stress on both the primary and secondary resonant capacitors compared to the CLLC converter. Additionally, the applied voltage to the primary side of the transformer in the CLLLC converter is slightly lower than that of the CLLC converter, leading to reduced core losses in the transformer.

Furthermore, the CLLC converter operates at a lower switching frequency compared to the CLLLC converter under the same operating conditions, resulting in decreased switching losses. Lastly, the primary and secondary MOSFET switches in the CLLLC converter exhibit lower current stress than those in the CLLC converter, leading to reduced conduction losses.

2.5.2 Achieving the Maximum gain in forward mode

820 --> 470 V	5KW		8KW		10.5KW	
	CLLLC	CLLC	CLLLC	CLLC	CLLLC	CLLC
V_Crp(V)	163	269	215	275	292	309
V_Crs(V)	52	65	84	107	112	142
I_rp(A)	10	14	13	14	16	16
I_rs(A)	13	13	21	23	28	30
V_tr_p(V)	818	1048	945	1013	947	996
V_tr_s(V)	445	436	446	437	449	439
Fsw (KHz)	99	82	98	81	98	80
I_mos_P(A)	7	10	9	10	11	16
I_mos_S (A)	9	9	15	16	19	30

Table 2. Voltage and current stress of the components in CLLLC and CLLC converters in different operating point while achieving the maximum voltage in forward mode

In this specific operating point, it is observed that the CLLLC converter exhibits lower voltage and current stress on both the primary and secondary resonant capacitors compared to the CLLC converter. This reduction in stress results in lower power losses.

Moreover, the applied voltage to the primary side of the transformer in the CLLLC converter is lower than that of the CLLC converter, leading to decreased core losses.

While the CLLLC converter operates at a higher switching frequency compared to the CLLC converter, contributing to increased switching losses, it switches near the resonant frequency. This facilitates Zero Voltage Switching (ZVS) and results in fewer harmonics.

Lastly, the current stress experienced by the primary and secondary MOSFET switches in the CLLLC converter is lower than that of the CLLC converter, leading to reduced conduction losses.

2.5.3 Achieving the Maximum gain in reverse mode

470 --> 820 V	5KW		8KW		10.5KW	
	CLLLC	CLLC	CLLLC	CLLC	CLLLC	CLLC
V_Crp(V)	60	38	100	75	141	108
V_Crs(V)	45	29	63	53	85	73
I_rp(A)	6	6	10	10	13	14
I_rs(A)	19	20	26	30	33	37
V_tr_p(V)	801	782	803	793	787	796
V_tr_s(V)	505	493	514	511	825	525
Fsw (KHz)	171	270	164	218	155	203
I_mos_P(A)	4	4	7	7	9	10
I_mos_S (A)	13	14	18	21	23	26

Table 3. Voltage and current stress of the components in CLLLC and CLLC converters in different operating point while achieving the maximum voltage in reverse mode

At this specific operating point, certain differences in voltage and current stress between the CLLLC and CLLC converters are observed. The voltage stress in the CLLLC converter is higher compared to the CLLC converter, while the current stress remains slightly higher in the CLLC converter, resulting in comparatively higher power losses.

The applied voltage to the primary side of the transformer is approximately the same in both converters, making it difficult to definitively determine whether core losses are decreased or increased.

Although the switching frequency of the CLLLC converter is lower than that of the CLLC converter, resulting in less switching loss, it operates close to the resonant frequency, thereby reducing harmonics.

Lastly, the overall current in the primary and secondary MOSFET switches is slightly decreased in the CLLLC converter, leading to comparatively higher conduction losses on these switches.

2.5.4 Achieving the Minimum gain in reverse mode

270 --> 650 V	5KW		8KW		10.5KW	
	CLLLC	CLLC	CLLLC	CLLC	CLLLC	CLLC
V_Crp(V)	201	133	328	215	434	283
V_Crs(V)	114	89	168	128	225	165
I_rp(A)	10	9	17	15	23	19
I_rs(A)	23	24	35	35	46	46
V_tr_p(V)	581	619	565	619	583	633
V_tr_s(V)	329	319	320	321	326	328
Fsw (KHz)	78	110	77	109	77	109
I_mos_P(A)	7	6	12	10	23	14
I_mos_S (A)	16	17	24	35	46	32

Table 4. Voltage and current stress of the components in CLLLC and CLLC converters in different operating point while achieving the minimum voltage in reverse mode

In this operating point, it is observed that the voltage and current stress on the primary and secondary resonant capacitors in the CLLLC converter are higher compared to the CLLC converter, consequently leading to increased power loss. Additionally, the applied voltage to the primary winding of the transformer is lower in the CLLLC converter, resulting in reduced core loss. Although the switching frequency in the CLLLC converter is lower than that of the CLLC converter, the latter switches near resonance, exhibiting fewer harmonics in this particular operating point. Finally, while the current flowing through the primary side MOSFET switches in the CLLLC converter is generally lower in most operating points, it is comparatively higher in the secondary MOSFET switches.

3 Electromagnetics design

In this section, we will delve into the design of three key electromagnetic components. Specifically, we will discuss the design and construction of an inductor and a transformer with a predetermined value of magnetizing inductance for the CLLC converter, where the magnetizing inductance serves as the parallel inductor. Additionally, for the CLLLC converter, we will detail the design of an integrated transformer-inductor. In this design, the leakage inductance is utilized as the series resonant inductance, while the magnetizing inductance is tailored to function as the parallel inductor. It's important to note that due to confidentiality reasons, the detailed specifications of these components cannot be disclosed.

3.1 Core sizing

The **Area Product (A_p)** approach is a commonly utilized method for determining the size of transformer and inductor cores. Equation (7) and Equation (8) are employed in this context to calculate the size of the transformer and inductor cores, respectively.

$$A_p = A_w A_e = \frac{2 \times P_o}{K_f\ K_u\ f\ B_{max}\ J} \qquad (7)$$

$$A_p = A_w A_e = \frac{L\ I_{max}^2}{K_u\ B_{max}\ J} \qquad (8)$$

Where A_w is the window area in which the windings will be winded, A_e is the cross-sectional area of the core, P_o is the output power (KW), K_f is the waveform coefficient: 4 for square waves and 4.44 for sinusoidal ones, K_u is the utilization or filling factor, f is the minimum frequency, B_{max} is the maximum flux density, J is the maximum current density, L is the inductance, and I_{max} is the inductor's peak current.

Once the core size is calculated, referring to the core manufacturer a suitable size core can be selected. To optimize the power loss and volume the size and geometry of the core can also be customized. For the current design, the shape and size of the core for all three components are customized.

3.2 Number of turns

Equation (9) and (10) will be used to determine the number of turns for transformer and inductor respectively.

$$N_1 = \frac{V_p\ D}{2 B_{max}\ f\ A_e} \qquad (9)$$

$$N_{ind} = \sqrt{\frac{L}{A_L}} \qquad (10)$$

Where
N_1 is the number of the turns of transformers' primary winding. The secondary will be caluculated using the transformation ratio

V_p is the primary peak voltage, which is equal to

$$\frac{4}{\pi} V_{dc-max} = \frac{4}{\pi} \times 820 = 1044.1 V$$

D is the duty ratio which is equal to 0.5 in this design

N_{ind} is the turn number of the inductor

L is the inductance

A_L is the specific inductance value of the core provided by the supplier

3.3 Conductor sizing

The bare conductor diameter d_{bare} for the primary and secondary of the transformer, as well as for the inductor, is computed individually using the Equation (11).

$$d_{bare} = \sqrt{\frac{4 \ I_{RMS}}{\pi \ J}} \qquad (11)$$

Where

I_{RMS} is the worst case RMS current

3.4 Skin and proximity effects

Given that the intended converter operates at high frequencies, it is essential to consider skin and proximity effects. The skin depth is determined using Equation (12).

$$\delta = \sqrt{\frac{\rho_{copper \ @ \ 100°C}}{\pi \ \mu_0 \ \mu_r \ f_{sw}}} \qquad (12)$$

Where

$\rho_{copper \ @ \ 100°C}$ is the copper resistivity at 100°C

μ_0 is the vacuum or free space permeability

μ_r is the copper permeability

f_{sw} is the switching frequency

In high-frequency applications, it is crucial for the skin depth to be larger than the conductor diameter d_{bare}. If this condition is not met, it becomes imperative to utilize Litz wire to mitigate the skin effect effectively. Moreover, implementing interleaved winding techniques helps reduce the proximity effect, leading to a notable reduction in AC copper loss.

3.5 Gap length

The minimum length of the air gap g_{min} is determined using Equation (13).

$$g_{min} = N^2 \frac{\mu_0 \ A_e}{L} \qquad (13)$$

In Equation (13), N represents the number of turns for the inductor or, in the case of transformer design; it denotes the number of turns in the transformer's primary winding. For the inductor, L corresponds to the desired inductance value. For the transformer, L represents the magnetizing inductance value, while for the integrated transformer-inductor, L is the sum of the magnetizing inductance and sum of the primary and secondary leakage inductances.

4 Comparison

The designed transformer, inductor, and integrated transformer-inductor are subjected to simulation using Ansys Maxwell, a 3D FEM electromagnetic simulation tool. Various types of simulations, namely Magneto-static, Eddy Current, Transient and Electro-static, are conducted to validate the parameters.

Following each simulation, feedback from the results is utilized to refine and redesign the components as necessary. This iterative process ensures that the final designs meet the desired performance specifications. In the subsequent analysis, we will compare two topologies: one employing separate inductor and transformer components (CLLC), and the other utilizing an integrated transformer-inductor configuration (CLLLC).

4.1.1 Resistance

	CLLC			CLLLC	
	Transformer		Inductor	Integrated Transformer-Inductor	
	Primary	Secondary		Primary	Secondary
Rdc (mΩ)	34	38	15	35	39
Rac(mΩ) @75KHz	48	44	17	41	45
Rac(mΩ) @100KHz	57	52	18	45	50
Rac(mΩ) @200KHz	125	109	28	74	84
Rac(mΩ) @250KHz	176	151	36	96	109

Table 5. AC resistance of the separated and integrated transformer-inductor

The comparison reveals that the AC resistance of the integrated transformer-inductor is lower than that of the separated transformer and inductor configuration. Consequently, it is anticipated that the integrated transformer-inductor will experience lower copper losses compared to the separated components.

4.1.2 Core loss at nominal power

	CLLC	CLLLC

	Transformer	Inductor	Integrated Transformer-Inductor
Core loss (W) @75KHz	2.3	0.9	2.3
Core loss (W) @100KHz	6.4	2.5	6.4
Core loss (W) @200KHz	72	27	72
Core loss (W) @250KHz	158	60	158

Table 6. Core loss of the separated and integrated transformer-inductor

The separated transformer and inductor configuration exhibits higher core volume, leading to increased core losses. Conversely, integrating the inductor into the transformer has effectively reduced the core losses.

4.1.3 Parasitic parameters

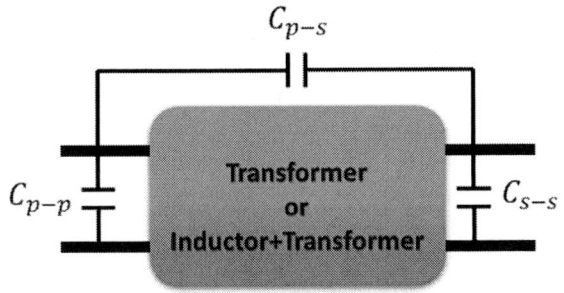

Figure 4. Parasitic elements in electromagnetics

Figure 4 illustrates the parasitic elements of the CLLC/CLLLC electromagnetic components, where C_{p-p} and C_{s-s} represent the primary and secondary winding parasitic capacitances, respectively, and C_{p-s} denotes the intertwining capacitance between the primary and secondary windings. Corresponding parasitic parameter values are presented in Table 7. Analysis of these parameters suggests that the EMI behavior of the CLLLC configuration is anticipated to outperform that of the CLLC.

	C_{p-p} (pF)	C_{s-s} (pF)	C_{p-s} (pF)
CLLC	2.03	1.27	257.79
CLLLC	1.5	2.65	64.6

Table 7. Parasitic capacitance values of CLLC and CLLLC electromagnetic components

4.1.4 Volume

	CLLC	CLLLC

	Transformer	Inductor	Integrated Transformer-Inductor
Total Core volume (mm^3)	67706	36989	68371
Total Copper volume (mm^3)	20056	5817	20476

Table 8. Core and copper volume used in the separated and integrated transformer-inductor

Total core volume reduction is:

$$\frac{V_{core-integrated}}{V_{core-seperated}} = (1 - \frac{68371.54419}{36989.61643+67706.66146}) \times$$

$$100 = 35.33\%$$

Total copper volume reduction is:

$$\frac{V_{copper-integrated}}{V_{copper-seperated}} = (1 - \frac{20476.779093}{5817.570889+20056.356137}) \times$$

$$100 = 20.85\%$$

4.1.5 Efficiency

Since the nature of the CLLC and CLLLC converters are different, the size of the resonant capacitance in each converter were different contributing to distinctive switching frequencies in each converter. Besides, the dedicated CLLC coolant system was not ideal one for CLLLC one. Thus, the generated heat by the electromagnetic components was not removed effectively resulting in an increased resistance and causing even more power loss and temperature rise on the winding in CLLLC converter. At boost stage, where the frequency difference in minimum, the archived efficiency is presented in Table 9.

	CLLC	CLLLC
Input voltage (V)	820	820
Output voltage (V)	464	465
Switching frequency (KHz)	80	99.6
Efficiency (%)	96.61	96.12

Table 9. CLLC and CLLLC Efficiency comparison in boost mode

However, the best way to compare these two scenarios is to refer to resistance, core loss and volume tables presented in the previous sections.

5   Conclusion

The conclusion of the paper presents a comparative analysis between two 11KW 400V converters, namely the CLLC and CLLLC converters. Throughout the investigation, several key observations were made:

In the CLLC converter, there was a slight shift in resonant frequency compared to the CLLLC converter. Variations in operating points revealed different voltage and current stresses on the components, making it challenging to definitively determine superiority from this aspect alone. The CLLLC converter demonstrated advantages in terms of resistance, core loss, and copper loss over the CLLC converter, primarily due to its smaller size. Notably, the CLLLC converter exhibited a reduced volume of electromagnetic components, leading to significant cost savings and higher power density compared to the CLLC counterpart. Additionally, the CLLLC converter showcased superior EMI behavior compared to the CLLC converter.

In summary, the findings suggest that the CLLLC converter offers several benefits over the CLLC converter, including reduced losses, cost savings, and improved EMI performance, thus making it a promising choice for high-power applications.

References

[1] Dhakar, AK, Soni, A., & Bansal, HO (2020, December). Design and Control of a Bi-directional CLLC Resonant Converter For Low voltage Energy Storage Systems. In 2020 IEEE 17th India Council International Conference (INDICON) (pp. 1-8). IEEE.

[2] Liu, Y., Du, G., Wang, X., & Lei, Y. (2019). Analysis and design of high-efficiency bidirectional GaN-based CLLC resonant converter. Energies, 12(20), 3859.

[3] Joševski, M., Korompili, A., & Monti, A. (2020). Modeling and Voltage Control of Bidirectional Resonant DC/DC Converter for Application in Marine Power Systems. IFAC-PapersOnLine, 53(2), 13056-13063.

[4] Instruments, T. (2018). Survey of resonant converter topologies. In 2018 Texas Instruments Power Supply Design Seminar SEM2300, SLUP (Vol. 376).

[5] Zahid, Z. U. (2015). Design, modeling and control of bidirectional resonant converter for vehicle-to-grid (V2G) applications (Doctoral dissertation, Virginia Tech).

[6] De, D., Klumpner, C., Rashed, M., Patel, C., Kulsangcharoen, P., & Asher, G. (2012, March). Achieving the desired transformer leakage inductance necessary in DC-DC converters for energy storage applications. In 6th IET International Conference on Power Electronics, Machines and Drives (PEMD 2012) (pp. 1-6). IET.

[7] Pavlovsky, M., De Haan, S. W. H., & Ferreira, J. A. (2006, June). Winding losses in high-current, high-frequency transformer foil windings with leakage layer. In 2006 37th IEEE Power Electronics Specialists Conference (pp. 1-7). IEEE.

[8] McLyman, C. W. T. (2004). Transformer and inductor design handbook. CRC press.

[9] Erickson, R. W., & Maksimovic, D. (2007). Fundamentals of power electronics. Springer Science & Business Media.

[10] Ansari, S. A., Davidson, J. N., & Foster, M. P. (2022, June). Fully-integrated transformer with asymmetric leakage inductances for a bi-directional resonant converter. In 11th International Conference on Power Electronics, Machines and Drives (PEMD 2022) (Vol. 2022, pp. 260-265). IET.

[11] Cougo, B., & Kolar, J. W. (2012, March). Integration of leakage inductance in tape wound core transformers for dual active bridge converters. In 2012 7th International Conference on Integrated Power Electronics Systems (CIPS) (pp. 1-6). IEEE.

[12] Steiger, U., & Mariethoz, S. (2010, September). Method to design the leakage inductances of a multiwinding transformer for a multisource energy management system. In 2010 IEEE Vehicle Power and Propulsion Conference (pp. 1-6). IEEE.

[13] Instruments, T. (2020). Bidirectional CLLLC resonant dual active bridge (dab) reference design for HEV/EV on-board charger. Texas Instruments reference design No. TIDM-02002. Accessed Oct, 26.

Dynamic Wireless Charging System Design for Extra-Urban Areas based on Resonant Inductive Power Transfer

Irene-Maria Torres-Alfonso[1], Carlos Costas-Sos[1], Juan M. Perie-Buil[1], José-Francisco Sanz-Osorio[2], Juan L. Villa[2], Oscar Garcia-Izquierdo[2]
[1] CIRCE Technological Centre, Spain
[2] CIRCE Institute, Spain

Corresponding author and speaker: Irene-Maria Torres-Alfonso, imtorres@fcirce.es

Abstract

This article outlines the hardware and control design of an LCC-S inductive power transfer system for dynamic wireless charging. The system enables variable power transfer within a defined misalignment range and without communication with the electric vehicle (EV) except detection. Interoperability and cost minimisation are also considered. The paper explores the design and control of the inductive power transfer (IPT) system and resonant topology, as well as the overall system structure, followed by considerations for implementation and electromagnetic emissions. The proposed system has been developed within the framework of H2020 project INCIT-EV.

Keywords: *Inductive Power Transfer (IPT), Wireless Power Transfer (WPT), Dynamic Wireless Power Transfer (DWPT), SPS-S (Series-Parallel-Series-Series), LCC-S Resonant inverters.*

Nomenclature:

AC	Alternating Current	ICNIRP	International Commission of Non-Ionizing Radiation Protection
AC/DC	Alternating current to Direct current converter, rectifier	IEC	International Electrotechnical Commission
AVC	Asymmetrical Voltage Cancellation	IPT	Inductive Power Transfer
CAN	Controller Area Network	ISO	International Organization for Standardization
DC/AC	Direct current to Alternating current converter, inverter	LCC-S	Inductor – Capacitor – Capacitor (Tx side) – series resonant topology (Rx side)
DWPT	Dynamic Wireless Power Transfer	PI	Proportional - Integral
EM	Electromagnetic	SPS-S	Series-Parallel-Series (Tx topology) – series (Rx topology)
EMC	Electromagnetic Compatibility	Rx	Receiver or secondary coil(s)
EMF	Electromagnetic Fields	SAE	Society of Automotive Engineers
EMI	Electromagnetic Interference	Tx	Transmitter or primary coil(s)
EV	Electric Vehicle	V2G	Vehicle To Grid
FEM	Finite Element Method	WPT	Wireless Power Transfer
FOD	Foreign Object Detection	ZVS	Zero voltage switching
HF	High Frequency		

1   Introduction

The growing EV market is being affected by the demand of more efficient and reliable battery charging methods. WPT systems emerge eliminating certain drawbacks of conventional conductive charging infrastructure. Based on magnetic coupling between transmitter and receiver coils, this technology requires no physical connection between EV and charging station, avoiding the consequent risk for users, and incorporating other desirable aspects such as automatic operation, safety in harsh climatic conditions, flexibility and interoperability [1]. WPT is now achieving power transfer ratings and efficiency comparable to conductive methods, and enables EV charging at any time and location, including while in motion through dynamic WPT.

DWPT technology significantly changes the framework of electric mobility limitations by providing unlimited driving range through its capacity to charge while driving [2]. Travel stress is avoided, charging waiting times are eliminated and battery capacity requirement is drastically reduced, impacting directly on EV weight and price [1], [3]. However, these advantages involve great challenges. Specifically related to dynamic charging functionality, issues such as internal communication and coils activation management based on EV movement, coils synchronization, misalignment control or thermal behavior on road integration must be faced. Common issues with static WPT are EMC and EMI mitigation, safety concerns related to FOD and the crucial aspect of interoperability between transmitter and receiver [4].

The main element of a DWPT charging station is the charging track. The road is fitted with underfloor transmitter (Tx) coils that are sequentially activated as the EV drives over them. Geometry, size, assembly of coils and distance between them affect their integration on the road and maintenance expenditures, EM emissions, power supply architecture, power transmission stability or control complexity [1]. According to [5], lumped, elongated pads and elongated rails is a common classification for track configuration.

Ferrites are commonly employed in charging track and Rx coils. Proper material selection, geometry and assembly design can enhance IPT efficiency and reduce EMF leakages, taking advantage of its capacity of guiding and concentrating the magnetic flux. Soft ferrites are frequently chosen for this application, serving as wiring core or located between wiring and shielding. However, incorporating ferrites can significantly increase costs, weight and complexity, as well as the addition of a heat source to the assembly [6][7]. On the other hand, the use of shielding on Rx coils is broadly extended in order to protect EV equipment and users from EMF leakages. Passive aluminium-plates-based shielding is preferred due to its lightweight and low cost. Shielding incorporation to both Tx and Rx systems has the main objective to reach EMC standards and protect users from unsafe EMF exposure. The heat generated on shielding due to Eddy currents and its impact on IPT efficiency must be considered in its use and design [7][8].

Misalignment and air-gap variation on IPT coils are aspects intrinsically present in WPT systems. They provoke changes in self-, leakage and especially in mutual inductances that can consequently detune the compensation networks, decrease power transmission and efficiency [9][10]. On DWPT, lateral misalignment is often critical. Coils design, resonant network topology or HF control methodology among others, define tolerance to misalignment. Advanced techniques such as automatic vehicle guidance, real-time roadside and EV inside information are frequently employed [11].

Compensation networks are necessary for both the Tx and Rx coils in magnetic resonance-based WPT systems to reduce the reactive rating of the HF converter, improve power transfer, reduce losses through soft switching operation and maximize efficiency. They enable the regulation of current at primary or supply side while voltage or power are regulated on the secondary or receiver side [7]. Primary side parallel-based topologies are suitable for DWPT systems due to their zero-coupling tolerance, which allows for a relatively high mutual inductance range and high efficiency at low coupling [12]. Hybrid or multi-resonant topologies are also commonly found on DWPT.

DWPT charging systems are subject to WPT standards which are currently under development. First guidelines were established by the American SAE, and J2954 version 4.0, released in 2022, is widely considered. European IEC 61980 includes the bases of SAE standards – power class classification, minimum efficiency without misalignment, maximum airgap, frequency range or communication standards -, adding guidelines for V2G capabilities and expanding frequency range for heavy-duty EVs. Concerns about EMI and EMF exposure are covered by ICNIRP recommendations, as of their 2010 version, for the applicable frequency range. On the other hand, ISO introduces interoperability in its 19363:2020 standard [13], [14].

The interoperability between Tx and Rx sides is defined in the standards through specific indicators related to compensation networks, magnetic couplers, and communication protocols. Further promotion is necessary for aspects such as impedance range definition, frequency range selection for systems above WPT4 (high power) avoiding radio frequency band and V2G functionality [15].

It is also relevant to note that road integration of DWPT systems represents a major challenge. Compatibility with pavements materials must be considered not only in terms of EM behavior, but regarding mechanical performance, which can be affected by the discontinuities associated with system integration or by low thickness of wearing course, consequence of airgap minimization. In addition, Joule effect produced by charging track performance affects inevitably to pavement [16].

Summarizing, this paper outlines a design proposal for a dynamic wireless charging station for the specific application of extra-urban areas. The document is structured as follows: after the introduction, section 2 focuses on the design, describing the IPT proposal and the overall topology of the system, emphasizing on the system modelling and introducing the control concept. Section 3 presents the results of the previous methods, including the charging track final approach, system modelling calculations and the analysis of its behaviour, the system response to control action and the proposed road integration.

2 Methodology

2.1 IPT design

The proposed system is conceived for extra-urban high-speed applications, opting for elongated pads Tx coils configuration, in contrast with Tx-Rx coils comparable areas commonly used in urban areas. Quasi-continuous charging tracks maximize medium flux linkage with Rx coil, and power transfer stability through reduction the number of transitions between Tx coils [1,13]. Additionally, the efficiency of the IPT system is not significantly affected by speed. This type of configuration requires simpler control algorithms and a reduced number of power modules per meter, but conversely, extra safety measurements are required due to the high stray magnetic fields generated during operation comparing to size-reduced Tx coils configurations. Excessive lengths can also result in elevated inductances, unmanageable voltages and high power losses [1]. A one-turn 10-m length geometry (Fig. 1) is determined to be the optimal compromise in this proposal.

Fig. 1 Primary and secondary coils proportion

A configuration without ferrites or shielding is chosen for Tx coils. The coreless configuration for elongated pads has several benefits, including a lower cost, lower insulation requirements due to lower self-inductance and less sensitivity to lateral misalignments for the same power rating [2][7]. Considering the DWPT operating frequency range, non-negligible high frequency phenomena such as skin and proximity effects are expected, increasing AC resistance [4]. Consequently, Litz wire is chosen as conductor with the aim of improving efficiency.

Although the presented charging track design is conceived to be interoperable in a certain range, the IPT system in the context of INCIT-EV project includes the Rx coil developed by project partner Vedecom. The pad consists of a multiturn Litz wire winding placed on a ferrite plane and counts with aluminum-based shielding which takes part of the EV's underside. It has a 1:10 proportion with proposed charging track (Fig. 1).

2.2 Overall system topology

The overall layout of the proposed DWPT charging station is illustrated in Fig. 2. The charging track comprises eight primary coils, covering a distance of 80 to 90 metres. An AC/DC stage establishes a common DC bus, to which eight DC/AC stages distributed in four cabinets, are connected. DC/AC stages consume power from AC/DC sequentially, ensuring Tx coils progressive activation. The AC/DC stage is based on a two-stage four-leg active rectifier, preceded by filtering and hardware protection, while the DC/AC stages are resonant full-bridge converters based on SPS-S resonant topology.

Fig. 2 Layout of DWPT charging station.

A notable benefit of the suggested system is the low density of electronic components required per metre of charging track, reducing cost and complexity compared to other common topologies. Considering a series resonant topology for the secondary side, a LCC topology is chosen for the primary side (Fig. 3), taking advantage of the intrinsic characteristics of the LC basic topology, such as its current source behavior, its high efficiency at low coupling and the safe response to misalignment related to the consequent

2505

impedance increase. This last property is especially convenient for DWPT, where the relative position of primary and secondary coils is inherently variable. Adding an extra capacitor to Tx coil branch reduces the size of series inductance, making the network more compact, and decreases the voltage on the parallel branch [7].

2.3 Resonant network modelling

Focusing on the magnetic coupling and its associated resonant networks, the proposed global topology can be defined as SPS-S (series-parallel-series-series, Fig. 3).

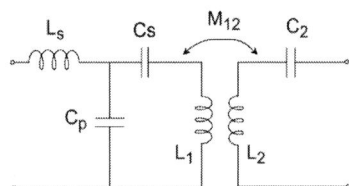

Fig. 3 SPS-S resonant topology

As part of the presented design, this study develops simplified circuit models for primary and secondary system resonant networks along with their associated mathematical expressions. The aim is to establish a reference method to calculate the primary side components values (L_s, C_s and C_p) meeting specific requirements which can be configured, while also allowing interoperability with secondary systems within a range.

The inductive coupling design is characterized by Tx and Rx coils leakage inductances (L_1 and L_2), mutual inductance (M_{12}) and its variation caused by the admissible lateral misalignment, track entrance and exit and transitions between Tx coils. The interdependence between Tx and Rx coils is the main principle of IPT. Equations 1 and 2 demonstrate how the current circulating through one coil induces a voltage on the other proportional to the mutual inductance between them.

$$V_{m2} = I_{m1} \cdot M_{12} \cdot w_{sw} \tag{1}$$

$$V_{m1} = I_{m2} \cdot M_{12} \cdot w_{sw} \tag{2}$$

Model's input and output parameters are defined as follows:

Inputs:

- IPT parameters: L_1, L_2, nominal M_{12}
- Coils' resistance: R_{L1}, R_{L2}, R_{Ls}
- Secondary resonant network components: C_2

- Range of displacement along the track and lateral misalignment: $[d_x, d_y]$
- Airgap or distance between coils: d_z
- Maximum rates: Tx side DC bus voltage range (V_{1DC_max} and V_{1DC_min}), maximum Tx side inverter output current (I_{1_lim}), maximum admissible induced voltage at Rx side (V_{m2_lim}), output power range (0 to P_{O_max})

Outputs:

- Tx side resonant components values: L_s, C_s and C_p
- Main Tx side circuit magnitudes: working frequency (f_{sw}), Tx coil current (I_{m1}), resonant circuit voltages and currents (V_{Cs_max}, V_{Cp_max}, I_{Cp_max}, I_{1_max})

2.4 Secondary side model

Modelling is approached defining first the secondary resonant network. In accordance with INCIT-EV secondary system topology, a series resonant topology is represented. A power control is assumed, as shown in Fig. 4.

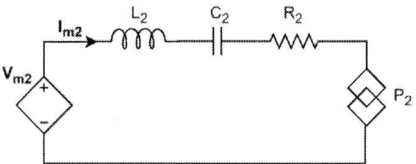

Fig. 4 Secondary side equivalent circuit

Secondary side induced voltage (V_{m2}) can be expressed as a function of Tx coil current (I_{m1}) and mutual inductance according to Eq. 1. On the other hand, active (P_2) and reactive power (Q_2) can be expressed as functions of the model inputs, induced voltage and secondary resonant network impedance (Z_2, Eq. 3 and 4). Maximum value is considered for active power (Eq. 3).

$$P_{2\_max} = \frac{V_{m2}^2}{2 \cdot Z_2} \tag{3}$$

$$Q_2^2 - \frac{V_{m2}^2}{Z_2} Q_2 + P_2^2 = 0 \tag{4}$$

2.5 Primary side model

Figure 5 illustrates the simplified model for the primary side. Tx coil current (I_{m1}) is defined in this proposal as primary side control variable and taken as phase reference for the second mesh variables.

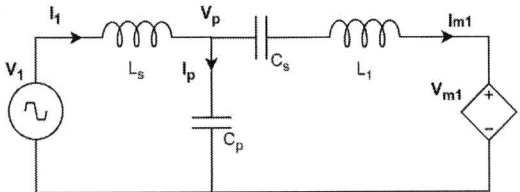

Fig. 5 Primary side equivalent circuit

Components of induced voltage at Tx coil can be approximated using Rx side active and reactive power as shown in Eq. 5. By solving the second mesh, voltage at parallel branch can be expressed as shown in Eq. 6; current is calculated as a function of parallel branch impedance (Eq. 7). Series resistance is taken into account in Tx coils.

The calculation of the inverter output current involves developing the node currents vectorial expression (Eq. 8), and voltage is expressed as a function of the voltages of the first mesh (Eq. 9). A phase reference swap to I_1 must be applied to V_{Cp} to incorporate it to Eq. 9. Additionally, resistive component is also taken into account for series inductance.

$$V_{m1x} = \frac{P_2}{I_{m1}}; \ V_{m1y} = \frac{Q_2}{I_{m1}} \qquad (5)$$

$$\overrightarrow{V_{Cp}} = \overrightarrow{V_{m1}} + \overrightarrow{I_{m1}} \cdot \left(j \cdot \omega_{sw} \cdot L_1 + R_1 + \frac{1}{j \cdot \omega_{sw} \cdot C_p}\right) \qquad (6)$$

$$\overrightarrow{I_{Cp}} = \overrightarrow{V_{Cp}} \cdot j \cdot \omega_{sw} \cdot C_p \qquad (7)$$

$$\overrightarrow{I_1} = \overrightarrow{I_{Cp}} + \overrightarrow{I_{m1}} \qquad (8)$$

$$\overrightarrow{V_1} = \overrightarrow{I_1}\left(j \cdot \omega_{sw} \cdot L_s + R_{L_s}\right) + \overrightarrow{V_{Cp}} \qquad (9)$$

2.6 Boundary conditions

Primary and secondary sides electronic stages impose some conditions that must be considered in the resonant network design.

Two factors restrict the value of I_{m1} set point. The first limit, expressed by 10, is determined by the Rx side electronic stages through the maximum admissible induced secondary voltage. In practice, I_{m1} set point is further restricted by the need to limit losses in Tx coils, in order to achieve a minimum efficiency on the IPT system (Ineq. 11). Apart from

that, I_{m1} is taken as phase reference for the second mesh variables.

$$I_{m1} < \frac{V_{m2\_lim}}{\omega_{sw} \cdot M_{12}} \qquad (10)$$

$$I_{m1} < \sqrt{\frac{P_{2\_max}}{R_{L_1}} \frac{1 - \eta_{min}}{\eta_{min}}} \qquad (11)$$

The operative range of the inverter output voltage is given by Ineq. 12 and 13, which are based on the chosen control modulation, outlined in section 2.9. Lower limit is related to ensure ZVS condition (Ineq. 12), and upper limit is determined by the maximum rms value (Ineq. 13). Both expressions consider the available DC bus voltage range. The inverter output current range is defined as designed condition too; maximum current is given by transistors, and minimum value is taken to ensure maximum power transfer under the condition of maximum inverter output voltage (Eq. 14). Last requirement defined is to have an inductive equivalent impedance at inverter output under any coupling condition considered in this proposal.

$$V_1 > \frac{2\sqrt{2} \cdot V_{1DC_{min}}}{4\pi} \qquad (12)$$

$$V_1 < \frac{2\sqrt{2} \cdot V_{1DC_{max}}}{\pi} \qquad (13)$$

$$I_1 > \frac{P_{2\_max}}{V_{1\_max}} \qquad (14)$$

2.7 Parametric analysis

A parametric sweep is defined to understand and validate the behavior of the designed system within the applicable range. Three types of variations are considered:

- Tx-Rx coils relative position

Displacement ranges on x and y axes are identified based on their quantifiable effect on leakage and, specially, on mutual inductances. Regarding longitudinal displacement, track entry, exit and transitions between Tx coils are considered. In practice, a COMSOL® model is created and executed for every combination of [x,y], obtaining values for L_1, L_2, M_{12} and mutual inductances between track coils (M_{13}, M_{34}).

- Components tolerance

The presented calculation applies a ±5% tolerance to nominal values of resonant networks components (L_s, C_p, C_s, L_1, L_2 and C_{s2}) to account for manufacturing tolerances. The model supports the introduction of a different tolerance percentage. Variations in these components values have a significant impact on the system's behavior due to its resonant nature.

- Output power

Maximum and minimum power demanded by Tx circuit are defined and taken into consideration to meet ZVS condition in the inverter and ensure I_{m1} controllability.

2.8 Control and coils activation

A crucial aspect for the design of a DWPT charging station is establishing a stable and fast-response control for the intrinsically broad range of M_{12}. Coupling varies during the EV displacement, as well as for different Rx systems. The design of the track coils activation sequence also plays a relevant role in the system response.

2.9 Control methodology

In this proposal, interoperability is approached from the point of view of control by establishing a constant current on Tx coils (I_{m1}) and acting over the secondary side through Rx coil induced voltage (Eq. 1). Defining working frequency as a constant, V_{m2} depends on coupling and I_{m1}. In that way, power is controlled by the Rx side.

Tx side control, executed at DC/AC stages, is based on asymmetrical voltage cancellation (AVC). This technique modifies the amplitude of the inverters output voltage fundamental component (V_1) acting on its duty cycle through the control angle α (Eq. 15). V_1 regulation is employed to control Tx coils current, designated as the process variable.

$$V_{1\_rms} = \frac{V_{DC}}{\pi}\sqrt{5 + 3\cos\alpha} \qquad (15)$$

The inverter output voltage (V_1) is designed as regulated variable. A PI controller is applied to it and defined by proportional and integral terms of the controlled variable, Tx coil current (I_{m1}). Equation 16 and Figure 6 illustrate the previous concept.

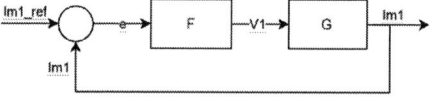

Fig. 6   Tx coils current feedback loop scheme

$$\Delta V_{1k} = K_i \cdot \Delta V_{1(k-1)} + K_p \cdot (I_{m1ref} - I_{m1}) \qquad (16)$$

2.10 Communication scheme and track coils activation

Communication between power stages is established via CAN bus connected in star topology. The AC/DC converter acts as master device, controlling DC/AC converters' sequential activation and deactivation. For their part, DC/AC converters send DC power and voltage references as well as generated errors to the AC/DC stage.

The consecutive activation of DC/AC stages starts after EV detection, established some meters before the first Tx coil. Once nominal current is established and power transfer to Tx coil is detected, the second DC/AC is activated, and so on.

3   Results

3.1   Charging track

As stated in section 2.1, a charging track based on elongated pads is defined in this approach. A rectangular geometry of 10m x 0.5m is determined as result of magnetic coupling optimization process, where the reference Rx coil is considered. 10 m length results from taking a compromise between power transfer continuity, leakage inductance and consequent coil voltage at nominal working conditions, as well as Joule losses limitation. 0.5 m width provides a magnetic flux distribution in lateral direction suitable for reference Rx system and defined misalignment (see section 3.3).

Considering coils' length, the one-turn configuration maintains inductance within a range that facilitates meeting standards frequency requirements for WPT, established in 79-90 kHz for light-duty vehicles by IEC 61980 [14]. In addition, this configuration enables a simple and cost-effective installation.

Litz wire electric design involves several considerations. Number of strands is determined considering rated current, Joule losses estimation and heat transmission paths involved in the proposed road integration (section 3.7). Strands diameter of 0.1 mm is identified as optimal for minimizing skin effect at the operating frequency range, and bunch design is also considered to minimize internal proximity effect. Additionally, the one-turn design allows reducing proximity effect between wires.

Infrastructure aspects are also taken into account in Litz wire structural characteristics. To enable its integration in asphalt, a 2.5 mm thick silicone jacket is defined, covering the thermal requirements associated with the asphalt pouring process. Silicone also provides desirable flexibility, resulting in a low turn radius that allows keeping design geometry, minimizing the distance between coils and facilitating wire's laying. The distance between coils is set as 15.2 cm (noted as "d" in Fig. 7).

3.2 Resonant network design and analysis

3.3 Inputs definition

Input values for system modelling are based on the following premises:

- Tx coils leakage inductance and resistance is based on design presented in section 2.1.
- Rx side resonant network parameters are associated with INCIT-EV reference system.
- Maximum inverter output voltage (V_1) and maximum induced voltage (V_{m2}) reflect the admissible range of power stages.

Table 1 displays the proposed values for input parameters. Fig. 7 can be used as reference for relative position ranges.

Input parameter	Value	Units
L_1	20.85	μH
L_2	141.39	μH
M_{12}	3.07	μH
R_{L1}	0.033	Ω
R_{L2}	0.08	Ω
C_2	33.5	nF
R_{Ls}	0.03	Ω
V_{m2_max}	750	V
P_{o_max}	30	kW
V_{1DC} range	[700,780]	V
I_{1_lim}	150	A
x range	[-17,7]	m
y range	±0.75	m
airgap	0.242	m

Table 1 Proposed values for modelling input parameters.

Fig. 7 Coordinates system defined for IPT coils.

3.4 Parametric sweeps

After defining the input parameters, Tx coils current set-point is calculated through the iterative script. Then, the execution of the parametric analysis generates a series of graphs to provide an intuitive understanding of both resonant circuits' behavior.

Regarding first defined variation, leakage and mutual inductances matrices considering x, y and z defined ranges are calculated using a COMSOL® model. Figure 8 is part of this analysis; x range corresponds to the first two Tx coils of the track, and y covers the lateral misalignment range.

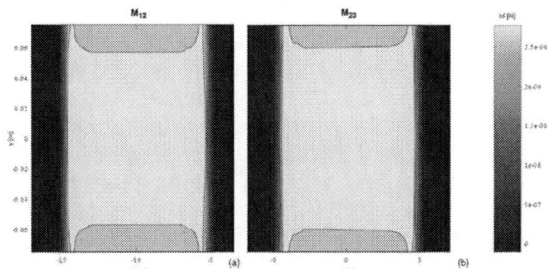

Fig. 8 Mutual inductance between 1st (a) and 2nd Tx coil and Rx coil in function of relative position (x) and lateral misalignment (y).

Apart from relative position variations, components tolerance is introduced in the parametric sweep when calculating circuit magnitudes for both resonant networks. Intermediate magnitudes of Rx circuit, such as induced voltage (V_{m2}), are calculated at working frequency (Fig. 9a). Rx side active power is also represented (Fig. 9b). Induced voltage possible values in x and y ranges are significantly below the limit imposed by Rx side power stages (see table 1). For rated Z_2 and established current at Tx coils, power transfer to Rx side is expected to reach the objective of 30 kW once Rx coil half length enters to Tx coil for the complete lateral misalignment range.

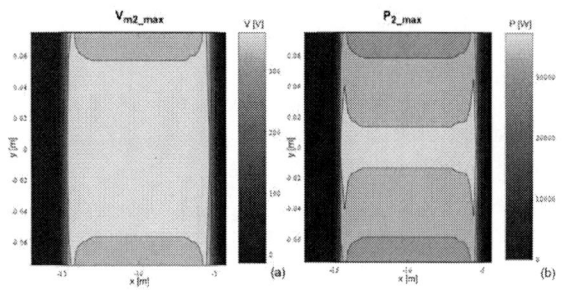

Fig. 9 Rx side induced voltage (a) and transferred power respect to relative position (b).

3.5 Outputs calculation and verification

Resonant network modelling presented in section 2.3 is solved through an iterative script based on a Montecarlo algorithm. Random values are assigned to output parameters, as well as done for input parameters dependent on defined variations. All possible combinations are covered for both data sets. System equations are solved at each iteration, and optimal combination in terms of imposed boundary conditions is returned.

The application of the iterative script to the defined input values generates the values presented in Table 2.

Output parameter	Value	Units
L_s	9	μH
C_s	265	nF
C_p	1.71	μF
I_{m1}	295	A
f_{sw}	72.6	kHz
V_{Ls_max}	595.99	V
V_{Cs_max}	2468.44	V
V_{Cp_max}	409.34	V
I_{1_max}	151.52	A

Table 2 Modelling results.

Output values are confirmed through a verification script. On it, model input parameters take a random value within the defined range on each iteration; 10^6 iterations are carried out. A boxplot per output variable for maximum and zero power transfer is generated. For instance, Fig. 10 represents the potential values of inverter output voltage and current fundamental components for the calculated Tx side compensation network considering the cases of maximum and zero transferred power to the Rx side. The majority of V_1's interquartile range falls within the range defined by upper and lower limits (Ineq. 12,13) for both cases of power transfer. With regard to I_1, the cases that do not guarantee the maximum power transfer correspond to relative positions of low or minimum coupling, such as track entrance and exit. Maximum limit, related to inverter maximum

ratings including a security factor, is not surpassed in the 98% of the cases.

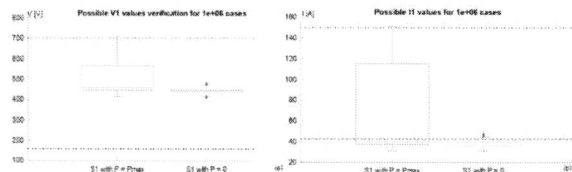

Fig. 10 DC/AC output voltage (a) and current (b) boxplot for maximum and zero output power.

3.6 Control simulation

The primary side IPT control is adjusted for the previously calculated primary side resonant network. Figures 11 and 12 show the variation of the main primary and secondary circuits variables considering the relative movement of the Rx system over two consecutive Tx coils and 0% of lateral misalignment. Primary side control establishes set-point current sequentially at Tx coils according to Rx coil coupling. Primary side inverters output voltage is regulated, and output current varies consequently (Fig. 11). On secondary side, induced voltage and current are kept under maximum ratings, and power control acts to achieve the required power transfer to the battery regulating load voltage (Fig. 12). It is also shown that calculated IPT system maximum efficiency reaches 90%, and the expected power transfer drop in the transition between Tx coils is close to 33%.

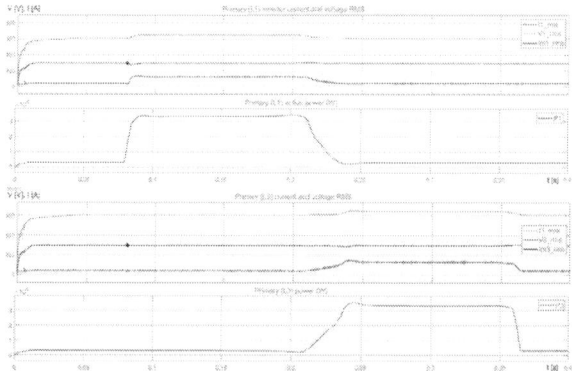

Fig. 11 Primary side main circuit variables evolution under control action.

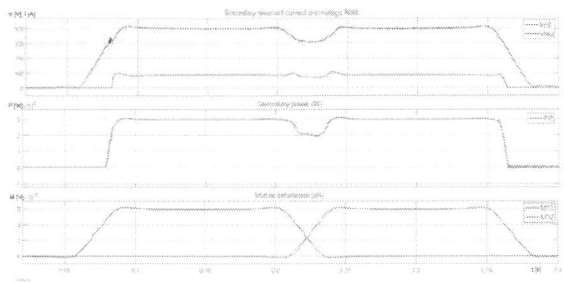

Fig. 12 Secondary side main circuit variables evolution under primary side control action

3.7 Road integration

The integration solution suggested in this proposal follows the recurrent principles of simplification and cost-reduction. The design assumes the installation of the charging track in an existing road. Micro-trenches adapted to Litz wire's diameter are proposed to be dug in the current pavement, providing a simple and effective way of positioning the coils' conductors. Trenches are filled and covered with 2 cm asphalt binder course, followed by a 6 cm wearing course (Fig. 13).

Fig. 13 Proposed road integration concept.

Airgap determined by IPT design counts with coils installation depth, established at 8 cm under the road surface.

3.8 Thermal behaviour

A FEM simulation model is created to estimate the thermal behavior of the proposed integration scheme. Asphalt concrete is considered to be surrounding the Litz wire; conduction thought asphalt and convection at wearing course surface are considered as heat transfer mechanisms. On the other hand, power losses of 180W/m are taken into consideration as heat source.

Simulations examine two scenarios in terms of ambient temperature: 15ºC and 40ºC. Continuous operation is assumed, which reflects a worse situation than the discontinuous operation regime associated with this application. Figure 14 illustrates the temperature evolution of wearing course accessible surface for operation periods of 1, 3 and 12h. It is observed that, for continuous operation periods shorter than 1 hour, temperature does not reach asphalt service threshold, established in 60 ºC, at any case.

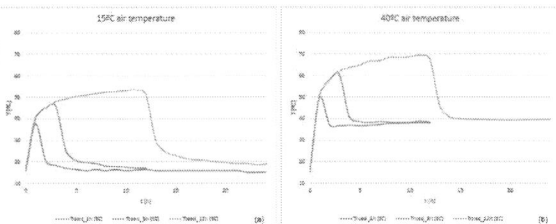

Fig. 14 Road surface temperature for 15ºC (a) and 40ºC air temperature conditions

3.9 EM emissions

Limits set by ICNIRP guidelines in terms of exposure to varying electrical and magnetic fields at working frequency range, covered by 2010 version, are considered in this proposal. A limit of 27uT is established for general public exposure to magnetic fields in the range of 3kHz – 100MHz [14] As described in section 3.1, shielding is provided by the EV in the presented design. Considering IPT system magnetic field distribution in the range of lateral misalignment, EV's users' exposure is maximum at centred position but remains below 23 uT and therefore below ICNIRP limit.

The most critical situation in the system arises along the active coil during a charging process in sections not covered by an EV. The system must be installed in extra-urban roads and in dedicated tracks, avoiding access to pedestrians and non-shielded EVs. Although not being covered by this design in the context of INCIT-EV, FOD safety measurements must be considered for the implementation of this proposal.

4 Conclusions

A versatile design method for dynamic wireless charging is presented. The proposal is oriented to extra-urban areas and roads with restricted access and applied to a particular case within the framework of INCIT-EV project. Its main advantages are its simplicity in terms of charging track structure and civil works requirements, as well as the significantly low number of electronic stages per track meter required. The prototype is being manufactured and currently under validation as demonstrator of INCIT-EV project (https://www.incit-ev.eu/), funded by the European Union (H2020-LC-GV-2019).

5 Acknowledgement

The authors acknowledge the contribution of Universidad de Zaragoza to Tx coils design, Vedecom for the design and supply of the project reference secondary system, as well as the rest of project formal partners that contributed to UC3's design. The research has been carried out in the framework of the IN-CIT-EV project (https://www.incit-ev.eu/), funded by the European Union under the Horizon 2020 Programme (H2020-LC-GV-2019).

References

[1] A. A. S. Mohamed, A. A. Shaier, H. Metwally, and S. I. Selem, "An Overview of Dynamic Inductive Charging for Electric Vehicles," *Energies (Basel)*, vol. 15, no. 15, Aug. 2022, doi: 10.3390/en15155613.

[2] G. Duarte, A. Silva, and P. Baptista, "Assessment of wireless charging impacts based on real-world driving patterns: Case study in Lisbon, Portugal," *Sustain Cities Soc*, vol. 71, p. 102952, Aug. 2021, doi: 10.1016/J.SCS.2021.102952.

[3] Lingxiao Xue *et al.*, "Design and Analysis of a 200 kW Dynamic Wireless Charging System for Electric Vehicles," Applied Power Electronics Conference, 2022. doi: https://doi.org/10.1109/APEC43599.2022.9773670.

[4] Y. S, N. R, J. Sathik Mohamed Ali, and D. Almakhles, "A Comprehensive Review of the On-Road Wireless Charging System for E-Mobility Applications," *Front Energy Res*, vol. 10, 2022, doi: 10.3389/fenrg.2022.926270.

[5] R. Tavakoli, E. M. Dede, C. Chou, and Z. Pantic, "Cost-Efficiency Optimization of Ground Assemblies for Dynamic Wireless Charging of Electric Vehicles," *IEEE Transactions on Transportation Electrification*, vol. 8, no. 1, pp. 734–751, Mar. 2022, doi: 10.1109/TTE.2021.3105573.

[6] E. K. E. I. Ahmed, "Effects of Ferromagnetic Cores in Wireless Power Transfer System for Charging Electric Vehicles," *Journal of Karary University for Engineering and Science*, 2022, [Online]. Available: https://api.semanticscholar.org/CorpusID:252932632

[7] X. Mou, D. T. Gladwin, R. Zhao, and H. Sun, "Survey on magnetic resonant coupling wireless power transfer technology for electric vehicle charging," *IET Power Electronics*, vol. 12, no. 12. Institution of Engineering and Technology, pp. 3005–3020, Oct. 16, 2019. doi: 10.1049/iet-pel.2019.0529.

[8] K. Wang, Z. Zuo, L. Sang, and X. Zhu, "Comprehensive Analysis for Electromagnetic Shielding Method Based on Mesh Aluminium Plate for Electric Vehicle Wireless Charging Systems," *Energies (Basel)*, 2022, [Online]. Available: https://api.semanticscholar.org/CorpusID:247041001

[9] L. Zhao, S. Ruddell, D. J. Thrimawithana, U. K. Madawala, and P. A. Hu, "A hybrid wireless charging system with DDQ pads for dynamic charging of EVs," *2017 IEEE PELS Workshop on Emerging Technologies: Wireless Power Transfer (WoW)*, pp. 1–6, 2017, [Online]. Available: https://api.semanticscholar.org/CorpusID:40221482

[10] M. Amjad, M. Farooq-i-Azam, Q. Ni, M. Dong, and E. A. Ansari, "Wireless charging systems for electric vehicles," *Renewable and Sustainable Energy Reviews*, vol. 167. Elsevier Ltd, Oct. 01, 2022. doi: 10.1016/j.rser.2022.112730.

[11] A. Azad, V. Kulyukin, and Z. Pantic, "Misalignment Tolerant DWPT Charger for EV Roadways with Integrated Foreign Object Detection and Driver Feedback System," in *2019 IEEE Transportation Electrification Conference and Expo (ITEC)*, 2019, pp. 1–5. doi: 10.1109/ITEC.2019.8790600.

[12] A. Mahesh, B. Chokkalingam, and L. Mihet-Popa, "Inductive Wireless Power Transfer Charging for Electric Vehicles-A Review," *IEEE Access*, vol. 9, pp. 137667–137713, 2021, doi: 10.1109/ACCESS.2021.3116678.

[13] G. Palani, U. Sengamalai, P. Vishnuram, and B. Nastasi, "Challenges and Barriers of Wireless Charging Technologies for Electric Vehicles," *Energies*, vol. 16, no. 5. MDPI, Mar. 01, 2023. doi: 10.3390/en16052138.

[14] A. Sagar *et al.*, "A Comprehensive Review of the Recent Development of Wireless Power Transfer Technologies for Electric Vehicle Charging Systems," *IEEE Access*, vol. 11. Institute of Electrical and Electronics Engineers Inc., pp. 83703–83751, 2023. doi: 10.1109/ACCESS.2023.3300475.

[15] K. Song *et al.*, "A Review on Interoperability of Wireless Charging Systems for Electric Vehicles," *Energies (Basel)*, vol. 16, p. 1653, Feb. 2023, doi: 10.3390/en16041653.

[16] B. Mazhoud *et al.*, "Pavement integration of an inductive charging system for electric vehicles. Results of the INCIT-EV project," *Transportation Engineering*, vol. 10, Dec. 2022, doi: 10.1016/J.TRENG.2022.100147.

Bidirectional Isolated 400-12V DC-DC Converter With Improved Power Density and Full-Range Operation for EV Applications

Héctor Sarnago[1], Ignacio Álvarez[1], and Oscar Lucía[1]

[1] Electronics Engineering and Communications Department. Universidad de Zaragoza, I3A, Spain.

Corresponding author: Óscar Lucía, olucia@unizar.es
Speaker: Óscar Lucía, olucia@unizar.es

Abstract

Efficient and high-performance dc-dc conversion is essential in modern electric vehicles to power key essential systems and allow for regenerative micro-grid operation. Usually, operating under a wide range of operating conditions, i.e. input/output voltage, is a challenging task, degrading the converter performance. In this paper, a bidirectional isolated 400-12 V dc-dc converter with improved power density and full-range operation for EV applications is proposed. The proposed novel topology significantly improves continuous-mode operation in the full input/output voltages range and improves power density thanks to the advantageous inductor and transformer implementation. The proposed 400-12 V converter has been designed and experimentally tested with a 100-A output current prototype.

1 Introduction

Transition towards electric transportation is a key step forward to achieve the Paris agreement and some of the UN Sustainable Development Goals [1, 2]. To achieve such aims, transportation electrification is a key pillar for sustainability. Consequently, electric vehicle (EV) technology development and adoption has grown exponentially in recent years. Nowadays, EV includes a number of key power converters [3], including inverters [4], on-board chargers [5-8] and dc-dc converters [9, 10] to manage the complete vehicle powertrain (Fig. 1).

Dc-dc conversion in EVs has significantly advanced in recent years to incorporate new function and higher performance and power densities. First designs where conceived to serve as a battery charger for the low-voltage battery. Later, dc-dc converters where also conceived to supply all the vehicle low-voltage subsystems and event o interact with local power sources such as photovoltaic modules or regenerative suspensions. The most aggressive design trends, consider even removing the low voltage battery, being the EV dc-dc converter in charge of maintaining the whole low-voltage nano-grid. In this context, the development of higher-performance and power density dc-dc converters has become essential for EVs development.

Classical dc-dc converter implementation includes the use of isolated resonant dc-dc converters [11, 12] or dual-active bridge [10, 13, 14] implementations. However, these designs suffer from poor power density and control issues when pushing the converter operation range boundaries under extreme voltage range conditions.

This paper proposes a novel integrated buck-boost dc-dc topology to achieve improved power device ratings due to the optimal usage of both the transformer and the SR devices by using a fixed-duty-cycle control, and consequently, improved power density due to the reduction of capacitive an inductive elements. The proposed converter features an advantageous magnetic component device placing and implementation to improve the power density. The proposed converter is able to operate in full-performance continuous operation in the 250-470 V input-voltage range and 8-16 V output voltage range with sustained 100 A output current.

The remainder of this paper is organized as follows. Section II details the proposed power converter, comparing it with state-of-the-art implementations. Section III shows the implemented experimental prototype and the main experimental results. Finally, Section IV summarizes the conclusions of this paper.

PCIM Europe 2024, 11– 13 June 2024, Nuremberg   DOI: 10.30420/566262352

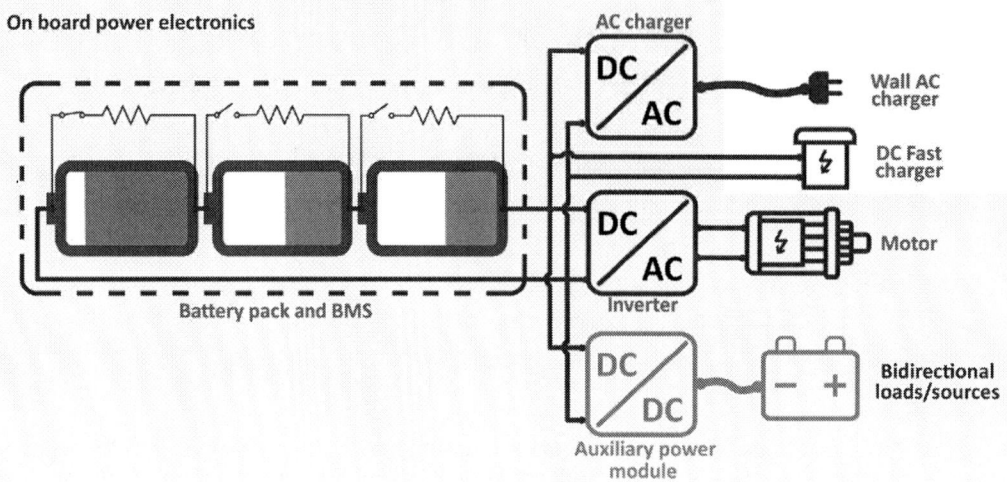

Fig. 1   EV power electronic architecture.

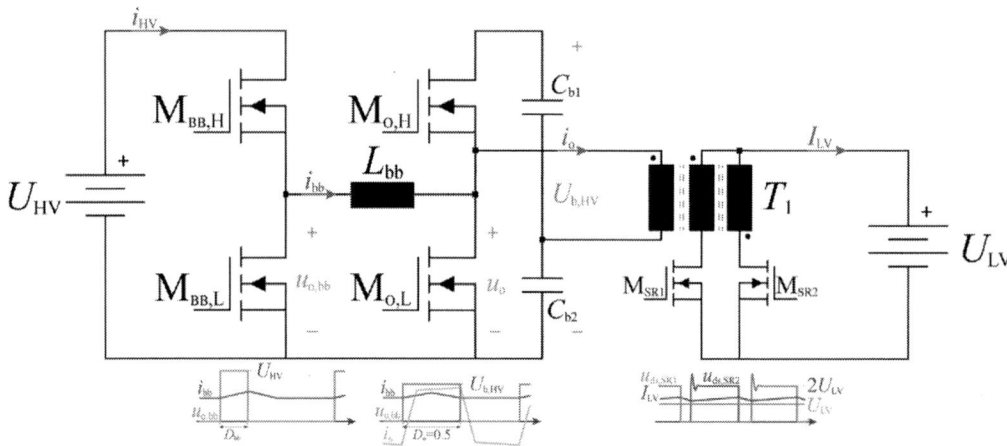

Fig. 2   Proposed converter and main waveforms.

2   Proposed power converter

The proposed converter (Fig. 2) features a full-bridge converter in the HV-side, where the left side is directly connected to the HV battery, while a flying voltage, $U_{b,HV}$, is established in the right side. The converter is completed by a DC transformer (DCX) structure, consisting of the dc-blocking capacitors, C_{b1}, C_{b2}, a transformer, T_1, and two synchronous rectifier FETS, M_{SR1} and M_{SR2}. As a result, a fixed duty cycle, D_o=0.5, is employed in v_o, while a variable duty cycle (Fig. 3) is used in synthesizing $v_{o,bb}$, to accommodate broader voltages variations of both the input and output voltages of the converter. Consequently, optimal utilization of the transformer is achieved, and, in contrast to conventional SR FETs, the average peak voltage is limited to $2U_{LV}$, enabling the use of low-voltage-rated devices. Furthermore, this approach offers

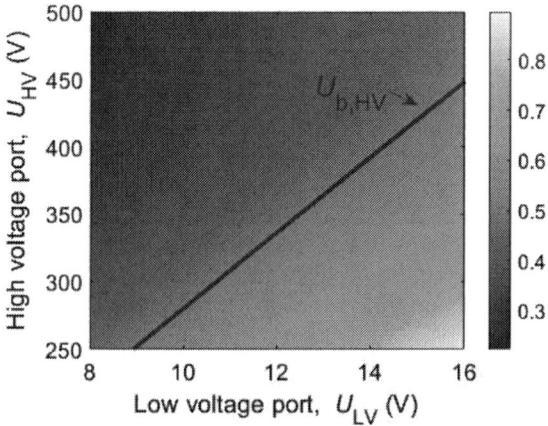

Fig. 3   Duty cycle map as a function of input and output voltages (b). Flying voltage, $U_{b,HV}$, is also included.

2514

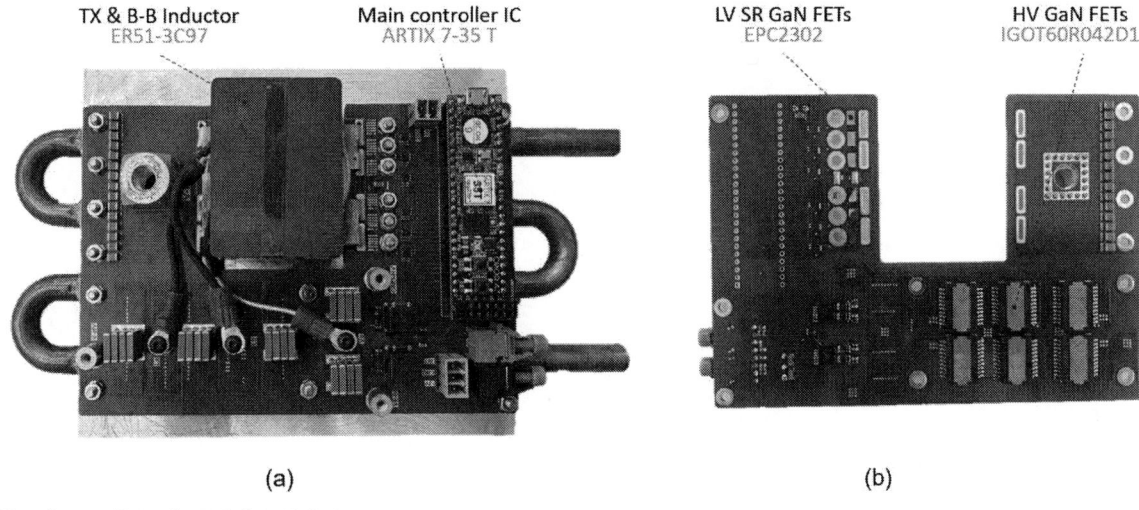

(a)   (b)

Fig. 4   Experimental prototype.

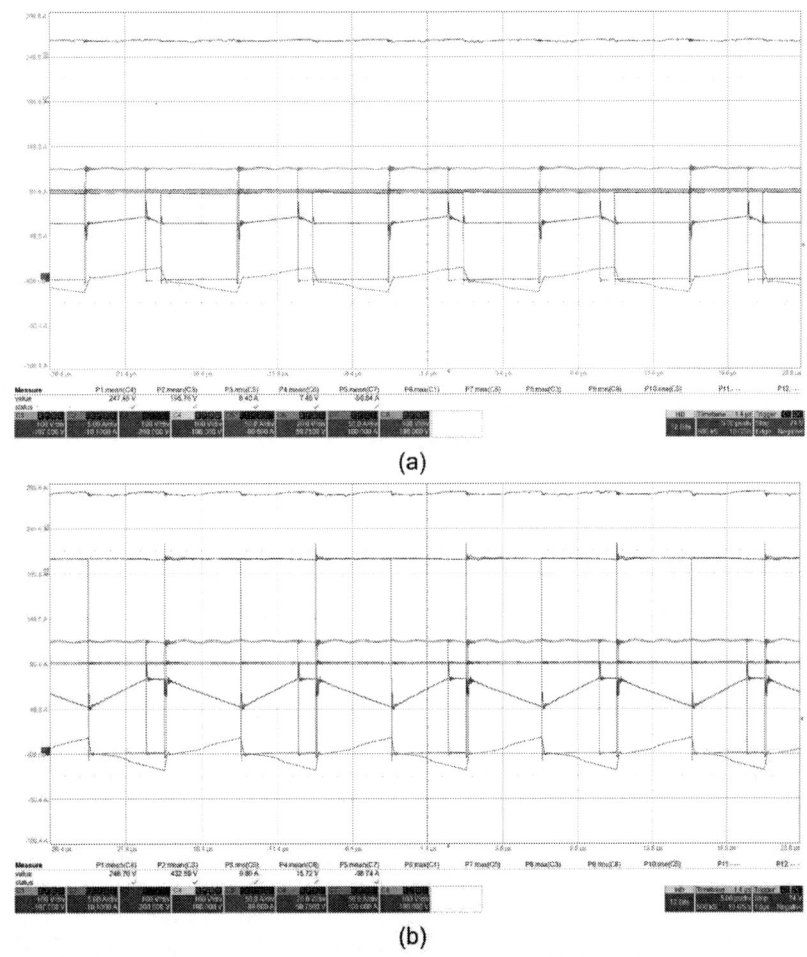

Fig. 5   Main experimental results at maximum output current (100 A): 250 V to 8 V (a), 250 V to 16 V (b). From top to bottom; LV-side voltage CH3, HV-side voltage CH4, flying bus voltage, $U_{b,HV}$, CH3, load current CH7, transformer current, i_o, CH5, output voltage, $u_{o,BB}$, CH1, and output voltage, u_o, CH8.

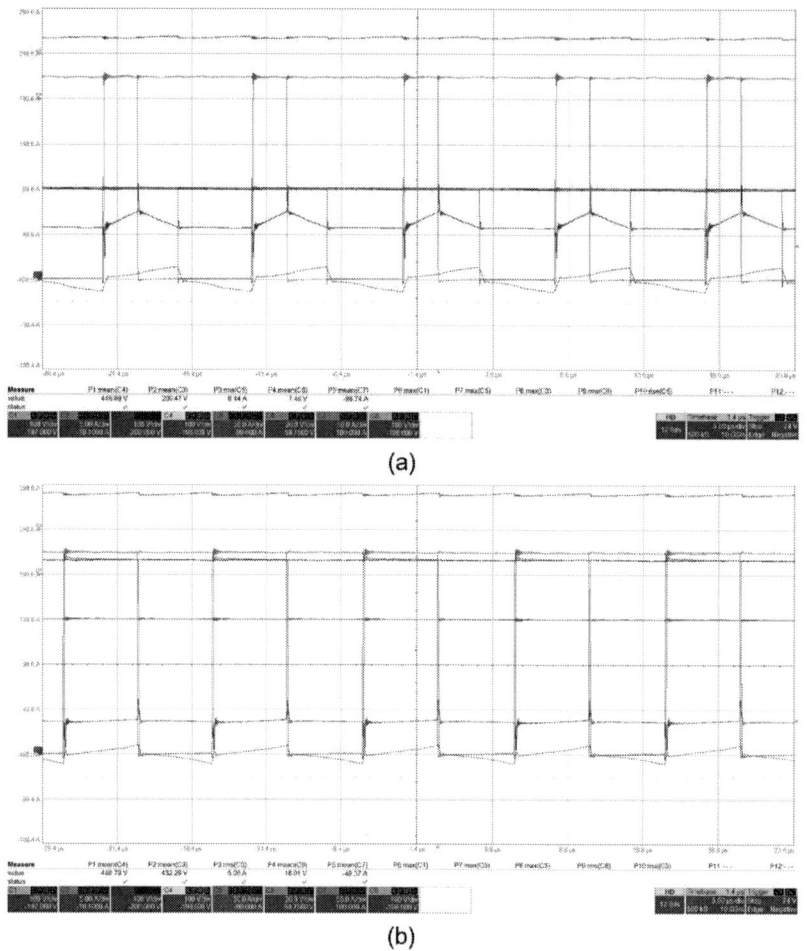

(a)

(b)

Fig. 6 Main experimental results at maximum output current (100 A): 450 V to 8 V (a) and. 450 V to 16 V (b). From top to bottom; LV-side voltage CH3, HV-side voltage CH4, flying bus voltage, $U_{b,HV}$, CH3, load current CH7, transformer current, i_o, CH5, output voltage, $u_{o,BB}$, CH1, and output voltage, u_o, CH8.

additional advantages compared to LLC or DAB implementations, such as fixed-frequency operation, a straightforward control scheme based on duty cycle, and minimal recirculating current.

3 Implementation and experimental results

In order to test the proposed topology and control strategy a 100-A output current 400-12 V dc-dc converter prototype has been designed and implemented. The implemented prototype is detailed in Fig. 4. It features GaN FETs IGOT60R042D1 devices in the HV-side full-bridge, whereas the synchronous rectifier FETs are composed of 3xEPC2302 GaN devices from EPC. Besides, the magnetic components have been implemented using an advantageous combined assembly to improve the power density, including in the same

component both L_{bb} inductor and T_1 transformer, using an ER51 core made of 3C97 material.

Finally, Fig. 5 and Fig. 6 shows the main experimental results at maximum output current (100 A) for the corner voltages of the converter. These results prove the feasibility of the proposed topology.

4 Conclusions

This paper has proposed a novel bidirectional isolated 400-12 V dc-dc converter with improved power density and full-range operation for EV applications. The proposed converter significantly improves continuous-mode operation in the full input/output voltages range and improves power density thanks to the advantageous inductor and transformer implementation. The proposed converter has been tested using a 100-A output current prototype, proving the feasibility of this proposal.

Acknowledgment

This work was partly supported by Projects TED2021-129274B-I00, CNS2023-144980, and PDC2023-145837-I00 co-funded by MICIU/AEI/10.13039/501100011033, by "ERDF A way of making Europe", by the "European Union NextGenerationEU/PRTR", and by the DGA-FSE.

References

[1] WEC, "A vision for a sustainable battery value chain in 2030. Unlocking the full potential to power sustainable development and climate change mitigation.," World Economic Forum. Global Battery Alliance., 2019.

[2] *Transforming our world : the 2030 Agenda for Sustainable Development,* U. G. Assembly, 2015.

[3] M. Yilmaz, and P. T. Krein, "Review of Battery Charger Topologies, Charging Power Levels, and Infrastructure for Plug-In Electric and Hybrid Vehicles," *IEEE Transactions on Power Electronics,* vol. 28, no. 5, pp. 2151-2169, 2013, doi: 10.1109/TPEL.2012.2212917.

[4] S. S. Williamson, A. K. Rathore, and F. Musavi, "Industrial Electronics for Electric Transportation: Current State-of-the-Art and Future Challenges," *IEEE Transactions on Industrial Electronics,* vol. 62, no. 5, pp. 3021-3032, 2015, doi: 10.1109/TIE.2015.2409052.

[5] A. Khaligh, and M. D. Antonio, "Global Trends in High-Power On-Board Chargers for Electric Vehicles," *IEEE Transactions on Vehicular Technology,* vol. 68, no. 4, pp. 3306-3324, 2019, doi: 10.1109/TVT.2019.2897050.

[6] H. L. Sarnago, O., "High Power Density On-Board Charger Featuring Power Pulsating Buffer," *IEEE Open Journal of Power Electronics,* vol. 5, pp. 162-170, 2024, doi: 10.1109/OJPEL.2024.3359271.

[7] H. Sarnago, O. Lucía, S. Chhawchharia, D. Menzi, and J. W. Kolar, "Novel Bidirectional Universal 1-Phase/3-Phase-Input Unity Power Factor Differential AC/DC Converter," *Electronics Letters,* vol. 59, no. 13, pp. 1-4, 2023, doi: https://doi.org/10.1049/ell2.12857.

[8] H. Sarnago, and O. Lucía, "Optimized EV ON-Board Charging Power Converter Using Hybrid DCX-Dab Topology," in *IEEE Applied Power Electronics Conference and Exposition*, 2024, pp. 1305-1309.

[9] P. He, and A. Khaligh, "Comprehensive Analyses and Comparison of 1 kW Isolated DC–DC Converters for Bidirectional EV Charging Systems," *IEEE Transactions on Transportation Electrification,* vol. 3, no. 1, pp. 147-156, 2017, doi: 10.1109/TTE.2016.2630927.

[10] H. Sarnago, and O. Lucía, "Bidirectional 400-12 V dc-dc Converter with Improved Dynamics and Integrated Transformer for EV Applications," in *IEEE Applied Power Electronics Conference and Exposition*, 2024, pp. 3081-3085.

[11] S. Zhao, A. Kempitiya, W. T. Chou, V. Palija, and C. Bonfiglio, "Variable DC-Link Voltage LLC Resonant DC/DC Converter With Wide Bandgap Power Devices," *IEEE Transactions on Industry Applications,* vol. 58, no. 3, pp. 2965-2977, 2022, doi: 10.1109/TIA.2022.3151867.

[12] X. Zhou *et al.*, "A High-Efficiency High-Power-Density On-Board Low-Voltage DC–DC Converter for Electric Vehicles Application," *IEEE Transactions on Power Electronics,* vol. 36, no. 11, pp. 12781-12794, 2021, doi: 10.1109/TPEL.2021.3076773.

[13] J. Tian, F. Wang, F. Zhuo, X. Cui, and D. Yang, "An Optimal Primary-Side Duty Modulation Scheme With Minimum Peak-to-Peak Current Stress for DAB-Based EV Applications," *IEEE Transactions on Industrial Electronics,* vol. 70, no. 7, pp. 6798-6808, 2023, doi: 10.1109/TIE.2022.3206698.

[14] I. Kougioulis, A. Pal, P. Wheeler, and M. R. Ahmed, "An Isolated Multiport DC–DC Converter for Integrated Electric Vehicle On-Board Charger," *IEEE Journal of Emerging and Selected Topics in Power Electronics,* vol. 11, no. 4, pp. 4178-4198, 2023, doi: 10.1109/JESTPE.2023.3276048.

PCIM Europe 2024, 11– 13 June 2024, Nuremberg   DOI: 10.30420/566262353

Gain optimization control method for CLLLC resonant converters under phase shift mode

Will Tai[1], Guangzhi Cui[2], Sheng-Yang Yu[3]

[1,2]Texas Instruments, China, [3]Texas Instruments, USA.

Corresponding author: Sheng-Yang Yu, seanyu@ti.com
Speaker: Sheng-Yang Yu, seanyu@ti.com

Abstract

Bi-directional capacitor-inductor-inductor-inductor-capacitor (CLLLC) resonant converter is widely used in applications such as energy storage systems and portable power stations to efficiently charge/discharge batteries. However, an efficient CLLLC resonant converter has a challenge of covering a wide battery voltage range with various load conditions. Especially at light load, output voltage tends to rise due to parasitic capacitance and could eventually go out of regulation. In this paper, phase shift control is introduced to main output voltage regulation at light load. In addition, a novel synchronous rectifier control method is proposed in this abstract to eliminate the nonlinearity effect caused by parasitic capacitance.

1   CLLLC design consideration

1.1   The basic working principle of the CLLLC converter

Bi-directional CLLLC resonant converter, shown in Fig. 1, is mainly used to charge and discharge batteries in energy storage systems (ESS). The magnetizing inductance of the transformer in the figure is L_m. L_{r1} and L_{r2} are the resonant inductances of the primary and secondary sides, respectively. L_{r1} and L_{r2} are the combined inductances of the transformer leakage inductance with discrete series inductance on each side, respectively. C_{r1} and C_{r2} are the resonant capacitors (and also serve as DC blocking capacitors) of the primary and secondary sides. $D_{S1} \sim D_{S8}$ are body diodes, and $C_{S1} \sim C_{S8}$ are junction capacitors of 8 switches.

Compared to conventional LLC resonant converters, bidirectional symmetric CLLLC resonant converters add an additional resonant inductor L_{r2} and a capacitor C_{r2} on the secondary side to make the structure and gain symmetrical. Detailed CLLLC resonant converter operation can be found in [1].

According to the circuit of the CLLLC resonant converter, there are two resonant frequencies in the CLLLC converter. Taking CLLLC forward operation as an example, one is the resonant frequency when L_{r1} and C_{r1}, L_{r2} and C_{r2} are resonant, which is called the first resonant frequency, and the other resonant frequency is the frequency

when L_{r1}, C_{r1} and L_m are resonant, which is called the second resonant frequency. The two resonant frequencies are as follows:

$$f_{r1} = 1/(2\pi\sqrt{L_{r1}C_{r1}}) \tag{1}$$

$$f_{r2} = 1/(2\pi\sqrt{(L_{r1}+L_m)C_{r1}}) \tag{2}$$

The traditional control method of CLLLC resonant converters is frequency control, that is, changing the frequency to obtain different gains. The traditional gain analysis method for resonant converters is first harmonic approximation (FHA)[2], Fig.2 is CLLLC resonant converter FHA equivalent circuit diagram without considering parasitic capacitance. As shown in the Fig. 3, the curve of frequency and gain in the ideal state is shown, where Q represents the quality factor and f_n represents the ratio of the switching frequency f_s and the resonant frequency f_{r1},

$$Q = \sqrt{L_{r1}C_{r1}}/R_{eq} \tag{3}$$

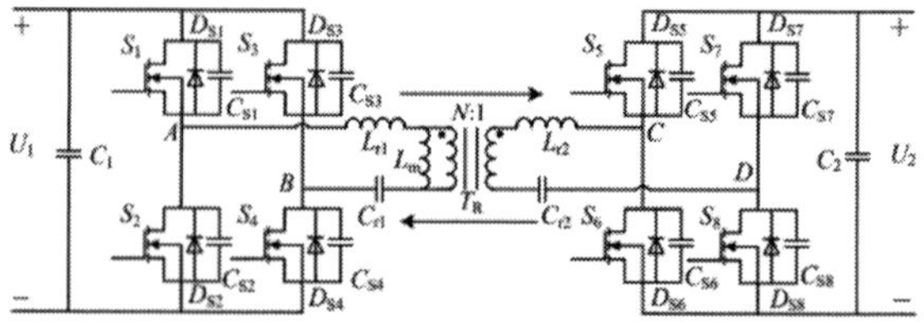

Fig. 1   CLLLC resonant converter topology

2518

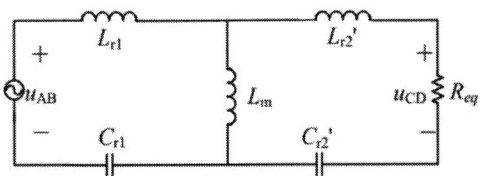

Fig. 2 CLLLC resonant converter FHA equivalent circuit diagram

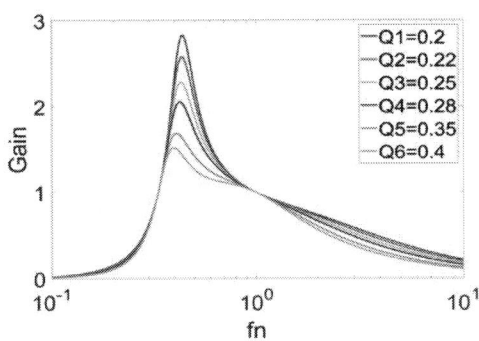

Fig. 3 CLLLC converter gain curve (Ideally)

1.2 Gain regulation problems in light load

Fig. 6 shows parasitic capacitance distribution of CLLLC resonant converter, this chapter will analyze the effect of parasitic capacitance of the transformer on the gain.

The parasitic parameters of a transformer mainly include parasitic capacitance and leakage inductance, and the equivalent circuit containing the parasitic parameters is shown in Fig. 7. In the figure, L_{l1} represents the primary side leakage inductance of the transformer, L_{l2} represents the secondary side leakage inductance, L_m is the magnetizing inductance of the transformer, C_p and C_s represent the inter-turn distributed capacitance of the primary and secondary sides, respectively, and C_{ps} can be expressed as the parasitic capacitance generated by the primary and secondary sides.

Since the bidirectional DCDC converter is used to charge and discharge the battery, and when the battery voltage is low, the battery charger is operating in a pre-charge state with low charging current to prolong battery life. Also, when the battery is nearly full charged, the charging current will become very small. In both conditions the system enters a light load.

Due to the wide input and output voltage range in ESS, we often ensure the switching frequency (f_s) at heavy load is around resonant frequency. Also, the inductance ratio of magnetizing inductor (L_m) and the series inductor (L_s) is generally set to be large for efficiency optimization. However, a large inductance ratio might result in output voltage out of regulation under light load conditions using variable frequency control.

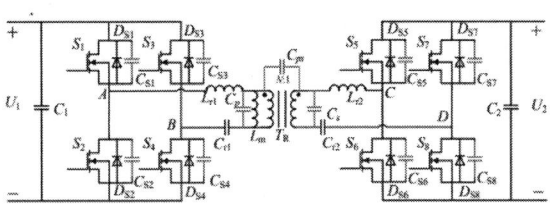

Fig. 6 Parasitic capacitance distribution of CLLLC resonant converter

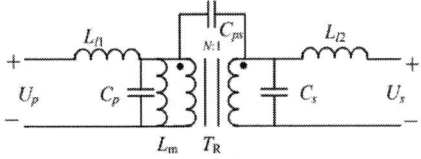

Fig. 7 High frequency transformer equivalent circuit

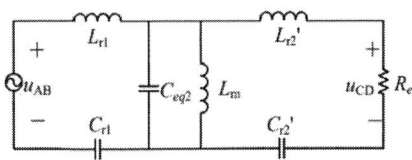

Fig. 8 Consider the distributed capacitance transformer equivalent circuit

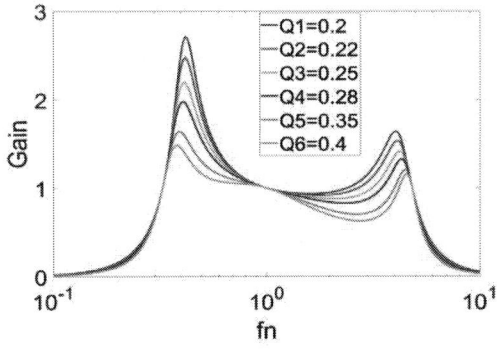

Fig. 9 CLLLC converter gain curve

2519

Moreover, the parasitic capacitance C_{ps} on the transformer could have a big effect on the gain of converter[3]. Fig. 8 shows the equivalent circuit considering the distributed capacitance on the transformer, where C_{eq2} is the distributed capacitance of the primary and secondary sides of the transformer equivalent to the primary side. After adopting FHA, we can get the curve of frequency and gain considering the parasitic capacitance, as shown in Fig. 9.

From this figure, we can see the parasitic capacitance of the transformer will lead to negative input/output voltage gain slope when $f_s > f_r$ – meaning we are not able to simply reduce gain by increasing f_s. In order to regulate output voltage under light load conditions, phase-shift control is introduced here.

Therefore, for the CLLLC resonant converter, the control scheme proposed in this paper is a phase-shift and varying-frequency hybrid control scheme. And in order to achieve a seamless switch, the control block diagram is shown in the Fig. 10.

The system uses the loop output to select either frequency modulation or phase shift mode, When the load is light and the gain needs to be reduced, the frequency rises and enters phase shift mode when reach the maximum frequency.

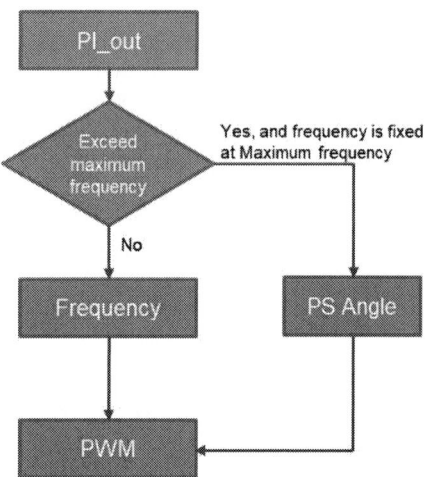

Fig. 10   The control block diagram in phase shift control

2   Phase-shift control in CLLLC converter

2.1   Phase shift basic operation

When a CLLLC converter operates in phase-shift control, f_s is fixed and the output voltage can be regulated by changing the phase shift angle φ between the two bridge legs[4]. The ideal waveform (without considering parasitic capacitance) of a phase-shift controlled CLLLC converter is shown in Fig. 11. By increasing φ, we are able to reduce the effective duty cycle and reduce input/output voltage gain, the relationship between the two is linear, as shown in Fig. 12.

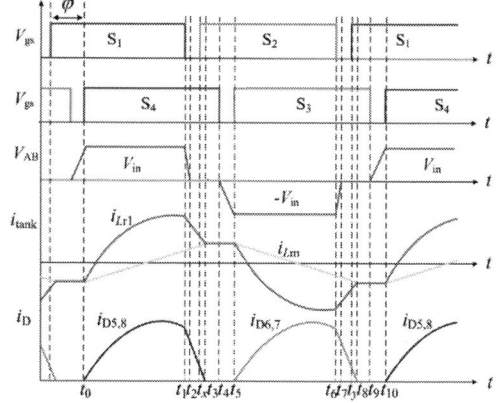

Fig. 11   Ideal waveforms under Phase Shift Mode($f_s > f_r$)

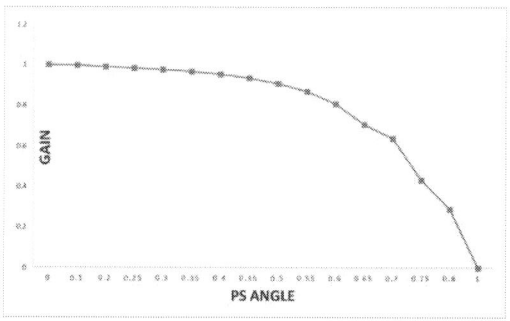

Fig. 12   Gain curve under Phase Shift Mode ideally

2.2   Problems caused by parasitic capacitance

Fig. 11 shows the ideal waveforms under Phase Shift mode, however if we consider parasitic capacitance such as the MOSFETs output capacitance (COSS) shown in Fig. 6, the tank current will oscillate with the output capacitors as shown in Fig. 13 compared to Fig. 14, and Fig. 15 is Experimental waveforms with C_{oss} under phase shift mode.

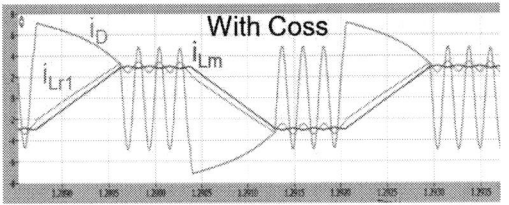

Fig. 13 Simulation waveforms with C_{OSS} under Phase shift mode (and open loop)

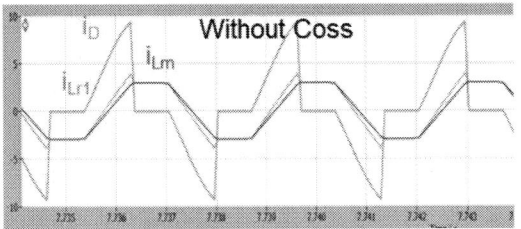

Fig. 14 Simulation waveforms without C_{OSS} under Phase shift mode (and open loop)

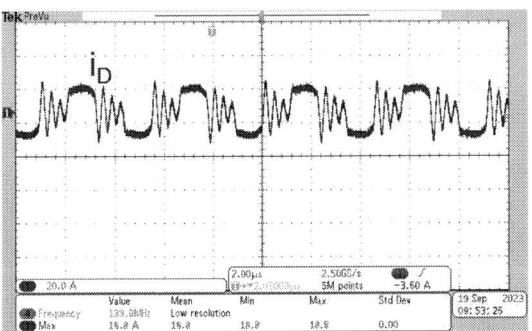

Fig. 15 Experimental waveforms with C_{OSS} under Phase shift mode (and open loop)

As the current initial condition at S1 and S2 turn off transients could be very different due to the oscillation current even with small φ difference. Taking Fig. 16 and Fig. 17 as an example, Ideally, a phase shift angle of 0.4 should transfer less energy than a phase shift of 0.35. But due to the oscillation current, initial current when phase shift angle is 0.4 is higher than it when phase shift angle is 0.35, which means that a phase shift angle of 0.4 will transfer more energy than a phase shift of 0.35.

In this condition, increasing φ is no longer necessary reducing the input/output voltage gain. A gain comparison of a CLLLC converter with and without considering MOSFET C_{oss} are plotted in Fig. 18. Fluctuation of the gain curve can be observed on the curve by considering MOSFET C_{oss} in the model. Therefore, φ might be adjusted to a wrong direction under closed loop control and result in large current spikes as shown in Fig. 19.

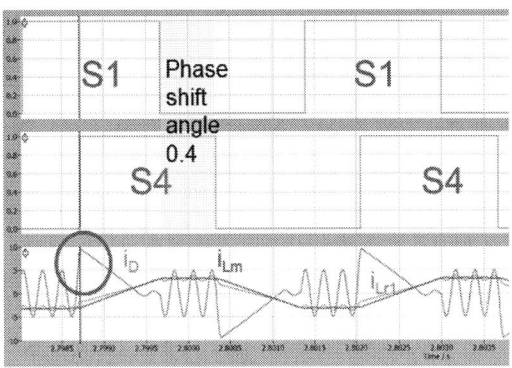

Fig. 16 Simulation waveforms when phase shift angle equals to 0.4

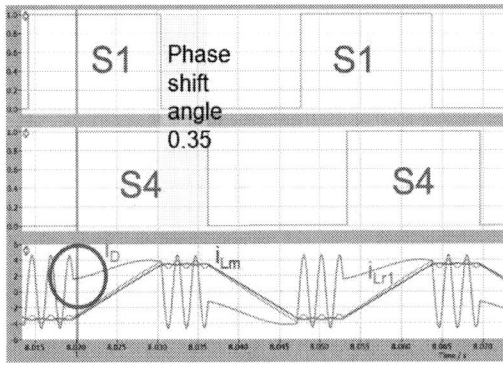

Fig. 17 Simulation waveforms when phase shift angle equals to 0.35

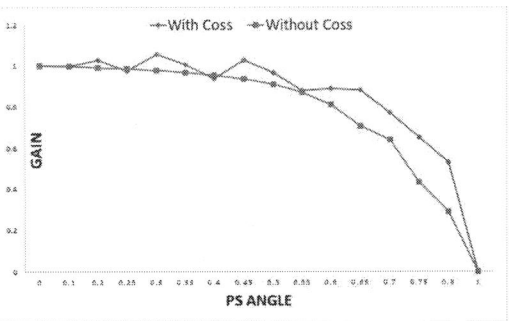

Fig. 18 Gain curve under Phase Shift Mode with and without C_{OSS}

Fig. 20 Waveforms with proposed SR control scheme

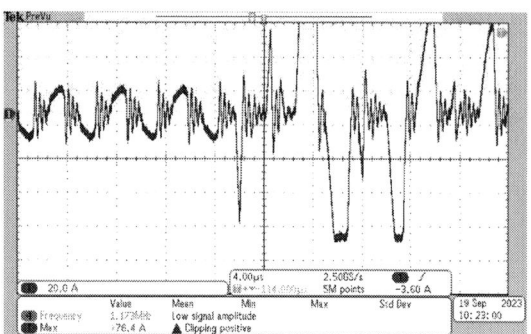

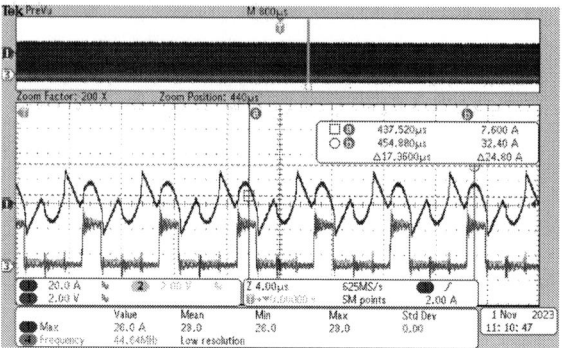

Fig. 19 Current spike under phase shift mode (and closed loop)

Fig. 21 Experimental waveforms with C_{OSS} under Phase shift mode (and close loop)

3 Solution for gain problems

3.1 SR control scheme

In order to address the current oscillation issue and to ensure a monotonic gain change, a synchronous rectifier (SR) control scheme is shown in Fig. 20. By ensuring either two upper or two lower SR switches to be turned on at the same time during the current oscillation period, the transformer secondary side winding is temporarily shorted and is not able to resonate with SR MOSFET C_{OSS}.

3.2 Test results

After adopting this method, the current waveforms become normal, as shown in Fig. 21. And by doing so, we are able to smooth the input/output voltage gain curve as shown in Fig. 22.

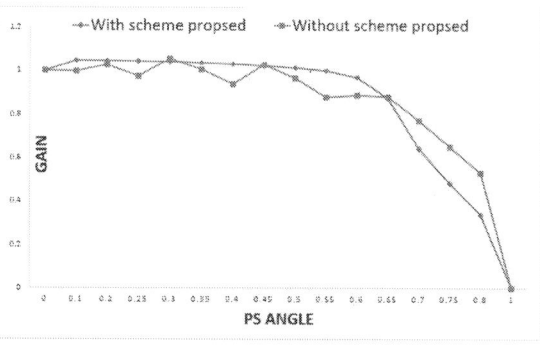

Fig. 22 Gain curve under Phase Shift Mode with and without scheme proposed

4 Summary

CLLLC resonant converter as a popular topology in ESS applications as it can offer soft-switching, high power density, and high efficiency. In order to maintain high converter efficiency and gain regulation ability, we often need to introduce control methods in addition to variable frequency control. A commonly used one is phase-shift control. As depicted in this paper, phase-shift control introduces current oscillation issue which becomes severe in the designs that have large C_{OSS}. A SR control method is proposed in this paper to address the current oscillation issue.

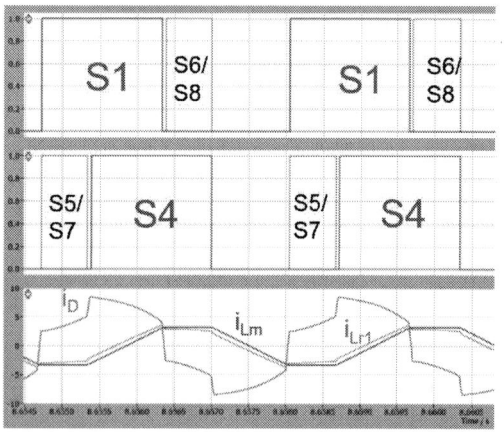

References

[1] J. -H. Jung, H. -S. Kim, M. -H. Ryu and J. -W. Baek, "Design Methodology of Bidirectional CLLC Resonant Converter for High-Frequency Isolation of DC Distribution Systems," in IEEE Transactions on Power Electronics, vol. 28, no. 4, pp. 1741-1755, April 2013

[2] K. Li et al., "Modeling and Hybrid Controller Design of CLLLC," 2019 IEEE 10th International Symposium on Power Electronics for Distributed Generation Systems (PEDG), 2019, pp. 168-172

[3] B. Lee, M. Kim, C. Kim, K. Park and G. Moon, "Analysis of LLC Resonant Converter considering effects of parasitic components," INTELEC 2009 - 31st International Telecommunications Energy Conference, Incheon, Korea (South), 2009, pp. 1-6.

[4] A. Safaee, M. Karimi-Ghartemani, P. K. Jain and A. Bakhshai, "Time-Domain Analysis of a Phase-Shift-Modulated Series Resonant Converter with an Adaptive Passive Auxiliary Circuit," in IEEE Transactions on Power Electronics, vol. 31, no. 11, pp. 7714-7734, Nov. 2016.

PCIM Europe 2024, 11– 13 June 2024, Nuremberg DOI: 10.30420/566262354

Analysis of Common and Split DC-Bus Interleaved H-Bridge Converters for High-Current Low-Ripple Applications

Bhavana Gudala [1,2], Riccardo Mandrioli [1], Vincenzo Cirimele [1], Gaetano Longo [2], Mattia Ricco [1]

[1] Department of Electrical, Electronic and Information Engineering, University of Bologna, Italy
[2] OCEM Power Electronics, Italy

Corresponding author: Riccardo Mandrioli, r.mandrioli@unibo.it
Speaker: Bhavana Gudala, bhavana.gudala2@unibo.it

Abstract

This paper investigates two possible configurations of a power converter with a generic number of parallel H-bridges in interleaved configuration, namely common DC-bus configuration (CDC) and split DC-bus configuration (SDC). Comparison of RMS currents in CDC and SDC are performed as a function of the number of H-bridges to analyze the effect of circulating currents. Normalization of the RMS current in CDC is proposed to decouple the effect of the circulating current from the output current. The analytical developments for both configurations are numerically validated considering a large set of parallel H-bridge and duty cycles.

1 Introduction

High-current, low-ripple applications, such as those used in nuclear fusion reactors, synchrotrons, and particle accelerators, demand substantial energy, often in the range of tens of megajoules [1]. In all these applications, coils serve critical functions related to magnetic field generation like particle beam manipulation and plasma confinement [2]. Generally, the coils are supplied with low- or medium-voltage, depending on their resistive component. Hence, currents are in the typical range of tens of kiloamperes. In the mentioned applications, when unloading the coils, the stored energy has to be transferred back to the DC-bus of the supplying power converter (PC) and this requires the ability of the PC to provide negative voltages [3]. In the technical literature, parallel H-bridge (HB) configurations have been widely acknowledged as effective solutions for leveraging low-power-rating switches in medium and high-power applications [4]. This is due to the inherent efficiency and cost-effectiveness of semi-conductor modules with low-power ratings compared to their high-power counterparts. Naturally, every semiconductor device has limitations in terms of voltage and current handling capabilities [5]. To overcome these constraints, one common practice is to parallel connect directly more HBs to effectively handle higher currents [6]. This approach

also enhances the system reliability and redundancy [7]. However, due to the negative thermal coefficient of the semiconductor devices (typically IGBTs and power diodes), the direct parallel connection of HBs may lead to thermal run-away [8]. Consequently, inductors are utilized to decouple the outputs of the HBs. With such architecture (see Fig. 1), pulse-width-modulation (PWM) carriers of the different HBs can be evenly shifted to achieve output current ripple cancellation in what is known as HBs interleaving [9]. This technique, often called phase-shift PWM, presents straightforward scalability and great implementation ease [10], [11].

From the source's point of view, in the aforementioned applications, the design of the power converter stage is based on two different topologies. One topology, shown in Fig. 1(a), considers all the HBs connected to a common DC-bus and therefore takes the name of common DC-bus configuration (CDC). The second topology, shown in Fig. 1(b), is a galvanically isolated version called split DC-bus configuration (SDC) which avoids the direct paralleling of input DC-buses. In the CDC topology, voltage imbalances caused by carrier phase-shift (interleaving operations) or power switch parasitics can cause the rise of circulating currents. These circulating currents can increase the current stress on the semiconductor devices and degrade the PCs efficiency [12].

Circulating currents have been the subject of numerous scientific studies. For instance, some authors

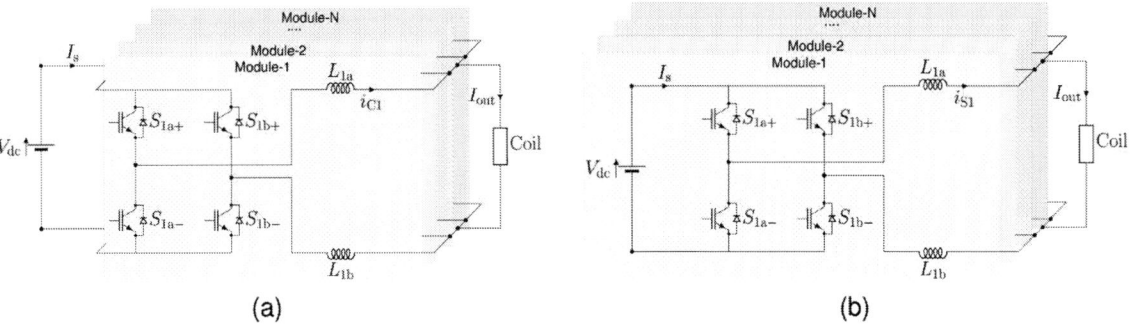

Fig. 1: Typical topologies for N-module interleaved HBs. (a) CDC topology. (b) SDC topology.

have proposed strategies to mitigate circulating currents in a generic number of parallel-connected HBs. These strategies include the implementation of specific control techniques such as the modified unipolar phase-shift PWM, as in [13], or the utilization of model predictive control with a sphere-decoding algorithm as proposed in [14]. Alternatively, authors in [15] have proposed a passive method for mitigation of circulating current in a single HB converter by using coupled inductors to increase the reactance in the circulating currents' path. Shifting the focus from mitigation strategies, certain studies concentrate on the analysis of circulating currents in different application contexts. For example, an analytical study in determining the circulating current within a power converter consisting of two parallel-connected three-phase rectifiers sharing a common DC-bus has been proposed in [16]. On the other hand, [17] presented an analytical evaluation of the arising circulating currents in configurations involving two parallel connected DC/DC buck converters, as well as for two parallel connected HBs. Notably, the analysis in [17] considers only two stages and the results can not be generalized for a larger number of parallel connected HBs as would be the case in high-current applications.

The present work focuses on analyzing the impact of circulating current in applications where a generic number of HBs are connected in parallel to a common DC-bus. This is achieved by developing the analytical formulations for the currents flowing into the interleaving HB (IHB) in both the CDC and SDC of the PC. The rest of the paper is structured as follows: section 2 provides background on both the SDC and CDC for the design of a PC, along with a recap of output current ripple. In section 3, the analysis of RMS currents in SDC, and CDC is presented. Section 4 presents simulation and analytical results. Finally, section 5 summarizes the conclusions drawn from the analysis.

2 Background on the Interleaved CDC and SDC Converters

To simplify the representation of the power converter for analytical purposes, all front-end components along with DC-bus (including protection devices, line frequency transformer, rectifier unit, and input filter components) are represented as a constant DC source V_{dc} as displayed in Fig. 1.

For the sake of comparing the SDC and CDC, the value of the inductances L_{na} and L_{nb}, represented in Fig. 1, are assumed to be the same and equal to a generic inductance L (i.e., $L_{na} = L_{nb} = L$). Legs are distinguished by means of the symbol n whose value ranges from 1 to N where N represents the number of HBs.

As can be seen from Fig. 1(b), as the number of HBs increases, the number of front-end components required in the SDC increases by a factor N, thereby increasing the overall footprint and cost of the PC. Conversely, in the CDC, the requirement for front-end components is reduced to only one. However, due to the lack of galvanic isolation between the HBs, this solution allows the existence of closed electrical paths among the N parallel connected HBs and the common DC-bus. The presence of these closed paths, along with the phase-shift difference in the carriers, can give rise to undesired currents, which circulate only among the HB modules without affecting the output current. Such currents are typically called circulating currents. These currents affect semiconductor devices and inductors (L_{na} and L_{nb}) resulting in increased current stress, operating temperatures, conduction losses, and ultimately a reduction in the PCs lifetime [18]. Whereas in SDC topology, even though the carrier signals are phase-shifted, the circulating current does not exist due to the presence of galvanic isolation provided by power transformers for each DC-buses.

Therefore, selecting a topology for a power converter is always a compromise in meeting operational requirements, reliability, volume, weight, and cost in the aforementioned applications.

2.1 Output Current Ripple Derivation

Similarly to multiphase interleaved DC/DC buck converters analyzed in [19], the peak-to-peak inductor current ripple of an interleaved HB converter can be expressed as:

$$\Delta i_{\rm rip} = \frac{V_{\rm dc}}{4Lf_{\rm s}}\delta\left(1 - \delta\right) , \tag{1}$$

where $f_{\rm s}$ is the switching frequency and δ is the HB gain ratio, whose value ranges from 0 to 1. In this work, the gain ratio δ is defined as the difference in the duty cycle of leg a and leg b for any n-th HB, i.e.:

$$\delta = \delta_{\rm na} - \delta_{\rm nb} , \tag{2}$$

where $\delta_{\rm na}$ equals to $0.5 + \delta/2$ and $\delta_{\rm nb}$ equals to $0.5 - \delta/2$. The relation between peak-to-peak output current ripple and δ of an interleaving HB converter at any operative condition can be expressed as in (3). The function $\mathrm{ceil}\left(N\delta\right)$ in (3) gives the closest highest integer of $N\delta$.

Equation (3) is valid for both the CDC and SDC topologies of the PC because the circulating current has no effect on the load. On the other hand, (1) is valid only in the case of SDC topology since the derived equation does not consider any contributions from circulating current. Therefore, to incorporate the contribution of circulating current in the CDC, RMS current equation for the inductor is formulated and compared with the RMS current in SDC in the following section.

3 Analysis of the Interleaved CDC and SDC Converter

In this section, analytical formulations are developed to analyze the RMS of HB current in both the SDC and CDC topology of the PC; in all the cases, N HB modules are considered. The RMS current is derived by taking advantage of orthogonality between the DC component resulting from the load contribution ($I_{\rm out}/N$) and the current ripple ($I_{\rm rip}$). As already specified in section 2.1, the non-zero-frequency components in SDC and CDC are different due to the presence of circulating currents in the CDC.

3.1 RMS Current of the IHB in the SDC

The RMS value of the inductor current in the leg a of the n-th HB $I_{\rm Sn}$ is calculated using the expression:

$$I_{\rm Sn} = \sqrt{\left(\frac{I_{\rm out}}{N}\right)^2 + \left(I_{\rm rip}\big|_{\rm SDC}\right)^2} \tag{4}$$

where $I_{\rm out}$ is the average output current, and $I_{\rm rip}\big|_{\rm SDC}$ is the RMS of the current ripple in the inductor of the SDC.

From (1), the RMS of the current ripple in the inductor of SDC can be written as:

$$I_{\rm rip}\big|_{\rm SDC} = \frac{\Delta i_{\rm rip}}{2\sqrt{3}} = \frac{V_{\rm dc}}{8\sqrt{3}Lf_{\rm s}}\delta\left(1 - \delta\right) . \tag{5}$$

Therefore, replacing (5) in (4) results into:

$$I_{\rm Sn} = \sqrt{\left(\frac{I_{\rm out}}{N}\right)^2 + \left(\frac{V_{\rm dc}}{8\sqrt{3}Lf_{\rm s}}\delta\left(1 - \delta\right)\right)^2} , \tag{6}$$

which represents the RMS current in the n-th HB module of the SDC. The RMS current derived in (6) can not be used for CDC topology of the PC because it lacks the circulating current component.

3.2 RMS Current of the IHB in the CDC

The approach used in section 3.1 cannot be used in CDC due to the presence of circulating currents. Therefore, to determine the RMS of the non-zero-frequency components present in CDC, an electrical equivalent circuit shown in Fig. 2 is considered. This circuit represents the connection of the n-th HB with the rest of the $N - 1$ HBs and the output which is nothing but the coil in the aforementioned applications. In Fig. 2, $V_{\rm na}$ represents the phase voltage of leg a in the n-th HB, $V_{\rm Lna}$ represents the mean voltage of the inductor present at leg a of the n-th HB, and $V_{\rm com}$ represents the common-mode voltage from leg a of n-th HB with respect to the output negative terminal. To derive relevant equations, Kirchhoff's voltage law is applied to the equivalent circuit represented in Fig. 2 obtaining:

$$V_{\rm na} = V_{\rm Lna} + V_{\rm out} + V_{\rm com} . \tag{7}$$

$$\Delta i_{\rm out} = \frac{V_{\rm dc}}{4Lf_{\rm s}}\left[\delta - \frac{\mathrm{ceil}\left(N\delta\right) - 1}{N}\right]\left\{1 - N\left[\delta - \frac{\mathrm{ceil}\left(N\delta\right) - 1}{N}\right]\right\} \tag{3}$$

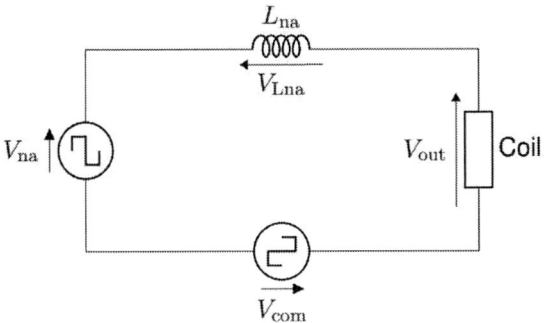

Fig. 2: Equivalent electrical schematic.

The mean voltage of leg a in the n-th HB can be expressed as the product of the DC-bus voltage and the leg's duty cycle:

$$V_{na} = \delta_{na} V_{dc} , \qquad (8)$$

The output voltage V_{out} in terms of the DC-bus voltage V_{dc} is given by the equation:

$$V_{out} = \delta V_{dc} . \qquad (9)$$

The average component of the voltage in the leg a of the n-th HB is zero, hence substituting the expressions of V_{na}, and V_{out}, in (7) yields:

$$V_{com} = \frac{V_{dc}}{2} (1 - \delta) . \qquad (10)$$

The RMS of the current ripple in the inductor of the CDC ($I_{rip}\big|_{CDC}$) is expressed as:

$$I_{rip}\big|_{CDC} = \frac{t_{on}}{2\sqrt{3}L} V_{Lna}\big|_{t_{on}} \qquad (11)$$

where, $V_{Lna}\big|_{t_{on}}$ is the voltage across the inductor in the leg a of the n-th HB during the conduction of switch S_{na+} in Fig. 1(a). This difference in this voltage with respect to SDC is precisely the source of circulating current. Thereby, substituting (8), (9) and, (10) in (7) at t_{on} it results in:

$$V_{Lna}\big|_{t_{on}} = \frac{V_{dc}}{2} (1 - \delta) , \qquad (12)$$

while the t_{on} of the inductor L_{na} in CDC is given by:

$$t_{on} = \frac{\delta_{na}}{f_s} . \qquad (13)$$

Replacing (12) and (13) in (11), the RMS of the current ripple in the inductor of the CDC results in:

$$I_{rip}\big|_{CDC} = \frac{V_{dc}}{8\sqrt{3}Lf_s} (1 - \delta^2) . \qquad (14)$$

Therefore, the total RMS value of the inductor current at leg a in the n-th HB in CDC (I_{Cn}) can be expressed as:

$$I_{Cn} = \sqrt{\left(\frac{I_{out}}{N}\right)^2 + \left(I_{rip}\big|_{CDC}\right)^2} , \qquad (15)$$

leading to:

$$I_{Cn} = \sqrt{\left(\frac{I_{out}}{N}\right)^2 + \left[\frac{V_{dc}}{8\sqrt{3}Lf_s} (1 - \delta^2)\right]^2} , \qquad (16)$$

which represents the RMS current in the n-th HB module of the CDC.

3.3 Comparison of RMS Current in the SDC and CDC

By comparing the RMS current equations derived in (6) and (16), it is evident that the DC component, given by the contribution of the load (i.e., I_{out}/N), remains the same. Whereas, the non-zero-frequency component, given by the harmonics, are different (i.e., $I_{rip}\big|_{CDC} \neq I_{rip}\big|_{SDC}$).

In order to partially decouple the computed RMS in the case of CDC from the influence of the output current I_{out}, a normalization factor I_{out}/N is applied. Specifically, the normalized current in the n-th HB of CDC is evaluated as:

$$I_C = \frac{I_{Cn}}{\frac{I_{out}}{N}} = N\frac{I_{Cn}}{I_{out}} . \qquad (17)$$

Equation (17) explains that in the CDC topology of a PC, as the number of HBs increases to provide a constant output current, the effect of circulating current becomes more dominant. This is due to the fact that when N increases the contribution of output current by each individual HB decreases, while the I_{rip} in CDC or SDC is invariant.

In the aforementioned applications, while choosing a topology for the PC, (17) provides insights on the contribution of circulating current in the HB modules, as the number of parallel connections increases for a single DC-bus for providing higher output currents, with the aim of reducing front-end components.

4 Results

Topologies depicted in Fig. 1(a) and Fig. 1(b) have been simulated on PLECS (Plexim GmBh) environment to validate the analytical results. Considered gain ratios for validation are $\delta = 0.3, 0.6$, and 0.9. Main system parameters are collected in Tab. 1.

Comparisons of currents for $N = 15$ HBs in CDC and SDC topologies of the PC are shown in Fig. 3. Specifically, Fig. 3(a), Fig. 3(c), and Fig. 3(e) represent the HB currents for two switching periods in CDC in case of $\delta = 0.3, 0.6$, and 0.9, respectively. Similarly, Fig. 3(b), Fig. 3(d), and Fig. 3(f) represent currents in case of SDC.

As previously mentioned in section 3.3, at a fixed number of HBs, the DC component in CDC and SDC remains the same, while the ripple components in the CDC are higher than in the SDC due to the contribution of circulating current. The comparison of plots in Fig. 3(a) and Fig. 3(b) for $\delta = 0.3$ reveals

Parameter	Symbol	Value	Unit
DC-bus voltage	V_{dc}	1	kV
Output current	I_{out}	10	kA
Interphase inductors	$L_{\mathrm{na}} = L_{\mathrm{nb}} = L$	25	μH
Coil resistance	R	0.1	Ω
Switching frequency	f_{s}	2	kHz
Number of HB modules	N	15	-

Tab. 1: Main system parameters used for the simulations.

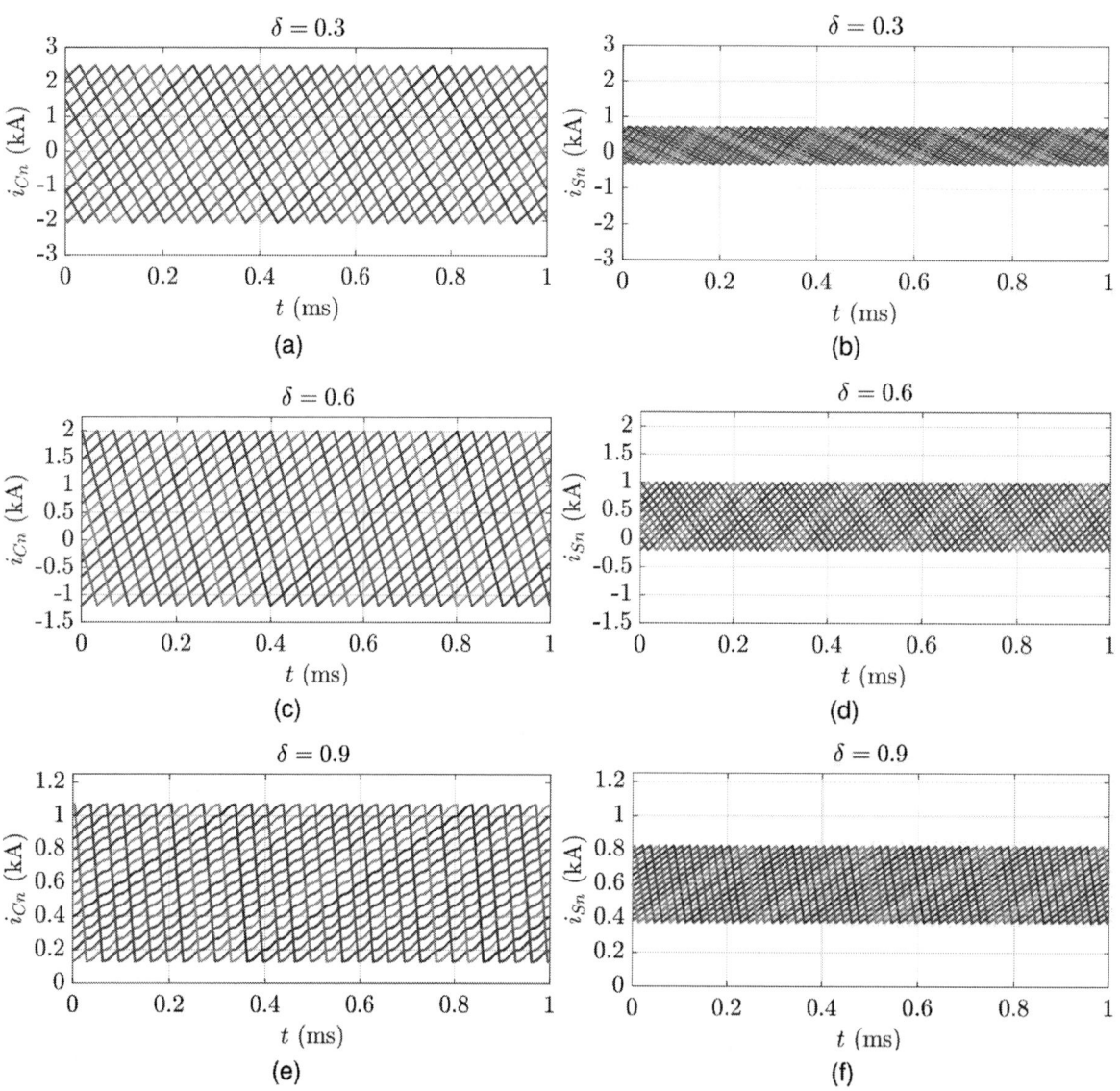

Fig. 3: IHB currents in case of:(a) CDC at $\delta = 0.3$, (b) SDC at $\delta = 0.3$, (c) CDC at $\delta = 0.6$, (d) SDC at $\delta = 0.6$, (e) CDC at $\delta = 0.9$ and, (f) SDC at $\delta = 0.9$.

that the mean current value is 200.0 A in both CDC and SDC. Whereas, the peak-to-peak current ripple is 1038 A in SDC while is 4520 A in CDC due to the presence of circulating currents. Similarly, at $\delta = 0.6$, from Fig. 3(c) and Fig. 3(d), the mean value of the current in both CDC and SDC is 400.0 A, but the peak-to-peak current ripple in SDC is 1188 A, while in CDC, it is 3177 A. Furthermore, in case of $\delta = 0.9$, from Fig. 3(e) and Fig. 3(f), the mean value of the current in both the CDC and SDC is 600.0 A, but the peak-to-peak current ripple in SDC is 449.9 A, while in CDC, it is 938.5 A. Peak-to-peak current ripple in the case of SDC fairly agree with (1).

To better understand the RMS currents in CDC and SDC with the variation of the number of HBs, Fig. 4 is presented. In particular, Fig. 4(a) presents the variation of RMS currents in SDC having N from 2 to 15 for $\delta = 0.3$, 0.6 and 0.9. Similarly, Fig. 4(b) presents the variation of RMS currents in CDC for the same range of N and the same set of δ values. Comparing these two plots, for instance, at $N = 7$ and $\delta = 0.3$, the RMS current in SDC is 0.5 kA, while in CDC, it is 1.4 kA. Similarly, at $N = 15$ and $\delta = 0.3$, the RMS current of HB in SDC is 0.3 kA, whereas in CDC it is 1.3 kA. These results suggest that the decay of RMS current in CDC does not follow a parabolic trend $(1/N)$. This is due to the fact that beyond a certain value of N (about $N = 6$), the circulating current predominates over the output current contribution. As foreseeable, lower values of gain ratio δ ensure lower RMS currents in the case of SDC (see Fig. 4(a)) regardless of the number of modules N. Higher values of gain ratio δ lead to higher output currents I_{out} and therefore higher RMS. This behavior holds for low values of N also in the case of CDC (see Fig. 4(b)). However, once N gets higher than 6, lower values of gain ratio δ lead to higher values of RMS. Such a trend inversion is attributable to the effect of the circulating currents.

Traces displayed in Figure 4(c) correspond to the normalization of RMS current (17) in the case of CDC as N varies from 2 to 15 for $\delta = 0.3$, 0.6 and 0.9. It is immediately clear that higher values of N are more affected by the circulating current. In particular, lower values of gain ratio δ cause higher values of circulating currents throughout the whole N diapason. Such a statement was not directly apparent from the non-normalized RMS of Fig. 4(b).

As visible from Fig. 5, the output current in both the CDC and the SDC configurations is unaffected by the circulating current. The latter statement holds regardless of gain ratio δ values.

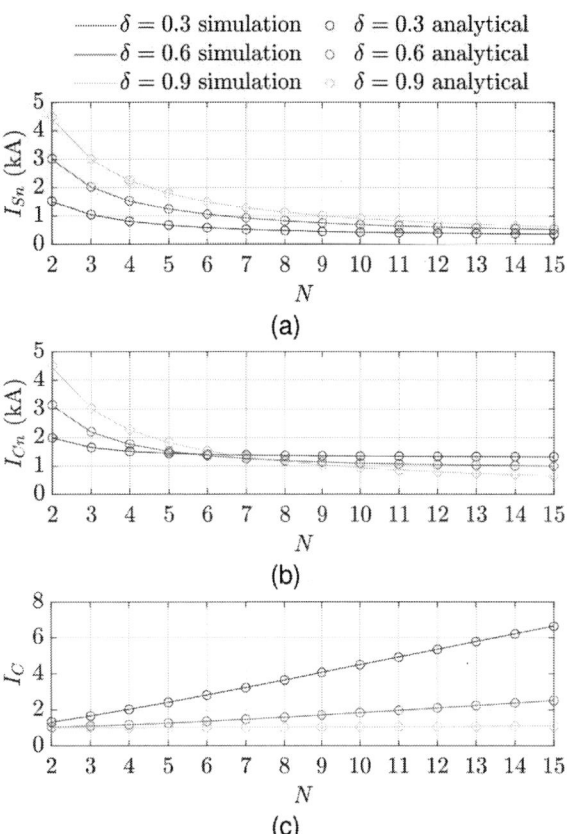

Fig. 4: Numerical validation of RMS current in IHB modules at $\delta = 0.3, 0.6,$ and 0.9: (a) in case of SDC; (b) in case of CDC; (c) normalization in case of CDC.

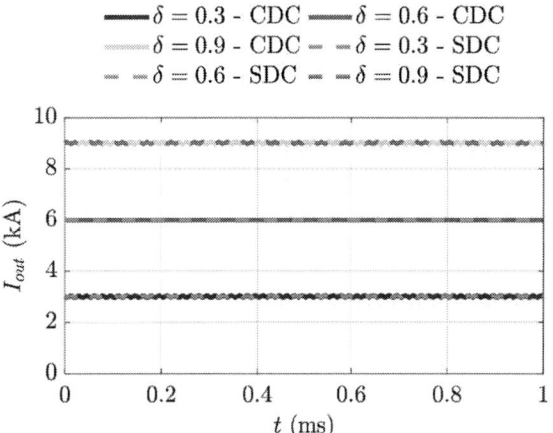

Fig. 5: Numerical validation of output current in case of SDC and CDC at $\delta = 0.3$, 0.6 and 0.9.

5 Conclusion

This paper has discussed the variation of current ripple in SDC and CDC of the PC. The analytical formula for the RMS currents in generic number of parallel connected HBs in CDC and SDC have been derived and compared. A normalized RMS current in CDC is proposed to decouple the effect of circulating currents from the output current. Eventually, these outcomes are validated using simulations for both CDC and SDC configurations. Current RMS values obtained numerically are in agreement with the analytical results.

Future work will define an algorithm for the automatic selection of the optimal topology considering the drawbacks of the circulating current and an increased number of components for an operational and cost-effective solution.

References

[1] A. Lampasi and S. Minucci, "Survey of electric power supplies used in nuclear fusion experiments," in *2017 IEEE International Conference on Environment and Electrical Engineering and 2017 IEEE Industrial and Commercial Power Systems Europe (EEEIC / I&CPS Europe)*, 2017, pp. 1–6. DOI: 10.1109/EEEIC.2017.7977851.

[2] L. Bottura, S. A. Gourlay, A. Yamamoto, and A. V. Zlobin, "Superconducting magnets for particle accelerators," *IEEE Transactions on Nuclear Science*, vol. 63, no. 2, pp. 751–776, 2016. DOI: 10.1109/TNS.2015.2485159.

[3] N. Wassinger, S. Maestri, R. Garcia Retegui, M. Funes, P. Antoszczuk, and S. Pittet, "Mosfet Selection for a 18kA Modular Power Converter for HI-Lhc Inner Triplet," *Available at SSRN 4484160*,

[4] A. de Paula Dias Queiroz, C. B. Jacobina, A. C. N. Maia, V. F. M. B. Melo, and I. da Silva, "Investigation of a single-phase multilevel inverter based on series/parallel-connected h-bridges," *IEEE Transactions on Industry Applications*, vol. 54, no. 5, pp. 4707–4716, 2018. DOI: 10.1109/TIA.2018.2839666.

[5] B. J. Baliga, "Trends in power semiconductor devices," *IEEE Transactions on electron Devices*, vol. 43, no. 10, pp. 1717–1731, 1996.

[6] D. Ma, W. Chen, and X. Ruan, "A review of voltage/current sharing techniques for series–parallel-connected modular power conversion systems," *IEEE Transactions on Power Electronics*, vol. 35, no. 11, pp. 12 383–12 400, 2020. DOI: 10.1109/TPEL.2020.2984714.

[7] Y. Huang and C. K. Tse, "Circuit Theoretic Classification of Parallel Connected DC–DC Converters," *IEEE Transactions on Circuits and Systems I: Regular Papers*, vol. 54, no. 5, pp. 1099–1108, 2007. DOI: 10.1109/TCSI.2007.890631.

[8] G. Konstantinou, G. J. Capella, J. Pou, and S. Ceballos, "Single-carrier phase-disposition pwm techniques for multiple interleaved voltage-source converter legs," *IEEE Transactions on Industrial Electronics*, vol. 65, no. 6, pp. 4466–4474, 2018. DOI: 10.1109/TIE.2017.2767541.

[9] R. Mandrioli, L. K. Pittala, V. Cirimele, M. Ricco, and G. Grandi, "Probabilistic approach for the study of neutral current ripple in split-capacitor inverters," in *2023 IEEE 17th International Conference on Compatibility, Power Electronics and Power Engineering (CPE-POWERENG)*, 2023, pp. 1–6. DOI: 10.1109/CPE-POWERENG58103.2023.10227446.

[10] H. Xiao and S. Xie, "A ZVS Bidirectional DC–DC Converter With Phase-Shift Plus PWM Control Scheme," *IEEE Transactions on Power Electronics*, vol. 23, no. 2, pp. 813–823, 2008. DOI: 10.1109/TPEL.2007.915188.

[11] T. Kohama and T. Ninomiya, "Automatic interleaving control for paralleled converter system and its ripple estimation with simplified circuit model," in *2007 7th Internatonal Conference on Power Electronics*, 2007, pp. 238–242. DOI: 10.1109/ICPE.2007.4692384.

[12] J.-W. Yang and H.-L. Do, "High-efficiency bidirectional dc–dc converter with low circulating current and zvs characteristic throughout a full range of loads," *IEEE Transactions on Industrial Electronics*, vol. 61, no. 7, pp. 3248–3256, 2014. DOI: 10.1109/TIE.2013.2279370.

[13] D. Verdugo, F. Rojas, J. Lillo, M. Diaz, J. Pereda, and G. Gatica, "Phase-shifted pulse width modulation with alternate zeros voltage for parallel connection of h-bridges for high-current low-voltage applications," in *IECON 2019 - 45th Annual Conference of the IEEE Industrial Electronics Society*, vol. 1, 2019, pp. 1950–1955. DOI: 10.1109/IECON.2019.8927568.

[14] C. Terlizzi, S. Bifaretti, and A. Lampasi, "Model predictive control with sphere-decoding algorithm for parallel-connected h-bridges," in *2022 IEEE Energy Conversion Congress and Exposition (ECCE)*, 2022, pp. 1–7. DOI: 10.1109/ECCE50734.2022.9947422.

[15] G. Zhu, B. A. McDonald, and K. Wang, "Modeling and analysis of coupled inductors in power converters," *IEEE Transactions on Power Electronics*, vol. 26, no. 5, pp. 1355–1363, 2011. DOI: 10.1109/TPEL.2010.2079953.

[16] M. Baumann and J. W. Kolar, "Parallel connection of two three-phase three-switch buck-type unity-power-factor rectifier systems with dc-link current balancing," *IEEE Transactions on Industrial Electronics*, vol. 54, no. 6, pp. 3042–3053, 2007. DOI: 10.1109/TIE.2007.907006.

[17] Y. Xia, M. Yu, Y. Peng, and W. Wei, "Modeling and analysis of circulating currents among input-parallel output-parallel nonisolated converters," *IEEE Transactions on Power Electronics*, vol. 33, no. 10, pp. 8412–8426, 2018. DOI: 10.1109/TPEL.2017.2777604.

[18] R. Bayerer, T. Herrmann, T. Licht, J. Lutz, and M. Feller, "Model for power cycling lifetime of igbt modules - various factors influencing lifetime," in *5th International Conference on Integrated Power Electronics Systems*, 2008, pp. 1–6.

[19] K. Drobnic, G. Grandi, M. Hammami, R. Mandrioli, M. Ricco, *et al.*, "An output ripple-free fast charger for electric vehicles based on grid-tied modular three-phase interleaved converters," *IEEE Transactions on Industry Applications*, vol. 55, no. 6, pp. 6102–6114, 2019. DOI: 10.1109/TIA.2019.2934082.

PCIM Europe 2024, 11– 13 June 2024, Nuremberg   DOI: 10.30420/566262355

Optimal Frequency Operating Points for Hybrid Switched Capacitor Converters and Lossless Current Sense Method

Roberto Rizzolatti [1], Simone Mazzer [1], Mario Ursino [1], Erik Medeossi[1], Ivan Seet[1], Stefano Saggini [2], Kevin Zufferli[2]

[1] Infineon Technologies AG, Austria
[2] University of Udine, Italy

Corresponding author: Roberto Rizzolatti, Roberto.Rizzolatti@infineon.com
Speaker: Simone Mazzer, Simone.Mazzer@infineon.com

Abstract

Data center electricity consumption has been increasing rapidly in the last few years as computation moved into cloud computing [1]. With the introduction of the 48 V power delivery architecture, powering digital loads (CPUs, GPUs, ASICs, . . .) is typically accomplished by means of a two-stages approach. Most commonly, the regulation is established at second-stage level, with an unregulated first stage. In general, the first stage is made with a fixed ratio dc-dc converter, such as LLC or Hybrid Switched Capacitor (HSC) converter. In this paper a novel optimal frequency feed-forward control and a lossless current sense methods are presented for HSC converters. The goal of the feed-forward method is to optimize the converter efficiency over the entire input voltage range, by tracking the optimal operating point. Experimental results for an HSC 5:1 750 W prototype in down-solution 18 x 47.5 x 7 mm are showing the effectiveness of the proposed approach.

1 Introduction

The escalating energy demands of modern data centers necessitate the development of innovative power conversion technologies capable of efficiently managing dynamic voltage fluctuations [2]. To achieve high peak efficiency, high power density intermediate-bus converters (IBCs) require soft-switching techniques such as zero-voltage switching (ZVS) and/or zero-current switching (ZCS) to minimize transistor switching losses. However, traditional soft-switching converter architectures as LLC [3], Switched Tank Converter (STC) [4] and Hybrid Switched Capacitor (HSC) converters [5] struggle to maintain desired circuit waveforms when power is reduced or input voltage deviates from nominal values. The main factor impacting the performance of these converters is the capacitance derating of class II multi-layer ceramic capacitors (MLCCs) when used as part of their resonant tanks, according to voltage and temperature operating conditions [6]. Capacitance derating reflects into deviation of the tank resonant frequency thus affecting converter operation. Regardless of these challenges, demonstrations of closed-loop control and accurate sensing that accounts for components mismatch and variation have demonstrated performance improvement for resonant type converters, such as STC [4]. In STC an accurate matching of the switching frequency with respect to the actual resonant frequency enables optimal transfer of the flying-capacitors charge (energy) to the output and reduced turn-off losses. To overcome main limitation coming from drifting of ceramic capacitor, a lock-in controller have been demonstrated [7] [8] for switched capacitor converter ZCS based, which adjust the switching frequency in a closed-loop fashion, thus accommodating any mismatch or drifts in the component values.

On the other hand, in HSC converters [5], where ZVS is ensured for all the MOSFETs, the optimal switching frequency does not necessarily match the tank resonant frequency. For instance, the converter can operate above the resonant frequency (hence loosing ZCS) while maintaining high-efficiency, having majority of the switching losses mitigated by the ZVS operation. Therefore, in HSC type converter, is not possible to rely on the ZCS information.

2532

In this work, a 5:1 HSC DCX converter is introduced, which utilizes the Optimal Frequency Operating Points without the need for additional networks or complicated controls, ensuring a flatter output resistance R_{out}. By leveraging this principle, it becomes possible to calculate the output current simply by measuring the input and output voltages, as the R_{out} keeps constant across the input voltage range.

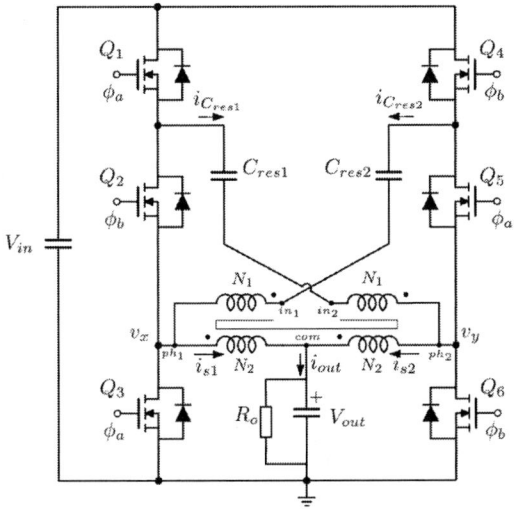

Fig. 1: HSC topology overview.

2 Key Features of Hybrid Switched Capacitor DCX Converters

HSC is a resonant DCX converter which comprises an interleaved flying-capacitor structure connected to a multi-tapped autotransformer (MTA) [5] . Topology is shown in Figure 1, with main waveform shown in Figure 2. Zero-voltage switching (ZVS) operation for all the FETs is primarily achieved by the transfer or energy forced by the autotransfomer magnetizing inductance. This is paramount important to consider since the converter can operate over the resonant frequency, keeping high efficiency and therefore low output and constant impedance, which is actually one of the main benefits highlighted in [5].

The ZVS transition is ideally wanted to be largely independent on the energy stored in the series resonant tank. While in general the validity of this statement depends on the amount of energy stored in the leakage inductance of the autotransformer, and how does this compare with the electrical energy stored in the semiconductors output capacitance, in practical cases it results to be especially true in

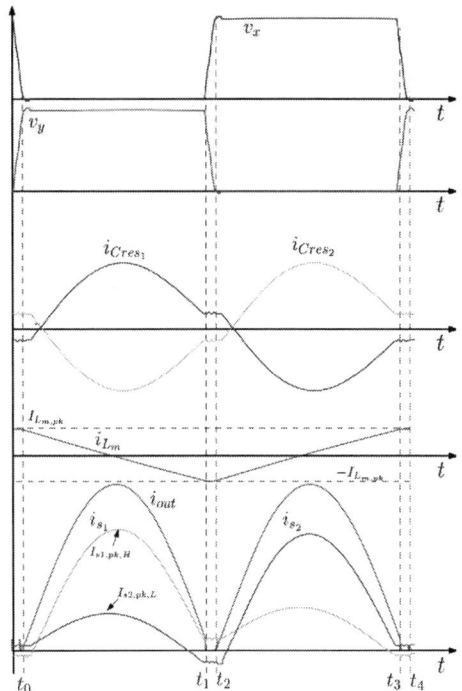

Fig. 2: HSC topology main waveforms.

all the implementations where the resonant tank quality factor Q is low (e.g., in a module implementation of the HSC converter). In fact, the lower the Q factor, the lower is the sensitivity against a non-zero current switching event, as the turn-off current varies within a narrow range at different resonant frequency (Figure 3).

However, when the resonant tank is characterized by an high quality factor Q, it is not guaranteed that the magnetic energy stored in the leakage inductance of the autotransformer will prevent the MOSFETs from deviating from the optimal ZVS trajectory (Figure 3).

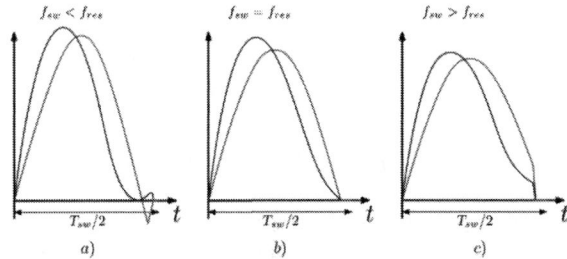

Fig. 3: MOSFET currents in the HSC converter, under low Q (blue) and high Q (red): a) below resonant frequency, b) at resonant frequency and c) above resonant frequency

The converter can be designed to always operate above resonant frequency, however - when class II dielectrics (e.g., X7R) ceramic capacitor are used - the high capacitance derating over input voltage and temperature operating conditions must be taken into account. X7R 50 V capacitors are typically adopted as part of the resonant tank of high-power dense HSC IBCs from 48 V input as they offer high volumetric capacitance density. However capacitance derating may cause the resonant frequency to increase (e.g., with input voltage), potentially leading to unintentional under-resonant operation. This could adversely affect both the converter's performance and/or violating the voltage rating of the MOSFETs. Indeed, the synchronous rectifiers may experience high over voltages when turning off with large currents. The degradation in performance can be reflected into an increase of the converter equivalent output resistance R_{out}: in fact, operating as a DCX, the HSC transfer function is accurately modeled by the following equation:

$$V_{out} = \frac{V_{in}}{4 + 2\frac{N_1}{N_2}} - R_{out}(T, \Delta) I_{out} \quad (1)$$

where the main contributors to the output resistance are respectively the MOSFETs and the multi-tapped autotransformer (MTA): $R_{out}(T, \Delta) = R_{out,MOS}(T, \Delta) + R_{out,MTA}(T, \Delta)$. It can be shown that, under the high-Q approximation and $f_{sw} = f_{res}$ the output resistance is given by a linear combination of terms (Equation 2), where R_w are the (ac) resistances associated with the transformer windings, R_{on} are the on-state resistances of the semiconductors, $n_{MTA} = 1 + N_1/N_2$ is the autotransformer turns ratio and $0 < D < 0.5$ is the duty cycle. Deviations in the switching frequency from the resonant frequency cause the actual output resistance to deviate from the ideal expression, in response to a different RMS to DC value of the rectified output current.

The output impedance is dependent on temperature T (via the copper and the semiconductors temperature coefficients) and distributed according to components mismatch Δ (i.e., due to fabrication tolerances). The converter output impedance is strictly connected to the overall converter efficiency.

For an high-efficiency converter:

$$\eta = \frac{1}{1 + \frac{P_{loss}}{P_{out}}} \approx 1 - \frac{P_{loss}}{P_{out}} \quad (3)$$

By neglecting all the sources of switching losses (including gate losses), the overall efficiency can be expressed solely in terms of R_{out}:

$$\eta \approx 1 - \frac{R_{out} I_{out}^2}{P_{out}} = 1 - \frac{R_{out}}{R} \quad (4)$$

where R is the load resistance. From Equation 4 is then clear that in a DCX converter, optimizing the efficiency is equivalent to optimizing the converter output impedance.

3 Flat R_{out} and lossless current sense method optimization for Hybrid Switched Capacitor DCX converters

To overcome the risk of operation below resonance and to optimize the DCX output resistance, a feed forward method adjusting the switching frequency is here proposed. In particular, it will be shown how the output resistance of an HSC IBC [5] from 48 V can be maintained relatively constant over the entire input voltage and temperature range by modulating the converter switching frequency. The ultimate goal of flattening the output resistance is twofold: other than optimizing the converter for best efficiency, a constant value of R_{out} also enables to implement an accurate, lossless current sense method. Differently from "lossless" approaches developed earlier [9], the method here proposed do not require to sacrifice autotransformer winding area.

The idea behind the novel method is to estimate the output current by the prior knowledge of the converter output impedance, together with measurements of the input and output voltages. By exploiting the resistive *droop* behavior of the HSC converter (Equation 1):

$$I_{out} = \frac{1}{R_{out}} \left[\frac{V_{in}}{4 + 2\frac{N_1}{N_2}} - V_{out} \right] \quad (5)$$

$$R_{out} = \frac{\pi^2}{8} \left(\frac{1}{2D} \right) \left[\frac{R_{w,p} + 2R_{on,MOS} + 4\left(R_{on,SR} + R_{w,s}\right)\left(1 + n_{MTA}^2\right)}{4\left(1 + n_{MTA}\right)^2} \right] \quad (2)$$

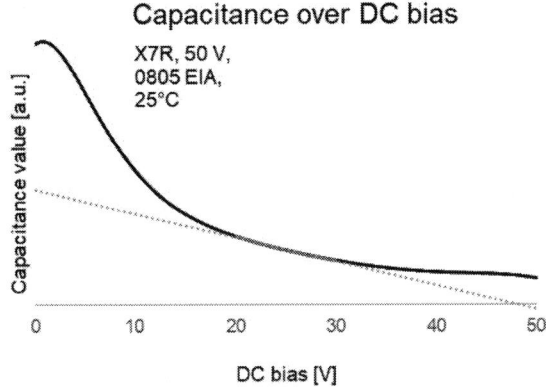

Fig. 4: Capacitance derating against DC bias of a commercially available MLCC with X7R dielectric, showing linear behavior in the range of interest.

the output current estimate can be thus obtained by sensing the input and output voltages and calculated cycle-by-cycle from a digital controller. Alternatively, the analog current information can be reconstructed with the aim of a differential amplifier. In both approaches, the accuracy of the estimation is dominated by the accurate knowledge of the converter output impedance. Large errors can occur if the actual converter impedance varies (with respect to the nominal value characterized e.g., at the nominal input voltage) as results of changing of the operating conditions of the converter, i.e., with input voltage and temperature.

In an attempt to control the converter impedance across the input voltage range, switching frequency is modulated in a feed forward fashion, solely based on the knowledge of the input voltage and of the class II resonant capacitor derating characteristic over voltage [10]. During the steady-state operation of the converter, the flying capacitors are subjected to a DC bias of:

$$V_{C_{res}} = \frac{V_{in}}{2}, \qquad 40\ V < V_{in} < 60\ V \qquad (6)$$

The capacitance over DC bias of commercially available X7R 50 V capacitors in EIA0805 is generally relatively linear in the range of interest of $V_{C_{res}}$ (Figure 4), such that is possible to write:

$$C_{res}(u) \approx C_{res}(25\ V)\left[1 - k(u - 25\ V)\right] \qquad (7)$$

with k being a coefficient resulting from a LSQ fit of the characteristic. Then the resonant frequency of the HSC converter results to be input voltage dependent:

$$f_{res}(u) = \frac{1}{2\pi\sqrt{L_r C_0}\sqrt{[1 - k(u - 25)]}} =$$
$$= \frac{f_{res,0}}{\sqrt{[1 - k(u - 25)]}} \qquad (8)$$

where $C_0 := C_{res}(25\ V)$. Being $k(u - 25) << 1$ then asymptotically:

$$f_{res}(V_{in}) \approx f_{res,0}\left[1 + \frac{k}{2}\left(\frac{V_{in}}{2} - 25\ V\right)\right] \qquad (9)$$

The HSC tank resonant frequency varies linearly with excellent approximation across the input voltage range of the converter. Thus it is possible to track this variation by feeding the V_{in} information into the PWM switching period register, according to the linear law in Equation 9.

Electrical properties of class II dielectrics are also known to be strongly dependent on temperature. However, when the resonant capacitors are subjected to fairly strong DC bias (e.g., in the range of one half of the rated voltage, like in Figure 4), the electric polarization tends to become almost insensitive to any sort of temperature variation, thus the value of the capacitance. This phenomena helps as is not needed to perform any further adjustment of the switching frequency to track resonant switching frequency deviations with temperature.

4 Experimental results

To validate the proposed approach, a 750 W 5:1 IBC fitting in OAM 2.0 form factor [11], and based on HSC topology has been built (Figure 5).

The converter has been lay-outed as a down-solution, with the planar MTA magnetic realized in PCB technology and placed as an SMD component. Q_1, Q_2, Q_4, and Q_5 are 1x IQE046N08LM5SC while Q_3 and Q_6 are 2x IQE008N03LM5SC. To realize the 5:1 down-conversion, the MTA is structured with half-turn primary windings $N_1 = 0.5$ and single-turn secondary windings $N_2 = 1$. Experimental peak efficiency (including driving and auxiliary

Tab. 1: HSC 5:1 750 W prototype specification.

Q_1, Q_4	2 × IQE046N08LM5SC – 80 V, 4.6 mΩ
Q_2, Q_5	2 × IQE046N08LM5SC – 80 V, 4.6 mΩ
Q_3, Q_6	4 × IQE008N03LM5SC – 30 V, 850 μΩ
Drivers	4 × 1EDN7550U, 2 × 2EDN7534G
Controller	XDP™ XDPP1100-Q024
SMD Autotransformer (MTA)	Sumida PS22-196
f_{sw}	250 kHz

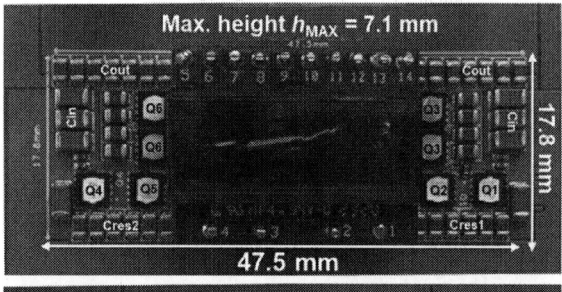

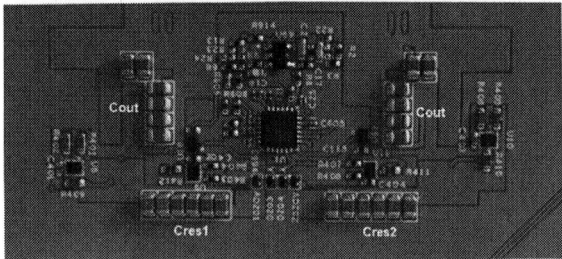

Fig. 5: Test board down-solution for OAM 2.0 form factor 5:1 fixed conversion 750 W (TOP/BOT layer views).

losses) is around 98.1% (at nominal input voltage 54 V), and according to the low R_{out}, keeps high up to full-load as shown in Figure 8. By implementing the frequency modulation based on input voltage feed-forward, the converter output impedance resulted in a quasi-constant value over the interesting input voltage range (44 V to 60 V). On the other hand, the implementation at constant frequency (i.e., 244 kHz) shows a clear monotonically increasing trend with the input voltage (Figure 6). By adjusting the switching frequency with the input voltage, the relative deviation from the nominal output impedance reduced from 15% to 5%, as shown in Figure 7.

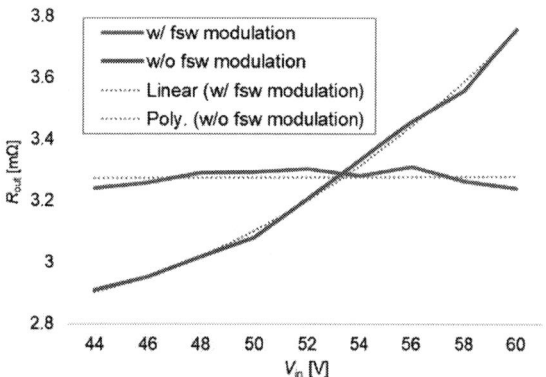

Fig. 6: HSC output impedance as function of the input voltage, at fixed switching frequency (red) and with variable switching frequency (blue).

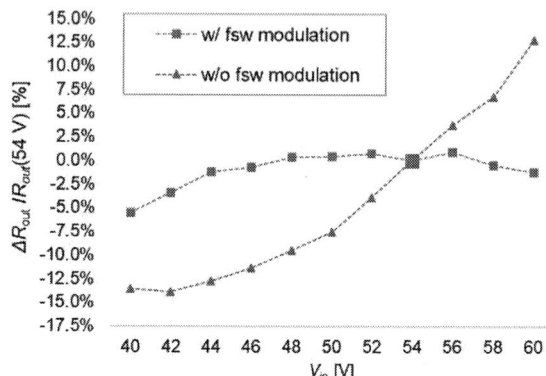

Fig. 7: Relative variation of the output resistance (across V_{in}) with respect to the nominal output resistance (taken at nominal $V_{in} = 54$ V).

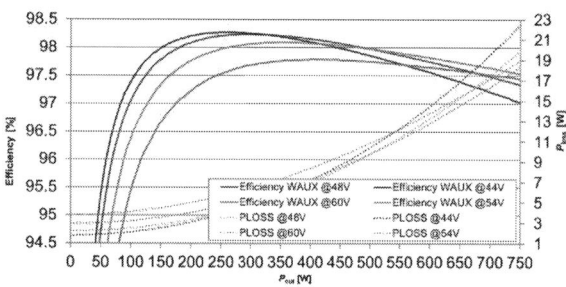

Fig. 8: Experimental efficiency with $P_{out} = 750$ W and $f_{sw} = 244$ kHz.

5 Conclusion

In this paper, a novel input-voltage feed-forward method is proposed to flatten the variation in output impedance of the HSC converter over the input voltage range, by adjusting the switching frequency. The capability to maintain constant impedance despite variations in input voltage and temperature has enabled a novel current sensing method. This method is based on the estimation of output current derived from the output impedance. The experimental results have shown that the proposed method allows for the estimation of the output resistance with an accuracy of 5%, which makes possible the implementation of high-accuracy, lossless current-sensing methods.

References

[1] Yole Intelligence, *Computing and ai technologies for data center 2022, market and technology report*, 2022.

[2] S. Oliver, *From 48 v direct to intel vr12.0: Saving 'big data' $500,000 per data center, per year*, Jul. 2012.

[3] G. Deng, Y. Sun, G. Xu, X. Chen, S. Xie, *et al.*, "Zvs analysis of half bridge llc-dcx converter considering the influence of resonant parameters and loads," in *2020 IEEE Energy Conversion Congress and Exposition (ECCE)*, 2020, pp. 1186–1190. DOI: 10.1109/ECCE44975.2020.9235371.

[4] S. Jiang, S. Saggini, C. Nan, X. Li, C. Chung, and M. Yazdani, "Switched tank converters," *IEEE Transactions on Power Electronics*, vol. 34, no. 6, pp. 5048–5062, 2019. DOI: 10.1109/TPEL.2018.2868447.

[5] R. Rizzolatti, C. Rainer, S. Saggini, and M. Ursino, "High density hybrid switched capacitor converter for data-center application," in *2021 IEEE Applied Power Electronics Conference and Exposition (APEC)*, 2021, pp. 1288–1293. DOI: 10.1109/APEC42165.2021.9487136.

[6] J. Xu, L. Gu, and J. Rivas-Davila, "Effect of class 2 ceramic capacitor variations on switched-capacitor and resonant switched-capacitor converters," *IEEE Journal of Emerging and Selected Topics in Power Electronics*, vol. 8, no. 3, pp. 2268–2275, 2020. DOI: 10.1109/JESTPE.2019.2951807.

[7] T. Urkin, G. Sovik, E. E. Masandilov, and M. M. Peretz, "Digital zero-current switching lock-in controller ic for optimized operation of resonant scc,"

IEEE Transactions on Power Electronics, vol. 36, no. 5, pp. 5985–5996, 2021. DOI: 10.1109/TPEL.2020.3029976.

[8] H. B. Sambo, Y. Zhu, T. Ge, N. M. Ellis, and R. C. N. Pilawa-Podgurski, "Autotuning of resonant switched-capacitor converters for zero current switching and terminal capacitance reduction," in *2023 IEEE Applied Power Electronics Conference and Exposition (APEC)*, 2023, pp. 1217–1224. DOI: 10.1109/APEC43580.2023.10131171.

[9] C. Rainer, R. Rizzolatti, S. Saggini, and M. Ursino, "Lossless current sensing method for hybrid switched capacitor converter," in *2021 IEEE Applied Power Electronics Conference and Exposition (APEC)*, 2021, pp. 934–938. DOI: 10.1109/APEC42165.2021.9487368.

[10] S. Coday, C. B. Barth, and R. C. Pilawa-Podgurski, "Characterization and modeling of ceramic capacitor losses under large signal operating conditions," in *2018 IEEE 19th Workshop on Control and Modeling for Power Electronics (COMPEL)*, 2018, pp. 1–8. DOI: 10.1109/COMPEL.2018.8460142.

[11] *Open Compute Project — opencompute.org*, https : / / www . opencompute . org / projects / open-accelerator-infrastructure, [Accessed 01-04-2024].

PCIM Europe 2024, 11– 13 June 2024, Nuremberg   DOI: 10.30420/566262356

Design and Testing of a 250 kW 50 kHz SiC-based Half-Bridge-Series-Resonant-Converter

Daniel Haake[1], Anton Gorodnichev[1], Matthias Klee[1], Fabian Schnabel[1], Matthias Buerger[2], Dirk Schekulin[3], Daniel Benner[3], Marco Jung[1,4]

[1] Fraunhofer Institute of Energy Economics and Energy System Technology IEE, Germany

[2] Infineon Technologies AG, Germany

[3] STS Spezial-Transformatoren-Stockach GmbH & Co. KG, Germany

[4] Power Electronics and Power System Laboratory (PEPS-Lab)
 Bonn-Rhein-Sieg University of Applied Sciences, Germany

Corresponding author: Daniel Haake, daniel.haake@iee.fraunhofer.de

Abstract

This paper presents the design of a 250 kW 50 kHz SiC-based Half-Bridge-Series-Resonant-Converter (HB-SRC) which was designed as a Power Block of a modular 1 MW Input-Series-Output-Parallel (ISOP) DC-DC converter. To achieve the switching frequency of 50 kHz even at the rated power, Silicon-Carbide (SiC) MOSFET power modules were used and the HB-SRC was operated in discontinuous conduction mode (DCM) for soft switching operation. A prototype of the HB-SRC was built and its performance was experimentally validated.

1 Introduction

Power electronic converters for high power applications are often built as modular systems using identical submodules, called Power Blocks (PB) in the following. In particular, such a modular structure is used for medium and high voltage systems such as solid-state transformers (SST) for DC and

AC grids, railway traction systems or electrified agricultural machinery and utility vehicles [1] – [4]. The number of PBs necessary depends on the voltage level of the system, the converter topology and power semiconductors used to design the PBs as well as the rated power of the complete system. However, when a DC-DC converter stage is needed, mostly derivatives of the basic series resonance converter (SRC) and dual active bridge (DAB) are used due to their high efficiency and power density [1] – [4]. To increase efficiency and power density further, SiC power semiconductors could be used.

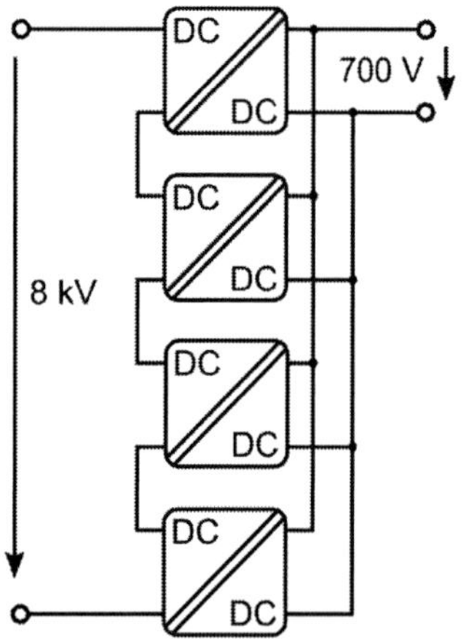

2 Converter Design

For an application of an electrified agricultural machinery, which is supplied with a medium voltage of 8 kV by a cable as described in [3], a 1 MW DC-DC converter with output voltage of 700 V was designed. For this application, high power density and high efficiency are crucial. To meet these requirements, the converter is built up of four identical Power Blocks (PB), which are connected in Input-Serial-Output-Parallel (ISOP), as shown in **Fig. 1**. Each PB of the modular DC-DC converter consists of a 250 kW HB-SRC with a switching frequency of 50 kHz and an input voltage of 2 kV (see **Table 1** for design parameters). In this application,

Fig. 1   1 MW ISOP DCDC converter structure [5]

2538

no regulation of the output voltage was required, therefore open-loop control of the converter system was used.

Table 1 Power Block design parameters

Symbol	Meaning	Value
	HB-SRC Parameter	
V_{in}	Nominal input voltage (MV-Side)	2 kV
V_{out}	Nominal output voltage (LV-Side)	700 V
$V_{\text{out,min}}$	Minimum output voltage	665 V
$I_{\text{out,max}}$	Maximum output current	376 A
	(at P_{out} and $V_{\text{out,min}}$)	
P_{out}	Nominal output power	250 kW
f_{sw}	Switching frequency	50 kHz
f_{res}	Resonance frequency	58 kHz
C_{MV}	Total input Capacitance	19.8 µF
C_{LV}	Output capacitance	480 µF
	Resonant Tank Parameter (ref. LV-Side)	
L_{σ}	Transformer leakage inductance	0.56 µH
L_{m}	Transformer magnetizing inductance	75 µH
C_{res}	Total resonance capacitance	12.87 µF
$N_{\text{LV}}{:}\,N_{\text{MV}}$	Transformer turn ratio	7:10

2.1 Topology

As mentioned before, the HB-SRC topology is used for the design of the PBs, shown in **Fig. 2**. In this application, the HB-SRC consists of a half bridge on the medium voltage side (MV-Side) with a split DC link, a medium frequency transformer (MFT) and the resonant tank and a full bridge on the low voltage side (LV-Side). The transformer is built as an integrated magnetic component, which means that it's leakage inductance L_{σ} is used as the resonance inductance of the HB-SRC. The converter operates as a DC transformer (DCX) with a fixed ratio between input and output voltage according to (1). To achieve this, the HB-SRC is operated in half-cycle discontinuous-conduction-mode (DCM) and the resonant tank must be designed carefully as described in [4]. In this operation mode, soft switching can be achieved over the whole operation range. The specific design of the resonant tank for this application is explained in detail in section 2.3.

$$V_{\text{out}} = \frac{N_{\text{LV}}}{N_{\text{MV}}} \cdot \frac{V_{\text{in}}}{2} \tag{1}$$

2.2 Power Block Concept

The requirements of the PB with a combination of a high input voltage of 2 kV, a high output power of 250 kW and a high switching frequency of 50 kHz led to several design challenges which had to be considered in the isolation coordination and the mechanical design of the PB, as follows.

2.2.1 Isolation Concept

In the targeted application of an electrified agricultural machinery, as mentioned above, (see [3] for details) the vehicle traction system was designed as an IT-system. All active conductors are isolated from vehicle ground so that the system could still operate under single isolation fault condition. Consequently, the isolation of all PBs in ISOP configuration should be able to withstand 4 kV in normal operation and a maximum of 8 kV under single isolation fault condition.

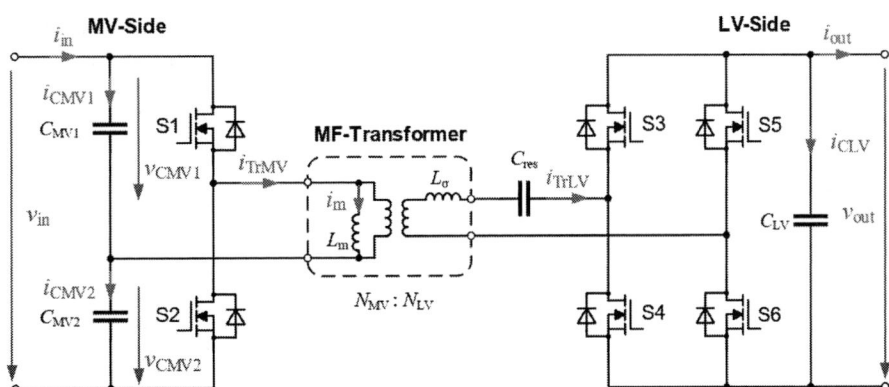

Fig. 2 HB-SRC topology used for Power Block design [6]

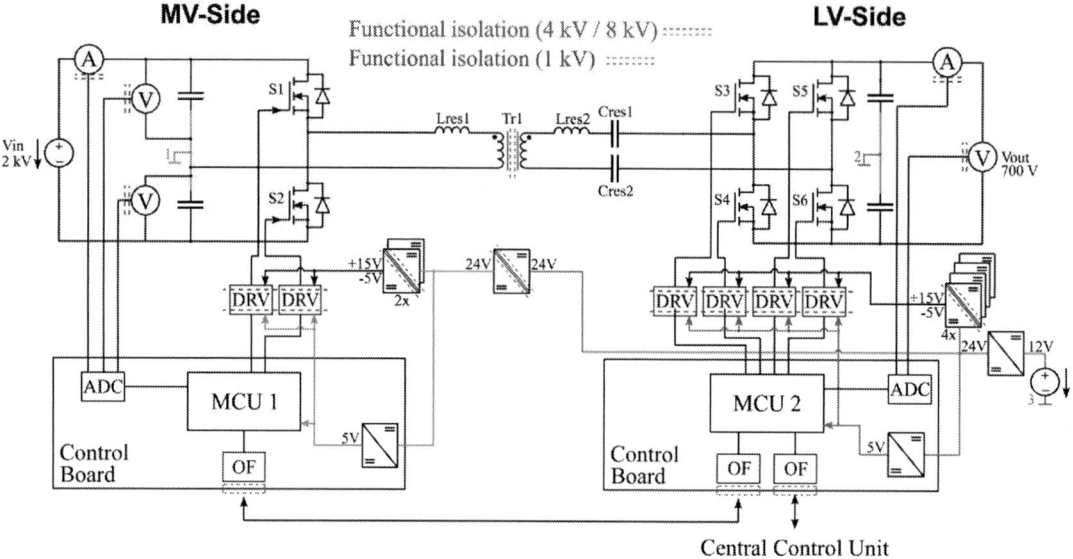

Fig. 3 Power Block isolation concept [3]

Therefore, the isolation concept shown in **Fig. 3** was utilized. For the auxiliary 24 V supplies of the MV-Side as well as the LV-Side converter, the respective midpoint potential of the capacitor voltage is used as refence potential ("local grounds", depicted in **Fig. 3** in orange and green). Both 24 V supply circuits are connected using an additional isolated DC-DC converter to ensure the required isolation between the LV-Side and MV-Side converters. Two separate control boards for measurement, control and fault handling are used for the MV-Side and LV-Side converters. These control boards are connected through optical fibers (OF) for communication and synchronization purposes. Using this isolation concept, the isolation effort was reduced significantly and the necessary number of components with a rated isolation voltage of 8 kV was minimized to only the MFT, 24V/24V DC-DC converter and OF, as marked red in **Fig. 3**. This approach also enables the use of commercially available standard components for inverters for industrial applications, e.q. gate drivers, driver supplies and sensors. For such components, especially on the LV-Side, only a lower isolation voltage rating of at least 1 kV must be considered (marked blue in **Fig. 3**). Such components usually are less expensive and smaller in size compared to components dedicated for MV applications. The isolation concept is described more detailed in [3].

2.2.2 Mechanical Design

Based on the isolation concept, the mechanical design of the PB was derived. The arrangement of the main components is depicted in **Fig. 4**. The MV-Side and LV-Side converter are physically separated by the transformer to ensure the required insolation. The power modules and capacitors are connected by specially designed busbars on both the MV-Side and the LV-Side. These busbars consist of laser-cut copper sheets to handle the high frequency currents in full load operation and to ensure the mechanical stability. As isolation between the copper sheets, laser-cut Mylar A foils are used, as shown in detail in **Fig. 5** and **Fig. 6**. The chosen design also minimizes the parasitic in-

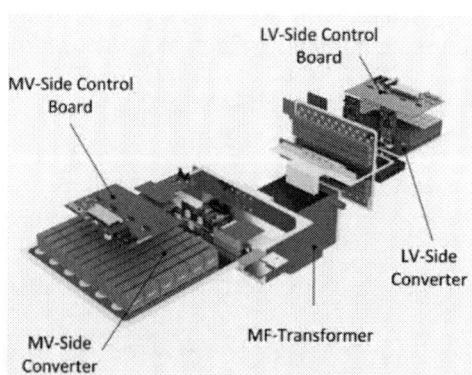

Fig. 4 Power Block of the DCDC-Converter (CAD)

ductance of the busbars. A parasitic inductance of 19 nH was measured for the MV-Side and a value of 10.4 nH was measured for the LV-Side of the converter using a Bode 100 impedance analyzer. Rectangular shaped copper busbars are used for the 2 kV terminals of the MV-Side as well as for the 700 V terminals of the LV-Side. These busbars are directly connected to the terminals of the power modules. The input and output DC current is carried by the respective busbars, while the high frequency AC current of the corresponding capacitors is carried by the busbars which connect those capacitors to the terminals of the power modules. By separating DC and high frequency AC current paths, the respective busbars could be optimized for either of these very different applications.

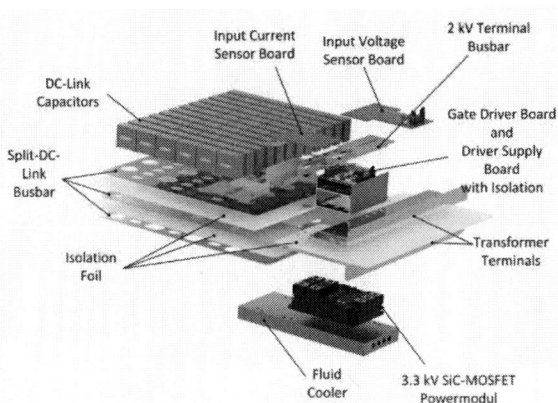

Fig. 5 Power Block MV-Side Converter (CAD exploded view)

The auxiliary circuits needed (see **Fig. 3**), such as voltage and current measurements, gate drivers and driver supplies etc. are designed as PCBs. They are integrated into the mechanical design of the busbars or connected to them by designated interfaces as depicted in **Fig. 4**. More detailed views for the MV-Side and the LV-Side are provided in **Fig. 5** and **Fig. 6**, respectively. 3D-printed parts are used for PCB mounting and also as additional isolation, where necessary.

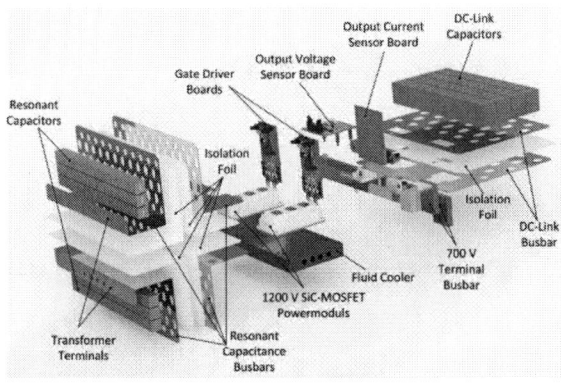

Fig. 6 Power Block LV-Side Converter [7] (CAD exploded view)

2.3 Resonant Tank Design

The series resonant circuit is comprised of a series connection of the internal leakage inductance L_σ of the MFT and an external resonant capacitance C_{res}. The resonance frequency can be calculated with (2).

$$f_{res} = \frac{1}{2\pi \cdot \sqrt{L_\sigma C_{res}}} \tag{2}$$

Since the resonant capacitance is located on the LV-Side of the converter, it must conduct the high LV-Side transformer current $I_{LV,max}$. The amplitude of the voltage over the resonant capacitance can be calculated according to (3).

$$\hat{V}_{res,max} = \frac{P_{out,max}}{\sqrt{2}\pi f_{sw} \cdot C_{res} \cdot V_{out,min}} \tag{3}$$

The amplitude of the voltage over the resonant capacitance $\hat{V}_{res}$ must always be smaller than the output voltage V_{out}. Therefore, $\hat{V}_{res,max} < V_{out,min}$ must apply. This leads to the following requirement for the dimensioning of the resonance capacitance.

$$C_{res} > \frac{P_{out,max}}{\sqrt{2}\pi f_{sw} \cdot V_{out,min}^2} \tag{4}$$

The resonant capacitance is split into two parts and is connected to both LV-ports of the transformer. To reduce the cost and to simplify the assembly, film capacitors were used to build the resonance capacitance. Several capacitors were connected in parallel to reduce the ESR and ESL values of the resonant capacitance and to achieve the required current rating of $I_{TrLV,max} = 450$ A, which was calculated using (5) according to [6].

$$I_{TrLV,RMS} = I_{out} \cdot \sqrt{\frac{\pi^2}{8} \cdot \frac{f_{res}}{f_{sw}}} \tag{5}$$

A good solution regarding the requirements can be achieved by using Kemet R76MI33901502 film capacitors for each part of the resonant capacitance. In total, 132 capacitors have been used. With 390 nF per capacitor and a configuration of two capacitor in series and 66 capacitors in parallel, a total capacitance of 12.87 µF has been achieved. The current rating of the resonant capacitance is calculated to be 726 A, which provides a safety margin of approx. 60 % with respect to a current rating of $I_{TrLV,max} = 450$ A. Furthermore, the requirement (4) is met. The leakage inductance can now be calculated with (2). With $f_{res} = 58$ kHz, the leakage inductance $L_\sigma = 585$ nH is obtained. Due to a very small value of L_σ, a tight coupling between the MV- and the LV-Side of the converter is achieved. However, special care must be taken to minimize the parasitic inductance of the resonant current loop since its parasitic inductance reduces the resonant frequency. Therefore, the leakage inductance is specified slightly less than the calculated value and amounts to L_σ = 560 nH.

The magnetizing inductance L_m is required to define the magnetizing current amplitude $\hat{i}_m$, which determines the discharging time of the output capacitances of the power modules. Using the datasheet values, the total charge $Q_{oss,total}$ stored in the output capacitances of the modules at V_{in} = 2 kV and V_{out} = 700 V of the power modules can be calculated to $Q_{oss,total}$ = 34.51 µC (referenced to the LV-Side).

The maximal discharging time is the dead time between the end of the resonance current pulse and the start of the next half cycle. It can be calculated according to (6).

$$T_{\text{discharge,max}} = \frac{1}{2} \cdot \left(\frac{1}{f_{sw}} - \frac{1}{f_{res}} \right) \qquad (6)$$

With the parameters given in the specification, $T_{\text{discharge,max}}$ = 1.38 µs is calculated. This allows the calculation of the magnetizing current according to (7).

Fig. 7   Liquid-cooled MF-Transformer

$$\hat{i}_m = \frac{Q_{oss,total}}{T_{\text{discharge,max}}} \qquad (7)$$

The obtained magnetizing current is 25 A. The maximum value of the magnetizing inductance is calculated following equation (8).

$$L_{m,max} = \frac{V_{out} \cdot \left(\frac{T_s}{2} - T_{\text{discharge,max}} \right)}{2 \cdot \hat{i}_m} \qquad (8)$$

This leads to $L_{m,max}$ = 120.7 µH. This inductance value is referenced to the LV-Side of the converter.

Considering that the discharging of the MV- and the LV-Side module capacitances occurs slightly time shifted, a smaller value of $L_{m,max}$ = 75 µH has been specified and implemented in the transformer shown in **Fig. 7**. Reducing the specified value of L_m provides a safety margin regarding the discharging of the output capacitances of the modules within $T_{\text{discharge,max}}$ and therefore enables zero voltage switching (ZVS).

2.4   MV-Side and LV-Side Capacitors

To select suitable capacitors for the realization of the MV-Side and LV-Side capacitances, it is necessary to determine the current stress during operation of the PB at an output power of 250 kW (at $I_{out,max}$). Therefore, equation (9) - (10) from [6] with the PB design parameters shown in **Table 1** were used (note that magnetizing current i_m is neglected in (9)). This leads to nominal currents of $I_{CMV,RMS}$ = 180 A for each of the capacitances on the MV-Side and $I_{CLV,RMS}$ = 247 A for the LV-Side capacitance. The maximum total capacitance value for the MV-Side was 20 µF according to the system specification due to safety requirements. This leads to a maximum capacitance of 40 µF. Therefore, 33 KEMET R75RW412050H4J capacitors with a nominal capacitance value of 1.2 µF, a current rating of 16.4 A and a voltage rating of 1250 V were used for each of the MV-Side capacitances.

The minimum capacitance value of the LV-Side capacitance C_{LV} = 157.2 µF was determined using (11) to achieve a maximum output voltage ripple of 1 %. To achieve the required current, voltage, and capacitance values, 48 WIMA DCP4N051006JD2KYSD capacitors with a nominal capacitance value of 10 µF, a current rating of 10 A and a voltage rating of 900 V were used.

$$I_{CMV,RMS} = I_{in} \cdot \sqrt{\frac{\pi^2}{4} \cdot \frac{f_{res}}{f_{sw}} - 1} \qquad (9)$$

$$I_{CLV,RMS} = I_{out} \cdot \sqrt{\frac{\pi^2}{8} \cdot \frac{f_{res}}{f_{sw}} - 1} \qquad (10)$$

$$C_{LV} = \frac{I_{out,max}}{2 \cdot \Delta V_{out}} \left[\frac{1}{f_{sw}} \sqrt{1 - \left(\frac{2}{\pi} \cdot \frac{f_{sw}}{f_{res}} \right)^2} - \frac{1}{f_{res}} + \frac{2}{\pi \cdot f_{res}} \arcsin\left(\frac{2}{\pi} \cdot \frac{f_{sw}}{f_{res}} \right) \right] \qquad (11)$$

2.5   Power Modules and Cooling

Due to the 2 kV input voltage, power modules with 3.3 kV SiC MOSFETs are used for the MV-Side half bridge of the PB. The used module is a customized variant of the Infineon module presented in [8] which are utilizing low inductive XHP™2

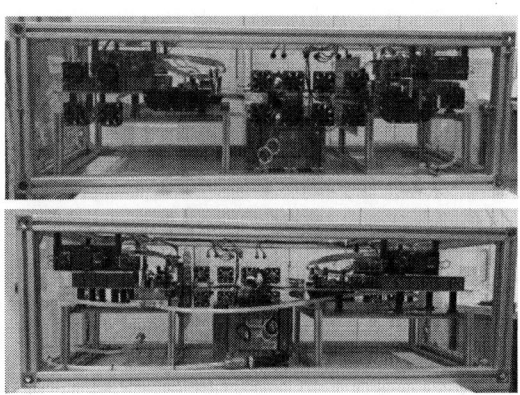

Fig. 8 Power Block Prototype

packages and 3.3 kV CoolSiC™ MOSFETs to achieve the switching frequency of 50 kHz. To reduce the output capacitance, the number of parallel chips for the high and low side switches of the module were reduced by a factor of 0.75. This facilitates soft switching of the HB-SRC at the cost of slightly increased conduction losses. For the realization of the LV-Side, two Infineon standard 62mm C-Series full bridge modules with 1.2 kV CoolSiC™ MOSFETs (FF2MR12KM1) are used.

Circuit simulations were carried out to determine the losses of the used power modules under full load conditions, resulting in overall losses of 1.02 kW for the MV-Side half bridge and 1.64 kW for the LV-Side full bridge. These values were used to determine the necessary cooling performance, i.e. the maximum permissible thermal resistance R_{th} of the heat sinks.

Targeting a maximum junction temperature of about 130 °C for the SiC semiconductors and assuming a maximum coolant temperature at the inlet of 65 °C results in a maximum R_{th} for the heat sink of approx. 39.8 K/kW for the MV-Side and 14 K/kW for the LV-Side, respectively. To ensure the required cooling performances for both the 3.3 kV SiC module on the MV-Side and the two 1.2 kV SiC modules on the LV-Side, two identical customized aluminium liquid coolers with an R_{th} of approx. 7 K/kW at a flow rate of 10 l/min were utilized.

2.6 Software and Synchronization

The control boards, one for the MV-Side and one for the LV-Side as depicted in **Fig. 3**, consist of two processing units each, an ARM-Processor and an FPGA. The ARM-Processor contains what is referred to as "Software", while the FPGA's code is referred to as "Firmware". The software is only implemented on one of the two control boards and is used to control both of them. It provides a simple interface allowing to configure both PWMs. The

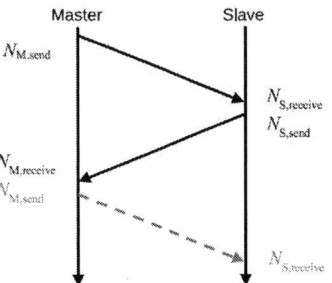

Fig. 9 Flowchart of the synchronization algorithm

$$N_{\text{offset}} = \frac{1}{2}\left(N_{\text{S,receive}} - N_{\text{M,send}} - N_{\text{M,receive}} + N_{\text{S,send}}\right) \quad (12)$$

firmware differs between the MV-Side and LV-Side converters. While they share the same safety functions, such as over-voltage and current detection, only one acts as a master and handles all communications.

For the PB to work, both converters need to keep their PWM signals synchronized. Typically, the oscillators used in the SoCs suffer from imperfections, resulting in a small frequency deviation. As the PWMs directly depend on them, the PWM itself will display a small deviation, causing the PWM signals of both converters to drift apart after some time. To prevent the PWM signals from drifting apart, a synchronization algorithm needs to be implemented. A slightly modified version of precision-time-protocol (PTP) was used. It allows to keep the PWM of both converters synchronized, while displaying a jitter of ±2 samples in relation to the base clock ($f_{\text{FPGA}} = 100$ MHz, $t_{\text{jitter}} = \pm 20$ ns). The PTP algorithm itself, given by the formula shown in [9], works by calculating the delay of the communication channel and the difference between the internal PWM counter of the master and slave. It uses a chain of messages in which the internal counter of the master and slave are compared during send and receive actions to isolate the channel delay from the counter offset. The main change done to the algorithm was that it only once calculates the channel delay according to (12) and only updates the counter offset every two transactions, as shown in **Fig. 9**.

3 Test Setup and Experimental Results

To evaluate the performance of the designed PB, two identical prototypes were built and operated in Back-to-Back configuration (B2B), as shown in **Fig. 10**. One PB prototype is shown in detail in **Fig. 8**. The complete test setup consists of the two PBs in B2B operation that are interconnected at their 2 kV terminals and a 700 V DC voltage source and a current sink with a rated power of 400 kW each are connected to the PB's 700 V terminals as depicted in **Fig. 10**.

For minimization of the switching losses, optimized pulse patterns according to [7] were used to provide ZVS of the PB's SiC MOSFETs. These pulse patterns also ensure minimal undesired current and voltage oscillations for the HB-SRC in DCM.

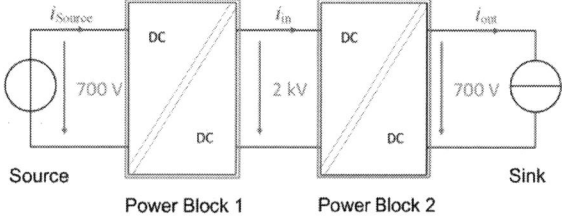

Fig. 10 Back-to-Back Test Setup [5]

In **Fig. 11**a – c, the measured waveforms the MFT LV-Side current and Drain-Source-Voltage of the MOSFETs S2 and S4 are shown, while the output power of PB2 is varied from 50 kW to the nominal 250 kW in steps of 50 kW. The LV-Side transformer current waveforms in **Fig. 11**a indicate that quasi zero current switching (ZCS) was achieved for the SiC MOSFETs of the LV-Side converter. The trapezoidal waveforms of the Drain-Source-Voltages of MOSFETs S2 and S4 depicted in **Fig. 11**b and c show that ZVS was achieved for both the MV-Side and LV-Side MOSFETs. These measurement results indicate that soft switching, ZVS for MV-Side and ZVS and ZCS for the LV-Side MOSFETs, is achieved over the entire operation range of the PB. This leads to minimal switching losses, which is the key feature to reach the specified 50 kHz switching frequency.

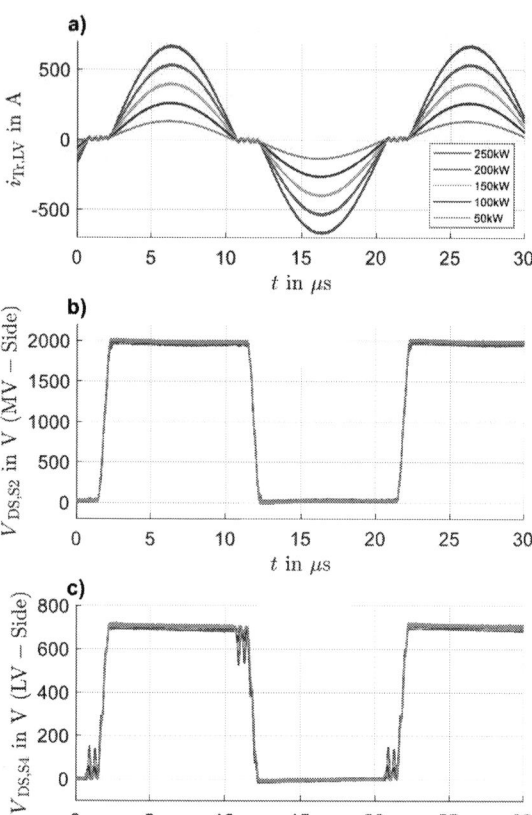

Fig. 11 Measured waveforms:
a) Transformer current (LV-Side)
b) Drains-Source-Voltage S2 (MV-Side)
c) Drains-Source-Voltage S4 (LV-Side)

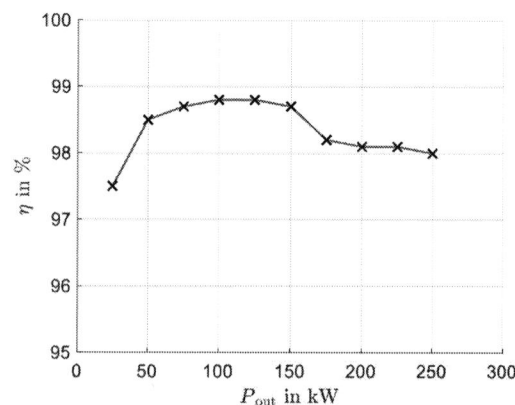

Fig. 12 Power Block efficiency over output power (power flow MV-Side to LV-Side)

In **Fig. 12,** the results of the efficiency measurements of PB2 are shown, which provided the energy flow from its MV-Side to the LV-Side. The efficiency measurement was carried out using a ZES Zimmer LMG671 power analyzer equipped with S-channels for high precision broadband measurements. Additionally, a precision wideband high voltage divider HST3-3 was used for the MV-Side voltage measurement and PCT1200 current sensors were used to measure input and output currents. At the nominal input and output voltages (see **Table 1**), the output power was varied in 25 kW steps between 25 kW and 250 kW.

At the rated output power of 250 kW, an efficiency of 98 % was achieved. The peak efficiency of 98.8 % was reached at partial load condition for output power between 100 kW and 125 kW.

Similar results with a slightly lower efficiency were achieved in [5]. Here, the same test setup was used, but with an oil-cooled MFT prototype with toroidal core and foil windings installed in PB2.

4   Conclusion

This paper presents the design of a Power Block for a 1 MW ISOP DC-DC converter. The Power Block consists of a 250 kW HB-SRC with a switching frequency of 50 kHz, an input voltage of 2 kV and an output voltage of 700 V. The HB-SRC was operating in DCM to achieve soft switching and power modules with 3.3 kV SiC MOSFETs were used on the MV-Side and 1.2 kV SiC MOSFETs on the LV-Side, respectively, to achieve high efficiency.

Due to the challenging specification of 8 kV input voltage of the 1 MW DC-DC converter built of four identical Power Blocks connected in ISOP, a special isolation concept was considered, which minimizes the number of components needed with a rated isolation voltage of 8 kV. This also enabled the usage of standard components with lower rated isolation voltage of at least 1 kV that are less expensive and smaller in size compared to components dedicated for MV applications.

The performance of the built Power Block prototypes was evaluated experimentally using two of them in Back-to-Back operation. An efficiency of 98 % was reached at the PB's rated output power of 250 kW and peak efficiency of 98.8 % at partial load condition for output power between 100 kW and 125 kW.

5   Acknowledgment

This work was funded by the German Federal Ministry for Economic Affairs and Climate Action BMWK under grant no. FKZ 03EN2014E through the research project "MUSiCel: Mobile Umrichter und Energieübertragungslösungen auf SiC Basis für elektrische, leistungsstarke Land- und Baumaschinen".

The authors express their gratitude to the funding agency for the opportunity to perform these investigations within the project "MUSiCel" and to the project partners for their assistance and continued cooperation in this (and other) projects.

Only the authors are responsible for the content of this publication.

References

[1]   L. Ferreira Costa: "Modular Power Converters for Smart Transformer Architectures", dissertation, University Kiel, 2019

[2]   H. Tu, H. Feng, S. Srdic, S. Lukic: "Extreme Fast Charging of Electric Vehicles: A Technology Overview", IEEE TRANSACTIONS ON TRANSPORTATION ELECTRIFICATION, VOL. 5, NO. 4, DECEMBER 2019

[3]   D. Tatusch, A. Gorodnichev, D. Haake, F. Schnabel, J. Friebe, M. Jung: "Hardware and control design considerations for a mobile 1 MW Input-Series Output-Parallel (ISOP) DC-DC converter in Medium Voltage range", IEEE Energy Conversion Congress and Expo 2021 (ECCE 2021)

[4]   G. Ortiz: "High-Power DC-DC Converter Technologies for Smart Grid and Traction Applications", dissertation, ETH Zürich, 2014

[5]   S. Lin, D. Haake, A. Gorodnichev, J. Friebe: „Design and Experimental Verification of an Oil-Cooled Medium-Frequency Transformer for a 250kW Half-Bridge Series Resonant Converter", International Power Electronics and Motion Control Conference 2024 (IPEMC 2024 ECCE Asia)

[6]   D. Haake, A. Gorodnichev, F. Schnabel, M. Jung: "Impact of Higher Current Harmonics on Component Current Stress and Conduction Losses of Half-Bridge-Series-Resonant-Converters in Discontinuous Conduction

Mode for High-Power Applications", European Conference on Power Electronics and Applications 2022 (EPE'22 ECCE Europe)

[7] A. Gorodnichev, D. Haake, M. Klee, M. Jung: "Switching Transition Analysis in a Subresonant Half-Bridge Series Resonant Converter for High Power and High Switching Frequency Applications", IEEE Energy Conversion Congress and Expo 2023 (ECCE 2023)

[8] M. Buerger, K.-H. Hoppe, K. Schraml, A. Wedi: "The New XHP2 Module Using 3.3 kV CoolSiC MOSFET and .XT Technology", International Exhibition and Conference for Power Electronics, Intelligent Motion, Renewable Energy and Energy Management 2023 (PCIM Europe 2023)

[9] IEEE Standards Association. Standard for a Precision Clock Synchronization Protocol for Networked Measurement and Control Systems; IEEE: Piscataway, NJ, USA, 2020; pp. 1–499

PCIM Europe 2024, 11–13 June 2024, Nuremberg   DOI: 10.30420/566262357

30kW – 97% Efficiency Isolated DC-DC Converter with Large Input Voltage Range Based on a BOOST-DAB Association

Olivier Martos[1-2], Jean-Jacques Huselstein[1], François Forest[1], Patrice Levron[2]

[1]INSTITUTE OF ELECTRONICS & SYSTEMS, Montpellier University, France

[2] EXAIL-ROBOTICS, France

Corresponding author: Jean-Jacques Huselstein, jean-jacques.huselstein@umontpellier.fr
Speaker: Jean-Jacques Huselstein, jean-jacques.huselstein@umontpellier.fr

Abstract

This paper presents a 30kW - 97% efficiency isolated DC-DC converter with fixed (200V or 400V) regulated output voltage and wide range (200V to 600V) input voltage.

After theoretical investigations on single stage isolated converters, our main orientation was to design a Dual Active Bridge (DAB) with fixed input and output voltages, to ensure galvanic insulation with very high efficiency. The fixed input voltage of the DAB is provided by an 8-phase interleaved BOOST converter. An original planar transformer structure has been designed for the DAB. Experimental results for the full structure are presented in the article.

1   Specifications and overall structure

The design of the final isolated converter must comply with the following specifications:

Output power:	P_{OUT} = **30kW**
Input voltage:	V_{IN} = **200 to 600V**
Output voltage:	V_{OUT} = **200 or 400V**
Switching frequency:	F_{SW} > **50kHz**
Efficiency:	η > **96%**

Table 1 Converter specifications

In this process of converter design, the choice of the topology is an important step. The search of high efficiency is primarily challenged by the significant variation on the input voltage. However, let's be clear about the fact that this efficiency level must be reached for V_{IN} > 300V. All isolated topologies are notably impacted by this aspect. Passive and active components will be strongly stressed on both current and voltage.

With that in mind, different single stage isolated converters could be used for this specification. Three isolated stages, widely used in this range of output power have been analyzed: the serial LLC converter, the ZVS phase shift full bridge and the DAB. All of them have the great advantage to allow soft switching for most of the power switches.

In this output power range and frequency, this switching behavior is mandatory to reach this efficiency level.

The DAB has been largely presented in the literature, specifically with adaptive and complex control to compensate for the effects of wide voltage variations (200V to 600V input voltage range four our application). A different path has been taken during this work. Best performances are reached for the DAB if the input/output voltage ratio is constant. This provides a significant increase in performance and reduce constraints of power switches and passive components.

Provide a stable input output voltage ratio to the isolated stage can be achieved with a pre-regulator, such as a BOOST converter. This is something feasible with the high increase in performance of power switches and has the advantage of significantly reduce complexity on all components (passive and active).

In order to choose the best-performing structure, the components constraints, qualitative criterions have been studied on each option mentioned. Semiconductors losses estimations, based on experimental results have been also added to the study. In this comparison, all options used same number of components, at same junction temperature and for both output voltage option [1].

Finally, parallelization of switching cells (instead of components) has also been a major choice in this study, to reduce as much as possible EMI interferences, losses, and integration complexity.

Losses computations have demonstrated the relevance of the BOOST-DAB association for this project. This one presenting the best performances in the nominal behavior (Fig. 1 & Fig. 2).

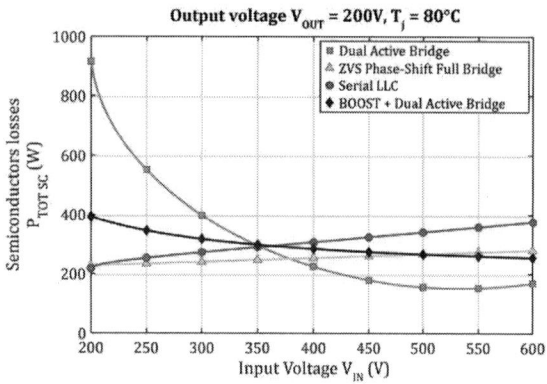

Fig. 1 Semiconductor losses comparison by topology for V_{OUT} = 200V

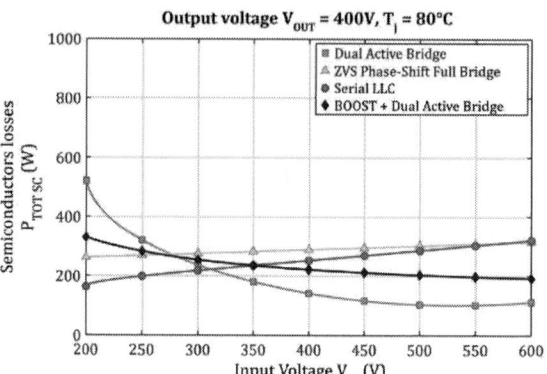

Fig. 2 Semiconductor losses comparison by topology for V_{OUT} = 400V

This last combination achieved the best overall calculated performance and led to a complete experimental realization with a non-isolated hard-switched 8-phase BOOST converter which provides the fixed input voltage of the second stage, the DAB (Fig. 3).

First, to reduce the DAB's current constraints and to be compatible with the input voltage range, the step-up function seemed to be the most efficient topology for this non-isolated stage. The intermediate voltage V_i between the BOOST and the DAB, based on theoretical estimations on transformer, performances of semiconductors, was fixed at 650V.

To ensure high efficiency, 8 interleaved cells were needed for the BOOST converter. Branch currents are significantly reduced by this "high" number of cells. To give numbers, at the worst operating point in terms of current (P_{OUT} = 30kW, V_{IN} = 200V), the average input current I_{IN} is about 150A. This configuration leads to an average current of 18.75A per leg or MOSFET. At the same time, this structure allows excellent filtering through the interleaving effect, a high reduction in the volume/mass of reactive components and a reduction of the switching frequency and losses.

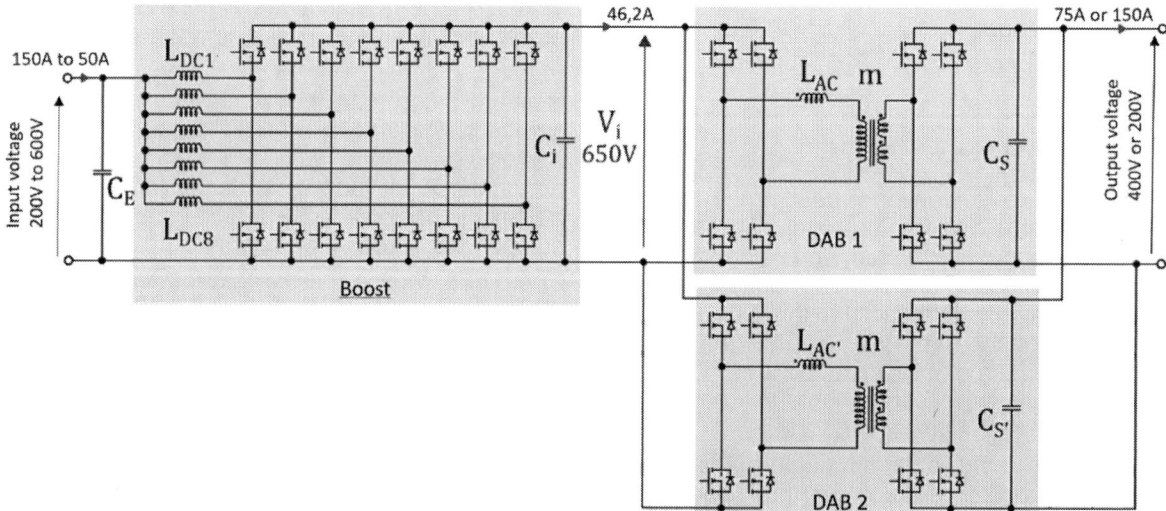

Fig. 3 BOOST-DAB association for a 30kW converter, 400V output voltage version

The total power of 30kW has been split between two 15kW-DAB structures with soft switching behavior and a switching frequency of 60kHz. A 15kW unit power made the construction of the transformers easier (1kg by transformer, windings and cores included) and enables a current level into the DAB bridges compatible with the available SiC MOSFETs capabilities. In addition, the interleaved control of the two DABs greatly reduces current ripples at the input and output of the full DAB stage.

Due to operating voltages (600-650V max in all converter stages), 1200V/14mΩ SiC MOSFETs (C3M0016120K) have been chosen for the entire converter. Those components enable both high switching speeds and very low conduction losses at this voltage level.

2 DAB transformers

2.1 Preliminary choices

Out of all the magnetic components in this converter, attention was first directed towards the transformer. After the specification analysis and examination of various isolated structures, the DAB operating at a fixed input/output voltage ratio appeared to be the most promising candidate. This operating configuration enables optimal transformer performance and facilitates its implementation.

As a reminder, the total power of the final magnetic component is 30kW. The decision between a single transformer or an array of scaled-down power transformers arises from a trade-off. It considers size constraints, estimated performance, and implementation complexity. The DAB behavior imposes on primary side a square wave with a 650V amplitude. On the secondary side, depending on the output version, 200 or 400V is applied.

To address this compromise, several aspects have been analyzed: the transformer technology, the shape of magnetic core, and the winding implementation.

Based on the laboratory's expertise in the field, as well as the demonstrated performance, the decision has been made to employ planar transformer. At equivalent product area, the planar technology is the one that allows for better heat dissipation [2]. Another advantage is the higher power density compared to classical technology.

To reach the required power, several cores of the largest model (E64) were assembled to obtain a larger winding surface.

This also has the advantage of reducing the relative effect of the coil heads as the number of magnetic circuits used increases. Several magnetic components can be used with this core shape. We chose the one presenting the lowest losses and the lowest temperature dependency, 3C95, with a minimum loss at 60°C (Fig. 4).

2.2 Transformer design

Transformer design was based on the use of advanced core and winding area product calculations with thermal constraints [3]. Iron losses were estimated using a broadband Steinmetz model. Copper losses were evaluated with a 1D Dowell model [4], with high frequency effects considered, supplemented by FEA simulations.

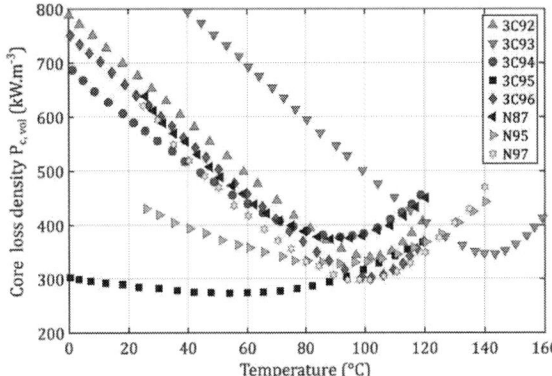

Fig. 4 Magnetic material losses, @0.2T, 100kHz

Analytically, the area product formula used for the design can be expressed as follows:

$$AP = P_{OUT} \cdot \frac{K_{P1} \cdot \frac{K_{w1}}{J_1} + K_{P2} \cdot \frac{K_{w2}}{J_2}}{\Delta B. F} \quad (1)$$

The area product gives an idea of the magnetic component volume. Obviously, the more the power P_{OUT}, the more the volume. Inversely with the frequency.

Some coefficients K_{P1}, K_{p2}, K_{w1} and K_{w2} are respectively related to the converter topology and the winding management. The two other parameters ΔB and $J_{1,2}$ indirectly represents the core P_c and the winding losses P_w.

Those two last parameters are, at the beginning, unknown and geometrically dependent. It's impossible to directly evaluate the equation (1). Maximum temperature rise ΔT evaluation is needed:

$$\Delta T = \frac{P_w + P_c}{H. A_{TH}} \quad (2)$$

Where H and A_{TH} are respectively the thermal exchange coefficient and surface area.

Winding losses are estimated with a 1D model based on Dowell formulas:

$$P_w = \rho \times \left(Vol_{w1}.J_1^2.g_1 + Vol_{w2}.J_2^2.g_2 \right) \quad (3)$$

Where ρ and $Vol_{w1,2}$ are respectively the conductor resistivity and the volume of the winding considered. g_1 and g_2 represents the AC vs DC resistance ratio based on current waveforms and winding management.

Core losses are on the other hand estimated with a modified Steinmetz model:

$$P_c = (K_{c1}F^{\alpha_1} + K_{c2}F^{\alpha_2}).\Delta B^{\beta_F}.Vol_c \qquad (4)$$

Where $K_{c1,2}$, $\alpha_{1,2}$ and β_F are magnetic material coefficients and Vol_c the volume of the core.

Because both side of the equation (1) are geometrically dependent, numerical approach is mandatory.

First, by imposing the frequency, a homothetic approach is used. It assumes that all the geometrical parameters evolve in a homothetic way. A core shape is imposed at the beginning, and geometry will evolve in order to respect imposed ΔT. This approach is the first step to have an idea of the final volume, more or less close to the initial magnetic circuit.

Secondly, the area product is imposed by imposing the magnetic circuit, and the frequency is the adjusting parameter.

After iteration, theoretical results of this design are summarized in the table below:

Power:	P = 15kW
Frequency:	F = 60kHz
Primary/Secondary turns:	$N_1 = 6$ / $N_2 = 4$
Primary/Secondary winding thickness	$e_1 = 200\mu m$ $e_2 = 300\mu m$
Maximum induction:	$\Delta B_{MAX} = 0.196T$
Current density:	$J_{1,2,\,RMS} = 7.9 Amm^{-2}$
Core losses:	$P_c = 24.1W$
Winding losses:	$P_w = 38.6W$
Temperature rise:	$\Delta T = 60°C$

Table 2 Transformer theoretical design

2.3 Transformer construction

The final design led to the construction of two 15 kW planar transformer units.

Regarding the winding construction, PCB technology with high copper thickness has been chosen. For the application, a maximum thickness of 210µm has been imposed by technical limitations on the market. This corresponds to the calculated primary optimal thickness but is 30% less than the calculated secondary optimal thickness. For the insulation, 350µm of FR4 have been used for each turn (and PCB layer).

Interleaved winding disposition between primary and secondary is represented in the Fig. 5. This arrangement limits the maximum magnetic field seen by the conductors and therefore minimizes the increase in apparent resistance at high frequencies due to the proximity effect.

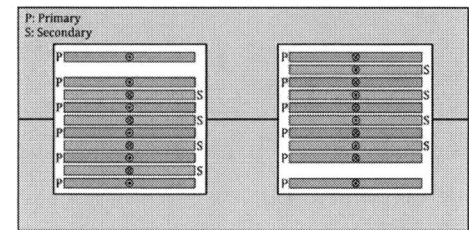

Fig. 5 Interleaved transformer's windings cross-section

The construction of the two nested windings from single-layer PCBs was obtained by a turn-by-turn interconnection by 200µm thick copper ribbon bridges brazed onto the PCBs (Fig. 6).

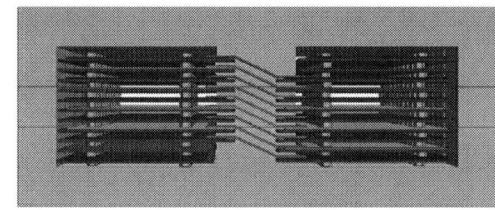

Fig. 6 Nested connections of primary and secondary turns. Soldered copper interconnections between the PCB layers.

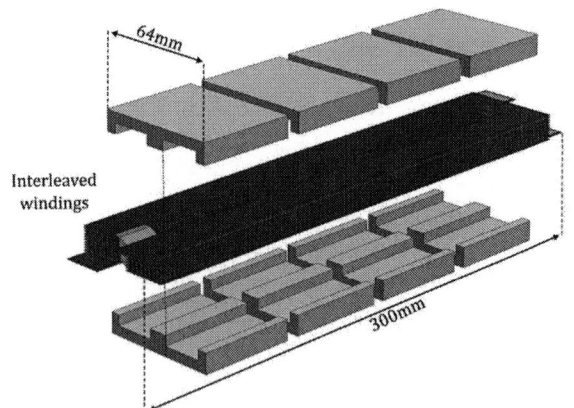

Fig. 7 Exploded view of the transformer

2.4 Transformers secondary windings for 200V and for 400V output versions

Our applications must include two output voltage versions. Either a 200V and 75A DC output per 15kW channel, or a 400V and 37.5A DC output per channel.

For the 400V version the secondary is formed by 4 winding turns and use one active bridge and one AC coil one the primary side.

For the 200V version the secondary is formed by two 2 turns windings which are parallelized indirectly with two active bridges on the DC side (Fig. 8). This allows a good balance between the currents in these two secondary windings.

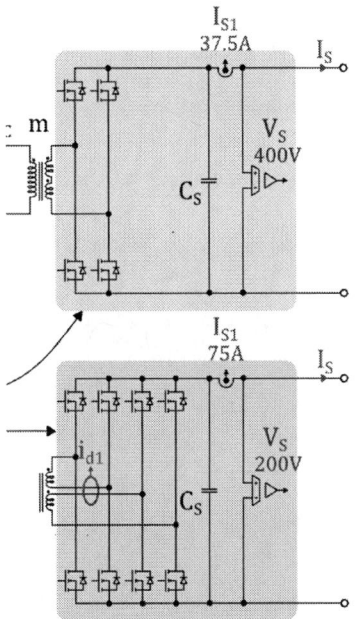

Fig. 8 400V DC output (top) and 200V output (bot)

Moreover, current unbalance between the two secondary windings could occurs. To allow natural balance for this 200V output version, without need of regulation, two AC inductors have been used on the secondary side of the DAB (9).

Our experimental tests have confirmed the excellent natural balancing of currents. The small secondary currents amplitudes imbalance essentially depends only on the eventual unbalance in the AC coil values.

Additionally, the total number of turns of the AC inductor is divided into two parts. The two half windings are each placed on one of the two legs of the bridge. It has been experimentally confirmed that this very considerably reduces the common mode currents during switching and the disturbances generated (to see further).

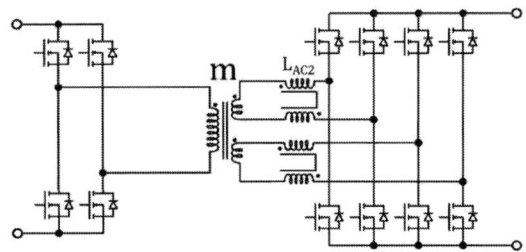

Fig. 9 200V output version with two 2 turns windings on transformer secondary and two AC coils with distributed windings

For the 400V output version, the AC inductance has been located on the primary side, to reduce current stress on this component and to facilitate construction.

Note: if we consider that the volume of an inductor is proportional to the storable energy, for identical DAB current variation durations, placing the AC inductor on the primary side or on the secondary side modifies the value of the inductor and its maximum current but not its volume.

2.5 Parasitic capacitances between transformer windings

Due to the large surface area of the windings and they interleaved arrangement the parasitic capacitance between primary and secondary is high. The effects of this parasitic capacitances have been encountered during experiments.

The square voltage applied by the primary bridge commutes between +650 V and -650 V. Fast voltage steps with an amplitude of 1300 V, together with the parasitic secondary primary capacitance, create large-amplitude common-mode current peaks and could generate large oscillations.

To reduce the capacitance, we can increase the gap thickness between the plates (PCB winding layers) and reduce the relative permittivity of the dielectric (use air instead of a polymer, $\varepsilon_r = 1$ instead of 4.4 for FR4).

This raised a dilemma between close-spaced winding (better thermal conduction and lower losses) and spaced winding (lower parasitic capacitances).

We therefore chose to separate the primary and secondary windings by layers of air and to use this air gap for forced convection cooling. To mechanically maintain the air gap between the PCBs, small mechanical spacers (3D printed) are placed between the PCBs for each turn (Fig. 10). That only represent 10% of the total area. In this way, most of the insulation material is air (measured total capacitance reduction from 1nF to

220pF) and forced air cooling is achieved inside the transformer itself. An air duct system has then been designed and 3D printed to achieve this cooling (Fig. 11). If the spacing increases slightly the copper losses, because of less efficient interleaving, global copper and core losses are reduced due to the temperature rise reducing (better copper conductivity).

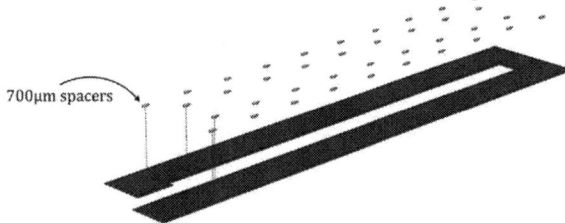

Fig. 10 Spacer insertion on each winding turn PCB layer

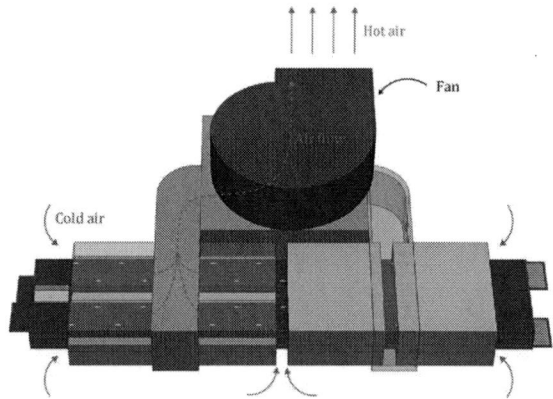

Fig. 11 Transformer assembly with cooling system

Four thermal sensors have been inserted inside the winding (two on primary and two on secondary winding) to measure the temperature rise during full power testing. At 16kW, the temperature of the windings only reached 36°C for a 22°C room temperature.

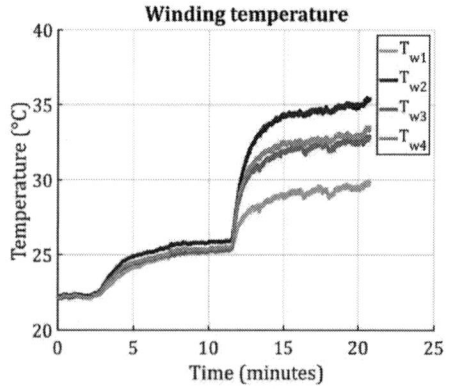

Fig. 12 Windings temperature during full power testing

These latest tests have validated the cooling method as well as the integration in the final prototype. Indeed, this cooling system can be centralized to extract all the sources of heat (power switches, inductances). In this way, this overcooled test presented before will be reduced to reach reasonable temperatures commonly encountered in this domain.

3 Tests and losses measurements

3.1 .Operation of the DAB stage

For illustration the Fig. 13 shows voltages and currents for the DAB operation for the 200V output voltage version at is nominal conditions. Since the voltage ratio is fixed and equal to the ratio of the number of turns in the transformer, currents are very close to the ideal form. Stresses on the switches are minimized, and the RMS values of the currents are at their minimum.

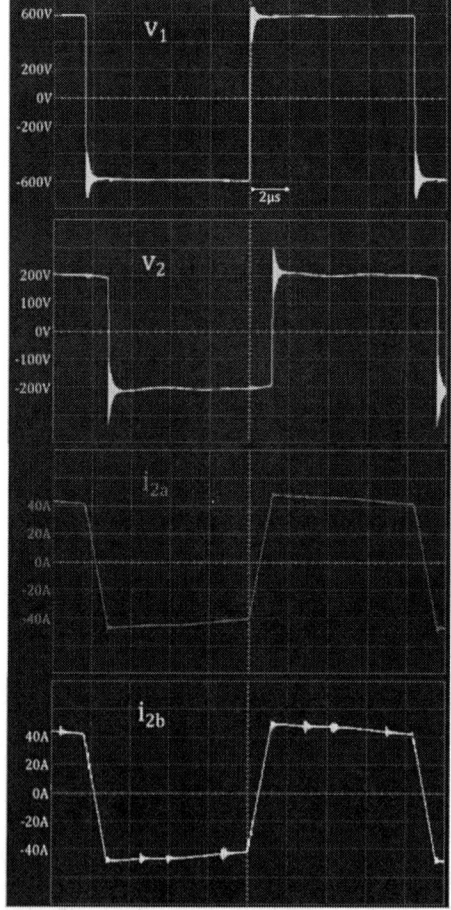

Fig. 13 DAB transformer operating: v_1: primary voltage, v_2: one of the secondary bridges voltage. i_{2a} and i_{2b}: currents on the booth secondary windings (showing the good current balance).

3.2 Opposition method for test and losses measurements for the 400V version

The experimental setup for evaluating the converter performance is illustrated in Fig. 14. This is a "load-free" configuration, where the system is looped on itself, and the DC power source only provide the total losses of the system and the required voltage levels [5].

$$P_{loss} = V_{IN} \times I_{loss} \quad (5)$$

$$\eta = \frac{V_{OUT} \times I_{OUT}}{V_{OUT} \times I_{OUT} + P_{loss}} \quad (6)$$

This is the most precise method to evaluate global losses of the converter. Because the entire converter can be halved, it's only necessary to test half the converter. At this efficiency level, the total power should not exceed 625W for 15kW.

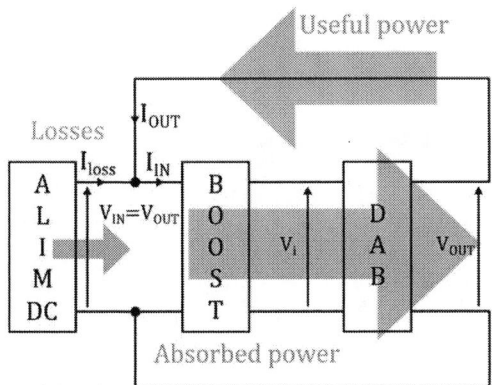

Fig. 14 Opposition method for no load nominal operating point

However, this configuration has the limitation to evaluate the losses for only one operating point, where the input voltage of the BOOST is equal to the output voltage of the DAB. This limits our estimation, but this is close to the worst operating point, with high current all over the converter. In this way, for other operating points with higher voltage, the losses should be lower, and the overall performance still validated.

Fig. 15 Full-bridge with its cooling system

This loss estimation has also been completed with separated losses estimations for all the other parts. First, all the power full bridges were mounted onto a dedicated cooling system (Fig. 15).

This configuration allowed us to fast implementation of the entire converter, but also helped us to estimate the related losses.

This losses estimation is based on the following assumption: most of the heat flux is dissipated through the heatsink. Fig. 16 illustrates this assumption and the simplified thermal model used for the losses estimation.

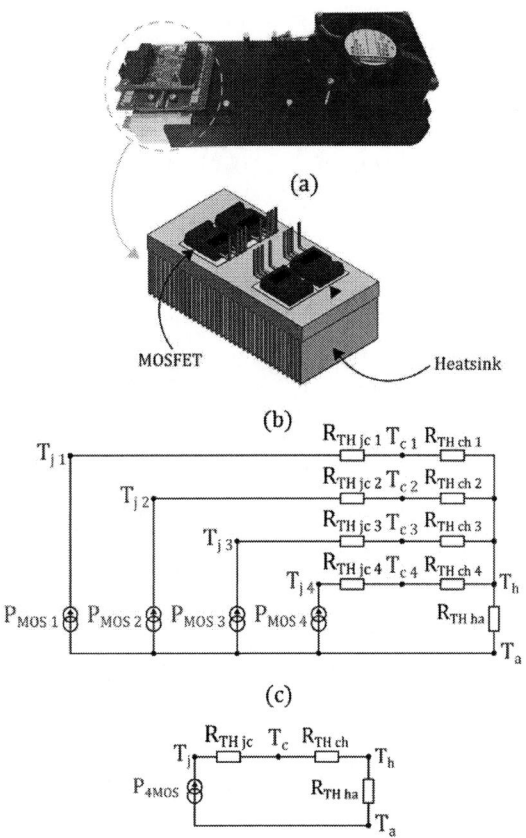

Fig. 16 (a): Full bridge with its cooling system, (b): detailed view of the heatsink, (c): simplified thermal schematic

An initial "calibrating" step, before any loss estimation has been done. Indeed, by applying a controlled DC current through the body diodes of the MOSFETs, all the losses (roughly 71W) are dissipated through the diodes. With the assumption explained before and with the measurement (see Fig. 17) of ΔT_{h-a}(19.3°C), the thermal resistance between the heatsink and the air $R_{TH\ h-a}$ has been evaluated at 0.27°C/W.

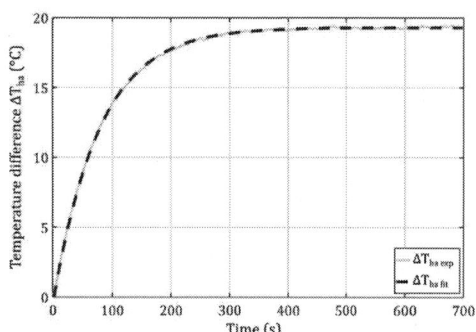

Fig. 17 Thermal measurement for the calibration

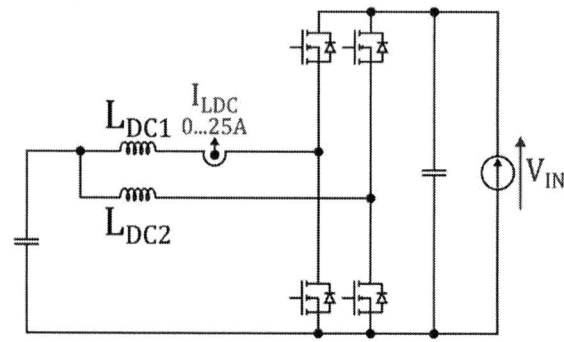

Fig. 18 DC inductances' losses evaluation test bench

During the tests, evaluating the temperature difference ΔT_{h-a} will allow us to evaluate the semiconductors' part of the global losses.

In this goal of evaluating each source of losses, the magnetic part still needs to be identified. A test bench has been used to evaluate separately the DC and AC inductances. The illustrative process of the test is shown on the Fig. 18.

This test bench allows to apply DC current with ripple like the nominal operating point. By neglecting the capacitor losses, and extracting the semiconductors' losses, then, the DC inductances' losses can be evaluated. With an operating DC current of 25A, losses for the two inductances have been estimated at roughly 50W.

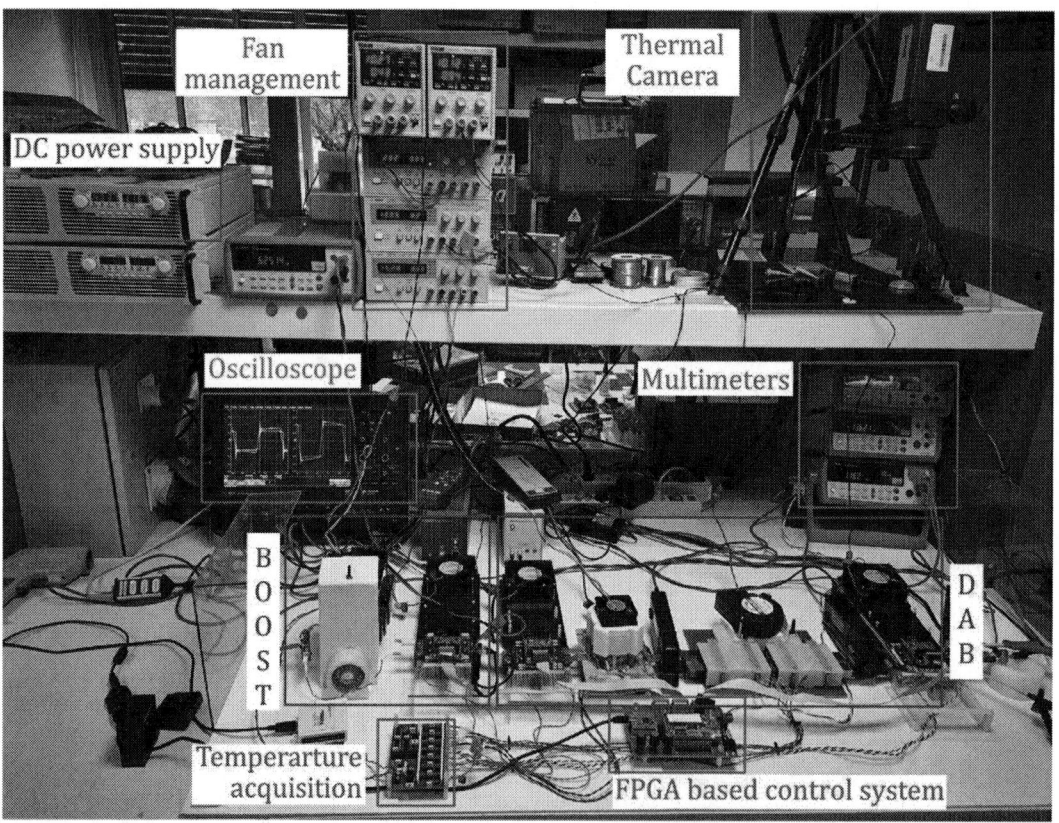

Fig. 19 Test bench for the converter: opposition method here with 400V input/output voltage (first converter version in open layout configuration).

A similar approach has been used for the AC inductances by applying a trapezoidal current, also similar to the operating point. A total of 63.5W has been estimated.

A complete test bench has been implemented, for a 15Kw / 400V V_{OUT} configuration (see Fig. 19). This configuration, with all the subsystems of the full converter, allowed us to demonstrate its technical feasibility. It facilitated all the electrical and thermal measurements.

A "built-in" version of this prototype has then been developed, on a more industrial vision. Further explanations will be provided on the section 4.

The electrical characteristics of the test can be resumed in the following table:

Input voltage:	$V_{IN} = \mathbf{400V}$
Input voltage:	
"Loss" current:	$I_{loss} = \mathbf{1.176A}$
Intermediate voltage:	$V_i = \mathbf{634V}$
Output voltage:	$V_{OUT} = \mathbf{400V}$
Output current:	$I_{OUT} = \mathbf{41A}$
Useful power:	$V_{OUT} \times I_{OUT} = \mathbf{16.4kW}$
Losses:	$P_{loss} = \mathbf{470W}$
Efficiency:	$\eta = \mathbf{97.22\%}$

Table 3 Losses evaluation for the full converter (DAB + BOOST association) – 400V version

With the thermal losses estimation explained before, an overview of the losses can be drafted (see 20). All the losses contributions have been evaluated by experiments, except for the transformer which represents the remaining part of the total losses, after subtraction of the other parts.

This test has proven the relevance of the BOOST+DAB association that respects the specifications, by exhibiting an overall efficiency of **97.22%** for an output power little higher than the nominal value and for the 400V version.

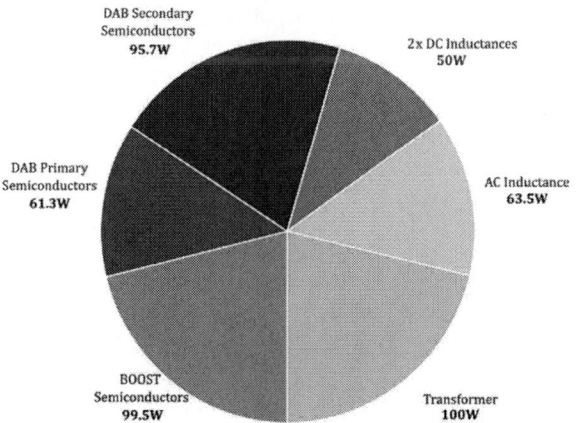

Fig. 20 BOOST-DAB association losses distribution, for 400V output voltage

For the 200V version, the feasibility tests have been done with the built-in version of the converter, explained in the next section.

4 Built-in version of the converter

4.1 Construction

At the end of this project, a built-in version has been developed. We decided to divide the full converter into two similar blocks, from a visual standpoint and from the perspective of their cooling systems. Those similarities allowed to fast implementation of the final prototype.

The first block integrates 30kW non-isolated DC-DC BOOST stage with 8 interleaved cells. Those are composed of SiC MOSFETs legs. The 8 DC inductors are mounted in the air box (only 4 represented in the share the same air flow used for the MOSFETs.

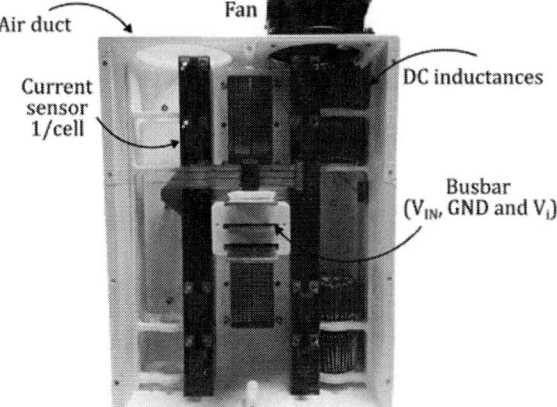

Fig. 21 Built-in version of the BOOST converter: inside view of the air box

The MOSFETs are controlled by a FPGA card ensuring current and voltage regulation and communicating via fast serial link with the DAB block control card.

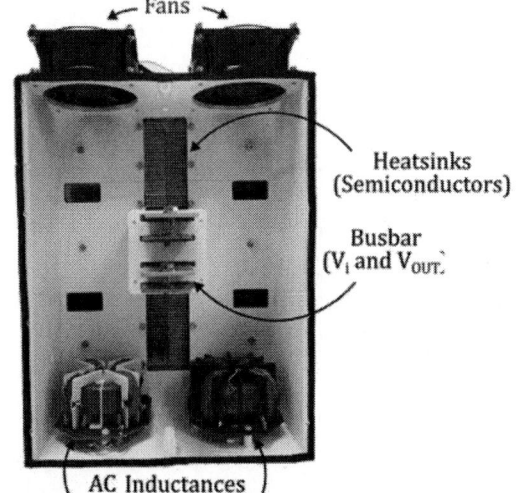

Fig. 22 Built-in version of the DAB converter: inside view of the air box

The second block integrates the isolated 30kW DAB stage, divided into two units of 15kW. Each primary is composed of one full bridge, made with 4 SiC MOSFETs. Each secondary is made up of one or two parallelized full bridges, depending on whether it's 400V and 200V output version respectively.

The magnetic part is made with two 15kW transformer, on each side of the block. A dedicated air duct allows to channelize the air between the transformer turns into the air box, sharing then the same cooling system with the MOSFETs and the AC inductors.

The MOSFETs are also controlled by a FPGA card providing close control as well as current and voltage regulation and communicating via a fast serial link with the BOOST block control card.

Fig. 23 Built-in version of the DAB converter: front view

4.2 Losses tests for the 200V version

Similar test done for the 400V version has been done. The electrical characteristics for this test can be resumed in the following table:

Input voltage:	V_{IN} = **204V**
"Loss" DC current:	I_{loss} = **3.59A**
Intermediate voltage:	V_i = **632V**
Output voltage:	V_{OUT} = **204V**
Output current:	I_{OUT} = **76.5A**
Useful power:	$V_{OUT} \times I_{OUT}$ = **15.6kW**
Losses:	P_{loss} = **733W**
Efficiency:	η = **95.5%**

Table 4 Losses evaluation for the full converter (DAB + BOOST association) – 200V version

This voltage configuration (input and output) is the worst operating point (see Table 1), due to higher current values. Even if the efficiency specified is not mandatory for this operating point, the result of this test is very close with a value of 95.5%.

Conclusion

The full converter was built for half the specifications total power (a single 15 kW DAB stage and therefore a single transformer). The BOOST stage was also halved (4 interleaved phases with 4 DC inductors). This system was tested at its full power of 16kW. Despite the cascading of two converters, the overall efficiency of the complete structure was measured higher than 97% with the 400V output and almost 96% with the 200V output version.

References

[1] **O. MARTOS**, PhD thesis « Étude d'un convertisseur DC/DC isolé de forte puissance, destiné à l'alimentation de bord de navires civils et militaires. ». Université de Montpellier, **2022**.

[2] **J.-S. N. T. MAGAMBO**, PhD thesis « Modélisaton et conception de transformateurs planar pour convertisseur de puissance DC/DC embarqué », Ecole Centrale de Lille, **2017**.

[3] **F. FOREST, E. LABOURE, T. MEYNARD, M. ARAB**, « Analytic design method based on homothetic shape of magnetic cores for high-frequency transformers », *IEEE Transactions on Power Electronics*, vol. 22, n° 5, Art. n° 5, **2007**

[4] **P. L. DOWELL**, « Effects of eddy currents in transformer windings », *Proceedings of the Institution of Electrical Engineers*, vol. 113, n° 8, Art. n° 8, **1966**,

[5] **F. FOREST, J. HUSELSTEIN, S. FAUCHER, M. ELGHAZOUANI, P. LADOUX, T. A. MEYNARD, F. RICHARDEAU ET C. TURPIN**, «Use of opposition method in the test of high-power electronic converters,» IEEE Transactions on Industrial Electronics, vol. 53, pp. 530-541, 2006.

PCIM Europe 2024, 11– 13 June 2024, Nuremberg   DOI: 10.30420/566262358

Full-Bridge Push-Pull Forward Dual Active Bridge DC-DC Converter

Sandip Guha Thakurta[1], Gean Sousa[1], Christos Leontaris[1], Marcelo Lobo Heldwein[1]

[1] Technische Universität München, Germany

Corresponding author: Sandip Guha Thakurta, sandipguha.thakurta@tum.de
Speaker: Gean Sousa, gean.sousa@tum.de

Abstract

This work introduces a full-bridge push-pull forward dual active bridge DC-DC converter designed exemplarily for on-board charging applications. The topology is derived from the conventional dual active bridge structure, which has a full-bridge configuration on the primary side. The key distinction lies on the secondary side, where a push-pull forward configuration is used instead of a full-bridge. This modification reduces the number of power MOSFETs, effectively minimizing the overall cost and complexity of the system. A combined phase-shift and frequency modulation scheme is utilized, which minimizes the overall losses in the converter and ensures soft-switching of all the power MOSFETs. Detailed insights into the converter topology, the modulation scheme, and its soft-switching range analysis are presented. To highlight the differences with the dual active bridge converter, analytical comparisons are also presented. The operation of the converter is illustrated using a hardware prototype, with a rated input voltage of 800 V, rated output voltage of 400 V, and rated power of 11 kW. The experimental results confirm the validity of the analytical descriptions.

1 Introduction

Although electric vehicles (EVs) have emerged as one of the key contributors to a cleaner and more sustainable future, their widespread adoption depends on the cost of the battery, performance, range, and an efficient and stable charging infrastructure. The research and development on EV chargers have garnered much attention in recent years. The charging system plays a pivotal role in determining the EV system's efficiency and performance [1], [2]. The onboard EV chargers (OBCs) are widely used in the automotive industry for their convenience. They allow the vehicle to be directly charged from the AC utility grid [3]. For such applications, bidirectional power flow between the EV and the grid proves beneficial, as the energy stored in the battery can be fed back to the grid, an isolated home, or some other loads during an emergency (V2X applications) [4]. One of the key components of a bidirectional OBC is an isolated bidirectional DC-DC converter (IBDC), which regulates the power flowing into and out of the EV battery unit. Therefore, a reliable power converter architecture is critical for an efficient and

reliable EV operation.

The dual active bridge (DAB) converter, first proposed in [5] and [6], has been widely adopted for EV on-board DC power conversion. It offers several benefits, including bidirectional power flow, galvanic isolation between the input and output, high efficiency, high power density, and zero voltage switching (ZVS) of all the power switches [7], [8], [9]. Conventional DAB converter utilizes two full-bridge (FB) structures on the primary and secondary sides, each requiring four switches. As a result, it leads to high component count, cost, and complexity. Consequently, research efforts have been dedicated to novel IBDC topologies that realize the benefits of DAB while reducing the number of power circuit components.

Several studies have integrated push-pull (PP) and push-pull forward (PPF) structures into the DAB topology, effectively minimizing the number of power semiconductors. A push-pull forward half-bridge converter topology is proposed in [10], which requires four MOSFETs in total. However, the converter utilizes single phase-shift (SPS) modulation, which leads to reduced degrees of optimization [7], [11]. A DC-DC converter utilizing

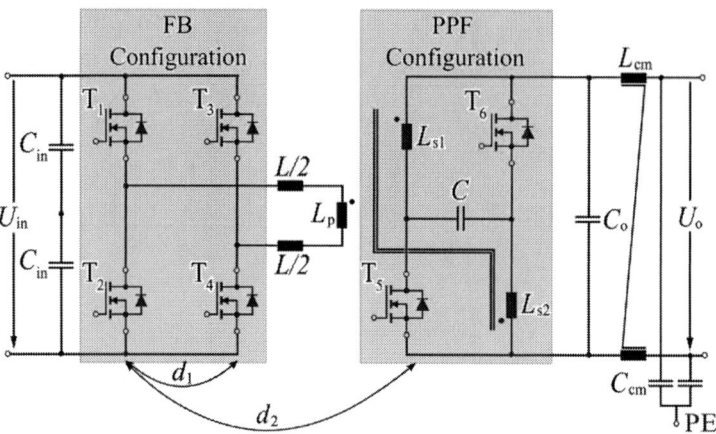

Fig. 1: The topology of the FBPPF-DAB converter.

a three-switch push-pull structure is presented in [12]. The arrangement uses an extended phase-shift (EPS) scheme at a fixed switching frequency. It reduces the current stress in the converter but fails to realize ZVS across the entire power range. Furthermore, a push-pull dual active bridge converter is introduced in [13]. It modifies the conventional DAB topology by integrating the PPF configuration on the primary side, keeping the FB configuration on the secondary side. Two external inductors are connected in series to the transformer's primary windings. While the topology minimizes losses, it is not optimal for buck mode operation [14]. The PPF structure clamps the MOSFET's drain-to-source voltages to twice the DC-link voltage during off state, necessitating high breakdown voltage MOSFETs on the PPF side (higher than 1.7 kV for 800 V DC-link voltage).

The topology of the FBPPF-DAB converter investigated in this work is shown in Fig. 1. The PPF configuration is implemented on the low-voltage secondary side to comply with the input and output voltage ratings. Considering the rated output voltage for this application is 400 V, this modification facilitates using high-performing 1.2 kV SiC MOSFETs. Compared to the conventional DAB topology, this arrangement reduces the number of required secondary switches by two. Consequently, it offers lower cost, reduced complexity, and eliminates losses associated with the two extra switches.

The FBPPF-DAB converter requires a three-winding transformer, with one primary and two secondary windings. A three-winding planar

transformer is employed to minimize the volume. While the transformer design involves complexity due to the additional secondary winding, it does not significantly increase the overall volume of the converter, still enabling a high power density. An external inductor is required as the leakage inductance of the planar transformer is insufficient to effectively regulate the power flow between the input and the output side. Equivalent inductance of L is introduced to the primary side and divided into two equal halves for enhanced design layout and symmetry. Additionally, a common-mode (CM) electromagnetic compatibility (EMC) filter at the output effectively suppresses the CM current from flowing into the load.

2 Analysis of the Converter under EPS Modulation

For the DAB converter, the power flow between the input and the output can be effectively controlled by adjusting the phase shift between the primary and secondary legs. A similar approach is implemented for the FBPPF-DAB topology. Figure 1 indicates the phase shifts between the individual legs of the converter. d_1 indicates the phase shift between the primary legs, between MOSFETs T_1 and T_4, while d_2 denotes the phase shift between the primary and the secondary, between MOSFETs T_1 and T_5. Neglecting the dead time intervals, MOSFETs T_2, T_3, and T_6 operates complementary to T_1, T_4, and T_5 respectively.

A simpler modulation scheme, SPS modulation, occurs when $d_1 = 0$. It has limited optimization

PCIM Europe 2024, 11– 13 June 2024, Nuremberg   DOI: 10.30420/566262358

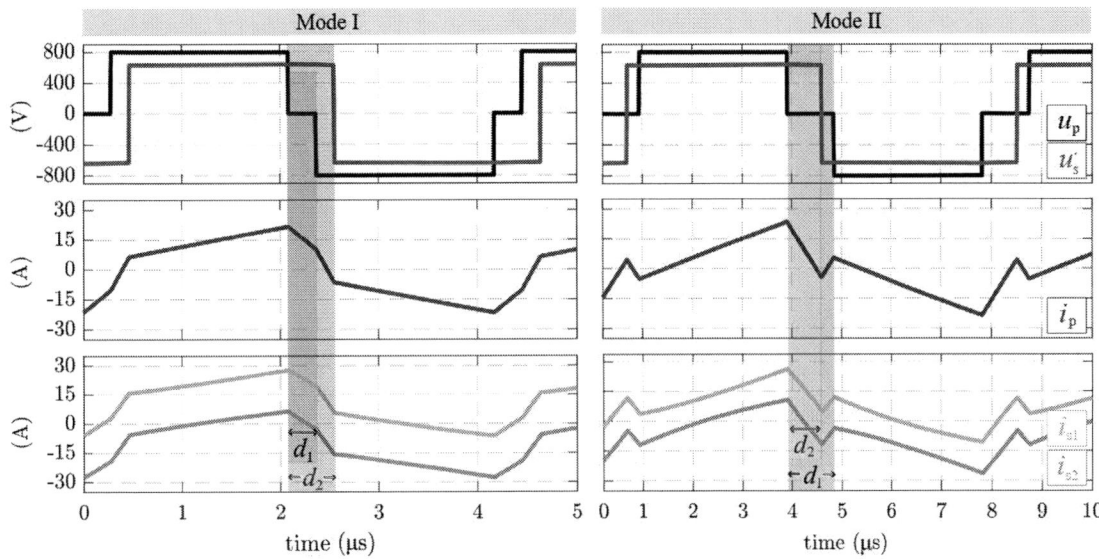

Fig. 2: The primary voltage (u_p), the reflected secondary voltage (u'_s), the primary current (i_p), and the secondary current (i_s1 and i_s2) waveforms for the operating conditions: (a) Mode I: U_in = 800 V, U_o = 400 V, P_o = 11 kW, and (b) Mode II: U_in = 800 V, U_o = 400 V, P_o = 6 kW.

prospects as d_2 is the only parameter that needs evaluation for a specific operating condition. Consequently, SPS modulation inherently shows higher currents and higher overall losses in the converter [11]. EPS modulation has more flexibility as both d_1 and d_2 can be evaluated for individual operating points defined by the input voltage U_in, output voltage U_o, and output power P_o. As one of the primary goals of the design is to improve the overall efficiency, the analysis of the converter is performed only under EPS modulation.

For a detailed analysis, the approach adopted in this study is similar to the one performed in [15] for the conventional DAB topology. The converter's operation is categorized into two modes based on the relationship between d_1 and d_2:

Mode I: $d_1 < d_2$.
Mode II: $d_1 > d_2$.

The ideal waveforms of the primary voltage (u_p), the secondary voltage reflected to the primary side (u'_s), the primary current (i_p), and the two secondary winding currents (i_s1 and i_s2) for each of these two modes are shown in Fig. 2. This division is beneficial, as it imposes a fixed form on the inductor current for each mode. Consequently, more accurate numerical expressions can be derived compared to some of the alternative approaches involving harmonic current analysis [16].

The transformer primary RMS current expressions for Mode I and Mode II are represented in Eqs. (1) and (2), respectively. The average output power

$$I_\mathrm{p-ModeI}^\mathrm{RMS} = \sqrt{\frac{\sqrt{\frac{N_\mathrm{p}}{N_\mathrm{s}}U_\mathrm{in}U_\mathrm{o}[4(d_1-d_2)^3+6(d_1-d_2)^2+2d_2^2(3-2d_2)]+d_1^2U_\mathrm{in}^2(2d_1-3)+\left(U_\mathrm{in}-\frac{N_\mathrm{p}}{N_\mathrm{s}}U_\mathrm{o}\right)^2}}{48L^2f_\mathrm{sw}^2}} \quad (1)$$

$$I_\mathrm{p-ModeII}^\mathrm{RMS} = \sqrt{\frac{\sqrt{\frac{N_\mathrm{p}}{N_\mathrm{s}}U_\mathrm{in}U_\mathrm{o}[6(d_1-d_2)^2+6d_2^2(1-2d_1)-4d_1^2(d_1-3d_2)]+d_1^2U_\mathrm{in}^2(2d_1-3)+\left(U_\mathrm{in}-\frac{N_\mathrm{p}}{N_\mathrm{s}}U_\mathrm{o}\right)^2}}{48L^2f_\mathrm{sw}^2}} \quad (2)$$

$$P_\mathrm{o-ModeI} = \left(\frac{N_\mathrm{p}U_\mathrm{in}U_\mathrm{o}}{N_\mathrm{s}4f_\mathrm{sw}L}\right)[2d_2(1-d_2)-d_1(1-2d_2)-d_1^2] \quad (3)$$

$$P_\mathrm{o-ModeII} = \left(\frac{N_\mathrm{p}U_\mathrm{in}U_\mathrm{o}}{N_\mathrm{s}4f_\mathrm{sw}L}\right)[2d_2-d_1(1+2d_2)+d_1^2] \quad (4)$$

2559

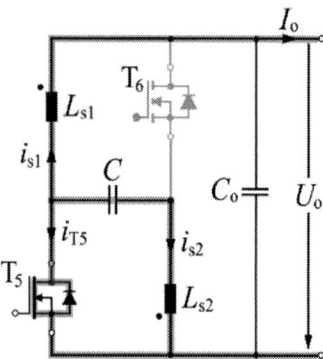

Fig. 3: The distribution of current in the secondary side when T_5 is on, and T_6 off.

expressions are provided in Eqs. (3) and (4) for Mode I and Mode II, respectively. The ratio N_p/N_s is the transformer's primary-to-secondary turns ratio. To derive the expressions of the MOSFET RMS current values and the transformer secondary winding RMS current values, a generic primary RMS current variable I_p^{RMS} has been used, as these expressions hold true for both Mode I and Mode II.

The primary MOSFET RMS current value can be derived from I_p^{RMS} as shown below.

$$I_{T_1}^{RMS} = \frac{I_p^{RMS}}{\sqrt{2}} \qquad (5)$$

Since the primary side of the converter is identical to the DAB, the currents and the resultant power losses in the individual components are also identical.

The significant difference lies in the transformer's secondary winding currents. As depicted in Fig. 2, each of the secondary winding currents has a DC offset, resulting from half of the average load current flowing through each of the windings. This DC current flows in the opposite direction with respect to the dot polarity through each of the secondary windings, preventing transformer core saturation. However, the overall RMS current in the secondary winding increases in the presence of this DC current, and the transformer is expected to have higher copper losses compared to the DAB. The instantaneous value of the secondary winding RMS currents, derived from the instantaneous primary current i_p, and the DC load current I_o, are represented in Eq. (6), and their RMS values are expressed in Eq. (7).

$$i_{s1} = \frac{N_p}{2N_s} i_p + \frac{I_o}{2}$$
$$i_{s2} = \frac{N_p}{2N_s} i_p - \frac{I_o}{2} \qquad (6)$$

$$I_{s1}^{RMS} = \sqrt{\left(\frac{N_p}{2N_s} I_p^{RMS}\right)^2 + \left(\frac{I_o}{2}\right)^2} \qquad (7)$$

To analyze the effect of the increased secondary winding current in the secondary MOSFETs, a specific condition is illustrated in Fig. 3, when T_5 is on, and T_6 off. The instantaneous current in the MOSFET T_5 can be represented by:

$$i_{T5} = -(i_{s1} + i_{s2})$$
$$i_{T5} = -\left(\frac{N_p}{N_s}\right) i_p \qquad (8)$$

$$I_{T5}^{RMS} = \frac{\left(\frac{N_p}{N_s}\right) I_p^{RMS}}{\sqrt{2}} \qquad (9)$$

Equation (8) shows the instantaneous current expression in the secondary MOSFETs, which resembles the current expressions of DAB. Their RMS current expression, given by Eq. (9), is also identical. The offset in the secondary winding currents is not reflected in the MOSFETs. Consequently, the losses contributed by the MOSFETs are effectively half compared to that of the DAB utilizing the same devices and for the same modulation scheme, as two fewer switches are utilized.

3 Soft-Switching Analysis and Range Estimation

One of the key benefits of the DAB converter is its ability to achieve ZVS for a wide range of operation without requiring additional circuit elements. Since the FBPPF-DAB topology is based on the DAB structure, it exhibits similar characteristics. The ZVS operation is extensively studied in the literature for the conventional DAB converter, and the criteria to ensure its attainment are well defined in [6], [17], [18], and [19]. In this work, a similar analysis is performed for the FBPPF-DAB topology. The turn-on of MOSFETs at zero voltage depends on the magnitude and direction of the series inductor current I_p. The orientation of the anti-parallel diodes of the individual MOSFETs

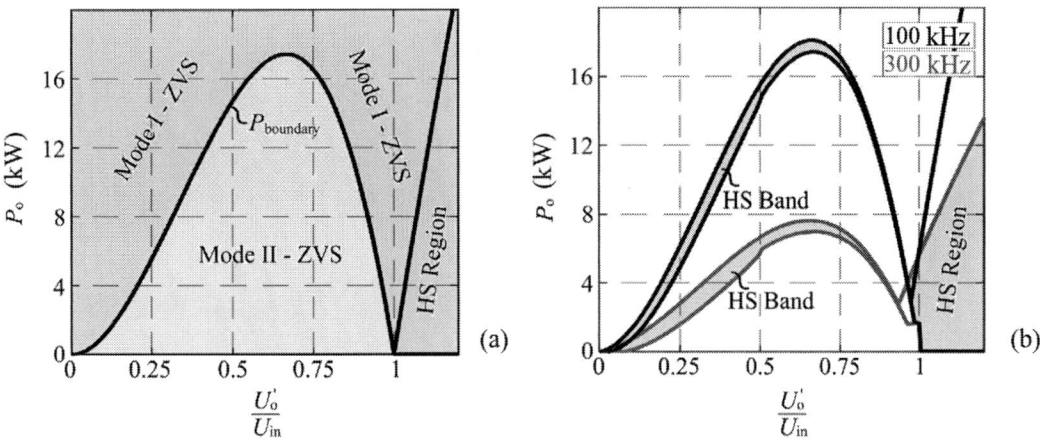

Fig. 4: The ZVS and HS regions of the FBPPF-DAB converter operating in Mode I and Mode II for: (a) f_{sw} = 100 kHz, neglecting the effect of C_{oss}, and (b) f_{sw} = 100 kHz and 300 kHz, considering the effect of C_{oss}.

determines the direction of the currents. For T_1, T_4, and T_5, the current must flow out of the dot terminal of the transformer. For T_2, T_3, and T_6, the current must enter the dot terminal.

The minimum inductor current criteria arise due to the presence of MOSFET output capacitor C_{oss}, which requires the inductor to store a minimum energy, facilitating charging and discharging of these capacitors whenever a transition occurs in any converter leg. To calculate the magnitudes of the minimum currents, the nonlinear capacitance C_{oss} is approximated with a linear capacitance $C_{Q,eq}(U_{ds})$, on the basis of total charge conservation principle. The expression of $C_{Q,eq}(U_{ds})$ is represented in Eq. (10) [20], [21].

$$C_{Q,eq}(U_{ds}) = \frac{\int_0^{U_{ds}} C_{oss}(v)\,dv}{U_{ds}} \qquad (10)$$

$$I_{\min}^{T_{12}} = U_{in}\sqrt{\frac{2C_{Q,eq}(U_{ds})\left(2\frac{U_o'}{U_{in}}-1\right)}{L}} \qquad (11)$$

$$I_{\min-\text{ModeI}}^{T34} = U_{in}\sqrt{\frac{2C_{Q,eq}(U_{ds})\left(2\frac{U_o'}{U_{in}}+1\right)}{L}}$$

$$\qquad\qquad (12)$$

$$I_{\min-\text{ModeII}}^{T34} = U_{in}\sqrt{\frac{2C_{Q,eq}(U_{ds})\left(1-2\frac{U_o'}{U_{in}}\right)}{L}}$$

The magnitude of the minimum inductor currents for the individual MOSFETs are defined, neglecting

the influence of transformer magnetizing inductance and winding capacitances [18]. For the MOSFETs T_1 and T_2, the current expression is given by Eq. (11). This equation holds true for operations in both Mode I and Mode II, as long as $2\frac{U_o'}{U_{in}} > 1$. For other cases, $I_{\min}^{T_{12}}$ can be zero. For T_3 and T_4, the current magnitude differs in Mode I and Mode II, as illustrated in Eq. (12). A minimum inductor current is always required in Mode I, while in Mode II, the magnitude criteria should only be satisfied when $2\frac{U_o'}{U_{in}} < 1$. Under the assumptions mentioned previously, MOSFETs T_5 and T_6 do not demand any current for ZVS. The capacitors fully charge and discharge even when there is no initial current in the inductor at the beginning of the resonant cycle. Although the minimum current criteria vary depending on the mode of operation and the input and output voltages, the current direction criteria must always be satisfied.

Figure 4(a) illustrates the range of ZVS operation of the converter operating in Mode I and Mode II, for a fixed U_{in} of 800 V and a variable U_o. U_o' is the output voltage reflected to the primary side. For this analysis, f_{sw} is held constant at 100 kHz, and the influence of dead time and the charging/discharging of MOSFET C_{oss} are neglected. The plot indicates that whenever U_o' is lower than U_{in}, all the MOSFETs exhibit ZVS across the entire range of P_o, if the appropriate mode is chosen. $P_{boundary}$ represents the power at the boundary between the Mode I and Mode II operation. Soft-switching is ensured in Mode

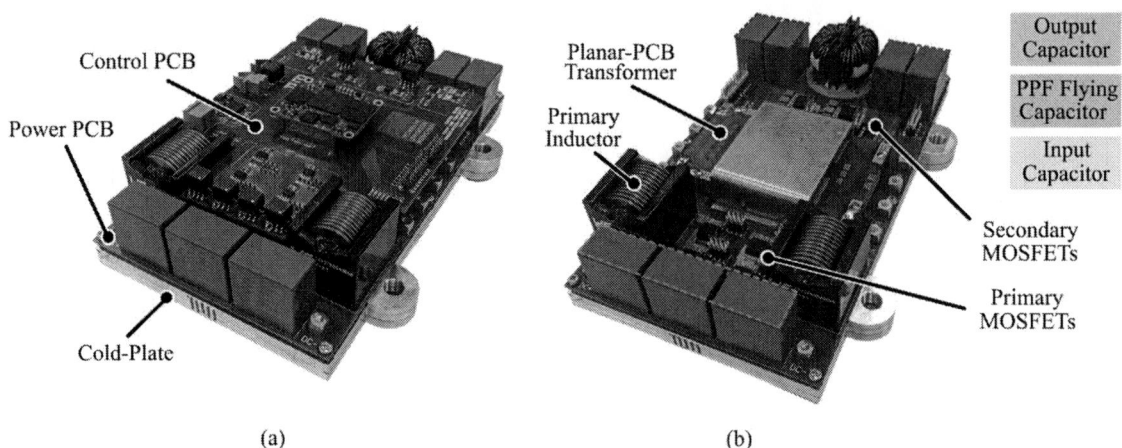

Fig. 5: (a) The complete hardware prototype, including the power board, the control board, and the cooling plate, (b) the individual components of the power board in accordance with the schematics of Fig. 1.

II when $P_o < P_{\text{boundary}}$, and in Mode I when $P_o > P_{\text{boundary}}$. However, when U'_o is higher than U_{in}, Mode II does not exhibit ZVS for some of the MOSFETs, resulting in a hard-switching (HS) region.

In addition to the HS region, the presence of C_{oss} narrows the ZVS range by introducing HS bands, representing the range of output power where neither Mode I nor Mode II can ensure ZVS. Varying the switching frequency shifts this band to different power ranges. Figure 4(b) illustrates these effects for $f_{\text{sw}} = 100$ and 300 kHz. As a result, as long as the $\frac{U'_o}{U_{\text{in}}}$ ratio is maintained below unity, soft-switching of all six MOSFETs can be ensured for a variable switching frequency operation. This study hence also considers the f_{sw} as an optimization parameter, along with d_1 and d_2. f_{sw} is varied between 100 and 300 kHz. The three modulation parameters are obtained by solving an optimization problem aimed at minimizing the overall converter losses at specific operating points. The details of this optimization procedure are beyond the scope of this paper.

4 Hardware Overview and Test Results

The overall converter system, with the power board, the control board, and the liquid cooling system, is illustrated in Fig. 5(a), and the individual components of the power board are represented in Fig. 5(b). The transformer has a metal casing

	Description	Value
T_{1-6}	MOSFETs	1.2 kV, 40 mΩ
		(C3M0040120J1)
L	Inductor	16 μH
N_p	Transf. prim. turns	11
N_s	Transf. sec. turns	7
C_{in},C,C_o	Capacitors	10 μF
L_{cm}	CM filter inductor	0.9 mH
C_{cm}	CM filter capacitor	20 nF

Tab. 1: The parameters of the FBPPF-DAB converter.

mounted on it for enhanced cooling. For this work, an FPGA has been used to generate the gate pulses. It is mounted on the control board alongside other analog signal conditioning circuitries. The converter has been tested for open-loop operation but incorporates all the necessary analog and sensing circuitry for the implementation of closed-loop output voltage and current control. The cold-plate is designed to implement forced liquid cooling by connecting it to an external cooling unit. The volumetric power density achieved with this system is 7.02 kW/dm³. The details of the main circuit components are listed in Tab. 1. The inductors shown in Fig. 5 were observed to have high core temperature and loss. They were replaced with two 8 μH inductors built using stacked cores (2xN87 E47/20/16, 7 turns).

The converter is tested for two different operating conditions, representing the operation in the two modes.

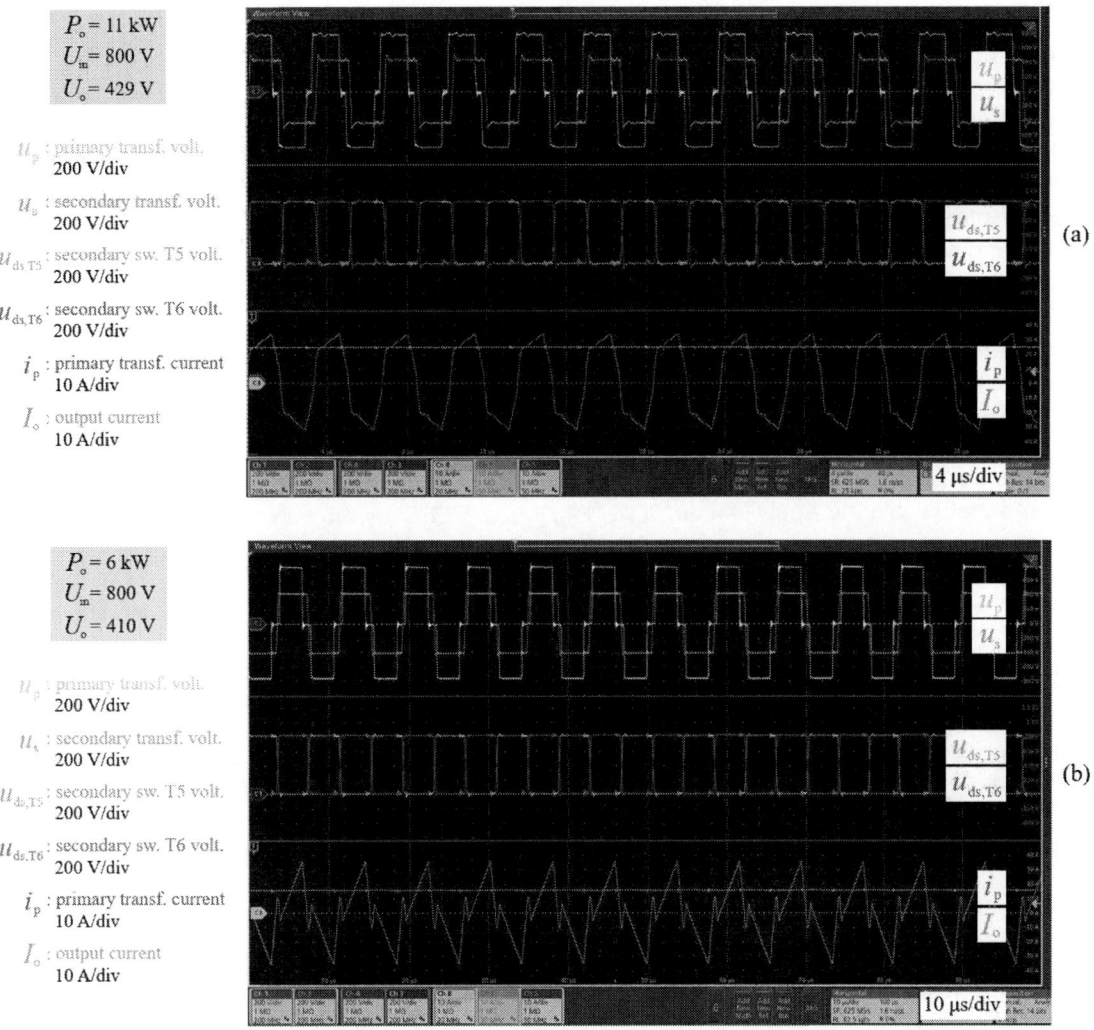

Fig. 6: The converter test results representing u_p, u_s, the secondary MOSFET drain-source voltages u_{ds_T5}, u_{ds_T6}, i_p, and the output current I_o for the operating conditions: (a) Mode I: U_{in} = 800 V, U_o = 429 V, P_o =11 kW, and (b) Mode II: U_{in} = 800 V, U_o = 410 V, P_o = 6 kW.

Figure 6(a) represents the operation of the converter in Mode I for U_{in} = 800 V, U_o = 429 V, P_o = 11 kW, and f_{sw} = 295 kHz. The measured efficiency of the converter is 98.12 %.

Figure 6(b) represents the operation of the converter in Mode II for U_{in} = 800 V, U_o = 410 V, P_o = 6 kW, and f_{sw} = 128 kHz. The efficiency of the converter is measured to be 98.04 %.

For both operating conditions, all the MOS-FETs were found to be operating under ZVS. To ensure this, the dead times of the individual legs are optimally adjusted for each operating point. The u_{ds} of T_5 and T_6 does not have any over-voltage.

5 Conclusion

In this paper, an 800/400 V, 11 kW FBPPF-DAB converter is presented for a step-down OBC application. The converter is shown to exhibit properties similar to those of the conventional DAB converter, with the additional benefits of lower component count, cost, and complexity. Consequently, it offers efficiency benefits as the number of power MOSFETs on the secondary side is reduced while the currents flowing through them remain the same as the DAB, reducing the overall MOSFET conduction losses. The complexity of the topology lies in the design of the three-winding transformers. The transformer is shown to exhibit higher RMS currents in the two secondary windings, resulting in higher losses.

The paper introduces the phase-shift modulation schemes and provides numerical descriptions of the converter currents under the EPS modulation, which can be used to optimize the performance. Also, a variable frequency operation has been shown to ensure soft-switching of all the MOSFETs. The hardware description and the test results, shown at two different operating conditions, prove the high power density of the system, high efficiency and match with the analytical descriptions. The ongoing work for this system includes the implementation of a closed-loop output voltage and current controller.

References

[1] S. S. G. Acharige, M. E. Haque, M. T. Arif, N. Hosseinzadeh, K. N. Hasan, and A. M. T. Oo, "Review of electric vehicle charging technologies, standards, architectures, and converter configurations," *IEEE Access*, vol. 11, pp. 41 218–41 255, 2023. DOI: 10.1109/ACCESS.2023.3267164.

[2] M. Yilmaz and P. T. Krein, "Review of battery charger topologies, charging power levels, and infrastructure for plug-in electric and hybrid vehicles," *IEEE Transactions on Power Electronics*, vol. 28, no. 5, pp. 2151–2169, 2013. DOI: 10.1109/ TPEL.2012.2212917.

[3] A. Khaligh and M. D'Antonio, "Global trends in high-power on-board chargers for electric vehicles," *IEEE Transactions on Vehicular Technology*, vol. 68, no. 4, pp. 3306–3324, 2019. DOI: 10.1109/TVT.2019.2897050.

[4] J. Yuan, L. Dorn-Gomba, A. D. Callegaro, J. Reimers, and A. Emadi, "A review of bidirectional on-board chargers for electric vehicles," *IEEE Access*, vol. 9, pp. 51 501–51 518, 2021. DOI: 10. 1109/ACCESS.2021.3069448.

[5] R. De Doncker, D. Divan, and M. Kheraluwala, "A three-phase soft-switched high-power-density dc/dc converter for high-power applications," *IEEE Transactions on Industry Applications*, vol. 27, no. 1, pp. 63–73, 1991. DOI: 10.1109/28.67533.

[6] M. Kheraluwala, R. Gascoigne, D. Divan, and E. Baumann, "Performance characterization of a high-power dual active bridge dc-to-dc converter," *IEEE Transactions on Industry Applica-*

tions, vol. 28, no. 6, pp. 1294–1301, 1992. DOI: 10.1109/28.175280.

[7] N. Hou and Y. W. Li, "Overview and comparison of modulation and control strategies for a non-resonant single-phase dual-active-bridge dc–dc converter," *IEEE Transactions on Power Electronics*, vol. 35, no. 3, pp. 3148–3172, 2020. DOI: 10.1109/TPEL.2019.2927930.

[8] F. Krismer and J. W. Kolar, "Accurate small-signal model for the digital control of an automotive bidirectional dual active bridge," *IEEE Transactions on Power Electronics*, vol. 24, no. 12, pp. 2756–2768, 2009. DOI: 10.1109/TPEL.2009.2027904.

[9] X. She, X. Yu, F. Wang, and A. Q. Huang, "Design and demonstration of a 3.6kv–120v/10kva solid state transformer for smart grid application," *2014 IEEE Applied Power Electronics Conference and Exposition - APEC 2014*, pp. 3429–3436, 2014.

[10] Z. Zhang, O. C. Thomsen, and M. A. E. Andersen, "Optimal design of a push-pull-forward half-bridge (ppfhb) bidirectional dc–dc converter with variable input voltage," *IEEE Transactions on Industrial Electronics*, vol. 59, no. 7, pp. 2761–2771, 2012. DOI: 10.1109/TIE.2011.2134051.

[11] H. Bai and C. Mi, "Eliminate reactive power and increase system efficiency of isolated bidirectional dual-active-bridge dc–dc converters using novel dual-phase-shift control," *IEEE Transactions on Power Electronics*, vol. 23, no. 6, pp. 2905–2914, 2008. DOI: 10.1109/TPEL.2008.2005103.

[12] Y. Lu, Q. Wu, Q. Wang, D. Liu, and L. Xiao, "Analysis of a novel zero-voltage-switching bidirectional dc/dc converter for energy storage system," *IEEE Transactions on Power Electronics*, vol. 33, no. 4, pp. 3169–3179, 2018. DOI: 10.1109/TPEL.2017. 2703949.

[13] A. Ali and Y. Liao, "Optimized control for modified push-pull dual active bridge converter to achieve wide zvs range and low current stress," *IEEE Access*, vol. 9, pp. 140 258–140 267, 2021. DOI: 10. 1109/ACCESS.2021.3117873.

[14] Y. Li, J. A. Anderson, M. Haider, J. Schäfer, J. Miniböck, *et al.*, "Optimal synergetic operation and experimental evaluation of an ultra-compact gan-based three-phase 10 kw ev charger," *IEEE Transactions on Transportation Electrification*, pp. 1–1, 2023. DOI: 10.1109/TTE.2023.3297502.

[15] A. Tong, L. Hang, G. Li, X. Jiang, and S. Gao, "Modeling and analysis of a dual-active-bridge-isolated bidirectional dc/dc converter to minimize rms current with whole operating range," *IEEE Transactions on Power Electronics*, vol. 33, no. 6, pp. 5302–5316, 2018. DOI: 10.1109/TPEL.2017. 2692276.

[16] J. Huang, Y. Wang, Z. Li, and W. Lei, "Unified triple-phase-shift control to minimize current stress and achieve full soft-switching of isolated bidirectional dc–dc converter," *IEEE Transactions on Industrial Electronics*, vol. 63, no. 7, pp. 4169–4179, 2016. DOI: 10.1109/TIE.2016.2543182.

[17] R. Steigerwald, R. De Doncker, and H. Kheraluwala, "A comparison of high-power dc-dc soft-switched converter topologies," *IEEE Transactions on Industry Applications*, vol. 32, no. 5, pp. 1139–1145, 1996. DOI: 10.1109/28.536876.

[18] Z. Shen, R. Burgos, D. Boroyevich, and F. Wang, "Soft-switching capability analysis of a dual active bridge dc-dc converter," in *2009 IEEE Electric Ship Technologies Symposium*, 2009, pp. 334–339. DOI: 10.1109/ESTS.2009.4906533.

[19] A. R. Rodríguez Alonso, J. Sebastian, D. G. Lamar, M. M. Hernando, and A. Vazquez, "An overall study of a dual active bridge for bidirectional dc/dc conversion," in *2010 IEEE Energy Conversion Congress and Exposition*, 2010, pp. 1129–1135. DOI: 10.1109/ECCE.2010.5617847.

[20] M. Kasper, R. M. Burkart, G. Deboy, and J. W. Kolar, "Zvs of power mosfets revisited," *IEEE Transactions on Power Electronics*, vol. 31, no. 12, pp. 8063–8067, 2016. DOI: 10.1109/TPEL.2016.2574998.

[21] D. Costinett, D. Maksimovic, and R. Zane, "Circuit-oriented treatment of nonlinear capacitances in switched-mode power supplies," *IEEE Transactions on Power Electronics*, vol. 30, no. 2, pp. 985–995, 2015. DOI: 10.1109/TPEL.2014.2313611.

PCIM Europe 2024, 11– 13 June 2024, Nuremberg   DOI: 10.30420/566262359

Symmetrical Operation of Four Channel Resonant Boost DC-DC Converters in Continuous Conduction Mode

János Hamar [1], Péter Stumpf [1], Kristóf Bándy [1], Róbert Orvai [1]

[1] Department of Automation and Applied Informatics, Faculty of Electrical Engineering and Informatics, Budapest University of Technology and Economics, Műegyetem rkp. 3, 1111 Budapest, Hungary

Corresponding author: János Hamar, Hamar.Janos@aut.bme.hu
Speaker: Kristóf Bándy, Bandy.Kristof@aut.bme.hu

Abstract

The steady-state operation of a four channel resonant step-up dc/dc converter is presented and analyzed in continuous current conduction mode (CCM). The converter is equipped with two input channels, four output channels, and facilitates zero voltage (ZVS) and/or zero current switching (ZCS), which help reducing the switching losses. The quasi-sinusoidal waveforms assist in improving the EMI characteristics. The relations among the input and output voltages, with the use of the control variables, are derived. The effect of the control variables on the operation is evaluated. The theoretical results are verified by computer simulations and experimental tests.

1   Introduction

Multichannel resonant DC/DC converters constitute a state-of-the-art power conversion technology that offers a number of advantages, including efficiencies of up to 96-99%. This is because they often operate at zero voltage switching (ZVS) and/or zero current switching (ZCS), enabling the minimization of the switching losses. Another advantage is the less electromagnetic interference (EMI), that they cause, due to the quasi-sinusoidal shape of voltages and currents, which contain less harmonics. Their operation at higher switching frequencies enables the usage of smaller circuit components, smaller size and weight, and consequently, higher power densities. These DC/DC converters are particularly well-suited for applications where multiple outputs are required, such as data center servers, telecommunications equipment, and industrial automation systems.

The four-channel, resonant boost converter was presented first in [1]. The investigations were limited to the operation in discontinuous conduction mode (DCM). In the current paper the operation in continuous conduction mode (CCM) is analysed.

2   Circuit Configuration and Operation

The circuitry and the basic operation is now briefly presented, based on [1]. The converter, shown in Fig. 1, includes controlled switches S_p and S_n in the positive (p) and negative (n) input channels, and two additional ones in output channels p_1, p_2 (S_{cp1}, S_{cp2}), and n_1, n_2 (S_{cn1}, S_{cn2}).

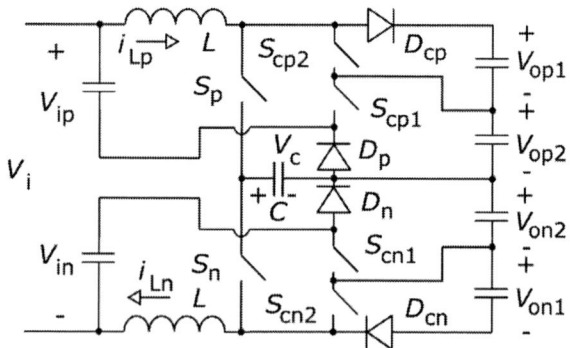

Fig. 1 Circuit configuration

The p and n channels are connected through a switched capacitor C, which formulates series resonating circuits with the p and n channel inductors L in different sub-periods. The resonant

PCIM Europe 2024, 11– 13 June 2024, Nuremberg   DOI: 10.30420/566262359

circuit has an angular frequency of $\omega = 2\pi f_r = 1/\sqrt{LC}$, a characteristic impedance of $Z = \sqrt{L/C}$, and a switching frequency $f_s=1/T_s$. It is supplied from input sources V_{ip} and V_{in} and has four output voltages V_{op1}, V_{op2}, V_{on1}, and V_{on2}. Each switching angle below is calculated as $\alpha_x = \omega t_x$.

The operation of the circuit is summarized in Fig. 2 and Fig. 3. The consecutive switching intervals are presented in Fig. 2, and the relating voltage and current waveforms in Fig. 3. The operation of the converter during each switching interval is as follows:

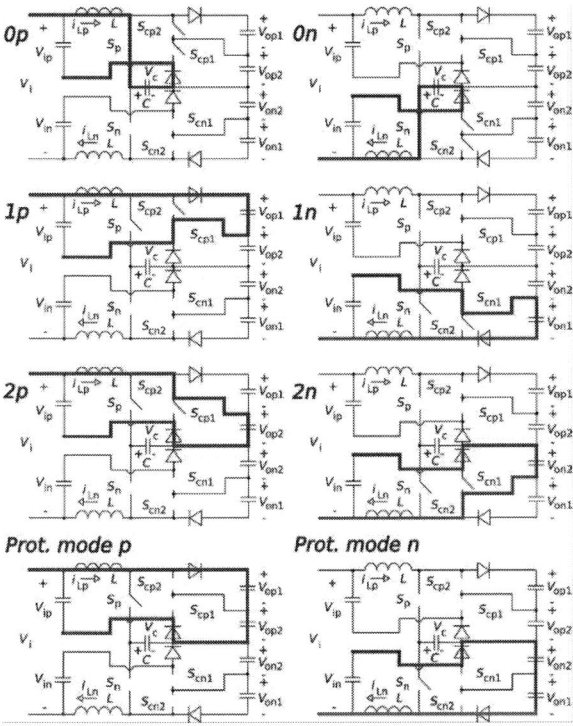

Fig. 2 Circuit operation intervals

In interval 0p ($0 < \omega t \le \alpha_p$), switch S_p is on, while switches S_{cp1} and S_{cp2} are off. The sinusoidal current $i_{Lp} = i_c$ is flowing in the circuit $V_{ip} - L - S_p - C - D_p$. The voltage of the switched capacitor $v_c(t)$ is sinusoidally increasing. At the end of the interval $i_{Lp}(\alpha_p) = i_c(\alpha_p) = I_{Lpa}$ and $v_c(\alpha_p) = V_{cp}$.

In interval 1p ($\alpha_p < \omega t \le \alpha_{cp}$), switch S_{cp1} is on, switches S_p and S_{cp2} are off. The inductor current $i_{Lp}(t)$ is linearly decreasing in circuit $L - D_{cp} - V_{op1} - S_{cp1} - V_{ip}$. The current $i_{Lp}(\alpha_{cp}) = I_{Lpb}$ at the end of this interval.

In interval 2p ($\alpha_{cp} < \omega t \le \omega T_s$), switch S_{cp2} is on, while S_p and S_{cp1} are off. The choke current $i_{Lp}(t)$ is further falling in the circuit $L - S_{cp2} - V_{op2} - D_p - V_{ip}$, and reaches $i_{Lp}(\omega T_s) = I_{Lpc}$ at the end. In continuous conduction mode (CCM), the current never reaches zero.

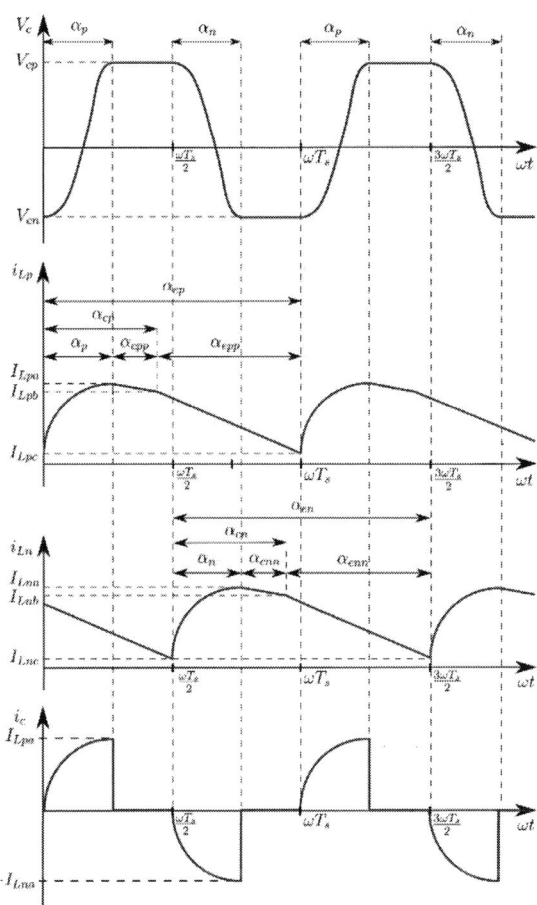

Fig. 3 Circuit operation waveforms

The construction of the converter is symmetrical, that is, each circuit component in channel n has its counterpart in channel n. They also have the same subperiods (0p→0n, 1p→1n, and 2p→2n), but the switching instants are shifted by a half period $\omega T_s /2$ in channel n.

The circuit can also operate in the so-called "protection mode". As discussed above, in interval 0p, the voltage of the switched capacitor $v_c(t)$ is increasing. If the voltage of capacitor C reaches the value of $v_c = V_{op1}+V_{op2}$ while $\omega t < \alpha_p$, the diodes D_p and D_{cp} will become forward biased and start conducting current in circuit $L - D_{cp} - V_{op1} - V_{op2} - D_p - V_{ip}$. Similarly, in channel n, when the voltage

2567

of the capacitor reaches the value of $v_c = -(V_{on1}+V_{on2})$, the diodes D_n and D_{cn} will become forward biased and start conducting current in circuit $L - V_{in} - D_n - V_{on2} - V_{on1} - D_{cn}$.

3 Steady-State Analysis of CCM Operation

The detailed steady state analysis of the symmetrical CCM operation is presented below, assuming that the circuit components in the p and n channels are identical.

3.1 Energy pulse equations

The energy balance of the converter is assessed for one switching period T_s.

In interval "0p" ($0 < \omega t \le \alpha_p$), the sinusoidal current $i_c = i_{Lp}$ is flowing in the series LC resonant circuit. The input energy, feeding the converter, is

$$w_{ip}^{(0p)} = \int_0^{\alpha_p/\omega} i_c v_{ip} dt = 2CV_c V_i \tag{1}$$

In symmetrical operation, the switched capacitor does not exchange energy between the input channels p, and n, that is

$$\Delta w_{Cp}^{(0p)} = \frac{C}{2}(V_{cp}^2 - V_{cn}^2) = 0 \tag{2}$$

In interval "1p" ($\alpha_p < \omega t \le \alpha_{cp}$), the input energy pulse is

$$w_{ip}^{(1p)} = \int_{\alpha_p/\omega}^{\alpha_{cp}/\omega} i_{Lp} v_{ip} dt = \frac{V_i V_o}{R f_s} \tag{3}$$

In interval "2p" ($\alpha_{cp} < \omega t \le \omega T_s$), it is similarly

$$w_{ip}^{(2p)} = \int_{\alpha_{cp}/\omega}^{T_s} i_{Lp} v_{ip} dt = \frac{V_i V_o}{R f_s} \tag{4}$$

The output energy, consumed by each load is

$$w_o = \frac{V_o^2}{R f_s} \tag{5}$$

The total input energy in one period is equal to the total energy consumed by the load resistors

$$2w_o = w_{ip}^{(0p)} + w_{ip}^{(1p)} + w_{ip}^{(2p)} \tag{6}$$

that is, summing up (1), (3), (4), and making it equal to the total output energy, furthermore expressing the voltage of the switched capacitor results in

$$V_c = \frac{V_o^2 - V_i V_o}{R C f_s V_i} \tag{7}$$

At the border of protection mode $V_c = 2V_o$. Substituting this condition into (7) yields

$$V_{o,prot} = V_i(1 + 2RCf_s) \tag{8}$$

3.2 Inductor current waveforms

The relations describing the time dependency of the choke current in the three above summarized intervals are as follows:

Interval "0p", when $0 < \omega t \le \alpha_p$

$$i_{Lp}^{(0p)} = i_c^{(0p)} = I_{Lpc}\cos(\omega t) + \frac{V_i + V_c}{Z}\sin(\omega t) \tag{9}$$

At the end of the "0p" interval

$$I_{Lpa} = I_{Lpc}\cos\alpha + \frac{V_i + V_c}{Z}\sin\alpha \tag{10}$$

Integrating (9) between 0 and α_p/ω yields

$$\int_0^{\frac{\alpha}{\omega}} i_{Lp}^{(0p)}(\omega t)\, dt = \frac{I_{Lpc}\sin(\alpha)}{\omega} + \frac{V_i + V_c}{Z\omega}[1 - \cos\alpha] =$$
$$= C(V_{cp} - V_{cn)} = 2CV_c \tag{11}$$

Let us now simplify it to

$$I_{Lpc} = \frac{V_c(1 + \cos\alpha) - V_i(1 - \cos\alpha)}{Z\sin\alpha} \tag{12}$$

Combining (12) with (10) in order to eliminate V_c, results in

$$I_{Lpa} = I_{Lpc} + \frac{2V_i}{Z}\tan\frac{\alpha}{2} \tag{13}$$

Interval "1p", when $\alpha_p < \omega t \le \alpha_{cp}$

$$i_{Lp}^{(1p)} = I_{Lpa} - \frac{V_o - V_i}{Z}(\omega t - \alpha) \tag{14}$$

At the end of the "1p" interval

$$I_{Lpb} = I_{Lpa} - \frac{V_o - V_i}{Z}(\alpha_c - \alpha) \tag{15}$$

Interval "2p", when $\alpha_{cp} < \omega t \le \omega T_s$

$$i_{Lp}^{(2p)} = I_{Lpb} - \frac{V_o - V_i}{Z}(\omega t - \alpha_c) \tag{16}$$

At the end of the "2p" interval

$$I_{Lpc} = I_{Lpb} - \frac{V_o - V_i}{Z}(\omega T_s - \alpha_c) \tag{17}$$

Substituting (15) into (17) gives

$$I_{Lpc} = I_{Lpa} - \frac{V_o - V_i}{Z}(\omega T_s - \alpha) \tag{18}$$

and combining (13) with (18) delivers the equation for the output voltage

$$V_o = V_i + \frac{2V_i \tan(\alpha/2)}{\omega/f_s - \alpha} \tag{19}$$

Due to the symmetrical operation, inductor L supplies the same energy to the loads in interval "1p" and "2p"

$$\frac{L}{2}(I_{Lpa}^2 - I_{Lpb}^2) = \frac{L}{2}(I_{Lpb}^2 - I_{Lpc}^2) \tag{20}$$

that is, I_{Lpb} will be the quadratic mean of I_{Lpa} and I_{Lpc}

$$I_{Lpb} = \sqrt{\frac{I_{Lpa}^2 + I_{Lpc}^2}{2}} \tag{21}$$

At the border between continuous and discontinuous conduction modes (CCM and DCM) $I_{Lpc} = 0$. Using (12) with this condition results in

$$V_c = V_i \tan^2 \frac{\alpha}{2} \tag{22}$$

Combining (7) with (19), and (22) gives the value of the load resistance when the operation is just at the border between CCM and DCM

$$R_{ccm,dcm} = Z \frac{2\pi - \alpha f_s/f_r + 2\tan(\alpha/2)f_s/f_r}{(2\pi - \alpha f_s/f_r)^2} \frac{4\pi}{\tan(\alpha/2)} \tag{23}$$

The operation is in CCM if $R < R_{ccm,dcm}$, and it is in DCM otherwise.

4 Discussion

Major relations are summarized below, focusing especially on the effect of the control variables to the operation. The output voltages can be controlled using the switching angle $\alpha = \alpha_p = \alpha_n$ and the switching frequency ($f_s = 1/T_s$). As an alternative to α, the clamping voltage of the switched capacitor $V_c = V_{cp} = -V_{cn}$ can also be used besides f_s. The symmetry between output channels p1 and p2 as well as n1 and n2 can be maintained by properly setting control variable α_c, or equivalently by controlling I_{Lb}.

Output voltage [see (19)]

$$V_o = V_i + \frac{2V_i \tan(\alpha/2)}{\omega/f_s - \alpha} \tag{24}$$

Clamping voltage of the capacitor [see (7)]

$$V_c = \frac{V_o^2 - V_i V_o}{RCf_s V_i} \tag{25}$$

Choke current at $\omega t = 0$ [see (12)]

$$I_{Lc} = \frac{V_c(1 + \cos\alpha) - V_i(1 - \cos\alpha)}{Z \sin\alpha} \tag{26}$$

Choke current at $\omega t = \alpha$ [see (13)]

$$I_{La} = I_{Lc} + \frac{2V_i}{Z}\tan\frac{\alpha}{2} \tag{27}$$

Choke current at $\omega t = \alpha_c$ [see (21)]

$$I_{Lb} = \sqrt{\frac{I_{La}^2 + I_{Lc}^2}{2}} \tag{28}$$

Switching angle α_c, set to maintain the symmetrical operation [see (15)]

$$\alpha_c = \alpha + \frac{(I_{La} - I_{Lb})Z}{V_o - V_i} \tag{29}$$

Border between CCM and DCM [see (23)]

$$R_{ccm,dcm} = Z \frac{2\pi - \alpha f_s/f_r + 2\tan(\alpha/2)f_s/f_r}{(2\pi - \alpha f_s/f_r)^2} \frac{4\pi}{\tan(\alpha/2)} \tag{30}$$

Border of protection mode ($V_c = 2V_o$) [see (8)]

$$V_{o,prot} = V_i(1 + 2RCf_s) \tag{31}$$

Based on the analysis, characteristic curves are developed (see in Fig. 4 and Fig. 5).

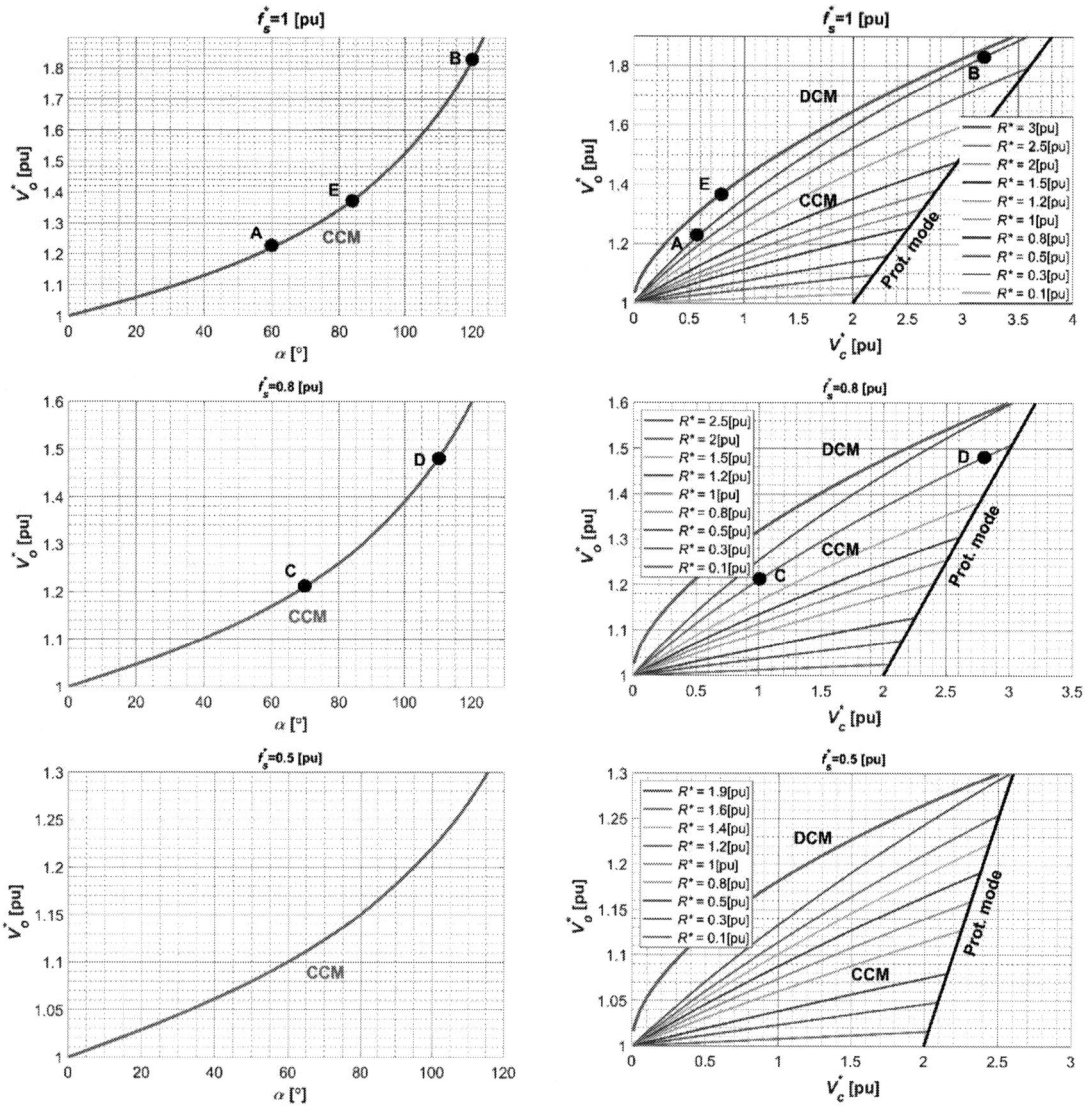

Fig. 4 Characteristic curves (V_o vs. α and V_o vs. V_c)

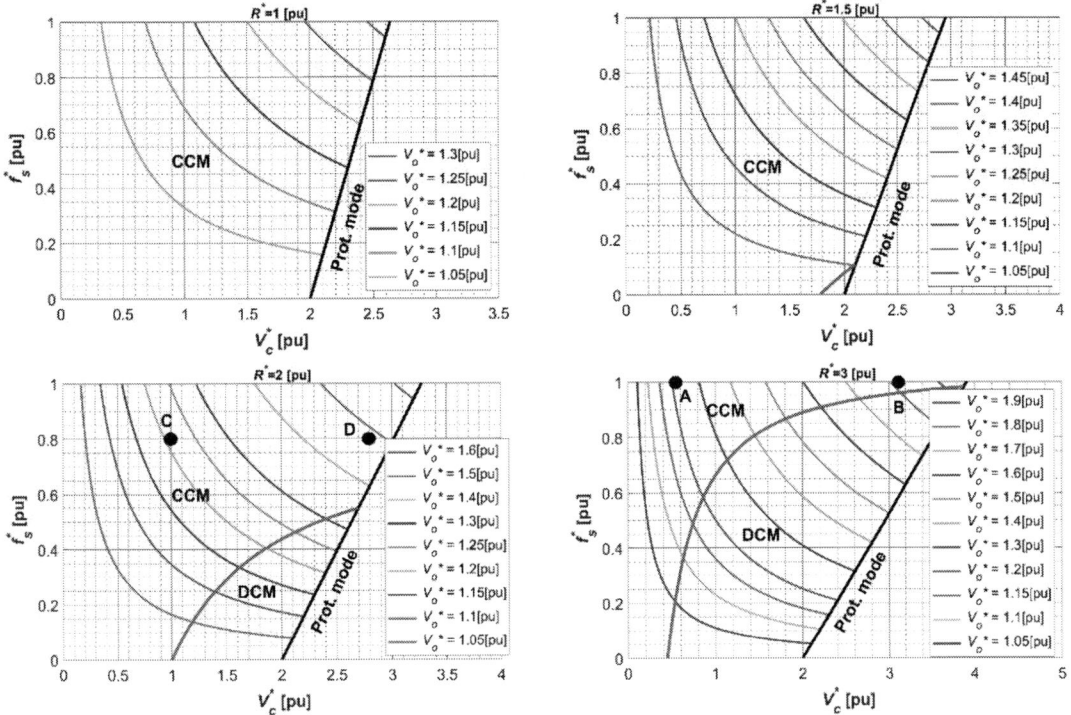

Fig. 5 Constant output voltage curves

Per-unit values were introduced, shown by stars (*). The unit is V_i at the voltages, Z at the resistances, V_i/Z at the currents, and f_r at the frequency. The examined CCM operation region is limited by the borders of protection mode and discontinuous conduction mode (DCM). The thick black curves indicate the border of protection mode, while the thick red curves highlight the border of DCM mode in the figures.

The major findings were as follows. In CCM operation, the output voltage is independent of the load resistances. The achievable output voltage is limited to $V_o / V_i < 2$. Above that, the operation becomes either discontinuous or the limit of protection mode is reached. By increasing α, the output voltage is monotonically increasing. The sensitivity of V_o against the variations of α, that is, the derivative $dv_o/d\alpha$ is growing, while α is being increased. The lowering switching frequency f_s causes a monotonically decreasing output voltage. The reduced switching frequency will take the operation mode closer to the DCM region. If using V_c instead of α, as a control variable, the load resistances will affect the output voltage so

that, the decrease in R causes the falling of V_o, assuming that in the meanwhile f_s and V_c are fixed.

Moving along any R^*=constant characteristic curve, while increasing V_c^*, the operation will reach either the border of DCM or the protection mode. When the protection mode is reached, the output voltage is the available highest, assuming constant load resistance and switching frequency. In these operation points, switches S_p and S_n work in zero voltage switching (ZVS) mode, facilitating the reduction of switching losses. In any operation point along the border between CCM and DCM, zero current switching (ZCS) is realised

5 Simulation and experimental results

To verify the results of the analysis, first, computer simulations were carried out. Results are summarized for operation points A, B, C, D, and E (see Table 1, Fig. 4, and Fig. 5). The circuit parameters were as follows:

$L = 33 \ \mu H$, $C = 0.33 \ \mu F$, $f_r = 48.23$ kHz, $C_i = 47 \ \mu F$, and $C_o = 200 \ \mu F$. The characteristic impedance was $Z = 10 \ \Omega$. The rated input voltages were $V_{ip} = V_{in} = 24V$. The simulations were carried out in a Matlab/Simulink environment using ideal active and passive circuit components in the model. The calculation and simulation results coincide and are summarized in Table 1.

	V_o [V]	V_c [V]	I_{La} [A]	I_{Lb} [A]	I_{Lc} [A]	R [Ω]	α [°]	α_c [°]	f_s [kHz]
A	29.28	13.53	3.73	2.72	0.96	30	60	169	48.23
B	43.85	75.95	8.54	6.04	0.23	30	120	192	48.23
C	29.07	24.10	5.12	3.83	1.76	20	70	216	38.58
D	35.55	67.20	8.13	5.82	1.28	20	110	225	38.58
E	32.86	19.05	4.28	3.02	0	40	83.4	164	48.23

Table 1 Simulation results in steady-state operation points A, B, C, D, and E

The capacitor voltage and inductor current waveforms are presented in Fig. 6… Fig. 10.

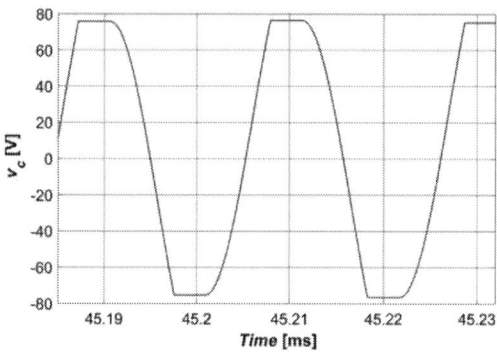

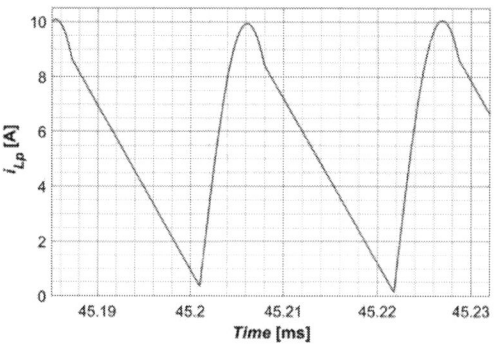

Fig. 7 Simulation results – Operation point B

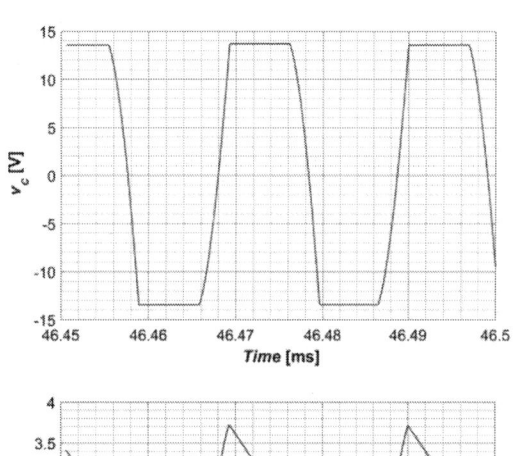

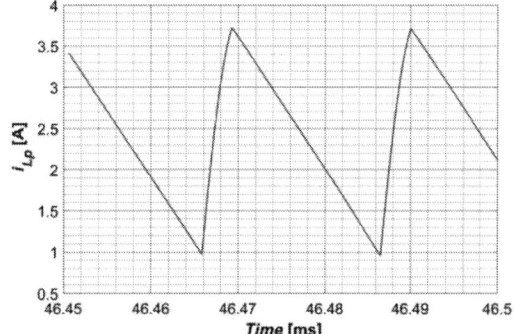

Fig. 6 Simulation results – Operation point A

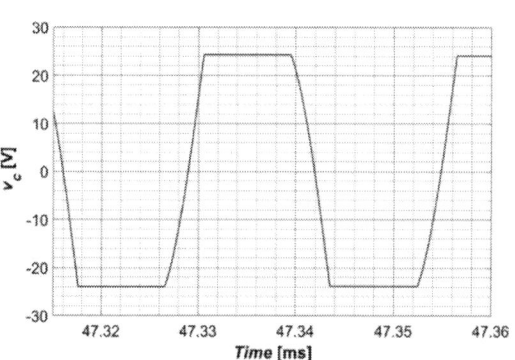

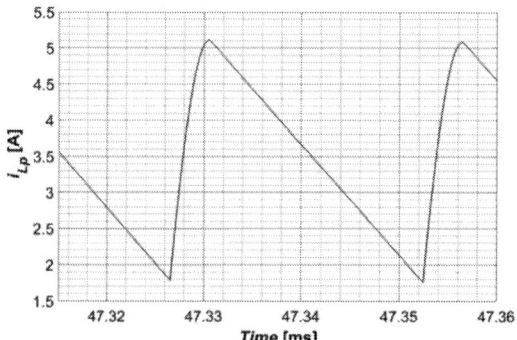

Fig. 8 Simulation results – Operation point C

PCIM Europe 2024, 11– 13 June 2024, Nuremberg   DOI: 10.30420/566262359

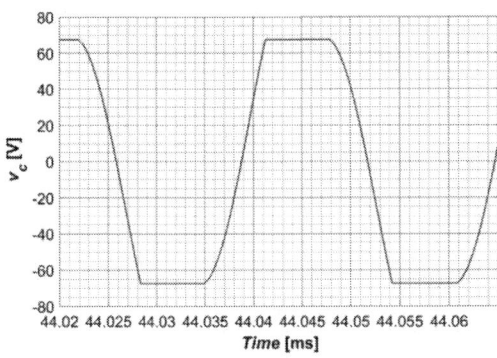

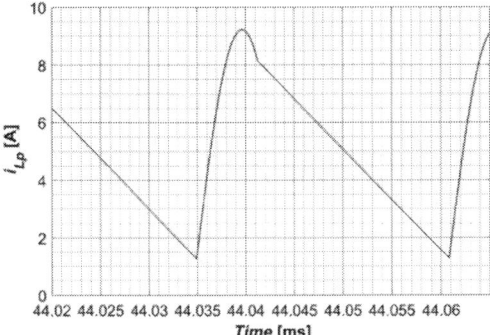

Fig. 9 Simulation results – Operation point D

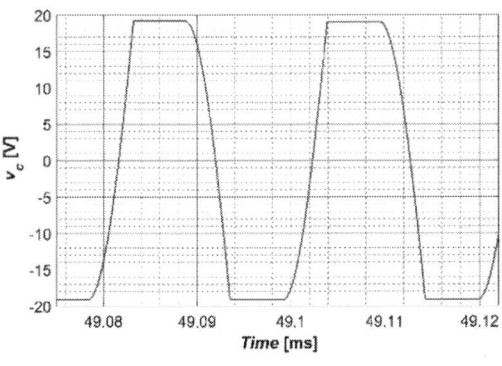

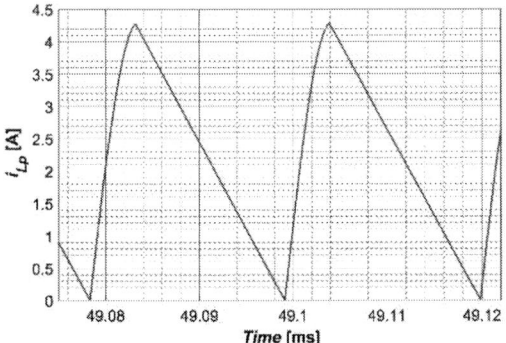

Fig. 10 Simulation results – Operation point E

The results were also confirmed by laboratory tests. In the experimental circuit, IPA083N10N MOSFETs and STTH802 diodes were applied as power switches. A TMS320F283790 Digital Signal Processor (DSP) is responsible for the open-loop control. The circuit parameters were the same as the simulation model parameters.

The measured capacitor voltage v_c and current i_c and inductor current i_{Lp} waveforms are presented in Fig.11… Fig.15 for operation points A, B, C, D and E. The value of the measured output voltages as well as the efficiencies are given in the captions of the figures.

The measurement results are in good agreement with the calculated and simulated values, which confirms the viability of the analytical results. As it can be seen there are slight differences between the calculated and measured output voltages and peak of inductor currents. These deviations are originated from the non-ideal circuit elements, like the finite resistance of the inductor, the voltage drop of real diodes, and the nonlinear behavior of the MOSFETs and diodes.

As it can be seen that, an efficiency value of around 92-94% can be achieved at this power level using the proposed topology.

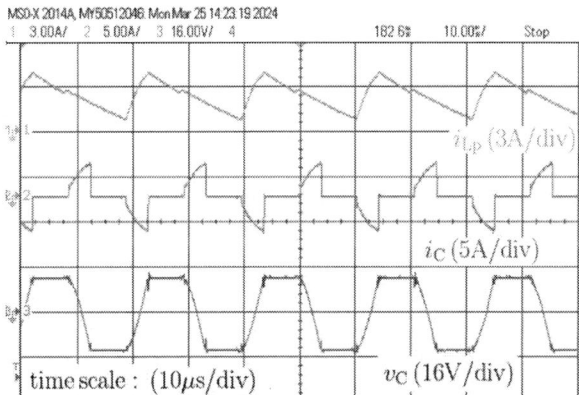

Fig. 11 Experimental results – Operation point A, measured output voltage and efficiency: V_o = 27.89V, η=92.1%

2573

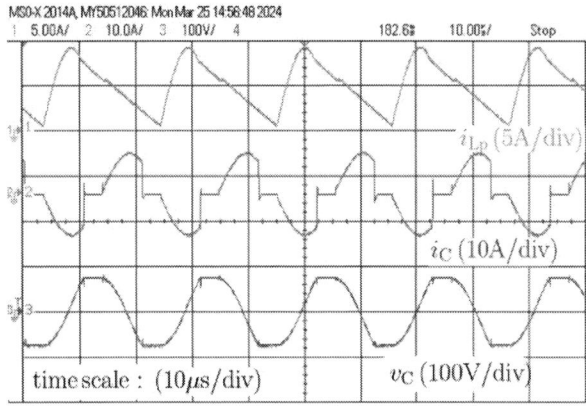

Fig. 12 Experimental results – Operation point B, measured output voltage and efficiency: V_o = 40.25V, η=93.6%

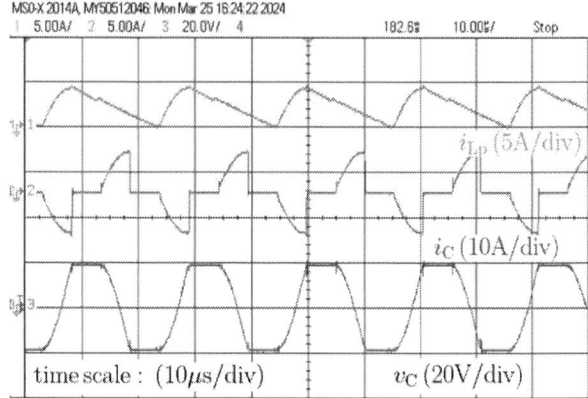

Fig. 15 Experimental results – Operation point D, measured output voltage and efficiency: V_o = 31.7 V, η=92.0%

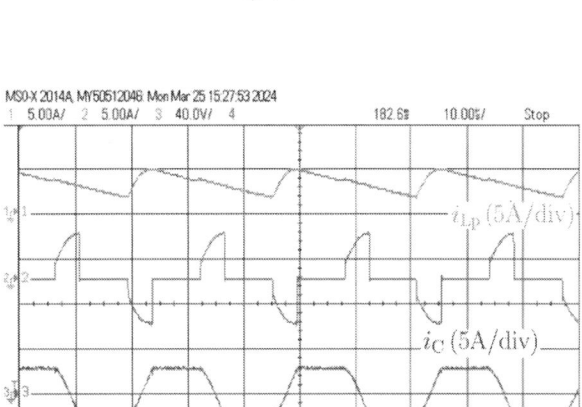

Fig. 13 Experimental results – Operation point C, measured output voltage and efficiency: V_o = 28 V, η=93.4%

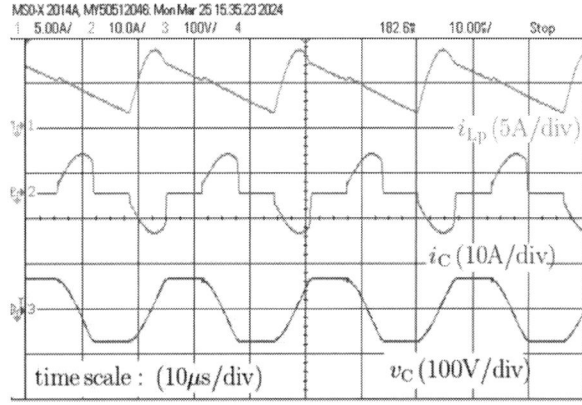

Fig. 14 Experimental results – Operation point D, measured output voltage and efficiency: V_o = 33.3 V, η=92.9%

6 Conclusions

The operation of a four channel resonant step-up dc-dc converter was analysed in continuous conduction mode (CCM). Relations, describing the CCM operation in steady state were presented for symmetrical conditions. Control variable pairs were selected to regulate the output voltage. On one hand the switching frequency, on the other hand either the switching angle α or the capacitor clamping voltage V_c can be applied for this purpose. The effect of the control variables on the output as well as the borders of the CCM-DCM operation and the protection mode were depicted. The simulation and test results confirmed and verified the theoretical considerations.

Acknowledgement

Supported by the ÚNKP-23-3-II-BME-60 New National Excellence Program of the Ministry for Culture and Innovation from the source of the National Research, Development and Innovation Fund. This work was supported by the National Research, Development, and Innovation Office under Grant FK 143429.

References

[1] J. Hamar and P. Stumpf, "New Four-Channel Resonant Boost DC/DC Converter," in IEEE Access, vol. 9, pp. 82335-82350, 2021, doi: 10.1109/ACCESS.2021.3086911.

[2] R. Kiguchi and Y. Nishida, "Cascaded Boost Converter to Achieve High Voltage Boost Rate - Conduction Loss Analysis," PCIM Europe 2019; International Exhibition and Conference for Power Electronics, Intelligent Motion, Renewable Energy and Energy Management, Nuremberg, Germany, 2019, pp. 1-7.

[3] J. Haruna, Y. Matano and H. Funato, "Efficiency Improvement for Diode-Clamped Linear Amplifier Using Unequally Divided Voltage Power Supply," in IEEE Transactions on Industry Applications, vol. 57, no. 3, pp. 2666-2672, May-June 2021, doi: 10.1109/TIA.2021.3065188.

[4] F. Kano, Y. Kasai, H. Kimura, K. Sagawa, J. Haruna and H. Funato, "Buck-Boost Type MPPT Circuit Suitable for Photovoltaic Generation of Vehicle Installation," 2018 International Power Electronics Conference (IPEC-Niigata 2018 -ECCE Asia), Niigata, Japan, 2018, pp. 2036-2041, doi: 10.23919/IPEC.2018.8507955.

[5] D. Menzi, F. Krismer, T. Ohno, J. Huber, J. W. Kolar and J. Everts, "Novel Bidirectional Single-Stage Isolated Three-Phase Buck-Boost PFC Rectifier System," 2023 IEEE Applied Power Electronics Conference and Exposition (APEC), Orlando, FL, USA, 2023, pp. 1936-1944, doi: 10.1109/APEC43580.2023.10131553.

[6] Z. Li, S. Dusmez and H. Wang, "A Novel Soft-Switching Secondary-Side Modulated Multioutput DC–DC Converter With Extended ZVS Range," in IEEE Transactions on Power Electronics, vol. 34, no. 1, pp. 106-116, Jan. 2019, doi: 10.1109/ TPEL.2018.2815718.

[7] Lee J-H, Park S-J, Lim S-K. Improvement of Multilevel DC/DC Converter for E-Mobility Charging Station. Electronics. 2020; 9(12):2037.

[8] X. Zhang and T. C. Green, "The Modular Multilevel Converter for High Step-Up Ratio DC–DC Conversion," in IEEE Transactions on Industrial Electronics, vol. 62, no. 8, pp. 4925-4936, Aug. 2015, doi: 10.1109/TIE.2015.2393846.

[9] G. Butti and J. Biela, "Novel high efficiency multilevel DC-DC boost converter topologies and modulation strategies," Proceedings of the 2011 14th European Conference on Power Electronics and Applications, Birmingham, 2011, pp. 1-10.

[10] A. B. Ponniran, K. Orikawa and J. Itoh, "Minimum Flying Capacitor for N-Level Capacitor DC/DC Boost Converter," in IEEE Transactions on Industry Applications, vol. 52, no. 4, pp. 3255-3266, July-Aug. 2016, doi: 10.1109/ TIA.2016.2555789.

Impact of Magnetics Tolerance on the Power Sharing of Parallel Dual-Output Phase-Shift Full-Bridge Converters

Lohith Kumar Pittala ©, Riccardo Barbone ©, Riccardo Mandrioli ©, Vincenzo Cirimele ©, Mattia Ricco ©

Department of Electrical, Electronic and Information Engineering, University of Bologna, Bologna, Italy.

Corresponding author: Riccardo Mandrioli, r.mandrioli@unibo.it
Speaker: Riccardo Mandrioli, r.mandrioli@unibo.it

Abstract

Understanding the effects of circuit element tolerances on power electronic devices is crucial for ensuring optimal performance and system reliability. Therefore, this paper explores the impact of magnetic tolerance in a parallel dual-output phase-shift full-bridge converter. Initially, a comprehensive analysis of power sharing among the individual legs on the converter's secondary side is provided. Taking into account ±20% uniform tolerances in inductance, a rigorous mathematical derivation of the probability density function for the converter's secondary-side leg power and for its partial-scale bridge power is provided. The outcomes of this derivation lead to a simplified triangular distribution for each leg, while an Irwin-Hall distribution for each bridge. These theoretical developments are further strengthened by numerical validation through extensive 30k PLECS simulations, incorporating randomized inductance parameters. The results of this validation process affirm the effectiveness of the adopted approach in accurately predicting power-sharing outcomes.

1 Introduction

Research on how circuit element tolerances affect power electronic device operating modes is becoming more important to guarantee the desired device performance [1]. Any electronic system's reliability is directly correlated with the precision of the tolerance analysis accomplished during the design phase. Essentially, uncertainties arising from construction procedures always impact the values of the physical parameters of commercial components [2]. Further uncertainties arise due to the fact that all parameters are contingent upon the specific operating conditions in which the components function, influenced by various factors such as frequency, temperature, aging, and others [3]. Authors in [4] stated that modifications in the circuit element values have a substantial effect on the power electronic device's working mode in both quantitative and qualitative ways. The analysis was performed by establishing a ±20% tolerance in inductance for commercially available inductors, which is roughly in line with the actual state of the circuit elements that are sold on the market.
Considering the goal of constraining the size of power electronic components to save costs, and

the need for uneven power flow between forward and reverse modes, especially in low-voltage distribution grids, the integration of isolated DC-DC converters featuring a partial-scale diode bridge and a partial-scale active bridge in parallel on the secondary side emerges as a promising solution [5], [6]. When two or more bridges are connected in parallel, ensuring adequate power sharing among them is essential [7]. However, factors such as environmental conditions, parasitic parameters, and tolerances in commercial components can impact power sharing between the bridges. These disparities may cause uneven thermal stress and component aging, ultimately affecting the lifetime and reliability of the system [8]. In order to ensure the robust performance of a converter, the aforementioned

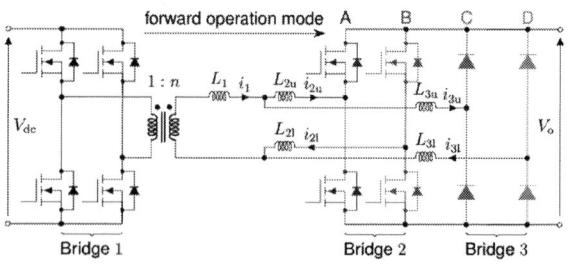

Fig. 1: Asymmetrical bidirectional DC/DC converter.

factors must be considered in a realistic design [9]. Therefore, it is important to investigate how magnetic component tolerances might affect the power sharing between the two partial-scale bridges when they are working in forward mode. This study could provide valuable insights, as any inaccuracies in power sharing might impact the choice of component sizes, highlighting the need for careful analysis in this area.

To carry out the tolerance analysis, it is common to characterize the uncertainties in statistical terms, such as resorting to probability distributions in various converters [10]–[12]. Accordingly, the present paper focuses on the asymmetrical bidirectional DC/DC (AB-DC/DC) topology investigated in [5], as illustrated in Fig. 1, emphasizing the significance of magnetic tolerance analysis.

The manuscript is structured as follows: Section 2 presents the background on the AB-DC/DC converter concerning forward power sharing. This section includes a comprehensive mathematical derivation of the analytical expression of the power probability density function (PDF) for the converter's secondary side legs and two partial-scale bridges. Section 3 encompasses the numerical validation, which was evaluated through 30k PLECS simulations with randomized inductance parameters. Finally, Section 4 concludes the paper by summarizing the findings and discussing potential future developments.

2 Power sharing under the forward operation mode

The analysis carried out in [5] has been considered in this paper. Under the forward operating mode, the secondary side active bridge is shut down and the conduction is relegated to body diodes; hence, the active bridge operates as a diode bridge. Consequently, the secondary side of the AB-DC/DC converter can be considered equivalent to a conventional phase-shift full-bridge (PSFB) converter with two parallel diode bridges. As a PSFB converter, the AB-DC/DC converter can operate in either discontinuous conduction mode (DCM) or continuous conduction mode (CCM), leading to the power flows expression:

$$P_o = \begin{cases} \frac{(1-m)V_{dc}^2}{L_{eq}f_{sw}}d^2 & \text{DCM: } d < \frac{m}{2} \\ \frac{mV_{dc}^2}{L_{eq}f_{sw}}\left(\frac{d}{2}(1-d) - \frac{m^2}{8}\right) & \text{CCM: } d \geq \frac{m}{2} \end{cases} \quad (1)$$

$$L_{eq} = L_1 + \frac{L_{2u}L_{3u}}{L_{2u} + L_{3u}} + \frac{L_{2l}L_{3l}}{L_{2l} + L_{3l}} \quad (2)$$

where d is the phase shift between the legs of the primary bridge, and m is the voltage gain defined as $(nV_o)/V_{dc}$. According to the notation in Fig. 1, the power theoretically processed by each leg (A, B, C, and D) can be predicted as:

$$P_A = \frac{P_o}{2}\left(\frac{L_{3u}}{L_{2u}+L_{3u}}\right), \quad P_B = \frac{P_o}{2}\left(\frac{L_{3l}}{L_{2l}+L_{3l}}\right),$$
$$P_C = \frac{P_o}{2}\left(\frac{L_{2u}}{L_{2u}+L_{3u}}\right), \quad P_D = \frac{P_o}{2}\left(\frac{L_{2l}}{L_{2l}+L_{3l}}\right). \quad (3)$$

From (3), it is evident that in ideal conditions the power distribution within each branch can be regulated by selecting inductance ratios. However, real magnetics can significantly drift from the nominal values. Hence, it is necessary to assume a probability distribution for the tolerances considered. In practice, however, the manufacturers neither provide this information on the data sheets, nor indicate if the uncertainty is provided according to a deterministic or probabilistic method [13]. In case the distribution is not known, then the uncertainty value due to a given contribution is taken according to the principle of maximum entropy leading to a uniform distribution. For the sake of simplicity, all the here considered contributions have the same tolerance, i.e., ±20%. It is more useful to refer to normalized powers, as they depend solely on inductances, which are the only parameters affected by uncertainty for the purposes of this analysis. Taking $P_o/2$ as the normalization basis, and based on (3), the following expressions for the powers shared by the converter legs are obtained:

$$N_A = \frac{L_{3u}}{L_{2u}+L_{3u}}, \quad N_B = \frac{L_{3l}}{L_{2l}+L_{3l}},$$
$$N_C = \frac{L_{2u}}{L_{2u}+L_{3u}}, \quad N_D = \frac{L_{2l}}{L_{2l}+L_{3l}}. \quad (4)$$

It is possible to demonstrate that the PDF related to the power flowing in each branch can be approximated with a triangular distribution (T-PDF), as validated in Section 3. However, it is useful to first go through the mathematical derivation of the analytical expression of the actual PDF (A-PDF), reported in the following paragraphs. After providing the derivation of the actual converter leg probability density function and its triangular approximation, given in Subsection 2.1, the approximated converter bridge probability density function is discussed in Subsection 2.2.

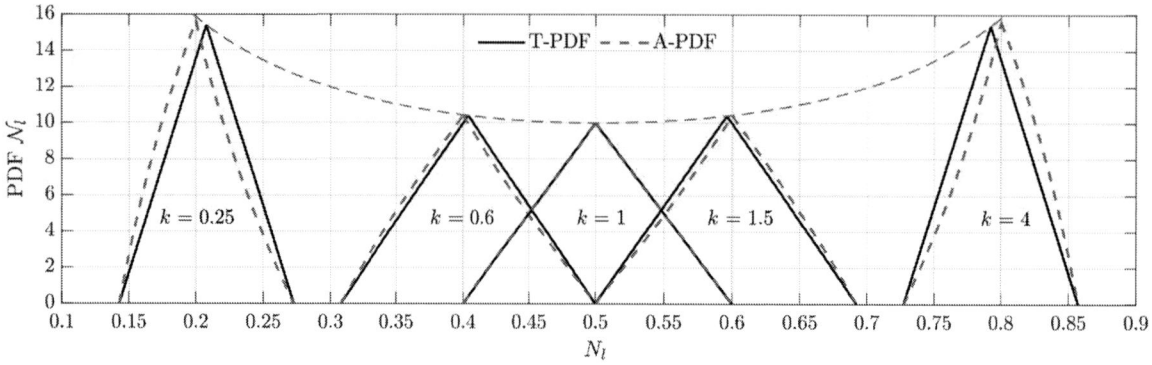

Fig. 2: Actual and approximated triangular PDF of the converter leg power with different k.

2.1 Converter Leg Power PDF

In the following, reference is made to the power shared by the leg A of the converter depicted in Fig. 1. As per (4), the relative power N_A is a function of the sole inductances, and presents a PDF denoted with $\mathcal{N}_A$. To make computation steps easier, the reciprocal of $\mathcal{N}_A$ is first determined:

$$\frac{1}{\mathcal{N}_A(x)} = 1 + \frac{\mathcal{L}_{2u}(x)}{\mathcal{L}_{3u}(x)} = 1 + \mathcal{L}_{2u}(x)\mathcal{R}_{3u}(x) , \quad (5)$$

where $\mathcal{L}_{2u}$ and $\mathcal{L}_{3u}$ are the uniform PDFs for inductances L_{2u} and L_{3u}, respectively, and are represented as:

$$\begin{aligned} \mathcal{L}_{2u}(x) &= \frac{1}{0.4L_{2u}} \quad x \in [0.8L_{2u}, 1.2L_{2u}] \\ \mathcal{L}_{3u}(x) &= \frac{1}{0.4L_{3u}} \quad x \in [0.8L_{3u}, 1.2L_{3u}] \end{aligned} . \quad (6)$$

On the other hand, $\mathcal{R}_{3u}$ is another PDF corresponding the reciprocal of $\mathcal{L}_{3u}$, and it is computed from (6) as:

$$\mathcal{R}_{3u}(x) = \frac{\mathcal{L}_{3u}(\frac{1}{x})}{x^2} = \frac{5}{2L_{3u}x^2} \quad x \in \left[\frac{5}{6L_{3u}}, \frac{5}{4L_{3u}}\right]. \quad (7)$$

Thanks to the procedure detailed in Appendix A, the PDF (5) is defined as:

$$\frac{1}{\mathcal{N}_A(x)} = \begin{cases} \frac{9k}{2} - \frac{2}{k(x-1)^2} & x \in \left[\frac{2}{3k}+1, \frac{1}{k}+1\right] \\ \frac{9}{2k(x-1)^2} - 2k & x \in \left[\frac{1}{k}+1, \frac{3}{2k}+1\right] \end{cases}, \quad (8)$$

where k is a function of the nominal inductances reported in Fig. 1, and it exhibits distinct expressions for the legs between the two secondary-side bridges. Concerning leg A, (8) is evaluated with $k = L_{3u}/L_{2u}$.

Finally, the PDF of the relative power $\mathcal{N}_l$ shared by

a generic leg l can be computed as:

$$\mathcal{N}_l(x) = \begin{cases} \frac{9}{2k(1-x)^2} - \frac{2k}{x^2} & x \in \left[\frac{2k}{3+2k}, \frac{k}{1+k}\right] \\ \frac{9k}{2x^2} - \frac{2}{k(1-x)^2} & x \in \left[\frac{k}{1+k}, \frac{3k}{2+3k}\right] \end{cases} \quad (9)$$

$$k = \begin{cases} \frac{L_{3u}}{L_{2u}} & \text{for A} \\ \frac{L_{3l}}{L_{2l}} & \text{for B} \end{cases} \quad \text{and} \quad k = \begin{cases} \frac{L_{2u}}{L_{3u}} & \text{for C} \\ \frac{L_{2l}}{L_{3l}} & \text{for D} \end{cases}.$$

Once the PDF of a random variable is available, one can determine both its expected value E and its standard deviation σ. Determining E and σ of the PDF reported in (9) might not be trivial, however, further considerations may be drawn by looking at Fig. 2. Fig. 2 shows the plots of $\mathcal{N}_l$, i.e., A-PDF, with different inductance ratios. It is evident that the shape of the A-PDF resembles that of a triangle, i.e., it can be approximated with a symmetric T-PDF. The parameters governing the resulting symmetric triangular distribution can be defined in relation to the factor k as:

$$\begin{aligned} &\text{Lower limit} \quad a = \frac{P_{nom}}{2} \frac{2k}{3+2k} \\ &\text{Upper limit} \quad b = \frac{P_{nom}}{2} \frac{3k}{2+3k} \quad \text{with} \\ &\text{Mode} \quad c = \frac{a+b}{2} \end{aligned} \quad (10)$$

$$k = \begin{cases} \frac{L_{3u}}{L_{2u}} & \text{for A} \\ \frac{L_{3l}}{L_{2l}} & \text{for B} \end{cases} \quad \text{and} \quad k = \begin{cases} \frac{L_{2u}}{L_{3u}} & \text{for C} \\ \frac{L_{2l}}{L_{3l}} & \text{for D} \end{cases}.$$

Based on the PDF given in (10), the expected value E^* and the standard deviation σ^* can be obtained as:

$$E^* = c , \qquad \sigma^* = \frac{b-a}{2\sqrt{6}} . \quad (11)$$

Looking at Fig. 2, one may see that the A-PDF in the $k = 1$ case perfectly matches the T-PDF. In the other cases (e.g., $k = 0.25$, 0.6, 1.5, 4, according to the figure), however, a slight deviation of the A-PDF from the approximated T-PDF may be observed. This discrepancy is all the more pronounced the further k moves away from unity.

2.2 Bridge Power PDF

The power processed by each bridge (the bridges are denoted as Bridge 2 and Bridge 3) can be determined as $P_2 = P_A + P_B$ and $P_3 = P_C + P_D$, and the normalized form can be obtained as $N_2 = N_A + N_B$ and $N_3 = N_C + N_D$. Considering that leg normalized powers are independent random variables, the PDF of the normalized bridge power $\mathcal{N}_b$ can be determined through convolution (see: [14]) as:

$$\begin{aligned} \mathcal{N}_2(x) &= \mathcal{N}_A(x) * \mathcal{N}_B(x) \\ \mathcal{N}_3(x) &= \mathcal{N}_C(x) * \mathcal{N}_D(x) \end{aligned}. \quad (12)$$

Assuming the same inductance ratio k for legs pertaining to the same bridge and taking advantage of T-PDF symmetry, it is possible to express (12) as an Irwin–Hall distribution of the fourth order (see: [15]):

$$\mathcal{N}_b(x) = \begin{cases} \frac{b-a}{12}z^3 & x \in [2a, a+c] \\ \frac{b-a}{12}(-3z^3+12z^2-12z+4) & x \in [a+c, 2c] \\ \frac{b-a}{12}(3z^3-24z^2+60z-44) & x \in [2c, b+c] \\ \frac{b-a}{12}(4-z)^3 & x \in [b+c, 2b] \end{cases}$$

$$\text{having} \qquad z = \frac{2(x-2a)}{b-a},$$

$$k = \frac{L_{3u}}{L_{2u}} = \frac{L_{3l}}{L_{2l}} \quad \text{for Bridge 2,}$$

$$k = \frac{L_{2u}}{L_{3u}} = \frac{L_{2l}}{L_{3l}} \quad \text{for Bridge 3.} \quad (13)$$

Based on the PDF given in (13), the expected value E^* and the standard deviation σ^* can be obtained as:

$$E^* = 2c, \qquad \sigma^* = \frac{b-a}{\sqrt{12}}. \quad (14)$$

As foreseeable, the expected value of the PDF associated with the bridge power is twice that of the leg power. Conversely, the standard deviation of the bridge PDF is $\sqrt{2}$ times higher than that of the legs.

Parameter	Symbol	Value	Unit
Input voltage	V_{dc}	200	V
Voltage gain	m	0.5	-
Turn ratio	n	1	-
Nominal output power	P_{nom}	1475	W
Switching frequency	f_{sw}	20	kHz
Case 1 Inductances	L_1	$38 \pm 20\%$	μH
	L_{2u}, L_{2l}	$15 \pm 20\%$	μH
	L_{3u}, L_{3l}	$10 \pm 20\%$	μH
Case 2 Inductances	L_1	$40 \pm 20\%$	μH
	L_{2u}, L_{2l}	$10 \pm 20\%$	μH
	L_{3u}, L_{3l}	$10 \pm 20\%$	μH
Case 3 Inductances	L_1	$34 \pm 20\%$	μH
	L_{2u}, L_{2l}	$40 \pm 20\%$	μH
	L_{3u}, L_{3l}	$10 \pm 20\%$	μH

Tab. 1: Numerical plant main system parameters.

3 Numerical validation

To validate the above considerations, the converter of Fig. 1 is simulated 30k times on PLECS (Plexim GmbH) environment considering uniformly distributed randomized inductances. In the analysis, three distinct cases (10k simulations each) were defined:

- **Case 1**: Bridge 2 (active) processes 40% of the nominal output power while Bridge 3 (passive) processes the remaining 60%.

- **Case 2**: Bridge 2 (active) processes 50% of the nominal output power while Bridge 3 (passive) processes the remaining 50%.

- **Case 3**: Bridge 2 (active) processes 20% of the nominal output power while Bridge 3 (passive) processes the remaining 80%.

Figure 3 illustrates the numerical validation of the normalized power distribution among each leg (A, B, C, D). The A-PDF, represented as a dashed red line and described by (9), perfectly aligns with the histogram data across all the cases. However, a slight deviation is noticeable between the T-PDF (solid line) and A-PDF in Case 1 and Case 3, as predicted in section 2.1. Furthermore, in Case 1, there is an additional power demand of $+25\%$ and $+15\%$ in each leg of Bridge 2 and Bridge 3, respectively. In Case 2, all legs experience a $+20\%$ additional power demand, whereas in Case 3, each leg of Bridge 2 and Bridge 3 faces a $+35\%$ and $+10\%$ additional power demand, respectively, in the worst-case scenario. This additional power demand necessitates the oversizing of silicon devices, potentially jeopardizing the converter's partial-scale feature.

As summarized in Table 2, the theoretical expected values E^* and standard deviations σ^* provided in (11) fairly match with the numerical acquisitions E and σ in Case 2. This alignment can be visually observed from the adequate fitting of the proposed T-PDF over the data histograms, as shown in Fig. 3. However, in Case 1 and Case 3, notable differences between the theoretical and numerical expected values and standard deviation values emerge due to the discrepancy introduced by the approximated T-PDF outlined in (10).

As detailed in Section 2.2, the power flow in each bridge is characterized by the Irwin-Hall distribution, obtained through convolution of the approximated T-PDFs relevant to each converter's leg. Thus, in Fig. 4, the normalized power distributions related to Bridge

2579

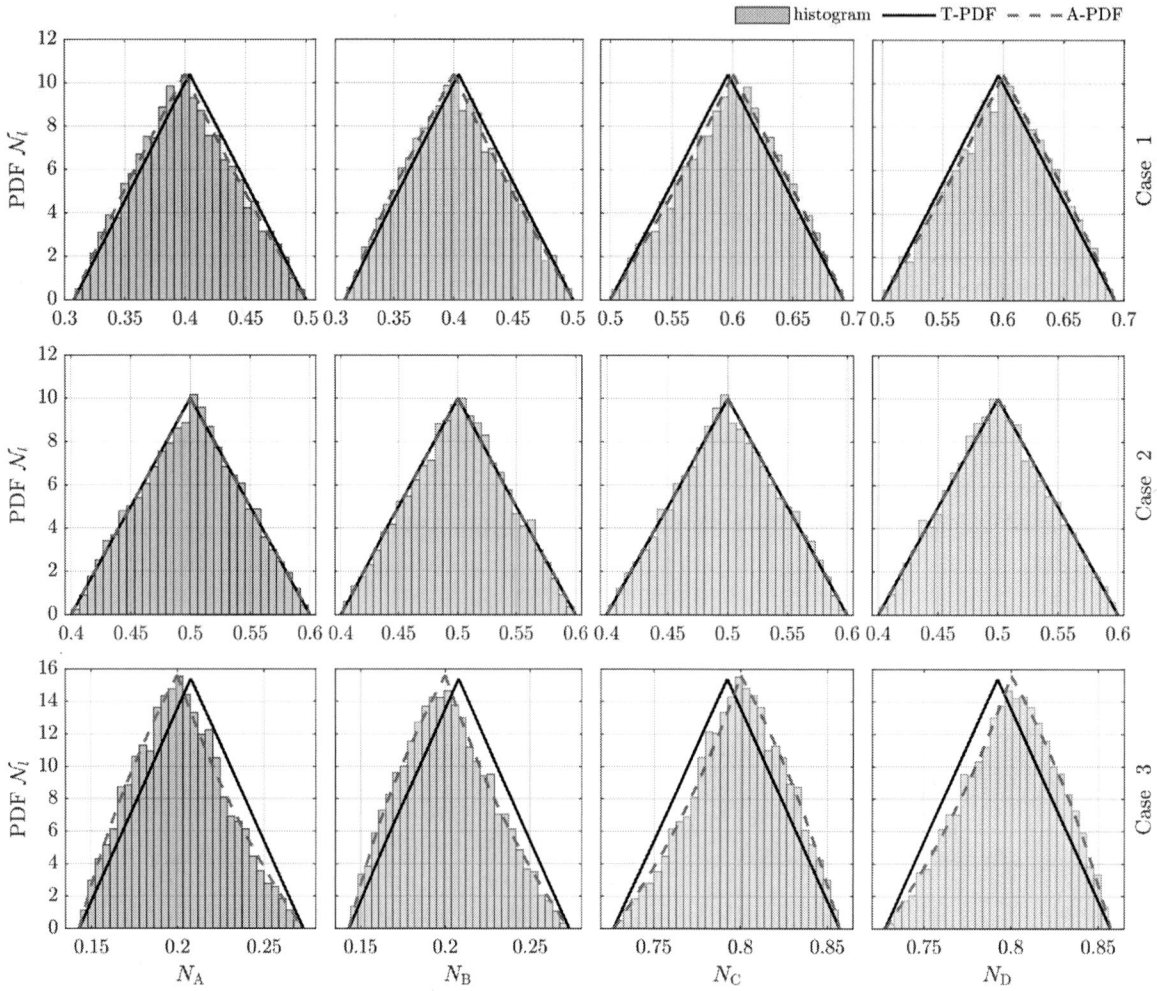

Fig. 3: Numerical power distribution together with actual and approximated triangular PDFs of the normalized leg power. Legs A, B, C, and D from left to right. Case 1, 2, and 3 from top to bottom.

		Leg A	B	C	D	Unit
Case 1	E^*	297.8	297.8	439.7	439.7	W
	E	295.5	295.8	442.0	442.8	W
	σ^*	28.95	28.95	28.95	28.95	W
	σ	29.17	29.19	29.17	29.19	W
	RMSD	5.646	5.660	6.790	6.742	mW
Case 2	E^*	368.7	368.7	368.7	368.7	W
	E	368.8	368.8	368.8	368.8	W
	σ^*	30.10	30.10	30.10	30.10	W
	σ	30.30	29.81	30.30	29.81	W
	RMSD	6.135	6.111	6.103	6.093	mW
Case 3	E^*	153.2	153.2	584.2	584.2	W
	E	148.3	148.1	589.2	589.4	W
	σ^*	19.55	19.55	19.55	19.55	W
	σ	19.14	19.41	19.14	19.41	W
	RMSD	2.878	2.900	6.290	6.248	mW

Tab. 2: Statistical analysis of measured and predicted powers in each leg.

		Bridge 2	3	Unit
Case 1	E^*	595.7	879.3	W
	E	591.2	883.7	W
	σ^*	40.94	40.94	W
	σ	41.42	41.42	W
	RMSD	8.938	11.63	mW
Case 2	E^*	737.5	737.5	W
	E	738.0	736.9	W
	σ^*	42.57	42.57	W
	σ	42.76	42.76	W
	RMSD	10.15	10.09	mW
Case 3	E^*	306.5	1168.5	W
	E	296.4	1178.6	W
	σ^*	27.64	27.64	W
	σ	27.49	27.49	W
	RMSD	4.544	12.02	mW

Tab. 3: Statistical analysis of measured and predicted powers on the secondary-side bridges.

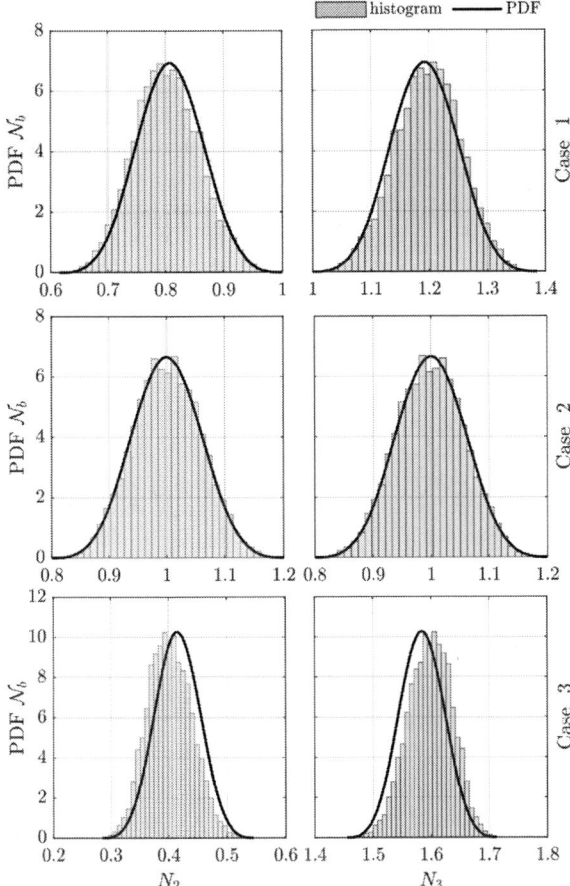

Fig. 4: Numerical power distribution together with approximated Irwin-Hall PDF of the normalized bridge power. Bridges 2 and 3 from left to right. Case 1, 2, and 3 from top to bottom.

2 and Bridge 3 are depicted. The PDF, represented by a solid line, aligns perfectly with the histogram data in Case 2. However, a slight deviation is noticeable in Case 1 and Case 3, attributed to the considered approximation. These discrepancies are also observed in the theoretical expected values E^* and standard deviations σ^* (14) compared to the numerical values, as shown in Table 3. Furthermore, the root mean square deviation (RMSD) between the measured power and the predicted one is provided for each leg in Table 2, and for each bridge in Table 3. As visible, RMSD absolute values present milliwatts order of magnitude. Such a deviation is in the range of 10 to 20 ppm and it can be almost exclusively attributed to numerical errors in all cases. This is compelling evidence that the theoretical developments (3) are in good match with the numerical validation, affirming the accuracy and reliability of the presented approach under non-ideal conditions.

4 Conclusion

This paper investigates the impact of magnetic tolerance on the power-sharing of a parallel dual-output PSFB converter. The proper selection of the inductors enables an effective control of the power-sharing among the bridges, as demonstrated successfully. Analytical determination of the power flowing through each leg yields an expression of the power-sharing ratio between the secondary-side bridges. Considering a ±20% tolerance in the inductances, the A-PDF of each leg power is derived and approximated to a T-PDF. Thanks to the convolution of the T-PDFs of the power flowing through the converter's leg for each bridge, the approximated power PDF in each bridge is shown to take on the shape of an Irwin-Hall distribution. Furthermore, theoretical developments are numerically validated across three distinct cases, with each case simulated 10k times. Simulation outcomes confirm accurate alignment of A-PDF with histogram data, with minor deviations in the T-PDF. Such deviations are also evident in the power distribution of each bridge. The precision of theoretical developments is further confirmed by the expected value and standard deviation obtained from simulations. Evaluation using RMSD reveals discrepancies at the milliwatt level between predicted and measured power considering inductance variations.

Future research will focus on investigating the active power-sharing control between partial-scale active and diode bridges. These advancements are aimed at reducing thermal stress on either bridge, thereby enhancing the overall reliability of the system.

Appendix A

The PDF resulting from the product of $\mathscr{L}_{2u}$ and $\mathscr{R}_{3u}$ inside (5), here indicated as $\mathscr{H}_A$, can be computed using the approach outlined in [16] as:

$$
\mathscr{H}_A(x) = \begin{cases} \displaystyle\int_{\alpha}^{\frac{x}{\gamma}} \frac{\mathscr{L}_{2u}(z)\mathscr{R}_{3u}\left(\frac{x}{z}\right)}{z}dz & \alpha\gamma \leq x \leq \beta\gamma \\[4mm] \displaystyle\int_{\frac{x}{\delta}}^{\beta} \frac{\mathscr{L}_{2u}(z)\mathscr{R}_{3u}\left(\frac{x}{z}\right)}{z}dz & \alpha\delta \leq x \leq \beta\delta \end{cases} \tag{15}
$$

having:

$$
\begin{aligned}
\alpha &= 0.8L_{2u} & \gamma &= \frac{1}{1.2L_{3u}} = \frac{1}{1.2kL_{2u}} \\
\beta &= 1.2L_{2u} & \delta &= \frac{1}{0.8L_{3u}} = \frac{1}{0.8kL_{2u}}
\end{aligned} \tag{16}
$$

Acknowledgments

Project funded under the National Recovery and Resilience Plan (NRRP), Mission 4 Component 2 Investment 1.3 - Call for tender No. 1561 of 11.10.2022 of Ministero dell'Università e della Ricerca (MUR); funded by the European Union – NextGenerationEU. Project code PE0000021, Concession Decree No. 1561 of 11.10.2022 adopted by Ministero dell'Università e della Ricerca (MUR), PE02 - NEST - CUP: J33C22002890007, Project title "Network 4 Energy Sustainable Transition – NEST".

References

[1] N. L. Hinov and T. H. Hranov, "Tolerance analysis of common transistor dc-dc converters," in *Proceedings of 25th International Conference Electronics*, 2021, pp. 1–6. DOI: 10.1109/IEEECONF52705.2021.9467442.

[2] M. Rimondi, R. Mandrioli, V. Cirimele, L. K. Pittala, M. Ricco, and G. Grandi, "Design of an integrated, six-phase, interleaved, synchronous dc/dc boost converter on a fuel-cell-powered sport catamaran," *Designs*, vol. 6, no. 6, 2022. DOI: 10.3390/designs6060113.

[3] A. Cirillo, N. Femia, and G. Spagnuolo, "An interval mathematics approach to tolerance analysis of switching converters," in *Proceedings of 27th Annual IEEE Power Electronics Specialists Conference*, vol. 2, 1996, 1349–1355 vol.2. DOI: 10.1109/PESC.1996.548757.

[4] N. Hinov, T. Hranov, and V. Dimitrov, "Tolerance analysis of resonant converters with parallel loaded capacitor," in *Proceedings of XXIX International Scientific Conference Electronics (ET)*, 2020, pp. 1–5. DOI: 10.1109/ET50336.2020.9238188.

[5] R. Zhu, F. Hoffmann, N. Vázquez, K. Wang, and M. Liserre, "Asymmetrical bidirectional dc–dc converter with limited reverse power rating in smart transformer," *IEEE Transactions on Power Electronics*, vol. 35, no. 7, pp. 6895–6905, 2020. DOI: 10.1109/TPEL.2019.2957407.

[6] S. Wu, *et al.*, "Topology and operation analysis of isolated dc/dc converters with bidirectional asymmetric power flow," in *Proceedings of 48th Annual Conference of the IEEE Industrial Electronics Society*, 2022, pp. 1–5. DOI: 10.1109/IECON49645.2022.9968702.

[7] B. M. H. Jassim, D. J. Atkinson, and B. Zahawi, "Modular current sharing control scheme for parallel-connected converters," *IEEE Transactions on Industrial Electronics*, vol. 62, no. 2, pp. 887–897, 2015. DOI: 10.1109/TIE.2014.2355813.

[8] K. Ma, M. Liserre, F. Blaabjerg, and T. Kerekes, "Thermal loading and lifetime estimation for power device considering mission profiles in wind power converter," *IEEE Transactions on Power Electronics*, vol. 30, no. 2, pp. 590–602, 2015. DOI: 10.1109/TPEL.2014.2312335.

[9] A. Diet, N. Couellan, X. Gendre, J. Martin, and J.-P. Navarro, "A statistical approach for tolerancing from design stage to measurements analysis," *Procedia CIRP*, vol. 92, pp. 33–38, 2020, 16th CIRP Conference on Computer Aided Tolerancing (CIRP CAT 2020). DOI: https://doi.org/10.1016/j.procir.2020.05.171.

[10] Y. Chen, A. Sangwongwanich, M. Huang, S. Pan, X. Zha, and H. Wang, "Failure risk assessment of grid-connected inverter with parametric uncertainty in lcl filter," *IEEE Transactions on Power Electronics*, vol. 38, no. 8, pp. 9514–9525, 2023. DOI: 10.1109/TPEL.2023.3274396.

[11] Y. Yang, H. Wang, A. Sangwongwanich, and F. Blaabjerg, "45 - design for reliability of power electronic systems," in *Power Electronics Handbook (Fourth Edition)*, M. H. Rashid, Ed., Fourth Edition, Butterworth-Heinemann, 2018, pp. 1423–1440. DOI: https://doi.org/10.1016/B978-0-12-811407-0.00051-9.

[12] R. Mandrioli, L. K. Pittala, V. Cirimele, M. Ricco, and G. Grandi, "Probabilistic approach for the study of neutral current ripple in split-capacitor inverters," in *Proceedings of IEEE 17th International Conference on Compatibility, Power Electronics and Power Engineering*, 2023, pp. 1–6. DOI: 10.1109/CPE-POWERENG58103.2023.10227446.

[13] V. Cirimele, *et al.*, "Uncertainty quantification for sae j2954 compliant static wireless charge components," *IEEE Access*, vol. 8, pp. 171489–171501, 2020. DOI: 10.1109/ACCESS.2020.3025052.

[14] M. Linford, "The gaussian-lorentzian sum, product, and convolution (voigt) functions used in peak fitting xps narrow scans, and an introduction to the impulse function.," *Vacuum Technology and Coating*, Jul. 2014. DOI: 10.1016/j.apsusc.2018.03.190.

[15] J. E. Marengo, D. L. Farnsworth, and L. Stefanic, "A geometric derivation of the irwin-hall distribution," *Int. J. Math. Math. Sci.*, vol. 2017, 3571419:1–3571419:6, 2017. DOI: 10.1155/2017/3571419.

[16] A. G. Glen, L. M. Leemis, and J. H. Drew, "Computing the distribution of the product of two continuous random variables," *Computational Statistics & Data Analysis*, vol. 44, no. 3, pp. 451–464, 2004. DOI: https://doi.org/10.1016/S0167-9473(02)00234-7.

PCIM Europe 2024, 11– 13 June 2024, Nuremberg   DOI: 10.30420/566262361

A Balancing Converter with Series Connected MOSFETs for ±700V Bipolar DC Grids

Sachin Yadav ©[1], Guangyao Yu ©[1], Zian Qin ©[1], Pavol Bauer ©[1]

[1] Delft University of Technology, The Netherlands

Corresponding author: Sachin Yadav, S.Yadav-1@tudelft.nl
Speaker: Sachin Yadav, S.Yadav-1@tudelft.nl

Abstract

This paper introduces a balancing converter designed for use in ±700 V bipolar DC grids. Integrating a buck-boost topology with triangular current modulation (TCM), the converter achieves zero-voltage switching (ZVS) turn-on across all switches, thereby enhancing efficiency and reducing other issues such as EMI emissions. Additionally, the utilization of series-connected MOSFETs allows for higher blocking voltages, broadening the converter's applicability. The study also explores how the switching frequency and duty cycle influence power flow, providing valuable insights for optimal operation. A prototype of the converter has been designed and tested, demonstrating a peak efficiency of 97.9% at load power of up to 5 kW. The findings showcase the converter's practical viability and mark a significant advancement in the design of balancing converters for bipolar DC grids.

1   Introduction

As global reliance on renewable energy sources intensifies, the demand for innovative power electronic solutions capable of integrating these resources into the power grid grows. DC grids, favored for their compatibility with inherently DC sources like batteries and photovoltaics, emerge as a promising avenue. Among them, bipolar DC grids stand out for their superior power flow capacity, reduced insulation voltage stresses, and enhanced voltage level diversity compared to their unipolar counterparts [1], [2]. However, the challenge of grid imbalance—prompted by differential loading across poles and neutral—calls for specialized converters to maintain equilibrium, thereby safeguarding grid equipment and minimizing losses.

In the Netherlands, bipolar DC grids adhere to standard voltages of ±350 V and ±700 V [3], with the latter capable of supporting up to 22.4 kW power flow at a standard pole conductor current of 16 A. Previous efforts, such as the series resonant balancing converter designed for ±350 V grids, highlighted innovations in passive component minimization and full-range Zero-Voltage Switching (ZVS) [4]. This work extends these advancements to the ±700 V bipolar DC (BiDC) grid, employing an in-verting buck-boost topology renowned for balancing functionalities but not yet fully explored for its ZVS potential through Triangular Current Modulation (TCM).

Addressing a gap in the literature, this paper introduces a novel balancing converter for the ±700 V BiDC grid. Key to this solution is the use of series-connected Silicon Carbide (SiC) MOSFETs, capable of withstanding voltages up to 1400 V, to accommodate high switching frequencies [5]. Incorporating TCM, the proposed converter achieves high efficiency while necessitating a duty cycle adjustment for ZVS across varying operational ranges—a consideration previously unaddressed. These innovations constitute the core contributions of our research.

The paper is structured as follows: Section 2 delves into the converter's operating principles and the analytical basis for the duty cycle adjustment essential for achieving ZVS. Section 3 outlines the converter's design. Experimental validations are presented in Section 4, followed by concluding remarks in Section 5.

2   Converter Operation

The schematic representation of the balancing converter is illustrated in Fig. 1. It depicts each switch's snubber circuit, identified by resistances ($R_{sn,Sx}$)

and capacitors ($C_{sn,Sx}$), which are crucial for transient suppression, and resistors ($R_{bl,Sx}$) that ensure static voltage balancing across the switches. The converter's input and output voltages are labeled as V_{L1} and V_{L2}, respectively. Additionally, the inductor's current and voltage are denoted by i_L and v_L. The parasitic resistances of the switches and the inductor are also included, represented by $R_{ds,Sx}$ and R_L, which are essential for a comprehensive analysis of the converter's performance.

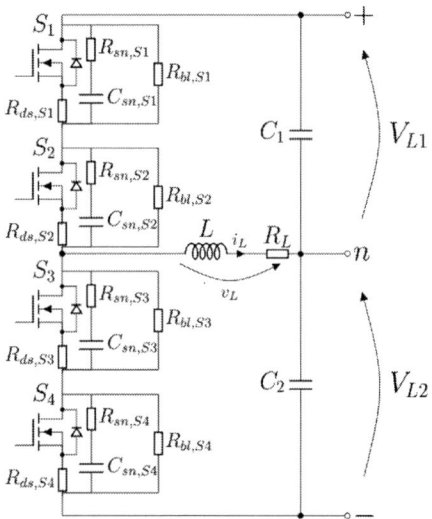

Fig. 1: Topology of the inverting buck-boost balancing converter including the parasitic resistances.

The TCM scheme is designed to modulate the switching frequency, thus controlling the inductor's current flow to match the output load current. This approach begins with establishing the switching frequency range, accounting for the converter's parasitic elements to adjust the duty cycle of the switches appropriately. The culmination of these considerations leads to the development of a cohesive power flow control strategy, detailed in the subsequent sections.

2.1 Switching Frequency Control

The operation of the converter, illustrated by the typical voltage and current waveforms of the inductor, alongside the switching pattern shown in Fig. 2, is analyzed. The direction of the inductor current, as depicted in Fig. 1, suggests a higher load is connected between the neutral and negative pole, assuming this specific load configuration.

The converter's operational phases are elucidated as follows:

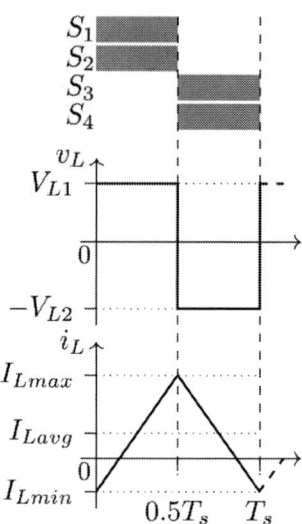

Fig. 2: Triangular current modulation scheme for the inverting buck-boost balancing converter.

– Activation of S_1 and S_2 imposes the + pole to neutral voltage (V_{L1}) across inductor L, facilitating energy storage from the + pole through inductor charging.

– The inductor current escalates to a peak (I_{Lmax}), determined by the load current ($I_{out} = I_{Lavg}$), as described by (1).

$$I_{Lmax} = 2I_{Lavg} - I_{Lmin} \qquad (1)$$

– Following the peak at I_{Lmax}, deactivation of S_1 and S_2 occurs. The inductor's inherent current continuity forces current through diodes in S_3 and S_4 during the interim, post capacitance discharging across S_3 and S_4, thereby enabling ZVS turn-on upon their activation.

– With S_3 and S_4 active, the inductor is subjected to the neutral to - pole voltage, causing a reduction in inductor current.

– The inductor current inverts, reaching I_{Lmin}, at which point S_3 and S_4 are deactivated. This action prepares S_1 and S_2 for subsequent ZVS turn-on, with I_{Lmin} further explored in the ensuing subsection.

Switching frequency control begins with establishing its operational boundaries. Converter power flow inversely correlates with switching frequency, diminishing to null at the equilibrium of positive and negative inductor current magnitudes, indicative of zero average current. The converter's maximal switching frequency is computed as per (2).

$$f_{sw,max} = \frac{2V_d}{L\,|I_{L,min}|} \quad (2)$$

Conversely, the minimal switching frequency is dictated by the peak output current capacity, ascertainable under balanced grid conditions ($V_{L1} = V_{L2} = V_d$), and quantified by (3).

$$I_{out,max} = \frac{P_{max}}{V_d} \quad (3)$$

In steady state, with balanced V_{L1} and V_{L2} equal to V_d, the optimal switching frequency ensuring achievement of I_{Lmin} for any given output current I_{out} (corresponding to I_{Lavg} in Fig. 2) is delineated by (4).

$$f_{sw} = \frac{V_d}{4L(I_{out} - I_{Lmin})} \quad (4)$$

2.2 Determination of Minimum Negative Inductor Current

In the framework of TCM, the discharging of the switch's output capacitance (C_{oss}) necessitates a specific minimum negative inductor current. Given the series configuration of the switches, it is imperative to consider the equivalent output capacitance ($C_{oss,eq}$) resulting from this arrangement. The requisite minimum inductor current ($I_{l,min}$) for efficiently discharging C_{oss} is mathematically established in (5).

$$I_{L,min} = V_d\sqrt{\frac{8}{3}\frac{C_{oss,eq}}{L}} \quad (5)$$

The transition dynamics of the output capacitances, during switching events, are effectively represented through an equivalent circuit model, as depicted in Fig. 3. The series alignment of two switches results in their capacitances also being in series, which consequently reduces the equivalent capacitance below that of a singular switch. Nonetheless, the influence of snubber capacitance is also crucial and must be incorporated into this analysis. A detailed exploration of this circuit model and its implications is provided in [6], offering insights into the nuanced behavior of these components under TCM conditions.

2.3 Duty Cycle Control

This study addresses the often-overlooked impact of parasitic resistances in the semiconductor

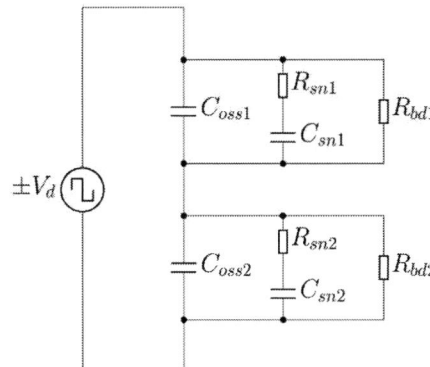

Fig. 3: Equivalent circuit during the charge or discharge of the switch capacitances.

switches and inductor on the operation of converters designed for grid balancing. Previous research frequently neglects these resistances, leading to suboptimal results in achieving ZVS turn-on for the switches [7]. A detailed examination of this oversight and its repercussions on a non-inverting buck-boost converter topology is presented in a recent publication [8]. In essence, disregarding parasitic resistances can result in the loss of ZVS turn-on capabilities with variations in switching frequency due to voltage drops across the MOSFET switches and inductor. To counteract this voltage drop, an adjustment in the duty cycle of the PWM signal is necessary. The formula for recalculating the duty cycle (D_{new}) to accommodate these considerations is specified in (6).

$$D_{new} = \frac{(V_{L1} + 2V_{L2}) - \left\{(2V_{L1} - V_{L2})^2 - 8I_{out}(V_{L1} + V_{L2})(2R_{ds,on} + R_L)\right\}^{\frac{1}{2}}}{4(V_{L1} + V_{L2})} \quad (6)$$

To simplify (6), we assume the converter operates in a steady state, allowing for a more straightforward duty cycle adjustment methodology.

$$D_{new} = \frac{3V_d - \sqrt{V_d^2 - 8I_{out}(2V_d)(2R_{ds,on} + R_L)}}{8V_d} \quad (7)$$

The necessity to adjust the duty cycle in response to fluctuating load currents introduces significant complexity into the converter's control strategy. Nonetheless, this adjustment is crucial for maintaining ZVS turn-on conditions. A method to streamline duty cycle management involves setting a fixed duty cycle, predetermined by the maximal expected output current (I_{out}). For the purposes of this design, a duty cycle of 0.504 was empirically determined to facilitate efficient power flow up to 5.5 kW, demonstrating satisfactory performance under varied load conditions.

2.4 Power Flow Control

This section elucidates a consolidated power flow control methodology, synthesizing previously discussed dependencies on switching frequency and duty cycle. The implemented control scheme is visually depicted in Fig. 4, providing a comprehensive overview of the system's operation.

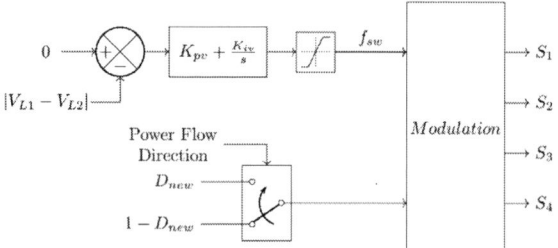

Fig. 4: Schematic of the power flow control scheme for the converter.

The core components of the power flow control mechanism are outlined as follows:

- Voltage Sensing: The input and output voltages (V_{L1} and V_{L2}) are monitored via voltage sensors. This data serves as a critical input for controlling the converter's operation.

- Switching Frequency Adjustment: The difference in sensed voltages is employed to calculate the appropriate switching frequency. Notably, the switching frequency is inversely related to this voltage difference, implying that a larger voltage discrepancy results in a reduced switching frequency which inadvertantly increases the output current. The frequency range is strategically limited, with an upper bound defined by conditions of zero output current and a lower threshold to facilitate maximum power transfer.

- Duty Cycle and Power Flow Direction: The direction of power flow is intricately linked to slight variations in the duty cycle, which, while proximate, is not precisely 0.5. Minute adjustments in the duty cycle can effectively reverse the direction of power flow. Consequently, a sophisticated algorithm is requisite for accurately

determining the power flow direction, enabling seamless transition between poles based on the duty cycle configuration.

By integrating voltage sensing with dynamic adjustments in switching frequency and duty cycle, this control strategy adeptly regulates the power flow, ensuring optimal converter performance across varying operational conditions.

3 Converter Design

The prototype of the designed ±700 V balancing converter is illustrated in Fig. 5. Central to this design is the implementation of independent and isolated gate driving circuits for each switch, ensuring precise control and operational safety. The gate drivers are controlled by inputs from the TI F28379D Launchpad microcontroller board. Additionally, the converter's architecture incorporates a high-efficiency inductor and a pair of DC link capacitors, critical for smoothing voltage fluctuations and enhancing the system's overall stability. Detailed specifications and operational parameters of the converter design are outlined in Table 1.

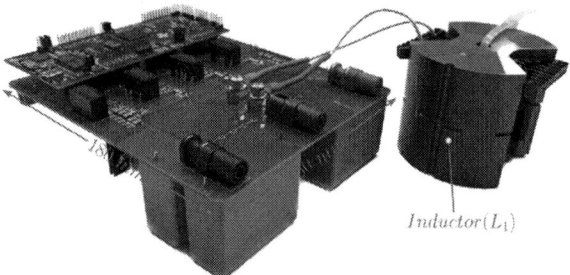

Fig. 5: Designed prototype of the balancing converter, including the inductor.

Tab. 1: Parameters of the designed converter.

Parameter	Value
MOSFET used	C3M0030090K
L_1 inductance	659 μH
DC link capacitance (C_1 & C_2)	20 μF
Switching frequency range	25 - 80 kHz
Snubber resistance	3.6 $k\Omega$
Snubber capacitance	270 pF
Static balancing capacitor	1.5 $M\Omega$

3.1 Switch Selection

The design choice for switch selection prioritizes both voltage handling capability and reliability. In this context, each switch, rated at 900V, is part of a strategic configuration where two switches are connected in series. This arrangement effectively doubles the voltage tolerance to 1800V, surpassing the requirement for a single 1700V switch. However, this solution introduces a significant challenge: ensuring equitable voltage sharing among the series-connected switches, an issue discussed above and well discussed in the literature [9], [10].

While various complex strategies have been proposed to address the voltage sharing conundrum [11], this work adopts a straightforward yet effective approach. We employ RC snubbers across each switch to facilitate proper voltage balancing. RC snubbers, recognized for their simplicity and robustness, provide a reliable solution to mitigate the risks associated with uneven voltage distribution across the switches. This method enhances the overall reliability of the converter's operation by ensuring that no single switch is subjected to voltage stresses beyond its rated capacity.

3.2 Snubber Circuit Considerations

The integration of a snubber circuit, composed of resistors ($R_{sn,Sx}$) and capacitors ($C_{sn,Sx}$), plays a pivotal role in achieving voltage balance across series-connected switches within the converter. The capacitor value, $C_{sn,Sx}$, is crucial for ensuring uniform voltage distribution. A larger $C_{sn,Sx}$ effectively reduces the voltage difference across the switches, facilitating more even stress distribution. However, this advantage comes with a trade-off: larger capacitance values necessitate greater discharge currents at the moment of switching to maintain ZVS conditions, complicating the ZVS achievement as outlined in [5].

Conversely, the resistor $R_{sn,Sx}$ serves to modulate the discharge current through $C_{sn,Sx}$. Selecting an overly high resistance value can lead to suboptimal voltage balancing and increased energy dissipation within the switches, as detailed in [6], [12]. Therefore, the design of the snubber circuit requires a careful balance: optimizing $C_{sn,Sx}$ to ensure voltage uniformity across switches without excessively hindering the ZVS capability, and tuning $R_{sn,Sx}$ to control the discharge current without inviting undue losses.

This delicate balancing act underscores the necessity of precise component selection and circuit design to mitigate potential inefficiencies and ensure the converter's operational integrity.

4 Results and Discussion

4.1 Test Setup

The configuration of the experimental test setup is depicted in Fig. 6. This setup comprises two converters, labeled V_{s1} and V_{s2}, which collectively emulate the bipolar DC grid. To simulate a 100 m cable within this grid, resistive and inductive components, R_{line} and L_{line}, are incorporated, effectively mimicking the electrical characteristics of a real cable. Additionally, a load converter is connected between the neutral and - pole.

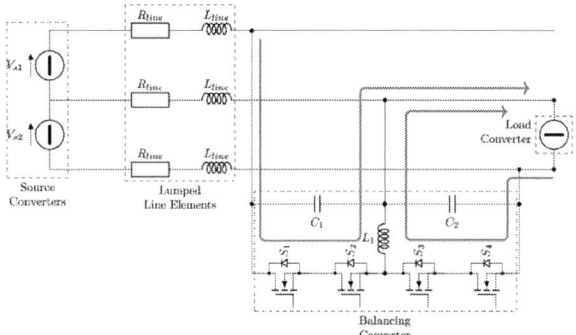

Fig. 6: Test setup used for the experiments and the load current flow shown by red arrows.

4.2 Voltage Measurement

At first consideration, measuring voltages across semiconductor switches might appear straightforward; however, the reality is far more complex. The off resistance of the SiC MOSFETs used in our setup exhibits significant variability, ranging from a few megaohms to tens of megaohms. This variability can lead to uneven voltage distribution across the switches during their off phases unless corrective measures are implemented. To address this challenge, static balancing resistors, typically in the range of hundreds of kiloohms, are employed to compensate for and harmonize the differences in switch resistance.

For the practical measurement of voltage across these switches, a differential voltage probe is commonly utilized. This probe integrates an RC circuit, which is connected between the measurement point and ground, ensuring that the measurement is not influenced by ground loop errors. The components of this RC circuit, specifically the resistors (R) and capacitors (C), are matched in value for both the positive (+) and negative (-) connections to maintain measurement integrity. Typically, for probes designed to measure up to

1000 volts, the resistance value is set around 4 MOhms [13], a parameter we confirmed through direct resistance measurements between probe pins and ground with a multimeter. However, when connected across the switches, the probe's high impedance can alter the voltage reading. This measurement discrepancy and potential solutions have been thoroughly discussed and addressed in the literature [13], [14].

4.3 Evaluation of ZVS Turn-On Performance

This subsection presents the ZVS turn-on performance of the converter under various load conditions. Oscilloscope outputs, captured at load powers of 770 W, 2.4 kW, and 4.8 kW, are depicted in Figures 7a, 7b, and 7c respectively. Additionally, the switching frequencies associated with these operational points are 50 kHz, 35 kHz, and 24 kHz, respectively, with the converter consistently operated at a duty cycle of 0.504 across all tests.

The oscilloscope captures demonstrate key parameters including the inductor current and the gate-to-source (V_{GS}) and drain-to-source (V_{DS}) voltages for switches S_3 and S_4. Notably, the detailed analysis reveals that both switches achieve ZVS well before the gate voltage is activated, highlighting the efficacy of the ZVS mechanism. Furthermore, the oscilloscope outputs indicate almost equal voltage sharing between switches S_3 and S_4. Given the symmetric design of the converter, it is reasonable to infer that similar ZVS performance and voltage sharing characteristics are attained for switches S_1 and S_2 as well.

This performance analysis confirms the converter's ability to achieve efficient ZVS turn-on across a range of operational conditions, underscoring the design's effectiveness in managing dynamic load scenarios while maintaining optimal switching efficiency.

4.4 Methodology for Efficiency Evaluation

Efficiency evaluation is pivotal in assessing the performance of the balancing converter, which facilitates power transfer between the poles of a bipolar DC grid. To determine the system's efficiency, we measure three distinct power values: $P1$, the power absorbed by the load converter; $P2$, the power output from source V_{s1}; and $P3$, the power output from source V_{s2}. These measurements allow us to cal-

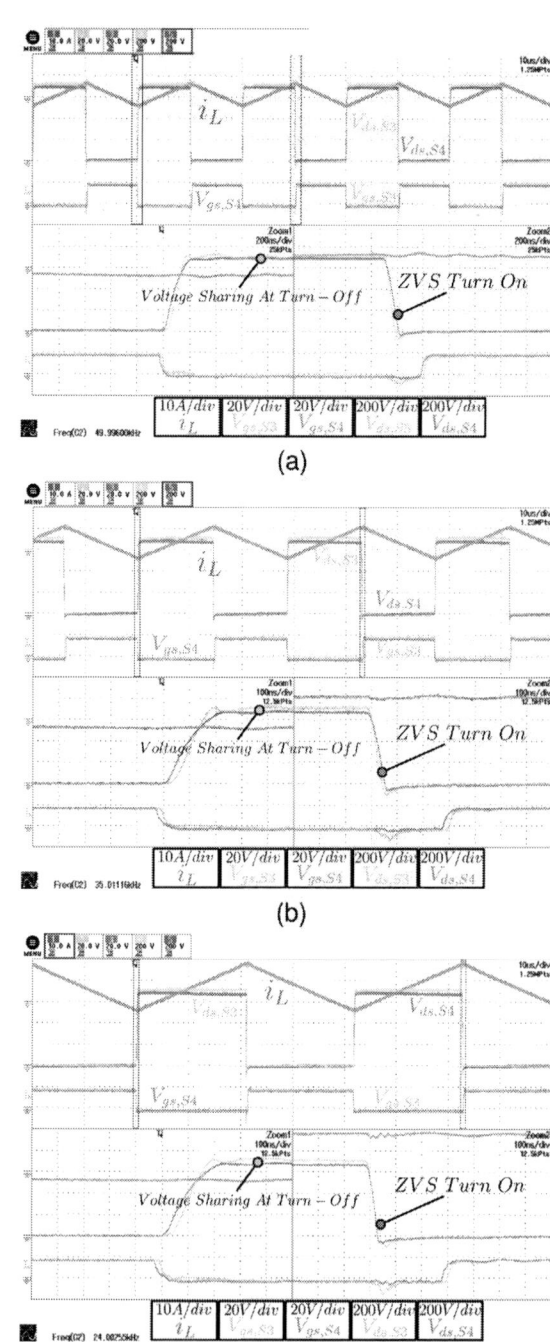

Fig. 7: Experimental results showing the oscilloscope output to verify ZVS turn-on of the switches.

culate the system efficiency (η_{sys}) as follows:

$$\eta_{sys} = \frac{P1}{P2 + P3},\qquad(8)$$

where η_{sys} represents the ratio of power utilized by the load to the total power supplied by the sources. Corresponding to the oscilloscope results above, the power analysis results are provided in Figures 8a, 8b, and 8c.

The Yokogawa WT500 power analyzer was employed to ascertain the system's efficiency, with the experimental results depicted in Figure 9. This figure illustrates the system efficiency across various load powers, demonstrating the prototype converter's performance under different operational conditions.

It is crucial to differentiate between system efficiency, as shown in Figure 9, and converter efficiency. The latter can be derived from the system efficiency through additional calculations, detailed by the equation (9). Given that the balancing converter operates on a partial power processing principle, its efficiency inherently falls below that of the system's overall efficiency. Therefore, a comprehensive analysis of both converter and system efficiency is essential for a full understanding of the converter's performance.

$$\eta_{conv} = 1 - 2(1 - \eta_{sys})\qquad(9)$$

5 Conclusion

In conclusion, this paper has detailed the design, operational principles, and control strategy of a balancing converter employing the buck-boost topology for enhanced voltage handling in bipolar DC grids. A significant innovation introduced in this work is the utilization of series-connected SiC MOSFET switches, specifically engineered to extend the converter's voltage endurance. Through experimentation, the converter demonstrated good voltage balancing across the switch pairs, underlining the effectiveness of the proposed design.

The experimental analysis confirmed an outstanding system efficiency of 98.93% at an output load power of 4.8 kW, showcasing the converter's operational efficiency and its potential for large-scale implementation in renewable energy systems and DC grids. These results not only validate the converter's design but also position it as a viable solution for improving the integration and management of renewable energy sources within DC distribution networks.

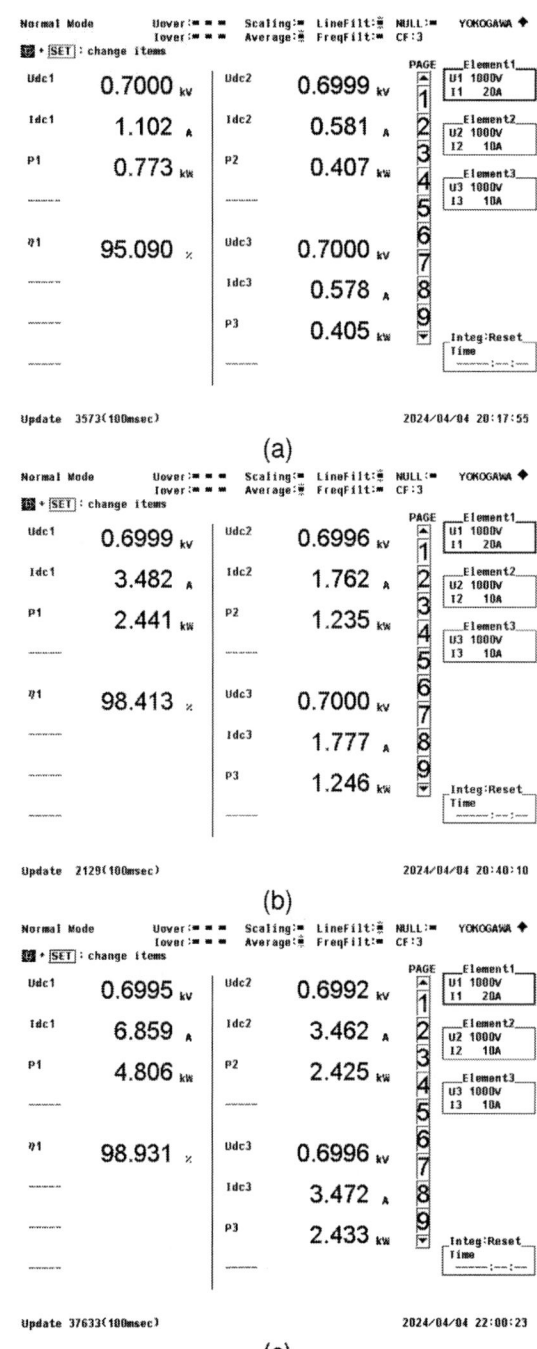

Fig. 8: Experimental results showing the power analyzer output showing the system efficiency.

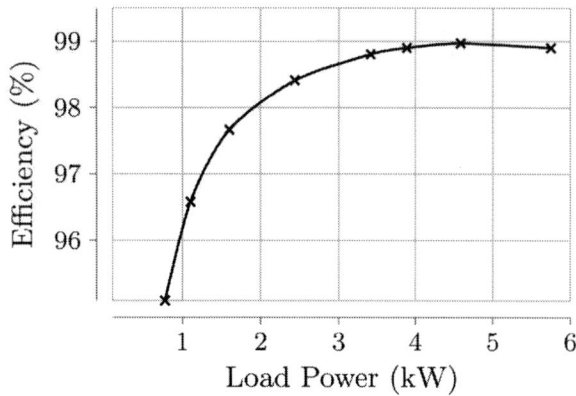

Fig. 9: Experimentally measured system efficiency with the designed converter prototype at various load powers.

Future work could explore the scalability of this converter design for higher power applications and its adaptability to varying grid conditions. Additionally, further research into optimizing the control strategy for dynamic load conditions could enhance the converter's efficiency and reliability, making it an even more compelling solution for modern power systems.

References

[1] S. Yadav, Z. Qin, and P. Bauer, "Bipolar DC grids on ships: Possibilities and challenges," *e & i Elektrotechnik und Informationstechnik*, Jun. 2022. DOI: 10.1007/s00502-022-01036-x.

[2] G. Van den Broeck, S. De Breucker, J. Beerten, J. Zwysen, M. Dalla Vecchia, and J. Driesen, "Analysis of three-level converters with voltage balancing capability in bipolar DC distribution networks," in *2017 IEEE Second International Conference on DC Microgrids (ICDCM)*, Nuremburg, Germany: IEEE, Jun. 2017, pp. 248–255. DOI: 10.1109/ICDCM.2017.8001052.

[3] Nederlands Normalisatie-instituut, "Dc-installaties voor laagspanning," Nederlands Normalisatie-instituut, Delft, The Netherlands, Nederlandse PraktijkRichtlijn 9090:2024, 2024, Replaces NPR 9090:2018; NPR 9090:2022 Ontw.

[4] S. Yadav, Z. Qin, and P. Bauer, "A series resonant balancing converter for bipolar dc grids on ships," in *2022 24th European Conference on Power Electronics and Applications (EPE'22 ECCE Europe)*, IEEE, 2022, pp. 1–8.

[5] P. Sinha, S. Yadav, Z. Qin, and P. Bauer, "Comparative analysis of series connected mosfets with single switch for zvs turn on converter topology," in *2023 25th European Conference on Power Elec-*

tronics and Applications (EPE'23 ECCE Europe), IEEE, 2023, pp. 1–7.

[6] P. Sinha, "Design of balancing converter for bipolar dc grids using series connected mosfet switches," 2022.

[7] V. Fernao Pires, A. Cordeiro, C. Roncero-Clemente, S. Rivera, and T. Dragicevic, "DC-DC Converters for Bipolar Microgrid Voltage Balancing: A Comprehensive Review of Architectures and Topologies," *IEEE Journal of Emerging and Selected Topics in Power Electronics*, pp. 1–1, 2022. DOI: 10.1109/JESTPE.2022.3208689.

[8] G. Yu, S. Yadav, J. Dong, and P. Bauer, "Revisiting the reverse switched current of buck, boost and buck-boost converters in voltage-mode tcm-zvs control considering parasitic resistances," *IEEE Transactions on Power Electronics*, 2024.

[9] C. Liu, Z. Zhang, Y. Liu, Y. Si, M. Wang, and Q. Lei, "Design guidelines of current source gate driver for series connected SiC MOSFETs," in *2020 IEEE Energy Conversion Congress and Exposition (ECCE)*, Detroit, MI, USA: IEEE, Oct. 2020, pp. 3803–3810. DOI: 10.1109/ECCE44975.2020.9236282.

[10] R. Wang, A. B. Jørgensen, W. Liu, H. Zhao, Z. Yan, and S. Munk-Nielsen, "Voltage balancing of series connected sic mosfets with adaptive-impedance self-powered gate drivers," *IEEE Transactions on Industrial Electronics*, 2022.

[11] L. F. Alves, P. Lefranc, P.-O. Jeannin, and B. Sarrazin, "Advanced voltage balancing techniques for series-connected sic-mosfet devices: A comprehensive survey," *Power Electronic Devices and Components*, p. 100 055, 2023.

[12] R. Chen, C. Li, H. Fang, R. Lu, C. Li, *et al.*, "Analysis and design for medium voltage dual active bridge converter based on series-connected sic mosfets," *IEEE Transactions on Power Electronics*, 2023.

[13] M. Grubmüller, B. Schweighofer, and H. Wegleiter, "Development of a differential voltage probe for measurements in automotive electric drives," *IEEE transactions on industrial electronics*, vol. 64, no. 3, pp. 2335–2343, 2016.

[14] W. Zhao, M. G. Niasar, P. Vaessen, and G. Rietveld, "Voltage sharing improvement methods in series-connected mosfets for future grid high voltage applications," in *23rd International Symposium on High Voltage Engineering*, 2023.

PCIM Europe 2024, 11– 13 June 2024, Nuremberg   DOI: 10.30420/566262362

Optimization and Design of Low-Voltage and High-Current Point-of-Load Converter under 48V Bus Architecture

Jiajia Guan [1] , Jin Wen [1] , Shuangxi Zhu [1] , Zongheng Wu [1] , Cai Chen [1] , Yong Kang [1]
Yue Wu [2], Zhipeng He [2]

[1] School of Electrical and Electronic Engineering, Huazhong University of Science and Technology, China

[2] State Key Laboratory of HVDC, Electric Power Research Institute, CSG, China

Corresponding author: Cai Chen, caichen@hust.edu.cn
Speaker: Jiajia Guan, jiajiaguan@hust.edu.cn

Abstract

The data center power supply architecture is changing from 12V to 48V, and the Point-of-Load (PoL) converter is facing the problem of how to achieve a high step-down ratio and high efficiency under high current. Based on the two-stage architecture, this paper proposed a method to improve the efficiency of the converter by optimizing the switching frequency and the phase number of Buck. Finally, a 48V-1V/200W prototype was designed based on the optimization results. Under the rated conditions, the prototype achieves peak efficiency of 91.7%, closed-loop control error less than 1% and power density of 52W/in3.

1 Induction

As the power level of the data center increases, the loss of the busbar in the traditional 12V architecture increases exponentially. To reduce the transmission path loss, the 12V architecture is gradually changing to the 48V architecture [1] - [4]. As shown in Fig. 1, Point-of-Load (PoL) needs to realize 48V to 1V power conversion. Therefore, how to realize the high step-down ratio and high efficiency of the PoL converter under high current has become a problem [5].

To realize the power conversion from 48V to 1V, current research focuses on quasi-single-stage architecture and two-stage architecture. The two-stage architecture has the characteristics of high reliability and easy control. The front stage uses a switched capacitor converter or LLC converter to realize step-down, and the latter stage uses Buck to realize voltage regulation [6]. There has been research on the use of switched capacitor topology in the front stage. The switched capacitor topology is used to achieve a 4:1 transformation ratio, and the rear stage uses multi-phase interleaved Buck to achieve the voltage stabilization function. The advantage of using switched capacitor topology in

the front stage is that it does not require inductive components and has high power density. However, it also has the disadvantage that it can only achieve a fixed transformation ratio and has low design freedom. The advantage of using LLC is that soft switching can be achieved within the load range, and the bus voltage can be adjusted by designing the transformer ratio and frequency regulation [7] – [10].

The front stage works near the resonance point, so the efficiency is high; the latter stage uses multi-

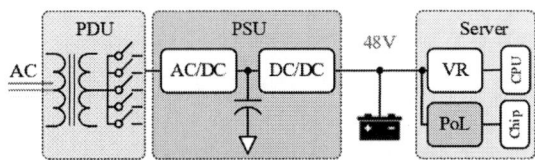

Fig. 1 48V bus architecture.

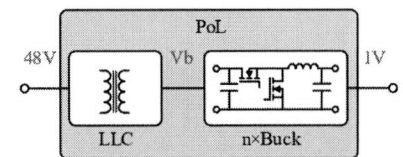

Fig. 2 Two-stage architecture of PoL converter.

phase interleaved Buck to reduce the output current ripple and the volume of the output capacitor. Although the topology types in the two-stage architecture are fixed, there are still many points worthy of attention: the selection of the intermediate bus voltage, the number of interleaved phases of the Buck, etc.

This paper is based on the two-stage topology, as shown in Fig. 2, to optimize the efficiency of the PoL converter under 48V to 1V/200A application. This paper first establishes the loss model of the converter, then optimizes the switching frequency and the number of Buck parallel phases based on the loss model, and finally designs a prototype to verify the optimization results.

2 Loss Model of the Two-Stage Topology

To optimize the design of the PoL converter, it is necessary to establish its loss model. The loss can be broken down into the following categories: switching device losses, drive losses, magnetic component losses, capacitor equivalent series resistance (ESR) losses, and other losses.

The loss of switching devices includes on-state loss and switching loss. LLC can realize soft switching and the lower switch of Buck can realize zero-voltage turn-on. Therefore, their switching

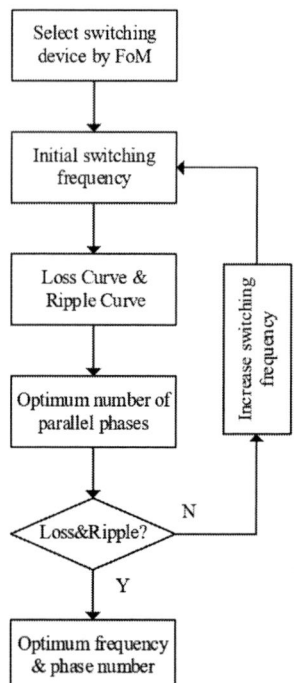

Fig. 3 Optimization process of frequency and number of parallel phases.

loss can be ignored. The RMS current (I_{LLC}) of the LLC device can be obtained under the rated condition, and the on-state loss of the LLC device can be calculated based on Ohm's law (1), where R_{on_LLC} is the on-state resistance of the LLC device. According to the output current (I_o), the number of parallel phases (k) and the on-state resistance (R_{on_B}), the on-state loss of the upper switch and the lower switch of the Buck converter can be calculated (2) (3) respectively, where D is the duty cycle of the upper switch, r is the current ripple coefficient of the inductor.

$$P_{con\_LLC} = I_{LLC}{}^2 R_{on\_LLC} \tag{1}$$

$$P_{con\_T} = (\frac{I_o}{2k}(D\left(1+\frac{r^2}{12}\right))^{0.5})^2 R_{on\_B} \tag{2}$$

$$P_{con\_B} = (\frac{I_o}{2k}((1-D)\left(1+\frac{r^2}{12}\right))^{0.5})^2 R_{on\_B} \tag{3}$$

The turn-on loss and turn-off loss of the upper switch can be calculated according to (4) (5) respectively. When ignoring the effect of dead time, the turn-off loss of the lower switch can be obtained by (6), where V_b is the intermediate bus voltage, tr is the turn-on time, t_f is the turn-off time, and fs is the switching frequency. The driver loss of the switching device can be calculated according to (7), where V_{drive} is the driver voltage, and Q_g is the gate charge.

$$P_{sw\_on\_T} = \frac{I_o V_b t_r f_s(1 - 0.5r)}{2k} \tag{4}$$

$$P_{sw\_off\_T} = \frac{I_o V_b t_f f_s(1 + 0.5r)}{2k} \tag{5}$$

$$P_{sw\_off\_B} = \frac{I_o V_b t_f f_s(1 - 0.5r)}{2k} \tag{6}$$

$$P_{drive} = V_{drive} Q_g f_s \tag{7}$$

Magnetic component losses include transformer losses and chopper inductor losses. The chopper inductor uses a commodity inductor, and its loss can be obtained from the datasheet. Transformer losses include winding losses and core losses. Winding loss needs to calculate its DC resistance (R_{dc}) first, then calculate its AC resistance (R_{ac}) according to Dowell's formula, and finally calculate

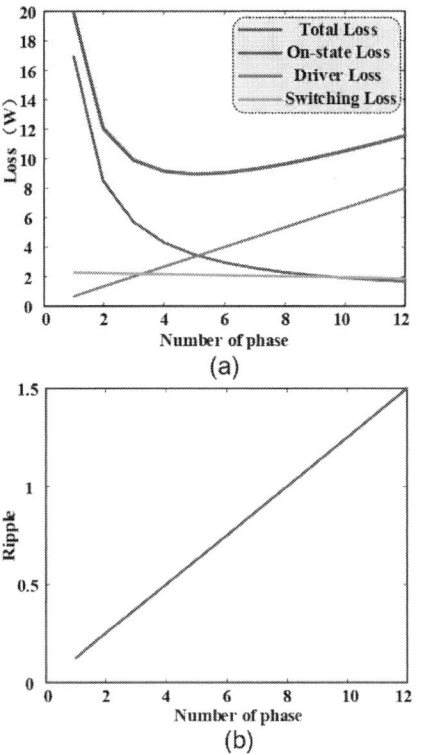

(a)

(b)

Fig. 4 Optimization Results. (a) Number of parallel phases. (b) Current ripple.

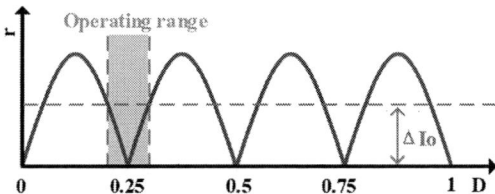

Fig. 5 Total output current ripple.

its winding loss under high-frequency current according to Ohm's law. Core loss can be calculated according to Steinmtz's equation. The ESR loss of the capacitor can be calculated according to the datasheet.

3 Optimization of Switching Frequency and Number of Parallel Phases

To reduce the size and loss of the chopper inductor, the current ripple ratio is limited to not more than 0.5, and the range of the number of parallel phases is [1,12]. The optimization process of parallel phase number and frequency is shown in Fig.

3. After selecting the optimal device based on the figure-of-merit (FoM), the device loss curve and current ripple curve under different interleaved parallel phase numbers can be drawn according to the loss model at the initial lower switching frequency. If the optimum point at this frequency does not meet the requirements of loss and ripple, increase the switching frequency until the optimal switching frequency and the number of parallel phases that meet the requirements of loss and ripple are found.

The optimization results are shown in Fig.4. When the switching frequency is 300kHz, four-phase interleaving can meet the ripple requirement of 0.5 with the lowest loss. In addition, this design also has the characteristics of low current ripple, and the output current ripple after four-phase interleaving is shown in Fig.5. It can be seen that after adopting four-phase interleaving, low output current ripple can be achieved in the full input voltage range. Furtherly, zero output current ripple can be achieved under the rated condition (V_i=48V). Since low ripple output can be achieved across the input voltage range, the size of the output filter capacitor can be reduced and the system power density can be improved. Therefore, this paper uses LLC cascaded four-phase interleaved Buck as the final topology, as shown in Fig.6. The secondary side of LLC uses a winding parallel structure to cope with large output current.

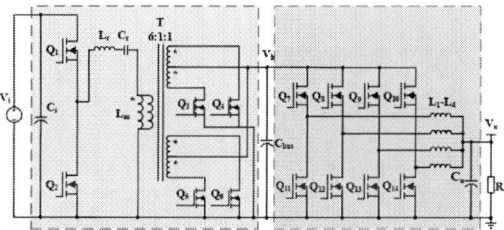

Fig. 6 The topology of this paper.

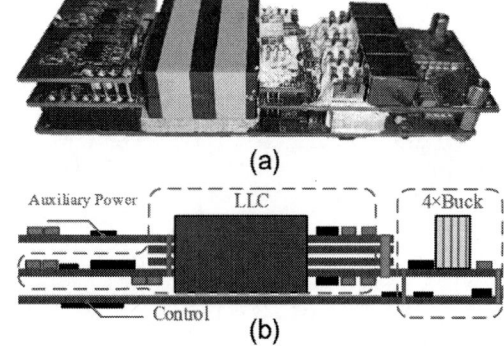

(a)

(b)

Fig. 7 PoL converter. (a) Prototype. (b) Structure.

	Input voltage	40 V-60 V
	Switching frequency	380 kHz
	Resonant capacitor	1.7 µF
LLC	Resonant inductor	150 nH
	Magnetizing inductor	4 µH
	Transformer ratio	6:1:1
	Device	GS61008T& IQE006NE2LM5
	Input voltage	3.33 V-5 V
	Output voltage	1 V
Buck	Switching frequency	300 kHz
	Inductance	100 nH
	Device	IQE006NE2LM5

Table 1 Parameters of the prototype.

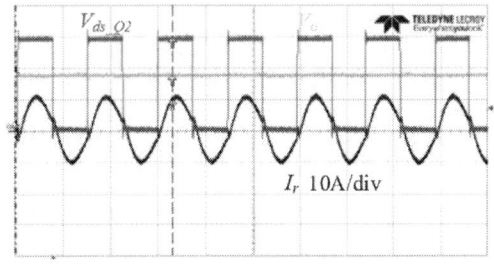

Fig. 9 Full-load under 60V input.

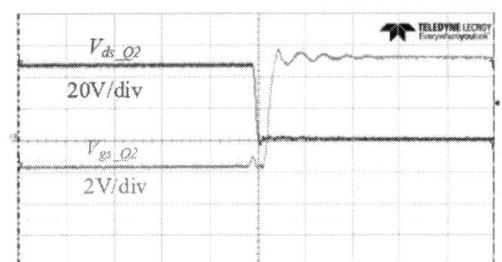

Fig. 10 Soft-switching waveform of LLC.

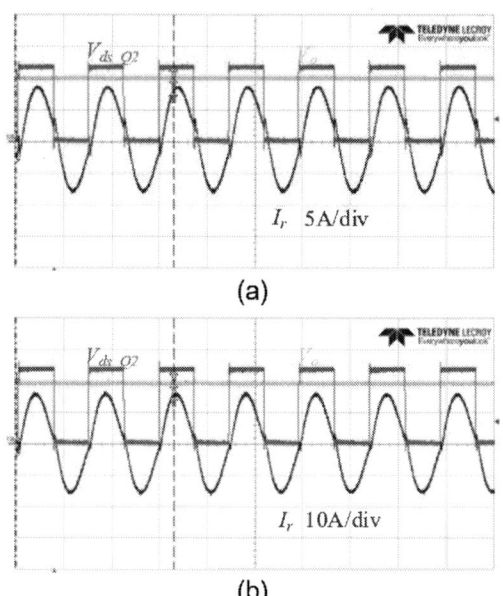

Fig. 8 Experimental waveform under 48V input. (a) Half-load. (b) Full-load.

4 Prototype and Experimental Results

According to the optimization results, the prototype of the final design is shown in Fig. 7. The size of the prototype is 106mm*33mm*18mm, the full load power is 200W, and the power density of 52W/in³ can be achieved. Table 1 is the detailed design parameters of the prototype, where the resonant inductance is replaced by the leakage inductance of the transformer. The prototype adopts a multi-board stacked structure to fully utilize the space in the thickness direction and improve power density. In addition, the control board and

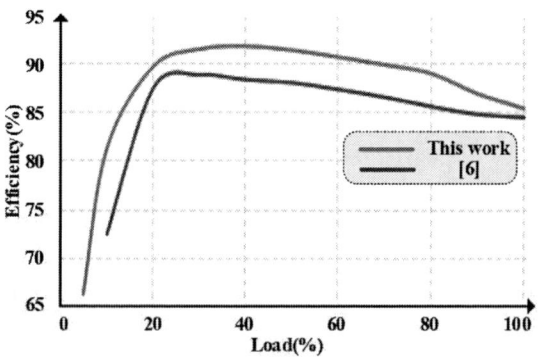

Fig. 11 Efficiency curve under rated condition.

power board are separated to reduce the impact of the power loop on the control loop.

Fig. 8(a) is the half-load waveform under 48V input, and Fig. 8(b) is the full-load waveform under 48V input, where I_r is the resonant current. Fig. 9 is the full-load waveform under 60V input, and Fig. 10 is the soft switching waveform of the LLC converter under rated conditions. It can be seen that the prototype can achieve a constant output voltage of 1V when the input voltage changes. Fig. 11 is a comparison of the efficiency curves of this work and [6]. The peak efficiency of this work is 91.7% and the full-load efficiency is 85.3%, both higher than [6].

5 Conclusion

This paper proposed a method to improve the efficiency of the PoL converter by optimizing the switching frequency and the number of Buck parallel phases and designed a 48V-1V/200W prototype. The prototype can achieve a peak efficiency of 91.7%, a full load efficiency of 85.3%, and a power density of $52W/in^3$, which verified the proposed optimization method.

References

[1] D. Kim, J. He, and D. G. Figueroa, "48V Power Delivery to Grantley Reference Board," 2016.

[2] M. H. Ahmed, C. Fei, F. C. Lee and Q. Li, "Single-Stage High-Efficiency 48/1 V Sigma Converter With Integrated Magnetics," in IEEE Transactions on Industrial Electronics, vol. 67, no. 1, pp. 192-202, Jan. 2020.

[3] M. Ahmed, C. Fei, F. C. Lee and Q. Li, "High-efficiency high-power-density 48/1V sigma converter voltage regulator module," 2017 IEEE Applied Power Electronics Conference and Exposition (APEC), Tampa, FL, USA, 2017, pp. 2207-2212.

[4] M. H. Ahmed, C. Fei, V. Li, F. C. Lee and Q. Li, "Startup and control of high efficiency 48/1V sigma converter," 2017 IEEE Energy Conversion Congress and Exposition (ECCE), Cincinnati, OH, USA, 2017, pp. 2010-2016.

[5] A. Fiore, Q. Huang and A. Q. Huang, "Loss Model and Output Impedance Analysis of a 48V-to-1V High Current Point-of-Load Converter," 2020 IEEE Energy Conversion Congress and Exposition (ECCE), Detroit, MI, USA, 2020, pp. 938-942

[6] Y. Chen, H. Cheng, D. M. Giuliano and M. Chen, "A 93.7% Efficient 400A 48V-1V Merged-Two-Stage Hybrid Switched-Capacitor Converter with 24V Virtual Intermediate Bus and Coupled Inductors," 2021 IEEE Applied Power Electronics Conference and Exposition (APEC), Phoenix, AZ, USA, 2021, pp. 1308-1315.

[7] J. Wen et al., "An Iterative-Based Dead-Time Compensation Method for Integrated Interleaved Boost-LLC Converter," 2023 IEEE Energy Conversion Congress and Exposition (ECCE), Nashville, TN, USA, 2023, pp. 3341-3348.

[8] Z. Wu, Z. Wang, T. Liu, W. Xu, C. Chen and Y. Kang, "High Efficiency and High Power Density Partial Power Regulation Topology With Wide Input Range," in IEEE Transactions on Power Electronics, vol. 38, no. 2, pp. 2074-2091, Feb. 2023.

[9] Bo Yang, F. C. Lee, A. J. Zhang and Guisong Huang, "LLC resonant converter for front end DC/DC conversion," APEC. Seventeenth Annual IEEE Applied Power Electronics Conference and Exposition (Cat.No.02CH37335), 2002, pp. 1108-1112 vol.2.

[10] F. Zhu, X. Lou and Q. Li, "An Improved Hybrid-Coupled Inductor Structure With Flux Reduction and Integrated Controllable Coupling Function," in IEEE Transactions on Power Electronics, vol. 39, no. 1, pp. 1103-1114, Jan. 2024.

PCIM Europe 2024, 11– 13 June 2024, Nuremberg   DOI: 10.30420/566262363

Interleaved Boost Converter Efficiency and Power Density Model for Active and Passive Component Design

Damien Lemaitre[1] , Jacques Ecrabey[1]

[1] Univ. Grenoble Alpes, CEA, Liten, France

Corresponding author: Damien Lemaitre, damien.lemaitre@cea.fr
Speaker: Damien Lemaitre, damien.lemaitre@cea.fr

Abstract

Interleaved boost converter (IBC) is used in hydrogen vehicle and solar applications. This converter requires a precise model. However, some expressions remain unknown in the state of the art, particularly for discontinuous conduction mode. In this article, the sizing and loss equations for switches, diodes and inductors are described for continuous and discontinuous conduction modes. A simplified method using component libraries for optimal IBC sizing is proposed. The design of a 1.6 kW IBC is carried out. The resulting prototype meets the specifications.

Introduction

Modelling a converter makes its design easier and more efficient. In this article, the converter studied is the interleaved boost (IBC) with uncoupled inductors, as defined in section 1.1.

IBC is widely studied in hydrogen vehicle and solar applications [1]. Key sizing steps are described in the scientific literature. The expression for the input current ripple ΔI_{IN} is described in [1], [2], [3], [4]. Articles [1], [4] also discuss the expression of the output voltage ripple ΔV_{out} in continuous conduction mode (CCM) and discontinuous conduction mode (DCM). The IBC loss model in DCM is described in [3], [5].

The aim of this article is firstly to review the state of the art of known expressions required for IBC sizing. Expressions of root mean square (RMS) currents in DCM components are then defined, which to the author's knowledge are not in the literature. Finally, a simplified sizing method for the IBC is proposed in order to determine the optimum switching frequency f_{sw}, number of legs N_{leg} and components for a given specification.

To achieve these objectives, the operation of the IBC is first described in section 1.

The design equations and the loss model are then discussed in section 2. The IBC sizing method is described in detail in section 3, applied to an example in section 4, and then a prototype is developed and tested in section 5.

The low ripple ΔI_{IN} obtained with the IBC is the main advantage of this topology. For this reason, the example discussed in sections 5 and 6 will focus on the optimal choice of inductors in order to meet the specifications.

1 The Interleaved Boost Converter

This part introduces the notations used to describe the IBC. Its operating principle is then described. The objectives of the IBC model are then discussed.

1.1 Notations

As illustrated in Fig 1 below, the N_{leg} IBC is the combination of N_{leg} boosts electrically connected in parallel, with N_{leg} an integer greater than one. Each leg consists of an inductor and a switching leg, composed of a switch T_k and a diode D_k. We can also use switches instead of diodes for synchronous rectification or when reversibility is required.

PCIM Europe 2024, 11– 13 June 2024, Nuremberg   DOI: 10.30420/566262363

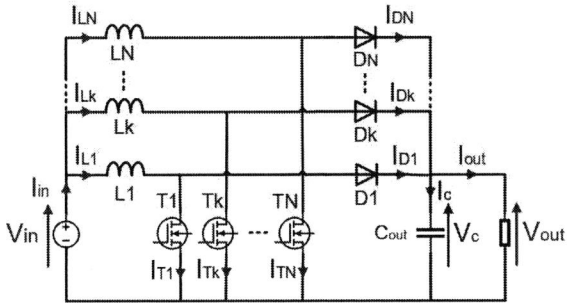

Fig 1 Electrical circuit of the interleaved boost converter (IBC).

The input and output voltages and currents are noted $v_{in}(t), v_{out}(t)$ and $i_{in}(t), i_{out}(t)$. Their mean values are noted $V_{in}, V_{out}, I_{in}, I_{out}$.

The ripple of the input current $i_{in}(t)$ is noted ΔI_{in}. The ripple of the output voltage $v_{out}(t)$ is noted ΔV_{out}.The voltage and current of the output filter capacitor C_{out} are noted $v_C(t)$ and $i_C(t)$.

The voltages and currents of the inductance L_k, the switch T_k , the diode D_k switches of the k-leg are respectively noted $v_{Lk}(t)$, $v_{Tk}(t)$, $v_{Dk}(t)$ et $i_{Lk}(t)$, $i_{Tk}(t)$, $i_{Dk}(t)$. The current ripple in the inductance L_k is noted ΔI_{Lk}.

1.2 Operation of the Boost Converter

The IBC is a voltage boosting DC/DC converter. To explain how the IBC works, it is therefore necessary to understand how the conventional boost works. Fig 2 below describes the two boost states in CCM.

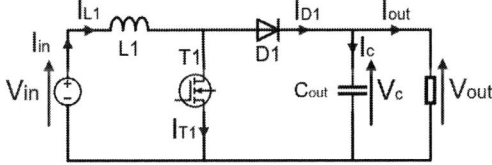

Fig 2 Electrical diagram of the boost converter.

The first state shown in Fig 3 (a) is defined by the ON state of the switch T_1 over the time $[0; \alpha T]$, where α is the duty cycle between 0 and 1. The L_1 inductor stores energy. Diode D_1 is blocked.

The second state shown in Fig 3 (b) occurs when switch T_1 is commanded to turn OFF, i.e. over the time $[\alpha T; T]$ .The diode D_1 conducts and the energy stored in the inductor is transferred to the load.

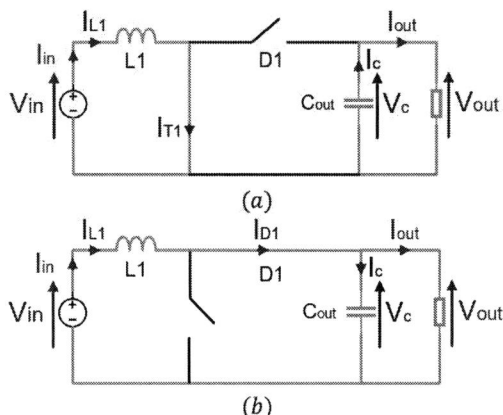

(a)

(b)

Fig 3 Electrical diagram of the boost converter in CCM during phase 1 (a) and during phase 2 (b). Current flow is shown in yellow.

The waveforms obtained for the currents in the inductor $i_{L1}(t)$, the switch $i_{T1}(t)$ and the diode $i_{D1}(t)$ are shown in Fig 4 below. It shows that the current $i_L(t)$ in the inductor never cancels out, which is characteristic of CCM.

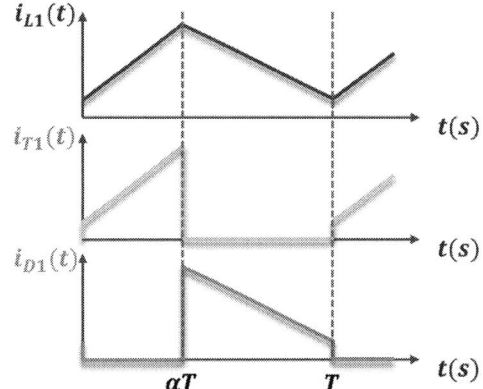

Fig 4 Currents in inductor L1, switch T1 and diode D1 of the boost operating in CCM.

A distinction is made between CCM and DCM according to the value of the inductor current ripple ΔI_L compared to the current in the inductor I_L defined by equation (1).

$$I_L = \langle i_L(t) \rangle = \frac{I_{in}}{N_{leg}} \tag{1}$$

CCM corresponds to the case described by equation (2) and illustrated by Fig 5 (a) below.

$$2\Delta I_L \leq \frac{I_{in}}{N_{leg}} \tag{2}$$

Otherwise, the boost operates in DCM. The current in the inductor then cancels out at a time noted βT and illustrated in Fig 5 (b).

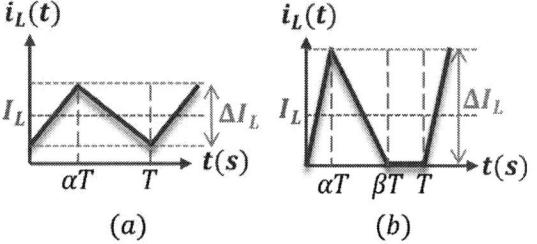

Fig 5 IBC inductor current in CCM (a) and DCM (b).

DCM operation results in a third state during the $[\beta T; T]$ period, with $i_L(t)$ equal to zero. We have explained the boost operation. The next section focuses on the IBC.

1.3 Operation of the IBC

The combination of several boosts in parallel then gives the IBC structure as shown in Fig 1 above. The main advantage of the IBC is that the input current ripple ΔI_{in} can be reduced by increasing the number of legs N_{leg}.

For example, in the case of a three-leg IBC in CCM, a typical shape of the currents $i_{L1}(t)$, $i_{L2}(t)$, and $i_{L3}(t)$ of the leg inductors is shown in Fig 6.

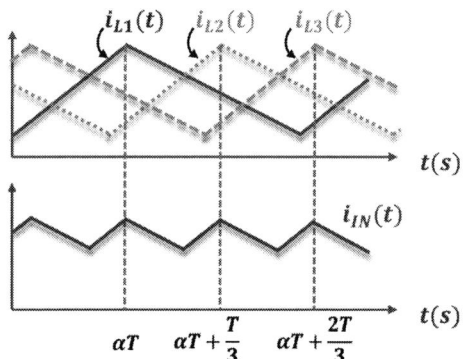

Fig 6 Example of the waveforms of the currents in the L1, L2 and L3 inductors of a three-legs IBC.

The principle is to phase-shift these three currents using switch control in order to reduce ΔI_{in}, as illustrated on Fig 6. The current $i_{in}(t)$ is the sum of the currents in the leg inductances of the IBC.

2 Converter Sizing Model

This section describes the equations used to size IBC components in CCM and DCM. Their voltage and current constraints are defined.

The expressions for the leg inductances and the output capacitor are defined.

2.1 IBC Sizing Model

The quantities with the index X_{Max} represent the maximum stresses of the variable "X". The voltage withstand of switches T_k, diodes D_k inductors L_k and output capacitor are defined by equations (3), (4) and (5). The maximum current withstand is defined by equations (6) and (7). Note that the expression for the current withstand depends on the conduction mode.

$$V_{Tk_{Max}} = V_{Dk_{Max}} = V_{out_{Max}} \tag{3}$$

$$V_{L_{Max}} = \begin{cases} V_{out} - V_{in} & if \ V_{out} > 2V_{in} \\ V_{in} & if \ V_{out} \le 2V_{in} \end{cases} \tag{4}$$

$$V_{CoutMax} = V_{out} + \frac{\Delta V_{out}}{2} \tag{5}$$

$$I_{Lk_{Max}} = I_{Tk_{Max}} = I_{Dk_{Max}} \tag{6}$$

$$I_{Lk_{Max}} = \begin{cases} \dfrac{I_{in}}{N_{leg}} + \dfrac{\Delta I_L}{2} & CCM \\ \Delta I_L & DCM \end{cases} \tag{7}$$

The average currents of switches I_{Tkav}, diodes I_{Dkav} and inductors I_{Lkav} are defined by equations (8), (9), and (10).

$$I_{Lk_{av}} = \frac{I_{in}}{N_{leg}} \tag{8}$$

$$I_{Tk_{av}} = \alpha \frac{I_{in}}{N_{leg}} \tag{9}$$

$$I_{Dk_{av}} = (1 - \alpha) \frac{I_{in}}{N_{leg}} \tag{10}$$

The expression of the RMS current of the components differs between CCM and DCM. In CCM, the RMS currents of switches I_{Tkrms} and diodes I_{Dkrms} are defined by equations (11) and (12). The equation (13) gives the expression of $I_{Tk_{min}}$ and $I_{Dk_{min}}$.

$$I_{Tk_{rms}} = \sqrt{\alpha \left(I_{Tk_{min}} \Delta I_L + \frac{\Delta I_L^2}{3} + I_{Tk_{min}}^2 \right)} \tag{11}$$

$$I_{Dk_{rms}} = \sqrt{\frac{V_{in}}{V_{out}} \left(I_{Dk_{min}} \Delta I_L + \frac{\Delta I_L^2}{3} + I_{Dk_{min}}^2 \right)} \tag{12}$$

$$I_{Tk_{min}} = I_{Dk_{min}} = \frac{I_{in}}{N_{leg}} - \frac{\Delta I_L}{2} \qquad (13)$$

The RMS currents in DCM of switches T_k and diodes D_k are defined by equations (14) and (15).

$$i_{Tk_{rms}} = \sqrt{\frac{2.\Delta I_L.I_{out}.(V_{out} - V_{in})}{3V_{in}N_{leg}}} \qquad (14)$$

$$i_{Dk_{rms}} = \sqrt{\frac{2.I_{out}\Delta I_L}{3.N_{leg}}} \qquad (15)$$

The RMS current I_{Lkrms} in the L_k inductors is defined in CCM and DCM by the equation (16):

$$I_{Lk_{rms}} = \sqrt{i_{Tk_{rms}}^2 + i_{Dk_{rms}}^2} \qquad (16)$$

2.1.1 Inductance Sizing

The value of the leg inductance in CCM is:

$$L_{CCM} = \frac{(V_{out} - V_{in})V_{in}}{f_{sw}V_{out}\Delta I_L} \qquad (17)$$

The value of the leg inductance in DCM is:

$$L_{DCM} = \frac{2P_{out}(V_{out} - V_{in})}{N_{leg}f_{sw}V_{out}\Delta I_L^2} \qquad (18)$$

The expression of the input current ripple ΔI_{IN} in CCM as a function of the duty cycle and generalized to N_{leg} IBC is given in [3]. This allows us to deduce the ΔI_L expression:

$$\Delta I_L = \frac{\Delta I_{in}N_{leg}\alpha(1 - \alpha)}{(N_{leg}\alpha - x)(x + 1 - N_{leg}\alpha)} \qquad (19)$$

$$\frac{x}{N_{leg}} < \alpha < \frac{x + 1}{N_{leg}}, x = 0, 1, ..., N_{leg} - 1 \qquad (20)$$

The expression of ΔI_{IN} in DCM is given for the case $N_{leg} = 3$ in [4]. The expression of ΔI_{IN} generalised to N legs does not appear in scientific literature. Thus, in CCM, the minimum value of the leg inductor is given by equation (21). This is obtained from equations (17) and (19).

$$L_{Min} = \frac{(V_{out} - V_{in})V_{in}}{f_{sw}V_{out}\Delta I_{in}\dfrac{N_{leg}\alpha(1 - \alpha)}{(N_{leg}\alpha - x)(x + 1 - N_{leg}\alpha)}} \qquad (21)$$

Note that the value of L depends on the duty cycle α. Knowing the minimum and maximum duty cycle α_{Min} and α_{Max} of a given application, L_{Min} can be sized more precisely.

The expressions for the output voltage ripple in CCM and DCM are given in [4]. The expression for the output capacitor is derived directly from this.

The electrical constraints of the components, as well as inductor and output capacitor values are now defined. Section 2.2 below discusses the IBC loss model.

2.2 IBC Loss Model

2.2.1 Inductor Losses

The losses P_L in the inductor are composed of the losses in the magnetic core P_{core} and the losses in the winding P_{wind} (22) :

$$P_L = P_{core} + P_{wind} \qquad (22)$$

The losses in the magnetic core P_{core} are given by the Steinmetz formula (23).

$$P_{core} = Kf_{sw}^\alpha B_{pp}^\beta \qquad (23)$$

This expression has the advantage of estimating core losses from datasheet data. Note that this formula considers sinusoidal magnetic flux density excitation with no DC component. However, the DC component affects the value of the Steinmetz parameters [6]. This may explain the difference between the estimated and actual losses of the magnetic core. The improved General Steinmetz Equation iGSE can also be used [7].

The losses in the inductance P_{wind} are composed of losses due to the skin effect P_{skin} and losses due to the proximity effect P_{prox}.

$$P_{wind} = P_{skin} + P_{prox} \qquad (24)$$

The P_{wind} loss model is described for a round wire and for a Litz wire in [8], [9], [10].

2.2.2 Diode Losses

The diode losses P_{Dk} are composed of conduction losses P_{Dkcond} and reverse recovery losses P_{Dkrr} define in equation (26) and (27) [11].

$$P_{Dk} = P_{Dk_{cond}} + P_{Dk_{rr}} \qquad (25)$$

$$P_{Dkcond} = V_{Do}I_{Dav} + R_D I_{D_{rms}}^2 \qquad (26)$$

$$P_{Dkrr} = \frac{1}{4} Q_{rr} V_{Drr} \qquad (27)$$

With the diode on-state resistance R_D, the diode on-state zero-current voltage V_{Do}, the voltage across the diode during reverse recovery V_{Drr}, and the reverse recovery charge Q_{rr}.

2.2.3 MOSFET Losses

Losses P_{Tk} in switches T_k define in equation (28) is composed of conduction losses P_{Tkcond} and switching losses P_{Tksw} define in (29) and (30):

$$P_{Tk} = P_{Tk_{cond}} + P_{Tk_{sw}} \qquad (28)$$

$$P_{Tk_{cond}} = R_{DSon} I_{Tkrms}^2 \qquad (29)$$

$$P_{Tksw} = P_{Tkon} + P_{Tkoff} \qquad (30)$$

P_{Tkcond} depends on drain-source on-state resistance R_{DSon} and on the RMS switch current I_{Tkrms}. The switching losses of switches T_k are composed of switch-on and switch-off losses, defined by equations (31) and (32).

$$P_{Tkon} = f_{sw}E_{on}(I_{ds})K_{on} \qquad (31)$$

$$P_{Tkoff} = f_{sw}E_{off}(I_{ds})K_{off} \qquad (32)$$

Where E_{on} and E_{off} are the energies lost when the switch T_k opens and closes. The expressions for the coefficients K_{on} and K_{off} are given below.

$$K_{on} = \frac{V_{ds}}{V_{dsth}} \frac{E_{on}(R_g)}{E_{on}(R_{gth})} \frac{E_{on}(T_j)}{E_{on}(T_{jth})} \qquad (33)$$

$$K_{off} = \frac{V_{ds}}{V_{dsth}} \frac{E_{off}(R_g)}{E_{off}(R_{gth})} \frac{E_{off}(T_j)}{E_{off}(T_{jth})} \qquad (34)$$

V_{ds} is the drain-to-source voltage, R_g the gate resistor, T_j the junction temperature. The quantities with the index X_{th} represent the theoretical value of the variable "X" considered in the datasheet for calculating switching energies.

All the parameters are shown in the datasheet. The only missing data is the expression of the $E_{on}(I_{ds})$ and $E_{off}(I_{ds})$ functions. The equations can be obtained by fitting the curves from the datasheet, as shown in Fig 7 below.

An approximation by a third-order equation is recommended, the coefficients $A_{on}, B_{on}, C_{on}, D_{on}$, $A_{off}, B_{off}, C_{off}, D_{off}$ are thus introduced:

$$E_{on} = A_{on}I_{DS}^3 + B_{on}I_{DS}^2 + C_{on}I_{DS} + D_{on} \qquad (35)$$

$$E_{off} = A_{off}I_{DS}^3 + B_{off}I_{DS}^2 + C_{off}I_{DS} + D_{off} \qquad (36)$$

To obtain these coefficients, the curve fitting can be carried out, for example, using the online software [12].

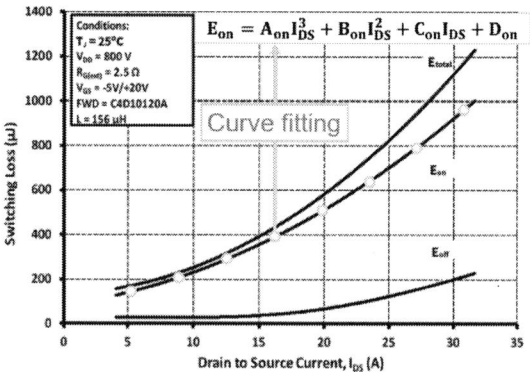

Fig 7 Example of curve fitting of the Eon, function of drain to source current I_{DS} for the C2M0080120D MOSFET.

The losses in the switches, diodes and inductances of the IBC can thus be calculated in CCM and DCM. The choice of the inductance and output capacitor values is dealt with in the next section.

In this section, the principle of the IBC sizing algorithm is first described. Then the set-up stage is detailed. The final selection of components by the algorithm is then discussed.

3 IBC Sizing Method

3.1 Principle

This method has three objectives. The first one is to comply with the ripple constraints on ΔI_{in} and ΔV_{out} defined in the specifications. The second one is to define the best values of N_{leg}, f_{sw}, and ΔI_L to find the optimal solution between the converter's efficiency and volume.

The third objective is to select the components from the libraries set up beforehand. Fig 8 below shows the process method to follow for an optimal IBC sizing.

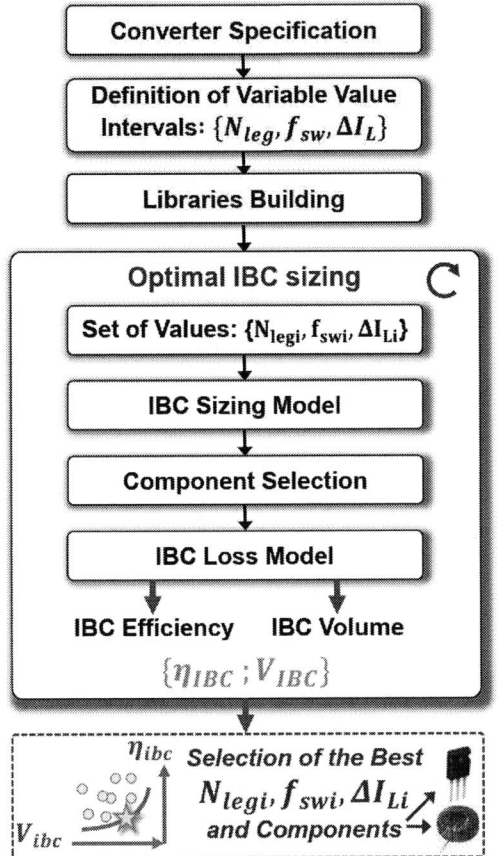

Fig 8 Method description for optimal IBC sizing.

The converter specifications are the rated output power P_{out}, the rated input and output voltages V_{in} and V_{out}, the switching frequency f_{sw}, the minimum and maximum duty cycle α_{Min} and α_{Max}, the maximum ripples of the input current ΔI_{in} and output voltage ΔV_{out}.

Then, several sets of parameters $\{N_{leg}, f_{sw}, \Delta I_L\}$ can provide interesting dimensioning. We therefore need to define the limits of these variables. Yet, as shown in the IBC sizing model section 3.1, component current constraints depend on the number of legs N_{leg} and ΔI_L. It is therefore preferable to have at least one switch, diode and inductor that is neither undersized nor oversized for each value of N_{leg}. This is to ensure that the optimisation is as relevant as possible. If the switching frequency range is large, the same consideration on components with regard to frequency can be applied.

The following section describes how to build component libraries

3.2 Libraries Building Example

The objective of the library of a component is to be able to provide it as input to the IBC loss model in order to obtain the losses, and the volume.

Let us consider the MOSFET library as an example. Each MOSFET must contain the following information: V_{dsMax}, I_{dsMax}, R_{DSon}, $E_{on}(R_g)$, $E_{on}(R_{gth})$, $E_{on}(T_j)$, $E_{on}(T_{jth})$, A_{on}, B_{on}, C_{on}, D_{on}, A_{off}, B_{off}, C_{off}, D_{off}. The volume $V_{cooling}$ of the active component cooling system is often considerable, hence it is advisable to specify it. Likewise, the losses $P_{cooling}$ associated with the energy it consumes.

The same principle must be followed with the diode library and inductor library, incorporating the characteristics involved in its loss model as described in section 2.2.

It is also possible to fix the choice of some components, while allowing multiple alternatives for the others. This scenario is applied in the section 4.

3.3 Component Selection

The selection of components can be carried out in several ways. The chosen method is to retain all components meeting the electrical constraints calculated during the sizing model of the IBC, as illustrated in Fig 8. This approach increases computation time but ensures no solution is omitted as a result. The method described here is applied in the following section. The resulting outcome will be implemented and experimentally tested in section 5.

4 Application Case

The IBC specifications to be designed are described in section 4.1. The influence of the inductor current ripple ΔI_L is then outlined in section 4.2. The selection of the optimal sizing of the IBC is carried out in section 4.3 among all the solutions.

4.1 Specifications

In this example, the study focuses on the inductance sizing.

The objective is to determine if the described IBC model allows meeting the set constraint on the input current ripple ΔI_{in}. The optimal point respecting the specifications will be selected.

The fast prototyping platform Imperix [13] will be used to implement the converter once sized. It consists of commercial switching arms, each composed of two MOSFETs C2M0080120D. Therefore, the choice of switches and diodes is predetermined. Table 1 presents the specifications considered for the IBC sizing.

P_{out}	1600 W
V_{in}	400 V
V_{out}	500V − 700 V
α_{Min}	0.2
α_{Max}	0.45
ΔI_{in}	$< 10\%$ I_{in}
f_{sw}	[10kHz; 30kHz; 50kHz; 70kHz; 90kHz]
N_{leg}	[1; 2; 3; 4]

Table 1   Specifications considered for the IBC sizing in the application case studied.

The volume of the switches, heatsinks, fans and output capacitors of the Imperix legs is known. With the inductor volume known, the converter's power density can be calculated directly. However, if the Imperix modules allow rapid production of the converter, there are oversized for the present specifications. Consequently, it is more relevant to consider the inductors volume only, what is done in this study.

Note that the input voltage V_{in} is fixed at 400V while the output voltage V_{out} can vary between 500V and 700V, implying duty cycle ratios α_{Min} and α_{Max} of 0.2 and 0.45 respectively.

In the current scenario, the inductances are sized for each set of values of the variables $N_{leg}, f_{sw}, \Delta I_L$ using an algorithm. The inductance library is thus replaced by a library of ETD magnetic cores of whose sizes are ETD 29, 34, 39, 44, et 59.

4.2   Influence of the Current Ripple

The maximum imposed on the input current ripple ΔI_{in} limits the allowable current ripple ΔI_L in the inductors. Due to the interleaving of legs, the limit on ΔI_L relaxes as the number of legs increases. This is illustrated in Fig 9 below. The losses Fig 9 (a) and the volume of the IBC Fig 9 (b) are plotted as a function of ΔI_L, for N_{leg} ranging from 1 to 4.

The switching frequency f_{sw} considered in this case is 50 kHz. Solutions meeting the constraint $\Delta I_{in} < 10\%$ I_{in} are located in the green regions, marked with the "√" indicator. Other solutions do not meet this condition and are therefore not acceptable.

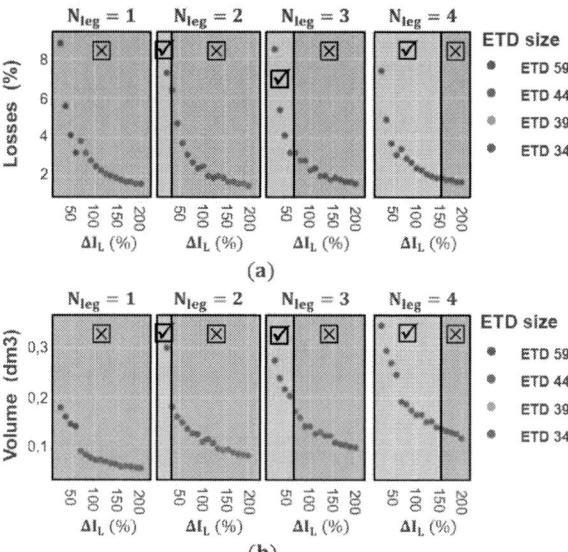

Fig 9   IBC Losses (a) and inductor volume (b) of the IBC as a function of the current ripple in the inductors ΔI_L (in percentage of the current inductor I_L), the number of legs N_{leg}, and the magnetic core constituting the inductors. Only solutions within the region marked by the "√" indicator meet the condition: $\Delta I_{in} < 10\%$ I_{in}. The switching frequency is 50 kHz.

Fig 9 shows that no solution exists for $N_{leg} = 1$. In this case, the ripple ΔI_L is equal to ΔI_{in}, imposing a low inductor current ripple ΔI_L: $\Delta I_L < 10\%$ I_{in}. This requires a high inductance value L, implying a peak induction in the core greater than the tolerated value. Fig 9 shows that increasing ΔI_L leads to a reduction in the losses of the IBC as well as the total inductors volume reduction. This allows a reduction in the inductance value. Consequently, it enables to select smaller magnetic cores and/or cores with fewer turns.

In addition, to increase ΔI_L reduces the value of the magnetic field seen by the core and the turns, consequently reducing losses. Increasing the number of legs allows to increase ΔI_L while still respecting the limit on ΔI_{in}. This explains why the optimal solution, with the selected criteria, is achieved for $N_{leg} = 4$.

The maximum inductor current ripple allowable is 150% in this case, as illustrated in Fig 9.

Fig 10 (a) and Fig 10 (b) below show the IBC losses and inductors volume as a function of the current ripple ΔI_L for $N_{leg} = 4$ and a switching frequency f_{sw} of 50 kHz.

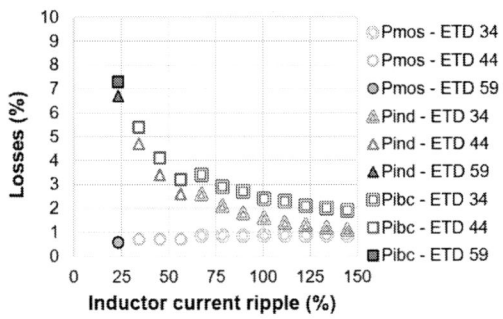

(a)

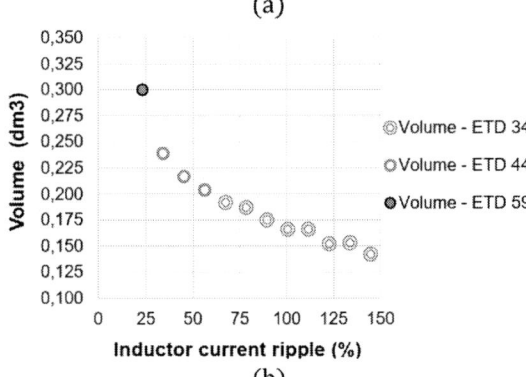

(b)

Fig 10 IBC losses (a) and IBC inductor volume (b) as function of the inductor current ripple ΔI_L (in percentage of the current inductor I_L). The switching frequency is 50 kHz and the leg number is 4.

Fig 10 (a) and Fig 10 (b) clearly demonstrate the influence of ΔI_L on the choice of the magnetic core, thereby impacting the IBC losses and the inductors volume.

The influence of the current ripple ΔI_L and the legs number N_{leg} on the IBC losses and inductors volume. We will now see all the solutions with the switching frequency as an additional variable.

4.3 Selected Solution

Fig 11 below illustrates the various IBC designs that meet the specifications. The variables are the legs number N_{leg}, the switching frequency f_{sw} and the inductor current ripple ΔI_L.

The converter losses are plotted as a function of the inductors volume. The values of variables N_{leg}

and f_{sw} are classified in Fig 11 (a). The ETDs used are distinguished in Fig 11 (b).

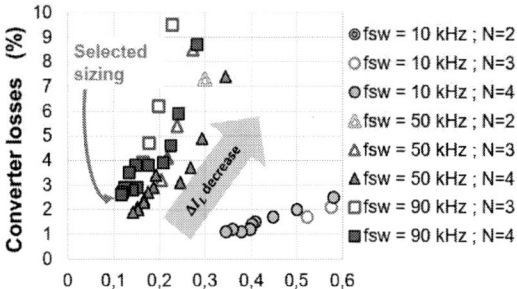

(a)

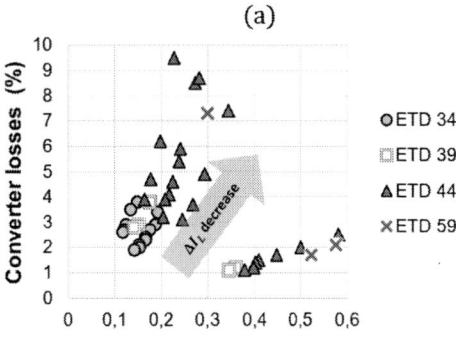

(b)

Fig 11 Converter losses as function of inductors volume with solutions classified by switching frequency and number of legs (a), then sorted by the size of ETDs used (b).

The selected IBC sizing is indicated in Fig 11 (a) and detailed in Table 2 below. The specifications of the required inductors are detailed in Table 3 below.

f_{sw}	N_{leg}	ETD	ΔI_L
90 kHz	4	ETD 34	145 %
P_{mos}	P_{ind}	V_{ind}	η_{ibc}
1,5 %	1,2 %	117 cm^3	97,3 %

Table 2 Sizing parameters of the optimal IBC as indicated in Fig 11 (a).

L	air gap	R_{wire}	N_t
1.33 mH	0.77 mm	0.25 mm	95

Table 3 Characteristics of the four inductors of the optimal IBC in Fig 11 (a).

Now that the IBC optimal sizing is known, the objective of the next section is to compare these theoretical results with those measured on the prototype.

5 Experimental Results

In this part, the IBC with the characteristics detailed in Table 2 and Table 3 above is realised and tested.

5.1 Converter Making

Fig 12 below shows the inductors made with the characteristics of Table 3. With these inductors and the switching legs of the Imperix PEB8024 modules, the four-leg IBC is made and illustrated in Fig 13.

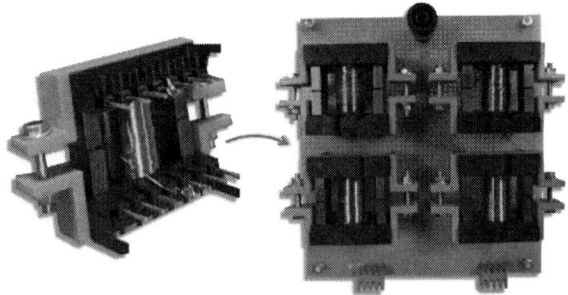

Fig 12 Optimised inductance on the left. Inductance of the 4 IBC legs on the right.

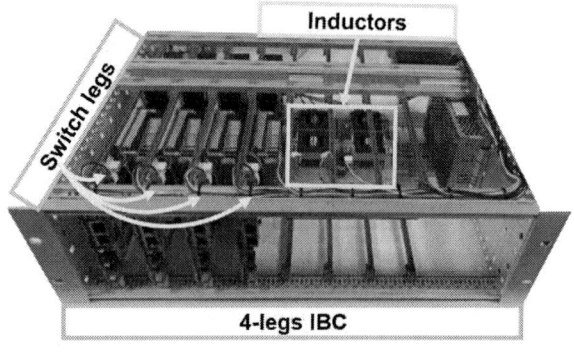

Fig 13 IBC made with Imperix switching legs and the sized inductors.

5.2 Measures

The inductance currents of the four legs of the IBC and the input and output currents are measured using the CP031 current probe. HVD3106 and ADP305 differential probes measure input and output voltages respectively.

The ripple of the input current is measured on a Wavesurfer 3024 oscilloscope using cursors.

To calculate efficiency, input and output voltages are measured using Keysight 34450A multimeters. The input and output currents are measured with Fluke 287 multimeters.

The input and output powers are then calculated, enabling the efficiency to be deduced. The converter test bench is shown in Fig 14.

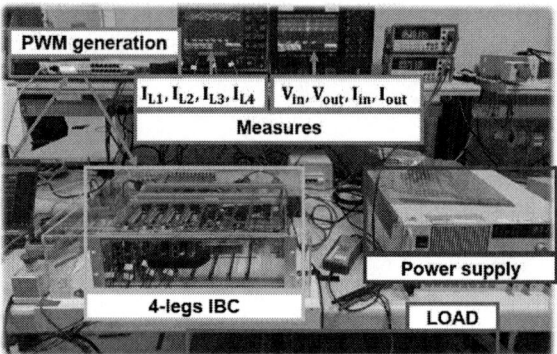

Fig 14 IBC test bench

The specifications considered imply a variable output voltage from 500 V to 700 V. For this reason, tests n°1 and n°3 are carried out close to these operating points. Test n°2 is an intermediate point. Table 4 below shows the characteristics measured for these three cases.

	Test n°1	Test n°2	Test n°3
V_{IN} (V)	400.67	377.94	379.30
I_{IN} (A)	4.02	4.03	4.03
V_{out} (V)	490.50	550.12	670.52
I_{out} (A)	3.21	2.68	2.19
P_{in} (W)	1610.69	1523.10	1527.82
P_{out} (W)	1574.51	1474.32	1468.44
P_{IBC} (W)	36.18	48.78	59.38
η_{IBC} (%)	97.75	96.80	96.11
α	0.27	0.40	0.52
ΔI_{IN} (A)	0.240	0.246	0.202
ΔI_{IN} (%)	5.97	6.11	5.01

Table 4 Measurements of characteristics of the 1.6 kW IBC characteristics for three operating points.

The estimated efficiency at operating point n°3 is 97.3%, compared to 96.11%. This results in a difference of 18W. Output capacitors losses not included in the model can be partly responsible of this difference. In addition, Imperix switching legs operate at a low current considering their characteristics. The losses for a switch current of 1 A are extrapolated from Fig 7, as manufacturer data are not available for currents below 4 A.

The ripple of the input current ΔI_{IN} is lower than 10% of I_{IN} in all three cases. The specifications are thus respected.

Conclusion

The sizing of the interleaved boost converter (IBC) as well as its loss model was addressed in continuous (CCM) and discontinuous (DCM) conduction mode. The proposed IBC sizing method based on component libraries was applied. The 1.6 kW prototype realized complies with the specifications including the constraint on the ripple of the input current ΔI_{in}. In the case studied, only inductors libraries were used. If the switches and output capacitors being imposed, the approach is the same with these components.

The maximum ripple ΔI_{in} measured is 6.11% of the continuous value I_{in}, less than the 10% set in specifications.

Outlook

The addition to the presented model of the volume and losses of the input and output filters, including EMC filters and the component cooling system, is a continuation of this work.

Acknowledgment

Thanks to the "Recherches Technologiques de Base" programme of the French Agency for National Research (ANR) and "CARNOT" program which supported this work with Carnot funding.

References

[1] G. R. Chandra Mouli, J. H. Schijffelen, P. Bauer, et M. Zeman, « Design and Comparison of a 10-kW Interleaved Boost Converter for PV Application Using Si and SiC Devices », *IEEE J. Emerg. Sel. Top. Power Electron.*, vol. 5, n° 2, p. 610-623, juin 2017, doi: 10.1109/JESTPE.2016.2601165.

[2] S. Kascak, M. Prazenica, M. Jarabicova, et R. Konarik, « Four Phase Interleaved Boost Converter: Theory and Applications », vol. 13, 2018.

[3] H. Wang, « Design and control of a 6-phase Interleaved Boost Converter based on SiC semiconductors with EIS functionality for Fuel Cell Electric Vehicle », phdthesis, Université Bourgogne Franche-Comté, 2019. Consulté le: 20 juin 2023. [En ligne]. Disponible sur: https://theses.hal.science/tel-02185678

[4] G.-Y. Choe, J.-S. Kim, H.-S. Kang, et B.-K. Lee, « An Optimal Design Methodology of an Interleaved Boost Converter for Fuel Cell Applications », *J. Electr. Eng. Technol.*, vol. 5, n° 2, p. 319-328, juin 2010, doi: 10.5370/JEET.2010.5.2.319.

[5] D.-D. Tran, S. Chakraborty, Y. Lan, M. E. Baghdadi, et O. Hegazy, « NSGA-II-Based Codesign Optimization for Power Conversion and Controller Stages of Interleaved Boost Converters in Electric Vehicle Drivetrains », *Energies*, vol. 13, n° 19, Art. n° 19, janv. 2020, doi: 10.3390/en13195167.

[6] J. Muhlethaler, J. Biela, J. W. Kolar, et A. Ecklebe, « Core Losses Under the DC Bias Condition Based on Steinmetz Parameters », *IEEE Trans. Power Electron.*, vol. 27, n° 2, p. 953-963, févr. 2012, doi: 10.1109/TPEL.2011.2160971.

[7] K. Venkatachalam, C. R. Sullivan, T. Abdallah, et H. Tacca, « Accurate prediction of ferrite core loss with nonsinusoidal waveforms using only Steinmetz parameters », in *2002 IEEE Workshop on Computers in Power Electronics, 2002. Proceedings.*, Mayaguez, Puerto Rico: IEEE, 2002, p. 36-41. doi: 10.1109/CIPE.2002.1196712.

[8] K. Umetani, S. Kawahara, J. Acero, H. Sarnago, O. Lucia, et E. Hiraki, « Analytical Formulation of Copper Loss of Litz Wire With Multiple Levels of Twisting Using Measurable Parameters », *IEEE Trans. Ind. Appl.*, vol. 57, n° 3, p. 2407-2420, mai 2021, doi: 10.1109/TIA.2021.3063993.

[9] R. P. Wojda et M. K. Kazimierczuk, « Winding Resistance and Power Loss of Inductors With Litz and Solid-Round Wires », *IEEE Trans. Ind. Appl.*, vol. 54, n° 4, p. 3548-3557, juill. 2018, doi: 10.1109/TIA.2018.2821647.

[10] C. R. Sullivan et R. Y. Zhang, « Analytical model for effects of twisting on litz-wire losses », in *2014 IEEE 15th Workshop on Control and Modeling for Power Electronics (COMPEL)*, Santander: IEEE, juin 2014, p. 1-10. doi: 10.1109/COMPEL.2014.6877187.

[11] Infineon, « MOSFET Power Losses Calculation Using the DataSheet Parameters ».

[12] « Curve fitting software ». [En ligne]. Disponible sur: https://plotdigitizer.com/app

[13] « Power electronic solutions for teaching, research and the industry - imperix ». [En ligne]. Disponible sur: https://imperix.com/solutions/

PCIM Europe 2024, 11– 13 June 2024, Nuremberg   DOI: 10.30420/566262364

Evaluation of a Hybrid Power Switch Based on Trench Clustered IGBT and SiC MOSFET

Alireza Sheikhan[1], E. M. Sankara Narayanan[1]

[1] The University of Sheffield, United Kingdom

Corresponding author: Alireza Sheikhan, asheikhan1@sheffield.ac.uk
Speaker: Alireza Sheikhan, asheikhan1@sheffield.ac.uk

Abstract

This paper reports performance of a hybrid power switch (HPS) based on parallel arrangement of a 1.2 kV silicon (Si) field stop trench clustered IGBT (FS-TCIGBT) and a 4H-Silicon Carbide (SiC) MOSFET of equivalent rating. The device is aimed to deliver optimum performance over a wide range of load conditions in terms of static and dynamic power losses while maintaining a low cost-to-performance ratio. The HPS can operate in unipolar or bipolar regime depending on load current and operating temperature which enables the device to offer a low on-state losses regardless of load condition. The results show that the HPS effectively combines high current capability of TCIGBT and switching performance of SiC in a single package.

1   Introduction

Motor drives require robust power semiconductors capable of delivering maximum efficiency over a wide range of load conditions. This is particularly important as the source of energy is limited. Wide bandgap semiconductors such as SiC are promising in terms of switching performance and power losses which is attributed to their superior material properties. The drift region, which is the main contributor to the on-state resistance in high voltage devices, can be made thinner than that of Si due to the high critical electric field strength and wide bandgap nature of 4H-SiC. Wide bandgap FETs also have extremely fast switching transitions due to the absence of bipolar charge carriers which results in lower switching energy than silicon MOS-bipolar counterparts [1,2]. However, despite advancements in manufacturing and fabrication technologies, the associated costs are still significantly higher when compared with Si counterparts. These costs increase non-linearly with higher current requirements. Si insulated gate bipolar transistors (IGBTs), on the other hand, have long established domination in high power applications. IGBTs are significantly cheaper than SiC and inherently suitable for high current operations due to their low voltage drop achieved through conductivity modulation. IGBTs offer superior cost-to-performance ratio, which is due to the availability of low-cost Si wafers. Despite the clear advantage of IGBTs in high current applications, their low current performance is

hindered by a 0.7 V p-n junction voltage drop at RT. The switching performances of IGBTs are also limited because of the injected carriers due to the bipolar action. The injected carriers which reduce the drift region resistance, need to be removed or recombined in order for the device to turn-off. The removal process of these excess carriers causes a tail current. IGBTs also suffer from dynamic avalanche, which limit high current density operation [3,4]. Therefore, the development trend in IGBTs has been focused toward improving these trade-off issues. One of the recent innovations in this area is the introduction of the field stop (FS) trench clustered IGBT (TCIGBT), which is shown in Fig. 1 [5,6]. TCIGBT is aimed as a more efficient chip-for-chip replacement of IGBTs. One of the key features of TCIGBT is the utilization of thyristor mode of conduction to reduce the on-state voltage drop. The addition of a highly conducting N well region over a floating P well layer in an IGBT creates a MOS controlled thyristor action. The resulting PNP transistor involving (P well/N well and P base), is designed in such a way that the N well layer is punched through at a certain anode voltage and the voltage across the gate is therefore clamped [7]. This mechanism protects the trench gate from high electric fields. The design also includes a PMOS formed by the p base-N-well-p well along the side walls of the gates to enable efficient hole diversion and removal of DA effects. The PMOS is turned on with increase in the N well potential, which is akin to applying a negative gate bias to the PMOS gate.

PCIM Europe 2024, 11– 13 June 2024, Nuremberg		DOI: 10.30420/566262364

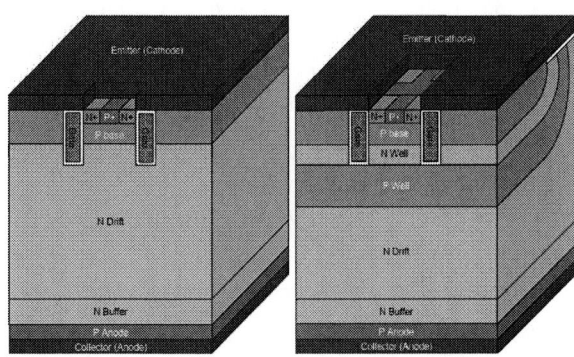

Fig. 1 Cross-sectional view of the trench IGBT (left) and trench CIGBT (right).

Consequently, holes are extracted by the electric field and dynamic avalanche issues are avoided, which hinders high dV/dt and high current density operation in conventional IGBTs [4]. To take advantage of both SiC and Si, most electric vehicles (EVs) utilize a combination of both technologies. A remedy for high cost of SiC is to use less chip area at the expense of higher conduction losses, inferior thermal conductivity, and less current handling capability. Another approach is to use a hybrid power switch (HPS) that can fit for a wide range of load demand without compromising conduction losses or switching performance. In this paper, a hybrid power switch based on the latest TCIGBT and SiC MOSFET technologies is presented for the first time. The device characteristics, working mechanism and switching performance are experimentally demonstrated and analyzed. Additionally, a comparison of characteristics between the TCIGBT and IGBT is presented.

1.1 Hybrid Power Switch Concept

One of the early reports on hybrid combinational transistors is given in [8]. Several studies to this date have reported various hybrid device configurations using IGBTs and MOSFETs [1][9-12]. HPS's are formed by a parallel combination of unipolar and bipolar devices as shown in Fig. 2. HPS have a unique bi-mode operation that enables them to maintain a low conduction loss profile over a wide range of load currents. Under low load conditions, the device operates in unipolar mode and the load is entirely handled by the MOSFET. Meanwhile, under high load conditions, the HPS switches to bipolar mode which diverts excess current to the IGBT. The boundary between bipolar and unipolar mode is determined by the on-state resistance of the MOSFET and bipolar on-set voltage. There are

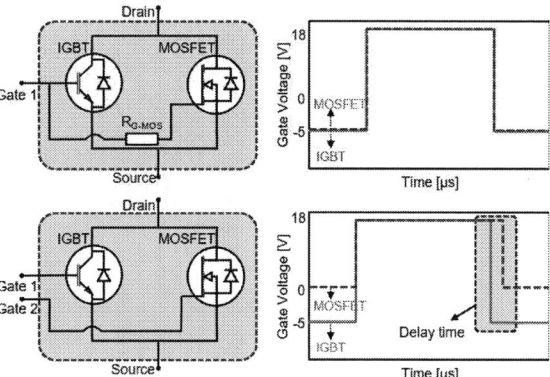

Fig. 2 Hybrid power switch configurations (left) and optimal gate drive approach (right).

two HPS gate configurations; (1) single gate, where the IGBT and MOSFET gates are connected. (2) dual gate, where gates are separated and independently controlled. This is because IGBTs generally have longer switching delay times than MOSFETs which is particularly troublesome during the turn-off as it results in higher switching losses in a single gate arrangement. However, with an intelligent control, a turn-off delay can be added in such a way that the MOSFET turns off after the IGBT in the dual gate mode. The benefit of this approach is that the load current is switched under the unipolar mode with the minimum switching losses. In this mode, the IGBT experiences a zero-voltage switching and does not contribute to switching losses. A turn-on delay is not always necessary as the MOSFET switches on first.

2 Experimental Setup

A 1200 V FS TCIGBT and a 1200 V SiC MOSFET (IMW120R030M1H – CoolSiC – Infineon [13]) with a current rating of 50 A each were selected for the HPS. For comparison purposes, a FS IGBT (IKW50N120CS7 – IGBT7 – Infineon [14]) with similar rating is also selected. The static characteristics of individual devices were measured using a curve tracer. The switching performance of the HPS were analyzed using an inductive switching test bench as shown in Fig. 3. The load inductor (L_{Load}) is 240 µH with 1200 V freewheeling SiC Schottky diode (D_{FWD}). The turn-on and turn-off switching performances were evaluated at V_{DC} of 600 V and I_{Load} of 50 A at different gate resistances R_{G1} to analyze the dV/dt controllability. The positive gate bias is set at 18 V. A turn-off gate bias of 0 V and -5 V are applied to the MOSFET and IGBT respectively.

2607

PCIM Europe 2024, 11– 13 June 2024, Nuremberg DOI: 10.30420/566262364

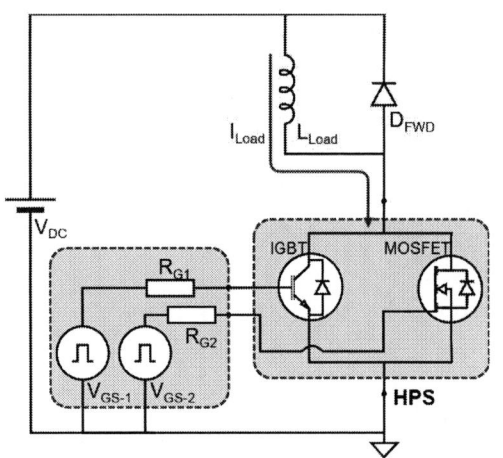

Fig. 3 Inductive switching test setup.

The switching losses were extracted from the measured data by integrating the dissipated power over switching periods as given by Eq. (1).

$$E_{SWT} = E_{On} + E_{Off} = \int V_{Drain} \cdot I_{Drain} \, dt \quad (1)$$

3 Device Characteristics

3.1 On-State Characteristics.

The on-state output I-V characteristics of the HPS were measured under pulse mode to prevent self-heating as shown in Fig. 4 and Fig. 5.

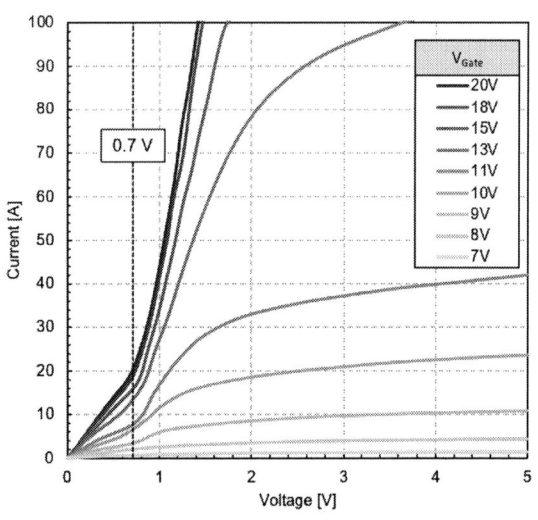

Fig. 4 Measured output IV characteristics of the HPS at different gate voltages and 25˚C. The pulse width is 250 µs.

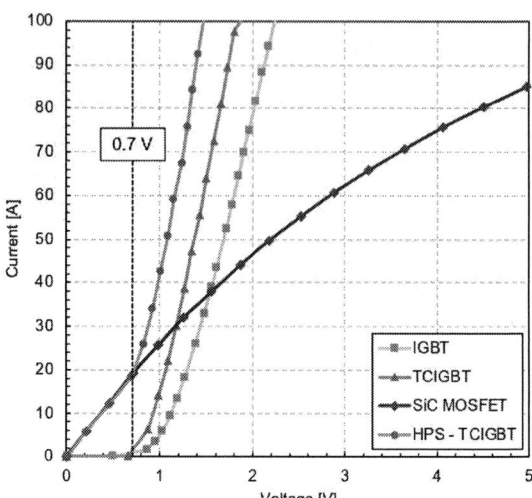

Fig. 5 Comparison of measured output IV curves at V_GS of 18 V and at 25˚C.

In this HPS configuration the boundary between bipolar and unipolar mode is ~20 A at RT. Below this boundary, the entire load current is conducted in unipolar mode through the MOSFET. The excess current above 20 A is handled by the TCIGBT as the voltage drop across the device is high enough to overcome the bipolar on-set voltage of 0.7 V at RT. With the increase in temperature, the boundary condition also changes because the bipolar on-set voltage decreases allowing more current to be diverted to the TCIGBT.

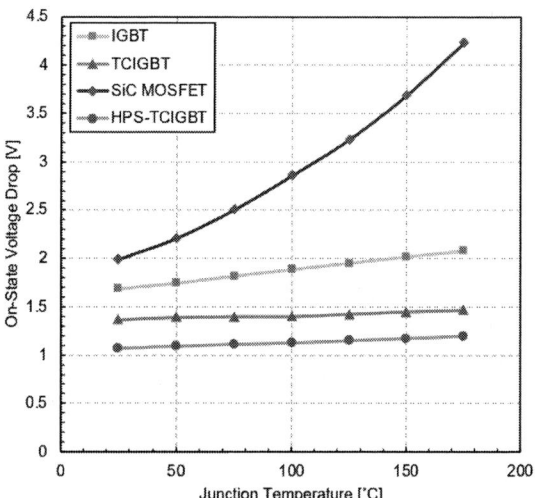

Fig. 6 Measured on-state voltage drop as a function of junction temperature at 50 A. The gate voltage is 18 V.

2608

This mechanism allows the HPS to maintain low on-state losses under high temperature/current conditions. It is important to note that the TCIGBT achieves lower on-state voltage drop compared with the IGBT for the same given power rating and active area making it highly suitable for the HPS. The on-state voltage of the HPS and its constituent components were extracted at different junction temperatures as shown in Fig. 6.

The HPS exhibits a mild positive temperature coefficient of on-state voltage drop similar to the TCIGBT. In contrast, the on-state voltage of the SiC MOSFET increases rapidly with the temperature in such a way that the on-state voltage is twice as high as RT. Therefore, a pure SiC solution will result in significant power losses at elevated operating temperatures while conducting its nominal current.

4 Switching Performance

4.1 Turn-On Switching

Figure 7 shows the turn-on switching waveforms of the HPS at different gate resistances at RT. In this experiment, no delay is added and both gate signals are applied at the same time. The MOSFET naturally turns-on first before the IGBT and thus the turn-on transition is handled in unipolar mode. The turn-on dV/dt and voltage fall time are presented in Fig. 8. As it can be seen the HPS offers a similar dV/dt and fall time figures to the SiC MOSFET which confirms the unipolar turn-on.

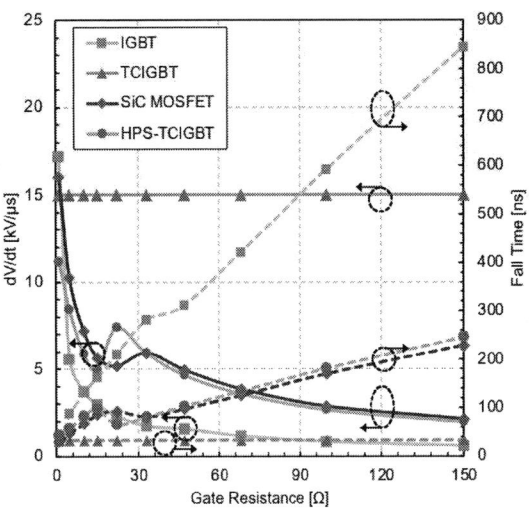

Fig. 8   Measured turn-on dV/dt and voltage fall time as function of gate resistance at 25°C.

As previously stated, in TCIGBTs, the turn-on is governed by its inherent thyristor action. The current gain of the thyristor constituents determine the turn-on performance and the influence of the gate resistance is nominal which explains the flat dV/dt response. A more detailed analysis of the TCIGBT turn-on behavior is explained in [15].

The turn-on losses were extracted from the measured switching waveforms as illustrated in Fig. 9. Similar to the turn-off, the turn-on power losses are

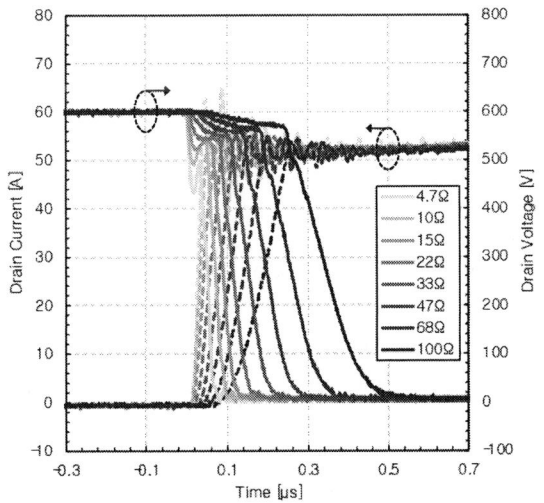

Fig. 7   Measured turn-on switching waveforms of the HPS at different gate resistances and 25°C.

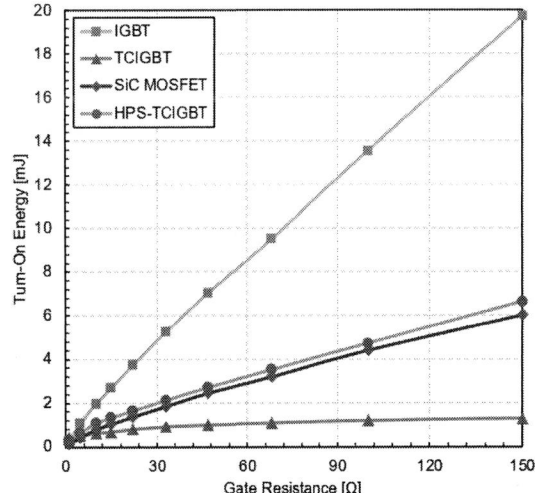

Fig. 9   Measured turn-on energy losses at different gate resistances and 25°C.

comparable to the SiC MOSFET because of the unipolar switching transition. The IGBT exhibits higher turn-on losses due to its slower dV/dt. Meanwhile, the switching energy in TCIGBT is almost constant regardless of the gate resistance.

4.2 Turn-Off Switching

The turn-off switching performance of the HPS is evaluated at different gate resistances at RT as shown in Fig. 10. In this test a delay of 2 µs is added to the MOSFET to ensure that it turns-off after the TCIGBT.

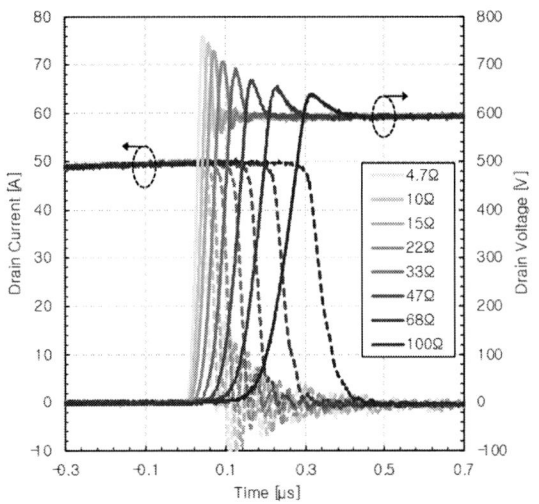

Fig. 10 Measured turn-off switching waveforms of the HPS at different gate resistances and 25˚C.

In the turn-off, no tail current can be observed which indicates a unipolar switching. Therefore, a higher switching speed can be achieved which results in lower losses. In several applications such as variable speed drive the dV/dt need to be limited to meet the application requirements and adhere to electromagnetic interference (EMI) requirements [16]. The turn-off dV/dt and voltage rise time were extracted for each device as shown in Fig. 11. The dV/dt can be effectively controlled by the gate resistor to fulfil various application requirements such as motor drives. At low gate resistance a plateau can be observed in dV/dt of the IGBT. This plateau is caused by the occurrence of a dynamic avalanche which limits the high-speed operation. The corresponding turn-off losses were extracted from the measured switching waveforms as shown in Fig. 12. Since the HPS switches in unipolar mode, the switching losses are similar to the SiC MOSFET. Meanwhile the HPS offers double the current handling capacity than the MOSFET. It is worth pointing out

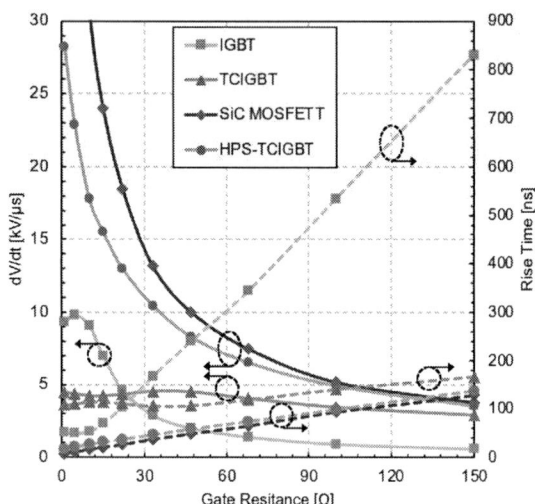

Fig. 11 Measured turn-off dV/dt and voltage rise time as function of gate resistance at 25˚C.

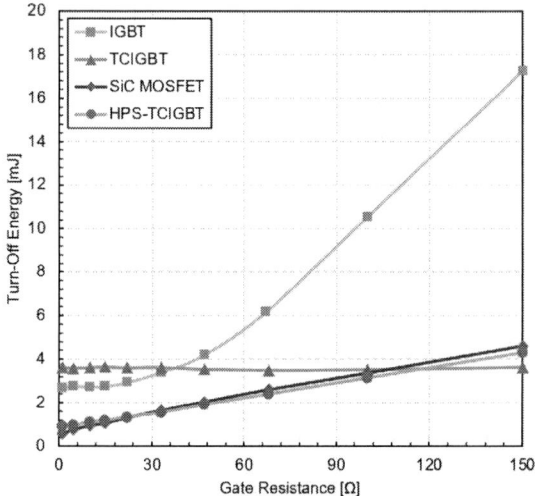

Fig. 12 Measured turn-off energy losses at different gate resistances and 25˚C.

that the TCIGBT power losses do not change with the gate resistor because of its nearly flat dV/dt. This is because of effective use of PMOS.

5 Conclusion

In this paper, a hybrid power switch using a FS-TCIGBT and SiC MOSFET have been evaluated experimentally. The device characteristics, operating principle and switching performance were analyzed. Based on the results, the HPS offers low switching losses similar to a SiC MOSFET while offering almost double the current handling capability. The bi-mode characteristics of the HPS allows the device to maintain a low power loss profile over a wide range of load current and operating temperatures. The results also show that the utilization of thyristor mode in TCIGBT contribute to improved on-state characteristics compared with the IGBT. Additionally, the unique self-clamping mechanism protect the gates from high electric fields and enables the device to operate high current densities without the risk of dynamic avalanche. Such characteristics makes this HPS configuration a highly suitable candidate for power electronic applications requiring flexibility and optimal performance particularly in the form of intelligent power modules for a wide range of applications, including drives.

6 Acknowledgements

The authors thank Eco semiconductors Limited for providing 1.2 kV FS-TCIGBT samples to undertake this study.

References

[1] A. Sheikhan and E. M. S Narayanan, "Evaluation of Characteristics and Turn-off dV/dt Controllability of 1.2 kV SiC Si Hybrid Power Switch," in *International Seminar on Power Semiconductors*, Prague, 2023.

[2] A. Sheikhan, E. M. S. Narayanan, H. Kawai, S. Yagi and H. Narui, "Evaluation of turn-off dV/dt controllability and switching characteristics of 1.2 kV GaN polarisation superjunction heterostructure field-effect transistors," *Japanese Journal of Applied Physics*, vol. 62, p. 064502, 2023.

[3] J. Lutz and R. Baburske, "Dynamic avalanche in bipolar power devices," *Microelectronics Reliability*, vol. 52, no. 3, pp. 475-481, 2012.

[4] P. Luo, S. N. E. Madathil, S.-I. Nishizawa and W. Saito, "Evaluation of Dynamic Avalanche Performance in 1.2-kV MOS-Bipolar Devices," *IEEE Transactions on Electron Devices*, vol. 67, no. 9, pp. 3691-3697, 2020.

[5] H. Y. Long, M. R. Sweet, M. M. D. Souza and E. M. S. Narayanan, "Next Generation 1200V Trench CIGBT for High Voltage Applications," in *International Symposium on Power Semiconductor Devices & IC's*, Hong Kong, 2015.

[6] K. Vershinin, M. Sweet, L. Ngwendson, J. Thomson, P. Waind, J. Bruce and E. M. S. Narayanan, "Experimental Demonstration of a 1.2kV Trench Clustered Insulated Gate Bipolar Transistor in Non Punch Through Technology," in *2006 IEEE International Symposium on Power Semiconductor Devices and IC's*, Naples, 2006.

[7] O. Spulber, M. Sweet, K. Vershinin, C. Ngw, L. Ngwendson, J. Bose, M. M. D. Souza and E. M. S. Narayanan, "A novel trench clustered insulated gate bipolar transistor (TCIGBT)," *IEEE Electron Device Letters*, vol. 21, no. 12, pp. 613-615, 2000.

[8] D. Y. Chen and S. A. Chin, "Bipolar-FET Combinational Power Transistors for Power Conversion Applications," in *Fifth International Telecommunications Energy Conference*, Tokyo, 1983.

[9] M. Rahimo, F. Canales, R. A. Minamisawa, C. Papadopoulos, U. Vemulapati, A. Mihaila, S. Kicin and U. Drofenik, "Characterization of a Silicon IGBT and Silicon Carbide MOSFET Cross-Switch Hybrid," *IEEE Transactions on Power Electronics*, vol. 30, no. 9, pp. 4638 - 4642, 2015.

[10] G. Ortiz, C. Gammeter, J. W. Kolar and O. Apeldoorn, "Mixed MOSFET-IGBT bridge for high-efficient Medium-Frequency Dual-Active-Bridge converter in Solid State Transformers," in *Workshop on Control and Modeling for Power Electronics*, Salt Lake City, 2013.

[11] P. Luo, H. Y. Long, M. R. Sweet, M. M. D. Souza and E. M. S. Narayanan, "Analysis of a Clustered IGBT and Silicon Carbide MOSFET Hybrid Switch," in *26th International Symposium on Industrial Electronics*, Edinburgh, 2017.

[12] C. R. Mueller, "Design and Analysis of a Low-Inductive Power-Semiconductor Module with SiC T-MOSFET and Si IGBT in Parallel Operation," in *PCIM Europe*, Nuremberg, 2017.

[13] Infineon, "Datasheet: IMW120R030M1H," Infineon, 2020.

[14] Infineon, "Datasheet: IKW50N120CS7," Infineon, 2021.

[15] S. N. E. Madathil, P. Luo, W. Saito and S.-I. Nishizawa, "Performance Comparison of Scaled IGBTs and CIGBTs," in *Solid State Devices and Materials conference*, 2021.

[16] H. Akagi, "Influence of high dv/dt switching on a motor drive system: a practical solution to EMI issues," in *International Symposium on Power Semiconductor Devices and ICs*, Kitakyushu, 2004.

PCIM Europe 2024, 11– 13 June 2024, Nuremberg   DOI: 10.30420/566262365

Contributions for Building Blocks for Normally-off 650V GaN-on-Si Power Integrated Circuits

Plinio Bau, Sebastian Gaviria-Duque, Thanh Hai Phung, Dominique Bergogne, Bernard Bancal
Wise-Integration, France
Corresponding author: Plinio Bau, Plinio.bau@wise-integration.com

Abstract

To increase power density, reliability and easy use, different circuits can be integrated in the same GaN-on-Si 650V die for power management. Input compatibility circuit and UVLO (undervoltage lockout) are essential circuits needed when integrating gate driver with a power transistor in the same die. This paper presents the characterization measurements of an UVLO and two different designs for input compatibility circuits with the objective to obtain one design with good process compensation for high efficiency mass production yields.

1   Introduction

In the domain of power electronics, the continuous pursuit of enhancing semiconductor device performance has been a driving force behind groundbreaking innovations. The emergence of Gallium Nitride (GaN) technology offers superior characteristics in terms of higher breakdown voltages, lower on-state resistance, and faster switching speeds compared to traditional silicon-based power transistors [1].

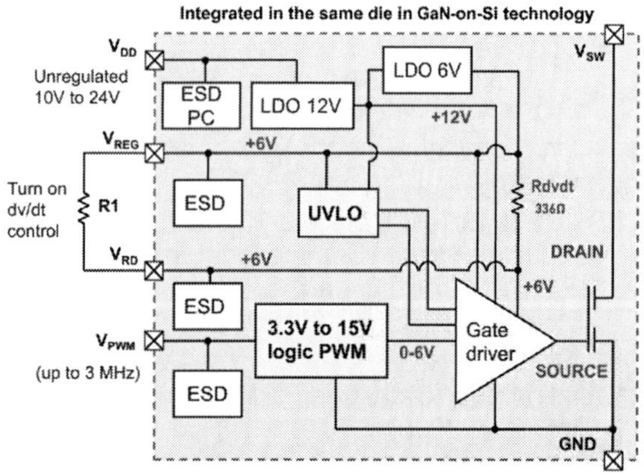

Fig. 1. Overall gate driver integrated with GaN HEMT (High Electron Mobility Transistor). The focus of this work is sharing technical design details of the UVLO, and the input compatibility circuits monolithically integrated in a same die with a power transistor.

GaN-based power devices, particularly 650V enhancement-mode (e-mode) GaN transistors, have garnered significant attention due to their potential to revolutionize numerous applications, ranging from power supplies and motor d

rives to electric vehicles and renewable energy systems. However, the successful operation of GaN devices heavily relies on efficient gate driver circuits. With a monolithic integration the GaN technology allows designers to use their full capabilities while mitigating inherent challenges such as high frequency switching and increased EMI due to high dv/dt's. Integrated gate drivers play a significant role in maximizing the performance of these power ICs, offering opportunities to achieve superior switching speeds, and therefore reduced switching energy and power dissipation [2-4].

Fig. 2. Photography of the chip with gate driver and auxiliary functions monolithically integrated in 650V pGaN AlGaN/GaN HEMT technology

The challenges in designing analog circuits for GaN technology are compounded by the inherent charge trapping effects, necessitating meticulous consideration of dynamic trapping and de-trapping mechanisms, which can significantly impact device characteristics, reliability, and overall circuit performance.
In order to obtain high efficiency yield in IC production, the IC design should have compensation methods for process fabrication [5-7]. A corner analysis simulation will evaluate a circuit behavior in extreme values to analyze perfor-

mance parameters if the design is acceptable and would obtain a high fabrication yield. The following methodologies are used to reduce fabrication dispersion:

- Common centroid layout.
- Design not using the minimum size devices.
- Use of dummy structures.
- Circuit branches to compensate for manufacturing (process) variation.

Fig. 1 and **Fig. 2** present a power GaN transistor with gate driver and auxiliary functions. This work will focus on UVLO and input compatibility circuits. It will discuss the challenges of analog design for GaN technology, and the measured values for those blocks and overall system will be presented.

2 UVLO in p-GaN Technology

2.1 General

Under-Voltage Lockout (UVLO) circuits act as crucial protective measures in power electronics, monitoring IC power supply rail voltages to prevent system activation or operation when levels dip below predefined thresholds. Essential for ensuring reliable performance, particularly in power converters, UVLO circuits guarantee controlled start-up and shutdown sequences, safeguarding against undervoltage scenarios that could compromise system integrity or cause damage. Designing undervoltage lockout (UVLO) circuits for Gallium Nitride (GaN) technology presents specific challenges due to the unique characteristics of GaN devices [8-13].

2.2 Schematic and Simulation

The UVLO topology integrating a differential pair for input amplification is a basic configuration in power electronics, ensuring accurate undervoltage detection despite process dispersion. This setup amalgamates the precision of UVLO functionality with the signal amplification capacity of a differential pair. By capitalizing on the differential pair's inherent ability to amplify small input signals, this topology offers a simple and reliable solution in detecting undervoltage conditions. Its resilience against process dispersion is verified by simulation at corners (SS, TT and FF) and ensures consistent and accurate operation, fortifying the safety and reliability of power electronic systems by providing a robust shield against inadequate supply voltages.

This topology was calculated to have less sensitivity to process dispersion from multiple DC

and transient simulations. **Fig. 3** below presents the simplified schematics without the dummy structures for simplicity. The dummy structures are a well-known option to reduce mismatches for CMOS technology. The voltage reference circuit (that generates V_{REF} node voltage) is composed by 12V GaN transistors M1, M2, M3 and R1 and is well described in [14]. This reference voltage generator is PVT (process, voltage and temperature) compensated to provide a stable value of 2.81V. Notice transistors M1 and M2 are normally-on (also called depletion-mode transistors) and M3 is a normally-off (also called enhanced-mode) GaN transistor. Two power supply rail nodes are required for this circuit. The power supply rail nodes are approximately 6V and 12V and change slightly with PVT changes. It was observed by simulations that sensing the 12V power supply node makes the output more insensitive to process dispersion compared to sensing the 6V supply rail. The voltage reference circuit is well described in [14]. Transistors M4 and resistor R4 make the hysteresis behavior required for a stable operation.

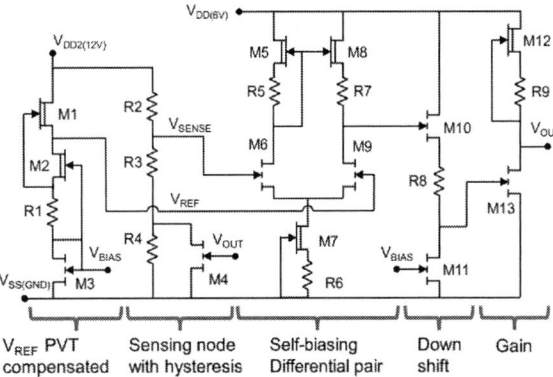

Fig. 3. UVLO undervoltage lockout circuit for GaN-on-Si technology with dGaN (normally-on) and eGaN (normally-off) devices.

Table 1 illustrates the anticipated simulated values for voltage threshold parameters, providing an overview of the expected outcomes derived from post layout simulation analysis. Those values are for junction temperature of Tj=25°C and Tj=150 °C.

Table 1 Post layout simulated values of the UVLO (TT)

Parameter	Symbol	25°C	150°C	Unit
Positive-going Threshold Volt.	V_{UVLO+}	8.44	9.17	V
Negative-going Threshold Volt.	V_{UVLO-}	7	7.49	V
Hysteresis	V_{HYS}	1.45	1.68	V
Current Consumption	I_{DD}	240	104	µA

Monte Carlo simulation is a powerful tool used in CMOS technology to understand and analyze sensitivity of circuit parameters versus manufacturing process variation. By modeling statistical variability, this method helps to predict how differences in fabrication, materials, and device structures affect circuit performance. It allows engineers to foresee potential variations in performance due to process fluctuations and material imperfections. With Monte Carlo simulations, designers can optimize circuit designs to ensure reliability, even amidst manufacturing uncertainties. This approach aids in creating GaN-based devices and systems that maintain consistent operation across different environments and manufacturing conditions.
The simulated values with Monte Carlo analysis are presented below in **Fig. 4** for positive-going threshold value and in **Fig. 5** for negative-going threshold value.

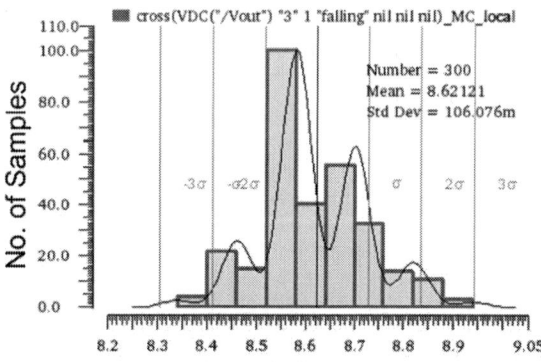

Simulated UVLO Threshold value V_{UVLO+} [V]

Fig. 4. Simulated positive-going threshold UVLO value as a function of process dispersion with Monte Carlo method. This is a simulation with local only variation.

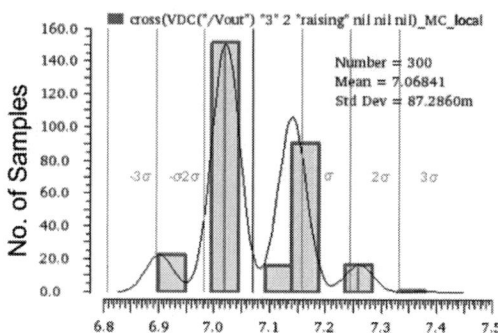

Simulated UVLO Threshold value V_{UVLO-} [V]

Fig. 5. Simulated fabrication dispersion of negative-going threshold UVLO value.

2.3 Experimental Results

The threshold values measured for 22 prototypes are presented in **Fig. 6** below. Both the positive-going (V_{UVLO+}) and negative-going (V_{UVLO-}) are plotted in the same image for comparison. The methodology to obtain those values is presented in **Fig. 7** below using a very low frequency triangular waveform of 1KHz. This method consists of obtaining the value observed in the V_{DD2} circuit node when the output of the UVLO (V_{OUT}) switches on or off.
Fig. 8 is the positive-going threshold value as a function of the negative-going threshold value ordered by ascending order. In this analysis we can observe the tendency of the hysteresis values.

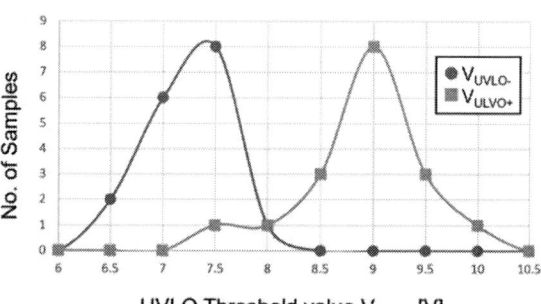

UVLO Threshold value V_{UVLO} [V]

Fig. 6. Measured threshold for UVLO values for 22 prototypes from same manufactured lot. The simulated values (dotted lines) were 7V and 8.5V for the negative-going and positive-going threshold respectively.

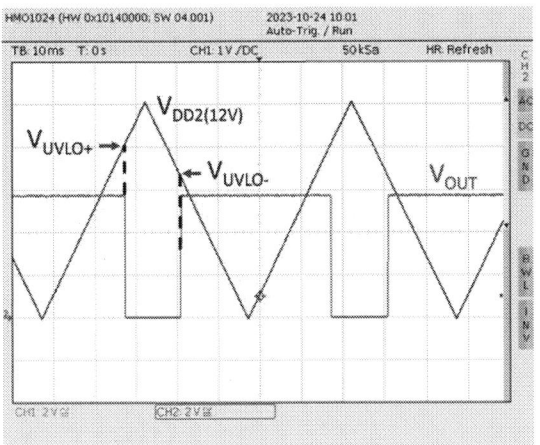

Fig. 7. Measured response to a triangular wave to obtain UVLO threshold values.

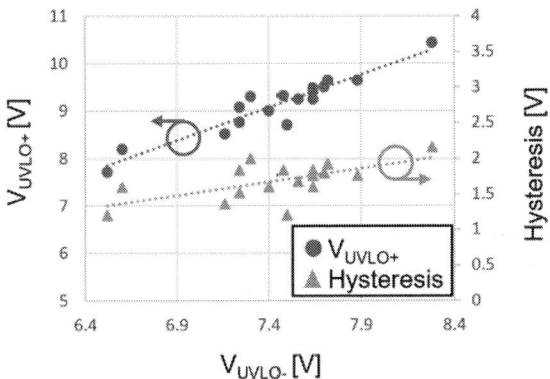

Fig. 8. Measured UVLO threshold value ordered by ascending order to analyze the minimum and maximum hysteresis.

The average and standard deviation for the threshold values for the UVLO are compared between simulated and measured in the table below. The parameters below are simulated for 300 parts and measured for 17 parts.

Table 2 UVLO Parameters comparison

Parameter	Simulated		Measured	
	AVG	StDev	AVG	StDev
UVLO-	7.06	0.08	7.44	0.414
UVLO+	8.62	0.10	9.12	0.605
Hyster.	1.45	-	1.68	0.254

It is observed that there is a good correlation between simulated and measured, the differences can be attributed to simulation models not very mature for this technology.

3 Input Compatibility Circuits

Input compatibility circuits serve as critical components within electronic systems, facilitating seamless interaction between various devices with distinct signal characteristics. These circuits play a role in ensuring the compatibility of input signals from different sources, addressing voltage level disparities, impedance mismatches, or signal format variations. Often implemented using level shifters, buffers, or interface modules, input compatibility circuits act as the interface gateway, translating and conditioning incoming signals to match the requirements of components or systems. For 650V pGaN technology the optimal value to switch on the gate of the power device is 6V. This work proposes 2 different circuits to act as level shifters to shift a PWM signal. The incoming source can be a microcontroller operating in 3.3V logic or can be a signal as high as a power CMOS gate driver with 0 to 15V. See **Fig. 9**.

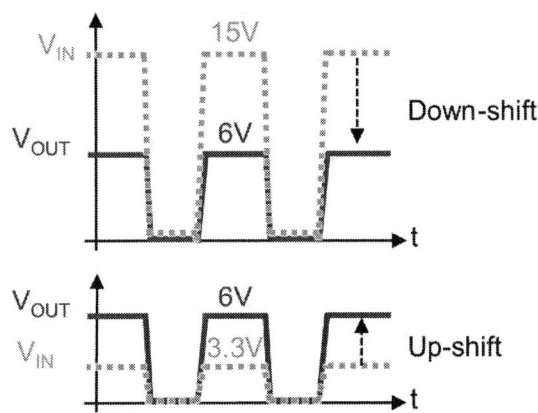

Fig. 9. Input compatibility circuits work as level shifters performing up-shift or down-shift according to the input signal level. The objective is to obtain a square wave of 0 to 6V in the output.

The schematics of 2 different circuits are presented in the next sections.

3.1 Schmitt Trigger Like Circuit

3.1.1 General

Schmitt trigger circuits stand as indispensable components within electronic systems, renowned for their pivotal role in signal conditioning and noise immunity. Leveraging hysteresis, these circuits exhibit a unique switching behavior, ensuring robustness against input noise and signal fluctuations. By employing positive feedback, Schmitt triggers possess dual threshold levels, enabling them to convert erratic, noisy signals into well-defined digital outputs. This characteristic makes them adept in applications where signal debouncing, waveform shaping, or threshold-based switching are paramount.

3.1.2 Schematic and Simulation

Fig. 10 presents this work's proposed GaN technology adapted Schmitt trigger. Transistor M1 is depletion-mode used for clamping the input voltage to a maximum 6V for the gate of M8.

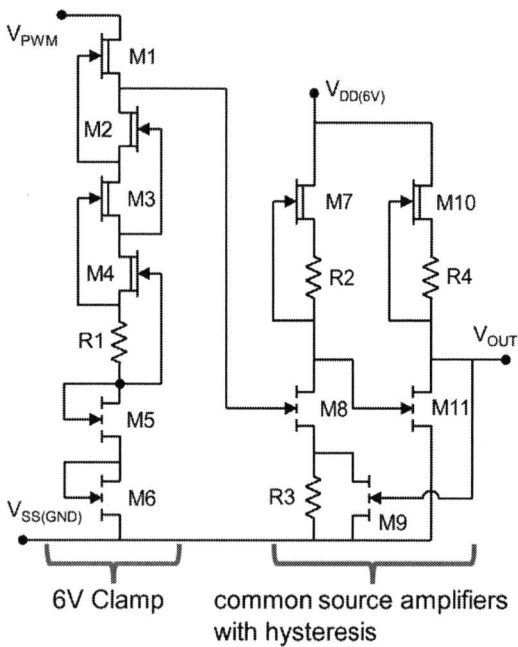

Fig. 10. A clamp circuit followed by a Schmitt Trigger like circuit adapted to GaN technology. This schematic contains normally-on, normally-off GaN transistors and 2DEG resistors.

Transistors M2 to M6 compose a self-biasing branch for M1. Transistor M9 add the desired hysteresis effect.

The simulated DC characteristics of the circuit is presented in Fig. X below. The main part of this circuit is switching M8 above its threshold value, that can be observed in the second curve of **Fig. 11** ($V_{G8}-G_{S8}$).

Notice the biasing current of each of the main branches is around 36 µA. The input current is around 10 µA at 25°C for typical process corner. Extensive simulations are performed for temperature and corners (SS and FF) to make sure the hysteresis is always above 400 mV.

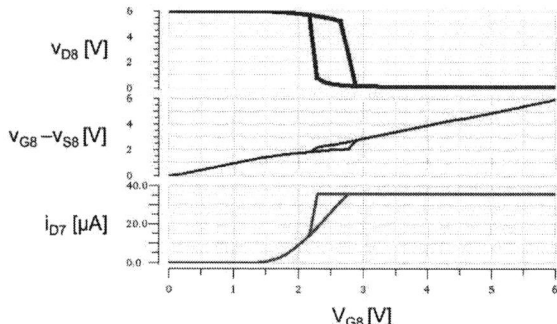

Fig. 11. DC Simulation DC for GaN adapted Schmitt trigger circuit. This circuits switches when gate-to-source voltage of transistor M7 is above the threshold value.

3.1.3 Experimental Results

Measured data for 15 dies are presented in **Fig. 12** below. To obtain the threshold values, a triangular square wave of very low frequency of 1Hz is applied and the value is obtained when V_{DS} signal switches on or off respectively. It is possible to observe the minimum measured hysteresis value above 0.4V.

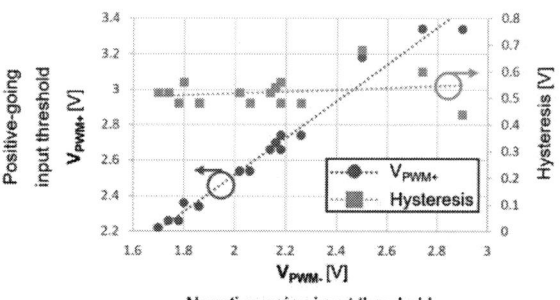

Negative-going input threshold

Fig. 12. Measured threshold values of Schmitt trigger GaN technology adapted circuit for 15 parts and arranged by ascending order of positive-going threshold value.

3.2 Common Gate Amplifier Circuit

Common gate amplifier circuits are integral components in power electronics, renowned for their high input impedance, voltage gain, and robust signal conditioning capabilities. These circuits efficiently amplify low-level signals while presser-ving linearity, making them valuable in power conversion systems. Their ability to provide substantial current gain contributes to enhanced signal control and stability in power converters, such as switched-mode power supplies and motor drives.

3.2.1 Schematic and Simulation

The proposed circuit is a common gate amplifier where the input signal arrives in the source and not the gate of the GaN transistor M7.
This makes it possible to switch on M7 with an input signal below Vth and improves reliability because a signal above 6V would not damage the gate of this transistor.
To compensate for process variation and mismatch, transistors M4 and M7 (**Fig. 13**) work in pairs as well as transistors M1 and M5. The biasing point of the gate of M7 is added by a Vth value by M4. Those two transistors are symmetric in chip's layout for reduced spatial dispersion. If the threshold value in M4 is for example close to a slow (SS) corner, its effect is compensated by M7 (also fabricated with physical properties close to a slow corner SS) and the same compensation effect is made for M1 and M5. They both contain a transistor in parallel to work as active diode with less voltage drop when in forward operation.

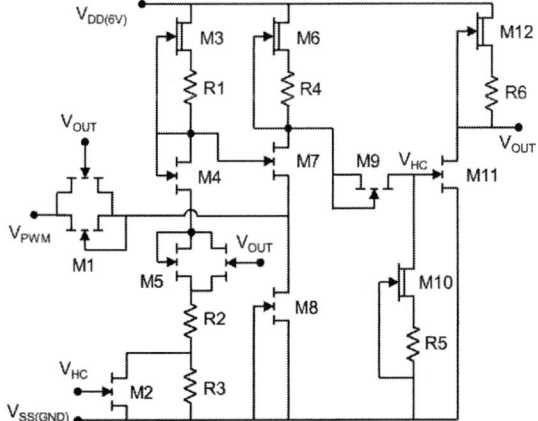

Fig. 13. Common gate amplifier schematic. This circuit is an input compatibility circuit that allows PWM signal from either 0 to 3.3V logic up to 0 to 15V and will output a PWM signal from 0 to 6V.

Fig. 14 contains static DC simulation for the main nodes of the circuits. Notice it switches the output when the gate-to-source voltage of M7 crosses its threshold value. Transient simulations are also performed extensively to observe the dynamic effects caused by capacitive current from the transistors' capacitances.

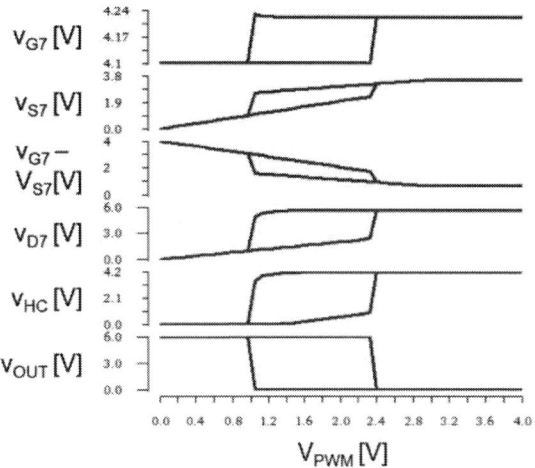

Fig. 14. Simulated DC waveforms for common gate amplifier circuit made for input compatibility in GaN technology.

3.2.2 Experimental Results

Measured data for 13 dies are presented in **Fig. 15**. It is possible to observe the minimum measured hysteresis value above 0.6V.

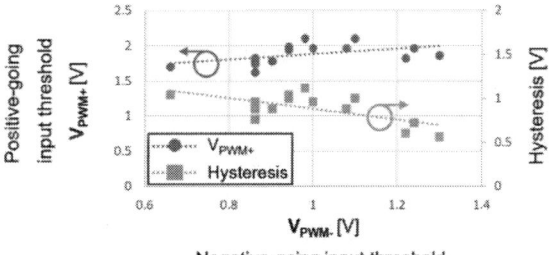

Fig. 15. Measured threshold values of common gate amplifier circuit trigger GaN technology adapted circuit for 6 parts and arranged by ascending order of positive-going threshold value.

By mitigating signal integrity issues and maintaining proper signal levels, these circuits contribute significantly to the overall robustness, reliability, and interoperability of complex electronic systems across diverse operational environments.

4 Experimental Results in High Voltage Operation

To validate the performance of those circuits embedded in the power chip test the system in a high voltage system level operation the circuits switch in a high voltage DC bus of 200V volts, and the measure is presented below in **Fig. 16**.

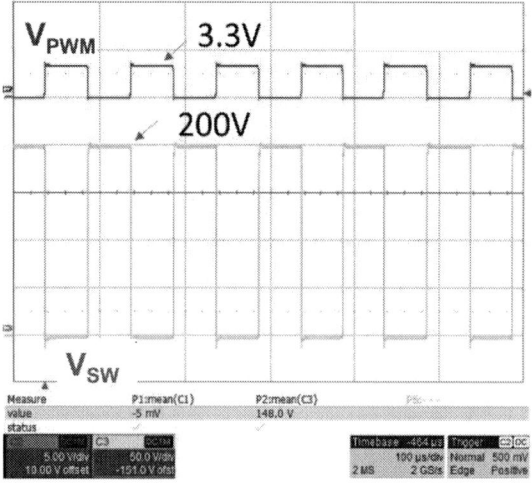

Fig. 16. Validation of operation in conditions close to the required application.

The monolithic power transistor with gate driver switching with 12V in the PWM signal in ZVS is presented in **Fig. 17**.

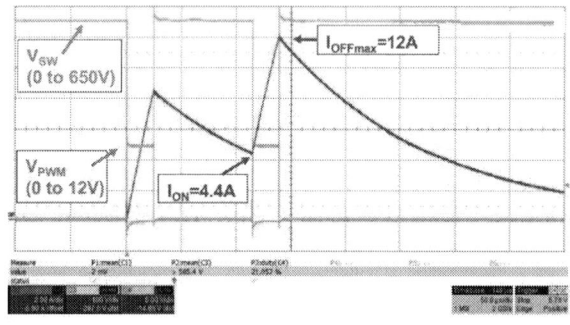

Fig. 17. Validation of the system in double pulse test in hard switching at 650V. The input compatibility circuit converts a PWM signal of 12V to a signal of 6V that is optimal for GaN transistors.

The die is also tested in soft switching (also known as ZVS) at 1MHz and the measure is presented below. In this condition the power devices in a half-bridge configuration heat up to only 83°C after 30 minutes of continuous operation. See **Fig. 18**.

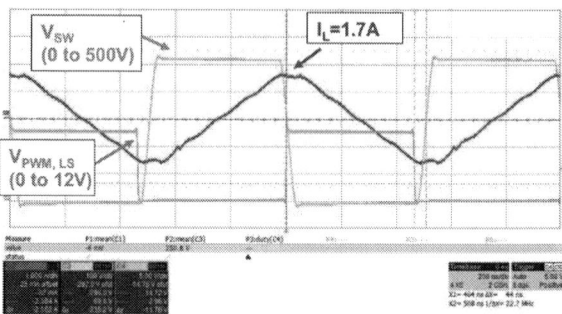

Fig. 18. Measure to validate the system in soft switching (ZVS) with I_{peak} = 1.7A at 500V (dV/dt =10V/ns).

5 Conclusion

This paper has presented one UVLO and two input compatibility circuits monolithically integrated into one 650V power e-GaN HEMT. Their most important parameters measured show the circuits work as expected. The characterization and measured parameters of analog circuits constructed on AlGaN/GaN normally-off technology offers valuable insights for analog designers. The aim of this work is to empower designers with technical information to help build better circuits for mass production with the intricate physical effects inherent in this technology, aiding in more informed and efficient circuit designs. Addressing the challenges posed by charge trapping-induced threshold voltage instability, fabrication dispersion, and mismatch, this study highlights the significance of overcoming these hurdles for optimized circuit performance. Through a campaign of characterization, it is evaluated the manufacturing dispersion. This paper contributes to research endeavors to develop designs resilient to process variations, aiming for better yields in large-scale production.

6 Reference

[1] Scrimizzi, F. et al. "The GaN Breakthrough for Sustainable and Cost-Effective Mobility Electrification and Digitalization". *Electronics, 12*, 1436. 2023.

[2] Xiangdong Li et al., "Integration of GaN analog building blocks on p-GaN wafers for GaN ICs", *Journal of Semiconductors*, Volume 42, Number 2, 2021.

[3] H. Xu, G. Tang, J. Wei, Z. Zheng and K. J. Chen, "Monolithic Integration of Gate Driver and Protection Modules With P-GaN Gate Power HEMTs," in *IEEE Transactions on Industrial Electronics*, vol. 69, no. 7, pp. 6784-6793, July 2022.

[4] Urmimala Chatterjee et al., "A fully integrated half-bridge driver circuit in All-GaN GAN-IC technology" *Solid-State Electronics*, Volume 207, 2023.

[5] A. K. Sharma et al., "Common-Centroid Layouts for Analog Circuits: Advantages and Limitations," *2021 Design, Automation & Test in Europe Conference & Exhibition (DATE)*, Grenoble, France, pp. 1224-1229, 2021.

[6] S. Dongaonkar, S. P. Mudanai and M. D. Giles, "From Process Corners to Statistical Circuit Design Methodology: Opportunities and Challenges," in *IEEE Transactions on Electron Devices*, vol. 66, no. 1, pp. 19-27, Jan. 2019.

[7] K. Samperi, S. Pennisi, F. Pulvirenti and G. Palmisano, "1-mS constant-Gm GaN transconductor with embedded process compensation," *2022 17th Conference on Ph.D Research in Microelectronics and Electronics (PRIME)*, Villasimius, SU, Italy, pp. 73-76, 2022.

[8] Cho, K. "An Ultra-Low Quiescent Current Under-Voltage Lockout Circuit for a High-Voltage Gate Driver IC". *Electronics, 9*, 1729, 2020.

[9] W. Guo et al., "Under voltage lockout circuit design for enhanced GaN HEMT drive," *2019 3rd International Conference on Electronic Information Technology and Computer Engineering (EITCE)*, Xiamen, China, pp. 671-675, 2019.

[10] H. Wu et al. "Design of a Low Temperature Drift UVLO Circuit with Base Current Compensation," *2019 IEEE International Conference on Electron Devices and Solid-State Circuits (EDSSC)*, Xi'an, China, pp. 1-3, 2019.

[11] W. Guo, S. Du, C. Bai, L. Lei and Y. Zhu, "Design of a Low Temperature Drift Undervoltage Lockout Circuit-Used for GaN FET Power Driver IC," *2018 15th China International Forum on Solid State Lighting: International Forum on Wide Bandgap Semiconductors China (SSLChina: IFWS)*, Shenzhen, China, pp. 1-4, 2018.

[12] M. -H. Cho, et al. "Development of undervoltage lockout (UVLO) circuit configurated Schmitt trigger," *2015 International SoC Design Conference (ISOCC)*, Gyeongju, Korea (South), pp. 59-60, 2015.

[13] Xiaozong, Huang et al. "A Proposed Under-voltage Lockout of Compensated Temperature Coefficient Threshold Voltage without Comparator," *Journal of Theoretical and Applied Information Technology*, Vol. 44 No.1 2012.

[14] P. Bau, S. Gavira-Duque, F. Rothan, C. Reymond and D. Bergogne, "Voltage Reference and Zero Current Detector Monolithically Integrated on p-GaN Technology Designed for Process Corners Compensation," *2023 IEEE Applied Power Electronics Conference and Exposition (APEC)*, Orlando, FL, USA, 2023.

[15] C. Reymond, L. Mistre, P. Bau, T. H. Phung and D. Bergogne, "Digital GaN 300W AC/DC Power Supply," *PCIM Europe 2023; International Exhibition and Conference for Power Electronics, Intelligent Motion, Renewable Energy and Energy Management*, Nuremberg, Germany, pp. 1-6, 2023.

[16] P. Bau, S. Hariharan, S. Gaviria-Duque and D. Bergogne, "Digital Control for Efficient Switching Over a Wide Range of Supply Voltages," *PCIM Europe 2023; International Exhibition and Conference for Power Electronics, Intelligent Motion, Renewable Energy and Energy Management*, Nuremberg, Germany, pp. 1-6, 2023.

PCIM Europe 2024, 11– 13 June 2024, Nuremberg   DOI: 10.30420/566262366

New Bidirectional Asymmetric High Voltage TVS (Transient Voltage Suppressor) diode

Boris Rosensaft[1], Martin Schulz[2], Xingchong Gu[3]

[1] Littelfuse, Germany

[2] Littelfuse Europe GmbH, Germany

[3] Littelfuse Inc., Wuxi

Corresponding author: Boris Rosensaft, brosensaft@littelfuse.com
Speaker: Boris Rosensaft, brosensaft@littelfuse.com

Abstract

Transient Voltage Suppressor (TVS) diodes are semiconductor devices with a well-defined breakdown voltage. Once this voltage is exceeded, the TVS-diode changes from blocking into conducting mode and thus limits the voltage in a paralleled device, TVS-diodes absorb the energy contained in the overvoltage event to protect power semiconductors from destructive voltage overshoots [1].
A further use is in protecting power semiconductors like IGBTs [2]. Here, the so-called active clamping is used to prevent damage by exceeding tolerable voltage levels. However, to achieve suitably high voltages of 1200 V, several devices need to be connected in series which results in some drawbacks including space consumption, device tolerances and cost. This work presents the design of monolithically integrated bidirectional high-voltage TVS-diode dedicated to the active clamping function.

1   State-of-the-art

Transient Voltage Suppressor (TVS) diodes protect sensitive circuit nodes from single and time-limited overvoltage faults. Such TVS diodes are widely used in modern high-voltage IGBT circuits for active clamping, connected between the collector und the gate of an IGBT as illustrated in Fig. 1.

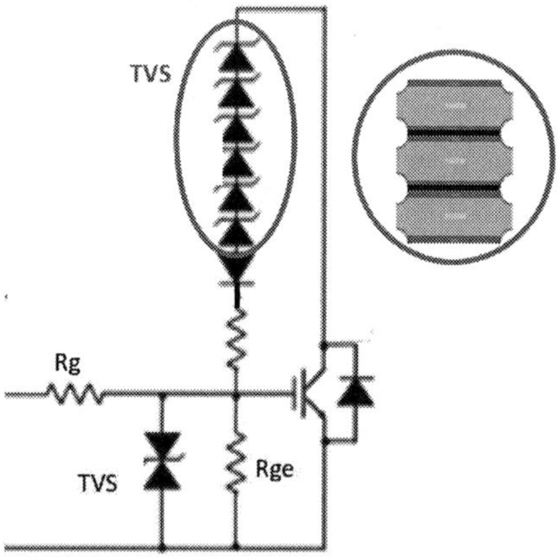

These TVS diodes need to fulfill restrictive requirements such as a high breakdown voltage with low deviation and low temperature drift as well as a high robustness in the clamping mode with a stable clamping voltage.

To guarantee safe operation, a low leakage current is a prerequisite. A too high leakage current could charge the gate during normal operation, leading to a complete loss of control.

Currently, two or more low-voltage TVS diodes are stacked inside a package to achieve the desired clamping voltage in a range higher than 500V.

Such series connection is both, costly and complicated in regards of packaging. Furthermore, the conventional low-voltage TVS diodes with MESA or with guard ring edge-termination are less suitable for high voltage TVS applications [3]. The disadvantage is, that these types of edge terminations have a large impact on breakdown voltage.

In contrast, the current work leads to a monolithically integrated, bidirectional, asymmetric high-voltage TVS diode which meets all the necessary high voltage requirements.

Fig. 1 Existing solution to set up active clamping for an IGBT

2 Design idea and operating principle

The basic idea for the new device is to combine non-punch-through (NPT) P+N-P+ structures with an isolation diffusion edge termination in a monolithic device. The cross-sectional view of the structure is given in Fig. 2.

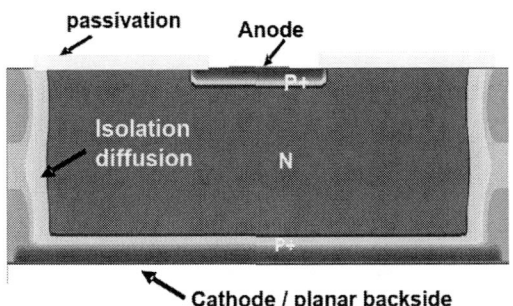

Fig. 2 Structure for the new bidirectional asymmetric HV TVS

The use of the isolation diffusion enables a low deviation of the breakdown voltage in the clamping direction. This becomes possible because the maximum electric field is now located in the silicon bulk. The use of the PNP structure ensures the negative dynamic resistance in clamping direction and consequently low power losses during the clamping events.

The asymmetric behavior can be seen in the measured I-V-characteristic depicted in Fig. 3. The I-V characteristic comprises two blocking directions. The forward blocking direction can be used in the IGBT active clamping mode.

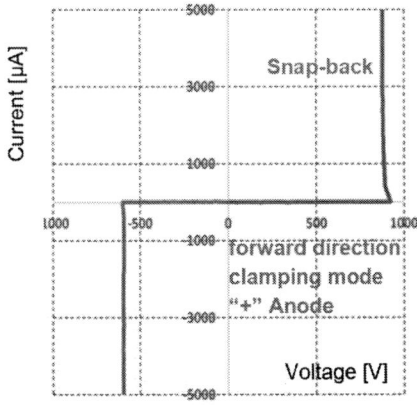

Fig. 3 Measured I-V-characteristic of a first sample, 25°C

This blocking direction features a very stable and precise breakdown voltage of about 930V. Due to the PNP structure, the device has a negative dynamic resistance in this direction, the so-called snap-back. This ensures fast-recharging of the IGBT's gate-emitter capacity and reduces the power loss during a clamping event. The reverse blocking direction has a breakdown voltage of 600V.

Figures 4 and 5 explain the physical advantages of the new HV TVS diode compared to the MESA terminated TVS diode. Both, the new and the MESA terminated TVS designs were simulated for the voltage range of 930V.

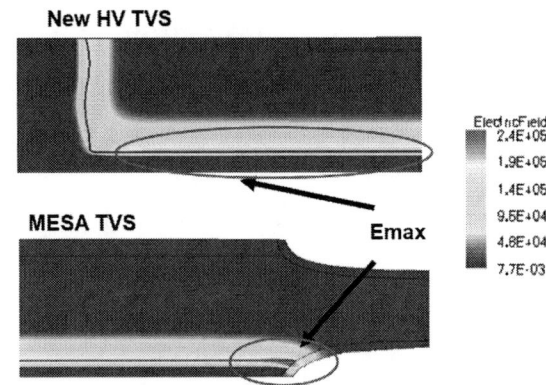

Fig. 4 Simulated electric field distribution @2A in avalanche mode, Vz+=930V, 25 °C

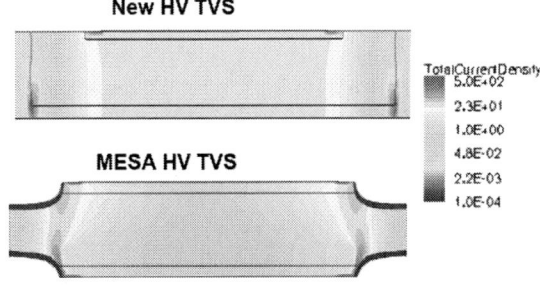

Fig. 5 Simulated total current density distribution @2A in avalanche mode, Vz+=930V, 25 °C

Figures 4 and 5 depict a simulated E-field- and current density distribution in clamping mode. A physical advantage of the new device is that the main current flows in the silicon bulk, as the maximum electric field is also located in the silicon bulk. This ensures high precision and process stability for the breakdown voltage, which only depends on the specific resistance of the silicon bulk.

In case of the MESA terminated device, the maximum electric field is located at the chip edge, so the breakdown voltage is very sensitive to edge design and process deviations.

Other advantages of the new TVS diode include a low temperature coefficient of the breakdown voltage and a planar backside. A planar backside simplifies the assembly process and reduces the thermal resistance.

The new HV TVS device can be manufactured based on the existing Littelfuse technologies.

A disadvantage of the initial HV TVS diode design was an increased leakage current in hot conditions due to the PNP structure, which also applies to other PNP designs. To mitigate this problem, a new technological method was proposed and tested.

3 Device characterization

The first batches of the new HV TVS diode were produced and characterized at Littelfuse. The graph in Fig. 6 presents a typical statistical distribution of the breakdown voltage Vz+.

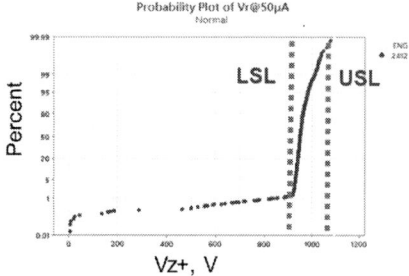

Fig. 6 Probability plot of forward breakdown voltage Vz+ @50µA, 25°C

The lower and upper specification limits should be defined regarding the targeted application. To ensure high-precision clamping, the chips can be divided into several small breakdown voltage ranges, as can be seen in Fig. 7.

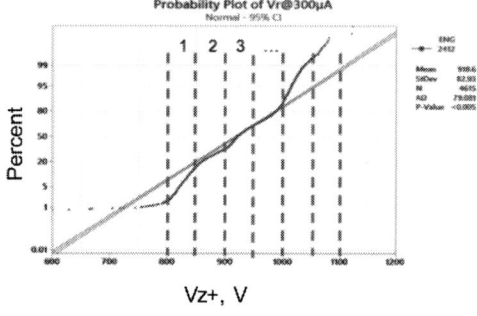

Fig. 7 Example of grouping TVS diodes into high-precision forward voltage ranges

To evaluate the temperature stability of the new HV TVS chip in avalanche mode, the temperature coefficient of the breakdown voltage in the temperature range between 25°C and 150°C was determined. For comparison, a stacked diode of the same voltage range was used, which consists of two standard low voltage TVS diodes. The new diode features a slightly lower temperature coefficient of the forward breakdown voltage of approx. 0.08 %/K as summarized in Fig. 8.

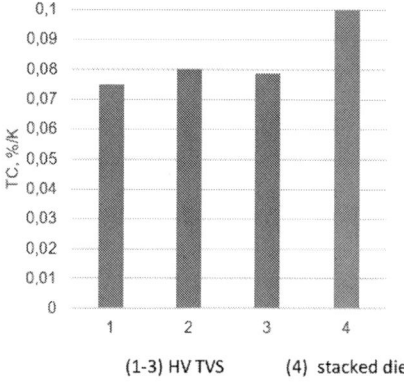

Fig. 8 Averaged temperature coefficient (TC) of the forward breakdown voltage Vz+, TC measured for 3 wafer batches in a temperature range between 25°C and 150°C

The leakage current at elevated temperatures of the new TVS diode is slightly higher than that of the stacked diode. The main reason for this is the fact that the diodes in the series connection are only biased with part of the total clamping voltage. To reduce the leakage current in hot condition, a technological method was proposed which leads to a controlled reduction in the carrier diffusion length and corresponding reduction in the transistor gain of the PNP structure. With this method, the hot leakage current can be reduced by about 50 %, as can be seen in the diagram in Fig. 9.

At the same time, the influence of the method on the breakdown voltage is very small.

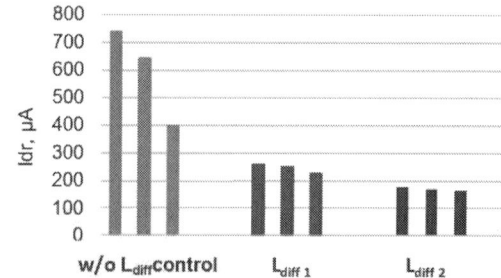

Fig. 9 Forward leakage current (Idr) reduced using a method of carrier diffusion length control, Idr measured on 3 wafer batches, 150°C, 800V

The new diodes were also tested for surge current (Ipp) in comparison to the stacked diode of the same voltage range. The test result is summarized in Fig. 10.

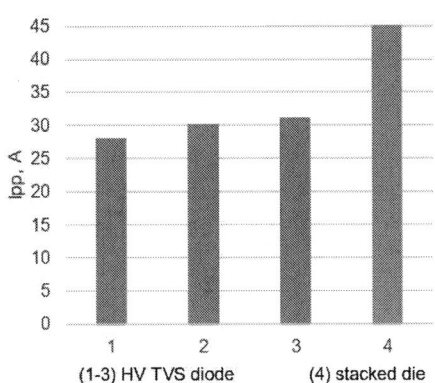

Fig. 10 Maximum surge current I_{pp} 8/20 µs measured for 3 wafer batches, 25°C

The new TVS diode achieves a surge current value of approx. 30 A. As the stacked diode has a higher heat capacity, it allows a higher peak value of approx. 45 A

4 IGBT active clamping test at room temperature

Fig. 11 pictures the equivalent circuit of the test setup.

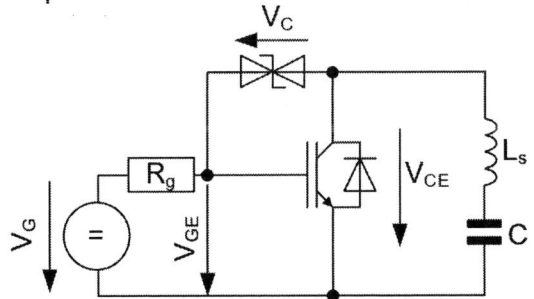

Fig. 11 Circuit used for the active clamping test

The TVS sample is connected between collector and gate of the IGBT. The gate-driver is imaged as voltage source V_G which is connected to the gate through the external resistance R_g of 10 Ohm. To provoke an overvoltage event, a stray inductance of 2 µH and a DC-link capacitor of 100 µF are used.

Fig. 12 demonstrates the waveforms measured during the active clamping test.

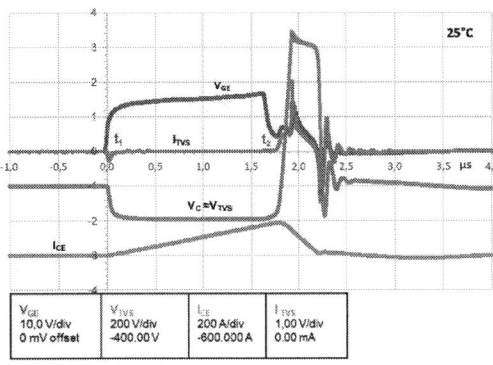

Fig. 12 Active clamping waveforms using the new HV TVS diode, 25°C

There is IGBT turn-on at the point t_1 which causes the collector current I_{CE} to increase. At the point t_2 the IGBT is turned off. This causes a decrease of the collector current with a high current change rate dI/dt and consequently overshooting of the collector-emitter voltage due to the stray inductance. Once the voltage overshoot exceeds the breakdown voltage of the TVS-diode, the TVS begins to conduct a current I_{TVS} which flows into the IGBT's gate and partially in the external resistance Rg.

Though the gate-driver provides a turn-off signal, the current from the TVS-diode charges the IGBT's gate and in turn forces the IGBT to turn on again. In turn, the collector-emitter-voltage collapses and the TVS-diode returns to blocking mode.

This way, a feedback-loop exists that achieves two effects:

1) the overvoltage across the IGBT is limited to a safe value of 1090 V in the test.
2) during turn-off, the current change rate di/dt is reduced which in turn helps limiting the overvoltage.

The test also proves, that the leakage current is low enough to prevent unwanted turn-on of the IGBT during normal mode of operation.

5 Thermal simulation of a temperature operation area

In a potential module structure, the TVS diode should be placed close to the IGBT to reduce the parasitic inductance in the gate-loop. As a result, the device is strongly thermally coupled to the IGBT and the freewheeling diode. In turn, the TVS-diode receives heat from these active elements through the heat spreading [4].

The results of a thermal simulation of the IGBT module with two IGBTs and corresponding two HV TVS diodes in Fig. 13 depicts the temperature distribution in the module in the case of normal operation without overvoltage in the collector circuit.

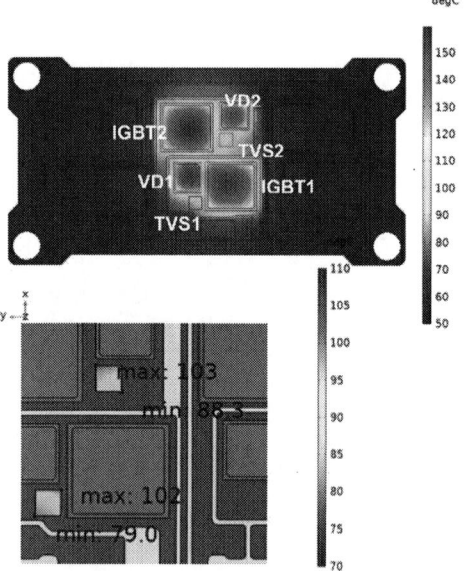

Fig. 13 Thermal simulation: normal operation, power loss each IGBT 400V, each freewheeling diode 160 W, both TVS-diodes in off-state

According to the simulation results, the diodes TVS1 and TVS2 reach temperatures of 102 °C and 103°C respectively, while the temperatures of both IGBTs is 150°C. The slightly higher temperature of TVS2 diode is due to its unfavourable position, as it receives additional heat from IGBT1 and VD1.

The case of overvoltage faults was also thermally simulated. In this case, additional power losses occur in the TVS diode as it conducts a higher current during the event. Fig. 14 depicts the resulting temperature rise of the TVS-diode over time during the active clamping event. The relative temperature increase is very small with about 2 K only, as the active clamping event only lasts a few hundred nanoseconds.

Based on the thermal simulation results, the operating temperature range between -40° and 105°C was defined and investigated further.

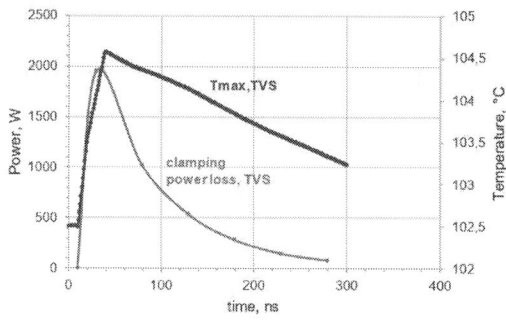

Fig. 14 Thermal simulation: Temperature increase of the diode TVS2 in Fig. 13 during an active clamping event

6 IGBT active clamping at high and low temperature limits. Temperature drift of HV TVS clamping parameters

The behaviour of the HV TVS diode in active clamping mode at -40°C and 105°C was investigated. The investigation served to observe the new HV TVS diode in the defined temperature operating range. In addition, the important drift parameters of the diode, which are of great importance for the application, were evaluated.

Fig. 15 includes the switching waveforms of the diode's current and voltage during the clamping process for the minimum limit temperatures of -40°C.

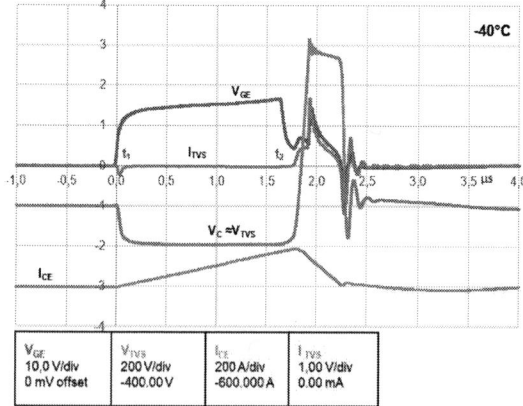

Fig. 15 Active clamping waveforms during clamping, -40°C

Being a silicon-based technology, it was expected that the breakdown voltage of the diode decreases at low temperatures. This behaviour is similar to the way, an IGBT's blocking voltage drops at such low temperature levels. At -40°C, the overvoltage was limited to 1030 V.

With growing TVS-diode's temperature, a growth in breakdown voltage is expected. Test with elevated temperatures of 105°C ended up with a clamping voltage of 1150 V as seen in Fig. 16, making this diode technology an efficient match for IGBTs.

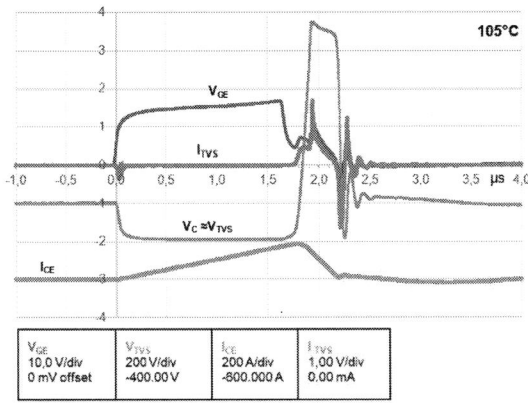

Fig. 16 Waveforms during an active clamping event, 105°C

The new HV TVS diode proved to be fully functional for both temperature limits. As expected, a positive temperature drift of the clamping voltage and a dependence of the maximum diode current on the temperature were determined.

Fig. 17 reveals the positive drift of the clamping voltage with temperature.

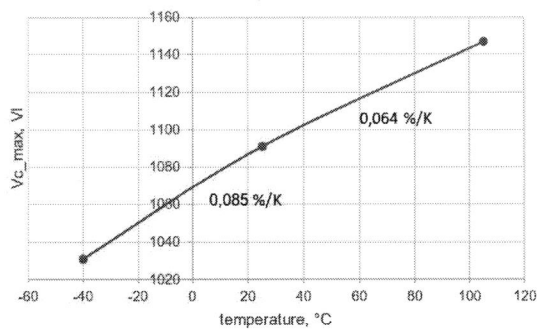

Fig. 17 Maximum clamping voltage V_{cmax} versus junction temperature of HV TVS diode

The diagram can be separated in two parts, contemplating the operation below and above 25°C.

For the part below room temperature and down to -40°C, a temperature coefficients of the maximum clamping voltage $V_{c\,max}$ of 0.085 %/K is obtained. For the high temperature range from 25°C to 105°C, 0.064 %/K can be taken from the graph respectively.

For the range from -40°C to 25°, the temperature coefficient of $V_{c\,max}$ is slightly higher compared to the one measured for the temperature range between 25°C and 105°C.

Further investigation towards higher temperatures is currently ongoing.

A further important parameter is the TVS-diode's forward current. This is the current that later charges the IGBT's gate. This value also changes with temperature. However, it may not change to a value that would not allow charging the gate fast enough.

Fig. 18 demonstrates the temperature dependence of the diode's maximum current during the active clamping event.

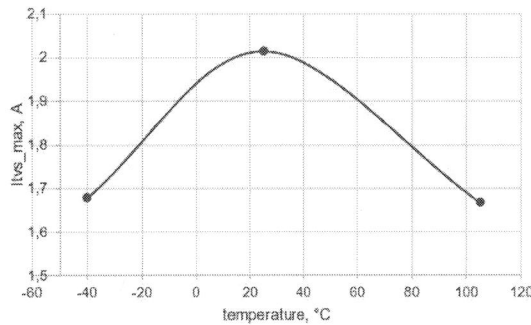

Fig. 18 Temperature dependence of the maximum current delivered by the HV TVS-diode in the clamping mode

This dependency should be taken into account when designing the IGBT circuit with active clamping. Individual testing is necessary to ensure, that the current remains high enough to activate the clamping sequence. This mostly depends on the IGBT's gate capacity and is of growing importance when handling high-power IGBTs or a larger number of paralleled dice.

7 Summary

The new design concept of a monolithic bidirectional asymmetric High Voltage TVS Diode was proposed and modeled using finite element simulation. The HV TVS diode has several advantages over the current high voltage TVS design, which consists of several stacked low-voltage diodes.

The most important advantages include easier fabrication of the new monolithic element and simple assembly process. The backside of the new HV TVS diode is planar. The first samples of the new TVS device were fabricated based on the existing Littelfuse chip technology, characterized and tested. The first results revealed that the new HV TVS diode is a fully functional element with outstanding performance, featuring low temperature drift of the maximum clamping voltage and low power losses in clamping mode. The active clamping function of the new TVS diode was also successfully tested in the junction temperature range from -40° to 105°C.

8 References

[1] TVS/Zener, Theory and Design Considerations, Handbook, HBD854/D Rev. 0, Jun–2005

[2] Mingfang Chen, Zhichao Xiong, Yongxi Zhang, Enxiao Zhu, Yuying Zhao and Zunbo Ma, *IGBT Overvoltage Protection Based on Dynamic Voltage Feedback and Active Clamping*, Appl. Sci. 2023

[3] Zeng Jianfei, Cai Yingda, *TVS Diode and Assembly having asymmetric breakdown voltage*, European Patent Application, EP 3832720A1

[4] Ahmet Mete Muslu; Ryan Wong; Vanessa Smet; Yogendra Joshi, *Heat Spreading and Heat Removal Needs of a Novel Power Electronics Package with Integrated Cooling*, 2021 20th IEEE Intersociety Conference on Thermal and Thermomechanical Phenomena in Electronic Systems (iTherm)

PCIM Europe 2024, 11– 13 June 2024, Nuremberg   DOI: 10.30420/566262367

ISO247: High Performance Ceramic based Advanced Isolated Discrete Package to Fully Exploit the Advantages of SiC MOSFET

Sachin Shridhar Paradkar[1], Francois Perraud[1], Aalok Bhatt[2]

[1]Littelfuse Europe GmbH, Germany
[2]Formerly, Littelfuse Europe GmbH, Germany

Corresponding author: Sachin Shridhar Paradkar, sparadkar@littelfuse.com

Abstract

In power electronic applications, achieving electrical isolation between a discrete semiconductor package and the heatsink often results in an increase in the thermal resistance junction-to-heatsink R_{thJH}, limiting the current carrying capacity of the packaged device. Littelfuse's innovative isolated discrete ISO247 package directly addresses this issue. The ISO247 with high performance ceramic enhances power and current density while providing inherent isolation and remaining footprint compatible with standard TO-247 package. This paper compares ISO247 and TO-247 devices carrying the same 1200 V SiC MOSFET. Thermal resistance, junction temperature and power handling capability are analyzed to demonstrate the superiority of the ISO247 package for SiC power devices. Thermal measurements indicate that selecting the ISO247 is optimal for minimizing chip junction temperature and thermal resistance junction-to-heatsink, potentially leading to increased application power output and cost savings at the system-level.

1   ISOPLUS™ – The Revolution in Discrete Isolation Technique

ISOPLUS™ is the family of internally isolated discrete power semiconductor devices, first pioneered by IXYS-Littelfuse in 2003. The ISO247 belongs to the ISOPLUS™ product family and adheres to the JEDEC TO-247AD outline, ensuring pin compatibility with the standard TO-247 package. The ISO247 displayed in Fig. 1, offers several key advantages:

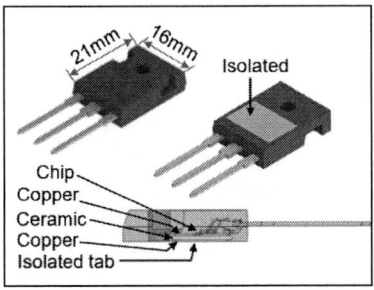

Fig. 1   ISO247 internal construction

• High performance ceramic-based active metal brazing (AMB) substrate offers inherent isolation, higher thermal conductivity and reduced thermal resistance junction-to-heatsink [1].

• Isolation voltage rating of 2.5 kV AC, 1 minute or 3 kV AC, 1 second.

• Higher temperature and fast power cycling withstand capability compared to standard TO-247 ascribed to the matched coefficient of thermal expansion (CTE) for SiC chip and AMB substrate [2].

• Increased power density and simplified thermal management.

• Reduced EMI attributed to the small chip-to-heatsink stray capacitance.

The primary motivation to use SiC MOSFETs in an application is to achieve improved efficiency, high power density and high frequency operation enabling a compact design. Maximizing chip performance becomes imperative to offset the high cost associated with SiC. Conventional TO-247 packages often pose challenging limitations stemming from thermal management, resulting in a suboptimal utilization of the SiC chip. The ISO247 package addresses this drawback and presents a revolutionary solution simplifying the way design engineers shall address system level performance, integration, and assembly needs. The ISO247 is available in a variety of technologies such as Si/SiC MOSFETs, IGBTs and Diodes with different voltage classes ranging from 70 V to 1600 V.

2 ISO247 – High Performance Ceramic based Discrete Isolated Package

The ISO247 package was engineered by IXYS-Littelfuse in 2003 based on direct bonding copper (DBC) substrates with aluminum oxide (Al_2O_3) ceramic. The increasing adoption of silicon carbide in power electronics is driven by the demand for increased power density, longer lifetimes, and elevated operating temperatures. Designers are compelled to choose advanced substrates for power semiconductor packages to fully exploit the benefits offered by SiC. In the context of isolated power semiconductor packages, the ceramic material plays a critical role in influencing thermal resistance, thereby impacting the power density of the packaged semiconductor device. Littelfuse has developed the ISO247 package with advanced high performance silicon nitride (Si_3N_4) ceramic, specifically tailored to meet the demanding requirements of SiC MOSFET-based applications. Si_3N_4's impressive bending strength, high fracture toughness, and high thermal conductivity make it an ideal choice for substrates in electrically isolated wide band gap (WBG) power semiconductor packages [3].

The bonding of Si_3N_4 ceramic substrates deviates from conventional DBC methods. While the DBC process is exclusively compatible with intermediate oxide-ceramic layers, the metallization process employed in Si_3N_4 substrate for Littelfuse ISO247 package utilizes hybrid active metal brazing (H-AMB) technology. This approach integrates the advantages of both the sputtering and AMB processes, constituting a two-step metallization procedure. In the initial phase, the ceramic substrate undergoes a sputtering process with an active metal filler layer. Subsequently, in the second stage, the copper layer is brazed onto the sputtered substrate at a temperature of approximately 850°C. The utilization of H-AMB on the ceramic substrate results in superior thermal conductivity, high reliability, a void free bonding surface and an improved cost per amp ratio.

Distinguished by its internal isolation using high performance ceramic, the ISO247 package is designed to significantly reduce the thermal resistance R_{thJC} while offering isolation, and enhanced power and current densities. This stands in stark contrast to the TO-247 package, which relies on an external isolation method. Although ISO247 and TO-247 packages share identical outer dimensions and pinout configurations, their internal structures and mounting approach exhibit notable distinctions as depicted in Fig. 2. The TO-247 device necessitates external isolation during its attachment to the heatsink. In contrast, the ISO247 device relies exclusively on thermal interface material for its mounting to the heatsink [4].

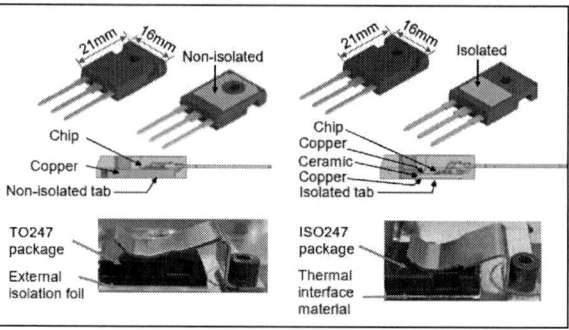

Fig. 2 Internal construction and mounting difference between TO-247 and ISO247 discrete packages.

3 Thermal Performance Comparison between ISO247 and TO-247 based SiC MOSFETs

The standard TO-247 package has an electrically conductive mounting tab, which is typically at the drain potential. It is generally desirable to electrically isolate the device mounting tab from the heat sink due to safety concerns and the desire to mount multiple discrete devices on the same heat sink frame. The utilization of an external thermally conductive, electrically isolating foil between the semiconductor package and the heatsink has become a widely adopted approach in the industry for this purpose. Nonetheless, employing external isolation entails significant drawbacks: increased thermal resistance, diminished power and current handling capacity, complex thermal management, and substantial assembly efforts. The mentioned penalties, particularly the issue of reduced power handling capabilities, become unacceptable, especially in cases where WBG semiconductors like SiC-MOSFETs are utilized. The Si_3N_4 ceramic based ISO247 offers improved overall thermal resistance and power handling capability compared to a standard TO-247 discrete with an external isolation foil, thereby aiding in maintaining SiC chips cooler at a given DC current.

Thermal measurements were conducted using a 1200 V 25 mΩ SiC MOSFET chip packaged in ISO247 and TO-247 packages to evaluate the performance advantages of the Littelfuse advanced ISO247 package. Thermal measurements were executed using the cooling curve method in accordance with IEC 60747-8 [5], with measurement set-up detailed in reference [6]. The measurements included various packaging configurations, as summarized in Table 1.

Device No., Type	MOSFET chip	Device isolated from heatsink by:
Device 1 TO-247		External isolation foil with thermal conductivity 1.8 W/mK
Device 2 TO-247	1200 V, 25 mΩ SiC	External isolation foil with thermal conductivity 6.5 W/mK
Device 3 ISO247		Internal isolation with high performance Si3N4 ceramic

Table 1 Devices for thermal measurement comparison featuring SiC MOSFETs

For the thermal measurements of the TO-247 devices, external isolation foils with thermal conductivity values of 1.8 W/mK and 6.5 W/mK were utilized between the package and the heatsink to replicate real-world applications. The ISO247 device employed a thermal paste between the package and the heat sink. The devices were mounted on temperature controlled, water-cooled heatsinks maintained at a constant temperature of 30°C. The thermal data obtained for all three devices were subsequently compared, with a focus on parameters such as thermal resistance R_{thJH} and junction temperature T_{vj} under identical heating current I_H conditions. These comparisons are visualized in Fig. 3 and Fig. 4.

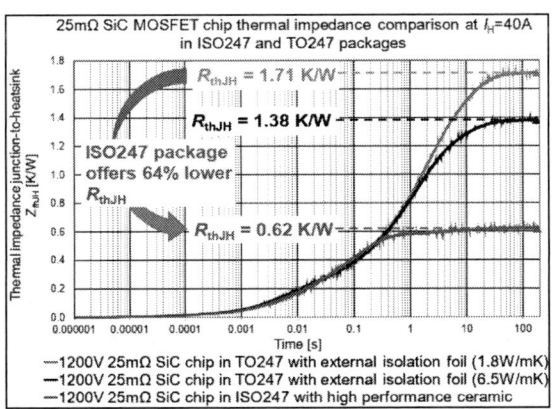

Fig. 3 Thermal impedance measurement comparison between ISO247 and TO-247 discrete devices.

As evident from Fig. 3, the ISO247 with high performance ceramic improves the steady state thermal resistance R_{thJH}, by 64% compared to the TO-247 devices with external isolation carrying the same SiC chip. This directly translates into an increased power handling potential and lower chip temperature at a given current. As depicted in Fig. 4, the SiC chip in the ISO247 package with advanced ceramic stays up to 60°C cooler when compared to the SiC chips in the TO-247 device with external isolation at I_H=40 A. This results in lower temperature swing between the junction and heatsink, ΔT_{JH}, at the given heating power. The ISO247 with high performance ceramic exhibits nearly a 53% reduction in temperature swing ΔT_{JH} compared to the standard TO-247 discrete, significantly enhancing the ISO247 device's lifetime and, consequently, the reliability in the application.

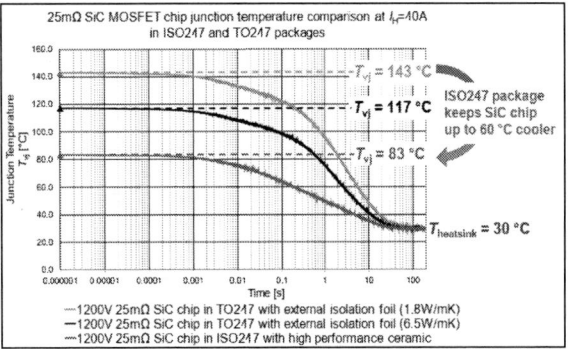

Fig. 4 Junction temperature comparison between ISO247 and TO-247 discrete devices.

However, it is important to note that the practical limit for T_{vj} within applications is typically up to 130°C to ensure safe operation. Therefore, a performance comparison of the ISO247 device to the TO-247 Device 2, utilizing a higher quality and more expensive 6.5 W/mK isolation foil offers a more realistic comparison. As depicted in Fig. 3 and Fig. 4, the ISO247 with high performance ceramic improves thermal resistance R_{thJH} by 55% and offers a 39% lower temperature swing ΔT_{JH}.

4 Increasing Application Power Output using the ISO247

To demonstrate the improvement in application power output using the ISO247, thermal measurements with a heating current I_H resulting in a chip temperature T_{vj} of 130°C was applied to different packages, all containing the same 25 mΩ SiC chip. A junction temperature T_{vj} of 130°C is selected as most real-world applications are designed to operate with chip temperatures $T_{vj} \leq 130°C$. The results from the thermal measurements have been summarized in Table 2.

Device and isolation type	T_{vj} [°C]	I_H [A]	R_{thJH} [K/W]	P_{DJH} [W]
TO-247 with 1.8 W/mK isolation foil	130	38.8	1.71	59
TO-247 with 6.5 W/mK isolation foil	130	42.2	1.38	73
ISO247 with high performance ceramic	130	50.8	0.62	160

Table 2 Thermal measurement results at $T_{vj}=130°C$

The observations can be interpreted to clearly understand the advantages offered by the ISO247 device at the application level. It is observed that the SiC chip in the TO-247 package with external isolation foil reached a junction temperature T_{vj} of 130°C with I_H=38.8 A. The same SiC chip in the ISO247 with high performance ceramic reached a junction temperature T_{vj} of 130°C with I_H=50.8 A. This translates to nearly 30% increase in the current carrying capacity of the SiC chip in the ISO247 device to reach the same junction temperature T_{vj} of 130°C. As demonstrated by the data presented in Table 2, when considering the SiC chip operating at a junction temperature of 130°C, it becomes apparent that the ISO247 package shows a remarkable 170% improvement in power handling capacity compared to the TO-247 solution.

These measurements show that the ISO247 with advanced ceramic takes 30% higher current to reach a chip temperature of 130°C. The exceptional thermal performance exhibited by the advanced ISO247 package unleashes the potential for enhancing power density and output power in the end application. Upgrading a given application with a DC-link voltage of 800 V originally designed for 20 kW from 1200 V 25 mΩ SiC MOSFETs in TO-247 package with external isolation foil to the same SiC MOSFETs in high performance ceramic-based ISO247 packaging solution could potentially increase the DC power output of this system to 30 kW. This represents a substantial 50% increase in DC power output, as visually depicted in Fig. 5.

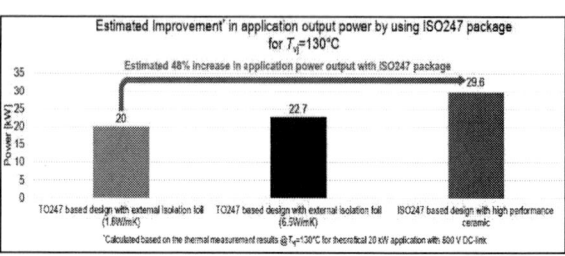

Fig. 5 Estimated increase in application power output by using ISO247.

5 Summary

The growing prevalence of WBG devices demands innovative advancements in packaging technology to fully harness the advantages offered by WBG semiconductors. Littelfuse's ISO247 is a unique isolated package specifically designed to address the rigorous requirements of SiC-based applications while remaining compatible with the standard TO-247 footprint.

From the thermal measurement comparison between the ISO247 and TO-247 packages, it has been established that the Littelfuse ISO247 with high performance Si_3N_4 ceramic offers a remarkable 64% reduction in thermal resistance R_{thJH} and 53% reduction in temperature swing ΔT_{JH}. Consequently, the SiC MOSFET chip in the ISO247 package remains up to 60°C cooler at the same DC power. This significantly improves overall device lifetime and application reliability. It also offers the possibility to increase application power output by 50% considering the 20 kW application example.

In simpler terms, due to the improved thermal resistance R_{thJH} and power dissipation P_{DJH} of the ISO247 package with high performance ceramic, engineers can choose higher $R_{DS(on)}$ chips for a given application power rating. This presents a significant cost-saving opportunity at the system level. Employing the ISO247 in power-electronic applications reduces mounting efforts, enables space-saving, decreases overall thermal resistance, and increases power density, all while simplifying thermal design. The advanced ISO247 package developed by Littelfuse is positioned to transform the approach to discrete semiconductor isolation within applications utilizing WBG semiconductors.

References

[1] M. Goetz, N. Kuhn et al., 'Comparison of Silicon Nitride DBC and AMB Substrates for different applications in power electronics', Nuremberg, Germany, PCIM Europe 2023.

[2] S. Yang, B. Yang et al., 'Selection of interfacial metals for Si3N4 ceramics by the density functional theory', Chemical Physics Letters, Volume 763, 2021, ISSN 0009-2614.

[3] M. Goetz, N. Kuhn et al., 'Silicon Nitride Substrates for Power Electronics', Article in Bodo's Power Systems, pp. 40-41, May 2018.

[4] Application Note: 'ISOPLUS™: Isolated Discrete Power Semiconductors', www.littelfuse.com, access on 29-12-2023.

[5] IEC Standard: 'IEC 60747-8:Semiconductor Devices – Discrete Devices – Part 8: Field-effect Transistors', Edition 3.0, 2010-12.

[6] A. Bhatt, U. Kulsoom, F. Perraud, M. Schulz, L. Gant, 'ISOPLUS - SMPD: An Advanced Isolated Packaging to Fully Exploit the Advantages of SiC MOSFETs', Nuremberg, Germany, PCIM Europe 2023, DOI: 10.30420/566091117, 2023.

PCIM Europe 2024, 11–13 June 2024, Nuremberg   DOI: 10.30420/566262368

Impact of Current Ripple Reduction Using High Switching Frequencies on PMSM Efficiency

Jannik Fuchs-Gade[1,2], Guilherme Bueno Mariani[1]

[1] Infineon Technologies, Austria
[2] Alpen Adria Universität, Klagenfurt

Corresponding author: Jannik Fuchs-Gade, Jannik.Gade@Infineon.com
Speaker: Jannik Fuchs-Gade, Jannik.Gade@Infineon.com

Abstract

With the advent of WBG technologies (SiC, GaN) as well as switching optimized Si, the penalty for increased switching frequencies is less significant opening up new possibilities for highly efficient motor drives. This paper discusses the impact of high switching frequencies on motor efficiency, based on experiments conducted on a permanent magnet synchronous motor (PMSM) under various load and speed conditions. The results show a benefit of going to higher switching frequencies across various operating points. Additionally, this paper presents a methodology based on FEM (Finite Element Method) to help understand the mechanism of losses in the motor.

1   Introduction

In recent years, energy efficiency has become a prominent concern due to the importance of mitigating carbon emissions. One of the key areas that can benefit from increased energy efficiency is motor drives which are extensively employed, not only in industrial settings, but also in major and small home appliances [1],[2],[3].

Current ripple is often associated with torque ripple and other undesirable phenomena linked to the control of electrical machines. However, mitigating current ripple, achieved through higher switching frequencies or the selection of inverter topology, can also lead to improved motor efficiency [4][5]. Emerging semiconductor technologies such as wide band-gap semiconductors and switching optimized MOSFETs present opportunities to operate motor drive inverters at higher switching frequencies without accruing significantly higher switching losses [1][4][5]. Operating at higher frequencies directly contributes to the reduction of current ripple while at the same time offering several advantages for motor drive applications[4][5].

One significant advantage is the reduction of high-frequency losses in the motor. Consequently, it is crucial to evaluate system losses using these new semiconductor technologies to determine potential efficiency gains and to assess which semiconductor technology is best suited for specific applications. Moreover, the advent of new motors with lower inductance necessitates operation at higher switching frequencies, as the effects of current ripple or insufficient control bandwidth would lead to significant issues [1], [5].

Different approaches for high-frequency motor losses have been proposed. In [6], [7], and [8] the focus was on eddy currents losses, while [9] delved into soft magnetic material losses. In [10] the focus was on low frequency rotating losses of a PMSM using FEA which builds on work done in [11] and [12].

This paper presents experimental results using synchronized high and low-frequency measurement paths to quantify the losses of a PMSM using an automated testbench. As future work, a FEM based methodology is proposed, which relies on FEM simulations combined with analytical equations. The methodology extends the work done in [10] to account for high frequency losses.

In this paper, Section 2 details the experimental setup used to measure system and motor HF (High Frequency) losses. Section 3 presents the system losses and the motor HF losses. Section 4 introduces the proposed method for evaluating HF losses using FEM simulations, and finally, in section 5, a conclusion is provided.

2 Experimental Setup

The measurements were carried out on an automated testbed consisting of a motor-generator setup which can be broken into two parts: **a)** system loss measurement and **b)** high frequency measurement. The aim of the system loss measurement is to capture motor and inverter losses during steady state operation at the selected operating point. Figure 1 shows the full setup used for the measurements.

The figure displays the blue area which represents the system loss measurement. This setup comprises two main measurements. The first part consists of a mechanical power measurement utilizing an *HBM T21WM* torque sensor, which also measures the speed of the motor. The electrical power measurement is performed by *LEM IN100* current sensors, measuring the phase currents and the current at the output of the DC source. The voltage measurement is directly carried out by the *Yokogawa WT5000* power analyzer, which also captures the mechanical input, compiling all the power measurements.

The orange area in figure 1 illustrates the HF loss measurement setup. The purpose of this setup is to capture losses at the timescale of the PWM, enabling the study of the effects of current ripple. Care was taken to minimize ringing in the measurement by ensuring a short distance between the inverter and motor. This setup includes *TCP0030A* current sensors for measuring the phase currents, *TCP4971* voltage probes to measure the phase voltage, and a *Tektronix MSO5B* oscilloscope for capturing the measurements at the desired scale. All the data captured by the oscilloscope will be post-processed in *MATLAB* to obtain the HF motor losses.

The motor setup comprises the DUT (Device Under Test) *Anaheim BLY343S* PMSM, chosen for its low inductance on the winding. The DUT imposes the torque on the shaft, while a *T-motor U15 II* is used as a generator to control the speed on the shaft.

The motor is driven by an inverter using *200V OptiMOS 6^{TM}*. This specific device was chosen for its optimized switching, considering the study's high-frequency operation. The inverter also incorporates *XENSIV^{TM} TLI4971* for phase current measurement and *EiceDRIVER ^{TM} 2EDF7235K* gate driver for driving the MOSFETs. Figure 2 shows an image of the inverter.

Fig. 2: B6 inverter used for the measurements.

3 Motor Loss Measurements

3.1 Procedure

The measurement procedure is as follows:

```
initFoc ()

waitForSteadyStateTemperature()

for w_gen in (−3000, 2000, 3000)
    setGeneratorSpeedRpm(w_gen)
    steadyStateSpeedDelay()
    for i_d in (0 : 2 : 18)
        setDutIdCurrentInAmp(i_d)
        steadyStateSpeedDelay()
        for f_s in (20, 40, 50, 80, 100)
            setDutSwitchingFrequencykHz(f_s)
            delay()
            triggerMeasurement()
        endfor
    endfor
endfor
```

Prior to testing the motor temperature is observed using an infrared camera. Once the steadystate temperature is achieved the testing is started. The switching frequency is changed in the inner loop to minimize its effect on the motor temperature and thus the efficiency measurement itself.

3.2 High Frequency Losses Evaluation

In order to evaluate the high frequency effect on the motor losses it is necessary to separate the losses into HF and LF components. This is done since the power generated by the high frequency components do not produce any torque, [5]-[4] and can therefore be considered fully as losses. The process of separation of the losses into HF and LF relies on the split of the phase current and voltage into HF and LF components. This is done by a series of filters applied on the current and voltage measurement. Figure 3 shows the result of phase current splitting into HF an LF components. The

PCIM Europe 2024, 11– 13 June 2024, Nuremberg   DOI: 10.30420/566262368

System Loss Measurement

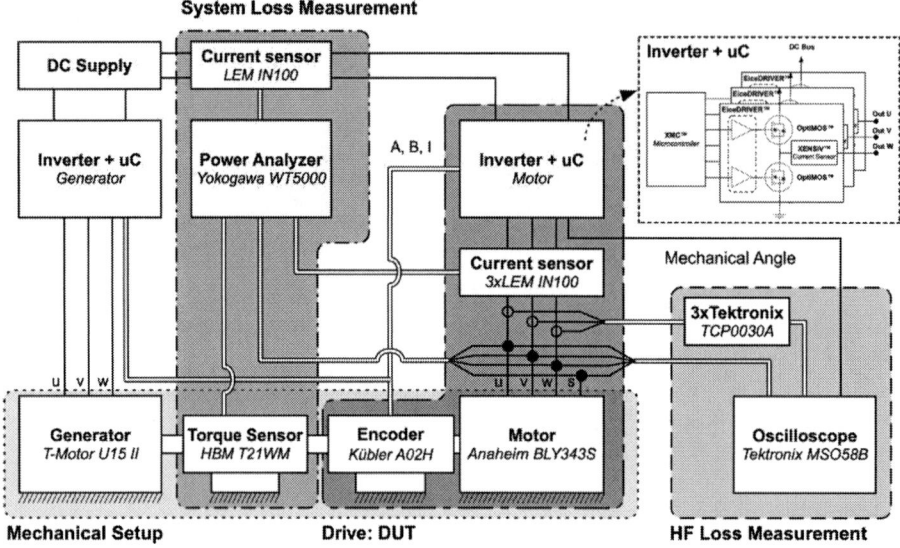

Fig. 1: Experimental setup depicting the system (blue) and the high-frequency measurement (orange) domains.

same process is adopted on the phase voltage.

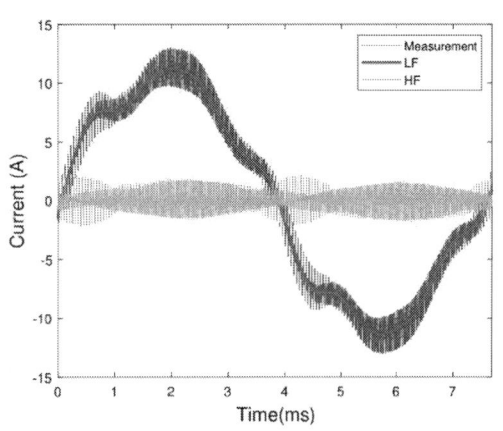

Fig. 3: Separation of the phase current into its low and high frequency components.

The ringing effect on the current measurement was negligible confirming the hypothesis cited in [5] that the ringing effect is related to the cable length and a side effect of the reflection on the cables. Another important note is that this effect can also be related to the bus voltage level. Figure 4 shows the effect of the switching frequency on the HF component of the current in which the envelope of the current is reduced as a function of frequency.

Similar to [5] -[1] the switching frequency, as expected, directly impacts the ripple of the current. With higher switching frequency the ripple becomes smaller. Once the current and the voltage are split

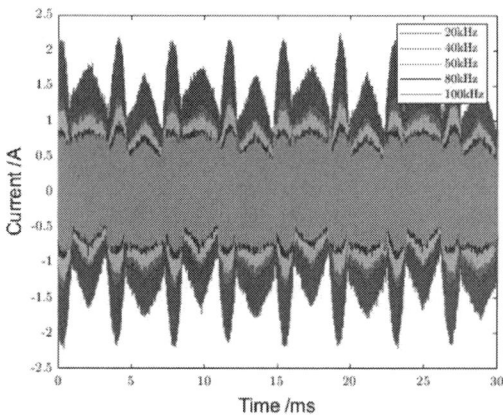

Fig. 4: High frequency content of the phase current.

into HF and LF the HF losses are calculated as:

$$P_{HF} = V_{A_{HF}}I_{A_{HF}} + V_{B_{HF}}I_{B_{HF}} + V_{C_{HF}}I_{C_{HF}}$$

HF motor losses are directly linked to the switching frequency, which is expected since the ripple is reduced with the switching frequency. On the other hand, inverter losses increase with switching frequency due to the presence of more switching events. This implies that there is a trade-off between motor HF losses and inverter losses in regards to the switching frequency.

In this paper an evaluation of an optimum switching frequency is offered. However, this kind of analysis should be adopted for other parameters e.g. deadtime, control etc. This evaluation is depicted in

2634

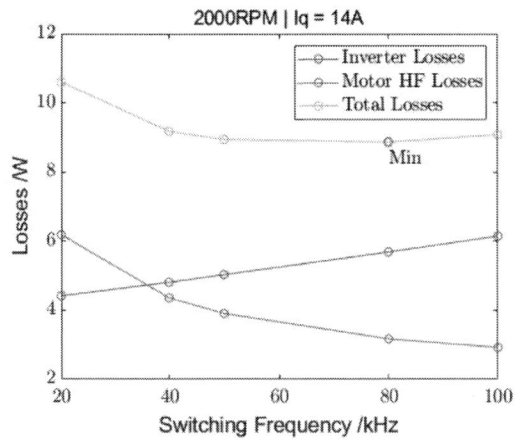

Fig. 5: Inverter and high frequency motor losses with respect to switching frequency.

figure 5 where a summary of the results for the experimental tests is given. The tests were carried out in a way to account for temperature variations between switching frequencies.

Figure 5 shows a trade-off between motor HF losses and inverter losses. To attain the lowest proportion of losses relative to the output power, the motor drive system should operate at a switching frequency of 80kHz when running at a nominal speed (2000 RPM) and at 1.24 Nm load (260W). The optimal frequency of 80 kHz is much higher than normal frequency for motor drive applications. Normally for this kind of applications the switching frequency is in the range of (5-20 kHz). New technologies of switching optimized MOSFETS as Optimos™ 6, CoolGAN™, CoolSIC™ allows to achieve these higher switching frequencies at a lower power loss, therefore shifting the system to work at higher switching frequencies. The optimum operational switching frequency shown here is defined by a specific set of conditions e.g. load, speed, temperature etc.. It is important to mention that this optimal switching frequency will change depending on the operational point of the machine.

3.3 Motor Loss Evaluation

Figure 6 shows the motor efficiency measured using the LF (System Loss Measurement in blue) path of the measurement setup seen in figure 1.

Here the motor efficiency is obtained using the power analyzer which measures the mechanical and electrical power of the motor. Across the three measured speeds there is a consistent widening of the high-efficiency range of the motor when increasing the switching frequency. This is especially true

where $I_q \geq 6A$ which can be seen by the widening of the yellow region as the switching frequency increases. The largest increase in efficiency can be seen in the range between 20kHz to 40kHz after which the impact of increasing the switching frequency on motor losses becomes lower.

4 Evaluating HF Losses Using FEA (Future Work)

4.1 Issues of HF Losses in FEA

In [10] the author post processes a series of static FEA simulations to calculate the rotating losses of a PMSM in combination with high frequency core loss data and other material constants. Simplifications have been made to neglect the effects of PWM and PWM induced current ripple.

According to the results, the stator core and magnet losses are dominant at higher speeds while resistive losses dominate at lower speeds.

Combining FEA simulations, which typically focus on motor timescales, with that of inverter simulations, which typically focus on switching behavior, can be a time consuming endeavor. In order to accurately model switching behavior, and therefore the losses, requires simulation steps in nano or pico second scale. In order to avoid computing a field solution using FEA in each of these timesteps a different methodology is here proposed.

4.2 Extending FEA to HF Loss Simulation

The methodology shown in figure 7 can be described as the following:

1. A phase current measurement is obtained by measurement or simulation using representative impedance to reproduce current ripple.

2. A frequency analysis is performed to split the frequency content into three groups: fundamental, harmonics, and noise; latter of which is removed.

3. The fundamental frequency is used in accordance with [10] but is in addition used as a frozen permeability operating point.

4. These points, denoted as $n = 0, n = 1, ..., n = k$, where k is the number of total DC operating points, serve as which serves as DC operating points for an AC sweep consisting of the selected harmonics.

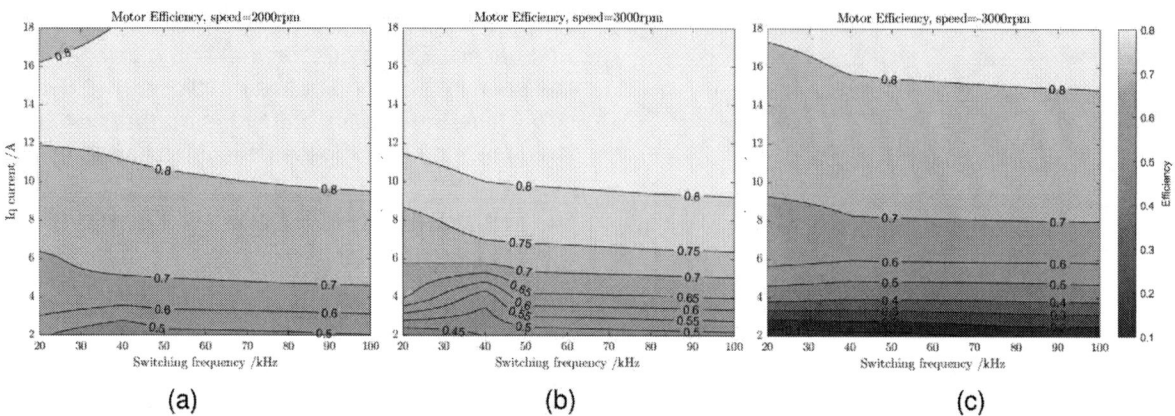

Fig. 6: System efficiency as a function of switching frequency at various Iq currents at **a)** 2000rpm, **b)** 3000rpm and **c)** -3000rpm. Note that the high efficiency region (in yellow) is widened as the switching frequency increases.

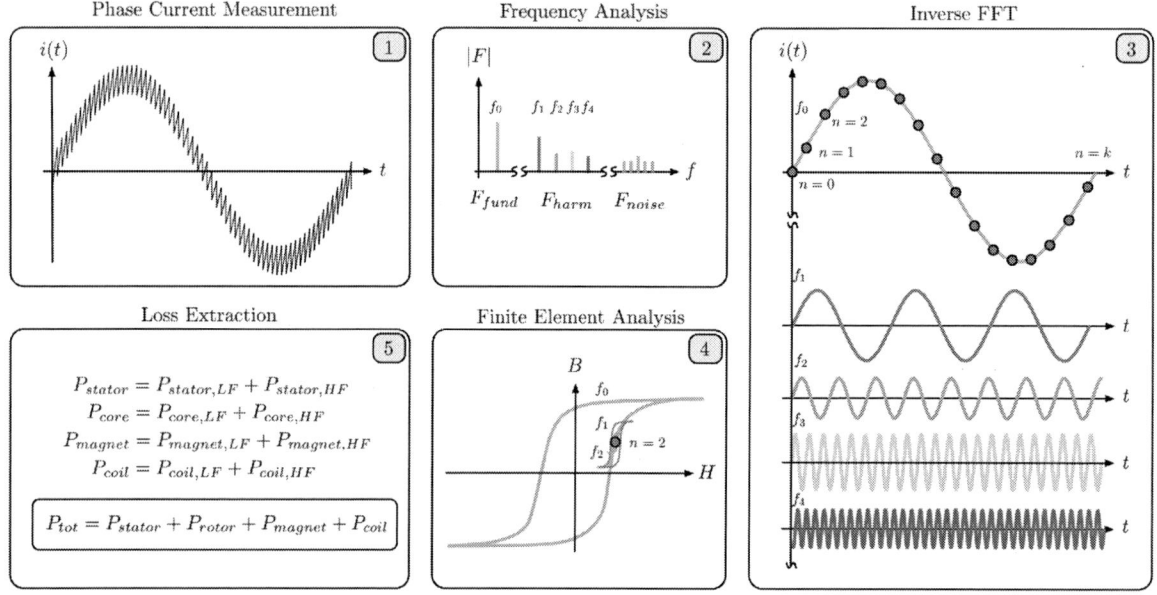

Fig. 7: Methodology for finite element based PMSM loss extraction.

5. The power losses for the low frequency losses are now augmented using the losses from the AC sweeps.

The main challenges with the approach is to obtain core loss data above 1kHz which typically isn't provided by manufacturers.

5 Conclusion

This paper underscores the importance of evaluating motor drive losses at the system level rather than focusing on motor or inverter losses in isola-

tion. A trade-off in the choice of switching frequency is imperative as the losses shift from the motor to the inverter at higher switching frequencies.

While it is observed that losses in the inverter escalate with higher switching frequencies, there is a significant reduction in motor losses as the high efficiency region is widened. This presents a compelling case for finding the optimal switching frequency for the inverter and motor as this can lower the overall system efficiency.

In addition to the insights provided, it is imperative to underscore the significance of evaluating other critical aspects at the system level in future research. As the field of motor drive systems continues to evolve, it becomes increasingly clear that a comprehensive evaluation of factors such as dead time, control methods, and modulation is essential to gain a holistic understanding of system performance and efficiency.

The advent of advanced semiconductor technologies, such as MOSFETs with optimized switching characteristics and WBG materials like GaN and SiC, but also new Si technologies such as Optimos^{TM} 6 has made the prospect of operating at higher switching frequencies increasingly feasible. Drawing from the experimental findings, it is evident that an optimal switching frequency emerges for a specific operational point of the machine. Notably, the trade-off between inverter losses and motor HF losses is optimized at 80 kHz, a frequency significantly higher than the typical operational switching frequency of motor drives. This underscores the potential for substantial gains in efficiency through strategic selection of switching frequencies, enabled by cutting-edge semiconductor technologies. Additionally, this paper has introduced a FEA based methodology for analyzing HF losses within the motor. This analytical approach serves as a foundation for future developments, as it offers the potential to comprehensively understand the intricacies of loss mechanisms within the motor. The ongoing refinement and expansion of this analysis will be invaluable in advancing our understanding of motor drive efficiency and in guiding the development of future motor drive systems.

References

[1] A. K. Morya, M. C. Gardner, B. Anvari, L. Liu, A. G. Yepes, et al., "Wide bandgap devices in ac electric drives: Opportunities and challenges," *IEEE Transactions on Transportation Electrification*, vol. 5, no. 1, pp. 3–20, 2019. DOI: 10.1109/TTE.2019.2892807.

[2] L. Chang, M. Alvi, W. Lee, J. Kim, and T. M. Jahns, "Efficiency optimization of pwm-induced power losses in traction drive systems with ipm machines using wide bandgap-based inverters," *IEEE Transactions on Industry Applications*, vol. 58, no. 5, pp. 5635–5649, 2022. DOI: 10.1109/TIA.2022.3178979.

[3] L. Chang, W. Lee, T. M. Jahns, and K. Rahman, "Investigation and prediction of high-frequency iron loss in lamination steels driven by voltage-source inverters using wide-bandgap switches,"

IEEE Transactions on Industry Applications, vol. 57, no. 4, pp. 3607–3618, 2021. DOI: 10.1109/TIA.2021.3075647.

[4] N. Voyer, G. Bueno-Mariani, A. Besri, V. Quemener, Y. Okamoto, and A. Satake, "High frequency modelling of permanent magnet synchronous machine," in *2018 8th International Electric Drives Production Conference (EDPC)*, 2018, pp. 1–6. DOI: 10.1109/EDPC.2018.8658275.

[5] G. Bueno Mariani, E. Alfawy, F. Wernegger, and J. Gade, "Wide band gap inverters and high-frequency effects on motor drives," in *PCIM Europe 2023; International Exhibition and Conference for Power Electronics, Intelligent Motion, Renewable Energy and Energy Management*, 2023, pp. 1–6. DOI: 10.30420/566091301.

[6] D. Ishak, Z. Zhu, and D. Howe, "Eddy-current loss in the rotor magnets of permanent-magnet brushless machines having a fractional number of slots per pole," *IEEE Transactions on Magnetics*, vol. 41, no. 9, pp. 2462–2469, 2005. DOI: 10.1109/TMAG.2005.854337.

[7] F. Z. Zhou, J. X. Shen, and W. Z. Fei, "Influence on rotor eddy-current loss in high-speed pm bldc motors," in *Proceedings of the 41st International Universities Power Engineering Conference*, vol. 2, 2006, pp. 734–738. DOI: 10.1109/UPEC.2006.367576.

[8] K. Yamazaki and A. Abe, "Loss investigation of interior permanent-magnet motors considering carrier harmonics and magnet eddy currents," *IEEE Transactions on Industry Applications*, vol. 45, no. 2, pp. 659–665, 2009. DOI: 10.1109/TIA.2009.2013550.

[9] G. von Pfingsten, S. Steentjes, A. Thul, T. Herold, and K. Hameyer, "Soft magnetic material degradation due to manufacturing process: A comparison of measurements and numerical simulations," in *2014 17th International Conference on Electrical Machines and Systems (ICEMS)*, 2014, pp. 2018–2024. DOI: 10.1109/ICEMS.2014.7013817.

[10] D. C. Meeker. "Rotating losses in a surface mount permanent magnet motor." (2017), [Online]. Available: https://www.femm.info/wiki/SPMLoss (visited on 04/09/2024).

[11] Z. Q. Z. D. Ishak and D. Howe, "Comparison of models for estimating magnetic core losses in electrical machines using the finite-element method," 2009.

[12] G. von Pfingsten, S. Steentjes, A. Thul, T. Herold, and K. Hameyer, "Eddy-current loss in the rotor magnets of permanent-magnet brushless machines having a fractional number of slots per pole," 2005.

PCIM Europe 2024, 11– 13 June 2024, Nuremberg   DOI: 10.30420/566262369

Maximizing Cost-Efficiency in Electric Drivetrains: A SiC/Si Fusion Switch Approach

Matthias Ippisch[1], Tomas Reiter[1], Michael Niendorf[2], Waldemar Jakobi[2], Mark Münzer[1]

[1] Infineon Technologies AG, Neubiberg, Germany
[2] Infineon Technologies AG, Warstein, Germany

Corresponding author: Matthias Ippisch, matthias.ippisch@infineon.com
Speaker: Matthias Ippisch, matthias.ippisch@infineon.com

Abstract

In order to further drive electrification of cars, there is a critical need to optimize cost-efficiency in electric drivetrain components. This paper investigates the potential of integrating silicon insulated-gate bipolar transistors (IGBTs) and diodes with silicon carbide (SiC) metal-oxide-semiconductor field-effect transistors (MOSFETs) in parallel configuration to address this imperative. The hybrid concept emerges from the fact that SiC, while offering improved light load efficiency, is more expensive due to higher manufacturing complexity and energy demand, while silicon IGBTs have cost-efficient high load capability. The focus of this research is set on ease of use, i.e. only one gate pin is used. To aid the optimization of this fusion drivetrain, analytical equations for the inverter losses are presented and used for analyzing cost-performance in the Worldwide Harmonized Light Vehicles Test Cycle (WLTC). A HybridPACK™ Drive G2 Fusion prototype is built up and measurements on static and dynamic performance are presented, revealing a good match of Infineon's CoolSiC™ and EDT3 chip technology for parallel operation.

1   Introduction

Silicon carbide MOSFETs are increasingly applied in automotive inverters due to their advantage in terms of inverter efficiency. This efficiency increase is due to the faster switching transients and unipolar conduction mechanism leading to improved light load efficiency. However, due to higher manufacturing and fabrication costs, this benefit outweighs the higher semiconductor content cost mainly in high performance cars benefitting from a system cost reduction. Therefore, the majority of automotive inverters is still using the mature silicon IGBT technology. The slower switching speed is thereby outweighed by the bipolar forward characteristic meeting power requirements at a lower price point. Moreover, the full potential of silicon carbide MOSFETs in automotive inverter applications isn't always realized due to switching speed restrictions imposed by EMC and load demands. Their unipolar nature renders them more susceptible to ringing compared to their bipolar counterparts. Thus, the pursuit of finding a sweet spot between cost and performance, particularly under high and light load conditions, by means of parallelizing two technologies has become a focal point of research. This

endeavor isn't entirely novel, with investigations dating back to 1993 utilizing silicon MOSFETs and IGBTS [1]. With the advent of silicon carbide technology in recent years, unipolar SiC MOSFETs have entered competition with Si IGBTS across automotive inverter voltage classes (750 V and 1200 V). This has spurred inquiries into paralleling these devices to balance cost and performance, especially under light load conditions. Notably, light load performance plays an important role as drive cycles are emphasizing this operation. Exemplarily, in Fig. 1 the WLTC is depicted (lower plot). The respective torque requirement and the accumulated probability thereof is depicted in the upper figure, underlining the light load focus for the drivetrain evaluation. Investigations are thereby mostly focusing on the turn-on and turn-off sequence of the two devices, introducing two-driver or specialized one-driver concepts. In [2–4] the MOSFET is switching in ZCS and ZVS condition and benefits of the hybrid implementation are gained through the unipolar output characteristic and the improved switching of the IGBT due to different charge carrier concentrations as well as a superior short-circuit robustness. In [5–8] the turn-off signal of the MOSFET is delayed in order to benefit from the reduced turn-off losses of the SiC

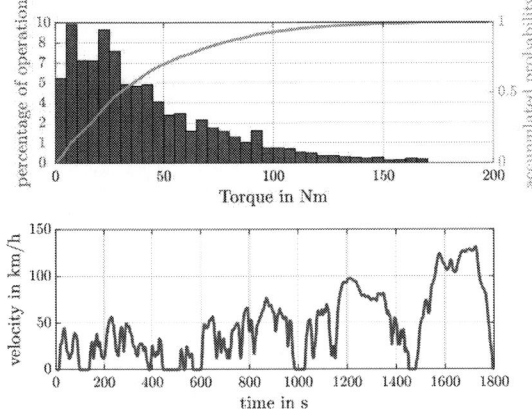

Fig. 1 WLTC drive cycle and exemplary torque requirements.

technology. Reference [9] combines different delays depending on the switched current within one output period. This is a balance between avoiding overcurrent stress in the SiC device and best possible switching losses. In this publication the focus of investigation is on the simplest possible realization of a mixed technology parallelization using the fusion switch like a single-technology device and applying the same gate signal to the respective pins. Therefore, the implementation, in contrast to e.g. [10], does not require specialized gate driving circuits or multiple gate signals and thus existing systems could be easily upgraded to a fusion implementation. In section 2 the on-state behavior of the fusion switch is revisited and the prototype of a fusion module is introduced. Moreover, analytical equations for the conduction losses in inverter operation are derived. In section 3 dynamic measurements are shown including switching loss and short circuit behavior. Then, in section 4 WLTC drive cycle investigations are performed. Finally, the paper is concluded in section 5.

2 On-state behavior

2.1 Basic considerations

The basic idea of a switch combining SiC and Si technology stems from their static output characteristic behavior. Due to its bipolar behavior the Si IGBT/Diode solution enables high output powers at a comparatively low price point. A drawback however emerges at low currents, where the forward voltage drop leads to higher losses compared to the SiC solution. The latter features ohmic behavior and therefore at light load losses are small. However, to be able to conduct full peak power, the SiC solution is costlier. The output characteristic of a fusion switch can be calculated with

$$J_{\text{Fusion}} = J_{\text{SiC}}\xi + J_{\text{Si}}(1 - \xi), \tag{1}$$

Where J_{Fusion}, J_{SiC} and J_{Si} denote the current densities of the fusion switch, silicon carbide MOSFET and the Silicon IGBT or Diode, depending on the respective operation. The ratio of silicon carbide to total area (SiC and either IGBT or Diode) is represented by ξ:

$$\xi = \frac{A_{\text{SiC}}}{A_{\text{Si}} + A_{\text{SiC}}}. \tag{2}$$

With a linear approximation of measured output characteristics of IGBT/Diode and MOSFET it can be written as

$$J_{\text{Fusion}} = \begin{cases} \dfrac{v_{ds}\xi}{r_{\text{SiC}}A_{\text{SiC}}} & \text{for} \quad v_{ds} < v_0 \\ \dfrac{v_{ds} - v_0}{r_{\text{Si}}A_{\text{Si}}}(1 - \xi) + \xi\dfrac{v_{ds}}{r_{\text{SiC}}A_{\text{SiC}}} & \text{for } v_{ds} \geq v_0 \end{cases}, \tag{3}$$

with v_0 denoting the knee-voltage of the IGBT/Diode and $r_{\text{Si}}A_{\text{Si}}$ and $r_{\text{SiC}}A_{\text{SiC}}$ describing the area normalized differential resistances of the respective technologies. For the design of the fusion switch the current sharing between the two devices is of major interest. With the simplified output characteristic as derived in (3) the current share of the SiC MOSFET can be calculated to

$$\frac{I_{\text{SiC}}}{I_{\text{Fusion}}} = \begin{cases} 1 & \text{for} \quad v_{ds} < v_0 \\ \dfrac{\xi}{\left(1 - \dfrac{v_0}{v_{ds}}\right)\dfrac{r_{\text{SiC}}A_{\text{SiC}}}{r_{\text{Si}}A_{\text{Si}}}(1 - \xi) + \xi} & \text{for } v_{ds} \geq v_0 \end{cases}. \tag{4}$$

For low currents only the SiC MOSFET is conducting current and helps improve light-load efficiency. For higher currents the current share of the MOSFET drops asymptotically to the value defined by the area ratio and technology parameters

$$\lim_{v \to \infty} \frac{I_{\text{SiC}}}{I_{\text{Fusion}}} = \frac{\xi}{\dfrac{r_{\text{SiC}}A_{\text{SiC}}}{r_{\text{Si}}A_{\text{Si}}}(1 - \xi) + \xi}. \tag{5}$$

2.2 Fusion prototype

A prototype of a HybridPACK™ Fusion module is built up to investigate the performance. Note that this prototype is not optimized in terms of module layout and chip sizes. Available chips and DCB layouts of other single-technology modules are used. A photograph of the module is depicted in Fig. 2. It is a six-pack module and consists of one

Fig. 2 HybridPACK™ Drive G2 Fusion prototype (**layout not optimized**).

IGBT chip of 160 mm² and two diode chips, each 59 mm², in parallel to three second generation 750 V CoolSiC™ chips with a total area of around 60 mm² per switch. In Fig. 3 the output characteristic of the fusion module is depicted along with the output characteristic of a FS1150 IGBT module with 320 mm² IGBT area and a FS01 full silicon-carbide module with a total area of 196 mm². Thereby the measured values as well as the linear approximation (see Eq. (3)) is shown. Also, the calculated values for the fusion module with Eq. (1) match well with the measured output characteristic. It can be seen that the fusion module has substantially lower losses in the light load region compared to a silicon module while high load performance is only slightly worse. With an area of 160 mm² Si IGBT and 60 mm² SiC MOSFET the fusion prototype has a ξ of around 0.27 for first quadrant operation.

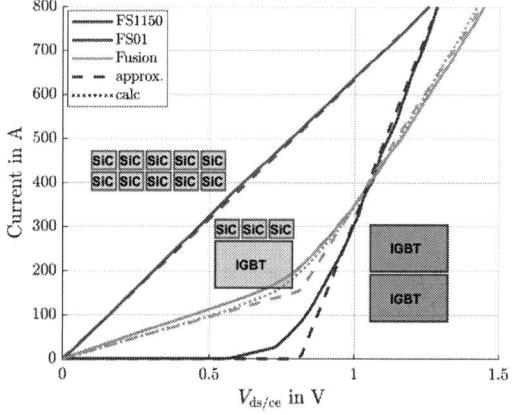

Fig. 3 Output characteristic of Full-SiC (FS01), Full-Si (FS1150) and Fusion prototype ($T_i = 25°C$).

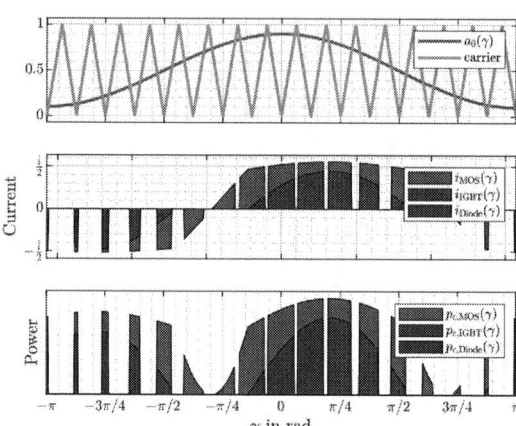

Fig. 4 Exemplary high-side switch waveforms for Fusion SPWM.

2.3 Inverter conduction losses

In order to be able to judge about the losses in inverter operation, the inverter conduction losses shall be derived in analytical form. The calculation is based on the simplified output characteristic as shown in section 2 (Eq. (3)). They are calculated according to [11] with

$$\overline{P}_c = \frac{1}{2\pi} \int_0^{2\pi} \overline{p}_c(\gamma)d\gamma, \tag{7}$$

where $\overline{p}_c(\gamma)$ denotes the averaged conduction loss of the high-side switch within a pulse period. Assumptions hereby include sinusoidal pulse width modulation (SPWM) and a high pulse ratio. Figure 4 shows the respective waveforms of a fusion inverter with a pulse ratio of 16. At the top the reference signal along with the carrier signal is depicted. In the middle the current waveforms are shown for a power factor of $\cos(\varphi) = 0.8$. It can be seen that corresponding to the output characteristic for low currents only the MOSFET is conducting the current (see Eq. (4)). At an angle from the current zero crossing,

$$\rho = \text{asin}\left(\frac{v_0}{r_{\text{SiC}}\hat{i}}\right), \tag{8}$$

the IGBT/Diode starts conducting current in its respective half-wave. Then the current waveform of the MOSFET flattens out as current is shared between the two devices. The bottom figure shows the respective power dissipation $p_c(\gamma)$ in the devices underlining the resistive behavior of the MOSFET. Analytical results of the calculation are summarized in Table 1. Negative signs are valid for the negative half-wave of the current, and therefore MOSFET/Diode operation. The constants are defined as

$$C_1 = \frac{r_{\text{SiC}}}{4\pi(r_{\text{SiC}} + r_{\text{Si}})^2}; \quad C_2 = \frac{r_{\text{Si}}}{4\pi(r_{\text{SiC}} + r_{\text{Si}})^2}. \quad (9)$$

Note that the formulas collapse to the well-known MOSFET inverter formulas for $\rho = \pi/2$ and to the IGBT/Diode formulas for $\rho = 0$ as expected.

3 Dynamic behaviour

For experimental validation of the dynamic behavior of the fusion technology the HybridPACK™ Drive G2 Fusion prototype (see Fig. 2) is measured in a double pulse test setup at different temperatures. A comprehensive overview over the switching transients at a current of around 550 A of the Full-Si, Full-SiC and Fusion module is given in Fig. 7. Thereby characterization is performed on the same setup with identical stray inductance. For each module gate resistors are adjusted to limit overvoltages in the system in the defined operating conditions. In terms of EMI and ringing the silicon solution shows a soft switching behavior, that can be attributed to the limited turn-off speed as well as technology specific improvements and the Si diode stored charge that leads to a soft turn-on transition. The SiC module however shows higher oscillations mainly driven by the faster turn-off speed, lacking tail current and the snappier body diode at turn-on. These oscillations and overvoltages limit the switching speed especially during turn-on, thus preventing to gain full loss reduction

potential of WBG technology. The turn-off waveform of the fusion module is very similar to the silicon module. However, the tail current is lower and therefore the resulting turn-off losses are reduced. This effect is already discussed in [3] and attributed to the reduced charge carrier concentration in the IGBT. The difference in the turn-on waveforms however is even more striking. The Fusion module shows again a similar characteristic like the silicon module, however a reduction in current peak is visible due to the reduced diode size. Compared to the SiC module the reverse recovery charge of the Si diode supports a smooth transition of the current despite a much faster turn-on speed, thus allowing a reduction of switching losses. In the next section the turn-off switching is investigated in more detail. [12]

3.1.1 Turn-off switching

The turn-off switching and subsequent short circuit behavior is investigated in an application near setup depicted in Fig. 5. The respective gate driver configuration is shown in Fig 6. Thereby, Infineon's latest generation isolated gate driver EiceDriver™ 1EDI3035AS is used for switching the single gate.

$r_{\text{SiC}}\hat{\imath} < v_0$ MOSFET	$\overline{P}_{\text{c,SiC}}^{\pm} = \dfrac{r_{\text{SiC}}\hat{\imath}^2}{8} \pm \dfrac{r_{\text{SiC}}\hat{\imath}^2 M\cos(\varphi)}{3\pi}$	(10)
	$\overline{P}_{\text{c,SiC,1}}^{\pm} = \dfrac{r_{\text{SiC}}\hat{\imath}^2}{4\pi}\left[\rho - \dfrac{1}{2}\sin(2\rho)\right] \pm \dfrac{r_{\text{SiC}}\hat{\imath}^2 M\cos(\varphi)}{24\pi}[-9\cos(\rho) + \cos(3\rho) + 8]$	(11)
$r_{\text{SiC}}\hat{\imath} > v_0$ MOSFET	$\overline{P}_{\text{c,SiC,2}}^{\pm} = C_1\left\{v_o^2[\pi - 2\rho] + 4v_o r_{\text{Si}}\hat{\imath}\cos(\rho) + \hat{\imath}^2 r_{\text{Si}}^2\left[\dfrac{\pi}{2} - \rho + \dfrac{1}{2}\sin(2\rho)\right]\right\}$	(12)
	$\overline{P}_{\text{c,SiC,3}}^{\pm} = \pm C_1 M\cos(\varphi)\left\{2v_o^2\cos(\rho) + v_o r_{\text{Si}}\hat{\imath}[\pi - 2\rho + \sin(2\rho)] + \dfrac{\hat{\imath}^2 r_{\text{Si}}^2}{6}[9\cos(\rho) - \cos(3\rho)]\right\}$	(13)
	$\overline{P}_{\text{c,Si,1}}^{\pm} = \dfrac{-v_0^2}{2\pi(r_{\text{SiC}} + r_{\text{Si}})}\left[\dfrac{\pi}{2} - \rho \pm M\cos(\varphi)\cos(\rho)\right]$	(14)
$r_{\text{SiC}}\hat{\imath} > v_0$ IGBT/ Diode	$\overline{P}_{\text{c,Si,2}}^{\pm} = \dfrac{\hat{\imath} r_{\text{SiC}} v_0}{8\pi(r_{\text{SiC}} + r_{\text{Si}})}\{4\cos(\rho) \pm M\cos(\varphi)[\pi - 2\rho + \sin(2\rho)]\}$	(15)
	$\overline{P}_{\text{c,Si,3}}^{\pm} = C_2\left\{v_o^2[\pi - 2\rho] - 4v_o r_{\text{SiC}}\hat{\imath}\cos(\rho) + \hat{\imath}^2 r_{\text{SiC}}^2\left[\dfrac{\pi}{2} - \rho + \dfrac{1}{2}\sin(2\rho)\right]\right\}$	(16)
	$\overline{P}_{\text{c,Si,4}}^{\pm} = \pm C_2 M\cos(\varphi)\left\{2v_o^2\cos(\rho) + v_o r_{\text{SiC}}\hat{\imath}[\pi - 2\rho + \sin(2\rho)] + \dfrac{\hat{\imath}^2 r_{\text{SiC}}^2}{6}[9\cos(\rho) - \cos(3\rho)]\right\}$	(17)

Table 1 Analytical formulas for Fusion conduction loss in inverter operation using SPWM.

PCIM Europe 2024, 11– 13 June 2024, Nuremberg   DOI: 10.30420/566262369

Fig. 5 Setup for short-circuit and switching measurements.

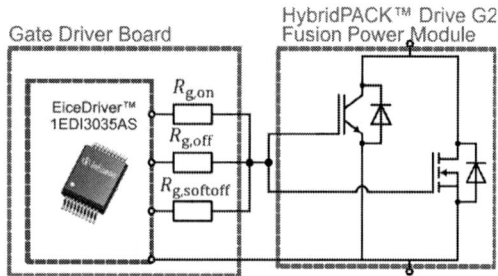

Fig. 6 Configuration of HybridPACK™ Drive G2 Fusion prototype.

Transient waveforms for a current of 600 A and a dc link voltage of 400 V can be seen in Fig. 8 underlining the smooth turn-off behavior of the Fusion switch. This shows a good match of Infineon's EDT3 750 V IGBT technology and second generation CoolSiC™ MOSFET trench technology in a

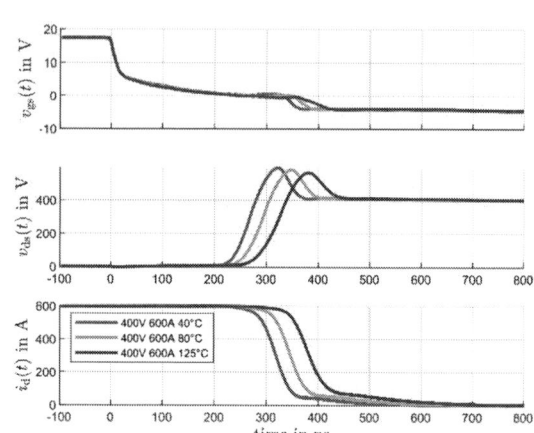

Fig. 8 Transient turn-off waveforms of Fusion module.

Fusion application. Higher temperature thereby has an influence on the commutation speed that can mostly be attributed to the IGBT switching behavior. In Fig. 9 the switching energies of the turn-off transition of the fusion switch are shown in dependency of the switched current for different temperatures. Thereby, a low temperature dependency for small currents can be observed, as the MOSFET takes the majority of the current and therefore the transient is mainly defined by the MOSFET commutation. At some point a change of slope is visible and energies start differing from each other for different temperatures. This is when the IGBT takes over current and contributes significantly to the turn-off transient. Note that this change of slopes matches well with the change of slopes in the output characteristic of the switch (see Fig. 3 for 25°C). This shape of turn-off losses

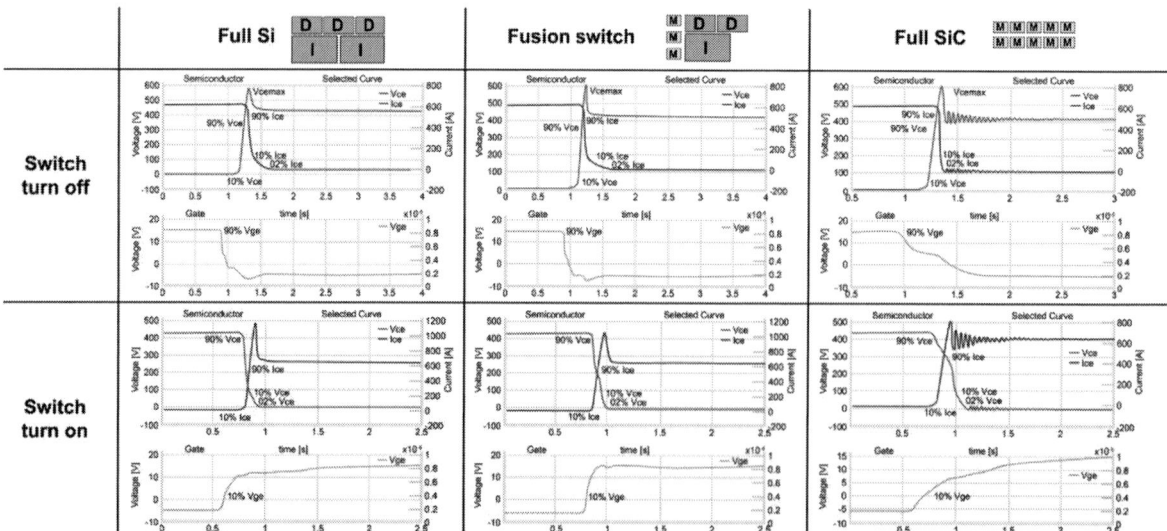

Fig. 7 Switching waveforms of fusion module, full SiC and full Si modules [12]

2642

PCIM Europe 2024, 11– 13 June 2024, Nuremberg   DOI: 10.30420/566262369

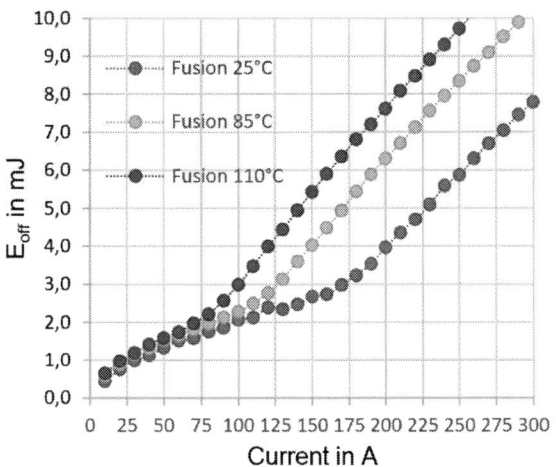

Fig. 9   Turn-off energies of Fusion module.

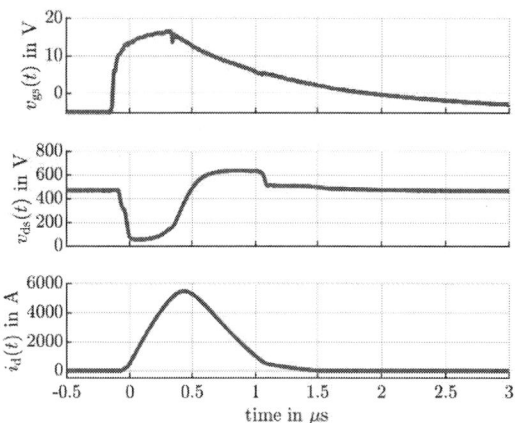

Fig. 10   Short circuit type 1 experimental results using EiceDriver™ 1EDI3035AS.

therefore again underlines the improved light-load efficiency of a fusion switch implementation similar to the improvement in the conduction losses. This is achieved by using a simple gate driver implementation as shown in Fig. 6 that does not differ from the implementation in Full-Si or Full-SiC modules. This is believed to be a key enabler for adopting fusion technology in the market.

3.1.2   Short circuit

Fulfilling functional safety goals for traction inverters usually requires the power devices to withstand short circuit conditions. Both technologies, SiC MOSFET and Si IGBTs individually, are short-circuit rugged. However, in combination, ringing or uneven current sharing might lead to early destruction. Therefore, it is mandatory to assess short circuit capability of a fusion switch in a wide range of conditions. The HybridPACK™ Drive G2 Fusion power module prototype is tested in a short circuit type 1 test according to AQG324 [13]. The measured waveform is shown in Fig. 10 at 470 V dc-link voltage. Like in the switching test, the isolated gate driver EiceDriver™ 1EDI3035AS with optimized DESAT detection and soft-turn-off feature is applied for this test (see Fig. 5 and Fig. 6). The supply voltage and turn-on resistor are set to 17.5 V and 0.5 Ω resulting in a fast turn-on slope. At 400 ns the gate driver detects the short circuit and activates the soft-turn-off feature. The soft turn-off resistor is 22 Ω in this experiment leading to a very smooth turn-off with minor turn-off overvoltage (compare maximum v_{ds} voltage with 750 V device rating). A short circuit time of 1.1 μs is achieved with about 1.5 J short circuit energy, which is uncritical for the applied Fusion power switches. The experiment demonstrates the ease of use of this Fusion configuration with latest standard gate drivers. Simplicity of the gate drive

unit as well as the ability to maintain the ruggedness of the system under short circuit conditions are clear benefits of a fusion switch approach in a simultaneous switching configuration. [12]

4   Drive cycle considerations

In order to judge about the system cost benefit of a fusion inverter design, Eqs. (10-17) derived in section 2.3 are used to evaluate the energy losses due to conduction of a compact car in the WLTC cycle (see Fig. 1) in a very simplified manner. The junction temperature is thereby kept constant at 80°C. Silicon carbide and silicon area are parameters sweeping from the Full-Si module to the Full-SiC module already investigated in this research (Ratio of IGBT area to overall silicon area is assumed 0.7) and the energy losses for the specific range are multiplied with the battery price. Together with the semiconductor costs the relative cost benefit on system level related to a Full-SiC solution with 200 mm² is calculated. The evaluation is depicted in Fig. 11, where areas that are not likely to be reached with the module layout are greyed out. It can be seen that for this configuration in terms of conduction losses the fusion switch offers system cost benefits up to 20% with regard to the Full-SiC solution in a wide range of semiconductor areas. By applying more silicon carbide area for a given silicon area the system costs decrease until a point where the increased cost of the silicon carbide area outweighs the decreasing conduction loss benefit. Moreover, it can be concluded that the silicon area does not have a big impact on conduction losses, except for low areas applied, in the WLTC drive cycle as expected. Therefore, the silicon area is a design parameter that should be designed to reach the desired power class of the inverter and meet thermal constraints in the relevant operating points. It should

2643

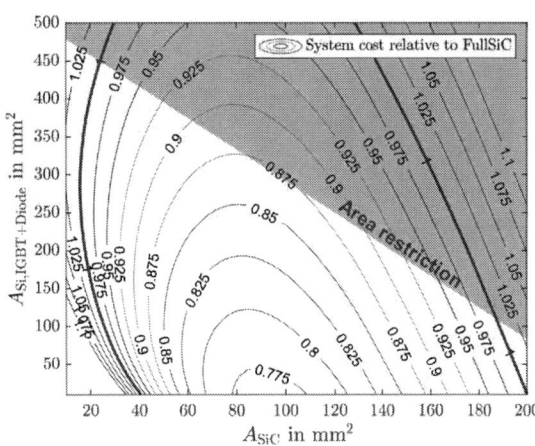

Fig. 11 Exemplary relative cost of Fusion solution with regard to Full-SiC solution ($T_j = 80°C$, $C_{bat} = 80€/kWh$) for a range of 200 km (conduction only)

be noted that range and battery price have a big influence on the system cost value proposition as already shown in [14]. Also, the calculation is based on the linearly approximated silicon output characteristic, thereby worsening the light load result for silicon (compare to Fig. 3).

5 Conclusion

This publication presents investigations on a Si/SiC mixed technology module using only one gate signal, the so-called Fusion module. This represents the simplest form of a combined switch circumventing specialized gate drivers and control strategies. Analytical results for the inverter conduction losses are presented. A non-optimized prototype of a HybridPACK™ fusion is investigated in terms of switching behavior. Thereby Infineon's second generation 750 V CoolSiC™ and 750V EDT3 IGBT technology are placed in parallel. Switching results reveal a good match between both technologies, resulting in clean switching waveforms and reduced switching losses. Moreover, investigations on the short circuit behavior do not show signs of both technologies mutually influencing their respective short circuit ruggedness. Furthermore, system cost benefits regarding conduction losses are calculated with the derived formulas revealing system cost benefits of Fusion technology with respect to the WLTC drive cycle.

Acknowledgements

Part of the work was funded within the frame of IPCEI ME/CT – Important Project of Common European Interest - by BMWK, the European Union - NextGenerationEU, the Bavarian State Ministry for Economic Affairs, Regional Development and Energy, Ministry of Economic Affairs, Climate Protection and Energy of the State of North Rhein-Westphalia.

References

[1] Y. Jiang, G. C. Hua, E. Yang and F. C. Lee, "Soft-switching of IGBTs with the help of MOSFETs in bridge-type converters," in *IEEE Power Electronics Specialist Conference - PESC*, Seattle, 1993.

[2] F. Kayser and H.-G. Eckel, "Event-Triggered Gate Drive for a 1.7kV Si-SiC Hybrid Switch with IGBT-like Short-Circuit Robustness," in *25th European Conference on Power Electronics and Applications (EPE'23 ECCE Europe)*, Aalborg, 2023.

[3] F. Kayser, R. Baburske , P. Brandt , U. Queitsch and H.-G. Eckel, "Hybrid Switch with SiC MOSFET and fast IGBT for High Power Applications," in *PCIM Europe digital days; International Exhibition and Conference for Power Electronics, Intelligent Motion, Renewable Energy and Energy Management*, 2021.

[4] F. Kayser, F. Pfirsch, F.-J. Niedernostheide, R. Baburske and H.-G. Eckel, "Novel Si-SiC hybrid switch and its design optimization path," *2022 IEEE 34th International Symposium on Power Semiconductor Devices and ICs (ISPSD)*, pp. 225-228, 2022.

[5] R. E. Mathieson, P. D. Judge and S. Finney, "Si/SiC Hybrid Switch for Improved Switching and Part-Load Performance," in *21st Workshop on Control and Modeling for Power Electronics (COMPEL)*, Aalborg, 2020.

[6] X. Song, L. Zhang and A. Q. Huang, "Three-Terminal Si/SiC Hybrid Switch," in *IEEE Transactions on Power Electronics*, 2020.

[7] X. Song, A. Q. Huang, M.-C. Lee and C. Peng, "High Voltage Si/SiC Hybrid Switch: An Ideal Next Step for SiC," in *IEEE 27th International Symposium on Power Semiconductor Devices & IC's (ISPSD)*, Hong Kong, 2015.

[8] A. Piccioni, "SiC MOSFET Assisted Si IGBT 1200 V Switch for 3-phase DC-AC Converters with Overload Condition," in *PCIM Europe 2023; International Exhibition and Conference for Power Electronics,*

Intelligent Motion, Renewable Energy and Energy Management, Nuremberg, 2023.

[9] Z. Peng, J. Wang , Z. Liu, Z. Li, Y. Dai, G. Zeng and Z. J. Shen, "A Variable-Frequency Current-Dependent Switching Strategy to Improve Tradeoff Between Efficiency and SiC MOSFET Overcurrent Stress in Si/SiC-Hybrid-Switch-Based Inverters," in *IEEE Transactions on Power Electronics*, 2021.

[10] STMicroelectronics, "A high-power inverter based on hybrid switch SIC+IGBT technology," 2024. [Online]. Available: https://www.st.com/content/dam/static-page/events/apec-2024/exhibitor-seminar-apec24-hybrid-gate-driver.pdf. [Accessed 11 04 2024].

[11] J. W. Kolar, H. Ertl and F. C. Zach, "Calculation of the passive and active component stress of three phase pwm converter systems with high pulse ratio," in *Power Electronics and Applications (EPE)*, 1989.

[12] M. Münzer, W. Jakobi, M. Niendorf, M. Ippisch and T. Reiter, "Range impact & peak power of power semiconductor fusion," in *37th International Electric Vehicle EVS37 Symposium*, Seoul, Korea, 2024.

[13] E. G. AQG324, "Qualification of Power Modules for Use in Power Electronics Converter Units in Motor Vehicles; Release no: 03.1/2021," in *online*.

[14] M. Walter and M.-M. Bakran, "Hybrid-Switch-Inverter - A New Approach Reducing the System Cost of the Electric Powertrain," *PCIM Europe 2023; International Exhibition and Conference for Power Electronics, Intelligent Motion, Renewable Energy and Energy Management,* pp. 1-10, 2023.

PCIM Europe 2024, 11– 13 June 2024, Nuremberg   DOI: 10.30420/566262370

Concise and Reliable SiC MOSFET Driver Circuits

Zhong Ye, Hailong Yang, Hongzi Yang

InventChip, China

Corresponding author: Zhong Ye, zhong.ye@inventchip.com.cn
Speaker: Zhong Ye, zhong.ye@inventchip.com.cn

Abstract

The high dv/dt of SiC MOSFET switching and the Miller effect can cause gate driver circuit loop ringing. Excessive ringing is the culprit for most SiC MOSFET failure in applications. This article presents a family of 6 and 8-pin SiC MOSFET driver ICs, which integrates the most needed features, including negative bias, desaturation protection and active Miller clamp. The driver ICs require very few external components and are able to drive paralleled MOSFETs with excellent dynamic current sharing. An 11kW compressor drive and a double-pulse board with 4 MOSFETs in parallel were designed to demonstrate the performance of the driver circuit.

1   Introduction

SiC MOSFETs have just been widely used in high voltage and high power applications. The speed of SiC MOSFET adoption is unprecedented and engineers are struggling to keep up with the new technology. Engineers with Si MOS and IGBT experience now have to work on a new territory of higher voltage, higher dv/dt, and even new topologies. IC industry, on the other hand, has created very few SiC MOSFET-specified drivers to facilitate the transition. Most engineers still have to use IGBT driver circuits to try their luck on SiC MOSFETs, but unfortunately the difficulty is significant. The drive circuit is still responsible to most of the MOSFET failure in applications. Compared with IGBTs, SiC MOSFETs can switch much faster and at higher switching frequencies. The high dv/dt and its related Miller effect can induce gate driver circuit loop ringing or voltage spikes. Some level's gate drive voltage disturbance is inevitable, but the disturbance with an excessive positive amplitude can mistrigger the MOSFETs and cause a large power loss or even a fatal shoot through since SiC MOSFETs have a relatively low Vth, especially at high temperature. Excessive negative spike, on the other hand, can cause MOSFET reliability issues. To tackle the problem, using a driver with three features of negative bias, active Miller clamp and or desaturation protection becomes necessary for a reliable driver circuit design[1,2,3]. Ideally, a driv-

er IC should integrate all the features, which are often seen in power module drivers. For a discrete MOSFET, such as in TO-247 package, a driver with a large pin count is mostly not feasible because of PCB space limitation. In terms of cost and driver circuit loop's length, a larger driver is not preferred either. To facilitate discrete MOSFET's drive circuit design, a family of low-pin count drivers was developed. The drivers integrate one or two of the three features mentioned above. By combining two different drivers in half bridge applications, all the three features can be realized.

To extend power levels, multiple MOSFETs are often connected in parallel. Current imbalance among the MOSFETs is the major concern for the application[4,5,6]. Static current sharing is determined by the MOSFETs' Ron while dynamic current is affected by the MOSFET Vth and drive circuit. Since SiC MOSFETs are positive temperature coefficient devices, static current sharing can be achieved naturally to a certain degree. However, Vth is negative temperature coefficient. A MOSFET with a lower Vth can hog the current at both turn-on and turn-off switching edges, which can result in more switching loss. The more the switching loss is, the higher the device temperature become, which in turn causes the Vth to decrease further and maybe device damage at the end. The forementioned drivers are so tiny and compact that the paralleled MOSFETS can be driven individually. By doing so, the MOSFETs are essentially decoupled each other and excellent dynamic current can be achieved even with significant Vth difference. This article introduces the family of drivers in section 2 and provides examples and test results of a compressor design and four MOSFET parallel operation in section 3 and 4. Conclusion is given in section 5.

2 A SiC MOSFET Driver Family

A family of SIC MOSFET drivers was designed to support the fast growing MOSFET applications. Three drivers are selected to showcase their features and special functions.

IVCR1401 is an 8-pin driver with an integrated -3.5V bias and desaturation protection. The driver requires just a single positive power supply. Its internal charge pump generates a negative 3.5V bias within 20us typically, which simplifies external circuit design and offers a reliable turn-off voltage right after the system is powered up. Based on the desaturation characteristic difference between SiC MOSFET and IGBT, IVCR1401 uses higher desaturation threshold voltage and higher constant current source for SiC MOSFET desaturation detection.

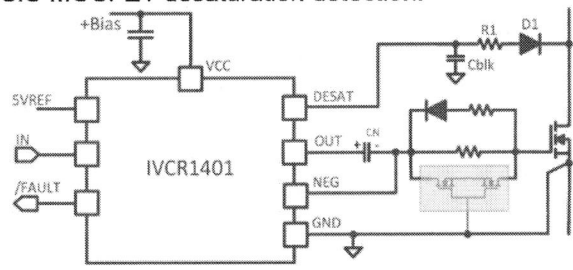

Fig. 1 IVCR1401 application diagram.

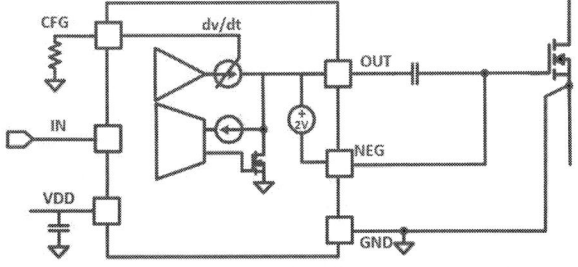

Fig. 2 IVCR1412 circuit and application diagram.

IVCR1412 is a compact 6-pin driver, which aims high density design and to drive paralleled MOSFETs. For SiC MOSFET-based bridge converters, gate drive circuity occupies significant PCB space. The 6-pin driver IVCR1412, as shown in Fig.2, eases the design. Since the driv-

er output is a constant current, which eliminate any gate drive resistors and shorten the gate drive loop. The output driving current can be programmed by CFG pin. Since the driver is very compact, it is suitable to drive paralleled MOSFETs by using one-on-one driving scheme, which effectively decouples the MOSFETs' gate drive circuits to eliminate the impact of the MOSFET Vth difference and achieve excellent dynamic current sharing. Session 4 will provide further details and test data.

For a bridge's top side MOSFET drive, an isolated gate driver is preferred. IVCO1412 was designed to provide an isolation and negative bias. It eases the drive circuit design and reduces the component count.

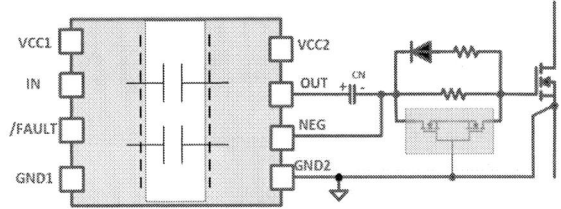

Fig. 3 IVCO1412 application diagram.

3 11kW Compressor Design

To reduce the size and weight and improve efficiency, Electrical Vehicles' air compressors use more and more SiC MOSFETs. A family of SIC MOSFET drivers was designed to ease the circuit design and PCB layout. IVCO1412 is the only available isolated driver which can utilize a bootstrap circuit to power the driver IC and generate a negative bias with internal charge pump for high side drive. The low side driver however can use non-isolated driver IVCR1401, which has an integrated negative bias and short circuit protection capability. Fig. 4 shows the circuit diagram of a high density air compressor design.

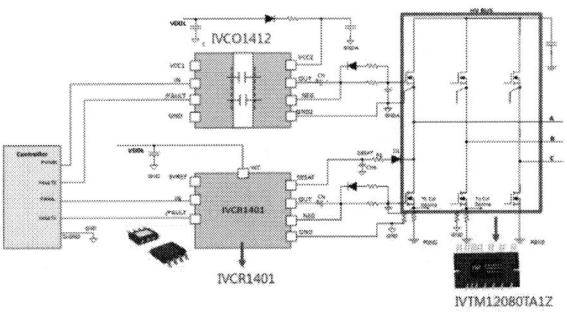

Fig. 4 Concise air compressor design.

It can be seen that very few external components are used to implement three-phase bridge drive. The following photo of Fig. 5 is an 11kw air compressor evaluation board design. A 1200V 80mohm three-phase transfer-mold module IVTM12080TA1Z is used for the power stage.

Fig. 5 11kW air compressor evaluation board.

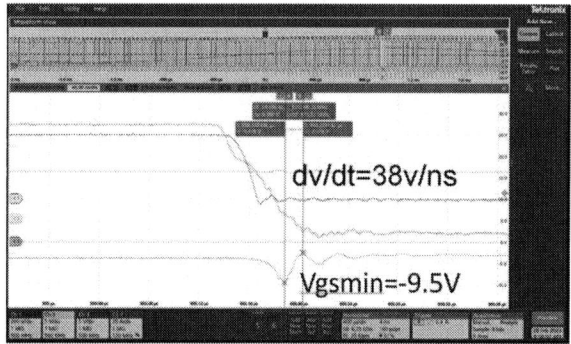

Fig. 6 Switching waveform of turning off at 25A.

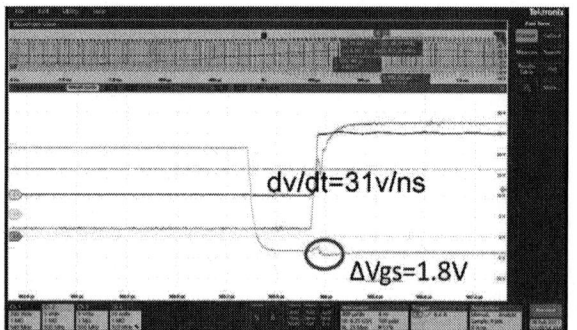

Fig. 7 Switching waveform of turning on at 25A.

The waveforms show that even under a dv/dt over 30V/ns the gate voltage spike is well below specifications.

4 Drive Circuit for Paralleled MOSFETs

MOSFETs in discrete packages, such as TO-247, are very popular and low cost. However, their power handling capability is limited. To extend the power range, the MOSFETs are often connected in parallel. Due to devices' parameter variation, paralleled MOSFETs often suffer from current balancing issues. Current balancing includes dynamic current balancing and static current balancing. The static current balancing is primarily determined by the MOSFETs' Ron, while the dynamic current balancing depends on the gate threshold voltage Vth and transconductance, etc.

In the application with multiple MOSFETs in parallel, adding a gate resistor to each MOSFET, as shown in Fig. 8, is a traditional method to achieve the MOSFETs' current balancing. However, this method actually can't decouple the MOSFETs completely. A larger Vth difference may still impair the current balancing. Adding additional gate resistor, on the other hand, makes the circuit more sensitive to the Miller effect. The location of the driver IC can also affect the current sharing since it is difficult to have even gate drive loop layout for all the MOSFETs. IVCR1412 with little mismatch from device to device was designed to provide one-on-one local drive and achieve excellent dynamic current balancing.

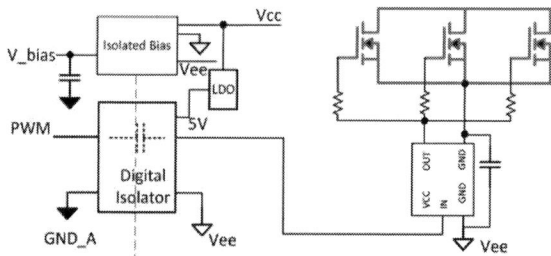

Fig. 8 A traditional method for current sharing.

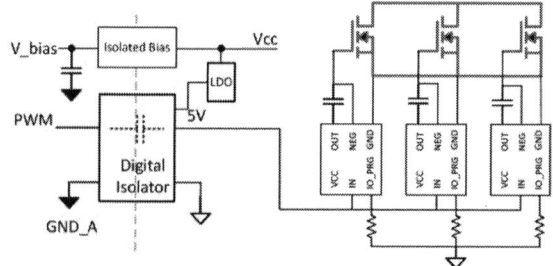

Fig.9 New current sharing control with IVCR1412.

Fig. 8 and 9 show the two different current sharing driver circuits. The IC-based driver circuit in Fig. 9 uses IVCR1412. The tests were conducted on two 4-paralleled-MOSFET double pulse boards with the different gate driver circuits.

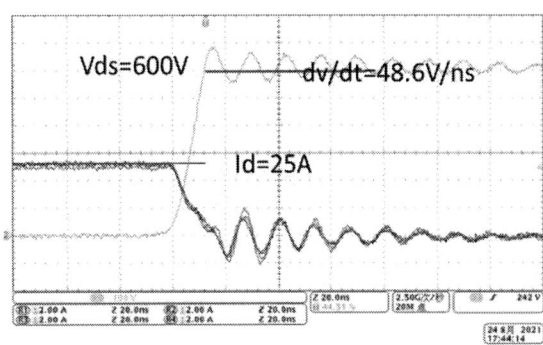

Fig.13 IC-based current sharing at turn off.

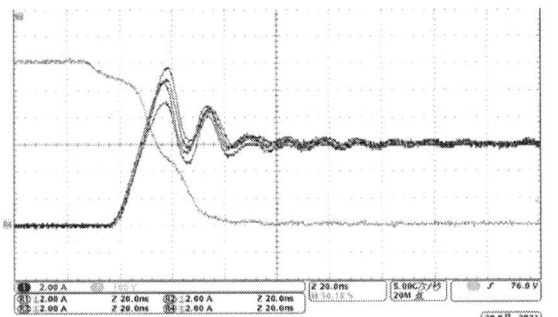

Fig. 10 Resistor-based current sharing at turn on.

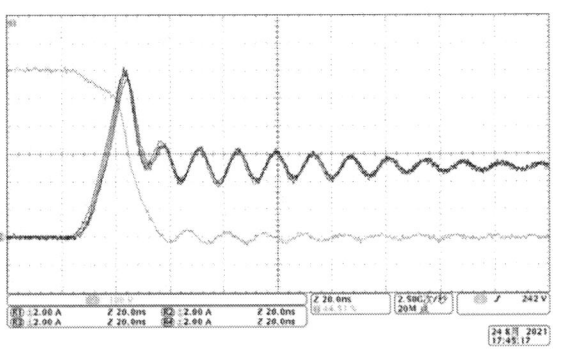

Fig. 11 IC-based current sharing at turn on.

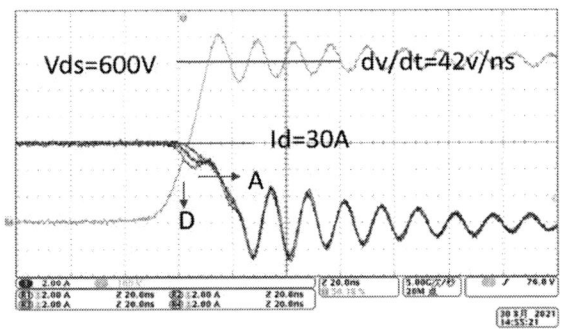

Fig. 12 Resistor-based current sharing at turn off.

The same set of the MOSFETs were used for the tests. The Vth differences were within 0.1V. It can be seen that the dynamic current balancing with resistor-based scheme still has some mismatching, while the current balancing with IC-based scheme can achieve almost perfect current balancing. When Vth difference increases, the current mismatching of the resistor-based scheme increases substantially, but the current balancing remains at the same excellent level for the IC-based scheme even at 0.6V Vth difference. It is evidenced that the IC (IVCR1412)-based driver can achieve much better dynamic current balancing at both turn on and turn off.

5 Conclusions

Almost all MOSFET applications require a negative bias. The driver ICs with integrated negative bias simplify the gate driver circuit and ease the system design. The family of products offers isolation and overcurrent Desat protection, which are necessary for a concise and reliable design. SiC MOSFETs with Vth variation can be paralleled and achieve excellent dynamic current sharing at both turn-on and turn-off edges with the specifically-designed gate driver IC IVCR1412.

References

[1] Qin H.; Wang D. " An overview of SiC MOSFET gate drivers," 2017 12th IEEE Conference on Industrial Electronics and Applications (ICIEA).

[2] L. Alves; P. Lefrance "Review on SiC-MOSFET Devices and Associated Gate Drivers," 2018 IEEE International Conference on Industrial Technology (ICIT)

[3] Julius Rice, John Mookken," SiC MOSFET Gate Drive Design Considerations," 2015 IEEE International Workshop on Integrated Power Packaging (IWIPP)

[4] Wolfspeed "Gate Drivers and Gate Driving with SiC MOSFETs," Oct. 2021 ,https://www.wolfspeed.com/knowledge-center/article/gate-drives-and-gate-driving-with-sic-mosfets/.

[5] H. Li, S. Zhao, X Wang, "Parallel Connection of Silicon Carbide MOSFETs—Challenges, Mechanism, and Solutions" IEEE Transactions on Power Electronics (Volume: 38, Issue: 8, August 2023).

[6] Y. He, X. Wang, S Shao, J. Zhang "Active Gate Driver for Dynamic Current Balancing of Parallel-Connected SiC MOSFETs ," IEEE Transactions on Power Electronics (Volume: 38, Issue: 5, May 2023).

PCIM Europe 2024, 11– 13 June 2024, Nuremberg   DOI: 10.30420/566262371

Artificial Intelligence Enhanced Resolver System for Automotive Traction Inverter Applications Based on AURIX TC4x

Mihail Jefremow[1], David Zipperstein[1], Juergen Schaefer[1], Arndt Voigtlaender[1]

[1] Infineon Technologies AG, Germany

Corresponding author: Mihail Jefremow, mihail.jefremow@infineon.com
Speaker: David Zipperstein, david.zipperstein@infineon.com

Abstract

Continuous performance increase due to scaling of semiconductor technology allows to utilize more artificial intelligence (AI) approaches on embedded devices. This paper focuses on an AI algorithm running on the automotive microcontroller of the new 28nm AURIX TC4x family to implement a novel resolver to digital converter solution. The proposed approach demonstrates how the AI is enhancing the commonly used hardware solution in the automotive industry by utilizing the new computation hardware inside the AURIX TC4x family. By doing so the safety compliance can be still guaranteed due to a redundant path approach and specific hardware features.

1   Introduction

This paper addresses the main traction inverter application, where the resolver system is responsible for motor position detection. The estimated angle of the resolver system plays a key role with a direct impact on overall inverter efficiency and motor torque [1]. The real time performance in this kind of applications is key and therefore the methods and approaches used in this paper are focused on real time execution and data processing. Automotive motor control is the process of regulating the speed, torque, and position of electric motors used in modern vehicles. There are many different motor position sensors that can be used, like encoder, resolver or hall-sensor. There are even sensorless approaches [3]. However, the resolver has proven to be the most robust, reliable and noise insensitive sensor, as described in [4]. In today's automotive motor control it is the most used feedback sensor in automotive applications. It works by using sets of coils in the stator, which are physically offset from each other by 90 degrees, and one coil in the rotor. The rotor coil is excited with a so-called carrier signal, so that amplitude-modulated (AM) sine and cosine waves are induced on the stator coils (please see left part of figure 1). These AM waves need to be demodulated, so that the angle and the speed of the motor can be obtained and used for the motor control.

The angle and speed calculation are usually done by a tracking loop observer.
This paper focuses on the AM signal demodulation part. There are several different possibilities how this can be achieved, like peak detection algorithms, undersampling, PLLs or tracking loops, as described in [4]. There are also machine learning based demodulation approaches which rely solely on artificial neural networks (ANN) using multilayer perceptron networks (MLP) [5-7].

2   State-of-the-art solution for automotive applications

Most common approach in today's automotive applications is to use the Delta-Sigma (DS)ADC in combination with a hardware filter chain similar as described in [2]. Instead of using tracking observer as in [2] the coherent demodulation technique utilizing the hardware of the AURIX family is used as shown in figure 1. Two DSADCs convert the modulated signals coming for the resolver. The pulse density modulated signal from the analog frontend of the DS ADC is filtered by the following CIC filter to create the digital representation of the oversampled analog signal [2].

PCIM Europe 2024, 11– 13 June 2024, Nuremberg DOI: 10.30420/566262371

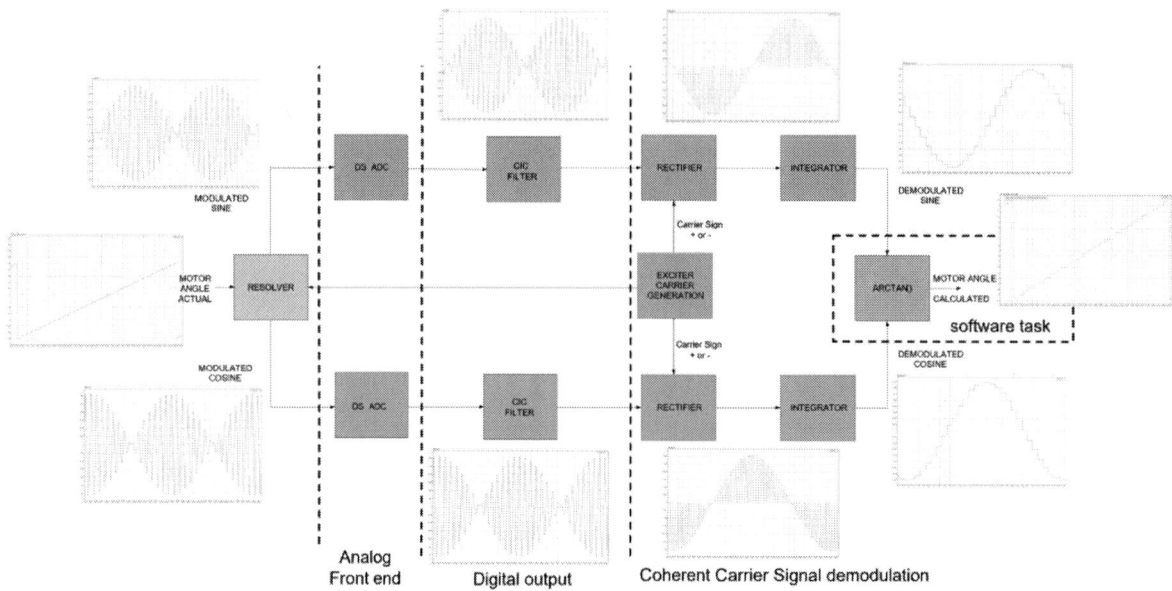

Fig. 1 State-of-the-art hardware solution for automotive resolver to digital converter

The coherent demodulation hardware part requires the so-called rectifier and the integrator. The rectifier is multiplying the digital output data either with +1 or -1 depending on the sign of the carrier signal, which is given the by the carrier generator exciting the resolver sensor.

The rectified signal is then integrated over a fixed period of samples (e.g. 50) to create one output data point of the demodulated signal (demodulated sine or demodulated cosine) as shown in fig.1. To finally obtain the digitally converted angle, the software task executed on main CPU needs to read out the demodulated data from the integrator

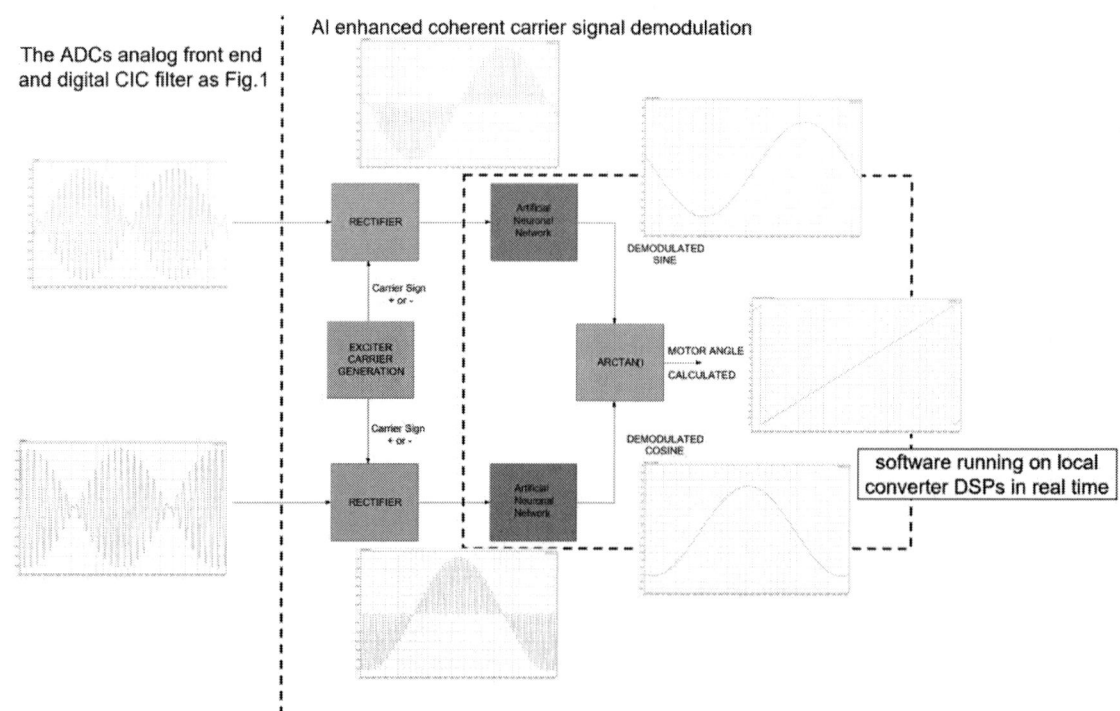

Fig. 2 Proposed AI enhanced automotive resolver to digital converter

2652

and calculate the arc tangent function. This resolver to digital solution has the big advantage of the instantaneous reaction on any speed change due to its open loop structure (no PLL or observer), but it needs a software task to read the integrator values and interpolate the value to the actual system time.

3 Proposed AI enhanced method

Implementation

Figure 2 shows the proposed enhancement of the hardware solution of figure 1 by artificial neuronal networks, which are executed on local converter DSPs (CDSPs) of AURIX TC4x family. This specific new CDSPs are directly located in the ADC cluster and can postprocess incoming ADC data in real time without any main CPU interaction.

The modulated signals coming from ADC hardware are converted and rectified first as in the previous solution. However instead of the hardware integrator an artificial neuronal network is used in form of an autoencoder to construct the demodulated sine and cosine waveforms. Each channel for sine and cosine is using one dedicated CDSP executing the pretrained autoencoder. Due to memory limitation (2kB for data and 2kB memory on each instance) of the CDSPs the ANN based on autoencoder structure is not as complex as it is mainly used for text or picture generation, but rather optimized on the current needs of the application.

Autoencoder structure

Figure 3 show the structure of the used autoencoder. The most resource efficient structure ob-

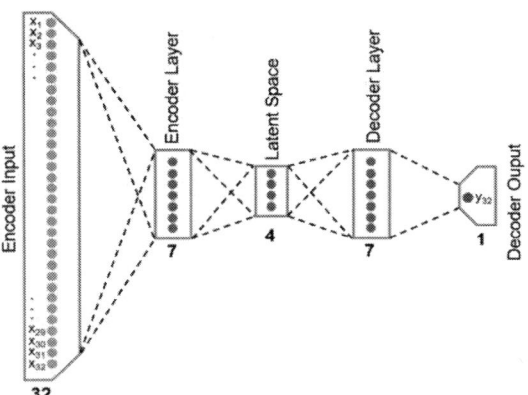

Fig. 3 Autoencoder structure

tained is taking the last 31 DSADC output samples plus the new incoming one (32 input samples in total) of the modulated signal as input and mapping them to 7 neurons of encoder layer and 4 neurons of latent space. To decode again 7 neurons were used, which then create one decoder output data point corresponding to the demodulated output data sample. However, during training the autoencoder was structured like a classical autoencoder with equal input and output dimensions (32 inputs, 32 outputs) to achieve the desired autoencoder properties. These are its inherent data compression and denoising property, which can be

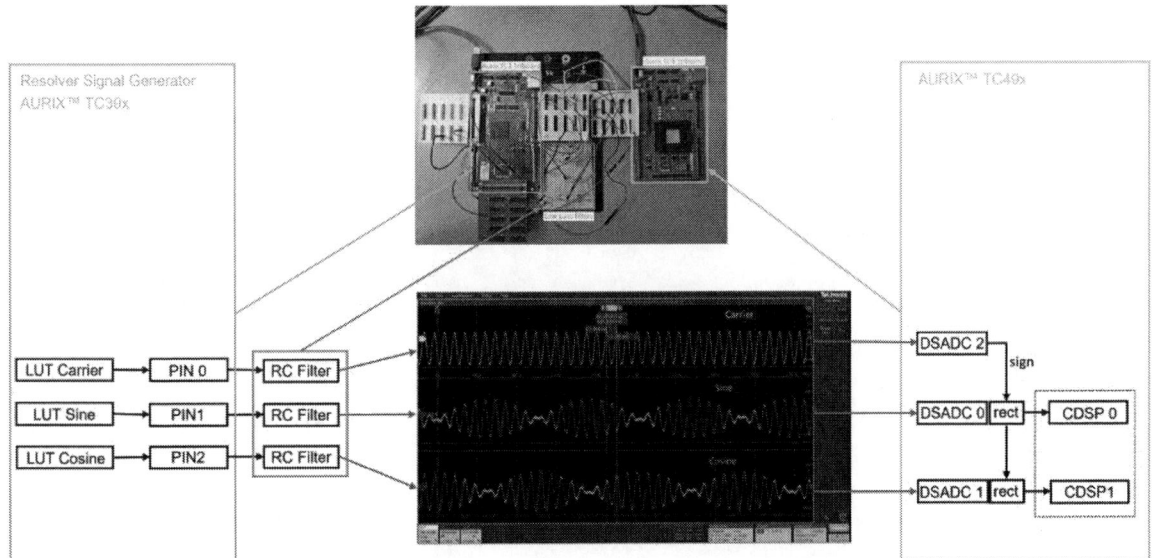

Fig. 4 Hardware in the loop (HiL) setup

seen like a function of a median filter and integrator together, with the big advantage to be trainable for any signal type.

Another very important point why the autoencoder was used instead of MLP structure as proposed by many papers [6,7], was due to the nature of the modulated signal which is becoming zero with each carrier zero cross. Therefore, in final application it cannot be guaranteed that the ADC data output will not be zero for the modulated sine and modulated cosine. MLP structure produced a zero in this case as output, which will create an unforeseen angle estimation error in the system, which is not acceptable in automotive applications requiring a certain ASIL level.

Test Setup

For training and test of the autoencoder setup the hardware in the loop (HiL) setup was created, which can be seen in Figure 4. It uses the previous generation AURIX TC39x to generate resolver signals: carrier, sine and cosine. The corresponding PWM look up tables (LUT) are transferred to the digital output pins. The corresponding pin voltages are low pass filtered by external RC filters to obtain the analog waveforms as shown in the oscilloscope snapshot.

The outputs of the low pass filters serve as inputs for the Delta-Sigma ADCs (DSADC) of the TC49x chip utilizing the proposed AI enhanced method. In this HiL setup the carrier signal was generated externally not by the carrier signal generator of the

TC49x as shown in figures 1 and 2. Therefore the DSADC channel 2 is used only as sign source feeding the rectifiers of the sine DSADC 0 and cosine DSADC 1 channel respectively. This setup allows a very flexible test environment without usage of specialized hardware or any motor bench hardware.

Figure 5 shows the measured rectified signals coming from the DSADC hardware and the processed data from the autoencoder structure executed on both CDSPs (each for cosine and sine signals). It can be clearly observed that the autoencoder structure can remove the high frequency carrier component from the input signal, delivering a clean envelope signal corresponding to the demodulated sine and cosine signals. These measurements also show a speed jump from 12000 rpm to 24000 rpm (electrical speed). The autoencoder solution copes well with speed change without any artifact in the output waveform.

4   Performance Comparison

To verify the performance of the proposed AI enhanced solution as shown in the figure 2 versus the state-of-the-art hardware solution from the figure 1, the HiL was used for the same chip of TC4x AURIX family. The Delta-Sigma ADC of the TC4x family has 14bit effective resolution (higher than 86dB signal-to-noise-ratio) and the modulator stage runs at 40MHz. TC4x supports both methods, the conventional one with integrator after the

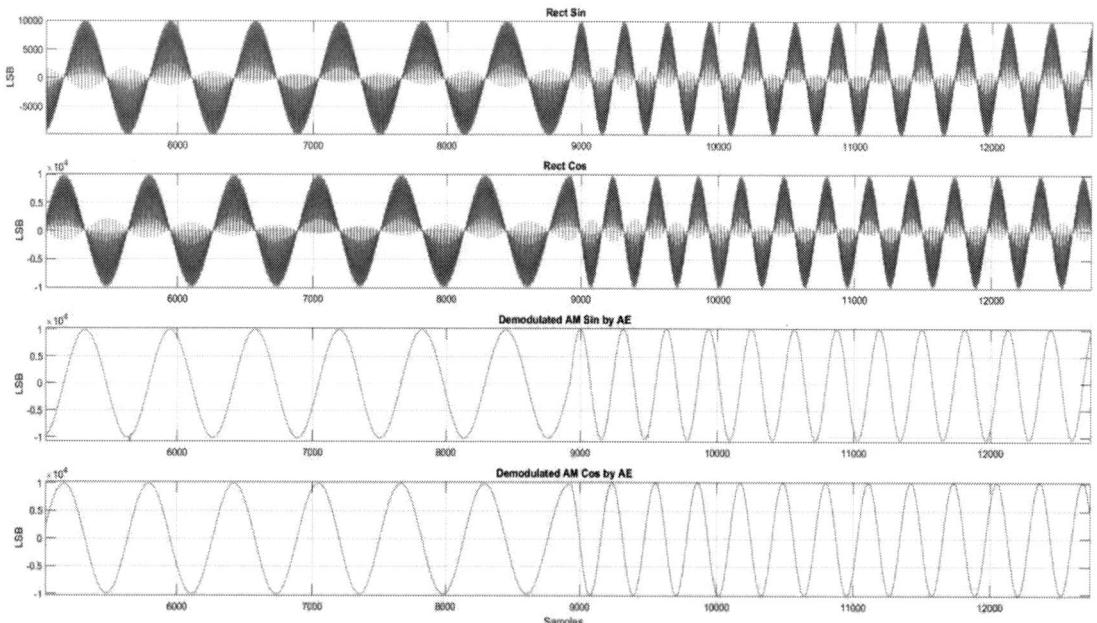

Fig. 5   Measurement results of the proposed AI enhanced method

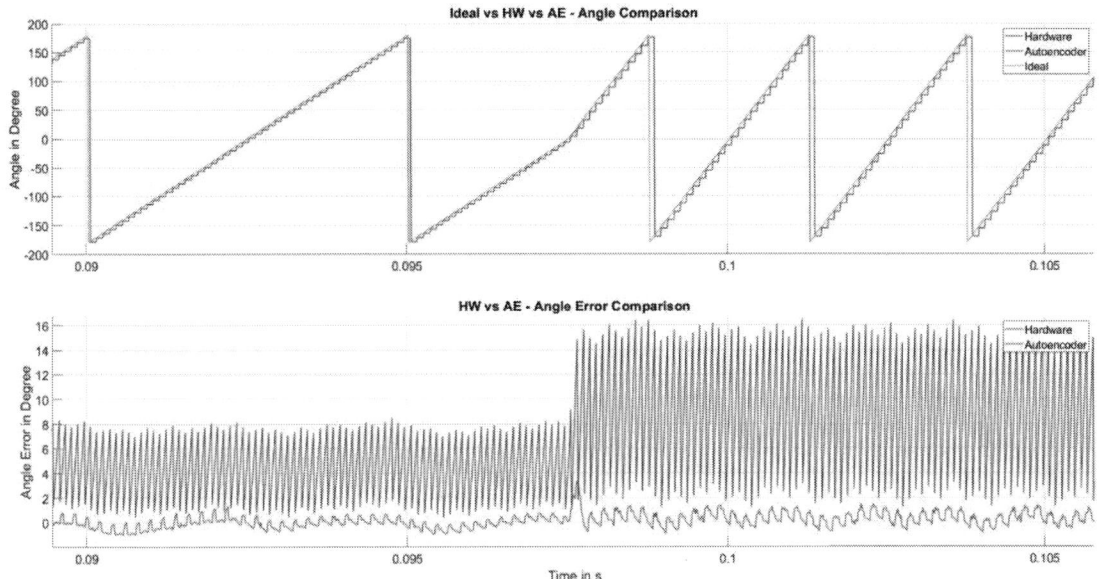

Fig.6 Measurement results state-of-the-art vs AI enhanced method

rectifier and the proposed one with dedicated DSPs executing the autoencoder structure. By doing so the two methods can be compared on the same ADC hardware allowing the pure comparison of the two algorithms without any other hardware or software dependances.

Figure 6 shows the obtained measurement results from both solutions using the same TC4x chip. The absolute angle error vs a reference input to resolver is shown for rotation speed jump from 12000 rpm to 24000 rpm.

It can be clearly observed that the previous pure hardware solution produces significantly higher angle error due to low output data rate. The hardware solution can run only at the carrier speed, which is 10kHz in this case. Because the integrator stage needs to accumulate the incoming ADC samples for the full carrier period to eliminate the high frequency carrier.

The proposed AI enhanced solution can run at 125kHz (more than 12x faster), as there is no need for long accumulation time of incoming samples.

The higher data rate is also beneficial for the settling time, because the autoencoder needs only 32 input samples from CIC filter (4µs) to reach steady state condition. The additional Delta-Sigma ADC hardware delay is 12µs due to the decimation factor of 320 in this case (16µs settling time in total). In contrast the state-of-the-art solution

needs in total 53µs to settle (more than x3 longer compared to AI solution). Although the additional ADC hardware delay is only 3µs, but the integrator

needs to accumulate for 50 samples (one 10kHz carrier period).

5 Conclusion and Summary

This paper proposed an AI enhanced resolver to digital solution utilizing the local converter DSPs of AURIX TC4x family together with dedicated Delta-Sigma ADC hardware.

The deployment of artificial intelligence algorithm in the converter DSPs (CDSP) made it possible to outperform the state-of-the-art resolver to digital converter solution.

The CDSP offers real time AI capabilities without any additional load on the main CPUs providing more MIPS on system level.

The measurement results obtained on the same chip using identical ADC hardware and software configuration stack for both methods clearly indicate the advantage of the proposed solution.

Table 1 summarizes the experimental results obtained on the HiL. The AI enhanced method outperforms the conventional resolver to digital converter solution due to its higher output data rate and almost instantaneous settling time. This is important for braking and acceleration scenarios of the electric vehicle. The low settling time is achieved by using the autoencoder structure with 32 input samples. The higher output data rate reflects in better angle accuracy, which is key for xEV inverter efficiency and torque accuracy.

The proposed AI enhanced method is also compliant with ASIL-D safety requirements, because the AURIX TC4x offers to run both solutions in parallel in fault tolerant time intervals for plausibility checks. E.g. using the previous HW based solution as monitor channel and the AI enhanced method as mission channel to benefit from better accuracy and faster settling time.

	State-of-the-art	AI enhanced
Output data rate in [kHz]	10	125
Settling time in [µs]	53	16
Mean angle error in degree (no interpolation)	4.2	0.45
Standard Deviation Angle Error in degree (no interpolation)	6.5	0.6

Table 1   Measurement results summary

References

[1] S. Gopalakrishnan, L. Hao, C. Namuduri, K. Rahman, A. Omekanda and C. Freitas, "Impact of position sensor accuracy on the performance of propulsion IPM drives," 2015 IEEE International Electric Machines & Drives Conference (IEMDC), Coeur d'Alene, ID, USA, 2015, pp. 946-952

[2] Krah, Jens & Schmirgel, Heiko & Albers, Marcel , "FPGA Based Resolver to Digital Converter Using Delta-Sigma Technology", 2006 PCIM EUROPE

[3] G. Boztas and O. Aydogmus, "ANN-Based Observer for Controlling a SynRM," 2018 International Conference on Artificial Intelligence and Data Processing (IDAP), Malatya, Turkey, 2018, pp

[4] M. Nemec and V. Ambrožič, "Comparison of different RDC techniques," 2015 9th International Conference on Compatibility and Power Electronics (CPE), Costa da Caparica, Portugal, 2015, pp. 373-377, doi: 10.1109/CPE.2015.7231104.

[5] P. Gaur et al., "A novel method for extraction of speed from resolver output using neural network in vector control of PMSM," India International Conference on Power Electronics 2010 (IICPE2010), New Delhi, India, 2011, pp. 1-7, doi: 10.1109/IICPE.2011.5728159.

[6] R. Celikel, ANN based angle tracking technique for shaft resolver, (2019), https://doi.org/10.1016/j.measurement.2019.106910.

[7] M. KhajueeZadeh, M. Emadaleslami and Z. Nasiri-Gheidari, "A High-Accuracy Two-Stage Deep Learning-Based Resolver to Digital Converter," 2022 13th Power Electronics, Drive Systems, and Technologies Conference (PEDSTC), Tehran, Iran, Islamic Republic of, 2022, pp. 71-75, doi: 10.1109/PEDSTC53976.2022.9767427.

PCIM Europe 2024, 11– 13 June 2024, Nuremberg   DOI: 10.30420/566262373

Multifunctional Grid Manager Topology with Configurable Output

S.J.C. Koning[1], D.C. Zuidervliet[1], P.J. van Duijsen[1]

[1] THUAS, DC-Lab, Delft, The Netherlands

Corresponding author: Peter van Duijsen, p.j.vanduijsen@hhs.nl
Speaker: Peter van Duijsen, p.j.vanduijsen@hhs.nl

Abstract

In this paper two multi-functional Grid Managers are presented. They are capable of controlling the output voltage, or bidirectional power flow. The grid manager replaces the traditional fuse-box in an AC grid, but includes much more functionality. This includes voltage regulation and power flow control, utilizing droop control. Next to droop control, there is also the requirement to disconnect active producers, such as solar panels and batteries, in case of a grid fault. The Grid Manager includes short-circuit and earth-leakage protection. Pre-charging and inrush current limitation are implemented on the outputs of the Grid Manager. The functionality of the Grid Manager is demonstrated experimentally in a Tiny-House, where an islanding low voltage DC grid is implemented.

1 Introduction

The Grid Manager is the converter between the DC Grid [1] and the prosumers, see Fig. 1.
The grid manager controls the power flow, to and from the prosumers in a DC grid [2]–[4]. Among

the prosumers are passive (LED) consumers, active (phone and laptop chargers) consumers, active (solar panels) producers, but also active (battery charging/discharging) prosumers.

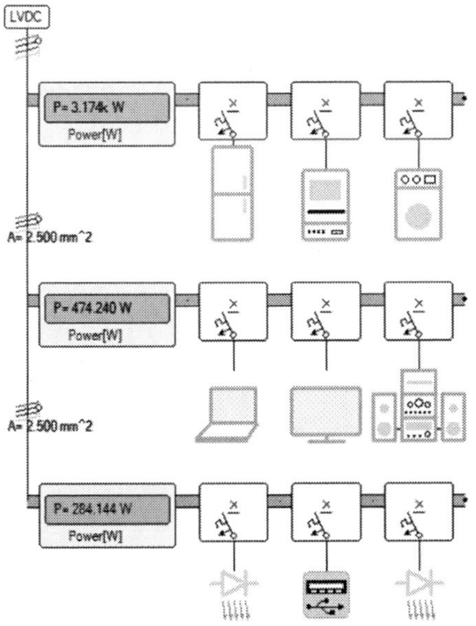

In Fig. 1, three Grid Managers connected to a DC grid are shown. The upper Grid Manager regulates the power flow into a kitchen [5], [6]. The middle Grid Manager supplies power to consumer electronics like phone chargers and television sets. The third Grid Manager supplies LED lighting and USB outlets.

This paper introduces a Grid Manager in a stand-alone self sustaining grid. This Grid Manager is capable to control the bidirectional power flow and output voltage. Different topologies are used for the Grid Manager. The producers and prosumers use bidirectional Dual Active Half Bridges to isolate the sources in case of a fault. Synchronous buck converters are used for passive and active consumers. Both can regulate power flow but the Dual Active Half Bridge has additional features that are necessary for a Grid Manager.

The topologies found in a Grid Manager are described in section 2, together with simulations with calculated wave forms. Section 3 states the protection and safety measures needed in a Grid Manager. Finally, an use case is demonstrated in section 4, where a Tiny House in a rural area is used to implement the grid manager.

Fig. 1: Three Grid Managers, each having power metering, supplying kitchen appliances, consumer electronics and LED lighting and USB outlets.

2 Topology

There are two typical topologies for the Grid Manager, the synchronous buck converter and the Dual Active Half Bridge [DAHB], see Fig. 2. Both share the same power electronics components, where the DAHB introduces an extra switching power leg. The DAHB is capable of isolating the output from the DC Grid. This is important in case of a fault in the prosumer or in case of a fault in the DC Grid. The Synchronous buck converter is not able to isolate the output from the DC Grid, and in case of a fault in the DC Grid, an active prosumer could still feed into the DC Grid via the anti-parallel diode in the Synchronous Buck converter, see Fig. 2b.

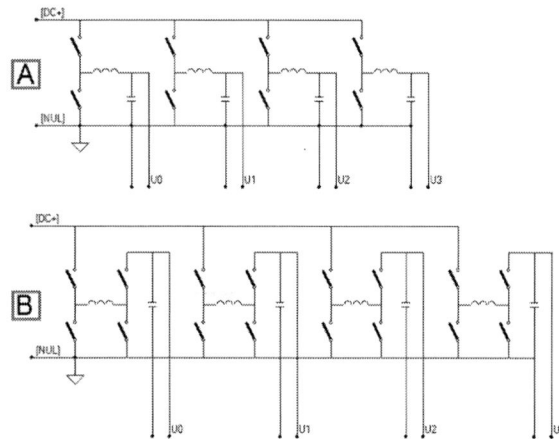

Fig. 2: Grid Managers. A: Synchronous Buck for loads. B: DAHB for active prosumers and producers.

2.1 Synchronous Buck

The synchronous buck converter is selected in the grid manager, whenever the output voltage is lower than the DC link voltage.

The origin of the spikes, as revealed by the simulation, are important to understand. Only then the design of the grid manager can be such, that these spikes are reduced in the real design. Experimental testing of the grid manager still reveal spikes in the waveforms, as can be seen in Fig. 5. In this figure the voltage ripple on the output is visible and is around $200mV$ top-top. The measurement of inductor current using a sense resistor, mostly shows ringing. This is caused by the switching of the mosfets connected to one side of the inductor. The voltage on the switching node between the two Mosfets, see node V_{out} in Fig. 4, is shown as the lower red trace in Fig. 5. Since the U4L in the Grid Manager [8] has a optimized PCB design, the ring-

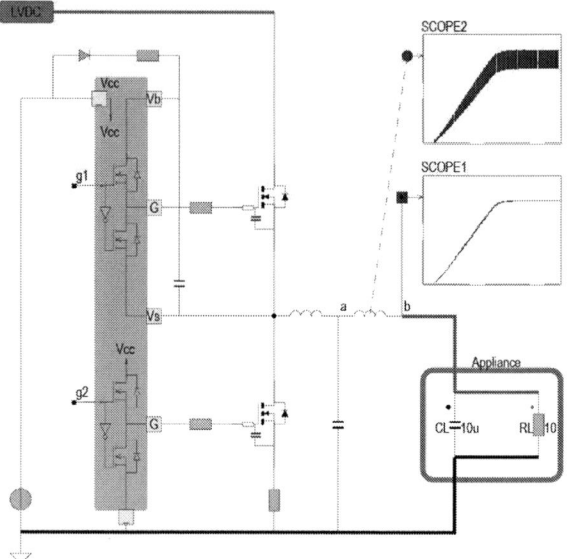

Fig. 3: Simulation in Caspoc of a synchronous buck converter pre-charging a passive load [7].

ing on that node is minimized, as can be seen by the red trace in Fig. 5.

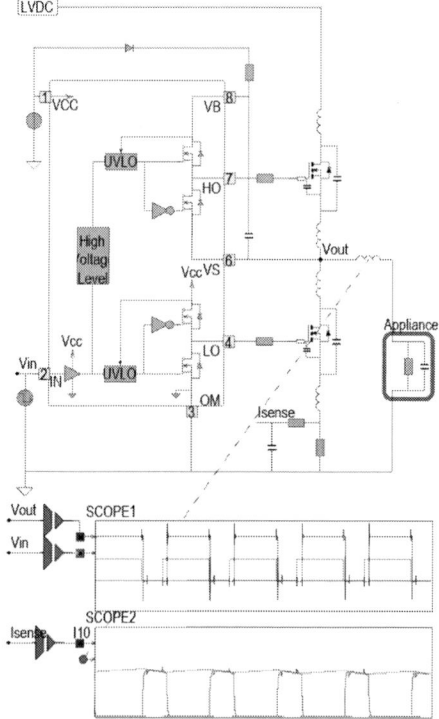

Fig. 4: Detailed simulation in Caspoc [7], of all parasitic components in the output of the Grid Manager. Scope1:Control signal and voltage on he switching node Vout. Scope2:Output current and measured current beneath the low-side Mosfet

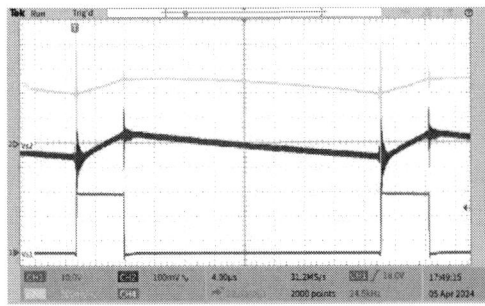

Fig. 5: Measurement from the output stage of the grid manager [timebase:$4\mu s$/div], from top to bottom: Ripple on the output voltage[200mV/div], Scaled inductor current, differential measurement over a shunt resistor, Voltage on the switching node[10V/div].

Typical examples are passive loads such as chargers for mobile applications and lighting. These applications require a constant output voltage and a limit on the output current. All of these can be realized by the synchronous buck converter. For the design care has to be taken on the switching node between the two Mosfets. As can be seen in Fig. 4 in scope 5, there is ringing due to the parasitic components between the source of the high-side Mosfet and the drain of the low-side Mosfet.

2.2 Dual Active Half-Bridge

The Dual Active Halve-Bridge [DAHB] can control either current or voltage. If the input is connected to a DC Grid with a constant voltage and the output is connected to a component with fixed voltage level, for example a battery, only the current flow between the input and output can be regulated. It can be regulated bidirectionally. If a passive load is connected to the output, the output voltage level can be controlled. Also pre-charging of the output appliance can be implemented and regulated by the DAHB. In Fig. 6, a DAHB is shown, which supplies power from the DC grid to a passive load.

To prevent an inrush current, pre-charging is applied. Scope1 in Fig. 6 shows the build up of the output voltage and scope2 in Fig. 6 shows the current through the appliance.

3 Protection

Protection and safety are two important functions in the grid manager [9]. Especially active prosumers require protection as well as a possibility to completely disconnect the prosumer from the DC grid. The principle an application of short circuit detection [8], [9], earth leakage detection as well a soft-start, to prevent inrush currents are the extra functions that can be implemented into the Grid Manager.

Over-current, short-circuit and earth-leakage protection is implemented via a current sensor in the output of the grid manager. In Fig. 7 a short circuit detection based on the Rate of Change of Current [RoCoC] is shown.

3.1 Short Circuit Protection

A short circuit can greatly increase the current flowing through a converter to hundreds of amperes in the matter of nanoseconds. An active protection unit implemented in a Grid Manager can detect and cut off a current source, such as a battery, in the same matter of seconds. This can prevent arcs and explosions from happening in a Grid Manager and connected components.

3.2 Earth Leakage Detection

Current can flow in unintended directions in case insulation is broken of a cable, especially if it makes contact with metal case. Active protection can measure the input and output currents at the output of the Grid Manager and ensures these currents are the same. In case these currents are not is there an earth leakage. The output is then disconnected before current can flow in unintended places.

3.3 Soft-Start

Because of the surplus amount of capacitors used in power converters and appliances are inrush currents a common occurrence in DC Grids. A soft-start is thus needed to increase the output voltage from zero with limited and controlled input current. This prevents large amounts of current to flow through the Grid Manager when a new component is connected.

4 Example: Tiny House

In the Tiny-House, a multi-functional Grid Manager is implemented, that is capable of controlling the output voltage or bidirectional power flow, see Fig. 10 [11].

PCIM Europe 2024, 11– 13 June 2024, Nuremberg   DOI: 10.30420/566262373

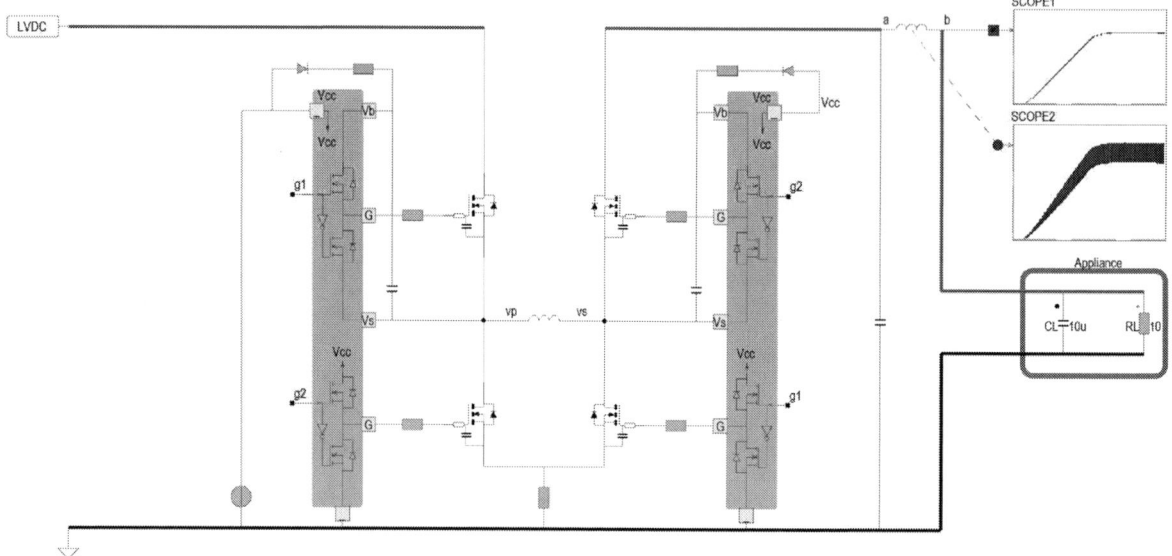

Fig. 6: Simualtion in Caspoc if the DAHB with soft-start capability soft-starting an appliance [7]

The low-voltage level is a Save Extra Low Voltage [SELV] of maximum $48v$. The grid manager replaces the traditional fuse-box in an AC grid, but includes much more functionality. Next to droop

control, there is also the requirement to disconnect active producers, such as solar panels and batteries, in case of a grid fault.

The experimental setup for testing control function of the grid manager is shown in Fig. 8. The power electronics board is the U4L, also being used for educational purposes [8]. The power electronics in this experimental setup is the same as is implemented in the Tiny-House. The resistors shown in Fig. 8 are replaced by the applications. The droop controllers inside the microcontroller, are directly controlling the output current.

In Fig. 9 the waveforms from the experimental setup are shown.

The Tiny House is a standalone grid with con-

Fig. 7: Detection of short circuit via Rate-Of-Change-Of-Current[RoCoC].

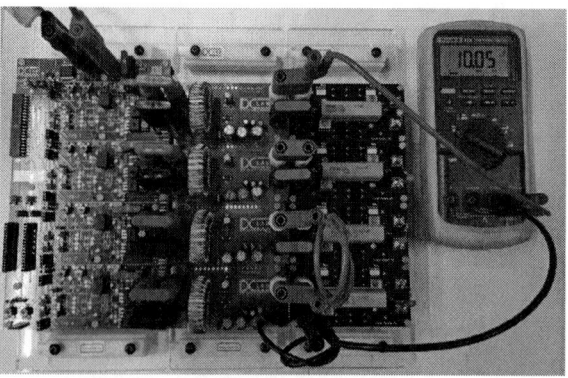

Fig. 8: Experimental setup of a Grid Manager to test the control functions [10]

2660

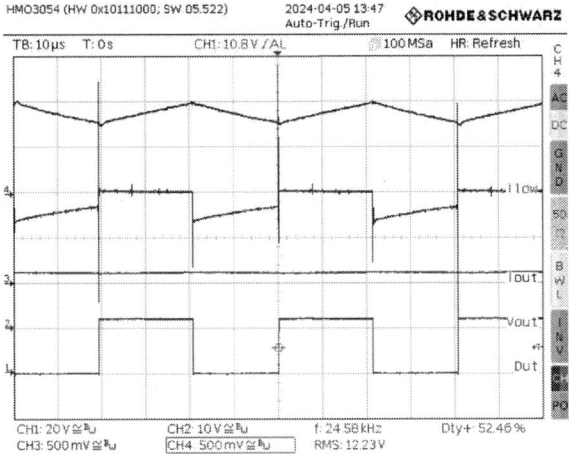

Fig. 9: Output voltage from the Grid Manager, and filter currents. From top to bottom: Output voltage ripple[$500mV/DIV$,Purple], Measured switched inductor current over a sense resistor[$500mV/DIV$,green], Constant output voltage[$10V/DIV$,blue], Voltage at the switch-node[$500mV/DIV$,darkgreen]

sumers, in the form of LED and low power chargers, producers, in the form of PV and prosumers, in the form of battery storage. All different components are connected together through a DC bus. The Grid managers connect to this bus in the form of a decentral grid. A combination of different Grid Managers can be applied because of the different type of connected components.

In each Grid Manager, a droop characteristic is included, which specify how much power flow can occur depending on the bus voltage [12]–[14]. This method of communication through the DC Bus voltage is called DC Bus Signaling [15], [16]. The different Grid Managers can apply power to the grid without knowing by what component is connected to another Grid Manager.

The separation of a very fast analog control for the voltage and power level, from the digital control in an Arduino [17] or C2000 [18] for functionality, improves the reliability of the grid manager. Real-time monitoring can be applied through the digital controller to monitor the power flow through the Grid manager [19]. Possible faults can also be shown through the same way. The installer can set the output parameters of each supply according to specification of the loads connected. The analog control can take these parameters and actively apply them as a reliable control system independent of the digital controller.

The real time monitoring also opens the possibility to monitor the solar yield, state of charge and power consumption of loads, seen in Fig. 10. Yield and storage can be checked to ensure that energy is available through night and winter. The digital controller or the user can then determine to limit the usage of appliances to preserve energy. It can also let the user know that the installation has to be upgraded with more producers or prosumers to make more capacity available if the option is available.

Conclusion

Two types of Grid Manager are presented. The synchronous buck converter can be applied when supplying passive loads. The DAHB has additional functionality, since the outputs can be switched of. This can prevent active prosumers to feed back into the DC grid, when there is a fault. An use case is also presented in the form of a standalone grid where the grid manager is implemented with consumers, producers and prosumers.

References

[1] B. Wunder, L. Ott, J. Kaiser, Y. Han, F. Fersterra, and M. März, "Overview of different topologies and control strategies for dc micro grids," in *2015 IEEE First International Conference on DC Micro-grids (ICDCM)*, 2015, pp. 349–354. DOI: 10.1109/ICDCM.2015.7152067.

[2] Y. Ito, Y. Zhongqing, and H. Akagi, "Dc micro-grid based distribution power generation system," in *The 4th International Power Electronics and Motion Control Conference, 2004. IPEMC 2004.*, vol. 3, 2004, 1740–1745 Vol.3.

[3] D. Boroyevich, R. Burgos, L. Arnedo, and F. Wang, "Synthesis and integration of future electronic power distribution systems," in *2007 Power Conversion Conference - Nagoya*, 2007, K-1-K–8. DOI: 10.1109/PCCON.2007.372910.

[4] D. Boroyevich, I. Cvetković, D. Dong, R. Burgos, F. Wang, and F. Lee, "Future electronic power distribution systems a contemplative view," in *2010 12th International Conference on Optimization of Electrical and Electronic Equipment*, 2010, pp. 1369–1380. DOI: 10.1109/OPTIM.2010.5510477.

[5] P. van Duijsen, J. Woudstra, and D. Zuidervliet, "Requirements on power electronics for converting kitchen appliances from ac to dc," in *2019 International Conference on the Domestic Use of Energy (DUE)*, 2019, pp. 190–197. DOI: ISBN: 978-0-6399647-3-7.

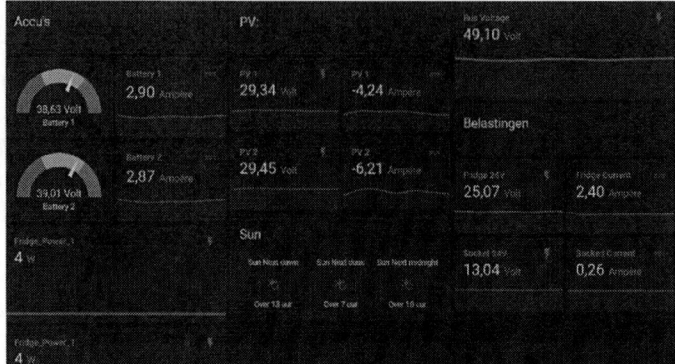

Fig. 10: Grid Manager and Dashboard for a Tiny House, containing solar, battery and consumers, but no connection to the main AC grid [10].

[6] A. Spaans, D. Zuidervliet, and P. van Duijsen, "Droop control in dc grids for kitchen appliances to avoid power congestion," in *2022 8th International Conference on Control, Decision and Information Technologies (CoDIT)*, vol. 1, 2022, pp. 791–796. DOI: 10.1109/CoDIT55151.2022.9804126.

[7] Simulation-Research. "Caspoc: Simulation and animation." (2024), [Online]. Available: https://www.caspoc.com.

[8] P. Van Duijsen and D. Zuidervliet, "Laboratory setup for teaching dc grid droop control and protection," in *2021 44th International Convention on Information, Communication and Electronic Technology (MIPRO)*, 2021, pp. 1587–1592. DOI: 10.23919/MIPRO52101.2021.9596738.

[9] P. J. van Duijsen and D. C. Zuidervliet, "Structuring, controlling and protecting the dc grid," in *2020 International Symposium on Electronics and Telecommunications (ISETC)*, 2020, pp. 1–4. DOI: 10.1109/ISETC50328.2020.9301065.

[10] DC-Lab. "Dc power laboratory." (2024), [Online]. Available: www.dc-lab.org.

[11] A. Benaissa, D. Zuidervliet, and P. van Duijsen, "Stand-alone dc nano-grid for a tiny house with droop control," in *2023 3rd International Conference on Electrical, Computer, Communications and Mechatronics Engineering (ICECCME)*, 2023, pp. 1–6. DOI: 10.1109/ICECCME57830.2023.10252427.

[12] J. M. Guerrero, J. C. Vásquez, and R. Teodorescu, "Hierarchical control of droop-controlled dc and ac microgrids — a general approach towards standardization," in *2009 35th Annual Conference of IEEE Industrial Electronics*, 2009, pp. 4305–4310. DOI: 10.1109/IECON.2009.5414926.

[13] X. Lu, J. M. Guerrero, K. Sun, and J. C. Vasquez, "An improved droop control method for dc microgrids based on low bandwidth communication with dc bus voltage restoration and enhanced current sharing accuracy," *IEEE Transactions on Power Electronics*, vol. 29, no. 4, pp. 1800–1812, 2014. DOI: 10.1109/TPEL.2013.2266419.

[14] L. Ott, Y. Han, B. Wunder, J. Kaiser, F. Fersterra, *et al.*, "An advanced voltage droop control concept for grid-tied and autonomous dc microgrids," in *2015 IEEE International Telecommunications Energy Conference (INTELEC)*, 2015, pp. 1–6. DOI: 10.1109/INTLEC.2015.7572406.

[15] J. Bryan, R. Duke, and S. Round, "Decentralized generator scheduling in a nanogrid using dc bus signaling," in *IEEE Power Engineering Society General Meeting, 2004.*, 2004, 977–982 Vol.1. DOI: 10.1109/PES.2004.1372983.

[16] J. Schonberger, R. Duke, and S. Round, "Dc-bus signaling: A distributed control strategy for a hybrid renewable nanogrid," *IEEE Transactions on Industrial Electronics*, vol. 53, no. 5, pp. 1453–1460, 2006. DOI: 10.1109/TIE.2006.882012.

[17] Arduino. "Arduino nano, [online] available:" (2024), [Online]. Available: https://store.arduino.cc/arduino-nano.

[18] Texas-Instruments. "C2000, [online] available:" (2024), [Online]. Available: https://www.ti.com/c2000.

[19] D. Termoshuizen, D. Zuidervliet, and P. van Duijsen, "Configurable educational dc system trainer combining universal code and hardware," in *2023 8th International Symposium on Electrical and Electronics Engineering (ISEEE)*, 2023, pp. 90–95. DOI: 10.1109/ISEEE58596.2023.10310330.

PCIM Europe 2024, 11– 13 June 2024, Nuremberg   DOI: 10.30420/566262375

CO$_2$ Footprint of Medium Voltage DC Solid State Transformer

Adriana Campos[1], Astrid Jasi[1], Konstantin Vershinin[1], Piotr Dworakowski[1]

[1] SuperGrid Institute, France

Corresponding author: Adriana Campos, adriana.campos@supergrid-institute.com
Speaker: Adriana Campos, adriana.campos@supergrid-institute.com

Abstract

Power converters are a key technology to support the massive integration of Renewable Energy Sources and it is therefore important to assess their environmental impact. This work uses the Life Cycle Assessment methodology as a base to assess the environmental impact of a DC Solid State Transformer. The carbon emissions were estimated for the DC SST with different design choices such as the operating frequency, number of semiconductor devices in parallel and the energy mix. The use phase is the biggest contributor of emissions for all cases, but it is likely to reduce with the decarbonation of electricity. Considering monetary value of the CO$_2$ emissions with the carbon tax, an optimal frequency is found for the DC SST. The economic weight of these emissions is expected to increase in the future.

1   Introduction

In order to stay in line with the 1.5°C pathway for 2050, it is vital to reduce the global carbon emissions. One likely scenario to do this, is the massive deployment of Renewable Energy Sources (RES) in the electric system [1]. Power electronic converters are a key interfacing technology for RES and thus they are expected to play a major role in the electric network of the future.

This future network is likely to have an increased presence of high and medium voltage DC links for some applications such as large solar power plants, the electrified railway infrastructure, ships and for reinforcement of the transmission and distribution network with point-to-point links [2]. For these applications, the isolated DC-DC converter, also called DC Solid State Transformer (DC SST) is a necessary technology to further develop the network. It is therefore important to assess its environmental impact, as part of an integral strategy to reduce carbon emissions.

The Life Cycle Assessment (LCA) methodology allows to account for the environmental impacts of a product throughout its life stages, from the extraction of raw materials, manufacturing, transportation, use and end of life. Different impacts are considered, some important ones are global warming, resource depletion, eutrophication and human toxicity. The global warming, usually measured in Global Warming Potential (GWP) represents the effect of a greenhouse gas with respect to carbon dioxide and it is expressed in CO$_2$-equivalent

weight [3]. This remains one of the most relevant impact categories of the LCA methodology, with a special focus in our society.

The importance of the environmental assessment of power electronic components and the use of LCA methodologies have already been pointed out. In [4] and [5] it is highlighted that the future design of power electronic equipment will have to shift towards Circular-Economy-Compatible (CEC) products. This would be done in practice by increasing the reusability and recyclability of products, by adding the environmental impact of the devices as a new dimension in a multi-objective optimization and by updating the manufacturer datasheets to include relevant environmental footprint data.

The re-use of electronic equipment can also contribute to increase the access to energy sources in developing countries. An environmental analysis of reuse components of a solar energy system was done in [6], and it was compared to the conventional solution without reused components. It was found that the reuse solution had a reduction of 30% on the global warming impact.

The evolution of the environmental impact of low power solar PV inverters rated from 2.5 to 20 kW was analyzed in [7]. For this, a Life Cycle Inventory (LCI) of solar inverters was done. It was found that the printed board assembly had the biggest impact for climate change. Furthermore, the printed board related emissions were found to be percentually higher for the newer 2.5 kW solar inverters compared to the old ones.

2663

The environmental impact of power electronics in electrified passenger cars has also been studied. In [8] the carbon emissions of the electric powertrain and power inverter unit of the vehicles was assessed for onshore manufacturing within the UK, compared to other key European and international regions. It was found that the GWP associated with production remained relatively low for the inverter, regardless of the region. The high-voltage battery accounted for most of the emissions in this phase. In the use phase, the Battery Electric Vehicle (BEV) was found to produce the lowest emissions with a European mix among the considered vehicles. Overall, the motor and inverter amounted for 12-17% of the total GWP.

The difference between Silicon (Si) and Silicon Carbide (SiC) semiconductors in terms of LCA was explored in [9]. A quantification of the production and use phases of both type of semiconductor devices was done. It was found that the energy savings from the efficiency gains of SiC devices are several times greater than the extra energy input needed to produce SiC vs. Si devices, when used in a solar PV inverter application.

In [10] the LCA of a 150 kW power electronics inverter was performed. The inverter is connected to a 450 V DC bus and operates 10000 hours in 15 years. The significant environmental impact of power electronics was highlighted, and a product eco-design and eco-optimization were presented as necessary to reduce its GWP. The manufacturing and use phases were found to be the most important in terms of environmental impact.

This article presents a methodology to perform the LCA of a DC SST. The methodology is applied to a Dual Active Bridge (DAB) DC-DC converter topology [11]. The calculation of the carbon footprint was done for different design choices, such as the operating frequency, the number of semiconductor devices in parallel or series. The goal was to understand how these factors influence the GWP considering a PV load profile. Finally, a techno-economic assessment of the carbon emissions is done. To the best knowledge of the authors, this type of analysis has not been done in the literature.

This article is organized as follows. First, the LCA methodology used to assess the environmental impact of the DC SST is described. Second, the case study and design specification for the DAB converter is presented. Finally, the CO_2 footprint of the converter is shown for different design choices, including the Medium Frequency Transformer (MFT) and the economic analysis is presented.

2 Methodology to assess the environmental impact of DC SST

The LCA methodology as presented in the ISO 14040 consists of four steps [12]:
i) Goal and scope definition
ii) Life Cycle Inventory (LCI)
iii) Impact Assessment
iv) Interpretation

The goal and scope definition includes the motivation of the study, the methods to be used, the system boundaries and the functional unit. The LCI details the inputs (materials and energy) and outputs (products, emissions) of each process. In the impact assessment, the environmental impacts of the product system are quantified for each impact category. There are nine impact categories, but this work focuses on the GWP. Finally, during the life cycle interpretation, the results are assessed, and potential environmental issues are identified. This allows to propose changes in the process to reduce the environmental impact. It is in this step that LCA provides an environmental parameter to be considered during the design and optimization of a product.

2.1 Goal and scope definition

The goal of the study is to determine how different design choices affect the carbon footprint of a DC SST. The scope of the analysis includes three phases of the life cycle of the converter as shown in Fig. 1, manufacturing, transportation and use. The raw material acquisition and end of life related emissions are excluded in this article.

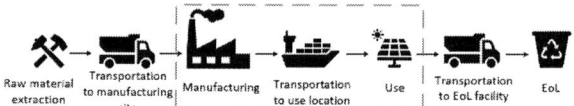

Fig. 1 LCA boundaries considered in this article

2.2 Life Cycle Inventory and Impact Assessment

The computed CO_2 footprint of the DC SST depends on the LCI of the converter during the manufacturing, transportation and use phases. As shown in Fig. 2, a detailed DC SST design is needed for the overall assessment. This design will allow to quantify the materials and weight of each subcomponent, as well as the efficiency of the DC SST and its total volume and weight. The load profile at which the converter is expected to operate is also needed for the analysis and the resulting CO_2 footprint will be greatly dependent on the application. Similarly, the resulting carbon

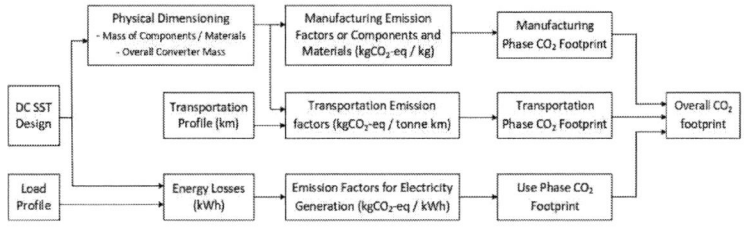

Fig. 2 CO₂ impact assessment of DC SST

emissions are directly dependent on the energy mix considered. Finally, a hypothesis about the transportation profile is needed to calculate the emissions corresponding to the transportation phase. With all these inputs, the carbon emissions for each phase can be computed with the corresponding emission factors.

2.2.1 Manufacturing phase

To account for the materials needed, a list of typical DC SST components is shown in Table 1. The associated CO_2 footprint from the manufacturing of these parts was taken from the literature.

Table 1 Typical components DC SST and its GWP data

Part	Subpart	GWP	Source
Power module	SiC chip	90 tonCO₂/m²	[4]
	Baseplate (Al)	5.9 kgCO₂/kg	[4], [10]
Film capacitors	-	47 kgCO₂/kg	[4]
MFT	Litz wires (Cu)	3.8 kgCO₂/kg	[13], [14]
	Ferrite	1.3 kgCO₂/kg	[15]
Heat sink (Al)	-	27.9 kgCO₂/kg	[4]
Busbar (Cu)	-	10.9 kgCO₂/kg	[4], [10]
PCB	(six layer)	56 kgCO₂/m²	[4], [8]

2.2.2 Transportation phase

Different modes of transportation are possible depending on the manufacturing site and the use site. Different emission factors are presented in Table 2 for transportation by ship, railway or road freight.

Table 2 CO₂ emissions for different transportation modes [8]

Mode	gCO₂-eq/tonne km
Road freight - diesel	200
Rail freight - electric	25
Rail freight - diesel	50
Container ship - ocean	25

It can also be seen that the final emissions will depend on the total distance. Assuming that the final destination is France, two possible manufacturing sites are considered, China and Germany. For the case of China, the estimated intercontinental distance is around 17000 km and the intracontinental

distance of around 1400 km. For the case of Germany an intracontinental distance of around 700 km is considered. With these hypotheses it is possible to find the emission factors for the different scenarios shown in Table 3.

Table 3 Emission factors of different transportation scenarios

Scenario - origin	Footprint (kgCO₂-eq/tonne)
China (Ship + rail electric)	461
China (Ship + rail diesel)	496
China (Ship + road diesel)	704
Europe (Road diesel)	139
Europe (Rail electric)	17
Europe (Rail diesel)	35

2.2.3 Use phase

For the use phase, the carbon footprint is measured with the DC SST losses throughout its lifetime. These losses are highly dependent on the losses of semiconductor devices and MFT, the load profile for which the converter operates and the energy mix that is used to power the converter. Table 4 presents the emission factors for three regions of interest, France, Europe and the world in 2022 as reported in [16], [17], [18]. As it can be seen, France has the lowest emission factor among these regions thanks to the large percentage of nuclear energy generated in the region. It is clear that a converter operating in France will have much lower emissions in the use phase than a converter in other European regions or other regions in the world, where the percentage of fossil energy sources is more significant.

Table 4 Mix emission factors per region in 2022

Region	gCO₂-eq/kWh
France	75
Europe	277
World	438

2.2.4 Load profile

For the load profile, a typical solar PV inverter profile has been considered. The operating power shown as a percentage of the nominal power and the duration of each power is presented in Table 5. The converter is assumed to operate 47% of time, following the intermittence of PV generation.

Table 5 Load profile - solar PV inverter [19]

Power, %	Duration, %
5	3
10	6
20	13
30	10
50	48
100	20

As it can be seen, excluding nighttime, the converter is expected to operate only 20% of the time at nominal power, therefore it is important to consider low power points for the DC SST design, in particular for the selection of the leakage inductance and the design of the MFT.

3 Case study and design

The application considered in this article consists of a MVDC collection network at 10 kV connected to the utility grid, the DC SST which interfaces the MVDC network with low voltage bus at 1.5 kV that is connected to a PV array, as shown in Fig. 3.

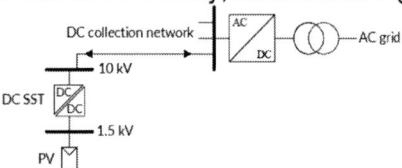

Fig. 3 System considered for the case study

3.1 Converter specification

The topology of the DC SST considered for the case study is a DAB converter [11] composed of two single-phase full bridges and an MFT, as shown in Fig. 4. As it can be seen the converter interfaces an LVDC network with an MVDC collection network and the converter can transfer power in either direction. On the MV bridge, a series connection of semiconductor devices is required in order to withstand the voltage on the MV bus terminals. The MFT is represented as an ideal transformer with a leakage inductance L_{lk} in series.

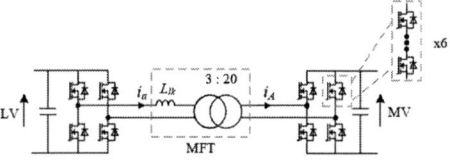

Fig. 4 DC SST considered for the case study

The specification for the converter is presented in Table 6. A variation of 10% on the MV bus is considered for nominal operation, and the LV bus voltage is assumed to be controlled at 1.5 kV. A maximum voltage ripple of 5% is considered to size the

DC bus capacitors. The nominal power of the converter is 250 kW. For the semiconductor devices, SiC MOSFETs have been selected to have a good performance of the converter at high frequencies.

Table 6 Specification for DC SST

Nominal voltage LV bus	1.5 kV
Nominal voltage MV bus	10 kV ± 10%
Max voltage ripple	5%
Nominal power	250 kW
Operating frequency	5 kHz, 10 kHz or 20 kHz
Device LV bridge	Hitachi 5SFG-0500X330100 SiC MOSFET 3.3 kV, 500 A Power module
Device MV bridge	GeneSiC – G2R50MT33K SiC MOSFET 3.3 kV, 44 A Discrete device
Transportation profile	China – ship + rail electric
Mission profile	Solar PV
Lifetime	20 years
Use location (energy mix)	France (2022)

Two design variations are of interest in terms of impact on the carbon footprint, the first one is the frequency and the second one is the number of devices in parallel. For the frequency variation, three different values are considered, 5 kHz, 10 kHz and 20 kHz. These frequencies are interesting since they allow a significant size reduction of the transformer. Regarding the number of devices connected in series or parallel, connecting more devices in series allows to increase the voltage in the DC bus, in this case, only six 3.3 kV devices are needed working at 1.7 kV to withstand the 10 kV on the MV bus, and only one device is enough to connect to the LV bus. On the other hand, connecting devices in parallel reduces the current flowing through each device, which allows to potentially reduce the semiconductor losses.
The lifetime of the converter is 20 years, and it is meant to be used in France.

3.2 Converter design

The modulation technique chosen for the DAB converter is the single-phase shift (SPS). With this modulation, both bridges operate with a duty cycle of 50% and the power transfer between the bridges is controlled only through the phase shift δ between them.

In terms of design choices, there is a degree of freedom that must be optimized for each of the proposed frequencies: the choice of the leakage inductance and the corresponding nominal phase shift. The relation between them is directly proportional. A higher leakage inductance will require a higher nominal phase shift to deliver the nominal power and vice versa. There are however important limitations that must be considered for both of these parameters. First, there are control limitations related to the phase shift δ. The upper limit is determined by the maximum power transfer, which happens at 90°. For the lower limit it is proposed to utilize 4 µs which corresponds to 200 steps of the FPGA running at 50 MHz. This number of steps provides sufficient resolution to achieve a smooth power sweep from 0 to 100%.

For the leakage inductance, the physical limitations of the MFT must be considered. Very small leakage inductances may not be feasible, while very high values will make the design complex and will affect the performance of the MFT [20]. The choice of the leakage inductance does not only affect the size and efficiency of the MFT, but it also has an impact on the semiconductor losses, as a higher leakage inductance will result in a higher reactive power (and current). By choosing a small leakage inductance, the reactive power (current) can be minimized, as shown in Fig. 5.

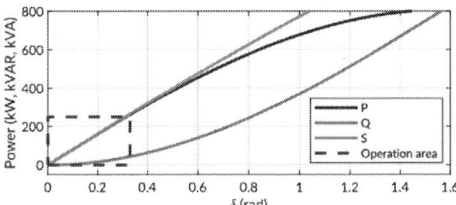

Fig. 5 P, Q and S for $L_{lk} = 36\,\mu H$ and $10\,kHz$

The leakage inductance and phase shift have another impact in the semiconductor losses related to the soft switching operation of the converter. One of the advantages of the DAB converter resides in its Zero Voltage Switching (ZVS) operation under certain operating conditions. During ZVS, there are no turn-on losses in the semiconductor devices. It is therefore desirable to remain within the conditions that allow ZVS. This is illustrated in Fig. 6, where the desired ZVS condition is in the white area. When the converter operates in the green or blue area, one of the bridges will lose ZVS, presenting higher losses.

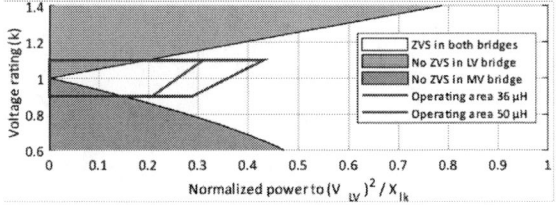

Fig. 6 ZVS zones of a 10 kHz DAB

As it can be seen, the ZVS zones are represented in terms of the power normalized to V_{LV}^2/X_{lk} and

the voltage rating. This figure shows how the ZVS zones are dependent on the leakage inductance: a higher leakage inductance will result in a higher nominal normalized power. This means in practice that with higher leakage inductance values, the converter will operate in ZVS in a larger range of power. For voltage ratings different from 1, the ZVS is lost in one of the bridges at low powers. It is therefore important to consider the load profile and specially the low power points.

To consider this, the semiconductor losses have been computed for different leakage inductances at a given frequency. The results are shown in Fig. 7. As it can be seen, for lower leakage inductance values the total losses are higher due to the loss of ZVS when k is different from 1.

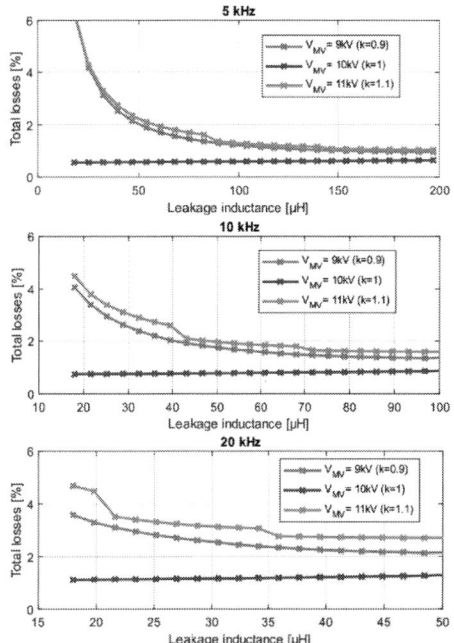

Fig. 7 Semiconductor losses vs. L_{lk} at 5, 10 and 20 kHz

3.3 MFT specification

Knowing the semiconductor losses with different leakage inductances for the considered load profile, it is possible to specify a target range for the specification of the MFT. Table 7 shows the specified range of the leakage inductance considering the control constraints and semiconductor losses and keeping in mind the goal of minimizing the L_{lk} value to reduce the reactive current.

Table 7 Target range for L_{lk} at different operating frequencies

f_{op}	$L_{lk,min}$	$L_{lk,max}$	Target range L_{lk}
5 kHz	36 µH	450 µH	50 µH < L_{lk} < 100 µH
10 kHz	36 µH	225 µH	36 µH < L_{lk} < 50 µH
20 kHz	36 µH	113 µH	36 µH < L_{lk} < 50 µH

As it can be seen, the target range for 10 kHz and 20 kHz is the same. This is mainly due to the control constraints, since lower target leakage inductance values would be expected with increasing frequency, as it is the case between the optimal range of 5 kHz and 10 kHz.

Other general specifications for the MFT are presented in Table 8. The turn ratio is specified as the voltage ratio between the LV and MV buses, with a tolerance of +5% and -7%, chosen to minimize losses. The efficiency of the MFT is to be greater than 99.5% computed with the European Efficiency [19]. The maximum ambient temperature considered is 40°C.

Table 8 General specifications for MFT

Turn ratio	3:20 +5% / -7%
Efficiency η_{EU}	>99.5%
Max ambient temperature	40°C

3.4 MFT design choices

Table 9 presents the technological choices taken for the design of the MFT. A parallel-series core type transformer has been chosen to help achieve the high turn ratio of the transformer and to split the high LV current between two windings. For the material, ferrite is to be used and only standard cores of the I93/28/30 assembly have been considered. For the transformer windings, Litz wires are to be used and for the cooling, forced convection with a speed of 4 m/s has been considered. It is important to mention that for the cooling, a thermal margin of at least 20°C is assumed, with a maximum temperature set to 80°C. This hypothesis remains conservative and accounts for unexpectedly high losses or low cooling performance.

Table 9 MFT technological choices

Transformer type	Core type parallel-series
Core material	Ferrite 3C97 Standard core I93/28/30
LV winding	Single layer Litz wire Nomex covering
MV winding	Single layer Litz wire Nomex covering
Cooling	Forced convection 4 m/s

Several MFT designs can be obtained with the hypotheses and design choices presented above. These designs will vary in efficiency, volume and cost of the transformer, and all of these parameters will be related to the carbon emissions of the MFT. The actual choice of the final design is a multiparameter optimization and the optimal value depends on the constraints such as space, weight, cost and emissions.

4 CO₂ footprint results

4.1 MFT CO₂ footprint calculation

For the choice of the leakage inductance of the MFT, it is of interest to see how the carbon emissions change with the spectrum of MFT designs. These emissions will be related to the MFT size, weight and efficiency. In general, bigger and more efficient transformers are more expensive, which is why representing the variation of the carbon footprint with cost seems relevant. To compute the material cost of the MFT, the values shown in Table 10 have been considered. Only the materials for the core and the windings are accounted for in the calculation. Other manufacturing costs and margins have not been considered, as these costs would be equivalent for the different MFTs and they would only shift the data. For comparison purposes, considering the material cost only does not affect the conclusions.

Table 10 MFT material cost assumptions

Ferrite 3C97	~44 €/kg
Litz wire	~30 €/kg

A Pareto front has been obtained for each frequency, and shown in Fig. 8. As it can be seen, the Pareto fronts are fairly discontinuous due to the selection of standard cores only. This is more noticeable at lower frequency.

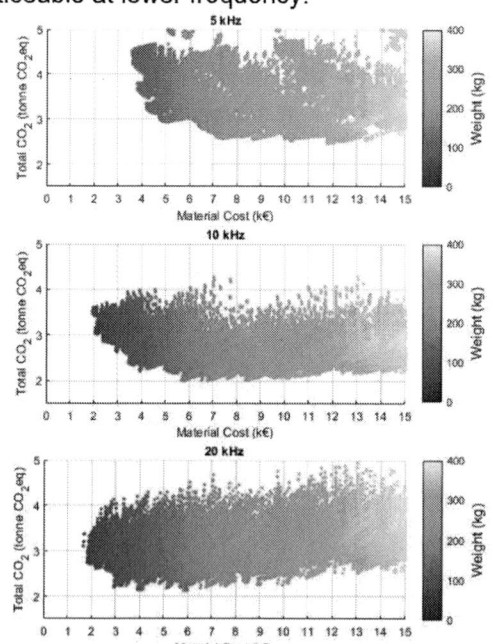

Fig. 8 MFT CO₂ pareto fronts for 5, 10 and 20 kHz

In terms of the CO₂ footprint variation with the frequency, the designs for 10 kHz are better than those at 5 kHz, with lower emissions and cost. The designs at 20 kHz have slightly higher emissions as the cost and weight increase. This is due to the relatively high leakage inductance values limited by the control constraints.

A comparison can be made for the same carbon footprint at three different frequencies. The point 3 tonnes CO_2eq is chosen as it has feasible designs for the three cases. Table 11 presents the design results for this point. As it can be seen, the material cost decreases as the frequency increases. The total weight of the MFT decreases greatly when passing from 5 kHz to 10 kHz, but the weight for the 10 and 20 kHz designs is very similar. Similarly, taking a closer look at the carbon emissions, a significant reduction occurs for the manufacturing and transport phases between 5 and 10 kHz, but the results are very similar for 10 and 20 kHz. In any case, it is clear that the use phase, i.e., the accumulated losses along the lifetime, is the most significant contribution to the total emissions for all the designs. Considering the MFT alone, higher frequency is preferred because it limits both the CO_2 emissions and the material cost.

Table 11 MFT designs with ~3 tonnes CO_2eq

	5 kHz	10 kHz	20 kHz
L_{lk}	53.3 µH	36.6 µH	43.4 µH
$\eta_{EU,\ MFT}$	99.62%	99.61%	99.65%
Weight	122 kg	56 kg	51 kg
CO_2 emissions (kgCO_2-eq)	M*: 330 Tr*: 56 U*: 2615 Tot*: 3001	M: 157 Tr: 26 U: 2837 Tot: 3020	M: 149 Tr: 23 U: 2816 Tot: 2989
Material cost	4.8 k€	2.1 k€	1.9 k€

* M: Manufacturing, Tr: Transportation, U: Use, Tot: Total

It is also interesting to compare different designs at low-cost points to see if higher frequencies are still preferred. Table 12 presents the design results for this case. Once again, the material cost of the MFT decreases as the frequency increases. Moreover, the overall carbon footprint also decreases as the frequency increases. Similarly to the previous case, the weight decreases drastically when passing from 5 to 10 kHz, but the decrease is less noticeable when choosing the 20 kHz design. The use phase remains the most important in terms of carbon emissions. Thus, also in this case, choosing a higher frequency would seem preferable to optimize the cost and emissions of the MFT.

Table 12 MFT designs minimising the cost (low cost)

	5 kHz	10 kHz	20 kHz
L_{lk}	67.3 µH	35.8 µH	36.0 µH
$\eta_{EU,\ MFT}$	99.57%	99.61%	99.67%
Weight	97 kg	54 kg	48 kg
CO_2 emissions	M*: 254	M: 148	M: 138
(kgCO_2-eq)	Tr*: 45 U*: 3342 Tot*: 3641	Tr: 25 U: 2872 Tot: 3045	Tr: 22 U: 2593 Tot: 2753
Material cost	3.8 k€	2.1 k€	1.8 k€

M: Manufacturing, Tr: Transportation, U: Use, Tot: Total

However, it is necessary to assess the overall emissions of the converter, to determine the weight of the transformer emissions and to see if at the DC SST scale, the higher frequency is indeed the most optimal one.

4.2 CO_2 footprint of DC SST

Taking the designs obtained for the low-cost point of the MFT presented in Table 12, the methodology was implemented to compute the overall carbon footprint of the DC SST. The low-cost designs were chosen over the constant emission ones (Table 11) as they illustrate very well the change in size and weight with a frequency increase, as well as the related cost and carbon emission reduction. Table 13 shows the European efficiency of the DC SST for the selected leakage inductances at the different frequencies. As it can be seen, the overall efficiency of the DC SST decreases at higher frequencies. This is due to the semiconductor losses, which are greater than the MFT losses.

Table 13 European efficiency of the DC SST

	5 kHz	10 kHz	20 kHz
L_{lk}	67.3 µH	35.8 µH	36.0 µH
$\eta_{EU,\ DC\ SST}$	99.16%	99.02%	98.64%

The results for the carbon emissions of the DC SST are presented in Fig. 9. As it can be seen, the use phase is the main source of carbon emissions for the converter along its lifetime. As expected, the total emissions increase with the frequency due to higher semiconductor losses. The manufacturing and transportation-related emissions are zoomed in to see the detail as they are very small compared to the use-related emissions. As it can be seen, both the manufacturing and transportation emissions decrease as the frequency increases. This is related to the size and weight reduction in the MFT and the capacitors of the converter as the frequency increases.

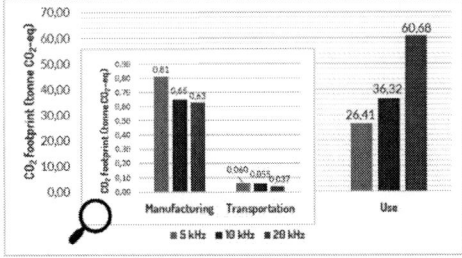

Fig. 9 CO_2 footprint of DC SST for different frequencies

Fig. 10 shows the detailed breakdown of the contribution of each subsystem of the converter (MFT, LV and MV bridges) in the carbon emissions of each phase. As expected, the emissions related to the semiconductor losses, and particularly the switching losses in both bridges are much more significative than the emissions related to the MFT for the three frequencies. Taking this into account, a reasonable criterion to select a good MFT design is the cost, since its related emissions only account for a little part of the overall ones. It can also be seen that the manufacturing and transportation-related emissions are only a small percentage of the overall emissions, so the DC SST losses should be optimized in priority.

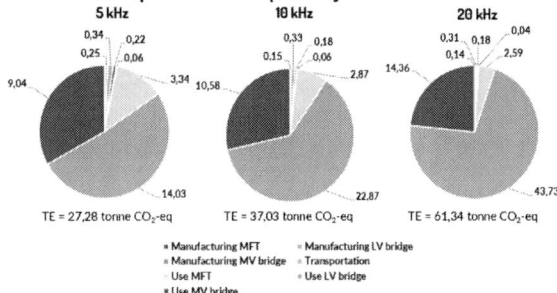

Fig. 10 Breakdown of carbon emissions in tonne CO_2-eq

4.3 Effect of the parallel connection of MOSFETs on the CO_2 footprint

One of the possible ways to optimize the semiconductor losses without changing the devices is to make a parallel connection of devices as shown in Fig. 11. This connection allows to reduce the current flowing through each device, which can result in smaller overall losses than with a single device.

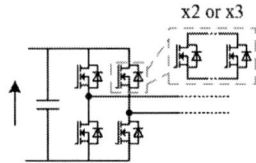

Fig. 11 Parallel connection of devices on bridge

Taking the 10 kHz design as a base, a parallel connection of MOSFETs is examined at the LV bridge. The resulting emissions are shown in Table 14 and Fig. 12. As it can be seen, by connecting 2 or 3 parallel devices on the LV bridge, the corresponding use-phase emissions (coming from the semiconductor losses) can be greatly reduced. An overall 25% reduction is observed when passing from one to two parallel devices, and a 47% reduction is observed with three parallel devices. As expected, the manufacturing emissions increase due to the higher number of semiconductor devices, but this increase remains small compared

to the overall emissions. If the cost of the extra semiconductor devices remains acceptable, a parallel connection of these is an interesting option to optimize the losses over the lifetime of the converter.

Table 14 CO_2 emissions for devices in parallel, tonne CO_2eq

Devices in parallel ‖	1 in ‖	2 in ‖	3 in ‖
Manufacturing	0.7	0.9	1.1
Transportation	0.1	0.1	0.1
Use	36	27	19
Total	37	28	21

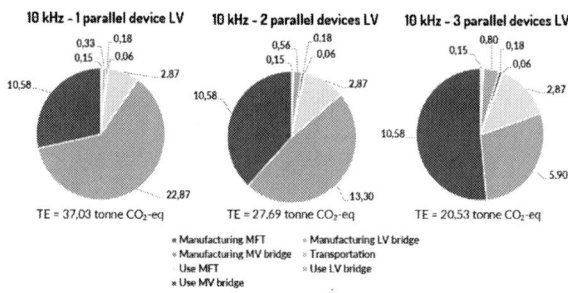

Fig. 12 Emissions with different number of parallel devices

4.4 Effect of the energy mix on the CO_2 footprint

The results presented so far have been computed considering the French energy mix of 2022. This energy mix is relatively "clean" in terms of carbon emissions, as almost 70% of the energy comes from nuclear plants and another 20% comes from RES such as hydro, wind and solar energy. Despite the low carbon emission factors of these energy sources, the use phase contributes the most to the overall carbon emissions of the converter. It is interesting to see how much the use-related emissions of the DC SST are affected if a different energy mix is used.

In Fig. 13, the use emissions of each subsystem are presented considering the French, European and world energy mixes for the 10 kHz design. Not surprisingly, the emissions are much higher when the European or world energy mixes are considered. The emissions obtained with the European mix are 4 times greater than those obtained with the French mix, while the emissions with the world mix are more than 6 times greater. Overall, the use phase increases from being 98% of the total losses with the French mix to 99.5% and 99.7% with the European and world mixes, respectively, within the LCA's boundaries. The total emissions increase from 37 to 146 and 241 tonne CO_2-eq, respectively.

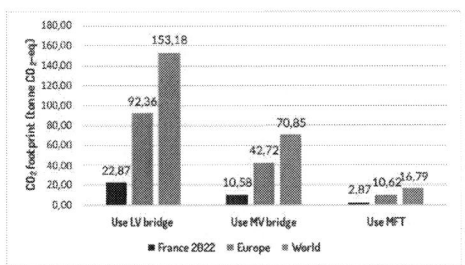

Fig. 13 Use emissions at 10 kHz for different energy mixes

4.5 Techno-economic assessment

The results presented in the previous sections show that when considering the carbon emissions of the DC SST, it is better not to go too high in frequency, since the semiconductor losses increase significantly which is reflected in the CO_2 emissions of the use phase. It was also shown that the material cost of the MFT tends to decrease as the frequency increases, thanks the reduction in materials needed. The semiconductor devices are the major cost in the bridges, and in the current design the same devices were used at all frequencies leading to the same semiconductor cost. This means that the total cost of the DC SST decreases at higher frequencies. The carbon emissions and the material cost seem to go in opposite directions, so it is interesting to see if there is an economic motivation to reduce the carbon emissions.

The carbon tax rate in European countries in March 2023 ranged from 15 to 120 €/tCO2eq [21], with an average value of 55 €/tCO2eq. Taking a look at the three designs presented in section 4.2, the costs related to the MFT materials and the CO_2 tax are presented in Fig. 14 considering the average carbon tax. The total cost is the sum of the MFT material cost and the carbon tax. As it can be seen, for the chosen designs, the optimal frequency for the converter is 10 kHz, while the designs for 5 kHz and 20 kHz seem equivalent in monetary terms.

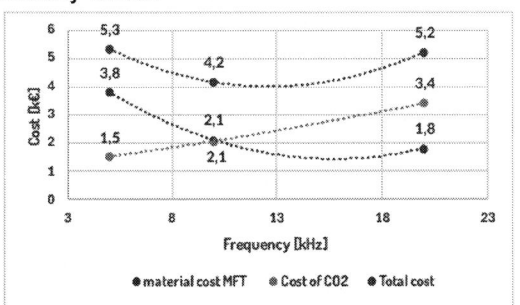

Fig. 14 Cost vs. frequency for DC SST, average CO_2 tax

In the future, the carbon tax rates are likely to increase, so the total carbon emissions will have more and more an economic impact on the DC SST design. Fig. 15 shows the costs for the DC SST considering the maximum carbon tax rate of 120 €/tCO2eq is considered. It can be seen that the 10 kHz design is still the optimal one. However, for this case, the designs for 5 and 20 kHz are no longer equivalent. The 20 kHz presents the highest total cost due to the higher tax on the emissions, which shows that in the future, the carbon emissions will have a significant weight on the optimization process of the DC SST.

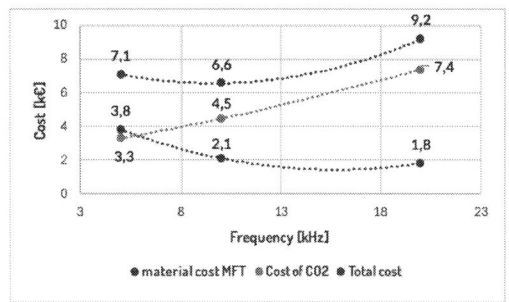

Fig. 15 Cost vs. frequency for DC SST, max CO2 tax

5 Conclusions

In this paper and LCA-based methodology to evaluate the carbon footprint of DC SSTs was used to estimate the CO_2 emissions of a DAB converter used in a solar PV application. A detailed design was done for three frequencies including the MFT specification, and three Pareto fronts were obtained. Overall, it was found that the use phase is responsible for more than 96% contribution to the CO_2 footprint of the DC SST. For the MFT design it was found that a higher frequency was overall preferred as it presents lower emissions and lower material costs. However, at the level of the DC SST semiconductor losses provide the main contribution and these increase with the switching frequency. Based on the TEA analysis, the optimal frequency is 10 kHz and in the future, the weight of the emissions will likely become more and more important in economic terms, making it necessary to add this environmental factor in the optimization of the DC SST design.

6 Acknowledgements

The authors would like to thank Alexis Fouineau and Martin Guillet for their valuable support in writing this article.

The authors would like to acknowledge FOR²EN-SICS project funded under European Union's Horizon Europe research and innovation programme under grant agreement No 101075672.

References

[1] IRENA, 'World Energy Transitions Outlook 2023: 1.5°C Pathway'. [Online]. Available: https://www.irena.org/Publications/2023/Jun/World-Energy-Transitions-Outlook-2023

[2] P. Le Métayer et al., 'Break-even distance for MVDC electricity networks according to power loss criteria', in 2021 23rd European Conference on Power Electronics and Applications (EPE'21 ECCE Europe), Sep. 2021, pp. 1–9. doi: 10.23919/EPE21ECCEEurope50061.2021.9570416.

[3] 'Suggestions for updating the Product Environmental Footprint (PEF) method'.

[4] F. Musil, C. Harringer, A. Hiesmayr, and D. Schoenmayr, 'How Life Cycle Analyses are Influencing Power Electronics Converter Design', in PCIM Europe 2023; International Exhibition and Conference for Power Electronics, Intelligent Motion, Renewable Energy and Energy Management, May 2023, pp. 1–9. doi: 10.30420/566091368.

[5] J. W. Kolar, 'Net-Zero-CO2 by 2050 is NOT Enough!', in 2023 25th European Conference on Power Electronics and Applications (EPE'23 ECCE Europe), Sep. 2023, pp. 1–2. doi: 10.23919/EPE23ECCEEurope58414.2023.10264345.

[6] B. Kim, C. Azzaro-Pantel, M. Pietrzak-David, and P. Maussion, 'Life cycle assessment for a solar energy system based on reuse components for developing countries', Journal of Cleaner Production, vol. 208, pp. 1459–1468, Jan. 2019, doi: 10.1016/j.jclepro.2018.10.169.

[7] L. Tschümperlin, P. Stolz, and R. Frischknecht, 'Life cycle assessment of low power solar inverters (2.5 to 20 kW)', treeze Ltd., commissioned by Swiss Federal Office of Energy (SFOE), Uster, CH, 2016.

[8] C. Antoniou et al., 'Life cycle analysis of power electronics and electric machines for future electrified passenger cars: Powertrain Systems for Net-Zero Transport conference, 2021', Powertrain systems for net-zero transport - Proceedings of the 2021 Powertrain Systems for Net-zero Transport Conference, 2021, pp. 315–332, 2022, doi: 10.1201/9781003219217-18.

[9] S. Schmidt, S. Glaser, and M. L. Makoschitz, 'A "life cycle thinking" approach to assess differences in the energy use of SiC vs. Si power semiconductors: e·nova 2021, International Conference', e-Nova 2021, 2021.

[10] B. Baudais, H. Ben Ahmed, G. Jodin, N. Degrenne, and S. Lefebvre, 'Life Cycle Assessment of a 150 kW Electronic Power Inverter', Energies, vol. 16, no. 5, Art. no. 5, Jan. 2023, doi: 10.3390/en16052192.

[11] R. W. A. A. De Doncker, D. M. Divan, and M. H. Kheraluwala, 'A three-phase soft-switched high-power-density DC/DC converter for high-power applications', IEEE Transactions on Industry Applications, vol. 27, no. 1, pp. 63–73, Jan. 1991, doi: 10.1109/28.67533.

[12] K.-M. Lee and A. Inaba, 'Life Cycle Assessment: Best Practices of International Organization for Standardization (ISO) 14040 Series', APEC. [Online]. Available: https://www.apec.org/publications/2004/02/life-cycle-assessment-best-practices-of-international-organization-for-standardization-iso-14040-ser

[13] A. Nordelöf, 'A scalable life cycle inventory of an automotive power electronic inverter unit—part II: manufacturing processes', Int J Life Cycle Assess, vol. 24, no. 4, pp. 694–711, Apr. 2019, doi: 10.1007/s11367-018-1491-3.

[14] R. Krishnan and K. R. M. Nair, 'Carbon Footprint of Transformer and the Potential for Reduction of CO2 Emissions', in 2019 IEEE 4th International Conference on Technology, Informatics, Management, Engineering & Environment (TIME-E), Nov. 2019, pp. 138–143. doi: 10.1109/TIME-E47986.2019.9353301.

[15] P. Gómez, D. Elduque, C. Pina, and C. Javierre, 'Influence of the Composition on the Environmental Impact of Soft Ferrites', Materials (Basel), vol. 11, no. 10, p. 1789, Sep. 2018, doi: 10.3390/ma11101789.

[16] '"Data Page: Carbon intensity of electricity generation", part of the following publication: Hannah Ritchie, Pablo Rosado and Max Roser (2023) - "Energy". Data adapted from Ember, Energy Institute.'

[17] 'eCO2mix - CO2 emissions per kWh of electricity generated in France'. [Online]. Available: http://www.rte-france.com/en/eco2mix/co2-emissions

[18] 'Carbon Dioxide Emissions From Electricity - World Nuclear Association'. [Online]. Available: https://www.world-nuclear.org/information-library/energy-and-the-environment/carbon-dioxide-emissions-from-electricity.aspx

[19] 'Le rendement européen d'un onduleur photovoltaïque'. [Online]. Available: https://www.photovoltaique.guidenr.fr/informations_techniques/onduleur-photovoltaique/rendement-europeen.php

[20] A. Fouineau, M.-A. Raulet, M. Guillet, F. Sixdenier, and B. Lefebvre, 'A Medium Frequency Transformer Design Tool with Methodologies Adapted to Various Structures', in 2020 Fifteenth International Conference on Ecological Vehicles and Renewable Energies (EVER), Monte Carlo, Monaco, Sep. 2020. doi: 10.1109/EVER48776.2020.9243104.

[21] 'Global carbon taxes by country 2023', Statista. [Online]. Available: https://www.statista.com/statistics/483590/prices-of-implemented-carbon-pricing-instruments-worldwide-by-select-country/

PCIM Europe 2024, 11– 13 June 2024, Nuremberg   DOI: 10.30420/566262376

Thermo-Electrical Analysis and Performance: A Comparative Study between Modular and Discrete Approaches

Stefano Orlando[1], Daniela Cavallaro[1], Marco Papaserio[1], Ludovica Longo[1], Alessandra Cascio[1], Domenico Nardo[2]

[1] STMicroelectronics, Italy
[2] STMicroelectronics, Germany

Abstract

In this paper a comparison between discrete and modular approaches is carried out. The evaluation is divided into four main steps: mechanical comparison, package parasitic extraction and switching performance evaluation, power losses evaluation inside a 22kW CLLC converter and, finishing with a thermal analysis, to figure out the max power dissipation to not exceed max allowable Tj for each packages.

1   Introduction

Nowadays, the power conversion sector is going through a period of strong modernization linked with the changes of how the electrical power is generated and managed.

Moreover, the demand for electricity increases and feed several different loads with different specifics and requirements.

This modernization leads to a redesign of all converters inside the power conversion stage. The requirements for these modern applications can be summarized as: high power density, increasing power efficiency, better thermal management, lighter and smaller power converters and in some cases the converters need to implement bidirectionality for vehicle to grid (V2G) and vehicle to load (V2L).

Semiconductors are becoming critical to achieve this target of efficiency and compactness associated with the new power conversion topologies.

To fulfill these new requirements, both the technology of power devices and packages are subject to disruptive changes.

Therefore, traditional silicon devices are updated to the wide band gap (WBG) technologies and innovative packages, such as top side cooling surface mounted device (SMD) and power module are necessary to target the demand of power density and power level.

The combination of new technologies and new packages enabling better thermal management and help lowering both the weight and the size of power conversion components.

2   Mechanical comparison

The first comparison is related to the physical dimension of the two packages, reported in table 1.

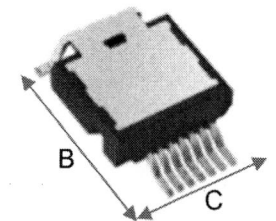

Fig. 1   HU3PAK, discrete package

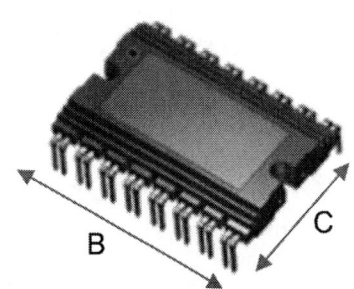

Fig. 2   DMT-32, power module, with full bridge configuration

Starting from the area of the single package, it was calculated the total space on the PCB for different topologies, full bridge and 3 phase B6.

2673

Package dimensions		
	HU3PAK	**DMT-32**
A (mm)	3.5	3.3
B (mm)	18.58	44
C (mm)	14	27.4
Area (B*C) (mm^2)	260.12	1205.6

Table 1 Physical dimension of the two solutions, discrete and power module

For the power module the area is always fixed, because it is possible to have it with different internal layouts; while for the HU3PAK is different and changes with the number of devices required by the topology.

PCB Area for different topologies		
	HU3PAK	**DMT-32**
Full bridge configuration (4 devices)	1040 mm^2	1205.6 mm^2
3ph B6 (six devices)	1561 mm^2	1205.6 mm^2

Table 2 PCB area occupied for different topologies considering the packages dimension

Considering, only the package dimensions, without any mechanical or physical constrains it possible to highlight that:

- For full bridge topology, HU3PAK allows to use a smaller area compared to DMT-32 by 13.73%
- For 3ph B6 topology, HU3PAK use a higher area compared to DMT-32 by 29.5%

Thus, when mechanical constraints are considered the difference between the two approaches for the full bridge case is reduced while for the 3ph B6 it is higher.

3 Parasitic elements evaluation

On top of thermal analysis, we carried out a parasitic extraction of both packages. The evaluation of parasitic electromagnetic inductances within the devices was conducted using the ANSYS Q3D simulation tool. This tool is a specialized solver for calculating RLGC parameters (resistance, inductance, capacitance and conductance). These parameters are critical for deriving the equivalent circuit and Spice model based on a given geometry.

To perform this analysis, the simulation tool examines individual paths within the device. This evaluation requires a precise definition of paths and nets representing the set of conductive objects separated by insulating materials or the surrounding nonconductive substance. Each net must include a source (where the current originates) and a sink (where the current is received).

For this study, the geometries of the two devices were imported. The electrical properties of the materials, such as relative permittivity, relative permeability and bulk conductivity, were assigned to the relevant layers within the devices.

The evaluation of parasitic elements was conducted by focusing on the four main paths present for the discrete: gate, source, drain and Kelvin connections while for DMT32, that is a full-bridge configuration module, only one leg was analyzed, and the paths DC+ HS Drain, HS Source LS Drain, and LS Source DC- were assessed. The terminations selected to serve as source and termination point are different between the power module and the discrete device. For the discrete device, the terminations are represented by the terminal surfaces that attach to the PCB. In the case of the module, however, the side surfaces of the terminals that will be inserted into the PCB are considered as terminations.

Once the 3D model is ready, it is essential to define the boundary conditions for the simulation. This includes choosing the frequency range over which the analysis will be performed. For this evaluation, a frequency of 1 MHz was selected. This frequency is commonly used as a standard reference because it provides a balanced representation of resistance and inductance values, facilitating comparison between simulation results and those derived from empirical tests. Table 3 shows the results of these evaluations.

Parasitic inductance at 1MHz			
Discrete HU3PAK		Power module DMT32	
L_{gate}	7.3 nH	$L_{DC+\_DrainHS}$	7.77 nH
L_{source}	2.9 nH	$L_{sourceHS\_DrainLS}$	3.03 nH
L_{drain}	1.6 nH	$L_{sourceLS\_DC-}$	7.9 nH
L_{kelvin}	8 nH	L_{gate}	10.9 nH

Table 3 Parasitic inductance evaluation for both devices at 1MHz

Comparative analysis of the parasitic components in the two packages under investigation indicates that the integrated module has a total loop inductance of 18.7 nH. For the discrete package, however, it is not sufficient to sum up the simulated values of the parasitic elements. In fact, the additional inductance supplied by the PCB must also be considered. With an assumed distance of at least 0.5 cm between the high-side (HS) and low-side (LS) devices, the inductance contributed by the PCB is estimated to be approximately 12.7725 nH. Consequently, the overall circuit inductance for the discrete package exceeds 21.77 nH. The inductance loop of both packages is shown in fig. 3.

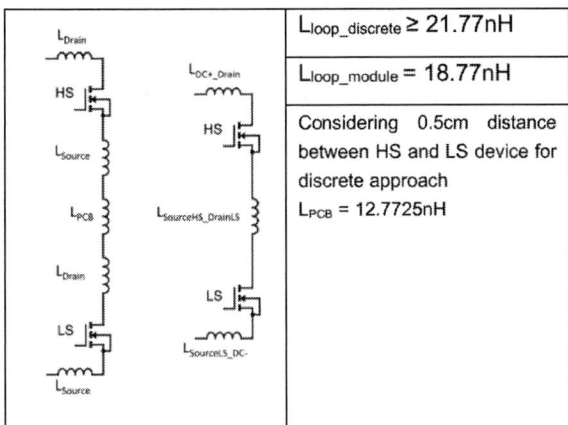

	$L_{loop\_discrete} \geq 21.77nH$
	$L_{loop\_module} = 18.77nH$
	Considering 0.5cm distance between HS and LS device for discrete approach $L_{PCB} = 12.7725nH$

Fig. 3 Loop inductance for both discrete and module approach

The extracted parasitic elements are used in the next section to evaluate switching performances. Switching performance.

4 Switching performance analysis

After the evaluation of the parasitic elements of the two packages under analysis, the impact of the different stray inductance is analyzed in association with the switching performances, at both turn-on and turn-off.

Figures 4 and 5 show respectively turn-on for the discrete and for the power module. The turn-on is performed at the same conditions for both DUTs, in terms of external gate resistance, current level and bus voltage.

The under-analysis parameter during the turn-on is the drain source current, especially the current peak, as reported in table 3, due to the lower parasitic inductances the total current peak for DMT32 is smaller, about 28.6%, than discrete case, this will help lowering the turn-on losses.

While for the turn-off the investigated waveforms is the drain source voltage. Again, thanks to the lower parasitic inductance in the power loop, the overvoltage associated with the stray inductance is lower for DMT32.

In this case the reduction of overvoltage is about 15%.

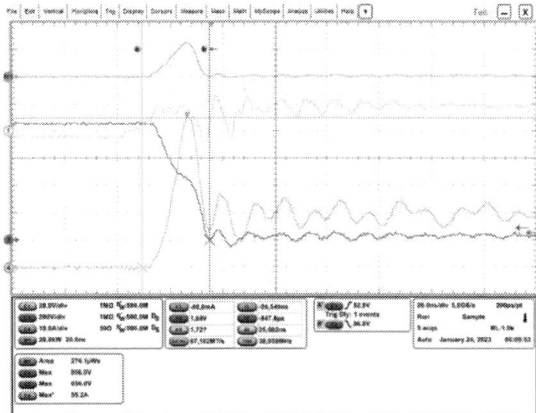

Fig. 4 Turn-on for discrete. In green I_{DS}, purple V_{DS}, yellow V_{GS}

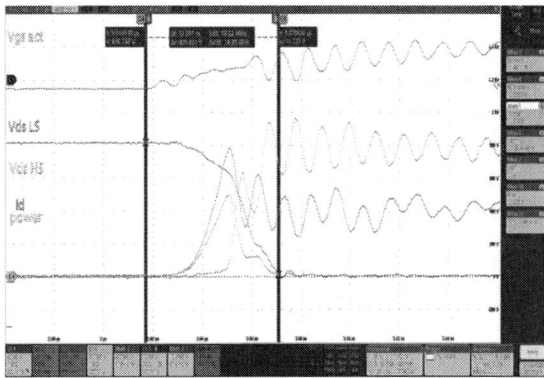

Fig. 5 Turn-on for power module. In orange I_{DS}, dark green V_{DS}, yellow V_{GS}

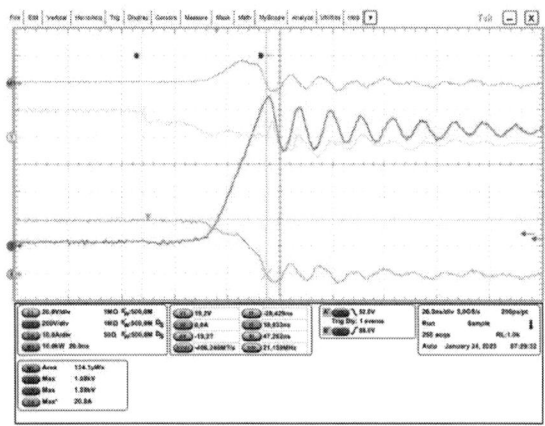

Fig. 6 Turn-off for discrete. In green I_{DS}, purple V_{DS}, yellow V_{GS}

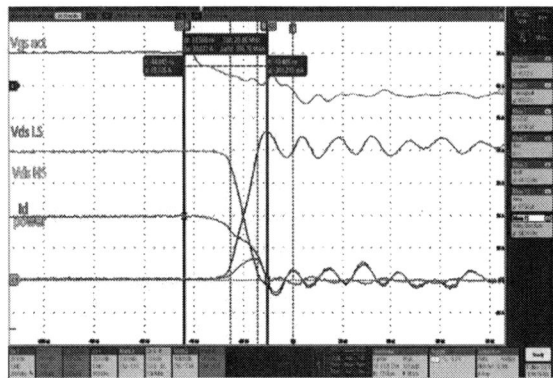

Fig. 7 Turn-off for power module. In orange I_{DS}, dark green V_{DS}, yellow V_{GS}

Turn-on		
	Discrete	Module
dv/dt (V/ns)	30.3	24.4
Current peak(A)	55.2	39.4
Turn-off		
	Discrete	Module
dv/dt (V/ns)	48.7	64.14
Overshoot (V)	1080	917

Table 3 Switching performances summary for both turn-on and turn-off

Lowering the overvoltage gives the designer more room to optimize the gate driving, and to use smaller external gate resistance to optimize the efficiency of the commutation by the reduction of the switching times and thus switching energies.

5 Power losses

After the evaluation of the switching performances an applicative analysis was carried out to find the max power rating of the application associated with the max power losses sustained by both packages.

A DCDC stage of an OBC, with a CLLC configuration is chosen and depending on the approach, discrete or modular, the DUTs are 4 HU3PAK for discrete while 1 DMT-32 for the modular one. For both packages the same SiC chip is used a 1200V – 40 mΩ MOSFET, reported in table 5.

In table 4 are reported the main parameters of the converter.

Vinput (V)	850
Vout (V)	800
Cr (nF)	60.392
Lr (µH)	12.94
Lm (µH)	103.52
fres (kHz)	180
Pmax (kW)	22
Dead time (ns)	200

Table 4 CLLC parameters

Where Cr and Lr are respectively the resonant capacitor and inductor of resonant tank and Lm is the magnetizing inductance.

$R_{DSONtyp}$ (mΩ)	40
V_{SD} (V)	4
Q_g (nC)	52
C_{oss} (pF)	57
C_{rss} (pF)	6
C_{iss} (pF)	1510
R_g (Ω)	6.8
V_{GSon} (V)	18
V_{GSoff} (V)	-5

Table 5 SiC MOSFET parameters

With the following equations, equation 1 to equation 6, the power losses at 22kW are evaluated.

$$P_{cond} = I_{RMS\_mos}^2 \cdot R_{DS(on)} \cdot \alpha \qquad \text{Eq. (1)}$$

$$P_{swoff} = (E_{off} - E_{Oss}) \cdot f_{SW} \qquad \text{Eq. (2)}$$

$$P_{diode} = I(mean\_diode) \cdot V_{SD} \qquad \text{Eq. (3)}$$

$$P_{driver} = Q_g \cdot V_{GS} \cdot f_{SW} \qquad \text{Eq. (4)}$$

$$I_{OFF\_mos} = \frac{n \cdot V_{OUT} \cdot T_R}{4 \cdot L_m} + \frac{V_{in} \cdot 2 \cdot t_{dead}}{4 \cdot (L_m + L_r)} \qquad \text{Eq. (5)}$$

$$E_{off} = I_{OFF\_mos} \cdot V_{DS} \cdot t_{fall} \cdot 0.5 \qquad \text{Eq. (6)}$$

The power losses for the single device are equal to 35.96W while for the full stage, 4 devices, the power losses are equal to 143.84W.

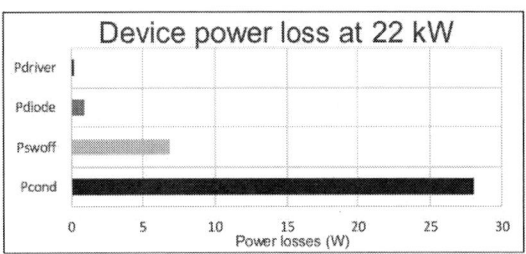

Fig. 8 Power losses breakdown for one device in the CLLC stage at 22kW

6 Thermal analysis

The simulation trials described in this paper have been performed on HU3PAK, a discrete device housing one SiC MOSFET chip, and ACEPACK DMT32 FULLBRIDGE ®, a module that contains four SiC MOSFET chips in full-bridge configuration mounted on a DBC.

A Finite Element Analysis (FEA) was carried out by simulating the thermal behavior of the module to evaluate thermal performance of both simulated solutions, to evaluate the maximum power that can be sustained by each one.

Both devices are attached to the same aluminum heatsink (13cm*15cm*1cm), used as reference, by a thermal interface material (TIM) with a conductivity of 3 W/mK. The material stacks of the devices, shown in figure 9, are the same, except for the DBC layer in the module. For this reason, the thickness of the TIM is 50 μm for DMT-32, while to provide insulation for HU3PAK without DBC, a TIM thickness of 600 μm is considered.

	HU3PAK	DMT32
Th TIM (μm)	600	50
λ TIM (W/m·K)	3.0	3.0
P$_{die}$ (W)	36	60
T$_{jmax}$ (°C)	150.32	150.980
T$_{javg}$ (°C)	145.6	141.546
T$_{case\ max}$ (°C)	136.18	108.48
T$_{heatsink\ max}$ (°C)	56.613	76.03
Rth $_{j-case}$ (°C/W)	0.262	0.551
Rth $_{j-heatsink}$ (°C/W)	2.472	1.092

Table 6 Main output and thermal parameters used in FEM analysis

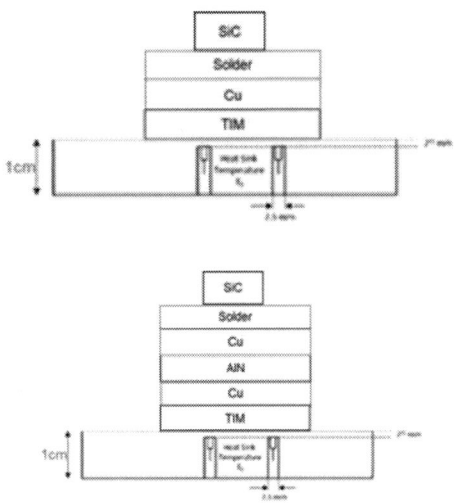

Fig. 9 Stack of materials of a) HU3PACK and b) DMT32

On the backside of heatsink a heat transfer coefficient of 10000 W/m²K has been fixed, considering a reference temperature of 50°C. Moreover, the main parameter for thermal simulation trials are thermal properties, especially Heat Capacitance, Density and Thermal Conductivity, for each material involved in the system.

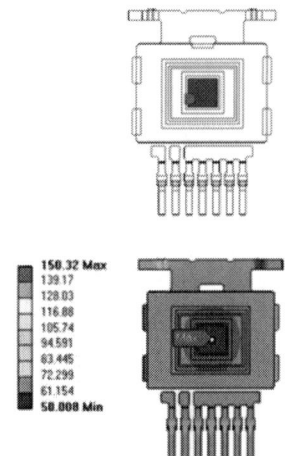

Fig. 10 HU3PAK thermal map at max power, 36W

Heat is assumed to be generated on the active volume of the die, and it is applied in terms of heat generated (power per unit volume) on this part (P_{die})

To investigate mutual thermal interaction between the dice, Power has been applied to all MOSFETs

in order to reach a 150°C maximum temperature junction.

By equation 7 and 8 we estimated the Rthjcase and Rthj-sink, where Tjavg is the average junction temperature of hottest die. T_{case} is referred to the maximum temperature on the contact surface between TIM layer and Cu layer. T_{sink} refers to the maximum heatsink temperature, taken 2mm from the top heatsink surface in contact with TIM layer, to reproduce experimental test.

$$Rth_{(j-case)} = \frac{Tjavg - Tcasemax}{P\_appl} \qquad \text{Eq. (7)}$$

$$Rth_{(j-heatsink)} = \frac{Tjavg - Theatsinkmax}{P\_appl} \qquad \text{Eq. (8)}$$

The max power losses to not exceeding the max Tj are: 36W for discrete and 240W for power module (60W per Die in the Module).

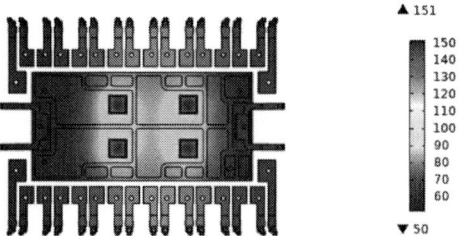

Fig. 11 DMT-32 thermal map at max power, 240W (60W per die)

7   Conclusions

In this paper a comparison between discrete and modular approaches was carried out. The paper analyzed the thermal performance of discrete and modular approaches, to find the limit of both solutions.

In the selected conditions the max power dissipation for the discrete package is 36W while for the power module the maximum is 60W (for each die). From the power losses analysis in a 22kW CLLC was found the device power losses at full load, equal to 35.96W, thus it is possible to point of the following considerations: both approaches are possible for this power rating but, HU3PAK works close to its limit so with high thermal stress and a more accurate thermal design is needed.

Differently DMT32 works far from its thermal limit thus a more relaxed thermal design is possible or is possible to shrink heatsink dimension.

References

[1] D. Nardo, et al., "Top-side cooling packages: disrupting technology to boost power density and performance in high-end power conversion systems" PCIM, 2023

[2] G. Mauromicale et al., "Modeling and Thermal Analysis of Cooling Solutions for High Voltage SMD Packages," PCIM Europe digital days 2021; International Exhibition and Conference for Power Electronics, Intelligent Motion, Renewable Energy and Energy Management, 2021.

[3] G. Bazzano, D. Cavallaro, R. Greco, A. Raffa, P. Veneziano "3D electro-thermal simulation of multilayer power MOSFET structure under electro-thermal stress" PCIM Europe digital days 2015; International Exhibition and Conference for Power Electronics, Intelligent Motion, Renewable Energy and Energy Management, 2015.

[4] D. Jo, S. Jang, K. Lee, T. Yim, B. Lee, J. Lee "Thermal Performance Comparison between New 650V Automotive Smart Power Module and Conventional Automotive Discrete IGBT" PCIM Europe 2019.

PCIM Europe 2024, 11– 13 June 2024, Nuremberg   DOI: 10.30420/566262377

Impact of Parameter Spread in Parallel-Operated SiC MOSFETs for Hard-Switching Conversion

Andrea Piccioni[1], Niklas Seltner[1]

[1] Infineon Technologies Austria AG, Austria

Corresponding author: Andrea Piccioni, andrea.piccioni@infineon.com
Speaker: Andrea Piccioni, andrea.piccioni@infineon.com

Abstract

This paper examines the impact of parameter spread in parallel-operated SiC MOSFETs during high switching frequency and hard switching operations. Seventy samples of a 1200V $60m\Omega$ rated SiC device in a TO-247 4-pin package were selected for the study. These samples were characterized for screening threshold voltage $V_{GS(th)}$ and transfer characteristics $I_D(V_{GS})$. The study focused on the dynamic performance imbalance caused by unequal current sharing, specifically during the turn-on event. This imbalance in current sharing arising from parameter spread was investigated through a double-pulse test and mathematical modeling of the phenomena. On top, this study provides recommendations for mitigating dynamic current imbalances. Finally, a temperature assessment showing the effects of imbalanced energies in continuous testing for a half-bridge DC-DC conversion is also presented.

1 Introduction

The demand for higher efficiency and high-switching frequency applications has led to a substantial shift towards the adoption of SiC MOSFETs in power electronic applications. SiC MOSFETs offer unmatched advantages, including operation under higher voltage and temperature, making them a promising alternative to traditional silicon-based devices [1][2][3]. One of the key challenges in achieving high current ratings in SiC MOSFETs lies in the limitations imposed by the manufacturing process. These limitations result in the production of SiC MOSFET chips with current ratings capped at 150 A consequently restricting the current rating of the single chip or devices available. As a result, parallel connection of multiple discrete devices or multi-chips in power modules is a common approach for achieving higher current ratings. The parallel operation of SiC MOSFETs is susceptible to current imbalances. This current imbalance is influenced by a variety of parameter spreads. It can lead to an electro-thermal imbalance that will eventually diminish the reliability of the system [4] [5]. Several studies in the realm of chip characterization and sorting have identified threshold voltage variations in SiC MOSFET devices that can signifi-

cantly impact current distribution in paralleled units. The spread in transfer characteristics also plays a role in determining the dynamic performance and current sharing characteristics of paralleled SiC MOSFETs [6][7][8]. This paper aims to comprehensively explore the challenges posed by parameter spread in the parallel operation of SiC MOSFETs and propose potential solutions to mitigate their effects. By delving into the factors affecting the current sharing, this paper provides insights into the limitations and opportunities associated with parallel configuration of SiC MOSFETs for high-power applications. The study was divided into several stages: a database of 70 devices was created by screening the two main parameters influencing the dynamic current unbalance ($V_{GS(th)}$ and $I_D(V_{GS})$) influencing dynamic current unbalance. This provided a guideline for possible worst-case scenarios designers may have to consider while designing paralleled SiC MOSFETs. Then selective double-pulse tests were conducted to establish the relation between parameter spreads and turn-on energy disparity. Finally, through mathematical modeling, the effect of energy disparity between the devices under tests (DUT) was translated into thermal imbalances in applications. This clarifies the effect of internal lot variations on worst-case electro-thermal imbalances in real application scenario.

PCIM Europe 2024, 11– 13 June 2024, Nuremberg   DOI: 10.30420/566262377

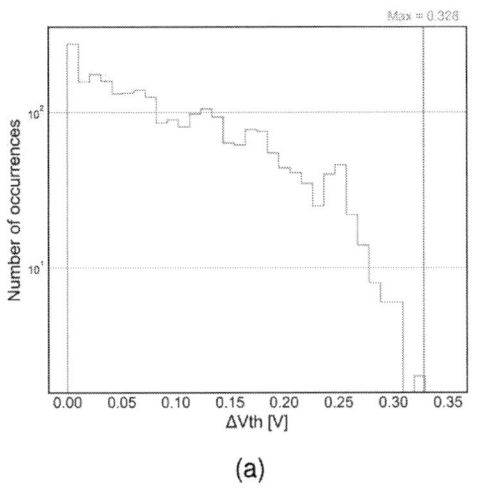 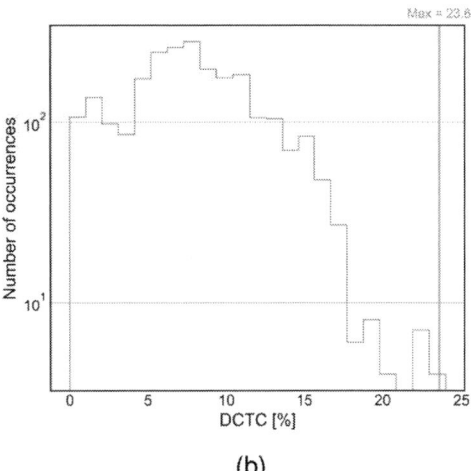

(a) (b)

Fig. 1: Statistical representation via the number of occurrence of (a) $\Delta V_{GS(th)(i,j)}$ and (b) $DCTC_{\%(i,j)}$

2 Devices characterization

In this paper, the screening of the parameters that influence current imbalance is limited to the $V_{GS(th)}$ and the $I_D(V_{GS})$, because the static current imbalance caused by $R_{DS(ON)}$ variation is compensated by the significant positive temperature coefficient (PTC). The devices are 70 commercially available new IMZ120R060M1H. They were characterized, and consequently, the analysis is a matrix approach with 2380 possible cases (70 x 70, symmetrical with the diagonal that represents the self-comparison). Assuming DUT*i* and DUT*j* as a couple of devices under examination, the figures of merit (FoMs) to evaluate their ability to parallel are the difference of threshold voltages $\Delta V_{GS(th)(i,j)}$ (1) and the distance coefficient of transfer characteristics (DCTC) which represents the average distance between the transfer curve between the two DUTs $DCTC_{\%(i,j)}$ (2):

$$\Delta V_{GS(th)(i,j)} = |V_{GS(th)(i)} - V_{GS(th)(j)}| \qquad (1)$$

$$DCTC_{\%(i,j)} = \begin{cases} NaN & V_{GS} < V_{GS(th)} \\ \frac{100}{m}\sum_k^m \frac{|I_{D(i,k)} - I_{D(j,k)}|}{I_{D(i,k)} + I_{D(j,k)}} & V_{GS} \geq V_{GS(th)} \end{cases}$$
$$(2)$$

Here, m is the discretization factor of the transfer curves, specifically, the sweep points of V_{GS}, and was kept constant at 90. $I_{D(i,k)}$ and $I_{D(j,k)}$ represent the values of drain current under the same gate voltage for transfer curves of DUT*i* and DUT*j*. The device characterization is at room temperature

since it has been demonstrated that the variation of $V_{GS(th)}$ and $I_D(V_{GS})$ manifests poor sensitivity to junction temperature [6]. After all the attributes from the 70 samples were organized in a data frame, a statistical analysis of the $\Delta V_{GS(th)(i,j)}$ and $DCTC_{\%(i,j)}$ was performed. Figure 1a shows the $\Delta V_{GS(th)(i,j)}$, statistically and in maximum value (worst-case scenario). Figure 1b shows the transfer characteristics $DCTC_{\%(i,j)}$ statistically and in maximum value (worst-case scenario). Figure 2 represents the population in a 2-D graph, allowing visualization of the population across the two FOMs.

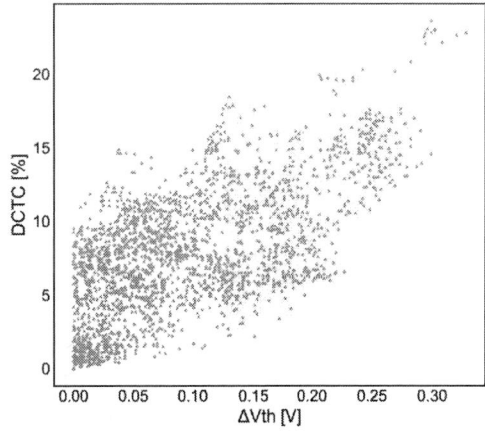

Fig. 2: Combinations of $\Delta V_{GS(th)(i,j)}$ and $DCTC_{\%(i,j)}$ in a 2-D plot for the entire population of samples

2681

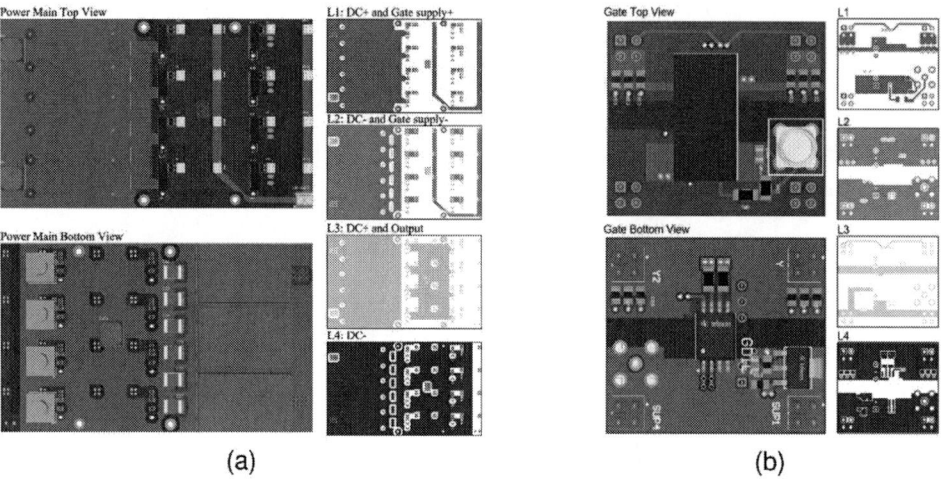

Fig. 3: (a) Realistic and fabrication view of the main half-bridge power board (b) Realistic and fabrication view of the gate driver daughter board (not to scale)

3 Layout symmetry verification

The methodology for estimating dynamic energy imbalance was explored through a double-pulse test using a half-bridge board that could accommodate up to four devices in parallel for each switching cell, as shown in Fig.3a. The center-left (L) and center-right (R) slots were employed for the study. To meet the requirements of bandwidth and rise time, this study employed SDN shunt resistors positioned on the low side of the half-bridge leg from T and M Research manufacturer. Figure 3b shows the gate driver daughter board used to drive the two switches in the main power board with one gate driver. The swallow-tailed source-sink connection in the setup is noteworthy. It ensures gate loop symmetry which is crucial for measuring dynamic energy imbalance accurately. In fact, to decouple the impact of variations in semiconductor parameters, it's essential to ensure symmetry in the test setup [9]. The symmetry of the layout between

the left and right spots was confirmed through an ANSYS® Q3D Extractor® analysis initially (results in Table 1). The symmetry was further validated by an overlap of the double pulse test waveform using the cross-transposition method with samples placed on the left (L) and right (R) spots. Figure 4 shows identical behavior after swapping positions ($\Delta E_{on(L)-(R)} = 2\mu J$), indicating that the position of the DUTs does not affect the results, thanks to the symmetry of the layout. Additionally, all the pins of the TO247-4 packages were cut to the same lengths to ensure consistency in the measurements.

Tab. 1: Parasitic inductance extraction

From	To	L_{1MHz} [nH]
High side drain (left)	DC+	13.96
High side drain (right)	DC+	13.68
High side source (left)	OUT	17.22
High side source (right)	OUT	17.14
Low side drain (left)	OUT	17.15
Low side drain (right)	OUT	17.11
Low side source (left)	DC-	36.87
Low side source (right)	DC-	37.02

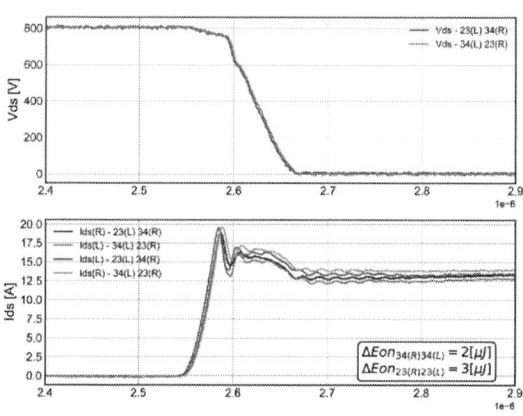

Fig. 4: Cross-transposition method with double pulse test results with samples placed on the left (L) and right (R) spots

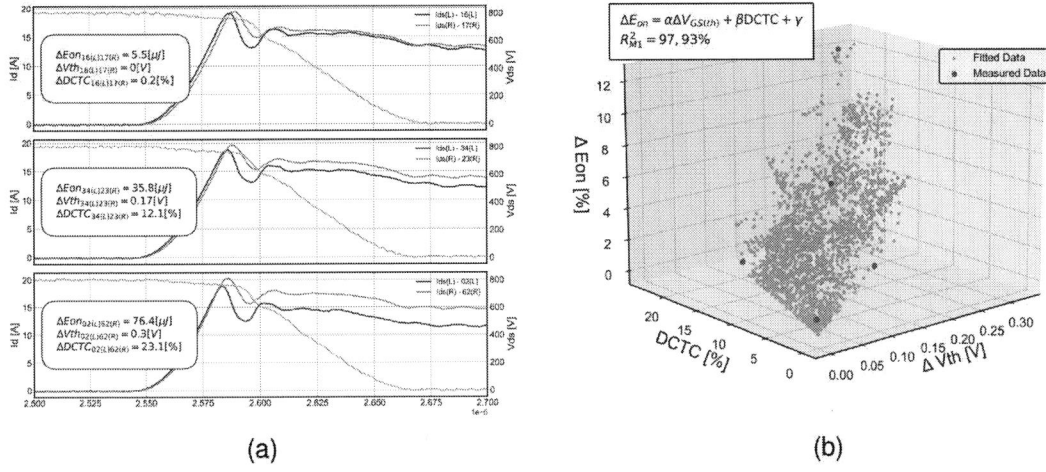

Fig. 5: (a) Measured turn-on data for three samples combinations (b) Function fitting ΔE_{on} ($\Delta V_{\text{GS(th)}}$, DCTC$_\%$)

4 Turn-on energy imbalance

Given the substantial number of sample combinations, achieving full double pulse test coverage was not practical. In similar studies, researchers often employ routine SPICE simulations or conduct a reasonable number of double pulse tests. In this study, a selective approach was adopted to pick sample combinations and perform DPTs, enabling the creation of reliable mathematical modeling through function fitting of the turn-on energy imbalance ΔE_{on}. The crucial aspect was to test combinations that allowed for sensitivity analysis of the influence of the two main parameters $\Delta V_{\text{GS(th)}}$ and DCTC$_\%$, as well as their combination, essentially conducting a corner analysis for the given population. Figures 5a and 5b outline the process for extracting the key parameter for evaluating the energy disparity ΔE_{on} from $\Delta V_{\text{GS(th)}}$ and DCTC$_\%$. The test conditions were chosen to be as close as possible to real-world applications, reflecting the hard-switching DC-DC boost converter (see Tab. 2). To prevent the influence of reverse recovery the test is performed using two equal (forward voltage screened) 40 A rated SiC Schottky barrier diodes for commutation on the high side. Note that the gate charging phase fastness significantly impacts the dynamic loss imbalance. Consequently, the ΔE_{on} can be significantly reduced by decreasing the value of the gate resistance. Figure 6 shows how decreasing the common gate resistance in the gate driver daughter board ($R_{\text{g(common)}}$) mitigates the ΔE_{on} worst-case scenario (Sample$_{02}$ and Sample$_{62}$). However, the designer should carefully select the appropriate R_g values that also fulfill the *dI/dt* and *dV/dt* requirements [10].

Tab. 2: Double pulse test conditions

Parameter	Unit	Value
V_{DC}	[V]	800
$I_{\text{(tot)}}$	[A]	26
V_{GS}	[V]	+18/-2
$R_{\text{g(single)}}$	[Ω]	14.2
$R_{\text{g(common)}}$	[Ω]	33

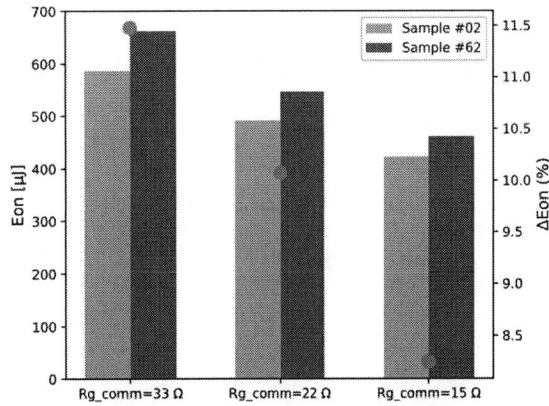

Fig. 6: Influence of the gate charging speed impacting the ΔE_{on} in magnitude and percentage

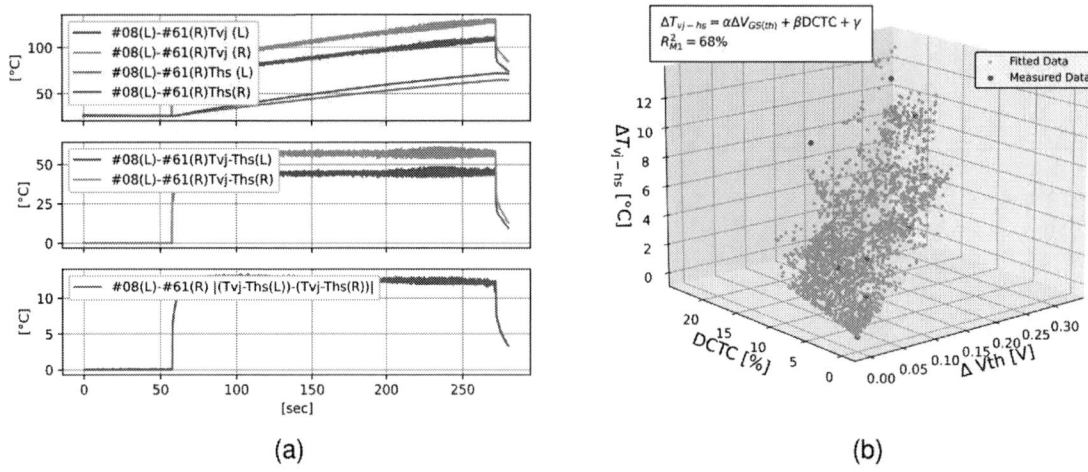

Fig. 7: (a) T_{vj} and $T_{heatsink}$ acquisition under application conditions (b) Function fitting $\Delta T_{vj\text{-}hs(i,j)}(\Delta V_{GS(th)}, DCTC_\%)$

5 Temperature imbalance under application conditions

To verify the application impacts of the parameter spreads, the board shown in Fig.3a was equipped with two separate heatsinks for the left and right spots, avoiding thermal cross-coupling effects. The devices were screw-mounted to the heatsinks via an AL_2O_3 pad. The mold compound was removed enabling real junction temperature measurement (T_{vj}) via thermal camera. The topology comprised a DC-DC boost converter switching at 65 kHz with a duty cycle of 0.5 to boost 400V to 800V (gate condition see Tab.2). Figure 7a shows the thermal camera acquisition for the 17 A load current, with DUT_{08} and DUT_{61} at the low side while being under parallel operation. With the same methodology used for ΔE_{on} modeling, Fig. 7b shows the function fitting for differences in temperature between junction and heatsink comparing the DUT_i and the DUT_j. In summary, the parameter spread of the 70 screened devices brings in the above-mentioned application conditions a worst-case scenario temperature difference of $\sim 12.5°C$.

6 Conclusion

This study presents a comprehensive investigation into the dynamic energy imbalance in parallel-operated SiC MOSFET converters, focusing on the impact of parameter spread on the performance of these devices. The methodology employed in this research, ranging from the estimation of dy-namic energy imbalance through a double-pulse test to real-world application temperature imbalances, provides a thorough understanding of the challenges and solutions in this field. The significance of maintaining a narrow parameter distribution is underscored by the results obtained in this study. It is particularly evident in the case of the turn-on energy imbalance, where a narrow distribution of parameters leads to improved performance and reduced disparity in energy loss. This study emphasizes the importance of a fast gate charging phase to achieve reliable performance in high-power, paralleled SiC-based converters. However, designers should select the appropriate gate resistance values carefully, taking into account the dI/dt and dV/dt requirements. Additional mitigation can be achieved by taking advantage of the thermal coupling effect, for example, using a common thermal interface material as a heat spreader and a common heatsink. If not enough, proper screening and clustering of the devices is recommended for applications in which the end-usage scenario demands close to zero imbalances. This approach can erase the problem of uneven temperature distribution and improve the overall performance of the high-power, paralleled SiC-based converters. The methodology employed in this study, combined with the findings, offers a comprehensive guide for designing and optimizing these devices, paving the way for better performance and reliability in high-power applications.

References

[1] A. Anthon, Z. Zhang, M. A. E. Andersen, D. G. Holmes, B. McGrath, and C. A. Teixeira, "The benefits of SiC MOSFETs in a t-type inverter for grid-tie applications," *IEEE Transactions on Power Electronics*, vol. 32, no. 4, pp. 2808–2821, Apr. 2017. DOI: 10.1109/TPEL.2016.2582344.

[2] M.-A. Ocklenburg, M. Dohmen, X.-Q. Wu, and M. Helsper, "Next generation DC-DC converters for auxiliary power supplies with SiC MOSFETs," in *2018 IEEE International Conference on Electrical Systems for Aircraft, Railway, Ship Propulsion and Road Vehicles & International Transportation Electrification Conference (ESARS-ITEC)*, Nottingham: IEEE, Nov. 2018, pp. 1–6. DOI: 10.1109/ESARS-ITEC.2018.8607463.

[3] A. K. Morya, M. C. Gardner, B. Anvari, L. Liu, A. G. Yepes, *et al.*, "Wide bandgap devices in AC electric drives: Opportunities and challenges," *IEEE Transactions on Transportation Electrification*, vol. 5, no. 1, pp. 3–20, Mar. 2019. DOI: 10.1109/TTE.2019.2892807.

[4] A. Borghese, M. Riccio, A. Fayyaz, A. Castellazzi, L. Maresca, *et al.*, "Statistical analysis of the electrothermal imbalances of mismatched parallel SiC power MOSFETs," *IEEE Journal of Emerging and Selected Topics in Power Electronics*, vol. 7, no. 3, pp. 1527–1538, Sep. 2019. DOI: 10.1109/JESTPE.2019.2924735.

[5] A. Borghese, M. Riccio, A. Castellazzi, L. Maresca, G. Breglio, and A. Irace, "Statistical electrothermal simulation for lifetime prediction of parallel SiC MOSFETs and modules," in *2020 2nd IEEE International Conference on Industrial Electronics for Sustainable Energy Systems (IESES)*, Cagliari, Italy: IEEE, Sep. 2020, pp. 383–386. DOI: 10.1109/IESES45645.2020.9210690.

[6] J. Ke, Z. Zhao, P. Sun, H. Huang, J. Abuogo, and X. Cui, "Chips classification for suppressing transient current imbalance of parallel-connected silicon carbide MOSFETs," *IEEE Transactions on Power Electronics*, vol. 35, no. 4, pp. 3963–3972, Apr. 2020. DOI: 10.1109/TPEL.2019.2934739.

[7] Y. Liu, X. Dai, X. Jiang, Z. Zeng, F. Qi, *et al.*, "A new screening method for alleviating transient current imbalance of paralleled SiC MOSFETs," in *2020 IEEE 1st China International Youth Conference on Electrical Engineering (CIYCEE)*, Wuhan, China: IEEE, Nov. 1, 2020, pp. 1–6. DOI: 10.1109/CIYCEE49808.2020.9332642.

[8] J. Ke, Z. Zhao, P. Sun, H. Huang, J. Abuogo, and X. Cui, "New screening method for improving transient current sharing of paralleled SiC MOSFETs," 2018.

[9] J. Chen, H. Peng, Z. Cheng, X. Liu, Q. Xin, *et al.*, "A novel power loop parasitic extraction approach for paralleled discrete SiC MOSFETs on multilayer PCB," *IEEE Journal of Emerging and Selected Topics in Power Electronics*, vol. 9, no. 5, pp. 6370–6384, Oct. 2021. DOI: 10.1109/JESTPE.2021.3071494.

[10] N. Oswald, P. Anthony, N. McNeill, and B. H. Stark, "An experimental investigation of the tradeoff between switching losses and EMI generation with hard-switched all-si, si-SiC, and all-SiC device combinations," *IEEE Transactions on Power Electronics*, vol. 29, no. 5, pp. 2393–2407, May 2014. DOI: 10.1109/TPEL.2013.2278919.

PCIM Europe 2024, 11– 13 June 2024, Nuremberg   DOI: 10.30420/566262378

Assessment of the $R_{ds,on}$ of SiC MOSFET Dies Through Kelvin Wire Connection

Philipp Rehlaender[1], Klaus Neumaier[1], Kaone Bogopa[1], Lukas Richert[1], Sara Kuzmanoska[1]
[1] onsemi (Germany) GmbH, Germany

Corresponding author: Philipp Rehlaender, philipp.rehlaender@onsemi.com
Speaker: Philipp Rehlaender, philipp.rehlaender@onsemi.com

Abstract

The market for silicon-carbide modules has seen phenomenal growth in recent years and is expected to grow further. Tier 1 suppliers and OEMs are expanding into the market to develop their own modules to reduce costs. As a result, the market for bare-die SiC MOSFET bare dies is expected to witness significant growth. Correctly assessing the performance of the die, therefore, is of utmost importance for the producer and the module manufacturer. This paper compares and reviews the measurement process for singulated die measurements of the $R_{ds,on}$ and the characterization of packaged dies to outline the limitations when using a packaged die to assess the performance of a bare die. Finally, the method of using the Kelvin pin to assess the on-state resistance is introduced and the fidelity of this method is investigated. Finite-element simulations are provided to support this investigation showing the significant influence from the packaging on the $R_{ds,on}$.

1   Introduction

The electric vehicle market has seen phenomenal growth over the last few years and is expected to grow with a staggering 20 percent CAGR through 2030 [1]. The sale of electric vehicles is expected to surge in the coming years rising from 10.5 million in 2022 to 27 million in 2026 [2]. The surge in electric vehicles will cause an increasing demand in silicon-carbide (SiC) MOSFETs, which are a primary component of traction inverters, on-board chargers, and DC-DC converters. Nearly 90 % of all SiC usage is in the SiC modules in the main inverter [3]. The increasing demand for power semiconductors in electric vehicles is expected to reach 3 billion USD in 2025 and 10 billion USD in the next decade, resulting in a significant reshaping of the market [4]. Many Tier 1 suppliers or OEMs have started to develop custom modules to reduce costs and increase revenue [3, 4]. This results in an increasing demand for bare dies with many chip SiC suppliers forming strategic partnerships with those Tier 1 suppliers and the OEMs [5–11]. With the increase of bare die business, the SiC producer is required to evaluate the die in a test package to verify common parameters such as $R_{ds,on}$. Correctly assessing the $R_{ds,on}$ of the employed bare dies in the test package is obviously key in building a pareto-optimal module with an optimal trade-off between cost, power density and losses. Correctly assessing the $R_{ds,on}$, however, is not a trivial task.   The test engineer must ensure

that the pulse width is kept to a minimum to assess the $R_{ds,on}$ at the correct junction temperature and avoid self-heating from the power loss of the die. Furthermore, all parasitic resistances need to be excluded from the measurement and the package should be representative of the actual application. At room temperature, the $R_{ds,on}$ of the die is usually measured using a wafer prober. The die is contacted using several needles that press against the metallization of the die. To exclude the contact resistance, a Kelvin probe is placed both at the source contact and drain contact to achieve a four-wire measurement. With several needles pressing against the die, the current is also well distributed. A drawback of the wafer prober is, however, that the long wires connected to the power device analyzer introduce a significant parasitic inductance, which substantially increases the settling time of the current pulse. As a result, several hundred microseconds may be necessary to achieve a settled current level, which itself heats up the device (self-heating). The actual temperature of the die is thus higher than the temperature of the tester. At elevated test temperatures, this problem becomes much more critical since the $R_{ds,on}$ is highly temperature dependent, which increases the power loss inside the die while measuring the resistance. A further limitation results from the tester itself as the needles limit the current level at which the die can be tested.

PCIM Europe 2024, 11– 13 June 2024, Nuremberg   DOI: 10.30420/566262378

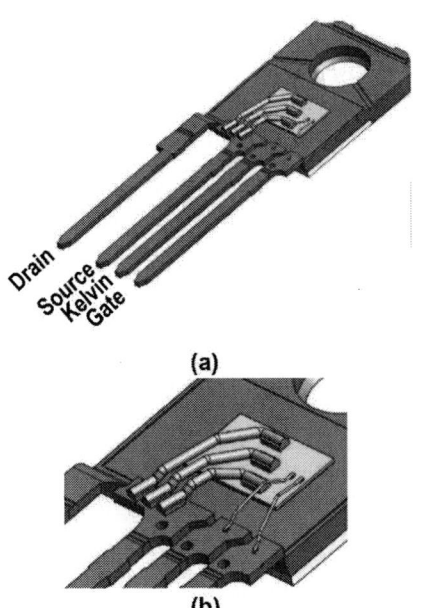

(a)

(b)

Fig. 1 Visualization of a die placed inside a TO-24-4L package with a Kelvin connection. (a) full package view, (b) zoom to the connection of the die with 3 source bond wires, a Kelvin bond wire and a gate bond wire.

As a result, for higher currents and reduced testing pulse widths, it is necessary to assess the $R_{ds,on}$ of the die via an evaluation package. The industry standard is to package the die inside a TO-247 package with three or four leads (drain, source, gate and optionally Kelvin). The package, depicted in Fig.1 (a), adds additional resistance to the die in the form of bond wires and leads. The Kelvin pin has become industry standard for wide-bandgap semiconductors to increase the switching speeds by providing a reference potential for the gate voltage without the influence from the voltage drop from the stray inductance and the current slope of the package. This enables faster switching speeds consequently resulting in reduced switching losses. The Kelvin pin is directly connected to the die (visualized in Fig. 1 (b)) using a bond wire. In the TO-247-4L the die attachment is commonly achieved through soldering while the source pads are attached using aluminum bond wires.

This contrasts with more sophisticated top side interconnections as only a small area is contacted and bond wires may result in reliability issues [12]. To further increase the current density in SiC inverters, more advanced packaging methods are required [13]. To increase the contact area and reduce the current spreading effect, many methods have been proposed such as ribbon bonding [14–16] where a large ribbon is placed on the source pad to increase the contact area. Double-sided

sintering has also been shown to increase the surface area and reduce contact resistance [17, 18]. Sintered [19] or soldered current buffers or a flip-chip structure [20] have also shown to be beneficial. A detailed investigation of more advanced methods was provided in [21] highlighting the extensive research process to develop more advanced topside interconnections for increased power density.

This paper analyzes the effect of the topside interconnection on the overall packaged $R_{ds,on}$. and analyzes the Kelvin connection as an appropriate measurement technique to assess the die $R_{ds,on}$. The paper is structured as follows: section 2 reviews the characterization of packaged semiconductors. Section 3 investigates the impact of the package through FEM analysis. Section 4 introduces the Kelvin wire as a tool to characterize the $R_{ds,on}$ of the actual die. The results are compared to singulated die measurements in section 5 before the paper is concluded.

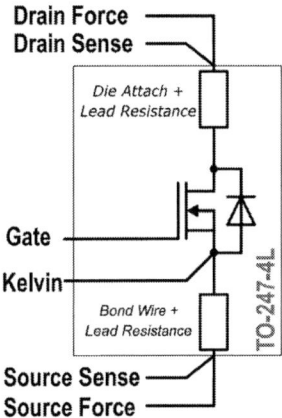

Fig. 2 Visualization of the equivalent circuit diagram of a MOSFET inside a package neglecting parasitic capacitance and inductance with separate force and sense connection for drain and source

2 Characterization of Packaged Semiconductors

When characterizing the $R_{ds,on}$ of the die, it is important to employ a four-wire measurement method with a separate force and sense connection. When characterizing discrete components, it is necessary to have two connections for both the source and drain lead, with one connection used for measuring the actual voltage on the device and the other connection used for forcing the current through the semiconductor. This is visualized in Fig. 2. In this setup, the Gate and Kelvin leads are used for biasing the gate of the MOSFET.

2687

Measurement results for a 1200 V semiconductor are shown in Fig. 3 showing the influence on $R_{ds,on}$ resulting from the self-heating associated with larger pulse widths. While the difference between 74 µs and 100 µs is insignificant for lower current values, a significant influence can be noted between 100 µs and 250 µs showing a small increase for room temperature and an almost 1.27 mΩ increase at 175°C.

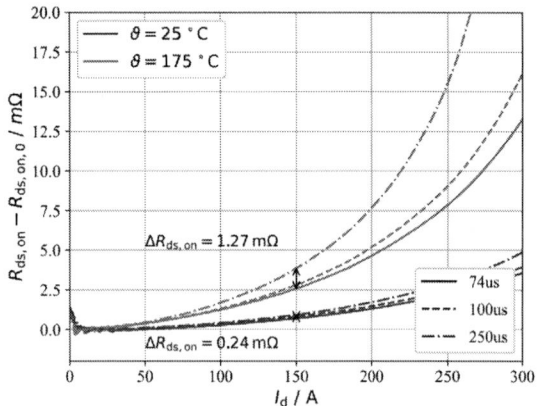

Fig. 3 $R_{ds,on}$ measurement showing the impact of the pulse width on the measured resistance. A significant self-heating can be observed for longer pulse widths. The plot is normalized to the low-current measurement to emphasize the impact.

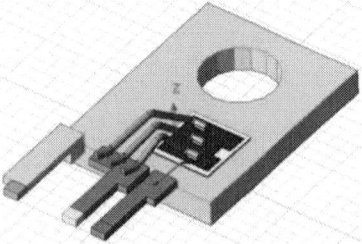

Fig. 4   Visualization of the 3D model of the TO-247-4L package with three bond wires and the Kelvin contact.

3   Analysis of the Package Influence Using FEM Simulations

To investigate the influence of the bonding and the package, the package was modeled in an FEM simulation. This allows for an investigation of the current spreading in the top metallization and its influence on the overall $R_{ds,on}$ of the device.

3.1   Methodology

Apart from the metallization, the bond wires, and the leads, which have been modelled explicitly, the die has been modeled using a solid body with the specific resistance calibrated to fit the measurement concept. The simulation model is depicted in

Fig. 4. Two models have been created for 25°C and 175°C, both calibrated to measurements. The metallization layers, pad, and VIA layer of the MOSFETs have been modeled corresponding to the same properties as the actual semiconductor.

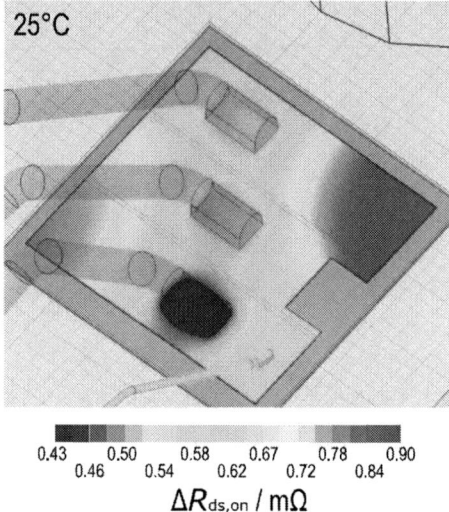

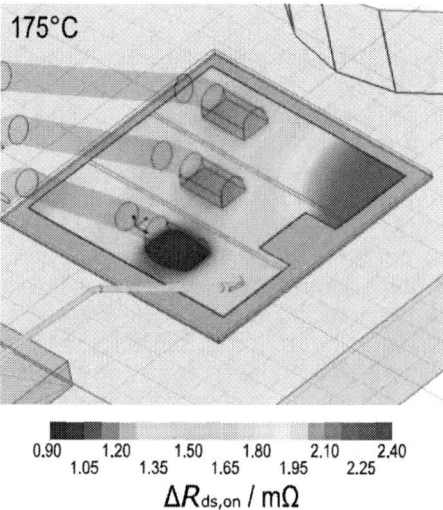

Fig. 5   FEM simulations of the die inside the package visualizing the equivalent resistance difference for a 100 A simulation at 25°C (top) and 175°C (bottom). The plot is normalized to the potential of the source lead.

3.2   Current Spreading in the Top Metallization

The results are shown for a current level of 100 A in Fig. 5 for 25°C and 175°C. To simulate the resistance difference to the source measurement method, $\Delta R_{ds,on}$ is shown as the difference resulting from a potential measurement contact. The source lead acts as the reference potential while the voltage on the top of the metallization was simulated. The difference in $R_{ds,on}$ has been extracted by the quotient of the potential of the top metallization and the applied current. The gate bond wire is not depicted. The results show a large influence of the top metallization on the overall resistance of the device. While the closest bond wire of the source pads shows the lowest difference to the resistance measured at the source lead itself, the difference increases with the distance to the leads. A clear influence from the current spreading can be noticed.

4 $R_{ds,on}$ measurement using a Kelvin connection

To investigate the actual $R_{ds,on}$ of the die, Fig. 6 shows a setup where the source sense is connected to the Kelvin lead, enabling the source potential to be directly probed from the top of the die as shown in Fig. 1 and Fig. 5.

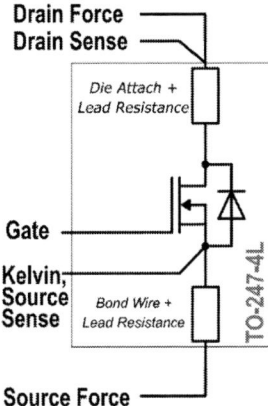

Fig. 6 Equivalent circuit diagram of the measurement circuit of a TO-247-4L package with the Kelvin connection utilized for the source sense connection.

The measured $R_{ds,on}$ of the Kelvin and Source probe is visualized in Fig. 7 and shows a significant decrease of approximately 1.1 mΩ at room temperature and an astonishing 2.5 mΩ at 175°C at a current level of 150 A, as depicted in Fig. 7. Furthermore, it is also evident that the reduced $R_{ds,on}$ value is almost constant over the entire current range and also independent from the applied pulse width, suggesting that the influence is not related to the actual $R_{ds,on}$ of the die but due to constant resistances in the form of bond wires and leads. The results differ from the simulation results in Fig. 5. While the measured $R_{ds,on}$ differs at room temperature by about 600 μΩ in the simulation, it differs by 1.08 mΩ in the actual measurement. At high temperature, it differs 1.48 mΩ in the simulation and 2.46 mΩ in the experiment. The difference between the source and Kelvin type are significantly dependent on the positioning of the Kelvin

bond wire. However, judging from the presented simulation results, the Kelvin wire's employed position captures the approximate average resistance difference well.

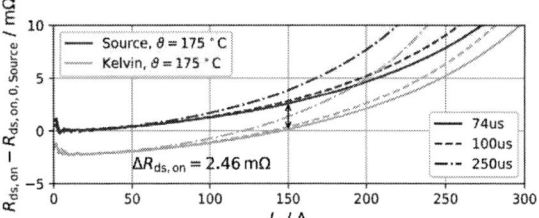

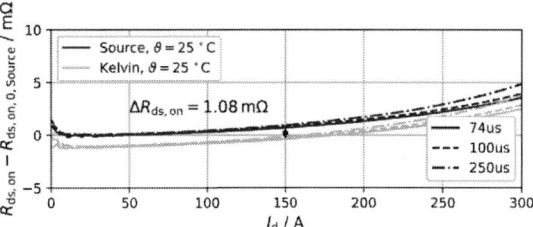

Fig. 7 (top) Measurement results for 175°C showing a consistent impact of about 2.5 mΩ across the current range, (bottom) measurement results for 25°C showing a consistent impact of about 1.1 mΩ across the current range.

5 Die $R_{ds,on}$ assessment with a singulated die measurement

Another measurement method for evaluating the die $R_{ds,on}$ is the use of a wafer prober or singulated die tester. Here the die is contacted using several needles pressing on the top metallization. A typical probe card is depicted in Fig. 8.

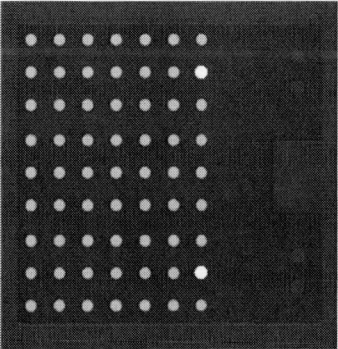

Fig. 8 Visualization of the singulated die probe connection

The green, white, and blue circles depict the source force, the source sense, and the Kelvin (used for the reference potential when biasing the gate) connections, respectively. For the simulation, the same assumptions as for the packaged device simulation were made with the same die calibrated to the packaged product as verified by experiments. The simulation results are given for the top metallization in Fig. 9 for a current of 100 A. The results show that the measurement location of the source sense is a good indicator for the actual source potential showing only a difference of around 140 µΩ compared to the actual source force potential. It is also visible that the $R_{ds,on}$ would be measured with a significant error of up to 0.66 mΩ if the Kelvin connections were to be used.

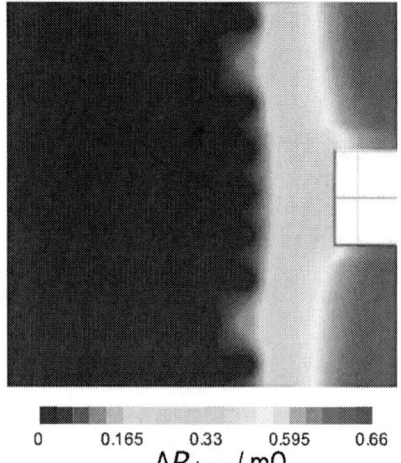

0   0.165   0.33   0.595   0.66

$\Delta R_{ds,on}$ / mΩ

Fig. 9 Visualization of the equivalent $R_{ds,on}$ difference of the die measured in singulated die using the probe connection of Fig. 8 for a temperature of 175°C.

	Ohmic loss based	Kelvin contact
TO-247-4L (WB)	-0.61 mΩ	-0.60 mΩ
Singulated die	-0.63 mΩ	-0.7 mΩ

Table 1: Simulated resistance differences (25°C) of package (relative to calibrated resistance of TO-247-4L with source leads) extracted at 100 A.

	Ohmic loss based	Kelvin contact
TO-247-4L (WB)	-1.18 mΩ	-1.48 mΩ
Singulated die	-1.31 mΩ	-1.47 mΩ

Table 2: Simulated resistance differences (175°C) of package (relative to calibrated resistance of TO-247-4L with source leads) extracted at 100 A.

A comparison of the singulated die and TO-247-4L simulations is provided in Table 1 and Table 2 for 175°C and 25°C respectively. The results show that the $R_{ds,on}$ derived from the singulated die simulation matches quite well the $R_{ds,on}$ of the Kelvin-type measurement. The source-type measurement on the other side shows a minimum error of 0.6 mΩ at 25°C and 1.18 mΩ at 175°C. A comparison of singulated die measurements, source and Kelvin type measurements is depicted in Table 3

showing the resistance difference at 175°C referenced to the source-type measurement method ($\Delta R_{ds,on} = R_{ds,S}^{meas} - R_{ds,X}^{meas}$) where $R_{ds,S}^{meas}$ is the resistance measured at the source leads. Since the devices could not been traced through the packaging process, the mean difference is shown in the table. The results show that the mean $R_{ds,on}$ of the singulated die measurement is 2.2 mΩ lower compared to the packaged product. If the $R_{ds,on}$ is measured with the Kelvin contact, the difference is in average 2.13 mΩ. Both measurements were performed at 100 A. It is important to note that the measurement setup is different resulting from differences in the parasitic inductance such that different pulse widths were used (380 µs for the packaged device and 250µs for the singulated die test). The test is, thus, influenced by the self-heating of the device. The results are remarkably similar for both tests and shows that the Kelvin-type measurement method can be a good way to estimate the $R_{ds,on}$ of the die.

6 Conclusion

When measuring the $R_{ds,on}$ of SiC MOSFETs on the die, one has to ensure that the self-heating is kept to a minimum by employing short measurement pulses. High-current and high-temperature measurements, however, usually cannot be achieved on die level because of material restrictions of the wafer prober. The engineer is forced to employ packaged products with additional parasitic resistances from the die attach, the bond wires and the leads. This paper showed that the $R_{ds,on}$ can be measured by using the Kelvin lead that is conventionally used for biasing the gate. The paper outlined a significant difference between source-lead measurement and Kelvin lead measurements suggesting a difference above 1 mΩ at room temperature and a difference of more than 2.5 mΩ at 175°C. The measurements were supported by FEM simulations which outlined a significant dependency on the position of the Kelvin bond wire.

	Singulated Die ($R_{ds,X}^{meas} = R_{ds,SD}^{meas}$)	TO-247-4L ($R_{ds,X}^{meas} = R_{ds,K}^{meas}$)
Resistance difference ($R_{ds,S}^{meas} - R_{ds,X}^{meas}$)	2.2 mΩ	2.13 mΩ

Table 3: Measured mean resistance difference at 175°C of the package with Kelvin lead and singulated die referenced to the measurement with source leads both measured at 100A.

References

[1] A. Brothers et al., *New silicon carbide prospects emerge as market adapts to EV expansion.* [Online]. Available: https://www.mckinsey.com/industries/semiconductors/our-insights/new-silicon-carbide-prospects-emerge-as-market-adapts-to-ev-expansion (accessed: Apr. 4 2024).

[2] BloombergNEF, *Electric Vehicle Outlook.* [Online]. Available: https://about.bnef.com/electric-vehicle-outlook/ (accessed: Apr. 4 2024).

[3] TrendForce Corporation, *Under the Hood: How is SiC Reshaping the Automotive Supply Chain?* [Online]. Available: https://www.trendforce.com/news/2023/06/28/under-the-hood-how-is-sic-reshaping-the-automotive-supply-chain/ (accessed: Apr. 4 2024).

[4] M. Slowick, *Automakers and Tier 1's Vie for a Slice of the SiC Pie.* [Online]. Available: https://www.electronicdesign.com/markets/automotive/article/21122006/automakers-and-tier-1s-vie-for-a-slice-of-the-sic-pie (accessed: Apr. 4 2024).

[5] J. Morra, *Cree Strikes Deal with GM for Silicon Carbide Chips, Changes Name.* [Online]. Available: https://www.electronicdesign.com/technologies/power/article/21177312/electronic-design-cree-strikes-deal-with-gm-for-silicon-carbide-chips-changes-name (accessed: Apr. 4 2024).

[6] G. Mason, *Vitesco Technologies develops silicon-carbide power module for EVs.* [Online]. Available: https://futurride.com/2023/09/14/vitesco-technologies-develops-silicon-carbide-power-module-for-evs/ (accessed: Apr. 4 2024).

[7] Magna International Inc., *onsemi and Magna Sign Strategic Agreements, Invest in Silicon Carbide Manufacturing for Growing Electric Vehicle Market.* [Online]. Available: https://www.magna.com/stories/news-press-release/2023/onsemi-and-magna-sign-strategic-agreements-invest-in-silicon-carbide-manufacturing-for-growing-electric-vehicle-market (accessed: Apr. 4 2024).

[8] I. Hübner, *STMicroelectronics-MOSFETs in Semikron-Power-Modules: SiC für künftige E-Fahrzeuge.* [Online]. Available: https://www.elektroniknet.de/automotive/elektromobilitaet/stmicroelectronics-mosfets-in-semikron-power-modules.196024.html (accessed: Apr. 4 2024).

[9] C. Hammerschmidt, *Infineon, Stellantis set for major SiC semiconductor supply contract.* [Online]. Available: https://www.eenewseurope.com/en/stellantis-to-secure-sic-manufacturing-capacity-from-infineon/ (accessed: Apr. 4 2024).

[10] Compound Semiconductor Magazine, Angel Business Comms. Ltd, *The growth of power SiC partnerships.* [Online]. Available: https://compoundsemiconductor.net/article/117486/The_growth_of_power_SiC_partnerships (accessed: Apr. 4 2024).

[11] BorgWarner Inc., *BorgWarner Expands Silicon Carbide Inverter Business with Major Global OEM.* [Online]. Available: https://www.borgwarner.com/newsroom/press-releases/2023/02/09/borgwarner-expands-silicon-carbide-inverter-business-with-major-global-oem (accessed: Apr. 4 2024).

[12] H. Luo, F. Iannuzzo, N. Baker, F. Blaabjerg, W. Li, and X. He, "Study of Current Density Influence on Bond Wire Degradation Rate in SiC MOSFET Modules," *IEEE J. Emerg. Sel. Topics Power Electron.*, vol. 8, no. 2, pp. 1622–1632, 2020, doi: 10.1109/JESTPE.2019.2920715.

[13] B. Shi et al., "A review of silicon carbide MOSFETs in electrified vehicles: Application, challenges, and future development," *IET Power Electronics*, vol. 16, no. 12, pp. 2103–2120, 2023, doi: 10.1049/pel2.12524.

[14] J. Helm, I. Dietz von Bayer, A. Olowinsky, and A. Gillner, "Influence of the surface properties of the connector material on the reliable and reproducible contacting of battery cells with a laser beam welding process," *Weld World*, vol. 63, no. 5, pp. 1221–1228, 2019, doi: 10.1007/s40194-019-00727-y.

[15] N. Marenco, M. Kontek, W. Reinert, J. Lingner, and M.-H. Poech, "Copper ribbon bonding for power electronics applications," in *2013 European Microelectronics Packaging Conference (EMPC)*, 2013, pp. 1–4.

[16] Heraeus Deutschland GmbH & Co. KG, *Aluminum Bonding Ribbon for Power Electronics.* [Online]. Available: https://www.heraeus.com/media/media/het/media_het/products_4/alu_thick_wire/alubond_pdfs/Flyer_Aluminium_Bonding_Ribbon_for_Power_Electronics.pdf (accessed: Apr. 4 2024).

[17] J. Rudzki, M. Becker, R. Eisele, M. Poech, and F. Osterwald, "Power Modules with Increased Power Density and Reliability Using Cu Wire Bonds on Sintered Metal Buffer Layers," in *Integrated Power Systems (CIPS), 2014 8th International Conference on*, 2014.

[18] Z. Ren, X. Guo, J. Fu, and Q. Lin, "A Double-sided Cooling SiC Power Module Applied to Electric Vehicles," in *2023 26th International Conference on Electrical Machines and Systems (ICEMS)*, 2023, pp. 5143–5148.

[19] M. Schaal, M. Klingler, and B. Wunderle, "Silver Sintering in Power Electronics: The State of the Art in Material Characterization and Reliability

Testing," in *2018 7th Electronic System-Integration Technology Conference (ESTC)*, 2018, pp. 1–18.

[20] S. Seal, M. D. Glover, and H. A. Mantooth, "3-D Wire Bondless Switching Cell Using Flip-Chip-Bonded Silicon Carbide Power Devices," *IEEE Trans. Power Electron.*, vol. 33, no. 10, pp. 8553–8564, 2018, doi: 10.1109/TPEL.2017.2782226.

[21] L. Wang, W. Wang, R. J. E. Hueting, G. Rietveld, and J. A. Ferreira, "Review of Topside Interconnections for Wide Bandgap Power Semiconductor Packaging," *IEEE Trans. Power Electron.*, vol. 38, no. 1, pp. 472–490, 2023, doi: 10.1109/TPEL.2022.3200469.

PCIM Europe 2024, 11– 13 June 2024, Nuremberg   DOI: 10.30420/566262379

Challenges in Scaling SiC Single-Chip Measurements to Corresponding Power Modules

Hao Wang[1], Pham Ha Trieu To [1], Felix Kayser[1], Florian Sawallich[1], Hans-Günter Eckel[1]

[1] University of Rostock, Germany

Corresponding author: Hao Wang, hao.wang@uni-rostock.de

Abstract

To experimentally model the electrical behaviour of high-power modules, it is common practice to scale based on measurements from a single chip. The traditional approach involves scaling commutation loop inductance L_σ to align the overshoot voltage between the scaled single chip and the module. Nevertheless, scaling only L_σ often is not enough to match transient behaviours. This is due to the impact of parasitic parameters like common source inductance L_{com} on transient behaviours. In this paper, a new scaling method is introduced, which takes additional parameters into consideration, offering a more comprehensive approach to aligning the transient behaviour.

1 Introduction

Wide band-gap semiconductors, such as SiC MOSFETs, are currently being developed in many power electronic applications and are in process of being developed to supersede the dominant silicon semiconductors due to SiC-MOSFETs offer notable advantages, including high switching speed and low conduction losses [1]. The transient behaviours in various applications are a primary concern, with critical factors being the setup parasitic parameters and chip physical properties. For cost-effective and low-risk investigation of module transient behaviors, it is advisable to scale the same SiC MOSFET chip with single-chip package to match the characteristics of the power module Properly scaled commutation loop inductance (L_σ) is crucial for matching overshoot voltage in both power single-chips and modules [2] [3].

However, scaling considerations should extend beyond L_σ, factors such as gate loop inductance (L_g) and common source inductance (L_{com}) must also be accounted for. The L_{com} is an important factor impacting switching behaviours. Due to the feedback of di/dt onto the gate voltage that will lead to a reduction of effective gate voltage on the chip, furthermore, the transient behaviours would be affected [4] [5]. Therefore, for a proper match, the difference of L_{com} between single-chip and power module should be considered. In a single-chip configuration, L_{com} is influenced by both package design and PCB layout, affecting the gate and load current loops. It is important to note that

package adjustments are not possible in single-chip configurations. However, the PCB layout can be used to adjust L_{com} values to match the switching behaviour of the module. This paper evaluates the impact of L_{com} on switching behaviours through double pulse test measurements, discussing overshoot voltage, di/dt, and gate voltage based on single-chip measurements. Transient behaviours in both single-chip and power module configurations are presented under well-scaled parasitic inductance conditions.

2 Scaling principle in single-chip and power module

2.1 Differences of experimental setups for single-chip and power module measurements

To configure an experimental setup for well-matched scaled single-chip measurement, a series of critical module parameters such as current change (di/dt), rate of voltage change (dv/dt), and overshoot voltage on the drain-source side should be mainly focused. The scaling aims to match the transient behavior of a single chip to that of the module, due to the same chip being applied in the single-chip and module package, the significant impact parameter is parasitic inductance. In the power modules with multiple chips in parallel, it is observed that the di/dt of current transients is n times higher than that of a single chip, leading to a proportional increase in voltage overshoot. To ensure equivalence in transient behavior, additional parasitic inductance is strategically introduced into a single-chip setup to adjust L_σ, L_{com} and L_g. The scaling formula outlined in Eq. (1), incorporates the scaling factor 'n', representing the number of chips within the module[2]. This formula shows the guide method for determining required values for L_σ, L_{com} and L_g at the single-chip, which should be n times higher than those in the module to achieve a comparable V_{gs} and V_{ds} overshoot.

$$V_{L_{module}} = V_{L_{single}} \qquad (1)$$

$$L_{module}\frac{di_{module}}{dt} = n * L_{module}\frac{di_{single}}{dt}$$

The corresponding experimental setups for the module and single-chip are illustrated in Fig. 1 and Fig. 2. In the power module setup, shown in Fig. 1, integrated inductances ($L_{\sigma\_module\_int}$, $L_{com\_module\_int}$, and $L_{gate\_module\_int}$) are accounted for within the module package, while $L_{\sigma\_module\_ext}$ is based on the DC link stray inductance.

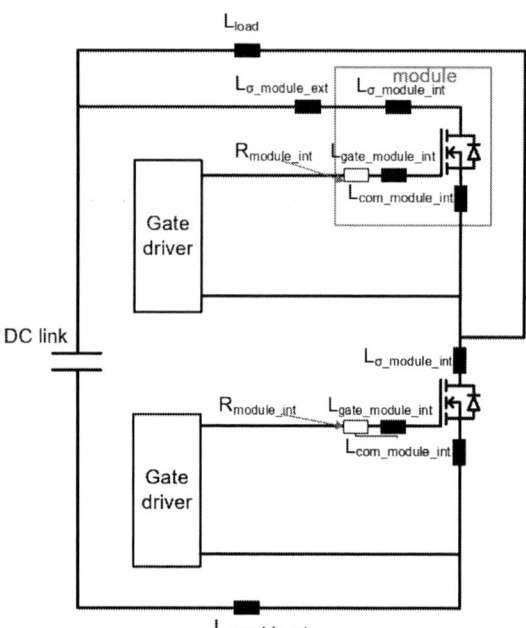

Fig. 1: Schematic of Module Showing Inductance Components and Double Pulse Setup

Recognizing the inherent lower parasitic inductance of single chip devices, additional parasitic inductance is incorporated in single chip measurements to match the n times higher parasitic inductance of the power module. Notably, due to the functional resemblance between

$L_{\sigma\_module\_int}$ and $L_{\sigma\_module\_ext}$, both are merged into a single scaled inductance. Fig. 2 illustrates the positioning of additional parasitic inductance in the single-chip setup. Taking into account that voltage drop across L_σ ($V_{L\sigma}$) is determined by the L_σ and di/dt, with R_g being one of the influence factors on di/dt, the matching of R_g should also be considered in the scaling process. Table 1 is presented to provide the clear scaling ratio from the single chip to the module to match in single-chip and module characterization.

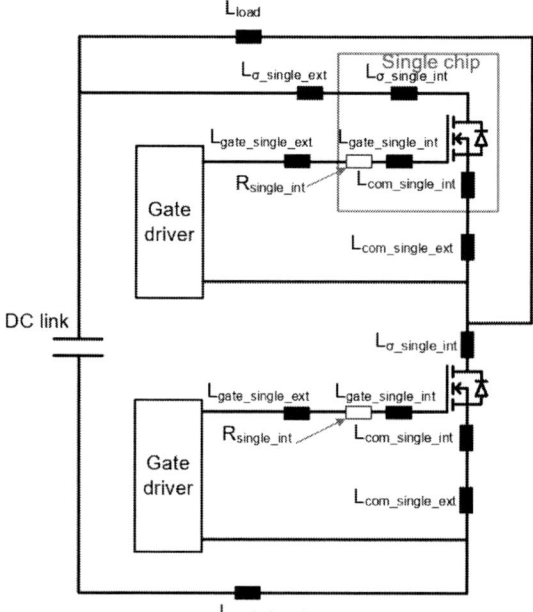

Fig. 2: Schematic of Single-Chip Showing Scaled Inductance Components and Double Pulse Setup

DUT	Single chip	module
Nominal voltage	V_{nom}	V_{nom}
Nominal current	I_{module_nom}/n	I_{module_nom}
R_g	$n*R_{g\_module}$	R_{g_module}
L_σ	$n*L_{\sigma\_module}$	L_{σ_module}
L_{com}	$n*L_{com\_module}$	L_{com_module}
L_{gate}	$n*L_{gate\_module}$	L_{gate_module}

Table 1: Comparing Setup and Inductance Configuration between Scaled Single-Chip and Modules

The conventional scaling method is mainly focused on adjusting L_σ in single-chip test setup to ensure that L_{σ_single} equal to n * L_{σ_module}, thereby achieving V_{dsmax} matching between single-chips and module, with the well scaled L_σ.

Eq.(2) and Eq.(3) represent the maximum voltage V_{dsmax} during turn-off for module and single-chip under suitable scaling, respectively. In addition to appropriately scaled L_σ in single-chip and module, the rate of di/dt should also be scaled accordingly. The $V_{gs}(t)$ plays a significant role in determining di/dt. Moreover, the presence of L_{com} affects $V_{gs}(t)$, consequently impacting V_{dsmax}. Therefore, it is imperative to consider L_{com} in the scaling analysis to accurately achieve V_{dsmax} equivalence between single chips and modules.

$$V_{dsmax\_module} = V_{DC} + L_{\sigma\_module} \frac{di_{module}}{dt} \quad (2)$$

$$V_{dsmax\_single} = V_{DC} + n * L_{\sigma\_module} \frac{di_{single}}{dt} \quad (3)$$

2.2 L_{com} effect

The L_{com} effect in transient behaviours is performed by the double pulse test to analyse the effect of L_{com} on the $V_{gs}(t)$. The simplified test setup is shown in Fig. 3 to focus on the L_{com} effect on the turn-off behaviour while the analysis of the L_g during turn-off is omitted. The chip gate source voltage ($V_{gs\_chip}(t)$) during turn-off can be expressed as Eq.(4). The $V_{gs\_chip}(t)$ can be simply rewritten as Eq.(5) because the $i_d \gg i_g$, thereby the voltage drop V_{Lcom} by gate current ($i_g(t)$) is rather smaller than that by drain-source current ($i_d(t)$) and can be neglected [6].

The $i_g(t)$ plays a key role in assessing the behaviour of the gate-source capacitance (C_{gs}). By establishing the correlation between $i_g(t)$ and L_{com}, a connection between L_{com} and the gate can be established. By integrating Eq.(5) and Eq.(6) with transconductance ($g_m = di_d/dv_{gs\_chip}(t)$), Eq.(7) is formulated, representing the relationship between $i_g(t)$ and L_{com}, which shows that an increase in L_{com} results in a reduction in $i_g(t)$, thereby extending switching period [6]. Consequently, an increase in the L_g setup corresponds to a decrease in overshoot voltage [7].

$$v_{gs\_chip}(t) = V_{GDU\_OFF} + R_g * i_g(t) \quad (4)$$
$$+ L_{com} \frac{d(i_d(t) - i_g(t))}{dt}$$

$$v_{gs\_chip}(t) = V_{GDU\_OFF} + R_g * i_g(t) + L_{com} \frac{di_d(t)}{dt} \quad (5)$$

$$v_{Lcom}(t) = L_{com} \frac{di_d(t)}{dt} \quad (6)$$
$$= L_{com} \frac{di_d(t)}{dv_{gs\_chip}(t)}$$
$$* \frac{dv_{gs\_chip}(t)}{dt}$$
$$= L_{com} * g_m * \frac{dv_{gs\_chip}(t)}{dt}$$
$$= L_{com} * g_m * \frac{i_g(t)}{C_{gs}}$$

$$i_g(t) = \frac{v_{gs\_chip}(t) - V_{GDU\_OFF}}{R_g + \frac{g_m * L_{com}}{C_{gs}}} \quad (7)$$

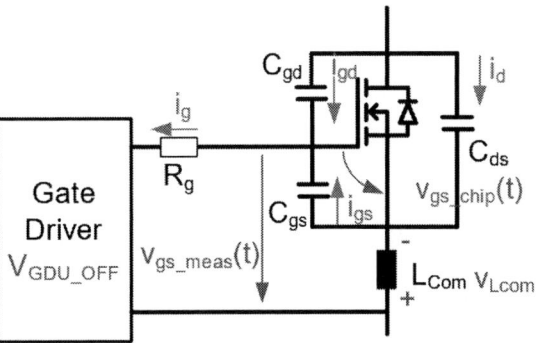

Fig. 3: A Simplified Schematic of Single-Chip MOSFET setup Demonstrating the Influence of L_{com} during Turn-Off

The corresponding simplified turn-on setup is shown in Fig. 4. The relationship between measurement gate voltage ($v_{gs\_meas}(t)$) and $i_g(t)$ during turn-on for L_{com} are shown in Eq.(8) and Eq.(9). It is similar to the turn-off scenario, an increase in L_{com} induces a decrease in i_g during turn-on, resulting in a slower turn-on process.

Obvious from Eq. (7) and Eq.(9), the L_{com} plays an essential role in transient behaviours impacting di/dt, overshoot voltage and switching period. The inherent differences in L_{com_int} arising from packaging distinctions between power modules and single-chip setups inevitably lead to variations in switching behaviour. To compensate for the different switching characteristics of different packages, an extra external L_{com} is added to the single-chip setup to achieve consistent switching dynamics across different package configurations.

$$v_{gs\_chip}(t) = V_{GDU\_ON} - R_g * i_g(t) - L_{com}\frac{di_d(t)}{dt} \quad (8)$$

$$i_g(t) = \frac{V_{GDU\_ON} - v_{gs\_chip}(t)}{R_g + \frac{g_m * L_{com}}{C_{gs}}} \quad (9)$$

L_{com_ext} name	L_{com_ext} value(nH)
$L_{com\_ext\_1}$	1.3
$L_{com\_ext\_2}$	6
$L_{com\_ext\_3}$	12

Table 2: different L_{com_ext} used in comparing single chip switching behaviour

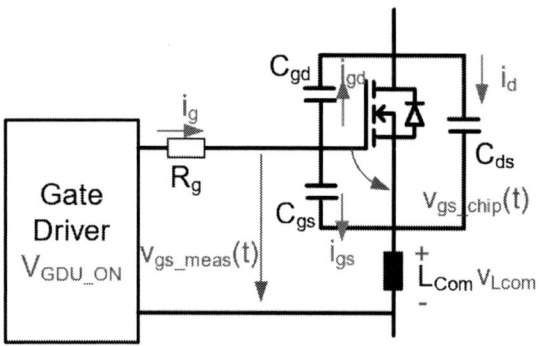

Fig. 4: A Simplified Schematic of Single-Chip MOSFET setup Demonstrating the Influence of L_{com} during Turn-On

3 Measurement Verification

3.1 Measurements on L_{com} effect

The double pulse test setup is used as a measurement platform as shown in Fig. 2. The single-chip is used to as switching DUT to assess the impact of L_{com} on switching performance and the gate control capability. A simple functional gate driver is utilized to control the MOSFET. In MOSFET mode on the low side, V_{GDU} values of -5V and +15V are used for turn-off and on the MOSFET. On the high side, the operation is in diode mode, with V_{GDU}=-5V applied during this mode. The $L_{com\_single\_ext}$ is incorporated into both the gate current and load current loops, thereby influencing $V_{gs}(t)$ and $V_{ds}(t)$ during the current commutation period.

A series of inductance values, shown as $L_{com\_single\_ext}$, are employed to investigate the influence of L_{com}. These inductance values are measured using an LCR measurement device (TELEDYNE T3LCR1300) and are presented in Table 2.

The L_σ is kept the same in the different L_{com_ext} measurement test setup due the maximum L_{com_ext} is considerably smaller than 1% of L_σ. As a result, its influence on the commutation loop is negligible and, making it suitable for neglect.

A series of figures are shown to illustrate the dynamic behavior of single-chip circuits, specifically focusing on the influence of L_{com} during turn-on and turn-off events.

Fig. 5 illustrates the turn-on curves in the low side (MOSFET mode). These measurements clearly indicate that an increase of L_{com_ext} leads to a decrease of di/dt during turn-on. It can be seen that the $V_{gs\_chip}(t)$ is influenced by L_{com}, resulting in a decrease, which matches to Eq.(8) and Eq.(9). In the high side (diode mode), in Fig. 6 a discernible trend is observed, a decrease in V_{rr_peak} with an increase in L_{com}. This can be explained as V_{rr_peak} is depending on the V_{DC}, di/dt and the L_σ. Maintaining a constant L_σ and DC link voltage, with the decrease in di/dt results in a proportional decrease in V_{rr_peak}.

In MOSFET mode, the $V_{gs\_meas}(t)$ exhibits an increase with the rise in L_{com}. This phenomenon is due to the inclusion of L_{com} in the gate measurement loop, consequently influencing the observed changes in V_{gs_meas} during MOSFET mode. The positive V_{Lcom} is included into the V_{gs_meas} (shown in Fig. 4), contributing to the observed increase as L_{com} grows. This indicates the impact of L_{com} on the $V_{gs\_meas}(t)$ in MOSFET mode.

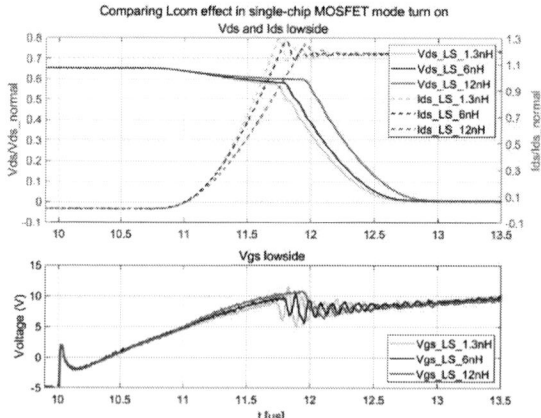

Fig. 5: The Influence of L_{com} on MOSFET Mode Turn-On Behaviour

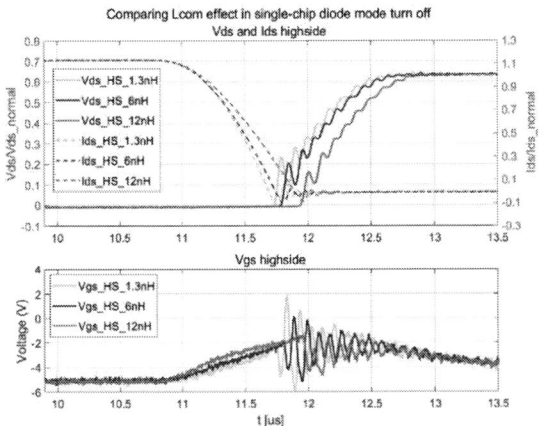

Fig. 6: The Influence of L_{com} on Diode Mode Turn-Off Behaviour

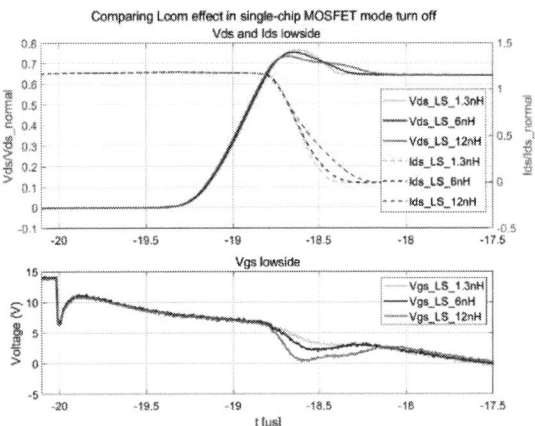

Fig. 7: The Influence of L_{com} on MOSFET Mode Turn-Off Behaviour

As depicted in Fig. 7, the turn-off curves reveal a consistent pattern. The presence of L_{com} results in lower di/dt, accompanied by a decrease in V_{peak} with an increasing L_{com}. However, the turn-off period lengthens due to the reduced I_g in the high L_{com} configuration. With an increase in L_{com}, the changes in V_{gs_meas} become more pronounced within the di/dt range. This amplification in $V_{gs\_meas}(t)$ variation can be attributed to the inclusion of V_{Lcom} in the V_{gs} measurement loop. Regarding the behavior of $V_{gs\_chip}(t)$ during turn-off, the measurement position on $V_{gs\_meas}(t)$ and $V_{gs\_chip}(t)$ is depicted in Fig. 3 and based on Kirchhoff's laws the mathematical formula is provided in Eq.(10), it reviews under the same L_{com} condition, $V_{gs\_chip}(t)$ is larger than the $V_{gs\_meas}(t)$ when the di/dt occurs during turn off.

$$v_{gs_{chip}}(t) = v_{gs_{meas}}(t) + L_{com}\frac{di_d(t)}{dt} \qquad (10)$$

The presented results underscore the important relationship between L_{com} and diverse dynamic parameters, highlighting the significance of L_{com} as a crucial parameter for scaling in both module and single-chip configurations.

3.2 Scaling measurements in single-chip and power module

The scaling function can be divided into 2 modes, differentiated by with or without of power module measurements. In the absence of module measurements, where no comparative results are available, scaling L_{com} appropriately becomes challenging. The recommended approach involves setting L_{com} to the smallest value to attain the maximum overshoot voltage, providing a safety margin for potential module measurements. The prioritization of module survival, considering energy loss, and ensuring a match in di/dt and dv/dt, becomes paramount.

On the other hand, when power module measurements are available, a comprehensive approach is needed to scale a single-chip to match the module switching behavior. The influential parameters that affect the scaling, namely R_g, L_σ, and L_{com}, should be matched with the corresponding parameters from the module.

This matching is crucial as these influential parameters directly impact the scaling evaluation parameters, including the rate of di/dt, rate of dv/dt, and overshoot voltage during switching events. Furthermore, variations in one influential parameter have the potential to affect one or more evaluation parameters. Consequently, carefully adjusting these influential parameters on the single-chip configuration becomes essential to match the module parameters.

The single-chip scaling method is shown in Fig. 8. During turn-on, L_{com} is initially set to its minimum, and the alignment of di/dt between the single-chip and the power module is achieved by adjusting the single-chip R_{gon}. Subsequently, L_σ is tuned to ensure alignment with the power module V_{ds}. The next step for a precise scaling is the adjustment of R_{goff} using turn-off transients. R_{goff} is determined based on the dv/dt of the rising voltage slope before a negative di/dt occurs. A well scaling standard for L_{com} is established by ensuring a match with the overshoot voltage between single-chip and module. Following this, the subsequent step involves checking and adjusting R_{gon} to align with the observed di/dt during the turn-on process. Sometimes, these steps may need to be repeated to ensure well-matched results between the single chip and module configurations.

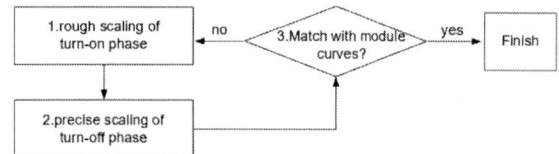

Fig. 8: Single-Chip Scaling Process for Matching with Power Module Measurement

To validate the proposed scaling method, measurements were conducted using identical SiC MOSFET chips, comparing both single-chip package and power module package containing n-times the number of single-chips. In the single chip measurement, the parasitic inductance, I_{ds} and V_{ds} are scaled according to Table 1, with the exception of R_g. Specifically, $R_{gon\_single-chip}$ is set to about 96% of $n^*R_{gon\_module}$ and $R_{goff\_single-chip}$ is adjusted to 92% of $n^*R_{goff\_module}$ to match di/dt and dv/dt.

The measurement probes and gate driver details are outlined in Table 3. and the comparative measurement figures are presented in Fig. 9 and Fig. 10, which the measurements between module and single-chip are similar.

Probe and gate driver	Module	Single-chip
$V_{ds\_highside}$	HVD3605	HVD3605
$V_{gs\_meas\_highside}$	HVFO108	HVFO108
GDU$_{highside}$	Same driver	
$V_{ds\_lowside}$	HVD3605	HVD3605
$V_{gs\_meas\_lowside}$	1:10 passive probe	HVFO108
GDU$_{lowside}$	Same driver	

Table 3: Measurement Probes and Gate Driver Inventory for Comparing Single-Chip and Power Module Measurement

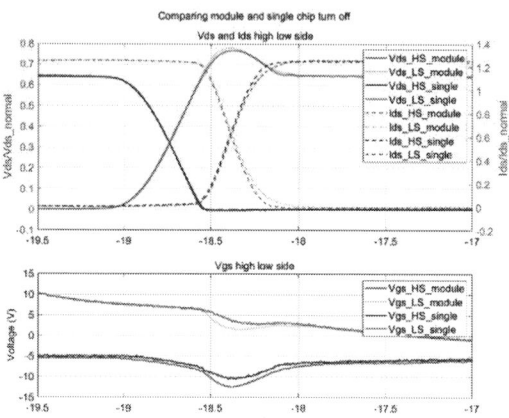

Fig. 9: Comparative of Turn-Off Characteristics: Module and Scaled Single Chip

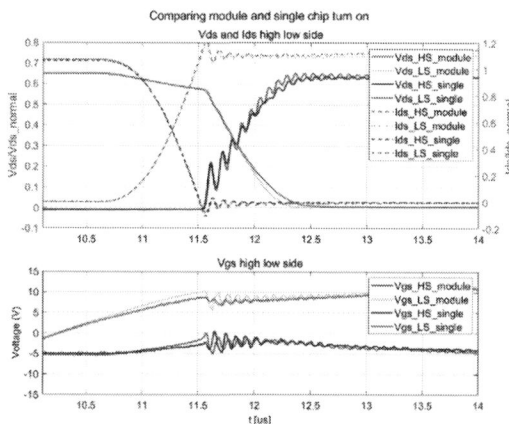

Fig. 10: Comparative of Turn-On Characteristics: Module and Scaled Single Chip

The effectiveness of the scaling process is assessed based on evaluation parameters. A comprehensive comparison of these ratios between the single-chip and module configurations is provided in Table 4. The difference is in 5%, which is in the acceptable margin.

Ratio of Single-chip/module			
Turn-on		Turn-off	
di_{rr}/dt	103%	di_{ds}/dt	95%
dv_{rr}/dt	103%	dv_{ds}/dt	96%
V_{rr_peak}	104%	V_{ds_peak}	98%

Table 4: Comparative of Switching Parameters between Scaled Single-Chip and Module

This difference indicates that factors such as parasitic capacitors or magnetic coupling may need to be taken into account for further refinement to achieve optimal matching between single-chip and module configurations [8] [9].

4 Conclusion

In this paper, the effect of L_{com} in scaled single-chip measurements is analysed and experimentally investigated, showing its impact on coupled load and gate loops. The coupling through L_{com} influences the charging and discharging of C_{gs} in the chip, subsequently affecting transient behaviors during switching. The variations in L_{com}, introduced by different package devices and setup layout, play a crucial role in scaling results. Unproper scaling of L_{com} can lead to significant differences in performance across various package devices.

To verify the significance of L_{com} scaling, a series of measurements are conducted with varied L_{com} values. The results demonstrate that, with well-scaled parasitic inductance, and well-adjusted R_g in both the single-chip and module configurations, the observed differences in switching behaviour are within an acceptable range of less than 5%.

However, even with properly adjusted L_{com}, the transient behavior on the scaled single-chip is not resulting in a perfect 100% matching on the module transient behavior. This observation highlights the complexity of scaling, indicating that factors such as parasitic capacitors and magnetic coupling should also be taken into account to provide a more thorough and precise representation of the system's behaviour. Further exploration and consideration of additional scaling factors are warranted to enhance the accuracy and reliability of the scaling process.

5 Acknowledgement

This work was funded by the German Federal Ministry for Economic Affairs and Climate Action.

6 References

[1] M. Nitzsche, C. Cheshire, M. Fischer, J. Ruthardt and J. Roth-Stielow, "Comprehensive Comparison of a SiC MOSFET and Si IGBT Based Inverter," PCIM Europe 2019; International Exhibition and Conference for Power Electronics, Intelligent Motion, Renewable Energy and Energy Management, Nuremberg, Germany, 2019, pp. 1-7.

[2] D. Wigger and H. . -G. Eckel, "Comparison of chip- and module-measurements with high power IGBTs and RC-IGBTs," Proceedings of the 2011 14th European Conference on Power Electronics and Applications, Birmingham, UK, 2011, pp. 1-8.

[3] R. W. Maier and M. -M. Bakran, "Switching SiC MOSFETs Under Conditions of a High Power Module," 2018 20th European Conference on Power Electronics and Applications (EPE'18 ECCE Europe), Riga, Latvia, 2018, pp. P.1-P.9.

[4] Stueckler F. and Vecino E., "Cool MOS C7 650V switch in a kelvin source configureation," Infineon Application note, 2013.

[5] B. Zojer, "A new gate drive technique for superjunction MOSFETs to compensate the effects of common source inductance," 2018 IEEE Applied Power Electronics Conference and Exposition (APEC), San Antonio, TX, USA, 2018, pp. 2763-2768, doi: 10.1109/APEC.2018.8341408.

[6] Z. Chen, "An inductive-switching loss model accounting for source inductance and switching loop inductance," 2014 IEEE Applied Power Electronics Conference and Exposition - APEC 2014, Fort Worth, TX, USA, 2014, pp. 497-504, doi: 10.1109/APEC.2014.6803355.

[7] H. Yilmaz, K. Owyang, P. O. Shafer and C. C. Borman, "Optimization of power MOSFET body diode for speed and ruggedness," in IEEE Transactions on Industry Applications, vol. 26, no. 4, pp. 793-797, July-Aug. 1990, doi: 10.1109/28.56007.

[8] D. N. Dalal et al., "Impact of Power Module Parasitic Capacitances on Medium-Voltage SiC MOSFETs Switching Transients," in IEEE Journal of Emerging and Selected Topics in Power Electronics, vol. 8, no. 1, pp. 298-310, March 2020, doi: 10.1109/JESTPE.2019.2939644.

[9] C. Martin, J. -L. Schanen, J. -M. Guichon and R. Pasterczyk, "Analysis of Electromagnetic Coupling and Current Distribution Inside a Power Module," in IEEE Transactions on Industry Applications, vol. 43, no. 4, pp. 893-901, July-aug. 2007, doi: 10.1109/TIA.2007.900453.

PCIM Europe 2024, 11– 13 June 2024, Nuremberg   DOI: 10.30420/566262380

Switching Performance Evaluation of High-Power 1.7 kV SiC MOS-FET Modules using a Common Busbar Design

Sebastian Neira ©[1], Mason Parker ©[1], Stephen J. Finney ©[1], Paul D. Judge ©[1]

[1] The University of Edinburgh, United Kingdom

Corresponding author: Sebastian Neira, s.neira@ed.ac.uk
Speaker: Sebastian Neira, s.neira@ed.ac.uk

Abstract

The high dI/dt present while switching Silicon Carbide (SiC) MOSFET modules, coupled with power-loop stray inductance results in increased voltage overshoots and oscillatory switching behaviour. Thus, good module packaging, DC-link capacitor selection and busbar design are fundamental to maximise the benefits of SiC technology. This paper presents an experimental investigation into the switching performance for three 1.7 kV SiC MOSFET half-bridge modules, examining the effects of different module packages and their connection to an optimised busbar. Results show that minor design changes significantly impacted the total loop inductance, with changes from ~30 nH to ~12 nH reflected in major improvements to the obtained switching performance metrics.

1 Introduction

Silicon Carbide (SiC) MOSFET technology has rapidly evolved in recent years to revolutionise high-power medium-voltage applications such as electrical transportation, renewable energy generation and energy storage systems [1]–[3]. Specifically, SiC modules show relevant benefits in these high-power density applications due to their ability to withstand higher operating temperatures and improved switching performance compared to Silicon (Si) IGBT modules [4]. Thus, different manufacturers have worked to increase switching speed and current carrying capabilities while reducing switching losses [5], [6]. However, the performance of SiC-based power converters is critically dependent on the parasitics of the commutation power loop, as higher switching speeds and current levels result in increased overshoot and oscillations in the switching waveforms [7]. Consequently, the design of power modules and busbars for SiC-based power converters with reduced commutation loop stray inductance and complying with current and voltage ratings is critical for maximising the benefits of using SiC technology [8].

This paper analyses the switching performance under datasheet recommended conditions for three different 1.7 kV SiC MOSFET half-bridge modules using a common busbar designed to fit the different packages with a low ESL decoupling capacitor. The studied modules are the CAS380M17HM3 and CAB650M17HM3 from Wolfspeed and the MSCSM170AM029CT6LIAG from Microchip. The modules present two packages with different stray inductance values between DC terminals. Thus, the busbar is designed to set a low commutation loop inductance for each package under analysis. Experimental double pulse tests allowed for obtaining critical parameters, such as switching energy losses, overshoots and overall switching speeds. Furthermore, the loop inductance added by different busbar connection designs is analysed using finite-element simulations and experimental validation in double-pulse tests. The obtained results show that switching performance can be significantly improved with minor mechanical modifications to the design of the busbar connection to the SiC modules.

2 SiC Half-Bridge Modules under Study

Table 1 summarises the main operational parameters of the three studied modules. The study considers three 1.7 kV SiC half-bridge modules using a 62 mm package with two different tab arrangements and stray inductance values. The first two modules are the CAS380M17HM3 and CAB650M17HM3

Parameter	Module 1	Module 2	Module 3
Current Rating	532 A (25 °C)	916 A (25 °C)	676 A (25 °C)
	406 A (90 °C)	694 A (90 °C)	538 A (80 °C)
Module Stray Inductance	4.9 nH	4.9 nH	3 nH
Anti-parallel Diodes	Schottky Diode	Body Diode	Schottky Diode
Internal Gate Resistance	1.23 Ω	0.62 Ω	0.79 Ω
External Gate Resistance (Datasheet Recommended)	0.0 Ω	1.5 Ω	0.5 Ω
Input Capacitance	47 nF (V_{ds}=1200 V)	97.3 nF (V_{ds}=1200 V)	39.6 nF (V_{ds}=1000 V)
Output Capacitance	2.6 nF (V_{ds}=1200 V)	2.3 nF (V_{ds}=1200 V)	1.8 nF (V_{ds}=1000 V)

Tab. 1: Parameters of the 1.7 kV modules under study. Module 1: CAS380M17HM3, Module 2: CAB650M17HM3 and Module 3: MSCSM170AM029CT6LIAG. (Name convention used for the rest of the article.)

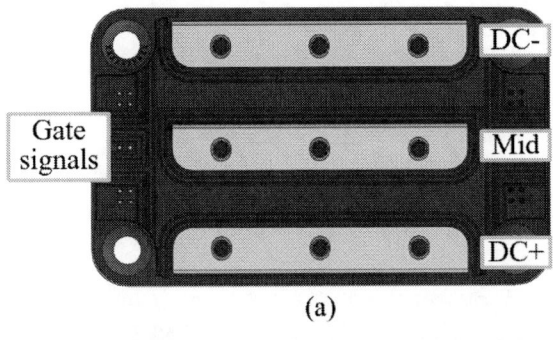

(a)

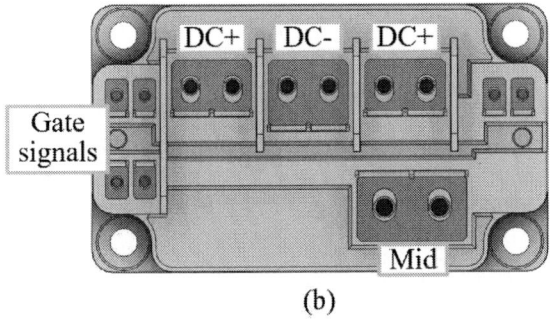

(b)

Fig. 1: Packages of modules under analysis. (a) HM3 package from Wolfspeed (Modules 1 and 2). (b) SP6LI package from Microchip (Module 3).

manufactured by Wolfspeed using the HM3 package (shown in Fig. 1(a)). This package presents a stray inductance between DC terminals of 4.9 nH and it uses a Silicon Nitride (Si_3N_4) power substrate. The CAS380M17HM3 module includes anti-parallel Schottky diodes and it has a nominal rated current of 380 A [9]. The CAB650M17HM3 uses MOSFET body diodes and it is rated for a nominal current of 650 A [10]. The third module under evaluation is the MSCSM170AM029CT6LIAG manufactured by Microchip using the SP6LI package

(shown in Fig. 1(b)). This package has a stray inductance of 3 nH between DC terminals and it uses an Aluminum Nitride (AlN) power substrate. The module includes anti-parallel Schottky diodes and it has a nominal current rating of 538 A [11].

3 Busbar Design

The busbar used in this study is designed to fit the two presented module packages with a low ESL decoupling capacitor (PowerRing 140 μF, ESL of less than 5 nH). The design aims to achieve a low parasitic inductance value on the commutation loop to avoid excessive overshoots and oscillatory switching behaviour. The laminated design considers using 3 mm thickness copper layers for the DC terminals with a 0.5 mm thickness polypropylene insulation layer between the conductors. The connection to the SiC modules varies with the different geometry of the terminals shown in Fig. 1. Moreover, the design also considers the capability of connecting current-sensing Rogowski coils to measure the drain current during switching transients.

Fig. 2(a) shows the designed busbar with the two analysed SiC half-bridge packages and the decoupling capacitor installed on the bottom side. The design for the HM3 modules considers a cutout between the DC terminals to allow the connection of the mid-point terminal to the load. Also, a slot in the middle of the DC tab was included to allow current measurement using two Rogowski coils, as shown in Fig. 2(b). The SP6LI package connection is performed with three tabs on the side of the busbar, allowing the straightforward connection of a current measurement coil (shown in Fig. 2(c)).

The 3D model of the busbar was used to analyse the stray inductance of the commutation loop us-

PCIM Europe 2024, 11– 13 June 2024, Nuremberg   DOI: 10.30420/566262380

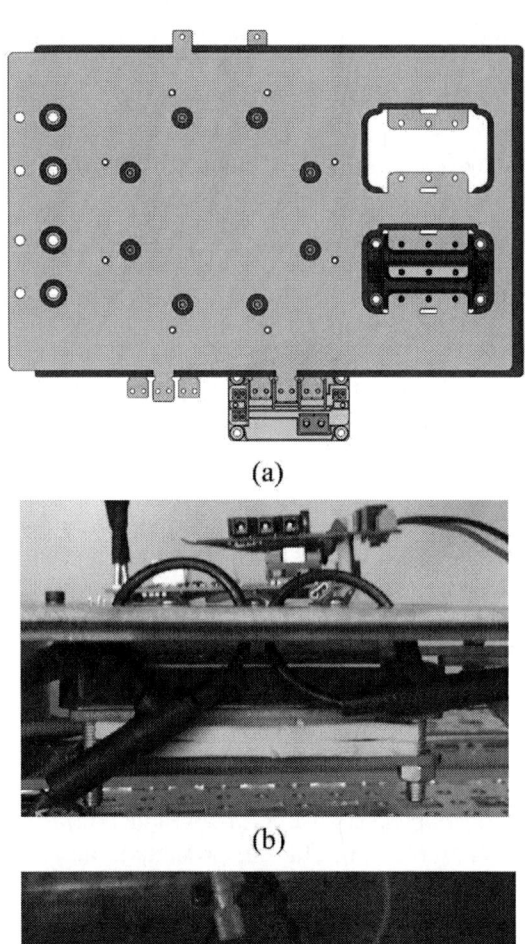

(a)

(b)

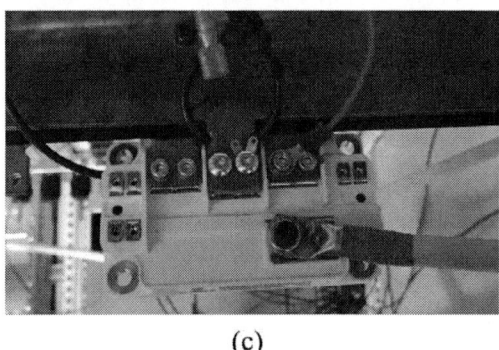

(c)

Fig. 2: Designed Busbar for connecting modules under study. (a) 3D model of busbar. (b) SP6LI package installed with Rogowski coil. (c) HM3 package installed with Rogowski coils.

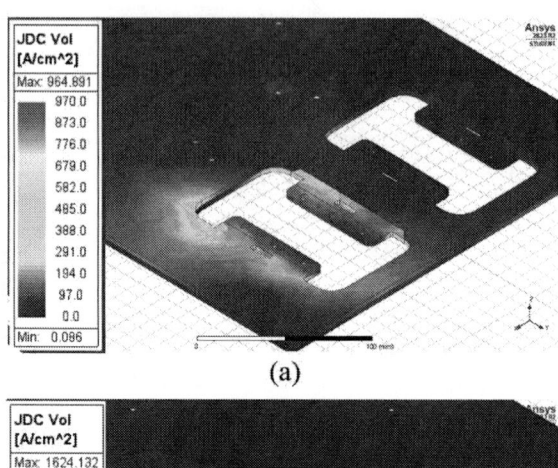

(a)

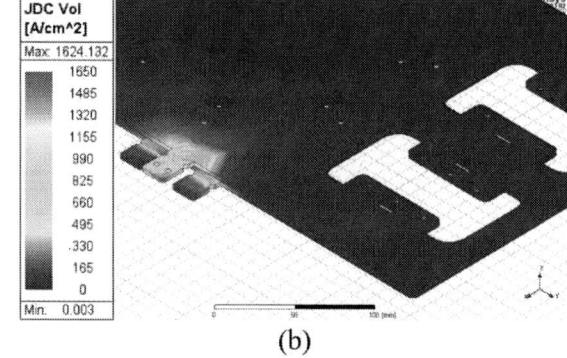

(b)

Fig. 3: Busbar current distribution in commutation loop for $I_{dc} = 400$ A. (a) HM3 Module. (b) SP6LI module.

nH for the SP6LI module, both calculated at 100 MHz. These inductance values account for the connection between each module and the decoupling capacitor. The total power loop inductance will also include the internal module stray inductance and the capacitor ESL.

The busbar-added inductance for the HM3 modules design is considerably higher than both the module and decoupling capacitor inductances (4.9 and <5 nH respectively). This increased value will lead to considerably higher magnitude overshoot and oscillations in the switching waveforms of modules 1 and 2. Thus, a re-design for the connection of this module was performed, eliminating the middle tab cutout to maximise the overlapping area between DC terminals. The new design considers two variants (A and B) shown in Fig. 4, where the difference is that the second variant eliminates the connection of the middle bolts maximizing the overlapped area between conductors. Moreover, the re-designed busbar includes different positions available to connect the HM3 module to analyse the effect of the location and orientation with respect to

ing Ansys Q3D Extractor software. Fig. 3 shows the current distribution for both modules connections considering a current of 400 A. The current distribution shows higher values at the module terminals, while density at the decoupling capacitor terminals is low due to having 8 tabs per DC terminal. The busbar inductance obtained from simulations is 21.8 nH for the HM3 module and 9.3

2702

PCIM Europe 2024, 11– 13 June 2024, Nuremberg   DOI: 10.30420/566262380

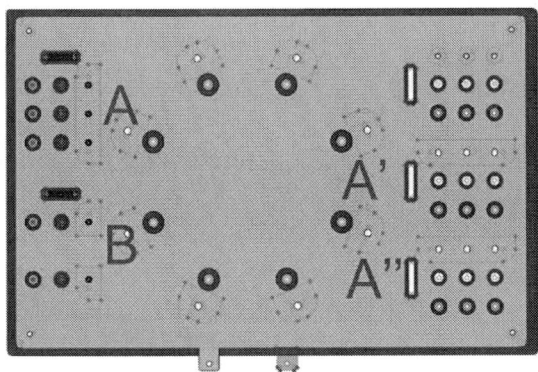

Fig. 4: Re-designed HM3 module connections. Variant A includes all mounting holes and Variant B omits middle holes. Variants A' and A" have the same design as A but with different location/orientation.

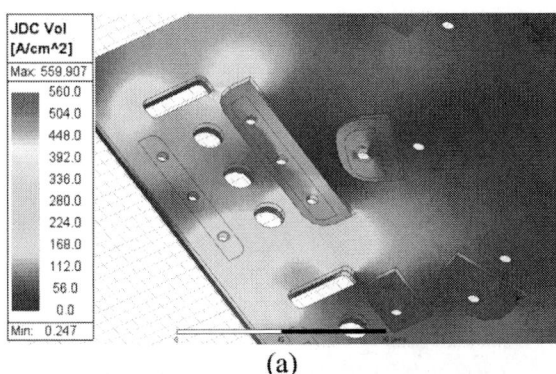

(a)

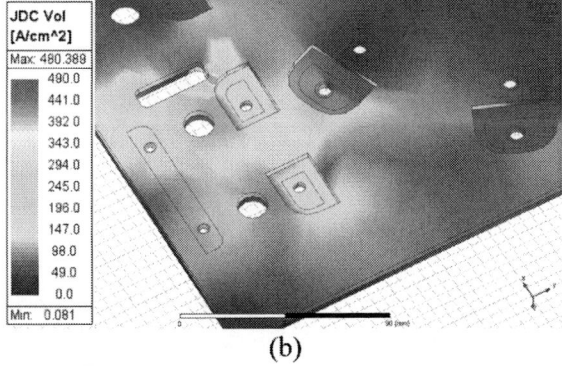

(b)

Fig. 5: Busbar current distribution in commutation loop for I_{dc} = 400 A. (a) HM3 Module variant A. (b) HM3 Module variant B.

the decoupling capacitor in the parasitic inductance (Variants A' and A"). All variants include a reduced-size cutout, only for connecting the gate signals and maintaining the voltage clearance distance to the busbar.

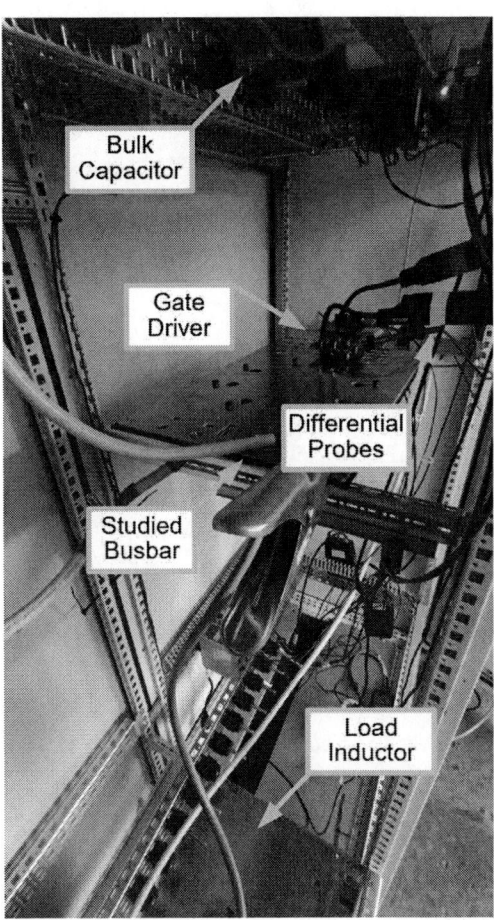

Fig. 6: DPTR with one of the studied busbars installed.

Fig. 5 displays the current distribution for variants A and B, showing that the new designs allow for the current to flow through the middle of the module. This change reduces the commutation loop current path, bringing the busbar inductance down to 7.8 nH for variant A and 10.6 nH for variant B at 100 MHz. Variants A' and A" show similar behaviour to A, with a calculated busbar inductance of 8.3 nH measured at the same frequency. Thus, eliminating the middle tab cutout region effectively reduces the commutation loop inductance for the HM3 modules by more than 50%, reaching similar values to the design for the SP6LI package.

4 Switching Performance Characterisation

This section presents the obtained results using a Double Pulse Test Rig (DPTR), shown in Fig. 6, developed for characterising modules switching up to 2 kV and 2 kA. Measurement of switching vari-

2703

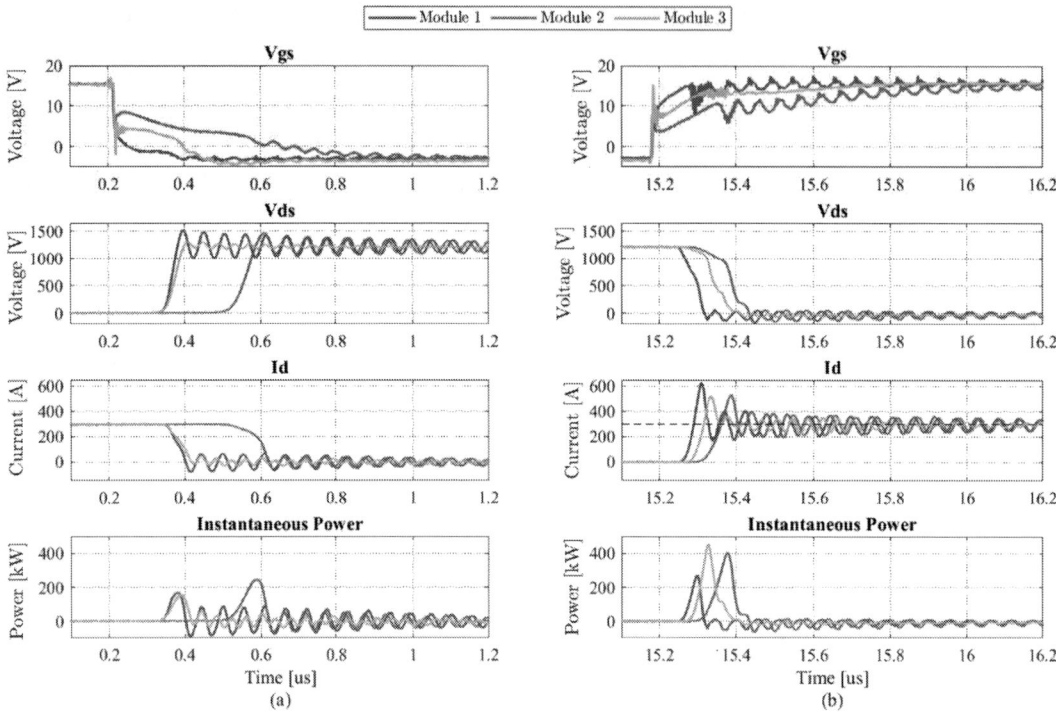

Fig. 7: Switching transients at 1200 V/300 A for studied modules using the initial busbar design in Fig. 2. (a) Turn-off. (b) Turn-on.

Parameter	Module 1	Module 2	Module 3
Turn-off Switching losses	6 mJ	14.3 mJ	6.3 mJ
Turn-off dv/dt	37.6 kV/μs	19.4 kV/μs	27.4 kV/μs
Turn-off di/dt	4.65 kA/μs	5.12 kA/μs	3.82 kA/μs
Turn-off Voltage Overshoot	1522 V (126.8%)	1462 V (121.8%)	1305 V (108.7%)
Turn-on Switching losses	7.6 mJ	20 mJ	20 mJ
Turn-on dv/dt	17.7 kV/μs	29.3 kV/μs	18 kV/μs
Turn-on di/dt	12.2 kA/μs	6.28 kA/μs	9.27 kA/μs
Oscillation frequency V_{ds}	17.95 MHz	18.81 MHz	24.52 MHz

Tab. 2: Performance indicators for switching transients at 1200 V/300 A. Module 1: CAS380M17HM3, Module 2: CAB650M17HM3 and Module 3: MSCSM170AM029CT6LIAG

ables was performed using two oscilloscopes: a Tektronix MSO64 1 GHz 25 GS/s for gate loop signals and a Teledyne LeCroy WaveRunner 604Zi 400 MHz 20 GS/s for power loop waveforms. Gate-source voltages were measured with 1 GHz IsoVu probes, while drain-source voltages were measured using HVD3605 200 MHz 6000 V differential probes. Drain switching current was measured using 50 MHz PEM Rogowski coils as shown in Figs. 2(b)-(c). Baseplate temperature was regulated using an Omrom Automation controller, enabling testing of modules at any temperature be-

tween 25 °C and 150 °C to emulate relevant operating conditions.

The double pulse tests were performed using a commercially available 1.7 kV gate driver from Wolfspeed (CGD1700HB3P-HM3) with an adapter board to connect with the SP6LI module. The tests considered the datasheet recommended external gate resistance for each module and gate-source voltage levels of 15/-4 V for on- and off-states.

Fig. 7 displays the turn-off and turn-on switching transitions at 1200 V/300 A for the three modules under analysis to compare their timing be-

PCIM Europe 2024, 11– 13 June 2024, Nuremberg   DOI: 10.30420/566262380

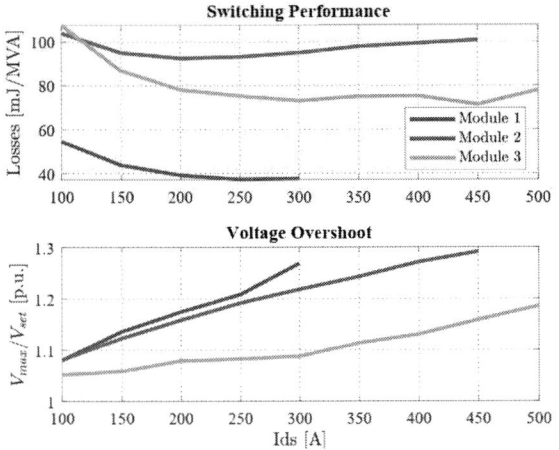

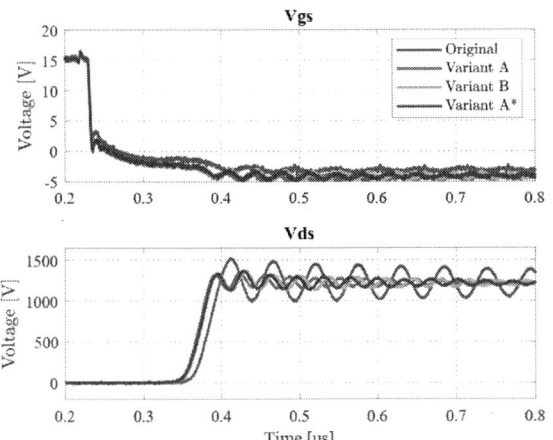

Fig. 8: Performance indicators at 1200V for the studied modules.

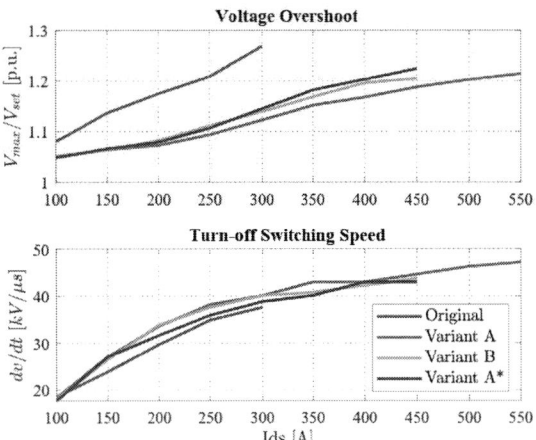

Fig. 10: Performance indicators at 1200V for the studied HM3 connection variants.

Fig. 9: Turn-off transient for different HM3 module busbar connection variants.

haviour. Results show that modules 1 and 3 achieve the highest turn-off dV/dt and turn-on dI/dt, in line with their lower gate charge requirement (displayed in Table 1). Table 2 summarises the main performance indicators for both switching transitions, where dv/dt and di/dt values are calculated with the 40-60% transition of the respective waveforms. Furthermore, Fig. 8 illustrates the normalised switching losses and voltage overshoot levels for a range of switching currents, showing consistently better switching performance for modules 1 and 3.

Both HM3 modules present considerably higher voltage overshoots compared to the SP6LI module due to the larger commutation loop inductance analysed in the previous section. These increased

overshoot levels prevent testing modules 1 and 2 at full capacity, as the peak values exceed the working isolation voltage for the gate driver (1500 V). The total loop inductance can be obtained using (1), considering the resonant frequencies in Table 2 and the output capacitance values for each module in Table 1 [12]. Thus, the calculated total commutation loop inductance is 30.3 nH for module 1, 31.1 nH for module 2 and 23.4 nH for module 3. An estimated value for the busbar added inductances is then obtained by subtracting the modules stray inductance and the decoupling capacitor ESL, which leads to a value of 20.3-21.2 nH for the HM3 modules and ~15.4 nH for the SP6LI package. The HM3 modules value closely matches the one obtained in Q3D simulations, while the SP6LI inductance has a larger variance of 65% respect to the simulated value. This difference can be partially explained due to the lack of data at 1200 V in the datasheet of module 3, which could lead to a variance in the assumed output capacitance.

$$L_{total} = \frac{1}{C_{oss}\omega_{res}^2} \quad (1)$$

The re-designed busbar for the HM3 modules shown in Fig. 4 was also implemented to validate the reduction in commutation loop inductance value. This busbar was tested with module 1, as it presents the highest switching speeds (thus generating the highest magnitude overshoot and oscillations). Fig. 9 shows a comparison of the turn-off voltages at 300 A for the original design and variants A, B and A'. All variants result in majorly

2705

decreased voltage overshoot values, with a maximum of 1380 V (115%) as opposed to the 1522 V from the original design. Furthermore, the dv/dt is also increased by roughly 10% compared to the initial case. Fig. 10 shows that the re-designed HM3 module connections enable testing the module 1 up to 550 A with limited overshoot levels and increased turn-off dv/dt values. The resulting total loop inductance calculates to 12.5 nH, using the same resonant frequency method. This value indicates that the added busbar inductance for the new variants is 2.6 nH, which greatly differs from the 7.8 nH value obtained from Q3D simulations. This difference can be partially explained due to unmodelled mutual inductances between the module and the busbar (as analysed in [13]), as now the current path goes on top of the module (shown in Fig. 5).

5 Conclusions

This paper presented a characterisation of the switching performance of three 1.7 kV SiC half-bridge modules focusing on the effect of the busbar design in the obtained metrics. Results indicate performance metrics in line with datasheet values for the three modules, achieving normalised losses between 40 and 100 mJ/MVA. Moreover, analysis of the total inductance of the commutation loop showed increased values for one of the packages under study (HM3 modules). Thus, a study of the busbar design considering finite-element simulations and experimental validation using a double-pulse test rig was performed. The obtained results showed that a minor mechanical re-design decreased the total commutation loop inductance from 30 nH to 12.5 nH. This improvement results in an overshoot reduction from 1522 V to 1380 V when switching the module at 1200 V/300 A. The overshoot reduction enabled testing the module at currents up to 550 A, with faster switching transients (turn-off dv/dt=48 kV/μs) while keeping the peak voltage under 1500 V.

Acknowledgement

The authors would like to acknowledge the support of Siemens Gamesa Renewable Energy.

References

[1] X. She, A. Q. Huang, O. Lucia, and B. Ozpineci, "Review of silicon carbide power devices and their applications," *IEEE Transactions on Industrial Electronics*, vol. 64, no. 10, pp. 8193–8205, 2017.

[2] A. Q. Huang, "Power semiconductor devices for smart grid and renewable energy systems," *Proceedings of the IEEE*, vol. 105, no. 11, pp. 2019–2047, 2017. DOI: 10.1109/JPROC.2017.2687701.

[3] K. Hamada, M. Nagao, M. Ajioka, and F. Kawai, "Sic—emerging power device technology for next-generation electrically powered environmentally friendly vehicles," *IEEE Transactions on Electron Devices*, vol. 62, no. 2, pp. 278–285, 2015. DOI: 10.1109/TED.2014.2359240.

[4] L. Zhang, X. Yuan, X. Wu, C. Shi, J. Zhang, and Y. Zhang, "Performance evaluation of high-power sic mosfet modules in comparison to si igbt modules," *IEEE Transactions on Power Electronics*, vol. 34, no. 2, pp. 1181–1196, 2019. DOI: 10.1109/TPEL.2018.2834345.

[5] S. Bontemps and L.-P. Doumergue, "Very low stray inductance, high frequency 1200 v 2 mohms full sic mosfet phase leg module," in *PCIM Europe 2018; International Exhibition and Conference for Power Electronics, Intelligent Motion, Renewable Energy and Energy Management*, 2018, pp. 1–8.

[6] A. H. Ismail, A. Al-Hmoud, Y. Zhao, A. Kumar, and K. Olejniczak, "A high-current 1.7 kv sic module enabling high efficiency, high power density renewable energy applications," in *PCIM Europe 2023; International Exhibition and Conference for Power Electronics, Intelligent Motion, Renewable Energy and Energy Management*, 2023, pp. 1–7. DOI: 10.30420/566091067.

[7] J. Chen, X. Du, Q. Luo, X. Zhang, P. Sun, and L. Zhou, "A review of switching oscillations of wide bandgap semiconductor devices," *IEEE Transactions on Power Electronics*, vol. 35, no. 12, pp. 13 182–13 199, 2020. DOI: 10.1109/TPEL.2020.2995778.

[8] H. Lee, V. Smet, and R. Tummala, "A review of sic power module packaging technologies: Challenges, advances, and emerging issues," *IEEE Journal of Emerging and Selected Topics in Power Electronics*, vol. 8, no. 1, pp. 239–255, 2020. DOI: 10.1109/JESTPE.2019.2951801.

[9] CAS380M17HM3 Data Sheet, Rev. 2, Jan. 2024.

[10] CAB650M17HM3 Data Sheet, Rev. 2, Jan. 2024.

[11] MSCSM170AM029CT6LIAG Data Sheet, Rev. A, Apr. 2021.

[12] S. Hu, R. Chen, X. Wu, M. Tahir, and Q. Yang, "Stray parameter extraction method based on high- frequency oscillation: An experimental study with theoretical and execution demo," *IEEE Transactions on Power Electronics*, vol. 38, no. 12, pp. 15870–15878, 2023. DOI: 10.1109/TPEL.2023.3309402.

[13] R. S. Krishna Moorthy, B. Aberg, M. Olimmah, L. Yang, D. Rahman, *et al.*, "Estimation, minimization, and validation of commutation loop inductance for a 135-kw sic ev traction inverter," *IEEE Journal of Emerging and Selected Topics in Power Electronics*, vol. 8, no. 1, pp. 286–297, 2020. DOI: 10.1109/JESTPE.2019.2952884.

PCIM Europe 2024, 11– 13 June 2024, Nuremberg   DOI: 10.30420/566262381

Characterizing the Switching Behavior of a 1.2 kV mixed SiC JFET and MOSFET Half Bridge

Tim Ringelmann[iD], Mark-M. Bakran[iD]
University of Bayreuth, Department of Mechatronics, Center of Energy Technology, Germany

Corresponding author: Tim Ringelmann, Tim.Ringelmann@uni-bayreuth.de
Speaker: Tim Ringelmann, Tim.Ringelmann@uni-bayreuth.de

Abstract

For cost savings and loss reduction in automotive inverters, SiC JFETs are promising devices compared to SiC MOSFETs. For this reason, the switching behavior of JFETs and MOSFETs is investigated and compared with each other. The normally-on behavior of the JFET, which is a disadvantage due to a phase short-circuit while the gate driver supply voltage fails, can be turned into an advantage by mixing each half bridge of a B6 inverter with a normally-off semiconductor. The mixed half bridge configuration allows an active short-circuit case in an automotive inverter, by itself. For this reason, the switching behavior and switching effects in a mixed JFET/MOSFET half bridge are investigated and compared with the uniform half bridges. To complete the switching characterization, a performance calculation is performed using a B6 inverter.

1   Introduction

One procedure for reducing the costs of automotive inverters is the reduction of the semiconductor die area A_{Die}. The area specific resistance $R_{DS(on)} \cdot A_{Die}$ is often used to differentiate the costs and losses of the inverter. In this context, a silicon carbide (SiC) junction field effect transistor (JFET) has a significant advantage over the state-of-the-art SiC metal oxide filed effect transistor (MOSFET) [1–3]. This is also illustrated by the forward and reverse conduction characteristics shown in Fig. 2. The Figure shows that the current density for a JFET is significantly lower compared to a MOSFET. In addition the JFET has no gate reliability problems according to [4]. Furthermore, the production costs of JFETs are lower due to their simple structure compared to the MOSFET, so the overall price of the semiconductor is expected to be lower [3]. This is one of the reasons why the switching behavior of 1.2 kV SiC JFETs is discussed in this paper compared to 1.2 kV SiC MOSFETs. However, a disadvantage of the JFET is the normally-on characteristic, which results in a half bridge (HB) short-circuit when the gate driver voltage fails. Combining the JFET and a MOSFET in a mixed HB, the normally-on behavior can be turned into an advantage. One possible topology of a mixed B6 inverter application is illustrated in Fig. 1.

Looking at the so called active short-circuit (ASC), a safety mechanism in automotive applications during gate driver outage, the mixed HB seems to be a promising topology [5]. The ASC prevents uncontrolled recuperation into the DC-link as well as a strong brake torque [5]. To perform an ASC, all high side switches (HSS) or all low side switches (LSS) must be switched in the on-state while the complementary switches are switched into the off-state [5]. In other words, the phases of the electric motor (EM) must be short-circuited [5]. To achieve this, a MOSFET B6 inverter requires a driver supply voltage. This is necessary due to the normally-off characteristic of the MOSFETs, which requires all HSS or LSS to be switched into the on-state. Typically, a redundant supply voltage is used, which is provided by the DC-link or the battery. In contrast to normally-off devices, normally-on devices do not require a driver-supply voltage to be in the on-state. Therefore, a mixed HB B6 inverter short-circuits the three phases of the inverter due to the normally-on behavior of the JFETs and prevents a HB short-circuit with the complementary MOSFETs, even without the redundant driver

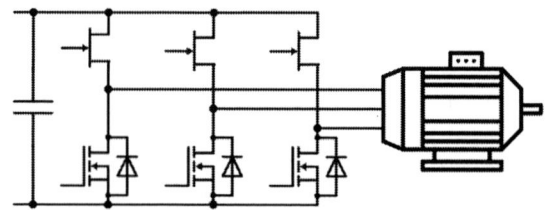

Fig. 1   Mixed JFET/MOSFET inverter application

2708

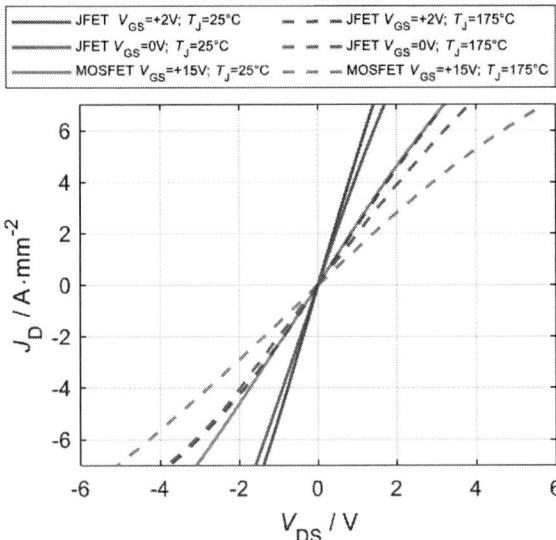

Fig. 2 Forward and reverse conduction characteristics [2]

supply voltage. This results in further potential for cost reduction. To summarize, the mixed JFET/MOSFET HB offers significant potential for cost optimization in automotive inverters. For this reason, this paper focusses on the switching behavior and the performance of mixed JFET/MOSFET HBs

2 Test Setup

The two devices under test (DUTs) being tested are a state-of-the-art 1.2 kV SiC JFET from the manufacturer *Qorvo* and a 1.2 kV SiC MOSFET from the manufacturer *Wolfspeed*. Their switch-

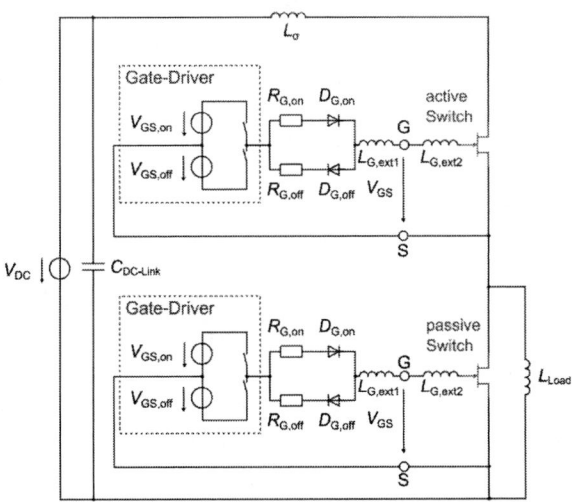

Fig. 3 Equivalent circuit of the hardware test setup

ing behavior is analyzed using a standard double pulse test bench as shown in Fig. 3. The DC-link voltage V_{DC} for this application is V_{DC}=800 V. To validate the switching behavior under application criteria, the stray inductance L_σ of the test setup was scaled according to the rules of [6]. L_σ is scaled to a modern inverter with a stray inductance of 10 nH and a rated current of 450 A. The switching behavior of the active switch (a.S.) and passive switch (p.S.) is analyzed on an inductive load L_{Load}. The gate driver applies the corresponding gate source voltage V_{GS} to the semiconductors depending on the switching sate (on/off). The turn-on and turn-off switching speed of the a.S. is controlled and can be adjusted by the external gate resistances $R_{G,on/off}$ and the diodes $D_{on,off}$. The a.S. gate resistances should be designed to the maximum rated voltage (1200 V) of the semiconductors, while the p.S. gate resistances are 0 Ω. Unfortunately, the application-scaled L_σ is so small that all gate resistances are designed to 0 Ω and the complete performance up to the device limit of 1200 V cannot be utilized (see chapter 4). In the equivalent circuit, the inductance $L_{G,ext}$ is used to model the gate-driver inductance.

3 Special Switching Effects of mixed Half Bridges

This chapter discusses the special switching effects of a mixed HB. As previously noted in [2, 7], the ratio of the gate drain to gate source charge Q_{GD}/Q_{GS} of the JFET is greater than one. If this ratio exceeds one, a semiconductor is susceptible to parasitic turn-on (PTon). As illustrated in Fig. 4 the JFET is susceptible to PTon, while the MOSFET is not. The PTon-effect is also addressed by [7, 8]. [2]

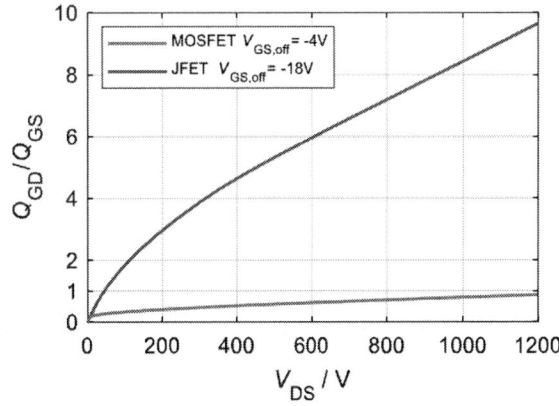

Fig. 4 Ratio of the gate-drain charge Q_{GD} to the gate-source charge Q_{GS} [2]

PCIM Europe 2024, 11– 13 June 2024, Nuremberg   DOI: 10.30420/566262381

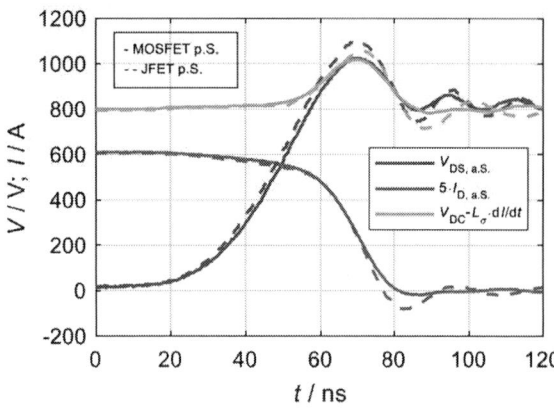

Fig. 5   Impact of the p.S. device structure on an a.S. turn-off event in a mixed HB with a MOSFET as a.S.

3.1 Passive Switch Turn-On Overvoltage

As already shown in [2], the Q_{GD}/Q_{GS} ratio causes a p.S. turn-on-overvoltage in a uniform JFET HB due to parasitic turn-off (PToff). This effect can be further demonstrated by the analysis of a mixed HB. In Fig. 5 the comparison of an a.S. turn-off event of a MOSFET for both JFET and MOSFET as p.S. is shown. To analyze the effect of the p.S. turn-on overvoltage the sum of the DC-link voltage V_{DC} and the inductive voltage increase $L_\sigma \cdot dI/dt$ is shown. As illustrated, the drain-source voltage of the a.S. $V_{DS,a.S.}$ is almost identical to $V_{DC}-L_\sigma \cdot dI/dt$ for a uniform MOSFET HB. The MOSFETs body diode reverse conduction voltage is negligible. For the mixed HB, the JFET's reverse conduction behavior (high knee voltage and PToff) results in a non-negligible turn-on overvoltage of the p.S., which is already visible during an a.S. turn-off event as previously mentioned in [2]. This results in a deviation of the $V_{DS,a.S}$ to $V_{DC}-L_\sigma \cdot dI/dt$ during an a.S. turn-off event. If the $R_{G,off,a.S.}$ is designed for the maximum rated overvoltage (1200 V), the increase in the a.S. turn-off overvoltage due to the p.S. turn-on overvoltage would lead to a slower switching process and correspondingly higher losses. To reduce the effect of the p.S. turn-on overvoltage a low impedance clamp circuit is introduced in [2]. The effect of the p.S. turn-on overvoltage in a mixed HB can be quantified by the turn-off energies of the a.S., illustrated in Fig. 12.

3.2 Reduction of the Parasitic Turn-On Behavior and its Influence

This section analyzes the PTon behavior due to the Q_{GD}/Q_{GS} ratio of the JFET. PTon affects both the a.S. turn-on energy and p.S. turn-off energy.

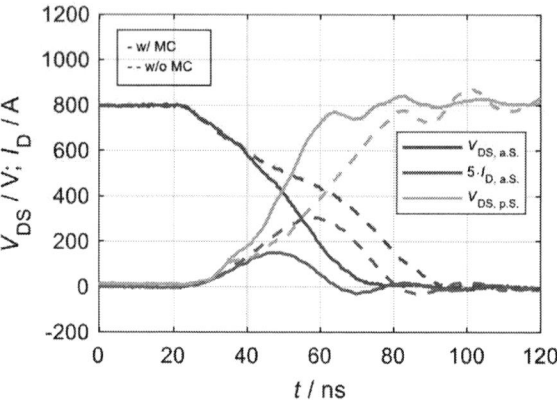

Fig. 6   Influence of the PTon on switching slopes for a JFET/JFET HB

The switching slopes of an a.S. turn-on event and a p.S. turn-off event at zero current switching is shown in Fig. 6, to visualize the PTon. Additionally, a measurement slope with a hardware adjustment (miller clamp) to reduce PTon is shown. The decrease in the turn-on energy of the a.S. and turn-off energy of the p.S. can be estimated as significant (refer to Fig. 21). To reduce PTon, a miller clamp (MC) printed circuit board (PCB) with low impedance, between the JFET and the MC, is used as a hardware adjustment in the double pulse test bench. The MC PCB is connected to the terminals G and S as illustrated in Fig. 3. According to [2], PToff can be reduced with a clamp circuit (in this case: a schottky diode D_{Clamp}, a ceramic capacitor C_{Clamp}, and a very low connection inductance L_{MC}). This clamp circuit is extended with a silicon (Si) low voltage (LV) MC MOSFET $T_{MC\text{-}MOSFET}$ to reduce PTon. If $T_{MC\text{-}MOSFET}$ is implemented, the clamp diode may be omitted due to the body diode of the MOSFET. This difference is not discussed further in this paper. The $T_{MC\text{-}MOSFET}$ is controlled by a MC-driver. To enhance comprehension of the MC, refer to Fig. 8 for the control signals of the semiconductors and the corresponding MC MOSFETs. The MC control signal typically follows the dashed yellow lines [9]. The time of PTon occurrence is at the red dotted vertical lines. To simplify the yellow control signals the blue ones are implemented.

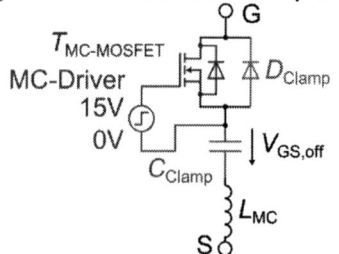

Fig. 7   Equivalent circuit of the used MC circuit

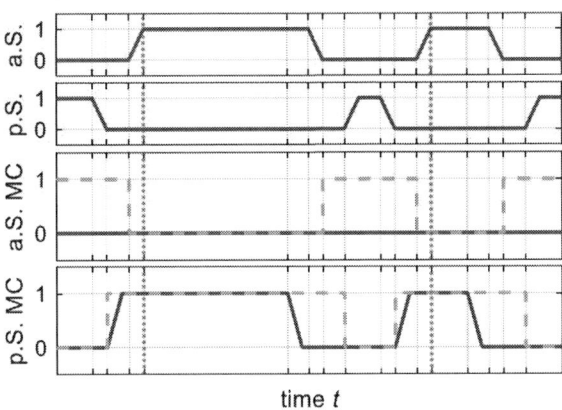

Fig. 8 Control signals of the used double pulse setup with MC (blue: control signals, yellow: regular MC control signals, red: time of PTon occurrence)

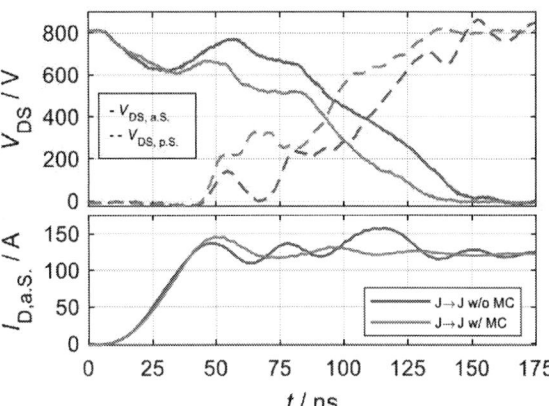

Fig. 10 Influence of the MC at a zero current a.S. turn-on and p.S. turn-off event for a uniform JFET HB

A reduction of the off-state gate source voltage reduces the PTon effect. To validate the MC, a uniform JFET HB with a MC and an off-state V_{GS} sweep of the p.S. is performed, as illustrated in Fig. 9. The JFET does not have reverse-recovery due to its unipolar structure, even in reverse conducting operation with a closed channel. Thus, this paper renames the reverse-recovery energy E_{rr}, commonly used for uniform MOSFET HBs, to E_{rr+PTO}. As shown, reducing the $V_{GS,off\ p.S.}$ further leads to a decrease of the reverse-recovery and PTon energy E_{rr+PTO} at low current. In summary, this plot shows that despite the implemented MC with $V_{GS,off\ p.S.}$=-18 V, there is still minimal PTon. Furthermore, the turn-on event of the a.S. is also influenced by the MC, as illustrated in Fig. 10. The comparison reveals a significant difference in the voltage slopes and current slopes, resulting in a remarkable reduction in the turn-off losses of the p.S. and a reduction in the turn-on losses of the active switch. To provide an overview of the different HB configurations, the following conven-

tion is introduced as an abbreviation and for better differentiation between individual HB configurations: "a.S.→p.S. with (w/) or without (w/o) MC". It defines which switch is the a.S., which is the p.S. and whether an MC is implemented.

The MC not only affects the turn-on event of the a.S., but also indirectly affects its turn-off event, as shown in Fig. 11. This may seem confusing at first, as the MC represented by the blue lines in Fig. 8, should not have any influence on the switching behavior except for the clamp behavior due to the clamp circuit. The influence shown in Fig. 11 is due to the additional capacitance (output capacitance C_{oss} of the MC MOSFET in series with the clamp capacitance C_{Clamp}) in parallel to the gate source capacitance of the DUT. The additional capacitance directly influences the dI/dt and therefore the voltage overshoot of the a.S. turn-off event. It is worth noting that this can be balanced with a lower $R_{G,off\ a.S.}$, but the $R_{G,off\ a.S.}$ is already 0 Ω due to the low application scaled stray inductance.

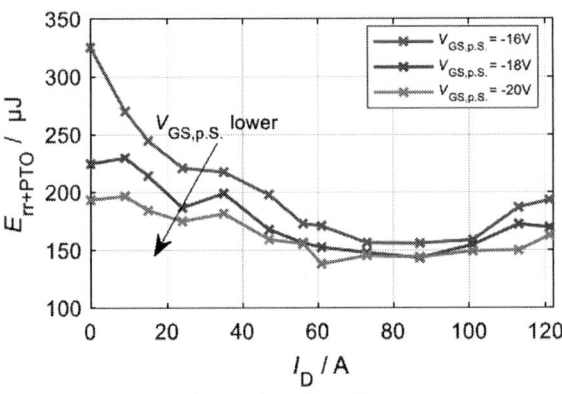

Fig. 9 Effect of reducing the off-state gate-source voltage of the p.S. on the p.S. turn-off energy for a uniform JFET HB with MC

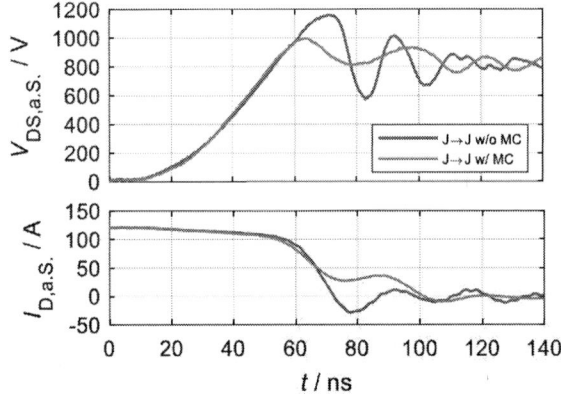

Fig. 11 Indirect effect of the MC on the a.S. turn-off event for a uniform JFET HB

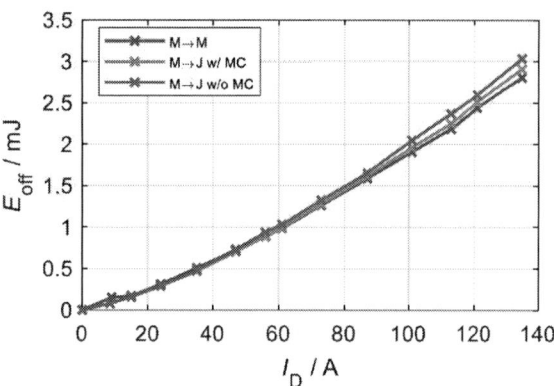

Fig. 12 Effect of the p.S. device and the MC on the a.S. turn-off energy in a mixed HB

The MC introduced in this paper is only used in combination with a JFET, not with a MOSFET. This is due to the JFET having a Q_{GD}/Q_{GS} ratio greater than one, while the MOSFET has a ratio smaller than one. To summarize the a.S. turn-off event and the MC influence in a mixed JFET/MOSFET HB, Fig. 12 illustrates the losses at different currents. As previously explained in chapter 3.1, the p.S. affects the turn-off losses of the active switch. However, this effect can be reduced by using a MC at the passive switch.

3.3 Influence of the Different Semiconductor Structure and Temperature

The switching behavior of the p.S. is influenced by the device structure. During the interlock or dead time, the MOSFET acts as a bipolar device due to its body diode, while the JFET acts as a unipolar device. Furthermore, the switching slopes differ during dead time [2, 10]. The reason for this is the relatively high "knee voltage" of the JFET compared to the MOSFET body diode threshold voltage $V_{G(th)}$ in the off-state [2].

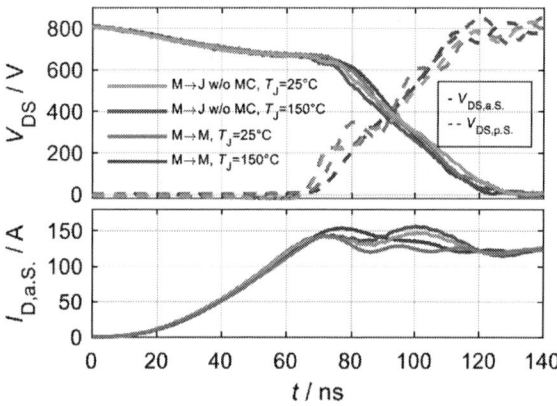

Fig. 13 Effect of the p.S. device and the temperature on an a.S. turn-on event

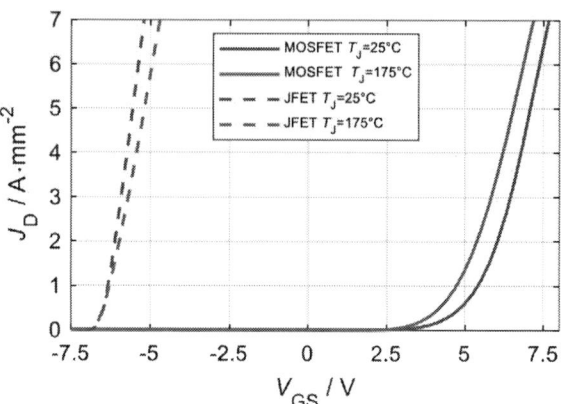

Fig. 14 Transfer characteristics of the JFET and MOSFET at different temperatures [2]

Furthermore, the behavior of the bipolar body diode exhibits increased reverse-recovery with rising temperature and current, resulting in an increase in E_{rr+PTO}. Due to the susceptibility of the JFET to PTon, it is challenging to differ between the effects of PTon and reverse-recovery when comparing the p.S. device. Therefore, the JFET with MC is also analyzed to reduce PTon. As depicted in Fig. 9, the JFET is not free of PTon despite the MC.

The bipolar nature of the MOSFET's body diode, in addition to the MOSFET transfer characteristics (see Fig. 14), is used to explain the a.S. turn-on events shown in Fig. 13. The PTon also influences the turn-on losses of the active switch. Due to the bipolar nature of the MOSFET's body diode, the $V_{DS,p.S.}$ slope occurs later in comparison to a unipolar p.S. device such as the JFET. Additionally, the MOSFET's transfer characteristic, including $V_{G(th)}$, is shifted to the left with increasing temperature [2]. Furthermore, the $R_{G,int}$ also increases with higher temperature, which has opposite effects in terms of dV/dt. Based on the measured slopes in Fig. 13, these two effects results in a small increase in switching speed. In addition, the current slopes for a MOSFET as p.S. illustrate the increasing reverese-recovery at higher temperatures. Furthermore, the JFET as p.S. has an increasing in PTon due to the higher $R_{G,int}$ of the p.S. at a higher temperature.

To summarize the switching slopes in Fig. 13, the turn-on energies of the a.S. are evaluated in Fig. 15, including all effects mentioned above. When switching a uniform MOSFET HB, the reverse-recovery effect dominates. When switching a MOSFET as a.S. against a JFET as p.S., there are again two opposite effects to consider: the PTon increases the losses, whereas the a.S. reduces the losses due to a faster switching. At low

PCIM Europe 2024, 11– 13 June 2024, Nuremberg   DOI: 10.30420/566262381

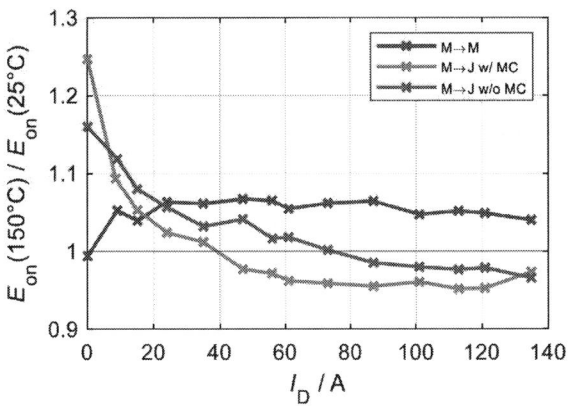

Fig. 15 Temperature influence on the a.S. turn-on energies for different p.S. devices

current, the PTon effect dominates. As the current increases, the dV/dt effect dominate.This is demonstrated by comparing the slope with and without MC. The crossover points with the horizontal line at 1 represents the trade-off between the PTon and the dV/dt effect. The MC reduces the PTon, shifting the crossover point with 1 to smaller load currents.

4 Switching Behavior

In this section, the switching behavior for different HB configurations will be discussed in detail. The different effects of mixed HBs are already explained in chapter 3. Therefore, the switching slopes are just shown as examples of the already explained effects. Due to space limitations, only the slopes at a temperature of 25 °C are shown. Furthermore, a comprehensive analysis of the switching energies indicates the impact of temperature variation from 25 °C to 150 °C on each HB, as well as the normalized switching energy of each HB to the uniform MOSFET HB at 25 °C.

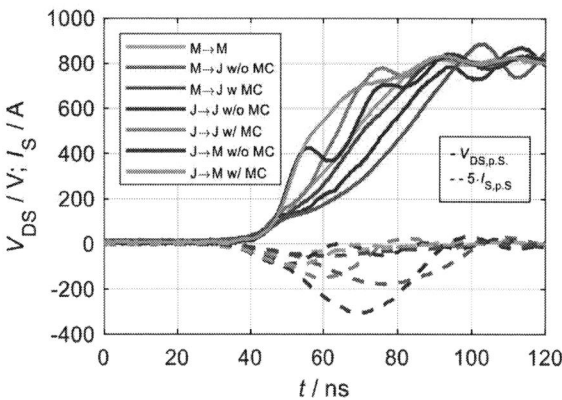

Fig. 17 Zero current switching: p.S. turn-off event

The first switching event analyzed is the a.S. turn-on event without load current which is shown in Fig. 16. Here, the PTon effect affects the voltage and current slopes. For a switching event without PTon, only capacitive losses would occur, as it is the case for the uniform MOSFET HB.

The second switching event, the p.S. turn-off event without load current, is shown in Fig. 17. The p.S. reacts to the a.S. turn-on event. This switching event is also influenced by the PTon.

Another switching event, the a.S. turn-off event with a current of 121 A, is shown in Fig. 18. One of the main effects in this plot is the difference in the p.S. turn-on overvoltage. The MC influences the dI/dt, which affects the a.S. turn-off overvoltage. In Addition, none of the HB configurations reach the maximum device voltage of 1200 V, even with a $R_{G,off\ a.S.}$ of 0 Ω. The p.S. turn-on losses are negligible, so these switching events are not discussed in this paper.

The forelast switching event considered is the a.S. turn-on event with a current of 121 A. This switching slope is again dominated by the PTon and the MC. The influence of the p.S. device

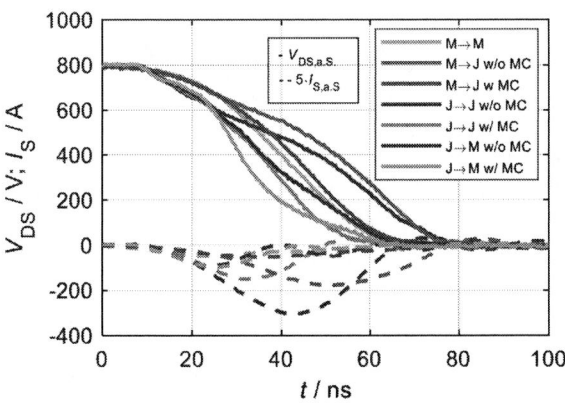

Fig. 16 Zero current switching: a.S. turn-on event

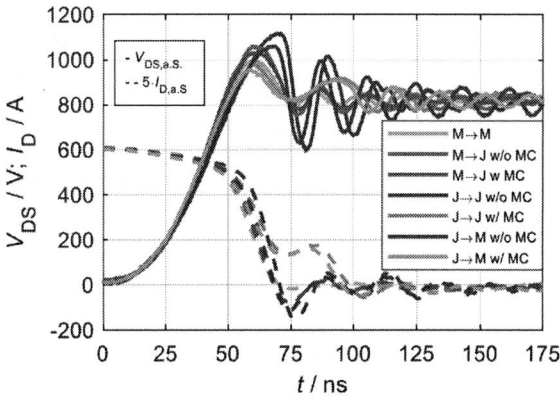

Fig. 18 Active switch turn-off event at 121 A

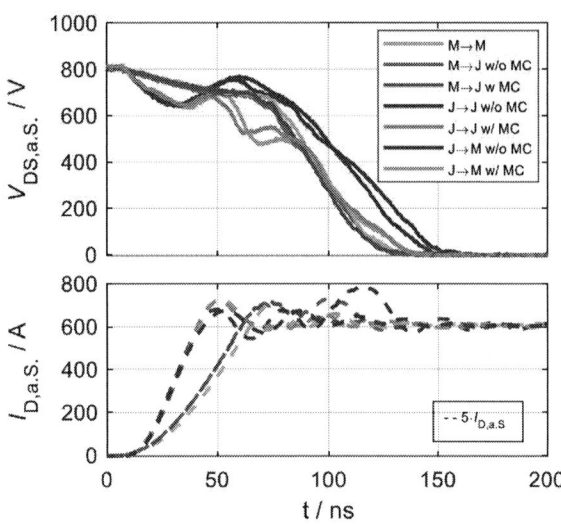

Fig. 19 Active switch turn-on event at 121 A

structure (bipolarity of the MOSFET body diode and unipolarity of the JFET in the dead time) is also detectable in the slopes.

Finally, the p.S. turn-off event at 121 A is analyzed in Fig. 20. The strong influence of the PTon also applies here, as well as the structure of the p.S. device (bipolarity of the MOSFET body diode and unipolarity of the JFET in the dead time) is also detectable in the slopes. As can be seen, the maximum device voltage of 1200 V is not reached in any configuration, despite a $R_{G,on\,a.S.}$=0 Ω. Therefore, the complete performance up to the device limit of 1200 V cannot be utilized.

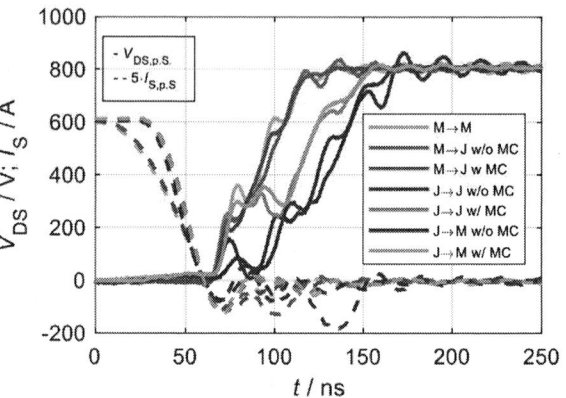

Fig. 20 Passive switch turn-off event at 121 A

The PTon is not always a disadvantage at an a.S. turn-on event or a p.S. turn-off event [11]. A small PTon can reduce the p.S. turn-off overvoltage and allow a faster switching of the a.S. turn-on event until the maximum rated device voltage is reached at the p.S. overvoltage [11]. This results in lower switching losses [11]. The stray inductance used for the switching events in this paper, is so small that the maximum device voltage is not even close to the device limit of 1200 V, despite a $R_{G,on\,a.S.}$ of 0 Ω. Therefore, a faster switching is not possible.

Finally, Fig. 21 shows the switching energies as an overall result of the switching behavior. To facilitate comparison with the state-of-the-art uniform MOSFET HB the switching energies at 25 °C are normalized to the switching energies from the uniform MOSFET HB (top side figures). In addition, Fig. 21 presents the temperature be-

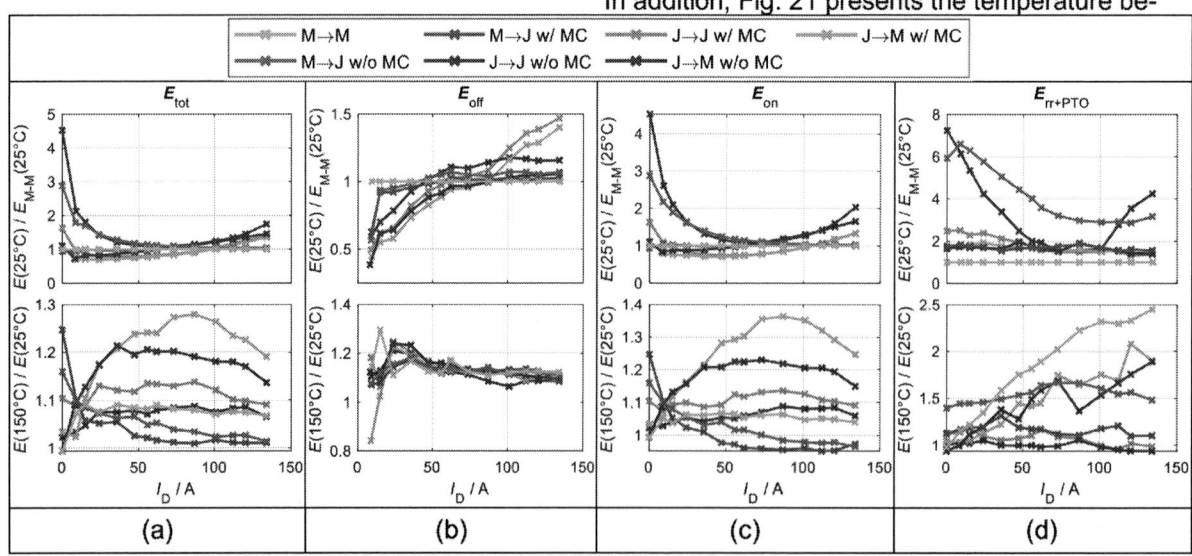

Fig. 21 Switching energies of the different HB configurations in comparison to the MOSFET HB and their temperature dependence to themselve; (a) a.S. total energy E_{tot}, (b) a.S. turn-off energy E_{off}, (c) a.S. turn-on energy E_{on}, (d) p.S. turn-off energy E_{rr+PTO}

Boundary condition	Value
Modulation m	1
Power factor $\cos(\varphi)$	1
DC-Link voltage V_{DC}	800 V
Electrical frequency f_{el}	200 Hz
Gate-source voltage $V_{GS,MOSFET/JFET}$	15 V / 2 V
Fluid temperature T_F	65 °C
Max. junction temperature $T_{J,max}$	150 °C
Area specific thermal resistance r_{th}	20 $\frac{K \cdot mm^2}{W}$
Apparent power S	200 kVA

Table 1   Standard boundary conditions for the performance analysis

havior of all HB configurations to themselves (bottom side figures). This enables individual evaluation of the temperature behavior of each HB configuration. The temperature influence is shown by the ratio of the switching energies at 25 °C to 150 °C. As illustrated in (a), the total energies E_{tot} ($E_{tot}=E_{on}+E_{off}$) of the HB configurations with MC are competitive to the uniform MOSFET HB. The temperature influence is dominated by the PTon/PToff and the p.S. device structure (bipolarity of the MOSFET body diode and unipolarity of the JFET in the dead time).

5   Performance

To conclude the discussion on the switching behavior, this chapter presents a performance calculation (sinus pulse width modulation) of a B6 inverter using the HB configurations analyzed in chapter 4. The boundary conditions used for the performance analysis are listed in Table 1.

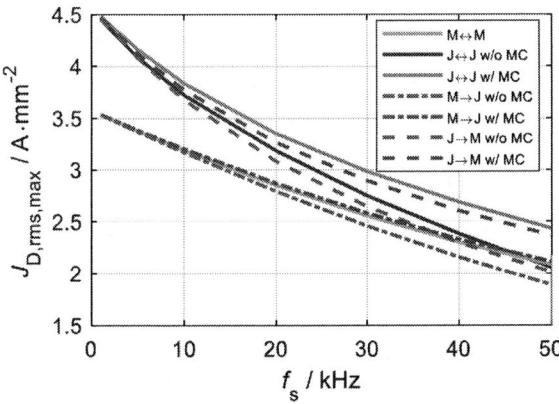

Fig. 22   Maximum current density of the a.S. at different f_s

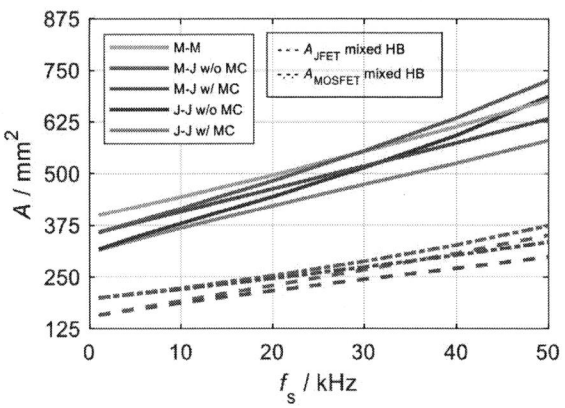

Fig. 23   SiC area of a B6 inverter at different f_s

First of all, the maximum current density $J_{D,rms,max}$ of an a.S. in a B6 inverter for different switching frequencies f_s are shown in Fig. 22. The calculation of $J_{D,rms,max}$, is based on a steady-state temperature condition, calculated with the conduction and switching losses in combination with the area specific thermal resistance r_{th} and a cooling fluid temperature T_F, at each current. $J_{D,rms}$ rises until the temperature reaches the maximum junction temperature $T_{J,max}$. The JFET HB with MC achieved the highest $J_{D,rms,max}$ due to the reduced PTon/PToff and the lower $R_{DS(on)} \cdot A_{Die}$ compared to the MOSFET HB. In all configurations with a MC the $J_{D,rms,max}$ is higher than without a MC. For higher f_s, the conduction losses are outweighed by the switching losses. If the JFET acts as a.S. in a uniform HB, the switching losses are higher compared to the MOSFET switching losses in a uniform HB. This leads to a greater decrease in the $J_{D,rms,max}$ with a JFET HB compared to the MOSFET HB for higher f_s. The basis for calculating the SiC area A of a B6 inverter is $J_{D,rms,max}$. A SiC area calculation for a B6 inverter with an apparent power S=200 kVA is shown in Fig. 23. In the case of mixed HBs, the SiC area can be divided into a JFET-area A_{JFET} and a MOSFET-area A_{MOSFET}. The sum of both areas results in the whole SiC area A of the analyzed inverter. The preferred use-case (shown in Fig. 1 with an additional MC for the JFETs) reduces the SiC area by 8 % compared to a state-of-the-art uniform MOSFET B6 inverter. As already mentioned in the introduction, reducing the SiC area results in lower inverter cost.

Another consideration is the load efficiency η, which is calculated and illustrated for a B6 inverter in Fig. 24. As illustrated, a MC significantly improves the partial load efficiency. The uniform JFET HB with MC performed the best at high loads due to the low $R_{DS(on)} \cdot A_{Die}$, which reduces the conduction losses. The uniform MOSFET HB

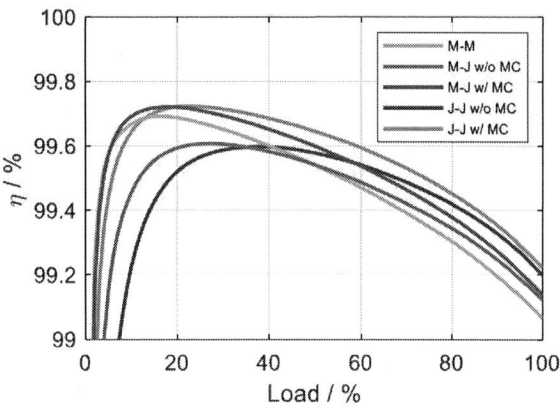

Fig. 24 Efficiency of a B6 inverter at different loads

outperforms the uniform JFET HB at low load due to its lower switching losses, which are more dominant than the conduction losses at low loads. The mixed JFET/MOSFET HB B6 inverter combines the advantages of the uniform HB B6 inverters. The mixed HB B6 inverter has lower switching losses at partial load than the uniform JFET B6 inverter and its conduction losses are smaller than those of the uniform MOSFET B6 inverter. For this reason, a small gap (5-20 % load) of the mixed HB B6 inverter resulted in a higher efficiency compared to the uniform HB B6 inverters.

Efficiencies of electric vehicles are often characterized using the worldwide harmonized light vehicle test cycle (WLTP). In this paper the analysis corresponds to a WLTP of an electric sport utility vehicle (SUV). The driving cycle is independent from the maximum apparent power of the inverter. Accordingly, the WLTP driving profile focuses

HB Configuration @ S = 200 kVA @ f_S = 10 kHz	$A_{B6} \cdot A^{-1}_{B6(M-M)}$ / %	η_{WLTP} / %
M − M	100	98.75
J − J w/o MC	85.6	97.97
J − J w/ MC	83.1	98.86
M − J w/o MC	93.7	98.35
M − J w/ MC	92.0	98.88

Table 2 Overview of the different B6 inverter configurations

more on the partial load of an inverter when the inverter's apparent power S is higher. For this reason, an MC should never be omitted when using a JFET (see Fig. 24). The efficiency of the WLTP η_{WLTP} for different S is shown in Fig. 25. The red, dashed line indicates the minimum $S_{WLTP\,Limit}$ required to handle the WLTP. When S is close to $S_{WLTP\,Limit}$, the overall losses are dominated by conduction losses. Therefore, the uniform JFET HB with MC got the best performance at the lowest S. However, for higher S, the partial load switching losses affect the driving cycle, resulting in the uniform MOSFET HB having better performance than the uniform JFET HB with MC at the highest visualized S. The mixed JFET/MOSFET B6 inverter with MC has a higher η_{WLTP} until 600 kVA compared to the state-of-the-art uniform MOSFET B6 inverter.

6 Conclusion

As described in this publication, the JFET has the potential to reduce losses due to its low $R_{DS(on)} \cdot A_{Die}$ and similar switching losses compared to a MOSFET HB. This paper investigates different switching effects of the uniform and of mixed JFET/MOSFET HBs in detail. Due to its Q_{GD}/Q_{GS} ratio, the JFET has PTon and PToff in contrast to the MOSFET. For this reason, the JFET HB results in higher switching losses compared to the MOSFET HB. To enhance competitiveness and minimize PTon losses compared to the uniform MOSFET HB, a miller clamp is introduced with a uniform JFET HB. The mixed JFET/MOSFET HB with MC is also competitive compared to a uniform MOSFET HB. Furthermore, a switching comparison (switching slopes and losses) is presented for different HB configurations. Finally, a performance calculation is performed to conclude the switching behavior. The performance analysis includes an investigation of the maximum current density, the total SiC area of a B6 inverter, the load efficiency, and a SUV WLTP efficiency.

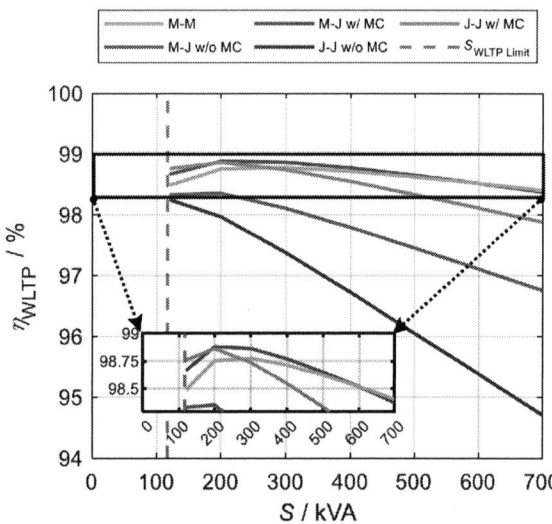

Fig. 25 WLTP efficiency at different apparent power

To summarize the performance of all B6 inverter configurations, Table 2 listed the B6 configuration chip areas A_{B6} normalized to the uniform MOSFET B6 inverter chip area $A_{B6(M-M)}$. In addition, the WLTP efficiency η_{WLTP} is listed for all B6 inverter configurations. The mixed HB with MC B6 inverter (see Fig. 1) is the targeted use case, which has a 8 % smaller SiC area than the uniform MOSFET B6 inverter and a 0.13 % better WLTP efficiency under the boundary conditions from Table 1 with $S = 200$ kVA and $f_s = 10$ kHz.

In conclusion, the alternative B6 inverter topology with a mixed HB of JFET with MC and MOSFET offers several advantages compared to the uniform MOSFET HB B6 inverter. The reduced SiC area leads to a cost-saving of the inverter. Furthermore, the topology of the HB presented (see Fig. 1) leads to an immediate ASC in the event of driver supply voltage failure, which is why a redundant driver supply voltage can be neglected.

Future investigations should compare the surge current robustness of 1.2 kV SiC JFETs to 1.2 kV SiC MOSFETs for the assessing an ASC event. Additionally, a driver concept should be developed to short-circuit the gate source contacts in the case of a driver supply voltage failure, to have a defined off-state of the normally-off and a defined on-state of the normally-on devices in a mixed HB. This results in defined control signals for the ASC.

Acknowledgement

This project was supported by the ZF Friedrichshafen AG. Many thanks go to the department Future Semiconductor, Bayreuth.

7 References

[1] A. Bhalla, X. Li, and J. Dodge, "Circuit Protection with SiC FETs in dual-gate configuration," in *PCIM Asia 2021; International Exhibition and Conference for Power Electronics, Intelligent Motion, Renewable Energy and Energy Management*, 2021, pp. 170–177.

[2] T. Ringelmann and M.-M. Bakran, "Characterization of the Static and Dynamic Behavior of a 1.2kV SiC JFET in Reverse Conduction," in *PCIM Europe 2023; International Exhibition and Conference for Power Electronics, Intelligent Motion, Renewable Energy and Energy Management*, 2023, pp. 815–823.

[3] X. Li, A. Bhalla, P. Alexandrov, and L. Fursin, "Study of SiC vertical JFET behavior during unclamped inductive switching," in *2013 Twenty-Eighth Annual IEEE Applied Power Electronics Conference and Exposition (APEC)*, 2013, pp. 2588–2592.

[4] X. Song and Y. Du, "Study of the SiC JFET Reverse Conduction and Reverse Blocking Characteristics," in *2019 IEEE Applied Power Electronics Conference and Exposition (APEC)*, 2019, pp. 2751–2756.

[5] T. Appel and A. Bieler, "Novel Method for Active Short Circuit (ASC) Tests of Power Module in Automotive Traction Application," in *2022 24th European Conference on Power Electronics and Applications (EPE'22 ECCE Europe)*, 2022, P.1-P.7.

[6] R. W. Maier and M. -M. Bakran, "Switching SiC MOSFETs Under Conditions of a High Power Module," in *2018 20th European Conference on Power Electronics and Applications (EPE'18 ECCE Europe)*, 2018, P.1-P.9.

[7] A. Maerz, R. Horff, T. Bertelshofer, M. Helsper, and M. Bakran, "Benchmarking of SiC JFET and SiC MOSFET modules for the application in medium power traction converters," in *PCIM Europe 2016; International Exhibition and Conference for Power Electronics, Intelligent Motion, Renewable Energy and Energy Management*, 2016, pp. 1–8.

[8] D. Heer, R. Bayerer, and D. Domes, "SiC-JFET in half-bridge configuration - parasitic turn-on at current commutation," in *PCIM Europe 2014; International Exhibition and Conference for Power Electronics, Intelligent Motion, Renewable Energy and Energy Management*, 2014, pp. 1–8.

[9] Infineon Technologies AG, "Datasheet EiceDRIVER 1ED34x1Mc12M Enhanced," Oct. 2021. Accessed: Mar. 20 2024. [Online]. Available:
https://www.infineon.com/dgdl/Infineon-1ED34x1Mx12M-DataSheet-v01_10-EN.pdf?fileId=5546d46274cf54d50174d97c2fb71f62

[10] C. Cai, W. Zhou, and K. Sheng, "Characteristics and Application of Normally-Off SiC-JFETs in Converters Without Antiparallel Diodes," *IEEE Transactions on Power Electronics*, vol. 28, no. 10, 2013, pp. 4850–4860, doi: 10.1109/TPEL.2012.2237417.

[11] P. Hofstetter, R. W. Maier, and M. Bakran, "Parasitic Turn-On of SiC MOSFETs – Turning a Bug into a Feature," in *PCIM Europe digital days 2020; International Exhibition and Conference for Power Electronics, Intelligent Motion, Renewable Energy and Energy Management*, 2020, pp. 1–7.

AUTHOR INDEX

Abbas, Khizra 764
Ackermann, Martin.............................1336
Aiello, Giuseppe 1217
Akbari, Saeed.....................................2094
Akturk, Akin 739
Alauzet, Louis....................................2811
Albert, Tianlong.................................1759
Alfonso, Irene Maria Torres................2503
Alfonzetti, Emanuela 1844
Allioua, Abdelmoumin........................2128
Ammar, Ahmed...................................1087
Appleby, Matthew...............................3276
Arai, Nobuhide 298
Araujo, Lucas.....................................1673
Arnaudov, Dimitar 2268
Askan, Kenan.....................................1545
Aspalter, Paul.....................................2258
Augustin, Tim 3086
Aunon, Fernando 1467
Ausseresse, Pierrick 1082
Austrup, Isabel...................................2956
Babaki, Amir......................................1227
Bagheribavaryani, Mohammadreza 1418
Baharizadeh, Mehdi............................378
Bai, Yeriel 1804
Baker, Nick 1923
Bándy, Kristóf..........................403, 2566
Barcelos, Renan Pillon.......................264
Barón, Kevin Muñoz 1978
Barth, Henry 2838
Basso, Christophe 3096
Bastawros, Adel 440, 1951
Batista, Emmanuel 2394
Baudais, Briac....................................3187
Behrendt, Stefan 361
Beiranvand, Hamzeh..........................1105
Beyerle, Raphael................................958
Bhatia, Tamanna 1259
Bicer, Ekin Alp 40
Bimmel, Luc 3206
Blechinger, Christoph 1717
Block, Marius 2217
Bockholt, Yannick..............................3334
Böhning, Lukas 2208
Boldyrjew-Mast, Roman.....................723
Bosnjic, Zlatko...................................1788
Boutry, Arthur 1878
Bouzerd, Souhila................................581

Branas, Christian................................2286
Brandl, Anja Katerina.........................1613
Breidenstein, Daniel...........................1634
Bürger, Matthias 863
Cairnie, Mark 599
Calmels, Alain 3305
Cammarata, Federica..........................1289
Campos, Adriana 2663
Cannone, Marco 502
Capobianco, Thomas Anthony 1168
Çay, Yunus...3247
Cepin, Simon......................................1051
Chaisakdanugull, Chanuch..................3067
Chatroux, Daniel 2278
Chatterjee, Bhaskar 774
Chen, Mengxing 424
Cherief, Wahid 1910
Cho, Wonjin Dylan.............................1046
Choo, Vin Loong 1775
Chorfi, Ilias 2175
Cinik, Sadik.......................................2453
Colak, Baris 490
Colomer, Pau 456
Conilh, Christophe 2227
Corbitt, Anna 135, 1123, 1821
Croston, Jose Andres Aguilar..............3150
Curbow, Austin 1475
Cusumano, Andrea 1627
Czerwenka, Philipp.................1139, 3034
Daire, Baptiste....................................3110
Dasch, Michael...................................1907
Davoodi, Hossein 1013
Debbadi, Karthik 2963
Deboy, Gerald 15
Dedew, Mohamed Lemine 34
Delaforge, Timothé 1797
Denk, Marco.......................................1192
Despesse, Ghislain 797
Diz, Sergio De Lopez..........................411
Do, Nguyen Nghia 1428
Dresel, Lars 2737
Du, Xinyuan 1987
Duijsen, Peter Van........1658, 2248, 2657, 3213
Dumollard, Yannick 1751
Dupont, Max 93
Dusmez, Serkan 383, 2334, 3060
Eichler, Felix 3020
Eyama, Takaaki 56

Fabian, Benjamin	190
Fenske, Florian	3390
Fey, Justin	1902
Fleck, Soenke	338
Förster, Nikolas	3237
Fotteler, Oleg	3328
Fräger, Lukas	926
Frank, Michael	754
Frank, Wolfgang	1770
Frei, Steffen	2478, 3007
Fuchs-Gade, Jannik	2632
Fuhrmann, Jan	1315
Gackowski, Bartosz	1504
Gandluru, Veera Bharath Chandra Reddy	2167
Gavin, Serge	1101
Gebhard, Thomas	1128
Gebhardt, Mathias	2769
Gellman, Ziv	608
Gendrin, Martin	909
Ghanbari, Alireza Ramezan	3175
Ghosh, Priyanka	1523
Gick, Sebastian	1264
Gioda, Alexis	3400
Girgin, Mehmet Oguz	3353
Giuffrida, Simone	248
Giuffrida, Vittorio	1065
Gleissner, Michael	2803
Goff, Gregoire Le	3160
Gomez, Antonio Miguel Munoz	625
Gottardo, Davide	2461
Gragger, Johannes	2104
Graham, Robert	1410
Groon, Fabian	3380
Groos, Gerhard	986
Guan, Jiajia	2591, 3395
Gudala, Bhavana	2524
Guiot, Eric	1604
Gunes, Ekrem R.	3221
Gupta, Gaurav	534
Gürlek, Yavuz	745
Haake, Daniel	2538
Haas, Tobias	2326, 3017
Haehre, Karsten	214
Haensel, Stefan	230
Hanf, Michael	351, 571
Harmand, Thomas	2138
Hasegawa, Kazunori	3002
Hauenschild, Philipp	1969
Hegarty, Timothy	1092
Hegde, Niranjan	1374
Heimler, Patrick	1955
Hellinger, Rolf	1

Hepp, Maximilian	3045
Herrera, Adolfo	1057
Herrmann, Clemens	731
Hertline, Joseph	1886
Herzog, Fabian	3136
Hirao, Takashi	1007
Hironaka, Yoichi	699
Hoffmann, Lennart	3264
Horat, Andreas	480
Hornbuckle, Malachi	2724
Hosseinzadehlish, Mana	1402, 1610
Hu, Jhih-Cheng	791
Huber, Jonas	254
Huerner, Andreas	681
Huselstein, Jean-Jacques	2547
Husev, Oleksandr	893
Igartuburu, Daniel San Laureano	2303
Imai, Ayano	180
Ippisch, Matthias	2638
Irifune, Hiroyuki	2028
Jahn, Simon	883
Jamal, Adeel	1346
Jappe, Tiago	2843
Jegal, Junhyeok	1590
Jha, Kunal	2930
Jia, Minli	2730
Jo, David	1732
Jones, Jeremy	1031
Kaiser, Jeremias	1538
Kampert, Erik	2342
Kanatzar, Paul	1361
Kangjia, He	62
Karout, Mohammed Amer	1835
Kasko, Igor	1991
Kato, Koji	1368
Kaufmann-Bühler, Marius	2400
Kawabata, Junya	2049
Keilmann, Robert	2972
Kempitiya, Asantha	497
Klever, Severin	1561
Knappstein, Lukas	1745
Knecht, Martin	3142
Koch, Jan-Niklas	2240
Koczy, Dawid	1651
Kohlhepp, Benedikt	2316
Koi, Kenichi	67
Kono, Hiroshi	2022
Kopischke, Ruben	2796
Körner, Patrick	615
Kragl, Robert	1385
Kreppel, Thomas	2416
Krigar, Tim	174

Kroics, Kaspars 510
Kugener, Jeff 3315, 3318
Kurukuru, Varaha Satya Bharath 875
Kuzmanoska, Sara 2745
Ladentin, Kevin 1964
Lambert, Adrien 1574
Langfermann, Sascha 1516
Lavery, Melanie 1485
Lee, Chih Hui 1152
Lee, Jongmu...1712
Lee, Kihyun 1724, 1737
Lemaitre, Damien 2596
Lenz, Travis 1352
Lenzen, Patrick 903
Leung, Wing Tai 74
Liao, Xinyuan 322
Lim, Alex 2937
Lindner, Lars 2370
Lippold, Florian 2981
Liu, Baihan 1072
Liu, Iris 1222
Liu, Yusi 3181
Lottis, Christian 3347, 3358
Lotz, Marc René 1457
Lu, Juncheng.. 19, 837
Lucia, Oscar 2448, 2513
Lutzen, Hauke 976
Lv, Jianwei 1872
Ma, Kwokwai 2778
Machtinger, Katharina 2119
Madloch, Sonja 369
Maheshwari, Ramkrishan 3118
Mai, Annette 284
Maier, Jannik 1642
Mandrioli, Riccardo 2576
Mannen, Tomoyuki...831
Mari, Jorge 843
Marie, Alexandre 2819
Martano, Emanuele 2874
Martínez, Alfonso 2359
Masuda, Akiyoshi 1018
Mauromicale, Giuseppe 2751
Mazzer, Simone 2162, 2532
McRae, Tim 2190, 2364, 3042
Medina-Garcia, Alfredo 1207
Meligy, Ahmed 1495
Menzel, Steffen..933
Merrouche, Abdennour 1113
Minamisawa, Renato Amaral 2036
Mirkovic, Nikola...2488
Mo, Xianghao 2386
Mochizuki, Yo..870

Mönch, Stefan 167
Mueller, Lukas 2425
Mühlfeld, Christian 3340
Muralikrishna, Ajay Krishna Voppu....................2886
Nachete, Idriss...2408
Nakako, Hideo 49
Nawaz, Muhammad 2013
Nehmer, Dominik 24
Neira, Sebastian 2700
Neuner, Matthias 3296
Nikiforidis, Ioannis 916, 2718
Nkembi, Armel Asongu 2469
Oberdieck, Karl 1828
O'Keeffe, Rosemary 1249
Olalla, David 1814
Ong, Shu Ee 2942
Orlando, Stefano 1434, 2673
Otori, Daichi...518
Otte, Raphael 2234
Ouhab, Merouane.................................589, 2948
Owzareck, Michael 3371
Palma, Marco 1568
Panchal, Pranav 315
Paradkar, Sachin Shridhar 2627
Patterson, Andrew 1330
Paul, Indrajit..1133
Peng, Hujun 665
Petzold, Tom 1158
Pham, Thanh-Toan 2786
Philippe, Antoine 1441, 1449
Phung, Thanh Hai 1555, 2612
Piccioni, Andrea 2680
Piepenbrock, Till 463
Poller, Tilo 2914
Porpora, Francesco 222
Pouresmaeil, Mobina 419
Prince, Aswathy M. 2850
Rabay, Battist 2082
Radix, Bryan 2112
Radomsky, Lukas 3286
Randerath, Joschka...3256
Raßmann, Rando 2350
Rauh, Michael 690
Rebenklau, Lars 2088
Reddy, Niranjan Suravarapu 2273
Rehlaender, Philipp 2686
Reimann, René 2377
Reiner, Richard 557
Reißenweber, Lukas 447, 525
Reitz, Niclas 1393
Ren, Linhao 1917
Ren, Xufu 803

Rendek, Karol	2831
Reymond-Laruina, Frédéric	103
Rezaeizadeh, Amin	1686
Ribarich, Tom	1254
Ribeiro, Kelly	2294
Rillo, Oriol Subirats	290
Ringelmann, Tim	2708
Rodrigues, Luis Alves	635
Rodriguez, Manuel Escudero	812
Rodruigez, Manuel Escudero	3077
Rosensaft, Boris	2620
Rudzki, Jacek	1942
Ruoff, Dominik	1277
Ruppert, Lukas	1703
Sakai, Junya	2006
Salomez, Florentin	2431
Samura, Koki	2764
Sankari, Rasched	197
Sawada, Takashi	161
Schindler, Stefanie	1147
Schindler, Tobias	140
Schmidhuber, Michael	2438
Schmidt, Matthias	397
Schmidt, Paul	1999
Schmitz, Laurids	2995
Schnell, Raffael	855
Schnitzler, Ruben	1851
Schulte, Felix	3364
Schulz, Martin	1077
Schwab, Stefan	343
Schwarz, Niklas	1200
Scuto, Alfio	2921
Seber, Elizabeth	888
Sekar, Ajith Kumar	2041
Sen, Gokhan	784
Seo, Hansol	1896
Sheikhan, Alireza	549, 2606
Shi, Sanbao	1212, 3029
Sifoune, Sarah	390
Singer, Mehyeddine	2309
Solomakha, Oleksandr	1174
Somarin, Hasan Mousavi	2494
Sos, Carlos Costas	205
Sousa, Gean	2557
Srikrishna, N. H	3269
Steenbock, Liska	3169
Steiner, Felix	1891
Stone, David A.	949
Subotic, Stefan	274
Sugie, Hisashi	1765
Sun, Qing	2791
Suzuki, Keita	2053

Syed, Hadiuzzaman	2758
Talits, Kevin	997
Tan, John Emmanuel	150
Tanikawa, Kohei	1272
Tarmoom, Ehab	942
Tekir, Bünyamin	672
Tengvall, Sebastian	1380
Thamm, Merlin	1532
Thekemuriyil, Tanya	645, 2986
Thirukoluri, Rajani Kumar	1865
Thomas, Mark	564
Thönnessen, André	1584
Tigira, Sandu	3130
To, Pham Ha Trieu	707
Tobler, Stefan	472
Tokorozuki, Takeshi	849
Torrisi, Marco	822
Tranchero, Maurizio	1322
Troudi, Rami	1185
Tuncay, Sebnem	1858
Uemura, Hirofumi	433
Ueno, Masaki	2909
Ugur, Abdulkerim	3229
Uhlemann, Andre	1283
Urbaneck, Daniel	2152
Varadarajan, Kamal	1598
Vemulapati, Umamaheswara Reddy	1025
Vinciguerra, Vincenzo	1039
Vobecky, Jan	1002
Vogelsberger, Markus	1307
Vogt, Michael	2866
Vuletic, Radovan	305
Walter, Michael	1297
Wang, Hamlin	2185
Wang, Hao	2693
Wang, Lei	1242
Wang, Lisheng	2060
Wang, Qilei	113
Wang, Rui	84
Wang, Yushi	966
Watanabe, Hiroki	3125
Weckbrodt, Julien	121
Wei, Frank	2146
Wei, Suhang	2074
Weihe, Sven	330
Wen, Jin	2182, 3103
Wessel, Wilfried	655
Wietschel, Martin	7
Wille, Christopher	128
Winkler, Paul	1511
Xie, Dong	2880
Xie, Luhong	1930

Yadav, Sachin..2583
Yan, Xingda 1680
Yan, Yiyang...1809
Ye, Yijun 2826
Ye, Zhong 1621, 2646
Yoshida, Satoshi...543
Yoshioka, Kentaro..2067
Yu, Renze...717
Yu, Sean... 1180, 2518
Yu, Sheng-Yang 2200
Zeng, Chenhang..1693
Zhang, Chi 1781
Zhang, Hongpeng 238
Zhang, Huaiyuan..2901
Zhang, Yi 1936
Zhao, Yue..3197
Zheng, Zexiang..2860
Zhu, Shiwu...2893
Zipperstein, David 2651
Zipprich, Robert...1664
Zocher, Markus 1233